Vector Calculus

Miroslav Lovrić

prepared for
Department of Mathematics
University of California, Santa Barbara

With additional material from
Howard Anton, Irl Bivens, Stephen Davis
Calculus Early Transcendentals Single Variable,
Chapter Ten: Infinite Series

Erwin Kreyszig
Advanced Engineering Mathematics, 9th Edition

BICENTENNIAL 1807 WILEY 2007 BICENTENNIAL

WILEY
CUSTOM SERVICES

Bicentennial Logo Design: Richard J. Pacifico

To order books or for customer service, please call 1(800)-CALL-WILEY (225-5945).

Printed in the United States of America.

ISBN 978-0-470-89578-8

Printed and bound by J & M Reproduction

10 9 8 7 6 5

Contents

From: Lovrić/Vector Calculus

From: Anton/Calculus 8e

From: Kreyszig/Advanced Engineering Mathematics 9e

Vector Calculus

BICENTENNIAL 1807 WILEY 2007 BICENTENNIAL

The Wiley Bicentennial–Knowledge for Generations

Each generation has its unique needs and aspirations. When Charles Wiley first opened his small printing shop in lower Manhattan in 1807, it was a generation of boundless potential searching for an identity. And we were there, helping to define a new American literary tradition. Over half a century later, in the midst of the Second Industrial Revolution, it was a generation focused on building the future. Once again, we were there, supplying the critical scientific, technical, and engineering knowledge that helped frame the world. Throughout the 20th Century, and into the new millennium, nations began to reach out beyond their own borders and a new international community was born. Wiley was there, expanding its operations around the world to enable a global exchange of ideas, opinions, and know-how.

For 200 years, Wiley has been an integral part of each generation's journey, enabling the flow of information and understanding necessary to meet their needs and fulfill their aspirations. Today, bold new technologies are changing the way we live and learn. Wiley will be there, providing you the must-have knowledge you need to imagine new worlds, new possibilities, and new opportunities.

Generations come and go, but you can always count on Wiley to provide you the knowledge you need, when and where you need it!

William J. Pesce
President and Chief Executive Officer

Peter Booth Wiley
Chairman of the Board

Vector Calculus

MIROSLAV LOVRIĆ
McMaster University

BICENTENNIAL 1807 WILEY 2007 BICENTENNIAL

JOHN WILEY & SONS. INC.

Publisher *Laurie Rosatone*
Assistant Editor *Shannon Corliss*
Editorial Assistant *Jeffrey Benson*
Freelance Editor *Jennifer Albanese*
Marketing Manager *Amy Sell*
Marketing Assistant *Tara Martinho*
Senior Production Editor *Ken Santor*
Senior Designer *Kevin Murphy*
Cover Design *David Levy*
Cover Photo *Dann Coffey / Getty Images, Inc.*
Illustrations *Erin Clements*
Media Editor *Stefanie Liebman*

This book was set in LaTeX by Techbooks, Inc. and printed and bound by R.R. Donnelley – Crawfordsville. The cover was printed by Phoenix Color Corporation.

This book is printed on acid-free paper. ♾

ISBN-13 978-0-471-72569-5
ISBN-10 0-471-72569-2

Printed in the United States of America

10 9 8 7 6 5 4 3 2 1

Preface

INTRODUCTION

This book provides a comprehensive and thorough introduction to ideas and major results in the theory of functions of several variables and modern vector calculus in two and three dimensions. Important concepts of calculus of real-valued functions of one variable (limit, continuity, derivative, differentiability, integral) are generalized to functions of several variables and to vector-valued functions. Attempts to generalize the definite integral result in a construction of path and surface integrals. Classical integration theorems of Green, Gauss and Stokes, which relate the (generalized) concepts of derivative and integral, preserve the spirit of the Fundamental Theorem of Calculus for real-valued functions of one variable.

The approach adopted in writing this text–easy to read and easy to understand narratives, numerous pictures and diagrams, clear explanations, large number of exercises and fully explained examples, and broad range and number of applications–makes the material suitable for a variety of audiences with a wide range of backgrounds and interests. Courses that could benefit from using this text include:

- Vector calculus course, usually taught to students who complete a full-year sequence in calculus of real-valued functions of one variable,
- Analysis or advanced calculus course that is a sequel to a two-semester (or three-term) first-year calculus course, or
- Engineering course on vector calculus.

The range of applications presented in the book conveys the importance of vector calculus in mathematics and beyond. Borrowed from a variety of disciplines far beyond physics and engineering, applications are also used to illustrate concepts and results that are discussed throughout the text. Presentation of theory is clear and rigorous, and yet not too technical nor dry.

APPROACH AND GOALS

We now outline guiding principles employed in creating this text, emphasizing major features of exposition of mathematics material.

Use a variety of approaches (algebraic, numeric, geometric) when presenting a concept or an idea, or when introducing a new object. For instance: differential operators (gradient, divergence, and curl) are given in algebraic form, using partial derivatives. Through a sequence of carefully designed examples, we build our understanding of their geometric properties, and relate them to properties of flows (vector fields). Numeric explorations (such as estimating divergence of a wind field in Example 4.59) further deepen our understanding. Finally, we witness the ways in which the three operators find their way into applications.

Convey excitement about the material; relate material to other disciplines. In the two sections, "World of Curves" in Chapter 3 and "World of Surfaces" in Chapter 7, we explore applications from an amazing array of disciplines: Bezier curves (used in design), catenary curve (suspension bridges), helices (a DNA molecule is a double helix!) and curves on surfaces. We learn how to parametrize a torus and Möbius strip, then study Archimedes' screw, model sea shells and animal horns, and identify surfaces used in modern architecture.

The harder the material, the larger the number of examples and illustrations. For instance, the section on the chain rule (which many students find quite challenging) contains detailed explanations, including comments on (often confusing) notation, as well as nine fully worked examples and over two dozen exercises for further practice.

Background material is reviewed as needed; new material is related to known material. Rather than starting right away with a new definition (for example, limit of a function of several variables), the text begins with a review of the limit from first-year calculus. On three pages, both numeric and algebraic aspects are discussed, preparing the reader for a generalization that will follow. The concept of closeness is related, in an example, to a (familiar) solution of a computer programming problem. The importance of this example lies in the fact that students will easily relate to it, and that will help them get a better grasp on the concept of closeness, which is essential for the definition of the limit. Likewise, as a prequel to double and triple integrals, Riemann sums and the definite integral of a real-valued function of one variable are reviewed.

Important concepts are revisited as often as possible, in a variety of contexts. For instance, paths and curves are introduced in Chapter 2, with the focus on algebraic features (parametrization). In Chapter 3 we continue our study, by exploring geometric concepts such as tangent and curvature, and introducing two fundamental physical quantities–velocity and acceleration. Flow lines establish a relationship between paths and vector fields (Chapter 4). Concepts related to integration along paths are discussed in Chapter 5. We revisit paths and curves in our discussion of parametrizations and properties of surfaces in Chapter 7.

Convey relevance and importance. The text is rich with applications coming from a wide range of disciplines (wind chill index, model of a hurricane, spirals, blood flow profile, to mention a few–aside from classical applications coming from physics and engineering). The emphasis is placed not on explaining details of the theory (although there is a bit of that), but on identifying physical (and other) quantities as mathematical objects and showing how to use calculus to manipulate them in meaningful ways. Thus, the reader will be prepared, when studying, say, physics or engineering, to proceed smoothly through the mathematical side and concentrate on understanding the physics (or engineering) of it. Several sections are entirely devoted to applications. For those interested in physics, we explore the power of vector calculus in electromagnetism and fluid flow.

▶ FOCUS ON CONCEPTUAL UNDERSTANDING AND MATHEMATICAL THINKING

This text provides students with an opportunity to build a clear, thorough, and deep understanding of ideas and concepts from calculus of functions of several variables. Mastering the material presented here will serve as a foundation for further study of mathematics and related disciplines. Let us repeat that the emphasis has been placed on depth of understanding, rather than on breadth of presentation topics.

One important step in building understanding is making sure that students attain a clear intuitive picture about the object(s) studied. To this end, the text offers exhaustive discussions and carefully crafted examples, clear and crisp statements of theorems and definitions, and–acknowledging that many of us are visual learners–numerous illustrations, graphs, and diagrams.

After being introduced, important concepts are revisited, placed in a variety of contexts, and shown in light of applications. Such an approach provides us with one of best opportunities to exercise deep learning. For instance, Implicit Function Theorem is discussed first in the special case of curves defined implicitly in Chapter 3. Then, in Chapter 4, a general version of the theorem is stated and illustrated in examples (that include implicitly defined surfaces). We witness the power of the Implicit Function Theorem as it gets applied to the change of variables in multiple integration calculations in Chapter 6, and in studying surfaces in Chapter 7.

▶ FOCUS ON DEVELOPMENT OF MATHEMATICAL SKILLS

It is impossible to fully master almost any topic in mathematics without adequate skills in symbolic (algebraic) manipulation. This text contains a large number of fully solved examples that are designed to illustrate formulas, algebraic methods and algorithms, and to give routine in technical intricacies of calculations.

Topics that students may find challenging (such as chain rule, change of variables in multiple integration, setting the limits of integration in multiple integrals, using classical integration theorems, or building parametrizations of curves and surfaces) are accompanied by a large number of solved examples.

Use of technology (graphing calculator, or mathematical software, such as Maple or Mathematica) is strongly encouraged, since–when done properly–it equips us with insights that might enhance our understanding of the material. To be more specific, consider an example. We will not ask a computer to do anything before we learn what the graph of a function of several variables is supposed to show, make sure we understand how to draw it by hand (at least in principle), and learn how computers generate such graphs (and the pitfalls of it!). All of the above accomplished, our attempts at interpreting a computer-generated "three-dimensional" graph or level curves of some surface will promote our understanding of functions of several variables and their properties.

No specific graphing calculator or software package is required.

▶ CONTENT AND ORGANIZATION

This book begins with a review of relevant topics from linear algebra: vectors, dot and cross products, matrices and determinants. This material is spiced with numerous applications, ranging from parametric equations of a line and center of mass to the work of a force and torque.

In Chapter 2 we study basic concepts of calculus of functions of several variables, and also get a first glimpse of vector-valued functions. In order to keep the text flowing, proofs of several theorems that are technical and not really revealing, such as differentiation theorems, have been moved to Appendix A at the back of the book.

Great care is taken with examples, comments and remarks in order to enhance our understanding of the material. For example, instead of insisting on proving differentiability using a fairly abstract definition, the book explains how to relate differentiability to a more intuitive, and geometric, hence visual, concept of a linear approximation. Or, instead of sweating out the proof of equality of mixed second partials (for an interested reader, actual sweating out is done in Appendix A), the book goes through a number of important situations, not only in vector calculus, but also in applications in electromagnetism and elsewhere, where this result is used.

Taylor formula for functions of several variables, extreme values and optimization (mainly for functions of two variables) are discussed in Chapter 4.

Chapter 3 and half of Chapter 4 are devoted to a study of two most popular classes of vector-valued functions: paths and vector fields. A vector-valued function of one variable is called a path, and its image, visualized as a geometric object, is a curve. We study various ways of constructing paths (so-called parametrizations) and investigate applications in the section "World of Curves," and elsewhere. We learn how to extract information such as length, velocity, and acceleration from a parametrization, and discover the close relationship between acceleration and curvature. Serret-Frenet formulas enable us to peek into the world of the differential geometry of curves.

A vector field is a function that assigns a vector to a point in a plane or in space (think of a river–every water molecule (a point) has its own direction of motion (i.e., a vector that is associated with it)–and you see a vector field). In chapter 4, with the help of partial derivatives, we define differential operators (curl and divergence) that we employ in our investigation of vector fields. In the overall spirit of the text, emphasis is placed on interpretations and applications of these operators in various contexts.

Building on intuitive understanding of implicitly defined curves in Chapter 3, we discuss the general version of Implicit Function Theorem in Section 4.7. Its major applications (change of variables in multiple integration and theory of surfaces) appear in Chapters 6 and 7.

Chapters 5–7 are devoted to generalizations of the definite integral to various regions in two and three dimensions. We learn how to integrate functions along curves, over two-dimensional regions, and over surfaces and three-dimensional solids. With the help of a parametrization, a path integral is reduced to the definite integral of a function of one variable.

The work of a force and the circulation of a vector field are presented as main applications. A vector field is called a gradient vector field if it is the gradient of some real-valued function (the most important example is a conservative force field). We end Chapter 5 with an investigation of remarkable properties of gradient vector fields with respect to integration.

Chapter 6 is devoted to the construction and techniques of evaluation of double integrals. Taking advantage of the machinery available, we end the chapter with a short study of triple integrals.

In Chapter 7 we investigate parametrizations of surfaces and define surface integrals. The section "World of Surfaces" contains some of many applications of surfaces. We also learn how to obtain more challenging parametrizations, such as a torus or a Möbius strip. Study of surface integrals leads into applications such as surface area and flux of a vector field. In the last section we unify various types of integration into a single concept and discuss further examples, properties, and physical applications.

The Fundamental Theorem of Calculus states that the definite integral of the derivative of a function depends not on the whole interval of integration but only on its endpoints (that is, on the boundary points of the interval). In Chapter 8 we investigate further the relation between the concepts of integration and differentiation. The results, contained in the theorems of Green, Gauss and Stokes (known as Classical Integration Theorems of vector calculus), are all variations on the same theme applied to different types of integration. Green's Theorem relates the path integral of a vector field along an oriented, simple closed curve in the xy-plane to the double integral of its derivative (to be precise, of the *curl*) over the region enclosed by that curve. Gauss' Divergence Theorem extends this result to closed surfaces and Stokes' Theorem generalizes it to simple closed curves in space.

Two sections in Chapter 8 are completely devoted to applications of vector calculus in electromagnetism and fluid flow. The emphasis is placed not on explaining the details of theory of electromagnetism or fluid flow, but on identifying physical quantities involved as mathematical objects and showing how to use calculus in manipulating them to obtain meaningful results. We also introduce a useful formalism of differential forms, the importance of which is understood best in the context of unification of Classical Integration Theorems.

▶ END-OF-CHAPTER SYNTHESIS AND REVIEW

Each chapter ends with a review section that is divided into four parts: *Chapter Summary, Review Questions, True/False Quiz*, and *Review Exercises and Problems*. *Chapter Summary* lists all concepts covered in the chapter, and is suitable for review or self-test of theory. *Review Questions* ask students to relate various concepts, rephrase or quote a definition or a theorem and discuss their statements and implications. On top of providing additional exercises, *Review Exercises and Problems* section includes harder, thought-provoking questions. On few occasions, certain theoretical concepts are introduced. *True/False Quiz* questions test conceptual understanding of the material.

The two appendices at the back of the book contain proofs of technical differentiation theorems, statements of several results that are used throughout the text, and answers to odd-numbered exercises.

▶ BALANCING PROCEDURAL AND CONCEPTUAL UNDERSTANDING

The structure of this book gives an opportunity to a course instructor to fine-tune the balance between the activities that promote conceptual and procedural understanding of the material. A course for engineers or physics majors might emphasize certain aspects a bit more than a course in introductory analysis for mathematics majors, and vice versa. Furthermore, the order of topics can be somewhat rearranged, and – in a few cases – certain sections can be omitted, without losing continuity.

There are different types of courses that will benefit from using this book: vector calculus for math or physics or engineering majors, advanced calculus, analysis, etc. Before outlining the core sections that a course instructor should consider, there are a few key features that might be helpful in planning the course.

The textbook is suitable for (advance) reading and practice assignments. For instance, the section Properties of Derivatives (Section 2.6) contains nine carefully explained examples of the chain rule that will help students fully understand all intricacies of the topic (that many might find challenging). Or, there is no need to lecture on techniques of evaluation of double integrals: moving from elementary, easy-to-do examples towards more challenging integrals, the material in Section 6.3 can be discussed in a tutorial, or studied as homework assignment.

Applications are picked based on students' interests. For instance, in Higher-Order Partial Derivatives (Section 4.1) physics or engineering majors will be interested in studying Example 4.8 (heat equation), or Example 4.11 (Korteweg-de Vries equation and soliton waves), whereas math majors might focus on the properties of Laplace's equation and harmonic functions (Example 4.10).

Depending on students' background preparation and course expectations, concepts and ideas can be presented at different levels. For instance: the divergence (Section 4.6) is discussed on an intuitive level (by looking at graphs of vector fields), and numerically (Example 4.59: Estimating Divergence of a Wind Field). Section 4.6 also contains a derivation of the relation between the divergence and total outflow in an easy-to-follow special case (with another special case appearing in the Exercises). A slightly more challenging general derivation, for time-dependent vector fields, is also given (in Section 8.6).

Key approach is learning by understanding. Readers of this book will discover numerous opportunities to deepen their conceptual grasp of the material; perhaps by working through a proof of a theorem, or by studying numerous examples that illustrate how a certain concept (or idea, technique, or algorithm) is used in a variety of situations.

It is assumed that this course will be taken after completing a full-year sequence in calculus of real-valued functions of one variable, using any of the standard calculus textbooks. Familiarity with some basic concepts in linear algebra and vectors is a prerequisite. Just in case this is lacking, relevant material is reviewed in Chapter 1.

List of core sections:

- Vector calculus course for students with interests in math or physics: The list includes sections that show various ways of working with mathematics content (from intuitive, numeric, geometric/visual reasoning, to theoretic) including selected applications: 2.1–2.8, Appendix A, 3.1–3.5, 4.1, 4.5–4.7, 5.1–5.4, 6.1–6.5, 7.1, 7.3–7.5, 8.1–8.5.
- Vector calculus course for engineering students includes material that provides clear, intuitive understanding, proficiency in numeric and algebraic calculations, and selected applications (sections marked *: theoretical intricacies can be de-emphasized, such as epsilon-delta arguments for limits and continuity, or abstract definition of differentiability): 2.1, 2.2, 2.3*, 2.4*, 2.5–2.8, 3.1–3.3, 4.1, 4.5–4.6, 5.1–5.4, 6.1–6.5, 7.1–7.4, 8.1–8.3, 8.6.
- Analysis course (level 2 or level 3, i.e., sophomore or junior level) that gives a rigorous treatment of multivariable calculus, i.e., emphasizes understanding and gives good theoretical foundations of the material (course title could be Analysis, Introductory Analysis, Honours Advanced Calculus, Advanced Calculus, etc.): 2.1–2.8, Appendix A, 3.1–3.3, 4.1–4.4, 4.6, 4.7, 5.1–5.4, 6.1, 6.2, 6.4, 6.5, 7.1–7.4, 8.1–8.4.
- Advanced calculus course (course title could be Calculus 3, Intermediate Calculus, Calculus and Analytic Geometry: usually the last course in the sequence of courses

offered under that name); first serious contact with functions of several variables (sections marked *: certain theoretical intricacies can be de-emphasized, such as epsilon-delta arguments for limits and continuity, or abstract definition of differentiability): 1.1–1.5 (brief review, if necessary), 2.1, 2.2, 2.3*, 2.4*, 2.5–2.8, 3.1–3.3, 4.1–4.4, 4.6, 5.1–5.4, 6.1–6.5, 7.1–7.4, 8.1–8.3.

Non-standard (fun!) courses that prefer intuitive, geometric, and numeric approaches to theoretical considerations, with focus on applications (sections marked *: many applications discussed could be used as starting points for independent projects):

- Vector calculus: 2.1, 2.2, 2.3 (limits from graphs, level curves), 2.4 (focus on linear approximation), 2.5, 2.6 (cases of functions of 2 or 3 variables and vector fields), 2.7, 3.1*, 3.2–3.4, 4.1*, 4.5*, 4.6, 5.1–5.4, 6.1–6.5 (de-emphasize calculations), 7.1, 7.2*, 7.3, 7.4, 8.1–8.3.
- Analysis: 2.1, 2.2, 2.3 (limits from graphs, level curves), 2.4 (focus on linear approximation), 2.5–2.7, 3.1*, 3.2, 3.3, 4.1*, 4.2–4.4, 4.6, 5.1–5.4 (de-emphasize calculations), 6.1–6.5 (de-emphasize calculations), 7.1, 7.2*, 7.3, 7.4, 8.1–8.3.

▶ ACKNOWLEDGEMENTS

First of all, I would like to thank the team at Wiley: Jennifer Albanese, Shannon Corliss, Jeff Benson, Laurie Rosatone, and Ken Santor, for their energy, enthusiasm and hard work invested in this project.

I am grateful to all reviewers whose valuable comments, criticism and suggestions helped me improve my manuscript: Robert Adams (University of British Columbia), William Allard (Duke University), Ivan Avramidi (New Mexico Tech), Michael Barbosu (State University of New York, Brockport), Scott Beaver (University of New Mexico), Thomas Bieske (University of Southern Florida), Adel Boules (University of North Florida), Brian Bradie (Christopher Newport University), Paul Campbell (Beloit College), Ricardo Carretero (San Diego State University), Margaret Cheney (Rensselaer Polytechnic Institute), Scott Crass (California State University, Long Beach), Joseph Fehribach (Worcester Polytechnic Institute), Clifton E. Ealy, Jr. (Western Michigan University), David Easley (Virginia), John Feroe (Vassar College), Viktor Ginzburg (University of California, Santa Cruz), Guillermo Goldsztein (Georgia Institute of Technology), Thomas Hoover (University of Hawaii), Tae-Chang Jo (New Mexico Institute of Mining and Technology), Matthias Kawski (Arizona State University), Yuji Kodama (Ohio State University), Srilal Krishnan (Iona College), Namyong Lee (Minnesota State University), Tanya Leise (Amherst College), Michelle LeMasurier (Hamilton College), Hans Lindblad (University of California, San Diego), Mark McKibben (Goucher College), Duy-Minh Nhieu (Georgetown University), Daniel Norman (Queen's University), Ralph Oberste-Vorth (Marshall University), Boris Okrun (University of Wisconsin, Milwaukee), Josef Paldus (University of Waterloo), Edgar Pechlaner (Simon Fraser University), Michael Raines (Western Michigan University), David Rollins (University of Central Florida), Kimberly Roth (Wheeling Jesuit University), David Royster (University of North Carolina at Charlotte), George Rublein (College of William and Mary), Ted Shifrin (University of Georgia), Plamen Simeonov (University of Houston, Downtown), Emilio Toro (University of Tampa), E.J. Janse von Rensburg (York

University), Nolan Wallach (University of California, San Diego), Peter Webb (University of Minnesota), Thomas Witelski (Duke University), and Xina Wu (University of Southern California).

For comments, suggestions, and help with exercises and solutions, I thank my colleagues, former students and friends: Whitney Black, John Blanchard, Henning Broge, Mark Chamberland, Derek DiFilippo, Daniel Egloff, Jean-Pierre Gabardo, Jeff Hopper, Fadil Khouli, Zdislav Kovarik, Maung Min-Oo, Lakshmi Narayani, Louis D. Nel, Rouset Shaiki-Adeh, and Willem Sluis.

Many thanks to others who have contributed in various ways and whose names or involvement I am not aware of.

Finally, I thank Erin Clements, for her work on illustrations and diagrams, for her faith in me, daily encouragement and advice. Big thank-you to my parents, Vilma and Ivan Lovrić, and the rest of my family.

▶ EPILOGUE

Like any book, this text is a communication between the reader and the author—and it would make me really happy if we can make it into a two-way communication. If you find an error, or have a comment, a question or a suggestion, or there is something you would like to discuss with me, please write me at lovric@mcmaster.ca.

I thank you for choosing to use this book and hope that you will like reading it, and that it will teach you some good, useful, and exciting math. It might convince you, perhaps, that math isn't that bad after all.

Miroslav Lovrić
McMaster University, December 2006

Contents

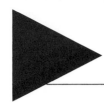

CHAPTER 1

Vectors, Matrices, and Applications

It suffices to use a single real number (together with a unit of measurement) to describe the average temperature on the surface of the Sun, the mass of a molecule, the distance between two cities, or the surface area of a lake. Quantities such as temperature, mass, distance, or area are called *scalars* or *scalar quantities*. On the other hand, the description, of a wind on a weather report contains not only its magnitude but the direction as well (e.g., "northwesterly"). The attractive force of a planet on a satellite is specified by its magnitude and direction ("radially toward the center of the planet"). Quantities such as force and motion are called *vectors* or *vector quantities*.

This chapter begins with the construction of rectangular coordinate systems, the definition of a vector, and a discussion of basic vector operations. Numerous applications, ranging from parametric equations of lines and planes to relative velocity, physical forces, and center of mass, are discussed in the following section. The two types of vector multiplication, the dot product and the cross product, are defined and their algebraic and geometric aspects studied in detail. Related examples introduce physical applications such as the work of a force, angular and tangential velocities of a rotating body and torque. Since matrices and determinants will appear in a number of situations in this book, we present a brief review covering only the relevant topics in a separate section.

1.1 VECTORS

In this section, we will review the definition and basic properties of vectors. Although we will concentrate on dimensions 2 and 3 (for reasons of convenience), all statements and results remain valid for vectors in n dimensions, where $n \geq 2$.

One way to identify points in a plane is to use a two-dimensional *Cartesian* (or *rectangular*) *coordinate system*: we first select a point of reference (usually denoted by O and called the *origin*) and two perpendicular number lines that intersect at O and are placed so that O represents the number zero for both of them. Their orientation (i.e., the direction of increasing values) is indicated by an arrow. The two lines (called the *coordinate axes*) are usually visualized as the horizontal line, called the *x-axis*, and the vertical line, called the *y-axis*. To describe the location of a point A, we have to specify two numbers: the (directed) distance from A to the x-axis (called the *y-coordinate* of A) and the (directed) distance from A to the y-axis (called the *x-coordinate* of A). "Directed distance" means that the

y-coordinates of points below the x-axis and the x-coordinates of points to the left of the y-axis are negative. We write $A(a_1, a_2)$, where a_1 is the x-coordinate and a_2 is the y-coordinate of A. A plane together with the coordinate system just constructed is called the *xy-plane* and denoted by $\mathbb{R}^2$ (sometimes by E^2). Using set notation, we can describe $\mathbb{R}^2$ as

$$\mathbb{R}^2 = \{(x, y) \mid x \in \mathbb{R}, y \in \mathbb{R}\},$$

where curly braces $\{\}$ are used to denote a set, (x, y) is called an ordered pair, the vertical bar is read "such that," and $\in$ means "belongs to" or "is an element of." In this new language, the formula written above can be translated as follows: $\mathbb{R}^2$ is the set of all ordered pairs (x, y) (that are visualized as points in the xy-plane) such that x and y are real numbers. Similarly,

$$\mathbb{R}^3 = \{(x, y, z) \mid x, y, z \in \mathbb{R}\}$$

describes all points in three-dimensional space. To give meaning to the numbers x, y, and z, we construct a three-dimensional *Cartesian* (or *rectangular*) *coordinate system* in much the same way as a two-dimensional system: choose a reference point O and three mutually perpendicular directed number lines that intersect at O (with O representing the number zero for all three of them). The three axes are called the *coordinate axes* and identified as the *x-axis*, the *y-axis*, and the *z-axis*; they are usually visualized as in Figure 1.1.

The coordinate axes define the three coordinate planes: the *xy-plane*, the *yz-plane*, and the *xz-plane*. The location of a point A in space can now be specified by listing the following three real numbers: the (directed) distance a_1 from A to the yz-plane (called the *x-coordinate* of A), the (directed) distance a_2 from A to the xz-plane (called the *y-coordinate* of A), and the (directed) distance a_3 from A to the xy-plane (called the *z-coordinate* of A). We say that the point A has coordinates (a_1, a_2, a_3) and write $A(a_1, a_2, a_3)$. Since it takes three real numbers to uniquely determine A, we say that the space is three-dimensional, and denote it (and we have done so already) by $\mathbb{R}^3$. Notice that the correspondence between the points and the coordinates is one-to-one. This means that every point in $\mathbb{R}^2$ ($\mathbb{R}^3$) is described by one ordered pair (triple) of real numbers and every ordered pair (triple) of real numbers represents one point in $\mathbb{R}^2$ ($\mathbb{R}^3$).

Analogously, we define

$$\mathbb{R}^n = \{(x_1, \ldots, x_n) \mid x_i \in \mathbb{R}, i = 1, \ldots, n\}.$$

In words, $\mathbb{R}^n$, $n \geq 2$, is the set of ordered n-tuples of real numbers, interpreted as points in n-dimensional space. Very soon we will come across another common interpretation of $\mathbb{R}^n$.

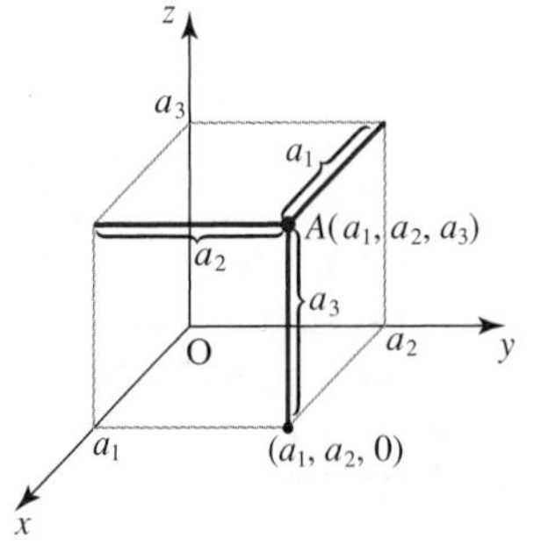

Figure 1.1 Three-dimensional rectangular coordinate system.

In the description of $\mathbb{R}^2$ and $\mathbb{R}^3$ (and, in general, $\mathbb{R}^n$ for $n \geq 1$) using the rectangular coordinate system, all coordinates are given the same meaning—that of a distance. However, there are other ways of describing the location of a point, such as using a combination of angles and distances, as in polar, cylindrical, or spherical coordinates. We will define polar coordinates in a moment, but will postpone the discussion of cylindrical and spherical coordinate systems (see Section 2.8).

Sometimes, more information than just distances is built into coordinates. For example, an ordered quadruple (x, y, z, t) (assume that t represents time) gives not only the location of a point in space, but also the time when something of interest occurred at that location.

A statement such as "the place you are looking for is three kilometers southwest of here" represents another common way of describing the location of a point. It uses *polar coordinates* that will now be constructed.

Choose a point in a plane, label it O (and name it the *pole*), and then choose a half-line starting at O (use an arrow to indicate its direction and call it the *polar axis*), as shown in Figure 1.2(a). The location of any point A in the plane is determined by the following two numbers: the distance r ($r \geq 0$) from O to A and the angle θ ($0 \leq \theta < 2\pi$) between the polar axis and the segment $\overline{OA}$. By convention, θ is measured in radians counterclockwise from the polar axis. We say that r and θ are the *polar coordinates* of A and write $A(r, \theta)$. The correspondence between the points and the polar coordinates is one-to-one, except in one case [the pole can be represented as $(0, \theta)$, for any θ, $0 \leq \theta < 2\pi$].

To compare polar and rectangular coordinate systems in a plane, we place the pole at the origin and the polar axis over the positive direction of the x-axis, as in Figure 1.2(b) (this is how the polar coordinate system is usually visualized). From $\cos\theta = x/r$ and $\sin\theta = y/r$, we obtain the formulas

$$x = r\cos\theta, \qquad y = r\sin\theta$$

for the Cartesian coordinates of a point if its polar coordinates are known. If x and y are known, then the formulas

$$r = \sqrt{x^2 + y^2}, \qquad \tan\theta = y/x, \quad 0 \leq \theta < 2\pi \tag{1.1}$$

give corresponding polar coordinates. Solving for θ [keeping in mind that the range of $\arctan(y/x)$ is between $-\pi/2$ and $\pi/2$, whereas the requirement for θ is $0 \leq \theta < 2\pi$], we get

$$\theta = \begin{cases} \arctan(y/x) & \text{if } x > 0 \text{ and } y \geq 0 \\ \arctan(y/x) + \pi & \text{if } x < 0 \\ \arctan(y/x) + 2\pi & \text{if } x > 0 \text{ and } y < 0 \end{cases} \tag{1.2}$$

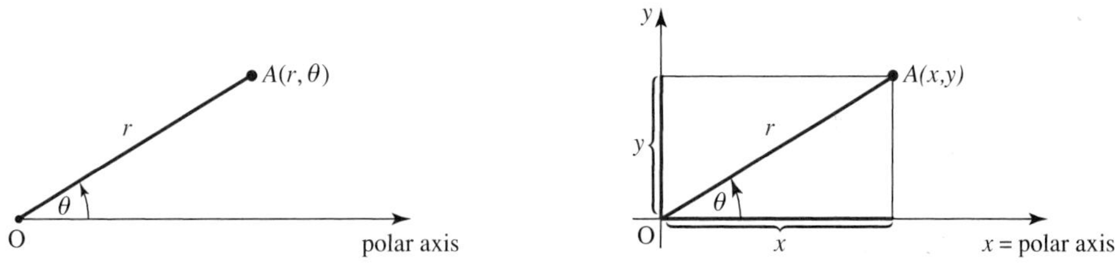

(a) Polar coordinates of a point. (b) Polar and Cartesian coordinates.

Figure 1.2 Polar coordinate system in a plane.

Furthermore, $\theta = \pi/2$ if $x = 0$ and $y > 0$ and $\theta = 3\pi/2$ if $x = 0$ and $y < 0$. If $x = 0$ and $y = 0$, then θ is not uniquely defined (i.e., we adopt a convention that the origin can be represented in polar coordinates as $r = 0$, with any value $0 \le \theta < 2\pi$).

Note that it is important to specify which coordinate system we are using. In other words, we have to use phrases such as "point $A(1, 1)$ in Cartesian coordinates" or "point $A(\sqrt{2}, \pi/4)$ in polar coordinates."

From now on (unless stated otherwise) we use Cartesian (rectangular) coordinate systems in $\mathbb{R}^2$ and $\mathbb{R}^3$.

Recall that the distance between two points $A(a)$ and $B(b)$ on a number line (points A and B are identified with the real numbers a and b) is given by $|b - a|$. We can generalize this formula to any dimension: the distance $d(A, B)$ between points $A(a_1, \ldots, a_n)$ and $B(b_1, \ldots, b_n)$ in $\mathbb{R}^n$ is given by

$$d(A, B) = \sqrt{(b_1 - a_1)^2 + \cdots + (b_n - a_n)^2}. \tag{1.3}$$

In low dimensions, this formula takes on the familiar forms

$$d(A, B) = \sqrt{(b_1 - a_1)^2 + (b_2 - a_2)^2},$$

if $A(a_1, a_2)$ and $B(b_1, b_2)$ are in $\mathbb{R}^2$, and

$$d(A, B) = \sqrt{(b_1 - a_1)^2 + (b_2 - a_2)^2 + (b_3 - a_3)^2},$$

if $A(a_1, a_2, a_3)$ and $B(b_1, b_2, b_3)$ are in $\mathbb{R}^3$. The proof of formula (1.3) for $n = 2$ and $n = 3$ uses the Pythagorean Theorem and is left as an exercise.

▶ EXAMPLE 1.1

Find the equation of the circle in $\mathbb{R}^2$ of radius r centered at $C(c_1, c_2)$, and the equation of the sphere in $\mathbb{R}^3$ of radius r whose center is located at the point $C(c_1, c_2, c_3)$.

SOLUTION

The circle consists of all points (x, y) in the xy-plane whose distance from $C(c_1, c_2)$ is constant and equal to r. Therefore, $\sqrt{(x - c_1)^2 + (y - c_2)^2} = r$, and hence the equation (square both sides)

$$(x - c_1)^2 + (y - c_2)^2 = r^2$$

represents the given circle. Similarly, we obtain

$$(x - c_1)^2 + (y - c_2)^2 + (z - c_3)^2 = r^2$$

for the equation of the sphere of radius r with its center at $C(c_1, c_2, c_3)$. ◀

A vector is a quantity characterized by both magnitude and direction. We will now give a precise definition.

DEFINITION 1.1 Vector in 2, 3, and *n* Dimensions

An *n-dimensional vector* (or a *vector in* $\mathbb{R}^n$) is an ordered n-tuple $\mathbf{v} = (v_1, \ldots, v_n)$ of real numbers, $n \ge 2$. In particular, a *two-dimensional vector* (or a vector in $\mathbb{R}^2$) is an ordered pair $\mathbf{v} = (v_1, v_2)$, and a *three-dimensional vector* (or a vector in $\mathbb{R}^3$) is an ordered triple $\mathbf{v} = (v_1, v_2, v_3)$. The real numbers $v_1, v_2, \ldots, v_n$ are called the *components* or the *coordinates* of $\mathbf{v}$. ◀

We often visualize a two-dimensional vector $\mathbf{v} = (v_1, v_2)$ as a line segment joining the origin and the point (v_1, v_2), with the direction (indicated by an arrow) from the origin toward (v_1, v_2). It is important to notice that the *same set* (namely $\mathbb{R}^2$) is viewed both as a set of points and as a set of vectors; that is, an ordered pair $(v_1, v_2) \in \mathbb{R}^2$ can be interpreted either as a point $A(v_1, v_2)$ or as a two-dimensional vector $\mathbf{v} = (v_1, v_2)$. Similarly, elements of $\mathbb{R}^3$ (or $\mathbb{R}^n$, $n \geq 2$ in general) are visualized either as points or as vectors in a three-dimensional (or n-dimensional) space. What we defined as a vector is sometimes called a *position vector* or a *directed line segment* (whose initial point is at the origin), or a *"bound" vector*. For convenience, we would like to have a vector that can be "moved around"; that is, that can "start" at any point, not necessarily at the origin. We now proceed by precisely defining this concept.

A *line segment* $\overline{AB}$ is the collection of points on the line joining A and B that lie between A and B (including the endpoints A and B). Once the direction has been specified (e.g., "from A to B"), we obtain the *directed line segment* $\overrightarrow{AB}$. In other words, $\overrightarrow{AB}$ is a line segment $\overline{AB}$ with the initial point (or the tail) A and the terminal point (or the tip) B. Now let $\mathbf{v} = (v_1, v_2)$ be a vector in $\mathbb{R}^2$ and pick any point $A(a_1, a_2)$. A *representative* of the vector $\mathbf{v}$ with initial point A is the directed line segment $\overrightarrow{AB}$, where B is the point $(a_1 + v_1, a_2 + v_2)$. Similarly, the directed line segment $\overrightarrow{AB}$, where $A(a_1, a_2, a_3)$ and $B(a_1 + v_1, a_2 + v_2, a_3 + v_3)$, is the representative of a vector $\mathbf{v} = (v_1, v_2, v_3)$ in $\mathbb{R}^3$ that starts at A. Figure 1.3 shows several representative directed line segments of a vector $\mathbf{v} = (v_1, v_2)$ in $\mathbb{R}^2$. The choice of a directed line segment that will represent a vector is determined from context. What we usually label as $\mathbf{v}$ is that chosen representative directed line segment. The representative of $\mathbf{v}$ that starts at the origin is called the *position vector*. Sometimes, we call representative directed line segments of a vector *"free" vectors*.

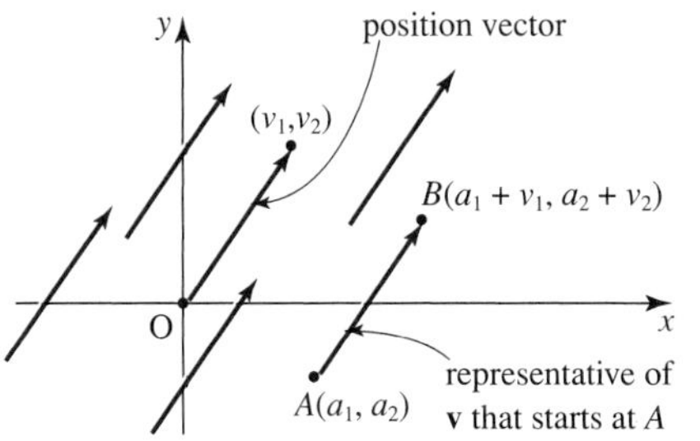

Figure 1.3 Several representatives of a vector $\mathbf{v} = (v_1, v_2)$ in $\mathbb{R}^2$.

Take two points $A(a_1, a_2, a_3)$ and $B(b_1, b_2, b_3)$ in $\mathbb{R}^3$ and consider the directed line segment $\overrightarrow{AB}$. Let us construct the vector $\mathbf{v} = (v_1, v_2, v_3)$ that is represented by $\overrightarrow{AB}$ (i.e., we want to find the "vector from A to B"). By definition, if $\overrightarrow{AB}$ represents $\mathbf{v}$, then $a_1 + v_1 = b_1$, $a_2 + v_2 = b_2$, and $a_3 + v_3 = b_3$; that is, $v_1 = b_1 - a_1$, $v_2 = b_2 - a_2$, and $v_3 = b_3 - a_3$, and therefore, $\mathbf{v} = (b_1 - a_1, b_2 - a_2, b_3 - a_3)$. Thus, the vector $\mathbf{v} \in \mathbb{R}^3$ represented by the directed line segment $\overrightarrow{AB}$, where the coordinates of A and B are $A(a_1, a_2, a_3)$ and $B(b_1, b_2, b_3)$, is given by $\mathbf{v} = (b_1 - a_1, b_2 - a_2, b_3 - a_3)$. By removing the last component, we obtain the corresponding statement for vectors in $\mathbb{R}^2$.

▶ EXAMPLE 1.2

This example illustrates the relationship between a vector and its representative directed line segments.

(a) Find the representative $\overrightarrow{AB}$ of the vector $\mathbf{v} = (3, 0, 4) \in \mathbb{R}^3$ that starts at the point $A(-2, 6, 2)$.

(b) Find the vector $\mathbf{v} \in \mathbb{R}^2$ that is represented by $\overrightarrow{AB}$, where $A = (3, 2)$ and $B = (-1, 4)$.

SOLUTION

(a) By definition, the representative of $\mathbf{v} = (3, 0, 4)$ is the directed line segment $\overrightarrow{AB}$, where $A = (-2, 6, 2)$. If $B = (b_1, b_2, b_3)$, then $b_1 = a_1 + v_1 = -2 + 3 = 1$, $b_2 = a_2 + v_2 = 6 + 0 = 6$, and $b_3 = a_3 + v_3 = 2 + 4 = 6$. Hence, $B = (1, 6, 6)$.

(b) From the discussion preceding this example, it follows that the vector $\mathbf{v} = (-1 - 3, 4 - 2) = (-4, 2)$ has the given directed line segment $\overrightarrow{AB}$ as its representative. ◀

The fact that all representatives of a vector are parallel translates of each other characterizes the Cartesian coordinate system. We will see in Section 2.8 that representatives of unit coordinate vectors in other coordinate systems depend on their location and no longer satisfy this property.

Consider a vector $\mathbf{v} = (v_1, v_2, v_3)$ and its representative directed line segment $\overrightarrow{AB}$, where $A = (a_1, a_2, a_3)$ and $B = (b_1, b_2, b_3)$ are points in $\mathbb{R}^3$ (this means that $v_1 = b_1 - a_1$, $v_2 = b_2 - a_2$, and $v_3 = b_3 - a_3$). The distance between A and B is computed by (1.3) to be

$$d(A, B) = \sqrt{(b_1 - a_1)^2 + (b_2 - a_2)^2 + (b_3 - a_3)^2} = \sqrt{v_1^2 + v_2^2 + v_3^2}.$$

We have just shown that the length of a representative $\overrightarrow{AB}$ of $\mathbf{v}$ does *not* really depend on the coordinates of the points A or B, but only on the components of $\mathbf{v}$. In other words, all representatives of $\mathbf{v}$ have the same length, equal to $\sqrt{v_1^2 + v_2^2 + v_3^2}$. Based on this observation, we will now define the *length* (or the *magnitude* or the *norm*) of a vector. It will be denoted by $|\mathbf{v}|$, or more often by $||\mathbf{v}||$.

DEFINITION 1.2 Length of a Vector

The length of a vector is equal to the length of any of its representatives. In particular, if $\mathbf{v} = (v_1, v_2)$ is a vector in $\mathbb{R}^2$, then $||\mathbf{v}|| = \sqrt{v_1^2 + v_2^2}$. If $\mathbf{v} = (v_1, v_2, v_3)$ is a vector in $\mathbb{R}^3$, then $||\mathbf{v}|| = \sqrt{v_1^2 + v_2^2 + v_3^2}$. ◀

The vectors $\mathbf{v} = (0, 0, 0) \in \mathbb{R}^3$ and $\mathbf{v} = (0, 0) \in \mathbb{R}^2$ are called the *zero vectors* and are denoted by $\mathbf{0}$. The representative of the zero vector starting at A is the (degenerate) directed line segment $\overrightarrow{AA}$. Clearly, $||\mathbf{0}|| = 0$.

DEFINITION 1.3 Addition of Vectors and Multiplication by Scalars

(a) The sum $\mathbf{v} + \mathbf{w}$ and the difference $\mathbf{v} - \mathbf{w}$ of two vectors $\mathbf{v} = (v_1, v_2)$ and $\mathbf{w} = (w_1, w_2)$ in $\mathbb{R}^2$ are the vectors given by $\mathbf{v} + \mathbf{w} = (v_1 + w_1, v_2 + w_2)$ and $\mathbf{v} - \mathbf{w} =$

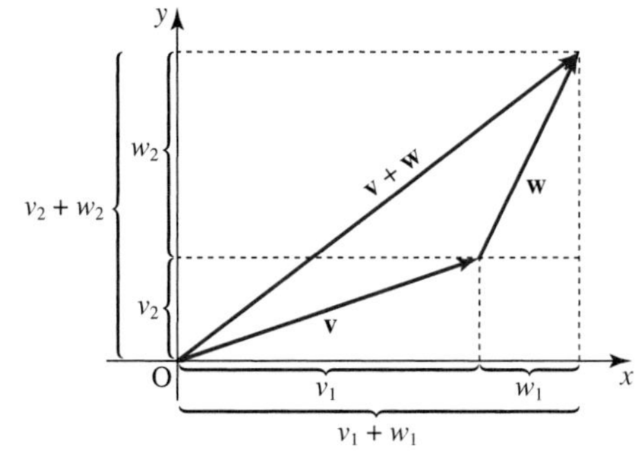

Figure 1.4 Triangle Law.

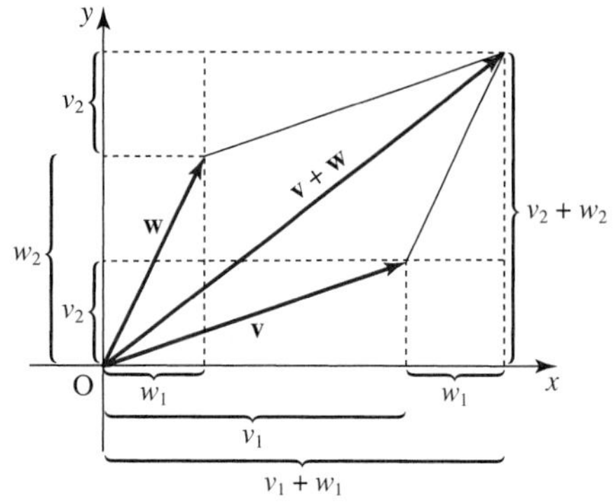

Figure 1.5 Parallelogram Law.

$(v_1 - w_1, v_2 - w_2)$. If $\mathbf{v} = (v_1, v_2, v_3)$ and $\mathbf{w} = (w_1, w_2, w_3)$ are in $\mathbb{R}^3$, then $\mathbf{v} + \mathbf{w} = (v_1 + w_1, v_2 + w_2, v_3 + w_3)$ and $\mathbf{v} - \mathbf{w} = (v_1 - w_1, v_2 - w_2, v_3 - w_3)$.

(b) If $\mathbf{v} = (v_1, v_2) \in \mathbb{R}^2$ and $\alpha \in \mathbb{R}$, then $\alpha\mathbf{v}$ is the vector in $\mathbb{R}^2$ defined by $\alpha\mathbf{v} = (\alpha v_1, \alpha v_2)$. If $\mathbf{v} = (v_1, v_2, v_3)$, then $\alpha\mathbf{v} = (\alpha v_1, \alpha v_2, \alpha v_3)$ for any real number α. ◀

The addition of vectors can be visualized as the Triangle Law or as the Parallelogram Law; see Figures 1.4 and 1.5. The details of the verification (similar triangles argument) are left as exercises.

THEOREM 1.1 Properties of Addition and Multiplication by Scalars

For all vectors $\mathbf{u}$, $\mathbf{v}$, and $\mathbf{w}$ in $\mathbb{R}^2$ (or, for all vectors $\mathbf{u}$, $\mathbf{v}$, and $\mathbf{w}$ in $\mathbb{R}^3$) and real numbers α and β, the following properties hold:

$$\mathbf{v} + \mathbf{w} = \mathbf{w} + \mathbf{v} \qquad \text{(commutativity)}$$

$$\mathbf{u} + (\mathbf{v} + \mathbf{w}) = (\mathbf{u} + \mathbf{v}) + \mathbf{w} \qquad \text{(associativity)}$$

$$\alpha(\mathbf{v} + \mathbf{w}) = \alpha\mathbf{v} + \alpha\mathbf{w} \qquad \text{(distributivity)}$$

$$(\alpha + \beta)\mathbf{v} = \alpha\mathbf{v} + \beta\mathbf{v} \qquad \text{(distributivity)}$$

$$(\alpha\beta)\mathbf{v} = \alpha(\beta\mathbf{v}).$$

If $\mathbf{0}$ denotes the zero vector, then $\mathbf{v} + \mathbf{0} = \mathbf{v}$. Finally, $1 \cdot \mathbf{v} = \mathbf{v}$. ◀

The product $(-1)\mathbf{v}$ is written as $-\mathbf{v}$. One could define the difference as $\mathbf{v} - \mathbf{w} = \mathbf{v} + (-\mathbf{w})$.

To show any of the identities, we have to write vectors in terms of their components, use the appropriate definitions of vector operations and properties of real numbers. For example, let us prove the first distributive law for vectors $\mathbf{v}$ and $\mathbf{w}$ in $\mathbb{R}^2$. Write $\mathbf{v} = (v_1, v_2)$ and $\mathbf{w} = (w_1, w_2)$ and compare the two sides. The left side is

$$\alpha(\mathbf{v} + \mathbf{w}) = \alpha((v_1, v_2) + (w_1, w_2)) = \alpha(v_1 + w_1, v_2 + w_2) = (\alpha(v_1 + w_1), \alpha(v_2 + w_2)).$$

Since the right side is

$$\alpha\mathbf{v} + \alpha\mathbf{w} = \alpha(v_1, v_2) + \alpha(w_1, w_2) = (\alpha v_1, \alpha v_2) + (\alpha w_1, \alpha w_2)$$
$$= (\alpha v_1 + \alpha w_1, \alpha v_2 + \alpha w_2),$$

it follows that the two sides are equal (by the distributive property of real numbers). The proof of the rest of the theorem is left as an exercise.

Theorem 1.1 shows that, as far as the operations of addition and multiplication by scalars are concerned, vectors behave in the same way as real numbers.

From the interpretation of the sum $\mathbf{v} + \mathbf{w}$ of $\mathbf{v}$ and $\mathbf{w}$ given by the Triangle Law (see Figure 1.4) and the fact that the sum of lengths of any two sides in a triangle is at least as large as the length of the third side, we get the *Triangle Inequality*

$$||\mathbf{v} + \mathbf{w}|| \le ||\mathbf{v}|| + ||\mathbf{w}||.$$

We say that the vectors $\mathbf{v}$ and $\mathbf{w}$ are *parallel* if there is a nonzero real number α such that $\mathbf{w} = \alpha\mathbf{v}$. The length of a vector $\alpha\mathbf{v}$ is computed to be

$$||\alpha\mathbf{v}|| = |\alpha|\; ||\mathbf{v}||.$$

If $\alpha > 0$, then $\mathbf{v}$ and $\alpha\mathbf{v}$ have the *same direction,* and if $\alpha < 0$, then they are of *opposite directions.* A vector whose length is 1 is called a *unit vector.* If $\mathbf{v}$ is a nonzero vector, then the vector $\mathbf{v}/||\mathbf{v}|| = (1/||\mathbf{v}||)\mathbf{v}$ is the unit vector in the same direction as $\mathbf{v}$. Constructing a unit vector $\mathbf{v}/||\mathbf{v}||$ from a nonzero vector $\mathbf{v}$ is sometimes called *normalizing a vector.*

A vector $\mathbf{v} = (v_1, v_2, v_3)$ in $\mathbb{R}^3$ can be written as

$$\mathbf{v} = (v_1, v_2, v_3) = v_1(1, 0, 0) + v_2(0, 1, 0) + v_3(0, 0, 1) = v_1\mathbf{i} + v_2\mathbf{j} + v_3\mathbf{k},$$

where $\mathbf{i} = (1, 0, 0)$, $\mathbf{j} = (0, 1, 0)$, and $\mathbf{k} = (0, 0, 1)$ are the *standard unit vectors* in $\mathbb{R}^3$. The set $\{\mathbf{i}, \mathbf{j}, \mathbf{k}\}$ is called the *standard basis of* $\mathbb{R}^3$ (the word *basis* refers to the fact that every vector in $\mathbb{R}^3$ can be expressed in terms of $\mathbf{i}$, $\mathbf{j}$, and $\mathbf{k}$). When the *order* of the basis vectors is specified (e.g., $\mathbf{i}$, $\mathbf{j}$, $\mathbf{k}$), we say that the space $\mathbb{R}^3$ is *oriented.* Similarly, if $\mathbf{v} \in \mathbb{R}^2$, then $\mathbf{v} = (v_1, v_2) = v_1(1, 0) + v_2(0, 1) = v_1\mathbf{i} + v_2\mathbf{j}$, where $\mathbf{i} = (1, 0)$ and $\mathbf{j} = (0, 1)$ are the standard unit vectors in $\mathbb{R}^2$. Vectors $\mathbf{i}$ and $\mathbf{j}$ (in that order) define an *orientation* of $\mathbb{R}^2$.

▶ EXAMPLE 1.3

Find the unit vector in the direction of $\mathbf{v} = \mathbf{i} - 2\mathbf{j} + \mathbf{k}$.

SOLUTION The length of $\mathbf{v}$ is $||\mathbf{v}|| = \sqrt{6}$, and therefore the vector

$$\frac{\mathbf{v}}{||\mathbf{v}||} = \frac{1}{\sqrt{6}}\mathbf{i} - \frac{2}{\sqrt{6}}\mathbf{j} + \frac{1}{\sqrt{6}}\mathbf{k}$$

is the required unit vector. ◀

▶ EXAMPLE 1.4

Find the vector $\mathbf{v}$ in $\mathbb{R}^2$ of length 4 whose direction makes an angle of $\pi/3$ radians (measured counterclockwise) with respect to the positive x-axis.

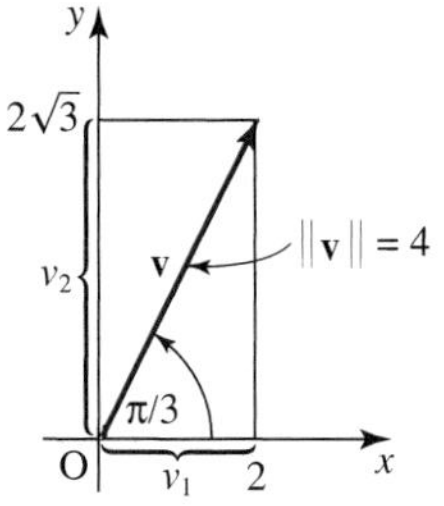

Figure 1.6 Vector **v** of Example 1.4.

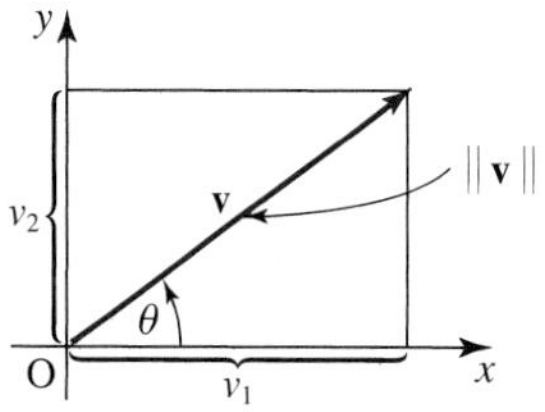

Figure 1.7 Quantities $||\mathbf{v}||$ and θ define the polar form of a vector.

SOLUTION See Figure 1.6. Since $||\mathbf{v}|| = 4$, it follows from $\cos(\pi/3) = v_1/||\mathbf{v}||$ that $v_1 = ||\mathbf{v}|| \cos(\pi/3) = 4(1/2) = 2$. Similarly, $v_2 = ||\mathbf{v}|| \sin(\pi/3) = 4(\sqrt{3}/2) = 2\sqrt{3}$. Hence, $\mathbf{v} = (2, 2\sqrt{3}) = 2\mathbf{i} + 2\sqrt{3}\mathbf{j}$. ◀

Let $\mathbf{v} = v_1\mathbf{i} + v_2\mathbf{j}$ be a nonzero vector in $\mathbb{R}^2$. Then $||\mathbf{v}|| \neq 0$, and (multiply and divide by $||\mathbf{v}||$)

$$\mathbf{v} = ||\mathbf{v}|| \left(\frac{v_1}{||\mathbf{v}||}\mathbf{i} + \frac{v_2}{||\mathbf{v}||}\mathbf{j} \right).$$

Now (see Figure 1.7) $v_1/||\mathbf{v}|| = \cos\theta$ and $v_2/||\mathbf{v}|| = \sin\theta$, where θ is the angle (measured counterclockwise) between the positive x-axis and the direction of $\mathbf{v}$. Consequently, $\mathbf{v}$ can be expressed in terms of $||\mathbf{v}||$ and θ as

$$\mathbf{v} = ||\mathbf{v}||(\cos\theta\mathbf{i} + \sin\theta\mathbf{j}).$$

This formula is called the *polar form of a vector,* since the components of $\mathbf{v}$ are expressed in polar coordinates with $r = ||\mathbf{v}||$. If $\mathbf{v}$ is a unit vector, then $||\mathbf{v}|| = 1$ and $\mathbf{v} = \cos\theta\mathbf{i} + \sin\theta\mathbf{j}$.

▶ EXERCISES 1.1

1. Find two vectors $\mathbf{v}$ and $\mathbf{w}$ in $\mathbb{R}^3$ such that $||\mathbf{v} + \mathbf{w}|| = ||\mathbf{v}|| + ||\mathbf{w}||$. Find two vectors $\mathbf{v}$ and $\mathbf{w}$ in $\mathbb{R}^2$ such that $||\mathbf{v} + \mathbf{w}|| < ||\mathbf{v}|| + ||\mathbf{w}||$, and find two other vectors $\mathbf{v}$ and $\mathbf{w}$ in $\mathbb{R}^2$ such that $||\mathbf{v} + \mathbf{w}|| < (||\mathbf{v}|| + ||\mathbf{w}||)/2$.

2. Show that $||\alpha\mathbf{v}|| = |\alpha|\ ||\mathbf{v}||$ for $\mathbf{v} \in \mathbb{R}^3$, and $\alpha \in \mathbb{R}$.

3. Show that $||\mathbf{v}|| = 0$ if and only if $\mathbf{v} = \mathbf{0}$ (in words, the only vector whose length is zero is the zero vector).

4. Find the Cartesian coordinates of points whose polar coordinates are $(0, \pi/2)$, $(10, \pi/2)$, $(2, 3\pi/4)$, $(1, \pi/6)$, and $(12, 3\pi/2)$.

5. Find the polar coordinates of points whose Cartesian coordinates are $(2, 2)$, $(-2, 2)$, $(2, -2)$, $(-2, -2)$.

6. Find the polar coordinates of points whose Cartesian coordinates are $(0, -3)$, $(1, \sqrt{3})$, $(-\sqrt{3}, -1)$, and $(-2, 0)$.

7. Using the Triangle Inequality, show that $||\mathbf{v} - \mathbf{w}|| \geq ||\mathbf{v}|| - ||\mathbf{w}||$ for $\mathbf{v}$ and $\mathbf{w}$ in $\mathbb{R}^2$ (or in $\mathbb{R}^3$). Give a geometric interpretation of this inequality.

8. Find two vectors $\mathbf{v}$ and $\mathbf{w}$ in $\mathbb{R}^2$ such that $||\mathbf{v} - \mathbf{w}|| = ||\mathbf{v}|| - ||\mathbf{w}||$.

9. Let $\mathbf{v}, \mathbf{w} \in \mathbb{R}^2$ be vectors such that $||\mathbf{v}|| = ||\mathbf{w}|| = 1$. What are the largest and the smallest possible values of $||\mathbf{v} + \mathbf{w}||$?

10. The sum of distances from the point (x, y) on an ellipse to the points $(-e, 0)$ and $(e, 0)$ (where $e > 0$) is constant and equal to ϵ. Use this characterization of the ellipse to show that its equation in the xy-plane is $x^2/a^2 + y^2/b^2 = 1$, where $a^2 = (\epsilon/2)^2$ and $b^2 = (\epsilon/2)^2 - e^2$.

Exercises 11 to 14: Find the length of the vector $\mathbf{v}$.

11. $\mathbf{v} = (0, 2, -1)$

12. $\mathbf{v} = \sin\theta\mathbf{i} + \cos\theta\mathbf{j} + \mathbf{k}$

13. $\mathbf{v} = \mathbf{w}/||\mathbf{w}||$, $\mathbf{w} = 3\mathbf{i} + 4\mathbf{k}$

14. $\mathbf{v} = \mathbf{w}/||\mathbf{w}||^2$, $\mathbf{w} = \mathbf{i} - \mathbf{j} + 2\mathbf{k}$

15. Using the Pythagorean Theorem, derive the formulas (1.3) for the distance between two points in $\mathbb{R}^2$ and $\mathbb{R}^3$.

16. Provide the details of the arguments that the formula for the sum of two vectors can be interpreted geometrically as the Triangle Law or as the Parallelogram Law.

17. Find the representatives of the vector $\mathbf{v} = (0, 2, -1)$ whose tails are located at $(0, 1, 1)$, $(0, 3, 0)$, $(8, 9, -4)$, and $(10, -1, 4)$.

18. Find the vector represented by the directed line segment $\overrightarrow{AB}$, where $A(3, 4)$ and $B(-1, 0)$. Find its representatives with tails located at $(0, 2)$, $(1, 1)$, and $(-4, -2)$.

19. Let $\mathbf{a} = 2\mathbf{i} - \mathbf{j} + \mathbf{k}$, $\mathbf{b} = \mathbf{k} - 3\mathbf{i}$, and $\mathbf{c} = 2\mathbf{i}$. Find vectors $\mathbf{a} - 2\mathbf{b}$, $\mathbf{a} - \mathbf{c}/||\mathbf{c}||$, $3\mathbf{a} + \mathbf{c} - \mathbf{j} + \mathbf{k}$ and the unit vector in the direction of $\mathbf{b} + 2\mathbf{a}$.

20. Find all vectors $\mathbf{v}$ of length 10 whose directions make an angle of $2\pi/3$ radians with respect to the positive x-axis.

21. Represent the vectors $-3\mathbf{i}$, $\mathbf{i}/2 - \mathbf{j}$, and $\mathbf{i} - 4\mathbf{j}$ in polar form.

22. Let $\mathbf{a} = 2\mathbf{i} + \mathbf{j}$ and $\mathbf{b} = -\mathbf{j} - 2\mathbf{k}$. Find the vector $\mathbf{x}$ such that $\mathbf{a} + 2(\mathbf{x} - \mathbf{b}) = 3\mathbf{x} + 2(\mathbf{a} - \mathbf{b})$.

23. Show that if $\mathbf{v} \in \mathbb{R}^3$ and $\alpha, \beta \in \mathbb{R}$, then $(\alpha + \beta)\mathbf{v} = \alpha\mathbf{v} + \beta\mathbf{v}$ and $(\alpha\beta)\mathbf{v} = \alpha(\beta\mathbf{v})$.

24. Explain how to construct geometrically the difference $\mathbf{v} - \mathbf{w}$ of two given vectors $\mathbf{v}$ and $\mathbf{w}$ in $\mathbb{R}^2$ (or in $\mathbb{R}^3$).

▶ 1.2 APPLICATIONS IN GEOMETRY AND PHYSICS

Vectors and vector operations introduced in the previous section will now be applied in a number of geometric and physical situations. We will start by computing equations of lines in $\mathbb{R}^2$ and $\mathbb{R}^3$. To make our exposition as clear as possible, we will not make a distinction between a vector and its representative. For example, when we say "a vector from A to B" or "a vector with initial point A" or "a tip of a vector $\mathbf{v}$," we are actually talking about the directed line segment (i.e., the "arrowed line") $\overrightarrow{AB}$ that is the representative of that vector. We should also keep in mind that it is often useful to identify vectors and points: to be more precise, a vector $\mathbf{v} = (v_1, v_2)$ in $\mathbb{R}^2$ can be thought of as a point (v_1, v_2) in a plane, and vice versa.

Let $\mathbf{v} = (v_1, v_2)$ be a nonzero vector in $\mathbb{R}^2$. Draw the line ℓ that contains the origin O and the point (v_1, v_2); see Figure 1.8(a) (in other words, the line ℓ "contains" $\mathbf{v}$, or the vector $\mathbf{v}$ "lies" on ℓ). Pick any point P on ℓ and consider the vector $\mathbf{w}$ determined by O and P.

Since $\mathbf{w}$ is parallel to $\mathbf{v}$ ($\mathbf{w}$ and $\mathbf{v}$ could fall in the same direction or in opposite directions), there is a real number t such that $\mathbf{w} = t\mathbf{v}$. In this way, for every point P on the line ℓ

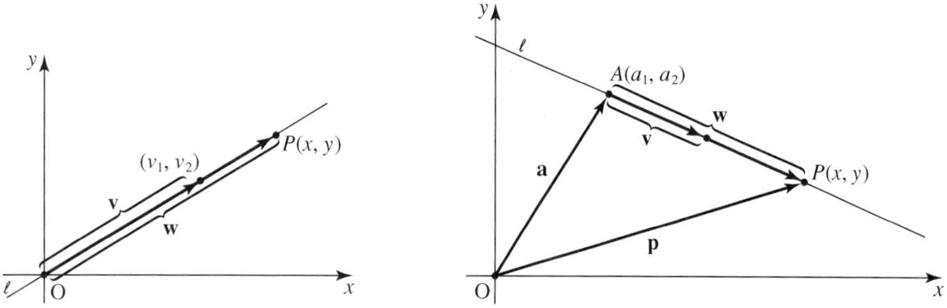

(a) ℓ goes through the origin. (b) ℓ goes through $A(a_1, a_2)$.

Figure 1.8 Line ℓ determined by a point and a vector.

we can find a unique scalar multiple $t\mathbf{v}$ of $\mathbf{v}$ whose tip is located at that point (the real number t is called a *parameter*). The line ℓ is "built" from all possible scalar multiples of $\mathbf{v}$:

$$\ell = \{t\mathbf{v} \mid t \in \mathbb{R}\}$$

(remember that we have identified a vector $t\mathbf{v}$ with its tip). Now pick a point $A(a_1, a_2)$ and a vector $\mathbf{v} = (v_1, v_2)$ and visualize $\mathbf{v}$ as its representative directed line segment that starts at A; see Figure 1.8(b). Let ℓ denote the line that contains A and whose direction is the same as $\mathbf{v}$, and let $P(x, y)$ be a point on it. By the Triangle Law, $\mathbf{p} = \mathbf{a} + \mathbf{w}$, where $\mathbf{p} = (x, y)$, $\mathbf{a} = (a_1, a_2)$, and $\mathbf{w}$ is the vector from A to P. Since $\mathbf{w}$ is parallel to $\mathbf{v}$, $\mathbf{w} = t\mathbf{v}$ for some $t \in \mathbb{R}$ and therefore $\mathbf{p} = \mathbf{a} + t\mathbf{v}$, $t \in \mathbb{R}$.

We have thus derived a *vector form of a parametric equation* of the line ℓ. As the parameter t takes on different values in $\mathbb{R}$, tips of vectors $\mathbf{p}$ describe points on the line. This equation is usually written as

$$\mathbf{l}(t) = \mathbf{a} + t\mathbf{v}, \quad t \in \mathbb{R},$$

or, in components as

$$\mathbf{l}(t) = (a_1 + tv_1, a_2 + tv_2), \quad t \in \mathbb{R},$$

or as

$$x = a_1 + tv_1, y = a_2 + tv_2, \quad t \in \mathbb{R}.$$

Any of the above forms is called a *parametric equation* (or *parametric equations*) *of a line.*

Similarly, we compute parametric equations of the line ℓ in $\mathbb{R}^3$ that contains a point $A(a_1, a_2, a_3)$ and whose direction is that of a vector $\mathbf{v} = (v_1, v_2, v_3)$ to be

$$\mathbf{l}(t) = \mathbf{a} + t\mathbf{v} = (a_1 + tv_1, a_2 + tv_2, a_3 + tv_3), \quad t \in \mathbb{R}.$$

The equation of a line can easily be memorized as "line equals point plus parameter times direction."

Note that we use ℓ to refer to a line viewed as a set of points, and $\mathbf{l}(t)$ to denote its parametric equation. Examples 1.7 through 1.9 will show that a single line has many representations in the form of parametric equations.

▶ EXAMPLE 1.5

Find a parametric equation of the line ℓ in $\mathbb{R}^3$ that passes through $(3, 2, -2)$ in the direction of the vector $\mathbf{i} - \mathbf{j} + 2\mathbf{k}$.

SOLUTION

An equation of this line is $\mathbf{l}(t) = \mathbf{a} + t\mathbf{v}$, where $\mathbf{a} = (3, 2, -2)$ and $\mathbf{v} = (1, -1, 2)$. Hence,

$$\mathbf{l}(t) = (3, 2, -2) + t(1, -1, 2) = (3 + t, 2 - t, -2 + 2t), \quad t \in \mathbb{R}$$

gives a possible form for the needed parametric equation. ◀

As our next example will illustrate, a parametric equation of a line establishes a one-to-one correspondence between points on the line and the values of the parameter.

▶ EXAMPLE 1.6

Consider the line ℓ in $\mathbb{R}^2$ given by $\mathbf{l}(t) = (2 - t, -1 + 3t)$, $t \in \mathbb{R}$.

Picking a value for t uniquely determines a point on the line. For instance, when $t = 2$, we get the point $\mathbf{l}(2) = (0, 5)$ on ℓ. Likewise, $\mathbf{l}(-3) = (5, -10)$ belongs to ℓ.

The correspondence between the values of t and the points on ℓ works the other way around as well: given a point P on ℓ, we can find a unique value of the parameter (call it t_0) such that $\mathbf{l}(t_0) = P$. For instance, take the point $P(-5, 20)$. Solving $2 - t = -5$ and $-1 + 3t = 20$ for t, we get $t = 7$, that is, $\mathbf{l}(7) = (-5, 20)$. Likewise, the point $(3, -4)$ is on ℓ, since $\mathbf{l}(-1) = (3, -4)$. ◀

▶ EXAMPLE 1.7

Find an equation of the line in $\mathbb{R}^2$ that contains $(1, 3)$ and $(0, -2)$.

SOLUTION

To determine an equation of a line, we need a point and a direction vector $\mathbf{v}$. Two points are given—choose one, say, $\mathbf{a} = (1, 3)$. The direction is that of the vector from $(1, 3)$ to $(0, -2)$. Consequently, $\mathbf{v} = (-1, -5)$ and an equation is

$$\mathbf{l}_1(t) = (1, 3) + t(-1, -5) = (1 - t, 3 - 5t), \quad t \in \mathbb{R}.$$

We made some choices along the way. Now let us recompute the equation using $(0, -2)$ as the point and keeping the same direction $\mathbf{v} = (-1, -5)$ [we could have taken the opposite direction $\mathbf{v} = (1, 5)$ as well]. This time,

$$\mathbf{l}_2(t) = (0, -2) + t(-1, -5) = (-t, -2 - 5t), \quad t \in \mathbb{R}.$$

The equations are not the same! For example, $\mathbf{l}_1(0) = (1, 3)$ and $\mathbf{l}_2(0) = (0, -2)$. However, $\mathbf{l}_2(-1) = (1, 3)$, that is, the point $(1, 3)$ does belong to both ℓ_1 and ℓ_2 (it is generated by different parameter values, but that is not important). Moreover, $\mathbf{l}_1(1) = (0, -2)$, so $(0, -2)$ also belongs to both ℓ_1 and ℓ_2. Since ℓ_1 and ℓ_2 have two points in common, they must be the same line!

To get a better feel for parametrizations, pick another point on the line ℓ_1, say, $\mathbf{l}_1(3) = (-2, -12)$. Now try to find the value of t so that $\mathbf{l}_2(t) = (-2, -12)$. Since $\mathbf{l}_2(t) = (-t, -2 - 5t) = (-2, -12)$, we get $t = 2$. Hence, $\mathbf{l}_2(2) = (-2, -12)$, so $(-2, -12)$ belongs to both lines.

Another way of proving that ℓ_1 and ℓ_2 represent the same line is to convert them to an explicit form: since $\mathbf{l}_1(t) = (1 - t, 3 - 5t)$, it follows that $x = 1 - t$ and $y = 3 - 5t$. Computing t from the first and substituting into the second equation, we get $y = 3 - 5(1 - x)$, that is, $y = 5x - 2$. Similarly, from $\mathbf{l}_2(t) = (-t, -2 - 5t)$, we get $x = -t$, $y = -2 - 5t$, and, after eliminating t, $y = -2 - 5(-x) = 5x - 2$. ◀

▶ EXAMPLE 1.8

Let us compute yet another form of parametric equations of the line in the previous example. Take $(1, 3)$ as the point and $\mathbf{v} = (1, 5)$ as the direction vector. By eliminating the parameter, compute an explicit equation of the line.

SOLUTION A parametric equation is

$$\mathbf{l}(t) = \mathbf{a} + t\mathbf{v} = (1, 3) + t(1, 5) = (1 + t, 3 + 5t), \quad t \in \mathbb{R}.$$

Eliminating t from $x = 1 + t,\ y = 3 + 5t$, we get $y = 3 + 5(x - 1)$ and $y = 5x - 2$. ◀

As we have just seen, the same line can be represented using parametric equations in many (actually, infinitely many) ways. Parametric equations of various curves will be studied in detail later.

▶ EXAMPLE 1.9

Find an equation of the line in $\mathbb{R}^3$ that contains $(1, 2, 0)$ and $(0, -2, 4)$.

SOLUTION As we have just seen, we might get different parametrizations, depending on our choices of the point and of the direction vector. However, they will all represent the same line. Choosing $(1, 2, 0)$ as the point and taking the direction to be determined by the vector from $(1, 2, 0)$ to $(0, -2, 4)$, we get

$$\mathbf{l}_1(t) = (1, 2, 0) + t(-1, -4, 4) = (1 - t, 2 - 4t, 4t), \quad t \in \mathbb{R}.$$

Using $(0, -2, 4)$ as the point and taking the direction to be the vector from $(0, -2, 4)$ to $(1, 2, 0)$, we get

$$\mathbf{l}_2(t) = (0, -2, 4) + t(1, 4, -4) = (t, -2 + 4t, 4 - 4t), \quad t \in \mathbb{R}$$

as another possible form of the parametric equations. ◀

Let $\mathbf{v}$ and $\mathbf{w}$ be nonzero, nonparallel vectors, visualized as directed line segments starting at the same point A. The point A and the tips of $\mathbf{v}$ and $\mathbf{w}$ determine the unique plane π, called the *plane (through A) spanned by* $\mathbf{v}$ *and* $\mathbf{w}$. We will now compute its parametric equation(s). Pick a point $P(x, y, z)$ on π and, drawing parallels to $\mathbf{v}$ and $\mathbf{w}$, construct the parallelogram whose diagonal is $\overline{AP}$, as shown in Figure 1.9.

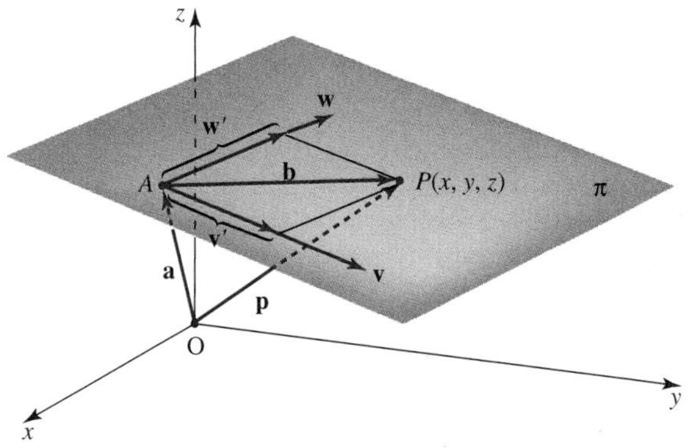

Figure 1.9 Plane through A spanned by $\mathbf{v}$ and $\mathbf{w}$.

The vector $\mathbf{b}$ that is represented by the directed line segment $\overrightarrow{AP}$ can be written as $\mathbf{b} = \mathbf{v}' + \mathbf{w}'$, where $\mathbf{v}'$ is parallel to $\mathbf{v}$ (and therefore $\mathbf{v}' = t\mathbf{v}$ for some real number t) and $\mathbf{w}'$ is parallel to $\mathbf{w}$ (hence $\mathbf{w}' = s\mathbf{w}$ for some s in $\mathbb{R}$). Consequently, $\mathbf{b} = t\mathbf{v} + s\mathbf{w}$, and since $\mathbf{p} = \mathbf{a} + \mathbf{b}$, it follows that the equation

$$\mathbf{p} = \mathbf{a} + t\mathbf{v} + s\mathbf{w}, \quad t, s \in \mathbb{R} \tag{1.4}$$

(where $\mathbf{a}$ is the vector from O to A) represents points in the plane π. The equation (or equations, if written in components) is (are) called *parametric equation(s) of the plane through* A, *spanned by* $\mathbf{v}$ *and* $\mathbf{w}$. This time, a single point is specified by assigning numerical values to *two* parameters, t and s.

▶ EXAMPLE 1.10

Find parametric equation(s) of the plane that contains the points $(0, 1, 0)$, $(2, 1, -3)$, and $(1, -1, 4)$.

SOLUTION We have to pick a point and two vectors. Depending on the choices we make, we will get different equations. However, the equation obtained from parametric equations by eliminating the parameters will be the same in all cases. Pick $(0, 1, 0)$ as the point, and let $\mathbf{v}$ be the vector from $(0, 1, 0)$ to $(2, 1, -3)$ [hence, $\mathbf{v} = (2, 0, -3)$] and let $\mathbf{w}$ be the vector from $(0, 1, 0)$ to $(1, -1, 4)$ [hence, $\mathbf{w} = (1, -2, 4)$]. It follows that the parametric equations

$$\mathbf{p} = \mathbf{a} + t\mathbf{v} + s\mathbf{w} = (0, 1, 0) + t(2, 0, -3) + s(1, -2, 4) = (2t + s, 1 - 2s, -3t + 4s),$$

for $t, s \in \mathbb{R}$ represent the given plane.

To obtain the equation in x, y, and z, we have to eliminate the parameters t and s from $x = 2t + s$, $y = 1 - 2s$ and $z = -3t + 4s$. From the second equation, $s = (1 - y)/2$. Substitute this into the first equation to get $x = 2t + (1 - y)/2$, and hence $t = (2x - 1 + y)/4$. Finally, substituting the expressions for t and s into the equation for z, we get $z = (-3)(2x - 1 + y)/4 + 4(1 - y)/2$, that is, $6x + 11y + 4z - 11 = 0$, or $z = (-6x - 11y + 11)/4$ as the explicit form. The reader is encouraged to try some other choices, compute the parametric equations, and check that they give the same explicit form.

Next, we will illustrate the use of vectors in various physical situations. Although we will concentrate on the dimension 2 (sometimes 3), all conclusions and arguments will remain valid for any dimension.

▶ EXAMPLE 1.11 Displacement (Position) and Relative Position

Suppose that we are located at the point $A(a_1, a_2)$ in the xy-plane and are walking toward $B(b_1, b_2)$ along a straight line. The vector $\mathbf{v} = (b_1 - a_1, b_2 - a_2)$, represented by the directed line segment $\overrightarrow{AB}$, gives the direction of our motion, and its magnitude $||\mathbf{v}||$ measures the distance we have to cover. The vector $\mathbf{v}$ is called the *displacement vector.* For example, suppose that $(2, 1)$ is the initial point and $(4, 4)$ the terminal point of our travel. The direction of our motion is given by $\mathbf{v} = (4 - 2, 4 - 1) = (2, 3)$, and the distance we have to cover is $||\mathbf{v}|| = \sqrt{4 + 9} = \sqrt{13}$ units.

Now suppose that a cat and a dog approach the same point H (see Figure 1.10), and that, at a particular moment, their displacements from H are given by the vectors $\mathbf{c}$ and $\mathbf{d}$. Their *relative position* at that moment is given by the vector $\mathbf{v}$ that joins the tails of $\mathbf{c}$ and $\mathbf{d}$. By the Triangle Law, $\mathbf{v} + \mathbf{d} = \mathbf{c}$, that is, the difference

$$\mathbf{v} = \mathbf{c} - \mathbf{d}$$

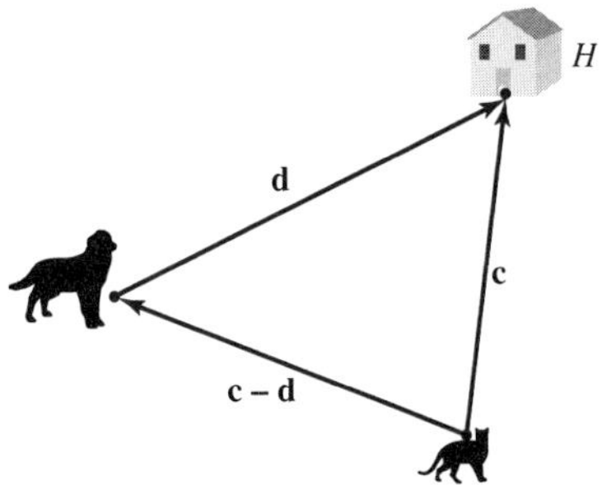

Figure 1.10 Relative position of the cat and the dog from Example 1.11.

is the vector giving their relative position (the distance between them is equal to $||\mathbf{v}||$). We could have taken $\mathbf{v}$ as the vector joining the tail of $\mathbf{d}$ with the tail of $\mathbf{c}$, in which case $\mathbf{v} = \mathbf{d} - \mathbf{c}$. Anyway, a relative position is always computed as the difference of displacement vectors. ◀

▶ EXAMPLE 1.12 Velocity and Speed

The velocity of an object moving with constant speed along a straight line can be represented as a vector: its direction is the direction of the motion and its magnitude is the speed.

(a) A particle moves along the line $y = 2x$ in the first quadrant away from the origin with speed 10 units/s. Find its position (i.e., the displacement) 2 s later.

(b) A particle initially located at $(3, -2, -1)$ starts moving along the line specified by the velocity $\mathbf{v} = \mathbf{i} + 2\mathbf{j} + \mathbf{k}$ units/s. Find its position 5 s later.

(c) A particle initially at $(1, 4)$ starts moving with speed 3 m/s in the direction of $3\mathbf{i} - 4\mathbf{j}$. Find its position 10 s later.

SOLUTION

(a) We have to find the velocity vector $\mathbf{v}$ first. Pick any point (other than the origin) on the line $y = 2x$, say, $(1, 2)$. Then $\mathbf{w} = \mathbf{i} + 2\mathbf{j}$ is a vector parallel to $\mathbf{v}$, but with magnitude $||\mathbf{w}|| = \sqrt{5}$. Since the speed is 10 units/s, that is, $||\mathbf{v}|| = 10$, we must take

$$\mathbf{v} = \frac{10}{\sqrt{5}}\mathbf{w} = \frac{10}{\sqrt{5}}\mathbf{i} + \frac{20}{\sqrt{5}}\mathbf{j} = 2\sqrt{5}\mathbf{i} + 4\sqrt{5}\mathbf{j}$$

(check: $||\mathbf{v}|| = |10/\sqrt{5}|\ ||\mathbf{w}|| = (10/\sqrt{5})\sqrt{5} = 10$). Using the formula

$$\textbf{displacement vector} = \textbf{velocity vector} \cdot \text{time}, \tag{1.5}$$

we finally see that the displacement $\mathbf{d}$ of the particle 2 s later is

$$\mathbf{d} = 2\mathbf{v} = \frac{20}{\sqrt{5}}\mathbf{i} + \frac{40}{\sqrt{5}}\mathbf{j} = 4\sqrt{5}\mathbf{i} + 8\sqrt{5}\mathbf{j},$$

that is, the particle is located at the point $(4\sqrt{5}, 8\sqrt{5})$.

(b) The equation of the line $\mathbf{l}(t)$ (that is the trajectory of the motion of the particle) is given by $\mathbf{l}(t) = \mathbf{a} + t\mathbf{v}$, where $\mathbf{a} = (3, -2, -1)$ and $\mathbf{v} = (1, 2, 1)$. Hence,

$$\mathbf{l}(t) = (3, -2, -1) + t(1, 2, 1) = (3 + t, -2 + 2t, -1 + t).$$

Now, $\mathbf{l}(0) = (3, -2, -1)$ is the initial position. Exactly 5 s later the particle is located at $\mathbf{l}(5) = (8, 8, 4)$.

Here is another way of solving this problem: given the velocity $\mathbf{v}$, the displacement vector (5 s later) of the particle is $\mathbf{d} = 5\mathbf{v} = (5, 10, 5)$. Since the particle is initially at $(3, -2, -1)$, its location 5 s later will be

$$\textbf{initial point} + \textbf{displacement vector} = (3, -2, -1) + (5, 10, 5) = (8, 8, 4).$$

(c) The velocity vector $\mathbf{v}$ of the particle has to be parallel to $\mathbf{w} = 3\mathbf{i} - 4\mathbf{j}$ (hence, $\mathbf{v} = \alpha\mathbf{w}$ for some real number α) and $||\mathbf{v}|| = 3$. Since $||\mathbf{w}|| = 5$, it follows that $\mathbf{v} = (3/5)\mathbf{w} = (3/5)(3\mathbf{i} - 4\mathbf{j}) = (9/5)\mathbf{i} - (12/5)\mathbf{j}$ is the velocity vector. The displacement 10 s later [see (1.5)] is $\mathbf{d} = 10\mathbf{v} = 18\mathbf{i} - 24\mathbf{j}$, and the position of the particle is

$$\mathbf{p} = \textbf{initial point} + \textbf{displacement vector} = (\mathbf{i} + 4\mathbf{j}) + (18\mathbf{i} - 24\mathbf{j}) = 19\mathbf{i} - 20\mathbf{j}.$$

As in (b), we could have computed the equation of the line of motion [with $\mathbf{a} = \mathbf{i} + 4\mathbf{j}$ and $\mathbf{v} = (9/5)\mathbf{i} - (12/5)\mathbf{j}$], thus obtaining

$$\mathbf{p}(t) = (\mathbf{i} + 4\mathbf{j}) + t\left(\tfrac{9}{5}\mathbf{i} - \tfrac{12}{5}\mathbf{j}\right) = \left(1 + \tfrac{9}{5}t\right)\mathbf{i} + \left(4 - \tfrac{12}{5}t\right)\mathbf{j},$$

and then substituted $t = 10$ to get the position $\mathbf{p}(10) = 19\mathbf{i} - 20\mathbf{j}$ of the particle 10 s later. ◀

▶ EXAMPLE 1.13

A wind is blowing in the northeast direction at a constant speed of 80 km/h. Find the vector in $\mathbb{R}^2$ that describes the velocity of the wind.

SOLUTION Assume that the y-axis points to the north and the x-axis to the east. In this case, the northeast direction is described by the angle of $\pi/4$ rad with respect to the positive x-axis. Using the polar form of a vector introduced in the previous section, we get $\mathbf{v} = ||\mathbf{v}||(\cos(\pi/4)\mathbf{i} + \sin(\pi/4)\mathbf{j}) = 80((\sqrt{2}/2)\mathbf{i} + (\sqrt{2}/2)\mathbf{j})$. Consequently, the vector $\mathbf{v} = 40\sqrt{2}\mathbf{i} + 40\sqrt{2}\mathbf{j}$ describes the velocity of the wind. ◀

▶ EXAMPLE 1.14 Relative Speed and Relative Velocity

Suppose that two cars, V and W, move along the same straight line with constant speeds v and w. At time t, V and W have covered the distances vt and wt, respectively, and the change in their (directed) distance as seen by the driver of V (i.e., measured from V to W) is $wt - vt$. It follows that the driver of V sees the car W moving with the speed (called the *relative speed of W with respect to V*)

$$\frac{wt - vt}{t} = w - v$$

(we have used the fact that speed = distance/time).

Now assume that the motions take place along the x-axis. The velocities of V and W are $\mathbf{v} = v\mathbf{i}$ and $\mathbf{w} = w\mathbf{i}$, and the *relative velocity of W with respect to V* is $w\mathbf{i} - v\mathbf{i} = \mathbf{w} - \mathbf{v}$.

In general, by analyzing each component of the velocity vectors $\mathbf{v}$ and $\mathbf{w}$ of V and W in this way, we will determine that the difference $\mathbf{w} - \mathbf{v}$ represents the relative velocity of W with respect to V [or the relative velocity of W as seen (experienced) by V]. ◀

▶ EXAMPLE 1.15 Physical Forces

A force of constant magnitude and direction can be thought of as a vector. Suppose that the vectors $\mathbf{F}_1, \ldots, \mathbf{F}_n$ define a system of n concurrent forces (i.e., all are applied at the same point A). The action of this system of n forces can be represented as the single force

$$\mathbf{F} = \mathbf{F}_1 + \cdots + \mathbf{F}_n$$

acting at A, called the *resultant force* of the system. If the resultant force is $\mathbf{0}$, we say that the system $\mathbf{F}_1, \ldots, \mathbf{F}_n$ of concurrent forces is in *equilibrium*.

Assume that the forces $\mathbf{F}_1$, $\mathbf{F}_2$, and $\mathbf{F}_3$ of magnitudes 2 N, 3 N, and 4 N (N denotes newton, a unit of force) are applied at the origin in the directions of $\pi/3$, $-\pi/2$, and $2\pi/3$ rad, respectively, with respect to the positive x-axis. Find the force $\mathbf{F}_4$ needed to keep the system $\mathbf{F}_1, \mathbf{F}_2, \mathbf{F}_3$ in equilibrium.

SOLUTION

We have to express $\mathbf{F}_1, \mathbf{F}_2, \mathbf{F}_3$ as vectors first. Using the polar form (see the end of Section 1.1), we get

$$\mathbf{F}_1 = ||\mathbf{F}_1|| \left(\cos \tfrac{\pi}{3}\mathbf{i} + \sin \tfrac{\pi}{3}\mathbf{j}\right) = 2\left(\frac{1}{2}\mathbf{i} + \frac{\sqrt{3}}{2}\mathbf{j}\right) = \mathbf{i} + \sqrt{3}\mathbf{j},$$

$$\mathbf{F}_2 = ||\mathbf{F}_2|| \left(\cos\left(-\tfrac{\pi}{2}\right)\mathbf{i} + \sin\left(-\tfrac{\pi}{2}\right)\mathbf{j}\right) = -3\mathbf{j}$$

and

$$\mathbf{F}_3 = ||\mathbf{F}_3|| \left(\cos \tfrac{2\pi}{3}\mathbf{i} + \sin \tfrac{2\pi}{3}\mathbf{j}\right) = 4\left(-\frac{1}{2}\mathbf{i} + \frac{\sqrt{3}}{2}\mathbf{j}\right) = -2\mathbf{i} + 2\sqrt{3}\mathbf{j}.$$

Forces $\mathbf{F}_1$, $\mathbf{F}_2$, and $\mathbf{F}_3$ are shown in Figure 1.11.

The resultant force of this system is

$$\mathbf{F} = \mathbf{F}_1 + \mathbf{F}_2 + \mathbf{F}_3 = (\mathbf{i} + \sqrt{3}\mathbf{j}) + (-3\mathbf{j}) + (-2\mathbf{i} + 2\sqrt{3}\mathbf{j}) = -\mathbf{i} + (-3 + 3\sqrt{3})\mathbf{j}.$$

Therefore, the force $\mathbf{F}_4 = \mathbf{i} - (-3 + 3\sqrt{3})\mathbf{j} = \mathbf{i} + (3 - 3\sqrt{3})\mathbf{j}$ has to be applied at the origin in order to keep the system $\mathbf{F}_1, \mathbf{F}_2, \mathbf{F}_3$ in equilibrium (see Figure 1.11).

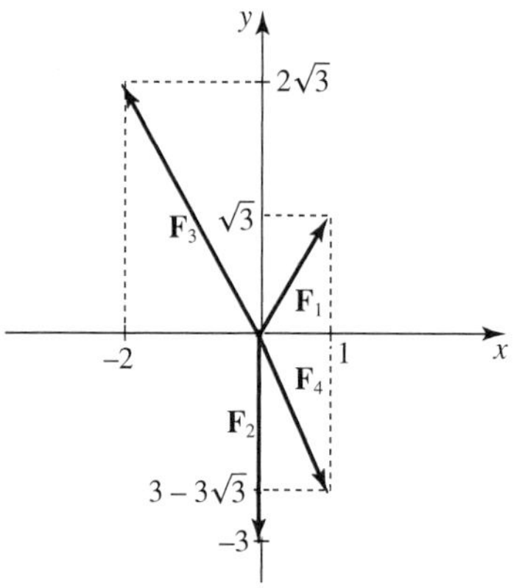

Figure 1.11 System of concurrent forces $\mathbf{F}_1$, $\mathbf{F}_2$, $\mathbf{F}_3$, $\mathbf{F}_4$. ◀

▶ EXAMPLE 1.16 Principle of Superposition

Coulomb's Law in electrostatics says that the electrostatic force on a charge q due to a charge Q is given by

$$\mathbf{F} = \frac{1}{4\pi\epsilon_0}\frac{Qq}{d(q, Q)^2}\mathbf{u},$$

where $d(q, Q)$ is the distance between q and Q, $\mathbf{u}$ is the unit vector in the direction from Q to q, and ϵ_0 is a constant [called the *permittivity of vacuum*; in the SI system it can be determined from $\epsilon_0 = 10^7/(4\pi c^2)$, where c is the speed of light in vacuum].

Now assume that $\mathbf{F}_1$ is the force on q due to a charge Q_1 with no other charges present and $\mathbf{F}_2$ is the force on q due to a charge Q_2 with no other charges present. The *Principle of Superposition* (which is one of the basic principles in electrostatics) states that the total force $\mathbf{F}$ on q when both Q_1 and Q_2 are present is the sum $\mathbf{F}_1 + \mathbf{F}_2$ of the forces $\mathbf{F}_1$ and $\mathbf{F}_2$. Let us explain this: with only Q_1 present, the force is $\mathbf{F} = \mathbf{F}_1$. When another charge Q_2 is included, it does not *change* the effect of the force $\mathbf{F}_1$ on q. Instead, its contribution to the total force is contained in a separate term $\mathbf{F}_2$ that is vectorially *added* to $\mathbf{F}_1$. ◀

▶ EXAMPLE 1.17 Center of Mass

For a particle of mass M located at the point (a_1, a_2, a_3), we define the three moments: the *moment about the xy-plane* is the scalar Ma_3, the *moment about the yz-plane* is the scalar Ma_1, and the *moment about the xz-plane* is the scalar Ma_2. Now assume that n particles of mass m_i are located at (x_i, y_i, z_i), $i = 1, \dots, n$, in space in such a way that their relative positions do not change (e.g., m_i could be the parts of a rigid body rotating in space).

When some aspects of the system $m_1, \dots, m_n$ are considered, it turns out that its behavior is the same as that of the single mass $M = m_1 + \cdots + m_n$ located at the point $C_M(x_M, y_M, z_M)$ called the *center of mass* (or the *centroid*), which is defined as follows: the moment of M about any plane is equal to the sum of the moments of m_i with respect to the same plane. Applying this definition to the xy-plane, we get

$$Mz_M = m_1 z_1 + \cdots + m_n z_n.$$

Similarly (considering moments about the xz-plane), we get

$$My_M = m_1 y_1 + \cdots + m_n y_n$$

and (considering moments about the yz-plane)

$$Mx_M = m_1 x_1 + \cdots + m_n x_n$$

(recall that $M = m_1 + \cdots + m_n$). The coordinates of the center of mass C_M are therefore

$$(x_M, y_M, z_M) = \left(\frac{m_1 x_1 + \cdots + m_n x_n}{M}, \frac{m_1 y_1 + \cdots + m_n y_n}{M}, \frac{m_1 z_1 + \cdots + m_n z_n}{M} \right).$$

The position vector $\mathbf{c}_M$ of the center of mass C_M is then

$$\begin{aligned}
\mathbf{c}_M &= x_M \mathbf{i} + y_M \mathbf{j} + z_M \mathbf{k} \\
&= \frac{1}{M}((m_1 x_1 + \cdots + m_n x_n)\mathbf{i} + (m_1 y_1 + \cdots + m_n y_n)\mathbf{j} + (m_1 z_1 + \cdots + m_n z_n)\mathbf{k}) \\
&= \frac{1}{M}(m_1(x_1\mathbf{i} + y_1\mathbf{j} + z_1\mathbf{k}) + \cdots + m_n(x_n\mathbf{i} + y_n\mathbf{j} + z_n\mathbf{k})) \\
&= \frac{1}{M}(m_1\mathbf{r}_1 + \cdots + m_n\mathbf{r}_n), \qquad (1.6)
\end{aligned}$$

where $\mathbf{r}_i = (x_i, y_i, z_i)$ is the position vector of the mass m_i, $i = 1, \dots, n$. ◀

▶ EXAMPLE 1.18

Three stones of mass 2 kg, 4 kg, and 3 kg are placed at (2, 1, 0), (−1, −1, 5), and (0, 0, 3), respectively. Where should a fourth stone of mass 1 kg be placed so that the center of mass of the system is at the origin?

SOLUTION Denote the masses by $m_1 = 2$, $m_2 = 4$, $m_3 = 3$, and $m_4 = 1$ and their positions by $\mathbf{r}_1 = (2, 1, 0)$, $\mathbf{r}_2 = (-1, -1, 5)$, $\mathbf{r}_3 = (0, 0, 3)$, and $\mathbf{r}_4 = (x_4, y_4, z_4)$. Using (1.6) with $M = 2 + 4 + 3 + 1 = 10$, we get

$$\begin{aligned}(0, 0, 0) &= \frac{1}{10}(2(2\mathbf{i} + \mathbf{j}) + 4(-\mathbf{i} - \mathbf{j} + 5\mathbf{k}) + 3(3\mathbf{k}) + 1(x_4\mathbf{i} + y_4\mathbf{j} + z_4\mathbf{k})) \\ &= \frac{1}{10}((x_4)\mathbf{i} + (-2 + y_4)\mathbf{j} + (29 + z_4)\mathbf{k}).\end{aligned}$$

Hence, $(x_4 = 0, y_4 = 2, z_4 = -29)$ is the location of the fourth stone. ◀

▶ EXAMPLE 1.19 Vectors on Meterorological Maps

Using arrows to represent vectors is a common practice in applications. However, there are situations—such as showing wind speed and direction on weather maps—where this approach has some disadvantages. For instance, to determine the magnitude of a vector, we need to compare its size to a scale (which needs to be given). Sometimes, a map gets cluttered with a large number of long arrows, or it might contain small arrows that are hard to read.

Several alternatives to using arrows have been used on meteorological maps; see Figure 1.12. A *hunting arrow* shows the direction of the wind, whereas its speed is determined from the number of "feathers." A *barb* is a simpified version, where the "feathers" are shown on only one side, and the arrow is dropped (presently—besides the common arrow—this is the most often used symbol). To represent a vector as a *whisker*, we indicate its initial point, use length to show magnitude, but drop the arrow (this representation, intuitively, resembles a wind sock that has been stretched out and blown downwind). ◀

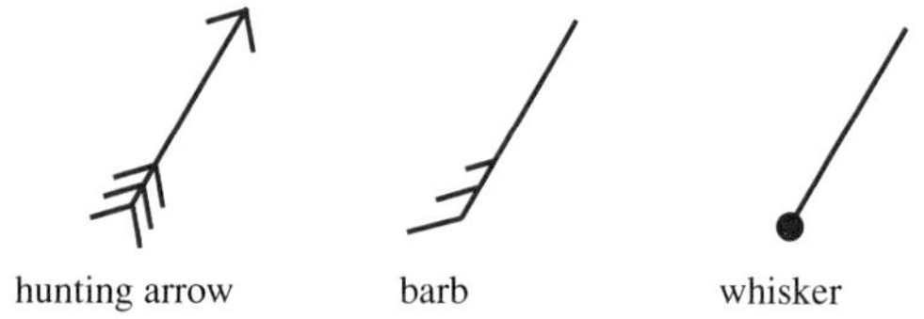

Figure 1.12 Representations of a vector used in meteorological maps.

▶ EXERCISES 1.2

1. Compute another form of a parametric equation of the line in Example 1.7: take $(1, 3)$ as the point and the multiple $3\mathbf{v}$ of $\mathbf{v} = (-1, -5)$ as the direction (clearly, $3\mathbf{v}$ and $\mathbf{v}$ have the same direction). Explain how to get infinitely many parametric equations of the line.

2. Find a parametric equation of the line that contains the points $(3, 2, 0)$ and $(0, -1, -1)$.

3. Find a parametric equation of the line that contains the points $(1, 1)$ and $(-2, 4)$. What values of the parameter should be used to describe the half-line starting at $(1, 1)$ and going through $(-2, 4)$? What values of the parameter should be used to describe the line segment between $(1, 1)$ and $(-2, 4)$, including both endpoints?

4. Compute a parametric equation of the line ℓ that contains the points $(2, -2, 0)$ and $(1, 1, 4)$. In this case, the parameter can be any real number. By restricting its values, describe the set of all points on the line segment defined by the two given points (that includes both of them).

5. Compute a parametric equation of the line ℓ that contains the points $(2, 1)$ and $(-1, 5)$. By restricting the values of the parameter, describe the half-line ℓ' with the initial point at $(2, 1)$ in the direction toward $(-1, 5)$.

6. Compute another set of parametric equations of the plane in Example 1.10: take $(1, -1, 4)$ as the point and the vectors from $(1, -1, 4)$ to $(2, 1, -3)$ and from $(1, -1, 4)$ to $(0, 1, 0)$ as spanning vectors. Compute the explicit form of the equation. Explain how to get infinitely many parametric equations of the plane.

7. Consider the rectangle that has three vertices located at $(1, 1)$, $(4, 1)$, and $(1, 2)$. By imitating parts of the construction of parametric equations of the plane, obtain a vector description of all points that belong to that rectangle (i.e., that are either inside the rectangle or on its boundary).

8. The tips of vectors $\mathbf{v} = (1, 4, 2)$ and $\mathbf{w} = (-2, 0, 3)$ (whose initial point is the origin) and the origin define a parallelogram (called the parallelogram spanned by $\mathbf{v}$ and $\mathbf{w}$). By adjusting the construction of parametric equations of the plane, obtain the vector description of all points that belong to this parallelogram.

9. Find the distance and relative position of two cars approaching the same intersection A if their displacement vectors from A are $\mathbf{i} + 3\mathbf{j} - \mathbf{k}$ and $\mathbf{j} + 2\mathbf{k}$.

10. A particle is ejected in the direction of $\mathbf{i} + 2\mathbf{j} - 2\mathbf{k}$ with speed 12 units/s. Find its position 1 min later.

11. A particle moves from the point $(3, 2, 4)$ with the constant velocity $3\mathbf{i} - \mathbf{k}$. When and where is it going to cross the xy-plane?

12. A particle moves from the point $(-2, 1)$ in the direction of $\pi/3$ rad with respect to the positive x-axis at a speed of 3 units/s. Find its velocity and location 10 s later.

13. Find the vector $\mathbf{F}$ that describes a force of magnitude 10 N acting on an object located at the origin having an angle of $\pi/6$ with respect to the positive x-axis.

14. A car A is moving north with speed 100 km/h and a car B is moving northeast with speed 80 km/h. Find the relative velocity of B as seen by the driver of A.

15. Find the center of mass of a system of four masses of 2 kg each located at the vertices of the parallelogram (one vertex is at the origin) spanned by $\mathbf{i} - 2\mathbf{j}$ and $\mathbf{i} + \mathbf{j}$.

16. Two objects of mass 2 kg each are placed at $(-1, 3)$ and $(1, -2)$. Where should a third object, of mass 3 kg, be placed so that the center of mass of the system occurs at the point $(1, 1)$?

17. Two stones of mass 2 kg each are placed at $(-1, 0)$ and $(1, -2)$. The third stone is placed at $(2, 1)$. Determine its mass from the requirement that the center of mass of this system be at $(1, 0)$.

18. Three forces of 3 N, 4 N, and 6 N are applied at the origin in the directions toward the points $(1, 1)$, $(0, -4)$, and $(10, 1)$, respectively. Find the resultant force of this system.

19. Three objects of mass 1 kg are placed at the vertices of an equilateral triangle with sides of length 2 m. Find the center of mass of the system.

20. Two cats walk away from each other along perpendicular paths with speeds of 10 km/h and 12 km/h. Find the relative velocity of the faster cat as seen by the slower cat.

21. Find the resultant force of a system of four forces, each of magnitude 5 N, applied at the origin at angles $\pi/10$, $\pi/5$, $\pi/2$, and $13\pi/10$ with respect to the positive x-axis.

▶ 1.3 THE DOT PRODUCT

We are going to define two more operations with vectors that produce meaningful quantities and are used in mathematics and various applications. In this section, we introduce the dot product, and in Section 1.5 we introduce the cross product.

DEFINITION 1.4 Dot Product

Let $\mathbf{v} = (v_1, \ldots, v_n)$ and $\mathbf{w} = (w_1, \ldots, w_n)$ be vectors in $\mathbb{R}^n$, $n \geq 2$. The *dot product* of $\mathbf{v}$ and $\mathbf{w}$ is the real number $\mathbf{v} \cdot \mathbf{w}$ defined by

$$\mathbf{v} \cdot \mathbf{w} = v_1 w_1 + \cdots + v_n w_n.$$

In particular, if $\mathbf{v}, \mathbf{w} \in \mathbb{R}^2$, then

$$\mathbf{v} \cdot \mathbf{w} = (v_1\mathbf{i} + v_2\mathbf{j}) \cdot (w_1\mathbf{i} + w_2\mathbf{j}) = v_1 w_1 + v_2 w_2,$$

and

$$\mathbf{v} \cdot \mathbf{w} = (v_1\mathbf{i} + v_2\mathbf{j} + v_3\mathbf{k}) \cdot (w_1\mathbf{i} + w_2\mathbf{j} + w_3\mathbf{k}) = v_1 w_1 + v_2 w_2 + v_3 w_3$$

if $\mathbf{v}$ and $\mathbf{w}$ are vectors in $\mathbb{R}^3$.

Let us emphasize that the dot product of two vectors is a *real number*. This is the reason why it is also called the *scalar product* (sometimes, the term *inner product* is used). Other notation for the dot product $\mathbf{v} \cdot \mathbf{w}$ includes angle brackets, $\langle \mathbf{v}, \mathbf{w} \rangle$, or (unfortunately) parentheses, $(\mathbf{v}, \mathbf{w})$. Applying the definition, we compute the dot products of the standard basis vectors $\mathbf{i}$, $\mathbf{j}$, and $\mathbf{k}$ to be $\mathbf{i} \cdot \mathbf{i} = 1, \mathbf{j} \cdot \mathbf{j} = 1, \mathbf{k} \cdot \mathbf{k} = 1, \mathbf{i} \cdot \mathbf{j} = \mathbf{j} \cdot \mathbf{i} = 0, \mathbf{i} \cdot \mathbf{k} = \mathbf{k} \cdot \mathbf{i} = 0$, and $\mathbf{j} \cdot \mathbf{k} = \mathbf{k} \cdot \mathbf{j} = 0$.

THEOREM 1.2 Properties of the Dot Product

Assume that $\mathbf{u}$, $\mathbf{v}$, and $\mathbf{w}$ are vectors in $\mathbb{R}^n$ (for $n \geq 2$), and α is a real number. The dot product is *commutative*

$$\mathbf{v} \cdot \mathbf{w} = \mathbf{w} \cdot \mathbf{v}$$

and *distributive* with respect to addition

$$\mathbf{u} \cdot (\mathbf{v} + \mathbf{w}) = \mathbf{u} \cdot \mathbf{v} + \mathbf{u} \cdot \mathbf{w}$$

and scalar multiplication

$$(\alpha\mathbf{u}) \cdot \mathbf{v} = \alpha(\mathbf{u} \cdot \mathbf{v}) = \mathbf{u} \cdot (\alpha\mathbf{v}).$$

Moreover,

$$\mathbf{0} \cdot \mathbf{v} = 0$$

($\mathbf{0}$ is the zero vector) and

$$\mathbf{v} \cdot \mathbf{v} = ||\mathbf{v}||^2.$$

If $\mathbf{v}$ and $\mathbf{w}$ are parallel, then

$$\mathbf{v} \cdot \mathbf{w} = ||\mathbf{v}||\ ||\mathbf{w}|| \tag{1.7}$$

if $\mathbf{v}$ and $\mathbf{w}$ have the same direction, and

$$\mathbf{v} \cdot \mathbf{w} = -||\mathbf{v}||\ ||\mathbf{w}|| \tag{1.8}$$

if they have opposite directions.

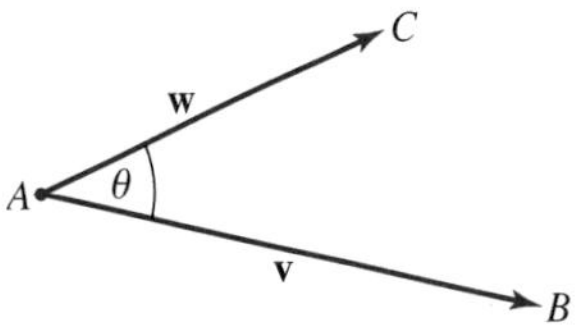

Figure 1.13 Angle between vectors.

The proof of the theorem is left as an exercise. All identities can be checked by writing the vectors involved in terms of their components, using the definitions of vector operations and the properties of real numbers. As an illustration, let us verify the last two statements for vectors $\mathbf{v} = (v_1, v_2, v_3)$ and $\mathbf{w} = (w_1, w_2, w_3)$ in $\mathbb{R}^3$ (the same proof works in any dimension; we choose $\mathbb{R}^3$ for convenience). Since $\mathbf{v}$ and $\mathbf{w}$ are parallel, there exists $\alpha \in \mathbb{R}$ such that $\mathbf{w} = \alpha\mathbf{v}$. Then $w_1 = \alpha v_1$, $w_2 = \alpha v_2$, and $w_3 = \alpha v_3$ and

$$\mathbf{v} \cdot \mathbf{w} = (v_1\mathbf{i} + v_2\mathbf{j} + v_3\mathbf{k}) \cdot (\alpha v_1\mathbf{i} + \alpha v_2\mathbf{j} + \alpha v_3\mathbf{k}) = \alpha(v_1^2 + v_2^2 + v_3^2).$$

We have used the distributivity (that allowed us to multiply each term in the first factor with each term in the second factor) and the results on dot products of standard basis vectors $\mathbf{i}$, $\mathbf{j}$, and $\mathbf{k}$. On the other hand (since $\sqrt{\alpha^2} = |\alpha|$),

$$||\mathbf{v}||\ ||\mathbf{w}|| = \sqrt{v_1^2 + v_2^2 + v_3^2} \cdot \sqrt{(\alpha v_1)^2 + (\alpha v_2)^2 + (\alpha v_3)^2} = |\alpha|\ (v_1^2 + v_2^2 + v_3^2).$$

If $\mathbf{v}$ and $\mathbf{w}$ have the same direction, then $\alpha > 0$, and consequently, $|\alpha| = \alpha$ and $\mathbf{v} \cdot \mathbf{w} = ||\mathbf{v}||\ ||\mathbf{w}||$. Otherwise, if $\mathbf{v}$ and $\mathbf{w}$ have opposite directions, then $\alpha < 0$, $|\alpha| = -\alpha$ and hence $\mathbf{v} \cdot \mathbf{w} = -||\mathbf{v}||\ ||\mathbf{w}||$.

Our next theorem gives a geometric interpretation of the dot product, from which a number of useful consequences follow. Before introducing the theorem, we have to define the angle between vectors. Let $\mathbf{v}$ and $\mathbf{w}$ be vectors in $\mathbb{R}^2$ or in $\mathbb{R}^3$ represented by the directed line segments $\overrightarrow{AB}$ and $\overrightarrow{AC}$ that start at the same point A; see Figure 1.13.

The angle θ between $\mathbf{v}$ and $\mathbf{w}$ is defined as the smaller of the two angles (in the plane through A spanned by $\mathbf{v}$ and $\mathbf{w}$) defined by the directed line segments $\overrightarrow{AB}$ and $\overrightarrow{AC}$. From our definition it follows that $0 \leq \theta \leq \pi$. If $\mathbf{v}$ and $\mathbf{w}$ are parallel, then $\theta = 0$ (if they have the same direction) or $\theta = \pi$ (if they have opposite directions).

THEOREM 1.3 Geometric Version of the Dot Product

Let $\mathbf{v}$ and $\mathbf{w}$ be vectors in $\mathbb{R}^2$ or $\mathbb{R}^3$. Then

$$\mathbf{v} \cdot \mathbf{w} = ||\mathbf{v}||\ ||\mathbf{w}||\ \cos\theta, \tag{1.9}$$

where θ is the angle between $\mathbf{v}$ and $\mathbf{w}$.

PROOF: If either $\mathbf{v} = \mathbf{0}$ or $\mathbf{w} = \mathbf{0}$ (or both), then both sides of (1.9) are zero. If $\theta = 0$, then $\mathbf{v}$ and $\mathbf{w}$ are parallel and have the same direction. In this case, $\cos\theta = 1$ and (1.9) follows from (1.7). If $\theta = \pi$, then $\mathbf{v}$ and $\mathbf{w}$ are parallel and have opposite directions, $\cos\theta = -1$ and (1.8) implies (1.9). Having verified all trivial cases, we must now consider the case $\mathbf{v} \neq \mathbf{0}$, $\mathbf{w} \neq \mathbf{0}$, and $0 < \theta < \pi$. Represent the vectors $\mathbf{v}$ and $\mathbf{w}$ by the directed line segments $\overrightarrow{AB}$ and $\overrightarrow{AC}$ starting at the same point A, as in Figure 1.14, and apply the Law of Cosines to the triangle ABC (the sides of the triangle are $\overline{AB} = ||\mathbf{v}||$, $\overline{AC} = ||\mathbf{w}||$, and $\overline{BC} = ||\mathbf{v} - \mathbf{w}|| =$

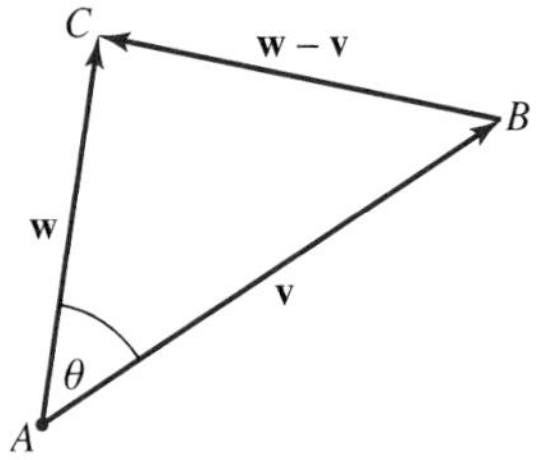

Figure 1.14 Triangle ABC from the proof of Theorem 1.3.

$||\mathbf{w}-\mathbf{v}||$):

$$||\mathbf{v}-\mathbf{w}||^2 = ||\mathbf{v}||^2 + ||\mathbf{w}||^2 - 2||\mathbf{v}||\ ||\mathbf{w}|| \cos\theta.$$

Using the properties of the dot product, we transform the left side as

$$||\mathbf{v}-\mathbf{w}||^2 = (\mathbf{v}-\mathbf{w})\cdot(\mathbf{v}-\mathbf{w}) = \mathbf{v}\cdot\mathbf{v} - \mathbf{w}\cdot\mathbf{v} - \mathbf{v}\cdot\mathbf{w} + \mathbf{w}\cdot\mathbf{w} = ||\mathbf{v}||^2 - 2\mathbf{v}\cdot\mathbf{w} + ||\mathbf{w}||^2.$$

Now substitute the right side of the previous equation into the Law of Cosines formula above to get

$$||\mathbf{v}||^2 - 2\mathbf{v}\cdot\mathbf{w} + ||\mathbf{w}||^2 = ||\mathbf{v}||^2 + ||\mathbf{w}||^2 - 2||\mathbf{v}||\ ||\mathbf{w}|| \cos\theta.$$

Thus, $-2\mathbf{v}\cdot\mathbf{w} = -2||\mathbf{v}||\ ||\mathbf{w}|| \cos\theta$ and

$$\mathbf{v}\cdot\mathbf{w} = ||\mathbf{v}||\ ||\mathbf{w}|| \cos\theta,$$

which is the formula (1.9).

In Theorems 1.4 and 1.5 we derive two important consequences of this theorem. Recall that two nonzero vectors **v** and **w** are called *orthogonal* (or *perpendicular*) if the angle between them is $\pi/2$.

THEOREM 1.4 Test for Orthogonality of Vectors

Let **v** and **w** be nonzero vectors in $\mathbb{R}^2$ or $\mathbb{R}^3$. Then $\mathbf{v}\cdot\mathbf{w} = 0$ if and only if **v** and **w** are orthogonal.

PROOF: From $\mathbf{v}\cdot\mathbf{w} = 0$ it follows that $||\mathbf{v}||\ ||\mathbf{w}|| \cos\theta = 0$, that is, $\cos\theta = 0$ (since $\mathbf{v}, \mathbf{w} \neq \mathbf{0}$ their magnitudes are nonzero as well) and hence $\theta = \pi/2$. Conversely, if $\theta = \pi/2$, then $\mathbf{v}\cdot\mathbf{w} = ||\mathbf{v}||\ ||\mathbf{w}|| \cos(\pi/2) = 0$.

DEFINITION 1.5 Orthonormal Set of Vectors

Vectors $\mathbf{v}_1, \ldots, \mathbf{v}_k$ (where $k \geq 2$) in $\mathbb{R}^n$, $n \geq 2$ are said to form an *orthonormal set* if they are of unit length and each vector in the set is orthogonal to the others.

Sometimes the vectors that form an orthonormal set are referred to as *orthonormal vectors*. The standard unit vectors **i**, **j**, **k** form an orthonormal set in $\mathbb{R}^3$. In Section 2.8 we will construct two more sets of orthonormal vectors in $\mathbb{R}^3$.

THEOREM 1.5 Angle Between Vectors

Let $\mathbf{v}$ and $\mathbf{w}$ be nonzero vectors in $\mathbb{R}^2$ or $\mathbb{R}^3$. Then

$$\cos\theta = \frac{\mathbf{v}\cdot\mathbf{w}}{||\mathbf{v}||\ ||\mathbf{w}||},$$

where θ is the angle between $\mathbf{v}$ and $\mathbf{w}$.

PROOF: All we have to do is to solve (1.9) for θ and notice that the denominator cannot be zero (since $\mathbf{v}$ and $\mathbf{w}$ are assumed to be nonzero vectors). ◀

▶ EXAMPLE 1.20

This example illustrates the use of Theorems 1.4 and 1.5.

(a) Find the angle between $\mathbf{v} = 4\mathbf{i} - \mathbf{j} + \mathbf{k}$ and $\mathbf{w} = \mathbf{j} + 3\mathbf{k}$ in $\mathbb{R}^3$.

(b) Show that the line ℓ_1 that goes through $(1, 0)$ and $(3, 4)$ is perpendicular to the line ℓ_2 that contains $(3, -1)$ and $(1, 0)$.

SOLUTION

(a) From $||\mathbf{v}|| = \sqrt{4^2 + (-1)^2 + 1^2} = \sqrt{18}$, $||\mathbf{w}|| = \sqrt{1+9} = \sqrt{10}$ and $\mathbf{v}\cdot\mathbf{w} = (4\mathbf{i} - \mathbf{j} + \mathbf{k})\cdot(\mathbf{j} + 3\mathbf{k}) = 2$, using Theorem 1.5, we get

$$\cos\theta = \frac{\mathbf{v}\cdot\mathbf{w}}{||\mathbf{v}||\ ||\mathbf{w}||} = \frac{2}{\sqrt{10}\sqrt{18}} = \frac{1}{3\sqrt{5}}$$

and $\theta = \arccos(1/3\sqrt{5}) \approx 1.42$ rad.

(b) The direction of ℓ_1 is that of the vector $\mathbf{v}' = (2, 4)$, and the direction of ℓ_2 is $\mathbf{v}'' = (-2, 1)$. Since their dot product $\mathbf{v}'\cdot\mathbf{v}'' = (2, 4)\cdot(-2, 1) = 0$, it follows that ℓ_1 and ℓ_2 are perpendicular. (It makes no difference if we take $(-2, -4)$ as the direction of ℓ_1 or $(2, -1)$ as the direction for ℓ_2.) ◀

We have already seen that every vector in $\mathbb{R}^2$ can be expressed in terms of the standard unit vectors $\mathbf{i}$ and $\mathbf{j}$. As a matter of fact, we can express a vector in $\mathbb{R}^2$ in terms of any pair of orthogonal vectors. That is the point of the next theorem.

THEOREM 1.6 Vector Expressed in Terms of Orthogonal Vectors

Let $\mathbf{v}$ and $\mathbf{w}$ be (nonzero) orthogonal vectors in $\mathbb{R}^2$ and let $\mathbf{a}$ be any vector in $\mathbb{R}^2$. Then

$$\mathbf{a} = a_{\mathbf{v}}\mathbf{v} + a_{\mathbf{w}}\mathbf{w},$$

where $a_{\mathbf{v}} = \mathbf{a}\cdot\mathbf{v}/||\mathbf{v}||^2$ is the component of $\mathbf{a}$ in the direction of $\mathbf{v}$ and $a_{\mathbf{w}} = \mathbf{a}\cdot\mathbf{w}/||\mathbf{w}||^2$ is the component of $\mathbf{a}$ in the direction of $\mathbf{w}$ (or in the direction orthogonal to $\mathbf{v}$).

PROOF: From linear algebra we know that every vector in $\mathbb{R}^2$ can be written as a linear combination (i.e., expressed in terms) of two mutually orthogonal vectors. Hence, $\mathbf{a} = a_{\mathbf{v}}\mathbf{v} + a_{\mathbf{w}}\mathbf{w}$ for some real numbers $a_{\mathbf{v}}$ and $a_{\mathbf{w}}$. Computing the dot product of $\mathbf{a} = a_{\mathbf{v}}\mathbf{v} + a_{\mathbf{w}}\mathbf{w}$ with $\mathbf{v}$, we get

$$\mathbf{a}\cdot\mathbf{v} = a_{\mathbf{v}}||\mathbf{v}||^2 + a_{\mathbf{w}}\mathbf{w}\cdot\mathbf{v}.$$

Since $\mathbf{v}$ and $\mathbf{w}$ are orthogonal, $\mathbf{w}\cdot\mathbf{v}=0$ and $\mathbf{a}\cdot\mathbf{v}=a_{\mathbf{v}}||\mathbf{v}||^2$. Hence, $a_{\mathbf{v}}=\mathbf{a}\cdot\mathbf{v}/||\mathbf{v}||^2$. Repeating the same argument (this time, computing the dot product of $\mathbf{a}$ and $\mathbf{w}$), we obtain the formula for the component $a_{\mathbf{w}}$ of $\mathbf{a}$. ◀

An analogous statement holds in three dimensions: if $\mathbf{u}$, $\mathbf{v}$, and $\mathbf{w}$ are nonzero and mutually orthogonal vectors (i.e., $\mathbf{u}\cdot\mathbf{v}=\mathbf{u}\cdot\mathbf{w}=\mathbf{v}\cdot\mathbf{w}=0$) and $\mathbf{a}\in\mathbb{R}^3$, then

$$\mathbf{a}=a_{\mathbf{u}}\mathbf{u}+a_{\mathbf{v}}\mathbf{v}+a_{\mathbf{w}}\mathbf{w},$$

where $a_{\mathbf{u}}=\mathbf{a}\cdot\mathbf{u}/||\mathbf{u}||^2$, $a_{\mathbf{v}}=\mathbf{a}\cdot\mathbf{v}/||\mathbf{v}||^2$, and $a_{\mathbf{w}}=\mathbf{a}\cdot\mathbf{w}/||\mathbf{w}||^2$.

▶ EXAMPLE 1.21

Check that $\mathbf{u}=(1,2,0)$, $\mathbf{v}=(2,-1,1)$, and $\mathbf{w}=(-2,1,5)$ are mutually orthogonal vectors and express $\mathbf{a}=(0,1,1)$ in terms of $\mathbf{u}$, $\mathbf{v}$, and $\mathbf{w}$.

SOLUTION

Since $\mathbf{u}\cdot\mathbf{v}=(1,2,0)\cdot(2,-1,1)=0$, $\mathbf{u}\cdot\mathbf{w}=(1,2,0)\cdot(-2,1,5)=0$, and $\mathbf{v}\cdot\mathbf{w}=(2,-1,1)\cdot(-2,1,5)=0$, the vectors $\mathbf{u}$, $\mathbf{v}$, and $\mathbf{w}$ are mutually orthogonal. Now $\mathbf{a}\cdot\mathbf{u}=(0,1,1)\cdot(1,2,0)=2$, $\mathbf{a}\cdot\mathbf{v}=(0,1,1)\cdot(2,-1,1)=0$, and $\mathbf{a}\cdot\mathbf{w}=(0,1,1)\cdot(-2,1,5)=6$; $||\mathbf{u}||=\sqrt{5}$, $||\mathbf{v}||=\sqrt{6}$, and $||\mathbf{w}||=\sqrt{30}$. Therefore,

$$\mathbf{a}=\frac{\mathbf{a}\cdot\mathbf{u}}{||\mathbf{u}||^2}\mathbf{u}+\frac{\mathbf{a}\cdot\mathbf{v}}{||\mathbf{v}||^2}\mathbf{v}+\frac{\mathbf{a}\cdot\mathbf{w}}{||\mathbf{w}||^2}\mathbf{w}=\tfrac{2}{5}\mathbf{u}+\tfrac{0}{6}\mathbf{v}+\tfrac{6}{30}\mathbf{w}=\tfrac{2}{5}\mathbf{u}+\tfrac{1}{5}\mathbf{w}.$$

◀

▶ EXAMPLE 1.22

Express the vector $\mathbf{a}=3\mathbf{i}+\mathbf{j}$ as the sum $\mathbf{a}=a_{\mathbf{v}}\mathbf{v}+a_{\mathbf{w}}\mathbf{w}$, where $\mathbf{v}$ and $\mathbf{w}$ are unit vectors, $\mathbf{v}$ is parallel to the line $y=x$, $\mathbf{w}$ is perpendicular to it, and both have positive $\mathbf{j}$ components.

SOLUTION

Picking two points on the line $y=x$, say, $(1,1)$ and $(2,2)$, we compute its direction to be $(1,1)=\mathbf{i}+\mathbf{j}$. The vector $\mathbf{v}$ has to be of unit length in that direction and hence $\mathbf{v}=(\mathbf{i}+\mathbf{j})/||\mathbf{i}+\mathbf{j}||=(\mathbf{i}+\mathbf{j})/\sqrt{2}$. The vector $\mathbf{w}$ has to be orthogonal to $\mathbf{v}$ and of unit length. So, first take any vector whose dot product with $\mathbf{v}$ is zero (and whose $\mathbf{j}$ component is positive) and then adjust its length. For example, $(-\mathbf{i}+\mathbf{j})\cdot(\mathbf{i}/\sqrt{2}+\mathbf{j}/\sqrt{2})=0$, so take $\mathbf{w}=(-\mathbf{i}+\mathbf{j})/||-\mathbf{i}+\mathbf{j}||=(-\mathbf{i}+\mathbf{j})/\sqrt{2}$. Although there are infinitely many vectors orthogonal to $\mathbf{v}$, only two (namely $\mathbf{w}$ and $-\mathbf{w}$) are of unit length; the requirement that its $\mathbf{j}$ coordinate be positive uniquely determines $\mathbf{w}$. Now $\mathbf{a}=a_{\mathbf{v}}\mathbf{v}+a_{\mathbf{w}}\mathbf{w}$, where $a_{\mathbf{v}}=\mathbf{a}\cdot\mathbf{v}/||\mathbf{v}||^2=(3\mathbf{i}+\mathbf{j})\cdot(\mathbf{i}/\sqrt{2}+\mathbf{j}/\sqrt{2})=4/\sqrt{2}=2\sqrt{2}$ (since $||\mathbf{v}||=1$) and $a_{\mathbf{w}}=\mathbf{a}\cdot\mathbf{w}/||\mathbf{w}||^2=(3\mathbf{i}+\mathbf{j})\cdot(-\mathbf{i}/\sqrt{2}+\mathbf{j}/\sqrt{2})=-2/\sqrt{2}=\sqrt{2}$ (since $||\mathbf{w}||=1$). Hence,

$$\mathbf{a}=2\sqrt{2}\left(\frac{1}{\sqrt{2}}\mathbf{i}+\frac{1}{\sqrt{2}}\mathbf{j}\right)-\sqrt{2}\left(-\frac{1}{\sqrt{2}}\mathbf{i}+\frac{1}{\sqrt{2}}\mathbf{j}\right)$$

is the required decomposition. ◀

We are going to give a geometric meaning to the vectors $a_{\mathbf{v}}\mathbf{v}$ and $a_{\mathbf{w}}\mathbf{w}$ in the expression $\mathbf{a}=a_{\mathbf{v}}\mathbf{v}+a_{\mathbf{w}}\mathbf{w}$, where $\mathbf{v}$ and $\mathbf{w}$ are orthogonal vectors in $\mathbb{R}^2$.

Take the representatives of $\mathbf{a}$ and $\mathbf{v}$ that start at the same point A, and draw the line ℓ through A in the direction $\mathbf{v}$; see Figure 1.15. The line perpendicular to ℓ that goes through the tip A' of $\mathbf{a}$ intersects ℓ at the point B. The vector represented by the directed line segment $\overrightarrow{AB}$ is called the *orthogonal projection* (or the *orthogonal vector projection*) of $\mathbf{a}$ onto $\mathbf{v}$, and is denoted by $proj_{\mathbf{v}}\mathbf{a}$. Its magnitude $||proj_{\mathbf{v}}\mathbf{a}||$ is called the *scalar projection.*

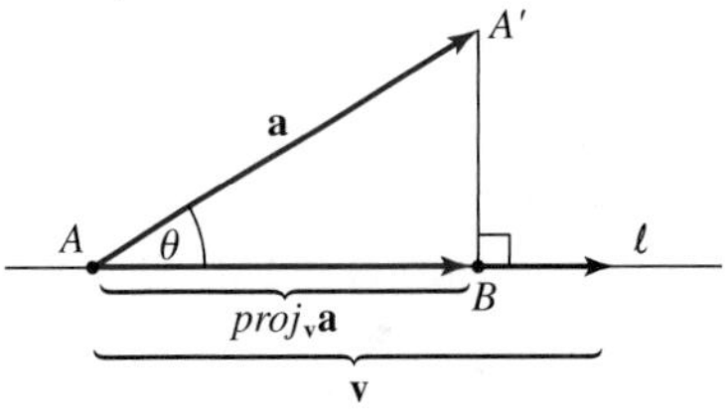

Figure 1.15 Projection of **a** onto **v**.

Let θ be the angle between the vectors $\mathbf{a}$ and $\mathbf{v}$. From the triangle ABA' we get $\cos\theta = ||proj_{\mathbf{v}}\mathbf{a}||/||\mathbf{a}||$, and hence $||proj_{\mathbf{v}}\mathbf{a}|| = ||\mathbf{a}||\cos\theta$. Combining this with the formula $\cos\theta = \mathbf{a}\cdot\mathbf{v}/||\mathbf{a}||\,||\mathbf{v}||$ of Theorem 1.5, we get the expression for the scalar projection

$$||proj_{\mathbf{v}}\mathbf{a}|| = ||\mathbf{a}||\cos\theta = ||\mathbf{a}||\,\frac{\mathbf{a}\cdot\mathbf{v}}{||\mathbf{a}||\,||\mathbf{v}||} = \frac{\mathbf{a}\cdot\mathbf{v}}{||\mathbf{v}||}. \tag{1.10}$$

To obtain the vector projection, we multiply the scalar projection by the unit vector $\mathbf{v}/||\mathbf{v}||$ in the direction of $\mathbf{v}$ (we need the direction of $\mathbf{v}$, but cannot multiply by $\mathbf{v}$ only as this would compromise the length of the projection). Hence,

$$proj_{\mathbf{v}}\mathbf{a} = \frac{\mathbf{a}\cdot\mathbf{v}}{||\mathbf{v}||}\,\frac{\mathbf{v}}{||\mathbf{v}||} = \frac{\mathbf{a}\cdot\mathbf{v}}{||\mathbf{v}||^2}\,\mathbf{v}. \tag{1.11}$$

Repeating this argument will yield

$$proj_{\mathbf{w}}\mathbf{a} = \frac{\mathbf{a}\cdot\mathbf{w}}{||\mathbf{w}||^2}\,\mathbf{w}.$$

Comparing these formulas with Theorem 1.6 we obtain $proj_{\mathbf{v}}\mathbf{a} = a_{\mathbf{v}}\mathbf{v}$ and $proj_{\mathbf{w}}\mathbf{a} = a_{\mathbf{w}}\mathbf{w}$. In other words, the decomposition $\mathbf{a} = a_{\mathbf{v}}\mathbf{v} + a_{\mathbf{w}}\mathbf{w}$ is the sum $\mathbf{a} = proj_{\mathbf{v}}\mathbf{a} + proj_{\mathbf{w}}\mathbf{a}$ of orthogonal projections of $\mathbf{a}$ onto $\mathbf{v}$ and $\mathbf{w}$ [keep in mind that $\mathbf{v}$ and $\mathbf{w}$ are (nonzero) orthogonal vectors].

Similarly, the decomposition $\mathbf{a} = a_{\mathbf{u}}\mathbf{u} + a_{\mathbf{v}}\mathbf{v} + a_{\mathbf{w}}\mathbf{w}$ in $\mathbb{R}^3$ is the sum of orthogonal projections of $\mathbf{a}$ onto $\mathbf{u}$, $\mathbf{v}$, and $\mathbf{w}$ [and $\mathbf{u}$, $\mathbf{v}$, and $\mathbf{w}$ are mutually orthogonal (nonzero) vectors].

▶ EXAMPLE 1.23

Compute the scalar and vector orthogonal projections of $\mathbf{a} = (2, 1, 4)$ onto $\mathbf{v} = (0, 2, 3)$.

SOLUTION

Since $\mathbf{a}\cdot\mathbf{v} = (2, 1, 4)\cdot(0, 2, 3) = 14$ and $||\mathbf{v}|| = \sqrt{13}$, it follows that

$$proj_{\mathbf{v}}\mathbf{a} = \frac{\mathbf{a}\cdot\mathbf{v}}{||\mathbf{v}||^2}\mathbf{v} = \tfrac{14}{13}(0, 2, 3) = (0, \tfrac{28}{13}, \tfrac{42}{13})$$

is the vector projection, and

$$||proj_{\mathbf{v}}\mathbf{a}|| = ||\tfrac{14}{13}\mathbf{v}|| = \tfrac{14}{13}\,||\mathbf{v}|| = 14/\sqrt{13}$$

is the scalar projection. ◀

▶ EXAMPLE 1.24 Work

Assume that a constant force **F** acts on an object located initially at $A(a_1, a_2)$ as the object moves along the straight line to the point $B(b_1, b_2)$. The *work* W done by the force **F** on the object is defined to be the product of the component of **F** in the direction of the motion (we now know that this is just the orthogonal projection) and the distance moved.

The distance moved is equal to the length $||\mathbf{d}||$ of the displacement vector $\mathbf{d} = (b_1 - a_1, b_2 - a_2)$. The component of **F** needed is the scalar projection $||proj_{\mathbf{d}}\mathbf{F}||$ of **F** onto **d**, and is given by [apply (1.10)] $\mathbf{F} \cdot \mathbf{d}/||\mathbf{d}||$; see Figure 1.16. Therefore,

$$W = \frac{\mathbf{F} \cdot \mathbf{d}}{||\mathbf{d}||} ||\mathbf{d}|| = \mathbf{F} \cdot \mathbf{d},$$

that is, the work is the dot product of the force and the displacement vectors.

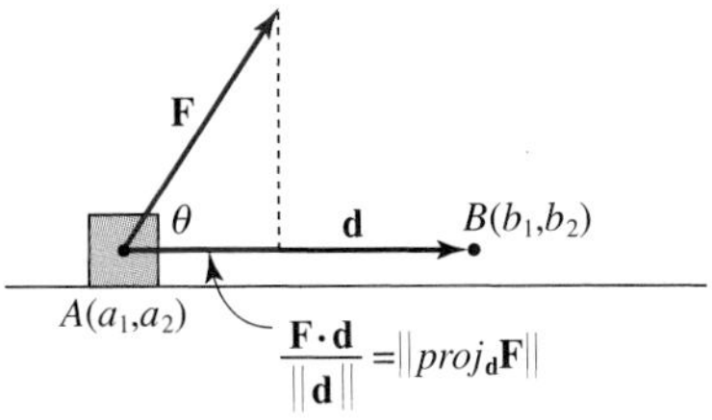

Figure 1.16 The work of a force **F** on an object is the dot product of **F** and the object's displacement vector **d**. ◀

▶ EXAMPLE 1.25

A force **F** of 5 N acts on an object as the object moves from the origin to the point (3, 2) along a straight line. In what direction with respect to the displacement should the force act to produce maximum work? Explain why **F** does zero work if applied in the direction perpendicular to the displacement.

SOLUTION According to the previous example, the work of **F** is $W = \mathbf{F} \cdot \mathbf{d} = ||\mathbf{F}||\ ||\mathbf{d}|| \cos\theta$, where $\mathbf{d} = 3\mathbf{i} + 2\mathbf{j}$ is the displacement vector and θ is the angle between **F** and **d**. Since $||\mathbf{F}|| = 5$ and $||\mathbf{d}|| = \sqrt{13}$, it follows that $W = 5\sqrt{13}\cos\theta$. The work is maximum when $\cos\theta = 1$, that is, when $\theta = 0$. Hence, maximum work is obtained by applying **F** in the direction of motion. Since $\cos(\pi/2) = 0$, the work is zero if **F** is applied in the direction perpendicular to the motion. ◀

▶ EXAMPLE 1.26 Equation of a Plane in Space

As an application of the dot product, we will compute the equation of the plane in space containing the point $A(x_0, y_0, z_0)$ and perpendicular to the vector $\mathbf{n} = (a, b, c)$ (here we depart from our standard use of the subscripts for the components of a vector; using a, b, and c instead will produce a familiar form of the equation). Visualize **n** as the directed line segment starting at A and pick any point $P(x, y, z)$ in the plane. The vector $\mathbf{v} = (x - x_0, y - y_0, z - z_0)$ defined by the directed line segment $\overrightarrow{AP}$ is perpendicular to **n**, and consequently $\mathbf{n} \cdot \mathbf{v} = 0$; that is,

$$(a, b, c) \cdot (x - x_0, y - y_0, z - z_0) = 0. \qquad (1.12)$$

[If a point $Q(x_1, y_1, z_1)$ does not belong to the plane, then $(a, b, c) \cdot (x_1 - x_0, y_1 - y_0, z_1 - z_0) \neq 0$.] Expression (1.12), written as

$$a(x - x_0) + b(y - y_0) + c(z - z_0) = 0,$$

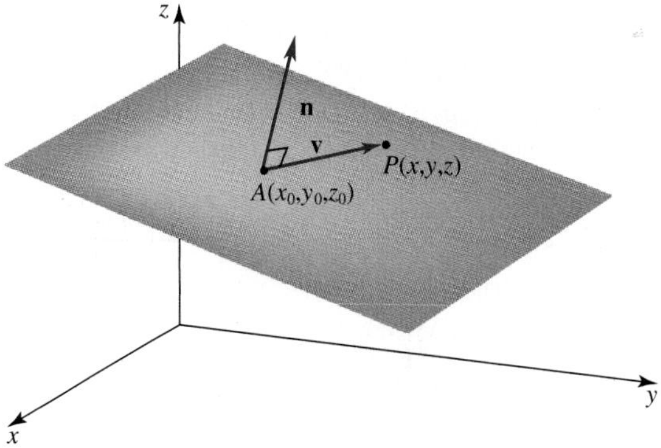

Figure 1.17 Plane through a point A with a normal vector $\mathbf{n}$.

represents the equation of the plane that contains the point (x_0, y_0, z_0) and is perpendicular to the vector $\mathbf{n} = (a, b, c)$. The vector $\mathbf{n}$ is called a *normal vector* to the plane; see Figure 1.17. ◀

From $a(x - x_0) + b(y - y_0) + c(z - z_0) = 0$, we get the usual form

$$ax + by + cz + d = 0 \tag{1.13}$$

of the equation of a plane in space (here, $d = -ax_0 - by_0 - cz_0$). In Exercises 28 to 31 below, we show how to convert parametric equation(s) of the plane (1.4) into the form (1.13), and vice versa.

▶ EXERCISES 1.3

1. Show that the dot product is commutative; that is, $\mathbf{v} \cdot \mathbf{w} = \mathbf{w} \cdot \mathbf{v}$ for $\mathbf{v}, \mathbf{w} \in \mathbb{R}^n$.

2. Explain why $\mathbf{u} \cdot (\mathbf{v} \cdot \mathbf{w}) \neq (\mathbf{u} \cdot \mathbf{v}) \cdot \mathbf{w}$ for vectors $\mathbf{u}$, $\mathbf{v}$, and $\mathbf{w}$ in $\mathbb{R}^2$ (or in $\mathbb{R}^3$).

3. Prove that, for vectors $\mathbf{u}$, $\mathbf{v}$, and $\mathbf{w}$ in $\mathbb{R}^3$ and any $\alpha \in \mathbb{R}$, $\mathbf{u} \cdot (\mathbf{v} + \mathbf{w}) = \mathbf{u} \cdot \mathbf{v} + \mathbf{u} \cdot \mathbf{w}$, $(\alpha\mathbf{u}) \cdot \mathbf{v} = \alpha(\mathbf{u} \cdot \mathbf{v})$, and $\mathbf{u} \cdot (\alpha\mathbf{v}) = \alpha(\mathbf{u} \cdot \mathbf{v})$.

4. Prove that $\mathbf{v} \cdot \mathbf{v} = ||\mathbf{v}||^2$ for any vector $\mathbf{v} \in \mathbb{R}^n$.

5. Show that if $\mathbf{u}$, $\mathbf{v}$, and $\mathbf{w}$ are nonzero and mutually orthogonal vectors in $\mathbb{R}^3$ and $\mathbf{a} \in \mathbb{R}^3$, then $\mathbf{a} = a_\mathbf{u}\mathbf{u} + a_\mathbf{v}\mathbf{v} + a_\mathbf{w}\mathbf{w}$, where $a_\mathbf{u} = \mathbf{a} \cdot \mathbf{u}/||\mathbf{u}||^2$, $a_\mathbf{v} = \mathbf{a} \cdot \mathbf{v}/||\mathbf{v}||^2$, and $a_\mathbf{w} = \mathbf{a} \cdot \mathbf{w}/||\mathbf{w}||^2$.

6. Find the angle between $2\mathbf{j} - \mathbf{k}$ and $\mathbf{i} + \mathbf{j} - 3\mathbf{k}$.

7. Find the angles in the triangle with vertices $(0, 3, 4)$, $(0, 3, 0)$, and $(12, 0, 5)$.

8. Determine whether the vectors $(3, 2, 4)$ and $(1, 2, 7)$ are parallel or orthogonal or neither. Do the same for $\mathbf{i} - \mathbf{j} - \mathbf{k}$ and $\mathbf{i} + \mathbf{j}$.

9. Let $\mathbf{v}$ and $\mathbf{w}$ be nonzero vectors in $\mathbb{R}^2$. Show that the vectors $\mathbf{w} - (\mathbf{v} \cdot \mathbf{w}/||\mathbf{v}||^2)\mathbf{v}$ and $\mathbf{v}$ are orthogonal.

10. Prove that $|\mathbf{v} \cdot \mathbf{w}| \leq ||\mathbf{v}||\ ||\mathbf{w}||$ for vectors in $\mathbb{R}^2$ (and then for vectors in $\mathbb{R}^3$). When is $|\mathbf{v} \cdot \mathbf{w}| = ||\mathbf{v}||\ ||\mathbf{w}||$? This important relation between the dot product of two vectors and their length is called the *Cauchy–Schwarz Inequality*.

11. Find the angles between the diagonal of a cube and its three adjacent edges.

12. Find the angle between the diagonals of the parallelogram, three of whose vertices are located at $(0, 2, -3)$, $(1, 1, 0)$, and $(-1, 0, 1)$.

13. Let $\mathbf{v}$ and $\mathbf{w}$ be nonzero vectors such that $||\mathbf{v}|| = ||\mathbf{w}||$ and let $\mathbf{a} = ||\mathbf{v}||\mathbf{w} + ||\mathbf{w}||\mathbf{v}$. Show that the angles between $\mathbf{v}$ and $\mathbf{a}$ and $\mathbf{w}$ and $\mathbf{a}$ are the same.

14. Find the work of the force of magnitude 10 N acting at an angle of $\pi/3$ on an object located at $(1, 1)$ as the object moves from $(1, 1)$ to $(3, -2)$.

15. Write the vector $\mathbf{i} + 2\mathbf{j}$ as the sum of its orthogonal projections onto $\mathbf{i} + \mathbf{j}$ and $\mathbf{i} - \mathbf{j}$.

16. Check that the vectors $(\mathbf{i} + \mathbf{j} - \mathbf{k})/\sqrt{3}$, $(\mathbf{i} - \mathbf{j})/\sqrt{2}$, and $(\mathbf{i} + \mathbf{j} + 2\mathbf{k})/\sqrt{6}$ form an orthonormal set.

17. Find an equation of the plane that contains the point $(0, -2, 2)$ and is orthogonal to the vector $(3, 0, -1)$.

18. Find an equation of the plane that contains the points $(0, -2, 3)$, $(1, 4, 3)$, and $(-1, 7, 0)$.

19. Find an equation of the plane that contains the point $(-2, 1, 4)$ and is perpendicular to the line $\mathbf{l}(t) = (2 + 3t, 1 - t, 7)$, $t \in \mathbb{R}$.

20. Fix a nonzero vector $\mathbf{v}$ in $\mathbb{R}^3$ and consider the dot products $\mathbf{v} \cdot \mathbf{u}$, where $\mathbf{u}$ is a unit vector in $\mathbb{R}^3$. Find all $\mathbf{u}$ such that $\mathbf{v} \cdot \mathbf{u}$ is the largest. Find all $\mathbf{u}$ such that $\mathbf{v} \cdot \mathbf{u}$ is the smallest. Describe geometrically all directions $\mathbf{u} \in \mathbb{R}^3$ that satisfy $\mathbf{v} \cdot \mathbf{u} = 0$.

21. Find the equation of the plane in $\mathbb{R}^3$ that contains the origin and is orthogonal to the line passing through the points $(3, 2, -1)$ and $(0, 0, 7)$.

22. Compute the scalar projection of the vector $3\mathbf{i} - 2\mathbf{j}$ onto the line passing through the points $(2, -1)$ and $(0, 4)$.

23. Show that the work done by a constant force $\mathbf{F}$ on an object that moves around any closed polygon is zero.

24. Find all unit vectors parallel to the yz-plane and perpendicular to the vector $\mathbf{i} + \mathbf{j} - 2\mathbf{k}$.

25. Check that the vectors $\mathbf{u} = \mathbf{i} + \mathbf{j} - 2\mathbf{k}$, $\mathbf{v} = 2\mathbf{j} + \mathbf{k}$, and $\mathbf{w} = 5\mathbf{i} - \mathbf{j} + 2\mathbf{k}$ are mutually orthogonal and express the vector $\mathbf{i}$ in terms of $\mathbf{u}$, $\mathbf{v}$, and $\mathbf{w}$.

26. The vectors $\mathbf{u} = (\mathbf{i} - 2\mathbf{j} + \mathbf{k})/\sqrt{6}$ and $\mathbf{v} = -(\mathbf{i} + \mathbf{j} + \mathbf{k})/\sqrt{3}$ are orthogonal and of unit length. Find all vectors $\mathbf{w}$ so that $\mathbf{u}$, $\mathbf{v}$, and $\mathbf{w}$ form an orthonormal set in $\mathbb{R}^3$.

27. Assume that $\mathbf{u}$, $\mathbf{v}$, and $\mathbf{w}$ are mutually orthogonal (nonzero) vectors in $\mathbb{R}^3$. Show that a nonzero vector in $\mathbb{R}^3$ orthogonal to all three of them does not exist.

28. A plane is given parametrically by $\mathbf{p} = \mathbf{a} + t\mathbf{v} + s\mathbf{w}$, where $t, s \in \mathbb{R}$, $\mathbf{p} = (x, y, z)$ is a point in the plane, and $\mathbf{a} = (a_x, a_y, a_z)$, $\mathbf{v} = (v_x, v_y, v_z)$, and $\mathbf{w} = (w_x, w_y, w_z)$ are vectors in $\mathbb{R}^3$ [see (1.4) in Section 1.2].

(a) Derive parametric equations in the form $x = a_x + tv_x + sw_x$, $y = a_y + tv_y + sw_y$, $z = a_z + tv_z + sw_z$.

(b) Eliminate the parameters s and t from the three equations in (a) to obtain the equation $Ax + By + Cz + D = 0$ [see Example 1.26 and formula (1.13)].

29. Convert the parametric equation $\mathbf{p} = (2, 0, -1) + t(0, -1, 1) + s(3, 0, 1)$ of the plane into the equation of the form $Ax + By + Cz + D = 0$. (*Hint*: See Exercise 28.)

30. Find parametric equation(s) of the plane [i.e., identify $\mathbf{a}$, $\mathbf{v}$, and $\mathbf{w}$ in (1.4)] given by $3x + y - z + 1 = 0$. (*Hint*: Since $\mathbf{v}$ and $\mathbf{w}$ belong to the plane, they must be perpendicular to its normal vector. There are many correct answers.)

31. Find parametric equation(s) of the plane [i.e., identify $\mathbf{a}$, $\mathbf{v}$, and $\mathbf{w}$ in (1.4)] given by $Ax + By + Cz + D = 0$.

▶ 1.4 MATRICES AND DETERMINANTS

In this section, we introduce two useful mathematical objects: matrices and determinants. For instance, the derivative of a function of several variables is defined (in Chapter 2) as a matrix, and the chain rule is expressed in terms of matrix multiplication. Determinants will appear in the analysis of extreme values of functions in Chapter 4, and also in the change of variables formula for multiple integrals in Chapter 6.

Our discussion will not be exhaustive, but rather will concentrate on those topics used in this book.

DEFINITION 1.6 Matrix

An $m \times n$ *matrix* A is a rectangular table of real numbers, arranged in m rows and n columns:

$$A = \begin{bmatrix} a_{11} & a_{12} & \dots & a_{1n} \\ a_{21} & a_{22} & \dots & a_{2n} \\ \vdots & \vdots & & \vdots \\ a_{m1} & a_{m2} & \dots & a_{mn} \end{bmatrix}.$$

An $m \times n$ matrix is also called a *matrix of type* $m \times n$ or of *order* $m \times n$. A matrix of type $n \times n$ is called a *square matrix of order* n.

The elements $a_{i1}, a_{i2}, \dots, a_{in}$ form the i*th row* of A, $i = 1, \dots, m$. The elements $a_{1j}, a_{2j}, \dots, a_{mj}$ form the j*th column* of A, $j = 1, \dots, n$. Sometimes, the ith row of A is written as a $1 \times n$ matrix

$$[\,a_{i1} \quad a_{i2} \quad \dots \quad a_{in}\,]$$

(and called a *row vector*) and the jth column of A is written as an $m \times 1$ matrix

$$\begin{bmatrix} a_{1j} \\ a_{2j} \\ \vdots \\ a_{mj} \end{bmatrix}$$

(and called a *column vector*). A 1×1 matrix $a = [a_{11}]$ is identified with its entry a_{11}. Matrices are usually denoted by uppercase letters A, B, C, etc. and their elements (or entries) by corresponding lowercase letters together with two subscripts that define their location in the matrix. For example, the element in the ith row and the jth column of A is denoted by a_{ij}. Sometimes, we use the notation $A = [a_{ij}]$.

For our purpose, we have to extend this (classical) definition of a matrix: we will allow the entries of a matrix to be real-valued functions of several variables, not only real numbers. All the properties that we will discuss will hold for these more general matrices (we still call them matrices, or matrices of functions).

EXAMPLE 1.27

Consider the 3×4 matrix

$$A = \begin{bmatrix} 9 & 0 & -5 & 5 \\ 2 & -5 & 9 & 1 \\ -4 & 0 & 2 & -7 \end{bmatrix}.$$

Let us list a few entries: $a_{12} = 0$, $a_{33} = 2$, $a_{21} = 2$, $a_{31} = -4$, $a_{24} = 1$, etc. The elements 2, -5, 9, and 1 form the second row and 5, 1, and -7 form the fourth column of A. We can also write them as matrices: the second row is the 1×4 matrix (or a row vector)

$$[2 \quad -5 \quad 9 \quad 1]$$

and the fourth column is the 3×1 matrix

$$\begin{bmatrix} 5 \\ 1 \\ -7 \end{bmatrix}$$

(also called a column vector).

The $m \times n$ matrix, all of whose entries are 0, is called a *zero matrix* (of type $m \times n$) and is denoted by 0. A matrix $A = [a_{ij}]$ of type $n \times n$ (i.e., A has the same number of rows and columns) is called a *square matrix of order* n. Its elements $a_{11}, a_{22}, \ldots, a_{nn}$ are said to form the *(main) diagonal* of A, and all other elements are called *off-diagonal elements*. The square matrix of order n whose diagonal elements are 1 and all off-diagonal elements are 0 is called an *identity matrix* (of order n) and is denoted by I or I_n (if its order needs to be specified).

Two matrices $A = [a_{ij}]$ and $B = [b_{ij}]$ are said to be *equal* (denoted by $A = B$) if and only if they are of the same type (say, $m \times n$) and all corresponding elements are equal; that is, $a_{ij} = b_{ij}$ for all $i = 1, \ldots, m$ and $j = 1, \ldots, n$.

DEFINITION 1.7 Elementary Operations with Matrices

Let $A = [a_{ij}]$ and $B = [b_{ij}]$ be matrices of type $m \times n$, and let α be a real number. The *sum* of A and B is the $m \times n$ matrix $C = A + B$ whose elements are defined by $c_{ij} = a_{ij} + b_{ij}$. The *difference* of A and B is the $m \times n$ matrix $C = A - B$ whose elements are defined by $c_{ij} = a_{ij} - b_{ij}$. The product of a scalar α and a matrix A is the $m \times n$ matrix $C = \alpha A$, where $c_{ij} = \alpha a_{ij}$.

In words, only matrices of the same type can be added or subtracted. The ijth entry in $A + B$ $(A - B)$ is the sum (difference) of the corresponding entries in A and B. In order to multiply a matrix by a real number α, we multiply all of its entries by α.

THEOREM 1.7 Properties of Matrix Addition and Multiplication by a Scalar

Assume that A, B, and C are matrices of type $m \times n$ and let α and β be real numbers. The addition of matrices satisfies $A + B = B + A$ *(commutativity)*, $(A + B) + C = A + (B + C)$ *(associativity)*, and, if 0 is the $m \times n$ zero matrix, then $A + 0 = A$. The *distributive* laws

$\alpha(A+B) = \alpha A + \alpha B$ and $(\alpha+\beta)A = \alpha A + \beta A$ hold for multiplication by scalars. Finally, $1 \cdot A = A$. ◀

By definition, $(-1) \cdot A$ is denoted by $-A$. One could define the difference $A - B$ as $A + (-B)$.

Theorem 1.7 states that, as far as addition (subtraction) and multiplication by scalars are concerned, matrices behave in the same way as real numbers. Proofs of all parts of Theorem 1.7 consist of checking that the corresponding entries in the matrices on both sides of the above identities are equal. For example, let us check the first distribution law: the ijth entry in the matrix on the left side is $\alpha(a_{ij} + b_{ij})$, and the corresponding entry on the right side is $\alpha a_{ij} + \alpha b_{ij}$. By the properties of real numbers, the two entries are equal.

▶ EXAMPLE 1.28

This example illustrates the matrix operations just defined.

(a) Let

$$A = \begin{bmatrix} 2\sin^2 x & x^2 \\ 1 & 0 \end{bmatrix} \quad \text{and} \quad B = \begin{bmatrix} -\cos^2 x & y^2 \\ xy & 2 \end{bmatrix}.$$

Compute $A - 2B + I_2$, where I_2 is the identity matrix of order 2.

(b) Let

$$A = \begin{bmatrix} 1 & 1 & 2 \\ 7 & -2 & 0 \end{bmatrix} \quad \text{and} \quad B = \begin{bmatrix} 0 & 2 & -1 \\ 6 & 2 & 4 \end{bmatrix}.$$

Find the 2×3 matrix X such that $3A + X = 2(B - X)$.

SOLUTION

(a) A straightforward computation gives

$$\begin{aligned} A - 2B + I_2 &= \begin{bmatrix} 2\sin^2 x & x^2 \\ 1 & 0 \end{bmatrix} - 2\begin{bmatrix} -\cos^2 x & y^2 \\ xy & 2 \end{bmatrix} + \begin{bmatrix} 1 & 0 \\ 0 & 1 \end{bmatrix} \\ &= \begin{bmatrix} 2\sin^2 x + 2\cos^2 x + 1 & x^2 - 2y^2 + 0 \\ 1 - 2xy + 0 & 0 - 4 + 1 \end{bmatrix} \\ &= \begin{bmatrix} 3 & x^2 - 2y^2 \\ 1 - 2xy & -3 \end{bmatrix}. \end{aligned}$$

(b) One way to solve for X is to substitute its entries and those of A and B into the given equation in order to obtain equations for the entries x_{ij} of X. There is an easier alternative: we will use the properties of matrix operations and the fact that $A = B$ implies $A + C = B + C$ (for any matrix C of the same size as A and B) and $\alpha A = \alpha B$ (for any real number α). Using the distributive property, we write the given equation as $3A + X = 2B - 2X$. Now add $2X$ to both sides and then subtract $3A$

from both sides to get $3X = 2B - 3A$. To find X, multiply both sides by $1/3$:

$$\begin{aligned} X &= \tfrac{1}{3}(2B - 3A) \\ &= \tfrac{1}{3}\left(\begin{bmatrix} 0 & 4 & -2 \\ 12 & 4 & 8 \end{bmatrix} - \begin{bmatrix} 3 & 3 & 6 \\ 21 & -6 & 0 \end{bmatrix}\right) \\ &= \begin{bmatrix} -1 & 1/3 & -8/3 \\ -3 & 10/3 & 8/3 \end{bmatrix}. \end{aligned}$$

Let us repeat that, as long as only addition, subtraction, and multiplication by scalars are involved, there is no difference between the properties of matrices and real numbers. In this case, we solved the matrix equation in exactly the same way as we would have solved an equation in real numbers.

DEFINITION 1.8 Matrix Multiplication

The *product* $C = AB$ of an $m \times n$ matrix A and an $n \times p$ matrix B is the $m \times p$ matrix $C = [c_{ij}]$, where

$$c_{ij} = a_{i1}b_{1j} + a_{i2}b_{2j} + \cdots + a_{in}b_{nj}, \tag{1.14}$$

$i = 1, \ldots, m$ and $j = 1, \ldots, p$.

The product AB of two matrices is defined only if the number of columns of the first factor A is equal to the number of rows of the second factor B. In that case, the product AB is the matrix C that has the same number of rows as A and the same number of columns as B. Let us look a bit more closely at formula (1.14) that gives the general element in the product matrix. The first factors in all terms are $a_{i1}, a_{i2}, \ldots, a_{in}$; that is, the elements of the ith row of A. Think of them as a row vector

$$[\, a_{i1} \quad a_{i2} \quad \ldots \quad a_{in} \,].$$

The second factors in all terms are $b_{1j}, b_{2j}, \ldots, b_{nj}$; that is, the elements of the jth column of B. Put them together in a column vector

$$\begin{bmatrix} b_{1j} \\ b_{2j} \\ \vdots \\ b_{nj} \end{bmatrix}$$

If we interpret both the row and column vectors as vectors in $\mathbb{R}^n$, we recognize the expression in (1.14) as their dot product! Therefore, the ijth entry in the product matrix AB is computed as the dot product of the ith row of A and the jth column of B.

▶ EXAMPLE 1.29

Let

$$A = \begin{bmatrix} 1 & 0 & 4 \\ 3 & 2 & -1 \end{bmatrix} \quad \text{and} \quad B = \begin{bmatrix} 4 & 1 \\ 7 & 6 \\ 0 & 0 \end{bmatrix}.$$

Compute (if defined) AB and BA.

SOLUTION The matrix A is of type 2×3 and B is of type 3×2. The product AB is defined (since the number of columns of $A =$ the number of rows of $B = 3$) and is a 2×2 matrix. The product BA is also defined (the number of columns of $B =$ the number of rows of $A = 2$), but is a 3×3 matrix. We have just detected the first big difference between real numbers and matrices. Since AB and BA are of different types, they certainly cannot be equal. Hence, in general, $AB \neq BA$ (and it gets worse, as we can easily imagine and will witness soon: in many cases one of the two products AB or BA will not be defined).

Let us compute $C = AB$. The entry c_{11} is the dot product of the first row of A and the first column of B:

$$c_{11} = \begin{bmatrix} 1 & 0 & 4 \end{bmatrix} \cdot \begin{bmatrix} 4 \\ 7 \\ 0 \end{bmatrix} = 1 \cdot 4 + 0 \cdot 7 + 4 \cdot 0 = 4$$

(although we use row and column vectors, we think of them as vectors in $\mathbb{R}^3$). Similarly,

$$c_{12} = \begin{bmatrix} 1 & 0 & 4 \end{bmatrix} \cdot \begin{bmatrix} 1 \\ 6 \\ 0 \end{bmatrix} = 1 \cdot 1 + 0 \cdot 6 + 4 \cdot 0 = 1,$$

$$c_{21} = \begin{bmatrix} 3 & 2 & -1 \end{bmatrix} \cdot \begin{bmatrix} 4 \\ 7 \\ 0 \end{bmatrix} = 3 \cdot 4 + 2 \cdot 7 + (-1) \cdot 0 = 26$$

and

$$c_{22} = \begin{bmatrix} 3 & 2 & -1 \end{bmatrix} \cdot \begin{bmatrix} 1 \\ 6 \\ 0 \end{bmatrix} = 3 \cdot 1 + 2 \cdot 6 + (-1) \cdot 0 = 15.$$

Hence,

$$AB = \begin{bmatrix} 4 & 1 \\ 26 & 15 \end{bmatrix}.$$

Similarly, we compute

$$D = BA = \begin{bmatrix} 4 & 1 \\ 7 & 6 \\ 0 & 0 \end{bmatrix} \cdot \begin{bmatrix} 1 & 0 & 4 \\ 3 & 2 & -1 \end{bmatrix} = \begin{bmatrix} 7 & 2 & 15 \\ 25 & 12 & 22 \\ 0 & 0 & 0 \end{bmatrix}.$$

Let us check a few entries in D. For example, d_{12} is obtained as the dot product of the first row of B and the second coulmn of A,

$$d_{12} = \begin{bmatrix} 4 & 1 \end{bmatrix} \cdot \begin{bmatrix} 0 \\ 2 \end{bmatrix} = 4 \cdot 0 + 1 \cdot 2 = 2.$$

Similarly,

$$d_{33} = \begin{bmatrix} 0 & 0 \end{bmatrix} \cdot \begin{bmatrix} 4 \\ -1 \end{bmatrix} = 0 \cdot 4 + 0 \cdot (-1) = 0,$$

$$d_{23} = \begin{bmatrix} 7 & 6 \end{bmatrix} \cdot \begin{bmatrix} 4 \\ -1 \end{bmatrix} = 7 \cdot 4 + 6 \cdot (-1) = 22,$$

etc. ◀

As we have already seen, the product of matrices is not commutative; that is, in general, $AB \neq BA$. However, our next theorem will show that some properties of ordinary multiplication are preserved.

THEOREM 1.8 Properties of Matrix Multiplication

Let A, B, and C be three matrices. Matrix multiplication is *associative*

$$(AB)C = A(BC)$$

and *distributive*

$$(A + B)C = AC + BC \qquad \text{and} \qquad A(B + C) = AB + AC,$$

whenever the operations on both sides of these identities are defined. ◀

The associativity property tells us that in order to compute the product of three matrices (if defined) ABC, we can either compute AB first and then multiply by C from the right, or compute BC first and then multiply by A from the left. Since the product of matrices is not commutative, we have to make the order of factors in the product precise. That is the reason why there are two distributive laws: one will not suffice, since neither one implies the other. The proofs are straightforward (the associativity part is fairly messy) and we will omit them here.

Next, we give a useful interpretation of one case of matrix multiplication. Let $A = [a_{ij}]$ be a square matrix of order n (i.e., of type $n \times n$) and let $\mathbf{v}$ be an $n \times 1$ matrix

$$\begin{bmatrix} v_1 \\ v_2 \\ \vdots \\ v_n \end{bmatrix}$$

($\mathbf{v}$ has only one column, so instead of using double subscripts $v_{11}, v_{21}, \ldots v_{n1}$ for its elements, we supress the second subscript). We think of $\mathbf{v}$ as a vector in $\mathbb{R}^n$. The product $A\mathbf{v}$ of A and $\mathbf{v}$ is an $n \times 1$ matrix that can again be visualized as a vector. In this way, we have defined a function (call it $\mathbf{F}_A$) that, for a given fixed matrix A, assigns the vector $A\mathbf{v}$ to a vector $\mathbf{v}$. The function $\mathbf{F}_A$ is called a *linear function* on $\mathbb{R}^n$ (see Exercise 32).

Visualizing $\mathbf{v}$ and $A\mathbf{v}$ as points rather than vectors, we get another interpretation: $\mathbf{F}_A$ maps points in $\mathbb{R}^n$ to points in $\mathbb{R}^n$ (i.e., "moves points in $\mathbb{R}^n$ around"; see Exercises 33 and 34).

There is one more interpretation: think of $\mathbf{v}$ as a point in $\mathbb{R}^n$ and think of $A\mathbf{v}$ as a vector in $\mathbb{R}^n$: thus, $\mathbf{F}_A$ assigns a vector in $\mathbb{R}^n$ to a point in $\mathbb{R}^n$. This function is an example of a vector field, called a *linear vector field.* Vector fields will be formally introduced in Section 2.1 and studied extensively in this book.

▶ EXAMPLE 1.30

Let

$$A = \begin{bmatrix} 3 & 2 \\ 1 & 1 \end{bmatrix}$$

and consider the linear map $\mathbf{F}_A$ just defined.

(a) Find the image of the vector $\mathbf{j}$ under the map $\mathbf{F}_A$.

(b) Compute $||\mathbf{F}_A(\mathbf{v})||$ if $\mathbf{v} = -2\mathbf{i} + 4\mathbf{j}$.

SOLUTION The matrix A is of type 2×2; that is, the map $\mathbf{F}_A$ is defined on vectors in $\mathbb{R}^2$.

(a) Think of $\mathbf{j}$ as a column vector $\begin{bmatrix} 0 \\ 1 \end{bmatrix}$. Then

$$\mathbf{F}_A(\mathbf{j}) = \mathbf{F}_A\left(\begin{bmatrix} 0 \\ 1 \end{bmatrix}\right) = \begin{bmatrix} 3 & 2 \\ 1 & 1 \end{bmatrix} \cdot \begin{bmatrix} 0 \\ 1 \end{bmatrix} = \begin{bmatrix} 2 \\ 1 \end{bmatrix}.$$

Hence, the image of $\mathbf{j}$ is the vector $\begin{bmatrix} 2 \\ 1 \end{bmatrix} = 2\mathbf{i} + \mathbf{j}$.

(b) Since

$$\mathbf{F}_A(-2\mathbf{i} + 4\mathbf{j}) = \mathbf{F}_A\left(\begin{bmatrix} -2 \\ 4 \end{bmatrix}\right) = \begin{bmatrix} 3 & 2 \\ 1 & 1 \end{bmatrix} \cdot \begin{bmatrix} -2 \\ 4 \end{bmatrix} = \begin{bmatrix} 2 \\ 2 \end{bmatrix} = 2\mathbf{i} + 2\mathbf{j},$$

it follows that $||\mathbf{F}_A(-2\mathbf{i} + 4\mathbf{j})|| = ||2\mathbf{i} + 2\mathbf{j}|| = \sqrt{8}$. ◀

Determinants

To any square matrix we can assign a real number, called the determinant of the matrix. Since we will need only the determinant of matrices of order 2 and 3, we will not give a general definition.

DEFINITION 1.9 Determinant of a Matrix

Let A be a square matrix of order 2. The *determinant* of A is the real number

$$\det(A) = \det \begin{bmatrix} a_{11} & a_{12} \\ a_{21} & a_{22} \end{bmatrix} = \begin{vmatrix} a_{11} & a_{12} \\ a_{21} & a_{22} \end{vmatrix}$$

defined by

$$\begin{vmatrix} a_{11} & a_{12} \\ a_{21} & a_{22} \end{vmatrix} = a_{11}a_{22} - a_{21}a_{12}.$$

If A is a square matrix of order 3, then its determinant $\det(A)$ is defined by

$$\begin{vmatrix} a_{11} & a_{12} & a_{13} \\ a_{21} & a_{22} & a_{23} \\ a_{31} & a_{32} & a_{33} \end{vmatrix} = a_{11}\begin{vmatrix} a_{22} & a_{23} \\ a_{32} & a_{33} \end{vmatrix} - a_{12}\begin{vmatrix} a_{21} & a_{23} \\ a_{31} & a_{33} \end{vmatrix} + a_{13}\begin{vmatrix} a_{21} & a_{22} \\ a_{31} & a_{32} \end{vmatrix},$$

and the 2×2 determinants are computed as above. ◀

The determinant of a 2×2 matrix is easy to compute: it is the difference of the product of the diagonal elements and the product of the off-diagonal elements. The determinant of a 3×3 matrix has three terms: each term is the product of the element a_{1j} from the first row and the jth column of A and the 2×2 determinant obtained from A by removing the first row and the jth column. The minus sign is placed in front of the a_{12} term, and then we add up all three terms.

▶ EXAMPLE 1.31

Compute the determinant of the matrix $\begin{bmatrix} 2 & 3 \\ 1 & 0 \end{bmatrix}$.

SOLUTION By definition,

$$\begin{vmatrix} 2 & 3 \\ 1 & 0 \end{vmatrix} = 2 \cdot 0 - 1 \cdot 3 = -3.$$

◀

▶ EXAMPLE 1.32

Compute

$$\begin{vmatrix} 3 & 2 & 0 \\ 1 & -7 & -2 \\ 6 & 0 & 6 \end{vmatrix}.$$

SOLUTION By definition, a determinant of order 3 is reduced to three order 2 determinants:

$$\begin{vmatrix} 3 & 2 & 0 \\ 1 & -7 & -2 \\ 6 & 0 & 6 \end{vmatrix} = 3\begin{vmatrix} -7 & -2 \\ 0 & 6 \end{vmatrix} - 2\begin{vmatrix} 1 & -2 \\ 6 & 6 \end{vmatrix} + 0\begin{vmatrix} 1 & -7 \\ 6 & 0 \end{vmatrix}$$
$$= 3(-42) - 2(18) + 0(42) = -162.$$

◀

We finish this section by listing properties of determinants. (They can all be checked by using the definition and are left as exercises.) Assume that A is a square matrix. Then

(a) If A contains a row or column of zeros, then $\det(A) = 0$.

(b) If B is a matrix obtained from A by multiplying *one* of its rows or columns by a real number α, then $\det(B) = \alpha \det(A)$.

(c) If B is a matrix obtained from A by interchanging two of its rows or columns, then $\det(B) = -\det(A)$.

(d) If A has two equal rows or columns, then $\det(A) = 0$.

(e) If $A = [a_{ij}]$, then its *transpose* A^t is defined by $A^t = [a_{ji}]$. In words, to transpose A, we turn its rows into columns and its columns into rows. The determinants of A and A^t are equal.

Assume that A and B are square matrices such that the product AB is defined. Then

(f) $\det(AB) = \det(A)\det(B)$.

▶ EXERCISES 1.4

Exercises 1 to 9: Let

$$A = \begin{bmatrix} 2 & -1 & 1 \\ 0 & -5 & 4 \end{bmatrix}, \qquad B = \begin{bmatrix} 0 & 5 \\ 4 & 0 \end{bmatrix}, \qquad C = \begin{bmatrix} -1 & 0 \\ 1 & 2 \\ 3 & -1 \end{bmatrix}$$

and let I_2 and I_3 denote the identity matrices of order 2 and 3. Compute the following expressions (or else state that they are not defined). The symbol X^2 denotes the product $X \cdot X$.

1. $2B - 16I_2$

2. $B^2 - 16I_2 + AC$

3. $CB - BA$

4. $I_3 - 3CA$

5. $AC + I_2$

6. $A(I_3 + CA)$

7. $C(BA)$

8. $(CA)B$

9. $(AC)^2 + 4B^2$

10. Let A, B, and C be matrices of the same type. Show that $A = B$ implies that $A + C = B + C$, and if α is a real number, then $A = B$ implies $\alpha A = \alpha B$.

Exercises 11 to 15: Let

$$A = \begin{bmatrix} 2 & -1 \\ 4 & 0 \end{bmatrix}, \qquad B = \begin{bmatrix} 10 & 1 \\ 0 & 0 \end{bmatrix}, \qquad C = \begin{bmatrix} 0 & 1 \\ 1 & 0 \end{bmatrix},$$

and let I_2 denote the identity matrix of order 2.

11. Solve the equation $3A - X = I_2 + 4(C - X)$ for X.

12. Solve the equation $X = 4BC - X$ for X.

13. Find the matrix X such that $AX = B$.

14. Find the images of the vector $4\mathbf{i} - 3\mathbf{j}$ under the linear maps $\mathbf{F}_A$, $\mathbf{F}_B$ defined in the text.

15. Describe in words the maps $\mathbf{F}_C$ and $\mathbf{F}_{I_2}$.

16. Consider the 2×2 matrices

$$A = \begin{bmatrix} 0 & 1 \\ 0 & 0 \end{bmatrix}, \qquad B = \begin{bmatrix} 4 & 7 \\ 0 & 0 \end{bmatrix}.$$

Show that $A^2 = A \cdot A$ and $A \cdot B$ are zero matrices. This is another big difference between matrices and real numbers: the product of nonzero matrices can be zero, whereas the product of two nonzero numbers is always nonzero. Find another pair A, B with this property.

Exercise 17 to 22: Compute the determinants of the following matrices.

17. $\begin{bmatrix} 3 & 4 \\ 4 & 3 \end{bmatrix}$

18. $\begin{bmatrix} -1 & 9 \\ 0 & 1 \end{bmatrix}$

19. $\begin{bmatrix} \cos\theta & \sin\theta \\ -r\sin\theta & r\cos\theta \end{bmatrix}$

20. $\begin{bmatrix} e^x & e^{-x} \\ e^{-x} & e^x \end{bmatrix}$

21. $\begin{bmatrix} 2 & 1 & 4 \\ 0 & 6 & 5 \\ 0 & 2 & 0 \end{bmatrix}$

22. $\begin{bmatrix} \cos t & 0 & \sin t \\ 0 & 1 & 0 \\ -\sin t & 0 & \cos t \end{bmatrix}$

23. Let A and B be 2×2 matrices. Prove that $\det(AB) = \det(A)\ \det(B)$. We know that, in general, $AB \neq BA$. However, is it true that $\det(AB) = \det(BA)$?

24. Show that if A contains a row or column of zeros, then $\det(A) = 0$.

25. Let A be a 2×2 matrix. Define the matrix B as follows: the first row of B is the first row of A, and the second row of B is the first row of A multiplied by a real number α. Show that $\det(B) = 0$.

26. Show that if B is a 2×2 or 3×3 matrix obtained from A by interchanging two of its rows or columns, then $\det(B) = -\det(A)$.

27. Let A be a 3×3 matrix and let B be the matrix obtained from A by switching the first and third rows of A and then by switching the second and third columns. What is the relation between $\det(B)$ and $\det(A)$?

28. Prove that if A is a square matrix of order 3 with two equal rows, then $\det(A) = 0$.

29. Show that if B is a 3×3 matrix obtained from A by multiplying one of its rows by a real number α, then $\det(B) = \alpha \det(A)$.

30. Show that $\det(A^t) = \det(A)$ if A is a 2×2 or 3×3 matrix.

31. Let A and B be 2×2 matrices. Show in an example that $(AB)^t \neq A^t B^t$. Prove that $(AB)^t = B^t A^t$.

32. Let A be an $n \times n$ matrix ($n \geq 2$) and let $\mathbf{F}_A$ be the linear function $\mathbf{F}_A(\mathbf{v}) = A\mathbf{v}$. Show that $\mathbf{F}_A(\alpha\mathbf{u} + \beta\mathbf{v}) = \alpha\mathbf{F}_A(\mathbf{u}) + \beta\mathbf{F}_A(\mathbf{v})$, for $\mathbf{u}, \mathbf{v} \in \mathbb{R}^n$ and $\alpha, \beta \in \mathbb{R}$ (this fact justifies the use of the word "linear").

33. Consider the matrices

$$A = \begin{bmatrix} 0 & 0 \\ 0 & 1 \end{bmatrix}, \qquad B = \begin{bmatrix} 2 & 0 \\ 0 & 2 \end{bmatrix}, \qquad C = \begin{bmatrix} 1 & 0 \\ 0 & -1 \end{bmatrix},$$

and the corresponding linear maps $\mathbf{F}_A, \mathbf{F}_B, \mathbf{F}_C : \mathbb{R}^2 \to \mathbb{R}^2$. Find a formula for each function, and give a geometric interpretation, that is, describe in words how they transform vectors.

34. Let $A = \begin{bmatrix} \cos\phi & -\sin\phi \\ \sin\phi & \cos\phi \end{bmatrix}$, where ϕ is a real number. Find a formula for $\mathbf{F}_A(\mathbf{v})$, where $\mathbf{v}$ is a vector in $\mathbb{R}^2$, and convince yourself that $\mathbf{F}_A$ is a rotation about the origin through the angle ϕ. (*Hint*: Compute the angle between $A\mathbf{v}$ and $\mathbf{v}$.)

35. Let A and B be $n \times n$ matrices ($n \geq 2$) and let $\mathbf{F}_A$ and $\mathbf{F}_B$ be the corresponding linear vector fields (or linear functions, as defined in this section). Express the linear vector fields corresponding to $A + B$, $2AB$, and $7B$ in terms of $\mathbf{F}_A$ and/or $\mathbf{F}_B$.

36. Let A and B be $n \times n$ matrices ($n \geq 2$) and let $\mathbf{F}_A$ and $\mathbf{F}_B$ be the corresponding linear functions defined in this section. Show that $\mathbf{F}_{AB} = \mathbf{F}_A \circ \mathbf{F}_B$, where $\circ$ denotes the composition of functions $\mathbf{F}_A$ and $\mathbf{F}_B$ and AB on the left side is the product of matrices A and B.

▶ 1.5 THE CROSS PRODUCT

The cross product of two vectors is an operation that assigns a vector (that is why it is sometimes called the *vector product*) to two given vectors. It is defined for vectors $\mathbb{R}^3$, and not for vectors in $\mathbb{R}^2$. There are generalizations to higher dimensions, but those will not be discussed here.

DEFINITION 1.10 Cross Product

The *cross product* of two vectors $\mathbf{v} = v_1\mathbf{i} + v_2\mathbf{j} + v_3\mathbf{k}$ and $\mathbf{w} = w_1\mathbf{i} + w_2\mathbf{j} + w_3\mathbf{k}$ is the vector $\mathbf{c} = \mathbf{v} \times \mathbf{w}$ in $\mathbb{R}^3$ defined by

$$\mathbf{c} = (v_2w_3 - v_3w_2)\mathbf{i} - (v_1w_3 - v_3w_1)\mathbf{j} + (v_1w_2 - v_2w_1)\mathbf{k}.$$ ◀

There is no need to memorize this formula. We will now present an easy way of computing the cross product with the help of the determinant. Construct a 3×3 determinant in the following way: put vectors **i**, **j**, and **k** in the first row, the components of **v** in the second row, and the components of **w** in the third row. Strictly speaking, this is not a determinant as we defined it in the previous section, since some of its entries are vectors. However, neglecting that fact and using the definition of the determinant, we obtain

$$\begin{vmatrix} \mathbf{i} & \mathbf{j} & \mathbf{k} \\ v_1 & v_2 & v_3 \\ w_1 & w_2 & w_3 \end{vmatrix} = \mathbf{i}\begin{vmatrix} v_2 & v_3 \\ w_2 & w_3 \end{vmatrix} - \mathbf{j}\begin{vmatrix} v_1 & v_3 \\ w_1 & w_3 \end{vmatrix} + \mathbf{k}\begin{vmatrix} v_1 & v_2 \\ w_1 & w_2 \end{vmatrix}$$
$$= (v_2w_3 - v_3w_2)\mathbf{i} - (v_1w_3 - v_3w_1)\mathbf{j} + (v_1w_2 - v_2w_1)\mathbf{k},$$

which is precisely the definition of the cross product!

▶ EXAMPLE 1.33

Let $\mathbf{v} = \mathbf{i} - 2\mathbf{k}$ and $\mathbf{w} = -2\mathbf{i} + 3\mathbf{j} - 4\mathbf{k}$. Compute $\mathbf{v} \times \mathbf{w}$ and $\mathbf{w} \times \mathbf{v}$.

SOLUTION We will form the cross product determinants and evaluate them:

$$\mathbf{v} \times \mathbf{w} = \begin{vmatrix} \mathbf{i} & \mathbf{j} & \mathbf{k} \\ 1 & 0 & -2 \\ -2 & 3 & -4 \end{vmatrix}$$
$$= \mathbf{i}\begin{vmatrix} 0 & -2 \\ 3 & -4 \end{vmatrix} - \mathbf{j}\begin{vmatrix} 1 & -2 \\ -2 & -4 \end{vmatrix} + \mathbf{k}\begin{vmatrix} 1 & 0 \\ -2 & 3 \end{vmatrix} = 6\mathbf{i} + 8\mathbf{j} + 3\mathbf{k}.$$

Similarly,

$$\mathbf{w} \times \mathbf{v} = \begin{vmatrix} \mathbf{i} & \mathbf{j} & \mathbf{k} \\ -2 & 3 & -4 \\ 1 & 0 & -2 \end{vmatrix}$$
$$= \mathbf{i}\begin{vmatrix} 3 & -4 \\ 0 & -2 \end{vmatrix} - \mathbf{j}\begin{vmatrix} -2 & -4 \\ 1 & -2 \end{vmatrix} + \mathbf{k}\begin{vmatrix} -2 & 3 \\ 1 & 0 \end{vmatrix} = -6\mathbf{i} - 8\mathbf{j} - 3\mathbf{k}.$$ ◀

In the previous example, $\mathbf{v} \times \mathbf{w} = -\mathbf{w} \times \mathbf{v}$. The next theorem confirms that this property holds in general.

THEOREM 1.9 Properties of the Cross Product

Let **u**, **v**, and **w**, be vectors in $\mathbb{R}^3$ and let α be any real number. The cross product satisfies $\mathbf{v} \times \mathbf{w} = -\mathbf{w} \times \mathbf{v}$ (*anticommutativity*), and $\mathbf{u} \times (\mathbf{v} + \mathbf{w}) = \mathbf{u} \times \mathbf{v} + \mathbf{u} \times \mathbf{w}$ and $(\mathbf{u} + \mathbf{v}) \times \mathbf{w} = \mathbf{u} \times \mathbf{w} + \mathbf{v} \times \mathbf{w}$ (*distributivity* with respect to the sum). Moreover, $\mathbf{v} \times \mathbf{v} = \mathbf{0}$ ($\mathbf{0}$ is the zero vector in $\mathbb{R}^3$) and $\alpha(\mathbf{v} \times \mathbf{w}) = (\alpha\mathbf{v}) \times \mathbf{w} = \mathbf{v} \times (\alpha\mathbf{w})$. ◀

PROOF: The proofs of all parts of Theorem 1.9 consist of writing the vectors involved in terms of components and using relevant definitions. As an illustration, let us prove that

$$\mathbf{u} \times (\mathbf{v} + \mathbf{w}) = \mathbf{u} \times \mathbf{v} + \mathbf{u} \times \mathbf{w}.$$

Let $\mathbf{u} = (u_1, u_2, u_3)$, $\mathbf{v} = (v_1, v_2, v_3)$, and $\mathbf{w} = (w_1, w_2, w_3)$. Then

$$\begin{aligned}
\mathbf{u} \times (\mathbf{v} + \mathbf{w}) &= \begin{vmatrix} \mathbf{i} & \mathbf{j} & \mathbf{k} \\ u_1 & u_2 & u_3 \\ v_1 + w_1 & v_2 + w_2 & v_3 + w_3 \end{vmatrix} \\
&= \mathbf{i} \begin{vmatrix} u_2 & u_3 \\ v_2 + w_2 & v_3 + w_3 \end{vmatrix} - \mathbf{j} \begin{vmatrix} u_1 & u_3 \\ v_1 + w_1 & v_3 + w_3 \end{vmatrix} + \mathbf{k} \begin{vmatrix} u_1 & u_2 \\ v_1 + w_1 & v_2 + w_2 \end{vmatrix} \\
&= (u_2(v_3 + w_3) - u_3(v_2 + w_2))\mathbf{i} - (u_1(v_3 + w_3) - u_3(v_1 + w_1))\mathbf{j} \\
&\quad + (u_1(v_2 + w_2) - u_2(v_1 + w_1))\mathbf{k}.
\end{aligned}$$

The right side is computed to be

$$\begin{aligned}
\mathbf{u} \times \mathbf{v} + \mathbf{u} \times \mathbf{w} &= \begin{vmatrix} \mathbf{i} & \mathbf{j} & \mathbf{k} \\ u_1 & u_2 & u_3 \\ v_1 & v_2 & v_3 \end{vmatrix} + \begin{vmatrix} \mathbf{i} & \mathbf{j} & \mathbf{k} \\ u_1 & u_2 & u_3 \\ w_1 & w_2 & w_3 \end{vmatrix} \\
&= \mathbf{i} \begin{vmatrix} u_2 & u_3 \\ v_2 & v_3 \end{vmatrix} - \mathbf{j} \begin{vmatrix} u_1 & u_3 \\ v_1 & v_3 \end{vmatrix} + \mathbf{k} \begin{vmatrix} u_1 & u_2 \\ v_1 & v_2 \end{vmatrix} \\
&\quad + \mathbf{i} \begin{vmatrix} u_2 & u_3 \\ w_2 & w_3 \end{vmatrix} - \mathbf{j} \begin{vmatrix} u_1 & u_3 \\ w_1 & w_3 \end{vmatrix} + \mathbf{k} \begin{vmatrix} u_1 & u_2 \\ w_1 & w_2 \end{vmatrix} \\
&= (u_2 v_3 - u_3 v_2)\mathbf{i} - (u_1 v_3 - u_3 v_1)\mathbf{j} + (u_1 v_2 - u_2 v_1)\mathbf{k} \\
&\quad + (u_2 w_3 - u_3 w_2)\mathbf{i} - (u_1 w_3 - u_3 w_1)\mathbf{j} + (u_1 w_2 - u_2 w_1)\mathbf{k}.
\end{aligned}$$

Clearly, both sides are equal.

► EXAMPLE 1.34

Find all cross products of the standard unit vectors **i**, **j**, and **k**.

SOLUTION Since the cross product of a vector with itself is a zero vector, we get $\mathbf{i} \times \mathbf{i} = \mathbf{0}$, $\mathbf{j} \times \mathbf{j} = \mathbf{0}$, and $\mathbf{k} \times \mathbf{k} = \mathbf{0}$. From the definition it follows that

$$\mathbf{i} \times \mathbf{j} = \begin{vmatrix} \mathbf{i} & \mathbf{j} & \mathbf{k} \\ 1 & 0 & 0 \\ 0 & 1 & 0 \end{vmatrix} = \mathbf{i} \begin{vmatrix} 0 & 0 \\ 1 & 0 \end{vmatrix} - \mathbf{j} \begin{vmatrix} 1 & 0 \\ 0 & 0 \end{vmatrix} + \mathbf{k} \begin{vmatrix} 1 & 0 \\ 0 & 1 \end{vmatrix} = \mathbf{k},$$

and then, by anticommutativity, $\mathbf{j} \times \mathbf{i} = -(\mathbf{i} \times \mathbf{j}) = -\mathbf{k}$. Similarly, $\mathbf{k} \times \mathbf{i} = \mathbf{j}$, $\mathbf{i} \times \mathbf{k} = -\mathbf{j}$ and $\mathbf{j} \times \mathbf{k} = \mathbf{i}$, $\mathbf{k} \times \mathbf{j} = -\mathbf{i}$.

One way to memorize these results is to write down the sequence **i j k i j** and read it from left to right: the cross product of two adjacent vectors is the next one in the sequence.

Consider the expression $\mathbf{u} \cdot (\mathbf{v} \times \mathbf{w})$, where $\mathbf{u} = (u_1, u_2, u_3)$, $\mathbf{v} = (v_1, v_2, v_3)$, and $\mathbf{w} = (w_1, w_2, w_3)$ are vectors in $\mathbb{R}^3$; it is the dot product of vectors $\mathbf{u}$ and $\mathbf{v} \times \mathbf{w}$, and therefore a scalar. Let us compute it: since

$$\mathbf{v} \times \mathbf{w} = \begin{vmatrix} \mathbf{i} & \mathbf{j} & \mathbf{k} \\ v_1 & v_2 & v_3 \\ w_1 & w_2 & w_3 \end{vmatrix} = (v_2w_3 - v_3w_2)\mathbf{i} - (v_1w_3 - v_3w_1)\mathbf{j} + (v_1w_2 - v_2w_1)\mathbf{k},$$

it follows that

$$\begin{aligned} \mathbf{u} \cdot (\mathbf{v} \times \mathbf{w}) &= (u_1\mathbf{i} + u_2\mathbf{j} + u_3\mathbf{k}) \cdot ((v_2w_3 - v_3w_2)\mathbf{i} - (v_1w_3 - v_3w_1)\mathbf{j} + (v_1w_2 - v_2w_1)\mathbf{k}) \\ &= u_1(v_2w_3 - v_3w_2) - u_2(v_1w_3 - v_3w_1) + u_3(v_1w_2 - v_2w_1). \end{aligned}$$

Once again, with the help of determinants we will be able to make sense of this expression. Form a 3×3 determinant whose rows consist of components of vectors $\mathbf{u}$, $\mathbf{v}$, and $\mathbf{w}$ (in that order). Then

$$\begin{vmatrix} u_1 & u_2 & u_3 \\ v_1 & v_2 & v_3 \\ w_1 & w_2 & w_3 \end{vmatrix} = u_1(v_2w_3 - v_3w_2) - u_2(v_1w_3 - v_3w_1) + u_3(v_1w_2 - v_2w_1),$$

and hence,

$$\mathbf{u} \cdot (\mathbf{v} \times \mathbf{w}) = \begin{vmatrix} u_1 & u_2 & u_3 \\ v_1 & v_2 & v_3 \\ w_1 & w_2 & w_3 \end{vmatrix}$$

(this is a "real" determinant, unlike the one coming from the cross product). We call the expression $\mathbf{u} \cdot (\mathbf{v} \times \mathbf{w})$ the *scalar triple product* of $\mathbf{u}$, $\mathbf{v}$, and $\mathbf{w}$.

We now turn to the geometric side of the cross product.

THEOREM 1.10 Geometric Properties of the Cross Product

Let $\mathbf{v}$ and $\mathbf{w}$ be vectors in $\mathbb{R}^3$. Then

(a) The cross product $(\mathbf{v} \times \mathbf{w})$ is a vector orthogonal to both $\mathbf{v}$ and $\mathbf{w}$.

(b) The magnitude of $\mathbf{v} \times \mathbf{w}$ is given by $||\mathbf{v} \times \mathbf{w}|| = ||\mathbf{v}|| \; ||\mathbf{w}|| \; \sin\theta$, where θ denotes the angle between $\mathbf{v}$ and $\mathbf{w}$. ◀

PROOF:

(a) The dot product $\mathbf{v} \cdot (\mathbf{v} \times \mathbf{w})$ of $\mathbf{v}$ and $\mathbf{v} \times \mathbf{w}$ is the scalar triple product of $\mathbf{v}$, $\mathbf{v}$, and $\mathbf{w}$. It is computed as the determinant whose first and second rows contain the components of $\mathbf{v}$ and the third row consists of components of $\mathbf{w}$. Since this determinant has two equal rows, it is equal to zero. (See Exercise 28 in Section 1.4.)

Hence, $\mathbf{v} \cdot (\mathbf{v} \times \mathbf{w}) = 0$ and, consequently, $\mathbf{v} \times \mathbf{w}$ is orthogonal to $\mathbf{v}$. An analogous argument proves that $\mathbf{v} \times \mathbf{w}$ is orthogonal to $\mathbf{w}$.

(b) Using the definition of the cross product and the formula for the length of a vector, we get

$$||\mathbf{v} \times \mathbf{w}||^2 = (v_2w_3 - v_3w_2)^2 + (v_1w_3 - v_3w_1)^2 + (v_1w_2 - v_2w_1)^2.$$

Now add zero, written as $v_1^2w_1^2 + v_2^2w_2^2 + v_3^2w_3^2 - v_1^2w_1^2 - v_2^2w_2^2 - v_3^2w_3^2$, to the right side and rearrange terms as follows:

$$\begin{aligned}||\mathbf{v} \times \mathbf{w}||^2 &= v_1^2w_1^2 + v_1^2w_2^2 + v_1^2w_3^2 + v_2^2w_1^2 + v_2^2w_2^2 + v_2^2w_3^2 + v_3^2w_1^2 + v_3^2w_2^2 + v_3^2w_3^2 \\ &\quad - (v_1^2w_1^2 + v_2^2w_2^2 + v_3^2w_3^2 + 2v_1w_1v_2w_2 + 2v_1w_1v_3w_3 + 2v_2w_2v_3w_3).\end{aligned}$$

Combine the first nine and the last six terms to obtain

$$\begin{aligned}||\mathbf{v} \times \mathbf{w}||^2 &= (v_1^2 + v_2^2 + v_3^2)(w_1^2 + w_2^2 + w_3^2) - (v_1w_1 + v_2w_2 + v_3w_3)^2 \\ &= ||\mathbf{v}||^2||\mathbf{w}||^2 - (\mathbf{v} \cdot \mathbf{w})^2.\end{aligned}$$

Using Theorem 1.3, we simplify the right side as follows:

$$||\mathbf{v}||^2||\mathbf{w}||^2 - (\mathbf{v} \cdot \mathbf{w})^2 = ||\mathbf{v}||^2||\mathbf{w}||^2 - ||\mathbf{v}||^2||\mathbf{w}||^2 \cos^2\theta = ||\mathbf{v}||^2||\mathbf{w}||^2 \sin^2\theta.$$

Therefore,

$$||\mathbf{v} \times \mathbf{w}||^2 = ||\mathbf{v}||^2||\mathbf{w}||^2 \sin^2\theta.$$

Computing the square root of both sides ($\sqrt{\sin^2\theta} = |\sin\theta| = \sin\theta$, since $0 \le \theta \le \pi$, and therefore $\sin\theta \ge 0$), we arrive at the formula in (b). ◀

We can now describe geometrically the cross product $\mathbf{c} = \mathbf{v} \times \mathbf{w}$. By the previous theorem, $\mathbf{c}$ is orthogonal to both $\mathbf{v}$ and $\mathbf{w}$ and its length is given by $||\mathbf{c}|| = ||\mathbf{v}||\ ||\mathbf{w}||\ \sin\theta$. This narrows down our search to two candidates (see Figure 1.18), $\mathbf{c}_1$ and $\mathbf{c}_2$, where $\mathbf{c}_2 = -\mathbf{c}_1$ and $||\mathbf{c}_1|| = ||\mathbf{c}_2||$. Consider an example first: let $\mathbf{v} = \mathbf{i}$ and $\mathbf{w} = \mathbf{j}$; then $\mathbf{c} = \mathbf{i} \times \mathbf{j} = \mathbf{k}$, and therefore $\mathbf{c}_1$ should represent $\mathbf{i} \times \mathbf{j}$. This example illustrates the following general rule. The direction of $\mathbf{v} \times \mathbf{w}$ is determined by the *"right-hand rule"*: place your right hand in the direction of $\mathbf{v}$, and curl your fingers from $\mathbf{v}$ to $\mathbf{w}$ through the angle θ (remember that θ is the smaller of the two angles formed by the lines with directions $\mathbf{v}$ and $\mathbf{w}$). Your thumb then points in the direction of $\mathbf{v} \times \mathbf{w}$. Applying this rule to vectors $\mathbf{v}$ and $\mathbf{w}$ of Figure 1.18, we see that $\mathbf{c}_1 = \mathbf{v} \times \mathbf{w}$. Another common interpretation is the following: when driving an ordinary screw, turning the screwdriver from $\mathbf{v}$ to $\mathbf{w}$ advances the screw in the direction of $\mathbf{v} \times \mathbf{w}$.

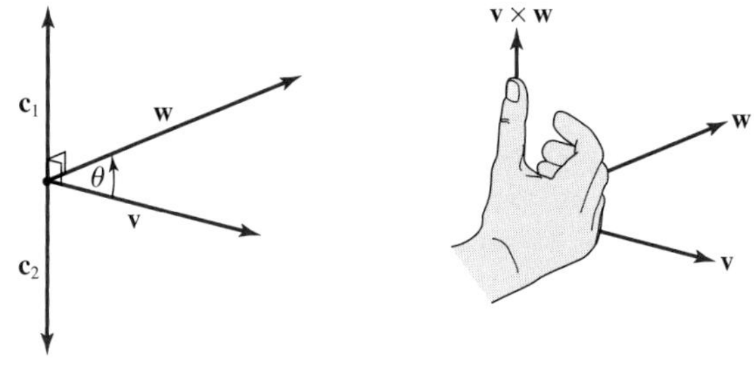

Figure 1.18 Candidates for the cross product of $\mathbf{v}$ and $\mathbf{w}$. By the "right-hand rule," $\mathbf{c}_1 = \mathbf{v} \times \mathbf{w}$.

▶ EXAMPLE 1.35

Find all unit vectors orthogonal to both $\mathbf{v} = \mathbf{i} + \mathbf{k}$ and $\mathbf{w} = \mathbf{j} - 2\mathbf{k}$.

SOLUTION The vector

$$\mathbf{c} = \mathbf{v} \times \mathbf{w} = \begin{vmatrix} \mathbf{i} & \mathbf{j} & \mathbf{k} \\ 1 & 0 & 1 \\ 0 & 1 & -2 \end{vmatrix} = -\mathbf{i} + 2\mathbf{j} + \mathbf{k}$$

is orthogonal to both $\mathbf{v}$ and $\mathbf{w}$. Now, normalize it to make it of unit length: $\mathbf{c}/||\mathbf{c}|| = (-\mathbf{i} + 2\mathbf{j} + \mathbf{k})/\sqrt{6}$. This is one desired vector. There is another one, $-(-\mathbf{i} + 2\mathbf{j} + \mathbf{k})/\sqrt{6}$, in the opposite direction. ◀

▶ EXAMPLE 1.36 Velocity Vector of a Rotating Body

Assume that a particle P rotates with constant angular speed w (units could be radians per second, e.g.) about an axis ℓ in space (its trajectory, thus, is a circle of radius a). The information on the motion of P is summarized by the *angular velocity vector* $\mathbf{w}$: its magnitude $w = ||\mathbf{w}||$ is the angular speed and its direction is parallel to the axis of rotation and determined by the right-hand rule (when the fingers of the right hand are curled in the direction of rotation, the thumb points in the direction of $\mathbf{w}$).

Assume (for convenience) that ℓ contains the origin O and visualize $\mathbf{w}$ as the directed line segment starting at O, as in Figure 1.19. Denote by π the plane through O spanned by $\mathbf{w}$ and $\mathbf{p}$, where $\mathbf{p}$ is the vector represented by $\overrightarrow{OP}$. The *tangential velocity vector* $\mathbf{v}$ is defined as the vector orthogonal to this plane, with magnitude $||\mathbf{v}|| = ||\mathbf{w}||\, a$ [since arc = angle · radius, it follows that (tangential) speed = arc/time = (angle/time) · radius = (angular) speed · radius], whose direction indicates the direction of motion. Let θ be the angle between $\mathbf{w}$ and $\mathbf{p}$. From $\sin\theta = a/||\mathbf{p}||$, we get $a = ||\mathbf{p}|| \sin\theta$ and hence $||\mathbf{v}|| = ||\mathbf{w}||\, ||\mathbf{p}|| \sin\theta$. Since $\mathbf{v}$ is orthogonal to $\mathbf{w}$ and $\mathbf{p}$, by the definition of $\mathbf{w}$ (the right-hand rule) it follows that $\mathbf{v} = \mathbf{w} \times \mathbf{p}$. In words, the tangential velocity vector is the cross product of the angular velocity vector and the position vector of a particle.

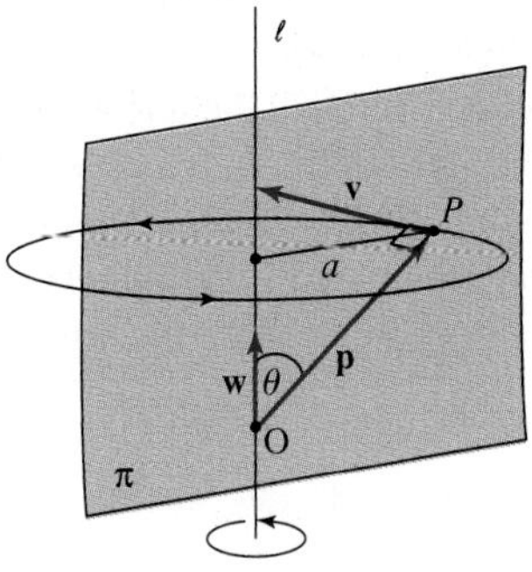

Figure 1.19 Angular and tangential velocity vectors.

◀

▶ EXAMPLE 1.37 Torque

Suppose that a constant force $\mathbf{F}$ acts at a point $A(x, y, z)$ on a rigid body M that is fixed at the origin (and capable of rotating only). The *torque* $\mathbf{T}$ (with respect to the origin) is defined as the cross product $\mathbf{T} = \mathbf{r} \times \mathbf{F}$ of the position vector $\mathbf{r} = x\mathbf{i} + y\mathbf{j} + z\mathbf{k}$ of A and the force vector $\mathbf{F}$. By

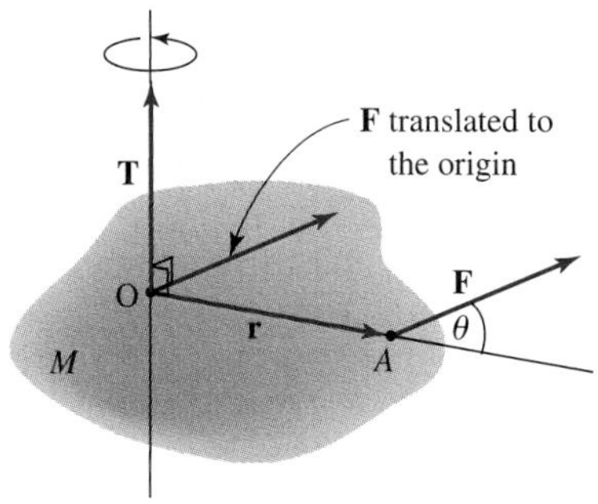

Figure 1.20 Torque of a force from Example 1.37.

Theorem 1.10, $\mathbf{T}$ is orthogonal to both $\mathbf{r}$ and $\mathbf{F}$ and its magnitude is $||\mathbf{T}|| = ||\mathbf{r}||\ ||\mathbf{F}||\ \sin\theta$, where θ is the angle between the force applied and the position vector of A (see Figure 1.20). If $\mathbf{F}$ acts parallel to $\mathbf{r}$, there will be no rotation ($\mathbf{F}$ just tries to pull A in the radial direction away from O) and the torque is zero. If the torque is nonzero, $\mathbf{F}$ tends to cause a rotation whose axis is parallel to $\mathbf{T}$. This is why we say that torque measures the tendency of a force to cause the rotation of a rigid body.

THEOREM 1.11 Test for Parallel Vectors

Nonzero vectors $\mathbf{v}$ and $\mathbf{w}$ in $\mathbb{R}^3$ are parallel if and only if $\mathbf{v} \times \mathbf{w} = \mathbf{0}$.

PROOF: Nonzero vectors $\mathbf{v}$ and $\mathbf{w}$ are parallel if and only if the angle θ between them is 0 or π. In either case, $\sin\theta = 0$, and therefore $||\mathbf{v} \times \mathbf{w}|| = ||\mathbf{v}||\ ||\mathbf{w}||\ \sin\theta = 0$, and hence $\mathbf{v} \times \mathbf{w} = \mathbf{0}$. Conversely, if $\mathbf{v} \times \mathbf{w} = \mathbf{0}$, then $||\mathbf{v} \times \mathbf{w}|| = 0$ and (since $||\mathbf{v}||, ||\mathbf{w}|| \neq 0$) it follows that $\sin\theta = 0$. Hence, $\theta = 0$ or $\theta = \pi$.

Let $\mathbf{v}$ and $\mathbf{w}$ be nonparallel vectors in $\mathbb{R}^3$ represented by the directed line segments $\overrightarrow{AB}$ and $\overrightarrow{AC}$, respectively, that start at the same point, as shown in Figure 1.21. By drawing parallels to $\overline{AB}$ and $\overline{AC}$, we obtain the parallelogram $ABDC$, called the *parallelogram spanned by* $\mathbf{v}$ *and* $\mathbf{w}$. Let h be its height drawn from C perpendicular to $\overline{AB}$, and denote the angle between $\mathbf{v}$ and $\mathbf{w}$ by θ. Then $\sin\theta = h/||\mathbf{w}||$, so $h = ||\mathbf{w}||\sin\theta$, and the area of the parallelogram is equal to $||\mathbf{v}||\ h = ||\mathbf{v}||\ ||\mathbf{w}||\sin\theta$, which is precisely the magnitude of $\mathbf{v} \times \mathbf{w}$! We have thus demonstrated the following theorem.

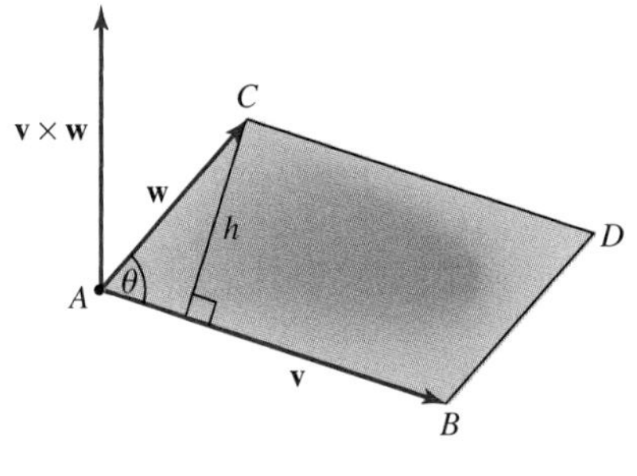

Figure 1.21 The magnitude of the cross product of $\mathbf{v}$ and $\mathbf{w}$ equals the area of the parallelogram spanned by $\mathbf{v}$ and $\mathbf{w}$.

THEOREM 1.12 Area of the Parallelogram Spanned by Two Vectors

Let **v** and **w** be nonzero, nonparallel vectors in $\mathbb{R}^3$. The magnitude $||\mathbf{v} \times \mathbf{w}||$ is the real number equal to the area of the parallelogram spanned by **v** and **w**. ◀

Since **v** and **w** are nonparallel, $0 < \theta < \pi$ and therefore $\sin\theta > 0$; it follows that $||\mathbf{v}||\,||\mathbf{w}||\sin\theta > 0$; this is reasonable, since areas are positive.

▶ EXAMPLE 1.38

Compute the area of the parallelogram in the xy-plane with three of its vertices located at $(1, 1)$, $(3, 2)$, and $(0, 2)$.

SOLUTION Labeling the vertices as points $A(1, 1)$, $B(3, 2)$, and $C(0, 2)$, we compute the vectors determined by the directed line segments $\overrightarrow{AB}$ and $\overrightarrow{AC}$ to be $\mathbf{v} = 2\mathbf{i} + \mathbf{j}$ and $\mathbf{w} = -\mathbf{i} + \mathbf{j}$. Now think of **v** and **w** as vectors in $\mathbb{R}^3$ whose **k** components are zero (a useful trick). The area of the parallelogram in question is

$$||\mathbf{v} \times \mathbf{w}|| = \left|\left| \begin{vmatrix} \mathbf{i} & \mathbf{j} & \mathbf{k} \\ 2 & 1 & 0 \\ -1 & 1 & 0 \end{vmatrix} \right|\right| = ||3\mathbf{k}|| = 3.$$

In the solution to this problem, we made choices in labeling the points, and hence in the vectors that span the parallelogram. The reader is invited to check that other combinations (producing possibly different spanning vectors) give the same result. ◀

Now let **u**, **v**, and **w** be nonzero vectors in $\mathbb{R}^3$ (represented by directed line segments with the same initial point) such that **v** and **w** are not parallel (so that they span a parallelogram) and such that **u** does not belong to the plane spanned by **v** and **w**.

Construct the parallelepiped with adjacent sides **u**, **v**, and **w** (i.e., *spanned* by **u**, **v**, and **w**) as shown in Figure 1.22, and consider the scalar triple product

$$\mathbf{u} \cdot (\mathbf{v} \times \mathbf{w}) = ||\mathbf{u}||\,||\mathbf{v} \times \mathbf{w}||\cos\theta,$$

where θ is the angle between **u** and $\mathbf{v} \times \mathbf{w}$ ($0 \le \theta \le \pi$ and $\theta \ne \pi/2$). By the previous theorem, $||\mathbf{v} \times \mathbf{w}||$ is the area of the parallelogram spanned by **v** and **w**. If $\theta < \pi/2$, the height h of the parallelepiped is $h = ||\mathbf{u}||\cos\theta$; see Figure 1.22(a). If $\theta > \pi/2$, then $h = ||\mathbf{u}||\cos(\pi - \theta) = -||\mathbf{u}||\cos\theta$; see Figure 1.22(b). In either case, $h = ||\mathbf{u}||\,|\cos\theta|$, and consequently, the absolute value

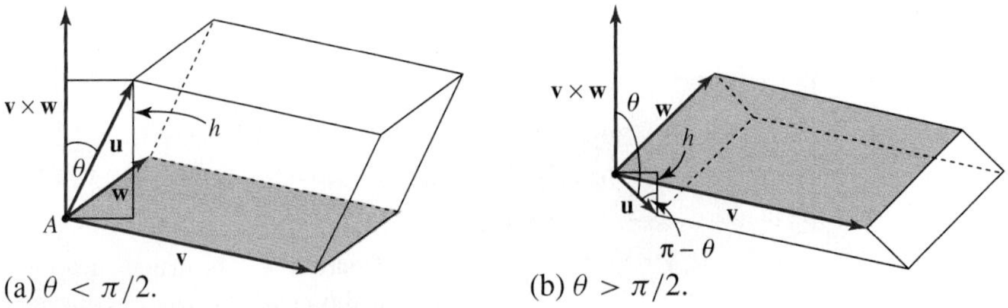

(a) $\theta < \pi/2$. (b) $\theta > \pi/2$.

Figure 1.22 Parallelepiped spanned by **u**, **v**, and **w**.

$$|\mathbf{u} \cdot (\mathbf{v} \times \mathbf{w})| = ||\mathbf{v} \times \mathbf{w}|| \, ||\mathbf{u}|| \, |\cos\theta|$$

represents the volume of the parallelepiped spanned by **u**, **v**, and **w**.

▶ EXAMPLE 1.39

Find the volume of the parallelepiped spanned by **i**, $\mathbf{i} - 2\mathbf{j}$, and $\mathbf{i} - \mathbf{j} + \mathbf{k}$.

SOLUTION The scalar triple product of the given vectors is

$$\mathbf{i} \cdot ((\mathbf{i} - 2\mathbf{j}) \times (\mathbf{i} - \mathbf{j} + \mathbf{k})) = \begin{vmatrix} 1 & 0 & 0 \\ 1 & -2 & 0 \\ 1 & -1 & 1 \end{vmatrix} = -2.$$

The volume of the parallelepiped is 2.

▶ EXAMPLE 1.40 k-cross Operator in Meteorology

In meteorology, a special cross product, called **k**-*cross,* is sometimes used to describe horizontal movements of air. Given a vector **a** in the xy-plane, the vector $\mathbf{k} \times \mathbf{a}$ (**k** is the unit vector in the direction of the z-axis) is obtained by rotating the vector **a** through 90° counterclockwise. We interpret $\mathbf{k} \times \mathbf{a}$ as the action of the operator (called **k**-cross) on **a**. For instance, **k**-cross transforms **i** to **j** and $-\mathbf{j}$ to **i**. Likewise, $\mathbf{k} \times (\mathbf{k} \times \mathbf{a}) = -\mathbf{a}$.

A mathematical model that is sometimes used to describe motions in Earth's atmosphere states that the wind speed **v** satisfies $\mathbf{v} = A\,(\mathbf{k} \times \mathbf{p})$, where A is a positive constant and **p** denotes the direction of the largest increase in horizontal air pressure (see Example 2.92 in Section 2.7 for details). In other words, the wind blows in the direction 90° counterclockwise from the largest increase in horizontal air pressure.

▶ EXERCISES 1.5

1. Let $\mathbf{v} = 2\mathbf{i} - 3\mathbf{k}$ and $\mathbf{w} = 4\mathbf{j} + \mathbf{i} + \mathbf{k}$. Compute $\mathbf{v} \times \mathbf{w}$, $(\mathbf{v} + \mathbf{w}) \times \mathbf{k}$, $(2\mathbf{v} - \mathbf{w}) \times (\mathbf{v} + \mathbf{w})$, and $\mathbf{i} \cdot (\mathbf{v} \times \mathbf{w})$.

2. Prove that for any two vectors **v** and **w** in $\mathbb{R}^3$, $\mathbf{v} \times \mathbf{w} = -\mathbf{w} \times \mathbf{v}$. Use this fact to prove that $\mathbf{v} \times \mathbf{v} = \mathbf{0}$.

3. Show that $(\mathbf{u} + \mathbf{v}) \times \mathbf{w} = \mathbf{u} \times \mathbf{w} + \mathbf{v} \times \mathbf{w}$ for **u**, **v**, and **w** in $\mathbb{R}^3$.

4. Prove that $\alpha(\mathbf{v} \times \mathbf{w}) = (\alpha\mathbf{v}) \times \mathbf{w}$, for any real number α and $\mathbf{v}, \mathbf{w} \in \mathbb{R}^3$.

5. Find all unit vectors orthogonal to both $\mathbf{i} + \mathbf{j} - 2\mathbf{k}$ and **k**.

6. Describe all vectors orthogonal to both $4\mathbf{i} + \mathbf{k}$ and $\mathbf{j} - \mathbf{i} + 2\mathbf{k}$. Find all unit vectors orthogonal to the given vectors.

7. Find vectors **u**, **v**, and **w** such that $(\mathbf{u} \times \mathbf{v}) \times \mathbf{w} \neq \mathbf{u} \times (\mathbf{v} \times \mathbf{w})$. In other words, show that the cross product is not associative. (*Hint*: Try standard unit vectors **i**, **j**, and **k**.)

8. Find the volume of the parallelepiped spanned by **i**, $\mathbf{i} + \mathbf{k}$, and $\mathbf{i} - \mathbf{j} + \mathbf{k}$.

9. Find the volume of the parallelepiped spanned by $3\mathbf{i} + \mathbf{j} - \mathbf{k}$, $-\mathbf{i} - \mathbf{j}$, and $\mathbf{j} + \mathbf{k}$.

10. Using the properties of the determinant, show that $\mathbf{u} \cdot (\mathbf{v} \times \mathbf{w}) = \mathbf{v} \cdot \mathbf{w} \times \mathbf{u} = \mathbf{w} \cdot (\mathbf{u} \times \mathbf{v})$, and $\mathbf{u} \cdot (\mathbf{v} \times \mathbf{w}) = -\mathbf{u} \cdot (\mathbf{w} \times \mathbf{v})$.

11. Writing vectors in terms of components, prove that $\mathbf{a} \times (\mathbf{b} \times \mathbf{c}) = (\mathbf{a} \cdot \mathbf{c})\mathbf{b} - (\mathbf{a} \cdot \mathbf{b})\mathbf{c}$, for **a**, **b**, and **c** in $\mathbb{R}^3$.

12. Use the identity from Exercise 11 to prove that $\mathbf{a} \times (\mathbf{b} \times \mathbf{c}) + \mathbf{b} \times (\mathbf{c} \times \mathbf{a}) + \mathbf{c} \times (\mathbf{a} \times \mathbf{b}) = \mathbf{0}$ for vectors **a**, **b**, and **c** in $\mathbb{R}^3$.

13. Let us go back to Example 1.38. Draw the parallelogram spanned by the vectors **v** and **w** chosen in the solution. Now choose another pair of vectors determined by the given points, draw the parallelogram that they span, and compute its area.

14. Let **u**, **v**, and **w** be nonzero vectors in $\mathbb{R}^3$. Show that **u** lies in the plane spanned by **v** and **w** if and only if $\mathbf{u} \cdot (\mathbf{v} \times \mathbf{w}) = 0$. Use this fact to check that the vectors $2\mathbf{i} + \mathbf{j}$, $-\mathbf{i} + 2\mathbf{k}$, and $3\mathbf{i} + 2\mathbf{j} + 2\mathbf{k}$ lie in the same plane.

15. Using the statement of Exercise 14, determine whether the points $(1, 0, 0)$, $(1, 2, -1)$, $(0, 2, -4)$, and $(2, -1, 0)$ lie in the same plane.

16. Find the area of the triangle with vertices $(0, 2, 1)$, $(3, 3, 3)$, and $(-1, 4, 2)$.

17. Use the cross product to show that the points $(1, 0, 0)$, $(-3, 2, 2)$, and $(3, -1, -1)$ lie in the same line.

18. Find an equation of the plane that passes through $(3, 2, -1)$ and contains the line $\mathbf{l}(t) = (-t, 2, 3 + t)$, $t \in \mathbb{R}$.

19. Find an equation of the plane that contains the lines $\mathbf{l}_1(t) = (-t, 1, 1 + t)$, $t \in \mathbb{R}$ and $\mathbf{l}_2(t) = (3t, 1, 1 - t)$, $t \in \mathbb{R}$.

20. Find an equation of the plane that passes through the origin and contains the points $(1, -2, -3)$ and $(0, 4, 1)$.

21. Find an equation of the plane that contains the points $(3, 0, 1)$, $(0, 4, -2)$, and $(2, -3, 0)$.

22. Describe geometrically all solutions $\mathbf{x} \in \mathbb{R}^3$ of the equation $\mathbf{x} \times \mathbf{a} = \mathbf{b}$, where **a** and **b** are given nonzero vectors in $\mathbb{R}^3$.

23. Find the (perpendicular) distance between the point $(3, 2, 0)$ and the plane through the origin spanned by $\mathbf{v} = (2, 4, 0)$ and $\mathbf{w} = (0, 1, -1)$.

24. Find the torque **T** of a force of 10 N acting parallel to the xy-plane at an angle of $\pi/6$ rad with respect to the positive x-axis, at the point $(3, -2, -2)$ on a rigid body M in $\mathbb{R}^3$.

25. Let M be a sphere of radius 1 whose center is fixed at the origin (so that M is capable of rotating only). Describe the points on M where the torque of $\mathbf{F} = 2\mathbf{k}$ has the largest and the smallest magnitude.

26. A rigid body is rotated about the z-axis at an angular speed of 5π rad/s. Find the angular velocity vector and the tangential velocity vector at the point $(1, 2, 1)$.

27. A disk in the xy-plane of radius 1, centered at the origin, is rotated counterclockwise as seen from above about the z-axis at an angular speed of 10 rad/s. Identify the points on the disk where the magnitude of the tangential velocity vector is the largest and the smallest.

28. Show that $\mathbf{k} \times (\mathbf{k} \times \mathbf{a}) = -\mathbf{a}$ for any vector **a** in the xy-plane. Also, prove that the **k**-cross operator preserves the length of horizontal vectors, that is, $||\mathbf{k} \times \mathbf{a}|| = ||\mathbf{a}||$ for all vectors **a** that belong to the xy-plane.

▶ CHAPTER REVIEW

CHAPTER SUMMARY

- **Vectors.** Cartesian coordinates, polar coordinates, length of a vector, unit vector, addition of vectors, and multiplication by scalars.

- **Applications of vectors.** Finding equations of lines and planes, displacement vector, velocity and speed, forces and superposition of forces, center of mass.
- **Dot product.** Properties, geometric interpretation, test for orthogonality of vectors, angle between vectors, orthonormal set of vectors, vector expressed in terms of orthogonal vectors, work.
- **Cross product.** Properties, geometric interpretation, area of the parallelogram spanned by two vectors, volume of the parallelepiped spanned by three vectors, torque.
- **Matrices and determinants.** Elementary operations with matrices, matrix multiplication, linear map, properties of determinants.

REVIEW

Discuss the following questions.

1. Give an analytic and a geometric description of a vector in $\mathbb{R}^3$. Explain how to identify vectors in $\mathbb{R}^3$ and points in $\mathbb{R}^3$.

2. Define a center of mass and explain how to find it. If one object is placed in the first quadrant, and the other in the third quadrant, is it possible for their center of mass to lie in the fourth quadrant?

3. Explain how to find an equation of the plane in $\mathbb{R}^3$ that contains three given points.

4. Is it true that the cross product of two unit vectors is a unit vector? Find an example to show that the cross product is not associative.

5. We defined a vector in $\mathbb{R}^n$ as an ordered n-tuple of real numbers. What other objects have been (for various reasons) identified as vectors?

6. State how the dot product of two vectors relates to their length.

7. What is the largest number of elements in an orthonormal set of vectors in $\mathbb{R}^2$, $\mathbb{R}^3$ (and in $\mathbb{R}^n$, in general)? Find a formula that expresses a vector $\mathbf{a} \in \mathbb{R}^3$ in terms of three orthonormal vectors $\mathbf{u}$, $\mathbf{v}$, and $\mathbf{w}$ in $\mathbb{R}^3$.

8. Explain how to find a vector projection of a vector onto a given vector.

9. Write down the formula for the angle between two vectors. State the method(s) that you can use to check whether two vectors are parallel or not.

10. Consider two unit vectors $\mathbf{v}$ and $\mathbf{w}$ in $\mathbb{R}^3$. When does their cross product have a maximum magnitude? Minimum magnitude? When is it zero?

11. List the differences between matrix multiplication and ordinary multiplication of real numbers. Find two 2×2 nonzero matrices A and B such that $AB = 0$.

12. Explain how to use vectors and vector operations to determine whether a given line and a given plane are parallel ("parallel" means that they do not intersect each other).

TRUE/FALSE QUIZ

Determine whether the following statements are true or false. Give reasons for your answer.

1. The vector $\mathbf{v} = (1/\sqrt{3}, -1/\sqrt{3}, 1/\sqrt{3})$ is a unit vector.

2. The vectors $(1/\sqrt{2}, -1/\sqrt{2}, 0)$, $(0, 0, 1)$, and $(-1/\sqrt{2}, 1/\sqrt{2}, 0)$ form an orthonormal set of vectors in $\mathbb{R}^3$.

3. Equations $y = -2x + 3$ and $\mathbf{l}(t) = (t - 1, -2t + 4)$, $t \in \mathbb{R}$ represent the same line.

4. The center of mass of four objects placed on the line $y = x$ lies somewhere on the line $y = x$.

5. If $\mathbf{v}$ and $\mathbf{w}$ are parallel vectors, then $\mathbf{v} \cdot \mathbf{w} = ||\mathbf{v}||\,||\mathbf{w}||$.

6. If $\mathbf{v} \cdot \mathbf{w} < 0$, then the angle between $\mathbf{v}$ and $\mathbf{w}$ is greater than π.

7. The triangle with vertices at the points $(1, 1, 1)$, $(0, -1, 1)$, and $(-2, -1, 4)$ is a right triangle.

8. If $\mathbf{v}$ and $\mathbf{w}$ are vectors in $\mathbb{R}^3$, the expression $(\mathbf{w} \cdot \mathbf{v})(\mathbf{v} \times \mathbf{w})$ is a scalar.

9. If A and B are nonzero square matrices, then $\det(AB) \neq 0$.

10. If the determinant of a matrix A is zero, then at least one entry of A must be zero.

11. If $A = \begin{bmatrix} 1 & 0 \\ 0 & 2 \end{bmatrix}$, then $A^2 = \begin{bmatrix} 1 & 0 \\ 0 & 4 \end{bmatrix}$ and $A^3 = \begin{bmatrix} 1 & 0 \\ 0 & 8 \end{bmatrix}$.

12. The dot product satisfies $\mathbf{w} \cdot \mathbf{v} = \mathbf{v} \cdot \mathbf{w}$, for all vectors $\mathbf{v}$ and $\mathbf{w}$.

13. The cross product of two nonzero perpendicular vectors is a nonzero vector.

REVIEW EXERCISES AND PROBLEMS

1. Let $\mathbf{v}$ and $\mathbf{w}$ be nonzero vectors. Prove that $||\mathbf{v} - \mathbf{w}||^2 + ||\mathbf{v} + \mathbf{w}||^2 = 2(||\mathbf{v}||^2 + ||\mathbf{w}||^2)$ and give a geometric interpretation.

2. Let $\mathbf{v}$ and $\mathbf{w}$ be nonzero vectors in $\mathbb{R}^3$, where $\mathbf{v} \neq \mathbf{w}$. Show that $\mathbf{v}$ and $(\mathbf{v} \times \mathbf{w}) \times \mathbf{v}$ are orthogonal. Prove that

$$\mathbf{w} = \frac{\mathbf{v} \cdot \mathbf{w}}{||\mathbf{v}||^2}\,\mathbf{v} + \frac{(\mathbf{v} \times \mathbf{w}) \times \mathbf{v}}{||\mathbf{v}||^2}.$$

3. Let $\mathbf{v} \in \mathbb{R}^3$ be a nonzero vector. The direction angles of $\mathbf{v}$ are the angles α, β, and γ $(0 \leq \alpha, \beta, \gamma \leq \pi)$ between $\mathbf{v}$ and the positive x-axis, y-axis, and z-axis, respectively. Express the so-called *direction cosines* $\cos\alpha$, $\cos\beta$, and $\cos\gamma$ in terms of $\mathbf{v}$ and the unit vectors $\mathbf{i}$, $\mathbf{j}$, and $\mathbf{k}$, and show that $\cos^2\alpha + \cos^2\beta + \cos^2\gamma = 1$. Express a nonzero vector $\mathbf{v}$ in $\mathbb{R}^3$ in terms of its norm and its direction cosines.

4. Prove the Triangle Inequality using the Cauchy–Schwarz Inequality $|\mathbf{v} \cdot \mathbf{w}| \leq ||\mathbf{v}||\,||\mathbf{w}||$ of Exercise 10 in Section 1.3 and the properties of the dot product. [*Hint*: Start by expanding $||\mathbf{v} + \mathbf{w}||^2 = (\mathbf{v} + \mathbf{w}) \cdot (\mathbf{v} + \mathbf{w})$.]

5. Find the distance from the point $(2, 3, -4)$ to the plane $x - 2y - z = 11$.

6. Find the area of a triangle with vertices $(1, -1, 0)$, $(2, 0, 1)$, and $(-1, -2, -3)$.

7. Let A be a square matrix of order 2. Show that the fact that A^2 $(A^2 = AA)$ is a zero matrix does not imply that A is a zero matrix. Prove that if AA^t is a zero matrix, then A must be a zero matrix.

8. Prove that $4\mathbf{v} \cdot \mathbf{w} = ||\mathbf{v} + \mathbf{w}||^2 - ||\mathbf{v} - \mathbf{w}||^2$ for any two vectors in $\mathbb{R}^2$ or in $\mathbb{R}^3$.

9. Compute an equation of the plane Π that contains the points $(1, 2, 0)$, $(2, 0, 4)$, and $(-1, -1, 2)$. Find an equation representing all planes perpendicular to Π that contain the point $(1, 2, 0)$.

10. Let D be the 3×3 matrix whose only nonzero elements lie on the main diagonal. Is it true that $DA = AD$ for any 3×3 matrix A?

11. Show that the parallelepiped spanned by $\mathbf{a} = \mathbf{i} - \mathbf{j}$, $\mathbf{b} = 2\mathbf{i} - 3\mathbf{j}$, and $\mathbf{c} = \mathbf{k}$ has the same volume as the parallelepiped spanned by $\mathbf{a}$, $\mathbf{b}$, and $\mathbf{a} + \mathbf{c}$. Explain why this is true using a geometric argument.

12. Find all vectors of length 7 that are perpendicular to the plane determined by the points $(3, 0, -4)$, $(4, 6, -2)$, and $(-1, 0, -1)$.

13. Let A be a 3×3 matrix and let α and β be two real numbers. Define the matrix B as follows: the first row of B is the first row of A, the second row of B is the second row of A, and the third row of B is the sum of the first row of A multiplied by α and the second row of A multiplied by β. Show that $\det(B) = 0$. Give a geometric reason why $\det(B) = 0$.

14. Show that the (perpendicular) distance from the point (x_0, y_0) to the line $ax + by + c = 0$ in $\mathbb{R}^2$ is given by $d = |ax_0 + by_0 + c|/\sqrt{a^2 + b^2}$.

15. Let A be a 2×2 matrix and let $\mathbf{F}_A$ be the corresponding linear function $\mathbf{F}_A(\mathbf{v}) = A\mathbf{v}$, $\mathbf{v} \in \mathbb{R}^2$.

(a) Find a nonzero matrix A and a nonzero vector $\mathbf{v}$ such that $\mathbf{F}_A(\mathbf{v}) = \mathbf{0}$.

(b) Assume that $\det(A) \neq 0$. Prove that $\mathbf{F}_A(\mathbf{v}) = \mathbf{0}$ implies that $\mathbf{v} = \mathbf{0}$.

16. Let $\mathbf{a}$ and $\mathbf{b}$ be nonzero vectors in $\mathbb{R}^3$. Find all vectors $\mathbf{x} \in \mathbb{R}^3$ such that $\mathbf{x} \times \mathbf{a} = \mathbf{b}$ and $\mathbf{x} \cdot \mathbf{a} = 1$. Use geometric arguments to determine the number of solutions.

17. A car A, initially located at the point $(10, 0)$ (assume that the units are km), moves in the negative direction of the x-axis (i.e., toward the origin) with the constant speed of 50 km/h. A car B, initially located at the origin, moves in the positive direction of the y-axis with the constant speed of 80 km/h.

(a) Find their positions at time t (hours).

(b) Find an expression for the distance between A and B in terms of t.

(c) Plot the distance as a function of t.

(d) When is the distance between A and B the smallest? What is that distance?

(e) When will the distance between A and B be 200 km?

CHAPTER 2

Calculus of Functions of Several Variables

The aim of this chapter is to generalize concepts of calculus of real-valued functions of one variable (limit, continuity, and, in particular, differentiability) to real-valued and vector-valued functions of several variables. Since the properties of continuity and differentiability of a vector-valued function depend on those of its components (which are real-valued), our study will focus on real-valued functions.

This chapter opens with the definition and examples of various functions of several variables. Different ways of visualizing such functions are discussed, the emphasis being not so much on the ways of constructing graphs but on understanding what they represent. In other words, we will discuss the ways to "read" the information on the function contained in its visual interpretation. The discussion of the limit is presented on a more intuitive level and covers only those aspects that will be needed in this book. A full and rigorous approach would require a lot of time, space, and patience (on the part of both the writer and the reader), and is certainly beyond the scope of this book. The definition of continuity is a straightforward generalization of the definition for real-valued functions of one variable, and the results presented in this section are the extensions of known results for such functions.

In this chapter, we cover and immediately employ several basic concepts related to curves: parametrization, graph, velocity, and tangent. A thorough study of curves is deferred to Chapter 3.

An exhaustive discussion of derivatives and differentiability is given. In order to preserve the theorem "differentiability implies continuity," we need a new definition of differentiability for functions of several variables. An interpretation (linear approximation) will make a rather abstract definition more transparent and understandable. We proceed by discussing properties of differentiable functions, in particular the chain rule. We learn how to use derivatives (partial derivatives, directional derivatives, gradient) to obtain valuable information about how functions change. Finally, in the last section, we introduce cylindrical and spherical coordinate systems.

► 2.1 REAL-VALUED AND VECTOR-VALUED FUNCTIONS OF SEVERAL VARIABLES

We need three pieces of data in order to describe a function: two sets (called the domain and the range of a function) and a rule that assigns to each element of the domain a unique element in the range. In this section, we present a definition and investigate examples of functions of several variables. We introduce a special class of such functions, called vector fields, that will be studied extensively in this book. In Section 2.5, we study paths, yet another important class of vector-valued functions.

A number of functions introduced in this section will appear in various contexts later in this book.

DEFINITION 2.1 Real-Valued and Vector-Valued Functions

A function whose domain is a subset U of $\mathbb{R}^m$, $m \geq 1$, and whose range is contained in $\mathbb{R}^n$ is called a *real-valued function of m variables* if $n = 1$, and a *vector-valued function of m variables* if $n > 1$.

Real-valued functions are also called *scalar-valued* or just *scalar functions*. A *vector function* is a synonym for a vector-valued function. We use the notation $f\colon U \subseteq \mathbb{R}^m \to \mathbb{R}$ (the symbol $\subseteq$ denotes a subset) to describe a scalar function, and $\mathbf{F}\colon U \subseteq \mathbb{R}^m \to \mathbb{R}^n$ (with $n > 1$) to describe a vector function. Instead of listing variables as an m-tuple of coordinates $(x_1, \dots, x_m)$, we take advantage of a vector notation $\mathbf{x} = (x_1, \dots, x_m)$. For functions with a low number of variables, we use $\mathbf{x} = (x, y)$, $\mathbf{x} = (x, y, z)$ and the like, instead of naming the variables by subscripts.

Let us rephrase the definition: a scalar function assigns a unique *real number* $f(\mathbf{x}) = f(x_1, \dots, x_m)$ to each element $\mathbf{x} = (x_1, \dots, x_m)$ in its domain U, whereas a vector function assigns a unique *vector* (or *point*, depending on the interpretation) $\mathbf{F}(\mathbf{x}) = \mathbf{F}(x_1, \dots, x_m) \in \mathbb{R}^n$ to each $\mathbf{x} = (x_1, \dots, x_m) \in U$ [the symbol $\in$ is read "in" or "belong(s) to"]. We write $\mathbf{F}(x_1, \dots, x_m) \in \mathbb{R}^n$ as

$$\mathbf{F}(x_1, \dots, x_m) = (F_1(x_1, \dots, x_m), \dots, F_n(x_1, \dots, x_m))$$

or as $\mathbf{F}(\mathbf{x}) = (F_1(\mathbf{x}), \dots, F_n(\mathbf{x}))$, where $F_1, \dots, F_n$ are the *components* (or the *component functions*) of $\mathbf{F}$. Notice that $F_1, \dots, F_n$ are *real-valued* functions of $x_1, \dots, x_m$, defined on U. Sometimes, instead of writing $\mathbf{F}(x, y, z) = (F_1(x, y, z), F_2(x, y, z))$ for a function $\mathbf{F}\colon U \subseteq \mathbb{R}^3 \to \mathbb{R}^2$, we will use vector notation $\mathbf{F}(x, y, z) = F_1(x, y, z)\mathbf{i} + F_2(x, y, z)\mathbf{j}$.

On a few occasions we will take advantage of matrix notation in order to keep track of the components of a function; for example,

$$\mathbf{F}(x, y, z) = \begin{bmatrix} F_1(x, y, z) \\ F_2(x, y, z) \end{bmatrix}$$

for $\mathbf{F}\colon U \subseteq \mathbb{R}^3 \to \mathbb{R}^2$.

▶ EXAMPLE 2.1 Important Functions of Several Variables

A function of the form $f(x, y) = ax + by + c$, where a, b, and c are constants (i.e., real numbers), is called a *linear function* (of two variables). Its domain is $U = \mathbb{R}^2$. In general, a linear function of n variables is defined on $\mathbb{R}^n$ by the formula $f(x_1, \dots, x_n) = a_1x_1 + a_2x_2 + \cdots + a_nx_n + b$, where $a_1, \dots, a_n$ and b are constants.

The *distance function* $f(x, y, z) = \sqrt{x^2 + y^2 + z^2}$ measures the distance from the point (x, y, z) to the origin. It is a real-valued function of three variables defined on $U = \mathbb{R}^3$. A *projection*

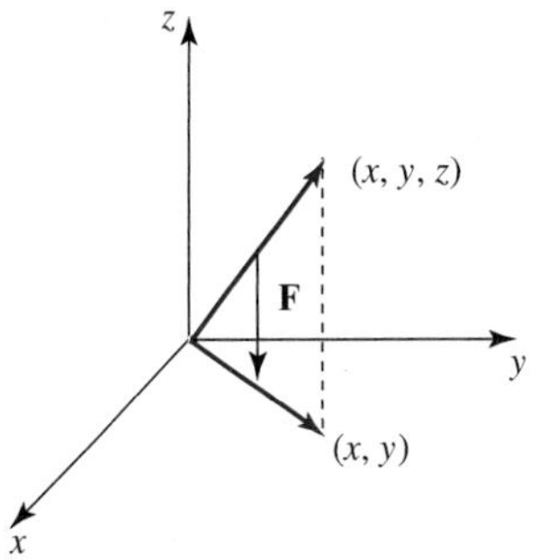

Figure 2.1 Projection function $\mathbf{F}: \mathbb{R}^3 \to \mathbb{R}^2$, given by $\mathbf{F}(x, y, z) = (x, y)$.

$\mathbf{F}(x, y, z) = (x, y)$ is a vector-valued function of three variables that assigns to every vector (x, y, z) in $\mathbb{R}^3$ its projection (x, y) onto the xy-plane; see Figure 2.1. ◀

▶ EXAMPLE 2.2

The component functions of the projection function $\mathbf{F}(x, y, z) = (x, y)$ from Example 2.1 are $F_1(x, y, z) = x$ and $F_2(x, y, z) = y$. Real-valued functions $F_1(x, y, z) = e^x - yz$ and $F_2(x, y, z) = y - 2$ are components of the vector-valued function $\mathbf{F}: \mathbb{R}^3 \to \mathbb{R}^2$ given by $\mathbf{F}(x, y, z) = (e^x - yz, y - 2)$.

The function $\mathbf{U}(\mathbf{r}) = \mathbf{r}/||\mathbf{r}||$, where $\mathbf{r} = x\mathbf{i} + y\mathbf{j} + z\mathbf{k}$, assigns to each vector $\mathbf{r} \neq \mathbf{0}$ the unit vector $\mathbf{r}/||\mathbf{r}||$ in the direction of $\mathbf{r}$. Sometimes, we write $\mathbf{U}$ in components as

$$\mathbf{U}(x, y, z) = \left(\frac{x}{\sqrt{x^2 + y^2 + z^2}}, \frac{y}{\sqrt{x^2 + y^2 + z^2}}, \frac{z}{\sqrt{x^2 + y^2 + z^2}} \right).$$

The domain of $\mathbf{U}$ is the set of all nonzero vectors in $\mathbb{R}^3$. Since $||\mathbf{U}(\mathbf{r})|| = 1$ for all $\mathbf{r} \neq \mathbf{0}$, the range of $\mathbf{U}$ consists of all unit vectors in $\mathbb{R}^3$. Identifying a vector with its terminal point (and placing its initial point at the origin), we visualize the range of $\mathbf{U}$ as a sphere in $\mathbb{R}^3$ of radius 1. ◀

▶ EXAMPLE 2.3

The function $f(x, y) = y \ln x$ is a real-valued function of two variables whose domain U consists of all pairs (x, y) such that $x > 0$. Using the curly braces {} to denote a set and a vertical bar | to say "such that," we write $U = \{(x, y) | x > 0\}$. ◀

▶ EXAMPLE 2.4

Find the domain and the range of the function

$$f(x, y) = \frac{x - 3y^2}{x^2 - y}.$$

SOLUTION

The domain U is a subset of $\mathbb{R}^2$, since f is a function of two variables. The domain is determined by the requirement that the denominator $x^2 - y$ be nonzero. Consequently, U consists of all points (x, y) in the xy-plane except those lying on the parabola $y = x^2$. Using set notation, we write $U = \{(x, y) | y \neq x^2\}$.

The collection of values of f for all $(x, y) \in U$ forms the range of f. Notice that $f(0, y) = 3y$ if $y \neq 0$; in other words, any real number $c \neq 0$ can be obtained as a value of f, because $f(0, c/3) = c$. Since $f(3, 1) = 0$, the range of f consists of all real numbers. ◀

The domain of a vector-valued function is the common domain of its components. For instance, the domain of the projection $\mathbf{F}$ of Example 2.1 consists of all vectors in $\mathbb{R}^3$. Let us consider another example.

▶ EXAMPLE 2.5

Find the domain U of the function $\mathbf{F}: U \subseteq \mathbb{R}^2 \to \mathbb{R}^3$ defined by

$$\mathbf{F}(x, y) = \left(\frac{y}{x^2 + y^2}, \frac{x}{x^2 + y^2}, \sqrt{xy} \right).$$

SOLUTION The domains of the first two components consist of all pairs (x, y) of real numbers, except $(0, 0)$. The function $\sqrt{xy}$ is defined whenever $xy \geq 0$; that is, when $x \geq 0$ and $y \geq 0$ or when $x \leq 0$ and $y \leq 0$. Therefore, U consists of all pairs (x, y) such that either both x and y are positive or 0, or both are negative or 0, but such that x and y are not simultaneously 0. In other words, U consists of all points in the first and third quadrants (including the x-axis and the y-axis), with the origin removed. ◀

▶ EXAMPLE 2.6 Cobb–Douglas Production Function in Economics

The function $P(L, K) = bL^{\alpha}K^{1-\alpha}$ (where b and α are constants, $b > 0$ and $0 < \alpha < 1$) has been used to model production, that is, the total value (in dollars) of all goods produced in a year. The variable K represents the amount of capital invested, and L is the monetary value of labor (i.e., the total number of person-hours in a year). The function $P(L, K)$ could represent the production of an individual company, or could model the entire economy of a country.

In 1928, Cobb and Douglas used the function $P(L, K) = 1.01L^{0.75}K^{0.25}$ to model the growth of U.S. economy (here, $b = 1.01$ and $\alpha = 0.75$). This model, called the *Cobb–Douglas production function,* has since been used successfully in a variety of situations. Note that the domain of P is the set $\{(L, K) \mid L \geq 0, K \geq 0\}$, and its range is given by $P(L, K) \geq 0$.

In Exercise 18 we explore one property of the function $P(L, K)$. ◀

▶ EXAMPLE 2.7 Wind Chill Index

When we feel cold, it is because our body senses the temperature of our skin (which is in contact with the surrounding air). If it is windy, we lose the heat faster, and thus feel that it is colder that it actually is (judging solely by the temperature). This sensation of coldness can be quantified using the *wind chill index.*

Let T denote the air temperature (in °C) and let v be the wind speed, measured in km/h. In 2001, Environment Canada used the function

$$W(T, v) = 13.12 + 0.6215T - 11.37v^{0.16} + 0.3965Tv^{0.16}$$

to model the wind chill index. For instance, $W(-5, 30) = -13$. This means that, when the air temperature is -5°C and the wind blows at 30km/h, it *feels* to us as if the temperature is -13°C.

See Exercises 19 and 20 for several properties of the wind chill index. ◀

▶ EXAMPLE 2.8 Pressure in an Ideal Gas

The pressure in kilopascals (kPa) in an ideal gas can be viewed as the scalar field $P: U \subseteq \mathbb{R}^2 \to \mathbb{R}$ given by $P(T, V) = RnT/V$, where V is the volume (in liters) occupied by the gas, T is the absolute temperature (in °K), n is the number of moles of gas, and $R = 8.314$ J mol^{-1} K^{-1} is the universal gas constant. ◀

DEFINITION 2.2 Vector Field

A *vector field in the plane* is a vector-valued function $\mathbf{F}\colon U \subseteq \mathbb{R}^2 \to \mathbb{R}^2$, defined on a subset $U \subseteq \mathbb{R}^2$. A *vector field in space* is a vector-valued function $\mathbf{F}\colon U \subseteq \mathbb{R}^3 \to \mathbb{R}^3$. In general, a function $\mathbf{F}\colon U \subseteq \mathbb{R}^m \to \mathbb{R}^m$ is called a *vector field* (or, a *vector field on* U).

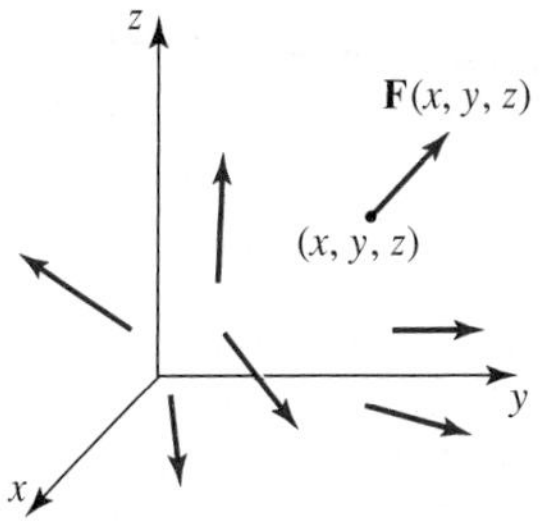

Figure 2.2 Vector field in $\mathbb{R}^3$.

Although the domain and range of a vector field belong to the same set $\mathbb{R}^m$, we often visualize them in a different way: elements of the domain are thought of as points, whereas the elements of the range are viewed as vectors. For example, a vector field on $\mathbb{R}^3$ is a function that assigns a *vector* $\mathbf{F}(x, y, z)$ to every *point* (x, y, z) in the three-dimensional space $\mathbb{R}^3$. This interpretation helps us graph a vector field: its value $\mathbf{F}(x, y, z)$ at (x, y, z) is represented as a vector $\mathbf{F}(x, y, z)$ whose initial point is (x, y, z), as shown in Figure 2.2.

Similarly, a vector field $\mathbf{F}$ on $\mathbb{R}^2$ assigns the vector $\mathbf{F}(x, y) \in \mathbb{R}^2$ to a point (x, y) in its domain [and (x, y) is taken as the initial point of $\mathbf{F}(x, y)$].

The diagrams representing vector fields $\mathbf{F}(x, y) = (-y, x) = -y\mathbf{i} + x\mathbf{j}$ in Figure 2.3(a) and $\mathbf{G}(x, y, z) = (x/\sqrt{x^2 + y^2}, y/\sqrt{x^2 + y^2}, 0)$ in Figure 2.3(b) were constructed in that

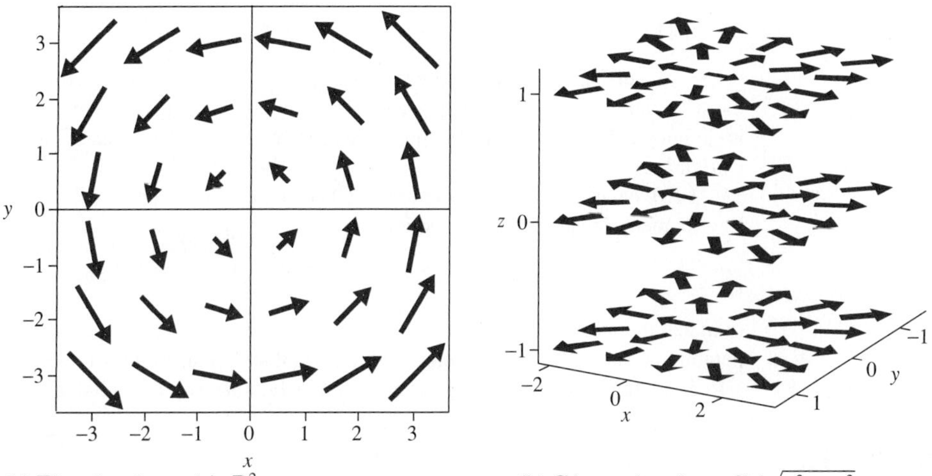

(a) $\mathbf{F}(x, y) = (-y, x)$ in $\mathbb{R}^2$.

(b) $\mathbf{G}(x, y, z) = (x, y, 0)/\sqrt{x^2 + y^2}$ in $\mathbb{R}^3$.

Figure 2.3 Visual representations of vector fields.

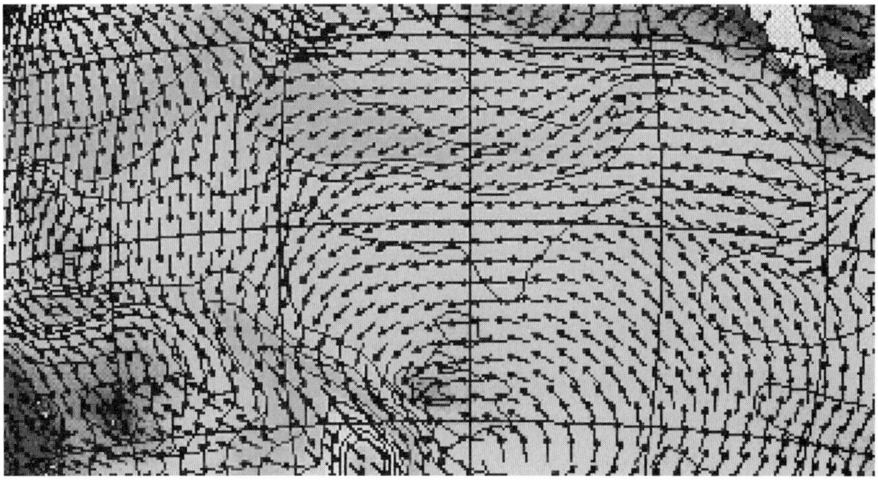

Figure 2.4 Surface wave motion in the Pacific Ocean.

way. Note that all vectors in Figure 2.3(b) should have the same length (equal to 1). It is a common practice (especially when depicting three-dimensional vector fields) to adjust lengths of vectors, so that we get a better feel of their position in space and so that the resulting diagram looks less cluttered. In Figure 2.4, we see how a vector field is used to depict the motions of surface waves in the Pacific Ocean.

We can think of a vector field as a "field of arrows," with one "arrow" emerging from every point. By analogy to this definition, a scalar function (i.e., an assignment of a real number to each point) is sometimes called a *scalar field.*

There are situations when it is more convenient to visualize a vector-valued function $\mathbf{F}: U \subseteq \mathbb{R}^m \to \mathbb{R}^m$ as a mapping of points. This useful way of thinking about $\mathbf{F}$ will be discussed and used in later sections, for example, in the change of variables technique (for integration) and in the definitions of path and surface. Yet another interpretation of maps $\mathbf{F}: \mathbb{R}^m \to \mathbb{R}^m$ (for $m = 3$) in the context of transformation of coordinates will be discussed in the last section of this chapter.

▶ EXAMPLE 2.9

Some rotary motions in the plane can be illustrated using the vector field $\mathbf{F}: \mathbb{R}^2 \to \mathbb{R}^2$, given by $\mathbf{F}(x, y) = (-y, x)$; see Figure 2.3(a). The corresponding unit vector field

$$\mathbf{F}(x, y)/||\mathbf{F}(x, y)|| = (-y/\sqrt{x^2 + y^2}, x/\sqrt{x^2 + y^2})$$

is defined on $U = \mathbb{R}^2 - \{(0, 0)\}$ (the symbol "$-$" denotes that U is obtained from $\mathbb{R}^2$ by removing the origin; it is read "without"). Vector fields $\mathbf{F}(x, y)$ and $\mathbf{F}(x, y)/||\mathbf{F}(x, y)||$ will appear again in various examples and applications. ◀

▶ EXAMPLE 2.10 Gravitational Force and Gravitational Potential

Imagine that a mass M is placed at the origin of the coordinate system. According to *Newton's Law of Gravitation,* the force of attraction or the *gravitational force* on a small mass m located at

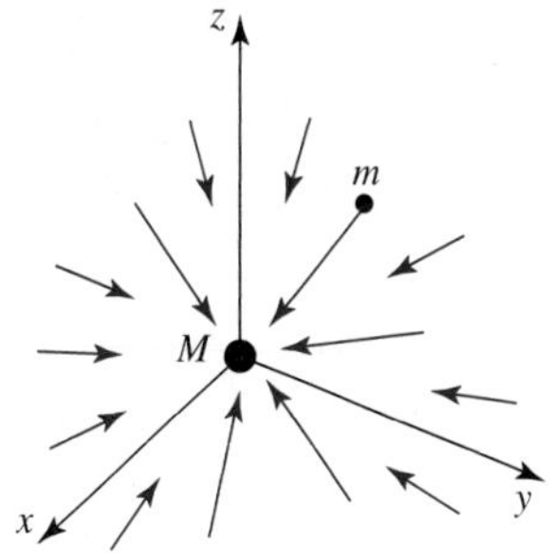

Figure 2.5 Gravitational force field $\mathbf{F}(\mathbf{r}) = -\frac{GMm}{||\mathbf{r}||^2}\frac{\mathbf{r}}{||\mathbf{r}||}$.

$(x, y, z) \neq (0, 0, 0)$ is given by

$$\mathbf{F}(\mathbf{r}) = -\frac{GMm}{||\mathbf{r}||^2}\frac{\mathbf{r}}{||\mathbf{r}||}, \tag{2.1}$$

where $\mathbf{r} = x\mathbf{i} + y\mathbf{j} + z\mathbf{k}$, $||\mathbf{r}|| = \sqrt{x^2 + y^2 + z^2}$ and $G = 6.67 \cdot 10^{-11}$ Nm^{2}kg^{-2} is the *gravitational constant*. The function $\mathbf{F}(\mathbf{r})$ can be thought of as a vector field (called the *gravitational force field*), whose value at a point $\mathbf{r} \neq \mathbf{0}$ is given by (2.1); see Figure 2.5. The *gravitational potential* is the scalar field

$$V(\mathbf{r}) = -\frac{GMm}{||\mathbf{r}||} \tag{2.2}$$

defined on $U = \mathbb{R}^3 - \{(0, 0, 0)\}$. The relation between (2.1) and (2.2) will be established in Section 2.4 (see Example 2.40 and the comment that follows). ◀

▶ EXAMPLE 2.11 Electrostatic Force and Electrostatic Potential

The electrostatic force on a charge q placed at a point $\mathbf{r} = (x, y, z) \neq (0, 0, 0)$ due to a charge Q located at the origin is given by *Coulomb's Law*

$$\mathbf{F}(\mathbf{r}) = \frac{1}{4\pi\epsilon_0}\frac{Qq}{||\mathbf{r}||^2}\frac{\mathbf{r}}{||\mathbf{r}||}, \tag{2.3}$$

where (in the SI system) $1/4\pi\epsilon_0 = c^2 \cdot 10^{-7} = 8.99 \cdot 10^9$ Nm^{2}C^{-2} (C denotes coulomb, a unit of charge, N is the newton, a unit of force, and m is the meter), and c denotes the speed of light in a vacuum. The constant ϵ_0 is called the *permittivity of vacuum*. The *electrostatic force field* is a vector field whose value at a point $\mathbf{r}$ in $\mathbb{R}^3 - \{(0, 0, 0)\}$ is given by (2.3). The scalar field defined by

$$V(\mathbf{r}) = \frac{1}{4\pi\epsilon_0}\frac{Qq}{||\mathbf{r}||} \tag{2.4}$$

is called the *electrostatic potential*. In Exercise 33 in Section 2.4, we will derive the relation between $\mathbf{F}(\mathbf{r})$ and $V(\mathbf{r})$. ◀

Note the similarity between the forces given by (2.1) and (2.3)—both are inverse-square-law forces—but there is also an important difference. Gravitational masses M and m are always positive, so the gravitational force is always attractive. The interaction between charges of opposite sign is attractive, but it is repulsive if charges of the same sign are involved. Experiments confirm that (2.3) remains valid even if the charge q moves with large velocities (and the charge Q remains at rest).

In physics, the potential functions introduced in formulas (2.2) and (2.4) are called the *gravitational potential energy* and the *electrostatic potential energy*.

▶ EXAMPLE 2.12 Electrostatic and Magnetic Fields

Assume that the charge q is at rest. The *electrostatic field* $\mathbf{E}(\mathbf{r})$ at the point $\mathbf{r} \neq \mathbf{0}$ due to a charge Q placed at the origin is given by the equation

$$\mathbf{E}(\mathbf{r}) = \frac{1}{4\pi\epsilon_0} \frac{Q}{||\mathbf{r}||^2} \frac{\mathbf{r}}{||\mathbf{r}||}.$$

By comparing this equation with (2.3), we can see that the force exerted by charge Q on another charge q located at $\mathbf{r}$ can be written as $\mathbf{F}(\mathbf{r}) = q\mathbf{E}(\mathbf{r})$.

Alternatively, suppose that there is no charge Q and the charge q is acted on by a *magnetic field* $\mathbf{B}(\mathbf{r})$. Then the force that the magnetic field exerts on q is $\mathbf{F}(\mathbf{r}) = q\mathbf{v} \times \mathbf{B}(\mathbf{r})$, where $\mathbf{v}$ is the velocity of q. Notice that the electrostatic force depends only on the location of q, not on how q is moving when it is at a particular location. However, the magnetic force on q depends on both its location and its velocity; in particular, note that the magnetic force on a charge at rest ($\mathbf{v} = \mathbf{0}$) is zero.

If both electrostatic *and* magnetic fields act on a charge q, then the total force acting on q (called the *electromagnetic* or *Lorentz force*) is

$$\mathbf{F}(\mathbf{r}) = q\Big(\mathbf{E}(\mathbf{r}) + \mathbf{v} \times \mathbf{B}(\mathbf{r})\Big). \tag{2.5}$$

This is an extension of the Principle of Superposition introduced in Example 1.16. ◀

▶ EXAMPLE 2.13

Motion of the surface of a fluid can be described as a function $\mathbf{F}: U \subseteq \mathbb{R}^3 \to \mathbb{R}^2$, where $\mathbf{F}(x, y, t)$ denotes the velocity vector of the fluid at a point (x, y) at time t; see Figure 2.6. The domain U consists of ordered triples (x, y, t), where (x, y) denotes the coordinates of points on the surface of the fluid and $t \geq 0$ represents time. ◀

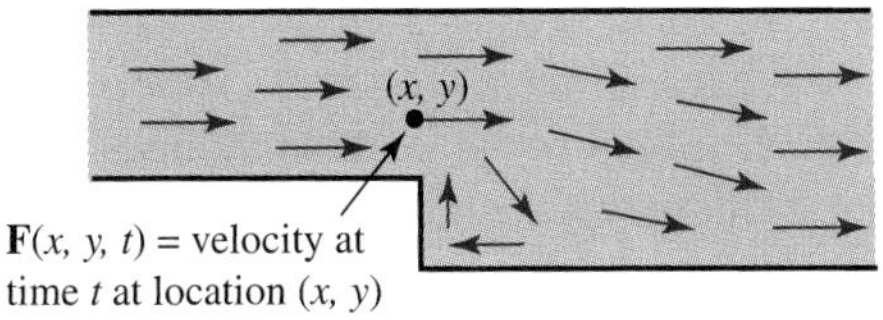

Figure 2.6 Motion of a fluid through a channel.

▶ EXAMPLE 2.14 Model of a Hurricane

Consider the fluid flow whose velocity at a point (x, y) is given by

$$\mathbf{H}(x, y) = \frac{(-x - y)\mathbf{i} + (x - y)\mathbf{j}}{x^2 + y^2}.$$

At a point (x, y), the speed of the flow is

$$\|\mathbf{H}(x, y)\| = \left(\frac{(-x-y)^2}{(x^2+y^2)^2} + \frac{(x-y)^2}{(x^2+y^2)^2}\right)^{1/2} = \left(\frac{2(x^2+y^2)}{(x^2+y^2)^2}\right)^{1/2} = \frac{\sqrt{2}}{\sqrt{x^2+y^2}}.$$

If $x^2 + y^2 = r^2$ [i.e., if a point (x, y) belongs to the circle of radius r centered at the origin], then $\|\mathbf{H}\| = \sqrt{2}/r$. Thus, we conclude that the speed of the fluid is constant on circles centered at the origin. Note that the speed is inversely proportional to the radius; thus, it increases the closer we get to the origin. Figure 2.7 shows the vector field $\mathbf{H}$ in the neighborhood of the origin (to keep the picture clear, velocity vectors close to the origin—due to their large magnitude—are not shown). Sometimes, vector fields such as $\mathbf{H}$ are used to model the flow of air around and within a hurricane.

The flow $\mathbf{H}$ can be written as $\mathbf{F}_1 + \mathbf{F}_2$, where $\mathbf{F}_1(x, y) = (-y\mathbf{i} + x\mathbf{j})/(x^2 + y^2)$ and $\mathbf{F}_2(x, y) = -(x\mathbf{i} + y\mathbf{j})/(x^2 + y^2)$. See Exercises 33 and 34 for several properties of $\mathbf{F}_1$ and $\mathbf{F}_2$.

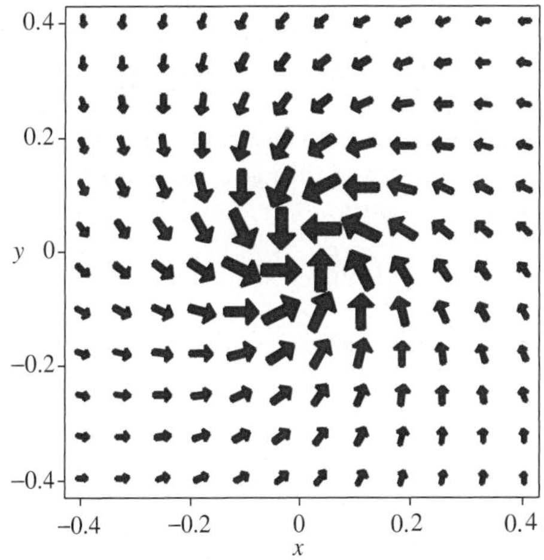

Figure 2.7 Vector field $\mathbf{H}$ (scaled for clarity) of Example 2.14.

In all examples we have presented so far, we visualized a vector field $\mathbf{F}$ by drawing the vector $\mathbf{F}(\mathbf{x})$ so that its initial point is located at $\mathbf{x}$. An alternate way of visualizing a vector field consists of drawing the slope lines of $\mathbf{F}$; see Exercise 32. This method conveys only the information on the direction of $\mathbf{F}$ (i.e., no arrows), and, sometimes, its magnitude, unlike the one we have discussed in this section, which shows all of the direction, orientation, and magnitude of $\mathbf{F}$. In certain situations we describe a vector field by drawing its flow lines, as explained in Section 4.5.

▶ EXERCISES 2.1

1. What are the component functions of the vector-valued function $\mathbf{G}(\mathbf{r}) = 2\|\mathbf{r}\|\mathbf{r}$? Write down the components of the gravitational force field $\mathbf{F}(\mathbf{r})$ given by (2.1).

2. Write a formula for a real-valued function f of three variables whose value at (x, y, z) is inversely proportional to the square of the distance from (x, y, z) to the origin, and is such that $f(1, 2, 3) = 3$.

Exercises 3 to 6: Find and sketch (or describe in words) the domain of the functions given below.

3. $f(x, y) = x(x^2 + y^2 - 1)^{-1}$

4. $f(x, y, z) = 5/xyz$

5. $f(x, y) = \arctan(x/y)$

6. $\mathbf{F}(r, \theta) = (r\cos\theta, r\sin\theta)$

7. What is the range of the linear function $f(x, y) = 3x + y - 7$? Discuss how the range of $f(x, y) = ax + by + c$ depends on the values of a, b, and c.

Exercises 8 to 17: Find the domain and range of the functions given below.

8. $f(x, y, z) = 3/(x + y)$

9. $f(x, y, z) = e^{-(x^2+y^2+z^2)}$

10. $f(u, v, t, z) = \tan(u + v) + t^3z^3$

11. $f(x, y) = 3x^2/(x^2 + y^2)$

12. $f(x, y) = \sqrt{4 - 4x^2 - y^2}$

13. $\mathbf{F}(x, y) = (\ln x, (\ln y)^2)$

14. $f(x, y) = \sqrt{xy}$

15. $f(x, y) = |x| + 2|y|$

16. $f(x, y) = \arctan(x^2 + y^2)$

17. $f(x, y) = y/|x|$

18. Show that the Cobb–Douglas production function $P(L, K) = bL^{\alpha}K^{1-\alpha}$ defined in Example 2.6 satisfies $P(mL, mK) = mP(L, K)$, where $m > 0$. Interpret this identity.

19. Consider the wind chill index given in Example 2.7.

(a) Show that $W(0, 0) = 13.12$, and compute $W(T, 0)$ for $T = 10, -10$, and -20. Explain why these results do not correspond to what we would expect a wind chill index to show.

(b) From (a) it follows that the points of the form (T, v), with $v = 0$, cannot belong to the domain of W. Define a possible domain for W.

(c) Compute $W(-5, v)$ and show that it is a decreasing function of v. Explain why this fact makes sense.

(d) Compute $W(T, 20)$ and show that it is an increasing function of T. Explain why this fact makes sense.

20. An alternate model for the wind chill index is given by

$$W_1(T, v) = 91 + (0.44 + 0.325\sqrt{v} - 0.023v)(T - 91),$$

where T denotes the temperature (in degrees Fahrenheit) without wind and v is the wind speed in mph, $5 \le v \le 45$.

(a) Compute $W_1(0, v)$, $5 \le v \le 45$, and show that it is a decreasing function. Explain why this fact makes sense.

(b) Compute $W_1(T, 10)$ and show that it is an increasing function. Explain why this fact makes sense.

21. Describe geometrically the range of the function $\mathbf{F}\colon \mathbb{R}^2 \to \mathbb{R}^2$ defined by $\mathbf{F}(\alpha, \beta) = (\sin\alpha, \cos\beta)$.

22. Write a formula for a vector field in two dimensions whose direction at (x, y) makes an angle of $\pi/4$ with respect to the positive x-axis and whose magnitude at (x, y) equals the distance from (x, y) to the origin.

23. Write a formula for a unit vector field in three dimensions whose direction at (x, y, z) is parallel to the line joining (x, y, z) and $(1, 2, -2)$ and points away from $(1, 2, -2)$.

Exercises 24 to 31: Sketch the vector field $\mathbf{F}$ in $\mathbb{R}^2$. Describe, in words, what each vector field looks like.

24. $\mathbf{F}(\mathbf{x}) = \mathbf{x}$

25. $\mathbf{F}(\mathbf{x}) = \mathbf{x}/||\mathbf{x}||$

26. $\mathbf{F}(\mathbf{x}) = \mathbf{x} - \mathbf{x}_0$, where $\mathbf{x}_0 = (1, 2)$

27. $\mathbf{F}(\mathbf{x}) = (\mathbf{x} - \mathbf{x}_0)/||\mathbf{x} - \mathbf{x}_0||$, where $\mathbf{x}_0 = (1, 2)$

28. $\mathbf{F}(x, y) = \frac{1}{2}\mathbf{i} + \frac{1}{2}\mathbf{j}$

29. $\mathbf{F}(x, y) = x^2\mathbf{i} + \mathbf{j}$

30. $\mathbf{F}(x, y) = \mathbf{i} + 2\mathbf{j}/y$

31. $\mathbf{F}(x, y) = x\mathbf{i} - y\mathbf{j}$

32. An alternate way of visualizing a vector field $\mathbf{F}$ consists of drawing the slope lines. Slope lines show the direction of $\mathbf{F}$ only, and ignore its orientation (and sometimes its magnitude). In Figure 2.8 identify the vector fields $\mathbf{F}_1(x, y) = (\sin x, \sin y)$ and $\mathbf{F}_2(x, y) = (x, -y)$ and $\mathbf{F}_3(x, y) = (-x, -y)$. Explain your answer.

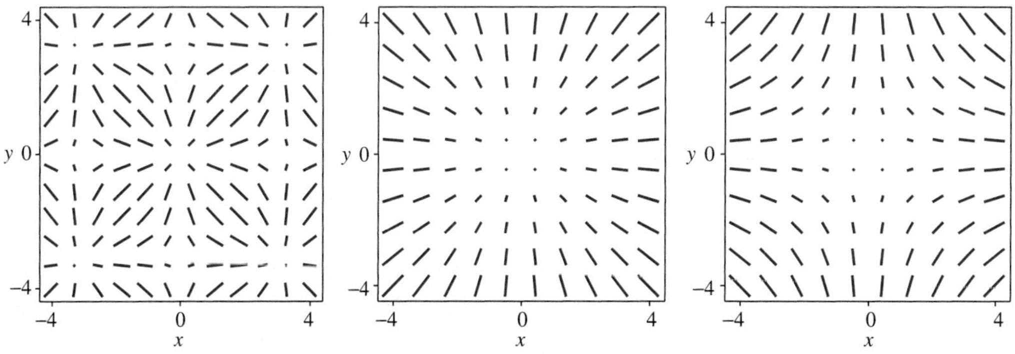

Figure 2.8 Slope line representation of three vector fields.

33. Assume that the vector field $\mathbf{F}_1(x, y) = (-y\mathbf{i} + x\mathbf{j})/(x^2 + y^2)$ represents a flow (i.e., the velocity) of a fluid. Prove that

(a) At every point $(x, y) \neq (0, 0)$, the flow is tangent to the circle centered at the origin that goes through (x, y).

(b) The speed of the fluid [i.e., $||\mathbf{F}_1(x, y)||$] at (x, y) is inversely proportional to the distance between (x, y) and the origin. In other words, the flow $\mathbf{F}_1$ along a fixed circle centered at the origin has constant speed, which is inversely proportional to the radius of the circle.

(c) Make a rough sketch of the flow $\mathbf{F}_1$. Convince yourself that $\mathbf{F}_1$ represents a counterclockwise motion. The type of motion modeled by $\mathbf{F}_1$ is sometimes called *counterclockwise vortex flow*.

34. Consider a motion of a fluid whose velocity is given by $\mathbf{F}_2(x, y) = -(x\mathbf{i} + y\mathbf{j})/(x^2 + y^2)$. Show that

(a) The speed of the fluid [i.e., $||\mathbf{F}_2(x, y)||$] is constant along the circles centered at the origin.

(b) The speed of the fluid at (x, y) is inversely proportional to the distance between (x, y) and the origin.

(c) Make a rough sketch of the flow $\mathbf{F}_2$ to convince yourself that the motion is directed toward the origin. The type of motion modeled by $\mathbf{F}_2$ is sometimes called *uniform sink flow*.

▶ 2.2 GRAPH OF A FUNCTION OF SEVERAL VARIABLES

The graph of a real-valued function $y = f(x)$ of one variable is a curve in the xy-plane. Each point (x, y) on that curve carries two pieces of data: the value x of the independent variable and the corresponding value $y = f(x)$ of the function. Alternatively, we can describe the graph of f as the set

$$\text{Graph}(f) = \{(x, y) | y = f(x) \text{ for some } x \in U\} \subseteq \mathbb{R}^2,$$

where $U \subseteq \mathbb{R}$ is the domain of f. Generalizing this description, we obtain the following definition.

DEFINITION 2.3 Graph of a Real-Valued Function of Two Variables

The *graph* of a real-valued function $f\colon U \subseteq \mathbb{R}^2 \to \mathbb{R}$ of two variables is the set

$$\text{Graph}(f) = \{(x, y, z) \mid z = f(x, y) \text{ for some } (x, y) \in U\} \subseteq \mathbb{R}^3,$$

where $U \subseteq \mathbb{R}^2$ denotes the domain of f.

Thus, both the values of the variables (x, y) in the domain and the corresponding value $z = f(x, y)$ of the function are encoded as a point on the graph. In general, the graph of a function of two variables is a *surface* in space; see Figure 2.9.

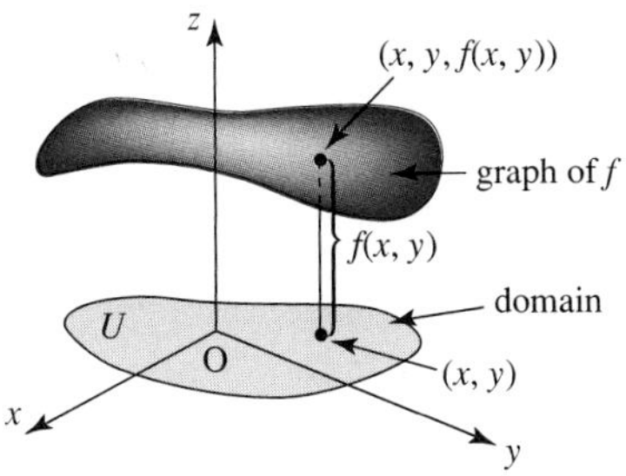

Figure 2.9 Graph of a function $z = f(x, y)$ is a surface in $\mathbb{R}^3$.

Analogously, the graph of a real-valued function $f\colon U \subseteq \mathbb{R}^m \to \mathbb{R}$ of m variables is the set

$$\text{Graph}(f) = \{(x_1, \ldots, x_m, y) | y = f(x_1, \ldots, x_m) \text{ for some } (x_1, \ldots, x_m) \in U\} \subseteq \mathbb{R}^{m+1}.$$

To graph a real-valued function $y = f(x)$, we need two dimensions (i.e., two coordinate axes): one for x and one for y. The graph of a real-valued function $z = f(x, y)$ has to be placed in three dimensions: two coordinate axes are needed for the domain (i.e., for the independent variables x and y) and the third one for the values of f (i.e., for the range). To construct a graph of $w = f(x, y, z)$, we would need four dimensions, so there are limitations on the visual representation of functions, even if they have a small number of variables.

However, there is another way of visualizing graphs that uses two dimensions to represent the graph of the function $z = f(x, y)$ of two variables. It consists of drawing *level curves* (or *contour curves*), and uses two-dimensional data to obtain three-dimensional information. This idea has been used quite often. In technical documentation, instead of drawing a complicated three-dimensional object, we draw its projections onto different planes (thus the two-dimensional drawings convey information on the three-dimensional features of the object in question). A contour curve on topographic maps indicates points of the same (usually, integer-valued) elevation. Not only can we find the elevation (or at least, approximate elevation) at various locations (and thus form a mental three-dimensional image of the hill), but we can also draw various conclusions: for example, the closer the contour curves are, the steeper the hill; the further apart they become, the smaller the slopes are. Figure 2.10 shows some (computer-generated) contour curves of part of the Himalayan Mountains north of Mount Everest.

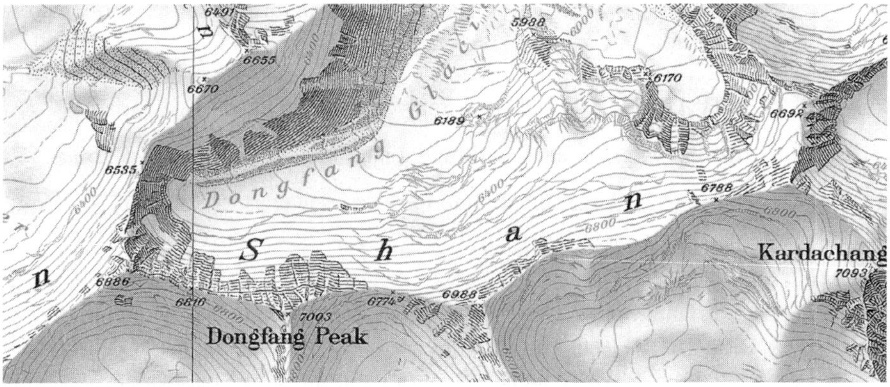

Figure 2.10 Contour curves of part of the Himalayan Mountains.

On a weather forecast map, contour curves are used to label points with the same air temperature. We can also draw isobars, by connecting points with the same barometric pressure. Looking at the isobars, we obtain information on the wind: for instance, the closer the isobars are (provided that they are plotted with equal difference values), the stronger the wind.

Contour curves are known under different names that usually have a prefix *iso-* or *equi-* [meaning "(of) equal" in Greek and Latin, respectively]. We have already mentioned isobars. Isotherms are the contour curves of the temperature function and isomers are the curves connecting points having the same monthly or seasonal precipitation. Equipotential curves are curves along which a potential function is constant.

To visualize a function $w = f(x, y, z)$ of three variables, we can use a three-dimensional analogue of level curves, namely the *level surfaces* (or *contour surfaces*). Figure 2.11 shows several isothermal surfaces for air temperature. Later on we will show that the equipotential surfaces for the gravitational potential of a point mass [see (2.2) in

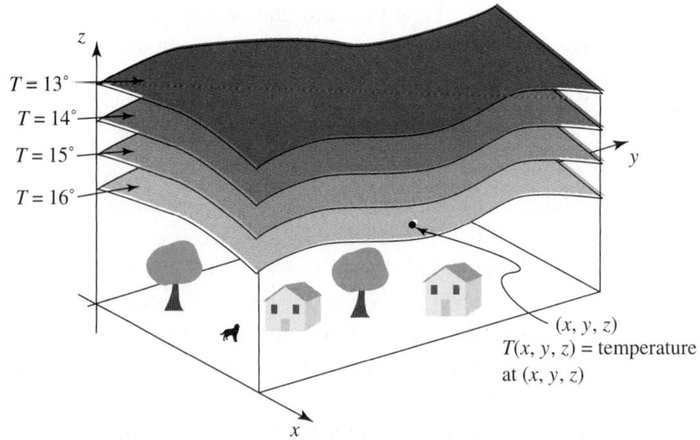

Figure 2.11 Isothermal surfaces of the temperature $T(x, y, z)$.

Example 2.10], and, consequently, for the electrostatic potential of a point charge [see (2.4) in Example 2.11] are concentric spheres.

DEFINITION 2.4 Level Set

Let $f: U \subseteq \mathbb{R}^m \to \mathbb{R}$ be a real-valued function of m variables and let $c \in \mathbb{R}$. The *level set of value* c is the set of all points in the domain U of f on which f has a constant value; that is,

$$\text{Level set of value } c = \{(x_1, \dots, x_m) \in U \mid f(x_1, \dots, x_m) = c\}.$$

In particular, for $m = 2$ the level set

$$\{(x, y) \in U \subseteq \mathbb{R}^2 \mid f(x, y) = c\}$$

is called a *level curve (of value c)* or a *contour curve (of value c)*, and for $m = 3$ the level set

$$\{(x, y, z) \in U \subseteq \mathbb{R}^3 \mid f(x, y, z) = c\}$$

is called a *level surface* or a *contour surface (of value c)*. In other words, a level curve of value c is the curve that contains all points where the value of a function of two variables is equal to c. Similarly, a level surface of value c contains all points where a function of three variables is constant and equal to c. By definition, level curves and level surfaces are always contained in the domain of a function. Sometimes, we do not draw a level curve of value c in the xy-plane , but in the plane $z = c$. In that case, we refer to it as the *level curve "lifted" to the surface*, or as the *cross-section* at (height) $z = c$. See Figures 2.13 and 2.16.

Level curves in the xy-plane , such as those shown in Figures 2.12, 2.15, 2.18, or 2.19, are said to form a *contour diagram* of the given function.

EXAMPLE 2.15

Describe the contour diagram of the linear function $f(x, y) = 3x - 2y + 1$.

SOLUTION From $3x - 2y + 1 = c$ it follows that $y = 3x/2 + (1 - c)/2$. Thus, the level curves of $f(x, y)$ are parallel lines (of slope $3/2$). The line corresponding to a value c crosses the y-axis at $(1 - c)/2$; see Figure 2.12.

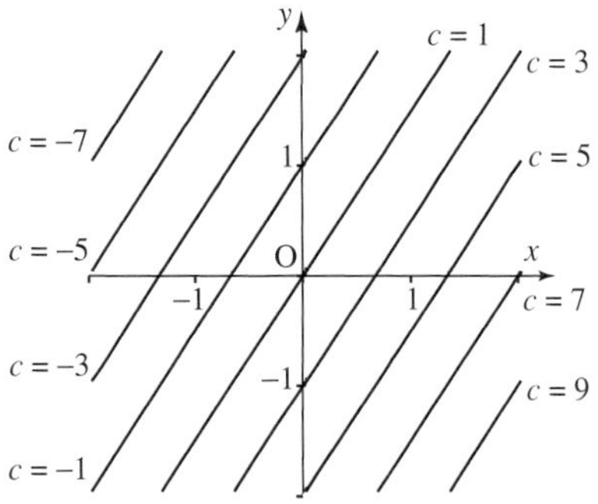

Figure 2.12 Contour diagram of the linear function $f(x, y) = 3x - 2y + 1$.

We notice that each time we increase the value of c by 2, the y-intercept moves down 1 unit. In other words, lines in Figure 2.12 are evenly spaced. See Exercise 12 for a generalization of this fact.

▶ EXAMPLE 2.16

Describe the level curves of $z = 9 - x^2 - y^2$.

SOLUTION

Setting $z = c$, we get $9 - x^2 - y^2 = c$ and $x^2 + y^2 = 9 - c$.

Since $x^2 + y^2 \geq 0$, there are no level curves for $9 - c < 0$, that is, for $c > 9$. In the case $c = 9$, the level curve consists of a single point, and for $c < 9$, it is a circle of radius $\sqrt{9 - c}$. Figure 2.13 shows several cross-sections identified by the value of c and the three-dimensional graph of the surface.

Here is another way of thinking about level curves. Imagine that we have a scanning device capable of detecting horizontal cross-sections, placed parallel to the xy-plane, and that we move it up and down. Suppose that we start scanning from the point 10 units above the xy-plane. As the scanner moves downward, it does not show anything until it reaches $z = 9$, when a single point is detected. From that moment on, the circles (of radius $\sqrt{9 - c}$) appear on the scanner, and their size keeps increasing as we continue moving the scanner further down. Hence, we can say that the surface in question is built of circles (smaller circles are placed on top of bigger ones) that shrink to a point when $z = 9$.

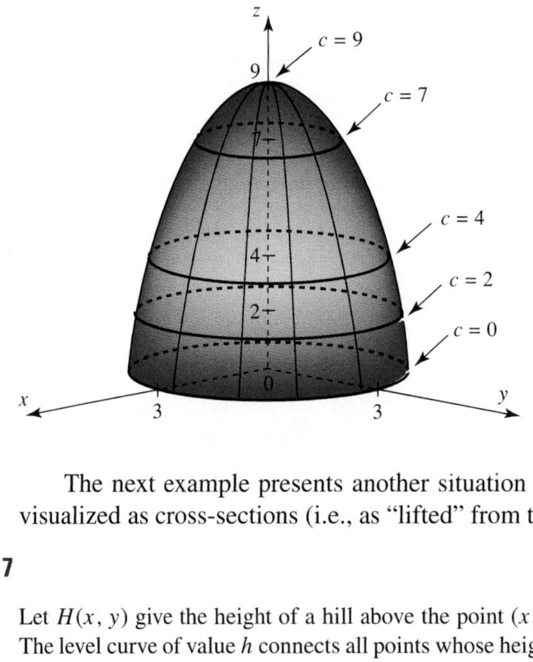

Figure 2.13 Several cross-sections of the surface $z = 9 - x^2 - y^2$.

The next example presents another situation when the level curves are conveniently visualized as cross-sections (i.e., as "lifted" from the xy-plane to the surface).

▶ EXAMPLE 2.17

Let $H(x, y)$ give the height of a hill above the point (x, y) in the xy-plane that represents sea level. The level curve of value h connects all points whose height above sea level is h; see Figure 2.14.

▶ EXAMPLE 2.18

Compute the level curves of the function $f(x, y) = e^{-x^2-2y^2}$.

SOLUTION

The fact that e^x is always positive implies that there are no level curves for $c \leq 0$. If $c > 0$, then $e^{-x^2-2y^2} = c$ gives $x^2 + 2y^2 = -\ln c$.

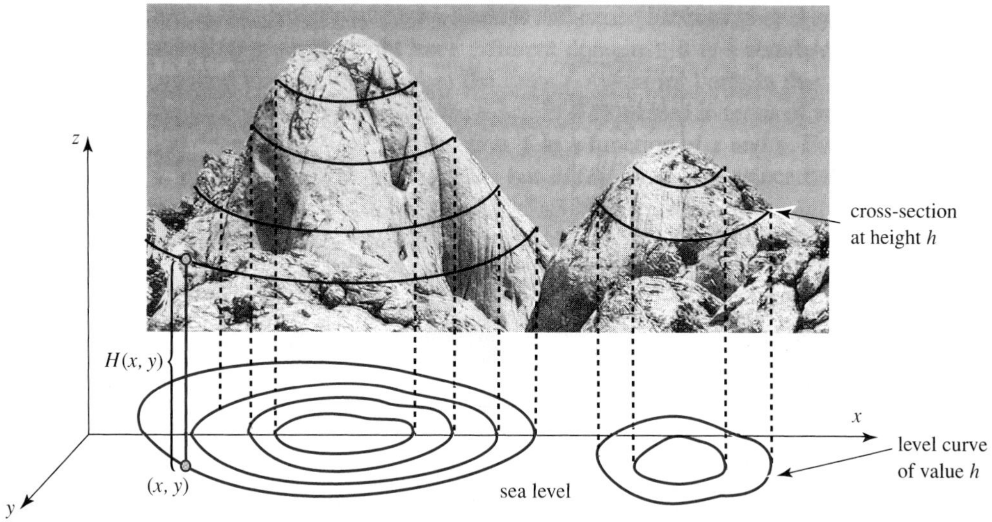

Figure 2.14 Cross-sections and corresponding level curves of Example 2.17.

Moreover, since $x^2 + 2y^2 \geq 0$, there are no level curves when $\ln c > 0$; that is, when $c > 1$. The level curve of value $c = 1$ consists of a single point (the origin, since $x^2 + 2y^2 = -\ln 1 = 0$ implies $x = y = 0$). It follows that the whole graph is contained in the region above the xy-plane and below the plane $z = 1$. For $0 < c < 1$ (for such a value of c, $-\ln c$ is positive), the level curve $x^2 + 2y^2 = -\ln c$ is the ellipse

$$\frac{x^2}{-\ln c} + \frac{y^2}{(-\ln c)/2} = 1$$

with semi axes $\sqrt{-\ln c}$ and $\sqrt{(-\ln c)/2}$; see Figure 2.15.

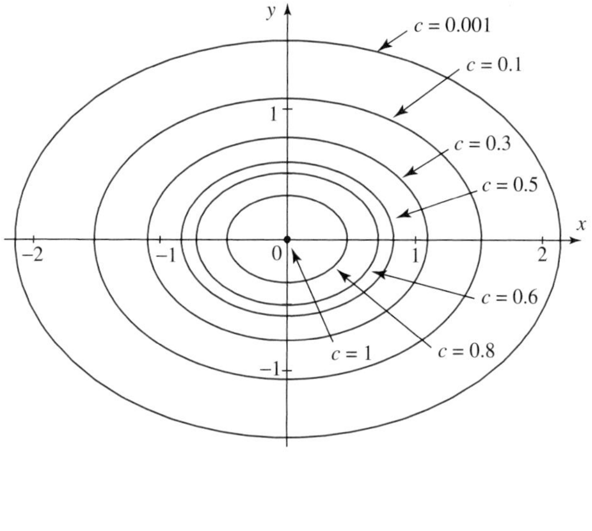

Figure 2.15 Level curves of the function $f(x, y) = e^{-x^2-2y^2}$.

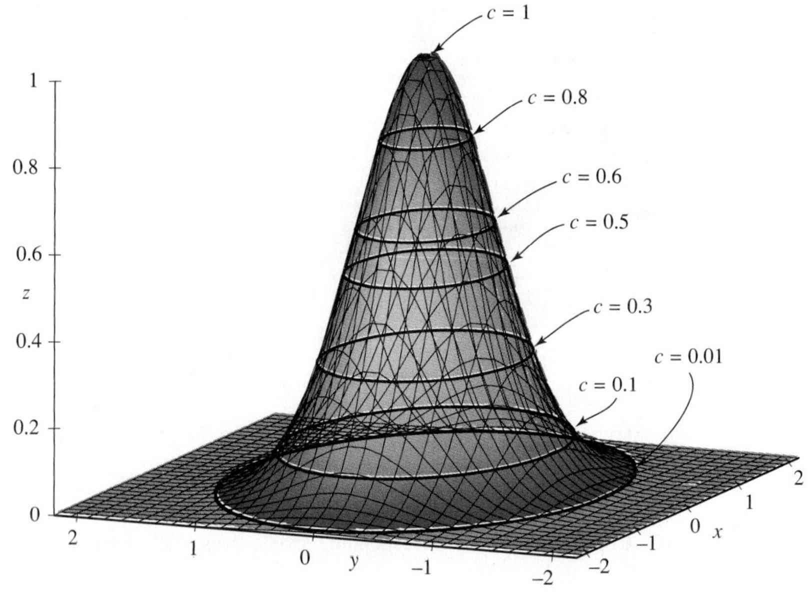

Figure 2.16 Graph of $f(x, y) = e^{-x^2-2y^2}$ with cross-sections corresponding to the level curves in Figure 2.15.

It follows that the surface $z = e^{-x^2-2y^2}$ is built of ellipses whose size increases as we move from $z = 1$ toward $z = 0$. Figure 2.16 shows a computer-generated plot of the function, together with the cross-sections corresponding to the level curves in Figure 2.15. ◀

▶ EXAMPLE 2.19

Find the level curves of $f(x, y) = 1 - (x^2 + y^2 - 1)^2$.

SOLUTION Setting $f(x, y) = c$ for a constant c, we get $1 - (x^2 + y^2 - 1)^2 = c$ and

$$x^2 + y^2 = 1 \pm \sqrt{1 - c}. \tag{2.6}$$

There are no level curves if $1 - c < 0$, that is, for $c > 1$. If $c = 1$, the level curve is the circle $x^2 + y^2 = 1$ of radius 1. If $0 < c < 1$, both $1 + \sqrt{1 - c}$ and $1 - \sqrt{1 - c}$ are positive and the corresponding level curve consists of two circles centered at the origin of radii $\sqrt{1 + \sqrt{1 - c}}$ and $\sqrt{1 - \sqrt{1 - c}}$. When $c = 0$, (2.6) reduces to $x^2 + y^2 = 1 \pm 1$, that is, the level curve of value 0 consists of the circle $x^2 + y^2 = 2$ *and* the point $(0, 0)$ (obtained from $x^2 + y^2 = 0$). If $c < 0$, then $1 + \sqrt{1 - c}$ is positive but $1 - \sqrt{1 - c}$ is negative, so this time the level curve consists only of the circle of radius $\sqrt{1 + \sqrt{1 - c}}$. Figure 2.17 shows the surface and several cross-sections. ◀

▶ EXAMPLE 2.20

Figure 2.18 shows a contour diagram of the wind chill index $W(T, v)$, introduced in Example 2.7 in Section 2.1.

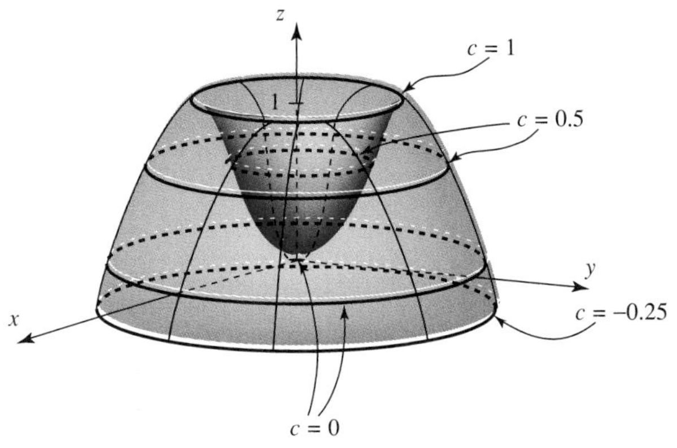

Figure 2.17 The surface $f(x, y) = 1 - (x^2 + y^2 - 1)^2$ and several cross-sections.

The points A, B, and C lie on the level curve of value -20. In other words, the air temperature of $-18°$C with a wind speed of 4 km/h (point A) feels the same as $-10°$C with the wind reaching 34 km/h (point B), or the same as the air temperature of $-8°$C with a wind speed of 60 km/h (point C).

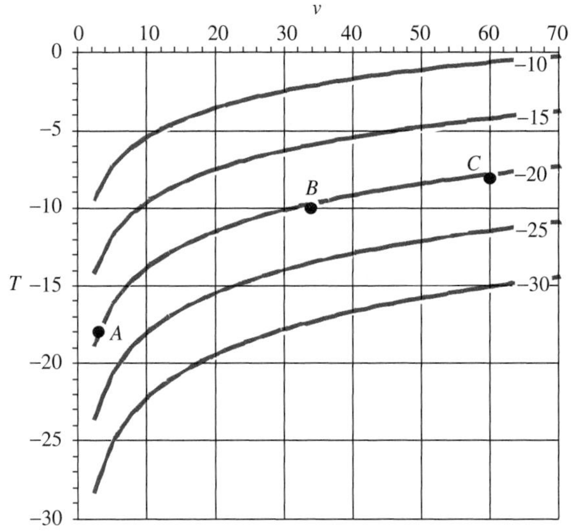

Figure 2.18 Contour curves of the wind chill index.

▶ EXAMPLE 2.21 Equipotential Curves of an Electric Dipole

An electric dipole is a configuration consisting of two point charges of the same magnitude q and of opposite signs placed at $(0, \ell/2)$ (the positive charge) and $(0, -\ell/2)$ (the negative charge), where $\ell > 0$. The potential at the point (x, y) due to the dipole is, for $\ell \to 0$ and $q \to \infty$, given by

$$\Phi(x, y) = \frac{1}{4\pi\epsilon_0} \frac{q\ell y}{(x^2 + y^2)^{3/2}},$$

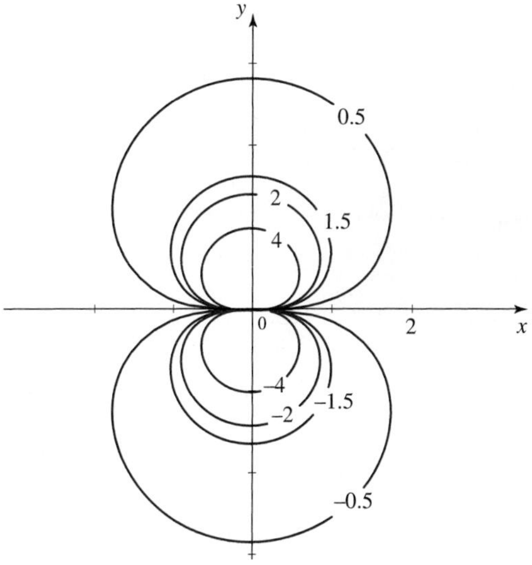

Figure 2.19 Equipotential curves of the electric dipole of Example 2.21.

or, in polar coordinates ($x = r\cos\theta$, $y = r\sin\theta$),

$$\Phi(r,\theta) = \frac{q\ell}{4\pi\epsilon_0}\frac{r\sin\theta}{(r^2)^{3/2}} = \frac{q\ell}{4\pi\epsilon_0}\frac{\sin\theta}{r^2}.$$

It follows that $\Phi(r,\theta) > 0$ if $0 < \theta < \pi$ (since then $\sin\theta > 0$) and $\Phi(r,\theta) < 0$ if $\pi < \theta < 2\pi$. The level curves (or equipotential curves) are given by

$$\Phi(r,\theta) = \frac{q\ell}{4\pi\epsilon_0}\frac{\sin\theta}{r^2} = c;$$

that is (after solving for r), $r = d\sqrt{\sin\theta/c}$, where $d = \sqrt{q\ell/4\pi\epsilon_0}$ and $0 < \theta < \pi$ if $c > 0$ and $\pi < \theta < 2\pi$ if $c < 0$. Several level curves (with $d = 2$) are drawn in Figure 2.19. ◀

▶ EXAMPLE 2.22

Find the level surfaces of the functions $f(x,y,z) = \ln(x^2+y^2+z^2)$ and $g(x,y,z) = 5x+y-z$.

SOLUTION In this case it is impossible to draw the graphs, but we can still describe them by analyzing level surfaces. For any $c \in \mathbb{R}$, $f(x,y,z) = c$ implies that $\ln(x^2+y^2+z^2) = c$, that is, $x^2+y^2+z^2 = e^c$. The level surfaces of f are spheres centered at the origin of radii $\sqrt{e^c}$.

From $5x+y-z = c$, it follows that the level surfaces of g are parallel planes [since they all have the same normal vector $(5, 1, -1)$]. ◀

▶ EXAMPLE 2.23 Equipotential Surfaces of a Gravitational Potential

Consider the gravitational potential (see Example 2.10) $V(\mathbf{r}) = -GMm/||\mathbf{r}||$. If $V(\mathbf{r}) = c$ (for some constant $c \neq 0$), then

$$-\frac{GMm}{||\mathbf{r}||} = -\frac{GMm}{\sqrt{x^2+y^2+z^2}} = c$$

(c has to be negative for the above to make sense), and therefore $x^2 + y^2 + z^2 = G^2M^2m^2/c^2$. In other words, level surfaces for this gravitational potential (also called equipotential surfaces) are spheres centered at the origin with radii $GMm/|c|$. ◀

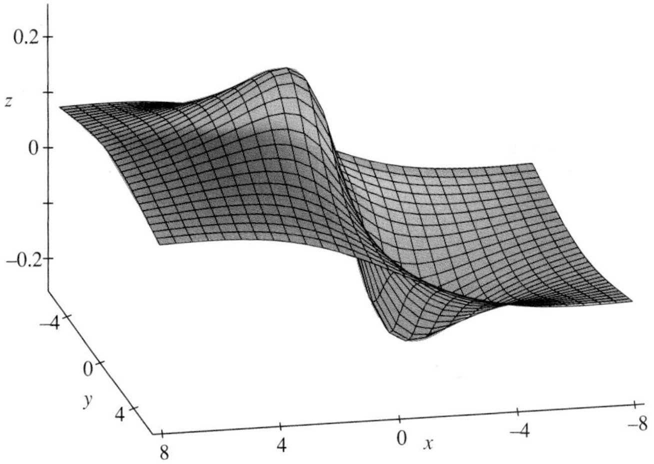

Figure 2.20 Plot of the function $f(x, y) = x/(x^2 + y^2 + 4)$.

How Computers Generate Plots of Functions

In order to request a plot of a function $y = f(x)$ of one variable, we have to input the formula for f and the interval where the plot will be constructed. The computer computes the values $y = f(x)$ at a certain number of equally spaced values x within the interval, plots the points (x, y) thus obtained, and joins them with straight-line segments. Some better graphing programs select extra points in those subintervals where there is a large difference in the values of the function to obtain a more precise plot. The same principle is used to plot paths in $\mathbb{R}^2$ and in $\mathbb{R}^3$.

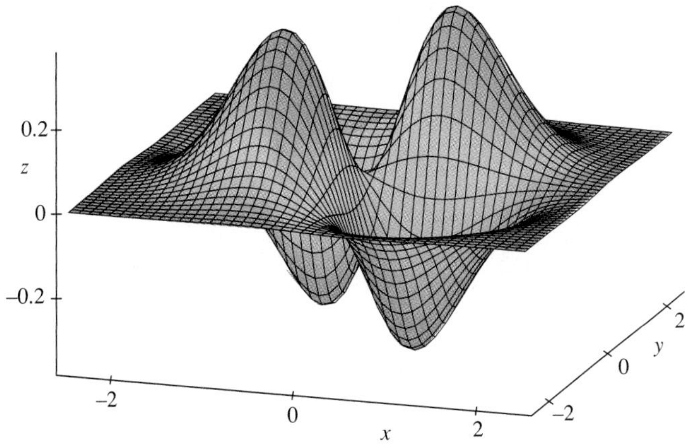

Figure 2.21 Plot of the function $f(x, y) = 2xye^{-x^2-y^2}$.

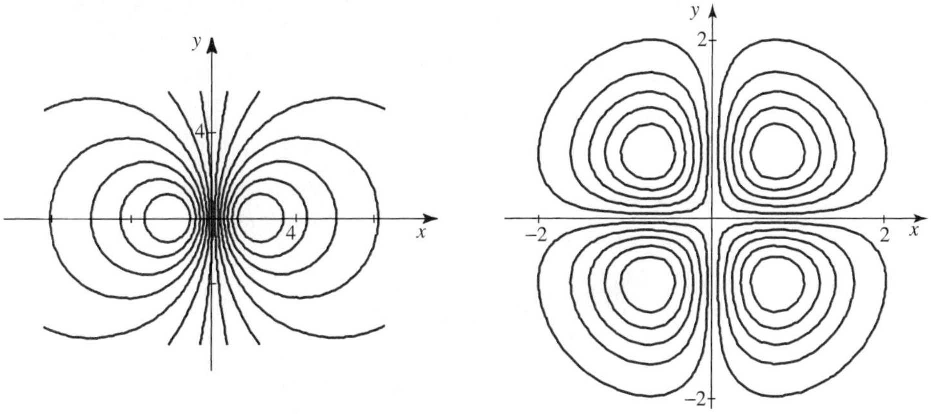

(a) $f(x, y) = x/(x^2 + y^2 + 4)$. (b) $f(x, y) = 2xye^{-x^2-y^2}$.

Figure 2.22 Contour diagrams of functions in Figures 2.20 and 2.21.

The plot of a function $z = f(x, y)$ of two variables is obtained in the following way: a computer chooses a number of equally spaced points in the given interval for x and in the given interval for y [thus obtaining a grid of elements (x, y) of the domain]. Then it computes the values $z = f(x, y)$, plots all points (x, y, z), and connects them with line segments, forming a "net" as shown in Figures 2.20, 2.21, and 2.23.

Figure 2.22 shows contour diagrams of functions $f(x, y) = x/(x^2 + y^2 + 4)$ and $f(x, y) = 2xye^{-x^2-y^2}$. We note that the former can be constructed quite easily with a

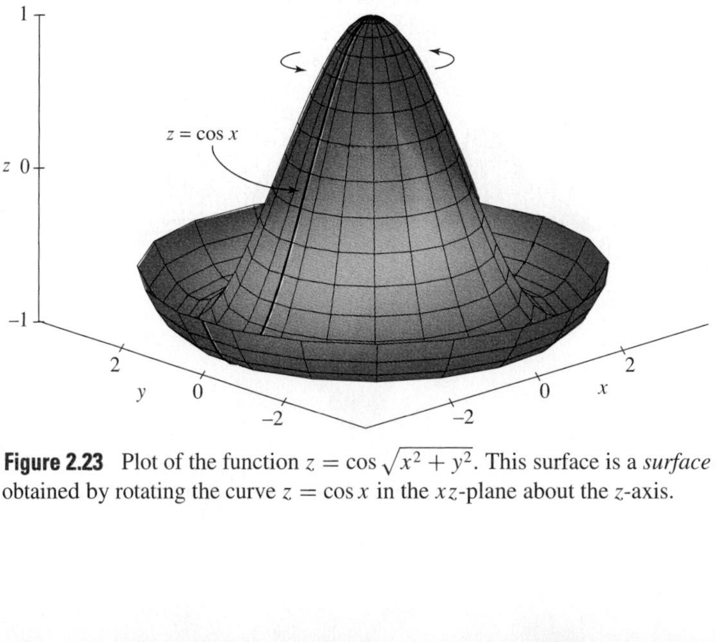

Figure 2.23 Plot of the function $z = \cos\sqrt{x^2 + y^2}$. This surface is a *surface of revolution*: it is obtained by rotating the curve $z = \cos x$ in the xz-plane about the z-axis.

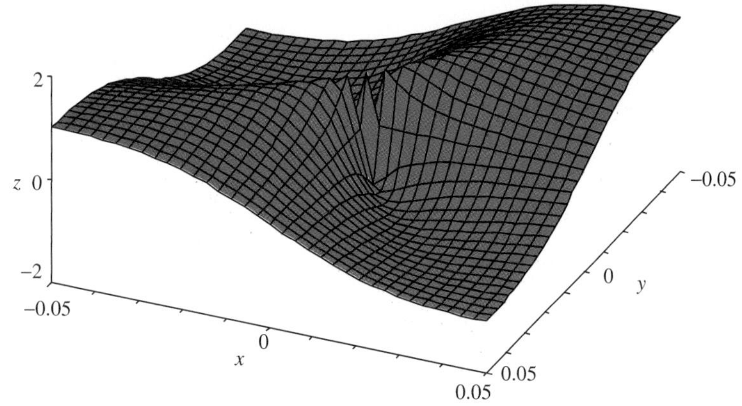

Figure 2.24 Near (0, 0), the plot of $f(x, y) = 2xy/(x^2 + y^2)$ is incorrect.

pencil and paper (see Exercise 41). However, without the use of a computer the contour diagram of $f(x, y) = 2xye^{-x^2-y^2}$ would be hard to obtain.

However, "seeing is not always believing" when considering computer-generated plots. Important features of the graph could be hidden or not shown at all. Figure 2.24 represents the function $f(x, y) = 2xy/(x^2 + y^2)$, and the graph of the function $g(x, y) = \arctan(0.2y/x)$ is shown in Figure 2.25. The complicated behavior of f near (0, 0) (we will see in the next section that f does not have a limit there) cannot be detected from the plot. The graph given in Figure 2.25 is incorrect: there should be a break along points on the y-axis, since g is not continuous there (this will also be discussed in the next section; see Figure 2.36).

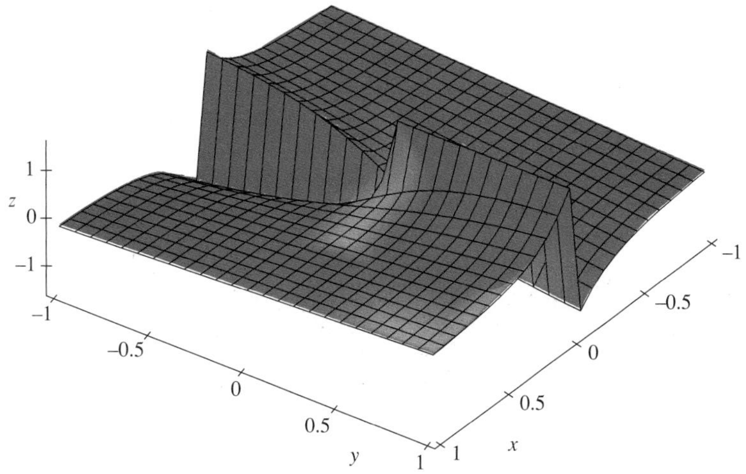

Figure 2.25 Incorrect plot of the function $f(x, y) = \arctan(0.2y/x)$.

▶ EXERCISES 2.2

1. A level set of a function $f\colon \mathbb{R}^2 \to \mathbb{R}$ does not have to be a curve (it is called a level curve anyway). We have already seen that a level set can degenerate to a point, or can be the union of two curves, as in Example 2.19. What are the "level curves" of the function $f(x, y) = 1$?

Exercises 2 to 11: Describe level curves of the functions given below.

2. $f(x, y) = 2x - 3y - 5$

3. $f(x, y) = 3 - x^2 - y^2$

4. $f(x, y) = -3y(x^2 + y^2 + 1)^{-1}$

5. $f(x, y) = e^{xy}$

6. $f(x, y) = \sqrt{x^2 - y}$

7. $f(x, y) = x/y$

8. $f(x, y) = (3x - y)/(x + 2y)$

9. $f(x, y) = y - \sin x$

10. $f(x, y) = 3x^2 - y^2 + 4$

11. $f(x, y) = x^2 - y^2$

12. Consider a linear function $f(x, y) = ax + by + d$, where $a \neq 0$ or $b \neq 0$. Show that

(a) Level curves of $f(x, y)$ are parallel lines.

(b) The distance between the lines corresponding to equal difference values is the same (i.e., the lines are equally spaced).

(c) Generalize (a) and (b) to a linear function of three variables.

13. Find formulas for linear functions whose contour diagrams are shown in Figure 2.26.

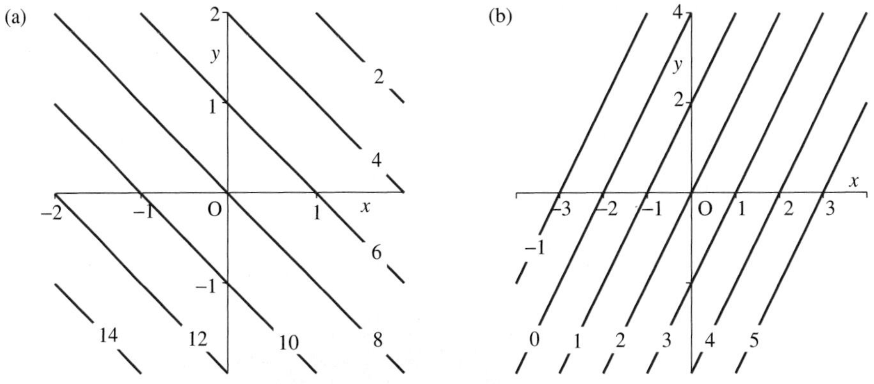

Figure 2.26 Contour diagrams of functions in Exercise 13.

14. Explain why is it not possible that level curves of two different values intersect each other.

15. Find the level curves of the function $f\colon \mathbb{R}^2 \to \mathbb{R}$ defined by

$$f(x, y) = \begin{cases} 1 - x^2 - y^2 & \text{if } x^2 + y^2 < 1 \\ 0 & \text{if } x^2 + y^2 \geq 1 \end{cases}$$

16. The function $T(x, y) = 50(1 + x^2 + 3y^2)^{-1}$ describes the temperature of a metal plate at the point $(x, y) \in \mathbb{R}^2$. Describe the level curves (isotherms) of T.

17. The voltage at a point (x, y) on a metal plate placed in the xy-plane is given by the function $V(x, y) = \sqrt{1 - 9x^2 - 4y^2}$. Sketch the equipotential curves (i.e., the curves of constant voltage).

18. Look at the contour diagram (Figure 2.18) of the wind chill index in Example 2.20. Identify the five points that lie on the intersection of the vertical line representing the wind speed of 40 km/h with the five contour curves. They appear to be evenly spaced. Is that true? Justify your answer.

Exercises 19 to 24: Describe the level surfaces of the functions of three variables given below.

19. $f(x, y, z) = y - x^2$

20. $f(x, y, z) = e^{x^2+y^2+z^2}$

21. $f(x, y, z) = 4x^2 + 4y^2 + z^2 + 1$

22. $f(x, y, z) = x - 2y + z + 3$

23. $f(x, y, z) = 3x + 4$

24. $f(x, y, z) = x^2 + y^2$

25. What is the graph of the equation:

(a) $x = a$ in $\mathbb{R}^2$? In $\mathbb{R}^3$?

(b) $y = b$ in $\mathbb{R}^2$? In $\mathbb{R}^3$?

26. Graph the following surfaces in $\mathbb{R}^3$:

(a) $y = ax + b$

(b) $z = ax + b$

(c) $z = ay + b$

27. Describe in words each graph (assume that $a > 0$).

(a) $x^2 + y^2 + z^2 = a^2$

(b) $(x - m)^2 + (y - n)^2 + (z - p)^2 = a^2$

Exercises 28 to 37: Sketch (and/or describe in words) the following surfaces in $\mathbb{R}^3$. In each case, describe the level curves and compute the traces in the xz- and the yz-planes (a trace is the intersection of the surface and a coordinate plane).

28. $x + 2y + 3z - 6 = 0$

29. $z = \sqrt{2 - x^2 - y^2}$

30. $x^2 + 3y^2 + 9z^2 = 9$

31. $x^2 + y^2 = 9$

32. $y - x^2 = 3$

33. $z = 4$

34. $z = x^2 + y^2$

35. $z = 2 - x^2 - y^2$

36. $z^2 = x^2 + y^2$

37. $z = \sqrt{x^2 + y^2}$

38. Consider the Cobb–Douglas production function $P(L, K) = 1.01L^{0.75}K^{0.25}$ of Example 2.6. Show that the level curve of $P(L, K)$ of value p is given by $K = cL^{-3}$, where $c = (p/1.01)^4$. Sketch a contour diagram of $P(L, K)$.

39. Consider the formula for the pressure of an ideal gas $p(T, V) = RnT/V$, introduced in Example 2.8 in Section 2.1. In the VT-coordinate system, what do level curves of p look like?

40. Draw a contour diagram of the function $f(x, y) = |y|$. Then sketch the graph of f.

41. Consider the function $f(x, y) = x/(x^2 + y^2 + 4)$.

(a) What is the level curve of value $c = 0$?

(b) Show that the level curve of value $c \neq 0$ is given by the equation $\left(x - \frac{1}{2c}\right)^2 + y^2 = \frac{1}{4c^2} - 4$. For which values of c does it represent a circle? A point?

(c) Sketch a contour diagram of f and compare the result with Figure 2.22.

42. Consider the function $f(x, y) = 2xy/(x^2 + y^2)$. (This exercise will be used in Example 2.26.)

(a) Complete the square to show that the level curve of value $c \neq 0$ is given by $(x - y/c)^2 + (1 - 1/c^2)y^2 = 0$.

(b) Prove that $f(x, y)$ has no level curves of value $c > 1$ or $c < -1$.

(c) Show that the level curve of value c, where $-1 \leq c \leq 1$, and $c \neq 0$, consists of the two lines $y = x(1/c \pm \sqrt{1/c^2 - 1})^{-1}$, from which the point $(0, 0)$ has been removed. What is the level curve of value $c = 0$?

43. A contour diagram of a linear function $f(x, y)$ consists of lines parallel to the x-axis. The contour curve of value c crosses the y-axis at $c/3$. Find $f(x, y)$.

44. Find a formula for a function $f(x, y, z)$ whose level set of value $c = 2$ is the surface $x^2 + y^3 - z = 4$.

45. Find a formula for a function $f(x, y, z)$ whose level surfaces are parallel planes with normal vector $\mathbf{n} = (2, 3, -4)$.

▶ 2.3 LIMITS AND CONTINUITY

Review: Limits of Functions of One Variable

A reader familiar with this material may wish to advance to the next subsection that introduces limits of functions of several variables.

Consider the function $f(x)$ defined by

$$f(x) = \begin{cases} x^2 + 1 & \text{if } x < 1 \\ 2x + 1 & \text{if } x \geq 1. \end{cases}$$

The graph of $f(x)$ is shown in Figure 2.27. The small empty circle means that the point $(1, 2)$ does not belong to the graph. The filled circle at $(1, 3)$ denotes the fact that $f(1) = 3$. Let us try to describe the behavior of f for values of x that are close to $a = 1$.

We start at, say, $x = 0.5$, and, as we walk along the x-axis toward $a = 1$, we compute the corresponding values of the function [keeping in mind that for $x < 1$, $f(x) = x^2 + 1$]. If $x = 0.5$, then $f(0.5) = 1.25$. If $x = 0.9$, then $f(0.9) = 1.81$. Similarly, $f(0.99) = 1.9801$, $f(0.999) = 1.998001$, etc. We see that the values of the function approach 2 as x gets closer and closer to 1. For example, if we need that $f(x)$ be closer to 2 than 1.9999, we can choose any x (substitute $y = 1.9999$ into $y = x^2 + 1$ and solve for x) such that $x > 0.99995$ (and $x < 1$). Let us check a few values: $f(0.99996) = 1.99992$, $f(0.99998) = 1.99996$, etc. The fact that f can be made as close to 2 as needed by choosing $x < 1$ close enough to 1 is written as $\lim_{x\to 1^-} f(x) = 2$, and is called the *left limit of $f(x)$ as x approaches* 1 or the *limit of $f(x)$ as x approaches* 1 *from the left* (the symbol $x \to 1^-$ denotes the fact that x gets closer and closer to 1, and $x < 1$).

A similar investigation shows that $f(x)$ can be made as close to 3 as needed by choosing $x > 1$ to be close enough to 1. This fact is written as $\lim_{x\to 1^+} f(x) = 3$, and represents the *right limit of $f(x)$ as x approaches* 1, or the *limit of $f(x)$ as x approaches* 1 *from the right* (the symbol $x \to 1^+$ is used to denote the fact that x gets close to 1 and $x > 1$). In a

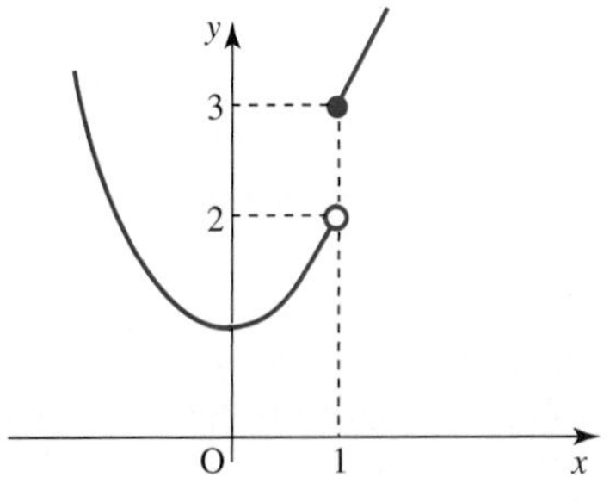

Figure 2.27 Graph of the function $f(x)$ defined above.

situation like this one, when different approaches give different values for the right and the left limits, we say that the *limit* or the *two-sided (or both-sided) limit* does not exist.

The left and right limits are also referred to as the *one-sided* limits. An analogous examination of the behavior of f near $a = 0$ [where $f(x) = x^2 + 1$] would show that $\lim_{x \to 0^+} f(x) = 1$ and $\lim_{x \to 0^-} f(x) = 1$. This time, the one-sided limits agree, and we say that the *two-sided (or both-sided) limit* of $f(x)$ as x approaches 0 exists and is equal to 1, and write $\lim_{x \to 0} f(x) = 1$.

Notice that in our investigation of limits, the value of f at $a = 1$ or at $a = 0$ did not play any role. Let us remember this important fact: when computing limits (as x approaches a), what matters is not the value of the function at $x = a$ (it may not even be defined) but the values of $f(x)$ "near" a (see Exercise 1).

Next, we give precise meaning to phrases like "x approaches a" and "$f(x)$ can be made as close to L as needed." Take an interval $(a - \delta, a + \delta)$ around a, where $\delta > 0$. The phrase "x approaches a" means that no matter what $\delta > 0$ is chosen, we can always find a value x that is inside $(a - \delta, a + \delta)$, and is such that $x \neq a$.

Recalling the fact that the absolute value $|x_1 - x_2|$ gives the distance between x_1 and x_2 on the x-axis, we can rephrase the above statement as follows: "x approaches a" means that no matter what δ is chosen, we can always find an $x \neq a$ whose distance from a is smaller than δ, that is, $|x - a| < \delta$. Since $x \neq a$, the distance between x and a cannot be zero (hence, $|x - a| > 0$) and we write the above as $0 < |x - a| < \delta$.

Here is one way of visualizing this process of "approaching." Assume that "x approaches 5," that is, let $a = 5$. Select a sequence of values for δ, like $\delta = 10^{-1}$, 10^{-2}, 10^{-3}, etc. This sequence defines a sequence of intervals of the form $(a - \delta, a + \delta)$; specifically, the intervals are (4.9, 5.1), (4.99, 5.01), (4.999, 5.001), etc. The phrase "x approaches 5" describes a process of selecting a number x from every interval in this sequence.

The phrase "$f(x)$ can be made as close to L as needed" means that, for any $\epsilon > 0$, the value $f(x)$ lies inside the interval $(L - \epsilon, L + \epsilon)$; that is, the distance $|f(x) - L|$ between $f(x)$ and L is less than ϵ.

Let us go back to our previous example and illustrate what we have just said. We claim that $\lim_{x \to 0} f(x) = \lim_{x \to 0}(x^2 + 1) = 1$ (i.e., $a = 0$ and $L = 1$). Take, for example, $\epsilon = 0.2$ and consider the interval $(L - \epsilon, L + \epsilon) = (0.8, 1.2)$. We should be able to find an interval $(a - \delta, a + \delta) = (-\delta, \delta)$ such that, no matter what nonzero x is selected from that interval, the corresponding value $f(x)$ lies inside (0.8, 1.2). From $|f(x) - 1| < \epsilon = 0.2$, we get $|x^2 + 1 - 1| < 0.2$, so $x^2 < 0.2$; hence, δ can be taken to be $\sqrt{0.2}$. In other words, for any $x \in (-\sqrt{0.2}, \sqrt{0.2})$, $f(x)$ is in (0.8, 1.2). The fact that the limit is 1 means that the above construction of the interval $(a - \delta, a + \delta) = (-\delta, \delta)$ can be carried out for *any* choice of $\epsilon > 0$.

In the same example, we demonstrated that $\lim_{x \to 1} f(x)$ does not exist. Let us think a bit about this: we will show that, for example, $L = 2$ cannot be the limit. Let $\epsilon = 0.1$; that is, consider the interval $(L - \epsilon, L + \epsilon) = (1.9, 2.1)$. Any interval $(a - \delta, a + \delta) = (1 - \delta, 1 + \delta)$, no matter how small, contains a number (call it x_0) to the right of 1. The corresponding value $f(x_0) = 2x_0 + 1$ is greater than 3 (since $x_0 > 1$) and certainly does not belong to (1.9, 2.1); see Figure 2.28. This violates the definition of the limit, and therefore $L = 2$ is not the limit of $f(x)$ as x approaches 1. A similar discussion would rule out any other real number as a candidate for the limit.

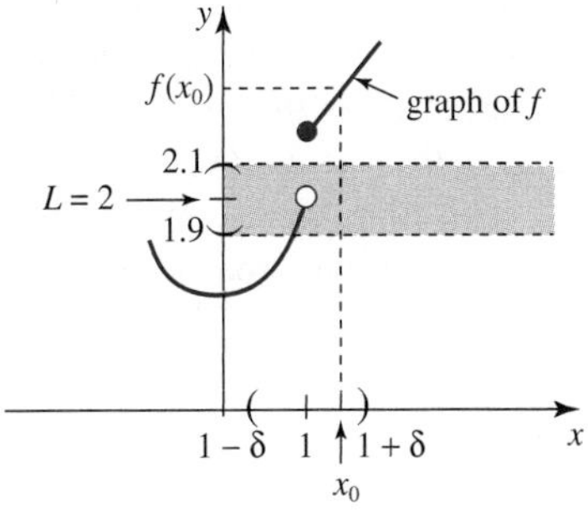

Figure 2.28 Limit of $f(x)$ as x approaches 1 does not exist.

▶ EXAMPLE 2.24 "Closeness" and "Approaching" in a Computer Programming Context

Let us relate this "closeness" and "approaching" to a problem sometimes encountered in a computer science class. Suppose that we have to find a real number x such that $x^2 = 2$ (the solutions are infinite nonperiodic decimal numbers $x = \pm\sqrt{2} = \pm 1.4142\ldots$). A bad computer program might try to solve the problem as follows:

```
choose an(other) x
if x^2 = 2 then done, else choose another x
```

The phrase "choose an(other) x" means that we have a way of selecting a new try for x based on the outcomes of previous passes through the loop (how this is done is not our concern). A program like this has a good chance of never ending! The computer might, for example, get $x^2 = 1.999999$ or $x^2 = 2.000001$ for some choices of x and continue trying with new choices, since neither of the two results is equal to 2. To fix the program, we change it to the following:

```
choose an(other) x
if x^2 is "close enough to 2" then done, else choose
  another x
```

In this case, the computer will stop when it hits a number x whose square is close to 2, for example, $x^2 = 1.999999$, and will return that x as a solution. Clearly, the result will be an approximation of the solution. If we require that x^2 be even "closer to 2," the approximation will be even better. More precisely, the program

```
let ϵ = 0.00001
choose an(other) x
if |x^2 − 2| < ϵ then done, else choose another x
```

will return an approximate solution. To find a better approximation, all we have to do is to choose a smaller interval (restrict the "tolerance"); that is, take, for example, $\epsilon = 10^{-10}$. ◀

DEFINITION 2.5 Limit of a Function of One Variable

A function $f\colon \mathbb{R} \to \mathbb{R}$ has limit L as x approaches a, in symbols $\lim_{x\to a} f(x) = L$, if and only if for any given number $\epsilon > 0$ there is a number $\delta > 0$ such that

$$0 < |x - a| < \delta \qquad \text{implies} \qquad |f(x) - L| < \epsilon. \quad ◀$$

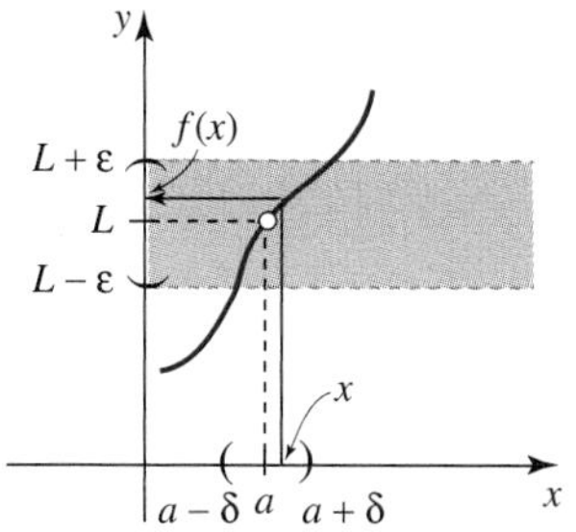

Figure 2.29 L is the limit of a function $y = f(x)$ as $x \to a$.

In other words, given the "tolerance" $\epsilon > 0$, we can find an interval $(a - \delta, a + \delta)$ around a ($\delta > 0$) so that for every $x \neq a$ inside $(a - \delta, a + \delta)$, the corresponding value $f(x)$ lies within the allowed "tolerance" interval $(L - \epsilon, L + \epsilon)$; see Figure 2.29.

Limits of Functions of Several Variables

We are on our way to generalizing the definition of a limit. Since the limit (and continuity) of a vector-valued function is based on the limits (continuity) of its components, it suffices to generalize limits to real-valued functions of several variables. In order to accomplish this task, we have to come up with an analogue of our concept of "closeness": we have to explain what is meant by phrases such as "$\mathbf{x} = (x, y)$ approaches $\mathbf{a} = (a, b)$" or "$\mathbf{x} = (x, y, z)$ approaches $\mathbf{a} = (a, b, c)$," etc. Since "closeness" was defined using open intervals, what we need now is their generalization to higher dimensions.

DEFINITION 2.6 Open Balls in $\mathbb{R}^m$

The *open ball* $B(\mathbf{a}, r) \subseteq \mathbb{R}^m$ with center $\mathbf{a} = (a_1, \ldots, a_m)$ and radius r ($r > 0$) is the set of all points $\mathbf{x}$ in $\mathbb{R}^m$ whose distance from a fixed point $\mathbf{a}$ is smaller than r. In symbols,

$$B(\mathbf{a}, r) = \{\mathbf{x} \in \mathbb{R}^m \mid ||\mathbf{x} - \mathbf{a}|| < r\},$$

where $\mathbf{x} = (x_1, \ldots, x_m)$ and $||\mathbf{x} - \mathbf{a}|| = \sqrt{(x_1 - a_1)^2 + \cdots + (x_m - a_m)^2}$.

For example, the open ball $B((1, 2), 3) \subseteq \mathbb{R}^2$ consists of all points in $\mathbb{R}^2$ whose distance from $(1, 2)$ is strictly smaller than 3; that is, $\sqrt{(x - 1)^2 + (y - 2)^2} < 3$. It is the inside of the circle of radius 3 centered at the point $(1, 2)$.

Similarly, the open ball $B((0, 0, 0), 2) \subseteq \mathbb{R}^3$ consists of the region inside the sphere of radius 2 centered at the origin. The open ball $B(3, 2) \subseteq \mathbb{R}$ contains all real numbers whose distance from 3 is less than 2; that is, $|x - 3| < 2$. It is the interval $(1, 5)$. The last example shows that, as subsets of $\mathbb{R}$, open balls coincide with open intervals. In particular, the statement $|x - a| < \delta$ (translated as "the distance from x to a is less than δ") can be written as $x \in B(a, \delta)$.

In the case of a function of one variable, the limit as $x \to a$ has been determined by investigating two special approaches, namely the right and left limits. If the two limits were equal, we said that the function had a (two-sided) limit, and its value $\lim_{x \to a} f(x)$ was equal to the common value of the one-sided limits. Consider now the function of two variables

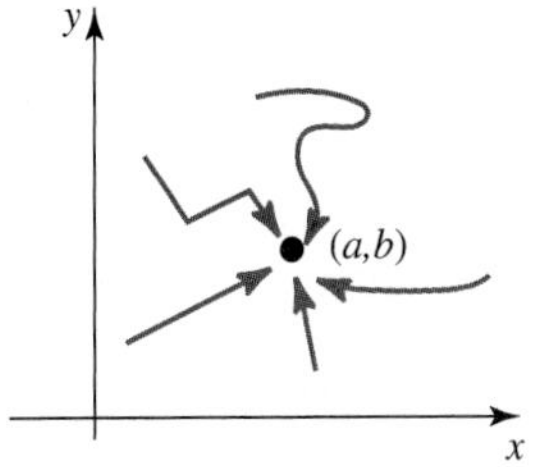

Figure 2.30 Point (a, b) can be approached in infinitely many ways.

(this is a choice of convenience: everything said holds for a function of any number of variables). It is impossible to investigate all possible ways of approaching a selected point $\mathbf{a} = (a, b)$ in $\mathbb{R}^2$: there are infinitely many of them! See Figure 2.30.

There is some good news here: the limit should be (and is) independent of the way we approach $\mathbf{a} = (a, b)$. Therefore, if two different approaches give two different candidates for the limit, we can be sure that the limit does not exist. However, if several approaches to (a, b) all give the same number, this does not prove anything yet: all it says is that, if the limit exists, it must be equal to that number.

We now imitate the one-variable case and define the limit of a function $f: U \subseteq \mathbb{R}^m \to \mathbb{R}$ as $\mathbf{x}$ approaches $\mathbf{a}$. The definition (applied to functions of two variables) says that if we can force the values $f(x, y)$ to move arbitrarily close to L as (x, y) gets close to (a, b), the function f has limit L; see Figure 2.31.

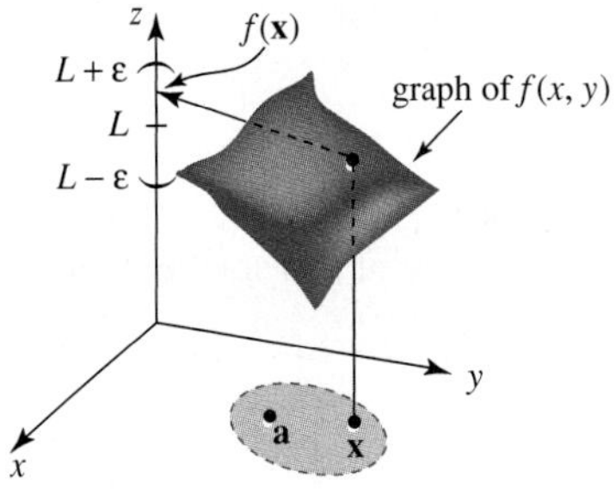

Figure 2.31 Limit of a function of two variables.

DEFINITION 2.7 Limit of a Real-Valued Function of Several Variables

Let $f: U \subseteq \mathbb{R}^m \to \mathbb{R}$ be a real-valued function of m variables. We say that the limit of $f(\mathbf{x}) = f(x_1, \ldots, x_m)$ as $\mathbf{x} = (x_1, \ldots, x_m)$ approaches $\mathbf{a} = (a_1, \ldots, a_m)$ is L, in symbols

$$\lim_{\mathbf{x}\to\mathbf{a}} f(\mathbf{x}) = L \quad \text{or} \quad \lim_{(x_1,\ldots,x_m)\to(a_1,\ldots,a_m)} f(x_1, \ldots, x_m) = L,$$

if and only if for every $\epsilon > 0$ there is a number $\delta > 0$ such that $\mathbf{x} \in U$ and

$$0 < ||\mathbf{x} - \mathbf{a}|| < \delta \quad \text{implies} \quad |f(\mathbf{x}) - L| < \epsilon. \tag{2.7}$$

The condition $||\mathbf{x} - \mathbf{a}|| > 0$ in (2.7) guarantees that $\mathbf{x} \neq \mathbf{a}$. Notice that there is no mention of the way $\mathbf{x}$ is supposed to approach $\mathbf{a}$. The requirement in (2.7) is that the *distance* $||\mathbf{x}-\mathbf{a}||$ between $\mathbf{x}$ and $\mathbf{a}$ becomes smaller and smaller and therefore the definition includes every possible path that brings $\mathbf{x}$ close to $\mathbf{a}$. A convenient way of thinking about the requirement that $||\mathbf{x}-\mathbf{a}|| < \delta$ is to consider a process of selecting a point $\mathbf{x}$ (of course, $\mathbf{x}$ has to belong to the domain U of the function) from the sequence of open balls (all centered at $\mathbf{a}$) that shrink in size (their radii δ becoming smaller and smaller).

Now let $\mathbf{x}=(x,y)$, $\mathbf{a}=(a,b)$, and consider the expression

$$||\mathbf{x}-\mathbf{a}|| = \sqrt{(x-a)^2+(y-b)^2}. \tag{2.8}$$

If $\mathbf{x}\to\mathbf{a}$, then the distance $||\mathbf{x}-\mathbf{a}||$ can be made as small (i.e., as close to 0) as needed. But that means the square root, and hence the expression $(x-a)^2+(y-b)^2$, can be made as close to 0 as needed. Since both summands are positive, it follows that each of $x-a$ and $y-b$ can be made as close to 0 as needed (but only one can actually be made equal to 0). In other words, if $\mathbf{x}\to\mathbf{a}$, then $x\to a$ and $y\to b$. An analogous statement holds in general; that is, if $\mathbf{x}\to\mathbf{a}$, where $\mathbf{x}=(x_1,\ldots,x_m)$ and $\mathbf{a}=(a_1,\ldots,a_m)$, then $x_1\to a_1,\ldots,x_m\to a_m$ or

$$\lim_{\mathbf{x}\to\mathbf{a}} x_i = a_i \tag{2.9}$$

for every $i=1,\ldots,m$.

DEFINITION 2.8 Interior and Boundary Points

A point $\mathbf{a}$ in U is called an *interior point* of U if there is an open ball centered at $\mathbf{a}$ that is completely contained in U. If every open ball centered at $\mathbf{a}$ (this time, $\mathbf{a}$ does not have to be in U) contains not only points in U but also points not in U, then we call $\mathbf{a}$ a *boundary point* of U. ◀

The dashed curves in Figure 2.32 indicate points that do not belong to the set U or to the open balls shown. The points on the "unbroken" curves belong to U. An interior point always belongs to the set. However, a boundary point of U may or may not belong to U, as shown in Figure 2.32(b).

The definition of the limit applies to both interior and boundary points. In approaching a boundary point, we have to make sure that in the process of picking $\mathbf{x}$ from shrinking balls, we always pick only those $\mathbf{x}$ that belong to the domain U of the function. For instance, in computing the limit of $f(x,y)=x^2\ln y$ as (x,y) approaches $(1,0)$, we can use any real value for x, but the y values need to be restricted to $y>0$.

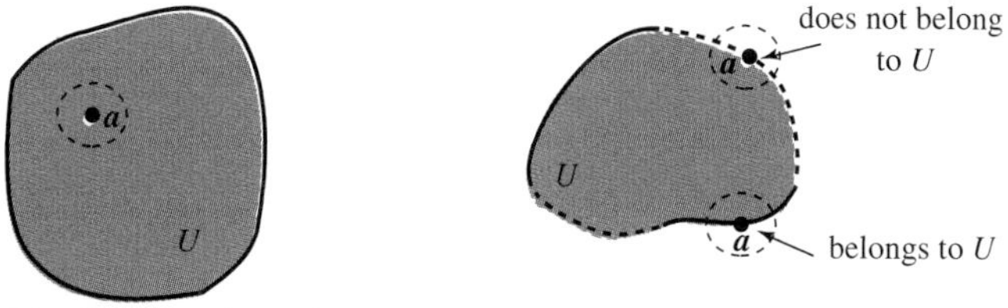

(a) $\mathbf{a}$ is an interior point. (b) Each $\mathbf{a}$ is a boundary point.

Figure 2.32 Interior and boundary points of a set.

▶ EXAMPLE 2.25

Several level curves of the function $f(x, y) = 3x^2y/(x^2 + y^2)$ are drawn in Figure 2.33. It looks like the values $f(x, y)$ approach 0 as (x, y) approaches $(0, 0)$. Show that this is indeed true, that is, use Definition 2.7 to prove that $\lim_{(x,y)\to(0,0)} f(x, y) = 0$.

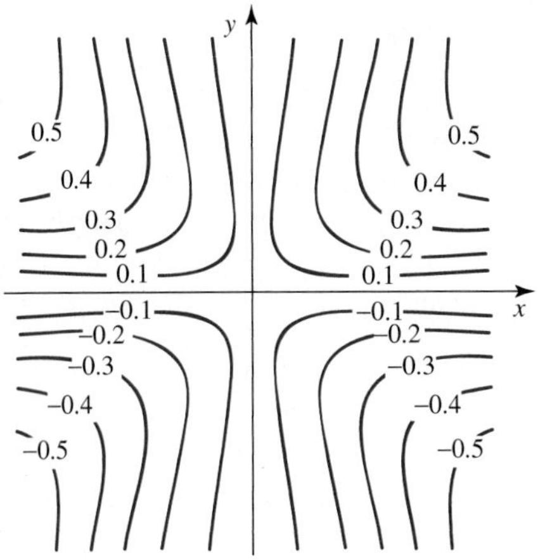

Figure 2.33 Level curves of $f(x, y)$ in Example 2.25.

SOLUTION

Pick any $\epsilon > 0$. We have to find $\delta > 0$ so that (using Definition 2.7 with $L = 0$)

$$|f(x, y) - 0| = \left|\frac{3x^2y}{x^2 + y^2} - 0\right| = \frac{3x^2}{x^2 + y^2}|y| < \epsilon,$$

whenever (x, y) satisfies $0 < \sqrt{x^2 + y^2} < \delta$. The latter is the statement $0 < ||\mathbf{x} - \mathbf{a}|| < \delta$, rewritten using the formula (2.8) with $\mathbf{x} = (x, y)$ and $\mathbf{a} = (0, 0)$. The inequality $x^2 \le x^2 + y^2$ implies that $x^2/(x^2 + y^2) \le 1$ and thus

$$|f(x, y) - 0| = \frac{3x^2}{x^2 + y^2}|y| \le 3|y|.$$

So, we need to find $\delta > 0$ that will guarantee $3|y| < \epsilon$. The inequality $\sqrt{x^2 + y^2} < \delta$, combined with $y^2 \le x^2 + y^2$, gives $|y| = \sqrt{y^2} \le \sqrt{x^2 + y^2} < \delta$. Therefore, if we take $\delta = \epsilon/3$, we will get

$$|f(x, y) - 0| = \frac{3x^2}{x^2 + y^2}|y| \le 3|y| < 3\delta = 3\,\frac{\epsilon}{3} = \epsilon.$$

We just showed that, given $\epsilon > 0$, if we take $0 < \sqrt{x^2 + y^2} < \delta = \epsilon/3$, then $|f(x, y) - 0| < \epsilon$. This, by definition, means that $\lim f(x, y) = 0$ as $(x, y) \to (0, 0)$. ◀

▶ EXAMPLE 2.26

In Section 2.2 we noticed that the computer-generated plot of the function $f(x, y) = 2xy/(x^2 + y^2)$ (see Figure 2.24) could not explain what happens to $f(x, y)$ near $(0, 0)$.

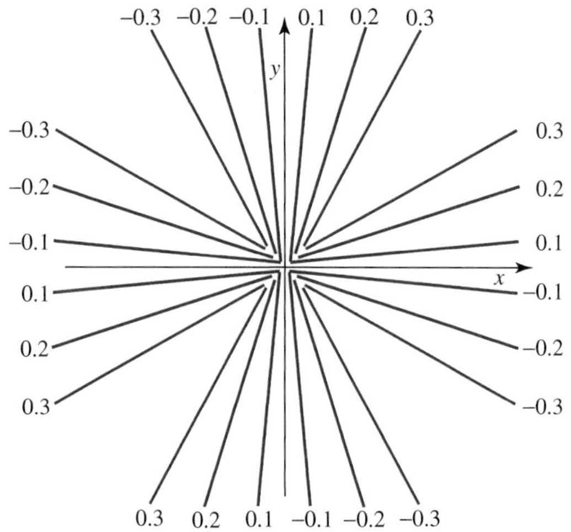

Figure 2.34 Level curves of the function $f(x, y)$ of Example 2.26.

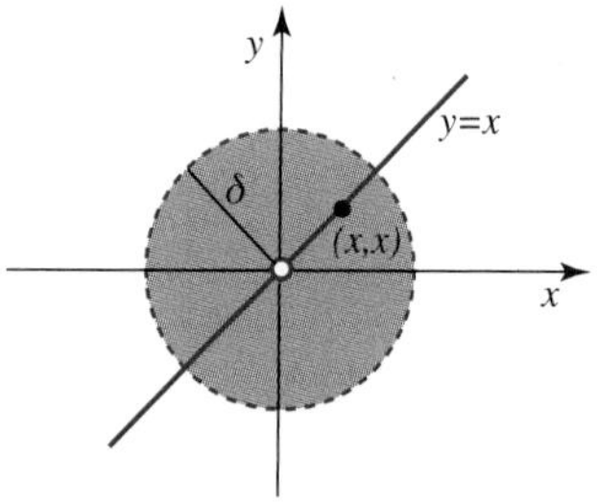

Figure 2.35 Any ball of radius $\delta > 0$ contains some points (x, x), $x \neq 0$.

Using limits, we will now describe the behavior of $f(x, y)$ near (and at) the origin. Figure 2.34 shows the contour diagram of $f(x, y) = 2xy/(x^2 + y^2)$ near $(0, 0)$. The level curves are lines (see Exercise 42 in Section 2.2 for a proof) that "collide" at $(0, 0)$, suggesting that $f(x, y)$ does not have a limit there.

Let us show that $L = 0$ cannot be the limit of $f(x, y)$ as (x, y) approaches $(0, 0)$. Pick an interval around $L = 0$, say, $(-\epsilon, \epsilon)$, for some $0 < \epsilon < 1$. Consider *any* open ball of radius $\delta > 0$ centered at the origin. No matter how small the ball is, it will always contain points that belong to the line $y = x$, for $(x, y) \neq (0, 0)$; see Figure 2.35.

At any point on this line [except at $(0, 0)$], the value of $f(x, y)$ is 1. So, any ball centered at $(0, 0)$ will contain points (x, y) for which $f(x, y)$ does not fall into the interval $(-\epsilon, \epsilon)$. Thus, $L = 0$ is not the limit. In a similar way (see Exercise 4), we can show that no real number is the limit of $f(x, y)$ as (x, y) approaches $(0, 0)$.

Now we present an alternate version (usually, easier to do) of the proof that $f(x, y)$ does not have a limit at $(0, 0)$. The idea lies in the discussion preceding the definition of the limit—all we have to do is to show that different ways of approaching $(0, 0)$ give different results. Let us approach $(0, 0)$ along the x-axis: then $y = 0$, and hence

$$\lim_{(x,y)\to(0,0)} \frac{2xy}{x^2 + y^2} = \lim_{x\to 0} \frac{0}{x^2} = \lim_{x\to 0} 0 = 0.$$

Now let us walk along the y-axis towards $(0, 0)$: in this case $x = 0$, and

$$\lim_{(x,y)\to(0,0)} \frac{2xy}{x^2 + y^2} = \lim_{y\to 0} \frac{0}{y^2} = \lim_{x\to 0} 0 = 0.$$

No luck! Moreover, we have not shown that the limit is zero, since there are infinitely many ways of approaching the origin and we would have to check each one of them. Now look at the approach along the line $y = x$:

$$\lim_{(x,y)\to(0,0)} \frac{2xy}{x^2 + y^2} = \lim_{x\to 0} \frac{2x^2}{x^2 + x^2} = \lim_{x\to 0} 1 = 1.$$

Consequently, the limit of $f(x, y)$ as $(x, y) \to (0, 0)$ does not exist. ◀

▶ EXAMPLE 2.27

Show that

$$\lim_{(x,y)\to(0,0)} \frac{y^2}{\sqrt{x^2+y^2}} = 0.$$

SOLUTION We will show that $y^2/\sqrt{x^2+y^2} \geq 0$ gets smaller and smaller as (x, y) moves closer and closer to $(0, 0)$. More precisely, for any $\epsilon > 0$, we will prove that

$$\left| \frac{y^2}{\sqrt{x^2+y^2}} - 0 \right| < \epsilon,$$

whenever $||(x, y) - (0, 0)|| = \sqrt{x^2+y^2} < \delta$, for some $\delta > 0$. Since $y^2 \leq x^2 + y^2$, we get

$$\frac{y^2}{\sqrt{x^2+y^2}} \leq \frac{x^2+y^2}{\sqrt{x^2+y^2}} = \sqrt{x^2+y^2}.$$

Thus, if we take $\delta = \epsilon$, we will obtain

$$\left| \frac{y^2}{\sqrt{x^2+y^2}} - 0 \right| \leq \sqrt{x^2+y^2} < \delta = \epsilon,$$

which establishes the desired inequality. ◀

▶ EXAMPLE 2.28

Show that

$$\lim_{(x,y)\to(0,0)} \frac{x^2y}{x^4+y^2}$$

does not exist.

SOLUTION Choose the approach $x = 0$ (that is, along the y-axis):

$$\lim_{(x,y)\to(0,0)} \frac{x^2y}{x^4+y^2} = \lim_{y\to 0} \frac{0}{y^2} = \lim_{y\to 0} 0 = 0.$$

Choose the approach $y = x$:

$$\lim_{(x,y)\to(0,0)} \frac{x^2y}{x^4+y^2} = \lim_{x\to 0} \frac{x^3}{x^4+x^2} = \lim_{x\to 0} \frac{x}{x^2+1} = 0.$$

Take a (general) line through the origin, $y = mx$ (by varying the values of m we get all lines through the origin, except the y-axis). Then

$$\lim_{(x,y)\to(0,0)} \frac{x^2y}{x^4+y^2} = \lim_{x\to 0} \frac{mx^3}{x^4+m^2x^2} = \lim_{x\to 0} \frac{mx}{x^2+m^2} = 0.$$

But we have not exhausted all possible approaches! (We have only exhausted all possible lines.) If we approach the origin along the parabola $y = x^2$, we get

$$\lim_{(x,y)\to(0,0)} \frac{x^2y}{x^4+y^2} = \lim_{x\to 0} \frac{x^4}{x^4+x^4} = \lim_{x\to 0} \frac{x^4}{2x^4} = \frac{1}{2},$$

and the proof is completed. ◀

Our next definition says that the limit of a vector-valued function is computed as the limit of its components. Let $\mathbf{F}\colon U \subseteq \mathbb{R}^m \to \mathbb{R}^n$ be a vector-valued function of m variables, defined on a set $U \subseteq \mathbb{R}^m$. $\mathbf{F}$ can be written as

$$\mathbf{F}(x_1, \ldots, x_m) = (F_1(x_1, \ldots, x_m), F_2(x_1, \ldots, x_m), \ldots, F_n(x_1, \ldots, x_m)), \tag{2.10}$$

where $F_i\colon \mathbb{R}^m \to \mathbb{R}$ is the ith *component of* $\mathbf{F}$, $i = 1, \ldots, n$. Let us emphasize that the components F_i are real-valued functions. In order to keep notation as simple as possible, we will use $\mathbf{x}$, instead of listing all variables $x_1, \ldots, x_m$; i.e., $\mathbf{x} = (x_1, \ldots, x_m)$. In particular, we will write $\mathbf{F}(\mathbf{x})$ instead of $\mathbf{F}(x_1, \ldots, x_m)$, and $\mathbf{F}(\mathbf{x}) = (F_1(\mathbf{x}), \ldots, F_n(\mathbf{x}))$ instead of (2.10).

DEFINITION 2.9 Limit of a Vector-Valued Function

Let $\mathbf{F}(\mathbf{x}) = (F_1(\mathbf{x}), \ldots, F_n(\mathbf{x}))$ be a vector-valued function of m variables, and let $\mathbf{a} = (a_1, \ldots, a_m)$ and $\mathbf{L} = (L_1, \ldots, L_n)$. We say that the function $\mathbf{F}(\mathbf{x})$ has limit $\mathbf{L}$ as $\mathbf{x}$ approaches $\mathbf{a}$, and write $\lim_{\mathbf{x}\to\mathbf{a}} \mathbf{F}(\mathbf{x}) = \mathbf{L}$, if and only if

$$\lim_{\mathbf{x}\to\mathbf{a}} F_1(\mathbf{x}) = L_1, \ldots, \lim_{\mathbf{x}\to\mathbf{a}} F_n(\mathbf{x}) = L_n.$$

In other words, the limit of a vector-valued function is computed componentwise:

$$\lim_{\mathbf{x}\to\mathbf{a}} \mathbf{F}(\mathbf{x}) = (\lim_{\mathbf{x}\to\mathbf{a}} F_1(\mathbf{x}), \ldots, \lim_{\mathbf{x}\to\mathbf{a}} F_n(\mathbf{x})),$$

provided that all limits on the right side (and those are limits of real-valued functions) exist. The computation of a limit can be simplified by the use of the limit laws and by the use of continuity—that will be discussed later in this section.

THEOREM 2.1 Limit Laws

Let $\mathbf{F}, \mathbf{G}\colon \mathbb{R}^m \to \mathbb{R}^n$, $f, g\colon \mathbb{R}^m \to \mathbb{R}$ and assume that $\lim_{\mathbf{x}\to\mathbf{a}} \mathbf{F}(\mathbf{x})$, $\lim_{\mathbf{x}\to\mathbf{a}} \mathbf{G}(\mathbf{x})$, $\lim_{\mathbf{x}\to\mathbf{a}} f(\mathbf{x})$ and $\lim_{\mathbf{x}\to\mathbf{a}} g(\mathbf{x})$ exist. Then

(a) $\lim_{\mathbf{x}\to\mathbf{a}}(\mathbf{F}(\mathbf{x}) + \mathbf{G}(\mathbf{x}))$ and $\lim_{\mathbf{x}\to\mathbf{a}}(\mathbf{F}(\mathbf{x}) - \mathbf{G}(\mathbf{x}))$ exist and

$$\lim_{\mathbf{x}\to\mathbf{a}}(\mathbf{F}(\mathbf{x}) \pm \mathbf{G}(\mathbf{x})) = \lim_{\mathbf{x}\to\mathbf{a}} \mathbf{F}(\mathbf{x}) \pm \lim_{\mathbf{x}\to\mathbf{a}} \mathbf{G}(\mathbf{x}).$$

(b) $\lim_{\mathbf{x}\to\mathbf{a}} f(\mathbf{x})g(\mathbf{x})$ and $\lim_{\mathbf{x}\to\mathbf{a}} c\mathbf{F}(\mathbf{x})$ (for any constant c) exist, and

$$\lim_{\mathbf{x}\to\mathbf{a}} (f(\mathbf{x})g(\mathbf{x})) = \left(\lim_{\mathbf{x}\to\mathbf{a}} f(\mathbf{x})\right)\left(\lim_{\mathbf{x}\to\mathbf{a}} g(\mathbf{x})\right) \quad \text{and} \quad \lim_{\mathbf{x}\to\mathbf{a}} (c\mathbf{F}(\mathbf{x})) = c \lim_{\mathbf{x}\to\mathbf{a}} \mathbf{F}(\mathbf{x}).$$

(c) If $\lim_{\mathbf{x}\to\mathbf{a}} g(\mathbf{x}) \neq 0$, then $\lim_{\mathbf{x}\to\mathbf{a}} f(\mathbf{x})/g(\mathbf{x})$ exists, and

$$\lim_{\mathbf{x}\to\mathbf{a}} \frac{f(\mathbf{x})}{g(\mathbf{x})} = \frac{\lim_{\mathbf{x}\to\mathbf{a}} f(\mathbf{x})}{\lim_{\mathbf{x}\to\mathbf{a}} g(\mathbf{x})}.$$

(d) For any $\mathbf{a} \in \mathbb{R}^m$ and any constant $\mathbf{c} \in \mathbb{R}^n$,

$$\lim_{\mathbf{x}\to\mathbf{a}} \mathbf{x} = \mathbf{a} \quad \text{and} \quad \lim_{\mathbf{x}\to\mathbf{a}} \mathbf{c} = \mathbf{c}.$$

In part (d) the symbol $\mathbf{c}$ denotes the function $\mathbf{F}\colon \mathbb{R}^m \to \mathbb{R}^n$ given by $\mathbf{F}(\mathbf{x}) = \mathbf{c}$ for all $\mathbf{x} \in \mathbb{R}^m$.

SKETCH OF PROOF: Rather than getting involved in arguments involving epsilons and deltas, we will provide more intuitive reasoning. Let $\lim_{\mathbf{x}\to\mathbf{a}} f(\mathbf{x}) = L$ and $\lim_{\mathbf{x}\to\mathbf{a}} g(\mathbf{x}) = M$, and consider part (a). We have to show that $f(\mathbf{x}) + g(\mathbf{x})$ can be made as close as needed to $L + M$ by selecting an $\mathbf{x}$ close enough to $\mathbf{a}$, $\mathbf{x} \neq \mathbf{a}$. Since $\lim_{\mathbf{x}\to\mathbf{a}} f(\mathbf{x}) = L$, it is possible to force $f(\mathbf{x})$ to fall as close as needed to L by requiring that $\mathbf{x}$ ($\mathbf{x} \neq \mathbf{a}$) belongs to a ball of a small enough radius δ. Similarly, $\lim_{\mathbf{x}\to\mathbf{a}} g(\mathbf{x}) = M$ means that $g(\mathbf{x})$ can be made as close as needed to M by taking $\mathbf{x}$ ($\mathbf{x} \neq \mathbf{a}$) from the inside of a ball of some small radius δ' centered at $\mathbf{a}$. Taking the smaller of the two balls, we can force both $f(\mathbf{x})$ and $g(\mathbf{x})$ to be as close as needed to L and M, thus making their sum $f(\mathbf{x}) + g(\mathbf{x})$ as close to $L + M$ as needed. Other properties are verified analogously. ◀

▶ EXAMPLE 2.29

Compute $\lim_{(x,y)\to(3,2)} (x^2 - 2 + xy^2)$.

SOLUTION

By the limit laws,

$$\begin{aligned}\lim_{(x,y)\to(3,2)} (x^2 - 2 + xy^2) &= \lim_{(x,y)\to(3,2)} x^2 - \lim_{(x,y)\to(3,2)} 2 + \lim_{(x,y)\to(3,2)} xy^2 \\ &= \left(\lim_{(x,y)\to(3,2)} x\right)\cdot\left(\lim_{(x,y)\to(3,2)} x\right) - \lim_{(x,y)\to(3,2)} 2 \\ &\quad + \left(\lim_{(x,y)\to(3,2)} x\right)\cdot\left(\lim_{(x,y)\to(3,2)} y\right)\cdot\left(\lim_{(x,y)\to(3,2)} y\right) = 9 - 2 + 12 = 19.\end{aligned}$$ ◀

Although we need limits to define and understand continuity and derivatives, we do not need to master (fortunately) technical intricacies involved in their computation (as we have seen in the calculus of functions of one variable, we rarely go all the way back to the limit definition of the derivative; instead, we use various formulas and properties, such as the quotient and the chain rules). Some technical issues involving limits are discussed in the exercises.

Continuity

Intuitively speaking, a function is continuous if its graph has no breaks. For a function of one variable this means that the curve (which is its graph) can be drawn on a piece of paper without lifting a pen. A bird flying describes a continuous function: it cannot happen that the bird disappears somewhere and reappears at some other location a moment later. One of the properties of continuous functions states that it is possible to predict their "short-term behavior." For example, assume that the air temperature at this moment is 18°C; a second later it could be 18.5°C or 17°C or 19°C; but it will not be −100°C.

On the other hand, having a glance at a traffic light (suppose that it is red) will not help us predict whether, a second later, it will still be red or will change to green. A traffic light's color is a discontinuous function. A hemisphere, or a plane, or the graphs given in Figures 2.20, 2.21, and 2.23, are graphs of continuous functions of two variables. However, the graph of $z = \arctan(0.2y/x)$ is "broken"; that is, f is not continuous at points where $x = 0$; see Figure 2.36.

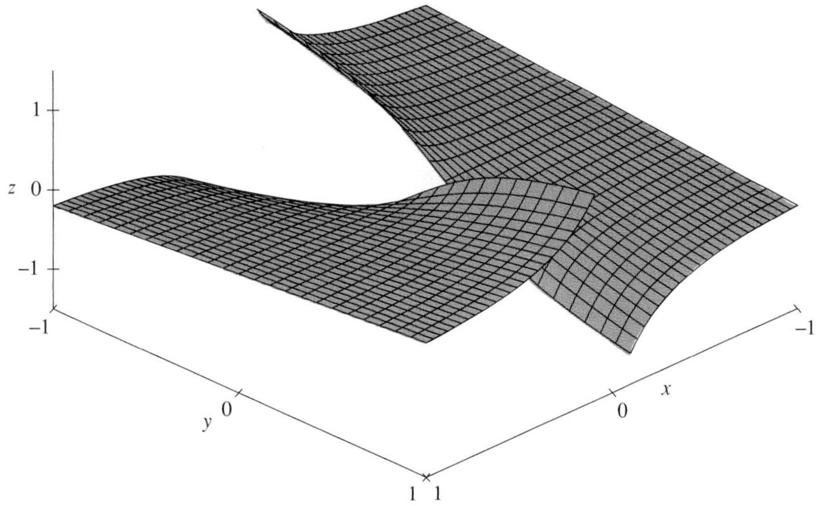

Figure 2.36 Graph of $z = \arctan(0.2y/x)$.

Recall that a function $f: \mathbb{R} \to \mathbb{R}$ is *continuous* at $x = a$ (see Figure 2.37) if and only if

(a) $\lim_{x\to a} f(x)$ exists,

(b) f is defined at a, and

(c) $\lim_{x\to a} f(x) = f(a)$.

We say that a function f is *continuous on an interval* (c, d) if it is continuous at every point a in (c, d). A function f is *continuous on a closed interval* $[c, d]$ if it is continuous on (c, d) and $\lim_{x\to c^+} f(x) = f(c)$ and $\lim_{x\to d^-} f(x) = f(d)$. We are *in* the interval $[c, d]$ and therefore can approach its endpoint c from the right only. Similarly, we can reach d from the left only. Any other point in $[c, d]$ can be reached from both sides; see Figure 2.38.

To understand this definition better, let us consider examples of functions that are not continuous at $x = a$.

The first function in Figure 2.39 does not have a limit as $x \to a$; hence, the condition (a) fails to hold [and (c) does not make sense]. For the second function, $\lim_{x\to a} f(x) = L$,

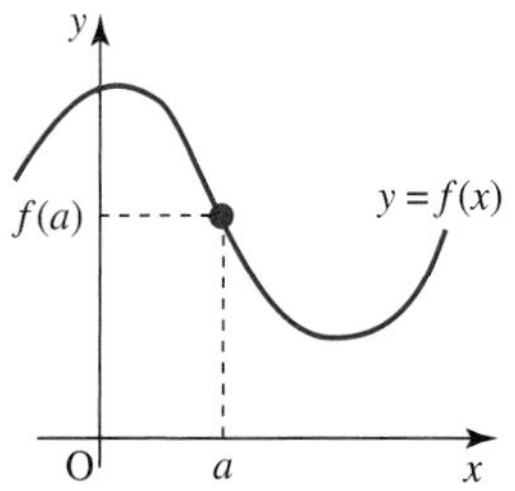

Figure 2.37 $f(x)$ is continuous at $x = a$.

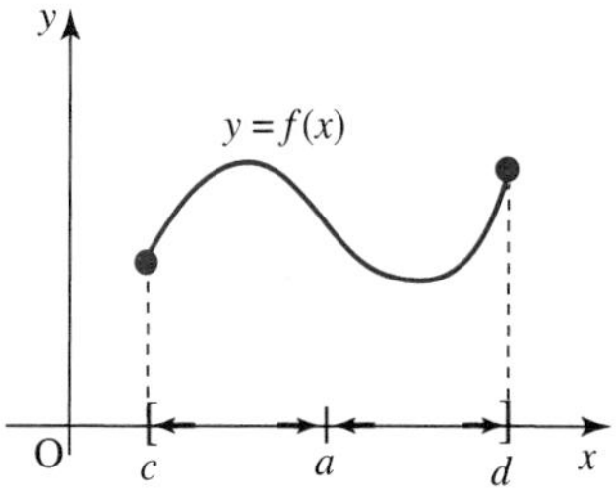

Figure 2.38 $f(x)$ is continuous on $[c, d]$.

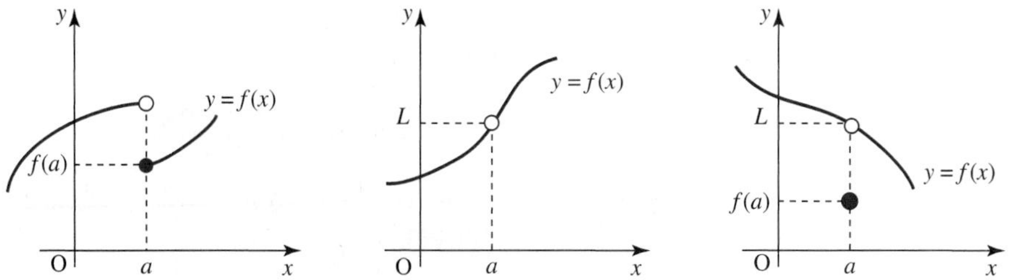

Figure 2.39 All three functions are not continuous at $x = a$.

but $f(a)$ is not defined; hence, (b) fails to hold [and again, (c) does not make sense]. Finally for the last graph in Figure 2.39, $\lim_{x\to a} f(x) = L$ and $f(a)$ is defined, but the two numbers are not equal. Condition (c) is not satisfied.

The following functions of one variable are continuous *at all points where they are defined*: $f(x) = c$, where c is a constant; $f(x) = x^n$, where n denotes any real number; polynomials and rational functions; $f(x) = e^x$, $f(x) = a^x$, for $a > 0$; $f(x) = \ln x$, $f(x) = \log x$; trigonometric and hyperbolic functions and their inverses, the absolute value function $|x|$, etc.

DEFINITION 2.10 Continuity of Functions of Several Variables

A function $f\colon U \subseteq \mathbb{R}^m \to \mathbb{R}$ is *continuous at* $\mathbf{x} = \mathbf{a}$ if and only if

(a) $\lim_{\mathbf{x}\to\mathbf{a}} f(\mathbf{x})$ exists,

(b) f is defined at $\mathbf{a}$, and

(c) $\lim_{\mathbf{x}\to\mathbf{a}} f(\mathbf{x}) = f(\mathbf{a})$.

We say that f is *continuous on a set* U (or just f is *continuous*) if and only if it is continuous at all points in U. A vector-valued function $\mathbf{F} = (F_1, \ldots, F_n)\colon U \subseteq \mathbb{R}^m \to \mathbb{R}^n$ is *continuous* if and only if its components F_i, $i = 1, \ldots, n$, are continuous. ◀

In light of our definition of the limit of a vector-valued function, a function $\mathbf{F}$ is continuous at $\mathbf{a}$ if and only if $\lim_{\mathbf{x}\to\mathbf{a}} \mathbf{F}(\mathbf{x})$ exists, $\mathbf{F}(\mathbf{a})$ is defined and $\lim_{\mathbf{x}\to\mathbf{a}} \mathbf{F}(\mathbf{x}) = \mathbf{F}(\mathbf{a})$. Let us emphasize that, when testing continuity at boundary points of U (if U has any) by computing the limit as $\mathbf{x} \to \mathbf{a}$, we must approach $\mathbf{a}$ from *within* U: that is, $\mathbf{x} \to \mathbf{a}$ assumes that $\mathbf{x} \in U$ (in the case of one variable, we had to use one-sided limits).

THEOREM 2.2 Properties of Continuous Functions

Let $\mathbf{F}, \mathbf{G}\colon U \subseteq \mathbb{R}^m \to \mathbb{R}^n$ $(n \geq 1)$ and $f, g\colon U \subseteq \mathbb{R}^m \to \mathbb{R}$ be continuous at $\mathbf{a} \in U$. Then

(a) the functions $\mathbf{F} \pm \mathbf{G}$, defined by $(\mathbf{F} \pm \mathbf{G})(\mathbf{x}) = \mathbf{F}(\mathbf{x}) \pm \mathbf{G}(\mathbf{x})$, are continuous at $\mathbf{a}$.

(b) the function $c\mathbf{F}$, defined by $(c\mathbf{F})(\mathbf{x}) = c\mathbf{F}(\mathbf{x})$, is continuous at $\mathbf{a}$.

(c) the function fg, defined by $(fg)(\mathbf{x}) = f(\mathbf{x})g(\mathbf{x})$, is continuous at $\mathbf{a}$.

(d) the function f/g, defined by $(f/g)(\mathbf{x}) = f(\mathbf{x})/g(\mathbf{x})$, is continuous at $\mathbf{a}$, if $g(\mathbf{a}) \neq 0$.

PROOF: To prove any of the above statements, all we have to do is to rewrite the corresponding limit statement. For example, let us prove part (c). Start with the definition of the product, use the "limit of the product law" (cf. (b), Theorem 2.1) and the assumption that f and g are continuous, thus getting

$$\lim_{\mathbf{x}\to\mathbf{a}}(fg)(\mathbf{x}) = \lim_{\mathbf{x}\to\mathbf{a}} f(\mathbf{x})g(\mathbf{x}) = \left(\lim_{\mathbf{x}\to\mathbf{a}} f(\mathbf{x})\right)\left(\lim_{\mathbf{x}\to\mathbf{a}} g(\mathbf{x})\right) = f(\mathbf{a})g(\mathbf{a}).$$

By the definition of the product (read in the opposite direction), $f(\mathbf{a})g(\mathbf{a}) = (fg)(\mathbf{a})$, and we are done.

Consider the function $pr_i\colon \mathbb{R}^m \to \mathbb{R}$, defined by $pr_i(x_1, \ldots, x_m) = x_i$; it extracts the ith component x_i from the list $(x_1, \ldots, x_m)$ of variables, and is called a *projection*. For example, $pr_2(x, y) = y$, $pr_1(x, y, z) = x$, $pr_2(x, y, z) = y$, etc. Now $\lim_{\mathbf{x}\to\mathbf{a}} pr_i(\mathbf{x}) = \lim_{\mathbf{x}\to\mathbf{a}} x_i$ by the definition of the projection, $\lim_{\mathbf{x}\to\mathbf{a}} x_i = a_i$ by (2.9), and $a_i = pr_i(\mathbf{a})$ again by the definition of projection read from right to left. In other words, $\lim_{\mathbf{x}\to\mathbf{a}} pr_i(\mathbf{x}) = pr_i(\mathbf{a})$, and the projection function is continuous. This means that, for example, functions such as $f(x, y) = x$, $f(x, y, z) = z$, etc., are continuous, viewed as functions of *several* variables.

THEOREM 2.3 Continuity of Composition of Functions

Let $\mathbf{F}\colon U \subseteq \mathbb{R}^m \to \mathbb{R}^n$ and $\mathbf{G}\colon V \subseteq \mathbb{R}^n \to \mathbb{R}^p$ be such that the range $\mathbf{F}(U)$ of $\mathbf{F}$ is contained in the domain V of $\mathbf{G}$, so that the composition $\mathbf{G} \circ \mathbf{F}$ is defined; see Figure 2.40. If $\mathbf{F}$ is continuous at $\mathbf{a}$ and $\mathbf{G}$ is continuous at $\mathbf{b} = \mathbf{F}(\mathbf{a})$, then $\mathbf{G} \circ \mathbf{F}$ is continuous at $\mathbf{a}$.

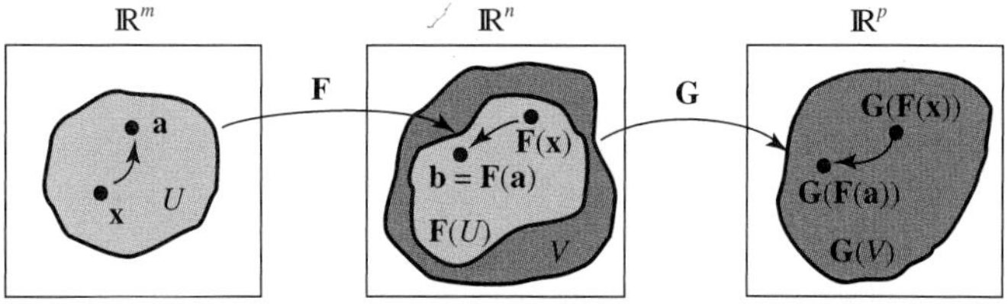

Figure 2.40 Composition of functions $\mathbf{G} \circ \mathbf{F}\colon U \subseteq \mathbb{R}^m \to \mathbb{R}^p$.

INTUITIVE PROOF: As $\mathbf{x}$ gets closer and closer to $\mathbf{a}$, the values $\mathbf{F}(\mathbf{x})$ get closer and closer to $\mathbf{F}(\mathbf{a})$, since $\mathbf{F}$ is continuous at $\mathbf{a}$. But now $\mathbf{G}$ is continuous at $\mathbf{b} = \mathbf{F}(\mathbf{a})$, and since $\mathbf{F}(\mathbf{x})$ gets closer and closer to $\mathbf{b} = \mathbf{F}(\mathbf{a})$, the values of $\mathbf{G}$, that is, $\mathbf{G}(\mathbf{F}(\mathbf{x}))$, get closer and closer to $\mathbf{G}(\mathbf{F}(\mathbf{a}))$.

The following functions of two variables are continuous *at all points where they are defined:* $f(x, y) = c$, where c is a constant; $f(x, y) = x$, $f(x, y) = y$ (these are the orthogonal projections onto the x-axis and onto the y-axis, respectively); $f(x, y) = x^n$, $f(x, y) = y^n$, where n denotes any real number. Therefore, polynomials and rational functions are continuous (whenever the denominator is not equal to zero), as is a composition involving any of the functions listed here with any function from the one-variable list. A list analogous to this one could be made for functions of m variables.

▶ EXAMPLE 2.30

Show that the function $\mathbf{F}(x, y, z) = (\sin x, x^2 + y^2, e^{xyz})$ is continuous for all $(x, y, z) \in \mathbb{R}^3$.

SOLUTION We have to analyze the components of $\mathbf{F}$. The first component $F_1(x, y, z) = \sin x$ is the composition of the projection $(x, y, z) \mapsto x$ and the trigonometric function $x \mapsto \sin x$, both of which are continuous. Hence, F_1 is continuous. The component F_2 is a polynomial and hence continuous. The function F_3 is continuous as it is the composition of the polynomial $(x, y, z) \mapsto xyz$ and the exponential function. ◀

▶ EXAMPLE 2.31

Show that the function

$$f(x, y) = \begin{cases} \dfrac{\cos(x^2 + y^2) - 1}{x^2 + y^2} & \text{if } (x, y) \neq (0, 0) \\ 0 & \text{if } (x, y) = (0, 0) \end{cases}$$

is continuous on $\mathbb{R}^2$.

SOLUTION The function $(x, y) \mapsto x^2 + y^2$ is a polynomial, and hence its composition with the cosine function is continuous. The numerator is continuous as it is the difference of continuous functions [the function $(x, y) \mapsto 1$ is a constant function, and hence continuous]. Since the denominator is continuous and nonzero except when $(x, y) = (0, 0)$, it follows that $f(x, y)$ is continuous at all points $(x, y) \neq (0, 0)$. It remains to check the point $(0, 0)$: by Definition 2.10, it suffices to show that $\lim_{(x,y)\to(0,0)} f(x, y) = f(0, 0) = 0$. To compute the limit, substitute $u = x^2 + y^2$; then $u \to 0$ (since both $x \to 0$ and $y \to 0$) and

$$\lim_{(x,y)\to(0,0)} \frac{\cos(x^2 + y^2) - 1}{x^2 + y^2} = \lim_{u\to 0} \frac{\cos u - 1}{u} = \lim_{u\to 0} \frac{-\sin u}{1} = 0,$$

by L'Hôpital's rule. Hence, f is also continuous at $(0, 0)$. ◀

▶ EXAMPLE 2.32

Find all points of discontinuity of the function

$$f(x, y) = \begin{cases} \dfrac{x^2 y}{x^4 + y^2} & \text{if } (x, y) \neq (0, 0) \\ 0 & \text{if } (x, y) = (0, 0) \end{cases}$$

SOLUTION The function $f(x, y) = x^2y/(x^4 + y^2)$ is continuous at all points except possibly at the origin (namely, it is a quotient of continuous functions with a nonzero denominator). It was shown in Example 2.28 that $\lim_{(x,y)\to(0,0)} f(x, y)$ does not exist, and consequently, $f(x, y)$ is not continuous at $(0, 0)$. ◀

▶ EXERCISES 2.3

1. In all three cases shown in Figure 2.41, $\lim_{x\to a} f(x)$ is equal to L. Describe the differences in terms of the behavior of $f(x)$ at a.

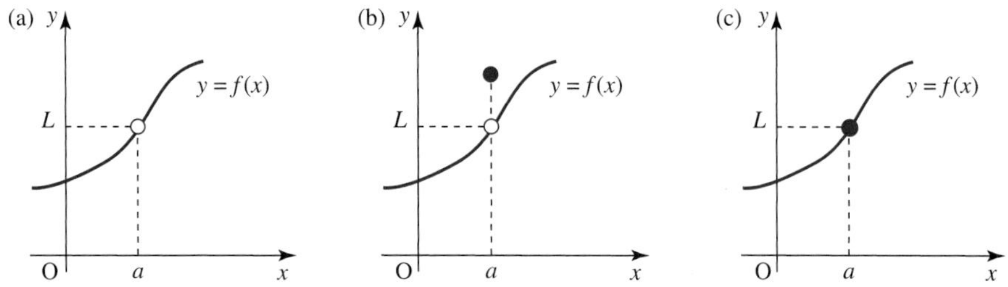

Figure 2.41 Functions from Exercise 1.

2. Consider the function $f\colon \mathbb{R}^2 \to \mathbb{R}$ defined by $f(x, y) = x^2y^3$. Find the radius of an open ball $B((0, 0), r)$ centered at the origin with the property that $|x^2y^3| < 0.005$, if $(x, y) \in B((0, 0), r)$. *Hint*: Find a and b such that $-a \le x \le a$ and $-b \le y \le b$ imply $|x^2y^3| < 0.005$ first. (What region in the xy-plane is represented by $-a \le x \le a$ and $-b \le y \le b$?)

3. Consider the function $f\colon \mathbb{R}^2 \to \mathbb{R}$ defined by $f(x, y) = e^{-(x^2+y^2)}$. Find an open ball $B((0, 0), r)$ (i.e., find its radius) such that, whenever $(x, y) \in B((0, 0), r)$, f satisfies $|f(x, y) - 1| < 0.01$.

4. Consider the function $f(x, y) = 2xy/(x^2 + y^2)$ of Example 2.26.

(a) Show that $L = 1/2$ cannot be the limit of $f(x, y)$ at $(0, 0)$. (*Hint*: Consider points on the x-axis.)

(b) Show that no number $L_o \neq 0$ can be the limit of $f(x, y)$ at $(0, 0)$.

5. Figure 2.42 shows level curves of a function $f\colon \mathbb{R}^2 \to \mathbb{R}$ whose limit at $(0, 0)$ is 3. Draw a ball $B((0, 0), r_1)$ such that $|f(x, y) - 3| < 0.04$ for every $(x, y) \in B((0, 0), r_1)$. Find another ball $B((0, 0), r_2)$ such that for every $(x, y) \in B((0, 0), r_2)$, $|f(x, y) - 3| < 0.01$. Assume that the values of f in the region between two level curves are between the values of f on those level curves. For example, the value of f at a point in the region between level curves of values 2.92 and 2.96 cannot be 4 or -2, but has to fall between 2.92 and 2.96.

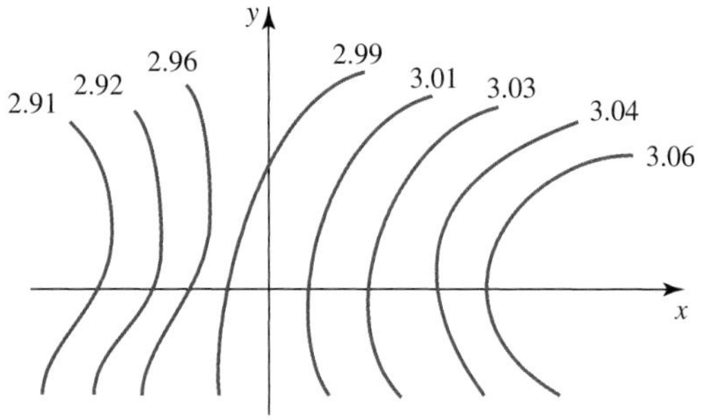

Figure 2.42 Level curves of the function f of Exercise 5.

6. Show that $\lim_{(x,y)\to(-1,2)} (y - 2)/(x + 1)$ does not exist.

7. Show that $\lim_{(x,y)\to(0,0)} \arctan(0.2x/y)$ does not exist.

Exercises 8 to 13: Find the limit of $f(x, y)$ as $(x, y) \to (0, 0)$, if it exists.

8. $f(x, y) = (x^3y - xy^3 - x)/(1 - xy)$

9. $f(x, y) = 1 - y - e^{-x^2-y^2} \cos x$

10. $f(x, y) = \sin(3x - 2y + xy)/(3x - 2y + xy)$ (*Hint*: Introduce a new variable $u = 3x - 2y + xy$ and reduce the limit to the one-variable case.)

11. $f(x, y) = (x + y)e^{-1/(x+y)}$

12. $f(x, y) = xy(x^2 + y^2)^{-1/2}$ (*Hint*: Switch to polar coordinates.)

13. $f(x, y) = x^2y/(x^2 + y^2)$

Exercises 14 to 20: Show that the limit of $f(x, y)$, as $(x, y) \to (0, 0)$, does not exist.

14. $f(x, y) = \dfrac{xy}{(x^2 + y^2)^{3/2}}$

15. $f(x, y) = \dfrac{2x^2 - y^2}{x^2 + 2y^2}$

16. $f(x, y) = \dfrac{3x^2y + 6xy + 19y^2}{x^2 + 4y^2}$

17. $f(x, y) = \dfrac{2xy}{2x^2 + y^2}$

18. $f(x, y) = \dfrac{x^2y}{x^4 + y^2}$

19. $f(x, y) = \dfrac{x^3y}{x^6 + y^2}$

20. $f(x, y) = \dfrac{xy}{x^3 + y^3}$ (*Hint*: Use polar coordinates, simplify and let $r \to 0$.)

21. Evaluate $\lim_{(x,y,z)\to(0,0,0)} \dfrac{xyz}{x^3 + y^3 + z^3}$, if it exists.

22. Prove that the function $\dfrac{\sin x \cos y}{x^2 + y^2 + 1}$ is continuous for every $(x, y) \in \mathbb{R}^2$.

23. Find all points where the function $\mathbf{F}(x, y, z) = (x/(x^2 + y^2 + z^2), y/(x^2 + y^2 + z^2))$ is not continuous.

24. Identify the domain of the function $f(x, y) = (\ln x)^2 + \ln y^2$. Explain why f is continuous at all points in its domain.

25. Find all points of discontinuity of the function $f(x, y) = (1 + \cos^2 x)(3 - \sin x \cos x)^{-1}$.

26. Show that the function $f\colon \mathbb{R}^m \to \mathbb{R}$, defined by $f(\mathbf{x}) = ||\mathbf{x}||$, is continuous for all $\mathbf{x} \in \mathbb{R}^m$. Find $\lim_{\mathbf{x}\to\mathbf{a}} ||\mathbf{x}||$.

27. Find all $\mathbf{x}$, where the function $f\colon \mathbb{R}^m \to \mathbb{R}$, defined by $f(\mathbf{x}) = (\mathbf{x} - \mathbf{x}_0)/||\mathbf{x} - \mathbf{x}_0||$, $\mathbf{x}_0 \in \mathbb{R}^m$, is not contunuous.

Exercises 28 to 30: Determine whether or not the limit of $f(x, y)$ as $(x, y) \to (0, 0)$, exists. If possible, define $f(0, 0)$ so as to make f continuous at $(0, 0)$.

28. $f(x, y) = \dfrac{\sin(3x^2 + y^2)}{x^2 + 2y^2}$

29. $f(x, y) = \dfrac{xy^3}{x^2 + y^6}$

30. $f(x, y) = \dfrac{\cos(x^2 + y^2) - 1}{x^2 + y^2}$

31. Find all interior and boundary points of the set $U = \{(x, y) \mid xy \neq 0\}$.

32. Find all interior and boundary points of the set $U = \{(x, y) \mid 1 < x^2 + y^2 \leq 2\}$. What boundary points belong to U?

33. Let $\mathbf{a}$ be a fixed vector in $\mathbb{R}^m$, and let $f\colon \mathbb{R}^m \to \mathbb{R}$ be a function defined by ($\cdot$ denotes the dot product) $f(\mathbf{x}) = \mathbf{x} \cdot \mathbf{a}$. Show that f is continuous at all $\mathbf{x}$ in $\mathbb{R}^m$.

34. Define a vector-valued function $\mathbf{F}\colon \mathbb{R}^3 \to \mathbb{R}^3$ by $\mathbf{F}(\mathbf{x}) = \mathbf{x} \times \mathbf{a}$, where $\mathbf{a}$ is a fixed vector in $\mathbb{R}^3$. Find all points where $\mathbf{F}$ is continuous. Find all points where $\mathbf{G}(\mathbf{x}) = \mathbf{x} \times \mathbf{a}/||\mathbf{x} \times \mathbf{a}||$ is continuous.

Exercises 35 to 37: Compute the limit, if it exists, of the function $\mathbf{F}(x, y)$, as (x, y) approaches (a, b). If possible, define $\mathbf{F}(a, b)$ so as to make it continuous at (a, b).

35. $\mathbf{F}(x, y) = \left(\dfrac{y \sin x}{x}, ye^x\right)$, $(a, b) = (0, 2)$

36. $\mathbf{F}(x, y) = \left(\dfrac{x}{\sqrt{x^2 + y^2}}, \dfrac{y}{\sqrt{x^2 + y^2}}\right)$, $(a, b) = (0, 0)$

37. $\mathbf{F}(x, y) = \left(\sin(x + y), \dfrac{\cos y - 1}{xy}, e^{xy}\right)$, $(a, b) = (1, 0)$

Exercises 38 to 40: Compute the limit, if it exists, of the function $f(x, y)$ as (x, y) approaches (a, b). If possible, define $f(a, b)$ so as to make it continuous at (a, b).

38. $f(x, y) = 3x \sin x + y^2 \ln (x - 2y)$, $(a, b) = (2, 1)$

39. $f(x, y) = \dfrac{\sin^2 (xy - 2)}{xy - 2}$, $(a, b) = (-1, -2)$

40. $f(x, y) = \tan (x^2 + y)$, $(a, b) = (0, \pi/2)$

▶ 2.4 DERIVATIVES

Using limits and continuity we can detect only some important properties of a function. To obtain more information, we make use of another powerful concept: the derivative of a function. For example, the graph of the function $f(x) = e^{-x^2}$ has no breaks [continuity information] and the line $y = 0$ is its horizontal asymptote [limit information]. With the help of the derivative [$f'(x) = -2xe^{-x^2}$] we can say much more: $f(x)$ is increasing for $x \leq 0$ and decreasing for $x \geq 0$; it has a maximum at $x = 0$, etc. Moreover, we can examine *how* $f(x)$ changes (recall that the derivative represents the rate of change): since $f'(-2) = 4e^{-4} \approx 0.0732$ and $f'(-1/2) = e^{-1/4} \approx 0.7788$, it follows that $f(x)$ increases much faster near $-1/2$ than near -2. Similarly, $f'(1) = -2e^{-1} \approx -0.7358$ and $f'(3) = -6e^{-9} \approx -0.0007$ imply that the function f decreases much faster (i.e., loses more per unit change in x) near 1 than near 3.

Recall that the derivative $f'(x)$ of a function $f(x)$ is defined as a limit of difference quotients

$$f'(x) = \lim_{h \to 0} \frac{f(x + h) - f(x)}{h},$$

provided that the limit exists. The number $f'(x_0)$ is the slope of the tangent to the graph of $f(x)$ at the point $(x_0, f(x_0))$.

The function $f'(x)$ is defined on open intervals (a, b) contained in the domain of $f(x)$. We say that the derivative, and, consequently, the tangent, do not exist at "ends" $x = a$ and $x = b$ of a graph. Similarly, the derivative of a function of several variables will be defined on special subsets in the domain: they are called open sets.

DEFINITION 2.11 Open Sets in $\mathbb{R}^m$

A set $U \subseteq \mathbb{R}^m$ is *open in* $\mathbb{R}^m$ if and only if all of its points are interior points. ◀

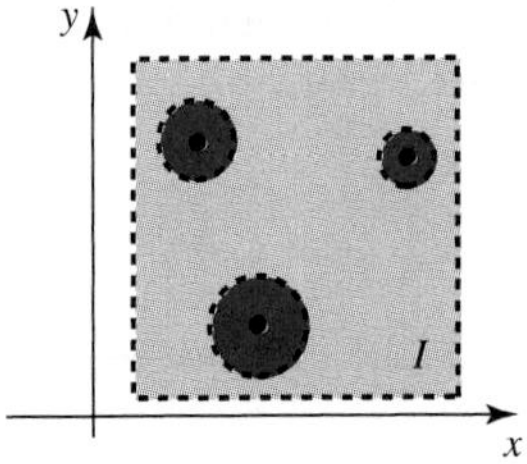

Figure 2.43 The inside of a square is an open set in $\mathbb{R}^2$.

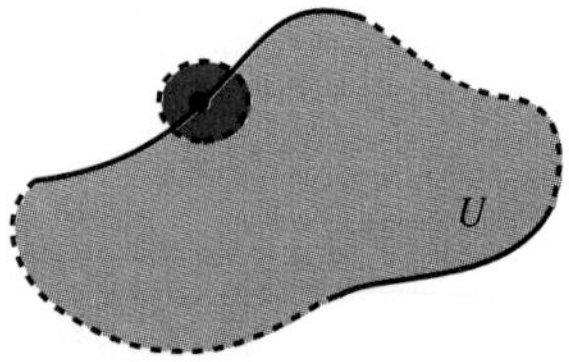

Figure 2.44 A set that contains a boundary point cannot be open.

In other words, a set $U \subseteq \mathbb{R}^m$ is open in $\mathbb{R}^m$ if and only if for any point $\mathbf{a} \in U$ there is an open ball centered at $\mathbf{a}$ that is completely contained in U. For example, the inside I of a square (boundary segments not included) is open in $\mathbb{R}^2$: no matter what point in I is chosen, there is always a small open ball that contains it and is contained in I. Clearly, the balls must get smaller and smaller as we approach the edges; see Figure 2.43.

The inside of a circle is an open set in $\mathbb{R}^2$; therefore, the use of the adjective "open" in the definition of the open ball has been justified. All of $\mathbb{R}^2$, or the upper half-plane $\{(x, y) | y > 0\}$ are open in $\mathbb{R}^2$. The first octant without the coordinate planes or the inside of a cube are open sets in $\mathbb{R}^3$.

The interval $(1, 2)$ is open in $\mathbb{R}$ (hence the name open interval). Consider the following two cases as illustration: pick a number in $(1, 2)$, say 1.8; the open ball $(1.7, 1.9)$ contains it, and is contained in $(1, 2)$. Pick, say, 1.9995; the open ball $(1.9992, 1.9998)$ satisfies the requirement of the definition: $1.9995 \in (1.9992, 1.9998) \subseteq (1, 2)$.

On the other hand, if a set U contains any of its boundary points, then it cannot be open: any ball centered at a boundary point, no matter how small, will always contain points outside of U, as shown in Figure 2.44. For example, the interval $[1, 2]$ is not open in $\mathbb{R}$. The set $\{(x, y) | x^2 + y^2 \leq 1\}$ (that contains the circle $x^2 + y^2 = 1$ and the region inside it) is not open in $\mathbb{R}^2$.

Partial Derivatives

We will start our presentation of the derivative by defining a partial derivative of a real-valued function. Throughout this section U denotes an open set.

Consider a function $f(x, y)$ and pick a point (a, b) in its domain U. In order to investigate how f changes at (a, b), we need to specify a direction in which the variables change (for instance, we feel an increase in the temperature as we approach a heater, and a decrease as we walk away from it). In this section, we study two special rates of change, defined by the direction of the coordinate axes. The rates of change in arbitrary directions are discussed in Section 2.7.

The function $g(x) = f(x, b)$ describes the values of f at the points on the line parallel to the x-axis that goes through (a, b). Note that g is a function of one variable. The rate of change of $f(x, y)$ at (a, b) in the direction of the x-axis is given by

$$g'(a) = \lim_{h \to 0} \frac{g(a + h) - g(a)}{h} = \lim_{h \to 0} \frac{f(a + h, b) - f(a, b)}{h},$$

if the limit exists. The expression on the right side is called the *partial derivative of f with respect to x* at (a, b), and is denoted by $(\partial f/\partial x)(a, b)$. In a similar way, we obtain the formula

$$\frac{\partial f}{\partial y}(a, b) = \lim_{h\to 0} \frac{f(a, b+h) - f(a, b)}{h}$$

for the *partial derivative of f with respect to y* at (a, b). Letting the point (a, b) vary, we obtain the functions

$$\frac{\partial f}{\partial x}(x, y) = \lim_{h\to 0} \frac{f(x+h, y) - f(x, y)}{h} \tag{2.11}$$

and

$$\frac{\partial f}{\partial y}(x, y) = \lim_{h\to 0} \frac{f(x, y+h) - f(x, y)}{h} \tag{2.12}$$

which represent *partial derivatives of f with respect to x and y.*

▶ EXAMPLE 2.33

Compute $(\partial f/\partial y)(2, 0)$ and $(\partial f/\partial x)(x, y)$ for $f(x, y) = (x-3)^2 e^y$.

SOLUTION Following the definition, we get

$$\frac{\partial f}{\partial y}(2, 0) = \lim_{h\to 0} \frac{f(2, h) - f(2, 0)}{h} = \lim_{h\to 0} \frac{e^h - 1}{h} = \lim_{h\to 0} e^h = 1$$

(note that we used L'Hôpital's rule to calculate the limit). Using (2.11),

$$\begin{aligned}\frac{\partial f}{\partial x}(x, y) &= \lim_{h\to 0} \frac{f(x+h, y) - f(x, y)}{h} = \lim_{h\to 0} \frac{(x+h-3)^2 e^y - (x-3)^2 e^y}{h} \\ &= \lim_{h\to 0} \frac{(h^2 + 2xh - 6h)e^y}{h} = \lim_{h\to 0}(h + 2x - 6)e^y = (2x-6)e^y.\end{aligned}$$

Next, we generalize partial derivatives to functions of any number of varibles.

DEFINITION 2.12 Partial Derivative

Let $f\colon U \subseteq \mathbb{R}^m \to \mathbb{R}$ be a real-valued function of m variables $x_1, \ldots, x_m$, defined on an open set U in $\mathbb{R}^m$. The *partial derivative* of f with respect to x_i (or with respect to the ith variable, $i = 1, \ldots, m$) is a real-valued function $\partial f/\partial x_i$ of m variables, defined by

$$\frac{\partial f}{\partial x_i}(x_1, \ldots, x_m) = \lim_{h\to 0} \frac{f(x_1, \ldots, x_i + h, \ldots, x_m) - f(x_1, \ldots, x_i, \ldots, x_m)}{h},$$

provided that the limit exists.

In other words, $\partial f/\partial x_i$ can be obtained by regarding all variables except x_i as constants, and applying standard rules for differentiating functions of one variable (in this case, the variable is x_i). If that is not possible, Definition 2.12 has to be used, as in Example 2.35.

Other commonly used symbols for partial derivatives $\partial f/\partial x_i$ are f_{x_i}, f_i, $D_{x_i} f$ and $D_i f$. If a function has a low number of variables, we use $\partial f/\partial x$, f_x, $D_1 f$ or $D_x f$ for the partial derivative of f with respect to x; similarly, the symbols $\partial f/\partial y$, f_y, $D_2 f$ or $D_y f$ denote the partial derivative of f with respect to y, etc.

▶ EXAMPLE 2.34

Let $f(x, y, z) = e^{xy} \sin(y^2 + z^2)$. Compute $\partial f/\partial x$, $\partial f/\partial y$ and $\partial f/\partial z$.

SOLUTION Regarding y and z as constants, we obtain

$$\frac{\partial f}{\partial x} = e^{xy} y \cdot \sin(y^2 + z^2) = ye^{xy} \sin(y^2 + z^2).$$

Similarly,

$$\frac{\partial f}{\partial y} = e^{xy} x \cdot \sin(y^2 + z^2) + e^{xy} \cos(y^2 + z^2) \cdot 2y$$
$$= e^{xy}(x \sin(y^2 + z^2) + 2y \cos(y^2 + z^2))$$

and

$$\frac{\partial f}{\partial z} = e^{xy} \cdot \cos(y^2 + z^2) \cdot 2z.$$

▶ EXAMPLE 2.35

Compute $(\partial f/\partial x)(x, y)$ for $f(x, y) = (x^4 + y^4)^{1/3}$.

SOLUTION The partial derivative

$$\frac{\partial f}{\partial x}(x, y) = \frac{1}{3}(x^4 + y^4)^{-2/3} \cdot 4x^3 = \frac{4x^3}{3(x^4 + y^4)^{2/3}} \tag{2.13}$$

is defined at all points (x, y) except at the origin. In order to compute $(\partial f/\partial x)(0, 0)$, we use Definition 2.12 or (2.11):

$$\frac{\partial f}{\partial x}(0, 0) = \lim_{h \to 0} \frac{f(h, 0) - f(0, 0)}{h} = \lim_{h \to 0} \frac{(h^4)^{1/3} - 0}{h} = \lim_{h \to 0} h^{1/3} = 0.$$

Therefore, $(\partial f/\partial x)(x, y)$ is given by (2.13) if $(x, y) \neq (0, 0)$, and by $(\partial f/\partial x)(0, 0) = 0$.

To get a better feel for partial derivatives we investigate several functions $f(x, y)$ of two variables. The partial derivative $(\partial f/\partial x)(x, y)$ represents the rate of change of f at (x, y) with respect to x when y is held fixed. A similar interpretation can be given to $(\partial f/\partial y)(x, y)$ [and for that matter, to any partial derivative of a function of any number of variables].

▶ EXAMPLE 2.36

The function $T(x, y) = 33e^{-x^2-2y^2}$ describes the air temperature at a location (x, y). Suppose that we start walking away from the origin along the x-axis in the positive direction. What rate of change in temperature do we experience at the moment when we reach the point $(1, 0)$? Compute $T_y(1, 2)$ and give a physical interpretation.

SOLUTION To answer the first question we have to compute $(\partial T/\partial x)(1, 0)$ (we are walking along the x-axis, so that $y = 0$, and it does not change). Using the chain rule, we get

$$\frac{\partial T}{\partial x} = 33e^{-x^2-2y^2}(-2x) = -66xe^{-x^2-2y^2},$$

and $(\partial T/\partial x)\,(1,0) = -66e^{-1} \approx -24.3$. So, we feel that the air is cooling down (the derivative is negative) as we pass through $(1,0)$. Similarly (the vertical bar is read "evaluated at"),

$$\frac{\partial T}{\partial y}(1,2) = -132ye^{-x^2-2y^2}\Big|_{(1,2)} = -264e^{-9} \approx -0.03.$$

In words, at the moment we reach the point $(1,2)$ on our walk along the vertical line $x = 1$ in the direction of the positive y-axis, we feel a very small decrease in temperature. ◀

▶ EXAMPLE 2.37

A contour diagram of a function $f(x, y)$ is shown in Figure 2.45. Estimate the values $(\partial f/\partial x)(10, 1)$ and $(\partial f/\partial y)(10, 1)$.

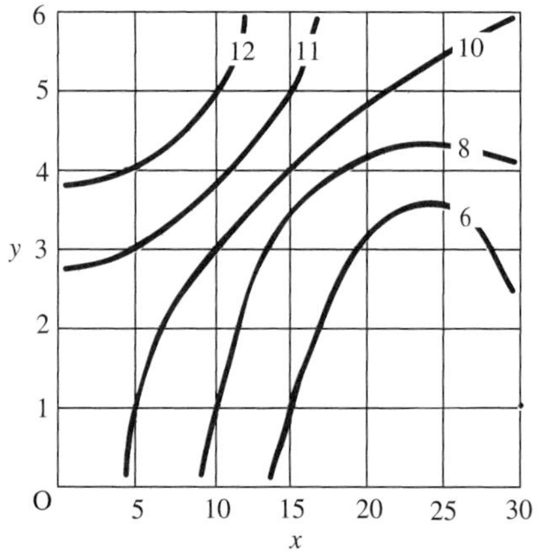

Figure 2.45 Contour diagram of the function $f(x, y)$ of Example 2.37.

SOLUTION To estimate partial derivatives, we use difference quotients; see (2.11) and (2.12). The point $(10, 1)$ lies on the contour of value 8. Moving 5 units to the right, we notice that $F(15, 1) = 6$. So, f decreases by 2 units as x increases by 5 units (and y is kept fixed at 1), and thus $(\partial f/\partial x)(10, 1) \approx -2/5$.

To compute $(\partial f/\partial y)(10, 1)$, we notice that, as y increases from 1 to 3 (with x kept at 10), f increases by 2 units (from 8 to 10). Thus, $(\partial f/\partial y)(10, 1) \approx 2/2 = 1$. ◀

Next, we discuss a geometric interpretation of partial derivatives. Consider a function $z = f(x, y)$ and pick a point (a, b) in its domain. The intersection of the graph of f and the vertical plane $y = b$ is the curve **c** that contains the point $(a, b, f(a, b))$ on the surface. Its equation is $z = f(x, b)$ (z is now a function of one variable) and the partial derivative $(\partial f/\partial x)(a, b)$ is equal to the slope of the tangent to that curve at $(a, b, f(a, b))$; see Figure 2.46.

▶ EXAMPLE 2.38

Compute $(\partial f/\partial x)(2, 1)$ for $f(x, y) = 2x^2 + 3xy - y^2$. Find the curve that is the intersection of the graph of $z = f(x, y)$ and the plane $y = 1$ and compute the slope of the tangent to that curve at $x = 2$. Give a geometric interpretation of $(\partial f/\partial x)(2, 1)$.

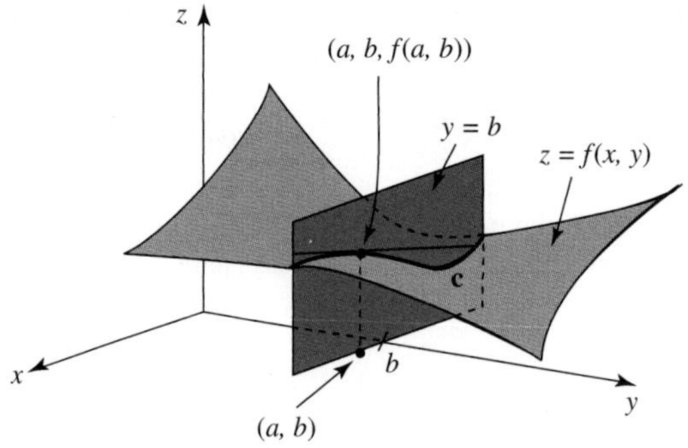

Figure 2.46 Partial derivative is the slope of a tangent.

SOLUTION The partial derivative is $(\partial f/\partial x)(2, 1) = (4x + 3y)|_{(2,1)} = 11$.

Substitute $y = 1$ into $z = 2x^2 + 3xy - y^2$ to get the curve $z = 2x^2 + 3x - 1$ in the plane $y = 1$. The slope of its tangent line at $x = 2$ is $z'(2) = (4x + 3)|_{x=2} = 11$. The partial derivative of f with respect to x at the point $(2, 1)$ equals the slope of the tangent line at $x = 2$ to the curve that is the intersection of the surface $z = 2x^2 + 3xy - y^2$ and the plane $y = 1$. ◀

Derivative of a Function of Several Variables

Let $\mathbf{F}$ be a vector-valued function $\mathbf{F}\colon U \subseteq \mathbb{R}^m \to \mathbb{R}^n$. Recall that $\mathbf{F}$ can be written in terms of its components as

$$\mathbf{F}(x_1, \ldots, x_m) = (F_1(x_1, \ldots, x_m), F_2(x_1, \ldots, x_m), \ldots, F_n(x_1, \ldots, x_m)),$$

or as $\mathbf{F}(\mathbf{x}) = (F_1(\mathbf{x}), F_2(\mathbf{x}), \ldots, F_n(\mathbf{x}))$, where $\mathbf{x} = (x_1, \ldots, x_m)$. In words, we can describe a vector-valued function $\mathbf{F}$ using n real-valued functions of m variables.

By $D\mathbf{F}(\mathbf{x})$ we denote the $n \times m$ matrix of partial derivatives of the components of $\mathbf{F}$ evaluated at $\mathbf{x}$ (provided that all partial derivatives exist at $\mathbf{x}$). Thus,

$$D\mathbf{F}(x) = \begin{bmatrix} \frac{\partial F_1}{\partial x_1}(\mathbf{x}) & \frac{\partial F_1}{\partial x_2}(\mathbf{x}) & \cdots & \frac{\partial F_1}{\partial x_m}(\mathbf{x}) \\ \frac{\partial F_2}{\partial x_1}(\mathbf{x}) & \frac{\partial F_2}{\partial x_2}(\mathbf{x}) & \cdots & \frac{\partial F_2}{\partial x_m}(\mathbf{x}) \\ \vdots & \vdots & & \vdots \\ \frac{\partial F_n}{\partial x_1}(\mathbf{x}) & \frac{\partial F_n}{\partial x_2}(\mathbf{x}) & \cdots & \frac{\partial F_n}{\partial x_m}(\mathbf{x}) \end{bmatrix}. \tag{2.14}$$

The matrix $D\mathbf{F}(\mathbf{x})$ has n rows and m columns (the number of rows is the number of component functions of $\mathbf{F}$, and the number of columns equals the number of variables). The ith row consists of partial derivatives of the ith component F_i of $\mathbf{F}$ with respect to all variables $x_1, \ldots, x_m$, evaluated at $\mathbf{x}$. The ith column is the matrix

$$\frac{\partial \mathbf{F}}{\partial x_i}(\mathbf{x}) = \mathbf{F}_{x_i}(\mathbf{x}) = \begin{bmatrix} \frac{\partial F_1}{\partial x_i}(\mathbf{x}) \\ \frac{\partial F_2}{\partial x_i}(\mathbf{x}) \\ \vdots \\ \frac{\partial F_n}{\partial x_i}(\mathbf{x}) \end{bmatrix},$$

that consists of partial derivatives of the component functions $F_1, \ldots, F_n$ with respect to the same variable x_i, evaluated at $\mathbf{x}$.

▶ EXAMPLE 2.39

Let $\mathbf{F}: \mathbb{R}^3 \to \mathbb{R}^4$ be given by $\mathbf{F}(x, y, z) = (e^{x+yz}, x^2 + 1, \sin(y + z), 4y)$. The components of $\mathbf{F}$ are $F_1(x, y, z) = e^{x+yz}$, $F_2(x, y, z) = x^2 + 1$, $F_3(x, y, z) = \sin(y + z)$, and $F_4(x, y, z) = 4y$. The matrix $D\mathbf{F}(x, y, z)$ is given by

$$D\mathbf{F}(x, y, z) = \begin{bmatrix} e^{x+yz} & ze^{x+yz} & ye^{x+yz} \\ 2x & 0 & 0 \\ 0 & \cos(y+z) & \cos(y+z) \\ 0 & 4 & 0 \end{bmatrix}.$$

The second column of $D\mathbf{F}(x, y, z)$ is equal to

$$\frac{\partial \mathbf{F}}{\partial y}(x, y, z) = \begin{bmatrix} ze^{x+yz} \\ 0 \\ \cos(y+z) \\ 4 \end{bmatrix},$$

which is the matrix of derivatives of component functions with respect to y. ◀

Let us consider several special cases. If $f(x): \mathbb{R} \to \mathbb{R}$, then $Df(x)$ is a 1×1 matrix whose entry is the derivative of (the only component) f with respect to (the only variable) x. Hence, $Df(x)$ is the usual derivative $f'(x)$.

Assume that $f(\mathbf{x}): U \subseteq \mathbb{R}^m \to \mathbb{R}$ is a real-valued function of m variables. Then $Df(\mathbf{x})$ is the $1 \times m$ matrix

$$Df(\mathbf{x}) = \left[\frac{\partial f}{\partial x_1}(\mathbf{x}) \quad \frac{\partial f}{\partial x_2}(\mathbf{x}) \quad \cdots \quad \frac{\partial f}{\partial x_m}(\mathbf{x}) \right],$$

whose only row consists of partial derivatives of f with respect to all variables $x_1, \ldots, x_m$, evaluated at $\mathbf{x} = (x_1, \ldots, x_m)$. Interpreted as a vector, $Df(\mathbf{x})$ is called the *gradient* of f at $\mathbf{x}$, and is denoted by *grad* $f(\mathbf{x})$ or $\nabla f(\mathbf{x})$. We will study the gradient in detail in Section 2.7.

Let $\mathbf{c}$: $[a, b] \to \mathbb{R}^n$ be a vector-valued function of one variable (we use t rather than x or x_1 to denote the independent variable). In this case, $D\mathbf{c}(t)$ is the $n \times 1$ matrix

$$D\mathbf{c}(t) = \begin{bmatrix} x_1'(t) \\ x_2'(t) \\ \vdots \\ x_n'(t) \end{bmatrix},$$

whose column consists of the derivatives of the components $x_1, \ldots, x_m$ of $\mathbf{c}$ with respect to t [since x_i are functions of one variable, we use x_i' instead of $\partial x_i/\partial t$ to denote the derivative]. Evaluated at a point t_0 and interpreted as a vector, $D\mathbf{c}(t_0)$ is called the *tangent vector* [provided that $D\mathbf{c}(t_0) \neq \mathbf{0}$], or the *velocity vector* of $\mathbf{c}$, and is denoted by $\mathbf{c}'(t_0)$.

The function $\mathbf{c}$ is called a *path* in $\mathbb{R}^n$. We study paths and related concepts (tangent vectors, velocity, etc.) in Section 2.5, and also in Chapter 3.

▶ EXAMPLE 2.40 Gradient of the Gravitational Potential

Consider the gravitational potential function

$$V(x, y, z) = -\frac{GMm}{\sqrt{x^2 + y^2 + z^2}}$$

discussed in Example 2.10. Compute its gradient $\nabla V(x, y, z)$.

SOLUTION By the chain rule,

$$\frac{\partial V}{\partial x} = \frac{1}{2} GMm(x^2 + y^2 + z^2)^{-3/2} \cdot 2x.$$

The partial derivatives $\partial V/\partial y$ and $\partial V/\partial z$ are computed in the same way. Hence,

$$\nabla V(x, y, z) = DV(x, y, z) = \begin{bmatrix} \frac{\partial V}{\partial x} & \frac{\partial V}{\partial y} & \frac{\partial V}{\partial z} \end{bmatrix}$$

$$= \begin{bmatrix} \frac{GMm}{(x^2 + y^2 + z^2)^{3/2}} x & \frac{GMm}{(x^2 + y^2 + z^2)^{3/2}} y & \frac{GMm}{(x^2 + y^2 + z^2)^{3/2}} z \end{bmatrix}.$$

Rewriting ∇V as a vector, we obtain (write $\mathbf{r} = x\mathbf{i} + y\mathbf{j} + z\mathbf{k}$)

$$\nabla V(x, y, z) = \frac{GMm}{(x^2 + y^2 + z^2)^{3/2}}(x\mathbf{i} + y\mathbf{j} + z\mathbf{k}) = \frac{GMm}{||\mathbf{r}||^3}\mathbf{r}.$$

Note that $GMm\mathbf{r}/||\mathbf{r}||^3$ is the negative of the gravitational force field. ◀

Example 2.40 shows that $\mathbf{F} = -\nabla V$, where $\mathbf{F}$ is the gravitational force field. In general, a force field $\mathbf{F}$ satisfying this formula is called *conservative*, and the scalar function V is the *potential function.* We will study properties of conservative fields and potential functions in Section 5.4.

Derivative and Differentiability

DEFINITION 2.13 Differentiability of a Vector-Valued Function

A vector-valued function $\mathbf{F} = (F_1, \ldots, F_n)\colon U \subseteq \mathbb{R}^m \to \mathbb{R}^n$, defined on an open set $U \subseteq \mathbb{R}^m$, is *differentiable at* $\mathbf{a} \in U$ if

(a) all partial derivatives of the components $F_1, \ldots, F_n$ of $\mathbf{F}$ exist at $\mathbf{a}$, and

(b) the matrix of partial derivatives $D\mathbf{F}(\mathbf{a})$ of $\mathbf{F}$ at $\mathbf{a}$ satisfies

$$\lim_{\mathbf{x}\to\mathbf{a}} \frac{||\mathbf{F}(\mathbf{x}) - \mathbf{F}(\mathbf{a}) - D\mathbf{F}(\mathbf{a})(\mathbf{x}-\mathbf{a})||}{||\mathbf{x}-\mathbf{a}||} = 0, \tag{2.15}$$

where $||.||$ in the numerator denotes the length in $\mathbb{R}^n$, and $||.||$ in the denominator is the length in $\mathbb{R}^m$. ◀

DEFINITION 2.14 Derivative of a Vector-Valued Function

If a vector-valued function $\mathbf{F}$ satisfies the conditions (a) and (b) of Definition 2.13, then the matrix $D\mathbf{F}(\mathbf{a})$ of partial derivatives given by (2.14) is called the *derivative of* $\mathbf{F}$ *at* $\mathbf{a}$. ◀

The subtractions in the numerator of (2.15) take place in $\mathbb{R}^n$: clearly, $\mathbf{F}(\mathbf{x})$ and $\mathbf{F}(\mathbf{a})$ are in $\mathbb{R}^n$; the third term is the product of the $n \times m$ matrix $D\mathbf{F}(\mathbf{a})$ and the vector (viewed as an $m \times 1$ matrix) $\mathbf{x} - \mathbf{a}$, and is therefore an $n \times 1$ matrix, that is, an element of $\mathbb{R}^n$.

Let us look more closely at condition (b) in Definition 2.13. Assume that $m = n = 1$; that is, consider the function $f\colon \mathbb{R} \to \mathbb{R}$ (a real-valued function of one variable can be considered as a special case of a general vector-valued function if n is allowed to equal 1).

Then $Df(x) = f'(x)$ and the statement (b) reads (the symbol $||.||$ is replaced by the absolute value, since all terms involved are real numbers)

$$\lim_{x\to a} \frac{\left|f(x) - \left[f(a) + f'(a)(x-a)\right]\right|}{|x-a|} = 0. \tag{2.16}$$

The expression

$$L_a(x) = f(a) + f'(a)(x-a)$$

appearing in the numerator of (2.16) is called the *linear approximation* or the *linearization* of f at a. Geometrically, L_a represents the equation of the line tangent to the graph of f at a [it is written in point-slope form: the point is $(a, f(a))$, and $f'(a)$ is the slope].

Since the limit in (2.16) is zero and the denominator goes to zero, it follows that the numerator $|f(x) - L_a(x)|$ has to approach zero as well. Consequently, (2.16) states that $L_a(x)$ approaches $f(x)$ as x approaches a; that is, *near* a the functions $f(x)$ and $L_a(x)$ have approximately the same value. This is not really important: as a matter of fact, *any* line that goes through $(a, f(a))$ satisfies this property. However, formula (2.16) says a lot more than that: if we rewrite it as

$$\lim_{x\to a} \left| \frac{f(x) - f(a) - f'(a)(x-a)}{x-a} \right| = \lim_{x\to a} \left| \frac{f(x) - f(a)}{x-a} - f'(a) \right| = 0,$$

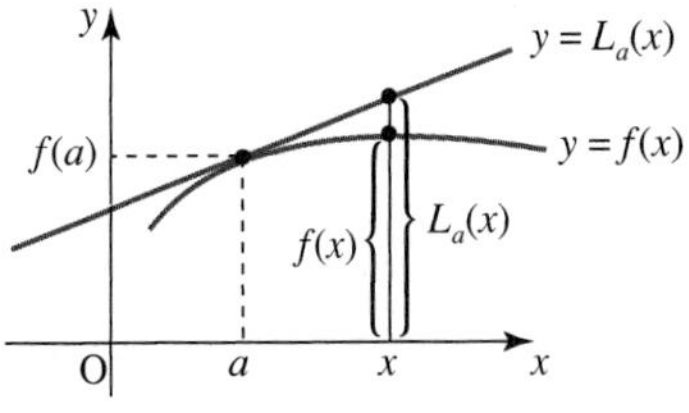

Figure 2.47 The tangent as a linear approximation.

we see that the *slopes* of $f(x)$ and $L_a(x)$ [recall that the slope of $L_a(x)$ is $f'(a)$] must approach each other. And that is true *only* for the tangent line.

In other words, $L_a(x)$ is a *good approximation* to $f(x)$ near a; that is, the tangent line is a *good approximation* to the curve $y = f(x)$ near a; see Figure 2.47.

▶ EXAMPLE 2.41

Let $f(x) = xe^{2x}$. Its linearization at $a = 1$ is $[f'(x) = e^{2x} + 2xe^{2x}]$

$$L_1(x) = f(1) + f'(1)(x-1) = e^2 + 3e^2(x-1).$$

Take a point near $a = 1$, say, $x = 1.0001$. Then $L_1(1.0001) = 7.3912728$ approximates the value of the function $f(1.0001) = 7.3912731$.

Clearly, the closer the number x is to 1, the better the approximation. For values of x that are far from 1, the linear approximation does not make any sense. For example, $L_1(0) = -2e^2 = -14.778112$, whereas $f(0) = 0$. ◀

Figure 2.48(a) shows the graph of the function $f(x)$ from the previous example on the interval [0.5, 1.5]. As we zoom in on the graph (in Figure 2.48(b), $f(x)$ is shown on the interval [0.9, 1.1], and in Figure 2.48(c) we used [0.95, 1.05]), we see that it looks like a straight line.

The fact that $f(x)$ is differentiable at $x = 1$ means that, as we continue zooming in on its graph around $x = 1$, it will resemble, closer and closer, a straight line. That straight line is the linear approximation (the tangent!) of $f(x)$ at $x = 1$. This property is sometimes called a *local linearity*. For instance, assuming that the curve in Figure 2.48(c)

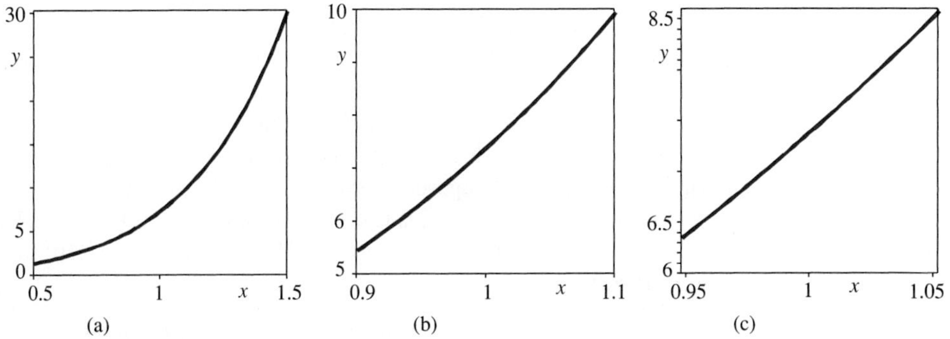

Figure 2.48 Zooming in on the graph of $f(x)$ near $x = 1$.

is a straight line, we compute its slope [using the endpoints (0.95, 6.35) and (1.05, 8.6)] to be $(8.6 - 6.35)/(1.05 - 0.95) = 22.5$. This is an approximation of the slope ($3e^2 \approx 22.17$) of the linear approximation of $f(x)$ [and of $f(x)$] at $x = 1$.

Next, we discuss another special case, that of a function $f\colon \mathbb{R}^2 \to \mathbb{R}$ (i.e., $m = 2$ and $n = 1$). In that case [with the notation $\mathbf{x} = (x, y)$ and $\mathbf{a} = (a, b)$],

$$Df(\mathbf{a})(\mathbf{x} - \mathbf{a}) = \begin{bmatrix} \frac{\partial f}{\partial x}(a, b) & \frac{\partial f}{\partial y}(a, b) \end{bmatrix} \cdot \begin{bmatrix} x - a \\ y - b \end{bmatrix}$$
$$= \frac{\partial f}{\partial x}(a, b) \cdot (x - a) + \frac{\partial f}{\partial y}(a, b) \cdot (y - b),$$

and hence (2.15) reads

$$\lim_{(x,y)\to(a,b)} \frac{|\, f(x, y) - L_{(a,b)}(x, y)\,|}{\sqrt{(x - a)^2 + (y - b)^2}} = 0, \tag{2.17}$$

where

$$L_{(a,b)}(x, y) = f(a, b) + \frac{\partial f}{\partial x}(a, b) \cdot (x - a) + \frac{\partial f}{\partial y}(a, b) \cdot (y - b) \tag{2.18}$$

is the *linear approximation* or the *linearization* of f at (a, b). It is a *good approximation* of f near (a, b) in the sense that the values of f and $L_{(a,b)}$ for points *near* (a, b) are almost the same; see Figure 2.49. This property does not make $L_{(a,b)}(x, y)$ special. What makes it unique (among all linear functions) is the requirement that it must satisfy (2.17). For a proof of this fact, see Exercise 45.

As an illustration, consider $f(x, y) = 1 - x^2 - 2y^2$. Its linearization at $\mathbf{a} = (1, 1)$ is

$$L_{(1,1)}(x, y) = f(1, 1) + (-2x)|_{(1,1)} \cdot (x - 1) + (-4y)|_{(1,1)} \cdot (y - 1) = 4 - 2x - 4y.$$

Take a point near (1, 1), for instance, $\mathbf{x} = (0.96, 1.02)$. The value of the function $f(\mathbf{x}) = f(0.96, 1.02) = 1 - (0.96)^2 - 2(1.02)^2 = -2.0024$ is approximated by the value of its linearization $L_{(1,1)}(\mathbf{x}) = L_{(1,1)}(0.96, 1.02) = -2$.

Geometrically, linear approximation represents the equation of a plane in $\mathbb{R}^3$ (in the previous example, $z = 4 - 2x - 4y$). This plane has the point $(a, b, f(a, b)) = (a, b, L_{(a,b)}(a, b))$ in common with the graph of f (see Figure 2.49) and is a unique plane that satisfies (2.17). A plane with these properties is called a *tangent plane*. It is defined by

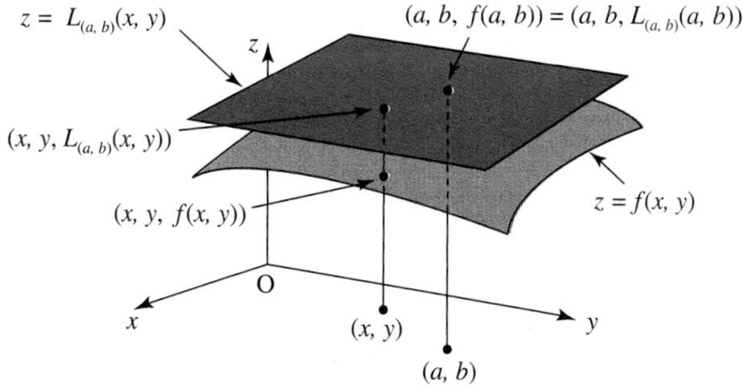

Figure 2.49 Linear approximation $L_{(a,b)}(x, y)$ of a function $f(x, y)$ at (a, b).

the equation

$$z = L_{(a,b)}(x, y) = f(a, b) + \frac{\partial f}{\partial x}(a, b) \cdot (x - a) + \frac{\partial f}{\partial y}(a, b) \cdot (y - b). \tag{2.19}$$

Based on the two examples we have considered, we say in general that the *linear approximation* $L_{\mathbf{a}}(\mathbf{x})$ of $\mathbf{F}(\mathbf{x})$ at $\mathbf{a}$ or the *linearization* of $\mathbf{F}(\mathbf{x})$ at $\mathbf{a}$, given by

$$L_{\mathbf{a}}(\mathbf{x}) = \mathbf{F}(\mathbf{a}) + D\mathbf{F}(\mathbf{a})(\mathbf{x} - \mathbf{a}), \tag{2.20}$$

is a *good approximation* of $\mathbf{F}$ near $\mathbf{a}$. Hence, according to Definition 2.13, a vector-valued function $\mathbf{F}$ is differentiable at $\mathbf{a}$ if and only if all partial derivatives of its components exist at $\mathbf{a}$ and its linearization at $\mathbf{a}$ is a *good approximation* in the sense just explained. (Another special case, that of a function $\mathbf{F}: \mathbb{R}^2 \to \mathbb{R}^2$, will be studied in Section 6.4.)

▶ EXAMPLE 2.42

Compute the equation of the plane tangent to the graph of the function $f(x, y) = \arctan(xy)$ at the point $(1, 1)$.

SOLUTION

From

$$\frac{\partial f}{\partial x} = \frac{y}{1 + x^2y^2} \quad \text{and} \quad \frac{\partial f}{\partial y} = \frac{x}{1 + x^2y^2},$$

we get $(\partial f/\partial x)(1, 1) = 1/2$ and $(\partial f/\partial y)(1, 1) = 1/2$. Since $f(1, 1) = \pi/4$, the equation of the tangent plane is $z = \frac{\pi}{4} + \frac{1}{2}(x - 1) + \frac{1}{2}(y - 1)$, that is, $2x + 2y - 4z + \pi - 4 = 0$. ◀

▶ EXAMPLE 2.43

Let $f(x, y) = 13 - x^2 - y^2$. Suppose that we use the linear approximation $L_{(1,2)}(x, y)$ of $f(x, y)$ at $(1, 2)$ to approximate the value of f at a point (x, y) near $(1, 2)$. Is this an overestimate or an underestimate of $f(x, y)$?

SOLUTION

The level curves $f(x, y) = 13 - x^2 - y^2 = C$ are circles of radius $\sqrt{13 - C}$ for $C < 13$. The intersections of the graph of f with the xz-plane and the yz-plane are the parabolas $z = 13 - x^2$ and $z = 13 - y^2$, both of which are concave down. In other words, the graph of f is a surface built of circles, smaller ones placed on top of the larger ones in such a way that the vertical cross-sections are parabolas. The surface is concave down, so the tangent plane must lie above it. Hence, the value of the linear approximation at a point near $(1, 2)$ is larger than the value of the function. The estimate is an overestimate. ◀

Recall that a function $f(x, y)$ of two variables is differentiable at (a, b) if and only if $(\partial f/\partial x)(a, b)$ and $(\partial f/\partial y)(a, b)$ exist and $f(x, y)$ satisfies (2.17). In analogy with functions of one variable, we say that $f(x, y)$ is differentiable if it is *locally linear* at (a, b); that is, zooming in on its graph near (a, b) will make it look like a plane.

▶ EXAMPLE 2.44

Determine whether the function $f(x, y) = \sqrt{x^2 + y^2}$ is differentiable at the origin.

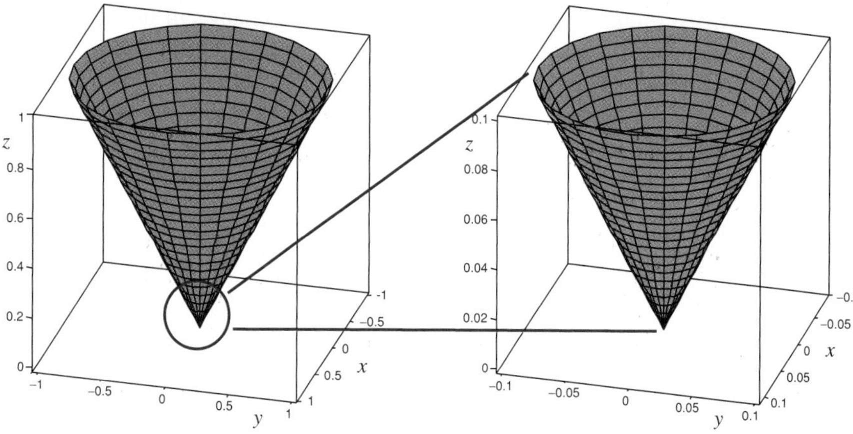

Figure 2.50 Graph of $f(x, y) = \sqrt{x^2 + y^2}$ with a zoom-in near the origin.

SOLUTION The graph of $f(x, y)$ is a cone whose vertex is at the origin; see Figure 2.50. Zooming in on the graph near the origin, we see that the vertex remains, that is, the graph does not flatten. In other words, local linearity does not hold at the origin. Thus, we believe that f is not differentiable at the origin.

To confirm our intuitive reasoning, we compute

$$\frac{\partial f}{\partial x}(0, 0) = \lim_{h \to 0} \frac{f(h, 0) - f(0, 0)}{h} = \lim_{h \to 0} \frac{\sqrt{h^2} - 0}{h} = \lim_{h \to 0} \frac{|h|}{h}.$$

As $h \to 0^+$, $|h|/h \to 1$, and as $h \to 0^-$, $|h|/h \to -1$; that is, the limit of $|h|/h$ as $h \to 0$ does not exist. So, $(\partial f/\partial x)(0, 0)$ does not exist and thus f is not differentiable at the origin. (See Exercise 46 for an alternate proof.)

Differential

We now discuss another interpretation of the linear approximation formula. Choose a point (a, b) in the domain of a differentiable function $f: \mathbb{R}^2 \to \mathbb{R}$. Measure the value of f at (a, b) and then move to a nearby point $(x, y) = (a + \Delta x, b + \Delta y)$ ("nearby" means that Δx and Δy are small). We would like to compare the value of f at this point with its initial value $f(a, b)$. In other words, we would like to compute or estimate the change (sometimes called the error) Δf in f, defined by $\Delta f = f(x, y) - f(a, b) = f(a + \Delta x, b + \Delta y) - f(a, b)$.

Since $f(x, y) \approx L_{(a,b)}(x, y)$, where $L_{(a,b)}(x, y)$ is the linear approximation of f at (a, b), it follows that

$$f(x, y) \approx f(a, b) + f_x(a, b)(x - a) + f_y(a, b)(y - b),$$

and therefore,

$$f(x, y) - f(a, b) \approx f_x(a, b)(x - a) + f_y(a, b)(y - b). \tag{2.21}$$

The right side of this approximate equality is equal to $f_x(a, b)\Delta x + f_y(a, b)\Delta y$ and is denoted by df (it is called the *differential* of f). The left side in (2.21) is the change (or the error) Δf, and hence $\Delta f \approx df$; that is,

$$\Delta f \approx f_x(a, b)\Delta x + f_y(a, b)\Delta y.$$

This formula says that we can estimate the change in the function Δf in terms of the change (or the error) Δx in the variable x and the change (or the error) Δy in the variable y. Analogous expressions can be obtained for a function of any number of variables.

The quantities Δx and Δy are usually replaced by dx and dy, and we write the differential of the function $f(x, y)$ at (a, b) as

$$df = f_x(a,b)dx + f_y(a,b)dy. \tag{2.22}$$

▶ EXAMPLE 2.45 Barometric Formula

The *barometric formula* (also known as *exponential atmosphere*) describes how the pressure P of the air changes with altitude and temperature. It is given by the formula

$$P(z,T) = P_0 e^{-Mg_0 z/RT},$$

where P_0 is the air pressure at sea level, $M = 0.029$ g is the mass of one mole of air, g_0 is the acceleration due to gravity (at sea level), R is the universal gas constant (see Example 2.8), T is the absolute temperature (in degrees Kelvin), and z is the height (in kilometers) above Earth's surface. We compute the differential of P as

$$dP = -\frac{Mg_0P_0}{RT}e^{-Mg_0z/RT}dz + \frac{Mg_0zP_0}{RT^2}e^{-Mg_0z/RT}dT.$$

The coefficient $-(Mg_0P_0/RT)e^{-Mg_0z/RT}$ of dz is negative since an increase in height (with the temperature kept fixed) will decrease air pressure.

The coefficient $(Mg_0zP_0/RT^2)e^{-Mg_0z/RT}$ of dT is positive since an increase in temperature (with no change in height) will cause an increase in air pressure.

Although this is a simplified model (it assumes that T does not change with altitude), it is fairly accurate to heights of about 140 km. In parts of the atmosphere where the temperature drops with the height (for instance, $0 \leq z \leq 15$ and $50 \leq z \leq 80$), we conclude that the above model gives an overestimate for the air pressure. ◀

Differentiability and Continuity

In the theory of functions of one variable, one proves that if a function f has a derivative, then it is continuous. The analogous statement ("differentiability implies continuity") also holds for functions of more than one variable; see Theorem 2.4. However, a function whose partial derivatives exist might not be continuous, as the following example shows.

▶ EXAMPLE 2.46

Define $f: \mathbb{R}^2 \to \mathbb{R}$ by

$$f(x,y) = \begin{cases} \dfrac{xy}{x^2+y^2} & \text{if } (x,y) \neq (0,0) \\ 0 & \text{if } (x,y) = (0,0) \end{cases}.$$

By definition,

$$\frac{\partial f}{\partial x}(0,0) = \lim_{h\to 0}\frac{f(h,0)-f(0,0)}{h} = \lim_{h\to 0}\frac{0-0}{h} = 0$$

and

$$\frac{\partial f}{\partial y}(0,0) = \lim_{h\to 0}\frac{f(0,h)-f(0,0)}{h} = \lim_{h\to 0}\frac{0-0}{h} = 0.$$

Hence, both partial derivatives exist at (0, 0). On the other hand, the limit of f as (x, y) approaches (0, 0) does not exist. To prove this, we will show that two different ways of reaching (0, 0) yield two different results. Walking along the y-axis toward (0, 0), we get

$$\lim_{x=0,y\to 0}\frac{xy}{x^2+y^2} = \lim_{y\to 0}\frac{0}{y^2} = \lim_{y\to 0} 0 = 0.$$

Approaching (0, 0) along the line $y = x$, we get

$$\lim_{(x,y)\to(0,0)}\frac{xy}{x^2+y^2} = \lim_{x\to 0}\frac{x^2}{x^2+x^2} = \lim_{x\to 0}\frac{1}{2} = \frac{1}{2},$$

and therefore the limit of $f(x, y)$ as (x, y) approaches (0, 0) does not exist. Consequently, f cannot be continuous at (0, 0).

Therefore, the mere existence of partial derivatives does not imply the continuity of a function. However, an extra assumption will fix this problem (see Theorem 2.4).

► EXAMPLE 2.47

Show that the function $f(x, y)$ defined in Example 2.46 is not differentiable at (0, 0).

SOLUTION

In Example 2.46 we obtained $(\partial f/\partial x)(0,0) = 0$ and $(\partial f/\partial y)(0,0) = 0$; thus, the condition on the existence of partial derivatives of f holds.

Next, we check (2.17). If f had a linear approximation at (0, 0), it would have to be

$$L_{(0,0)} = f(0,0) + \frac{\partial f}{\partial x}(0,0)(x-0) + \frac{\partial f}{\partial y}(0,0)(y-0) = 0.$$

The limit in (2.17) is then equal to

$$\lim_{(x,y)\to(0,0)}\frac{\left|\dfrac{xy}{x^2+y^2}-0\right|}{\sqrt{x^2+y^2}} = \lim_{(x,y)\to(0,0)}\frac{|xy|}{(x^2+y^2)^{3/2}}.$$

Using the approach $y = x$, we get

$$\lim_{(x,y)\to(0,0)}\frac{|xy|}{(x^2+y^2)^{3/2}} = \lim_{x\to 0}\frac{x^2}{(2x^2)^{3/2}} = \lim_{x\to 0}\frac{1}{2^{3/2}x}.$$

Since the limit on the right side does not exist, it follows that (2.17) does not hold, and thus f is not differentiable at (0, 0).

THEOREM 2.4 Differentiable Functions Are Continuous

Let $\mathbf{F}\colon U \subseteq \mathbb{R}^m \to \mathbb{R}^n$ be a vector-valued function and let $\mathbf{a} \in U$. If $\mathbf{F}$ is differentiable at $\mathbf{a}$, then it is continuous at $\mathbf{a}$.

In other words, a function that is not continuous at **a** cannot be differentiable there either. Thus, the conclusion that we arrived at in Example 2.46 implies that f is not differentiable at (0, 0) (this is an alternative to the proof presented in Example 2.47).

Theorem 2.4 is the correct generalization of the one-variable case: namely, if the components of **F** have partial derivatives *and* the derivative $D\mathbf{F}$ is a good approximation of **F**, then **F** is continuous. The proof of the theorem is presented in Appendix A.

A function whose partial derivatives exist might not be differentiable (see Examples 2.46 and 2.47). In other words, the existence of partial derivatives does not imply differentiability. However, if all partial derivatives are continuous, the implication is valid, as the following theorem shows.

THEOREM 2.5 Continuity of Partial Derivatives Implies Differentiability

Let $\mathbf{F}: U \subseteq \mathbb{R}^m \to \mathbb{R}^n$ be a vector-valued function with components $F_1, \ldots, F_n: U \subseteq \mathbb{R}^m \to \mathbb{R}$. If all partial derivatives $\partial F_i/\partial x_j$ $(i = 1, \ldots, n,\ j = 1\ldots, m)$ are continuous at **a**, then **F** is differentiable at **a**. ◀

Proving the differentiablity of a function using Definition 2.13 is usually fairly complicated. This theorem gives a more convenient alternative: all we have to do is to check that all partial derivatives exist and are continuous at the point(s) in question.

The proof of Theorem 2.5 is given in Appendix A.

DEFINITION 2.15 Function of Class C^1

A function whose partial derivatives exist and are continuous is said to be *continuously differentiable,* or *of class* C^1. ◀

The definitions and theorems we have stated could be visually represented in a diagram; see Figure 2.51 (containment means implication; i.e., functions contained in one "box" have properties defining any other "box" that contains it).

Let us identify a few facts from the diagram. Functions of class C^1 (those are in the smallest "box") are differentiable (those functions are in the larger box)—that is the statement of Theorem 2.5. Differentiable functions are continuous (Theorem 2.4). If a function has partial derivatives, it might not be differentiable (that was the conclusion of Examples 2.46 and 2.47). Not every differentiable function is of class C^1. A function can be continuous, but its partial derivatives might not exist (see Example 2.44), etc.

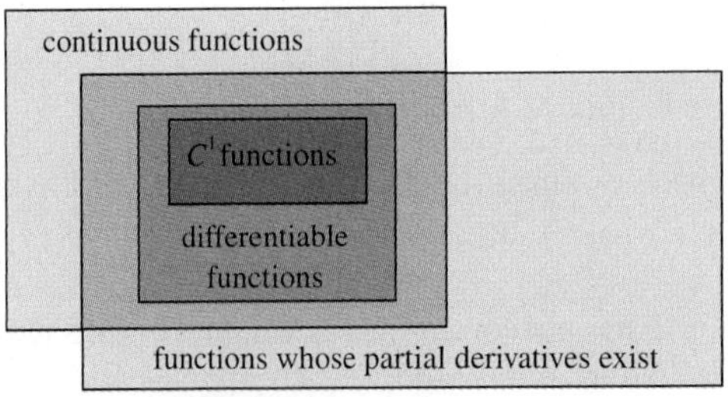

Figure 2.51 Continuity, differentiability, and partial derivatives.

▶ EXERCISES 2.4

Exercises 1 to 6: Determine which of the following sets are open.

1. $U = \{(x, y) \mid 2 < x^2 + y^2 < 3\} \subseteq \mathbb{R}^2$

2. $U = \{(x, y, z) \mid x \geq 0\} \subseteq \mathbb{R}^3$

3. $U = \{(x, y) \mid x + y = 2\} \subseteq \mathbb{R}^2$

4. $U = \{(x, y) \mid x + y < 2\} \subseteq \mathbb{R}^2$

5. $U = \{(x, y, z) \mid xyz > 0\} \subseteq \mathbb{R}^3$

6. $U = \{(x, y, z) \mid x \neq 0, y > 0\} \subseteq \mathbb{R}^3$

7. Consider the function $f(x, y)$ whose contour diagram is shown in Figure 2.45.

(a) Determine the sign of $(\partial f/\partial x)(5, 3)$.

(b) Which of the two numbers, $(\partial f/\partial x)(10, 3)$ or $(\partial f/\partial x)(10, 5)$, is larger?

8. Draw a contour diagram of a function $f(x, y)$ that satisfies $(\partial f/\partial x)(x, y) > 0$ and $(\partial f/\partial y)(x, y) < 0$ for all (x, y).

Exercises 9 to 18: Find the indicated partial derivatives.

9. $f(x, y) = x^y + y \ln x$; f_x, f_y

10. $f(x, y, z) = xe^{yz^2}$; f_x, f_y, f_z

11. $f(x, y, z) = \ln(x + y + z^2)$; f_x, f_z

12. $f(x, y) = \arctan(x/y)$; f_x, f_y

13. $f(x, y) = e^{xy} \cos x \sin y$; f_x, f_y

14. $f(x, y, z) = x\sqrt{y\sqrt{z}}$; f_x, f_y, f_z

15. $f(x_1, \ldots, x_m) = \sqrt{x_1^2 + \cdots + x_m^2}$; $\partial f/\partial x_i$, $i = 1, \ldots, m$

16. $f(x_1, \ldots, x_m) = e^{x_1 \cdots x_m}$; $\partial f/\partial x_i$, $i = 1, \ldots, m$

17. $f(x, y) = \int_0^x te^{-t^2} dt$; f_x, f_y

18. $f(x, y) = \int_{\ln y}^0 (t + 1)^2 \, dt$; f_x, f_y

Exercises 19 to 22: The function $z(x, y)$ is defined in terms of two differentiable real-valued functions f and g of one variable. Compute z_x and z_y.

19. $z = f(x) + g(y)$

20. $z = f(x)g(y)$

21. $z = f(x)/g(y)$

22. $z = f(x)^{g(y)}$

23. A hiker is standing at the point $(2, 1, 11)$ on a hill whose shape is given by the graph of the function $z = 14 - (x - 3)^2 - 2(y - 2)^4$. Assume that the x-axis points east and the y-axis points north. In which of the two directions (east or north) is the hill steeper?

24. The volume of a certain amount of gas is determined by $V = 0.12TP^{-1}$, where T is the temperature and P is the pressure. Compute and interpret $\partial V/\partial P$ and $\partial V/\partial T$ when $P = 10$ and $T = 370$.

25. Consider the function $f(x, y) = -xe^{-x^2-2y^2}$.

(a) Compute $f_y(2, 3)$.

(b) Find the curve that is the intersection of the graph of f and the vertical plane $x = 2$ and compute the slope of its tangent at $y = 3$.

(c) Using (a) and (b), give a geometric interpretation of $f_y(2, 3)$.

26. Let $u(x, y, t) = e^{-2t} \sin(3x) \cos(2y)$ denote the vertical displacement of a vibrating membrane from the point (x, y) in the xy-plane at the time t. Compute $u_x(x, y, t)$, $u_y(x, y, t)$, and $u_t(x, y, t)$ and give physical interpretations of your results.

Exercises 27 to 31: Compute the derivative of the function $\mathbf{F}$ at the point $\mathbf{a}$.

27. $\mathbf{F}(x, y) = (y, x, 11)$, $\mathbf{a} = (0, 0)$

28. $\mathbf{F}(x, y) = (e^{xy}, x^2 + y^2)$, $\mathbf{a} = (a_1, a_2)$

29. $\mathbf{F}(x, y, z) = (\ln(x^2 + y^2 + z^2), 2xy + z)$, $\mathbf{a} = (1, 1, 0)$

30. $\mathbf{F}(x, y) = (x/\sqrt{x^2 + y^2}, y/\sqrt{x^2 + y^2})$, $\mathbf{a} = (a_1, a_2) \neq (0, 0)$

31. $f(x, y, z) = ||x\mathbf{i} + y\mathbf{j} + z\mathbf{k}||^2$, $\mathbf{a} = (a_1, a_2, a_3)$

32. Compute $\nabla f(2, 1, -1)$ if $f(x, y, z) = xy \ln(z^2 + xy)$.

33. The electrostatic force field $\mathbf{F}(\mathbf{r})$ and the electrostatic potential $V(\mathbf{r})$ were defined in Example 2.11. Show that $\mathbf{F}(\mathbf{r}) = -\nabla V(\mathbf{r})$. Compare with Example 2.40.

34. Let $f(x, y, z) = xyz(x^2 + y^2 + z^2)^{-2}$. Compute $\nabla f(x, y, z)$ for $(x, y, z) \neq (0, 0, 0)$.

35. Define $f: \mathbb{R}^3 \to \mathbb{R}$ by $f(\mathbf{x}) = ||\mathbf{x}||$. Find $\nabla f(\mathbf{x})$ and state its domain.

Exercises 36 to 42: Find the linear approximation of the function f at the point $\mathbf{a}$.

36. $f(x, y) = e^{-x^2-y^2}$, $\mathbf{a} = (0, 0)$

37. $f(x, y) = \ln(3x + 2y)$, $\mathbf{a} = (2, -1)$

38. $f(x, y) = xy(x^2 + y^2)^{-1}$, $\mathbf{a} = (0, 1)$

39. $f(x, y) = x^2 - xy + y^2/2 + 3$, $\mathbf{a} = (3, 2)$

40. $f(x, y, z) = \ln(x^2 - y^2 + z)$, $\mathbf{a} = (3, 3, 1)$

41. $f(x, y, z) = \sqrt{x^2 + y^2 + z^2}$, $\mathbf{a} = (0, 1, 1)$

42. $f(x, y) = \int_x^y e^{-t^2} dt$, $\mathbf{a} = (1, 1)$

43. Verify that $xy(x + y)^{-1} \approx \frac{6}{5} + \frac{9}{25}(x - 2) + \frac{4}{25}(y - 3)$, for (x, y) sufficiently close to $(2, 3)$.

44. Prove that $\ln(2x^2 + 3y - 4) \approx 4x + 3y - 7$, for (x, y) sufficiently close to $(1, 1)$.

45. Assume that $f(x, y)$ is differentiable at (a, b) and let $\overline{L}(x, y) = f(a, b) + m(x - a) + n(y - b)$ be a linear function that satisfies (2.17), that is,

$$\lim_{(x,y)\to(a,b)} \frac{\left| f(x, y) - \overline{L}(x, y) \right|}{\sqrt{(x - a)^2 + (y - b)^2}} = 0.$$

(a) Substitute $y = b$ into the above formula to show that $m = (\partial f/\partial x)(a, b)$.

(b) Prove that $n = (\partial f/\partial y)(a, b)$ and conclude that $\overline{L}$ must be equal to the linear approximation $L_{(a,b)}$.

46. Consider the function $f(x, y) = \sqrt{x^2 + y^2}$ (see Example 2.44) and assume that it has a linear approximation $L_{(0,0)}(x, y)$ at $(0, 0)$.

(a) Explain why $L_{(0,0)}(x, y) = mx + ny$ for some real numbers m and n.

(b) Use (2.17) to show that f is differentiable at the origin if and only if

$$\lim_{(x,y)\to(0,0)} \left(1 - \frac{mx + ny}{\sqrt{x^2 + y^2}} \right) = 0.$$

(c) Use the approach $x \to 0$ and $y = 0$ to show that the above limit is not equal to 0. Conclude that f is not differentiable at the origin.

Exercises 47 to 51: Approximate the value of the given expression and compare it (except in Exercise 51) with the calculator value.

47. $\sqrt{0.99^3 + 2.02^3}$

48. $-0.09\sqrt{4.11^3 - 14.98}$

49. $7.95 \ln 1.02$

50. $\sin(\pi/50)\cos(49\pi/50)$

51. $\int_{0.995}^{1.02} e^{-t^2} dt$

Exercises 52 to 55: Compare the values of Δf and df.

52. $f(x, y) = x^2 - xy + 2y^2 + 1$, $(a, b) = (0, 1)$, $\Delta x = 0.01$, $\Delta y = 0.2$

53. $f(x, y) = e^x - ye^y$, $(a, b) = (0, 1)$, $\Delta x = 0.3$, $\Delta y = 0.01$

54. $f(x, y) = x^3 + xy + y^3$, $(a, b) = (-2, 1)$, $(x, y) = (-2.05, 0.9)$

55. $f(x, y, z) = x^2y - xyz + z^3$, $(a, b, c) = (1, 2, -1)$, $\Delta x = -0.02$, $\Delta y = 0.01$, $\Delta z = 0.02$

56. Estimate the maximum possible error in computing $f(x, y) = x \cos y$, where $x = 2$ and $y = \pi/3$, with maximum possible errors $\Delta x = 0.2$ and $\Delta y = 0.1$.

57. The pressure in an ideal gas is given by $P(T, V) = RnT/V$; see Example 2.8. Compute the differential of P and explain the signs of the coefficients of dT and dV.

58. Consider the Cobb–Douglas function $P(L, K) = bL^\alpha K^{1-\alpha}$ discussed in Example 2.6. Compute the differential dP and explain the signs of the coefficients of dL and dK.

59. About how accurately can the volume of a cylinder be calculated from the measurements of its height and radius that are in error by 1.5%?

60. The dimensions of a closed rectangular box are measured as 20, 50, and 120 cm, respectively, with a possible error of 0.4 cm in each dimension. Estimate the maximum error in computing the volume and the surface area of the box.

61. The length and the width of a rectangle are measured with a possible error of 2% in length and 3% in width. Approximate the error in computing the area of the rectangle.

62. Let $f(x, y) = 2x^2y^3$. Estimate the change in the function f if x increases by 3% and y increases by 2%.

63. Find the equation of the tangent plane to the graph of the function $z = 6 - x^2 - y^2$ at the point $(1, 2, 1)$.

64. Find the equation of the tangent plane to the surface $z = 3xy/(x - 2y)$ at the point $(3, 1, 9)$. Check whether the tangent plane contains the origin.

65. Define the function $f\colon \mathbb{R}^2 \to \mathbb{R}$ by

$$f(x, y) = \begin{cases} y \ln (x^2 + y^2) & \text{if } (x, y) \neq (0, 0) \\ 0 & \text{if } (x, y) = (0, 0) \end{cases}.$$

Show that f_x is defined for all (x, y), but that f_x is not continuous at $(0, 0)$.

66. Define the function $f\colon \mathbb{R}^2 \to \mathbb{R}$ by

$$f(x, y) = \begin{cases} \dfrac{xy^2}{x^2 + y^4} & \text{if } (x, y) \neq (0, 0) \\ 0 & \text{if } (x, y) = (0, 0) \end{cases}.$$

(a) Is f continuous at $(0, 0)$?

(b) Compute the linear approximation (if it exists) at $(0, 0)$.

(c) Is f_x continuous at $(0, 0)$?

(d) Is f differentiable?

67. Show that the function $\mathbf{F}(x, y) = (x + y^2, 2xy)$ is differentiable at $(0, 0)$.

68. Show that the function $f(x, y) = (xy)^{1/5}$ is not differentiable at $(0, 0)$.

69. Consider the function

$$f(x, y) = \begin{cases} \ln(x^2 + y^2) & \text{if } (x, y) \neq (0, 0) \\ 0 & \text{if } (x, y) = (0, 0) \end{cases}.$$

(a) Is f differentiable at $(0, 0)$?

(b) Is it possible to conclude from (a) that f is continuous at $(0, 0)$?

(c) Is f continuous at $(0, 0)$?

▶ 2.5 PATHS AND CURVES IN $\mathbb{R}^2$ AND $\mathbb{R}^3$

A trajectory of a moving object, a sound wave, a current in an electric circuit, the conversion between degrees Fahrenheit and Celsius, or the dependence of air pressure on altitude can be visually represented as curves in a plane or in three-dimensional space. Various measurement instruments such as oscilloscopes, heart-beat monitors, computers, and other devices display their data in the form of curves, which are more convenient and easier to interpret than a listing of thousands of numbers. The graph of a real-valued function $y = f(x)$ is a curve. The equation $f(x, y) = 0$ represents a curve described in a slightly different way (it is given "implicitly").

Continuing with vector-valued functions, we now introduce a new way of defining a curve and study its properties (that we will find useful in subsequent sections). We will resume our investigation of curves in Chapter 3. Concepts relevant to integration along curves are discussed in Chapter 5.

We are going to restrict our study to $\mathbb{R}^2$ and $\mathbb{R}^3$, although all statements (except those involving cross products) hold in any dimension.

DEFINITION 2.16 Path and Curve

A *path* in $\mathbb{R}^3$ (or $\mathbb{R}^2$) is a function $\mathbf{c}$: $[a, b] \to \mathbb{R}^3$ (or $\mathbb{R}^2$), whose domain is a subset $[a, b] \subseteq \mathbb{R}$. The image of $\mathbf{c}$ is called a *curve* in $\mathbb{R}^3$ (or $\mathbb{R}^2$). The function $\mathbf{c}$ is also known as a *parametrization* (or *parametric representation* or *parametric equation*) of the curve. ◀

According to the definition, a path is a function, whereas a geometric object in $\mathbb{R}^3$ (or $\mathbb{R}^2$) that is the image of that function is called a curve. In other words, a path or parametrization (the two are synonyms) represents an analytic way of describing a curve. We will soon witness that a single curve can have infinitely many parametrizations, not all of them characterized by the same properties. The reasons why the distinction between a path and a curve is needed will surface in Chapter 3 and in sections on integration along paths.

On a few occasions we will use the term "curve" to refer to both notions since the context will keep the meaning clear. For example, if we talk about the composition of curves or velocity, we think of a function; on the other hand, the statement "curves are orthogonal to each other" refers to a curve as a geometric object. Likewise, we will use the same notation for both the path and the corresponding curve.

Sometimes, it is useful to extend the domain $[a, b]$ in the definition of a path $\mathbf{c}$ so that $a = -\infty$ or $b = \infty$, or both (i.e., intervals $(-\infty, b]$, $[a, \infty)$ or $(-\infty, \infty) = \mathbb{R}$ are allowed as the domain of $\mathbf{c}$). This will enable us to describe, for example, lines (such as the tangent to a curve) as paths in a plane or in space. The variable of $\mathbf{c}$ is denoted by t and is often referred to as time. In components, we can represent a curve in $\mathbb{R}^2$ as

$$\mathbf{c}(t) = (x(t), y(t)), \qquad t \in [a, b],$$

and a curve in $\mathbb{R}^3$ as

$$\mathbf{c}(t) = (x(t), y(t), z(t)), \qquad t \in [a, b],$$

where $x(t)$, $y(t)$, and $z(t)$ are real-valued functions of t.

▶ EXAMPLE 2.48

The curve $\mathbf{c}_1$ parametrized by $\mathbf{c}_1(t) = (t\cos t, t\sin t)$, $t \in [0, 3\pi]$ in $\mathbb{R}^2$ has been drawn in Figure 2.52. Figure 2.53 shows the plot of the curve $\mathbf{c}_2$ in space given by $\mathbf{c}_2(t) = (\cos t, \sin t, \cos 4t)$, $t \in [0, 2\pi]$.

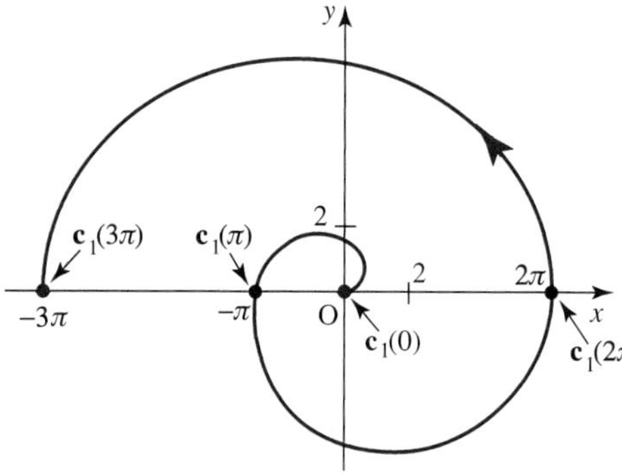

Figure 2.52 Curve $\mathbf{c}_1$ parametrized by $\mathbf{c}_1(t) = (t\cos t, t\sin t)$, $t \in [0, 3\pi]$.

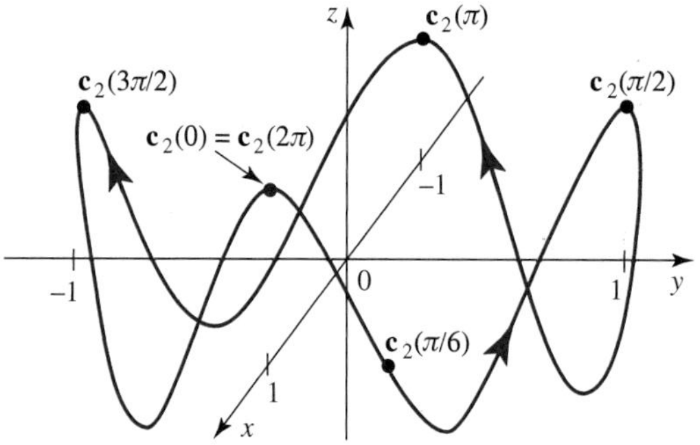

Figure 2.53 Curve $\mathbf{c}_2$ parametrized by $\mathbf{c}_2(t) = (\cos t, \sin t, \cos 4t)$, $t \in [0, 2\pi]$. ◀

A parametric representation of a curve gives a sense of orientation, as explained in the following definition.

DEFINITION 2.17 Orientation

Let $\mathbf{c}(t)$: $[a, b] \to \mathbb{R}^3$ (or $\mathbb{R}^2$) be a path. The point $\mathbf{c}(a)$ is called the *initial point*, and we call $\mathbf{c}(b)$ the *terminal point* of $\mathbf{c}$. The initial and the terminal points are called the *endpoints* of $\mathbf{c}$. The direction corresponding to increasing values of t gives the *positive orientation*, whereas the opposite direction defines the *negative orientation* of $\mathbf{c}$. ◀

If the domain of $\mathbf{c}$ includes $-\infty$ or ∞ (or both), one (or both) endpoints are not defined. The orientation is indicated in the graph by an arrow; see Figures 2.52 and 2.53. According to the definition, the positive orientation is the direction from the initial point toward the terminal point (if defined).

▶ EXAMPLE 2.49

For the path $\mathbf{c}_1(t)$ in Figure 2.52, the initial point is $\mathbf{c}(0) = (0, 0)$, and the terminal point is $\mathbf{c}(3\pi) = (-3\pi, 0)$. The arrows indicate the positive orientation (that can also be described as the counterclockwise orientation). The path $\mathbf{c}_2(t)$ of Figure 2.53 has the same point $\mathbf{c}(0) = \mathbf{c}(2\pi) = (1, 0, 1)$ as its initial and terminal points. To determine the orientation, compute the values of $\mathbf{c}$ at increasing values of t; for example, $\mathbf{c}(\pi/6) = (\sqrt{3}/2, 1/2, -1/2)$, $\mathbf{c}(\pi/4) = (\sqrt{2}/2, \sqrt{2}/2, -1)$, etc. The positive orientation is given by the direction from $(1, 0, 1)$ to $(\sqrt{3}/2, 1/2, -1/2)$, then to $(\sqrt{2}/2, \sqrt{2}/2, -1)$, etc., as indicated in the graph. ◀

▶ EXAMPLE 2.50 Parametric Representation of a Line and a Line Segment

A parametric representation of the line segment joining the points $A = (a_1, a_2, a_3)$ and $B = (b_1, b_2, b_3)$ in $\mathbb{R}^3$ is given by $\mathbf{c}(t) = \mathbf{a} + t\mathbf{v}$, $t \in [0, 1]$, where $\mathbf{a} = (a_1, a_2, a_3)$ and $\mathbf{v} = (b_1 - a_1, b_2 - a_2, b_3 - a_3)$. This parametrization was discussed at the beginning of Section 1.2. The initial point is $A = \mathbf{c}(0)$ and the terminal point is $B = \mathbf{c}(1)$. In coordinates,

$$\mathbf{c}(t) = (a_1 + t(b_1 - a_1), a_2 + t(b_2 - a_2), a_3 + t(b_3 - a_3)), \qquad t \in [0, 1].$$

A line going through $A = (a_1, a_2, a_3)$ in the direction $\mathbf{v} = (v_1, v_2, v_3)$ is represented as

$$\mathbf{c}(t) = \mathbf{a} + t\mathbf{v} = (a_1 + tv_1, a_2 + tv_2, a_3 + tv_3), \qquad t \in \mathbb{R}.$$

As "time" t increases (positive orientation), the point $\mathbf{c}(t)$ moves along the line, away from A, in the direction of $\mathbf{v}$. The direction of movement corresponding to decreasing time (negative orientation) corresponds to movement in the direction of $-\mathbf{v}$. ◀

▶ EXAMPLE 2.51 Parametrization of a Circle and an Ellipse

The curve represented parametrically as

$$\mathbf{c}(t) = (a\cos t, a\sin t), \qquad t \in [0, 2\pi]$$

(where $a > 0$) is the circle in $\mathbb{R}^2$ of radius a centered at the origin ($x(t) = a\cos t$, $y(t) = a\sin t$, and hence $x(t)^2 + y(t)^2 = a^2$); see Figure 2.54. The parameter t represents the angle between the x-axis and the position vector $\mathbf{c}(t)$. The initial point is $\mathbf{c}(0) = (a, 0)$ and the terminal point is $\mathbf{c}(2\pi) = (a, 0) = \mathbf{c}(0)$. Thinking of $\mathbf{c}$ as a trajectory of a moving object, we see that it takes the object 2π units of time to complete one full revolution and come back to its initial position. The positive orientation (i.e., direction of increasing t) corresponds to counterclockwise motion along the circle. The path

$$\mathbf{c}(t) = (o_1 + a\cos t, o_2 + a\sin t) = (o_1, o_2) + (a\cos t, a\sin t), \qquad t \in [0, 2\pi]$$

represents the circle centered at $O = (o_1, o_2)$ of radius a. The ellipse $x^2/a^2 + y^2/b^2 = 1$ (with semi-axes $a, b > 0$) can be parametrized as $\mathbf{c}(t) = (a\cos t, b\sin t)$, $t \in [0, 2\pi]$. ◀

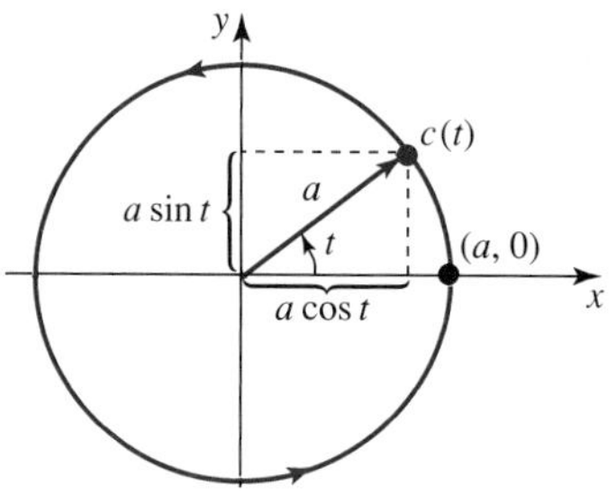

Figure 2.54 Circle $\mathbf{c}(t) = (a\cos t, a\sin t)$, $t \in [0, 2\pi]$.

▶ EXAMPLE 2.52 Parametrization of the Graph of $y = f(x)$

The graph of a real-valued function $f\colon [a,b] \to \mathbb{R}$ of one variable defined on an interval $[a,b] \subseteq \mathbb{R}$ is a curve that can be parametrized as $\mathbf{c}(t) = (t, f(t))$, $t \in [a,b]$.

For example, a parametric representation of the graph of $y = x^2$ on $[0,2]$ is given by $\mathbf{c}(t) = (t, t^2)$, $t \in [0,2]$. Similarly, $\mathbf{c}(t) = (t, te^{-t})$, $t \in [0,\infty)$, represents the graph of $y = xe^{-x}$ for $x \geq 0$. ◀

Let us for a moment go back to the previous example and compare the two ways of describing the parabola in question. Although both descriptions $y = x^2$, $x \in [0,2]$, and $\mathbf{c}(t) = (t, t^2)$, $t \in [0,2]$, do produce (geometrically) the same curve, there are differences. To understand them better, suppose that the curve $\mathbf{c}$ represents the motion of an object. Parametric representation conveys a lot more information than just the geometric curve: for example, the initial point of the motion is $\mathbf{c}(0) = (0,0)$, and the terminal point is $\mathbf{c}(2) = (2,4)$. Since $(t_1, t_1^2) \neq (t_2, t_2^2)$ for $0 \leq t_1, t_2 \leq 2$ and $t_1 \neq t_2$, the object moves along the parabola *from* $(0,0)$ *to* $(2,4)$, keeping the same direction all the time (i.e., it does not move back and forth; compare with Example 2.54).

From $\mathbf{c}(t) = (t, t^2)$ we can read off the location of the object at *any* time t, $0 \leq t \leq 2$. On the other hand, the graph of $y = x^2$ produces the trajectory of the object without showing any details of the motion. Later in this section, and also in Chapter 3, we will learn how to extract a lot more information from a parametric representation. For example, we will be able to measure the curvature, or determine the acceleration of a motion.

It is important to notice that the values of the parameter t are not built into the graph. In other words, if we select a point on the curve, we cannot read off the value of t that produced it. To somewhat remedy this deficiency, besides plotting a point on the curve, we also indicate the corresponding value of t, as shown in Figure 2.55.

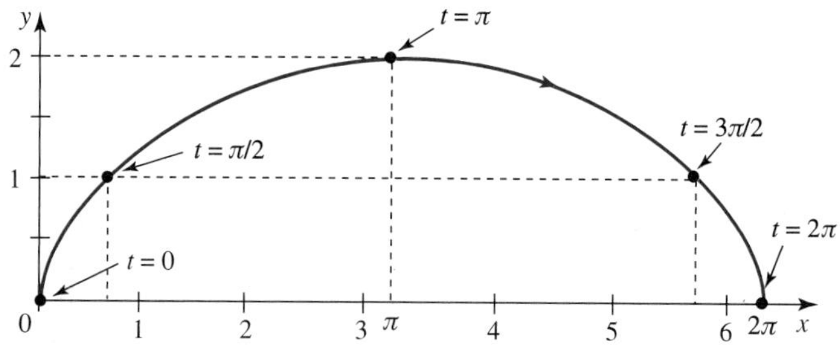

Figure 2.55 Graph of the path $\mathbf{c}(t) = (t - \sin t, 1 - \cos t)$, $t \in [0, 2\pi]$.

▶ EXAMPLE 2.53

Sketch the graph of the function $\mathbf{c}$: $[0, 3] \to \mathbb{R}^2$ given by $\mathbf{c}(t) = (t^2, 1 - t)$.

SOLUTION We interpret the values of $\mathbf{c}(t)$ as the coordinates $x(t) = t^2$ and $y(t) = 1 - t$ of a point in $\mathbb{R}^2$. If $t = 0$, then $x(0) = 0$ and $y(0) = 1$. For $t = 1$, we get $x(1) = 1$ and $y(1) = 0$, and similarly, $x(2) = 4$, $y(2) = -1$ and $x(3) = 9$, $y(3) = -2$. Continuing this process and connecting all points thus obtained produce a curve that is the graph of $\mathbf{c}$, see Figure 2.56 (alternative graphing techniques will be discussed later).

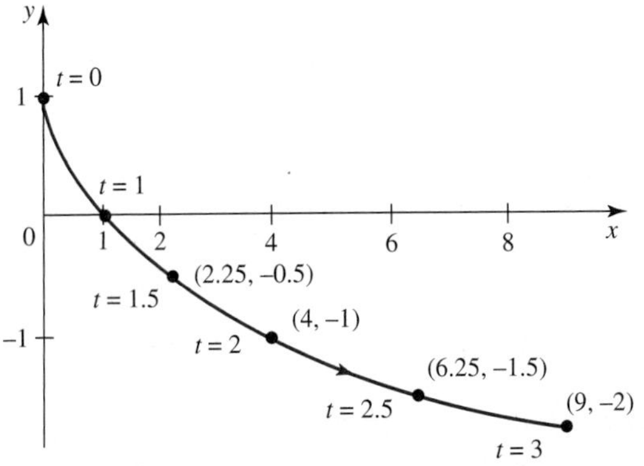

Figure 2.56 Graph of the path $\mathbf{c}(t) = (t^2, 1 - t)$ for $t \in [0, 3]$.

▶ EXAMPLE 2.54

Consider motions defined by the paths $\mathbf{c}_1(t) = (t, t^2)$, $t \in [-1, 1]$, and $\mathbf{c}_2(t) = (\sin t, (\sin t)^2)$, $t \in [-\pi/2, 5\pi/2]$. Note that both represent the graph of the function $y = x^2$, $-1 \le x \le 1$.

From $\mathbf{c}_1(-1) = \mathbf{c}_2(-\pi/2) = (-1, 1)$ and $\mathbf{c}_1(1) = \mathbf{c}_2(5\pi/2) = (1, 1)$, we conclude that both paths have the same initial and terminal points. Since $\mathbf{c}_1(t_1) \neq \mathbf{c}_1(t_2)$ for $t_1 \neq t_2$, it follows that $\mathbf{c}_1$ describes the motion along the graph of $y = x^2$ from $(-1, 1)$ to $(1, 1)$ without retracing any parts of it (i.e., $\mathbf{c}_1$ keeps the same direction all the time).

The motion of an object given by $\mathbf{c}_2$ is different. As t changes from $-\pi/2$ to $\pi/2$, $\sin t$ changes from -1 to 1 (and is one-to-one). Thus, the object moves from $(-1, 1)$ to $(1, 1)$. For $\pi/2 \le t \le 3\pi/2$, $\sin t$ decreases from 1 to -1, and so the object moves back along $y = x^2$ from $(1, 1)$ to $(-1, 1)$. Finally, when $3\pi/2 \le t \le 5\pi/2$, $\sin t$ increases from -1 to 1, and so the object moves back to $(1, 1)$.

▶ EXAMPLE 2.55

In Example 2.51 we showed that the path $\mathbf{c}(t) = (\cos t, \sin t)$, $t \in [0, 2\pi]$ represents the circle $x^2 + y^2 = 1$. There are other parametrizations: for instance, $\mathbf{c}_1(t) = (\cos 2t, \sin 2t)$, $t \in [0, \pi]$ satisfies $x(t)^2 + y(t)^2 = 1$ and, since $\mathbf{c}_1(0) = \mathbf{c}_1(\pi) = (1, 0)$, it describes the whole circle. Similarly, we can check that $\mathbf{c}_2(t) = (\sin (t + 3), \cos (t + 3))$, $t \in [0, 2\pi]$, or $\mathbf{c}_3(t) = (-\cos (t/4), \sin (t/4))$, $t \in [0, 8\pi]$, represent the same circle.

As a matter of fact, a curve has infinitely many parametrizations (in terms of Definition 2.16, we say that there are infinitely many *paths* that have the same image; i.e., parametrize the same curve). That is the reason why we made a distinction between a path and a curve. It is worth repeating that a curve is a geometric object and a path is a way of describing it algebraically in terms of a parameter. Parametrizations need not look alike: for example,

$$\mathbf{c}_4(t) = ((4 \cos t + \sin t)/\sqrt{17}, (4 \sin t - \cos t)/\sqrt{17}), \qquad t \in [0, 2\pi]$$

is another representation of the circle $x^2 + y^2 = 1$.

▶ EXAMPLE 2.56

Let $\mathbf{c}$ be the part of the curve $y = 2x^4$ between $(-1, 2)$ and $(1, 2)$. Write down several parametrizations of $\mathbf{c}$.

SOLUTION

As in Example 2.52, we can take $x = t$. Then $y = 2x^4 = 2t^4$, and we obtain the parametrization $\mathbf{c}_1(t) = (t, 2t^4)$, $t \in [-1, 1]$. There is no reason why we have to choose $x = t$. Try $x = mt$, $(m \neq 0)$; then $y = 2x^4 = 2m^4t^4$ and $\mathbf{c}_2(t) = (mt, 2m^4t^4)$, $t \in [-1/|m|, 1/|m|]$. (In defining an interval $[a, b]$, we have to make sure that $a \le b$; that is why we used the absolute value.) We already have infinitely many parametrizations, one for each nonzero value of m.

Let us list a few more parametrizations [of course, in every case $x(t)$ and $y(t)$ have to satisfy $y(t) = 2x(t)^4$, and the endpoints of the interval for the parameter must give $(-1, 2)$ and $(1, 2)$]: $\mathbf{c}_3(t) = (mt + 1, 2(mt + 1)^4)$, $t \in [-2/m, 0]$ (works for $m > 0$), then $\mathbf{c}_4(t) = (t^{1/3}, 2t^{4/3})$, $t \in [-1, 1]$, or $\mathbf{c}_5(t) = (\tan t, 2\tan^4 t)$, $t \in [-\pi/4, \pi/4]$, etc. On the other hand, $\mathbf{c}_6(t) = (t^2, 2t^8)$, $t \in [-1, 1]$, parametrizes the part of the parabola in the first quadrant only: for example, no value of t gives $(-1, 2)$. ◀

In Chapter 3, and in subsequent chapters, we will learn that there are significant differences between parametrizations. Not every parametrization of a curve can be used to compute its length. Some parametrizations will be more suitable as trajectories of the motion than others. A path integral will be defined for a special class of parametrizations, etc.

▶ EXAMPLE 2.57

The parametrizations (paths)

(a) $\mathbf{c}_1(t) = (2\cos t, 2\sin t)$, $t \in [0, \pi]$,

(b) $\mathbf{c}_2(t) = (-2\cos t, 2\sin t)$, $t \in [0, \pi]$,

(c) $\mathbf{c}_3(t) = (2\cos(3t), 2\sin(3t))$, $t \in [0, \pi/3]$,

(d) $\mathbf{c}_4(t) = (-2\cos(t/4), 2\sin(t/4))$, $t \in [0, 4\pi]$,

represent the same curve. Identify the curve and describe the differences between the parametrizations.

SOLUTION

In all four cases, $x(t)^2 + y(t)^2 = 4$ and $y(t) \ge 0$. Next, we compute the endpoints for all paths: $\mathbf{c}_1(0) = (2, 0)$, $\mathbf{c}_1(\pi) = (-2, 0)$, $\mathbf{c}_2(0) = (-2, 0)$, $\mathbf{c}_2(\pi) = (2, 0)$, $\mathbf{c}_3(0) = (2, 0)$, $\mathbf{c}_3(\pi/3) = (-2, 0)$, $\mathbf{c}_4(0) = (-2, 0)$, and $\mathbf{c}_4(4\pi) = (2, 0)$. Consequently, the curve in question is the semicircle of radius 2 (centered at the origin) in the upper half-plane with the endpoints $(2, 0)$ and $(-2, 0)$. Paths $\mathbf{c}_1$ and $\mathbf{c}_3$ are oriented counterclockwise, whereas $\mathbf{c}_2$ and $\mathbf{c}_4$ [having initial points at $(-2, 0)$ and terminal points at $(2, 0)$] are oriented clockwise. Now view t as time and interpret the interval for t as the total time needed to complete the motion along the curve. The motion along $\mathbf{c}_3$ is the fastest, and along $\mathbf{c}_4$ the slowest. Motions along $\mathbf{c}_1$ and $\mathbf{c}_2$ are completed in π units of time. ◀

▶ EXAMPLE 2.58

Sketch the curve $\mathbf{c}(t) = (3\cos t, 3\sin t, t)$, $t \in [0, 2\pi]$.

SOLUTION

Since $x = 3\cos t$ and $y = 3\sin t$, it follows that $x^2 + y^2 = 9$, which means that the curve lies on the surface of the cylinder of radius 3 whose axis is the z-axis. Its projection onto the xy-plane (take $z = 0$) is the circle of radius 3 (centered at the origin) oriented counterclockwise. As time t increases, z-coordinates of points on c increase from 0 to 2π. The initial point is $\mathbf{c}(0) = (3, 0, 0)$ and the terminal

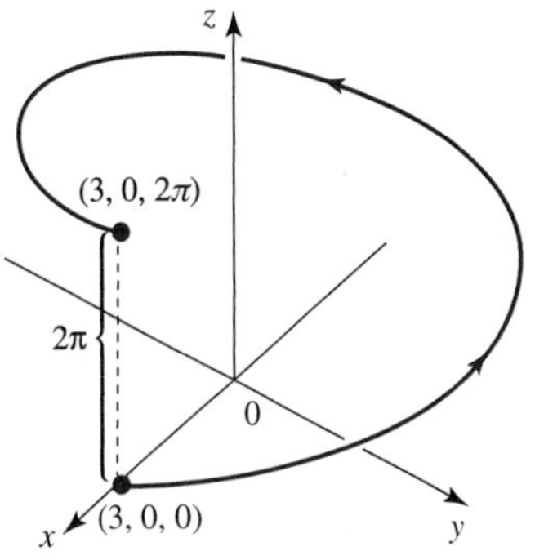

Figure 2.57 The graph of $\mathbf{c}(t) = (3\cos t, 3\sin t, t)$, $t \in [0, 2\pi]$ is a helix of "pitch" 2π.

point is $(3, 0, 2\pi)$. The curve is obtained in the following way: as we move the point along the circle of radius 3 counterclockwise, we simultaneously increase its height (at a constant rate) from 0 to 2π. The curve thus obtained is called a helix; see Figure 2.57. ◀

DEFINITION 2.18 Continuous, Differentiable, and C^1 Paths and Curves

A path (or a parametrization) $\mathbf{c}$: $[a, b] \to \mathbb{R}^2(\mathbb{R}^3)$ is *continuous* if and only if its component functions $x(t)$ and $y(t)$ (or $x(t)$, $y(t)$ and $z(t)$) are continuous on $[a, b]$. If the component functions of $\mathbf{c}(t)$ are differentiable (respectively C^1), then $\mathbf{c}(t)$ is called a *differentiable* (respectively C^1) path.

A curve is called *continuous* (*differentiable*, C^1) if among all of its parametrizations there is at least one that is continuous (differentiable, C^1). ◀

Let us clarify the meaning of the statements "$f(t)$ is continuous on $[a, b]$" and "$f(t)$ is differentiable on $[a, b]$." Continuity and differentiability are defined in terms of a limit: a function $f(t)$ is continuous at $t_0 \in (a, b)$ if and only if $\lim_{t\to t_0} f(t)$ exists and equals $f(t_0)$; it is differentiable at $t_0 \in (a, b)$ if and only if $\lim_{h\to 0}(f(t_0 + h) - f(t_0))/h$ exists. These definitions apply to any t_0 that lies inside the interval (a, b). To define continuity and differentiability at the endpoints a and b, all we have to do is to replace the (two-sided) limits with the appropriate one-sided limits in such a way that the endpoints are always approached from within the interval. For example, the function $f(t)$ is continuous at $t = b$ if and only if $\lim_{t\to b^-} f(t)$ exists and equals $f(b)$; it is differentiable at $t = a$ if and only if $\lim_{h\to 0^+}(f(a + h) - f(a))/h$ exists.

Recall that a real-valued function of one variable is called C^1 if its derivative is continuous. All curves that have appeared in this section are continuous and differentiable. When we state that "a curve $\mathbf{c}$ is differentiable," we mean to say that some parametric representation of that curve (usually also denoted by $\mathbf{c}$) is differentiable.

The parametrization $\mathbf{c}_4(t) = (t^{1/3}, 2t^{4/3})$, $t \in [-1, 1]$ of Example 2.56 is not differentiable at 0 [the derivative of $x(t)$ is $x'(t) = t^{-2/3}/3$, and hence not defined at 0]. Nevertheless, the parametrizations $\mathbf{c}_1$ (and $\mathbf{c}_2$, $\mathbf{c}_3$, and $\mathbf{c}_5$) are differentiable and C^1, and hence the curve that is the graph of $y = x^4$ on $[-1, 1]$ is differentiable and C^1. The graph of $y = |x|$, $x \in [-1, 1]$ is an example of a continuous, nondifferentiable curve: $|x|$ has a "corner" at

$x = 0$, and consequently, does not have a tangent (i.e., the derivative) there. As a matter of fact, to prove nondifferentiability, we would have to show that no parametrization of $|x|$ is differentiable. We prefer to rely on our intuitive reasoning at this moment.

Let us mention an issue related to notation. Consider the parametrizations $\mathbf{c}_1(t) = (t, 2t^4)$, $t \in [-1, 1]$ and $\mathbf{c}_5(t) = (\tan t, 2\tan^4 t)$, $t \in [-\pi/4, \pi/4]$ of Example 2.56. Strictly speaking, we should have used different symbols for the parameters, since t in $\mathbf{c}_1$ and $\mathbf{c}_5$ is not the same. For example, $\mathbf{c}_1(\pi/4) = (\pi/4, \pi^4/128)$, but $\mathbf{c}_5(\pi/4) = (1, 2)$. However, as t in $\mathbf{c}_1$ changes from -1 to 1, $\mathbf{c}_1$ describes the same curve as $\mathbf{c}_5$ (when its t changes from $-\pi/4$ to $\pi/4$). Beware of this common practice so that it will not become a source of confusion.

Let $\mathbf{c}(t) = (x(t), y(t), z(t))$ be a differentiable path in $\mathbb{R}^3$. The derivative $\mathbf{c}'(t_0) = D\mathbf{c}(t_0)$ of $\mathbf{c}$ at t_0 is the 3×1 matrix

$$\begin{bmatrix} dx/dt \\ dy/dt \\ dz/dt \end{bmatrix}_{\text{at } t=t_0} = \begin{bmatrix} x'(t_0) \\ y'(t_0) \\ z'(t_0) \end{bmatrix}$$

that can be interpreted as the vector $\mathbf{c}'(t_0) = x'(t_0)\mathbf{i} + y'(t_0)\mathbf{j} + z'(t_0)\mathbf{k}$ in $\mathbb{R}^3$. We visualize $\mathbf{c}'(t_0)$ as a vector whose initial point is located at $\mathbf{c}(t_0)$, as shown in Figure 2.58. To further explore this geometric interpretation, we rewrite $\mathbf{c}'(t_0)$ in the limit form:

$$\begin{aligned} \mathbf{c}'(t_0) &= x'(t_0)\mathbf{i} + y'(t_0)\mathbf{j} + z'(t_0)\mathbf{k} \\ &= \lim_{h\to 0} \frac{x(t_0+h) - x(t_0)}{h}\mathbf{i} + \lim_{h\to 0} \frac{y(t_0+h) - y(t_0)}{h}\mathbf{j} + \lim_{h\to 0} \frac{z(t_0+h) - z(t_0)}{h}\mathbf{k} \\ &= \lim_{h\to 0} \frac{(x(t_0+h)\mathbf{i} + y(t_0+h)\mathbf{j} + z(t_0+h)\mathbf{k}) - (x(t_0)\mathbf{i} + y(t_0)\mathbf{j} + z(t_0)\mathbf{k})}{h} \\ &= \lim_{h\to 0} \frac{\mathbf{c}(t_0+h) - \mathbf{c}(t_0)}{h}. \end{aligned}$$

The vector $(\mathbf{c}(t_0 + h) - \mathbf{c}(t_0))/h$, being parallel to $\mathbf{c}(t_0 + h) - \mathbf{c}(t_0)$, falls in the direction of the secant line joining $\mathbf{c}(t_0 + h)$ and $\mathbf{c}(t_0)$; see Figure 2.58. As $h \to 0$, the point $\mathbf{c}(t_0 + h)$ slides along the curve toward $\mathbf{c}(t_0)$ and the secant line approaches its limit position, the tangent line at $\mathbf{c}(t_0)$. Hence, $\mathbf{c}'(t_0)$ [if $\mathbf{c}'(t_0) \neq \mathbf{0}$] represents the direction of the line tangent to $\mathbf{c}$ at $\mathbf{c}(t_0)$.

A parametric equation of the tangent line (recall that "line = point plus parameter times vector") is given by $\mathbf{l}(t) = \mathbf{c}(t_0) + t\mathbf{c}'(t_0)$, $t \in \mathbb{R}$. This argument justifies the terminology introduced in our next definition.

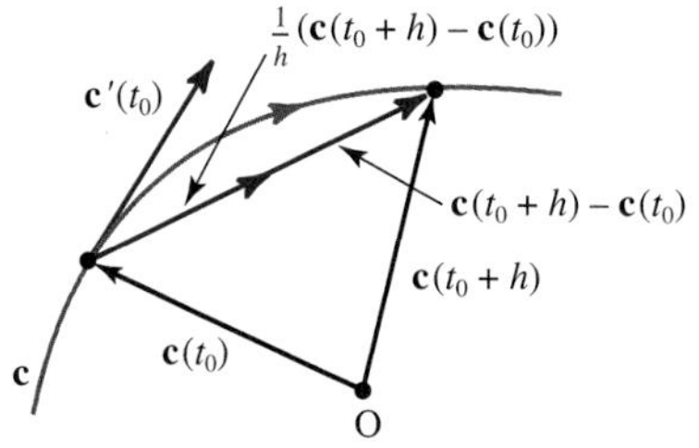

Figure 2.58 $\mathbf{c}'(t_0)$ is in the direction of the tangent line to $\mathbf{c}$ at a point $\mathbf{c}(t_0)$.

DEFINITION 2.19 Tangent Vector and Tangent Line

Let $\mathbf{c}$ be a differentiable path in $\mathbb{R}^2$ or $\mathbb{R}^3$. The vector $\mathbf{c}'(t_0)$ is called a *tangent vector* to $\mathbf{c}$ at $\mathbf{c}(t_0)$. The line tangent to a curve $\mathbf{c}$ [that is represented by a path $\mathbf{c}(t)$] at $\mathbf{c}(t_0)$ is given by $\mathbf{l}(t) = \mathbf{c}(t_0) + t\mathbf{c}'(t_0)$, $t \in \mathbb{R}$, provided that $\mathbf{c}'(t_0) \neq \mathbf{0}$. ◀

Now suppose that a path $\mathbf{c}(t)$ describes the trajectory of a moving object (and t represents time). Since $\mathbf{c}(t_0 + h) - \mathbf{c}(t_0) =$ (position at time $t_0 + h$) $-$ (position at time t_0), that is,

$$\frac{\mathbf{c}(t_0 + h) - \mathbf{c}(t_0)}{h} = \frac{\text{displacement vector}}{\text{time}},$$

the limit (as time h approaches 0) gives the *instantaneous velocity vector*.

DEFINITION 2.20 Velocity, Speed, and Acceleration

Let $\mathbf{c}(t) = (x(t), y(t), z(t))$ be a differentiable path in $\mathbb{R}^3$. The *velocity* $\mathbf{v}(t)$ at time t is given by the vector-valued function

$$\mathbf{v}(t) = \mathbf{c}'(t) = (x'(t), y'(t), z'(t)).$$

The *speed* is the real-valued function

$$||\mathbf{v}(t)|| = \sqrt{(x'(t))^2 + (y'(t))^2 + (z'(t))^2},$$

which is the length of the velocity vector. The *acceleration* $\mathbf{a}(t)$ at time t is given by

$$\mathbf{a}(t) = \mathbf{v}'(t) = \mathbf{c}''(t) = (x''(t), y''(t), z''(t)),$$

provided that $\mathbf{c}$ is twice differentiable. ◀

Usually, we visualize velocity and acceleration as vectors whose tails are located at the point $\mathbf{c}(t)$ on the curve. Example 2.60 will serve as an illustration.

▶ EXAMPLE 2.59

Using the notion of speed, we can now confirm our somewhat intuitive reasoning in Example 2.57. Since $||\mathbf{c}_1'(t)|| = ||\mathbf{c}_2'(t)|| = 2$, $||\mathbf{c}_3'(t)|| = 6$ and $||\mathbf{c}_4'(t)|| = 1/2$, it follows (since all parametrizations have constant speed) that the parametrization $\mathbf{c}_3$ is the fastest, $\mathbf{c}_4$ is the slowest, and $\mathbf{c}_1$ and $\mathbf{c}_2$ have the same speed. ◀

This example shows how speed can be used as a way of describing the differences between parametrizations of the same curve.

▶ EXAMPLE 2.60

Assume that the function $\mathbf{c}(t) = (t \sin t, t \cos t, t)$, $1 \leq t \leq 2$, represents the motion of an object in $\mathbb{R}^3$. The matrix

$$D\mathbf{c}(t) = \mathbf{c}'(t) = \begin{bmatrix} \sin t + t \cos t \\ \cos t - t \sin t \\ 1 \end{bmatrix}$$

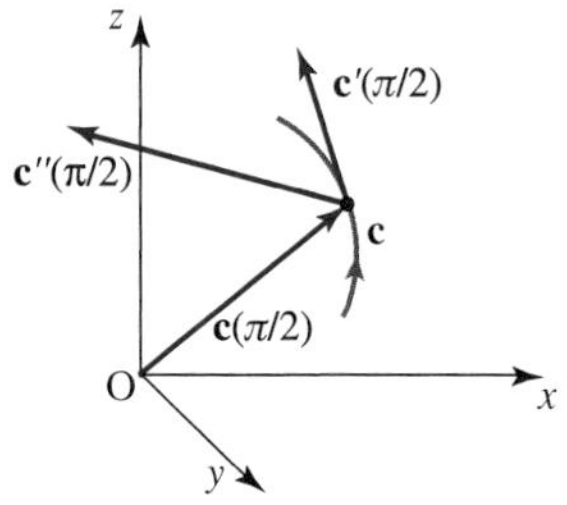

Figure 2.59 Position, velocity, and acceleration vectors of Example 2.60.

(thought of as a vector) gives the velocity of the object. For example, when $t = \pi/2$, the object is located at the point $\mathbf{c}(\pi/2) = (\pi/2, 0, \pi/2)$. Its velocity vector at that moment is computed to be

$$D\mathbf{c}(\pi/2) = \mathbf{c}'(\pi/2) = \begin{bmatrix} 1 \\ -\pi/2 \\ 1 \end{bmatrix},$$

and the acceleration is

$$\mathbf{c}''(\pi/2) = \begin{bmatrix} 2\cos t - \sin t \\ -2\sin t - t\cos t \\ 0 \end{bmatrix}_{at\ t=\pi/2} = \begin{bmatrix} -\pi/2 \\ -2 \\ 0 \end{bmatrix}.$$

We visualize the velocity $\mathbf{c}'(\pi/2)$ and the acceleration $\mathbf{c}''(\pi/2)$ as vectors whose tails are located at the point $\mathbf{c}(\pi/2)$ on the curve; see Figure 2.59. ◀

We will continue our study of tangents, velocity, and acceleration in Chapter 3.

▶ EXERCISES 2.5

Exercises 1 to 9: Find a parametric representation of the given curve:

1. The line segment in $\mathbb{R}^3$ joining the points $(3, 1, -2)$ and $(0, 5, 0)$

2. The line in $\mathbb{R}^2$ going through the point $(3, 2)$ in the direction of the vector $(-1, 1)$

3. The circle in $\mathbb{R}^2$ centered at the origin, of radius $\sqrt{5}$

4. The circle in the plane $z = 4$ centered at the point $(0, 0, 4)$, of radius 4

5. The ellipse in $\mathbb{R}^2$ with semiaxes of length 3 (in the x-direction) and 1 (in the y-direction) whose center is located at the point $(-2, -1)$

6. The graph of $f(x) = 3x^2 - 2$ in $\mathbb{R}^2$ for $-3 \le x \le 2$

7. The graph of $x - y^2 = 1$ in $\mathbb{R}^2$ for $0 \le y \le 1$

8. The graph of $3x^2 + y^3 = 1$ in $\mathbb{R}^2$ for $x, y \ge 0$

9. The graph of $x^{2/3} + y^{2/3} = 1$ in $\mathbb{R}^2$

10. What curve is represented by $\mathbf{c}(t) = (\cos t, \cos^2 t)$, $t \in \mathbb{R}$?

Exercises 11 to 15: Find an equation of the curve $\mathbf{c}(t)$ in a Cartesian coordinate system and sketch it, indicating its endpoints (if any) and orientation.

11. $\mathbf{c}(t) = (t - 3)\mathbf{i} + t^2\mathbf{j}$, $t \in [0, 2]$

12. $\mathbf{c}(t) = 3\sin 2t\mathbf{i} + 3\cos 2t\mathbf{j}$, $t \in [0, \pi/2]$

13. $\mathbf{c}(t) = (2\cosh t, 2\sinh t)$, $t \in \mathbb{R}$ (*Hint*: $\cosh^2 t - \sinh^2 t = 1$)

14. $\mathbf{c}(t) = (t^3, t^9)$, $t \in [0, 3]$

15. $\mathbf{c}(t) = (t, e^{3t})$, $t \in [0, \ln 2]$

16. Identify the curve parametrized by $c(t) = (2 - t, 1 + t, t)$, $t \in \mathbb{R}$. If t is replaced by

(a) $-t$ (b) t^2 (c) t^3 (d) e^t

what does the resulting parametrization represent?

Exercises 17 to 25: Sketch (or describe in words) the curve $\mathbf{c}(t)$, indicating its endpoints and orientation.

17. $\mathbf{c}(t) = (\cos t, \sin t, 3)$, $t \in [0, 2\pi]$

18. $\mathbf{c}(t) = (\cos t, \sin t, t)$, $t \in [0, 3\pi]$

19. $\mathbf{c}(t) = (\cos t, \sin t, t^3)$, $t \in [0, \pi]$

20. $\mathbf{c}(t) = (t, \cos t, \sin t)$, $t \in [0, 10\pi]$

21. $\mathbf{c}(t) = (t, \arctan t)$, $t \in [-1, 1]$

22. $\mathbf{c}(t) = \left(1+t^{-1}\right)\mathbf{i} + \left(1 - t^{-1}\right)\mathbf{j}$, $t \in [1, 2]$

23. $\mathbf{c}(t) = \left(t + t^{-1}\right)\mathbf{i} + \left(t - t^{-1}\right)\mathbf{j}$, $t \in [1, 2]$

24. $\mathbf{c}(t) = 4\mathbf{i} + (5 + 2\cos t)\mathbf{j} + (1 + 2\sin t)\mathbf{k}$, $t \in [0, 6\pi]$

25. $\mathbf{c}(t) = (e^{t/4}\sin t, e^{t/4}\cos t)$, $t \in [0, 2\pi]$

26. The following parametrizations have the same image. Describe their differences.

(a) $\mathbf{c}_1(t) = (t, t^2)$, $t \in [-1, 1]$

(b) $\mathbf{c}_2(t) = (\sin t, \sin^2 t)$, $t \in [-\pi/2, \pi/2]$

(c) $\mathbf{c}_3(t) = (\sin t, \sin^2 t)$, $t \in [-\pi/2, 3\pi/2]$

(d) $\mathbf{c}_4(t) = (t^{1/3}, t^{2/3})$, $t \in [-1, 1]$

(e) $\mathbf{c}_5(t) = (2t/\sqrt{1+t^2}, 4t^2/(1+t^2))$, $t \in [-1/\sqrt{3}, 1/\sqrt{3}]$

27. Check that the following parametrizations have the same image; that is, that they represent the same curve. Discuss their differences in terms of their speeds and orientations. Find two more parametrizations with the same image as the curves in (a)–(d).

(a) $\mathbf{c}_1(t) = (2\sin t, 2\cos t)$, $t \in [0, 2\pi]$

(b) $\mathbf{c}_2(t) = (2\cos t, 2\sin t)$, $t \in [0, 2\pi]$

(c) $\mathbf{c}_3(t) = (2\sin 3t, 2\cos 3t)$, $t \in [0, 2\pi]$

(d) $\mathbf{c}_4(t) = (-2\cos(t/2), 2\sin(t/2))$, $t \in [0, 4\pi]$

28. Show that the path $\mathbf{c}(t) = (t^{1/3}, 2t^{2/3})$, $t \in [-1, 1]$ is not differentiable. Identify the curve that is the image of $\mathbf{c}$ and prove that it is differentiable.

29. Write down a parametrization of the line $y = 2x$ in $\mathbb{R}^2$ that is not differentiable.

30. The curve $\mathbf{c}(t) = (t^2, 1/t)$, $t > 0$, represents the position of an object in the xy-plane. Find its velocity and acceleration at $t = 2$, $t = 1$, and $t = 1/10$. Describe what happens (in terms of magnitudes of the velocity and the acceleration) as t approaches 0.

Exercises 31 to 34: Consider the parametrization $\mathbf{c}(t)$ of the curve $y = x^3$, $-1 \le x \le 1$. Determine whether the parametrization is continuous, differentiable or C^1.

31. $\mathbf{c}(t) = (t^{1/3}, t)$, $t \in [-1, 1]$

32. $\mathbf{c}(t) = (2\tan t, 8\tan^3 t)$, $t \in [-\pi/4, \pi/4]$

33. $\mathbf{c}(t) = (t|t|, t^3|t|^3)$, $t \in [-1, 1]$

34. $\mathbf{c}(t) = (e^t - 2, (e^t - 2)^3)$, $t \in [0, \ln 3]$

35. Let $\mathbf{c}(t) = (te^t, (1-t)e^t, e^t)$, $t \in [0, 1]$ describe the position of an object. Find its velocity and acceleration.

36. Show that the parametrization $\mathbf{c}_4(t) = ((4\cos t + \sin t)/\sqrt{17}, (4\sin t - \cos t)/\sqrt{17})$, $t \in [0, 2\pi]$ of Example 2.55 represents the circle $x^2 + y^2 = 1$. Identify the initial and the terminal points of $\mathbf{c}_4(t)$.

Exercises 37 to 41: The vector function $\mathbf{c}(t)$ represents the trajectory of a moving object in $\mathbb{R}^2$ or in $\mathbb{R}^3$. Compute the velocity, speed, and acceleration.

37. $\mathbf{c}(t) = (1 + t^3, t^{-1}, 2)$

38. $\mathbf{c}(t) = e^t \cos t\mathbf{i} + e^t \sin t\mathbf{j} + t\mathbf{k}$

39. $\mathbf{c}(t) = e^{2t} \sin(2t)\mathbf{i} + e^{2t} \cos(2t)\mathbf{j}$

40. $\mathbf{c}(t) = (\cosh t, \sinh t, t)$

41. $\mathbf{c}(t) = (t^{1/2}, t, t^{3/2})$

2.6 PROPERTIES OF DERIVATIVES

After presenting the definition of a derivative, the calculus of functions of one variable proceeds by proving theorems that relate the derivatives of combinations of two functions (such as the sum, the product, or the composition) to the derivatives of the functions themselves. For example, the product rule formula $(fg)' = f'g + fg'$ expresses the derivative of the product of f and g in terms of f and g and their derivatives f' and g'. Although it is always possible to use the definition to find the derivative of a function, the computation is usually (technically) hard and quite lengthy. The differentiation rules provide a significantly easier alternative. We start by generalizing these rules to functions of several variables.

THEOREM 2.6 Properties of Derivatives

(a) Assume that the functions $\mathbf{F}, \mathbf{G}: U \subseteq \mathbb{R}^m \to \mathbb{R}^n$ are differentiable at $\mathbf{a} \in U$. Then the sum $\mathbf{F} + \mathbf{G}$ and the difference $\mathbf{F} - \mathbf{G}$ are differentiable at $\mathbf{a}$ and

$$D(\mathbf{F} \pm \mathbf{G})(\mathbf{a}) = D\mathbf{F}(\mathbf{a}) \pm D\mathbf{G}(\mathbf{a}).$$

(b) If the function $\mathbf{F}: U \subseteq \mathbb{R}^m \to \mathbb{R}^n$ is differentiable at $\mathbf{a} \in U$ and $c \in \mathbb{R}$ is a constant, then the product $c\mathbf{F}$ is differentiable at $\mathbf{a}$ and

$$D(c\mathbf{F})(\mathbf{a}) = cD\mathbf{F}(\mathbf{a}).$$

(c) If the real-valued functions $f, g: U \subseteq \mathbb{R}^m \to \mathbb{R}$ are differentiable at $\mathbf{a} \in U$, then their product fg is differentiable at $\mathbf{a}$ and

$$D(fg)(\mathbf{a}) = g(\mathbf{a})Df(\mathbf{a}) + f(\mathbf{a})Dg(\mathbf{a}).$$

(d) If the real-valued functions $f, g: U \subseteq \mathbb{R}^m \to \mathbb{R}$ are differentiable at $\mathbf{a} \in U$, and $g(\mathbf{a}) \neq 0$, then their quotient f/g is differentiable at $\mathbf{a}$ and

$$D\left(\frac{f}{g}\right)(\mathbf{a}) = \frac{g(\mathbf{a})Df(\mathbf{a}) - f(\mathbf{a})Dg(\mathbf{a})}{g(\mathbf{a})^2}.$$

(e) If the vector-valued functions $\mathbf{v}, \mathbf{w}: U \subseteq \mathbb{R} \to \mathbb{R}^n$ are differentiable at $a \in U$, then their dot (scalar) product $\mathbf{v} \cdot \mathbf{w}$ is differentiable at a and

$$(\mathbf{v} \cdot \mathbf{w})'(a) = \mathbf{v}'(a) \cdot \mathbf{w}(a) + \mathbf{v}(a) \cdot \mathbf{w}'(a).$$

(f) If the vector-valued functions $\mathbf{v}, \mathbf{w}: U \subseteq \mathbb{R} \to \mathbb{R}^3$ are differentiable at $a \in U$, their cross (vector) product $\mathbf{v} \times \mathbf{w}$ is differentiable at a and

$$(\mathbf{v} \times \mathbf{w})'(a) = \mathbf{v}'(a) \times \mathbf{w}(a) + \mathbf{v}(a) \times \mathbf{w}'(a).$$

Algebraic operations on the right sides of formulas (a)–(d) are matrix operations. The sum and difference of two matrices appear in (a), (c), and (d) [the matrices are of type

$n \times m$ in (a), and of type $1 \times m$ in (c) and (d)]. The product of a scalar and a matrix appears in (b), (c), and (d) [the fraction in (d) is the product of the scalar $1/g(\mathbf{a})^2$ and the matrix $g(\mathbf{a})Df(\mathbf{a}) - f(\mathbf{a})Dg(\mathbf{a})$]. Using ∇ to denote the gradient, we can rewrite (c) and (d) as

$$\nabla(fg)(\mathbf{a}) = g(\mathbf{a})\nabla f(\mathbf{a}) + f(\mathbf{a})\nabla g(\mathbf{a})$$

and

$$\nabla\left(\frac{f}{g}\right)(\mathbf{a}) = \frac{g(\mathbf{a})\nabla f(\mathbf{a}) - f(\mathbf{a})\nabla g(\mathbf{a})}{g(\mathbf{a})^2}.$$

If $\mathbf{v}$ and $\mathbf{w}$ are vector-valued functions of one variable (that is usually denoted by t), then their dot (or scalar) product is a real-valued function that assigns to every t the real number $\mathbf{v}(t) \cdot \mathbf{w}(t)$. Therefore, the derivative on the left side of (e) is the derivative of a real-valued function of one variable [hence the notation $()'$ instead of D]. Each term on the right side is a dot product of two vectors in $\mathbb{R}^n$. This time, $()'$ denotes the derivative of a vector-valued function of one variable (also called the velocity). All derivatives in (f) are derivatives of vector-valued functions of one variable. The left side is the derivative of the function that assigns a cross product of vectors $\mathbf{v}(t)$ and $\mathbf{w}(t)$ to every t. Since the cross product is defined only in $\mathbb{R}^3$, both $\mathbf{v}$ and $\mathbf{w}$ must have values in $\mathbb{R}^3$.

The proofs of statements (a)–(d) are analogous to the proofs of corresponding statements in the one-variable case. If we write vectors $\mathbf{v}$ and $\mathbf{w}$ in terms of their components, we can reduce the proofs of (e) and (f) again to the one-variable case. For completeness, the proofs are given in Appendix A.

▶ EXAMPLE 2.61

Let $f(x, y, z) = xy + e^z$ and $g(x, y, z) = y^2 \sin z$. Compute $D(fg)(0, 1, \pi)$.

SOLUTION

By the product rule (the vertical bar is read "evaluated at"),

$$\begin{aligned} D(fg)(0,1,\pi) &= g(0,1,\pi)D(f)(0,1,\pi) + f(0,1,\pi)D(g)(0,1,\pi) \\ &= y^2 \sin z|_{(0,1,\pi)}[y \quad x \quad e^z]|_{(0,1,\pi)} + (xy + e^z)|_{(0,1,\pi)}[0 \quad 2y\sin z \quad y^2\cos z]|_{(0,1,\pi)} \\ &= 0[1 \quad 0 \quad e^\pi] + e^\pi[0 \quad 0 \quad -1] = [0 \quad 0 \quad -e^\pi]. \end{aligned}$$

Alternatively, we compute the product $(fg)(x, y, z) = xy^3 \sin z + y^2 e^z \sin z$ first, and then differentiate

$$\begin{aligned} D(fg)(0,1,\pi) &= [y^3 \sin z \quad 3xy^2 \sin z + 2ye^z \sin z \quad xy^3 \cos z + y^2 e^z \sin z + y^2 e^z \cos z]\Big|_{(0,1,\pi)} \\ &= [0 \quad 0 \quad -e^\pi]. \end{aligned}$$

◀

▶ EXAMPLE 2.62

Let $\mathbf{v}(t) = t\mathbf{i} + \sin t\mathbf{j} + \cos t\mathbf{k}$ and $\mathbf{w} = 3t\mathbf{i} + 2\mathbf{k}$. Compute $(\mathbf{v} \cdot \mathbf{w})'(t)$ directly (i.e., by first computing the dot product and then differentiating) and check your result by using the product rule (e) from Theorem 2.6.

SOLUTION

We compute the dot product of $\mathbf{v}$ and $\mathbf{w}$ to be

$$(\mathbf{v} \cdot \mathbf{w})(t) = \mathbf{v}(t) \cdot \mathbf{w}(t) = (t\mathbf{i} + \sin t\mathbf{j} + \cos t\mathbf{k}) \cdot (3t\mathbf{i} + 2\mathbf{k}) = 3t^2 + 2\cos t,$$

and thus, $(\mathbf{v}\cdot\mathbf{w})'(t) = 6t - 2\sin t$. Since $\mathbf{v}'(t) = \mathbf{i} + \cos t\mathbf{j} - \sin t\mathbf{k}$ and $\mathbf{w}'(t) = 3\mathbf{i}$, we get

$$\begin{aligned}\mathbf{v}'(t)\cdot\mathbf{w}(t) + \mathbf{v}(t)\cdot\mathbf{w}'(t) &= (\mathbf{i} + \cos t\mathbf{j} - \sin t\mathbf{k})\cdot(3t\mathbf{i} + 2\mathbf{k}) + (t\mathbf{i} + \sin t\mathbf{j} + \cos t\mathbf{k})\cdot(3\mathbf{i})\\ &= 6t - 2\sin t.\end{aligned}$$

▶ EXAMPLE 2.63

Compute $\nabla(f/g)(x, y, z)$ if $f(x, y, z) = -x^2y^2$ and $g(x, y, z) = 2yz$.

SOLUTION Using the quotient rule, we get

$$\begin{aligned}\nabla(f/g)(x, y, z) &= \frac{g(x, y, z)\nabla f(x, y, z) - f(x, y, z)\nabla g(x, y, z)}{g(x, y, z)^2}\\ &= \frac{2yz[-2xy^2 \quad -2x^2y \quad 0] + x^2y^2[0 \quad 2z \quad 2y]}{4y^2z^2}\\ &= \frac{[-4xy^3z \quad -2x^2y^2z \quad 2x^2y^3]}{4y^2z^2}\\ &= \begin{bmatrix}\frac{-xy}{z} & -\frac{x^2}{2z} & \frac{x^2y}{2z^2}\end{bmatrix}.\end{aligned}$$

In the last step, the matrix in the numerator was multiplied by the function $1/4y^2z^2$.

We could have computed $\nabla(f/g)$ in the previous example without using the quotient rule: since $(f/g)(x, y, z) = -x^2y^2/2yz = -x^2y/2z$, it follows that

$$\nabla(f/g)(x, y, z) = \begin{bmatrix}\frac{-xy}{z} & -\frac{x^2}{2z} & \frac{x^2y}{2z^2}\end{bmatrix}.$$

However, in certain situations it will be impossible to avoid using the rules (a)–(f) from Theorem 2.6 (see Examples 2.64 and 2.65).

▶ EXAMPLE 2.64 Motion of an Object on the Surface of a Sphere

Assume that an object moves in space so that its distance from the origin O remains constant; that is, $||\mathbf{r}(t)|| = c$, where $\mathbf{r}(t)$ is the position vector of the object and $c > 0$. In other words, the object moves along the surface of the sphere with radius c centered at the origin. Now $||\mathbf{r}(t)||^2 = c^2$ is also constant and hence $(d/dt)||\mathbf{r}(t)||^2 = 0$ and (by the product rule)

$$0 = \frac{d}{dt}||\mathbf{r}(t)||^2 = \frac{d}{dt}(\mathbf{r}(t)\cdot\mathbf{r}(t)) = \left(\frac{d}{dt}\mathbf{r}(t)\right)\cdot\mathbf{r}(t) + \mathbf{r}(t)\cdot\left(\frac{d}{dt}\mathbf{r}(t)\right) = 2\mathbf{r}(t)\cdot\mathbf{v}(t),$$

where $\mathbf{v}t = d\mathbf{r}(t)/dt$ is the velocity of the object at time t. Hence, $\mathbf{r}(t)\cdot\mathbf{v}(t) = 0$, so that either $\mathbf{v}(t) = 0$ (which means that the object is at rest), or the velocity vector is always orthogonal to the position vector $\mathbf{r}(t)$; see Figure 2.60.

The converse of the above statement is true as well: if the object moves so that $\mathbf{r}(t)\cdot\mathbf{v}(t) = 0$, then the computation above (read from right to left) shows that $d||\mathbf{r}(t)||^2/dt = 0$; that is, $||\mathbf{r}(t)|| =$ constant. Consequently, the object moves on the surface of a sphere.

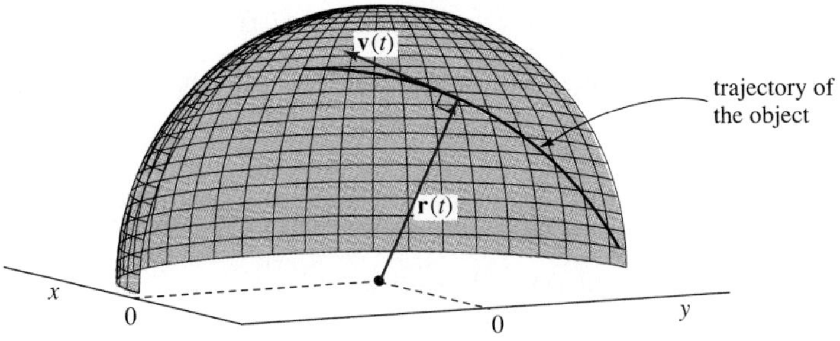

Figure 2.60 For motion on a sphere centered at the origin, the velocity vector is always perpendicular to the position vector.

▶ EXAMPLE 2.65 Kinetic Energy of a Charged Particle in a Magnetic Field

Consider the motion of a particle [described by the vector function $\mathbf{r}(t)$] of mass m and charge q in a constant magnetic field $\mathbf{B}$, with no electric field present. The electromagnetic force [see formula (2.5), with $\mathbf{E} = \mathbf{0}$], $\mathbf{F}(\mathbf{r}(t)) = q(\mathbf{v}(t) \times \mathbf{B})$ and Newton's Second Law, $\mathbf{F}(\mathbf{r}(t)) = m\mathbf{a}(t) = m\mathbf{v}'(t)$, imply that $m\mathbf{v}'(t) = q(\mathbf{v}(t)) \times \mathbf{B})$. Show that the kinetic energy $K(t) = m||\mathbf{v}(t)||^2/2$ is constant in time.

SOLUTION

We will show that the derivative of $K(t)$ is zero:

$$\begin{aligned}\left(\frac{1}{2}m||\mathbf{v}(t)||^2\right)' &= \frac{1}{2}m(\mathbf{v}(t)\cdot\mathbf{v}(t))' = \frac{1}{2}m(\mathbf{v}(t)\cdot\mathbf{v}'(t) + \mathbf{v}'(t)\cdot\mathbf{v}(t))\\ &= m\mathbf{v}(t)\cdot\mathbf{v}'(t) = m\left(\mathbf{v}(t)\cdot\frac{q}{m}(\mathbf{v}(t)\times\mathbf{B})\right)\\ &= q(\mathbf{v}(t)\cdot(\mathbf{v}(t)\times\mathbf{B}))\end{aligned}$$

(the product rule was used in the first line, and Newton's Second Law was used in the second line). By definition of the vector product, the vector $\mathbf{v}(t) \times \mathbf{B}$ is perpendicular to $\mathbf{v}(t)$ and hence their scalar product $\mathbf{v}(t) \cdot (\mathbf{v}(t) \times \mathbf{B})$ is zero.

Note that the fact $m||\mathbf{v}(t)||^2/2$ is constant implies that the speed $||\mathbf{v}(t)||$ of the particle is constant. ◀

The statement of our next theorem is a generalization of the one-variable chain rule.

THEOREM 2.7 Chain Rule

Suppose that $\mathbf{F}\colon U \subseteq \mathbb{R}^m \to \mathbb{R}^n$ is differentiable at $\mathbf{a} \in U$, U is open in $\mathbb{R}^m$, $\mathbf{G}\colon V \subseteq \mathbb{R}^n \to \mathbb{R}^p$ is differentiable at $\mathbf{F}(\mathbf{a}) \in V$, V is open in $\mathbb{R}^n$, and $\mathbf{F}(U) \subseteq V$ (so that the composition $\mathbf{G} \circ \mathbf{F}$ is defined). Then $\mathbf{G} \circ \mathbf{F}$ is differentiable at $\mathbf{a}$ and

$$D(\mathbf{G} \circ \mathbf{F})(\mathbf{a}) = D\mathbf{G}(\mathbf{F}(\mathbf{a})) \cdot D\mathbf{F}(\mathbf{a}),$$

where $\cdot$ denotes matrix multiplication. ◀

This theorem states that the derivative of the composition $\mathbf{G} \circ \mathbf{F}$ at a point $\mathbf{a}$ in U can be computed as a matrix product of the derivative of $\mathbf{G}$ [evaluated at $\mathbf{F}(\mathbf{a})$] and the derivative of $\mathbf{F}$ [evaluated at $\mathbf{a}$]. One easily checks that the matrices on both sides of the chain rule formula are of the same type: since $\mathbf{G} \circ \mathbf{F}\colon U \subseteq \mathbb{R}^m \to \mathbb{R}^p$, $D(\mathbf{G} \circ \mathbf{F})$ is a $p \times m$ matrix.

The right side is a product of the $p \times n$ matrix $D\mathbf{G}(\mathbf{F}(\mathbf{a}))$ and the $n \times m$ matrix $D\mathbf{F}(\mathbf{a})$, and is hence a $p \times m$ matrix. The proof of this theorem is given in Appendix A.

We now study examples, to gain experience in working with the chain rule.

▶ EXAMPLE 2.66

Let $\mathbf{F}$: $\mathbb{R}^2 \to \mathbb{R}^3$ be given by $\mathbf{F}(x, y) = (x^3 + y, e^{xy}, 2 + xy)$ and let $\mathbf{G}$: $\mathbb{R}^3 \to \mathbb{R}^2$ be given by $\mathbf{G}(u, v, w) = (u^2 + v, uv + w^3)$. Compute $D(\mathbf{G} \circ \mathbf{F})(0, 1)$.

SOLUTION By the chain rule,

$$D(\mathbf{G} \circ \mathbf{F})(0, 1) = D\mathbf{G}(\mathbf{F}(0, 1)) \cdot D\mathbf{F}(0, 1) = D\mathbf{G}(1, 1, 2) \cdot D\mathbf{F}(0, 1).$$

The derivatives of F and G are computed to be

$$D\mathbf{F}(0, 1) = \begin{bmatrix} 3x^2 & 1 \\ ye^{xy} & xe^{xy} \\ y & x \end{bmatrix}_{at\ (0,1)} = \begin{bmatrix} 0 & 1 \\ 1 & 0 \\ 1 & 0 \end{bmatrix}$$

and

$$D\mathbf{G}(1, 1, 2) = \begin{bmatrix} 2u & 1 & 0 \\ v & u & 3w^2 \end{bmatrix}_{at\ (1,1,2)} = \begin{bmatrix} 2 & 1 & 0 \\ 1 & 1 & 12 \end{bmatrix},$$

so that

$$D(\mathbf{G} \circ \mathbf{F})(0, 1) = \begin{bmatrix} 2 & 1 & 0 \\ 1 & 1 & 12 \end{bmatrix} \cdot \begin{bmatrix} 0 & 1 \\ 1 & 0 \\ 1 & 0 \end{bmatrix} = \begin{bmatrix} 1 & 2 \\ 13 & 1 \end{bmatrix}.$$

To check the result, compute the composition

$$\begin{aligned}(\mathbf{G} \circ \mathbf{F})(x, y) &= \mathbf{G}(\mathbf{F}(x, y)) = \mathbf{G}(x^3 + y, e^{xy}, 2 + xy) \\ &= ((x^3 + y)^2 + e^{xy}, (x^3 + y)e^{xy} + (2 + xy)^3),\end{aligned}$$

to obtain the function $\mathbf{G} \circ \mathbf{F}$: $\mathbb{R}^2 \to \mathbb{R}^2$. Its derivative $D(\mathbf{G} \circ \mathbf{F})$ is a 2×2 matrix

$$D(\mathbf{G} \circ \mathbf{F})(x, y) = \begin{bmatrix} 6(x^3 + y)x^2 + ye^{xy} & 2(x^3 + y)^2 + xe^{xy} \\ 3x^2e^{xy} + (x^3 + y)ye^{xy} + 3(2 + xy)^2y & e^{xy} + xye^{xy} + 3(2 + xy)^2x \end{bmatrix},$$

and therefore,

$$D(\mathbf{G} \circ \mathbf{F})(0, 1) = \begin{bmatrix} 1 & 2 \\ 13 & 1 \end{bmatrix}.$$

◀

▶ EXAMPLE 2.67

The composition $(f \circ \mathbf{c})(t)$ of f: $\mathbb{R}^2 \to \mathbb{R}$, $f(x, y) = x^2 + 2y^2$, and $\mathbf{c}$: $\mathbb{R} \to \mathbb{R}^2$, $\mathbf{c}(t) = (e^t, te^t)$, is a real-valued function of one variable. Compute $(f \circ \mathbf{c})'(0)$.

SOLUTION By the chain rule,

$$(f \circ \mathbf{c})'(0) = D(f \circ \mathbf{c})(0) = Df(\mathbf{c}(0)) \cdot D\mathbf{c}(0) = Df(1, 0) \cdot D\mathbf{c}(0).$$

The function f is a real-valued function of two variables, so its derivative (also called the gradient) is a 1×2 matrix $Df(x, y) = [2x \quad 4y]$. Hence, $Df(1, 0) = [2 \quad 0]$. The function $\mathbf{c}$ (also called a path in $\mathbb{R}^2$) is a function of one variable, and its derivative $D\mathbf{c}(t) = \mathbf{c}'(t)$ is the 2×1 matrix

$$D\mathbf{c}(t) = \begin{bmatrix} e^t \\ e^t + te^t \end{bmatrix}.$$

Consequently,

$$D\mathbf{c}(0) = \begin{bmatrix} 1 \\ 1 \end{bmatrix}$$

and

$$(f \circ \mathbf{c})'(0) = Df(1,0) \cdot D\mathbf{c}(0) = [2 \quad 0] \cdot \begin{bmatrix} 1 \\ 1 \end{bmatrix} = 2.$$

We check this by direct computation: since $(f \circ \mathbf{c})(t) = f(\mathbf{c}(t)) = f(e^t, te^t) = e^{2t} + 2t^2e^{2t}$, it follows that $(f \circ \mathbf{c})'(t) = 2e^{2t} + 4te^{2t} + 4t^2e^{2t}$, and, consequently, $(f \circ \mathbf{c})'(0) = 2$. ◀

▶ EXAMPLE 2.68

Consider the composition $f \circ \mathbf{c}$, where $f = f(x, y, z)$: $\mathbb{R}^3 \to \mathbb{R}$, and $\mathbf{c}$: $\mathbb{R} \to \mathbb{R}^3$ is given by $\mathbf{c}(t) = (x(t), y(t), z(t))$. Assume that both f and $\mathbf{c}$ are differentiable. Then

$$(f \circ \mathbf{c})(t) = f(\mathbf{c}(t)) = f(x(t), y(t), z(t)),$$

and, by the chain rule,

$$D(f \circ \mathbf{c})(t) = Df(\mathbf{c}(t)) \cdot D\mathbf{c}(t) = \begin{bmatrix} \dfrac{\partial f}{\partial x} & \dfrac{\partial f}{\partial y} & \dfrac{\partial f}{\partial z} \end{bmatrix}_{at\ \mathbf{c}(t)} \cdot \begin{bmatrix} \dfrac{\partial x}{\partial t} \\ \dfrac{\partial y}{\partial t} \\ \dfrac{\partial z}{\partial t} \end{bmatrix}_{at\ t}, \tag{2.23}$$

so that

$$D(f \circ \mathbf{c})(t) = \frac{\partial f}{\partial x}(\mathbf{c}(t))\frac{\partial x}{\partial t}(t) + \frac{\partial f}{\partial y}(\mathbf{c}(t))\frac{\partial y}{\partial t}(t) + \frac{\partial f}{\partial z}(\mathbf{c}(t))\frac{\partial z}{\partial t}(t),$$

or (dropping the notation for the dependence on a point)

$$D(f \circ \mathbf{c})(t) = \frac{\partial f}{\partial x}\frac{dx}{dt} + \frac{\partial f}{\partial y}\frac{dy}{dt} + \frac{\partial f}{\partial z}\frac{dz}{dt}. \tag{2.24}$$

We have replaced the partial derivative notation $\partial x/\partial t$, $\partial y/\partial t$, and $\partial z/\partial t$ by dx/dt, dy/dt, and dz/dt, since x, y, and z are functions of one variable (we could have used x', y', and z' instead). The 1×3 matrix in (2.23) is the gradient of f evaluated at $\mathbf{c}(t)$, and the 3×1 matrix is the derivative $\mathbf{c}'(t)$. Hence, (2.24) can be written as

$$D(f \circ \mathbf{c})(t) = \nabla f(\mathbf{c}(t)) \cdot \mathbf{c}'(t), \tag{2.25}$$

where the multiplication on the right side is interpreted either as a matrix multiplication, or as a dot product if both matrices $\nabla f(\mathbf{c}(t))$ and $\mathbf{c}'(t)$ are viewed as vectors in $\mathbb{R}^3$. ◀

Note that the calculations in the previous example can easily be extended to any number of variables. Thus, (2.25) holds for any differentiable function $f : \mathbb{R}^n \to \mathbb{R}$ and any differentiable path $\mathbf{c}$: $\mathbb{R} \to \mathbb{R}^n$.

▶ EXAMPLE 2.69

Assume that $f = f(x, y)\colon \mathbb{R}^2 \to \mathbb{R}$ is a differentiable function.

(a) Let $g_1(t) = f(t, t^2)$. Find $g_1'(t)$.

(b) Let $g_2(t) = f(t, f(t, t^2))$. Compute $g_2'(t)$. Assuming that $f(1, 1) = 2$, find $g_2'(1)$.

SOLUTION

(a) Note that $g_1 = f \circ \mathbf{c}$, where $\mathbf{c}(t) = (t, t^2)$. Using (2.25), we obtain

$$g_1'(t) = (f \circ \mathbf{c})'(t) = \nabla f(\mathbf{c}(t)) \cdot \mathbf{c}'(t) = \nabla f(t, t^2) \cdot (1, 2t).$$

Using D_1 and D_2 to denote the partial derivatives of f with respect to its first and second variables, we rewrite the above as

$$g_1'(t) = (D_1 f(t, t^2), D_2 f(t, t^2)) \cdot (1, 2t) = D_1 f(t, t^2) + 2t D_2 f(t, t^2).$$

Alternatively, we compute the derivative directly from $g_1(t) = f(t, t^2)$ using a variant of (2.24) for a function of two variables ("f with respect to its first variable times first variable with respect to t, plus f with respect to its second variable times second variable with respect to t"):

$$g_1'(t) = D_1 f(t, t^2)(t)' + D_2 f(t, t^2)(t^2)' = D_1 f(t, t^2) + 2t D_2 f(t, t^2).$$

(b) Proceeding as above, we start with the chain rule (2.24) for two variables:

$$g_2'(t) = D_1 f(t, f(t, t^2))(t)' + D_2 f(t, f(t, t^2))(f(t, t^2))'$$

Using (a), we obtain

$$g_2'(t) = D_1 f(t, f(t, t^2)) + D_2 f(t, f(t, t^2))(D_1 f(t, t^2) + 2t D_2 f(t, t^2)).$$

Thus, when $t = 1$,

$$g_2'(1) = D_1 f(1, 2) + D_2 f(1, 2)(D_1 f(1, 1) + 2D_2 f(1, 1)).$$

◀

▶ EXAMPLE 2.70

Let $f\colon \mathbb{R}^3 \to \mathbb{R}$, and let $\mathbf{G}\colon \mathbb{R}^3 \to \mathbb{R}^3$ be given by

$$\mathbf{G}(x, y, z) = (u(x, y, z), v(x, y, z), w(x, y, z)).$$

Assume that f and $\mathbf{G}$ are differentiable. Define $h\colon \mathbb{R}^3 \to \mathbb{R}$ by $h = f \circ \mathbf{G}$, that is,

$$h(x, y, z) = (f \circ \mathbf{G})(x, y, z) = f(\mathbf{G}(x, y, z)) = f(u(x, y, z), v(x, y, z), w(x, y, z)).$$

Compute $\partial h/\partial x$, $\partial h/\partial y$, and $\partial h/\partial z$ using the chain rule.

SOLUTION

The derivative of h (which is a function of x, y, and z) is given by the 1×3 matrix

$$Dh = \begin{bmatrix} \dfrac{\partial h}{\partial x} & \dfrac{\partial h}{\partial y} & \dfrac{\partial h}{\partial z} \end{bmatrix}.$$

Using the symbols $D_1 f$, $D_2 f$, and $D_3 f$ to denote the partial derivatives of f with respect to its variables, we write $Df = [D_1 f \quad D_2 f \quad D_3 f]$. By the chain rule,

$$Dh = Df \cdot D\mathbf{G} = [D_1 f \quad D_2 f \quad D_3 f] \cdot \begin{bmatrix} \frac{\partial u}{\partial x} & \frac{\partial u}{\partial y} & \frac{\partial u}{\partial z} \\ \frac{\partial v}{\partial x} & \frac{\partial v}{\partial y} & \frac{\partial v}{\partial z} \\ \frac{\partial w}{\partial x} & \frac{\partial w}{\partial y} & \frac{\partial w}{\partial z} \end{bmatrix},$$

and therefore,

$$\frac{\partial h}{\partial x} = D_1 f \frac{\partial u}{\partial x} + D_2 f \frac{\partial v}{\partial x} + D_3 f \frac{\partial w}{\partial x},$$

$$\frac{\partial h}{\partial y} = D_1 f \frac{\partial u}{\partial y} + D_2 f \frac{\partial v}{\partial y} + D_3 f \frac{\partial w}{\partial y},$$

and

$$\frac{\partial h}{\partial z} = D_1 f \frac{\partial u}{\partial z} + D_2 f \frac{\partial v}{\partial z} + D_3 f \frac{\partial w}{\partial z}. \tag{2.26}$$

Using u, v, and w for the variables of f, we write $D_1 f = \partial f / \partial u$, $D_2 f = \partial f / \partial v$, and $D_3 f = \partial f / \partial w$ and hence,

$$\frac{\partial h}{\partial x} = \frac{\partial f}{\partial u}\frac{\partial u}{\partial x} + \frac{\partial f}{\partial v}\frac{\partial v}{\partial x} + \frac{\partial f}{\partial w}\frac{\partial w}{\partial x}, \tag{2.27}$$

with similar expressions for $\partial h / \partial y$ and $\partial h / \partial z$. ◀

Example 2.70 shows us how to write partial derivatives of f in two different ways. As another exercise in notation, let us compute $\partial h / \partial x$ if $h(x, y) = f(x^2 + y^2, yz, e^x + y)$. We want the partial derivatives of f with respect to its variables, but cannot use expressions like $\partial f / \partial(x^2 + y^2)$ or $\partial f / \partial(yz)$. One approach to solving this notational difficulty is to introduce new variables $u = x^2 + y^2$, $v = yz$, $w = e^x + y$, write $h = f(u, v, w)$, and then use $\partial f / \partial u$, $\partial f / \partial v$, and $\partial f / \partial w$ for partial derivatives. Thus,

$$\frac{\partial h}{\partial x} = \frac{\partial f}{\partial u} \cdot 2x + \frac{\partial f}{\partial v} \cdot 0 + \frac{\partial f}{\partial w} \cdot e^x.$$

Alternatively, using the "D_i" notation for partial derivatives, we write

$$\frac{\partial h}{\partial x} = D_1 f \cdot 2x + D_2 f \cdot 0 + D_3 f \cdot e^x,$$

without explicitly mentioning the names of variables.

Let us make note of a notational convention commonly used in expressions involving the chain rule. Assume that $f = f(x, y)$ is a real-valued function of two variables x and y—take, for example, $f(x, y) = x^2 - 2y^2$ and let $x = u^2 + v^2$ and $y = uv$. Then

$$f(x, y) = f(u^2 + v^2, uv) = (u^2 + v^2)^2 - 2(uv)^2 = u^4 + v^4,$$

so f becomes a function of (new) variables u and v. This process is called *change of variables* and will be discussed in more detail later (e.g., as a technique in integration). It can be described as a composition of functions $h = f \circ \mathbf{P}$, where $\mathbf{P}$: $\mathbb{R}^2 \to \mathbb{R}^2$ is defined by $\mathbf{P}(u, v) = (u^2 + v^2, uv)$. Let us check this:

$$h(u, v) = (f \circ \mathbf{P})(u, v) = f(\mathbf{P}(u, v)) = f(u^2 + v^2, uv) = u^4 + v^4.$$

Although, strictly speaking, f and h are two different functions (they depend on different variables and, in general, might have different domains), it is a standard practice (especially in applied mathematics) to use the same notation for both. In this context, $f(u, v)$ denotes the function $h(u, v)$, that is, the function f expressed in terms of variables u and v, and $f(x, y)$ denotes, as usual, the function f as a function of x and y. For example, from $f(x, y) = x^2 - 2y^2$, we get $\partial f/\partial x = 2x$, but $\partial f/\partial u$ is not zero, since it does not refer to $f(x, y) = x^2 - 2y^2$ but to $f(u, v) = u^4 + v^4$. Hence, $\partial f/\partial u = 4u^3$.

With this convention in mind, we write the chain rule formula (2.27) as

$$\frac{\partial f}{\partial x} = \frac{\partial f}{\partial u}\frac{\partial u}{\partial x} + \frac{\partial f}{\partial v}\frac{\partial v}{\partial x} + \frac{\partial f}{\partial w}\frac{\partial w}{\partial x}.$$

(f on the left represents f viewed as a function of x, y, and z, and f on the right side is f viewed as a function of u, v, and w.)

▶ EXAMPLE 2.71

Let $f(u, v, w) = u^2 + v^3 e^w$, where $u = \sin(x + y + z)$, $v = x^2 e^y$, and $w = z$. Compute partial derivatives $\partial f/\partial x$, $\partial f/\partial y$, and $\partial f/\partial z$.

SOLUTION Using the convention just adopted and formula (2.27), we get

$$\frac{\partial f}{\partial x} = \frac{\partial f}{\partial u}\frac{\partial u}{\partial x} + \frac{\partial f}{\partial v}\frac{\partial v}{\partial x} + \frac{\partial f}{\partial w}\frac{\partial w}{\partial x}$$

[f on the left side is $f(x, y, z) = (\sin(x + y + z))^2 + (x^2 e^y)^3 e^z$, and f on the right side is $f(u, v, w) = u^2 + v^3 e^w$]. Thus,

$$\begin{aligned}\frac{\partial f}{\partial x} &= 2u \cdot \cos(x + y + z) + 3v^2 e^w \cdot 2xe^y + v^3 e^w \cdot 0\\ &= 2\sin(x + y + z)\cos(x + y + z) + 6x^5 e^{3y+z}.\end{aligned}$$

The last line was obtained by substituting the expressions for u, v, and w. Similarly,

$$\begin{aligned}\frac{\partial f}{\partial y} &= \frac{\partial f}{\partial u}\frac{\partial u}{\partial y} + \frac{\partial f}{\partial v}\frac{\partial v}{\partial y} + \frac{\partial f}{\partial w}\frac{\partial w}{\partial y} = 2u \cdot \cos(x + y + z) + 3v^2 e^w \cdot x^2 e^y + v^3 e^w \cdot 0\\ &= 2\sin(x + y + z)\cos(x + y + z) + 3x^6 e^{3y+z},\end{aligned}$$

and

$$\begin{aligned}\frac{\partial f}{\partial z} &= \frac{\partial f}{\partial u}\frac{\partial u}{\partial z} + \frac{\partial f}{\partial v}\frac{\partial v}{\partial z} + \frac{\partial f}{\partial w}\frac{\partial w}{\partial z} = 2u \cdot \cos(x + y + z) + 3v^2 e^w \cdot 0 + v^3 e^w \cdot 1\\ &= 2\sin(x + y + z)\cos(x + y + z) + x^6 e^{3y+z}.\end{aligned}$$

◀

▶ EXAMPLE 2.72 Partial Derivatives on a Surface

Let $f = f(x, y, z)\colon \mathbb{R}^3 \to \mathbb{R}$ and $z = g(x, y)\colon \mathbb{R}^2 \to \mathbb{R}$ be differentiable functions [recall that the graph of $z = g(x, y)$ is a surface in $\mathbb{R}^3$]. The function $w = w(x, y) = f(x, y, g(x, y))\colon \mathbb{R}^2 \to \mathbb{R}$ computes the value of f at the points $(x, y, g(x, y))$, which belong to the (surface which is the) graph of $z = g(x, y)$. The function w is called the *restriction* of f to the surface $z = g(x, y)$. Compute $\partial w/\partial x$ and $\partial w/\partial y$.

SOLUTION View w as the composition $w = f \circ \mathbf{G}$, where the function $\mathbf{G}: \mathbb{R}^2 \to \mathbb{R}^3$ is defined by $\mathbf{G}(x, y) = (x, y, g(x, y))$. By the chain rule, $Dw = Df \cdot D\mathbf{G}$, where

$$Dw = \begin{bmatrix} \dfrac{\partial w}{\partial x} & \dfrac{\partial w}{\partial y} \end{bmatrix}$$

and

$$Df \cdot D\mathbf{G} = \begin{bmatrix} \dfrac{\partial f}{\partial x} & \dfrac{\partial f}{\partial y} & \dfrac{\partial f}{\partial z} \end{bmatrix} \cdot \begin{bmatrix} 1 & 0 \\ 0 & 1 \\ \dfrac{\partial g}{\partial x} & \dfrac{\partial g}{\partial y} \end{bmatrix}.$$

Computing the product $Df \cdot D\mathbf{G}$ and comparing to Dw, we get

$$\frac{\partial w}{\partial x} = \frac{\partial f}{\partial x} + \frac{\partial f}{\partial z}\frac{\partial g}{\partial x},$$

and

$$\frac{\partial w}{\partial y} = \frac{\partial f}{\partial y} + \frac{\partial f}{\partial z}\frac{\partial g}{\partial y}$$

Alternatively, we could have applied the first formula in (2.26) to $w = f(x, y, z)$, where $z = g(x, y)$, to get

$$\frac{\partial w}{\partial x} = D_1 f \frac{\partial x}{\partial x} + D_2 f \frac{\partial y}{\partial x} + D_3 f \frac{\partial z}{\partial x}.$$

Realizing that $\partial x/\partial x = 1$, $\partial y/\partial x = 0$, and $\partial z/\partial x = \partial g/\partial x$, we can rewrite it as

$$\frac{\partial w}{\partial x} = D_1 f + D_3 f \frac{\partial g}{\partial x}.$$

Finally, using x, y, and z as the variables of f and z to replace $g(x, y)$, we write

$$\frac{\partial w}{\partial x} = \frac{\partial f}{\partial x} + \frac{\partial f}{\partial z}\frac{\partial z}{\partial x}.$$

◀

▶ EXAMPLE 2.73 Polar Coordinates

Let $x = r\cos\theta$, $y = r\sin\theta$, and let $f = f(x, y)$ be a differentiable function. Express $\partial f/\partial r$ and $\partial f/\partial\theta$ in terms of $\partial f/\partial x$ and $\partial f/\partial y$.

SOLUTION We interpret the change from Cartesian coordinates to polar coordinates as the map $\mathbf{P}: \mathbb{R}^2 \to \mathbb{R}^2$ defined by $\mathbf{P}(r, \theta) = (r\cos\theta, r\sin\theta)$, and consider the composition $h = f \circ \mathbf{P}$. Then

$$h(r, \theta) = f(\mathbf{P}(r, \theta)) = f(r\cos\theta, r\sin\theta);$$

that is, h is the function f "expressed in terms of polar coordinates." The chain rule applied to $h(r, \theta) = (f \circ \mathbf{P})(r, \theta)$ yields

$$Dh(r, \theta) = Df(\mathbf{P}(r, \theta)) \cdot D\mathbf{P}(r, \theta) = Df(x, y) \cdot D\mathbf{P}(r, \theta),$$

and in matrix notation,

$$\begin{bmatrix} \dfrac{\partial h}{\partial r} & \dfrac{\partial h}{\partial \theta} \end{bmatrix} = \begin{bmatrix} \dfrac{\partial f}{\partial x} & \dfrac{\partial f}{\partial y} \end{bmatrix} \cdot \begin{bmatrix} \cos\theta & -r\sin\theta \\ \sin\theta & r\cos\theta \end{bmatrix}.$$

Now, replacing $h(r,\theta)$ by $f(r,\theta)$, following the usual convention, we obtain

$$\frac{\partial f}{\partial r} = \frac{\partial f}{\partial x}\cos\theta + \frac{\partial f}{\partial y}\sin\theta,$$

and

$$\frac{\partial f}{\partial \theta} = \frac{\partial f}{\partial x}(-r\sin\theta) + \frac{\partial f}{\partial y} r\cos\theta.$$

◀

▶ EXAMPLE 2.74

Let $x = r\cos\theta$, $y = r\sin\theta$, and $f(x,y) = xe^{x^2+y^2}$. Find $\partial f/\partial r$ and $\partial f/\partial\theta$ directly, and then using the chain rule.

SOLUTION Since $f(x,y) = xe^{x^2+y^2} = r\cos\theta e^{r^2}$, we get $f(r,\theta) = r\cos\theta e^{r^2}$ (both functions are called f—recall the notational convention!) and hence $\partial f/\partial r = (e^{r^2} + 2r^2e^{r^2})\cos\theta$ and $\partial f/\partial\theta = -re^{r^2}\sin\theta$. Using the result of the previous example, we obtain

$$\begin{aligned}\frac{\partial f}{\partial r} &= \frac{\partial f}{\partial x}\cos\theta + \frac{\partial f}{\partial y}\sin\theta = (e^{x^2+y^2} + 2x^2e^{x^2+y^2})\cos\theta + 2xye^{x^2+y^2}\sin\theta\\ &= (e^{r^2} + 2r^2\cos^2\theta e^{r^2})\cos\theta + 2r^2e^{r^2}\sin\theta\cos\theta\sin\theta = (e^{r^2} + 2r^2e^{r^2})\cos\theta.\end{aligned}$$

The expression for $\partial f/\partial\theta$ is obtained similarly. ◀

Notice that in this case the direct computation was faster (and easier). However, there are situations where not only does the chain rule provide a more efficient way, but the direct computation cannot be applied at all; see Exercise 27.

▶ EXERCISES 2.6

1. Assume that g is a differentiable, real-valued function of two variables and let $f(x,y) = g(x^2 - y^2, y^2 - x^2)$. Prove that $x(\partial f/\partial y) + y(\partial f/\partial x) = 0$.

2. Assume that g is a differentiable real-valued function of one variable, such that $g(1) = 2$ and $g'(1) = 3$.

(a) If $f(x,y) = g(x) + g(x^2)g(y)$, find $(\partial f/\partial x)(x,y)$ and $(\partial f/\partial x)(1,1)$.

(b) If $f(x,y) = g(x)^{g(y)}$, find $(\partial f/\partial x)(1,1)$ and $(\partial f/\partial y)(1,1)$.

3. Find $g'(t)$ if $g(t) = f(t\sin t, t\cos t, t)$, where f is a differentiable function.

4. Assume that f is a differentiable function and let $g(t) = \sin(f(-t,t,2t))$. Find $g'(t)$.

Exercises 5 to 7: In each case, compute $(f \circ \mathbf{c})'(t)$ in two different ways: by computing the composition first and then differentiating, and by using formula (2.25).

5. $f(x,y) = x^2y$, $\mathbf{c}(t) = (\sin t, \cos t)$.

6. $f(x,y) = ye^{xy}$, $\mathbf{c}(t) = (t, \ln t)$.

7. $f(x,y,z) = xy + \cos(x^2 + z^2)$, $\mathbf{c}(t) = (t\sin t, t, t\cos t)$.

8. Assume that f is a differentiable function of two variables, and $D_1f(2,2) = -2$ and $D_2f(2,2) = 4$.

(a) Find $g'(2)$ if $g(x) = f(x,2)$.

(b) Let $g(x) = f(x, x)$. Find $g'(2)$.

(c) Let $g(x) = f(x^2, x^3)$. Find $g'(x)$.

9. Let $f(x, y) = g(x^2y, 2x + 5y, x, y)$, where g is a differentiable function of four variables. Find f_x and f_y.

10. Let $f\colon \mathbb{R}^2 \to \mathbb{R}^3$ be given by $f(x, y) = (h(x), g(y), k(x, y))$, where h, g, and k are differentiable functions of variables indicated. Find Df.

11. Let $F(x, y) = f(h(x), g(y), k(x, y))$, where $f\colon \mathbb{R}^3 \to \mathbb{R}$, and all functions involved are assumed to be differentiable. Find F_x and F_y.

12. Let $z = f(r)$, where $r = \sqrt{x^2 + y^2}$ and f is a differentiable function. Prove that $yz_x - xz_y = 0$ for all $(x, y) \neq (0, 0)$.

13. Let $f(x, y) = x^2 + xy$ and $g(x, y) = \ln x + \ln y$. Compute $\nabla(fg)(x, y)$ and $\nabla(f/g)(2, 2)$.

14. Let $\mathbf{G}(x, y) = (2xy, y^2 - x^2)$. Compute $D\mathbf{G}(x, y)$ and $D\mathbf{G}(3, 0)$.

15. Let $f(x, y, z) = x^2 + \sin(yz) - 3$. Find $D(f/x)(1, \pi, -1)$ and $D(x^2yf)(2, 0, 1)$.

16. Let $w = f(x, y, z)$, where $x = r\cos\theta$ and $y = r\sin\theta$. Find $\partial w/\partial r$, $\partial w/\partial\theta$, and $\partial w/\partial z$.

17. Let $w = f(x, y, z)$, where $x = \rho\sin\phi\cos\theta$, $y = \rho\sin\phi\sin\theta$, and $z = \rho\cos\phi$. Find $\partial w/\partial\rho$, $\partial w/\partial\theta$, and $\partial w/\partial\phi$.

18. Let $\mathbf{v}(t) = t\mathbf{i} + (t^2 + 1)\mathbf{j}$ and $\mathbf{w}(t) = \mathbf{i} - 2t\mathbf{j} + e^t\mathbf{k}$. Compute $(\mathbf{v}\cdot\mathbf{w})'(t)$ directly (i.e., by computing the dot product first and then differentiating) and then check your answer using the product rule.

19. Let $\mathbf{v}(t) = t^3\mathbf{i} + te^t\mathbf{k}$ and $\mathbf{w}(t) = -2t\mathbf{j}$. Compute $(\mathbf{v}\times\mathbf{w})'(t)$ directly (i.e., by computing the cross product first) and then check your answer using the product rule.

20. Let $\mathbf{u}(t) = \sin t\mathbf{i} + \cos t\mathbf{j} + t\mathbf{k}$, $\mathbf{v}(t) = \mathbf{i} + t\mathbf{j} + \mathbf{k}$, and $\mathbf{w}(t) = t^3(\mathbf{i} + \mathbf{j} + \mathbf{k})$. Compute $(\mathbf{u}\cdot(\mathbf{v}\times\mathbf{w}))'(t)$.

21. The function $\mathbf{F}\colon \mathbb{R}^2 \to \mathbb{R}^3$ is given by $\mathbf{F}(x, y) = (e^x, xy, e^y)$. Compute $D(g\circ\mathbf{F})(0, 0)$, where $g\colon \mathbb{R}^3 \to \mathbb{R}$ is given by $g(u, v, w) = uw + v^2$.

22. Let $f\colon \mathbb{R}^3 \to \mathbb{R}$ and $\mathbf{c}\colon \mathbb{R} \to \mathbb{R}^3$ be given by $f(x, y, z) = \sqrt{x^2 + y^2 + z^2}$ and $\mathbf{c}(t) = (\cos t, \sin t, 1)$. Compute $(f\circ\mathbf{c})'(t)$ and $(f\circ\mathbf{c})'(0)$.

23. Compute $\partial w/\partial x$ and $\partial w/\partial z$ if $w = f(x, y, z)$ and $y = g(x, z)$ are differentiable functions.

24. Let $w = \ln(r^2 + 1)$, where $r = \sqrt{x^2 + y^2}$. Find $\partial w/\partial y$.

25. Define a function $\mathbf{F}\colon \mathbb{R}^2 \to \mathbb{R}^2$ by $\mathbf{F}(\mathbf{x}) = A\cdot\mathbf{x}$, where A is a 2×2 matrix, and the dot indicates matrix multiplication. Compute $D\mathbf{F}(\mathbf{x})$. Prove that $\mathbf{F}$ is differentiable at any point $(a, b) \in \mathbb{R}^2$.

26. Let A and B be 2×2 matrices. Define $\mathbf{F}, \mathbf{G}\colon \mathbb{R}^2 \to \mathbb{R}^2$ by $\mathbf{F}(\mathbf{x}) = A\cdot\mathbf{x}$ and $\mathbf{G}(\mathbf{x}) = B\cdot\mathbf{x}$, where $\mathbf{x} \in \mathbb{R}^2$, and the dot indicates matrix multiplication. Find $D(\mathbf{G}\circ\mathbf{F})(\mathbf{x})$.

27. Let $f(x, y) = x^3y$, where $x^3 + tx = 8$ and $ye^y = t$. Find $(df/dt)(0)$.

28. In Examples 2.46 and 2.47 in Section 2.4, we studied the function

$$f(x, y) = \begin{cases} \dfrac{xy}{x^2 + y^2} & \text{if } (x, y) \neq (0, 0) \\ 0 & \text{if } (x, y) = (0, 0) \end{cases}.$$

Let $\mathbf{c}(t) = (t, t^2)$.

(a) Compute the composition $(f\circ\mathbf{c})(t)$ and show that $(f\circ\mathbf{c})'(0) = 1$.

(b) Show that $\nabla f(\mathbf{c}(0)) \cdot \mathbf{c}'(0) = 0$.

(c) Explain why the results in (a) and (b) do not contradict the chain rule formula (2.25).

29. Assume that $\mathbf{F}\colon \mathbb{R}^2 \to \mathbb{R}^2$ is a differentiable function, and define $f(\mathbf{x}) = \cos\left(||\mathbf{F}(\mathbf{x})||^2\right)$. Find $Df(\mathbf{x})$.

▶ 2.7 GRADIENT AND DIRECTIONAL DERIVATIVE

We now use gradient vector fields to investigate rates of change of real-valued functions of several variables. We will be able to obtain (in this section, and in Chapter 4) valuable information that will provide us with a deeper understanding of the behavior and properties of real-valued functions.

Recall that the derivative $Df(\mathbf{x})$ of a differentiable real-valued function $f(\mathbf{x})$ of m variables $\mathbf{x} = (x_1, \dots x_m)$ is a $1 \times m$ matrix

$$Df(\mathbf{x}) = \begin{bmatrix} \frac{\partial f}{\partial x_1}(\mathbf{x}) & \frac{\partial f}{\partial x_2}(\mathbf{x}) & \cdots & \frac{\partial f}{\partial x_m}(\mathbf{x}) \end{bmatrix}.$$

This matrix, interpreted as a vector in $\mathbb{R}^m$, is denoted by $\nabla f(\mathbf{x})$ or sometimes by *grad* $f(\mathbf{x})$. It is called the gradient of f at $\mathbf{x}$.

DEFINITION 2.21 Gradient of a Function of Two and Three Variables

Let $f\colon U \subseteq \mathbb{R}^2 \to \mathbb{R}$ be a differentiable function. The *gradient* of f is the vector field ∇f whose value at a point (x, y) in U is given by

$$\nabla f(x, y) = \left(\frac{\partial f}{\partial x}(x, y), \frac{\partial f}{\partial y}(x, y)\right).$$

The gradient of a differentiable function $f\colon U \subseteq \mathbb{R}^3 \to \mathbb{R}$ is the vector field

$$\nabla f(x, y, z) = \left(\frac{\partial f}{\partial x}(x, y, z), \frac{\partial f}{\partial y}(x, y, z), \frac{\partial f}{\partial z}(x, y, z)\right),$$

defined on the subset U of $\mathbb{R}^3$. ◀

▶ EXAMPLE 2.75

Let $f(x, y) = x^2 + x \ln y$. Find the gradient vector field ∇f and compute its value $\nabla f(2, 1)$ at the point $(2, 1)$.

SOLUTION By definition, $\nabla f(x, y) = (2x + \ln y, x/y)$. The value of ∇f at $(2, 1)$ is computed to be $\nabla f(2, 1) = (2x + \ln y, x/y)|_{(2,1)} = (4, 2)$. ◀

▶ EXAMPLE 2.76

Let $f(x, y, z) = \sqrt{x^2 + y^2 + z^2}$; that is, the function f measures the distance from the origin. Compute the gradient of f.

SOLUTION Using the definition, we obtain

$$\begin{aligned}\nabla f(x, y, z) &= \left(x(x^2 + y^2 + z^2)^{-1/2}, y(x^2 + y^2 + z^2)^{-1/2}, z(x^2 + y^2 + z^2)^{-1/2}\right) \\ &= (x^2 + y^2 + z^2)^{-1/2} \cdot (x, y, z) = \frac{1}{||\mathbf{r}||}\mathbf{r},\end{aligned}$$

where $\mathbf{r} = x\mathbf{i} + y\mathbf{j} + z\mathbf{k}$ is the position vector of the point (x, y, z). In words, the gradient of the distance from a point to the origin is the unit vector in the direction of that point. ◀

Recall that the partial derivatives $(\partial f/\partial x)(a, b)$ and $(\partial f/\partial y)(a, b)$ give the rates of change of the function $f(x, y)$ at the point (a, b) in the directions of the coordinate axes (i.e., in the directions given by the unit vectors $\mathbf{i}$ and $\mathbf{j}$). Next, we generalize partial derivatives, so that we will be able to determine how a given function changes in *any* direction (i.e., in the direction of an arbitrary unit vector in $\mathbb{R}^2$).

Assume that $f(x, y)$ is a function differentiable on an open set $U \subseteq \mathbb{R}^2$, and let $\mathbf{p} = (a, b) \in U$. We are going to compute the rate of change of $f(x, y)$ at $\mathbf{p}$ in the direction of a unit vector $\mathbf{u} = (u, v)$.

The expression $\mathbf{l}(t) = \mathbf{p} + t\mathbf{u} = (a + tu, b + tv)$ represents the line in $\mathbb{R}^2$ that goes through the point $\mathbf{p} = \mathbf{l}(0)$ and whose direction is given by the direction of $\mathbf{u} = (u, v)$. Compute the value $f(\mathbf{p} + t\mathbf{u})$ of the function f for each point $\mathbf{p} + t\mathbf{u}$ on $\mathbf{l}(t)$ that is in U.

The collection of all points $\mathbf{c}(t) = (a + tu, b + tv, f(\mathbf{p} + t\mathbf{u}))$ forms a curve on the surface that is the graph of f. Notice that $\mathbf{c}(t)$ belongs to the plane perpendicular to the xy-plane that crosses it along the line $\mathbf{l}(t)$; see Figure 2.61. Denote the point $\mathbf{c}(0) = (a, b, f(a, b))$ by P. We define the directional derivative $D_{\mathbf{u}} f(a, b)$ as the slope of the tangent to $\mathbf{c}(t)$ at P. Thus, $D_{\mathbf{u}} f(a, b)$ describes how f changes in the direction specified by the unit vector $\mathbf{u}$.

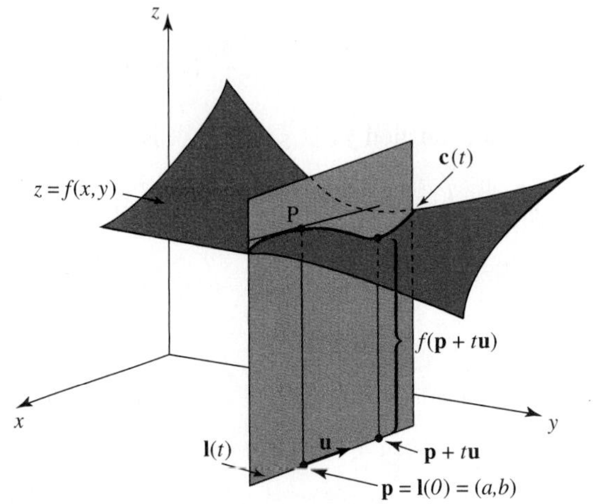

Figure 2.61 Directional derivative is the slope of the tangent.

DEFINITION 2.22 Directional Derivative

Let $f\colon U \subseteq \mathbb{R}^2 \to \mathbb{R}$ be a real-valued differentiable function. The *directional derivative* of f at the point $\mathbf{p} = (a, b)$ in the direction of the *unit* vector $\mathbf{u} = (u, v)$ is given by

$$D_{\mathbf{u}} f(a, b) = \frac{d}{dt} f(\mathbf{p} + t\mathbf{u})\Big|_{t=0}.$$

◀

Let us rephrase: what we are interested in are not the values of f everywhere in its domain, but only at the points on the line $\mathbf{l}(t)$ (specified by the given direction) that are in U; that is, we consider the so-called *restriction* of f to $\mathbf{l}(t)$. The directional derivative $D_{\mathbf{u}}f(a, b)$ gives the rate of change of that restriction at (a, b). If x and y represent distance along the coordinate axes, then the parameter t represents the distance along $\mathbf{l}(t)$ ($\mathbf{u}$ is a unit vector, and consequently, $\mathbf{p} + t\mathbf{u}$ is t units away from $\mathbf{p}$); thus, the directional derivative gives the rate of change with respect to *distance*.

From the limit definition of the derivative, it follows that

$$D_{\mathbf{u}}f(a, b) = \lim_{t \to 0} \frac{f(\mathbf{p} + t\mathbf{u}) - f(\mathbf{p})}{t} = \lim_{t \to 0} \frac{f((a, b) + t(u, v)) - f(a, b)}{t}. \tag{2.28}$$

The directional derivative of a function $f\colon U \subseteq \mathbb{R}^3 \to \mathbb{R}$ of three variables is defined and interpreted analogously.

▶ EXAMPLE 2.77

Compute the directional derivative of $f(x, y) = x^2 + 3xy$ in the direction of the vector $3\mathbf{i} + 4\mathbf{j}$ at the point $\mathbf{p} = (2, -1)$.

SOLUTION

First of all, we need a unit vector in the given direction: $\mathbf{u} = (3\mathbf{i} + 4\mathbf{j})/||3\mathbf{i} + 4\mathbf{j}|| = (3/5)\mathbf{i} + (4/5)\mathbf{j}$. By definition, $D_{\mathbf{u}}f(2, -1)$ is the derivative of the function

$$\begin{aligned} f(\mathbf{p} + t\mathbf{u}) &= f\left((2, -1) + t\left(\frac{3}{5}, \frac{4}{5}\right)\right) = f\left(2 + \frac{3}{5}t, -1 + \frac{4}{5}t\right) \\ &= \left(2 + \frac{3}{5}t\right)^2 + 3\left(2 + \frac{3}{5}t\right)\left(-1 + \frac{4}{5}t\right) = \frac{9}{5}t^2 + \frac{27}{5}t - 2 \end{aligned}$$

of *one* variable, evaluated at zero. Therefore,

$$D_{\mathbf{u}}f(2, -1) = \frac{d}{dt} f(\mathbf{p} + t\mathbf{u})\Big|_{t=0} = \frac{18}{5}t + \frac{27}{5}\Big|_{t=0} = \frac{27}{5}.$$

▶ EXAMPLE 2.78

Let $f\colon \mathbb{R}^2 \to \mathbb{R}$ be given by

$$f(x, y) = \begin{cases} \dfrac{x^2 y}{x^4 + y^2} & \text{if } (x, y) \neq (0, 0) \\ 0 & \text{if } (x, y) = (0, 0) \end{cases}$$

(a) Compute $D_{(1,0)}f(0, 0)$.

(b) Compute $D_{\mathbf{u}}f(0, 0)$, where $\mathbf{u} = (u, v)$ is a unit vector, and $v \neq 0$.

SOLUTION

(a) Using the limit version of the definition, that is, formula (2.28), we get

$$D_{(1,0)}f(0, 0) = \lim_{t \to 0} \frac{f((0, 0) + t(1, 0)) - f(0, 0)}{t} = \lim_{t \to 0} \frac{f(t, 0) - f(0, 0)}{t} = \lim_{t \to 0} \frac{0 - 0}{t} = 0.$$

(b) Similarly,

$$D_{\mathbf{u}} f(0,0) = \lim_{t\to 0} \frac{f((0,0)+t(u,v)) - f(0,0)}{t} = \lim_{t\to 0} \frac{f(tu,tv) - f(0,0)}{t}$$

$$= \lim_{t\to 0} \frac{1}{t} \frac{t^3u^2v}{t^4u^4 + t^2v^2} = \lim_{t\to 0} \frac{u^2v}{t^2u^4 + v^2} = \frac{u^2}{v}.$$

Thus, $D_{(u,v)} f(0,0) = u^2/v$, if $v \neq 0$. If $v = 0$, then $D_{(1,0)} f(0,0) = 0$, as shown in (a). ◀

The previous example shows how "pathological" a function of several variables can be. We showed that f has a derivative at the origin *in all directions,* but f is *not continuous* there [in Example 2.28 in Section 2.3 we showed that the limit of $x^2y/(x^4+y^2)$ as $(x,y) \to (0,0)$ does not exist].

Let us verify that the directional derivative is indeed a generalization of partial derivatives. From (2.28), with $\mathbf{u} = \mathbf{i} = (1,0)$, we get

$$D_{\mathbf{i}} f(a,b) = \lim_{t\to 0} \frac{f(a+t,b) - f(a,b)}{t} = \frac{\partial f}{\partial x}(a,b)$$

[recall defining formulas (2.11) and (2.12) for partial derivatives]. Similarly, $D_{\mathbf{j}} f(a,b) = (\partial f/\partial y)(a,b)$. For a function $f(x,y,z)$ of three variables, at a point $\mathbf{p} = (a,b,c)$, $D_{\mathbf{i}} f(\mathbf{p}) = (\partial f/\partial x)(\mathbf{p})$, $D_{\mathbf{j}}(\mathbf{p}) = (\partial f/\partial y)(\mathbf{p})$, and $D_{\mathbf{k}} f(\mathbf{p}) = (\partial f/\partial z)(\mathbf{p})$.

Next, we will obtain a workable description of the directional derivative. Having only the limit definition of the derivative could lead to technically involved and long computations.

THEOREM 2.8 Coordinate Description of the Directional Derivative

Let $f\colon U \subseteq \mathbb{R}^2 \to \mathbb{R}$ be a differentiable function, and $\mathbf{p} = (a,b) \in U$. Then

$$D_{\mathbf{u}} f(a,b) = \nabla f(a,b) \cdot \mathbf{u},$$

where $\mathbf{u} = (u,v)$ is a unit vector in $\mathbb{R}^2$. ◀

PROOF: Let $\mathbf{l}(t) = \mathbf{p} + t\mathbf{u} = (a+tu, b+tv)$; then $\mathbf{l}(0) = (a,b) = \mathbf{p}$ and $\mathbf{l}'(0) = \mathbf{u} = (u,v)$. Consider the composition $f(\mathbf{l}(t)) = f(\mathbf{p}+t\mathbf{u})$ of $\mathbf{l}$ and f. By the chain rule [see (2.25) in Section 2.6],

$$\frac{d}{dt} f(\mathbf{p}+t\mathbf{u}) = \frac{d}{dt} f(\mathbf{l}(t)) = \nabla f(\mathbf{l}(t)) \cdot \mathbf{l}'(t).$$

Consequently,

$$D_{\mathbf{u}} f(a,b) = \frac{d}{dt} f(\mathbf{p}+t\mathbf{u})\Big|_{t=0} = \nabla f(\mathbf{l}(0)) \cdot \mathbf{l}'(0) = \nabla f(a,b) \cdot \mathbf{u}.$$ ◀

This theorem actually states two facts: a differentiable function has a directional derivative in every direction, and it is computed as the dot product of the gradient of the function at the point in question and the unit vector in the desired direction. (Note that the converse is not true: the existence of all directional derivatives does not imply differentiability; see Example 2.78, and the comment following it.)

Note that Theorem 2.8 does not apply to Example 2.78, since the function $f(x, y)$ is not differentiable at (0, 0), and thus does not have a gradient there (despite having well-defined directional derivatives).

If f is a function of three variables, then

$$D_{\mathbf{u}} f(a, b, c) = \nabla f(a, b, c) \cdot \mathbf{u}, \tag{2.29}$$

where $\mathbf{u}$ is a unit vector in $\mathbb{R}^3$.

▶ EXAMPLE 2.79

Compute the rate of change of the function $f(x, y, z) = x^2 + y^2 + z^2$ in the direction of the vector $\mathbf{u} = (1, 1, 1)$ at the point $(1, -2, 0)$.

SOLUTION The requested rate of change is given by the directional derivative

$$D_{\mathbf{u}} f(1, -2, 0) = \nabla f(1, -2, 0) \cdot \frac{\mathbf{u}}{||\mathbf{u}||} = (2x, 2y, 2z)\Big|_{(1,-2,0)} \cdot \frac{1}{\sqrt{3}}(1, 1, 1)$$

$$= (2, -4, 0) \cdot (1/\sqrt{3}, 1/\sqrt{3}, 1/\sqrt{3}) = -2/\sqrt{3}.$$ ◀

Let us interpret the result of this exercise. Assume that we are located at the point $(1, -2, 0)$ in space. The fact that $D_{\mathbf{u}} f(1, -2, 0) = -2/\sqrt{3}$ means that, when we move one unit in the direction of the vector $(1, 1, 1)$, the function will change by approximately $-2/\sqrt{3}$ units.

It is because of this interpretation ("by approximately how many units does a function change if we walk one unit in the given direction") that we insist on using a unit vector in the definition of the directional derivative.

▶ EXAMPLE 2.80

The function $T(x, y) = 30e^{-x^2-y^2}$ gives the temperature in degrees Celsius at a location (x, y) in the plane (coordinates x and y are measured in meters). Compute $D_{\mathbf{u}} T(0, 1)$ if $\mathbf{u} = (\mathbf{i} - \mathbf{j})/\sqrt{2}$ and interpret your result.

SOLUTION Using Theorem 2.8 ($\mathbf{u}$ is unit vector), we get

$$\begin{aligned} D_{\mathbf{u}} T(0, 1) &= \nabla T(0, 1) \cdot (1/\sqrt{2}, -1/\sqrt{2}) \\ &= (-60xe^{-x^2-y^2}, -60ye^{-x^2-y^2})\Big|_{(0,1)} \cdot (1/\sqrt{2}, -1/\sqrt{2}) \\ &= (0, -60e^{-1}) \cdot (1/\sqrt{2}, -1/\sqrt{2}) = 60e^{-1}/\sqrt{2}. \end{aligned}$$

So, if we walk away from (0, 1) in the direction of $(\mathbf{i} - \mathbf{j})/\sqrt{2}$ (i.e., south-east) for 1 m, we will experience an increase in temperature of approximately $60e^{-1}/\sqrt{2} \approx 15.6$°C. ◀

▶ EXAMPLE 2.81 Directional Derivative of the Gravitational Potential

Consider the gravitational potential function $V(x, y, z) = -GMm/||\mathbf{r}||$ in $\mathbb{R}^3 - \{(0, 0, 0)\}$, where $\mathbf{r} = (x, y, z) = x\mathbf{i} + y\mathbf{j} + z\mathbf{k}$. The gradient of V is given by $\nabla V(x, y, z) = GMm\mathbf{r}/||\mathbf{r}||^3$ (see Example 2.40 in Section 2.4), and we compute the directional derivative in the unit direction $\mathbf{u}$ to be

$$D_{\mathbf{u}}V(x, y, z) = \nabla V(x, y, z) \cdot \mathbf{u} = \frac{GMm}{||\mathbf{r}||^3}\mathbf{r} \cdot \mathbf{u}.$$

Choose the point $(1, 0, 0)$; that is, let $\mathbf{r} = \mathbf{i}$. We compute $D_{\mathbf{i}}V(1, 0, 0) = GMm\mathbf{i} \cdot \mathbf{i} = GMm$. Similarly, $D_{-\mathbf{i}}V(1, 0, 0) = GMm\mathbf{i} \cdot (-\mathbf{i}) = -GMm$, $D_{\mathbf{j}}V(1, 0, 0) = GMm\mathbf{i} \cdot \mathbf{j} = 0$ and $D_{\mathbf{k}}V(1, 0, 0) = GMm\mathbf{i} \cdot \mathbf{k} = 0$. ◀

An interpretation of the results in the previous example will be given in Example 2.82.

THEOREM 2.9 Maximum Rate of Change of a Function

Let f be a differentiable function on $U \subseteq \mathbb{R}^2$ (or $\mathbb{R}^3$) and assume that $\nabla f(\mathbf{x}) \neq \mathbf{0}$ for $\mathbf{x} \in U$. The direction of the largest rate of increase in f at $\mathbf{x}$ is given by the vector ∇f. ◀

PROOF: Let $\mathbf{u}$ be a unit vector. By Theorem 2.8,

$$D_{\mathbf{u}}f(\mathbf{x}) = \nabla f(\mathbf{x}) \cdot \mathbf{u} = ||\nabla f(\mathbf{x})||\ ||\mathbf{u}|| \cos\theta = ||\nabla f(\mathbf{x})|| \cos\theta$$

(since $||\mathbf{u}|| = 1$), where θ denotes the angle between $\nabla f(\mathbf{x})$ and $\mathbf{u}$. Since $-1 \leq \cos\theta \leq 1$, $D_{\mathbf{u}}f(\mathbf{x})$ attains its largest value when $\cos\theta = 1$; that is, when $\theta = 0$. Consequently, maximum directional derivative $D_{\mathbf{u}}f(\mathbf{x})$ at $\mathbf{x}$ occurs in the direction parallel to the vector $\nabla f(\mathbf{x})$. ◀

When $\theta = 0$, $D_{\mathbf{u}}f(\mathbf{x}) = ||\nabla f(\mathbf{x})||$; that is, the magnitude of the largest rate of increase of f at $\mathbf{x}$ equals $||\nabla f(\mathbf{x})||$.

An argument similar to the one given in Theorem 2.9 shows that the function f decreases most rapidly in the direction opposite to the gradient (i.e., when $\cos\theta = -1$). The magnitude of that decrease equals $||\nabla f(\mathbf{x})||$ per unit length. From $D_{\mathbf{u}}f(\mathbf{x}) = \nabla f(\mathbf{x}) \cdot \mathbf{u}$, it follows that the rate of change in f in directions perpendicular to $\nabla f(\mathbf{x})$ is zero.

If $\nabla f(\mathbf{x}) = \mathbf{0}$, then $D_{\mathbf{u}}f(\mathbf{x}) = 0$ in all directions.

▶ EXAMPLE 2.82

We are now ready to interpret the results obtained in Example 2.81. Suppose that the mass m is located at the point $(1, 0, 0)$. The gradient of the potential is $\nabla V(1, 0, 0) = GMm\mathbf{i}$. The potential decreases most rapidly (at the rate GMm per unit distance) in the radial direction of $-\mathbf{i}$; that is, toward the mass M, located at the origin. The largest increase in the gravitational potential occurs in the (radial) direction away from the origin, at the rate GMm per unit distance. The rate of change of the potential in the $\mathbf{j}$ and $\mathbf{k}$ directions at $(1, 0, 0)$ is zero. These directions belong to the plane tangent to the level surface (so-called equipotential surface) through $(1, 0, 0)$; see Figure 2.62.

Note that the potential $V = -GMm/\|\mathbf{r}\|$ is always negative, so when it decreases it becomes large (in terms of magnitude). Thus, an accurate translation from mathematics to physics is: as we move away from the mass, the potential increases (mathematics), i.e., gets weaker (physics). As we move closer to the mass, the potential decreases (mathematics), i.e., grows stronger (physics).

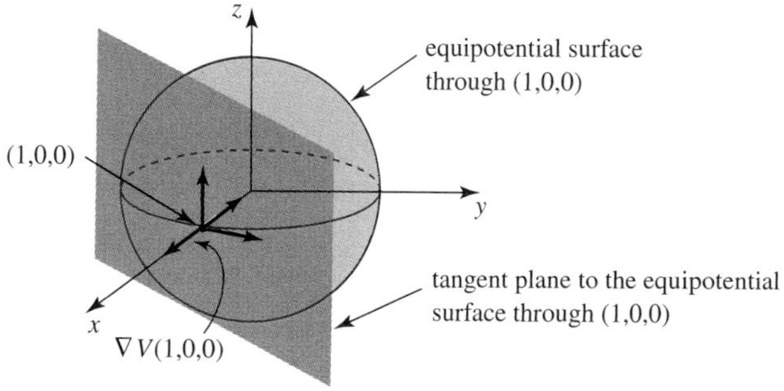

Figure 2.62 The gravitational potential at $(1, 0, 0)$ increases fastest in the direction away from the origin, perpendicular to the equipotential surface through $(1, 0, 0)$.

Note that $\mathbf{i}$ and $-\mathbf{i}$, that is, the directions of the most rapid increase and most rapid decrease of V, are perpendicular to the level surface of V through $(1, 0, 0)$. In Theorem 2.11 we will prove that this observation holds in general, for any differentiable function.

EXAMPLE 2.83

Let $f(x, y, z) = x^2 + y^2 + z^2$; that is, f is the distance squared from the origin. The function f increases most rapidly in the direction of its gradient $\nabla f(x, y, z) = (2x, 2y, 2z) = 2(x, y, z)$, that is, radially away from the origin. The maximum rate of change at a point (x, y, z) is $||\nabla f(x, y, z)|| = 2\sqrt{x^2 + y^2 + z^2}$.

Consider a C^1 function $f = f(x, y)\colon U \subseteq \mathbb{R}^2 \to \mathbb{R}$ and its level curve $\mathbf{c}$ given by $f(x, y) = C$, where C is a constant. This means that at all points on the level curve, the value of f is the same (and equal C), that is, $f(\mathbf{c}(t)) = C$ for all t, where $\mathbf{c}(t)$ is a parametrization of $\mathbf{c}$. Consequently, $(f(\mathbf{c}(t)))' = 0$ and since (by the chain rule) $(f(\mathbf{c}(t)))' = \nabla f(\mathbf{c}(t)) \cdot \mathbf{c}'(t)$, it follows that $\nabla f(\mathbf{c}(t)) \cdot \mathbf{c}'(t) = 0$. The conclusion of this argument is contained in our next theorem.

THEOREM 2.10 Gradient Vector in $\mathbb{R}^2$ Is Perpendicular to a Level Curve

Let $f\colon U \subseteq \mathbb{R}^2 \to \mathbb{R}$ be a differentiable function and let $\mathbf{c}(t)$ be a parametrization of its level curve. If $\nabla f(\mathbf{c}(t)) \neq \mathbf{0}$, then $\nabla f(\mathbf{c}(t))$ is perpendicular to the tangent vector $\mathbf{c}'(t)$.

In other words, the gradient vector (if nonzero) is always perpendicular to the level curves; see Figure 2.63.

To understand this fact better, let us consider a more intuitive argument. Pick a level curve of f of value c and a point (a, b) on it. Since f is differentiable (i.e., has the local linearity property), its level curves near (a, b) are approximated by parallel lines, as shown in Figure 2.64. Consider different paths from (a, b) to a level curve of value $c + \Delta c > c$ (the change in f is Δc). The largest rate of change of f will be obtained if we take the shortest path between the two level curves—which means that we need to move in the direction perpendicular to the level curves.

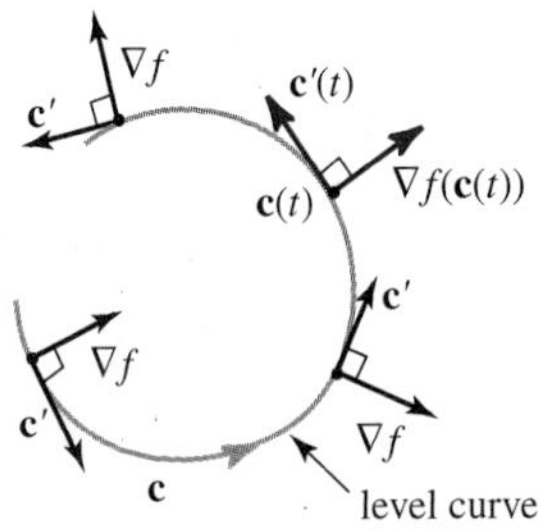

Figure 2.63 Gradient vector in $\mathbb{R}^2$ is perpendicular to a level curve.

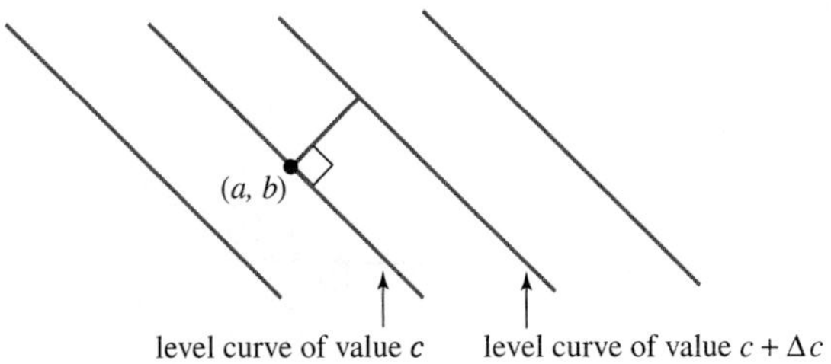

Figure 2.64 Contour diagram of f near a point (a, b).

▶ **EXAMPLE 2.84**

Compute a parametric equation of the line tangent to the graph of the equation $x^{2/3} + y^{2/3} = 1$ at the point $(1/\sqrt{8}, 1/\sqrt{8})$.

SOLUTION

Interpret the given curve as the level curve of the function $f(x, y) = x^{2/3} + y^{2/3}$ of value 1. By Theorem 2.10, the gradient

$$\nabla f(1/\sqrt{8}, 1/\sqrt{8}) = \left(\frac{2}{3x^{1/3}}, \frac{2}{3y^{1/3}}\right)\Bigg|_{(1/\sqrt{8},1/\sqrt{8})} = \frac{2\sqrt{2}}{3}\mathbf{i} + \frac{2\sqrt{2}}{3}\mathbf{j} = \frac{2\sqrt{2}}{3}(\mathbf{i}+\mathbf{j})$$

is perpendicular to the level curve in question. Consequently, the direction of the tangent is any vector perpendicular to $\mathbf{i}+\mathbf{j}$; for example, $\mathbf{i}-\mathbf{j} = (1, -1)$. So, a parametric equation of the tangent line is

$$\mathbf{l}(t) = (1/\sqrt{8}, 1/\sqrt{8}) + t(1, -1) = (1/\sqrt{8} + t, 1/\sqrt{8} - t), \qquad t \in \mathbb{R}.$$

We could have interpreted the equation $x^{2/3} + y^{2/3} = 1$ as $x^{2/3} + y^{2/3} - 1 = 0$ (i.e., as the level curve of $x^{2/3} + y^{2/3} - 1$ of value 0), or as $x^{2/3} + y^{2/3} + 6 = 7$ (i.e., as the level curve of $x^{2/3} + y^{2/3} + 6$ of value 7), or in many other ways. However, the answers would have been the same (the reader is encouraged to check this). ◀

Now assume that $f: U \subseteq \mathbb{R}^3 \to \mathbb{R}$ is differentiable and choose its level surface S defined by $f(x, y, z) = C$. Pick a point P in S and consider any curve $\mathbf{c}$ in S that passes through P [i.e., $\mathbf{c}(t) = P$ for some t]. Since the curve $\mathbf{c}$ belongs to the level surface S, it follows that $f(\mathbf{c}(t)) = C$. Continuing as before (see the text preceding Theorem 2.10), we see that $\nabla f(\mathbf{c}(t))$ is perpendicular to $\mathbf{c}'(t)$ at P for *any* curve $\mathbf{c}$ in S that contains P. In other words, $\nabla f(\mathbf{c}(t))$ (if nonzero) is perpendicular to all tangent vectors (at P) belonging to curves in S; consequently, it is perpendicular to the tangent plane to S at P (see Figure 2.65). This, by definition, means that $\nabla f(\mathbf{c}(t))$ is perpendicular to the level surface S.

THEOREM 2.11 Gradient Vector in $\mathbb{R}^3$ Is Perpendicular to a Level Surface

Let S be a level surface of a differentiable function $f: U \subseteq \mathbb{R}^3 \to \mathbb{R}$, and let (a, b, c) be a point in S. If $\nabla f(a, b, c) \neq \mathbf{0}$, then $\nabla f(a, b, c)$ is perpendicular (normal) to S at (a, b, c). ◀

We now use the above theorem to compute the equation of the plane tangent to the surface (which is the graph of) $z = f(x, y)$ at (a, b). Interpret the surface $z = f(x, y)$ as

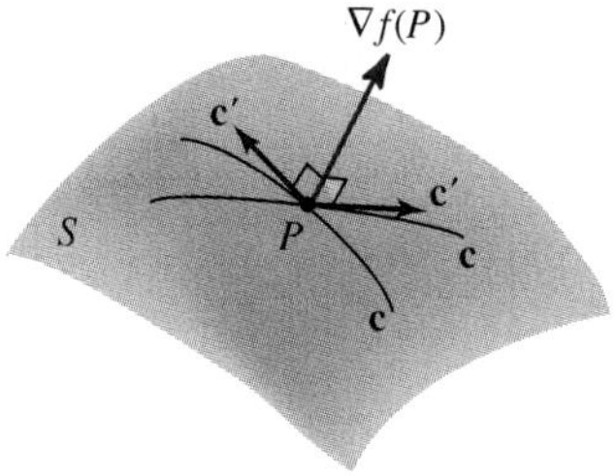

Figure 2.65 Gradient in $\mathbb{R}^3$ is perpendicular to a level surface.

the level surface $F(x, y, z) = 0$ of the function $F(x, y, z) = z - f(x, y)$ of value 0. By Theorem 2.11, the vector $\nabla F(a, b, c)$, where $c = f(a, b)$, is normal to the level surface (hence also normal to the tangent plane). If $\mathbf{x} = (x, y, z)$ is a point in the tangent plane other than $\mathbf{a} = (a, b, c)$, then $\nabla F(a, b, c)$ is perpendicular to $\mathbf{x} - \mathbf{a}$, and hence both $\nabla F(\mathbf{a}) \cdot (\mathbf{x} - \mathbf{a}) = 0$ and

$$\frac{\partial F}{\partial x}(a, b, c)(x - a) + \frac{\partial F}{\partial y}(a, b, c)(y - b) + \frac{\partial F}{\partial z}(a, b, c)(z - c) = 0$$

represent the plane tangent to the surface $z = f(x, y)$ (or $F(x, y, z) = 0$).

Since $F(x, y, z) = z - f(x, y)$, it follows that $\partial F/\partial x = -\partial f/\partial x$, $\partial F/\partial y = -\partial f/\partial y$, and $\partial F/\partial z = 1$, and we can rewrite the tangent plane equation as

$$z = c + \frac{\partial f}{\partial x}(a, b)(x - a) + \frac{\partial f}{\partial y}(a, b)(y - b).$$

This is precisely the equation we discussed in Section 2.4 in the context of the linear approximation of a function of several variables.

▶ **EXAMPLE 2.85**

Compute the equation of the plane tangent to the surface $z = x^2 + y^2 - 1$ at the point $(1, 0, 0)$.

SOLUTION

Think of the given surface as the level surface of $F(x, y, z) = z - x^2 - y^2 + 1$ of value 0. The normal to the surface is

$$\nabla F(1, 0, 0) = (-2x, -2y, 1)\Big|_{(1,0,0)} = (-2, 0, 1),$$

so the tangent plane is given by the equation $-2(x - 1) + 0(y - 0) + 1(z - 0) = 0$. Simplifying, we get $-2x + z + 2 = 0$. ◀

Let us gather together the properties of the gradient that we have discussed (and proved) so far in this section.

THEOREM 2.12 Properties of the Gradient

(a) If $\nabla f(\mathbf{x}) = \mathbf{0}$ at some point $\mathbf{x}$, then $D_{\mathbf{u}} f(\mathbf{x}) = 0$ in all directions.

(b) If $\nabla f(\mathbf{x}) \neq \mathbf{0}$, then there is a direction at $\mathbf{x}$ [given by $\nabla f(\mathbf{x})$], where f increases most rapidly. The magnitude of the most rapid increase is $||\nabla f(\mathbf{x})||$. The opposite direction, $-\nabla f(\mathbf{x})$, points in the direction of the most rapid decrease in f. The rate of change of f in the directions perpendicular to $\nabla f(\mathbf{x})$ is zero.

(c) If f is a differentiable function of two variables, then $\nabla f(\mathbf{x}) = \mathbf{0}$ (if nonzero) is perpendicular to the level curve of f that goes through $\mathbf{x}$. If the gradient vector $\nabla f(\mathbf{x})$ of a differentiable function of three variables is nonzero, then it is perpendicular to the level surface of f that contains $\mathbf{x}$.

We now explore a few examples to illustrate Theorem 2.12.

▶ EXAMPLE 2.86

Figure 2.66 shows contour curves on a topographic map [in other words, it is a contour diagram of the function $f(x, y)$ that measures the height above sea level].

(a) Explain why the vectors at points S and T could not represent the gradient of f.

(b) Explan why the vectors at points P, Q, and R could represent the gradient of f.

SOLUTION

(a) The vector at S is perpendicular to the level curve through S, but points in the wrong direction (toward a decrease, rather than toward an increase in f). The vector at T is not perpendicular to the level curve through T.

(b) All three vectors are perpendicular to corresponding level curves and point toward increasing values of f. Moreover, their relative size is correct: the contours $f(x, y) = 30$ and $f(x, y) = 40$ are closer together at R than at Q, and so the rate of change at R is larger. The vector at P is the smallest of the three vectors, because the change in height of 10 m requires the largest change in distance.

▶ EXAMPLE 2.87

Concentric circles in Figure 2.67 represent level curves of a function $f(x, y)$ whose graph is a cone with its vertex at the origin. Find $\nabla f(2, 0)$ and $\nabla f(1, 1)$.

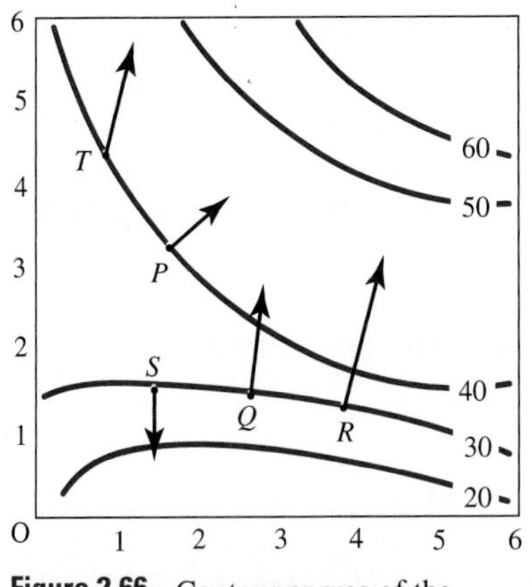

Figure 2.66 Contour curves of the height function of Example 2.86.

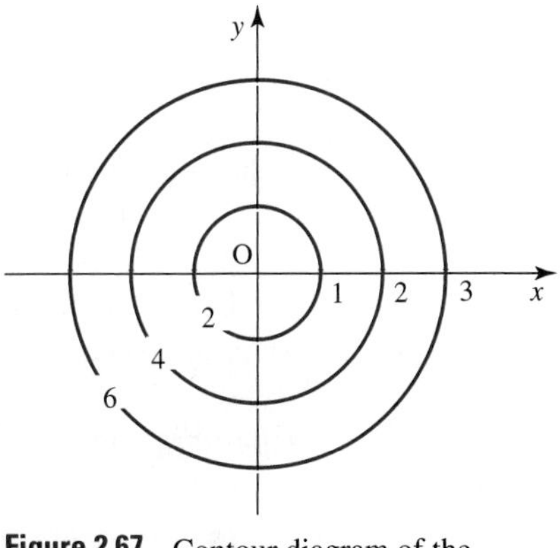

Figure 2.67 Contour diagram of the function in Example 2.87.

SOLUTION First note that all intersections of a cone with planes perpendicular to the xy-plane that contain the z-axis are lines of the same slope. In other words, $||\nabla f(x, y)||$ is the same at all points [except at $(0, 0)$, where the gradient is not defined]. Going from $(1, 0)$ to $(2, 0)$, the function changes by 2, and thus $||\nabla f(1, 0)|| = 2$. So, $||\nabla f(x, y)|| = 2$ for all $(x, y) \neq (0, 0)$.

Since $\nabla f(2, 0)$ must be perpendicular to the level curve $f(x, y) = 4$ and point toward increasing values of f, its direction is $\mathbf{i}$. Thus, $\nabla f(2, 0) = 2\mathbf{i}$. The unit direction perpendicular to the circle at $(1, 1)$ (and pointing toward increasing values of f) is $(\mathbf{i} + \mathbf{j})/\sqrt{2}$. It follows that $\nabla f(1, 1) = 2(\mathbf{i} + \mathbf{j})/\sqrt{2} = \sqrt{2}(\mathbf{i} + \mathbf{j})$.

▶ EXAMPLE 2.88

Let $f(x, y) = e^{-(x^2+y^2)}$. If $0 < C \leq 1$, then the level curves of f are circles $x^2 + y^2 = -\ln C$ of radius $\sqrt{-\ln C}$ centered at the origin. Level curve of value $C = 1$ collapses to a point (origin). There are no level curves of value $C > 1$ or $C \leq 0$.

Pick the level curve corresponding to $C = e^{-1}$ (which is the circle of radius 1) and the point $P(1/\sqrt{2}, 1/\sqrt{2})$ on it; see Figure 2.68. The gradient of f is $\nabla f(x, y) = -e^{-(x^2+y^2)}(2x, 2y)$. In particular, at P, $\nabla f(1/\sqrt{2}, 1/\sqrt{2}) = -e^{-1}(\sqrt{2}, \sqrt{2})$.

So, f increases most rapidly in the direction toward the origin. The maximum rate of change is $||\nabla f(1/\sqrt{2}, 1/\sqrt{2})|| = 2e^{-1}$. Since $\nabla f(1/\sqrt{2}, 1/\sqrt{2})$ points (radially) toward the origin, it must be perpendicular to the circle $x^2 + y^2 = 1$ (see Theorem 2.10).

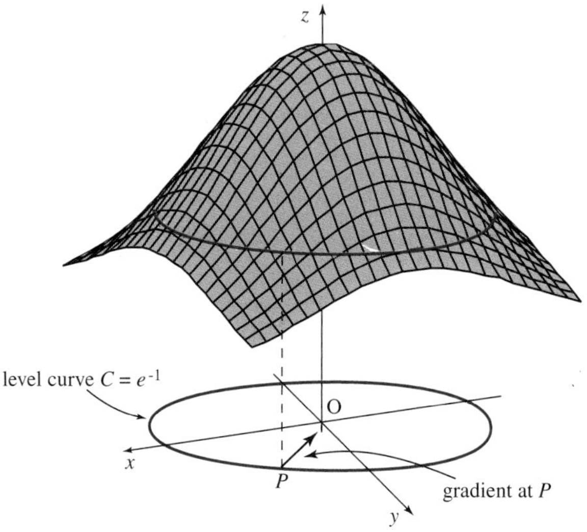

Figure 2.68 The surface and level curve of Example 2.88.

▶ EXAMPLE 2.89 Conservative Field and Its Potential Function

We have already seen that the gravitational force field can be obtained as the negative of the gradient of a scalar function V; that is, $\mathbf{F} = -\nabla V$ (see Example 2.40 in Section 2.4). In general, a vector field $\mathbf{F}$ with that property is called *conservative* and the corresponding scalar function V is the *potential function*. Examples of such fields are gravitational and electrostatic fields (see Section 2.1). Their properties with respect to integration (e.g., work) will be discussed in Chapter 5.

The gradient vector is perpendicular to level surfaces. Translated into the language of physics, this sentence states that the conservative force field $\mathbf{F} = -\nabla V$ is perpendicular to its *equipotential surfaces*. For example, the equipotential surfaces for the gravitational potential are spheres centered at the origin (see Example 2.23 in Section 2.2). The direction of the gradient vector field (i.e., the direction of the force) is radial (i.e., along lines through the origin), and clearly perpendicular to the spheres. ◀

▶ EXAMPLE 2.90 Heat Flux Vector Field

We know from experience that heat flows from regions of higher temperature toward regions of lower temperature. This process is described by the *heat flux vector field* $\mathbf{J} = -k\nabla T$, where $T = T(x, y, z)$ gives the temperature at a point (x, y, z) and $k > 0$ is a constant (called the *conductivity*). The direction of the gradient ∇T is the direction of the largest increase in temperature; the heat flows in the opposite direction; that is, in the direction $-\nabla T$ of the largest decrease in temperature. ◀

▶ EXAMPLE 2.91 Conservation of Energy

Assume that a particle of mass m moves along a curve $\mathbf{c}(t)$ in a conservative force field $\mathbf{F}$ (with potential energy function V). The quantity

$$E(t) = \tfrac{1}{2}m||\mathbf{c}'(t)||^2 + V(\mathbf{c}(t))$$

is called the *total energy*. It is the sum of the *potential energy* of the particle $V(\mathbf{c}(t))$ and its *kinetic energy* $m||\mathbf{c}'(t)||^2/2$. We are going to check the *Law of Conservation of Energy*, which states that the total energy is constant (i.e., energy only transforms from one form to another, but can neither be destroyed nor created). In the computation, Newton's Second Law $\mathbf{F}(\mathbf{c}(t)) = m\mathbf{c}''(t)$ will be used.

SOLUTION We will show that the derivative of the energy is zero. Using the product rule and the chain rule and writing $\|\mathbf{c}'(t)\|^2$ as $\mathbf{c}'(t) \cdot \mathbf{c}'(t)$, we get

$$\frac{d}{dt}E(t) = \frac{1}{2}m(\mathbf{c}'(t) \cdot \mathbf{c}'(t))' + (V(\mathbf{c}(t)))' = m\mathbf{c}'(t) \cdot \mathbf{c}''(t) + \nabla V(\mathbf{c}(t)) \cdot \mathbf{c}'(t),$$

and hence,

$$\frac{d}{dt}E(t) = (m\mathbf{c}''(t) + \nabla V(\mathbf{c}(t))) \cdot \mathbf{c}'(t).$$

The force $\mathbf{F}$ is conservative, so $\mathbf{F}(\mathbf{c}(t)) = -\nabla V(\mathbf{c}(t))$. Combining with Newton's Second Law, we get $m\mathbf{c}''(t) = -\nabla V(\mathbf{c}(t))$; that is, the first factor in the scalar product on the right side in $dE(t)/dt$ is zero. Hence, $dE(t)/dt = 0$, which implies that $E(t)$ is constant. ◀

▶ EXAMPLE 2.92 Model for Large-Scale Motions in Earth's Atmosphere

The mathematical model that is used to describe large-scale frictionless motions in Earth's atmosphere establishes the balance

$$\phi\,(\mathbf{k} \times \mathbf{v}) = -\rho^{-1}\nabla p \tag{2.30}$$

between the Coriolis force (left side) and horizontal pressure gradient force (right side). The positive constant ϕ is called the Coriolis constant, $\mathbf{v}$ gives wind direction and magnitude, $\rho > 0$ is the air density, and the function p represents horizontal air pressure.

The formula (2.30), written as $\nabla p = -\rho\phi\,(\mathbf{k} \times \mathbf{v})$, states that the pressure gradient vector ∇p is 90° clockwise from the wind [recall that the vector $\mathbf{k} \times \mathbf{v}$ is 90° counterclockwise from $\mathbf{v}$; see the $\mathbf{k}$-cross operator in Example 1.40 of Section 1.1].

Applying the $\mathbf{k}$-cross to (2.30), we get $\phi(\mathbf{k} \times (\mathbf{k} \times \mathbf{v})) = -\rho^{-1}(\mathbf{k} \times \nabla p)$, and $\phi\mathbf{v} = \rho^{-1}(\mathbf{k} \times \nabla p)$, since $\mathbf{k} \times (\mathbf{k} \times \mathbf{v}) = -\mathbf{v}$. This equation states that the wind blows in the direction 90° counterclockwise from the strongest increase in horizontal air pressure.

▶ EXAMPLE 2.93 Potential due to a Point Charge

The potential due to a point charge q located at $\mathbf{r}_0 = (x_0, y_0, z_0)$ in $\mathbb{R}^3$ is given by the formula

$$\Phi = \frac{1}{4\pi\epsilon_0}\frac{q}{||\mathbf{r} - \mathbf{r}_0||},$$

where ϵ_0 is a constant. The corresponding electrostatic field $\mathbf{E}$ at the point $\mathbf{r} = (x, y, z)$ is given by $\mathbf{E} = -\nabla\Phi$; that is,

$$\begin{aligned}
\mathbf{E} &= -\frac{q}{4\pi\epsilon_0}\nabla\left(\frac{1}{||\mathbf{r} - \mathbf{r}_0||}\right)\\
&= -\frac{q}{4\pi\epsilon_0}\nabla\left(((x - x_0)^2 + (y - y_0)^2 + (z - z_0)^2)^{-1/2}\right)\\
&= \frac{q}{4\pi\epsilon_0}\left((x - x_0)^2 + (y - y_0)^2 + (z - z_0)^2\right)^{-3/2}(x - x_0, y - y_0, z - z_0)\\
&= \frac{q}{4\pi\epsilon_0}\frac{\mathbf{r} - \mathbf{r}_0}{||\mathbf{r} - \mathbf{r}_0||^3}.
\end{aligned}$$

Let us clarify the use of the word "potential." In physics, one distinguishes between *potential* and *potential energy*. In general, potential = potential energy per unit "source." For example, electrostatic potential = electrostatic potential energy per unit charge; gravitational potential = gravitational potential energy per unit mass; etc. So if $\mathbf{F} = q\mathbf{E}$, ($\mathbf{F}$ is an electrostatic force and $\mathbf{E}$ the corresponding electrostatic field), then $\mathbf{F} = -\nabla V$, but $\mathbf{E} = -\nabla\Phi$, where $\Phi = V/q$. The function V is the electrostatic potential energy and Φ is the electrostatic potential. In math literature this distinction is somewhat blurred, and both the potential and potential energy are referred to as "potential."

The next example explains how to find a potential function for a given conservative vector field. (Not every vector field has a potential function. Necessary and sufficient conditions for the existence of potential functions will be discussed in Chapter 5.)

▶ EXAMPLE 2.94

The function $\mathbf{F}(x, y, z) = yz\mathbf{i} + (zx - 1)\mathbf{j} + xy\mathbf{k}$ defines a conservative vector field. Find its potential function $V(x, y, z)$.

SOLUTION From $\nabla V = -\mathbf{F} = -(yz, zx - 1, xy)$, it follows that $\partial V/\partial x = -yz$, $\partial V/\partial y = -xz + 1$, and $\partial V/\partial z = -xy$. Integrating the first equation with respect to x, we get

$$V(x, y, z) = -xyz + C(y, z),$$

where the "constant" $C(y, z)$ of integration might depend on y and z, since these two variables were viewed as constants. We managed to partially recover V. To compute $C(y, z)$, substitute the expression for $V(x, y, z)$ into $\partial V/\partial y$, thus obtaining

$$-zx + \frac{\partial C(y, z)}{\partial y} = -xz + 1,$$

which implies that $\partial C(y, z)/\partial y = 1$ and $C(y, z) = y + C(z)$, by integration with respect to y (the variable z was kept fixed, so the integration "constant" might still depend on z). Hence,

$$V(x, y, z) = -xyz + y + C(z).$$

Finally, substituting this expression into the equation for $\partial V/\partial z$, we get

$$-xy + C'(z) = -xy,$$

so that $C(z) = C$ after integrating with respect to z. (C is a real number, not a function any longer.) It follows that any function of the form

$$V(x, y, z) = -xyz + y + C$$

(where C is a real number) is a potential function for the given vector field. ◀

▶ EXERCISES 2.7

Exercises 1 to 5: Consider a contour diagram of a function $f(x, y)$ in Figure 2.69. Estimate the directional derivative $D_{\mathbf{v}} f(a, b)$ at the given point in the given direction.

1. $(a, b) = (2, 1)$, $\mathbf{v} = \mathbf{j}$

2. $(a, b) = (2, 1)$, $\mathbf{v} = -\mathbf{i} + \mathbf{j}$

3. $(a, b) = (3, 2)$, $\mathbf{v} = -2\mathbf{i} + \mathbf{j}$

4. $(a, b) = (3, 2)$, $\mathbf{v} = \mathbf{i}$

5. $(a, b) = (4, 1)$, $\mathbf{v} = -\mathbf{i}$

6. Consider the contour diagram in Figure 2.69. Draw gradient vectors at several points on the level curve $f(x, y) = 16$.

Exercises 7 to 11: Find the directional derivative of the function f at the point $\mathbf{p}$ in the direction of the vector $\mathbf{v}$.

7. $f(x, y) = e^{xy}(\cos x + \sin y)$, $\mathbf{p} = (\pi/2, 0)$, $\mathbf{v} = 2\mathbf{i} - \mathbf{j}$

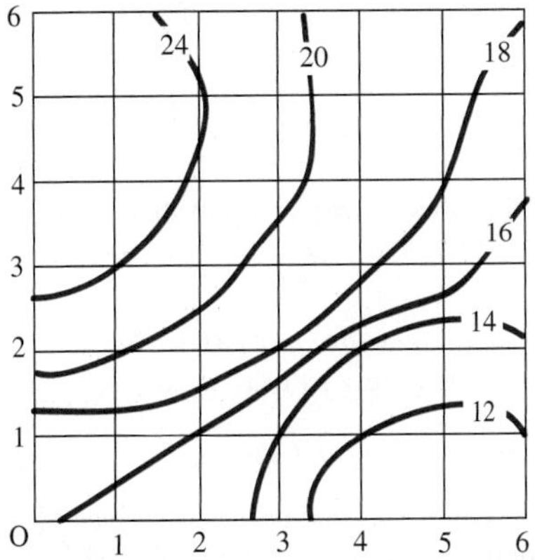

Figure 2.69 Contour diagram used in Exercises 1 to 6.

8. $f(x, y) = x^3y + 2x^2y^2 - xy^3$, $\mathbf{p} = (2, 3)$, $\mathbf{v} = (1, -1)$

9. $f(x, y, z) = e^{-x^2-y^2-z^2}$, $\mathbf{p} = (0, -1, 2)$, $\mathbf{v} = \mathbf{i} + \mathbf{j} + \mathbf{k}$

10. $f(x, y) = \arctan(y/x)$, $\mathbf{p} = (1, 1)$, $\mathbf{v} = \mathbf{i} - 4\mathbf{j}$

11. $f(x, y) = x \ln y^2 + 2y - 3$, $\mathbf{p} = (1, 2)$, $\mathbf{v} = 3\mathbf{i} + 4\mathbf{j}$

Exercises 12 to 14: Find the directional derivative of the function f at the point $\mathbf{p}$ in the direction given by the angle θ, measured from the positive direction of the x-axis in the counterclockwise direction.

12. $f(x, y) = xy^4 + x^2y^2 - 2$, $\mathbf{p} = (0, -1)$, $\theta = \pi/4$

13. $f(x, y) = e^{xy}$, $\mathbf{p} = (0, 1)$, $\theta = \pi/2$

14. $f(x, y) = \cos(2x + y)$, $\mathbf{p} = (2, -3)$, $\theta = -\pi/3$

15. Let $f: \mathbb{R}^2 \to \mathbb{R}$ be given by

$$f(x, y) = \begin{cases} \dfrac{2xy}{x^2 + y^2} & \text{if } (x, y) \neq (0, 0) \\ 0 & \text{if } (x, y) = (0, 0) \end{cases}.$$

Compute $D_{\mathbf{u}} f(0, 0)$, where $\mathbf{u} = (u, v)$ is a unit vector in $\mathbb{R}^2$.

16. Show that the function $f(x, y) = x^{1/3}y^{1/3}$ is continuous at $(0, 0)$ and has partial derivatives f_x and f_y at $(0, 0)$, but the directional derivative of f in any other direction does not exist.

Exercises 17 to 21: Determine the maximum rate of change of the function f at the point $\mathbf{p}$, and the direction in which it occurs.

17. $f(x, y) = \sec x \tan y$, $\mathbf{p} = (\pi/4, \pi/4)$

18. $f(x, y) = 2ye^x + e^{-y}$, $\mathbf{p} = (0, 0)$

19. $f(x, y, z) = xy^{-1} + yz^{-1} + zx^{-1}$, $\mathbf{p} = (1, 2, -1)$

20. $f(x, y, z) = \sqrt{xyz}$, $\mathbf{p} = (3, 3, 2)$

21. $f(x, y) = |xy|$, $\mathbf{p} = (3, -2)$

22. The temperature inside an object is given by $T(x, y, z) = 30(x^2 + y^2 + z^2)^{-1}$, at all points $(x, y, z) \neq (0, 0, 0)$.

(a) Find the rate of change of the temperature at the point $(1, 2, 0)$ inside the object in the direction toward the point $(2, -1, -1)$.

(b) Find the direction of the largest rate of increase in temperature at the point $(0, 0, 1)$.

(c) Find the direction of the most rapid decrease in temperature at a point (x, y, z) inside the object, if $(x, y, z) \neq (0, 0, 0)$.

23. The pressure $P(x, y)$ at a point $(x, y) \in \mathbb{R}^2$ on a metal membrane is given by the function $P(x, y) = 100e^{-x^2-2y^2}$.

(a) Find the rate of change of the pressure at the point $\mathbf{p} = (0, 1)$ in the direction $\mathbf{i} + \mathbf{j}$.

(b) In what direction away from the point $\mathbf{p}$ does the pressure increase most rapidly? Decrease most rapidly?

(c) Find the maximum rate of increase of pressure at $\mathbf{p}$.

(d) Locate the direction(s) at $\mathbf{p}$ in which the rate of change of pressure is zero.

24. Let $f(x, y) = e^x \cos(2x - y)$. Find the directional derivative of f at the point $(0, 1)$ in the direction of the line $y = 3x + 1$, for increasing values of x.

25. Consider the function $f(x, y) = 2xy$. In what directions at the point $(1, 2)$ is the directional derivative of f equal to 4?

26. Let $f(x, y) = 3x^2y - y^4 + 2$. Compute $\nabla f(2, 1)$. Identify all directions at the point $(2, 1)$ in which the rate of change of f is no larger than $||\nabla f(2, 1)||/2$.

27. Let $f(x, y)$ be a differentiable function. Identify all directions at a point (x, y) in which the rate of increase of f is at least 80% of the largest possible increase at that point.

28. The temperature produced by a source located at the origin is given by the formula $T(x, y, z) = e^{-(x^2+y^2+z^2)}$.

(a) Find the isothermal surfaces; that is, the surfaces on which the temperature is constant.

(b) What point is the warmest?

(c) What is the direction of the most rapid decrease in temperature at the point $(1, 2, -4)$?

(d) Describe the direction(s) at the point $(1, 2, -4)$ in which the rate of change of the temperature is no larger than $0.4e^{-21}$.

29. The direction of the gratest increase in height at some point on a hill, 15 m per 100 m, is toward the southwest. At this point, in the direction toward the west, how steep is the hill?

30. A climber is standing at a location $(2, 0, 2996)$ on a hill whose shape is given by $z = 3000 - x^2 + 2y^2 - xy$ meters.

(a) In which direction should the climber proceed if she wants to ascend the hill fastest?

(b) At what angle with respect to the horizontal is she climbing if she chooses the direction from (a)?

(c) Identify the curve along which the climber will walk if she decides to keep her altitude at 2996 m.

31. Assume that the function $f = f(x, y)$ has continuous partial derivatives. The directional derivative of f at $(0, 2)$ in the direction $\mathbf{i} + 2\mathbf{j}$ is 4 and the directional derivative of f at $(0, 2)$ in the direction $2\mathbf{i} - \mathbf{j}$ is 12. Find the directional derivative of f at $(0, 2)$ in the direction of the vector $3\mathbf{i} + 3\mathbf{j}$.

32. Consider the function $f(x, y) = 3x^2 - y^2 - 2$.

(a) Find the directional derivative of f in the direction of the tangent vector to the curve $\mathbf{c}(t) = (\cos t, t \sin t)$ at the point where $t = \pi/2$.

(b) Let $F(t) = f(\mathbf{c}(t))$. Compute $F'(\pi/2)$.

(c) What is the relation between the directional derivative in (a) and the rate of change $F'(\pi/2)$ in (b)?

Exercises 33 to 36: Let f and g be differentiable functions and let a, b, and n be constants. Prove the following identities.

33. $\nabla(af \pm bg) = a\nabla f \pm b\nabla g$

34. $\nabla(fg) = g\nabla f + f\nabla g$

35. $\nabla(f/g) = (g\,\nabla f - f\,\nabla g)/g^2$, at all points where g is not zero.

36. $\nabla f^n = nf^{n-1}\nabla f$

37. Find the acute angle between the surfaces $x^3y^3 - 3yz = 8$ and $x + 3z^2 = y^2 - 3$ at the point $(1, 2, 0)$.

Exercises 38 to 41: Find an equation of the tangent plane to the graph of the given surface at the point $\mathbf{p}$.

38. $x^2 - y^2 + z^2 = 2$, $\mathbf{p} = (0, 0, \sqrt{2})$

39. $\sin(xy) - 2\cos(yz) = 0$, $\mathbf{p} = (\pi/2, 1, \pi/3)$

40. $2e^{xyz} = 3$, $\mathbf{p} = (1, 1, \ln 1.5)$

41. $z^2 = \dfrac{4x - y}{x + y + 1}$, $\mathbf{p} = (1, 0, \sqrt{2})$

Exercises 42 to 46: Find a parametric equation of the line normal to the given surface at the point $\mathbf{p}$, and an equation of the plane tangent to the given surface at the same point.

42. $2x - 6y - z = -4$, $\mathbf{p} = (1, 2, -6)$

43. $(x-2)^2 + (y-3)^2 + z^2 - 4 = 0$, $\mathbf{p} = (2, 4, \sqrt{3})$

44. $xyz - 16 = 0$, $\mathbf{p} = (-2, 2, -4)$

45. $\ln(x^2 + y^2) - 2 = 0$, $\mathbf{p} = (0, e, 1)$

46. $e^x \cos y - e^x \cos z = 0$, $\mathbf{p} = (1, \pi/4, \pi/4)$

Exercises 47 to 50: Find the equations of the tangent line and the normal line to the given curve at the point $\mathbf{p}$.

47. $3x - 2y - 4 = 0$, $\mathbf{p} = (0, -2)$

48. $x^{3/2} + y^{3/2} = 1$, $\mathbf{p} = (0, 1)$

49. $e^x \sin y = 2$, $\mathbf{p} = (\ln 2, \pi/2)$

50. $y = x^{-2}$, $\mathbf{p} = (1/2, 4)$

51. Locate all points on the paraboloid $z = x^2 + y^2 - 5$ where the tangent plane is parallel to the plane $x + 3y - z = 0$.

52. Find an equation of the plane tangent to the sphere centered at $(2, 0, -1)$ at the point $(1, 3, 3)$.

53. Find unit normal vectors to the surface $\cos(xy) = e^z - 1$ at the point $(1, \pi/2, 0)$.

54. A particle is ejected from the surface $2x^2 + 4y^2 + z^2 = 16$ at the point $(2, 1, 2)$ in the direction perpendicular to the surface. Assuming that its trajectory is a straight line, and that it moves at a constant speed of 3 units/s, find its position 10 s later.

55. Show that a line normal to a sphere goes through its center.

56. Let $g(s, t) = f(s^2 t, s^2 - t^2)$, where f is a differentiable function. If $\nabla f(4, 3) = (3, -1)$, find $\nabla g(2, 1)$.

57. Check that the families of curves $xy = k$ and $x^2 - y^2 = m$ (k and m are constants) intersect orthogonally. Such families are sometimes called *orthogonal trajectories*.

58. Check that the families of curves $(x-k)^2 + y^2 = k^2$ and $x^2 + (y-m)^2 = m^2$ (k and m are constants) intersect orthogonally.

59. Show that the spheres $x^2 + y^2 + z^2 = 16$ and $(x-5)^2 + y^2 + z^2 = 9$ are orthogonal (i.e., corresponding normals at the points of intersection are orthogonal).

60. Show that $\nabla ||\mathbf{r}||^{-1} = -||\mathbf{r}||^{-3}\mathbf{r}$, where $\mathbf{r} = x\mathbf{i} + y\mathbf{j} + z\mathbf{k}$.

Exercises 61 to 65: The vector field $\mathbf{F}$ is conservative. Find its potential function (i.e., find V such that $\mathbf{F} = -\nabla V$).

61. $\mathbf{F}(x, y) = e^{xy}(1 + xy)\mathbf{i} + x^2 e^{xy}\mathbf{j}$

62. $\mathbf{F}(x, y) = (3x^2 - 3y^2)\mathbf{i} - 6xy\mathbf{j}$

63. $\mathbf{F}(x, y, z) = -(2xy^2 + 3x^2 z, 2x^2 y - z^3, x^3 - 3yz^2)$

64. $\mathbf{F}(x, y) = -y\cos(xy)\mathbf{i} - x\cos(xy)\mathbf{j}$

65. $\mathbf{F}(x, y) = (xy^2 + 3x^2 y)\mathbf{i} + (x^3 + yx^2)\mathbf{j}$

66. Define $f\colon \mathbb{R}^2 \to \mathbb{R}^2$ by $f(\mathbf{x}) = ||A\mathbf{x}||^2$, where A is a 2×2 matrix. Compute $\nabla f(\mathbf{x})$.

▶ 2.8 CYLINDRICAL AND SPHERICAL COORDINATE SYSTEMS

There are many ways of representing points in $\mathbb{R}^3$ other than using the Cartesian coordinates x, y, and z. Two commonly used sets of coordinates (e.g., in integration) are cylindrical and spherical coordinates.

DEFINITION 2.23 Cylindrical Coordinates r, θ, z

The *cylindrical coordinates* r, θ, z are related to Cartesian coordinates by

$$x = r\cos\theta, \quad y = r\sin\theta, \quad z = z, \tag{2.31}$$

where $0 \le \theta < 2\pi$, and $r \ge 0$.

In other words, cylindrical coordinates are a combination of polar coordinates in the xy-plane $\mathbb{R}^2$ and the z-axis (this is just a convention: we could have taken polar coordinates in the xz-plane and added the y-axis). They are normally used when the object involved exhibits symmetry with respect to an axis.

▶ EXAMPLE 2.95

The cylinder $x^2 + y^2 = a^2$ (whose axis of symmetry is the z-axis) has a particularly simple equation in cylindrical coordinates: $r = a$. The equation $r^2 = z^2$ in cylindrical coordinates represents the cone, which is given in Cartesian coordinates by $x^2 + y^2 = z^2$. The equation of the sphere $x^2 + y^2 + z^2 = a^2$ in Cartesian coordinates becomes $r^2 + z^2 = a^2$ when transformed into cylindrical coordinates.

The relations inverse to (2.31), expressing r, θ, and z in terms of Cartesian coordinates x, y, and z, are given by (1.1) and (1.2) in Section 1.1, and by $z = z$.

▶ EXAMPLE 2.96

This example illustrates the conversion between Cartesian and cylindrical coordinate systems.

(a) Find cylindrical coordinates of the points with Cartesian coordinates $A_1(2, 5, -2)$, $A_2(-1, 1, 3)$, and $A_3(0, -4, 2)$.

(b) The points B_1 and B_2 have cylindrical coordinates $B_1(2, \pi/4, 1)$ and $B_2(1, 3\pi/2, -4)$. Find their Cartesian coordinates.

(c) Express the equation of the double cone $4x^2 + 4y^2 = z^2$ in cylindrical coordinates.

(d) A surface has the equation $r^2 + z^2 = z$ in cylindrical coordinates. Convert the equation to Cartesian coordinates and identify the surface.

SOLUTION All we need are the conversion formulas (2.31) and (1.2).

(a) Let us find the cylindrical coordinates of A_1. Since $x = 2$ and $y = 5$, it follows that $r = \sqrt{2^2 + 5^2} = \sqrt{29}$ and $\arctan(y/x) = \arctan 2.5 \approx 1.19$ rad. Therefore, $\theta \approx 1.19$ and $(\sqrt{29}, 1.19, -2)$ are the cylindrical coordinates of A_1. For A_2, we get $r = \sqrt{2}$ and $\arctan(y/x) = \arctan(-1) = -\pi/4$. Therefore, $\theta = -\pi/4 + \pi = 3\pi/4$ and the cylindrical coordinates of A_2 are $(\sqrt{2}, 3\pi/4, 3)$. For A_3, we get $r = \sqrt{16} = 4$ and (since $x = 0$ and $y = -4 < 0$) $\theta = 3\pi/2$. Hence, A_3 has cylindrical coordinates $A_3(4, 3\pi/2, 2)$.

(b) The Cartesian coordinates of B_1 are $x = r\cos\theta = 2\cos(\pi/4) = \sqrt{2}$, $y = r\sin\theta = 2\sin(\pi/4) = \sqrt{2}$ and $z = 1$. For the point B_2, we get $x = 2\cos(3\pi/2) = 0$, $y = 2\sin(3\pi/2) = -1$, and $z = -4$.

(c) Since $x^2 + y^2 = r^2$, the equation of the double cone $4x^2 + 4y^2 = z^2$ in cylindrical coordinates is $4r^2 = z^2$.

(d) From $r^2 = x^2 + y^2$, we get $x^2 + y^2 + z^2 = z$ and (complete the square) $x^2 + y^2 + (z - 1/2)^2 = 1/4$. The surface in question is the sphere centered at $(0, 0, 1/2)$ of radius $1/2$.

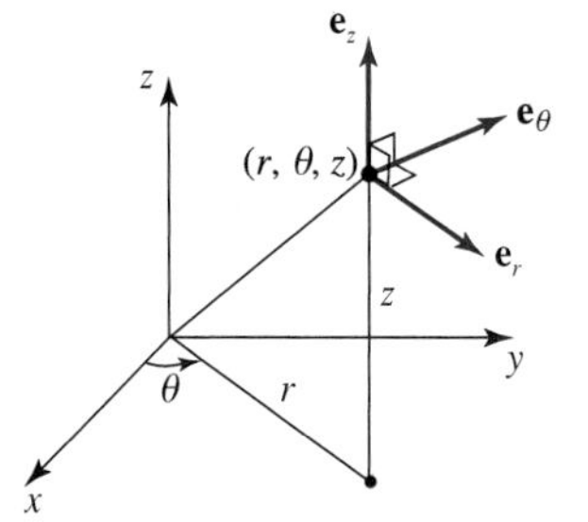

Figure 2.70 Cylindrical coordinates with unit vectors $\mathbf{e}_r$, $\mathbf{e}_\theta$, and $\mathbf{e}_z$.

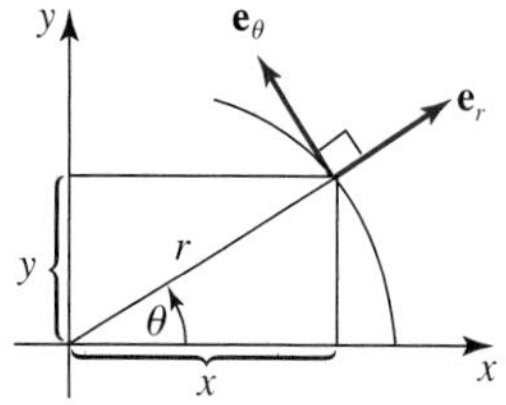

Figure 2.71 Unit vectors that correspond to polar coordinates r and θ.

The point $(1, 1, 1)$ in cylindrical coordinates is represented in Cartesian coordinates as $(\cos 1, \sin 1, 1) = (0.540, 0.841, 1)$. The point $(1, 1, 1)$ in Cartesian coordinates is represented as $(\sqrt{2}, \pi/4, 1) = (1.414, 0.785, 1)$ in cylindrical coordinates. Obviously, it is important to know what coordinates we are using—therefore, to avoid ambiguity, we must say "point $(1, 1, 1)$ in Cartesian coordinates" or "point $(1, 1, 1)$ in cylindrical coordinates." By convention, when no coordinate system has been specified, we use Cartesian coordinates.

Let us describe the unit vectors $\mathbf{e}_r$, $\mathbf{e}_\theta$, and $\mathbf{e}_z$ for the cylindrical coordinates r, θ, and z; see Figure 2.70. In general, a unit vector that corresponds to a coordinate function has a direction in which that coordinate increases while the remaining one(s) is (are) kept constant.

Since we took the z-axis from the Cartesian coordinate system, $\mathbf{e}_z = \mathbf{k}$. To find $\mathbf{e}_r$ and $\mathbf{e}_\theta$, we consider the polar coordinate system in the xy-plane. The vector $\mathbf{e}_r$ is the unit vector whose direction at a point (r, θ) is the direction in which r increases and θ remains constant; see Figure 2.71. It follows that

$$\mathbf{e}_r = \frac{\mathbf{r}}{||\mathbf{r}||} = \frac{x\mathbf{i} + y\mathbf{j}}{\sqrt{x^2 + y^2}} = \frac{x}{\sqrt{x^2 + y^2}}\mathbf{i} + \frac{y}{\sqrt{x^2 + y^2}}\mathbf{j} = \cos\theta\mathbf{i} + \sin\theta\mathbf{j}.$$

The unit vector $\mathbf{e}_\theta$ has to be perpendicular to $\mathbf{e}_r$ (take any vector in the xy-plane whose dot product with $\mathbf{e}_r$ is zero and divide your choice by its norm) and therefore $\mathbf{e}_\theta = \pm(\sin\theta\mathbf{i} - \cos\theta\mathbf{j})$. From Figure 2.71 it follows that, in the first quadrant, the direction of increasing values of θ corresponds to the negative $\mathbf{i}$ and the positive $\mathbf{j}$ components. Therefore, we must choose $\mathbf{e}_\theta = -\sin\theta\mathbf{i} + \cos\theta\mathbf{j}$, which you may verify is valid for all θ. Hence,

$$\begin{aligned} \mathbf{e}_r &= \cos\theta\mathbf{i} + \sin\theta\mathbf{j} \\ \mathbf{e}_\theta &= -\sin\theta\mathbf{i} + \cos\theta\mathbf{j} \\ \mathbf{e}_z &= \mathbf{k}. \end{aligned} \tag{2.32}$$

Vectors $\mathbf{e}_r$, $\mathbf{e}_\theta$, and $\mathbf{e}_z$ are called *orthonormal* (= orthogonal + normal), since they are perpendicular to each other and of unit length. The set $\{\mathbf{e}_r, \mathbf{e}_\theta, \mathbf{e}_z\}$ is called an *orthonormal basis of* $\mathbb{R}^3$. A vector $\mathbf{F}$ can be represented as $\mathbf{F} = F_r\mathbf{e}_r + F_\theta\mathbf{e}_\theta + F_z\mathbf{e}_z$, where

$$F_r = \mathbf{F}\cdot\mathbf{e}_r, \quad F_\theta = \mathbf{F}\cdot\mathbf{e}_\theta, \quad \text{and} \quad F_z = \mathbf{F}\cdot\mathbf{e}_z. \tag{2.33}$$

To verify this fact, compute the dot product of $\mathbf{F}$ with $\mathbf{e}_r$:

$$\mathbf{F}\cdot\mathbf{e}_r = F_r\mathbf{e}_r\cdot\mathbf{e}_r + F_\theta\mathbf{e}_\theta\cdot\mathbf{e}_r + F_z\mathbf{e}_z\cdot\mathbf{e}_r,$$

and use orthonormality ($\mathbf{e}_r\cdot\mathbf{e}_r = ||\mathbf{e}_r||^2 = 1$, $\mathbf{e}_\theta\cdot\mathbf{e}_r = 0$, and $\mathbf{e}_z\cdot\mathbf{e}_r = 0$) to get $F_r = \mathbf{F}\cdot\mathbf{e}_r$. The expressions for F_θ and F_z are checked analogously.

▶ EXAMPLE 2.97

Represent the vector field $\mathbf{F}(x, y, z) = xy\mathbf{i} + x^2z\mathbf{k}$ in cylindrical coordinates.

SOLUTION The components of $\mathbf{F}$ in cylindrical coordinates are computed from (2.33) by substituting (2.31) and (2.32):

$$\begin{aligned}
F_r &= \mathbf{F}\cdot\mathbf{e}_r = (r^2\cos\theta\sin\theta\mathbf{i} + r^2z\cos^2\theta\mathbf{k})\cdot(\cos\theta\mathbf{i} + \sin\theta\mathbf{j}) = r^2\cos^2\theta\sin\theta,\\
F_\theta &= \mathbf{F}\cdot\mathbf{e}_\theta = (r^2\cos\theta\sin\theta\mathbf{i} + r^2z\cos^2\theta\mathbf{k})\cdot(-\sin\theta\mathbf{i} + \cos\theta\mathbf{j}) = -r^2\cos\theta\sin^2\theta,\\
F_z &= \mathbf{F}\cdot\mathbf{e}_z = (r^2\cos\theta\sin\theta\mathbf{i} + r^2z\cos^2\theta\mathbf{k})\cdot\mathbf{k} = r^2z\cos^2\theta.
\end{aligned}$$

Therefore, $\mathbf{F}(r, \theta, z) = r^2\cos^2\theta\sin\theta\mathbf{e}_r - r^2\cos\theta\sin^2\theta\mathbf{e}_\theta + r^2z\cos^2\theta\mathbf{e}_z$. ◀

Solving equations (2.32) for unit vectors $\mathbf{i}$, $\mathbf{j}$, and $\mathbf{k}$, we get

$$\begin{aligned}
\mathbf{i} &= \cos\theta\mathbf{e}_r - \sin\theta\mathbf{e}_\theta\\
\mathbf{j} &= \sin\theta\mathbf{e}_r + \cos\theta\mathbf{e}_\theta\\
\mathbf{k} &= \mathbf{e}_z.
\end{aligned} \tag{2.34}$$

These formulas, combined with (2.31), provide an alternative way of expressing a vector in cylindrical coordinates: the vector field $\mathbf{F}$ of Example 2.97 can be written as

$$\begin{aligned}
\mathbf{F} = xy\mathbf{i} + x^2z\mathbf{k} &= r^2\cos\theta\sin\theta(\cos\theta\mathbf{e}_r - \sin\theta\mathbf{e}_\theta) + r^2z\cos^2\theta\mathbf{e}_z\\
&= r^2\cos^2\theta\sin\theta\mathbf{e}_r - r^2\cos\theta\sin^2\theta\mathbf{e}_\theta + r^2z\cos^2\theta\mathbf{e}_z.
\end{aligned}$$

▶ EXAMPLE 2.98 Velocity and Acceleration in Cylindrical Coordinates

Let $\mathbf{c}(t) = x(t)\mathbf{i} + y(t)\mathbf{j} + z(t)\mathbf{k}$ be a position vector of an object (in Cartesian coordinates). Recall that $\mathbf{v}(t) = \mathbf{c}'(t)$ is the velocity and $\mathbf{a}(t) = \mathbf{v}'(t)$ the acceleration. Compute the expressions for the velocity and acceleration in cylindrical coordinates $x = r\cos\theta$, $y = r\sin\theta$, $z = z$ [coordinate functions are now functions of time: $r = r(t)$, $\theta = \theta(t)$, and $z = z(t)$].

SOLUTION We first express the position vector $\mathbf{c}(t)$ in cylindrical coordinates (drop t to keep notation simple):

$$\begin{aligned}
\mathbf{c}(t) &= x\mathbf{i} + y\mathbf{j} + z\mathbf{k}\\
&= r\cos\theta(\cos\theta\mathbf{e}_r - \sin\theta\mathbf{e}_\theta) + r\sin\theta(\sin\theta\mathbf{e}_r + \cos\theta\mathbf{e}_\theta) + z\mathbf{e}_z\\
&= r\cos^2\theta\mathbf{e}_r - r\cos\theta\sin\theta\mathbf{e}_\theta + r\sin^2\theta\mathbf{e}_r + r\cos\theta\sin\theta\mathbf{e}_\theta + z\mathbf{e}_z\\
&= r\mathbf{e}_r + z\mathbf{e}_z.
\end{aligned}$$

The velocity is computed from $\mathbf{c}(t)$ using the product rule:

$$\mathbf{v}(t) = \frac{d}{dt}(\mathbf{c}(t)) = \frac{dr}{dt}\mathbf{e}_r + r\frac{d\mathbf{e}_r}{dt} + \frac{dz}{dt}\mathbf{e}_z + z\frac{d\mathbf{e}_z}{dt}.$$

The derivatives of unit vectors are computed from (2.32) using the chain rule (keep in mind that θ is a function of t):

$$\frac{d\mathbf{e}_r}{dt} = -\sin\theta\frac{d\theta}{dt}\mathbf{i} + \cos\theta\frac{d\theta}{dt}\mathbf{j} = (-\sin\theta\mathbf{i} + \cos\theta\mathbf{j})\frac{d\theta}{dt} = \frac{d\theta}{dt}\mathbf{e}_\theta,$$
$$\frac{d\mathbf{e}_\theta}{dt} = -\cos\theta\frac{d\theta}{dt}\mathbf{i} - \sin\theta\frac{d\theta}{dt}\mathbf{j} = -(\cos\theta\mathbf{i} + \sin\theta\mathbf{j})\frac{d\theta}{dt} = -\frac{d\theta}{dt}\mathbf{e}_r,$$

and $d\mathbf{e}_z/dt = d\mathbf{k}/dt = 0$. Substituting $d\mathbf{e}_r/dt$, and $d\mathbf{e}_z/dt$ into the expression for $\mathbf{v}(t)$, we get

$$\mathbf{v}(t) = \frac{dr}{dt}\mathbf{e}_r + r\frac{d\theta}{dt}\mathbf{e}_\theta + \frac{dz}{dt}\mathbf{e}_z. \tag{2.35}$$

The acceleration $\mathbf{a}(t) = \mathbf{v}'(t)$ is computed similarly: apply the product rule to $\mathbf{v}(t)$

$$\mathbf{a}(t) = \frac{d^2r}{dt^2}\mathbf{e}_r + \frac{dr}{dt}\frac{d\mathbf{e}_r}{dt} + \frac{dr}{dt}\frac{d\theta}{dt}\mathbf{e}_\theta + r\frac{d^2\theta}{dt^2}\mathbf{e}_\theta + r\frac{d\theta}{dt}\frac{d\mathbf{e}_\theta}{dt} + \frac{d^2z}{dt^2}\mathbf{e}_z + \frac{dz}{dt}\frac{d\mathbf{e}_z}{dt}$$

and use the expressions for the derivatives of unit vectors to get

$$= \frac{d^2r}{dt^2}\mathbf{e}_r + \frac{dr}{dt}\frac{d\theta}{dt}\mathbf{e}_\theta + \frac{dr}{dt}\frac{d\theta}{dt}\mathbf{e}_\theta + r\frac{d^2\theta}{dt^2}\mathbf{e}_\theta + r\frac{d\theta}{dt}\left(-\frac{d\theta}{dt}\right)\mathbf{e}_r + \frac{d^2z}{dt^2}\mathbf{e}_z$$
$$= \left(\frac{d^2r}{dt^2} - r\left(\frac{d\theta}{dt}\right)^2\right)\mathbf{e}_r + \left(r\frac{d^2\theta}{dt^2} + 2\frac{dr}{dt}\frac{d\theta}{dt}\right)\mathbf{e}_\theta + \left(\frac{d^2z}{dt^2}\right)\mathbf{e}_z.$$

Notice that $d\mathbf{e}_r/dt = (d\theta/dt)\mathbf{e}_\theta \neq \mathbf{0}$ and $d\mathbf{e}_\theta/dt = -(d\theta/dt)\mathbf{e}_r \neq \mathbf{0}$, whereas in Cartesian coordinates $d\mathbf{i}/dt = d\mathbf{j}/dt = d\mathbf{k}/dt = \mathbf{0}$.

DEFINITION 2.24 Spherical Coordinates ρ, θ, ϕ

The point $(x, y, z) \in \mathbb{R}^3$ is represented in *spherical coordinates* using the following data [$\mathbf{r} = x\mathbf{i} + y\mathbf{j} + z\mathbf{k}$ denotes the position vector of (x, y, z); see Figure 2.72]:

(a) Distance $\rho = ||\mathbf{r}|| = \sqrt{x^2 + y^2 + z^2} \geq 0$ from the origin.

(b) Angle θ ($0 \leq \theta < 2\pi$) in the xy-plane (measured counterclockwise) between the x-axis and the projection $\mathbf{r}_p$ of the position vector $\mathbf{r} = x\mathbf{i} + y\mathbf{j} + z\mathbf{k}$ onto the xy-plane.

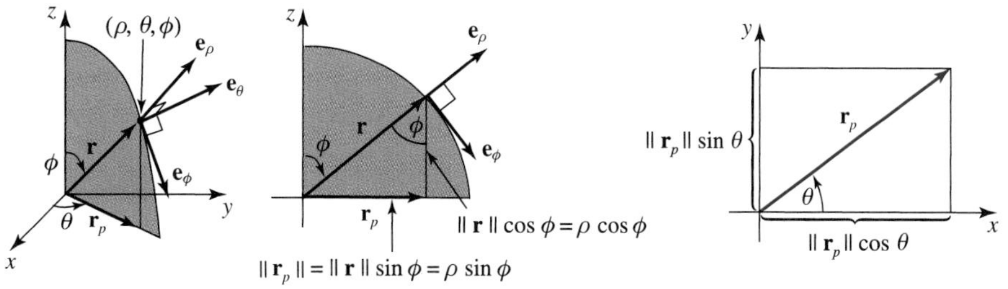

Figure 2.72 Spherical coordinate system and the corresponding unit vectors; view in the plane containing $\mathbf{r}_p$ and the z-axis; view in the xy-plane.

(c) Angle ϕ (in the plane containing the z-axis and the position vector $\mathbf{r}$, measured from the positive direction of the z-axis), $0 \le \phi \le \pi$. If a point lies on the z-axis, then $\phi = 0$ if $z \ge 0$ and $\phi = \pi$ if $z < 0$.

Since $\sin\phi = ||\mathbf{r}_p||/\rho$, i.e., $||\mathbf{r}_p|| = \rho\sin\phi$, it follows that

$$\begin{aligned} x &= ||\mathbf{r}_p||\cos\theta = \rho\sin\phi\cos\theta \\ y &= ||\mathbf{r}_p||\sin\theta = \rho\sin\phi\sin\theta \\ z &= \rho\cos\phi. \end{aligned} \tag{2.36}$$

The coordinates ρ, θ, and ϕ are called *spherical coordinates* in $\mathbb{R}^3$. They are customarily used in dealing with spherically symmetric objects (i.e., objects that are symmetric with respect to a point). For example (remember that $x^2 + y^2 + z^2 = ||\mathbf{r}||^2 = \rho^2$), the equation of the sphere $x^2 + y^2 + z^2 = a^2$ in Cartesian coordinates is $\rho = a$ in sphereical coordinates.

The inverse relations, expressing ρ, θ, and ϕ in terms of x, y, and z, are

$$\begin{aligned} \rho &= \sqrt{x^2 + y^2 + z^2} \\ \theta &= \begin{cases} \arctan(y/x) & \text{if } x > 0 \text{ and } y \ge 0 \\ \arctan(y/x) + \pi & \text{if } x < 0 \\ \arctan(y/x) + 2\pi & \text{if } x > 0 \text{ and } y < 0 \end{cases} \\ \phi &= \arccos(z/\rho). \end{aligned} \tag{2.37}$$

Note that the spherical coordinate θ is the same as the coordinate θ in polar (and cylindrical) coordinates; see Section 1.1, formulas (1.1) and (1.2), and the text following (1.2).

▶ EXAMPLE 2.99

This example illustrates the use of conversion formulas between Cartesian and spherical coordinate systems.

(a) Compute the spherical coordinates of the points whose Cartesian coordinates are $A_1(2, -2, 3)$ and $A_2(1, 1, -1)$.

(b) Spherical coordinates of a point in $\mathbb{R}^3$ are $(2, \pi/4, 2\pi/3)$. Find its Cartesian coordinates.

(c) Express the equation of the sphere $x^2 + y^2 + (z-1)^2 = 1$ in spherical coordinates.

SOLUTION

We make use of (2.36) and (2.37).

(a) The Cartesian coordinates of A_1 are $x = 2$, $y = -2$, and $z = 3$. Therefore, $\rho = \sqrt{x^2 + y^2 + z^2} = \sqrt{2^2 + (-2)^2 + 3^2} = \sqrt{17}$, $\arctan(y/x) = \arctan(-1) = -\pi/4$ so that $\theta = -\pi/4 + 2\pi = 7\pi/4$, and $\phi = \arccos(3/\sqrt{17}) \approx 0.76$ rad. It follows that, in spherical coordinates, $A_1(\sqrt{17}, 7\pi/4, 0.76)$. Similarly, the spherical coordinates of A_2 are $\rho = \sqrt{3}$, $\arctan(y/x) = \pi/4$ so that $\theta = \pi/4$, and $\phi = \arccos(-1/\sqrt{3}) \approx 2.19$ rad.

(b) The Cartesian coordinates of the point $(2, \pi/4, 2\pi/3)$ are $x = 2\sin(2\pi/3)\cos(\pi/4) = 2(\sqrt{3}/2)(\sqrt{2}/2) = \sqrt{6}/2$, $y = 2\sin(2\pi/3)\sin(\pi/4) = 2(\sqrt{3}/2)(\sqrt{2}/2) = \sqrt{6}/2$, and $z = 2\cos(2\pi/3) = 2(-1/2) = -1$.

(c) Recall that $x^2 + y^2 + z^2 = \rho^2$. From $x^2 + y^2 + (z-1)^2 = 1$, we get $x^2 + y^2 + z^2 - 2z = 0$ and hence $\rho^2 - 2\rho\cos\theta = 0$; that is, $\rho = 2\cos\theta$.

Let us compute the unit vectors e_ρ, e_θ, and e_ϕ for spherical coordinates. Vector e_ρ is the unit vector in the radial direction; hence,

$$\begin{aligned} e_\rho &= \frac{x\mathbf{i} + y\mathbf{j} + z\mathbf{k}}{\sqrt{x^2+y^2+z^2}} = \frac{\rho \sin\phi \cos\theta \mathbf{i} + \rho \sin\phi \sin\theta \mathbf{j} + \rho \cos\phi \mathbf{k}}{\rho} \\ &= \sin\phi \cos\theta \mathbf{i} + \sin\phi \sin\theta \mathbf{j} + \cos\phi \mathbf{k}. \end{aligned}$$

In order to find e_θ, we fix ρ and ϕ and consider the increase in θ in the (fixed) horizontal plane $z = \rho\cos\phi$. But this is just like θ in cylindrical coordinates (with r replaced by $\rho\sin\phi$; see Figure 2.72) and therefore $\mathbf{e}_\theta = -\sin\theta\mathbf{i} + \cos\theta\mathbf{j}$. From the definition, it follows that the increase in ϕ occurs in the direction perpendicular to $\mathbf{e}_\rho$ and $\mathbf{e}_\theta$. Hence,

$$\begin{aligned} \mathbf{e}_\phi &= \pm(\mathbf{e}_\rho \times \mathbf{e}_\theta) = \pm \begin{vmatrix} \mathbf{i} & \mathbf{j} & \mathbf{k} \\ \sin\phi\cos\theta & \sin\phi\sin\theta & \cos\phi \\ -\sin\theta & \cos\theta & 0 \end{vmatrix} \\ &= \pm(-\cos\phi\cos\theta\mathbf{i} - \sin\theta\cos\phi\mathbf{j} + \sin\phi\mathbf{k}). \end{aligned}$$

In the first octant, the $\mathbf{k}$ component of $\mathbf{e}_\phi$ has to be negative (see Figure 2.72). Since $\sin\phi > 0$ for points in the first octant, we have to choose the "$-$" sign above, therefore obtaining $\mathbf{e}_\phi = \cos\phi\cos\theta\mathbf{i} + \cos\phi\sin\theta\mathbf{j} - \sin\phi\mathbf{k}$. You may wish to verify that this relation is valid for all θ and ϕ. Thus, the orthonormal set of basis vectors for spherical coordinates is

$$\begin{aligned} \mathbf{e}_\rho &= \sin\phi\cos\theta\mathbf{i} + \sin\phi\sin\theta\mathbf{j} + \cos\phi\mathbf{k} \\ \mathbf{e}_\theta &= -\sin\theta\mathbf{i} + \cos\theta\mathbf{j} \\ \mathbf{e}_\phi &= \cos\phi\cos\theta\mathbf{i} + \cos\phi\sin\theta\mathbf{j} - \sin\phi\mathbf{k}. \end{aligned} \tag{2.38}$$

Repeating the argument presented in obtaining (2.33) in the case of cylindrical coordinates, we determine that a vector $\mathbf{F}$ can be represented in spherical coordinates as $\mathbf{F} = F_\rho\mathbf{e}_\rho + F_\theta\mathbf{e}_\theta + F_\phi\mathbf{e}_\phi$, where $F_\rho = \mathbf{F}\cdot\mathbf{e}_\rho$, $F_\theta = \mathbf{F}\cdot\mathbf{e}_\theta$ and $F_\phi = \mathbf{F}\cdot\mathbf{e}_\phi$. This method of decomposition works in $\mathbb{R}^2$ ($\mathbb{R}^3$) for any set of two (three) orthonormal vectors.

▶ EXAMPLE 2.100

Represent the vector field $\mathbf{F}(x, y, z) = z\mathbf{i} - x\mathbf{k}$ in spherical coordinates.

SOLUTION

In spherical coordinates, $\mathbf{F}(\rho, \theta, \phi) = F_\rho\mathbf{e}_\rho + F_\theta\mathbf{e}_\theta + F_\phi\mathbf{e}_\phi$, where

$$\begin{aligned} F_\rho &= \mathbf{F}\cdot\mathbf{e}_\rho = (z\mathbf{i} - x\mathbf{k})\cdot(\sin\phi\cos\theta\mathbf{i} + \sin\phi\sin\theta\mathbf{j} + \cos\phi\mathbf{k}) \\ &= (\rho\cos\phi\mathbf{i} - \rho\sin\phi\cos\theta\mathbf{k})\cdot(\sin\phi\cos\theta\mathbf{i} + \sin\phi\sin\theta\mathbf{j} + \cos\phi\mathbf{k}) \\ &= \rho\cos\phi\sin\phi\cos\theta - \rho\sin\phi\cos\theta\cos\phi = 0, \\ F_\theta &= \mathbf{F}\cdot\mathbf{e}_\theta = (\rho\cos\phi\mathbf{i} - \rho\sin\phi\cos\theta\mathbf{k})\cdot(-\sin\theta\mathbf{i} + \cos\theta\mathbf{j}) = -\rho\cos\phi\sin\theta, \end{aligned}$$

and

$$\begin{aligned} F_\phi &= \mathbf{F}\cdot\mathbf{e}_\rho = (\rho\cos\phi\mathbf{i} - \rho\sin\phi\cos\theta\mathbf{k})\cdot(\cos\phi\cos\theta\mathbf{i} + \cos\phi\sin\theta\mathbf{j} - \sin\phi\mathbf{k}) \\ &= \rho\cos^2\phi\cos\theta + \rho\sin^2\phi\cos\theta = \rho\cos\theta. \end{aligned}$$

It follows that $\mathbf{F}(\rho, \theta, \phi) = -\rho\cos\phi\sin\theta\mathbf{e}_\theta + \rho\cos\theta\mathbf{e}_\phi$. ◀

▶ EXAMPLE 2.101

Consider unit vectors in spherical coordinates and assume that $\rho = \rho(t)$, $\theta = \theta(t)$, and $\phi = \phi(t)$. Compute $d\mathbf{e}_\rho/dt$.

SOLUTION Using (2.28) and applying the product rule and the chain rule, we get

$$\begin{aligned}
\frac{d\mathbf{e}_\rho}{dt} &= \frac{d}{dt}(\sin\phi\cos\theta\mathbf{i} + \sin\phi\sin\theta\mathbf{j} + \cos\phi\mathbf{k}) \\
&= \cos\phi\frac{d\phi}{dt}\cos\theta\mathbf{i} - \sin\phi\sin\theta\frac{d\theta}{dt}\mathbf{i} + \cos\phi\frac{d\phi}{dt}\sin\theta\mathbf{j} + \sin\phi\cos\theta\frac{d\theta}{dt}\mathbf{j} - \sin\phi\frac{d\phi}{dt}\mathbf{k} \\
&= \frac{d\phi}{dt}(\cos\phi\cos\theta\mathbf{i} + \cos\phi\sin\theta\mathbf{j} - \sin\phi\mathbf{k}) + \frac{d\theta}{dt}(-\sin\phi\sin\theta\mathbf{i} + \sin\phi\cos\theta\mathbf{j}) \\
&= \frac{d\phi}{dt}\mathbf{e}_\rho + \sin\phi\frac{d\theta}{dt}\mathbf{e}_\theta.
\end{aligned}$$

◀

Notice that the derivative $d\mathbf{e}_\rho/dt$ of the unit vector $\mathbf{e}_\rho$ is a nonzero vector. We have observed the same phenomenon in the case of cylindrical coordinates. On the contrary, in Cartesian coordinates the derivatives of unit coordinate vectors are always zero.

The representatives of unit vectors $\mathbf{e}_r$ and $\mathbf{e}_\theta$ in cylindrical coordinates and $\mathbf{e}_\rho$ $\mathbf{e}_\theta$, and $\mathbf{e}_\phi$ in spherical coordinates starting at different points in $\mathbb{R}^3$ are *not* parallel translates of each other. The property that all representatives of a vector are parallel translates of each other holds only in Cartesian coordinate systems.

▶ EXERCISES 2.8

1. Find cylindrical coordinates of the points whose Cartesian coordinates are $(-4, 0, 0)$, $(0, 0, 3)$, $(0, 2, 4)$, and $(2, -3, -1)$.

2. Find spherical coordinates of the points whose Cartesian coordinates are $(-2, 0, 0)$, $(0, 4, 0)$, $(0, 0, 6)$, and $(4, 2, -3)$.

3. Describe geometrically the image of a cube of side a in the first octant whose faces are parallel to the coordinate planes under the mapping $T: \mathbb{R}^3 \to \mathbb{R}^3$ given in cylindrical coordinates by $T(r, \theta, z) = (2r, \theta + \pi, z)$.

4. Describe geometrically the image of an object in the first octant under the mapping $T: \mathbb{R}^3 \to \mathbb{R}^3$ given in spherical coordinates by $T(\rho, \theta, \phi) = (\rho, \theta, \phi + \pi/2)$. Repeat for the mappings $T(\rho, \theta, \phi) = (2\rho, \theta, \phi)$ and $T(\rho, \theta, \phi) = (\rho, \theta + \pi, \phi)$.

5. Express the equation of the paraboloid $z = 4 - x^2 - y^2$ and the equation of the plane $x + 2y - z = 0$ in both cylindrical coordinates and spherical coordinates.

6. Describe the surface whose equation in spherical coordinates is $\rho = 2\sin\phi$.

7. Describe the coordinate surfaces $r =$ constant, $\theta =$ constant, and $z =$ constant and coordinate curves for the cylindrical coordinate system. Coordinate curves are the intersections of coordinate surfaces.

8. Describe the coordinate surfaces $\rho =$ constant, $\theta =$ constant, and $\phi =$ constant and coordinate curves (i.e., the intersections of coordinate surfaces) for the spherical coordinate system.

9. Express the vectors **i**, **j**, and **k** in terms of the unit orthonormal vectors $\mathbf{e}_\rho$, $\mathbf{e}_\theta$, and $\mathbf{e}_\phi$ for spherical coordinates.

Exercises 10 to 14: Represent the vector field $\mathbf{F}(x, y, z)$ in both cylindrical and spherical coordinate systems.

10. $\mathbf{F}(x, y, z) = x\mathbf{i} - 2y\mathbf{j} + z\mathbf{k}$

11. $\mathbf{F}(x, y, z) = \mathbf{i} + \mathbf{j}$

12. $\mathbf{F}(x, y, z) = (x^2 + y^2)\mathbf{i} - \mathbf{k}$

13. $\mathbf{F}(x, y, z) = y\mathbf{i} - x\mathbf{j}$

14. $\mathbf{F}(x, y, z) = \mathbf{i} - x\mathbf{j} + \mathbf{k}$

15. Assume that spherical coordinates $\rho = \rho(t)$, $\theta = \theta(t)$, and $\phi = \phi(t)$ depend on time t and let $\{\mathbf{e}_\rho, \mathbf{e}_\theta, \mathbf{e}_\phi\}$ be the corresponding orthonormal vectors. Compute $d\mathbf{e}_\theta/dt$ and $d\mathbf{e}_\phi/dt$ ($d\mathbf{e}_\rho/dt$ was computed in Example 2.101).

16. Show that the position vector $\mathbf{r} = x\mathbf{i} + y\mathbf{j} + z\mathbf{k}$ is represented in spherical coordinates as $\mathbf{r} = \rho\mathbf{e}_\rho$. Find the expressions for the velocity and acceleration of a particle in spherical coordinates.

17. Let $\mathbf{r} = x\mathbf{i} + y\mathbf{j} + z\mathbf{k}$ be the position vector of a point. Define $d\mathbf{r} = dx\mathbf{i} + dy\mathbf{j} + dz\mathbf{k}$ and $ds^2 = d\mathbf{r} \cdot d\mathbf{r} = dx^2 + dy^2 + dz^2$, where dx, dy, and dz are the differentials of x, y, and z. The expression ds^2 is called the *square of the line element* (or the *metric*) in $\mathbb{R}^3$. Find its expression in cylindrical coordinates and spherical coordinates.

▶ CHAPTER REVIEW

CHAPTER SUMMARY

- **Functions.** Real-valued and vector-valued functions, linear functions, domain and range, vector field, graph, level curves and level surfaces, contour diagram.
- **Limit and continuity.** Using definition to compute limits, strategy for showing that limit does not exist, continuity, list of continuous functions.
- **Derivative.** Partial derivatives and directional derivatives and their geometric interpretation, derivative of a vector-valued function, properties of derivatives, chain rule, differential, relation between continuity and differentiability, linear approximation, local linearity, tangent line and tangent plane.
- **Gradient.** Geometric interpretation, relation to rates of change, relation to level curves and level surfaces, conservative vector field, potential function.
- **Paths and curves.** Distinction between a path and a curve, parametric representation of curves in $\mathbb{R}^2$ and $\mathbb{R}^3$, tangent vector, velocity, speed, and acceleration.
- **Cylindrical and spherical coordinate systems.** Conversion of coordinates of points and vectors, orthonormal set of vectors.

REVIEW

Discuss the following questions.

1. Define the domain and the range of a function. What assumption(s) on the domain and range of vector-valued functions $\mathbf{F}$ and $\mathbf{G}$ guarantee that the composition $\mathbf{G} \circ \mathbf{F}$ is defined?

2. Describe in words what a contour diagram of a linear function $f(x, y) = ax + by + c$ looks like. Is it possible that a surface, other than a plane, has the same contour diagram?

3. Let $f_1(x, y) = x^2 + y^2$, $f_2(x, y) = \sqrt{x^2 + y^2}$, and $f_3(x, y) = (x^2 + y^2)^2$. Explain why contour diagrams of all three functions consist of concentric circles. Describe the differences between the three contour diagrams. Identify the surfaces that are the graphs of f_1, f_2, and f_3.

4. Is it true that the composition of continuous functions is continuous? State the chain rule and all assumptions needed for the formula to hold. Write down the chain rule formula in the case of the composition $f \circ \mathbf{c}$ of a path $\mathbf{c}$ in $\mathbb{R}^3$ and a real-valued function f of three variables.

5. Explain what is meant by the statement "$\mathbf{x} = (x, y)$ approaches $\mathbf{a} = (a, b)$." Suppose that the limit of a function $f(x, y)$ as (x, y) approaches $(0, 0)$ along any straight line segment is 3. Is it true that the limit of the function is 3?

6. Sketch the graph of a function $y = f(x)$, defined on an interval $[a, b]$, that is continuous except at the boundary points $x = a$ and $x = b$. Let $D = \{(x, y) \mid -1 \leq x \leq 1, 0 \leq y \leq 2\}$. What is the boundary of D? Sketch the graph of a function $f(x, y)$: $D \to \mathbb{R}^3$ that is continuous at all points inside D but discontinuous at all boundary points.

7. We plan to calculate the area of a rectangular piece of land that has one side much larger than the other. Which dimension should we measure more carefully? Explain your answer.

8. Define the derivative $D\mathbf{F}$ of a function $\mathbf{F}$: $\mathbb{R}^m \to \mathbb{R}^n$. When is $\mathbf{F}$ differentiable? What is the size of the matrix $D\mathbf{F}$? Write down the definition in the special cases when $m = n = 1$, $m = 1$, $n > 1$ and $m > 1$, $n = 1$.

9. Define the linear approximation of a function f: $\mathbb{R}^2 \to \mathbb{R}$. What assumptions on f guarantee its existence?

10. We have seen in Section 1.4 that the product of two nonzero matrices can be a zero matrix. Suppose that $f(\mathbf{x})$: $\mathbb{R}^3 \to \mathbb{R}$ and $\mathbf{c}(t)$: $\mathbb{R} \to \mathbb{R}^3$ are differentiable functions, their derivatives Df and $D\mathbf{c}$ are nonzero but the product $Df(\mathbf{c}(t)) \cdot D\mathbf{c}(t)$ is zero for all t. Give a geometric explanation of the fact that the product of the two derivatives is zero. Find an example of such f and $\mathbf{c}$. Find an example of two functions $\mathbf{F}, \mathbf{G}$: $\mathbb{R}^2 \to \mathbb{R}^2$ such that $D\mathbf{F}$ and $D\mathbf{G}$ are nonzero 2×2 matrices but their product $D\mathbf{G} \cdot D\mathbf{F}$ is the zero matrix. What is (in your case) the composition $\mathbf{G} \circ \mathbf{F}$?

11. Define a path and a curve and describe the difference. What is a continuous path? Continuous curve? Is it possible for a continuous curve to have a discontinuous parametrization? Define a differentiable path and a differentiable curve. Write down a parametric representation of the line $y = 0$ that is not differentiable.

12. What is $D\mathbf{F}(x, y)$ if $\mathbf{F}(x, y) = (ax + by, cx + dy)$, and a, b, c, and d are constants?

13. State the relationship between the gradient and level curves or level surfaces. Give several examples that illustrate this relationship. Explain how to use a topographical map (i.e., a map of level curves of the height function) to climb to the top of a hill fastest.

14. Define a conservative vector field and a corresponding potential function. How many potential functions does a conservative vector field have?

TRUE/FALSE QUIZ

Determine whether the following statements are true or false. Give reasons for your answer.

1. The function $\mathbf{F}(x, y) = (x^2 y, \sin x, 1)$ is a vector field.
2. If a contour diagram of a function $f(x, y)$ consists of concentric circles, then its graph is a cone.
3. Level surfaces of a function $f(x, y, z) = ax + by + cz + d$, $a \neq 0$, are parallel planes.
4. If $\mathbf{F}$: $\mathbb{R}^2 \to \mathbb{R}^2$ is a differentiable vector field, then $D\mathbf{F}(1, 3)$ is a vector in $\mathbb{R}^2$.
5. The function $L(x, y) = 3x + y^2 - 2$ is a linear approximation of some function.
6. The function $f(x, y, z) = x^2 \ln(y + 4) - z^3$ is differentiable at $(3, 2, 0)$.
7. If $f(x, y, z)$ is a constant function, then $df = 0$.

8. Assume that $f(3, 1) = 2$ and $f_x(3, 1) = 4$. Then $f(4, 1) = 2 + 4 = 6$.

9. If $\nabla f(a, b) = \mathbf{i}$, then the rate of change of $f(x, y)$ at (a, b) in the direction of $\mathbf{j}$ is zero.

10. There is a function f such that $||\nabla f(a, b)|| = 3$ and $D_{\mathbf{u}} f(a, b) = 4$, where $\mathbf{u}$ is a unit vector.

11. The path $\mathbf{c}(t) = (\cos 2t, -\sin 2t)$, $t \in [0, \pi]$, is oriented clockwise.

12. If an object moves according to $\mathbf{c}(t) = (2\cos t, 2\sin t, 6t)$, $t \geq 0$, then its speed is constant.

13. Let $\mathbf{c}$ be a path in $\mathbb{R}^2$. If its speed is constant, then so is its velocity.

14. Let $g(t) = \ln f(t, -t)$, where f is a differentiable function. Then $g'(t) = 1/f(t, -t)$.

REVIEW EXERCISES AND PROBLEMS

1. Give a possible formula for a vector field $\mathbf{G}(x, y)$ whose magnitude decreases as its distance from the point (2, 3) increases.

2. Consider a Cobb–Douglas production function $P(L, K) = bL^\alpha K^\beta$, where $b > 0$ and $0 < \alpha$, $\beta < 1$. In Example 2.6 we discussed the special case where $\beta = 1 - \alpha$, that is, $\alpha + \beta = 1$. Assume that $m > 1$ and compute $P(mL, mK)$ in the case $\alpha + \beta < 1$. Explain why this situation is called *decreasing returns to scale.* Similarly, justify the term *increasing returns to scale* in the case $\alpha + \beta > 1$.

3. Using a graphing device, plot (in the *LK* coordinate system) the level curves of the Cobb–Douglas production function $P(L, K) = 1.01L^{0.75}K^{0.25}$ of values $p = 1.2$, $p = 1.4$, $p = 1.6$, and $p = 1.8$ (see Exercise 38 in Section 2.2). Fix a value for L, say, $L = 1.3$. Notice that, as we move vertically along $L = 1.3$, we distance between level curves increases. Interpret what this means for P. Explain the behavior of P as K is kept fixed.

4. Find a formula for a function $f(x, y, z)$ whose level surfaces are concentric spheres centered at $(-2, 4, 0)$. There are many correct answers.

5. Describe the level curves of the function $f(x, y) = 1 - (x^2 + y^2 - 9)^2$.

6. An object moves along the surface of the paraboloid $z = 10 - x^2 - 2y^2$ from the point $(-2, 3, -12)$ in such a way that the projection of its trajectory onto the *xy*-plane is a line parallel to the *x*-axis. Find the highest position of the object.

7. Determine whether or not the limit of $f(x, y) = xy^4/(x^2 + y^6)$ exists as $(x, y) \to (0, 0)$. If possible, define $f(0, 0)$ so as to make f continuous at $(0, 0)$. *Hint*: Switch to polar coordinates.

8. The period of a simple pendulum is given by $T = 2\pi\sqrt{\ell/g}$, where ℓ is its length and g is a constant. If ℓ is measured to be 1.2 m with an error of 0.03 m, g is taken to be 9.8 m/s^2 (thus, an error of no more than 0.02 m/s^2 is made), and π is taken to be 3.14 (an error of no more than 0.002), find an approximate value of T [your result should be in the form (value of T) $\pm$ error].

9. The temperature of a metal rod of length 4 at position x (where $0 \leq x \leq 4$) and at time t (with $t \geq 0$) is $T(x, t) = 30e^{-2t} \sin(\pi x/2)$.

(a) Find the rate of change of temperature with respect to position when $x = 3/2$ and $t = 2$. Sketch the cross-section for $t = 2$ and interpret your result.

(b) Find the rate of change of temperature with respect to time when $x = 3/2$ and $t = 2$. Sketch the cross-section for $x = 3/2$ and interpret your result.

10. The function $T(x, t) = a + e^{-ct} \sin(2x)$ (a and c are constants and $c > 0$) describes the temperature at time t and at the point x on a metal rod of length π, placed along the positive x-axis with one end at the origin.

(a) What is the initial temperature at the left end? At the right end?

(b) At time $t = t_0 \geq 0$, what is the warmest (coolest) point of the rod?

(c) Fix the point x_0 on the rod, $0 \leq x_0 \leq \pi/2$. When is the temperature going to reach its maximum value at that point? What happens when $t \to \infty$? Answer the same questions for the point x_0 with $\pi/2 \leq x_0 \leq \pi$.

(d) Locate the points on the rod where the temperature does not change.

11. Prove that if $|f(x, y)| \leq x^2 + y^2$ for all $(x, y) \in \mathbb{R}^2$, then f is differentiable at $(0, 0)$.

12. Consider the function $f(x, y) = (xy)^{4/5}$.

(a) Is f differentiable at $(0, 0)$?

(b) Is it possible to conclude from (a) that f is continuous at $(0, 0)$?

(c) Is f continuous at $(0, 0)$?

(d) Is f_x continuous at $(0, 0)$?

13. The force of gravity of a planet on an object of mass m at a distance r from the surface of the planet has magnitude $F(m, r) = mgR^2(R + r)^{-2}$, where g and R are constants. Find an expression for the time rate of change of F acting on a comet (whose mass m is changing) approaching the planet.

14. Let $f(x, y) = (x^3 + 3x^2y + e^y)\cos(x^4y^3)e^{yx^2\cos(1-x-y)}$. Find $(\partial f/\partial y)(0, 1)$. *Hint*: The computation of $(\partial f/\partial y)(x, y)$ with the use of the product rule and the chain rule is fairly complicated. Find a way to simplify your task.

15. Assume that $y = f(x)$ is a continuous function defined on an interval (a, b) and that $f(c) \neq 0$ for some $c \in (a, b)$. Show that there exists an interval $(a', b') \subseteq (a, b)$ such that $f(x) \neq 0$ for all $x \in (a', b')$ (in other words, a continuous function cannot have a nonzero value only at a single point). Formulate an analogous statement for a continuous function $f\colon U \subseteq \mathbb{R}^m \to \mathbb{R}$.

16. Assume that a function $f(x, y)$ is differentiable at (a, b) and let P be the tangent plane to the graph of $z = f(x, y)$ at (a, b).

(a) Use (2.19) in Section 2.4 to show that the upward pointing normal vector to P is given by $\mathbf{n} = (-(\partial f/\partial x)(a, b), -(\partial f/\partial y)(a, b), 1)$.

(b) Prove that $\cos\alpha = \left(1 + ((\partial f/\partial x)(a, b))^2 + ((\partial f/\partial y)(a, b))^2\right)^{-1/2}$, where α is the angle between $\mathbf{n}$ and the unit vector $\mathbf{k}$.

(c) By definition, the angle between two planes is the angle between their normal vectors. Thus, α is the angle between the tangent plane P and the xy-plane. Show that $\|\nabla f(a, b)\| = \tan\alpha$. In other words, $\|\nabla f(a, b)\|$ is the slope of the tangent plane at (a, b).

17. Sometimes, the parameter t can be eliminated from the parametric equation $\mathbf{c}(t) = (x(t), y(t))$ and an equation of the form $f(x, y) = 0$ is obtained [e.g., $\mathbf{c}(t) = (\cos t, \sin t)$ gives $x^2 + y^2 = 1$ after the elimination of t]. Express dy/dx and d^2y/dx^2 in terms of the derivatives of x and y with respect to t.

18. Prove that the equation of the tangent plane to the ellipsoid $x^2/a^2 + y^2/b^2 + z^2/c^2 = 1$ at the point (x_0, y_0, z_0) is $xx_0/a^2 + yy_0/b^2 + zz_0/c^2 = 1$.

19. Let $f\colon \mathbb{R}^3 \to \mathbb{R}$ be a differentiable function satisfying $f(t\mathbf{x}) = t^p f(\mathbf{x})$, where p is a constant and $t \in \mathbb{R}$. Prove that $\mathbf{x} \cdot \nabla f(\mathbf{x}) = pf(\mathbf{x})$ for every $\mathbf{x} \in \mathbb{R}^3$.

20. The shape of a hill corresponds to the graph of the function $z = 440 - 0.1x^2 - 0.4y^2$. A climber is located at the point $(10, 10, 390)$ on the hill.

(a) In which direction should the climber proceed in order to descend most rapidly?

(b) In which direction should the climber proceed in order to reach the top of the hill fastest? At what angle with respect to the horizontal is the climber climbing in that case?

(c) In which direction should the climber proceed in order to gain height at the rate of 10% (i.e., 1 m up for 10 m horizontal distance)?

21. Assume that $D_{(\mathbf{i}+\mathbf{j})/\sqrt{2}} f(\mathbf{a}) = 2$ and $D_{(\mathbf{i}-\mathbf{j})/\sqrt{2}} f(\mathbf{a}) = 5$, where $f\colon \mathbb{R}^2 \to \mathbb{R}$ is a differentiable function and $\mathbf{a}$ is a point in $\mathbb{R}^2$. Find $(\partial f/\partial x)(\mathbf{a})$ and $(\partial f/\partial y)(\mathbf{a})$.

22. Consider the function

$$f(x, y) = \frac{x^2 y + x y^3 + 2}{x^2 + y^2}.$$

(a) Find the equations of the level curves of f that go through the points $(1, 0)$, $(0.9, 0)$, $(0.95, 0)$, $(1.05, 0)$, and $(1.1, 0)$.

(b) Plot the level curves in (a) in the same coordinate system.

(c) Looking at your sketch in (b) determine whether the partial derivative $\partial f/\partial x$ at $(1, 0)$ is positive or negative. Estimate its value.

(d) Using rules for derivatives, find $(\partial f/\partial x)(1, 0)$ [thus checking your answers for (c)].

(e) For the function $g(x) = f(x, 0)$ of one variable, compute the slope at $x = 1$.

(f) What is the relation between the results of (d) and (e)?

(g) Looking at the plot of the level curves, determine which of the partial derivatives $(\partial f/\partial y)(1/2, 1)$ or $(\partial f/\partial x)(1/2, 1)$ is larger.

CHAPTER 3

Vector-Valued Functions of One Variable

In this chapter, we continue our study of paths that we started in Section 2.5. Recall that a path is a vector-valued function of one variable, and its image, visualized as a geometric object, is a curve. The study of vector fields, one other important class of vector-valued functions, begins in Chapter 4.

We start the chapter by investigating properties of a number of important curves that appear in a variety of applications (nature and environment, engineering, computer-aided design, physics, astronomy, medicine, etc.). Next, we present a special case of the Implicit Function Theorem, as it applies to curves in a plane.

We study numerous examples of paths (parametrizations) to learn how to extract valuable information, such as velocity, acceleration, or length. These concepts are then applied to the investigation of physical situations such as Kepler's Laws, Coriolis acceleration, or the motion of a projectile. Next, we investigate the relationship between the acceleration and geometric properties of a curve, namely its curvature. This investigation will lead us toward basic equations (the so-called Serret–Frenet formulas) of the differential geometry of curves in space.

▶ 3.1 WORLD OF CURVES

In this section, we study various curves, to show how they appear in applications, to further practice parametrizations, and because we will need some of them later in this book. We devote a significant amount of space to curves defined in alternate ways, such as curves defined by intersecting two surfaces, and curves defined implicitly. We state and discuss the Implicit Function Theorem in the context of curves in $\mathbb{R}^2$.

▶ EXAMPLE 3.1

All parametrizations of the circle that we studied in Examples 2.51 and 2.55 in Section 2.5 were based on the fact that $\sin^2 t + \cos^2 t = 1$. Consider a different parametrization

$$\mathbf{c}(t) = \left(\frac{2t}{1+t^2}, \frac{1-t^2}{1+t^2}\right), \qquad t \in R.$$

Find its image and describe the motion it represents.

SOLUTION Since

$$\left(\frac{2t}{1+t^2}\right)^2 + \left(\frac{1-t^2}{1+t^2}\right)^2 = \frac{t^4+2t^2+1}{(1+t^2)^2} = 1,$$

it follows that the image of $\mathbf{c}(t)$ is a part (or all) of the circle of radius 1 centered at the origin.

To understand how an object moves along the image of $\mathbf{c}(t)$, we plot the graphs of its coordinates; see Figure 3.1.

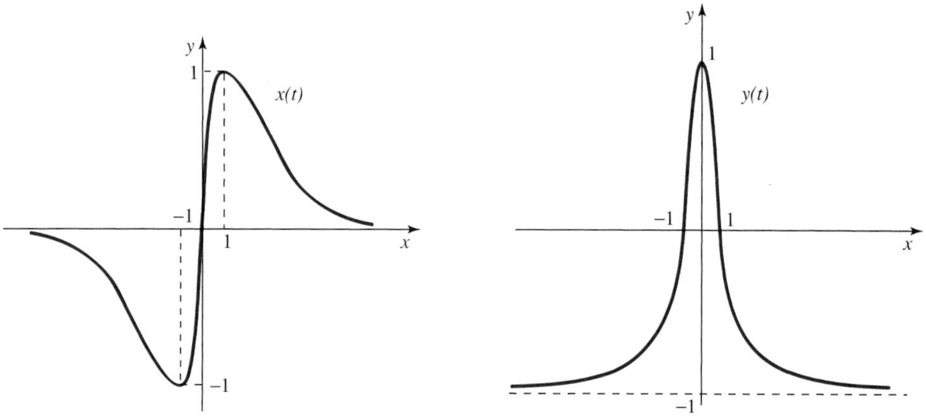

(a) $x(t) = 2t/(1+t^2)$.

(b) $y(t) = (1-t^2)/(1+t^2)$.

Figure 3.1 Graphs of coordinates of $\mathbf{c}(t)$.

Note that, since $t \in \mathbb{R}$, $\mathbf{c}(t)$ does not have endpoints. At $t \to -\infty$, $x(t) \to 0$ and $y(t) \to -1$. In words, moving backward, the object draws closer and closer to $(0, -1)$.

As t increases from $-\infty$ to -1, $x(t)$ decreases from 0 to -1 and $y(t)$ increases from -1 to 0. Thus, the motion is clockwise (in the third quadrant), and when $t = -1$, the object is at $(-1, 0)$. Looking again at the graphs in Figure 3.1, we see that, as t goes from -1 to 0, $x(t)$ increases from -1 to 0 and $y(t)$ increases from 0 to 1. Thus, the object continues its clockwise motion around the circle, reaching the point $(0, 1)$ at time $t = 0$.

Continuing our analysis in the same way, we realize that $\mathbf{c}(t)$ is a clockwise motion around the circle. As $t \to \infty$, $\mathbf{c}(t) \to (0, -1)$. The image of $\mathbf{c}(t)$ is the circle $x^2 + y^2 = 1$ without the point $(0, -1)$. ◀

▶ EXAMPLE 3.2 Cycloid

Fix a point P on the circumference of a wheel of radius 1. The curve that is traced out by P as the wheel rolls along the x-axis is called a *cycloid.* Find its parametric equation.

SOLUTION We pick the coordinate system so that the wheel is initially centered at $(0, 1)$ and P is at the origin. Figure 3.2 shows the initial location of $P$and several intermediate positions.

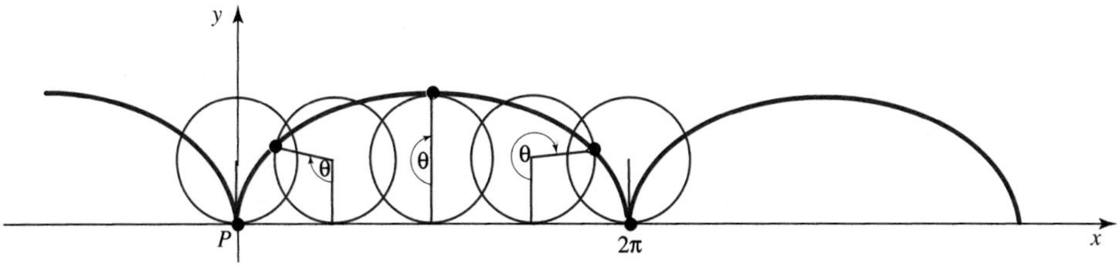

Figure 3.2 Mechanical way of obtaining a cycloid.

Let θ denote the angle of rotation (initially, $\theta = 0$). We need to find the coordinates of P as functions of θ; see Figure 3.3. (We use $|AB|$ to denote the length of the line segment $\overline{AB}$ joining A and B.)

Since the segment $\overline{PC}$ is a radius of the circle, $|\overline{PC}| = 1$. From the triangle PRC, we see that $|\overline{RC}| = \cos\theta$ and $|\overline{PR}| = \sin\theta$. Thus, $y = |\overline{QC}| - |\overline{RC}| = 1 - \cos\theta$. Note that the distance from O to Q is equal to the length of the arc from P to Q, that is, $|\overline{OQ}| = \theta$ (since θ is in radians). Thus, $x = |\overline{OQ}| - |\overline{PR}| = \theta - \sin\theta$. It follows that the parametrization of the cycloid is given by $\mathbf{c}(\theta) = (\theta - \sin\theta, 1 - \cos\theta)$, $\theta \in \mathbb{R}$.

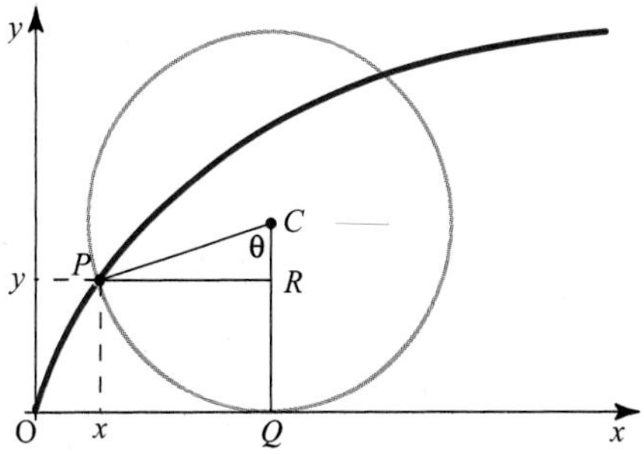

Figure 3.3 Computing the parametric equation of the cycloid. ◀

Galileo was among the first (of many) mathematicians who studied the cycloid. He contemplated using it in designing a bridge and attempted to find the area under its arch. Later, it was discovered that the cycloid solves two important problems: the *Brachistochrone problem* and the *Tautochrone problem.*

We are given two points A and B at different heights (assume that B is lower, and that it does not lie directly below A). The *Brachistochrone problem* consists of finding the curve from A to B along which a ball will slide in the shortest time, under the influence of gravity only. It can be proved that the inverted cycloid (i.e., the cycloid reflected with respect to the x-axis) is the optimal curve (for an investigation of the problem, see Exercise 17 in the Chapter Review section). To solve a *Tautochrone problem* means to find a curve with the following property: a ball placed anywhere on it will roll to the bottom in the same amount of time. In Exercise 18 in the Chapter Review, we prove that the inverted cycloid satisfies the desired property.

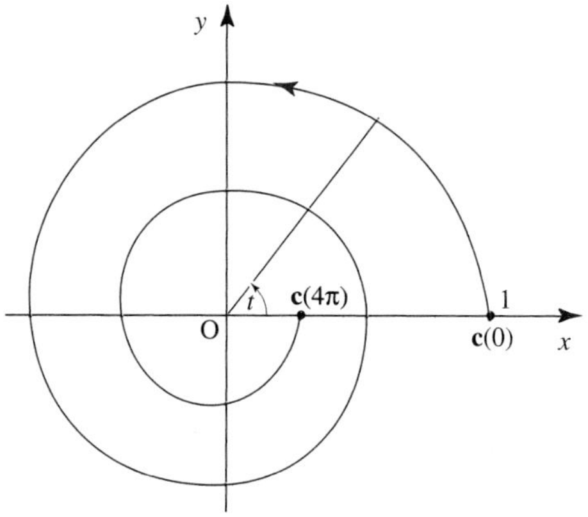

(a) Logarithmic, with $a = -0.1$, $t \in [0, 4\pi]$.

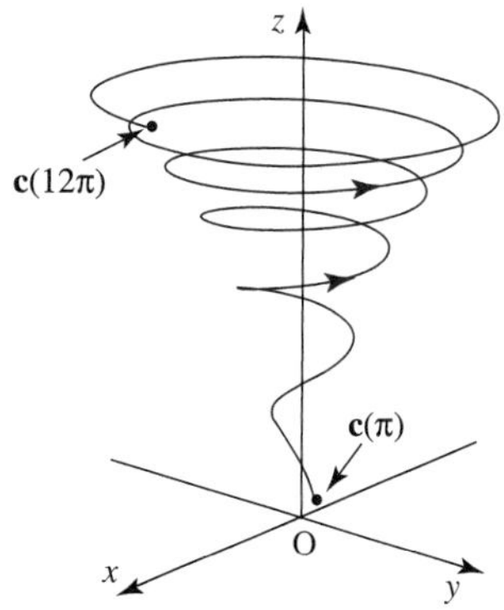

(b) $\mathbf{c}(t) = (t\cos t, t\sin t, \ln t)$, $t \in [\pi, 12\pi]$.

Figure 3.4 Spirals in two and three dimensions.

▶ EXAMPLE 3.3 Spirals

The curve given parametrically by $\mathbf{c}(t) = (e^{at}\cos t, e^{at}\sin t)$, $a \neq 0$, $t \geq 0$, is called a *logarithmic spiral;* see Figure 3.4(a). Computing $\|\mathbf{c}(t)\| = e^{at}$, we conclude that $\mathbf{c}(t)$ spirals inward [and approaches $(0, 0)$ as $t \to \infty$] if $a < 0$; for $a > 0$, it spirals outward. The parameter t represents the angle with respect to the x-axis.

From $x(t) = e^{at}\cos t$ and $y(t) = e^{at}\sin t$, we determine that $x^2 + y^2 = e^{2at}$. Switching to polar coordinates, and replacing t by θ, we get $r = e^{a\theta}$. In Section 4.5, we will see how the logarithmic spiral $r = e^{a\theta}$ is related to the flow of air near and within a hurricane.

In general, a spiral is an image of the path $\mathbf{c}(t) = (f(t)\cos t, f(t)\sin t)$, where $f(t) \geq 0$ is a continuous function, either increasing or decreasing for all t (see Exercises 28 and 29).

Spirals appear in nature quite often, for instance, as a growth form (sea shells, sunflower seeds, spider webs). Some galaxies in our universe are referred to as "spiral galaxies," due to their shape. The spiral is a common element in engineering design. Spirals in three dimensions, such as the one in Figure 3.4(b), are sometimes used to describe fluid flow. ◀

▶ EXAMPLE 3.4 Bézier Curves

In late 1950s, French mathematician and automobile designer Pierre Bézier introduced special types of curves (now known as *Bézier curves*), in order to carry out automobile design and calculations related to it. Bézier curves have since been used in a variety of applications, including computer-aided design, economics, and data analysis. Designers of typefaces (fonts) for computers and laser printers use Bézier curves. Bézier curves are included in every computer graphics program (and were used to create a number of images in this book).

A Bézier curve is defined using four control points $A(a_x, a_y)$, $B(b_x, b_y)$, $C(c_x, c_y)$, and $D(d_x, d_y)$. Its parametric equation is $\mathbf{c}(t) = (x(t), y(t))$, where

$$x(t) = a_x(1-t)^3 + 3b_x(1-t)^2 t + 3c_x(1-t)t^2 + d_x t^3 \tag{3.1}$$

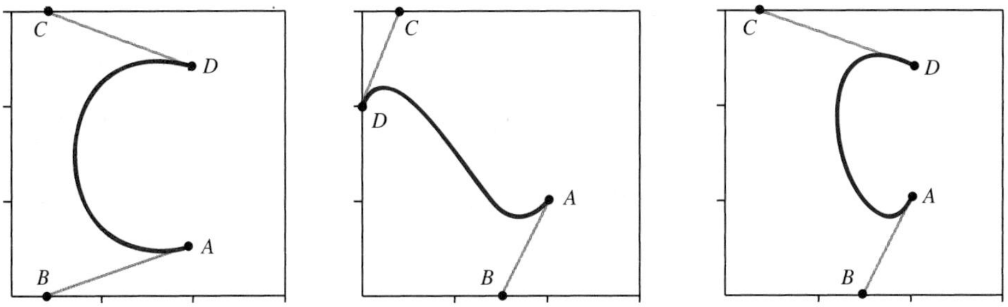

Figure 3.5 Bézier curves.

and

$$y(t) = a_y(1-t)^3 + 3b_y(1-t)^2 t + 3c_y(1-t)t^2 + d_y t^3, \tag{3.2}$$

and $0 \le t \le 1$. We notice that the coordinates of $\mathbf{c}(t)$ are polynomials of degree 3. Its initial point is $\mathbf{c}(0) = (a_x, a_y) = A$, and the terminal point is $\mathbf{c}(1) = (d_x, d_y) = D$. The remaining two points determine the direction of $\mathbf{c}(t)$ at its endpoints: since $\mathbf{c}'(0) = (-3a_x + 3b_x, -3a_y + 3b_y) = 3(b_x - a_x, b_y - a_y)$, it follows that the direction of the tangent to $\mathbf{c}(t)$ at $t = 0$ is given by the direction of the vector from A to B. Likewise (see Exercise 5 for details), the direction of $\mathbf{c}'(1)$ is the same as the direction of the vector from C to D.

Thus, to draw a Bézier curve, we specify initial and terminal points and the directions at the two endpoints. By changing the control points B and C, we obtain various curves; see Figure 3.5. In Exercises 5 and 6, we examine some properties of Bézier curves.

▶ EXAMPLE 3.5 Catenary Curve

The word "catenary" comes from a Latin word *catena* for "chain." Galileo believed that a hanging chain, suspended by its ends fixed at the same height and acted on by gravity only, would describe a parabola. Later, it was proved (under the assumptions that the chain is flexible, nonstretchable, and of constant density) that its shape is not the parabola, but another curve, named the *catenary curve* (however, Galileo was not completely wrong—see Exercise 32). The catenary curve is the image of $\mathbf{c}(t) = (t, a\cosh(t/a))$, $a > 0$, where $\cosh t = (e^t + e^{-t})/2$ is the hyperbolic cosine function; see Figure 3.6(a).

Cables in some suspension bridges are in the form of a catenary. Due to its structural stability, an inverted catenary is sometimes used to build arches; see Figure 3.6(b). The inverted catenary has an interesting property: if, for some reason, we would like to use a vehicle with square wheels, we would need to build a road in the shape of a sequence of inverted catenary curves.

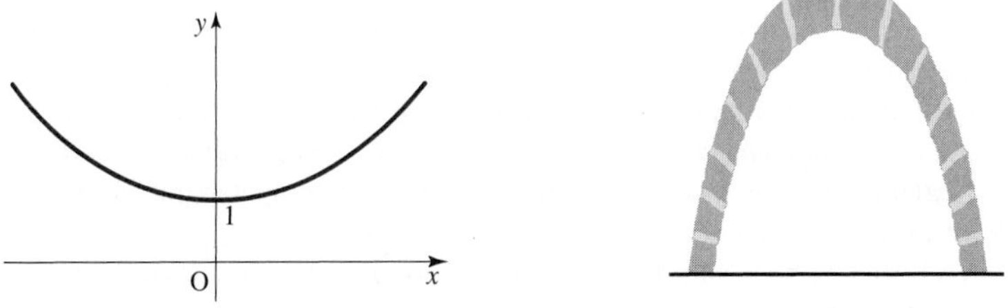

(a) $\mathbf{c}(t) = (t, \cosh t)$.

(b) Catenary arch.

Figure 3.6 Catenary curve.

▶ EXAMPLE 3.6 Lissajous Curves

Consider the path $\mathbf{c}(t) = (\cos 3t, \sin 4t)$, $t \in [0, 2\pi]$. Since $-1 \leq \cos 3t, \sin 4t \leq 1$, we conclude that the image of $\mathbf{c}(t)$ is a curve contained within the square in the xy-plane bounded by the lines $x = -1$, $x = 1$, $y = -1$, and $y = 1$. The curve $\mathbf{c}$ touches the line $x = 1$ ($\cos 3t = 1$ implies that $t = 0$, $2\pi/3$, or $4\pi/3$) at three points: $\mathbf{c}(0) = (1, 0)$, $\mathbf{c}(2\pi/3) = (1, \sqrt{3}/2)$, and $\mathbf{c}(4\pi/3) = (1, -\sqrt{3}/2)$. Similarly, we compute that $\mathbf{c}$ touches $x = -1$ at three points, and each of $y = 1$ and $y = -1$ at four points; see Figure 3.7(a). The initial and terminal points of $\mathbf{c}(t)$ are $\mathbf{c}(0) = \mathbf{c}(2\pi) = (1, 0)$. The curve is retraced every 2π units.

Figure 3.7(b) shows the image of the curve $\mathbf{c}(t) = (\cos 2t, \sin 6t)$, $t \in [0, \pi]$.

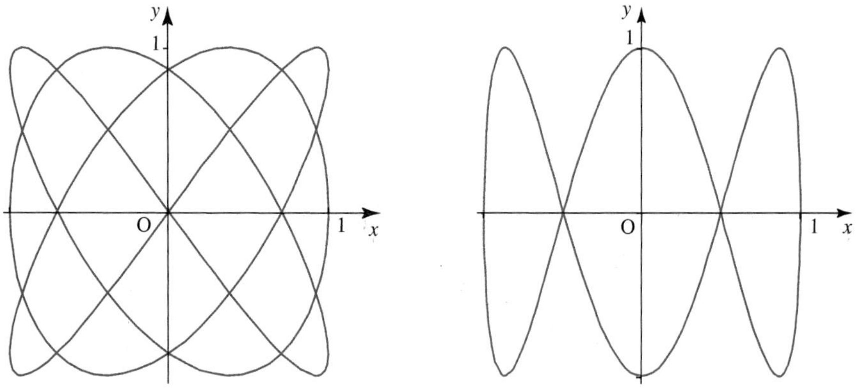

(a) $\mathbf{c}(t) = (\cos 3t, \sin 4t)$, $t \in [0, 2\pi]$.

(b) $\mathbf{c}(t) = (\cos 2t, \sin 6t)$, $t \in [0, \pi]$.

Figure 3.7 Lissajous curves.

In general, the image of a path $\mathbf{c}(t) = (\cos mt, \sin nt)$, where m and n are integers, is called a *Lissajous curve*. Lissajous curves were first studied in relation to vibrations (mechanical and acoustic) in various media. They have since been used in a number of applications in physics, astronomy, and elsewhere. In Exercises 7 through 9, we explore several properties of these curves.

▶ EXAMPLE 3.7

A common way of visualizing a path is by showing its image in $\mathbb{R}^2$ or in $\mathbb{R}^3$. Sometimes, it might be useful to indicate the speed of a path as well. Here is how it can be done: let $\mathbf{c}(t)$ be a path in $\mathbb{R}^2$ or in $\mathbb{R}^3$ defined on an interval $[a, b]$. Choose equally spaced values of t, say, $a = t_0 < t_1 < t_2 < \cdots < t_{n-1} < t_n = b$ and add the points $\mathbf{c}(t_i)$, $i = 0, \ldots, n$, to the plot. By looking at the points, we can get the speed: the closer the points, the smaller the speed; the further apart they become, the larger the speed.

Figure 3.8 shows plots by speed of the paths $\mathbf{c}_1(t) = (\cos t, \sin t)t \in [0, 2\pi]$, and $\mathbf{c}_2(t) = (\cos t^2, \sin t^2)\ t \in [0, \sqrt{2\pi}]$.

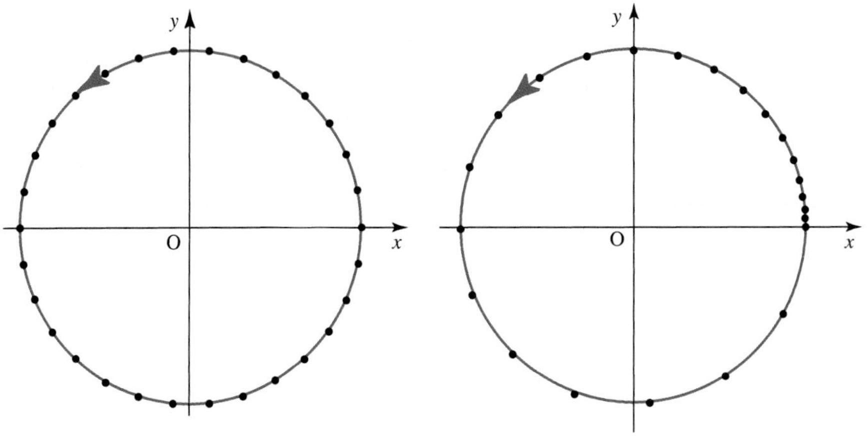

Figure 3.8 Paths $\mathbf{c}_1(t)$ (left) and $\mathbf{c}_2(t)$ (right) with an indication of their speed.

Curves on Surfaces and Curves as Intersections of Surfaces

In certain situations, a curve is limited to existing on a given surface. We now explore several cases of such curves.

▶ EXAMPLE 3.8 Helix

The image of the path $\mathbf{c}(t) = (a\cos t, a\sin t, bt)$, where $a > 0$ and $b > 0$, is called a *helix.* Sometimes, a more general path $\mathbf{c}(t) = (a\cos t, a\sin t, f(t))$, where f is a positive, continuous, and increasing function, is used to define a helix. In words, helix is a circular motion subjected to the change in height, which is determined by the function bt [or, in general, by $f(t)$]. Clearly, the image of $\mathbf{c}(t)$ lies on the cylinder of radius a whose axis of rotation is the z-axis. Figure 3.9(a) shows the helix $\mathbf{c}(t) = (\cos t, \sin t, 2t)$, $t \in [0, 6\pi]$. Note that the distance between consecutive windings (sometimes called the pitch of a helix) is constant, and equal to 4π.

Helix is one of the most useful shapes that we meet in everyday life: for instance, the binding that all kinds of screws and bolts provide is based on its shape. Archimedes, the famous Greek mathematician, used a helix to construct a device (called the Archimedes's screw) to force the water upward inside a tube (numerous variations of his invention are still widely used). A double helix (i.e., two copies of the same helix) represents the geometric shape of the DNA; see Figure 3.9(b). ◀

▶ EXAMPLE 3.9

A straightforward computation shows that the image of the path $\mathbf{c}(t) = (\cos t\cos 18t, \cos t\sin 18t, \sin t)$, $t \in [0, 2\pi]$, lies on the sphere $x^2 + y^2 + z^2 = 1$; see Figure 3.10(a).

Figure 3.10(b) shows a curve, called a *torus knot,* whose parametric equation is $\mathbf{c}(t) = ((3 + \cos 8t)\cos 7t, (3 + \cos 8t)\sin 7t, \sin 8t)$, for $t \in [0, 2\pi]$. The image of $\mathbf{c}(t)$ wraps around the surface of a torus (we will study the properties of the torus in Chapter 7).

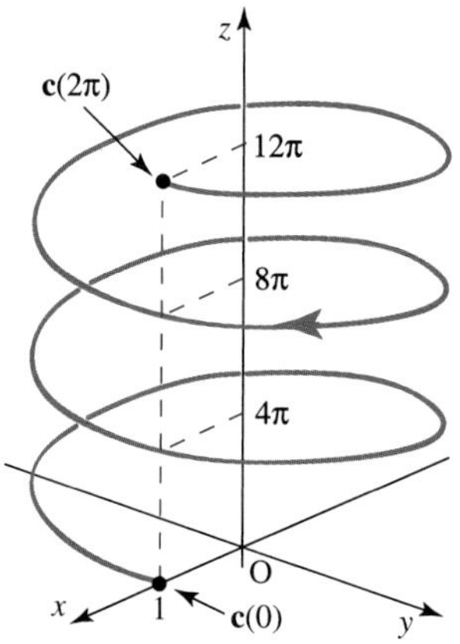

(a) Helix of constant pitch.

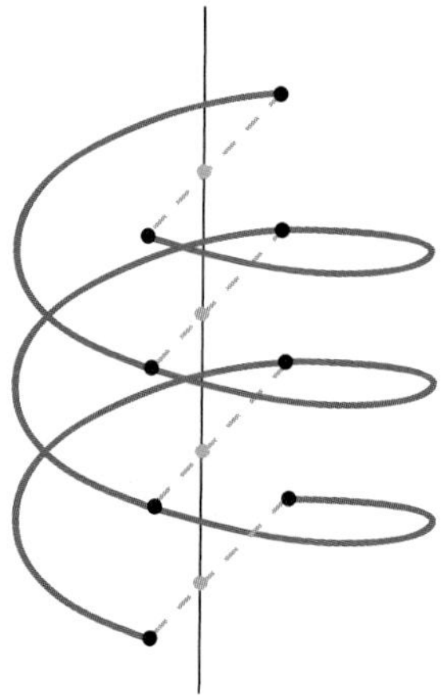

(b) Double helix of the DNA molecule.

Figure 3.9 Pictures of two helices.

Investigating properties of curves on surfaces is important not only in mathematics, but also in numerous applications, such as engineering, chemistry, and medicine.

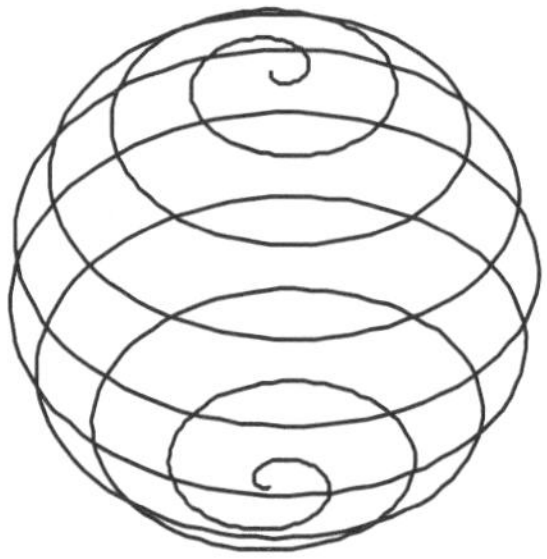

(a) Spiral on a sphere.

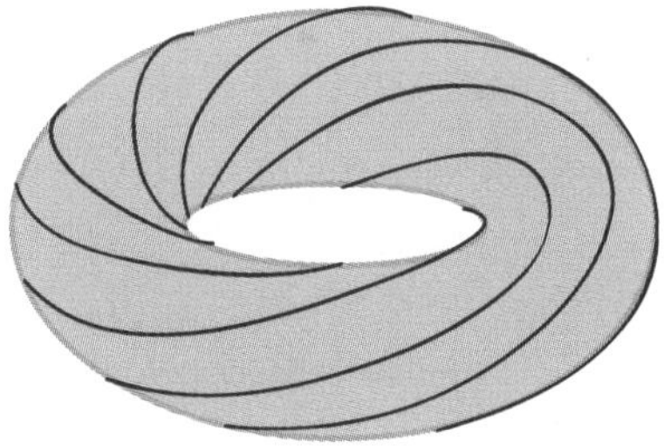

(b) Torus knot.

Figure 3.10 Curves on surfaces.

Next, we show how to parametrize curve(s) obtained by intersecting surfaces in three-dimensional space. We will apply this idea later, in computing line and surface integrals.

▶ **EXAMPLE 3.10**

Let $\mathbf{c}$ be the ellipse that is the intersection of the cylinder $x^2 + y^2 = 13$ and the plane $z = 2y$; see Figure 3.11. Find several parametrizations of $\mathbf{c}$.

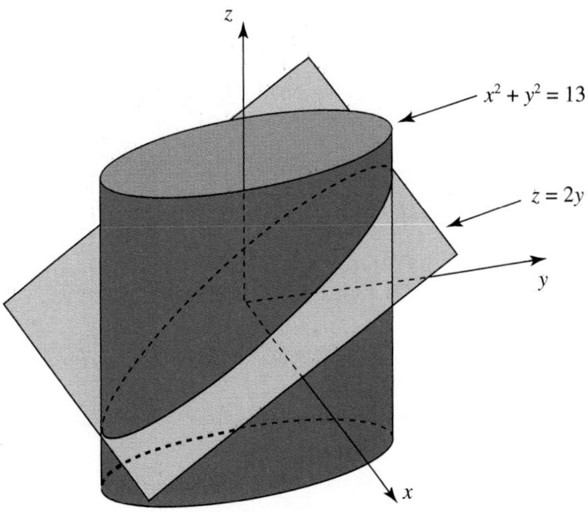

Figure 3.11 Intersection of $x^2 + y^2 = 13$ and $z = 2y$.

SOLUTION

We are looking for functions $\mathbf{c}(t) = (x(t), y(t), z(t))\colon [a, b] \to \mathbb{R}^3$ such that $x(t)^2 + y(t)^2 = 13$ and $z(t) = 2y(t)$.

Take, for example, $x(t) = \sqrt{13}\cos t$ and $y(t) = \sqrt{13}\sin t$; then $z(t) = 2\sqrt{13}\sin t$, and

$$\mathbf{c}_1(t) = (\sqrt{13}\cos t, \sqrt{13}\sin t, 2\sqrt{13}\sin t), \qquad t \in [0, 2\pi],$$

is a possible parametrization. Replacing t by mt, $m \neq 0$ (of course, adjusting the interval for t), we get infinitely many parametrizations. Apart from those, there are others, such as

$$x(t) = 2\cos t - 3\sin t, \quad y(t) = 3\cos t + 2\sin t, \quad z(t) = 6\cos t + 4\sin t,$$

where $t \in [0, 2\pi]$. ◀

▶ EXAMPLE 3.11

Find a parametric equation of the line that is the intersection of the planes $x + 2y - z - 4 = 0$ and $2x - y - z - 3 = 0$.

SOLUTION

Let $z = t$. Then $x + 2y = t + 4$ and $2x - y = t + 3$. Solving this system of two equations for x and y, we get $x = \frac{3}{5}t + 2$ and $y = \frac{1}{5}t + 1$, and the desired parametrization is

$$\mathbf{l}(t) = \left(\frac{3}{5}t + 2, \frac{1}{5}t + 1, t\right) = \left(\frac{3}{5}, \frac{1}{5}, 1\right)t + (2, 1, 0),$$

where $t \in \mathbb{R}$. In words, the intersection of the two given planes is the line going through the point $(2, 1, 0)$ whose direction is given by the vector $(3/5, 1/5, 1)$. See Exercise 17 for an alternate solution. ◀

▶ EXAMPLE 3.12

Find parametric equations of the curves that are obtained by intersecting the cylinder $x^2 + z^2 = 1$ with the sphere $x^2 + y^2 + z^2 = 4$.

SOLUTION We are looking for parametric equations of the form $\mathbf{c}(t) = (x(t), y(t), z(t))$. Since $x^2 + z^2 = 1$, we choose $x(t) = \cos t$ and $z(t) = \sin t$. Since $x^2 + y^2 + z^2 = 4$, it follows that $y^2 = 3$, and thus the intersection consists of two curves $\mathbf{c}_1(t) = (\cos t, \sqrt{3}, \sin t)$ and $\mathbf{c}_2(t) = (\cos t, -\sqrt{3}, \sin t)$, where $0 \le t \le 2\pi$. (Of course, we could have used any other parametric equation of the circle $x^2 + z^2 = 1$.)

Curves Defined Implicitly and Implicit Function Theorem

We study an important way of describing curves in the plane and discuss the theorem that will give us an insight into the properties of such curves.

DEFINITION 3.1 Implicitly Defined Curve

Assume that $F : \mathbb{R}^2 \to \mathbb{R}$ is a continuously differentiable function. The set of points (x, y) in the domain of F where $F(x, y) = 0$ is called an *implicitly defined curve* in $\mathbb{R}^2$.

Recall that F is continuously differentiable (also called of class C^1) if all its partial derivatives exist and are continuous.

Usually, we refer to the set of zeros of $F(x, y)$ as "the set $F(x, y) = 0$" or "the curve $F(x, y) = 0$." We have already met implicitly defined curves: to obtain level curves of a function $f(x, y)$ of two variables, we set $f(x, y) = c$ for various values of a constant c. In other words, the level curve of value c can be viewed as the set $F_c(x, y) = 0$, where $F_c(x, y) = f(x, y) - c$.

The curve defined implicitly by the function $F_1(x, y) = x^2 + y^2 - 1 = 0$ is the circle of radius 1 centered at the origin (note that F_1 is continuously differentiable). The curves defined implicitly by $F_2(x, y) = (x^2 + y^2)^2 - x^2 + y^2 = 0$ and $F_3(x, y) = y^2 - x^3 = 0$ are shown in Figure 3.12. We call the curve in Figure 3.12(a) a *lemniscate*.

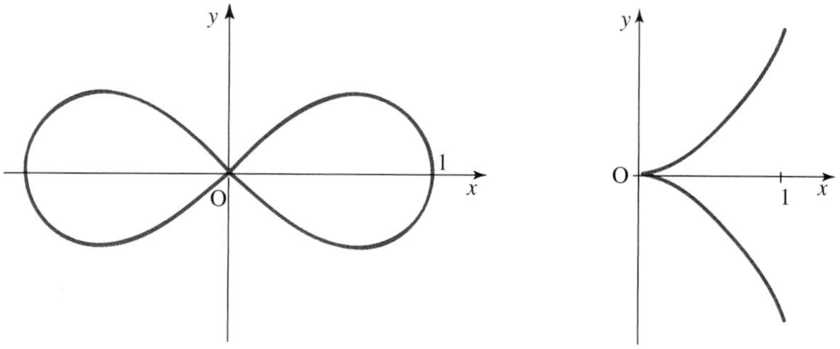

(a) $F_2(x, y) = (x^2 + y^2)^2 - x^2 + y^2 = 0$. (b) $F_3(x, y) = y^2 - x^3 = 0$.

Figure 3.12 Implicitly defined curves.

From $F(x, y) = 0$, we obtain curves with a wide variety of properties. For instance, a curve can intersect itself [as in Figure 3.12(a)], or appear to be nondifferentiable although it is

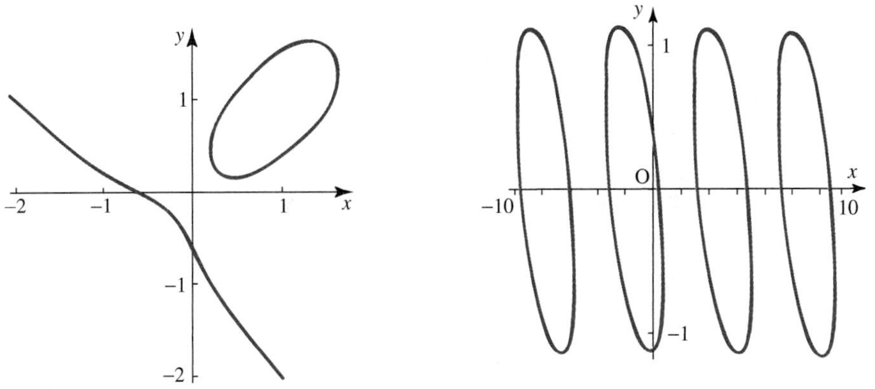

(a) $F_4(x, y) = x^3 + y^3 - 3xy + 0.1 = 0$. (b) $F_5(x, y) = y^2 + \sin(x + y) - 0.3 = 0$.

Figure 3.13 Curves defined implicitly by continuously differentiable functions.

defined using a differentiable function [as in Figure 3.12(b); the function $F_3(x, y) = y^2 - x^3$ is differentiable for all (x, y)].

Moreover, the set $F(x, y) = 0$ can be empty [for instance, if $F(x, y) = x^2 + y^2 + 1 = 0$], or a single point [if $F(x, y) = x^2 + y^2 = 0$], or can consist of several [Figure 3.13(a)] or infinitely many [Figure 3.13(b)] disconnected pieces.

When we write $y = f(x)$, for instance, as in $y = e^x$ or in $y = 3x \sin x$, we express y *explicitly* as a function of x. Solving $F_1(x, y) = x^2 + y^2 - 1 = 0$ for y, we get $y = f_1(x) = \sqrt{1 - x^2}$ and $y = f_2(x) = -\sqrt{1 - x^2}$. The two functions $f_1(x)$ and $f_2(x)$ are said to be *defined implicitly by* $F_1(x, y) = 0$.

One reason why we are thinking about solving for y is because the explicit form of a function is easier to work with. In general, solving $F(x, y) = 0$ is quite difficult, or, in many cases, impossible to do. For instance, solving $F_4(x, y) = x^3 + y^3 - 3xy + 0.1 = 0$ for y involves complicated expressions coming from the formula for solutions of a cubic equation. It is not possible to solve $F_5(x, y) = y^2 + \sin(x + y) - 0.3 = 0$ to obtain explicit formula(s) for y.

Applying implicit differentiation to $F_4(x, y) = x^3 + y^3 - 3xy + 0.1 = 0$, we get

$$3x^2 + 3y^2y' - 3y - 3xy' = 0, \tag{3.3}$$

and thus $y' = (y - x^2)/(y^2 - x)$. So, although we do not have an explicit formula for y, we can still compute its derivative—and, consequently, its tangent (i.e., linear approximation)—and thus obtain important information about the function.

Our next theorem will help us understand better what the curves coming from $F(x, y) = 0$ look like. Assuming certain properties of F, it will state that locally (i.e., when we focus on a small part of it) an implicitly defined curve does look like a graph of a function of one variable (however, we might not be able to find an explicit formula for that function). Furthermore, the way we computed the derivative in (3.3) can be applied to a general differentiable function.

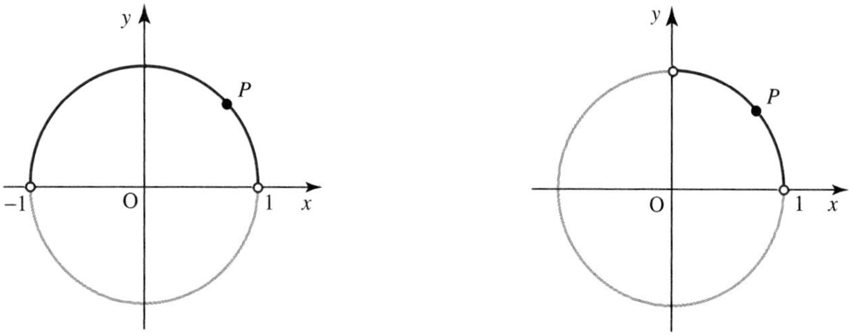

(a) $y = \sqrt{1 - x^2}$, $-1 < x < 1$, $y > 0$. (b) $y = \sqrt{1 - x^2}$, $0 < x < 1$, $0 < y < 1$.

Figure 3.14 Near P, circle $F_1(x, y) = x^2 + y^2 - 1 = 0$ is the graph of a function.

As we have already noted, the circle $F_1(x, y) = x^2 + y^2 - 1 = 0$ cannot be represented as the graph of a one-variable function *globally* [i.e., for *all* (x, y)]. However, once we pick a point on the circle—say, $P(1/\sqrt{2}, 1/\sqrt{2})$—we can describe a part of the circle near P as the graph of a function: for instance, $y = \sqrt{1 - x^2}$, for $-1 < x < 1$ and $y > 0$; see Figure 3.14(a). Of course, we could have used other intervals, such as $y = \sqrt{1 - x^2}$, for $0 < x < 1$ and $0 < y < 1$; see Figure 3.14(b).

Let us consider another example: pick the point $Q(\frac{1}{2}, -\frac{1}{2}\sqrt{-3 + 2\sqrt{3}})$ on the lemniscate $F_2(x, y) = (x^2 + y^2)^2 - x^2 + y^2 = 0$ (see Exercise 18). In Figure 3.15(a), we identified the curve $y = g(x)$ whose graph locally (i.e., near Q) is the lemniscate. In other words, the function $y = g(x)$, $0 < x < 1$, $y < 0$, shown in Figure 3.15(a), is a local solution to $F_2(x, y) = (x^2 + y^2)^2 - x^2 + y^2 = 0$ near Q. Incidentally, we can solve for y explicitly in this case: in Exercise 18 we show that

$$g(x) = -\frac{1}{2}\sqrt{-2 - 4x^4 + 2\sqrt{8x^2 + 1}}.$$

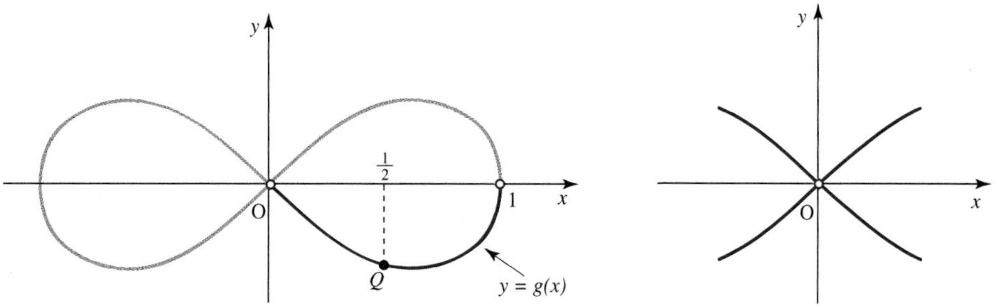

(a) Local solution near Q. (b) Graph near (0, 0).

Figure 3.15 The lemniscate $F_2(x, y) = (x^2 + y^2)^2 - x^2 + y^2 = 0$.

Note that $F_1(x, y) = x^2 + y^2 - 1 = 0$ cannot be solved for y (even locally) at the points $(\pm 1, 0)$. The same is true for the lemniscate at the points $(\pm 1, 0)$. At these points, tangent lines are perpendicular to the x-axis.

This fact gives us a useful insight: as long as we can draw the tangent $y = mx + b$ (m is a real number) to a curve $F(x, y) = 0$ at a point P, we will be able to represent the curve near P as the graph of a one-variable function $y = g(x)$ (recall that, near the point of tangency, the curve looks like its tangent).

Near the point $(0, 0)$ the lemniscate looks like a cross; see Figure 3.15(b). Thus, the lemniscate cannot be represented locally, near $(0, 0)$, as the graph of a function. From $F_2(x, y) = (x^2 + y^2)^2 - x^2 + y^2 = x^4 + 2x^2y^2 + y^4 - x^2 + y^2$, we get $\partial F_2/\partial x = 4x^3 + 4xy^2 - 2x$ and $\partial F_2/\partial y = 4x^2y + 4y^3 + 2y$; thus, at $(0, 0)$, both partial derivatives are zero.

Keeping these comments in mind, we now state (without proof) the theorem.

THEOREM 3.1 Implicit Function Theorem for Curves

Assume that the function $F\colon \mathbb{R}^2 \to \mathbb{R}$ has continuous partial derivatives, $F(x_0, y_0) = 0$ and $(\partial F/\partial y)(x_0, y_0) \neq 0$ at a point (x_0, y_0) in its domain. Then:

(a) There exist an open interval U containing x_0 and an open interval V containing y_0 such that there is a unique function $y = g(x)$ defined on U with values in V satisfying

$$F(x, g(x)) = 0$$

[i.e., $g(x)$ solves the equation $F(x, y) = 0$ locally near (x_0, y_0)].

(b) If x in U and y in V satisfy $F(x, y) = 0$, then they are related by $y = g(x)$.

(c) The function $y = g(x)$ is continuously differentiable (i.e., of class C^1) and its derivative is given by

$$g'(x) = -\frac{\dfrac{\partial F}{\partial x}(x, g(x))}{\dfrac{\partial F}{\partial y}(x, g(x))}, \tag{3.4}$$

whenever $(\partial F/\partial y)(x, g(x)) \neq 0$.

◀

The statement in (a) implies that the set $\{(x, y) \mid x \in U, y \in V \text{ and } y = g(x)\}$ is a subset of the set $\{(x, y) \mid x \in U, y \in V \text{ and } F(x, y) = 0\}$. In (b), the reverse containment is established, and thus, the set of points where $F(x, y) = 0$ is locally the same as the graph of the function $y = g(x)$.

For the circle $F_1(x, y) = x^2 + y^2 - 1$, we get $\partial F_1/\partial y = 2y$, and thus $(\partial F_1/\partial y)(\pm 1, 0) = 0$ [recall that, at $(\pm 1, 0)$, the circle has vertical tangents]. Likewise, for the lemniscate $F_2(x, y) = 0$, we get $\partial F_2/\partial y = 4x^2y + 4y^3 + 2y$, and so $(\partial F_2/\partial y)(\pm 1, 0) = 0$. Thus, we see that the assumption $(\partial F/\partial y)(x_0, y_0) \neq 0$ in the theorem helps eliminate vertical tangents.

In the discussion preceding the theorem, we considered the point $(1/\sqrt{2}, 1/\sqrt{2})$ (i.e., $x_0 = y_0 = 1/\sqrt{2}$ in the theorem) on the circle $F_1(x, y) = x^2 + y^2 - 1$. In Figure 3.14(b), we see that U is the interval $(0, 1)$ on the x-axis, and V is the interval $(0, 1)$ on the y-axis. For the lemniscate in Figure 3.15(a), $(x_0, y_0) = \left(\frac{1}{2}, -\frac{1}{2}\sqrt{-3+2\sqrt{3}}\right)$ and the function $g(x)$ is defined on $U = (0, 1)$, with values in, say, $V = (-1, 0)$.

The formula for the derivative given in (3.4) is easy to prove: we apply the chain rule to $F(x, g(x)) = 0$ to get

$$D_1F(x, g(x)) + D_2F(x, g(x))g'(x) = 0,$$

then replace D_1F by $\partial F/\partial x$ and D_2F by $\partial F/\partial y$, and solve for $g'(x)$.

The appearance of the minus sign in (3.4) can be explained geometrically; see Figure 3.16. Both components of the gradient $\nabla F(x, y)$ are positive. However, $g'(x) < 0$, since $g'(x)$ is the slope of the tangent line to $y = g(x)$ at (x, y).

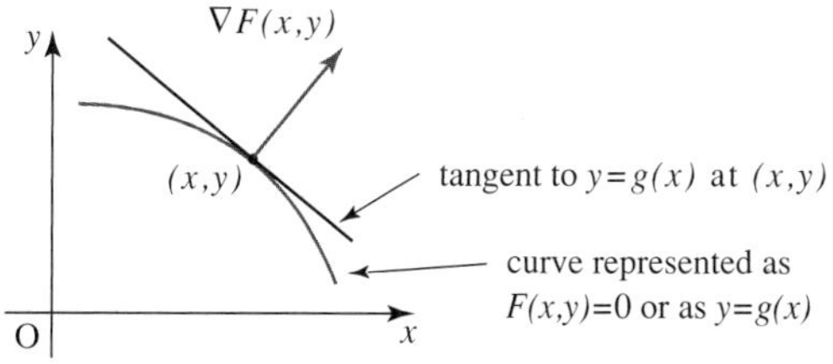

Figure 3.16 Geometry of the formula (3.4).

As an illustration of part (c) of the Implicit Function Theorem, consider the point $(x_0, y_0) = (0, -0.1^{1/3})$ on the curve $F_4(x, y) = x^3 + y^3 - 3xy + 0.1 = 0$ Since $F_4(0, -0.1^{1/3}) = 0$ and $(\partial F_4/\partial y)(0, -0.1^{1/3}) = 3y^2 - 3x\big|_{(0,-0.1^{1/3})} = 3(0.1^{3/2}) \neq 0$, the theorem states that the derivative of the local solution $y = g(x)$ at $(0, -0.1^{1/3})$ is given by

$$g'(x) = -\frac{\dfrac{\partial F_4}{\partial x}(x, g(x))}{\dfrac{\partial F_4}{\partial y}(x, g(x))} = -\frac{3x^2 - 3y}{3y^2 - 3x} = \frac{y - x^2}{y^2 - x};$$

compare it with (3.3). Thus, $g'(0) = -1/0.1^{1/3} \approx -2.154$ is the slope of the tangent to $y = g(x)$ at $(0, -0.1^{1/3})$.

Note that the roles of x and y in Theorem 3.1 can be reversed. To illustrate this point, use the circle $F_1(x, y) = x^2 + y^2 - 1 = 0$ again. The partial derivative $(\partial F/\partial y)(1, 0)$ is zero, but the partial derivative $(\partial F/\partial x)(1, 0) = 2$! Thus, the Implicit Function Theorem applies: near $(1, 0)$, we can solve for x explicitly in terms of y, $x = \sqrt{1 - y^2}$. Likewise, near $(-1, 0)$, $x = -\sqrt{1 - y^2}$. See Figure 3.17.

In this section, we discussed a special case of the Implicit Function Theorem, as it applies to functions $F\colon \mathbb{R}^2 \to \mathbb{R}$. In Chapter 4, we will quote a general form of the theorem. Another special case that applies to functions $F\colon \mathbb{R}^3 \to \mathbb{R}$ will be discussed in the context of surfaces in Chapter 7. In Chapter 6, we will use the Implicit Function Theorem in performing a change of variables in multiple integrals.

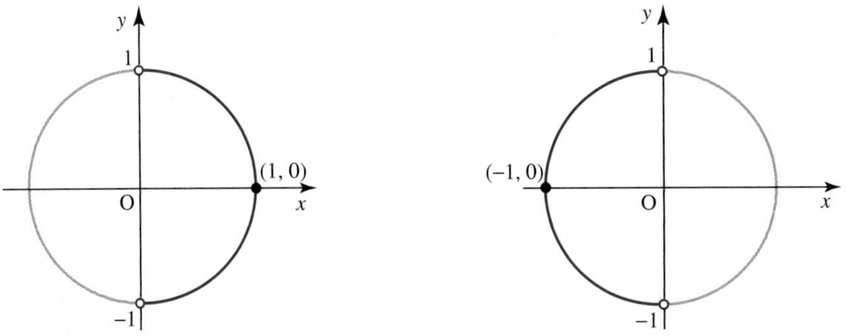

(a) $x = \sqrt{1 - y^2}$. (b) $x = -\sqrt{1 - y^2}$.

Figure 3.17 Local solutions near $(1, 0)$ and $(-1, 0)$.

▶ EXERCISES 3.1

1. Compute the velocity of the path $\mathbf{c}(t)$ in Example 3.1, and show that its speed is equal to $2/(1 + t^2)$. Identify all points where the speed is the largest.

2. Replace t by e^t in Example 3.1 to obtain the path $\mathbf{c}(t) = (2e^t/(1 + e^{2t}), (1 - e^{2t})/(1 + e^{2t}))$, $t \in \mathbb{R}$. What is its image? Describe the motion given by $\mathbf{c}(t)$, that is, discuss endpoints, orientation, and speed.

3. In Example 3.1 (see also Exercise 1), we discussed the parametrization of a circle that does not have a constant speed. Find another parametrization of the circle with nonconstant speed.

4. Glue together two concentric disks of radii $a > 0$ and $b > 0$ and fix a point P on the boundary of the disk of radius b. A generalized cycloid is the curve traced by P as the other disk is rolled without sliding along a straight line; see Figure 3.18 (the case $a < b$ is shown). Show that a parametric representation of this generalized cycloid is given by $\mathbf{c}(\theta) = (a\theta - b\sin\theta, a - b\cos\theta)$. Using a graphing device, plot $\mathbf{c}$ for various values of a and b.

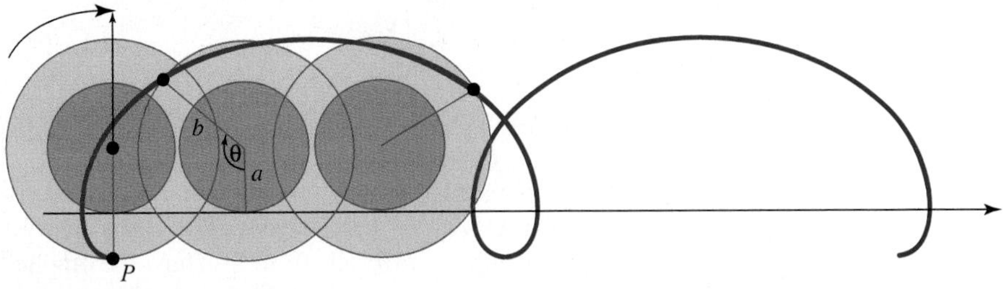

Figure 3.18 Cycloid, case $a < b$.

5. For a Bézier curve $\mathbf{c}(t)$ defined by (3.1) and (3.2), compute $\mathbf{c}'(t)$ and show that $\mathbf{c}'(0) = 3(b_x - a_x, b_y - a_y)$ and $\mathbf{c}'(1) = 3(d_x - c_x, d_y - c_y)$.

6. Consider the Bézier curve whose control points are $A(0, 0)$, $B(0, 1)$, $C(1, 1)$, and $D(1, 0)$.

(a) Find a parametric equation $\mathbf{c}(t) = (x(t), y(t))$ of the curve.

(b) Show that $0 \le x(t) \le 1$ and $0 \le y(t) \le 1$ whenever $0 \le t \le 1$. Conclude that the curve $\mathbf{c}(t)$ is contained within a square whose vertices are the control points. (This fact is true in general: a Bézier curve is contained inside a quadrilateral whose vertices are the control points.)

(c) Using a graphing device, plot the curve and indicate the control points.

(d) Examine how the curve changes if A and D are kept where they are, but B is moved to $(3, 0)$ and C is moved to $(1, 3)$.

7. Consider the path $\mathbf{c}(t) = (\cos 2t, \sin 3t)$, $t \in [0, 2\pi]$.

(a) Explain why the image of $\mathbf{c}(t)$ is contained in the square with vertices $(1, 1)$, $(-1, 1)$, $(-1, -1)$, and $(1, -1)$. At which points does the curve $\mathbf{c}$ touch the vertical line $x = 1$? $x = -1$? Horizontal line $y = 1$? $y = -1$?

(b) Using a graphing device, obtain a plot of the image of $\mathbf{c}(t)$. What are its initial and terminal points? Describe the motion of an object whose position is given by $\mathbf{c}(t)$, as t changes from 0 to 2π.

8. Using a graphing device, investigate and compare the images of $\mathbf{c}(t) = (\cos mt, \sin nt)$ when $m = 2$ and $n = 2, 5$, and when $m = 3$ and $n = 2, 4$.

9. Let $\mathbf{c}(t) = (\cos 2t, \sin 4t)$, $t \in [0, \pi]$.

(a) Eliminate t to find an implicit equation $F(x, y) = 0$ for $\mathbf{c}(t)$. Use it to show that $\mathbf{c}(t)$ is symmetric with respect to both the x-axis and y-axis.

(b) Find a local solution to $F(x, y) = 0$ of the form $y = g(x)$ near the point $(\sqrt{2}/2, 1)$.

(c) Plot the curve $\mathbf{c}(t)$ and indicate the solution $g(x)$ from (b) on it.

Exercises 10 to 13: Find a parametric equation of the given curve.

10. The intersection of the cylinder $x^2 + y^2 = 1$ and the plane $x + z = 1$ in $\mathbb{R}^3$

11. The intersection of the planes $x + y - z = 2$ and $2x - 5y + z = 3$ in $\mathbb{R}^3$

12. The intersection of the cylinder $y^2 + z^2 = 4$ and the surface $x = yz$ in $\mathbb{R}^3$

13. The intersection of the cylinder $(x + 2)^2 + (z - 2)^2 = 4$ and the plane $y = 3$ in $\mathbb{R}^3$

14. Find a parametric representation of the curve that is the intersection of the two paraboloids $z = 3 + 2x^2 + 4y^2$ and $z = 7 - x^2 - 2y^2$.

15. Find parametric representations of both curves that are the intersection of the cylinders $x^2 + z^2 = 2$ and $y^2 + z^2 = 2$ in $\mathbb{R}^3$.

16. Show that the intersection of the cylinder $(x - 1)^2 + y^2 = 1$ and the sphere $x^2 + y^2 + z^2 = 4$ can be parametrized by $\mathbf{c}(t) = (1 + \cos t, \sin t, 2 \sin t/2)$, $t \in [0, 2\pi]$.

17. We consider an alternate way of computing an equation of the line from Example 3.11.

(a) Compute the cross product of the vectors normal to the given planes. Explain why the vector you obtained is a direction vector for the line $\mathbf{l}$).

(b) Find coordinates of a point that belongs to both planes (and thus to $\mathbf{l}$).

(c) Use (a) and (b) to obtain a parametric equation of the line $\mathbf{l}$.

18. Consider the lemniscate $F_2(x, y) = (x^2 + y^2)^2 - x^2 + y^2 = 0$.

(a) Show that there are two points on the lemniscate whose x coordinate is $1/2$, namely $(\frac{1}{2}, \frac{1}{2}\sqrt{-3 + 2\sqrt{3}})$ and $(\frac{1}{2}, -\frac{1}{2}\sqrt{-3 + 2\sqrt{3}})$.

(b) Simplify $(x^2 + y^2)^2 - x^2 + y^2 = 0$ to get $y^4 + (2x^2 + 1)y^2 + x^4 - x^2 = 0$ and use the quadratic formula to obtain the solutions $y = \pm\frac{1}{2}\sqrt{-2 - 4x^4 + 2\sqrt{8x^2 + 1}}$ for y.

(c) Identify a local solution to $F_2(x, y) = 0$ near the point $(\frac{1}{2}, -\frac{1}{2}\sqrt{-3 + 2\sqrt{3}})$.

19. The point (1, 1.5066) lies on the graph of the equation $F_4(x, y) = x^3 + y^3 - 3xy + 0.1 = 0$; see Figure 3.13(a). Let $y = g(x)$ be the function defined implicitly by $F_4(x, y) = 0$ near (1, 1.5066). Find its linear approximation at $x = 1$.

20. Find local solutions of the equation $F_2(x, y) = (x^2 + y^2)^2 - x^2 + y^2 = 0$ near $(\pm 1, 0)$ in the form $x = g(y)$. [See Figure 3.12(a) for the graph of $F_2(x, y) = 0$.]

21. Assume that $F(x, y) = 0$ (where F is a C^1 function) defines y uniquely as a function of x. Knowing that $F(2, 4) = 0$ and $\nabla F(2, 4) = (3, -3)$, find $g'(2)$. Also, find the linear approximation of $g(x)$ at $x = 2$.

22. Let $y = g(x)$ be the function defined implicitly by $F(x, y) = \cos y - xy = 0$, near $(0, \pi/2)$. Find the linear approximation of $g(x)$ at $x = 0$.

23. Find examples of differentiable functions $F(x, y)$ such that the set $F(x, y) = 0$

(a) Is a single point $(1, -4)$.

(b) Consists of two parallel lines.

(c) Consists of two lines intersecting each other at $\pi/4$ rad.

(d) Is empty.

24. Let $\mathbf{c}(t) = (-3 + 2\cos t, 1 - 4\sin t)$, $t \in [0, 2\pi]$. Find a continuously differentiable function $F(x, y)$ such that the image of $\mathbf{c}(t)$ is the set $F(x, y) = 0$.

25. Show that $\mathbf{c}(t) = (3t/(1 + t^3), 3t^2/(1 + t^3))$, $t \in \mathbb{R}$, is a parametrization of the curve $x^3 + y^3 - 3xy = 0$.

26. Consider the points of intersection of the logarithmic spiral $\mathbf{c}(t) = (e^{at}\cos t, e^{at}\sin t)$, $t \geq 0$, with a ray emanating from the origin, see Figure 3.4(a). Show that these points are spaced according to a geometric progression.

27. Take a point $\mathbf{c}(t_0)$ on the logarithmic spiral given by $\mathbf{c}(t) = (e^{at}\cos t, e^{at}\sin t)$, and let $\mathbf{l}(t)$ be the line joining the origin and $\mathbf{c}(t_0)$. Compute the angle between the line $\mathbf{l}(t)$ and the spiral at $\mathbf{c}(t_0)$. What can you conclude from your answer?

28. Show that the speed of the spiral $\mathbf{c}(t) = (f(t)\cos t, f(t)\sin t)$ (assume that f is differentiable) is equal to $\sqrt{(f(t))^2 + (f'(t))^2}$.

29. Consider spirals $\mathbf{c}(t) = (f(t)\cos t, f(t)\sin t)$, where $f(t) = t$ $(t \geq 0)$, $f(t) = t^2$ $(t \geq 0)$, $f(t) = \ln t$ $(t > 1)$, and $f(t) = 1/t$ $(t > 0)$. Describe how the images of $\mathbf{c}(t)$ differ in terms of orientation and in the way in which they spiral.

30. Find a parametrization of a helix on a cylinder of radius 2 whose pitch is equal to 1.

31. An object, initially at the point (1, 0, 0), starts traveling around the cylinder $x^2 + y^2 = 1$ counterclockwise as seen from above, at a constant speed of 3 units/s. At all points, its trajectory makes an angle of 45° with respect to a horizontal plane. Find the position of the object 12 s later.

32. Eliminating t from the parametrization for a catenary curve, we obtain $y = f(x) = a\cosh(x/a)$ (note that $x = 0$, $y = a$ is its minimum). Show that the first two terms in MacLaurin expansion for $f(x)$ are at $a + x^2/2a$. This shows that a catenary curve can be approximated by a parabola. Thus, Galileo was not completely wrong in assuming that the hanging chain is a parabola.

33. A parameterization $\mathbf{c}(t) = (t, a\cosh bt)$, where a and b are real numbers, represents a catenary curve (that is more general than the one we studied in Example 3.5 and Exercise 32.) Find a catenary that is obtained by suspending a cable from supports 25 m high and 100 m apart, knowing that the lowest point on the cable is 10 m.

▶ 3.2 TANGENTS, VELOCITY, AND ACCELERATION

We continue our investigation of curves by studying concepts that are defined in terms of derivatives: velocity and acceleration.

Recall that the velocity of a differentiable path $\mathbf{c}(t)$ in $\mathbb{R}^2$ or $\mathbb{R}^3$ is given by $\mathbf{v}(t) = \mathbf{c}'(t)$, and its speed by the scalar function $\|\mathbf{v}(t)\|$. If $\mathbf{c}(t)$ is twice differentiable, we define the acceleration by $\mathbf{a}(t) = \mathbf{v}'(t) = \mathbf{c}''(t)$.

▶ EXAMPLE 3.13

Suppose that a particle, subjected to a force field, moves away from the origin along the path $\mathbf{c}(t) = (t, 2t^2, t^3)$, where t is the time in seconds. At time $t = 1$, the source of the force field is shut off, and the particle flies off along a tangent line, as shown in Figure 3.19. Find its position at $t = 3$.

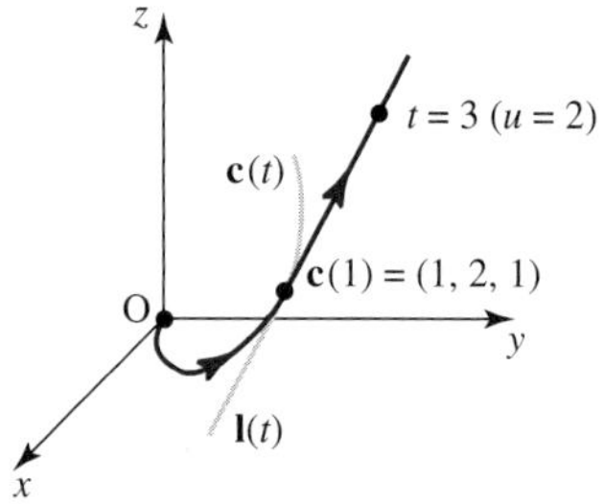

Figure 3.19 Path of the particle in Example 3.13.

SOLUTION At $t = 1$, the particle is located at $\mathbf{c}(1) = (1, 2, 1)$ and is moving with velocity

$$\mathbf{v}(1) = \mathbf{c}'(1) = (1, 4t, 3t^2)\Big|_{t=1} = (1, 4, 3).$$

The tangent line $\mathbf{l}(u)$ is now uniquely determined: it goes through the point $(1, 2, 1)$ and is parallel to the velocity vector $(1, 4, 3)$:

$$\mathbf{l}(u) = (1, 2, 1) + u(1, 4, 3), \qquad u \in \mathbb{R}.$$

The position of the particle at $t = 3$ is given by $\mathbf{l}(2)$ (i.e., when $u = 2$), since the particle already used 1 s to reach the point where it flew off its initial trajectory. Hence, the position of the particle 3 s after it started moving is $\mathbf{l}(2) = (3, 10, 7)$. ◀

▶ EXAMPLE 3.14

The position of a projectile fired from the origin at an angle of θ rad with respect to the positive x-axis and with the initial speed v_0 is given by

$$\mathbf{c}(t) = (v_0\cos\theta)t\mathbf{i} + ((v_0\sin\theta)t - gt^2/2)\mathbf{j},$$

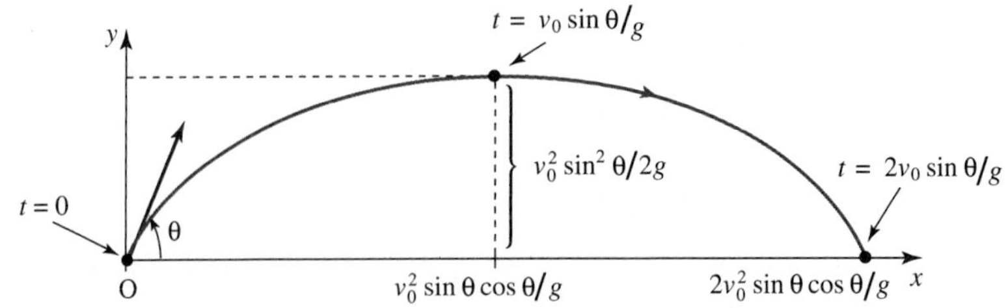

Figure 3.20 Path of a projectile in Example 3.14.

where t is time and g is the acceleration due to gravity; see Figure 3.20. (In Example 3.15 we will derive this formula.) Find

(a) The coordinates of the highest point on the trajectory, the time when the projectile reaches that point, and its velocity at that moment;

(b) The time when the projectile hits the ground (that is represented by the x-axis) and the velocity at that moment.

SOLUTION

(a) The height is given by the y-coordinate $y(t) = (v_0 \sin\theta)t - gt^2/2$ of $\mathbf{c}(t)$. The critical point of $y(t)$ is [set $y'(t) = v_0 \sin\theta - gt = 0$ and solve for t] $t = v_0 \sin\theta/g$. Since $y''(t) = -g < 0$, $y(t)$ has a maximum at $t = v_0 \sin\theta/g$. Hence, the projectile reaches its highest point when $t = v_0 \sin\theta/g$. The coordinates of the highest point are

$$\mathbf{c}(v_0 \sin\theta/g) = v_0 \cos\theta \frac{v_0 \sin\theta}{g}\mathbf{i} + \left(v_0 \sin\theta \frac{v_0 \sin\theta}{g} - \frac{g}{2}\frac{v_0^2 \sin^2\theta}{g^2}\right)\mathbf{j}$$
$$= \frac{v_0^2 \sin\theta\cos\theta}{g}\mathbf{i} + \frac{v_0^2 \sin^2\theta}{2g}\mathbf{j}.$$

Since $\mathbf{v}(t) = \mathbf{c}'(t) = v_0 \cos\theta\mathbf{i} + ((v_0 \sin\theta) - gt)\mathbf{j}$, it follows that $\mathbf{v}(v_0 \sin\theta/g) = v_0 \cos\theta\mathbf{i}$ is the velocity of the projectile at the moment it reaches the highest point of its trajectory.

(b) At the moment the projectile hits the ground, $y(t) = (v_0 \sin\theta)t - gt^2/2 = 0$ and $t > 0$ ($t = 0$ is the moment when the projectile is fired). Consequently, $t = 2v_0 \sin\theta/g$ is the moment when it hits the ground. Notice that this time is twice the time the projectile needs to reach its highest position. The velocity at the moment it hits the ground is $\mathbf{c}'(2v_0 \sin\theta/g) = v_0 \cos\theta\mathbf{i} - v_0 \sin\theta\mathbf{j}$. ◀

DEFINITION 3.2 Antiderivative of a Vector-Valued Function

Let $\mathbf{b}(t)$ and $\mathbf{c}(t)$ be two vector-valued functions of one variable. If $\mathbf{b}'(t) = \mathbf{c}(t)$, then the function $\mathbf{b}(t)$ is called an *antiderivative* (or an *indefinite integral*) of $\mathbf{c}(t)$, and is denoted by $\mathbf{b}(t) = \int \mathbf{c}(t)\,dt$. If $\mathbf{c}(t) = c_1(t)\mathbf{i} + c_2(t)\mathbf{j} + c_3(t)\mathbf{k}$, then

$$\mathbf{b}(t) = \int \mathbf{c}(t)\,dt = \left(\int c_1(t)\,dt\right)\mathbf{i} + \left(\int c_2(t)\,dt\right)\mathbf{j} + \left(\int c_3(t)\,dt\right)\mathbf{k}.$$ ◀

We use antiderivatives when we need to recover velocity from acceleration or position (path) from velocity, as in the following example.

▶ EXAMPLE 3.15

Suppose that a projectile is launched from the origin with initial speed v_0 at an angle of θ rad with respect to the positive x-axis. Assuming no forces or effects other than gravity are present (hence the acceleration is $\mathbf{a}(t) = -g\mathbf{j}$, $g \approx 9.80$ m/s^2), find the trajectory of the projectile.

SOLUTION

Since the acceleration is $\mathbf{a}(t) = -g\mathbf{j}$, it follows that the velocity is

$$\mathbf{v}(t) = \int \mathbf{a}(t)\,dt = \int -g\mathbf{j}\,dt = C_1\mathbf{i} + (-gt + C_2)\mathbf{j},$$

where C_1 and C_2 are constants. The initial condition [the data for $\mathbf{v}(0)$ is given in polar form] $\mathbf{v}(0) = v_0\cos\theta\mathbf{i} + v_0\sin\theta\mathbf{j}$ implies that $v_0\cos\theta\mathbf{i} + v_0\sin\theta\mathbf{j} = C_1\mathbf{i} + C_2\mathbf{j}$, that is, $C_1 = v_0\cos\theta$, $C_2 = v_0\sin\theta$, and hence,

$$\mathbf{v}(t) = v_0\cos\theta\mathbf{i} + (-gt + v_0\sin\theta)\mathbf{j}.$$

Another integration gives the position vector

$$\mathbf{c}(t) = ((v_0\cos\theta)t + D_1)\mathbf{i} + (-gt^2/2 + (v_0\sin\theta)t + D_2)\mathbf{j}.$$

Using $\mathbf{c}(0) = (0, 0)$, we get $D_1 = 0$, $D_2 = 0$, and thus,

$$\mathbf{c}(t) = (v_0\cos\theta)t\mathbf{i} + (-gt^2/2 + (v_0\sin\theta)t)\mathbf{j}.$$

This is precisely the formula used in Example 3.14. ◀

▶ EXAMPLE 3.16 Motion Around a Circle with Constant Speed

An object moves around the circle centered at the origin of radius R with constant speed s in the counterclockwise direction. Find a parametric representation of its trajectory.

SOLUTION

By analyzing Examples 2.57 and 2.59 in Section 2.5, we notice that the arguments of sine and cosine are involved in determining the speed. Therefore, we start with $c_1(t) = (R\cos\omega t, R\sin\omega t)$, $t \ge 0$, and determine the value of ω from the requirement that the speed be equal to s. From $\mathbf{c}_1'(t) = (-R\omega\sin\omega t, R\omega\cos\omega t)$ it follows that $\|\mathbf{c}_1'(t)\| = R\omega$, and therefore, $s = R\omega$ and $\omega = s/R$. Consequently, the parametrization

$$\mathbf{c}(t) = (R\cos(st/R), R\sin(st/R)), \qquad t \ge 0$$

has the required speed (see Figure 3.21). When $t = 0$, $\mathbf{c}(0) = (R, 0)$; for small positive values of t, both components of $\mathbf{c}(t)$ are positive; that is, $\mathbf{c}(t)$ is in the first quadrant. It follows that the curve $\mathbf{c}$ is oriented counterclockwise. ◀

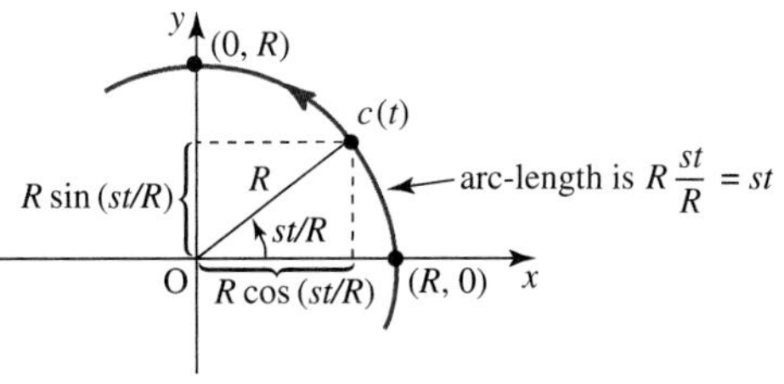

Figure 3.21 Motion around the circle with constant speed s.

From arc $=R\cdot$ angle (R is the radius), we see that speed $s=\text{arc}/\text{time}=R\cdot\text{angle}/\text{time}$. Since $s=R\omega$, it follows that $\omega=\text{angle}/\text{time}$. In other words, the scalar ω in Example 3.16 represents the angular speed.

Denote the angle measured counterclockwise from the positive x-axis by θ. From $\theta=\omega t$, it follows that $d\theta/dt=\omega$. Consequently, the sign of ω determines the orientation: if $\omega>0$, then $d\theta/dt>0$; that is, θ is increasing, and the curve is oriented counterclockwise. If $\omega<0$, then $d\theta/dt<0$ and the curve is oriented clockwise.

▶ EXAMPLE 3.17 Centripetal Acceleration and Centripetal Force

Compute the acceleration of a particle of mass m that moves along the circle of radius R with constant speed s.

SOLUTION Assuming that the center of the circular path is at the origin, we determine (see Example 3.16) that $\mathbf{c}(t)=(R\cos(s/R)t, R\sin(s/R)t)$ describes the position of the particle at time t. Hence, $\mathbf{v}(t)=(-s\sin(s/R)t, s\cos(s/R)t)$, and thus, the acceleration is

$$\mathbf{a}(t)=(-(s^2/R)\cos(s/R)t, -(s^2/R)\sin(s/R)t).$$

Since

$$\mathbf{a}(t)=-(s^2/R^2)(R\cos(s/R)t, R\sin(s/R)t)=-(s^2/R^2)\mathbf{c}(t),$$

it follows that the acceleration vector points in the direction opposite to the position vector $\mathbf{c}(t)$; that is, points toward the center of rotation. For this reason, $\mathbf{a}(t)$ is called a *centripetal acceleration,* and we call the corresponding force $\mathbf{F}=m\mathbf{a}(t)$ a *centripetal force*.

It is important to notice that although the speed s (scalar!) in this example is constant, the velocity $\mathbf{v}(t)$ (vector!) is not (the velocity vector changes its direction continuously), and therefore the acceleration is not zero. ◀

In the following example, we prove several formulas that we will find useful in calculations with paths.

▶ EXAMPLE 3.18

Let $\mathbf{c}(t)$ be a differentiable path such that $\mathbf{c}(t)\neq\mathbf{0}$ and $\mathbf{c}'(t)\neq\mathbf{0}$ for every t. Show that

(a) $\left(\|\mathbf{c}(t)\|^2\right)'=2\mathbf{c}(t)\cdot\mathbf{c}'(t)$.
(b) $(\|\mathbf{c}(t)\|)'=\mathbf{c}(t)\cdot\mathbf{c}'(t)/\|\mathbf{c}(t)\|$.
(c) $\left(\|\mathbf{c}(t)\|^{-1}\right)'=-\mathbf{c}(t)\cdot\mathbf{c}'(t)/\|\mathbf{c}(t)\|^3$.
(d) $(\mathbf{c}(t)/\|\mathbf{c}(t)\|)'=(-(\mathbf{c}(t)\cdot\mathbf{c}'(t))\,\mathbf{c}(t)+(\mathbf{c}(t)\cdot\mathbf{c}(t))\,\mathbf{c}'(t))/\|\mathbf{c}(t)\|^3$.

SOLUTION We drop t to keep the notation as simple as possible.

(a) Using the product rule and the fact that $\|\mathbf{c}\|^2=\mathbf{c}\cdot\mathbf{c}$, we obtain

$$\left(\|\mathbf{c}\|^2\right)'=(\mathbf{c}\cdot\mathbf{c})'=\mathbf{c}\cdot\mathbf{c}'+\mathbf{c}'\cdot\mathbf{c}=2\mathbf{c}\cdot\mathbf{c}'.$$

(b) Combining $\left(\|\mathbf{c}\|^2\right)'=2\|\mathbf{c}\|\,\|\mathbf{c}\|'$ (which is obtained from the chain rule) and (a), we get $2\mathbf{c}\cdot\mathbf{c}'=2\|\mathbf{c}\|\,\|\mathbf{c}\|'$, and hence, $\|\mathbf{c}\|'=\mathbf{c}\cdot\mathbf{c}'/\|\mathbf{c}\|$.

(c) From the chain rule and (b), it follows that

$$\left(\|\mathbf{c}\|^{-1}\right)' = -\|\mathbf{c}\|^{-2}\|\mathbf{c}\|' = -\frac{1}{\|\mathbf{c}\|^2}\frac{\mathbf{c}\cdot\mathbf{c}'}{\|\mathbf{c}\|} = -\frac{\mathbf{c}\cdot\mathbf{c}'}{\|\mathbf{c}\|^3}.$$

(d) Using the product rule and (c), we get

$$\left(\frac{\mathbf{c}}{\|\mathbf{c}\|}\right)' = \left(\|\mathbf{c}\|^{-1}\mathbf{c}\right)' = \left(\|\mathbf{c}\|^{-1}\right)'\mathbf{c} + \|\mathbf{c}\|^{-1}\mathbf{c}'$$

$$= -\frac{\mathbf{c}\cdot\mathbf{c}'}{\|\mathbf{c}\|^3}\mathbf{c} + \frac{1}{\|\mathbf{c}\|}\mathbf{c}' = \frac{-(\mathbf{c}\cdot\mathbf{c}')\,\mathbf{c} + \|\mathbf{c}\|^2\,\mathbf{c}'}{\|\mathbf{c}\|^3}.$$

▶ EXAMPLE 3.19 Kepler's First Law of Planetary Motion

Based on Newton's Second Law and Law of Gravitation we now derive a fundamental result related to motions of planets around a sun (actually, the law applies to a motion of any object subjected to the force of gravity of a larger object; for instance, the Moon or a satellite orbiting Earth, or a comet moving around some star).

Assume that a planet of mass m revolves around the Sun (of mass M); for convenience, we place the Sun at the origin of the coordinate system. By $\mathbf{r}(t)$ we denote the position vector of m; $\mathbf{v}(t)$ is its velocity and $\mathbf{a}(t)$ its acceleration. [To simplify the notation, we drop t and use $\mathbf{r}$ instead of $\mathbf{r}(t)$, $\mathbf{v}$ instead of $\mathbf{v}(t)$, etc.] Recall that Newton's Second Law of Motion states that $F = m\mathbf{a}$, and his Law of Universal Gravitation can be expressed as $\mathbf{F}(\mathbf{r}) = -(GMm/\|\mathbf{r}\|^3)\mathbf{r}$ (G is the gravitational constant; see Example 2.10 in Section 2.1).

First, we show that the orbit of the planet is confined to a plane. Equating the two forces, we obtain $m\mathbf{a} = -(GMm/\|\mathbf{r}\|^3)\mathbf{r}$, and thus,

$$\mathbf{a} = -\frac{GM}{\|\mathbf{r}\|^3}\mathbf{r}. \tag{3.5}$$

By the product rule,

$$(\mathbf{r}\times\mathbf{v})' = \mathbf{r}'\times\mathbf{v} + \mathbf{r}\times\mathbf{v}' = \mathbf{v}\times\mathbf{v} + \mathbf{r}\times\mathbf{a} = \mathbf{0}.$$

The first summand is zero since the cross product of a vector by itself is zero. The formula (3.5) implies that $\mathbf{a}$ and $\mathbf{r}$ are parallel, and thus $\mathbf{r}\times\mathbf{a} = \mathbf{0}$.

From $(\mathbf{r}\times\mathbf{v})' = \mathbf{0}$, we conclude that $\mathbf{r}\times\mathbf{v} = \mathbf{d}$, where $\mathbf{d}$ is a *constant* vector. Since $\mathbf{r}$ and $\mathbf{v}$ are not parallel (the planet is not trying to fly away from its sun!), their cross product is not zero, and thus $\mathbf{d} \neq \mathbf{0}$. Now from $\mathbf{r}\times\mathbf{v} = \mathbf{d}$, we conclude that $\mathbf{r} = \mathbf{r}(t)$ is perpendicular to $\mathbf{d}$ for *all* t. Thus, the orbit of m lies in the plane that contains the Sun and is perpendicular to the vector $\mathbf{d}$.

Next, we would like to obtain an equation for $\|\mathbf{r}\|$, so that we can recognize the shape of the trajectory of m. Start with

$$\mathbf{a}\times\mathbf{d} = \left(-\frac{GM}{\|\mathbf{r}\|^3}\mathbf{r}\right)\times\mathbf{d} = -\frac{GM}{\|\mathbf{r}\|^3}\mathbf{r}\times(\mathbf{r}\times\mathbf{v})$$

and use the formula from Exercise 11 in Section 1.5 to get

$$= -\frac{GM}{\|\mathbf{r}\|^3}((\mathbf{r}\cdot\mathbf{v})\,\mathbf{r} - (\mathbf{r}\cdot\mathbf{r})\,\mathbf{v}) = \frac{GM}{\|\mathbf{r}\|^3}(-(\mathbf{r}\cdot\mathbf{v})\,\mathbf{r} + (\mathbf{r}\cdot\mathbf{r})\,\mathbf{v})$$

and simplify using (d) in Example 3.18 (with $\mathbf{c} = \mathbf{r}$ and $\mathbf{c}' = \mathbf{v}$):

$$= \frac{GM}{\|\mathbf{r}\|^3}\|\mathbf{r}\|^3\left(\frac{\mathbf{r}}{\|\mathbf{r}\|}\right)' = GM\left(\frac{\mathbf{r}}{\|\mathbf{r}\|}\right)'. \tag{3.6}$$

On the other hand,

$$(\mathbf{v} \times \mathbf{d})' = \mathbf{v}' \times \mathbf{d} + \mathbf{v} \times \mathbf{d}' = \mathbf{a} \times \mathbf{d} + \mathbf{v} \times \mathbf{0} = \mathbf{a} \times \mathbf{d}, \tag{3.7}$$

since $\mathbf{d}$ is constant (so that $\mathbf{d}' = \mathbf{0}$). From (3.6) and (3.7), we get $(\mathbf{v} \times \mathbf{d})' = GM\,(\mathbf{r}/\|\mathbf{r}\|)'$. Integrating both sides with respect to t, we obtain the relation

$$\mathbf{v} \times \mathbf{d} = GM\frac{\mathbf{r}}{\|\mathbf{r}\|} + \mathbf{c}, \tag{3.8}$$

where $\mathbf{c}$ is a constant vector.

Now we choose a convenient coordinate system: let $\mathbf{d}$ point in the direction of the positive z-axis (so that the orbit of m lies in the xy-plane). We pick the direction of the x-axis to be determined by the (constant) vector $\mathbf{c}$ [from (3.8) we know that $\mathbf{c}$ lies in the xy-plane]. Thus, $\mathbf{c} = \|\mathbf{c}\|\,\mathbf{i}$.

To obtain an expression for $\|\mathbf{r}\|$, we need to eliminate $\mathbf{v}$ from (3.8). Computing the dot product of $\mathbf{r}$ with (3.8), we get

$$\mathbf{r} \cdot (\mathbf{v} \times \mathbf{d}) = GM\frac{\mathbf{r} \cdot \mathbf{r}}{\|\mathbf{r}\|} + \mathbf{r} \cdot (\|\mathbf{c}\|\,\mathbf{i}).$$

Now (see Exercise 10 in Section 1.5) $\mathbf{r} \cdot (\mathbf{v} \times \mathbf{d}) = \mathbf{d} \cdot (\mathbf{r} \times \mathbf{v}) = \mathbf{d} \cdot \mathbf{d} = \|\mathbf{d}\|^2$ and thus,

$$\|\mathbf{d}\|^2 = GM\frac{\|\mathbf{r}\|^2}{\|\mathbf{r}\|} + \|\mathbf{c}\|\,\|\mathbf{r}\|\,\cos\theta = \|\mathbf{r}\|\Big(GM + \|\mathbf{c}\|\,\cos\theta\Big),$$

where θ denotes the angle between $\mathbf{r}$ and $\mathbf{i}$. Finally,

$$\|\mathbf{r}\| = \frac{\|\mathbf{d}\|^2}{GM + \|\mathbf{c}\|\,\cos\theta} = \frac{\|\mathbf{d}\|^2/GM}{1 + (\|\mathbf{c}\|/GM)\cos\theta} \tag{3.9}$$

is the equation of the orbit of m (in polar coordinates; $\mathbf{i}$ is the polar axis). In Exercise 30 we show that (3.10) represents an ellipse with one focus at the origin. Thus, we have proved the Kepler's First Law, which states that the orbit of a planet rotating around the Sun is an ellipse, with the Sun at one focus.

The remaining two Kepler's Laws are discussed in Exercises 13 and 14 in the Chapter Review. ◀

▶ EXAMPLE 3.20 Geometric Interpretation of Derivatives

Let $\mathbf{F}\colon \mathbb{R}^2 \to \mathbb{R}^2$ be a differentiable function and let $\mathbf{c} = \mathbf{c}(t)\colon [a, b] \to \mathbb{R}^2$ be a differentiable curve. The derivative $\mathbf{c}'(t)$ can be interpreted as a vector tangent to $\mathbf{c}$ at the point $\mathbf{c}(t)$. The composition

$$\mathbf{d}(t) = \mathbf{F} \circ \mathbf{c}(t)\colon [a, b] \to \mathbb{R}^2$$

is again a differentiable curve in the plane, and $\mathbf{d}'(t)$ is its tangent vector at $\mathbf{d}(t)$; see Figure 3.22. By the chain rule,

$$\mathbf{d}'(t) = D\mathbf{F}(\mathbf{c}(t)) \cdot \mathbf{c}'(t);$$

that is, the tangent vector $\mathbf{d}'(t)$ is the image of the tangent vector $\mathbf{c}'(t)$ under the function $D\mathbf{F}$, which is the derivative of $\mathbf{F}$. Functions like $D\mathbf{F}$ that assign vectors to vectors have already been discussed in Section 1.4 (see Example 1.30 and the text preceding it).

To illustrate these ideas, consider this example: let $\mathbf{F}(x, y) = (\cos x, y)$ and take the curve $\mathbf{c}(t) = (t, \sin t)$, $t \in [0, 2\pi]$. The composition $\mathbf{d} = \mathbf{F} \circ \mathbf{c}$ is a curve

$$\mathbf{d}(t) = \mathbf{F}(\mathbf{c}(t)) = \mathbf{F}(t, \sin t) = (\cos t, \sin t), \qquad t \in [0, 2\pi].$$

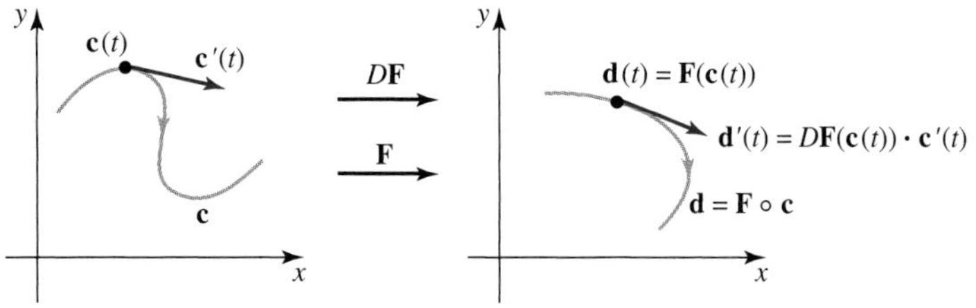

Figure 3.22 Function **F** maps points to points, and its derivative $D\mathbf{F}$ maps tangent vectors to tangent vectors.

Tangents to **c** and **d** are given by

$$\mathbf{c}'(t) = \begin{bmatrix} 1 \\ \cos t \end{bmatrix} \quad \text{and} \quad \mathbf{d}'(t) = \begin{bmatrix} -\sin t \\ \cos t \end{bmatrix}.$$

Our conclusion states that the tangent vector $\mathbf{d}'(t)$ is the image of $\mathbf{c}'(t)$ under $D\mathbf{F}$. To check this, we compute the derivative

$$D\mathbf{F}(x, y) = \begin{bmatrix} -\sin x & 0 \\ 0 & 1 \end{bmatrix}$$

and the image of $\mathbf{c}'(t)$ under $D\mathbf{F}$:

$$D\mathbf{F}(\mathbf{c}(t)) \cdot \mathbf{c}'(t) = D\mathbf{F}(t, \sin t) \cdot \mathbf{c}'(t) = \begin{bmatrix} -\sin t & 0 \\ 0 & 1 \end{bmatrix} \cdot \begin{bmatrix} 1 \\ \cos t \end{bmatrix} = \begin{bmatrix} -\sin t \\ \cos t \end{bmatrix}.$$

As an example, take $t = \pi$. Then the point $\mathbf{c}(\pi) = (\pi, 0)$ on the curve **c** gets mapped to the point $\mathbf{F}(\pi, 0) = (\cos\pi, 0) = (-1, 0)$, which belongs to **d** [since $\mathbf{d}(\pi) = (\cos\pi, \sin\pi) = (-1, 0)$]. Now

$$\mathbf{c}'(\pi) = \begin{bmatrix} 1 \\ \cos\pi \end{bmatrix} = \begin{bmatrix} 1 \\ -1 \end{bmatrix}$$

is the tangent vector to **c** at $\mathbf{c}(\pi) = (\pi, 0)$. The derivative of **F** at $\mathbf{c}(\pi)$ is

$$D\mathbf{F}(\mathbf{c}(\pi)) = D\mathbf{F}(\pi, 0) = \begin{bmatrix} 0 & 0 \\ 0 & 1 \end{bmatrix}.$$

The image of the tangent vector $\mathbf{c}'(\pi)$ under $D\mathbf{F}(\mathbf{c}(\pi))$ is

$$D\mathbf{F}(\mathbf{c}(\pi)) \cdot \mathbf{c}'(\pi) = \begin{bmatrix} 0 & 0 \\ 0 & 1 \end{bmatrix} \cdot \begin{bmatrix} 1 \\ -1 \end{bmatrix} = \begin{bmatrix} 0 \\ -1 \end{bmatrix}.$$

To check, we compute directly that $\mathbf{d}'(\pi) = (-\sin\pi, \cos\pi) = (0, -1)$. Hence, **F** mapped the point $(\pi, 0)$ to the point $(-1, 0)$, and its derivative $D\mathbf{F}(\pi, 0)$ at $(\pi, 0)$ mapped the tangent vector $\mathbf{c}'(\pi)$ to **c** at $(\pi, 0)$ to the tangent vector $\mathbf{d}'(\pi)$ of **d** at $(-1, 0)$.

Let us repeat the conclusion of the previous example: we interpret $\mathbf{F}\colon \mathbb{R}^2 \to \mathbb{R}^2$ (or $\mathbf{F}\colon \mathbb{R}^3 \to \mathbb{R}^3$) as a function that maps points on one curve to points on the image curve. In that case, the derivative $D\mathbf{F}$ maps the tangent vector at a point on the first curve to the tangent vector at the corresponding point on the image curve.

We conclude this section by exploring further applications of vector calculus concepts that we have developed so far.

▶ EXAMPLE 3.21 Coriolis Acceleration in $\mathbb{R}^2$

A plate in the shape of a disk of radius 1 rotates in the xy-plane with constant angular speed ω so that its center (located at the origin) coincides with the center of rotation. A particle starts moving from the center toward the edge, with the trajectory

$$\mathbf{r}(t) = t\,\mathbf{c}(t) = t\,(\cos(\omega t), \sin(\omega t)),$$

where $\mathbf{c}(t) = (\cos(\omega t), \sin(\omega t))$ describes the rotation of the boundary circle of the disk; see Figure 3.23. The function $\mathbf{r}(t)$ describes the motion relative to the x-axis and the y-axis that have their origin at the center of the disk and are fixed; that is, do not participate in the rotation (such a coordinate system is called an *inertial system*). In a coordinate system that rotates with the disk (imagine its axes painted on the disk), the motion of the particle occurs along a radial line ℓ.

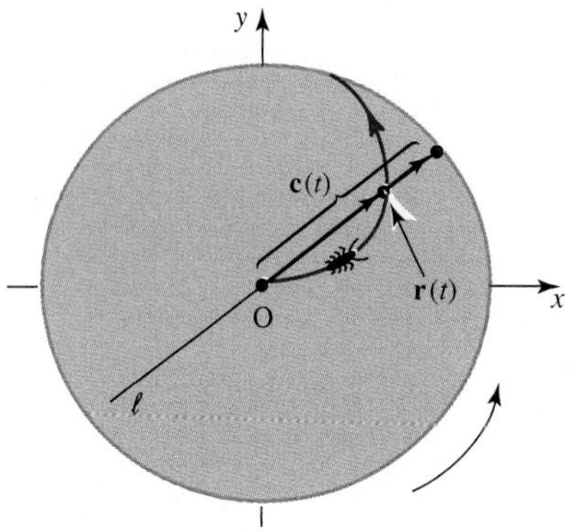

Figure 3.23 Trajectory of a particle in Example 3.21.

Another way to describe the trajectory $\mathbf{r}(t)$ is as follows: suppose that the disk is at rest and imagine a bug crawling on the surface of the disk from its center radially toward the edge. Now start rotating the disk and assume that the bug does not react to a sudden change, but continues crawling along its planned straight line. To the observer standing away from the rotating disk, the bug will appear to be crawling along $\mathbf{r}(t)$.

The velocity is computed to be $\mathbf{v}(t) = \mathbf{r}'(t) = \mathbf{c}(t) + t\,\mathbf{c}'(t)$, and the acceleration is

$$\mathbf{a}(t) = \mathbf{v}'(t) = 2\mathbf{c}'(t) + t\mathbf{c}''(t) = 2\mathbf{c}'(t) - \omega^2 t\mathbf{c}(t),$$

since $\mathbf{c}''(t) = -\omega^2 t\mathbf{c}(t)$. The term $-\omega^2\mathbf{c}(t)$ has appeared in Example 3.17—it is the *centripetal acceleration*. The contribution $2\mathbf{c}'(t)$ to the acceleration results from the interaction of the rotation of the disk and the movement of the particle. It is called the *Coriolis acceleration*, and it points in the direction of the velocity, as shown in Figure 3.24.

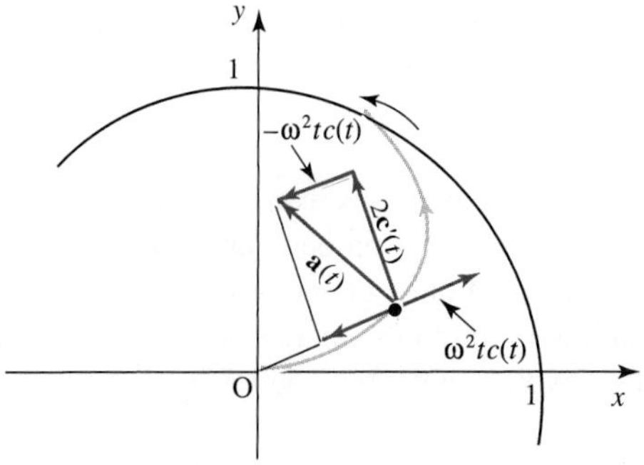

Figure 3.24 Acceleration $\mathbf{a}(t)$ is the sum of the Coriolis acceleration $2\mathbf{c}'(t)$ and the centripetal acceleration $-\omega^2\mathbf{c}(t)$. ◀

▶ EXAMPLE 3.22 Torque Equals the Rate of Change of Angular Momentum

Let $\mathbf{r}(t)$ be the position vector of a particle moving in $\mathbb{R}^3$. The *angular momentum* is defined as the vector

$$\mathbf{L}(t) = \mathbf{r}(t) \times \mathbf{p}(t),$$

where $\mathbf{p}(t) = m\mathbf{v}(t)$ is the momentum vector [m is the mass of the particle and $\mathbf{v}(t)$ is its velocity]. The *torque* of a force $\mathbf{F}(t) = m\mathbf{a}(t)$ exerted at the point $\mathbf{r}(t)$ is

$$\mathbf{T}(t) = \mathbf{r}(t) \times \mathbf{F}(t).$$

Since (with the use of the product rule for the cross product of vector functions)

$$\begin{aligned}\frac{d\mathbf{L}(t)}{dt} &= \frac{d}{dt}(\mathbf{r}(t) \times \mathbf{p}(t)) = \frac{d\mathbf{r}(t)}{dt} \times (m\mathbf{v}(t)) + \mathbf{r}(t) \times \frac{d(m\mathbf{v}(t))}{dt} \\ &= m\mathbf{v}(t) \times \mathbf{v}(t) + \mathbf{r}(t) \times m\frac{d\mathbf{v}(t)}{dt} \\ &= \mathbf{r}(t) \times m\mathbf{a}(t) = \mathbf{r}(t) \times \mathbf{F}(t) = \mathbf{T}(t),\end{aligned}$$

it follows [because $\mathbf{v} \times \mathbf{v} = \mathbf{0}$ for any vector $\mathbf{v}$ and $\mathbf{v}'(t) = \mathbf{a}(t)$] that the torque $\mathbf{T}(t)$ on a particle equals the rate of change of the particle's angular momentum. ◀

▶ EXERCISES 3.2

1. Compute the velocity and speed of the cycloid defined in Example 3.2 in Section 3.1. Identify points where the speed is the largest.

2. Show that the speed of the logarithmic spiral $\mathbf{c}(t) = (e^{at}\cos t, e^{at}\sin t)$, $a \neq 0$, $t \geq 0$, is equal to $e^{at}\sqrt{1+a^2}$. Compute the angle that $\mathbf{c}(t)$ makes with the velocity vector and give an interpretation of your answer.

3. Show that the speed of the helix $\mathbf{c}(t) = (a\cos t, a\sin t, bt)$, $a > 0$, $b > 0$, is constant. Compute the dot product $\mathbf{c}'(t) \cdot \mathbf{c}''(t)$ and give a physical interpretation.

4. Find the maximum speed of the projectile in Example 3.14 and the time when it is reached.

Exercises 5 to 9: Find the velocity $\mathbf{v}(t)$ and the position $\mathbf{c}(t)$ of a particle, given its acceleration $\mathbf{a}(t)$, initial velocity, and initial position.

5. $\mathbf{a}(t) = (-1, 1, 0)$, $\mathbf{v}(0) = (1, 2, 0)$, $\mathbf{c}(0) = (0, 2, 0)$

6. $\mathbf{a}(t) = -9.8\mathbf{k}$, $\mathbf{v}(0) = \mathbf{i} + \mathbf{j}$, $\mathbf{c}(0) = \mathbf{i} + 2\mathbf{j} - \mathbf{k}$

7. $\mathbf{a}(t) = (t, 1, 1)$, $\mathbf{v}(0) = (0, 1, 0)$, $\mathbf{c}(0) = (-2, 0, 3)$

8. $\mathbf{a}(t) = e^t(1, 0, 1)$, $\mathbf{v}(0) = (1, 0, -2)$, $\mathbf{c}(0) = (0, 1, 0)$

9. $\mathbf{a}(t) = t\mathbf{i} + t^2\mathbf{j} + t\mathbf{k}$, $\mathbf{v}(0) = 2\mathbf{j} - 3\mathbf{k}$, $\mathbf{c}(0) = 4\mathbf{i} + 2\mathbf{j} - 6\mathbf{k}$

10. Find a parametrization of the circle $x^2 + y^2 = 1$ of nonconstant speed. Find another parametrization $\mathbf{c}(t)$ such that $\|\mathbf{c}''(t)\|$ is nonconstant and $\|\mathbf{c}''(t)\| \neq 0$ for all t.

11. The position of a particle is given by $\mathbf{c}(t) = (t^{-1}, 1, t^2)$, where $t \in [1, 4]$. When and where does the particle reach its maximum speed?

12. The position of a particle is given by $\mathbf{c}(t) = (3e^{-t}\cos t, 3e^{-t}\sin t)$, where $t \in [1, 3]$. When and where does the particle reach its maximum speed?

13. A particle moves with acceleration $\mathbf{a}(t) = (3, 0, 1)$, where $0 \leq t \leq 12$. Assuming that the particle is initially located at the origin and its initial velocity is $(1, 3, 2)$, find the time needed for the particle to reach its highest position.

14. Prove that if a particle moves with constant speed, then its velocity and acceleration vectors are always perpendicular.

15. A projectile is fired from the origin with an initial speed of 700 m/s at an angle of elevation of 60°. Find the range of the projectile, the maximum height reached, the time needed to reach it, and the speed and the time of impact. Assume that no forces other than gravity act on the projectile.

16. An object is thrown upward from a point 10 m above the ground at an angle of 30° and with an initial speed of 100 m/s. Find a parametric equation of the path of the object. When does it reach its highest point? Where and when does it hit the ground? Assume that no forces other than gravity act on the object.

17. Find a parametrization of the line tangent to the ellipse $x^2 + 4y^2 = 3$ at the point where $x = \sqrt{3}$.

Exercises 18 to 21: Find a parametrization of the line tangent to the given curve $\mathbf{c}(t)$ at the point indicated.

18. $\mathbf{c}(t) = (2t, t^3, 0)$, at the point $(4, 8, 0)$

19. $\mathbf{c}(t) = (3\cos t, 3\sin t, 4t)$, at the point $(0, 3, 2\pi)$

20. $\mathbf{c}(t) = (t, t^2, t^3)$, at the point $(1, 1, 1)$

21. $\mathbf{c}(t) = (-\cosh t, 1 + \sinh t)$ at the point $(-1, 1)$

22. Let $\mathbf{c}(t) = (x(t), y(t))$ be a curve in $\mathbb{R}^2$ and let $\mathbf{c}(t_0) = (x_0, y_0)$. Find an equation and the slope of the tangent line at (x_0, y_0).

23. Let $\mathbf{c}(t) = (r(t)\cos\theta(t), r(t)\sin\theta(t))$ be the trajectory of a particle moving in $\mathbb{R}^2$. Show that its velocity and acceleration are given by the expressions (drop the notation for the dependence on t) $\mathbf{v} = r'(\cos\theta, \sin\theta) + r\theta'(-\sin\theta, \cos\theta)$ and $\mathbf{a} = (r'' - r(\theta')^2)(\cos\theta, \sin\theta) + (2r'\theta' + r\theta'')$ $(-\sin\theta, \cos\theta)$.

24. Show that if a particle moves along the spiral $(2e^t\cos t, 2e^t\sin t)$, then the angle between its position and velocity vectors is constant.

25. The curve $\mathbf{c}(t) = (e^{-t}\cos t, e^{-t}\sin t)$, $0 \leq t \leq 3\pi$, represents the trajectory of a particle moving in $\mathbb{R}^2$. Compute its velocity and find all points at which the velocity is horizontal or vertical.

26. Let $\mathbf{F}(x, y) = (-y, x)$. Compute the curve that is the image under $\mathbf{F}$ of $\mathbf{c}(t) = (\sin t, \cos t)$, $t \in [0, \pi]$. Describe in words the map $D\mathbf{F}$.

27. Define the map $\mathbf{F}\colon \mathbb{R}^2 \to \mathbb{R}^2$ by $\mathbf{F}(x, y) = (x^2y - x^3, ye^x - 2)$. Find the tangent vector to the image of the curve $\mathbf{c}(t) = (\sin t, t^2 - t)$ under $\mathbf{F}$ at $t = 0$.

28. Let $\mathbf{c}(t)$ be a curve such that $\mathbf{c}(0) = (1, 1)$ and $\mathbf{c}'(0) = (2, -1)$. Find the tangent vector to the image of $\mathbf{c}$ under the map $\mathbf{F} = (-y/\sqrt{x^2 + y^2}, x/\sqrt{x^2 + y^2})$ at $t = 0$.

29. Define the map $\mathbf{F}\colon \mathbb{R}^2 \to \mathbb{R}^2$ by $\mathbf{F}(\mathbf{x}) = A \cdot \mathbf{x}$, where A is a nonzero 2×2 matrix and the dot denotes matrix multiplication. Find the tangent vector to the image under $\mathbf{F}$ of the curve $\mathbf{c}(t)$ at $t = 0$ such that $\mathbf{c}(0) = (0, 0)$ and $\mathbf{c}'(0) = (c_1, c_2)$.

30. Assume that the curve given in polar coordinates by $r = A/(1 + B\cos\theta)$, where A and B are positive constants, represents the orbit of a planet revolving around the Sun.

(a) Convert the given equation to Cartesian coordinates and complete the square to obtain

$$\left(x + \frac{AB}{1 - B^2}\right)^2 + \frac{y^2}{1 - B^2} = \frac{A^2}{(1 - B^2)^2}.$$

Using the fact that the planet's orbit is a closed curve, argue that this equation represents an ellipse (i.e., $1 - B^2 > 0$).

(b) Show that (a) represents the ellipse $(x - x_0)^2/a^2 + y^2/b^2 = 1$, where $x_0 = -AB/(1 - B^2)$, $a^2 = A^2/(1 - B^2)^2$, and $b^2 = A^2/(1 - B^2)$.

(c) Recall that an ellipse is the set of points in the plane for which the sum of distances to two fixed points F_1 and F_2 is constant. The points F_1 and F_2 are called foci (plural of "focus") of the ellipse. For an ellipse of the form $x^2/a^2 + y^2/b^2 = 1$, the foci are located at $(\pm c, 0)$, where $c^2 = a^2 - b^2$ (if $a > b$) or $c^2 = b^2 - a^2$ (if $a < b$). Using the values a^2 and b^2 from (b), show that $c = AB/(1 - B^2)$. Conclude that the given equation represents an ellipse whose one focus is at the origin.

▶ 3.3 LENGTH OF A CURVE

Our next goal is to find a way of calculating the length of a curve. We will first define the length of a path (i.e., of a description of the curve using a parametrization). The length of a curve is defined to be equal to the length of a special class of paths, called smooth paths. The definition is independent of the choice of a smooth path (parametrization). This sounds reasonable: no matter how fast or how slow (or in what direction) we walk along the same curve, we always cover the same distance.

Let $\mathbf{c}(t) = (x(t), y(t))\colon [a, b] \to \mathbb{R}^2$ be a differentiable path in the xy-plane. Divide the interval $[a, b]$ into n subintervals $[a = t_1, t_2]$, $[t_2, t_3]$, ..., $[t_n, t_{n+1} = b]$ and join the corresponding points on the curve $\mathbf{c}(t_1)$, $\mathbf{c}(t_2)$, ..., $\mathbf{c}(t_{n+1})$ on $\mathbf{c}$ with straight-line segments $c_1, c_2,, \ldots, c_n$ (see Figure 3.25), thus forming a polygonal path p_n.

The lengths of $c_1, \ldots, c_n$ are $\ell(c_1) = \|\mathbf{c}(t_2) - \mathbf{c}(t_1)\|, \ldots, \ell(c_n) = \|\mathbf{c}(t_{n+1}) - \mathbf{c}(t_n)\|$, and the total length of p_n is $\ell(p_n) = \ell(c_1) + \cdots + \ell(c_n)$. As we keep increasing n, we obtain polygonal paths p_n that approximate the path $\mathbf{c}$ better and better. The length $\ell(\mathbf{c})$ of $\mathbf{c}$ is defined as the limit of lengths of those polygonal paths as $n \to \infty$. Now, for every $i = 1, \ldots, n$ (the symbol $\approx$ means "approximately equal"),

$$\begin{aligned}
\mathbf{c}'(t_i) &= x'(t_i)\mathbf{i} + y'(t_i)\mathbf{j} \\
&\approx \frac{x(t_{i+1}) - x(t_i)}{t_{i+1} - t_i}\mathbf{i} + \frac{y(t_{i+1}) - y(t_i)}{t_{i+1} - t_i}\mathbf{j} \\
&= \frac{(x(t_{i+1}) + y(t_{i+1}))\mathbf{i} - (x(t_i) + y(t_i))\mathbf{j}}{t_{i+1} - t_i} = \frac{\mathbf{c}(t_{i+1}) - \mathbf{c}(t_i)}{t_{i+1} - t_i}.
\end{aligned}$$

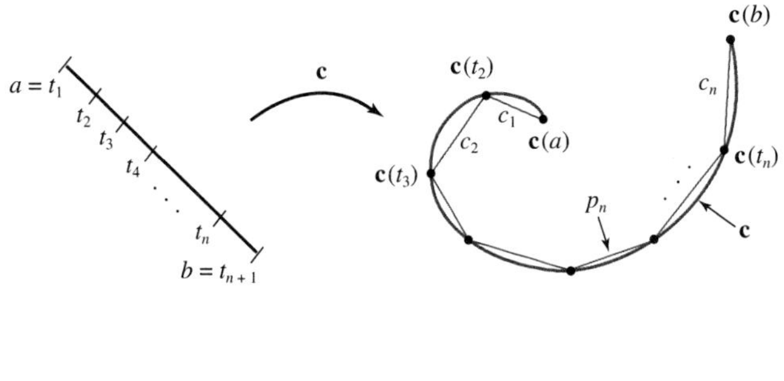

Figure 3.25 Polygonal path p_n approximates the path $\mathbf{c}$.

We viewed $x(t)$ and $y(t)$ as real-valued functions of one variable and used the fact that $x'(t_i)$ and $y'(t_i)$, interpreted as the slope of a tangent, can be approximated by the slopes of secant lines (our reasoning is a bit intuitive; see Exercise 3 for a fully rigorous argument). Thus,

$$\mathbf{c}'(t_i) \approx \frac{\mathbf{c}(t_{i+1}) - \mathbf{c}(t_i)}{t_{i+1} - t_i} \tag{3.11}$$

and hence, the length of the ith segment, c_i, is approximately equal to

$$\ell(c_i) = \|\mathbf{c}(t_{i+1}) - \mathbf{c}(t_i)\| \approx \|\mathbf{c}'(t_i)\|(t_{i+1} - t_i) = \|\mathbf{c}'(t_i)\|\Delta t_i, \tag{3.12}$$

where $\Delta t_i = t_{i+1} - t_i$. Consequently, the length of the polygonal path p_n

$$\ell(p_n) = \sum_{i=1}^{n} \ell(c_i) \approx \sum_{i=1}^{n} \|\mathbf{c}'(t_i)\|\Delta t_i \tag{3.13}$$

approximates the length of the curve. Now compute the limit as $n \to \infty$. By definition, the limit $\lim_{n\to\infty}\ell(p_n)$ on the left side is equal to the length $\ell(\mathbf{c})$ of the path $\mathbf{c}$. Since $n \to \infty$, $\Delta t_i \to 0$ and we recognize the limit on the right side as the definition of the definite integral of the function $\|\mathbf{c}'(t)\|$ over the interval $[a, b]$. This argument justifies the following definition. Recall that a function is called C^1 if its derivative is continuous.

DEFINITION 3.3 Length of a Path

Let $\mathbf{c}\colon [a, b] \to \mathbb{R}^2$ (or $\mathbb{R}^3$) be a C^1 path. The *length* $\ell(\mathbf{c})$ of $\mathbf{c}$ is given by

$$\ell(\mathbf{c}) = \int_a^b \|\mathbf{c}'(t)\|\, dt.$$

Before proceeding, let us consider an example.

▶ EXAMPLE 3.23

Compute the length of the paths $\mathbf{c}_1(t) = (\cos t, \sin t)$, $t \in [0, 2\pi]$ and $\mathbf{c}_2(t) = (\cos 3t, \sin 3t)$, $t \in [0, 2\pi]$ in $\mathbb{R}^2$.

SOLUTION By definition,

$$\ell(\mathbf{c}_1) = \int_0^{2\pi} \sqrt{(-\sin t)^2 + (\cos t)^2}\, dt = \int_0^{2\pi} dt = 2\pi$$

and

$$\ell(\mathbf{c}_2) = \int_0^{2\pi} \sqrt{(-3\sin 3t)^2 + (3\cos 3t)^2}\, dt = \int_0^{2\pi} 3\, dt = 6\pi.$$

Both $\mathbf{c}_1$ and $\mathbf{c}_2$ represent the same curve, the circle of radius 1. However, only the first parametrization gives its length. The path $\mathbf{c}_2$ traverses the circle three times, and hence its length is three times the length of the circle (in Exercise 4 we suggest a good way of thinking about this). Therefore, if we want to measure the length of a curve, we have to exclude some parametrizations, such as $\mathbf{c}_2$ in the previous example. This is one of the reasons why we made a fuss about the distinction between a path and a curve.

DEFINITION 3.4 Smooth Path (Smooth Parametrization)

A C^1 path (or a C^1 parametrization) $\mathbf{c}: [a, b] \to \mathbb{R}^2$ (or $\mathbb{R}^3$) is called *smooth* if $\mathbf{c}'(t) \neq \mathbf{0}$ for all $t \in (a, b)$, and if distinct points in (a, b) map to distinct points on the curve.

According to the definition, a smooth path can be closed [i.e., $\mathbf{c}(a) = \mathbf{c}(b)$ is allowed] but cannot intersect itself, be tangent to itself, or (partly or completely) retrace itself (like the path $\mathbf{c}_2$ of Example 3.23).

Notice that a smooth path has a nonzero velocity at every point.

EXAMPLE 3.24

Figure 3.26 shows the path $\mathbf{c}(t) = (\cos^3 t, \sin^3 t)$, $t \in [-\pi/2, \pi/2]$. It is C^1 and has no self-intersections. However, it is not smooth [and, looking at the point $(1, 0)$, we expect it not to be smooth] since $\mathbf{c}'(t) = (-3\cos^2 t \sin t, 3\sin^2 t \cos t)$ is zero when $t = 0$.

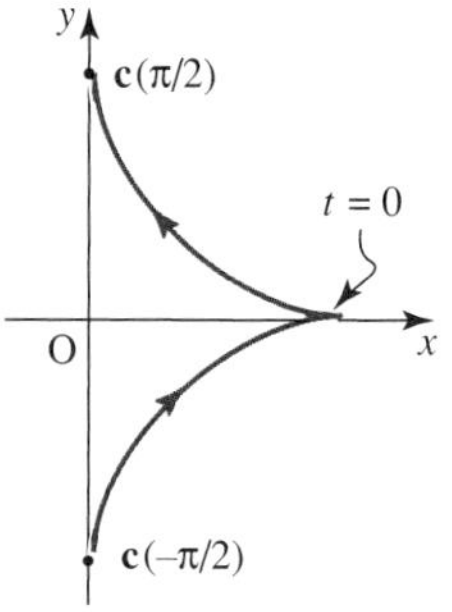

Figure 3.26 The path $\mathbf{c}(t) = (\cos^3 t, \sin^3 t)$, $t \in [-\pi/2, \pi/2]$ is not smooth.

EXAMPLE 3.25

The path $\mathbf{c}(\theta) = (\theta - \sin\theta, 1 - \cos\theta)$ of Example 3.2 in Section 3.1 is clearly continuously diferentiable (i.e., C^1). However, it is not smooth: since $\mathbf{c}'(\theta) = (1 - \cos\theta, \sin\theta)$, it follows that $\mathbf{c}'(2\pi k) = 0$ (k is an integer).

The path $\mathbf{c}(t) = (e^{at}\cos t, e^{at}\sin t)$, $a \neq 0$, $t \geq 0$, whose image is the logarithmic spiral [shown in Figure 3.4(a) in Section 3.1] is C^1. Since $\|\mathbf{c}'(t)\| = e^{at}\sqrt{1 + a^2} \neq 0$ for all t (see Exercise 2 in Section 3.2), and because $t_1 \neq t_2$ implies that $\mathbf{c}(t_1) \neq \mathbf{c}(t_2)$, it follows that $\mathbf{c}(t)$ is a smooth path.

The parametrizations of Lissajous curves shown in Figure 3.7 in Section 3.1 have self-intersections, and hence cannot be smooth.

Thinking of $\mathbf{c}$ as a trajectory of a motion, we translate the smoothness condition as the requirement that the *speed* be nonzero. At points where the speed is zero, the particle can make sharp turns (thus describing a nonsmooth trajectory). See Example 3.24 and the cycloid in Example 3.25 (and yet, the parametrizations are differentiable; see Exercise 5).

DEFINITION 3.5 Smooth Curve and Its Length

A curve $\mathbf{c}$ is called *smooth* if it has a C^1 parametrization that is smooth. The length $\ell(\mathbf{c})$ of $\mathbf{c}$ (also called the *arc-length*) is defined as the length of that smooth parametrization.

In order to show that a curve $\mathbf{c}$ is smooth, all we have to do is to find *one* smooth parametrization of $\mathbf{c}$. Its length is then equal to the length of the curve. As a consequence of the Change of Variables Theorem (in an integral), we show in Section 5.2 that the length of a curve is independent of the smooth parametrization that is used in the computation.

▶ EXAMPLE 3.26

Compute the length of the graph of the function $y = x^2/2$ for $0 \le x \le 1$.

SOLUTION The parametrization $\mathbf{c}(t) = (t, t^2/2)$, $t \in [0, 1]$, is C^1 (since its component functions are C^1). From $\mathbf{c}'(t) = (1, t)$ we see that $\mathbf{c}'(t) \neq \mathbf{0}$; since $\mathbf{c}(t_1) \neq \mathbf{c}(t_2)$ for $t_1 \neq t_2$, we conclude that $\mathbf{c}(t)$ is a smooth parametrization of the given curve. The desired length is (to integrate, we use a formula from a table of integrals)

$$\ell(\mathbf{c}) = \int_0^1 \sqrt{1+t^2}\, dt$$

$$= \frac{1}{2} t\sqrt{1+t^2} + \frac{1}{2} \ln(t + \sqrt{1+t^2})\Big|_0^1 = \frac{\sqrt{2}}{2} + \frac{1}{2}\ln(1+\sqrt{2}) \approx 1.148.$$ ◀

▶ EXAMPLE 3.27

Compute the length of the helix $\mathbf{c}(t) = (\cos t, \sin t, t)$ between $\mathbf{c}(0) = (1, 0, 0)$ and $\mathbf{c}(2\pi) = (1, 0, 2\pi)$ (that corresponds to one full revolution of its projection onto the xy-plane).

SOLUTION The given parametrization is smooth since it is C^1 (components are C^1 by inspection), $\|\mathbf{c}'(t)\| = \sqrt{(-\sin t)^2 + (\cos t)^2 + 1} = \sqrt{2} \neq 0$, and if $t_1 \neq t_2$, then $\mathbf{c}(t_1) \neq \mathbf{c}(t_2)$ (since, e.g., the z-coordinates are different). Consequently, the length (or the arc-length) of the helix in question is

$$\ell(\mathbf{c}) = \int_0^{2\pi} \sqrt{2}\, dt = 2\sqrt{2}\pi.$$

In this case, there is an alternative, more visual way to compute this. Cut the cylinder along the vertical line joining $(1, 0, 0)$ and $(1, 0, 2\pi)$ and unfold it, as shown in Figure 3.27. Then $\ell(\mathbf{c})$ is the length of the hypotenuse of the triangle with sides 2π (one side is the circumference of the circle that is the projection of the helix onto the xy-plane and the other side is the "pitch"; i.e., total increase in the z-coordinate from $t = 0$ to $t = 2\pi$).

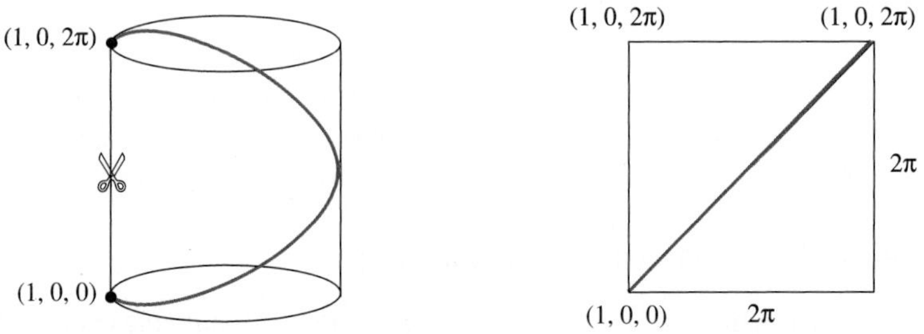

Figure 3.27 "Unfolding" the helix of Example 3.27.

◀

▶ EXAMPLE 3.28 Length of the Graph of $y = f(x)$

The graph of a C^1 function $f(x)\colon [a, b] \to \mathbb{R}$ can be parametrized as $\mathbf{c}(t) = (t, f(t))$, $t \in [a, b]$. The path $\mathbf{c}$ is C^1 (since f is C^1 by assumption), $\|\mathbf{c}'(t)\| = \sqrt{1 + (f'(t))^2} \neq 0$ for all t, and if $t_1 \neq t_2$, then $\mathbf{c}(t_1) = (t_1, f(t_1)) \neq (t_2, f(t_2)) = \mathbf{c}(t_2)$. Consequently, the path $\mathbf{c}(t)$ is smooth and its length (and hence the length of the graph of $y = f(x)$ on $[a, b]$) is

$$\ell(\mathbf{c}) = \int_a^b \sqrt{1 + (f'(t))^2}\, dt = \int_a^b \sqrt{1 + (f'(x))^2}\, dx.$$

In other words, we have shown that our definition agrees with the formula from the calculus of functions of one variable. ◀

THEOREM 3.2 The Shortest Distance Between Two Points

The shortest smooth curve connecting two given points in a plane is a straight line. ◀

PROOF: Take two points, A and B, and choose the coordinate system so that they both lie on the x-axis, having the coordinates $A = (a, 0)$ and $B = (b, 0)$. Choose any smooth curve $\mathbf{c}$ [think of it as of a graph of some continuously differentiable function $y = f(x)$] other than the straight-line segment $\mathbf{s}$ joining A and B (see Figure 3.28). Parametrizing $\mathbf{c}$ by $\mathbf{c}(t) = (t, f(t))$, we get

$$\ell(\mathbf{c}) = \int_a^b \sqrt{1 + (f'(t))^2}\, dt.$$

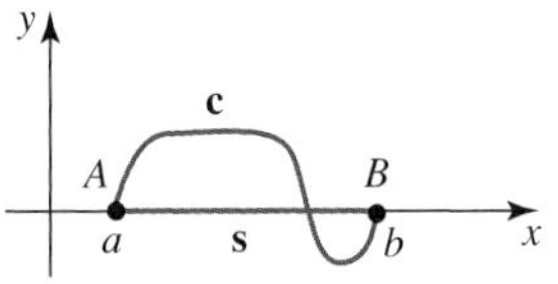

Figure 3.28 The length of $\mathbf{c}$ is larger than the length $b - a$ of $\mathbf{s}$.

We have to show that $\ell(\mathbf{c}) > \ell(\mathbf{s}) = b - a$. First of all, there exists t such that $f'(t) \neq 0$ [otherwise, if $f'(t) = 0$ for all t, then $f(t) =$ constant; i.e., the graph of f is the straight line]. But f' is continuous, so there exists a whole interval (a', b') containing t, where $f' \neq 0$; see Figure 3.29 (in other words, a continuous function f' cannot have a nonzero value at a single point in its domain; this is a consequence of the definition of continuity). It follows that the strict inequality $\sqrt{1 + (f'(t))^2} > 1$ holds on (a', b') [and clearly, $\sqrt{1 + (f'(t))^2} \geq 1$ everywhere else]. Therefore,

$$\begin{aligned}
\ell(\mathbf{c}) &= \int_a^b \sqrt{1 + (f'(t))^2}\, dt \\
&= \int_a^{a'} \sqrt{1 + (f'(t))^2}\, dt + \int_{a'}^{b'} \sqrt{1 + (f'(t))^2}\, dt + \int_{b'}^{b} \sqrt{1 + (f'(t))^2}\, dt \\
&> \int_a^{a'} 1\, dt + \int_{a'}^{b'} 1\, dt + \int_{b'}^{b} 1\, dt = \int_a^b 1\, dt = b - a = \ell(\mathbf{s});
\end{aligned}$$

that is, $\ell(\mathbf{c}) > \ell(\mathbf{s})$.

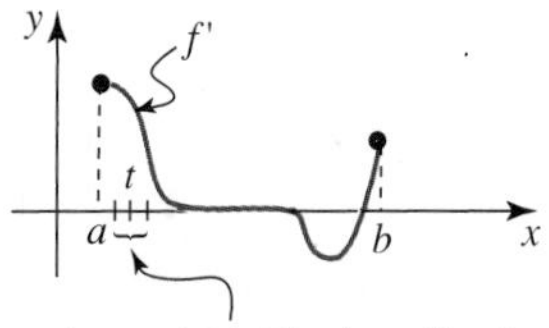

Figure 3.29 Possible graph of the derivative $f'(t)$.

DEFINITION 3.6 Arc-Length Function

Let $\mathbf{c}\colon [a, b] \to \mathbb{R}^2$ (or $\mathbb{R}^3$) be a C^1 path. The *arc-length function* $s(t)$ of $\mathbf{c}(t)$ is given by

$$s(t) = \int_a^t \|\mathbf{c}'(\tau)\| d\tau.$$

Geometrically, the arc-length function $s(t)$ measures the length of $\mathbf{c}$ from $\mathbf{c}(a)$ to the point $\mathbf{c}(t)$; that is, the distance traversed in time t. Clearly, $s(a) = \int_a^a \|\mathbf{c}'(\tau)\| d\tau = 0$ and $s(b) = \int_a^b \|\mathbf{c}'(\tau)\| d\tau = \ell(\mathbf{c})$. By the Fundamental Theorem of Calculus,

$$\frac{d}{dt}s(t) = \frac{d}{dt}\left(\int_a^t \|\mathbf{c}'(\tau)\| d\tau\right) = \|\mathbf{c}'(t)\|. \tag{3.14}$$

This makes sense: $ds(t)/dt$ is the rate of change of the arc-length (i.e., the distance) with respect to time, which is the speed $\|\mathbf{c}'(t)\|$.

The arc-length function can be used as a new parameter for $\mathbf{c}$; that is, a parametrization $\mathbf{c}(t)$ can be replaced by the parametrization $\mathbf{c}(s)$ (read the next two examples to see how this is done). Strictly speaking, we should use different symbols for the two parametrizations (although they represent the same curve, they are not equal as functions). However, this kind of sloppiness is common practice.

We say that the parametrization $\mathbf{c}(s)$ is the *parametrization by arc-length.*

▶ EXAMPLE 3.29 Parametrization by Arc-Length

Compute the arc-length function of the circle $\mathbf{c}(t) = (3\cos t, 3\sin t)$, where $t \in [0, 2\pi]$, and parametrize the circle by its arc-length.

SOLUTION

Since $\|\mathbf{c}'(t)\| = 3$, it follows that $s(t) = \int_0^t 3dt = 3t$ is the arc-length function. To express $\mathbf{c}$ in terms of s, solve $s = 3t$ for t and substitute the result into the given parametrization, thus getting the parametrization by arc-length

$$\mathbf{c}(s) = (3\cos(s/3), 3\sin(s/3)), \qquad s \in [0, 6\pi].$$

Since $0 \le t \le 2\pi$, it follows that $0 \le s = 3t \le 6\pi$.

The curve that corresponds to values of t from $t = 0$ to $t = \pi/2$ in the original parametrization $\mathbf{c}(t)$ is the part of the circle of radius 3 in the first quadrant; its length is $3\pi/2$. On the other hand, it takes $s = 3\pi/2$ units of "s-time" for the point on the path $\mathbf{c}(s)$ to reach (0, 3) from (3, 0). So in this case, the "s-time" needed is (numerically) the same as the distance traveled.

▶ EXAMPLE 3.30 Parametrization by Arc-Length

Consider the circular helix in $\mathbb{R}^3$ given by $\mathbf{c}(t) = (a\cos t, a\sin t, bt)$, $t \in [0, 2\pi]$, where $a, b > 0$. Compute its arc-length function and parametrize it by arc-length.

SOLUTION Since $\mathbf{c}'(t) = (-a\sin t, a\cos t, b)$, it follows that

$$s(t) = \int_0^t \sqrt{a^2+b^2}\,dt = \sqrt{a^2+b^2}\,t.$$

Consequently, $t = s/\sqrt{a^2+b^2}$ and the required parametrization is given by

$$\mathbf{c}(s) = \left(a\cos\frac{s}{\sqrt{a^2+b^2}}, a\sin\frac{s}{\sqrt{a^2+b^2}}, \frac{bs}{\sqrt{a^2+b^2}}\right), \qquad s \in [0, 2\pi\sqrt{a^2+b^2}].$$

Notice that we adjusted the interval (domain) of $\mathbf{c}(s)$.

The length of the helix is

$$\ell(c) = \int_0^{2\pi} \|\mathbf{c}'(t)\|\,dt = \int_0^{2\pi} \sqrt{a^2+b^2}\,dt = 2\pi\sqrt{a^2+b^2},$$

which is exactly how much "s-time" is needed to trace the reparametrized curve $\mathbf{c}(s)$.

The previous two examples have something in common: the time spent walking along a curve parametrized by arc-length is numerically equal to the distance covered. We will now demonstrate that this property is always true.

Take a C^1 path $\mathbf{c}(t)$, $t \in [a, b]$, and parametrize it by the arc-length function; that is, consider the parametrization $\mathbf{c}(s)$, $s \in [a', b']$. By definition,

$$\ell(\mathbf{c}(s)) = \int_{a'}^{b'} \|\mathbf{c}'(s)\|ds.$$

From the chain rule and (3.14), it follows that

$$\mathbf{c}'(t) = \frac{d\mathbf{c}(t)}{dt} = \frac{d\mathbf{c}(s)}{ds}\frac{d(s)}{dt} = \mathbf{c}'(s)\|\mathbf{c}'(t)\|.$$

Therefore, $\mathbf{c}'(s) = \mathbf{c}'(t)/\|\mathbf{c}'(t)\|$ [consequently, $\|\mathbf{c}'(s)\| = 1$] and

$$\ell(\mathbf{c}(s)) = \int_{a'}^{b'} 1ds = b' - a'.$$

In words, the length $b' - a'$ of a curve parametrized by its arc-length function is numerically equal to the time needed to trace it (which is equal to the length of the interval $[a', b']$). This should come as no surprise since $\|\mathbf{c}'(s)\| = 1$ (i.e., a curve parametrized by arc-length is traversed with a constant speed of 1).

The vector [defined whenever $\mathbf{c}'(t) \neq \mathbf{0}$]

$$\mathbf{T}(t) = \frac{\mathbf{c}'(t)}{\|\mathbf{c}'(t)\|} \tag{3.15}$$

is called the *unit tangent vector* to $\mathbf{c}$ at $\mathbf{c}(t)$. It is worth repeating that $\mathbf{T}(t) = \mathbf{c}'(s)$.

No matter what smooth parametrization of a curve we use, we always arrive at the same parametrization by arc-length. The fact that the unit tangent vector for any parametrization is equal to $\mathbf{c}'(s)$ means that although tangent vectors differ from parametrization to parametrization, the tangent *direction* is the same for all of them.

▶ EXAMPLE 3.31

Let $\mathbf{c}(t) = (t^2/2, t^3/3)$, $t \in [0, 2]$.

(a) Find the arc-length function s and parametrize $\mathbf{c}$ by the new parameter s.

(b) Check that $ds/dt = \|\mathbf{c}'(t)\|$.
(c) Verify that $\mathbf{T}(t) = \mathbf{c}'(s)$.

SOLUTION

(a) From $\mathbf{c}'(t) = (t, t^2)$, it follows that $\|\mathbf{c}'(t)\| = \sqrt{t^2 + t^4} = |t|\sqrt{1+t^2} = t\sqrt{1+t^2}$ (since $t \geq 0$) and

$$s = \int_0^t \|\mathbf{c}'(\tau)\|\, d\tau = \int_0^t \tau\sqrt{1+\tau^2}\, d\tau = \frac{1}{3}\,(1+\tau^2)^{3/2}\Big|_0^t = \frac{1}{3}(1+t^2)^{3/2} - \frac{1}{3}.$$

This computation gives us the arc-length parameter $s = (1+t^2)^{3/2}/3 - 1/3$. Solving for t, we get $3s + 1 = (1+t^2)^{3/2}$ and $t = \left((3s+1)^{2/3} - 1\right)^{1/2}$ [we take the positive value of the square root since t is positive by the definition of the domain of $\mathbf{c}(t)$]. Finally, the parametrization of $\mathbf{c}$ by the arc-length function s is

$$\mathbf{c}(s) = \left(\frac{1}{2}\left((3s+1)^{2/3} - 1\right), \frac{1}{3}\left((3s+1)^{2/3} - 1\right)^{3/2}\right).$$

(b) From $s = (1+t^2)^{3/2}/3 - 1/3$, it follows that

$$\frac{ds}{dt} = \frac{1}{3}\cdot\frac{3}{2}(1+t^2)^{1/2}\cdot 2t = t(1+t^2)^{1/2} = \|\mathbf{c}'(t)\|.$$

(c) The unit tangent vector is computed to be

$$\mathbf{T}(t) = \frac{\mathbf{c}'(t)}{\|\mathbf{c}'(t)\|} = \frac{(t, t^2)}{t\sqrt{1+t^2}} = \left(\frac{1}{\sqrt{1+t^2}}, \frac{t}{\sqrt{1+t^2}}\right).$$

The velocity of $\mathbf{c}(s)$ is

$$\begin{aligned}\mathbf{c}'(s) &= \left(\frac{1}{2}\cdot\frac{2}{3}(3s+1)^{-1/3}\cdot 3, \frac{1}{3}\cdot\frac{3}{2}\left((3s+1)^{2/3} - 1\right)^{1/2}\cdot\frac{2}{3}(3s+1)^{-1/3}\cdot 3\right)\\ &= (3s+1)^{-1/3}\left(1, \left((3s+1)^{2/3} - 1\right)^{1/2}\right)\\ &= (1+t^2)^{-1/2}(1, t) = \mathbf{T}(t),\end{aligned}$$

since $3s + 1 = (1+t^2)^{3/2}$ [that was computed in (a)]. ◀

▶ EXAMPLE 3.32

Let $\mathbf{c}(t) = (t, \cosh t)$, $t \in [0, 1]$, be a catenary curve (see Example 3.5 in Section 3.1). The arc-length function is

$$s(t) = \int_0^t \sqrt{1 + (\sinh\tau)^2}\, d\tau = \int_0^t \cosh\tau\, d\tau = \sinh t.$$

Hence, $t = \sinh^{-1} s$, so that the parametrization by arc-length is given by

$$\mathbf{c}(s) = (\sinh^{-1} s, \cosh(\sinh^{-1} s)), \qquad s \in [0, \sinh 1].$$

The tangent is computed to be [the derivative of $\sinh^{-1} x$ is $(1+x^2)^{-1/2}$]

$$\mathbf{c}'(s) = \left(\frac{1}{\sqrt{1+s^2}}, s\frac{1}{\sqrt{1+s^2}}\right),$$

which is a unit vector:

$$\|\mathbf{c}'(s)\| = \sqrt{\frac{1}{1+s^2} + \frac{s^2}{1+s^2}} = 1.$$

As shown earlier, the length of the curve from $\mathbf{c}(0)$ to $\mathbf{c}(s)$ equals s. For example, the length of the part of the curve between $\mathbf{c}(s = 0) = (0, 1)$ and $\mathbf{c}(s = 1) = (\sinh^{-1} 1, \cosh(\sinh^{-1} 1))$ is 1. ◀

▶ EXERCISES 3.3

1. Consider the curve parametrized by $\mathbf{c}(t) = (t, 1/t)$, $t \in [1, 2]$. Divide $[1, 2]$ into 5 subintervals of equal length, and sketch the curve $\mathbf{c}$ and polygonal path p_5 that approximates it. Approximate the length of p_5 using formula (3.13).

2. Consider the curve parametrized by $\mathbf{c}(t) = (t, e^{2t})$, $t \in [0, 1]$. Divide $[0, 1]$ into 4 subintervals of equal length, and sketch the curve $\mathbf{c}$ and polygonal path p_4 that approximates it. Approximate the length of p_4 using formula (3.13). Compare your approximation with the length of p_4 computed using the formula for the distance between two points.

3. Assume that $\mathbf{c}(t)\colon [a, b] \to \mathbb{R}^2$ is a differentiable path in $\mathbb{R}^2$, and consider the partition $[a = t_1, t_2], [t_2, t_3], \ldots, [t_n, t_{n+1} = b]$ of $[a, b]$, as in the beginning of the section. Let $\ell(c_i) = \|\mathbf{c}(t_{i+1}) - \mathbf{c}(t_i)\|$ be the length of the line segment from $\mathbf{c}(t_i)$ to $\mathbf{c}(t_{i+1})$.

(a) Show that $\ell(c_i) = \sqrt{(x(t_{i+1}) - x(t_i))^2 + (y(t_{i+1}) - y(t_i))^2}$.

(b) Apply the Mean Value Theorem from one-variable calculus to show that there exist t_i^*, t_i^{**} in $[t_i, t_{i+1}]$ such that $x(t_{i+1}) - x(t_i) = x'(t_i^*)\Delta t$ and $y(t_{i+1}) - y(t_i) = y'(t_i^{**})\Delta t$, where $\Delta t = t_{i+1} - t_i$. Conclude that $\ell(c_i) = \sqrt{(x'(t_i^*))^2 + (y'(t_i^{**}))^2}\,\Delta t$.

(c) Using (b), find a formula for the length of the polygonal path p_n [see (3.13)]. Compute the limit as $n \to \infty$ to obtain the formula from Definition 3.3.

4. Two students run around a circular track, given by $\mathbf{c}(t) = (50 \sin t, 50 \cos t)$, $t \in [0, 2\pi]$. Student A runs according to $\mathbf{c}_A(t) = (50 \sin(t/5), 50 \cos(t/5))$, $t \in [0, 30\pi]$, and student B according to $\mathbf{c}_B(t) = (50 \sin(t/4), 50 \cos(t/4))$, $t \in [0, 32\pi]$. Note that both A and B start and end at the point $(0, 50)$.

(a) What is the length of the track?

(b) Which student is running faster? Compute the distance covered by each student.

5. Consider the curve that is the graph of the function $y = x^{2/3}$ on $[-1, 1]$.

(a) Show that y is not differentiable at 0. Conclude that the parametrization $\mathbf{c}(t) = (t, t^{2/3})$, $t \in [-1, 1]$, is not differentiable.

(b) Show that $\mathbf{c}(t) = (\cos^3 t, \cos^2 t)$, $t \in [-\pi, \pi]$, is a differentiable parametrization of the given curve.

(c) Prove that the parametrization in (b) is not smooth.

6. Prove that the statement we made in Example 3.27 is true; that is, by "unfolding" the helix, we obtain a straight-line segment (see Figure 3.27).

Exercises 7 to 14: Find the length of the path $\mathbf{c}(t)$.

7. $\mathbf{c}(t) = (\sin 2t, \cos 2t)$, $t \in [0, \frac{\pi}{2}]$

8. $\mathbf{c}(t) = (2t^{3/2}, 2t)$, from $(0, 0)$ to $(2, 2)$

9. $\mathbf{c}(t) = e^t \cos t\mathbf{i} + e^t \sin t\mathbf{j}$, $0 \le t \le \pi$

10. $\mathbf{c}(t) = t^3\mathbf{i} + t^2\mathbf{j}$, $-2 \le t \le 1$

11. $\mathbf{c}(t) = ((1 + t), (1 + t)^{3/2})$, $t \in [0, 1]$

12. $\mathbf{c}(t) = (e^{2t}, e^{-2t}, \sqrt{8}t)$, $t \in [0, 1]$

13. $\mathbf{c}(t) = (2t - t^2)\mathbf{i} + \frac{8}{3}t^{3/2}\mathbf{j} + \mathbf{k}$, from $t = 1$ to $t = 3$

14. $\mathbf{c}(t) = \cos^2 t\mathbf{i} + \sin^2 t\mathbf{j}$, from $t = 0$ to $t = 2\pi$

15. Show that the length of a logarithmic spiral $\mathbf{c}(t) = (e^{at} \cos t, e^{at} \sin t)$, where $a < 0$ and $t \ge 0$, is finite.

16. Find the length of the catenary curve given by $\mathbf{c}(t) = (t, a \cosh(t/a))$, $a > 0$, $t \in [-a, a]$.

17. Is it true that the curve $y = 2 \sin x$, $x \in [0, 2\pi]$ is twice as long as $y = \sin x$, $x \in [0, 2\pi]$?

18. Let $r = f(\theta)$, $\theta \in [\alpha, \beta]$ be a representation of a path in polar coordinates. Show that its length is $\ell = \int_\alpha^\beta \sqrt{r^2 + (dr/d\theta)^2}\, d\theta$.

Exercises 19 to 24: Using Exercise 18, compute the length of the given curve.

19. $r = a\cos\theta$, $a > 0$, $0 \le \theta \le \pi/4$

20. $r = a\cos 2\theta$, $a > 0$, $0 \le \theta \le \pi/4$

21. $r = 3\theta^2$, $1 \le \theta \le 2$

22. $r = e^{2\theta}$, $0 \le \theta \le \ln 6$

23. $r = 1 + \sin\theta$, $-\pi/2 \le \theta \le \pi/2$

24. (cardioid) $r = 1 - \cos\theta$, $0 \le \theta \le \pi$

25. Find the arc-length function of $\mathbf{c}(t) = (t\sin 2t, t\cos 2t, 4t^{3/2}/3)$, $0 \le t \le 2\pi$.

Exercises 26 to 30: Consider the path $\mathbf{c}(t)$. Find its arc-length function $s(t)$ and reparametrize it by its arc-length.

26. $\mathbf{c}(t) = \sin 2t\mathbf{i} + \cos 2t\mathbf{j} + \frac{2}{3}t^{3/2}\mathbf{k}$, $t \in [-4, 4]$

27. $\mathbf{c}(t) = 5\cos t\mathbf{i} + 5\sin t\mathbf{j} + 12t\mathbf{k}$, $t \in [0, \pi/4]$

28. $\mathbf{c}(t) = (3 + 4t)\mathbf{i} + (5t - 1)\mathbf{j} + 11\mathbf{k}$, $t \in [-2, 1]$

29. $\mathbf{c}(t) = e^t\cos t\mathbf{i} + e^t\sin t\mathbf{j}$, $t \in [0, 1]$

30. $\mathbf{c}(t) = 2\cos^3 t\mathbf{i} + \cos 2t\mathbf{j} + 2\sin^3 t\mathbf{k}$, $t \in [\pi/4, \pi/2]$

31. Check that the parametrizations (a)–(c) represent the same curve. Compute the length in each case and compare your results.

(a) $\mathbf{c}_1(t) = (\cos 2t, \sin 2t, t)$, $t \in [0, \pi]$

(b) $\mathbf{c}_2(t) = (\cos t, \sin t, t/2)$, $t \in [0, 2\pi]$

(c) $\mathbf{c}_3(t) = (\cos 2t, -\sin 2t, -t)$, $t \in [-\pi, 0]$

32. Find the unit tangent vector to the curve $\mathbf{c}(t) = (e^{-t}\sin t, e^{-t}\cos t)$.

33. Find the unit tangent vector to the curve $\mathbf{c}(t) = (t^{-1} - t, t^{-1} + t, -1)$ at $t = 1$.

34. Consider the curve $\mathbf{c}$ parametrized by $\mathbf{c}(t) = (t^3, 3t^6 + 1)$, $t \in [-1, 1]$. This parametrization is not smooth (where?), and hence cannot be used to compute the length of $\mathbf{c}$. Find a smooth parametrization of $\mathbf{c}$ and compute its length.

35. Is the curve $\mathbf{c}(t) = (\sin t, 2\sin 2t)$, $t \in [0, 2\pi]$, smooth?

36. Compute the length of the curve $\mathbf{c}(t) = (\cos^3 t, \sin^3 t)$, $t \in [-\pi/2, \pi/2]$, of Example 3.24.

▶ 3.4 ACCELERATION AND CURVATURE

The acceleration of a particle moving in $\mathbb{R}^2$ or $\mathbb{R}^3$ can be written as a sum of two components: one is parallel to the motion, and the other orthogonal to it. We now study this decomposition and relate it to geometric features of a path, namely its curvature. We start with an example.

▶ EXAMPLE 3.33

Consider the motion $\mathbf{c}_1(t) = (1 + t^3, 2t^3 - 3)$ along a straight line and the motion $\mathbf{c}_2(t) = (a\cos(bt), a\sin(bt))$ around a circle of radius $a > 0$ (assume that $b \ne 0$). Compute the acceleration in both cases and describe its direction.

SOLUTION The path $\mathbf{c}_1(t)$ is indeed a line: we can rewrite it in a more recognizable form as $\mathbf{c}_1(t) = (1, -3) + t^3(1, 2)$. It follows that it goes through the point $(1, -3)$ and the direction is given by the vector $(1, 2) = \mathbf{i} + 2\mathbf{j}$. The velocity $\mathbf{v}_1(t) = \mathbf{c}_1'(t) = 3t^2(1, 2)$ is not constant (it *does* have a constant direction), and

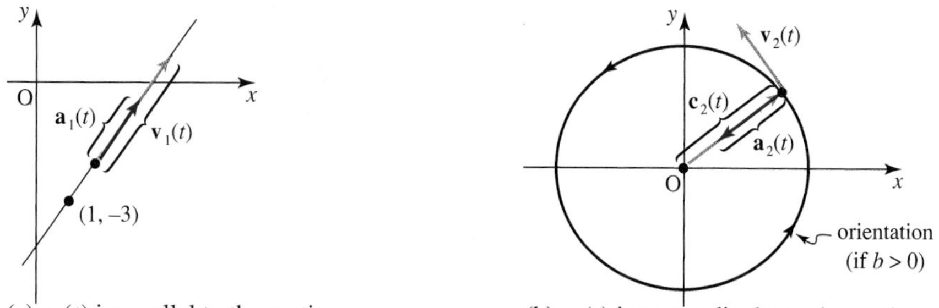

(a) $\mathbf{a}_1(t)$ is parallel to the motion. (b) $\mathbf{a}_2(t)$ is perpendicular to the motion.

Figure 3.30 Acceleration vectors for Example 3.33.

therefore, the acceleration $\mathbf{a}_1(t) = \mathbf{c}_1''(t) = 6t(1, 2)$ is a nonzero vector. The direction of $\mathbf{a}_1(t)$ is parallel to the motion; see Figure 3.30(a).

The velocity of $\mathbf{c}_2(t)$ is $\mathbf{v}_2(t) = \mathbf{c}_2'(t) = (-ab\sin(bt), ab\cos(bt))$, and its acceleration is $\mathbf{a}_2(t) = \mathbf{v}_2'(t) = (-ab^2\cos(bt), -ab^2\sin(bt))$. Since $\mathbf{a}_2(t) = -ab^2\mathbf{c}_2(t)$ [and $\mathbf{a}_2(t)\cdot\mathbf{v}_2(t) = 0$], it follows that the acceleration is perpendicular to the motion and points toward the center of the circle, as shown in Figure 3.30(b). ◀

The acceleration vectors for both $\mathbf{c}_1$ and $\mathbf{c}_2$ in the previous example are nonzero. The path $\mathbf{c}_1$ represents motion along a straight line with nonconstant speed (the speed increases as t increases), hence the nonzero acceleration (whose direction coincides with the direction of motion). The path $\mathbf{c}_2$ has a constant speed of $|ab|$ (units). However, the *velocity* changes its direction all the time, and thus, in this case the acceleration (which is now orthogonal to the motion) is also nonzero.

These two cases are the "extreme" cases. In general, an acceleration vector has a component in the direction of motion, and a component in the direction perpendicular to it. Our goal is to find this decomposition of an acceleration vector.

Let $\mathbf{c}(t)$ be a smooth C^2 path (we are going to compute the acceleration, and hence need two derivatives) describing the trajectory of a motion and let $s(t)$ be its arc-length function. Maybe the best way to think about $s(t)$ at this moment is to visualize it with the aid of

$$ds/dt = \text{rate of change of arc-length with respect to time} = \text{speed} = \|\mathbf{c}'(t)\|.$$

The velocity of $\mathbf{c}(t)$ is computed to be

$$\mathbf{v}(t) = \mathbf{c}'(t) = \|\mathbf{c}'(t)\| \frac{\mathbf{c}'(t)}{\|\mathbf{c}'(t)\|} = \frac{ds}{dt}\mathbf{T}(t),$$

where $\mathbf{T}(t) = \mathbf{c}'(t)/\|\mathbf{c}'(t)\|$ is the unit tangent vector (we need the smoothness assumption to guarantee that the denominator is nonzero).

Applying the product rule, we compute the acceleration vector

$$\begin{aligned}\mathbf{a}(t) &= \frac{d\mathbf{v}(t)}{dt} = \frac{d}{dt}\left(\frac{ds}{dt}\mathbf{T}(t)\right) = \frac{d}{dt}\left(\frac{ds}{dt}\right)\mathbf{T}(t) + \frac{ds}{dt}\left(\frac{d}{dt}\mathbf{T}(t)\right)\\ &= \frac{d^2s}{dt^2}\mathbf{T}(t) + \|\mathbf{c}'(t)\|\,\mathbf{T}'(t).\end{aligned} \tag{3.16}$$

The vector $\mathbf{T}(t)$ is of unit length; hence, $\|\mathbf{T}(t)\|^2 = \mathbf{T}(t) \cdot \mathbf{T}(t) = 1$. Differentiating $\mathbf{T}(t) \cdot \mathbf{T}(t) = 1$ with respect to t, we get

$$\mathbf{T}'(t) \cdot \mathbf{T}(t) + \mathbf{T}(t) \cdot \mathbf{T}'(t) = 2\mathbf{T}(t) \cdot \mathbf{T}'(t) = 0;$$

that is, $\mathbf{T}'(t)$ is always perpendicular to $\mathbf{T}(t)$. Consequently, formula (3.16) gives the decomposition of the acceleration vector into the tangential component

$$\mathbf{a}_T = \frac{d^2s}{dt^2}\,\mathbf{T}(t)$$

(called the *tangential acceleration*) and the normal (i.e., perpendicular to tangent) component

$$\mathbf{a}_N = \|\mathbf{c}'(t)\|\,\mathbf{T}'(t)$$

(called the *normal acceleration*). With the help of the chain rule, we obtain the expression for the normal component in terms of the arc-length parameter: since

$$\mathbf{T}'(t) = \frac{d}{dt}\mathbf{T}(t) = \left(\frac{d}{ds}\mathbf{T}(s)\right)\frac{ds}{dt} = \|\mathbf{c}'(t)\|\frac{d\mathbf{T}(s)}{ds},$$

it follows that

$$\mathbf{a}_N(s) = \|\mathbf{c}'(t)\|\,\mathbf{T}'(t) = \|\mathbf{c}'(t)\|^2\frac{d\mathbf{T}(s)}{ds}.$$

▸ EXAMPLE 3.34

Express the acceleration of the motion $\mathbf{c}(t) = 2\cos t\mathbf{i} + \sin t\mathbf{j}$, $t \in [0, 2\pi]$ as the sum of its normal and tangential components.

SOLUTION The velocity and acceleration of $\mathbf{c}$ are given by $\mathbf{c}'(t) = -2\sin t\mathbf{i} + \cos t\mathbf{j}$ and $\mathbf{a}(t) = \mathbf{c}''(t) = -2\cos t\mathbf{i} - \sin t\mathbf{j}$. The arc-length function of $\mathbf{c}(t)$ is

$$s(t) = \int_0^t \|\mathbf{c}'(\tau)\|\,d\tau = \int_0^t \sqrt{4\sin^2\tau + \cos^2\tau}\,d\tau,$$

and the unit tangent vector is computed to be

$$\mathbf{T}(t) = \frac{\mathbf{c}'(t)}{\|\mathbf{c}'(t)\|} = \frac{-2\sin t}{\sqrt{4\sin^2 t + \cos^2 t}}\mathbf{i} + \frac{\cos t}{\sqrt{4\sin^2 t + \cos^2 t}}\mathbf{j}.$$

Since $ds(t)/dt = \|\mathbf{c}'(t)\| = \sqrt{4\sin^2 t + \cos^2 t}$, it follows that

$$\frac{d^2s}{dt^2} = \frac{1}{2}\left(4\sin^2 t + \cos^2 t\right)^{-1/2}(8\sin t\cos t - 2\cos t\sin t) = \frac{3\sin t\cos t}{\sqrt{4\sin^2 t + \cos^2 t}},$$

and consequently, the tangential component of the acceleration is

$$\mathbf{a}_T = \frac{d^2s}{dt^2}\mathbf{T}(t) = \frac{3\sin t\cos t}{4\sin^2 t + \cos^2 t}(-2\sin t\mathbf{i} + \cos t\mathbf{j}).$$

The normal component $\mathbf{a}_N$ can be computed from either of the formulas we derived or directly from $\mathbf{a} = \mathbf{a}_T + \mathbf{a}_N$ since $\mathbf{a}$ and $\mathbf{a}_T$ are known. To avoid somewhat complicated computation of the derivatives, we choose the second approach:

$$\begin{aligned}\mathbf{a}_N &= \mathbf{a} - \mathbf{a}_T = -2\cos t\mathbf{i} - \sin t\mathbf{j} - \frac{3\sin t\cos t}{4\sin^2 t + \cos^2 t}(-2\sin t\mathbf{i} + \cos t\mathbf{j})\\ &= \frac{(-8\sin^2 t\cos t - 2\cos^3 t + 6\cos t\sin^2 t)\mathbf{i} + (-4\sin^3 t - \sin t\cos^2 t - 3\cos^2 t\sin t)\mathbf{j}}{4\sin^2 t + \cos^2 t}\\ &= \frac{-2\cos t(\cos^2 t + \sin^2 t)\mathbf{i} - 4\sin t(\sin^2 t + \cos^2 t)\mathbf{j}}{4\sin^2 t + \cos^2 t}\\ &= \frac{-2}{4\sin^2 t + \cos^2 t}(\cos t\mathbf{i} + 2\sin t\mathbf{j}).\end{aligned}$$ ◀

▶ EXAMPLE 3.35

Consider the path $\mathbf{c}(t) = (\cos t + t\sin t, -\sin t + t\cos t, t^2/2)$, $t \geq 0$. Find the normal and tangential components of its acceleration.

SOLUTION The velocity is $\mathbf{v}(t) = \mathbf{c}'(t) = (t\cos t, -t\sin t, t)$, and, thus, $\|\mathbf{c}'(t)\| = \sqrt{2}t$. The unit tangent vector is $\mathbf{T}(t) = \mathbf{c}'(t)/\|\mathbf{c}'(t)\| = (\cos t, -\sin t, 1)/\sqrt{2}$, and the acceleration is $\mathbf{a}(t) = \mathbf{c}''(t) = (\cos t - t\sin t, -\sin t - t\cos t, 1)$.

We do not need to compute the arc-length function, but only its derivatives: $ds/dt = \|\mathbf{c}'(t)\| = \sqrt{2}t$, and $d^2s/dt^2 = \sqrt{2}$. The acceleration can be written as $\mathbf{a}(t) = \mathbf{a}_T(t) + \mathbf{a}_N(t)$, where

$$\mathbf{a}_T(t) = \frac{d^2s}{dt^2}\mathbf{T}(t) = (\cos t, -\sin t, 1)$$

and

$$\mathbf{a}_N(t) = \|\mathbf{c}'(t)\|\,\mathbf{T}'(t) = \sqrt{2}t\,\frac{(-\sin t, -\cos t, 0)}{\sqrt{2}} = (-t\sin t, -t\cos t, 0).$$ ◀

Let us repeat what we have done so far: we have decomposed the acceleration vector $\mathbf{a}(t)$ as the sum $\mathbf{a}(t) = \mathbf{a}_T(t) + \mathbf{a}_N(t)$ of the component $\mathbf{a}_T(t) = (d^2s/dt^2)\mathbf{T}(t)$ in the direction of the motion [the term $d^2s/dt^2 = (d/dt)(ds/dt) = d\|\mathbf{c}'(t)\|/dt$ is the rate of change of the speed] and the component $\mathbf{a}_N(t) = \|\mathbf{c}'(t)\|\,\mathbf{T}'(t)$ perpendicular to the motion [$\mathbf{T}'(t)$ is the rate of change of the unit tangent vector]. In words, the tangential component of acceleration corresponds to changes in *speed*. The normal component of acceleration corresponds to changes in the *direction* of motion. We can now fully explain the results of Example 3.33: for motion along the line, there is no change in direction and hence the normal component of acceleration is zero. Motion along the circle has constant speed and, consequently, the tangential component of acceleration is zero.

These observations suggest that there is a connection between the acceleration of a moving particle and geometric properties (i.e., the "shape") of the curve representing the particle's trajectory. Our next goal is to investigate this connection.

Since the normal component of acceleration $\mathbf{a}_N$ is related to motion around a circle in Example 3.33, our idea is to find the circle that best approximates the motion at a given point on a curve. To solve this problem, we need the concept of curvature that will enable us to measure how fast a curve changes its direction. It will be defined to be independent of a particular parametrization, in terms of the unit tangent vector $\mathbf{T}$ and the arc-length function s.

DEFINITION 3.7 Curvature

Let **c** be a curve that is the image of a smooth C^2 path in $\mathbb{R}^2$ (or $\mathbb{R}^3$) parametrized by its arc-length s. The *curvature* $\kappa(s)$ of **c** at a point $\mathbf{c}(s)$ is given by

$$\kappa(s) = \left\| \frac{d\mathbf{T}(s)}{ds} \right\|,$$

where $\mathbf{T}(s)$ is the unit tangent vector.

In words, $\kappa(s)$ is the magnitude of the rate of change of the unit tangent vector $\mathbf{T}(s)$ (e.g., the unit tangent expressed in terms of s) with respect to the arc-length. By definition, $\kappa(s) \geq 0$. The application of the chain rule

$$\frac{d\mathbf{T}(s)}{ds} = \frac{d\mathbf{T}(t)}{dt}\frac{dt}{ds} = \mathbf{T}'(t)\frac{1}{\|\mathbf{c}'(t)\|} = \frac{\mathbf{T}'(t)}{\|\mathbf{c}'(t)\|}$$

gives the formula for the curvature

$$\kappa(t) = \left\| \frac{d\mathbf{T}(s)}{ds} \right\| = \frac{\|\mathbf{T}'(t)\|}{\|\mathbf{c}'(t)\|},$$

in terms of the parameter t instead of s (that is usually easier to use). Looking at this expression, we notice that the assumption about the smoothness of **c** is needed to guarantee that the denominator of $\kappa(t)$ is not zero.

▶ EXAMPLE 3.36 Curvature of a Line

Compute the curvature of the line $\mathbf{c}(t) = (at, bt)$, where $t \in \mathbb{R}$ and at least one of a or b is nonzero.

SOLUTION The unit tangent vector is

$$\mathbf{T}(t) = \frac{\mathbf{c}'(t)}{\|\mathbf{c}'(t)\|} = \frac{(a, b)}{\sqrt{a^2 + b^2}}.$$

Since $\mathbf{T}'(t) = \mathbf{0}$, it follows that the curvature of the line is $\kappa(t) = \|\mathbf{T}'(t)\|/\|\mathbf{c}'(t)\| = 0$. This certainly coincides with our intuitive understanding of curvature: as we move along the line, we do not change our direction.

▶ EXAMPLE 3.37 Curvature of a Circle

Compute the curvature of the circle $\mathbf{c}(t) = (a\cos t, a\sin t)$ of radius $a > 0$.

SOLUTION From $\mathbf{c}'(t) = (-a\sin t, a\cos t)$, we get the speed $\|\mathbf{c}'(t)\| = a$ and the unit tangent vector $\mathbf{T}(t) = \mathbf{c}'(t)/\|\mathbf{c}'(t)\| = (-\sin t, \cos t)$. The arc-length function of the circle is

$$s(t) = \int_0^t \|\mathbf{c}'(\tau)\|\, d\tau = \int_0^t a\, d\tau = at.$$

Hence, $t = s/a$, $\mathbf{T}(s) = (-\sin(s/a), \cos(s/a))$, and

$$\kappa(s) = \left\| \frac{d\mathbf{T}(s)}{ds} \right\| = \left\| \left(-\frac{\cos(s/a)}{a}, -\frac{\sin(s/a)}{a} \right) \right\| = \frac{1}{a}.$$

Alternatively, $\mathbf{T}'(t) = (-\cos t, -\sin t)$ and $\kappa(t) = \|\mathbf{T}'(t)\|/\|\mathbf{c}'(t)\| = 1/a$.

From our computation, it follows that the circle has the same curvature at all points (certainly not a surprise), equal to the reciprocal of its radius. Circles of smaller radii are "curved more"; that is, have larger curvature than big circles (that are "curved less"). Once again, the results correspond to our intuitive understanding of curvature. ◀

▶ EXAMPLE 3.38

Compute the curvature of the helix

$$\mathbf{c}(t) = a\cos t\mathbf{i} + a\sin t\mathbf{j} + bt\mathbf{k}, \qquad t \geq 0,$$

where $a, b > 0$.

SOLUTION

From $\mathbf{c}'(t) = -a\sin t\mathbf{i} + a\cos t\mathbf{j} + b\mathbf{k}$, we get $\|\mathbf{c}'(t)\| = \sqrt{a^2+b^2}$ and

$$\mathbf{T}(t) = -\frac{a\sin t}{\sqrt{a^2+b^2}}\mathbf{i} + \frac{a\cos t}{\sqrt{a^2+b^2}}\mathbf{j} + \frac{b}{\sqrt{a^2+b^2}}\mathbf{k}.$$

Consequently,

$$\mathbf{T}'(t) = -\frac{a\cos t}{\sqrt{a^2+b^2}}\mathbf{i} - \frac{a\sin t}{\sqrt{a^2+b^2}}\mathbf{j},$$

and $\|\mathbf{T}'(t)\| = a/\sqrt{a^2+b^2}$. So, the curvature $\kappa(t)$ at the point $\mathbf{c}(t)$ on the helix is

$$\kappa(t) = \frac{\|\mathbf{T}'(t)\|}{\|\mathbf{c}'(t)\|} = \frac{a/\sqrt{a^2+b^2}}{\sqrt{a^2+b^2}} = \frac{a}{a^2+b^2}.$$

It turns out that a helix, like a line or circle, has constant curvature. ◀

▶ EXAMPLE 3.39

Compute the curvature of the parabola $y = x^2$.

SOLUTION

Parametrize $y = x^2$ as $\mathbf{c}(t) = t\mathbf{i} + t^2\mathbf{j}$. Then $\mathbf{c}'(t) = \mathbf{i} + 2t\mathbf{j}$, $\|\mathbf{c}'(t)\| = \sqrt{1+4t^2}$, and the unit tangent vector is

$$\mathbf{T}(t) = \frac{\mathbf{c}'(t)}{\|\mathbf{c}'(t)\|} = \frac{1}{\sqrt{1+4t^2}}\mathbf{i} + \frac{2t}{\sqrt{1+4t^2}}\mathbf{j}.$$

Its derivative is computed to be

$$\mathbf{T}'(t) = -\frac{4t}{(1+4t^2)^{3/2}}\mathbf{i} + \frac{2}{(1+4t^2)^{3/2}}\mathbf{j}.$$

Consequently,

$$\|\mathbf{T}'(t)\| = \sqrt{\frac{16t^2+4}{(1+4t^2)^3}} = \sqrt{\frac{4}{(1+4t^2)^2}} = \frac{2}{1+4t^2}$$

and

$$\kappa(t) = \frac{\|\mathbf{T}'(t)\|}{\|\mathbf{c}'(t)\|} = \frac{2}{(1+4t^2)^{3/2}}.$$

Figure 3.31 shows the graph of the parabola $y = x^2$ and its curvature. Our intuition works again: the parabola is curved most at the vertex $\mathbf{c}(0) = (0, 0)$ [the curvature there is $\kappa(0) = 2$]. As we move

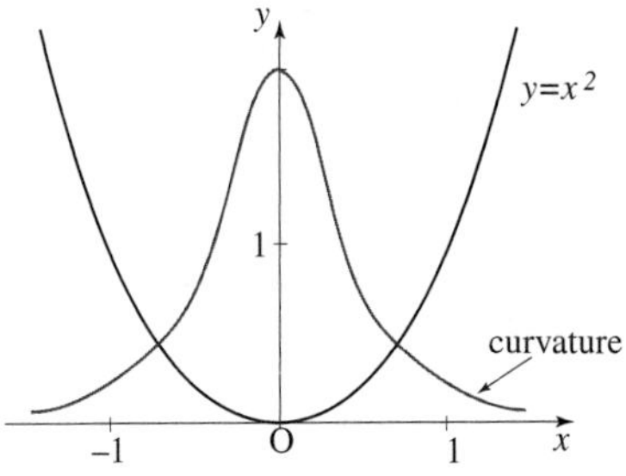

Figure 3.31 The parabola $y = x^2$ and its curvature function.

away from the vertex in either of the two directions, we notice that the parabola becomes curved less and less, so the graph of κ comes closer and closer to the x-axis. As t (or x) approaches $\pm\infty$, the curvature approaches zero. ◀

Let $\mathbf{c}(t)$ be a smooth C^2 path in $\mathbb{R}^2$ (or $\mathbb{R}^3$) and let $\mathbf{T}(t) = \mathbf{c}'(t)/\|\mathbf{c}'(t)\|$ be its unit tangent vector. Since $\|\mathbf{T}(t)\| = 1$, it follows that $\mathbf{T}'(t)$ is perpendicular to $\mathbf{T}(t)$ [see the computation immediately following (3.16)]. However, $\mathbf{T}'(t)$ might not be of unit length. But if $\mathbf{T}'(t) \neq \mathbf{0}$, we can define the *principal unit normal* (or just *unit normal*) *vector* $\mathbf{N}(t)$ by

$$\mathbf{N}(t) = \frac{\mathbf{T}'(t)}{\|\mathbf{T}'(t)\|}. \tag{3.17}$$

▶ EXAMPLE 3.40

Compute the principal unit normal vectors for the circle of Example 3.37 and the helix of Example 3.38.

SOLUTION

In the case of the circle, $\mathbf{T}'(t) = (-\cos t, -\sin t)$ is a unit vector, and hence, $\mathbf{N}(t) = \mathbf{T}'(t) = (-\cos t, -\sin t)$. The vector $\mathbf{N}(t)$ is orthogonal to $\mathbf{T}(t)$ and points toward the center of the circle (the origin); see Figure 3.32(a). For the helix, the unit normal $\mathbf{N}(t) = \mathbf{T}'(t)/\|\mathbf{T}'(t)\| = -\cos t\mathbf{i} - \sin t\mathbf{j}$ is the vector pointing toward the z-axis, as shown in Figure 3.32(b). ◀

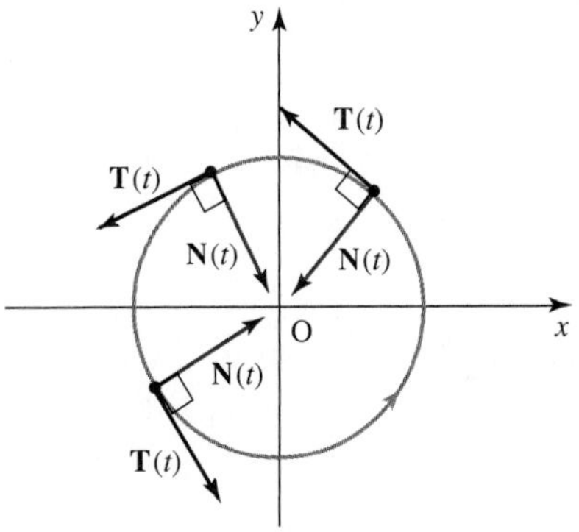

(a) To the circle of Example 3.37.

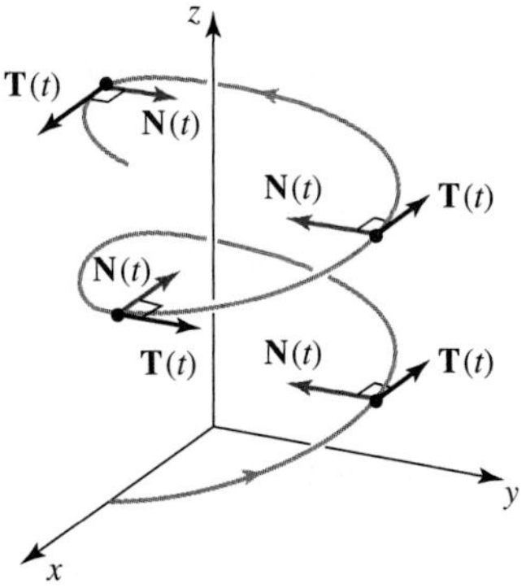

(b) To the helix of Example 3.38.

Figure 3.32 Principal unit normal vectors.

DEFINITION 3.8 Osculating Plane and Osculating Circle

Let $\mathbf{c}(t)$ be a smooth C^2 path such that $\mathbf{c}''(t) \neq \mathbf{0}$ [recall that smoothness guarantees that $\mathbf{c}'(t) \neq \mathbf{0}$]. The plane through $\mathbf{c}(t)$ spanned by the vectors $\mathbf{T}(t)$ and $\mathbf{N}(t)$ is called the *osculating plane* of $\mathbf{c}$ at $\mathbf{c}(t)$. The circle that lies in the osculating plane of $\mathbf{c}$ at $\mathbf{c}(t)$ on the side of $\mathbf{c}(t)$ toward which $\mathbf{N}(t)$ points, that has the point $\mathbf{c}(t)$ in common with the curve, whose tangent at $\mathbf{c}(t)$ is parallel to $\mathbf{c}'(t)$, and whose radius is $1/\kappa(t)$ [where $\kappa(t)$ is the curvature of $\mathbf{c}$ at $\mathbf{c}(t)$] is called the *osculating circle* (or the *circle of curvature*) of $\mathbf{c}$ at $\mathbf{c}(t)$; see Figure 3.33.

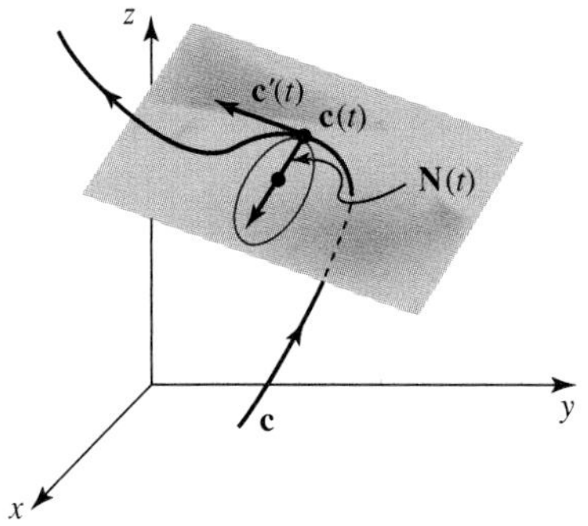

Figure 3.33 Osculating plane and osculating circle.

The osculating plane of a curve in $\mathbb{R}^2$ is $\mathbb{R}^2$. The osculating circle (the Latin word *osculum* means "kiss"), having the same tangent, normal, and curvature as the given curve, is the best approximation by a circle to the curve at a point $\mathbf{c}(t)$.

▶ EXAMPLE 3.41

Find the equation of the osculating plane at the point $\mathbf{c}(\pi) = (-a, 0, b\pi)$ for the helix $\mathbf{c}(t) = a\cos t\mathbf{i} + a\sin t\mathbf{j} + bt\mathbf{k}$, $t \geq 0$, where $a, b > 0$.

SOLUTION

The osculating plane goes through $\mathbf{c}(\pi) = (-a, 0, b\pi)$ and is spanned by the vectors (see Examples 3.38 and 3.40)

$$\mathbf{T}(\pi) = -\frac{a}{\sqrt{a^2+b^2}}\mathbf{j} + \frac{b}{\sqrt{a^2+b^2}}\mathbf{k}$$

and $\mathbf{N}(\pi) = \mathbf{i}$. The normal vector to the osculating plane is

$$\mathbf{T}(\pi) \times \mathbf{N}(\pi) = \begin{vmatrix} \mathbf{i} & \mathbf{j} & \mathbf{k} \\ 0 & -a/\sqrt{a^2+b^2} & b/\sqrt{a^2+b^2} \\ 1 & 0 & 0 \end{vmatrix} = \frac{b}{\sqrt{a^2+b^2}}\mathbf{j} + \frac{a}{\sqrt{a^2+b^2}}\mathbf{k},$$

and hence its equation is (recall the equation of the plane defined by a normal vector and a point discussed at the end of Section 1.3)

$$0(x+a) + \frac{b}{\sqrt{a^2+b^2}}(y-0) + \frac{a}{\sqrt{a^2+b^2}}(z-b\pi) = 0,$$

that is, $by + az - ab\pi = 0$.

▶ EXAMPLE 3.42

Find the curvature and osculating circle of the parabola $y = x^2$ at the point $(0, 0)$.

SOLUTION All computations that are needed here have been already done in Example 3.39. The osculating circle has radius of $1/\kappa = 1/2$, since the curvature of $y = x^2$ at the origin is $\kappa = 2$. It has to lie above (i.e., "inside") the parabola, since the normal $\mathbf{N}(0) = \mathbf{T}'(0)/\|\mathbf{T}'(0)\| = (0, 1) = \mathbf{j}$ points that way; see Figure 3.34. The center of the osculating circle has to lie on the line normal to the tangent at $(0, 0)$; that is, on the y-axis. It follows that the equation $x^2 + (y - 1/2)^2 = 1/4$ represents the desired osculating circle.

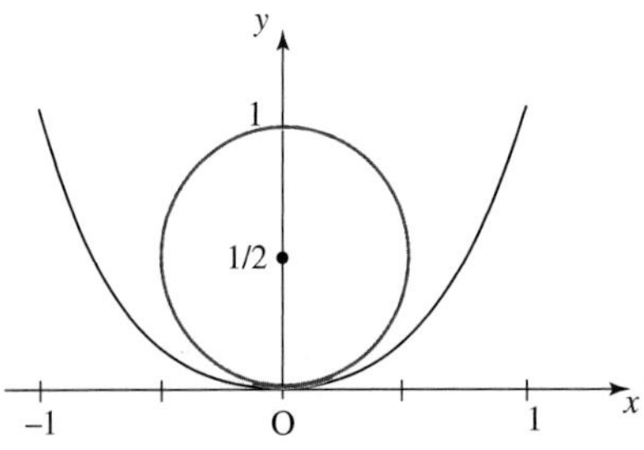

Figure 3.34 The parabola $y = x^2$ and its osculating circle at the origin.

◀

We are now ready to give a geometric interpretation of the normal acceleration $\mathbf{a}_N$. Start with $\mathbf{a}_N = \|\mathbf{c}'(t)\|\,\mathbf{T}'(t)$, and divide and multiply the right side by $\|\mathbf{c}'(t)\|\,\|\mathbf{T}'(t)\|$ to get

$$\mathbf{a}_N = \|\mathbf{c}'(t)\|^2 \frac{\|\mathbf{T}'(t)\|}{\|\mathbf{c}'(t)\|} \frac{\mathbf{T}'(t)}{\|\mathbf{T}'(t)\|}.$$

We recognize the first factor as the square of the speed $\|\mathbf{c}'(t)\| = ds/dt$. The second factor is the curvature $\kappa(t)$, and the third is the principal normal vector. Hence,

$$\mathbf{a}_N = \left(\frac{ds}{dt}\right)^2 \kappa(t)\mathbf{N}(t);$$

that is, the magnitude of the normal component of acceleration is the product of the square of the speed and the curvature.

▶ EXERCISES 3.4

Exercises 1 to 6: Find the tangential and normal components of acceleration for the motion of a particle described by its position vector $\mathbf{c}(t)$.

1. $\mathbf{c}(t) = (t^2, t, t^2)$

2. $\mathbf{c}(t) = (e^t, \sqrt{2}t, e^{-t})$

3. $\mathbf{c}(t) = 5t\mathbf{i} + 12\sin t\mathbf{j} + 12\cos t\mathbf{k}$

4. $\mathbf{c}(t) = 2t\mathbf{i} + 2\sin^2 t\mathbf{j} - 2\cos^2 t\mathbf{k}$

5. $\mathbf{c}(t) = (t - \sin t)\mathbf{i} + (1 - \cos t)\mathbf{j}$

6. $\mathbf{c}(t) = (\cos 3t, \sin 3t, 4t)$

7. Let $\mathbf{c}(t) = (x(t), y(t))$ be a smooth C^2 parametrization of a curve $\mathbf{c}$ in $\mathbb{R}^2$. Show that its curvature is $\kappa(t) = |x'(t)y''(t) - x''(t)y'(t)|/[(x'(t))^2 + (y'(t))^2]^{3/2}$.

8. Using the formula from Exercise 7, compute the curvature of $\mathbf{c}_1(t) = (t\sin t, t\cos t)$ and $\mathbf{c}_2(t) = (t, t^2)$.

9. Compute the curvature of $\mathbf{c}(t) = (2 - 2t^3, t^3 + 1)$, $t \in \mathbb{R}$. Identify the curve, thus checking your answer.

10. Compute the curvature of $\mathbf{c}(t) = (t, \sin t)$, $t \in \mathbb{R}$, and plot the curve and its curvature function using the same coordinate system. Identify the points (if any) where the curvature is zero and where it is largest.

11. Find the curvature of the plane curve $\mathbf{c}(t) = (t^2, 3 - t)$. Identify the point(s) where the curvature is largest. What happens as $t \to \infty$?

Exercises 12 to 17: For each parametrization, find the unit tangent and the unit normal vector, the curvature, and the normal component of acceleration.

12. $\mathbf{c}(t) = (\sin 2t, \cos 2t, 5t)$

13. $\mathbf{c}(t) = (e^t \sin t, 0, e^t \cos t)$

14. $\mathbf{c}(t) = (3 + 2t, -t, t - 3)$

15. $\mathbf{c}(t) = (t, \cos t, 1 - \sin t)$

16. $\mathbf{c}(t) = (1, 0, t^2/2)$

17. $\mathbf{c}(t) = (e^{-t} \cos t, e^{-t} \sin t, e^{-t})$

18. Find equations of the lines tangent and normal to the curve $\mathbf{c}(t) = (t^3/3 - t)\mathbf{i} + t^2\mathbf{j}$ at the point $(0, 3)$.

19. Find an equation of the osculating circle of $\mathbf{c}(t) = t^3\mathbf{i} + t\mathbf{j}$ at the point $(8, 2)$.

20. Find an equation of the osculating plane of the curve $\mathbf{c}(t) = (1/t + 1, 1/t - 1, t)$ at a point $\mathbf{c}(t_0)$, $t_0 \neq 0$.

21. Find an equation of the osculating plane of the helix $\mathbf{c}(t) = (2\cos t, 2\sin t, t)$ at the point $\mathbf{c}(\pi/2)$.

22. Prove that the curvature of the graph of a C^2 function $y = f(x)$ is given by the formula $\kappa(x) = |f''(x)|/(1 + (f'(x))^2)^{3/2}$. Show that this formula is a special case of the formulas in Exercises 7 and 28.

Exercises 23 to 26: Use the formula of Exercise 22 to solve the following problems.

23. Find the curvature of $y = x^2$ at a point (x_0, y_0).

24. Find the curvature of $y = x + \ln x$ at $(1, 1)$. What happens to the curvature as $x \to \infty$?

25. Where does the graph of $y = \ln x$ have maximum curvature?

26. Find the curvature of the graph of $y = \sin x$ at $(\pi/2, 1)$ and at $(\pi, 0)$.

27. Find the equation of the osculating plane of the curve $\mathbf{c}(t) = (t, 1 - t^2, 2t^2)$ at the point $\mathbf{c}(1)$.

28. Prove that the curvature of a smooth C^2 curve $\mathbf{c}(t)$ in $\mathbb{R}^3$ can be computed from the formula $\kappa(t) = \|\mathbf{c}'(t) \times \mathbf{c}''(t)\|/\|\mathbf{c}'(t)\|^3$. Show that the equation of Exercise 7 is a special case of this formula.

29. Find the equation of the osculating circle of the graph of $y = \sin x$ at $(\pi/2, 1)$.

30. Find the equation of the osculating circle of the graph of $y = e^x$ at $(1, e)$.

▶ 3.5 INTRODUCTION TO DIFFERENTIAL GEOMETRY OF CURVES

The material we have covered so far in this chapter shows that a large amount of information can be represented visually as a curve (or algebraically as its parametric representation). This is certainly a good reason to study curves in more depth. In this section, we only indicate one possible approach, the so-called differential geometry of curves. In Section 4.5, we will relate curves to vector fields by defining a flow line of a vector field. Concepts relevant to integration along paths will be discussed at the beginning of Chapter 5.

Consider a path $\mathbf{c}$ in $\mathbb{R}^3$. We will construct a coordinate system that "travels along $\mathbf{c}$"; that is, we will find continuous, mutually orthogonal unit vector fields defined at the points of $\mathbf{c}$. The motion along the curve can then be studied in terms of the changes of those three vectors as the point $\mathbf{c}(t)$ (that is their common initial point) slides along the curve.

Let us clarify one technical point first. We use $\mathbf{c}(t)$ to denote a representation of a curve in terms of a general parameter t and $\mathbf{c}(s)$ to denote the parametrization of $\mathbf{c}$ by its arc-length parameter s (strictly speaking, we should have used two different symbols). Recall that s is defined in terms of t by

$$s = \int_0^t \|\mathbf{c}'(\tau)\|\, d\tau. \tag{3.18}$$

However, we generally use the equivalent form $ds/dt = \|\mathbf{c}'(t)\|$; the equivalence is a consequence of the Fundamental Theorem of Calculus. The "prime" ($'$) notation denotes the derivative with respect to the corresponding variable. For example, $\mathbf{c}'(t) = d\mathbf{c}(t)/dt$ and $\mathbf{c}'(s) = d\mathbf{c}(s)/ds$ or $\mathbf{T}'(t) = d\mathbf{T}(t)/dt$ and $\mathbf{T}'(s) = d\mathbf{T}(s)/ds$, etc. To find the formula that relates the two derivatives, we use the chain rule. For example,

$$\mathbf{c}'(t) = \frac{d\mathbf{c}(t)}{dt} = \frac{d\mathbf{c}(s)}{ds}\frac{ds}{dt} = \mathbf{c}'(s)\,\|\mathbf{c}'(t)\|.$$

The part $d\mathbf{c}(t)/dt = (d\mathbf{c}(s)/ds)(ds/dt)$ is a case of the general chain rule sloppiness (as mentioned in Section 2.6): the left side is the derivative of $\mathbf{c}(t)$ with respect to t and the first factor on the right is the derivative of $\mathbf{c}$, viewed as a function of s, with respect to s.

The above computation shows that the velocity $\mathbf{c}'(s)$ of a path parametrized by its arc-length is always equal to the unit tangent vector [provided that, of course, $\mathbf{c}'(t) \neq \mathbf{0}$]; that is, $\mathbf{c}'(s) = \mathbf{c}'(t)/\|\mathbf{c}'(t)\| = \mathbf{T}(t)$ [the equation $\mathbf{c}'(s) = \mathbf{T}(t)$ actually means that, after replacing s by t [according to the defining identity (3.18)] on the left, or t by s on the right side, the two vectors are the same]. One more consequence of our computation: since $\mathbf{T}(t)$ is of unit length, it follows that $\|\mathbf{c}'(s)\| = 1$; that is, the speed of a path parametrized by its arc-length parameter is always constant and equal to 1.

Now let us return to our task of finding the three mutually orthogonal unit vectors. Part of the work has been done already: in the previous section we defined the unit tangent vector $\mathbf{T}(t) = \mathbf{c}'(t)/\|\mathbf{c}'(t)\|$ and the principal unit normal vector $\mathbf{N}(t) = \mathbf{T}'(t)/\|\mathbf{T}'(t)\|$. These formulas make sense only if $\mathbf{c}(t)$ is C^2 [to get $\mathbf{T}'$, we need two derivatives of $\mathbf{c}$], smooth [meaning that $\|\mathbf{c}'(t)\| \neq \mathbf{0}$, which implies that $\mathbf{T}(t)$ is defined] and such that $\mathbf{T}'(t) \neq \mathbf{0}$ [so that $\mathbf{N}(t)$ is defined].

Recall that we defined the curvature by $\kappa(s) = \|d\mathbf{T}(s)/ds\|$. It follows that

$$\frac{d\mathbf{T}(s)}{ds} = \|d\mathbf{T}(s)/ds\|\frac{d\mathbf{T}(s)/ds}{\|d\mathbf{T}(s)/ds\|} = \kappa(s)\mathbf{N}(s).$$

The vectors $\mathbf{T}(t)$ and $\mathbf{N}(t)$ define the osculating plane of $\mathbf{c}$ at $\mathbf{c}(t)$. In a neighborhood of $\mathbf{c}(t)$, the curve $\mathbf{c}$ looks like part of the circle (called the osculating circle) of radius equal to the reciprocal of the curvature centered along the line normal to $\mathbf{c}$ at $\mathbf{c}(t)$. The points on the curve that are near $\mathbf{c}(t)$ either lie in the osculating plane or are very close to it.

The properties of the cross product imply that the vector $\mathbf{B}(t) = \mathbf{T}(t) \times \mathbf{N}(t)$ is orthogonal to both the unit tangent vector and the principal unit normal vector and is of unit length. It is called the *binormal vector* to $\mathbf{c}$ at $\mathbf{c}(t)$. The vectors $\mathbf{T}(t)$, $\mathbf{N}(t)$, and $\mathbf{B}(t)$ form the right coordinate system at $\mathbf{c}(t)$ ("right" means that the third unit orthogonal vector is determined

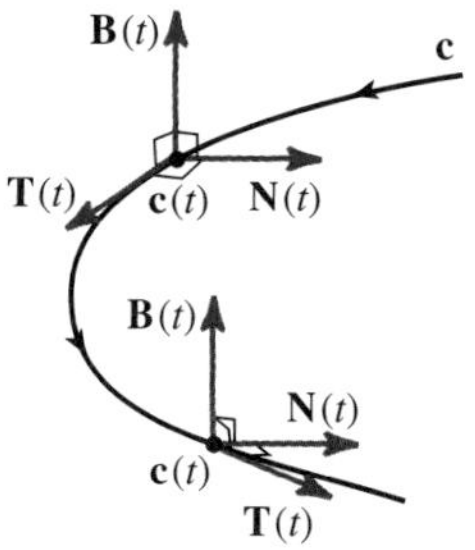

Figure 3.35 Frenet frame moving along a curve.

by the right-hand rule) that is called the *Frenet frame at* $\mathbf{c}(t)$, or the *TNB frame at* $\mathbf{c}(t)$. It is useful to think of a Frenet frame as a set of three unit, mutually orthogonal vectors that are locked in their positions. No matter how the frame changes along a curve, their mutual position remains the same; see Figure 3.35.

If a curve lies in a plane, its tangent and normal vectors lie in it as well—consequently, the binormal is a *constant* vector: it is a unit vector perpendicular to that plane.

▶ EXAMPLE 3.43

Find the Frenet frame for the helix $\mathbf{c}(t) = a\cos t\mathbf{i} + a\sin t\mathbf{j} + bt\mathbf{k}$, $t \geq 0$, where $a, b > 0$.

SOLUTION In Example 3.38, we computed the unit tangent vector to be

$$\mathbf{T}(t) = -\frac{a\sin t}{\sqrt{a^2+b^2}}\mathbf{i} + \frac{a\cos t}{\sqrt{a^2+b^2}}\mathbf{j} + \frac{b}{\sqrt{a^2+b^2}}\mathbf{k}.$$

Since

$$\mathbf{T}'(t) = -\frac{a\cos t}{\sqrt{a^2+b^2}}\mathbf{i} - \frac{a\sin t}{\sqrt{a^2+b^2}}\mathbf{j}$$

and $\|\mathbf{T}'(t)\| = a/\sqrt{a^2+b^2}$, it follows that the principal unit normal vector is

$$\mathbf{N}(t) = \frac{\mathbf{T}'(t)}{\|\mathbf{T}'(t)\|} = -\cos t\mathbf{i} - \sin t\mathbf{j}.$$

The binormal vector is the cross product of $\mathbf{T}$ and $\mathbf{N}$:

$$\mathbf{B} = \mathbf{T} \times \mathbf{N} = \begin{vmatrix} \mathbf{i} & \mathbf{j} & \mathbf{k} \\ -a\sin t/\sqrt{a^2+b^2} & a\cos t/\sqrt{a^2+b^2} & b/\sqrt{a^2+b^2} \\ -\cos t & -\sin t & 0 \end{vmatrix}$$

$$= \frac{b\sin t}{\sqrt{a^2+b^2}}\mathbf{i} - \frac{b\cos t}{\sqrt{a^2+b^2}}\mathbf{j} + \frac{a}{\sqrt{a^2+b^2}}\mathbf{k}.$$

The vectors $\mathbf{T}(t)$, $\mathbf{N}(t)$, and $\mathbf{B}(t)$ form the Frenet frame at the point $\mathbf{c}(t)$ on the helix. ◀

Now let us investigate how the Frenet (or TNB) frame changes along the curve. We already know that

$$\frac{d\mathbf{T}(s)}{ds} = \kappa(s)\mathbf{N}(s), \tag{3.19}$$

where $\kappa(s)$ is the curvature. Geometrically, (3.19) states that the tangent $\mathbf{T}(s)$ turns toward the normal $\mathbf{N}(s)$ at a rate $\kappa(s)$ [since there is no $\mathbf{B}$ component in (3.19), $\mathbf{T}(s)$ changes *only*

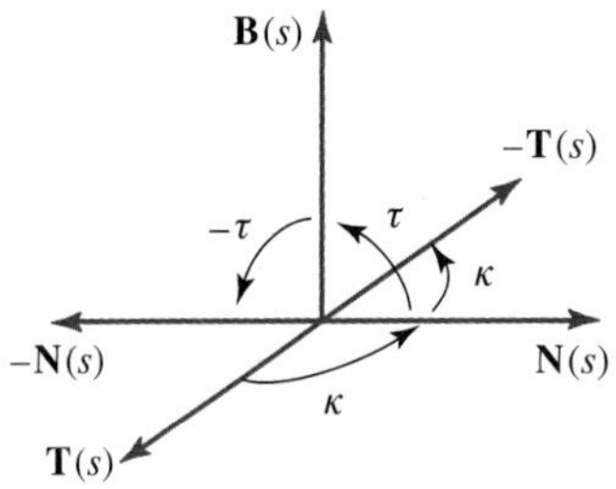

Figure 3.36 Changes of vectors in the Frenet (TNB) frame.

in the $\mathbf{N}(s)$ direction]. Since the mutual position of the three vectors is fixed, if $\mathbf{T}(s)$ turns toward $\mathbf{N}(s)$ at a rate $\kappa(s)$, then $\mathbf{N}(s)$ must turn toward $-\mathbf{T}(s)$ at the same rate.

But $\mathbf{N}(s)$ could also turn toward $\mathbf{B}(s)$ [e.g., $\mathbf{N}(s)$ could rotate about the direction of $\mathbf{T}(s)$]; see Figure 3.36. Therefore,

$$\frac{d\mathbf{N}(s)}{ds} = -\kappa(s)\mathbf{T}(s) + \tau(s)\mathbf{B}(s), \tag{3.20}$$

where the scalar $\tau(s)$ is called the *torsion* of the curve. Since $\mathbf{N}(s)$ turns toward $\mathbf{B}(s)$ at a rate $\tau(s)$, $\mathbf{B}(s)$ has to turn at the same rate toward $-\mathbf{N}(s)$. $\mathbf{B}(s)$ cannot turn toward $\mathbf{T}(s)$, since then $\mathbf{T}(s)$ would have to turn toward $-\mathbf{B}(s)$ [but $d\mathbf{T}(s)/ds = \kappa(s)\mathbf{N}(s)$ means that $\mathbf{T}(s)$ turns *only* toward $\mathbf{N}(s)$]. It follows that

$$\frac{d\mathbf{B}(s)}{ds} = -\tau(s)\mathbf{N}(s). \tag{3.21}$$

The formulas (3.19), (3.20) and (3.21) are called the *Serret–Frenet formulas* and describe the way the Frenet frame changes along a curve.

Computing the dot product of (3.21) with $\mathbf{N}(s)$ (remember that $\mathbf{N}$ is of unit length), we get the following formula:

$$\tau(s) = -\mathbf{N}(s) \cdot \frac{d\mathbf{B}(s)}{ds}$$

for the torsion. Computing the norms of both sides in (3.21) yields

$$\left\| \frac{d\mathbf{B}(s)}{ds} \right\| = |-\tau(s)|\, \|\mathbf{N}(s)\| = |\tau(s)|,$$

that is, $\tau(s) = \pm\|d\mathbf{B}(s)/ds\|$.

▶ EXAMPLE 3.44

Compute the torsion of a plane curve.

SOLUTION By construction, both $\mathbf{T}$ and $\mathbf{N}$ belong to the plane of the curve. Consequently, the binormal $\mathbf{B}$ is a constant vector (always unit and perpendicular to the plane) and therefore $\tau(s) = \pm\|d\mathbf{B}(s)/ds\| = 0$. ◀

If the binormal vector starts changing, it will force the curve out of the plane; in other words, the torsion (i.e., the rate of change of the binormal) measures how fast the curve moves away from its osculating plane.

▶ EXAMPLE 3.45

Compute the curvature and torsion of the helix $\mathbf{c}(t) = a\cos t\mathbf{i} + a\sin t\mathbf{j} + bt\mathbf{k}$, $t \geq 0$, $a, b > 0$, from the Serret–Frenet formulas (3.19) and (3.21). Using the results obtained, check the second Serret–Frenet formula (3.20).

SOLUTION The arc-length parameter of the helix is computed to be

$$s(t) = \int_0^t \|\mathbf{c}'(\tau)\| d\tau = \int_0^t \sqrt{(-a\sin\tau)^2 + (a\cos\tau)^2 + b^2}\, d\tau = \sqrt{a^2+b^2}\, t.$$

Using $t = s/\sqrt{a^2+b^2}$, we now express the unit tangent vector, the principal unit normal vector, and the binormal vector from Example 3.43 as

$$\mathbf{T}(s) = -\frac{a}{\sqrt{a^2+b^2}}\sin(s/\sqrt{a^2+b^2})\mathbf{i} + \frac{a}{\sqrt{a^2+b^2}}\cos(s/\sqrt{a^2+b^2})\mathbf{j} + \frac{b}{\sqrt{a^2+b^2}}\mathbf{k},$$

$$\mathbf{N}(s) = -\cos(s/\sqrt{a^2+b^2})\mathbf{i} - \sin(s/\sqrt{a^2+b^2})\mathbf{j},$$

and

$$\mathbf{B}(s) = \frac{b}{\sqrt{a^2+b^2}}\sin(s/\sqrt{a^2+b^2})\mathbf{i} - \frac{b}{\sqrt{a^2+b^2}}\cos(s/\sqrt{a^2+b^2})\mathbf{j} + \frac{a}{\sqrt{a^2+b^2}}\mathbf{k}.$$

According to the Serret–Frenet formula (3.19), the rate of change of the unit tangent vector

$$\frac{d\mathbf{T}(s)}{ds} = -\frac{a}{a^2+b^2}\cos(s/\sqrt{a^2+b^2})\mathbf{i} - \frac{a}{a^2+b^2}\sin(s/\sqrt{a^2+b^2})\mathbf{j}$$

is equal to $\kappa(s)\mathbf{N}(s)$, where $\kappa(s)$ is the curvature. Since

$$\frac{d\mathbf{T}(s)}{ds} = \frac{a}{a^2+b^2}\mathbf{N}(s),$$

it follows that $\kappa(s) = a/(a^2+b^2)$ (in Example 3.38 in the previous section, we obtained the same result using the definition of curvature). The third Serret–Frenet formula (3.21) states that the rate of change of the binormal

$$\frac{d\mathbf{B}(s)}{ds} = \frac{b}{a^2+b^2}\cos(s/\sqrt{a^2+b^2})\mathbf{i} + \frac{b}{a^2+b^2}\sin(s/\sqrt{a^2+b^2})\mathbf{j}$$

is equal to $-\tau(s)\mathbf{N}(s)$, where $\tau(s)$ is the torsion. Since

$$\frac{d\mathbf{B}(s)}{ds} = -\frac{b}{a^2+b^2}(-\cos(s/\sqrt{a^2+b^2})\mathbf{i} - \sin(s/\sqrt{a^2+b^2})\mathbf{j}),$$

it follows that $\tau(s) = b/(a^2+b^2)$. The left side in (3.20) is computed to be

$$\frac{d\mathbf{N}(s)}{ds} = \frac{1}{\sqrt{a^2+b^2}}\sin(s/\sqrt{a^2+b^2})\mathbf{i} - \frac{1}{\sqrt{a^2+b^2}}\cos(s/\sqrt{a^2+b^2})\mathbf{j}.$$

The right side

$$-\kappa(s)\mathbf{T}(s) + \tau(s)\mathbf{B}(s) = -\frac{a}{a^2+b^2}\left(-\frac{a}{\sqrt{a^2+b^2}}\sin(s/\sqrt{a^2+b^2})\mathbf{i}\right.$$

$$\left.+\frac{a}{\sqrt{a^2+b^2}}\cos(s/\sqrt{a^2+b^2})\mathbf{j} + \frac{b}{\sqrt{a^2+b^2}}\mathbf{k}\right)$$

$$+\frac{b}{a^2+b^2}\left(\frac{b}{\sqrt{a^2+b^2}}\sin(s/\sqrt{a^2+b^2})\mathbf{i}\right.$$

$$-\frac{b}{\sqrt{a^2+b^2}}\cos\,(s/\sqrt{a^2+b^2})\mathbf{j}+\frac{a}{\sqrt{a^2+b^2}}\mathbf{k}\Bigg)$$

$$=\left(\frac{a^2}{(a^2+b^2)^{3/2}}+\frac{b^2}{(a^2+b^2)^{3/2}}\right)\sin\,(s/\sqrt{a^2+b^2})\mathbf{i}$$

$$-\left(\frac{a^2}{(a^2+b^2)^{3/2}}+\frac{b^2}{(a^2+b^2)^{3/2}}\right)\cos\,(s/\sqrt{a^2+b^2})\mathbf{j}$$

$$=\frac{1}{\sqrt{a^2+b^2}}\sin\,(s/\sqrt{a^2+b^2})\mathbf{i}-\frac{1}{\sqrt{a^2+b^2}}\cos\,(s/\sqrt{a^2+b^2})\mathbf{j}$$

is equal to $d\mathbf{N}(s)/ds$, as predicted by the Serret–Frenet formula (3.20). ◀

▶ EXERCISES 3.5

1. Find formulas for the following three quantities: $d(\mathbf{T}(s)\cdot\mathbf{T}(s))/ds$, $(d\mathbf{c}(t)/dt)\cdot\mathbf{T}(t)$, and $d\mathbf{N}(s)/\,ds\cdot\mathbf{B}(s)$.

2. Show that the curve $\mathbf{c}(t)=(a\cos t, a\sin t, \sin t+\cos t+b)$, where a and b are any constants, is a plane curve.

3. Is it true that the acceleration of a particle moving along a path in $\mathbb{R}^3$ is always perpendicular to the binormal vector? Explain.

Exercises 4 to 7: Find the Frenet (TNB) frame.

4. $\mathbf{c}(t)=(e^t, 2e^t, 0)$, at $t=1$

5. $\mathbf{c}(t)=e^t\cos t\mathbf{i}+e^t\sin t\mathbf{k}$, at $t=\pi/2$

6. $\mathbf{c}(t)=(t\sin t, t\cos t, t)$, for any t

7. $\mathbf{c}(t)=(\sin t, \sin t, \sqrt{2}\cos t)$, for any t

8. Using the fact that the binormal vector $\mathbf{B}$ is of unit length and orthogonal to $\mathbf{T}$, show that $d\mathbf{B}/ds$ is parallel to $\mathbf{N}$ (s, as usual, denotes the arc-length parameter).

9. Show that $d\mathbf{N}/ds+\kappa\mathbf{T}$ is perpendicular to both $\mathbf{T}$ and $\mathbf{N}$, without using formula (3.20). Explain how one can use this fact to define the torsion of a curve.

10. By differentiating $\mathbf{B}=\mathbf{T}\times\mathbf{N}$, prove the third Serret–Frenet formula (3.21).

Exercises 11 to 14: Let $\mathbf{c}$ be a smooth C^2 curve (C^3 for Exercise 14) in $\mathbb{R}^3$. Use the Serret–Frenet formulas to prove the following results.

11. $\mathbf{c}''(t)=\dfrac{d^2s}{dt^2}\mathbf{T}(t)+\kappa(t)\left(\dfrac{ds}{dt}\right)^2\mathbf{N}(t)$

12. Show that if $\|\mathbf{c}(t)\|=1$, then $\mathbf{c}(t)=-\dfrac{1}{\kappa}\mathbf{N}-\dfrac{1}{\tau}\left(\dfrac{d\kappa^{-1}}{ds}\right)\mathbf{B}$. (*Hint*: Start by differentiating $\mathbf{c}(t)\cdot\mathbf{c}(t)=1$.)

13. $\mathbf{c}'(t)\times\mathbf{c}''(t)=\kappa(t)\left(\dfrac{ds}{dt}\right)^3\mathbf{B}(t)$

14. $\tau(t)=\dfrac{(\mathbf{c}'(t)\times\mathbf{c}''(t))\cdot\mathbf{c}'''(t)}{\|\mathbf{c}'(t)\times\mathbf{c}''(t)\|^2}$

15. Using Exercise 14, find the torsion of the helix $\mathbf{c}(t)=a\cos t\mathbf{i}+a\sin t\mathbf{j}+bt\mathbf{k}$, $t\geq 0$, $a, b>0$.

16. Show that the curve $\mathbf{c}(t)=(1+1/t, -t+1/t, t)$, $t>0$, lies in a plane.

17. Find the curvature and torsion of $\mathbf{c}(t)=at^2\mathbf{i}+2at\mathbf{j}$, where $a>0$ and $t\in\mathbb{R}$. At what point(s) on $\mathbf{c}(t)$ does the curvature attain its maximum?

18. Find the Frenet frame for the curve $\mathbf{c}(t)=(3t-t^3)\mathbf{i}+3t^2\mathbf{j}+(3t+t^3)\mathbf{k}$ at $\mathbf{c}(0)$.

19. Compute the curvature and torsion of the curve $\mathbf{c}(t)=(t+\cos t, t-\sin t, t)$ in $\mathbb{R}^3$.

▶ CHAPTER REVIEW

CHAPTER SUMMARY

- **Curves.** Parametrization of a curve, curves on surfaces, curves as intersections of surfaces, implicitly defined curves, Implicit Function Theorem.
- **Curves and applications.** Cycloid, spiral, Bézier curve, Lissajous curve, catenary curve, helix, centripetal acceleration, Kepler's Laws, Coriolis acceleration.
- **Path (parametrization).** C^1 path, smooth path, length of a path, arc-length function, parametrization by arc-length, acceleration, normal and tangential components of acceleration.
- **Geometry of curves.** Length of a curve, curvature, osculating plane, torsion, Frenet (TNB) frame.

REVIEW

Discuss the following questions.

1. Define a path and a curve and describe the difference. What is a continuous path? Differentiable path? C^1 path? Give an example of a path that is continuous, but not differentiable.

2. Write down a parametrization of the line $y = 0$ that is not differentiable. Write down a parametric representation of the line $y = 0$ that is differentiable, but not smooth.

3. State the Implicit Function Theorem. Identify all points on the graph of the equation $x^3 + x^2y - 3xy + y = 0$ where the theorem does not apply.

4. Explain how to obtain an equation of the line tangent to the graph of $F(x, y) = 0$ (where F is continuously differentiable) at some point (x_0, y_0) on it. Identify case(s) when the tangent does not exist.

5. Describe the relationship between the position, velocity, and acceleration vectors of motion along a circle with constant speed. Write down a parametrization of the circle $x^2 + y^2 = 1$ with nonconstant speed. Is the relation between the position, velocity, and acceleration vectors the same as in the constant-speed case?

6. Describe the geometric meaning of the derivative DF of a function $\mathbf{F}\colon \mathbb{R}^2 \to \mathbb{R}^2$. If two curves $\mathbf{c}_1(t)$ and $\mathbf{c}_2(t)$ intersect at the angle θ, is it true that their images under $\mathbf{F}$ intersect at the same angle? (If false, provide a counterexample.)

7. Let $\mathbf{c}(t)$ be a continuous path. Define the arc-length parameter s and consider the parametrization $\mathbf{c}(s)$ of $\mathbf{c}$ by its arc-length parameter. Explain the geometric/physical meaning of $ds/dt = \|\mathbf{c}'(t)\|$. Using the chain rule, show that $\|\mathbf{c}'(s)\| = 1$ and give a physical interpretation.

8. Define the unit tangent vector, the principal unit normal vector, and the binormal vector. Define the curvature and torsion of a curve. Explain the relationship between curvature and acceleration. Why do we use the arc-length parameter instead of a general parameter t in *defining* the curvature and torsion?

TRUE/FALSE QUIZ

Determine whether the following statements are true or false. Give reasons for your answer.

1. If $y = g(x)$ is a local solution of $F(x, y) = 0$, then $y' = (\partial F/\partial y)/(\partial F/\partial x)$.

2. The equation $x^3y^3 - x - 3y^2 + 12 = 0$ has a local solution of the form $y = g(x)$ near the point $(0, 2)$.

3. The curvature of the helix $\mathbf{c}(t) = (2\cos t, 2\sin t, 12t)$, $t \in \mathbb{R}$, is constant.

4. The curvature of a circle is proportional to its radius.

5. If $\|\mathbf{c}(t)\|$ is constant, then $\mathbf{c}(t)$ is perpendicular to $\mathbf{c}'(t)$.

6. If a curve lies in the xy-plane, then its torsion is zero.

7. The length of the path $\mathbf{c}(t) = (\cos 4t, \sin 4t)$, $t \in [-\pi, \pi]$, is 4π.

8. The intersection of the planes $x = 2$ and $y = 3$ is the line $\mathbf{l}(t) = (2, 3, t)$, $t \in \mathbb{R}$.

REVIEW EXERCISES AND PROBLEMS

Exercises 1 and 2: An object moves according to $\mathbf{c}(t) = (x(t), y(t))$, where the components $x(t)$ and $y(t)$ are shown in the graphs below. In each case, state the initial and terminal points of $\mathbf{c}(t)$ and describe the motion.

1.

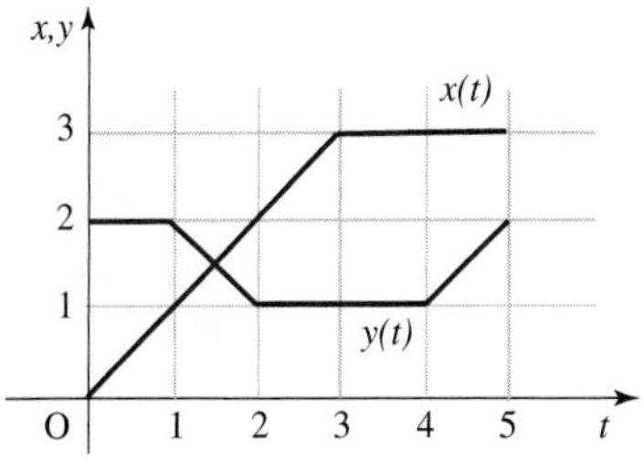

2.

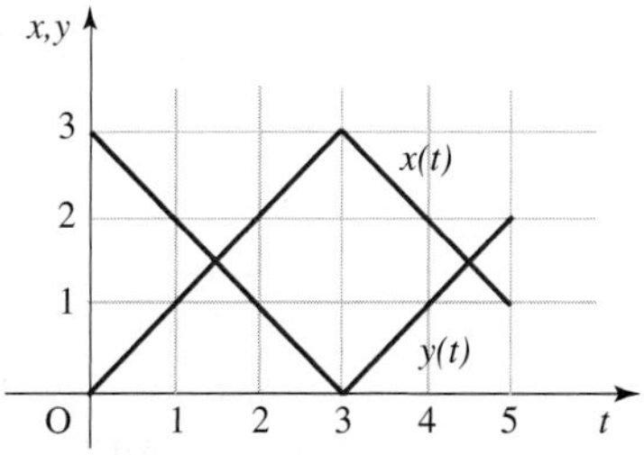

3. Find all points of intersection of the line $\mathbf{l}(t) = (3 - t, 2 + 2t, 4)$ and the cone $x^2 + y^2 = z^2$.

4. Find parametric representation of the curve that is the intersection of the spheres $x^2 + y^2 + z^2 = 1$ and $(x - 1)^2 + y^2 + z^2 = 1$.

5. Find a vector $\mathbf{W}$ such that the Serret–Frenet formulas can be written as $d\mathbf{T}/ds = \mathbf{W} \times \mathbf{T}$, $d\mathbf{N}/ds = \mathbf{W} \times \mathbf{N}$, and $d\mathbf{B}/ds = \mathbf{W} \times \mathbf{B}$.

6. Sometimes, the parameter t can be eliminated from the parametric equation $\mathbf{c}(t) = (x(t), y(t))$ and an equation of the form $f(x, y) = 0$ is obtained [e.g., $\mathbf{c}(t) = (\cos t, \sin t)$ gives $x^2 + y^2 = 1$ after the elimination of t]. Express dy/dx and d^2y/dx^2 in terms of the derivatives of x and y with respect to t.

7. Let $\mathbf{c}$ be a plane curve and let θ be the angle between the tangent to $\mathbf{c}$ at $\mathbf{c}(t)$ and the positive x-axis. Show that the curvature of $\mathbf{c}$ is equal to $|d\theta/ds|$, where s is the arc-length parameter of $\mathbf{c}$.

8. Assume that a curve $\mathbf{c}(s)$ is parametrized by its arc-length.
(a) Show that $\mathbf{c}'''(s) = -\kappa^2(s)\mathbf{T}(s) + \kappa'(s)\mathbf{N}(s) + \tau(s)\kappa(s)\mathbf{B}(s)$.
(b) Prove that the scalar triple product $\mathbf{c}'(s) \cdot (\mathbf{c}''(s) \times \mathbf{c}'''(s))$ is equal to $\kappa^2(s)\tau(s)$.

9. Show that if a particle is acted on by the radial force $\mathbf{F} = (f(r)\cos\theta, f(r)\sin\theta)$, where $r = r(t)$ and $\theta = \theta(t)$, then $r^2(d\theta/dt)$ is constant in time.

10. A particle moves along the trajectory $\mathbf{r}(t) = (\sin 3t, \cos 3t, 2t^{3/2})$, where $0 \le t \le \pi$.
(a) Show that $\mathbf{r}(t)$ lies on the surface of a cylinder of radius 1.
(b) Sketch the curve $\mathbf{r}(t)$.
(c) Find the velocity and speed of the particle.
(d) Find the distance traveled by the particle from $t = 0$ to $t = \pi$.
(e) Suppose that the particle flies off the given trajectory at $t = \pi$ and continues its motion, subject only to a constant gravitational force directed along the (negative) z-axis. Find its position π units of time later.

11. A particle moves in $\mathbb{R}^3$ along the path $\mathbf{c}(t) = (e^t \cos t, e^t \sin t, e^t)$. Compute its velocity. Does the particle travel with constant speed? Is the magnitude of its acceleration constant?

12. The center C of the small circle in Figure 3.37 travels counterclockwise around the circle of radius R, completing one full revolution in time T. The smaller circle (of radius r) rotates counterclockwise about its center, completing one full revolution in time $T/4$. Find a parametric representation of the curve traced out by the point P on the smaller circle. Assume that the initial point of P is on the x-axis, as shown.

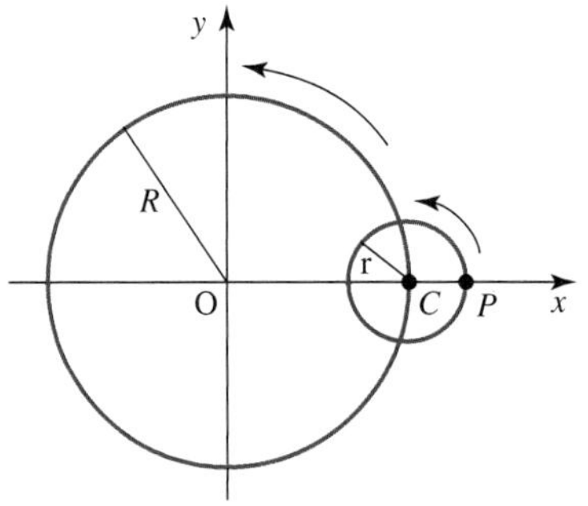

Figure 3.37 The motion described in Exercise 12.

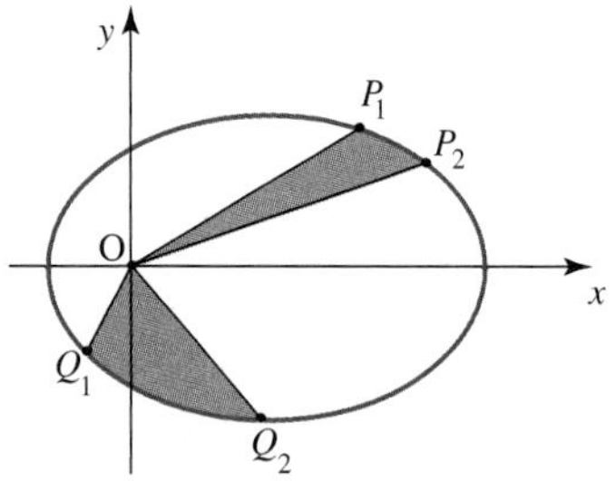

Figure 3.38 Area swept by the position vector of the planet, Exercise 13.

13. (Kepler's Second Law) Recall that Kepler's First Law states that the orbit of a planet rotating around the Sun is an ellipse. Using the setup and notation of Example 3.19 in Section 3.2, we now explore this rotation in more depth.

Assume that the planet takes the same amount of time to move along its elliptic orbit from P_1 to P_2 and from Q_1 to Q_2 (see Figure 3.38). We will prove that the shaded regions, that is, the areas of OP_1P_2 and OQ_1Q_2 are equal. In words, the areas swept by the position vector $\mathbf{r}(t) = (\|\mathbf{r}(t)\| \cos\theta(t), \|\mathbf{r}(t)\| \sin\theta(t))$ of the planet in equal times are equal (this is Kepler's Second Law).

(a) Denote by $A(t)$ the area swept by $\mathbf{r}(t)$ from time t_0 to time t. Show that $A'(t) = (\|\mathbf{r}(t)\|^2/2)\theta'(t)$.

(b) Recall that $\mathbf{d} = \mathbf{r} \times \mathbf{v}$ is a constant vector. Show that $\mathbf{d} = \|\mathbf{r}(t)\|^2\theta'(t)\mathbf{k}$. Computing the dot product of both sides by $\mathbf{k}$, prove that $\|\mathbf{d}\| = \|\mathbf{r}(t)\|^2\theta'(t)$.

(c) Combine (a) and (b) to obtain $A'(t) = \|\mathbf{d}\|/2$. Explain why this statement completes the proof of Kepler's Second Law.

14. (Kepler's Third Law) Let T be the time it takes the planet to complete one revolution around the Sun (T is called the period of the planet).

(a) Integrate the formula for $A'(t)$ from (c) in Exercise 13 to prove that $T = 2ab\pi/\|\mathbf{d}\|$. (Recall that the area of the ellipse with semiaxes a and b is $ab\pi$.)

(b) Using notation and part (b) of Exercise 30 in Section 3.2, show that $b^2/a = A = \|\mathbf{d}\|^2/GM$.

(c) Combining (a) and (b), prove that $T^2 = Ca^3$, where $C = 4\pi^2/GM$ is independent of the planet. The formula $T^2 = Ca^3$ represents Kepler's Third Law. State it in words.

15. Knowing that Earth completes one full revolution around the Sun in 365.25 days, use Kepler's Third Law to compute the largest distance between Earth and the Sun. The mass of the Sun is approximately $2 \cdot 10^{30}$kg, and the gravitational constant is $G = 6.67 \cdot 10^{-11}$ Nm^{2}kg^{-2}.

16. Consider the ellipse $x^2 + y^2/4 = 1$ in the xy-plane.

(a) Compute its curvature and torsion.

(b) Plot the curvature function. Where is it largest? smallest?

(c) Find the equation of the osculating circle at $(1, 0)$.

17. (Brachistochrone Problem) The *Brachistochrone* problem can be stated as follows: given two points, A and a lower point B (that is not directly below A), find the curve joining A and B along which a particle will slide in the shortest time, under the influence of gravity only. We will not attempt to solve the problem, but will rather compute times along different curves and compare them.

Given a parametrization $\mathbf{c}(t)$, $t \in [a, b]$ of a curve [such that $\mathbf{c}(a) = A$ and $\mathbf{c}(b) = B$], the time needed for the particle to slide from A to B is given by

$$T = \frac{1}{\sqrt{2g}} \int_a^b \sqrt{\frac{(x'(t))^2 + (y'(t))^2}{y}}\, dt.$$

Use a coordinate system in which the y-axis points downward, and let $A(0, 0)$ and $B(1, 1)$. Consider the following paths.

(a) $\mathbf{c}_1(t) = (t, t)$, $t \in [0, 1]$

(b) $\mathbf{c}_2(t) = ((t - \sin t)/\pi, (1 - \cos t)/2)$, $t \in [0, \pi]$

(c) $\mathbf{c}_3(t) = (\sin t + 1, \cos t)$, $t \in [3\pi/2, 2\pi]$

(d) $\mathbf{c}_4(t) = (t, \sin(\pi t/2))$, $t \in [0, 1]$

Check that all parametrizations give curves with initial point $A(0, 0)$ and terminal point $B(1, 1)$. Compute the lengths of the paths in (a)–(d). Plot all curves and compute T for each of them. Which one is the "fastest"?

Proof of the formula for T that we use in this exercise: we start with the Conservation of Energy principle $mv^2/2 = mgy$ (where m is the mass of the object moving at speed v, y is its height, and g is the acceleration of the gravity). Thus, $v = \sqrt{2gy}$, and since $v = ds/dt$, we get

$$T = \int_a^b dt = \int_a^b \frac{ds}{\sqrt{2gy}} = \frac{1}{\sqrt{2g}} \int_a^b \frac{\|\mathbf{c}'(t)\|}{\sqrt{y}}\, dt = \frac{1}{\sqrt{2g}} \int_a^b \sqrt{\frac{(x'(t))^2 + (y'(t))^2}{y}}\, dt.$$

18. (Tautochrone Problem) Consider the cycloid $\mathbf{c}(\theta) = (\theta - \sin\theta, 1 - \cos\theta)$, $0 \le \theta \le 2\pi$, and imagine it drawn in a coordinate system where the y-axis points downward (i.e., in the direction of the gravity). This way, the cycloid is inverted: its highest points are $\mathbf{c}(0) = (0, 0)$ and $\mathbf{c}(2\pi) = (2\pi, 0)$, and its lowest point is $\mathbf{c}(\pi) = (\pi, -2)$.

(a) Using the formula for T from Exercise 17, show that the time a ball needs to roll from the top of the cycloid to its bottom is $T = \pi/\sqrt{g}$.

(b) Now pick any other location $\mathbf{c}(\theta_0)$ (other than its lowest point, $\theta_0 = \pi$) on the cycloid. Using the formula for T from Exercise 17 [with y in the denominator replaced by $y - y(\theta_0)$], show that

$$T_{\theta_0} = \frac{1}{\sqrt{g}} \int_{\theta_0}^{\pi} \sqrt{\frac{1 - \cos\theta}{\cos\theta_0 - \cos\theta}}\, d\theta,$$

where T_{θ_0} is the time needed for a ball to roll from $c(\theta_0)$ to the lowest point $c(\pi)$.

(c) Using half-angle formulas, show that the integrand in (b) can be reduced to

$$\frac{\sin(\theta/2)}{(\cos^2(\theta_0/2) - \cos^2(\theta/2))^{1/2}}.$$

Use the substitution $u = \cos(\theta/2)/\cos(\theta_0/2)$ to integrate, and show that $T_{\theta_0} = \pi/\sqrt{g}$. Compare the result with (a).

CHAPTER 4

Scalar and Vector Fields

We open this chapter by discussing higher-order derivatives, for several reasons. In Section 4.2, we will construct a Taylor polynomial of a function of several variables. Certain properties of differential operators (divergence, gradient, and curl) that we study later in this chapter involve second-order partial derivatives. Moreover, second-order derivatives appear in numerous physical applications (some are discussed in the first section of this chapter, such as the wave equation or the heat equation).

Using Taylor's Theorem, we will be able to obtain better numeric approximations than those provided by the linear approximation. In particular, our study of quadratic approximations will help us analyze the extreme values of functions of several variables. The two problems we study are finding relative (or local) extreme values of a function on its domain and finding absolute (or global) extreme values of a function on a closed and bounded set contained in its domain. Next, we study the constrained optimization problem: finding extreme values of a differentiable function subject to certain conditions (or constraints) on its variables.

A useful way of visualizing a vector field (borrowed from fluid mechanics and electromagnetism) consists of drawing flow lines. If we think of a vector field as a velocity field of a fluid, then a flow line describes how a fluid particle moves under the influence of the field. In Section 4.6, we introduce and study two operations involving partial derivatives, divergence and curl, and use them to investigate the rate of change of a vector field. Next, we state (without proof) a general version of the Implicit Function Theorem. We have seen the importance of its special case in Section 3.1, where we studied curves defined by the equation $F(x, y) = 0$, for a C^1 function $F : \mathbb{R}^2 \to \mathbb{R}$. We will appreciate the power of this theorem when performing calculations involving change of variables in multiple integration in Chapter 6 and in studying surfaces in Chapter 7.

We close the chapter with an appendix that contains useful identities involving differential operators on functions and on vector fields.

► 4.1 HIGHER-ORDER PARTIAL DERIVATIVES

We study higher-order partial derivatives for several reasons. In Section 4.2, we construct quadratic approximation that will help us investigate extreme values of a function of several variables. Some important properties of differential operators (divergence, gradient, and curl) that we study later in this chapter involve second-order partial derivatives. Moreover, second-order derivatives appear in numerous physical applications (some are discussed in this section, such as the wave equation and the heat equation).

A real-valued function of m variables $x_1, \ldots, x_m$ is said to be of class C^1 if all of its partial derivatives $\partial f/\partial x_i$ are continuous. If, in turn, all $\partial f/\partial x_i$ have continuous partial derivatives with respect to all variables x_j, then f is called *twice continuously differentiable* or *of class* C^2. The "box" containing functions of class C^2 would be inside the smallest "box" (i.e., contained in the C^1 "box") in Figure 2.51. Similarly, we define a function of class C^k for any positive integer k.

To introduce notation for higher-order derivatives, we assume that f is a function of three variables x, y, and z. Recall that the first-order partial derivatives are denoted either by $\partial f/\partial x$, $\partial f/\partial y$, etc., or by f_x, f_y, etc., or by $D_1 f$, $D_2 f$, etc. Likewise,

$$\frac{\partial^2 f}{\partial x^2} = \frac{\partial}{\partial x}\left(\frac{\partial f}{\partial x}\right) \quad \text{or} \quad f_{xx} = (f_x)_x \quad \text{or} \quad D_{11} f = D_1(D_1 f)$$

$$\frac{\partial^2 f}{\partial z\,\partial y} = \frac{\partial}{\partial z}\left(\frac{\partial f}{\partial y}\right) \quad \text{or} \quad f_{yz} = (f_y)_z \quad \text{or} \quad D_{23} f = D_3(D_2 f)$$

$$\frac{\partial^2 f}{\partial y\,\partial z} = \frac{\partial}{\partial y}\left(\frac{\partial f}{\partial z}\right) \quad \text{or} \quad f_{zy} = (f_z)_y \quad \text{or} \quad D_{32} f = D_2(D_3 f),$$

etc. are *second-order partial derivatives* (also called *iterated partial derivatives*) of f. We can also form third-order, fourth-order, etc., partial derivatives. A function of two variables has four, and a function of three variables has nine second-order partial derivatives. However, as we will soon witness, not all of them are distinct if the function is of class C^2.

▶ EXAMPLE 4.1

Find all second-order partial derivatives of $f(x, y) = x^3 y + 3y^2 - x$.

SOLUTION

Differentiating $f_x = 3x^2 y - 1$ with respect to x and y, we get $f_{xx} = 6xy$ and $f_{xy} = 3x^2$. Similarly, from $f_y = x^3 + 6y$, we get $f_{yx} = 3x^2$ and $f_{yy} = 6$. ◀

▶ EXAMPLE 4.2

Compute all second-order partial derivatives of $f(x, y) = x^2 e^{2y}$.

SOLUTION

From $\partial f/\partial x = 2xe^{2y}$ and $\partial f/\partial y = 2x^2 e^{2y}$, we compute

$$\frac{\partial^2 f}{\partial x^2} = \frac{\partial}{\partial x}(2xe^{2y}) = 2e^{2y}, \qquad \frac{\partial^2 f}{\partial x\,\partial y} = \frac{\partial}{\partial x}(2x^2 e^{2y}) = 4xe^{2y},$$

$$\frac{\partial^2 f}{\partial y\,\partial x} = \frac{\partial}{\partial y}(2xe^{2y}) = 4xe^{2y}, \qquad \text{and} \qquad \frac{\partial^2 f}{\partial y^2} = \frac{\partial}{\partial y}(2x^2 e^{2y}) = 4x^2 e^{2y}.$$ ◀

Note that in the previous two examples, $f_{xy} = f_{yx}$ (or $\partial^2 f/\partial x\,\partial y = \partial^2 f/\partial y\,\partial x$). This is not a coincidence, but a consequence of the following theorem.

THEOREM 4.1 Equality of Mixed Partial Derivatives

Let f be a real-valued function of m variables $x_1, \ldots, x_m$ with continuous second-order partial derivatives (i.e., of class C^2). Then

$$\frac{\partial^2 f}{\partial x_i \partial x_j} = \frac{\partial^2 f}{\partial x_j \partial x_i},$$

for all $i, j = 1, \ldots, m$.

This theorem states that the order of computing derivatives of a function is irrelevant, as long as the function is "smooth enough" (i.e., of class C^2). This fact will be used in a number of situations, one of which is the interchanging of the gradient and the time derivative of a function. More precisely, consider a function $f(x, y, z, t)$ of four variables (where the variable t denotes time, and x, y, and z are the coordinates in the space $\mathbb{R}^3$). The gradient of f, interpreted as a vector, is written as

$$\nabla f = \frac{\partial f}{\partial x}\mathbf{i} + \frac{\partial f}{\partial y}\mathbf{j} + \frac{\partial f}{\partial z}\mathbf{k}$$

(partial derivatives are taken only with respect to "space" variables x, y, and z and not with respect to t). Then

$$\begin{aligned}
\nabla\left(\frac{\partial f}{\partial t}\right) &= \frac{\partial}{\partial x}\left(\frac{\partial f}{\partial t}\right)\mathbf{i} + \frac{\partial}{\partial y}\left(\frac{\partial f}{\partial t}\right)\mathbf{j} + \frac{\partial}{\partial z}\left(\frac{\partial f}{\partial t}\right)\mathbf{k} \\
&= \frac{\partial^2 f}{\partial x \partial t}\mathbf{i} + \frac{\partial^2 f}{\partial y \partial t}\mathbf{j} + \frac{\partial^2 f}{\partial z \partial t}\mathbf{k} = \frac{\partial^2 f}{\partial t \partial x}\mathbf{i} + \frac{\partial^2 f}{\partial t \partial y}\mathbf{j} + \frac{\partial^2 f}{\partial t \partial z}\mathbf{k} \\
&= \frac{\partial}{\partial t}\left(\frac{\partial f}{\partial x}\right)\mathbf{i} + \frac{\partial}{\partial t}\left(\frac{\partial f}{\partial y}\right)\mathbf{j} + \frac{\partial}{\partial t}\left(\frac{\partial f}{\partial z}\right)\mathbf{k} \\
&= \frac{\partial}{\partial t}\left(\frac{\partial f}{\partial x}\mathbf{i} + \frac{\partial f}{\partial y}\mathbf{j} + \frac{\partial f}{\partial z}\mathbf{k}\right) = \frac{\partial}{\partial t}(\nabla f).
\end{aligned}$$

The proof of Theorem 4.1 is presented in Appendix A.

▶ EXAMPLE 4.3

Compute f_{xzzy} if $f(x, y, z) = x^2 e^{y^2 - x^2} + x^2 y z^2 - \cos(x^2 + y^2)$.

SOLUTION We are asked to compute the fourth-order partial derivative of f with respect to x, z, z, and y. Since f consists of sums, products, and compositions of functions (polynomials, exponential function, cosine) that have as many derivatives as needed, it is certainly of class C^2 (as a matter of fact, it is of class C^k for any integer $k \geq 1$). Moreover, its partial derivatives are also of class C^2. Thus, by Theorem 4.1 we can compute the required partial derivatives in any order. Since the first and third terms of f do not depend on z, we differentiate with respect to z first: $f_z = 2x^2 yz$. Now $f_{zz} = 2x^2 y$, $f_{zzy} = 2x^2$, and, finally, $f_{zzyx} = 4x$. Thus, $f_{xzzy} = 4x$.

▶ EXAMPLE 4.4

In this example, we illustrate the use of the chain rule.

(a) Let $u(x, y) = e^x \sin y$, where $x = s^2$ and $y = st$. Compute $\partial u/\partial t$ and $\partial u/\partial s$.

(b) Let $u(x, y)$ be a twice continuously differentiable real-valued function and let $x = s^2$ and $y = st$. Compute $\partial u/\partial t$ and $\partial u/\partial s$.

(c) Compute all second partial derivatives of the function $u(x, y)$ defined in (b).

SOLUTION **(a)** One way to solve this problem is to immediately substitute expressions for x and y into $u(x, y)$ [thus getting $u(s, t) = e^{s^2} \sin(st)$] and then compute the derivatives. We choose another approach. Using the chain rule, we obtain

$$\frac{\partial u}{\partial t} = \frac{\partial u}{\partial x}\frac{\partial x}{\partial t} + \frac{\partial u}{\partial y}\frac{\partial y}{\partial t} = e^x \sin y \cdot 0 + e^x \cos y \cdot s = se^{s^2} \cos(st),$$

and

$$\frac{\partial u}{\partial s} = \frac{\partial u}{\partial x}\frac{\partial x}{\partial s} + \frac{\partial u}{\partial y}\frac{\partial y}{\partial s} = e^x \sin y \cdot 2s + e^x \cos y \cdot t = 2se^{s^2} \sin(st) + te^{s^2} \cos(st).$$

(b) As in (a), the chain rule gives

$$\frac{\partial u}{\partial t} = \frac{\partial u}{\partial x}\frac{\partial x}{\partial t} + \frac{\partial u}{\partial y}\frac{\partial y}{\partial t} = \frac{\partial u}{\partial x} \cdot 0 + \frac{\partial u}{\partial y} \cdot s = s\frac{\partial u}{\partial y},$$

and

$$\frac{\partial u}{\partial s} = \frac{\partial u}{\partial x}\frac{\partial x}{\partial s} + \frac{\partial u}{\partial y}\frac{\partial y}{\partial s} = \frac{\partial u}{\partial x} \cdot 2s + \frac{\partial u}{\partial y} \cdot t = 2s\frac{\partial u}{\partial x} + t\frac{\partial u}{\partial y}.$$

(c) Applying the product and chain rules, we get, from (b),

$$\begin{aligned}\frac{\partial^2 u}{\partial t^2} &= \frac{\partial}{\partial t}\left(\frac{\partial u}{\partial t}\right) = \frac{\partial}{\partial t}\left(s\frac{\partial u}{\partial y}\right) = \frac{\partial}{\partial t}(s)\frac{\partial u}{\partial y} + s\frac{\partial}{\partial t}\left(\frac{\partial u}{\partial y}\right)\\ &= 0 + s\left(\frac{\partial}{\partial x}\left(\frac{\partial u}{\partial y}\right)\frac{\partial x}{\partial t} + \frac{\partial}{\partial y}\left(\frac{\partial u}{\partial y}\right)\frac{\partial y}{\partial t}\right)\\ &= s\left(\frac{\partial^2 u}{\partial x \partial y} \cdot 0 + \frac{\partial^2 u}{\partial y^2} \cdot s\right) = s^2\frac{\partial^2 u}{\partial y^2},\end{aligned}$$

and

$$\begin{aligned}\frac{\partial^2 u}{\partial s \partial t} &= \frac{\partial}{\partial s}\left(\frac{\partial u}{\partial t}\right) = \frac{\partial}{\partial s}\left(s\frac{\partial u}{\partial y}\right) = \frac{\partial}{\partial s}(s)\frac{\partial u}{\partial y} + s\frac{\partial}{\partial s}\left(\frac{\partial u}{\partial y}\right)\\ &= 1 \cdot \frac{\partial u}{\partial y} + s\left(\frac{\partial}{\partial x}\left(\frac{\partial u}{\partial y}\right) \cdot 2s + \frac{\partial}{\partial y}\left(\frac{\partial u}{\partial y}\right) \cdot t\right)\\ &= \frac{\partial u}{\partial y} + 2s^2\frac{\partial^2 u}{\partial x \partial y} + st\frac{\partial^2 u}{\partial y^2}.\end{aligned}$$

Continuing in the same way (skipping details), we obtain

$$\frac{\partial^2 u}{\partial s^2} = 2\frac{\partial u}{\partial x} + 4s^2\frac{\partial^2 u}{\partial x^2} + 4st\frac{\partial^2 u}{\partial x \partial y} + t^2\frac{\partial^2 u}{\partial y^2}.$$

Since u is of class C^2, it follows that $\partial^2 u/\partial s \partial t = \partial^2 u/\partial t \partial s$. ◀

▶ EXAMPLE 4.5

Let $u(x, y)$ be a twice continuously differentiable function and let $x = e^s$, $y = e^t$. Show that

$$x^2 u_{xx} + y^2 u_{yy} + x u_x + y u_y = u_{ss} + u_{tt}.$$

SOLUTION By the chain rule, $u_s = u_x e^s + u_y \cdot 0 = u_x e^s$, and thus,

$$\begin{aligned} u_{ss} &= (u_x)_s\, e^s + u_x\, (e^s)_s = \left((u_x)_x\, e^s + (u_x)_y\, 0\right) e^s + u_x e^s \\ &= u_{xx} e^{2s} + u_x e^s = x^2 u_{xx} + x u_x. \end{aligned}$$

Replacing x by y and s by t (we can do this, due to symmetry) we get

$$u_{tt} = u_{yy} e^{2t} + u_y e^t = y^2 u_{yy} + y u_y.$$

The proof is now completed by adding the expressions for u_{ss} and u_{tt}. So, a complicated differential equation can be made simpler by using new, suitably defined variables s and t. ◀

Higher-order partial derivatives appear in a number of important applications. We will examine some of them in the examples that follow.

▶ EXAMPLE 4.6 One-Dimensional Wave Equation

Consider the motion of a vibrating string, such as a guitar string, and let the function $u(x, t)$ represent the vertical displacement at a location x, at time t; see Figure 4.1. It can be shown that $u(x, t)$ satisfies the *wave equation*

$$u_{tt} = c^2 u_{xx}, \tag{4.1}$$

where c is a constant.

Intuitively, this makes sense. The function $u(x, t)$, for fixed x, represents the vertical motion of the point in the string, and the partial derivative $u_{tt}(x, t)$ represents its acceleration at time t. The term $u_{xx}(x, t)$ determines the concavity of the string (at some fixed time t); the larger the concavity, the stronger the force that is trying to bring the string back into its equilibrium. Thus, (4.1) establishes the relation between the acceleration and concavity (i.e., the force; so, this is Newton's Second Law of Motion).

The wave equation (4.1) is used in the study of the propagation of sound (in various media), propagation of electromagnetic waves and electromagnetic radiation, vibrations of strings and drums, elasticity, etc. The equation appearing in this example is a special case of a general wave equation and is called the *one-dimensional* wave equation.

(a) Show that $u(x, t) = \sin x \sin(2t)$ satisfies the wave equation $u_{tt} - 4u_{xx} = 0$.

(b) Show that the wave equation $u_{tt} = c^2 u_{xx}$ transforms under the change of variables $v = x + ct$, $z = x - ct$ into the equation $u_{vz} = 0$.

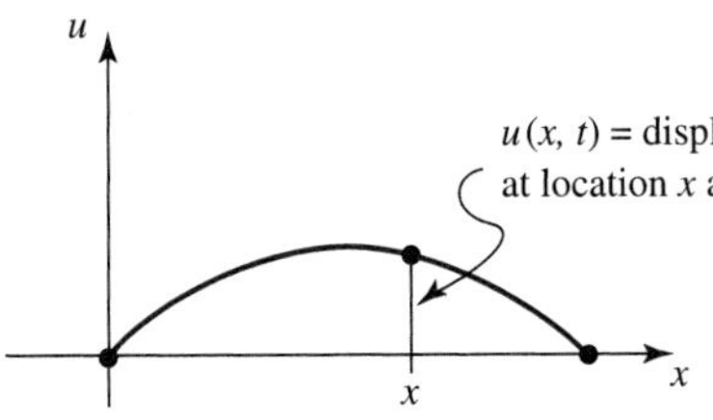

Figure 4.1 A snapshot of a vibrating string taken at a particular time t.

(c) Let ϕ and ψ be differentiable functions of one variable. Show that for any constant c the function $u = \phi(x - ct) + \psi(x + ct)$ is a solution of (4.1).

SOLUTION

(a) A straightforward computation gives $u_t = -2\sin x \cos(2t)$, $u_{tt} = -4\sin x \sin(2t)$, and $u_x = \cos x \sin(2t)$, $u_{xx} = -\sin x \sin(2t)$. Therefore, $u_{tt} - 4u_{xx} = 0$.

(b) Using subscripts to denote partial derivatives, we get, by the chain rule,

$$u_t = u_v v_t + u_z z_t = u_v c + u_z(-c) = cu_v - cu_z$$

and

$$\begin{aligned} u_{tt} &= c((u_v)_v v_t + (u_v)_z z_t) - c((u_z)_v v_t + (u_z)_z z_t) \\ &= c(u_{vv}c + u_{vz}(-c)) - c(u_{zv}c + u_{zz}(-c)) = c^2 u_{vv} + c^2 u_{zz} - 2c^2 u_{vz}. \end{aligned}$$

We used the equality of mixed partials; see Theorem 4.1. Furthermore,

$$u_x = u_v v_x + u_z z_x = u_v + u_z,$$

and thus,

$$u_{xx} = (u_v)_v v_x + (u_v)_z z_x + (u_z)_v v_x + (u_z)_z z_x = u_{vv} + 2u_{vz} + u_{zz}.$$

Substituting the expressions for u_{tt} and u_{xx} into (4.1), we get

$$0 = u_{tt} - c^2 u_{xx} = c^2 u_{vv} + c^2 u_{zz} - 2c^2 u_{vz} - c^2 (u_{vv} + 2u_{vz} + u_{zz}) = -4c^2 u_{vz},$$

that is, $u_{vz} = 0$.

(c) Abbreviating $z = x - ct$ and $v = x + ct$, we rewrite the function u as $u = \phi(z) + \psi(v)$. The functions ϕ and ψ are functions of one variable, so "prime" notation is used to denote their derivatives: ϕ' is the derivative of ϕ with respect to its variable z, and ψ' is the derivative of ψ with respect to v. Hence,

$$u_x = \phi' \cdot 1 + \psi' \cdot 1 = \phi' + \psi', \quad u_{xx} = \phi'' \cdot 1 + \psi'' \cdot 1 = \phi'' + \psi''$$

and

$$u_t = \phi'(-c) + \psi'(c) = -c\phi' + c\psi', \quad u_{tt} = -c\phi''(-c) + c\psi''(c) = c^2(\phi'' + \psi'').$$

It follows that $u_{tt} - c^2 u_{xx} = 0$. ◀

▶ EXAMPLE 4.7

We now analyze the vibrating motion $u(x,t) = \sin x \sin(2t)$, $0 \le x \le \pi$, of the string from Example 4.6(a). Figure 4.2 shows the position of the string at times $t = 0$, $t = 0.25$, $t = 0.5$, $t = 1.5$, and $t = 2$. We will evaluate first and second partial derivatives of $u(x,t)$ at $P(x = \pi/4, t = 0.5)$ and give an interpretation of the answers.

Since $u_x = \cos x \sin(2t)$, it follows that $u_x(P) = u_x(\pi/4, 0.5) = 0.595$. The function $u(x, 0.5) = \sin x \sin 1 = 0.841 \sin x$ represents the position of the string at time $t = 0.5$. Therefore, $u_x(P) = 0.595$ is the slope of the string at $x = \pi/4$ at the moment when $t = 0.5$. From $u_{xx} = -\sin x \sin(2t)$, it follows that $u_{xx}(P) = -0.595$. In words, the string is concave down at P.

From $u_t = 2\sin x \cos(2t)$, we compute $u_t(P) = u_t(\pi/4, 0.5) = 0.764$. The function $u(\pi/4, t) = \sqrt{2}\sin(2t)/2$ describes the vertical motion of the (fixed) point P in the wire. Consequently, the derivative $u_t(P) = 0.764$ is the speed of P at time $t = 0.5$. The fact that $u_t(P) > 0$

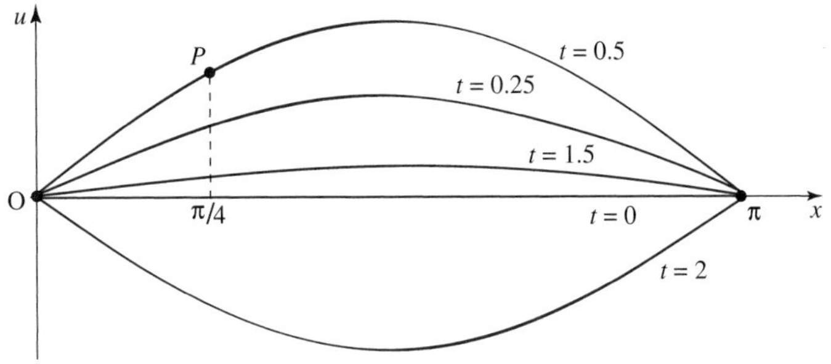

Figure 4.2 Position of the string at different times.

means that P is moving upward at the moment when $t = 0.5$. Using $u_{tt} = -4 \sin x \sin(2t)$, we get $u_{tt}(P) = -2.380$. Thus, the speed of P is decreasing at the moment when $t = 0.5$ (since its acceleration is negative).

Next, $u_{xt} = 2 \cos x \cos(2t)$, and $u_{xt}(P) = 0.764$. The function u_x represents the slope, and so the partial derivative u_{xt} gives the rate of change of the slope with respect to time. Since $u_{xt}(P) > 0$, we conclude that the slope at $x = \pi/4$ is increasing when $t = 0.5$. From $u_{tx} = 2 \cos x \cos(2t)$, we compute $u_{tx}(P) = 0.764$. Since the function u_t measures the velocity of the point P in the string, the derivative u_{tx} describes how velocity changes near $x = \pi/4$ at a fixed time t. The fact that $u_{tx}(P) > 0$ means that the points on the string to the immediate right of P move faster than P. ◀

▶ EXAMPLE 4.8 Heat Equation

Let $T(x, t)$ represent the temperature, at time t, at a point in a thin metal rod that is placed along the x-axis, with one end at the origin. Assume that, at some fixed time t, the temperature $T(x, t)$ is given by the graph in Figure 4.3(a). We will try to figure out whether the temperature at the point P in the rod is increasing or decreasing.

Pick the points x_1 and x_2 near P, one on each side of it (as shown), and consider the segment $[x_1, x_2]$ on the rod between x_1 and x_2. Since the heat flows from warmer regions toward colder regions, the direction of the flow is from right to left; that is, the heat will flow into the segment at x_2 and flow out at x_1.

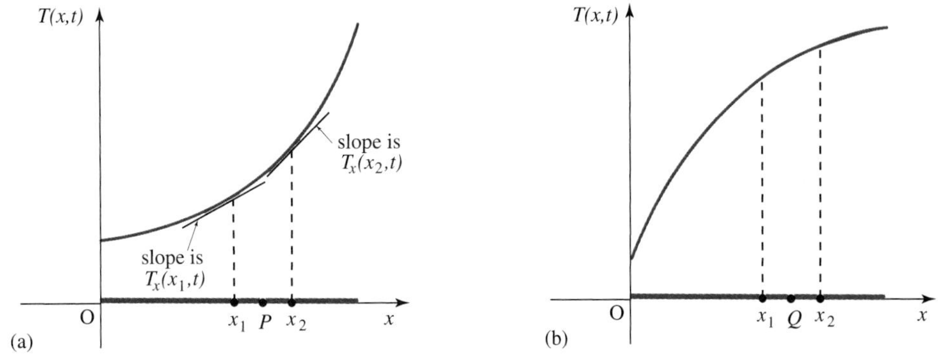

Figure 4.3 Temperature of the rod at a fixed time t.

Note that $T_x(x, t)$ measures how temperature changes with respect to position x at some fixed time t. According to Newton's Law of Cooling, the rate of heat flow is proportional to $T_x(x, t)$ (which seems logical: the larger the change in temperature, the stronger the flow; likewise, smaller changes in temperature will yield weaker flow). Since $T_x(x_2, t) > T_x(x_1, t)$, we conclude that more heat enters the segment $[x_1, x_2]$ than leaves it, that is, the inflow is larger than the outflow. We conclude that the temperature at P is increasing, and thus $T_t(P, t) > 0$. Note that at P, the graph of the function T is concave up, so $T_{xx}(P, t) > 0$. Reasoning in the same way, we conclude that in the case shown in Figure 4.3(b), $T_t(Q, t) < 0$. The graph is concave down, so $T_{xx}(Q, t) < 0$.

It follows that (at least in this case) T_t and T_{xx} have the same sign. It turns out that, in many situations, the two derivatives are actually proportional.

A function $T(x, t)$ is said to satisfy the *heat equation* if

$$\frac{\partial T}{\partial t} = \sigma \frac{\partial^2 T}{\partial x^2}, \tag{4.2}$$

where $\sigma > 0$ is a constant. The heat equation can be used to describe how heat flows in some medium (the constant σ characterizes that medium). Only one space variable is used, and hence (4.2) describes the heat flow in one dimension. An equation of the form

$$\frac{\partial T}{\partial t} = \sigma \left(\frac{\partial^2 T}{\partial x^2} + \frac{\partial^2 T}{\partial y^2} \right)$$

could describe the heat flow in a two-dimensional medium.

Studying heat equations helps us understand properties of the heat flow. As a consequence, by measuring temperature on a surface of a body (say, Earth), we can estimate the temperature in its interior.

The process of *diffusion* (for instance, the way that air pollutants spread in the atmosphere, or sugar dissolves in coffee) is also governed by (4.2) and its higher-dimensional analogues and generalizations (see Exercise 26). ◀

▶ EXAMPLE 4.9

Show that $T(x, t) = a + e^{-ct} \sin(kx)$ (a, c, and k are constants, $c > 0$, and $k \neq 0$) satisfies the heat equation (4.2) with $\sigma = c/k^2$.

SOLUTION

A straightforward computation gives

$$\frac{\partial T}{\partial x} = ke^{-ct} \cos(kx), \quad \frac{\partial^2 T}{\partial x^2} = -k^2 e^{-ct} \sin(kx), \quad \frac{\partial T}{\partial t} = -ce^{-ct} \sin(kx),$$

and therefore,

$$\frac{\partial T}{\partial t} = -ce^{-ct} \sin(kx) = \frac{-c}{k^2} k^2 e^{-ct} \sin(kx) = \frac{c}{k^2} \frac{\partial^2 T}{\partial x^2}.$$

In the last line, we divided and multiplied by k^2 so that we could recognize the expression for $\partial^2 T / \partial x^2$. ◀

▶ EXAMPLE 4.10 Laplace's Equation and Harmonic Functions

A function $f(x, y)$ is said to satisfy *Laplace's equation* if

$$\Delta f(x, y) = 0, \tag{4.3}$$

where Δf is the *Laplace operator* or the *Laplacian* of f, defined by

$$\Delta f = \frac{\partial^2 f}{\partial x^2} + \frac{\partial^2 f}{\partial y^2}.$$

A function that satisfies Laplace's equation is called *harmonic*. A generalized version of Laplace's equation, $\Delta f(x, y) = g(x, y)$, where $g(x, y)$ is some function, is usually called *Poisson's equation*.

Laplace's and Poisson's equations have been studied in relation to a variety of phenomena. In this book, we discuss a few, such as gravitational attraction (in Example 4.65 we show that gravitational potential satisfies Laplace's equation), electric fields (Example 4.53), or diffusion (Example 4.66).

(a) Show that the function $f(x, y) = \ln(x^2 + y^2)$ is harmonic for all $(x, y) \neq (0, 0)$.

(b) Show that $u(x, y) = e^{x^2} \sin(\sqrt{2}y)$ satisfies the equation $\Delta u = 4x^2 u$.

SOLUTION

(a) The partial derivatives are computed using the chain rule and quotient rule:

$$f_x = \frac{2x}{x^2 + y^2},$$

$$f_{xx} = \frac{2(x^2 + y^2) - 4x^2}{(x^2 + y^2)^2} = \frac{2y^2 - 2x^2}{(x^2 + y^2)^2}.$$

The function f is symmetric in x and y (i.e., by interchanging x and y in the formula for f, we get the same function). Consequently, f_{yy} can be obtained from the expression for f_{xx} by interchanging x and y. It follows that $\Delta f = f_{xx} + f_{yy} = 0$, and thus f is harmonic at all points, except at the origin.

(b) To compute the Laplacian, we need second partials:

$$\frac{\partial u}{\partial x} = 2xe^{x^2} \sin(\sqrt{2}y), \qquad \frac{\partial^2 u}{\partial x^2} = 2e^{x^2} \sin(\sqrt{2}y) + 4x^2 e^{x^2} \sin(\sqrt{2}y)$$

and

$$\frac{\partial u}{\partial y} = \sqrt{2}e^{x^2} \cos(\sqrt{2}y), \qquad \frac{\partial^2 u}{\partial y^2} = -2e^{x^2} \sin(\sqrt{2}y).$$

Hence,

$$\Delta u = \frac{\partial^2 u}{\partial x^2} + \frac{\partial^2 u}{\partial y^2} = 4x^2 e^{x^2} \sin(\sqrt{2}y) = 4x^2 u;$$

that is, u satisfies the given Poisson's equation. ◀

▶ EXAMPLE 4.11 Korteweg–de Vries Equation and Soliton Waves

Show that the function $u(x, t) = 2k^2 \cosh^{-2}(k(x - 4k^2 t))$, where $k \geq 0$, is a solution of the *Korteweg–de Vries equation* (sometimes abbreviated as the *KdV equation*)

$$u_t + 6uu_x + u_{xxx} = 0. \tag{4.4}$$

This equation is also called a *soliton equation* and describes, for example, the motion of a certain type of water wave in shallow water. The solution $u(x, t)$ is called a *soliton wave*. No detailed knowledge of hyperbolic functions is needed here; only the derivative formulas $(\cosh x)' = \sinh x$ and $(\sinh x)' = \cosh x$ and the basic identity $\cosh^2 x - \sinh^2 x = 1$ are used.

SOLUTION Denote $k(x - 4k^2 t)$ by A, so that $u(x, t) = 2k^2 \cosh^{-2} A$. We compute

$$\begin{aligned} u_t &= 2k^2(-2)\cosh^{-3}A \cdot \sinh A \cdot k(-4k^2) = 16k^5 \cosh^{-3}A \cdot \sinh A, \\ u_x &= 2k^2(-2)\cosh^{-3}A \cdot \sinh A \cdot k = -4k^3 \cosh^{-3}A \cdot \sinh A, \\ u_{xx} &= -4k^3\left(-3\cosh^{-4}A \cdot \sinh A \cdot k \sinh A + \cosh^{-3}A \cdot \cosh A \cdot k\right), \\ &= -4k^4\left(-3\cosh^{-4}A \cdot \sinh^2 A + \cosh^{-2}A\right). \end{aligned}$$

Thus (for details, see Exercise 21), $u_{xxx} = -4k^5\left(12\cosh^{-5}A \cdot \sinh^3 A - 8\cosh^{-3}A \cdot \sinh A\right)$, and we conclude that $u_t + 6uu_x + u_{xxx} = 0$.

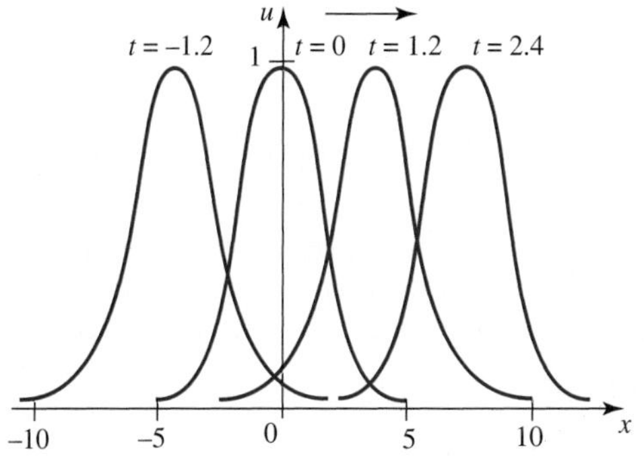

Figure 4.4 Soliton wave $u(x, t) = 2k^2 \cosh^{-2}(k(x - 4k^2 t))$ with $k = 0.7$, at four different times t.

Figure 4.4 shows the soliton wave $u(x, t)$ at four different times, and the graph of the function $u(x, t)$ with all positions of the wave for $-2 \le t \le 2$ is illustrated in Figure 4.5.

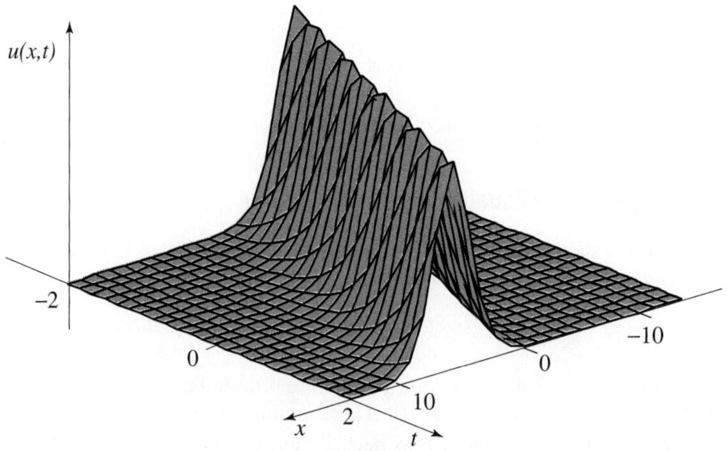

Figure 4.5 Plot of $u(x, t)$ shows how the soliton wave moves for $-2 \le t \le 2$. ◀

EXERCISES 4.1

Exercises 1 to 4: Looking at the level curves of a function $f(x, y)$, determine whether the partial derivatives $f_x(P)$, $f_y(P)$, $f_{xx}(P)$, $f_{xy}(P)$, and $f_{yy}(P)$ are positive, negative, or zero.

1.

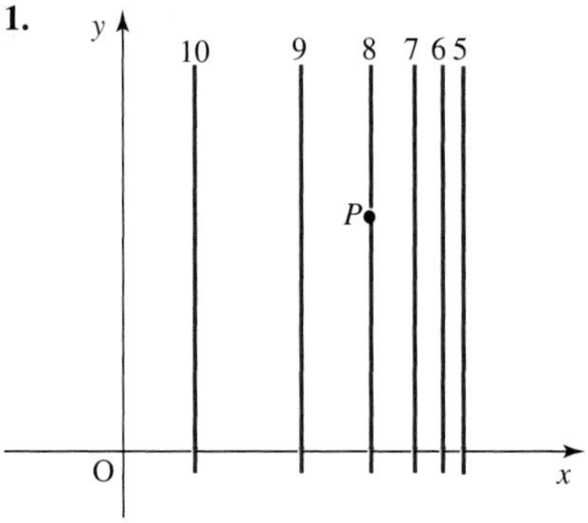

2.

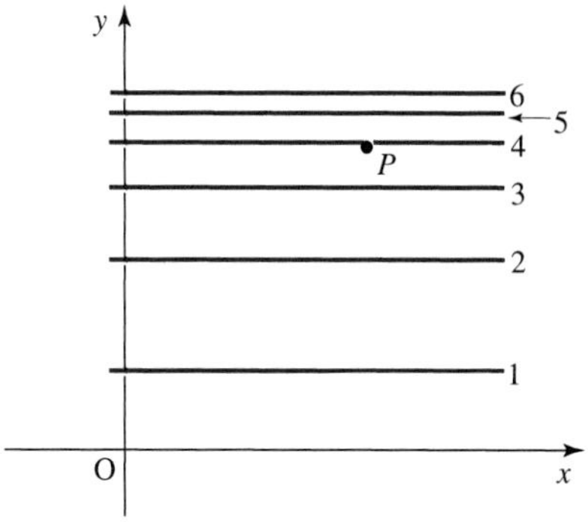

3.

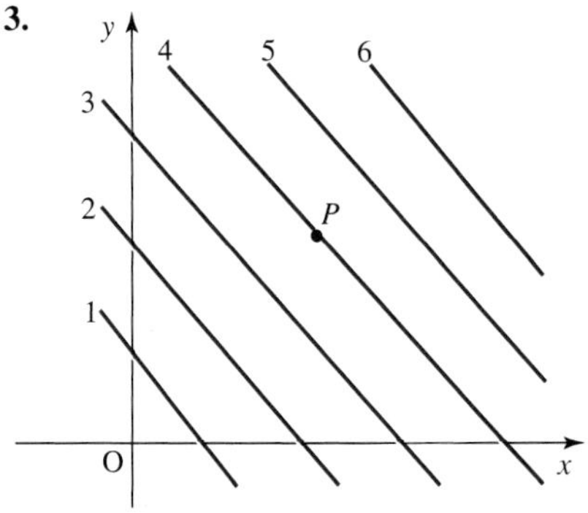

4.

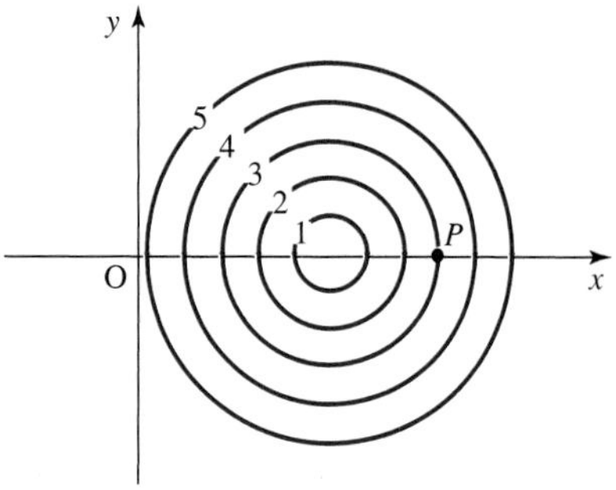

Exercises 5 to 13: Find the indicated second (or higher-order) partial derivatives of the given function.

5. $z = e^{xy} + \ln(x^2y^3)$; z_{xx}, z_{xy}, z_{yx}, z_{yy}

6. $z = x^y + (\ln y)^x$; z_{xx}, z_{xy}, z_{yx}, z_{yy}

7. $z = (x^2 + y^2)^{5/2}$; z_{xx}, z_{xy}, z_{yx}, z_{yy}

8. $z = x \arctan(y/x)$; z_{xx}, z_{xy}, z_{yx}, z_{yy}

9. $z = \sin^2(x + y)$; z_{xx}, z_{xy}, z_{yx}, z_{yy}

10. $z = f(x)g(y)$; z_{xx}, z_{xy}, z_{yx}, z_{yy} (f and g are differentiable real-valued functions)

11. $z = f(ax + by) + g(ax/y)$; z_{xx}, z_{xy}, z_{yx}, z_{yy} (f and g are differentiable real-valued functions of one variable and a and b are constants)

12. $z = e^{xy}$; z_{xx}, z_{xxx}, z_{xxxx}, z_{yyyy}

13. $w = y^3 \ln(x^2 + 3x + e^y) + x^3y^2z^4$; w_{xyzx}

14. A differential equation of the form $u_t = cu_{xx}$ where $u = u(x, t)$ and c is a constant, is called a *diffusion equation.*

(a) Show that $u(x, t) = e^{ax+bt}$ (a and b are constants) satisfies the diffusion equation with $c = b/a^2$.

(b) Show that $u(x, t) = t^{-1/2}e^{-x^2/t}$ satisfies the diffusion equation with $c = 1/4$.

15. Show that $z = xe^y + ye^x$ satisfies the equation $z_{xxx} + z_{yyy} = xz_{xyy} + yz_{yxx}$.

16. Explain why there is no C^2 function $f(x, y)$ such that $f_x(x, y) = e^x + xy$ and $f_y(x, y) = e^x + xy$.

17. How many different second-order partial derivatives does a C^2 function of 3 (and, in general, of m) variables have?

18. Show that $u(x, t) = \sin(x - ct) + \sinh(x + ct)$ (c is a constant) satisfies the wave equation $u_{tt} = c^2 u_{xx}$.

19. Assume that f and g are real-valued functions of one variable, and are of class C^2. Show that $u(x, y) = xf(x + y) + y\,g(x + y)$ satisfies the equation $u_{xx} - 2u_{xy} + u_{yy} = 0$.

20. Show that $z = e^{-ay}\cos(ax)$ (a is a constant) satisfies the equation $z_{xx} = az_y$.

21. Provide details of the calculation of u_{xxx} and of the proof that $u_t + 6uu_x + u_{xxx} = 0$ for the function $u(x, t) = 2k^2 \cosh^{-2}\left(k\left(x - 4k^2 t\right)\right)$ of Example 4.11.

22. Show that $z = x^4 - 6x^2y^2 + y^4$ satisfies Laplace's equation $z_{xx} + z_{yy} = 0$.

23. Show that the function $f(x, y) = \arctan(y/x)$ is harmonic at the points where it is defined.

24. Let $f(x, y, z)\colon \mathbb{R}^3 \to \mathbb{R}$ and $\mathbf{c}(t)\colon \mathbb{R} \to \mathbb{R}^3$ be C^2 functions. Using the chain rule, find $(f \circ \mathbf{c})''(t)$.

25. Show that the gravitational potential $V(x, y, z) = -GMm/\|\mathbf{r}\|$, where $\mathbf{r} = x\mathbf{i} + y\mathbf{j} + z\mathbf{k}$, satisfies Laplace's equation $V_{xx} + V_{yy} + V_{zz} = 0$, whenever $\mathbf{r} \neq \mathbf{0}$.

26. The concentration of a substance dissolving at a point x along a tube (i.e., we are looking at one-dimensional dispersion) is given by $C(x, t) = Me^{-x^2/4t}/\sqrt{4\pi Dt}$, where M is the mass of the substance, D is the diffusion coefficient, and t is time. Show that $C(x, t)$ satisfies (4.2). What is the value of σ?

27. The temperature at time t at a point x in a metal rod of length π placed along the x-axis with one end at the origin is given by $T(x, t) = 1 + e^{-t}\sin x$.

(a) What is the initial temperature at the ends of the rod? Initially, what is the warmest point on the rod?

(b) Show that $T(x, t)$ satisfies the heat equation $T_t = T_{xx}$.

(c) Sketch the graph of the temperature $T(x, t)$ at times $t = 0$, $t = 1$, and $t = 2$ in the same coordinate system. What happens as $t \to \infty$?

▶ 4.2 TAYLOR'S FORMULA

One reason why the linear approximation of a function is important is that it helps us find numeric approximations for the values of the function. Geometrically, linear approximation represents a tangent line (if a function depends on one variable) or a tangent plane (if it depends on two variables). Near the point of tangency, the graph of a function $y = f(x)$ looks like its tangent line. A tangent plane gives us a good idea of what the surface $z = f(x, y)$ looks like locally. Using Taylor's Theorem, we will obtain better numeric approximations than those provided by the linear approximation. Moreover, our study of quadratic approximations will help us analyze extreme values of functions of several variables.

One-Variable Taylor Formula

Assume that a function $y = f(x)$ has derivatives of all orders at a point x_0 in its domain. Recall that, by the Fundamental Theorem of Calculus,

$$\int_a^b F'(t)dt = F(b) - F(a), \tag{4.5}$$

for a continuously differentiable function F. Applying (4.5) to the function f, with $a = x_0$ and $b = x_0 + h$ (h is a real number), we get

$$f(x_0 + h) = f(x_0) + \int_{x_0}^{x_0+h} f'(t)\,dt.$$

Now we integrate by parts, using $u = f'(t)$ and $dv = dt$ (recall that $\int u dv = uv - \int v du$),

$$\begin{aligned} f(x_0 + h) &= f(x_0) + tf'(t)\Big|_{x_0}^{x_0+h} - \int_{x_0}^{x_0+h} tf''(t)\,dt \\ &= f(x_0) + (x_0 + h)f'(x_0 + h) - x_0 f'(x_0) - \int_{x_0}^{x_0+h} tf''(t)\,dt, \end{aligned}$$

rewrite the term $x_0 f'(x_0)$ as $(x_0 + h)f'(x_0) - hf'(x_0)$

$$= f(x_0) + hf'(x_0) + (x_0 + h)(f'(x_0 + h) - f'(x_0)) - \int_{x_0}^{x_0+h} tf''(t)\,dt,$$

and, in the third term, use (4.5) with $F = f'$

$$\begin{aligned} &= f(x_0) + f'(x_0)h + (x_0 + h)\int_{x_0}^{x_0+h} f''(t)\,dt - \int_{x_0}^{x_0+h} tf''(t)\,dt \\ &= f(x_0) + f'(x_0)h + \int_{x_0}^{x_0+h} (x_0 + h - t)f''(t)\,dt. \end{aligned} \tag{4.6}$$

Thus, we have obtained the *first-order Taylor formula*

$$f(x_0 + h) = T_1(x_0, h) + R_1(x_0, h), \tag{4.7}$$

where

$$T_1(x_0, h) = f(x_0) + f'(x_0)h \tag{4.8}$$

is the *first-order Taylor polynomial* and

$$R_1(x_0, h) = \int_{x_0}^{x_0+h} (x_0 + h - t)f''(t)\,dt \tag{4.9}$$

is the *first-order remainder*.

Replacing h by $x - x_0$ in (4.8), we get the usual form $T_1(x) = f(x_0) + f'(x_0)(x - x_0)$ of the first-order Taylor polynomial [which is, of course, equal to the linear approximation of $f(x)$ at x_0].

We would like to obtain an estimate for $R_1(x_0, h)$. Since t is in the interval $[x_0, x_0 + h]$, it follows that $|x_0 + h - t| \le |h|$. Moreover, the fact that f'' is continuous on $[x_0, x_0 + h]$ implies that f'' is bounded there, that is, $|f''(t)| \le M$ for some $M > 0$ and for all $t \in [x_0, x_0 + h]$. (Recall that a continuous function defined on a closed interval must have an absolute maximum and an absolute minimum and, thus, must be bounded.) It follows that

$$|R_1(x_0, h)| = \left| \int_{x_0}^{x_0+h} (x_0 + h - t) f''(t)\, dt \right|$$
$$\leq \int_{x_0}^{x_0+h} |x_0 + h - t|\, |f''(t)|\, dt \leq \int_{x_0}^{x_0+h} |h|\, M dt = |h|^2 M. \quad (4.10)$$

This estimate was carried out under the asumption that $h > 0$; the case $h < 0$ is done analogously (see Exercise 1). The inequality in (4.10) implies that

$$0 \leq \frac{|R_1(x_0, h)|}{|h|} \leq \frac{|h|^2 M}{|h|} = |h|\, M,$$

and thus,

$$\lim_{h \to 0} \frac{|R_1(x_0, h)|}{|h|} = 0.$$

In words, $|R_1(x_0, h)|$ is small compared to $|h|$ (i.e., goes to 0 faster than $|h|$), as $h \to 0$.

To conclude: if we approximate the value $f(x_0 + h)$ by computing $T_1(x_0, h) = f(x_0) + f'(x_0)h$ instead, the error $|R_1(x_0, h)| = |f(x_0 + h) - T_1(x_0, h)|$ (i.e., the absolute value of the difference between the value of the function and the value of its approximation using Taylor's formula) is small compared to $|h|$, in the sense that $|R_1(x_0, h)|/|h| \to 0$ as $h \to 0$. Let us illustrate this in an example.

▶ EXAMPLE 4.12

Compute the first-order Taylor formula for the function $f(x) = x + e^{-2x}$ at $x_0 = 0$. Use it to get an approximation of $f(0.075) = 0.075 + e^{-0.15}$ and give an estimate for the error of that approximation.

SOLUTION From $f(0) = 1$ and $f'(0) = -1$ [since $f'(x) = 1 - 2e^{-2x}$], we compute $T_1(0, h) = f(0) + f'(0)h = 1 - h$. Thus, we obtain the first-order Taylor formula

$$f(h) = T_1(0, h) + R_1(0, h) = 1 - h + R_1(0, h),$$

where $|R_1(0, h)|/|h| \to 0$ as $h \to 0$.

We need to approximate $f(0.075)$, so we substitute $h = 0.075$ into $T_1(0, h)$, and get $T_1(0, 0.075) = 1 - 0.075 = 0.925$. How close is this value to $f(0.075)$? To answer this, we need to estimate $R_1(0, h)$. From (4.9), we get

$$R_1(0, h) = \int_0^h (h - t) f''(t)\, dt = \int_0^h (h - t)\, 4e^{-2t}\, dt.$$

We proceed by imitating the way we obtained the estimate in (4.10), using the fact that $|e^{-2t}| \leq 1$ on the interval $[0, h]$,

$$|R_1(0, h)| \leq \int_0^h |h - t|\, |4e^{-2t}|\, dt \leq \int_0^h 4|h|\, dt = 4|h|^2.$$

So, when $h = 0.075$, we obtain $|R_1(0, 0.075)| \leq 4(0.075)^2 = 0.0225$. Indeed, the estimate that we got $[T_1(0, 0.075) = 0.925]$ differs from the value $f(0.075) = 0.075 + e^{-0.15} = 0.9357$ by 0.0107 (which is clearly smaller than our estimate of 0.0225 for the error). ◀

The above example, and the derivation that we did in (4.10), show how (4.9) is used to compute an estimate for the error we make when we replace a function by its first-order Taylor estimate (i.e., by its linear approximation).

The purpose of all this was not to rederive the formula for linear approximation. The process that we used (a combination of the Fundamental Theorem of Calculus and integration by parts) can now be repeated to yield higher-order estimates.

Recall the formula (4.6):

$$f(x_0+h) = f(x_0) + f'(x_0)h + \int_{x_0}^{x_0+h} (x_0+h-t) f''(t)\, dt. \tag{4.11}$$

Apply integration by parts with $dv = (x_0+h-t)dt$ and $u = f''(t)$, to get

$$\begin{aligned}\int_{x_0}^{x_0+h} (x_0+h-t) f''(t)\, dt &= -\frac{(x_0+h-t)^2}{2} f''(t)\Big|_{x_0}^{x_0+h} + \int_{x_0}^{x_0+h} \frac{(x_0+h-t)^2}{2} f'''(t)\, dt \\ &= \frac{h^2}{2} f''(x_0) + \frac{1}{2}\int_{x_0}^{x_0+h} (x_0+h-t)^2 f'''(t)\, dt.\end{aligned}$$

Substituting this expression into (4.11), we obtain the *second-order Taylor formula*

$$f(x_0+h) = f(x_0) + f'(x_0)h + \frac{f''(x_0)}{2}h^2 + \frac{1}{2}\int_{x_0}^{x_0+h} (x_0+h-t)^2 f'''(t)\, dt, \tag{4.12}$$

where

$$T_2(x_0, h) = f(x_0) + f'(x_0)h + \frac{f''(x_0)}{2}h^2 \tag{4.13}$$

is the *second-order Taylor polynomial* and

$$R_2(x_0, h) = \frac{1}{2}\int_{x_0}^{x_0+h} (x_0+h-t)^2 f'''(t)\, dt \tag{4.14}$$

is the *second-order remainder*. Replacing h by $x - x_0$ in (4.13), we get the usual form

$$T_2(x) = f(x_0) + f'(x_0)(x-x_0) + \frac{f''(x_0)}{2}(x-x_0)^2$$

of the second-order Taylor polynomial. Arguing analogously to the case of the first-order remainder, we can show that $|R_2(x_0, h)| \le M|h|^3/2$, where $|f'''(t)| \le M$ for all $t \in [x_0, x_0+h]$; see Exercise 2. Thus,

$$\lim_{h\to 0} \frac{|R_2(x_0, h)|}{|h|^2} = 0,$$

that is, $|R_2(x_0, h)|$ goes to zero faster than $|h|^2$, when $h \to 0$.

▶ EXAMPLE 4.13

Compute the second-order Taylor formula at $x_0 = 0$ for the function $f(x) = x + e^{-2x}$ of Example 4.12. Use the formula you obtained to approximate the value $0.075 + e^{-0.15}$, and give an estimate for the error.

SOLUTION The first three derivatives of $f(x) = x + e^{-2x}$ are $f'(x) = 1 - 2e^{-2x}$, $f''(x) = 4e^{-2x}$, and $f'''(x) = -8e^{-2x}$. Using (4.13) with $f(0) = 1$, $f'(0) = -1$, and $f''(0) = 4$, we get the second-order Taylor polynomial $T_2(0, h) = 1 - h + 2h^2$. The desired approximation is

$$T_2(0, 0.075) = 1 - 0.075 + 2(0.075)^2 = 0.936250.$$

From $|f'''(t)| \le 8$ (for $t \ge 0$), we obtain the estimate $|R_2(x_0, h)| \le 8|h|^3/2 = 4|h|^3$. In other words, using the second-order approximation $T_2(0, 0.075)$ for $0.075 + e^{-0.15}$, we made an error of no more than $4|h|^3 = 0.0016875$. [Indeed, $|f(0.075) - T_2(0, 0.075)| = |0.935708 - 0.936250| = 0.000542$.] Clearly, T_2 gave a better approximation than T_1. ◀

Continuing as before—applying integration by parts to (4.12)—we would obtain the third-order Taylor formula (see Exercise 3). Iterating this process, we arrive at the *nth-order Taylor formula.*

THEOREM 4.2 Taylor Formula for Functions of One Variable

Assume that the first $(n + 1)$ derivatives of $f(x)$ are continuous [i.e., $f(x)$ is of order C^k, for $0 \le k \le n + 1$]. Then

$$f(x_0 + h) = f(x_0) + f'(x_0)h + \frac{f''(x_0)}{2!}h^2 + \cdots + \frac{f^{(n)}(x_0)}{n!}h^n + R_n(x_0, h), \tag{4.15}$$

where

$$T_n(x_0, h) = f(x_0) + f'(x_0)h + \frac{f''(x_0)}{2!}h^2 + \cdots + \frac{f^{(n)}(x_0)}{n!}h^n$$

is the *nth-order Taylor polynomial.* The *nth-order remainder*

$$R_n(x_0, h) = \frac{1}{n!}\int_{x_0}^{x_0+h} (x_0 + h - t)^n f^{(n+1)}(t)\, dt$$

satisfies

$$\lim_{h\to 0} \frac{|R_n(x_0, h)|}{|h|^n} = 0.$$

◀

Replacing h by $x - x_0$, we get the usual form

$$T_n(x) = f(x_0) + f'(x_0)(x - x_0) + \frac{f''(x_0)}{2!}(x - x_0)^2 + \cdots + \frac{f^{(n)}(x_0)}{n!}(x - x_0)^n$$

of the nth-order Taylor polynomial. Often, as we have done it above, we drop x_0 from the notation, and use $T_n(x)$ instead of $T_n(x_0, x)$.

▶ EXAMPLE 4.14 Kinetic Energy in Special Relativity and in Classical Physics

According to the Special Theory of Relativity, the mass of a moving object depends on its speed according to the formula

$$m(v) = \frac{m_0}{\sqrt{1 - v^2/c^2}},$$

where v is the speed of the object, m_0 is its mass at rest (i.e., when $v = 0$), and c is the speed of light. We will compute the second-order Taylor formula for $m(v)$ at $v = 0$ (i.e., for small speeds), and use it to approximate the *kinetic energy* of the object, $K(v) = m(v)c^2 - m_0c^2$.

Starting with $m(v) = m_0(1 - v^2/c^2)^{-1/2}$, we compute

$$m'(v) = \frac{m_0 v}{c^2}\left(1 - \frac{v^2}{c^2}\right)^{-3/2}$$

and

$$m''(v) = \frac{m_0}{c^2}\left(1 - \frac{v^2}{c^2}\right)^{-3/2} + \frac{3m_0v^2}{c^4}\left(1 - \frac{v^2}{c^2}\right)^{-5/2}.$$

It follows that $m(0) = m_0$, $m'(0) = 0$, and $m''(0) = m_0/c^2$, and thus,

$$T_2(v) = m_0 + \frac{m_0}{2c^2}v^2.$$

Using $T_2(v)$ instead of $m(v)$ to compute the kinetic energy, we obtain

$$K(v) = T_2(v)c^2 - m_0c^2 = \left(m_0 + \frac{m_0}{2c^2}v^2\right)c^2 - m_0c^2 = \frac{1}{2}m_0v^2.$$

This calculation shows that for small speeds, kinetic energy $K(v)$ is approximately equal to the kinetic energy $m_0v^2/2$ from classical physics. ◀

Taylor's Theorem for Functions of Several Variables

Assume that $f: U \subseteq \mathbb{R}^m \to \mathbb{R}$ is a differentiable function of m variables. By Definition 2.13 in Section 2.4 (replace $\mathbf{x}$ by $\mathbf{x}_0 + \mathbf{h}$, $\mathbf{h} \in \mathbb{R}^m$, and $\mathbf{a}$ by $\mathbf{x}_0$), this means that

$$\lim_{\mathbf{h}\to\mathbf{0}} \frac{|f(\mathbf{x}_0 + \mathbf{h}) - f(\mathbf{x}_0) - \nabla f(\mathbf{x}_0)\mathbf{h}|}{\|\mathbf{h}\|} = 0.$$

(Recall that $Df = \nabla f$ for a real-valued function.) Thus, if we take

$$R_1(\mathbf{x}_0, \mathbf{h}) = f(\mathbf{x}_0 + \mathbf{h}) - f(\mathbf{x}_0) - \nabla f(\mathbf{x}_0)\mathbf{h},$$

we obtain the *first-order Taylor formula*

$$f(\mathbf{x}_0 + \mathbf{h}) = T_1(\mathbf{x}_0, \mathbf{h}) + R_1(\mathbf{x}_0, \mathbf{h}) \tag{4.16}$$

where [write $\mathbf{h} = (h_1, h_2, \ldots, h_m)$]

$$T_1(\mathbf{x}_0, \mathbf{h}) = f(\mathbf{x}_0) + \nabla f(\mathbf{x}_0)\mathbf{h} = f(\mathbf{x}_0) + \frac{\partial f}{\partial x_1}(\mathbf{x}_0)\,h_1 + \cdots + \frac{\partial f}{\partial x_m}(\mathbf{x}_0)\,h_m$$

and

$$\lim_{\mathbf{h}\to\mathbf{0}} \frac{|R_1(\mathbf{x}_0, \mathbf{h})|}{\|\mathbf{h}\|} = 0.$$

Thus, the first-order Taylor formula is just a restatement of the requirement that the function involved must be differentiable.

We now derive a second-order Taylor formula in the case of a function of two variables (later, we state the general case of a function of any number of variables).

Assume that $f(x, y)$ is of class C^2, and let $\mathbf{x}_0 = (x_0, y_0)$ and $\mathbf{h} = (h_1, h_2)$. Keeping $\mathbf{x}_0$ and $\mathbf{h}$ fixed, we define the function

$$F(t) = f(\mathbf{x}_0 + t\mathbf{h}) = f(x_0 + th_1, y_0 + th_2).$$

Note that $F(0) = f(\mathbf{x}_0) = f(x_0, y_0)$ and $F(1) = f(\mathbf{x}_0 + \mathbf{h}) = f(x_0 + h_1, y_0 + h_2)$. The function F depends on one variable, so we apply (4.12) to get

$$F(t_0 + h) = F(t_0) + F'(t_0)h + \frac{F''(t_0)}{2}h^2 + R_2(t_0, h).$$

In particular, when $t_0 = 0$ and $h = 1$,

$$F(1) = F(0) + F'(0) + \frac{F''(0)}{2} + R_2(0, 1), \tag{4.17}$$

where

$$R_2(0, 1) = R_2(\mathbf{x}_0, \mathbf{h}) = \frac{1}{2}\int_0^1 (1-t)^2 F'''(t)\,dt. \tag{4.18}$$

[Since the expression for $F'''(t)$ contains $\mathbf{x}_0$ and $\mathbf{h}$, we prefer to use $R_2(\mathbf{x}_0, \mathbf{h})$ to $R_2(0, 1)$ for the remainder.] Next, we compute $F'(0)$ and $F''(0)$. By the chain rule,

$$F'(t) = f_x(x_0 + th_1, y_0 + th_2)\,h_1 + f_y(x_0 + th_1, y_0 + th_2)\,h_2$$

and thus,

$$F'(0) = f_x(x_0, y_0)\,h_1 + f_y(x_0, y_0)\,h_2.$$

To compute $F''(t)$, we use the chain rule again:

$$\begin{aligned} F''(t) = {} & f_{xx}(x_0 + th_1, y_0 + th_2)\,h_1^2 + f_{xy}(x_0 + th_1, y_0 + th_2)\,h_1h_2 \\ & + f_{yx}(x_0 + th_1, y_0 + th_2)\,h_1h_2 + f_{yy}(x_0 + th_1, y_0 + th_2)\,h_2^2. \end{aligned}$$

Consequently,

$$F''(0) = f_{xx}(x_0, y_0)\,h_1^2 + 2f_{xy}(x_0, y_0)\,h_1h_2 + f_{yy}(x_0, y_0)\,h_2^2.$$

Substituting $F'(0)$ and $F''(0)$ back into (4.17), and keeping in mind that $F(0) = f(x_0, y_0) = f(\mathbf{x}_0)$ and $F(1) = f(x_0 + h_1, y_0 + h_2) = f(\mathbf{x}_0 + \mathbf{h})$, we get the following theorem.

THEOREM 4.3 Second-Order Taylor Formula for Functions of Two Variables

Assume that $f = f(x, y)$ has continuous second partial derivatives (i.e., is of class C^2), and let $\mathbf{x}_0 = (x_0, y_0)$ and $\mathbf{h} = (h_1, h_2)$. Then

$$f(\mathbf{x}_0 + \mathbf{h}) = T_2(\mathbf{x}_0, \mathbf{h}) + R_2(\mathbf{x}_0, \mathbf{h}),$$

where

$$\begin{aligned} T_2(\mathbf{x}_0, \mathbf{h}) = {} & f(x_0, y_0) + f_x(x_0, y_0)\,h_1 + f_y(x_0, y_0)\,h_2 \\ & + \tfrac{1}{2}\left(f_{xx}(x_0, y_0)\,h_1^2 + 2f_{xy}(x_0, y_0)\,h_1h_2 + f_{yy}(x_0, y_0)\,h_2^2\right) \end{aligned} \tag{4.19}$$

is the *second-order Taylor polynomial*. The *second-order remainder* $R_2(\mathbf{x}_0, \mathbf{h})$ is given by (4.18). As $\mathbf{h} \to \mathbf{0}$, $|R_2(\mathbf{x}_0, \mathbf{h})|/\|\mathbf{h}\|^2 \to 0$. ◀

Replacing h_1 by $x - x_0$ and h_2 by $y - y_0$, we get the usual form of the *second-order Taylor polynomial*

$$\begin{aligned} T_2(x, y) = {} & f(x_0, y_0) + f_x(x_0, y_0)(x - x_0) + f_y(x_0, y_0)(y - y_0) \\ & + \tfrac{1}{2}\left(f_{xx}(x_0, y_0)(x - x_0)^2 + 2f_{xy}(x_0, y_0)(x - x_0)(y - y_0)\right. \\ & \left. + f_{yy}(x_0, y_0)(y - y_0)^2\right) \end{aligned} \tag{4.20}$$

of $f(x, y)$ at (x_0, y_0). An estimate for the *second-order remainder* $R_2(\mathbf{x}_0, \mathbf{h})$ is derived in Exercise 16.

▶ EXAMPLE 4.15

Compute the second-order Taylor formula for the function $f(x, y) = \sin(x + y) + \cos(x - 3y)$ at the point $(x_0, y_0) = (0, 0)$.

SOLUTION

Clearly, $f(0, 0) = 1$. We now compute partial derivatives and evaluate at $(0, 0)$. From $f_x(x, y) = \cos(x + y) - \sin(x - 3y)$ and $f_y(x, y) = \cos(x + y) + 3\sin(x - 3y)$, we get $f_x(0, 0) = 1$ and $f_y(0, 0) = 1$. The second-order partial derivatives are

$$\begin{aligned} f_{xx}(x, y) &= -\sin(x + y) - \cos(x - 3y), \\ f_{xy}(x, y) &= -\sin(x + y) + 3\cos(x - 3y) \end{aligned}$$

and

$$f_{yy}(x, y) = -\sin(x + y) - 9\cos(x - 3y),$$

and thus, $f_{xx}(0, 0) = -1$, $f_{xy}(0, 0) = 3$, and $f_{yy}(0, 0) = -9$. The second-order Taylor formula at (0,0) can be written as $f(h_1, h_2) = T_2(h_1, h_2) + R_2(h_1, h_2)$, where

$$T_2(h_1, h_2) = 1 + h_1 + h_2 - \tfrac{1}{2}h_1^2 + 3h_1h_2 - \tfrac{9}{2}h_2^2$$

and $|R_2(h_1, h_2)|/(h_1^2 + h_2^2) \to 0$ as $(h_1, h_2) \to (0, 0)$.

To keep notation clear, we use $T_2(h_1, h_2)$ and $R_2(h_1, h_2)$ instead of $T_2((0, 0), (h_1, h_2))$ or $R_2((0, 0), (h_1, h_2))$, remembering that the Taylor formula has been calculated at $(0, 0)$.

Of course, $T_1(h_1, h_2) = 1 + h_1 + h_2$ is the first-order Taylor polynomial (linear approximation). Figure 4.6 shows level curves near $(0, 0)$ of f, its linear approximation T_1, and its second-order approximation T_2.

Clearly, T_2 approximates f near $(0, 0)$ better than T_1. To illustrate this algebraically, compare, for instance, the value $f(0.1, 0.2) = 1.173103$ with the values of the two approximations, $T_1(0.1, 0.2) = 1.3$ and $T_2(0.1, 0.2) = 1.175$. ◀

Substituting $h_1 = x - x_0 = x$ and $h_2 = y - y_0 = y$ into the formula for T_2, we obtain the usual form

$$T_2(x, y) = 1 + x + y - \tfrac{1}{2}x^2 + 3xy - \tfrac{9}{2}y^2$$

of the second-order Taylor polynomial of $f(x, y)$ near $(0, 0)$. An alternative way of computing T_2 is to use the one-variable Taylor series for $\sin x$ and $\cos x$ and discard all terms

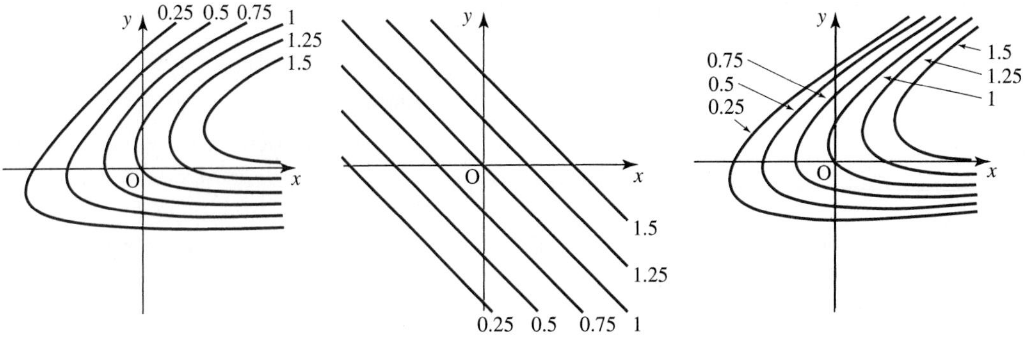

Figure 4.6 Level curves of f (left), its linear approximation (center), and its second-order Taylor approximation (right).

beyond the quadratic. Thus, from

$$\sin(x+y) + \cos(x-3y) = (x+y) - \frac{(x+y)^3}{3!} + \cdots$$
$$+1 - \frac{(x-3y)^2}{2!} + \frac{(x-3y)^4}{4!} + \cdots,$$

we get

$$T_2(x, y) = (x+y) + 1 - \frac{(x-3y)^2}{2!} = 1 + x + y - \tfrac{1}{2}x^2 + 3xy - \tfrac{9}{2}y^2.$$

▶ **EXAMPLE 4.16**

Find the first-order and second-order Taylor formulas for the function $f(x, y) = ye^{-x^2} + 2$ at $(1, 0)$.

SOLUTION The value of f at $(1, 0)$ is 2. From $f_x(x, y) = -2xye^{-x^2}$ and $f_y(x, y) = e^{-x^2}$, we get $f_x(1, 0) = 0$ and $f_y(1, 0) = e^{-1}$. Thus, the first-order Taylor formula is

$$f(h_1, h_2) = 2 + e^{-1}h_2 + R_1(h_1, h_2),$$

where $|R_1(h_1, h_2)|/\sqrt{h_1^2 + h_2^2} \to 0$ as $(h_1, h_2) \to (0, 0)$. The second-order partial derivatives $f_{xx}(x, y) = -2ye^{-x^2} + 4x^2ye^{-x^2}$, $f_{xy}(x, y) = -2xe^{-x^2}$, and $f_{yy}(x, y) = 0$, evaluated at the given point, give $f_{xx}(1, 0) = 0$, $f_{xy}(1, 0) = -2e^{-1}$, and $f_{yy}(1, 0) = 0$. It follows that

$$f(h_1, h_2) = 2 + e^{-1}h_2 - 2e^{-1}h_1h_2 + R_2(h_1, h_2),$$

where the remainder $R_2(h_1, h_2)$ satisfies $|R_2(h_1, h_2)|/(h_1^2 + h_2^2) \to 0$ as $(h_1, h_2) \to (0, 0)$.

Substituting $h_1 = x - 1$ and $h_2 = y$, we get

$$T_1(x, y) = 2 + e^{-1}y$$

and

$$T_2(x, y) = 2 + e^{-1}y - 2e^{-1}(x-1)y$$

for the first-order and second-order Taylor polynomials at $(1, 0)$. ◀

▶ EXAMPLE 4.17

Find the second-order Taylor polynomial of $f(x, y) = x + \ln(x - 2y)$ at the point (3, 1). Compare the linear and quadratic approximations of $f(3.15, 0.88)$.

SOLUTION

The derivatives of f are $f_x(x, y) = 1 + (x - 2y)^{-1}$, $f_y(x, y) = -2(x - 2y)^{-1}$, $f_{xx}(x, y) = -(x - 2y)^{-2}$, $f_{xy}(x, y) = 2(x - 2y)^{-2}$, and $f_{yy}(x, y) = -4(x - 2y)^{-2}$. Therefore, $f(3, 1) = 3$, $f_x(3, 1) = 2$, $f_y(3, 1) = -2$, $f_{xx}(3, 1) = -1$, $f_{xy}(3, 1) = 2$, and $f_{yy}(3, 1) = -4$, and it follows that

$$T_1(x, y) = 3 + 2(x - 3) - 2(y - 1)$$

and

$$T_2(x, y) = 3 + 2(x - 3) - 2(y - 1) - \tfrac{1}{2}(x - 3)^2 + 2(x - 3)(y - 1) - 2(y - 1)^2.$$

Letting $x = 3.15$ and $y = 0.88$, we get $T_1(3.15, 0.88) = 3.54$ and $T_2(3.15, 0.88) = 3.46395$ [the value of f is $f(3.15, 0.88) = 3.47930$]. ◀

Note that the second-order Taylor formula in Theorem 4.3 can be written as

$$\begin{aligned} f(\mathbf{x}_0 + \mathbf{h}) &= f(\mathbf{x}_0) + (f_x(\mathbf{x}_0), f_y(\mathbf{x}_0)) \cdot (h_1, h_2) \\ &\quad + \frac{1}{2}[\, h_1 \;\; h_2 \,] \begin{bmatrix} f_{xx}(\mathbf{x}_0) & f_{xy}(\mathbf{x}_0) \\ f_{yx}(\mathbf{x}_0) & f_{yy}(\mathbf{x}_0) \end{bmatrix} \begin{bmatrix} h_1 \\ h_2 \end{bmatrix} + R_2(\mathbf{x}_0, \mathbf{h}) \\ &= f(\mathbf{x}_0) + \nabla f(\mathbf{x}_0) \cdot \mathbf{h} + \tfrac{1}{2}\, \mathbf{h}^t Hf(\mathbf{x}_0)\, \mathbf{h} + R_2(\mathbf{x}_0, \mathbf{h}), \end{aligned} \tag{4.21}$$

where $\nabla f(\mathbf{x}_0)$ is the gradient of f at $\mathbf{x}_0$ [the second term is the dot product of vectors $\nabla f(\mathbf{x}_0)$ and $\mathbf{h}$], and $Hf(\mathbf{x}_0)$ is the matrix of second partial derivatives

$$Hf(\mathbf{x}_0) = \begin{bmatrix} f_{xx}(\mathbf{x}_0) & f_{xy}(\mathbf{x}_0) \\ f_{yx}(\mathbf{x}_0) & f_{yy}(\mathbf{x}_0) \end{bmatrix}, \tag{4.22}$$

called the *Hessian (matrix)* of f at $\mathbf{x}_0$. We represented $\mathbf{h}$ as the matrix $\begin{bmatrix} h_1 \\ h_2 \end{bmatrix}$, so that $\mathbf{h}^t = [\, h_1 \;\; h_2 \,]$. Both operations in $\mathbf{h}^t Hf(\mathbf{x}_0)\, \mathbf{h}$ are matrix multiplications. Formula (4.21) generalizes, in a straightforward way, to a function of m variables, $m \geq 1$. To summarize, consider the following theorem.

THEOREM 4.4 First-Order Taylor Formula for Functions of m Variables

Assume that $f: U \subseteq \mathbb{R}^m \to \mathbb{R}$ is differentiable at $\mathbf{x}_0 \in U$. Then

$$f(\mathbf{x}_0 + \mathbf{h}) = T_1(\mathbf{x}_0, \mathbf{h}) + R_1(\mathbf{x}_0, \mathbf{h}),$$

where

$$T_1(\mathbf{x}_0, \mathbf{h}) = f(\mathbf{x}_0) + \nabla f(\mathbf{x}_0) \cdot \mathbf{h},$$

and the remainder $R_1(\mathbf{x}_0, \mathbf{h})$ satisfies $|R_1(\mathbf{x}_0, \mathbf{h})|/\|\mathbf{h}\| \to 0$ as $\mathbf{h} \to \mathbf{0}$. ◀

Generalizing (4.22), we obtain the Hessian matrix of a function $f = f(x_1, x_2, \ldots, x_m)$ of m variables:

$$Hf(\mathbf{x}_0) = \begin{bmatrix} f_{x_1x_1}(\mathbf{x}_0) & f_{x_1x_2}(\mathbf{x}_0) & \ldots & f_{x_1x_m}(\mathbf{x}_0) \\ f_{x_2x_1}(\mathbf{x}_0) & f_{x_2x_2}(\mathbf{x}_0) & \ldots & f_{x_2x_m}(\mathbf{x}_0) \\ \vdots & \vdots & & \vdots \\ f_{x_mx_1}(\mathbf{x}_0) & f_{x_mx_2}(\mathbf{x}_0) & \ldots & f_{x_mx_m}(\mathbf{x}_0) \end{bmatrix}. \tag{4.23}$$

In order to obtain certain form of the remainder, we will need to assume that the function f in Theorem 4.4 is C^2 (see Exercise 19).

THEOREM 4.5 Second-Order Taylor Formula for Functions of m Variables

Assume that $f: U \subseteq \mathbb{R}^m \to \mathbb{R}$ has continuous second partial derivatives at $\mathbf{x}_0 \in U$. Then

$$f(\mathbf{x}_0 + \mathbf{h}) = T_2(\mathbf{x}_0, \mathbf{h}) + R_2(\mathbf{x}_0, \mathbf{h}),$$

where

$$T_2(\mathbf{x}_0, \mathbf{h}) = f(\mathbf{x}_0) + \nabla f(\mathbf{x}_0) \cdot \mathbf{h} + \tfrac{1}{2}\, \mathbf{h}^t Hf(\mathbf{x}_0)\, \mathbf{h}.$$

The second-order remainder $R_2(\mathbf{x}_0, \mathbf{h})$ satisfies $|R_2(\mathbf{x}_0, \mathbf{h})|/\|\mathbf{h}\|^2 \to 0$ as $\mathbf{h} \to \mathbf{0}$. ◀

The main reason why we developed second-order Taylor formula is to analyze extreme values of functions of several variables (see Section 4.3).

Formula (4.17) can be generalized to

$$F(1) = F(0) + F'(0) + \frac{F''(0)}{2!} + \cdots + \frac{F^{(n)}(0)}{n!} + R_n(0, 1).$$

Therefore, it is possible to compute nth-order Taylor formula (for $n \geq 1$) for a function of any number of variables (see Exercise 17). Since we will not use it in this book, we do not state it here. In Exercises 19 and 20, we derive formulas for remainders for the first-order and second-order Taylor formulas.

In Theorem 4.5 we assumed that f is C^2. However, in order to obtain certain forms of the remainder, we need to assume that f is C^3; see Exercises 16 and 20.

▶ EXERCISES 4.2

1. In computing the estimate for $R_1(x_0, h)$ in (4.10), we used the formula $|\int_{x_0}^{x_0+h} f(t)\,dt| \leq \int_{x_0}^{x_0+h} |f(t)|\,dt$. Explain why this formula works for $h \geq 0$ only. If $h < 0$, then $x_0 + h < x_0$, so we start the estimate by $|R_1(x_0, h)| = \left|\int_{x_0}^{x_0+h} (x_0 + h - t) f''(t)\,dt\right| = \left|\int_{x_0+h}^{x_0} (x_0 + h - t) f''(t)\,dt\right|$. Explain why this step is correct. Proceed as in (4.10) to complete the estimate.

2. Assuming that $|f'''(t)| \leq M$ for all $t \in [x_0, x_0 + h]$, prove that the second-order remainder (4.14) satisfies $|R_2(x_0, h)| \leq M|h|^3/2$. What condition(s) must f satisfy so that $|f'''(t)| \leq M$ for all $t \in [x_0, x_0 + h]$?

3. Apply integration by parts [with $u = f'''(t)$ and $dv = (x_0 + h - t)^2 dt$] to the formula (4.12) to obtain the third-order Taylor formula. Find an integral formula for the remainder $R_3(x_0, h)$, and show that $|R_3(x_0, h)|/|h|^3 \to 0$ as $h \to 0$.

4. Check that $T_3(x) = x - x^3/6$ is the third-order Taylor polynomial of $\sin x$ at $x_0 = 0$. Find an estimate for the error if $T_3(x)$ is used to compute $\sin x$ for $-0.1 \le x \le 0.1$.

5. Find the second-order Taylor polynomial for the function $f(x) = \sqrt{x}$ at $x_0 = 3$. Find an estimate for the error when $2 \le x \le 4$.

6. Find Taylor polynomials $T_2(x)$, $T_3(x)$, and $T_4(x)$ for the function $f(x) = e^x \cos x$ at $x_0 = 0$. Graph $f(x)$ and all three polynomials on $[-\pi/2, \pi/2]$.

Exercises 7 to 10: Find the second-order Taylor formula for the function $f(x)$ at the given point x_0. Give the remainder in integral form.

7. $f(x) = \sin x$, $x_0 = \pi/4$

8. $f(x) = \cos x$, $x_0 = \pi/3$

9. $f(x) = \ln x$, $x_0 = 4$

10. $f(x) = \sqrt{1 + x^2}$, $x_0 = 1$

11. Using the second-order Taylor polynomial, give an estimate for $\sin 0.1 + \cos 0.1$. Estimate the same expression using the third-order Taylor polynomial, and compare the two approximations.

12. Using the second-order Taylor polynomial, give an estimate for $0.087 \ln(1.087)$. Estimate the same expression using the first-order Taylor polynomial, and compare the two approximations.

13. Show that the equation $F(x, y) = y^3 - 4y + x^2 = 0$ defines y implicitly as a function of x, $y = g(x)$, near the point $x = 0$, $y = 2$. Find the second-order Taylor polynomial of $g(x)$ at $x = 0$.

14. Show that the equation $F(x, y) = \cos y - xy = 0$ defines implicitly, near the point $x = 0$, $y = \pi/2$, the function $y = g(x)$. Find the second-order Taylor polynomial of $g(x)$ at $x = 0$.

15. Find an approximation of $0.2e^{-0.2}$ using the second-order Taylor polynomial at $x_0 = 0$. Estimate the error term $R_2(0, 0.2)$.

16. Assume that $f = f(x, y)$ is of class C^3. Continuing the calculations preceding the statement of Theorem 4.3, obtain a formula for $F'''(t)$. Using (4.18), show that $R_2(\mathbf{x}_0, \mathbf{h}) = \frac{1}{2}\int_0^1 (t-1)^2 G(t)\,dt$, where $G(t) = f_{xxx}(\mathbf{x}_0 + t\mathbf{h})h_1^3 + 3f_{xxy}(\mathbf{x}_0 + t\mathbf{h})h_1^2 h_2 + 3f_{xyy}(\mathbf{x}_0 + t\mathbf{h})h_1 h_2^2 + f_{yyy}(\mathbf{x}_0 + t\mathbf{h})h_2^3$. Show that $|R_2(\mathbf{x}_0, \mathbf{h})| \le C\|\mathbf{h}\|^3$, where C is a positive constant.

17. Assume that $f = f(x, y, z)$ is of class C^2, and let $\mathbf{x}_0 = (x_0, y_0, z_0)$ and $\mathbf{h} = (h_1, h_2, h_3)$. Write down the formula for $T_2(\mathbf{x}_0, \mathbf{h})$. If $f = f(x_1, x_2, \ldots, x_m)$ is of class C^2, and $\mathbf{h} = (h_1, h_2, \ldots, h_m)$, show that $\mathbf{h}^t Hf(\mathbf{x}_0)\mathbf{h} = \sum_{i=1}^m \sum_{j=1}^m \frac{\partial^2 f}{\partial x_i \partial x_j}(\mathbf{x}_0)\, h_i h_j$. Write down the formula for $T_2(\mathbf{x}_0, \mathbf{h})$ in this case.

18. If $f = f(x_1, x_2, \ldots, x_m)$ is of class C^2, how many entries in its Hessian matrix $Hf(\mathbf{x}_0)$ are repeated?

19. Assume that $f: U \subseteq \mathbb{R}^2 \to \mathbb{R}$ is a C^2 function.

(a) Imitate the derivation of the formula (4.17) to obtain the first-order remainder $R_1(\mathbf{x}_0, \mathbf{h}) = R_1(0, 1) = \int_0^1 (1-t)F''(t)\,dt$. Show that $R_1(\mathbf{x}_0, \mathbf{h}) = \int_0^1 (1-t)G(t)\,dt$, where $G(t) = f_{xx}(\mathbf{x}_0 + t\mathbf{h})h_1^2 + 2f_{xy}(\mathbf{x}_0 + t\mathbf{h})h_1 h_2 + f_{yy}(\mathbf{x}_0 + t\mathbf{h})h_2^2$.

(b) Recall the Second Mean-Value Theorem for integrals: if g and h are continuous functions and $h(t) \ge 0$ on $[a, b]$, then $\int_a^b g(t)h(t)\,dt = g(c)\int_a^b h(t)\,dt$, where c is a number in $[a, b]$. Use this theorem to show that $R_1(\mathbf{x}_0, \mathbf{h}) = \frac{1}{2}\left(f_{xx}(\mathbf{c}_{11})h_1^2 + 2f_{xy}(\mathbf{c}_{12})h_1 h_2 + f_{yy}(\mathbf{c}_{22})h_2^2\right)$, where $\mathbf{c}_{11}$, $\mathbf{c}_{12}$, and $\mathbf{c}_{22}$ lie on the line joining $\mathbf{x}_0$ and $\mathbf{x}_0 + \mathbf{h}$.

(c) Generalize (b) to obtain the formula for $R_1(\mathbf{x}_0, \mathbf{h})$ in the case of a differentiable function of m variables.

20. Apply the Second Mean-Value Theorem for integrals [see (b) in Exercise 19] to the remainder from Exercise 16 to obtain a formula for $R_2(\mathbf{x}_0, \mathbf{h})$ for a function of two variables. Show that, if $f = f(x_1, x_2, \ldots, x_m)$ is of class C^3, and $\mathbf{h} = (h_1, h_2, \ldots, h_m)$, then $R_2(\mathbf{x}_0, \mathbf{h}) = \frac{1}{3!}\sum_{i=1}^{m}\sum_{j=1}^{m}\sum_{k=1}^{m} \frac{\partial^3 f}{\partial x_i \partial x_j \partial x_k}(\mathbf{c}_{ijk})\, h_i h_j h_k$, where all $\mathbf{c}_{ijk}$ lie on the line joining $\mathbf{x}_0$ and $\mathbf{x}_0 + \mathbf{h}$.

21. Derive the third-order Taylor formula for a C^4 function $f(x, y)$ of two variables.

Exercises 22 to 27: Find the second-order Taylor formula for the function f at the given point $\mathbf{x}_0$.

22. $f(x, y) = e^{-x} \sin y$, $(2, 0)$

23. $f(x, y) = x^2 + y^2 - 2xy + 1$, $(1, 1)$

24. $f(x, y) = \ln(x^2 + y^2 + 1)$, $(0, 1)$

25. $f(x, y) = \sin x + \sin 2y$, $(0, \pi/2)$

26. $f(x, y) = (x - 2)^2(y + 4)^2$, $(0, 0)$

27. $f(x, y) = (xy)^{-1}$, $(1, 2)$

28. Find the first-order and second-order Taylor polynomials of the function $f(x, y) = \arctan(xy)$ at $(1, 1)$. Compare the two approximations of $f(1.15, 0.93)$ with the value of the function.

29. Find the first-order and second-order Taylor polynomials of the function $f(x, y) = \sqrt{x + 4y - 1}$ at $(5, 3)$. Compare the two approximations of $f(4.9, 3.1)$ with the value of the function.

30. Compute linear and quadratic approximations of $f(x, y) = (x + y + 3)^{-1}$ at $(0, 0)$. Compare the values of the two approximations at $(0.1, 0.04)$ with the value $f(0.1, 0.04)$.

31. Find the second-order Taylor polynomial of the function $f(x, y) = y \sin x$ at $(0, 1)$ and use it to draw an approximation of the contour diagram of $f(x, y)$ near $(0, 1)$.

32. Find the second-order Taylor polynomial of the function $f(x, y) = ye^{x^2}$ at $(0, 0)$ and use it to draw an approximation of the contour diagram of $f(x, y)$ near $(0, 0)$.

Exercises 33 to 36: Use the quadratic approximation (i.e., the second-order Taylor formula) to give estimates for the following expressions.

33. $e^{0.03^2 - 0.95^2}$

34. $0.98 \ln 1.03$

35. $3.98 \arctan 0.02$

36. $\sin 0.96 \cos 0.04$

▶ 4.3 EXTREME VALUES OF REAL-VALUED FUNCTIONS

In the mid-18th century, French mathematician and astronomer Pierre Louis Moreau de Maupertius formulated a so-called "metaphysical principle" that could serve as a guiding mechanism driving the laws of nature. The principle states that every "action" of nature is actually an attempt to minimize or maximize a certain quantity. An animal, sleeping in the snow, curls up—in doing so, it minimizes the surface area of its body that is exposed to cold temperatures and loses the least amount of heat. A river, flowing down a mountain, follows the curve of steepest descent (provided that there are no obstructions in its way). A large number of celestial objects have the (approximate) shape of a ball—and a ball can be shown to possess a remarkable number of minimizing and maximizing properties. Moreover, a "bottom line" in economics is usually a synonym for a maximum profit or a minimum loss. Our daily actions sometimes follow "the line of least resistance."

In this section, we develop tools that will allow us to investigate extreme values of functions of several variables. For simplicity, we focus our investigation on functions of two variables. Most results we state hold, however, for a function of any number of variables. The two problems we study are finding relative (or local) extreme values of a function on its domain and finding absolute (or global) extreme values of a function on a closed and bounded set contained in its domain.

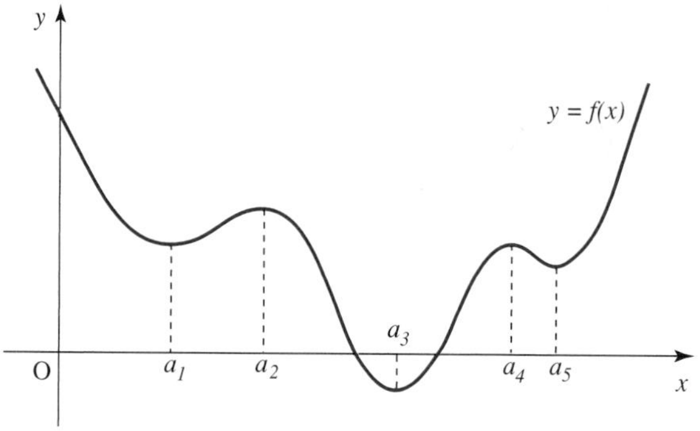

Figure 4.7 Extreme values of a function.

Extreme Values of Functions of One Variable

A function $y = f(x)$ has a local (or relative) maximum at x_0 if there is an open interval (c, d) containing x_0 such that $f(x) \leq f(x_0)$ for all x in (c, d). In other words, f has a local maximum at x_0 if f takes on no value larger than $f(x_0)$ at points near x_0. Similarly, $y = f(x)$ has a local (or relative) minimum at x_0 if there is an open interval (c, d) containing x_0 such that $f(x) \geq f(x_0)$ for all x in (c, d). A local minimum value and a local maximum value are called local (or relative) extreme values.

Let D be a subset (i.e., all or part) of the domain of the function $y = f(x)$. We say that f has an absolute (or global) maximum on D at x_0 if $f(x) \leq f(x_0)$ for all x in D. The number $f(x_0)$ is called the absolute (or global) maximum value of f on D. Similarly, f has an absolute (or global) minimum on D at x_0 if $f(x) \geq f(x_0)$ for all x in D. The number $f(x_0)$ is called the absolute (global) minimum value of f on D. An absolute minimum value and an absolute maximum value are called absolute (or global) extreme values.

For example, the function $y = f(x)$ defined for all $x \in \mathbb{R}$ shown in Figure 4.7 has a local maximum at a_2 and at a_4, and a local minimum at a_1, a_3, and a_5. It does not have an absolute maximum on $\mathbb{R}$. Its absolute minimum (D is taken to be $\mathbb{R}$) occurs at a_3; absolute minimum value is $f(a_3)$.

The function $y = \sin x$ (defined for all x; i.e., $D = \mathbb{R}$) has an absolute maximum value of 1, and it occurs at $x = \pi/2 + 2\pi k$ (k is an integer). Its absolute minimum value is -1; it occurs at infinitely many points $x = 3\pi/2 + 2\pi k$. The function $y = x^2$ does not have an absolute minimum on the set $D = (0, 2]$. It takes on the absolute maximum value of 4 at $x = 2$. The function $y = e^x$ does not have any local extreme values. It does not have any absolute extreme values on $D = \mathbb{R}$ either.

As the above examples show, not every function has extreme values. However, there is a situation in which the existence of extreme values is guaranteed.

THEOREM 4.6 Extreme Value Theorem

A continuous function $f(x)$ defined on a closed interval $[c, d]$ attains its (absolute) maximum value $f(a_1)$ and its (absolute) minimum value $f(a_2)$ at some points a_1 and a_2 in $[c, d]$.

This theorem is a "good news–bad news" theorem. It states that a continuous function has a minimum and a maximum on a closed interval, but it does not tell us how to find them. In finding extreme values, we make use of the following result.

THEOREM 4.7 Fermat's Theorem

If a function $y = f(x)$ has a local maximum or a local minimum at x_0 and if $f'(x_0)$ exists, then $f'(x_0) = 0$.

Note that the theorem does not say that if $f'(x_0) = 0$, then there must be a minimum or a maximum. For example, if $f(x) = x^3$, then $f'(0) = 0$; however, $x = 0$ is neither a minimum nor a maximum of x^3. What the theorem says is that, in finding extreme values of a function f, we must locate points where $f'(x_0) = 0$ [and also where $f'(x_0)$ does not exist]. A point in the domain of f where $f'(x_0) = 0$ or where $f'(x_0)$ does not exist is called a critical point (or a critical number) of f. The following theorem helps us examine critical points.

THEOREM 4.8 Second Derivative Test

Assume that f'' is continuous on an open interval containing x_0 and $f'(x_0) = 0$. If $f''(x_0) > 0$, then f has a local minimum at x_0. If $f''(x_0) < 0$, then f has a local maximum at x_0.

This theorem cannot be used for critical points x_0 in cases when $f'(x_0)$ does not exist [for instance, $f(x) = |x|$ does not have a derivative at $x_0 = 0$]. An alternative method, such as checking intervals of increase and decrease of f near x_0, must be employed.

Extreme Values of Functions of Two Variables

We now generalize these concepts and results to functions of two variables. Recall that a two-dimensional analogue of an open interval in $\mathbb{R}$ is an open ball (see Definition 2.6 in Section 2.3).

DEFINITION 4.1 Local and Absolute Extreme Values of $z = f(x, y)$

A function $z = f(x, y)$ has a *relative (local) maximum* at (x_0, y_0) if $f(x, y) \leq f(x_0, y_0)$ for all points (x, y) in some open ball centered at (x_0, y_0). The number $f(x_0, y_0)$ is called a *relative (local) maximum value.* Similarly, if $f(x, y) \geq f(x_0, y_0)$ for all points (x, y) in some open ball centered at (x_0, y_0), then $f(x, y)$ has a *relative (local) minimum* at (x_0, y_0). The number $f(x_0, y_0)$ is called a *relative (local) minimum value.*

Let D be a subset of the domain of $z = f(x, y)$. If $f(x, y) \leq f(x_0, y_0)$ for all points (x, y) in D, then $f(x, y)$ has an *absolute (global) maximum* on D at (x_0, y_0). The number $f(x_0, y_0)$ is called the *absolute (global) maximum value* of $f(x, y)$ on D. Similarly, $f(x_0, y_0)$ is the *absolute (global) minimum value* of $f(x, y)$ on D if $f(x, y) \geq f(x_0, y_0)$ for all points (x, y) in D.

Minimum and maximum values of a function are called *extreme values.*

By a commonly accepted abuse of the language, both a point (x_0, y_0) and the corresponding value $f(x_0, y_0)$ are referred to as a "minimum of f" (or a "maximum of f"). For instance, when we say "$(1, 2)$ is a relative maximum of f," we mean that the point $(1, 2)$ is a point where f attains its relative maximum, or that $f(1, 2)$ is a relative maximum value

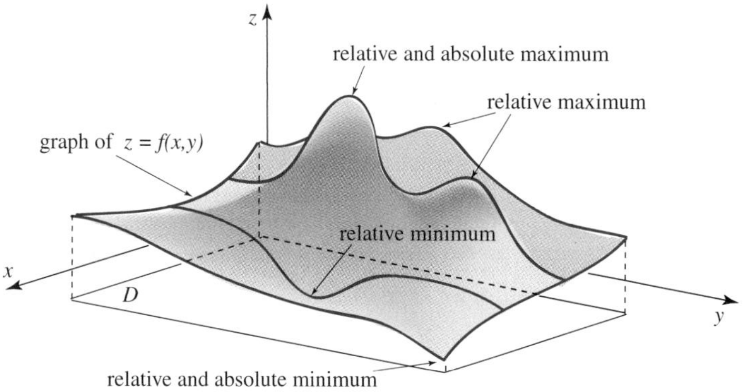

Figure 4.8 Extreme values of a function $z = f(x, y)$.

of f. Similarly, the statement "(x_0, y_0) is an absolute extreme point of f" means that f has an absolute extreme value (an absolute minimum or an absolute maximum) at (x_0, y_0).

It is always safe to say "relative minimum" or "relative maximum". However, if we want to use adjectives "absolute" or "global", we must say what the set D is. (So absolute extreme values are actually relative—but let us not get into that.) Take the set of all mountains on Earth, and consider the function that measures their height. Mount Everest is the absolute maximum (height: 8,848 m), and, for example, Mount McKinley (height: 6,194 m) is a relative maximum. If D is the set of all mountains in North America, then McKinley is the absolute maximum on D. However, Mount McKinley is not the absolute maximum on $\overline{D}$ = the set of all mountains in the Americas—Mount Aconcagua (height: 6,959 m) and some other mountains are higher than Mount McKinley.

Figure 4.8 shows the graph of a function $z = f(x, y)$, and its relative extreme values and absolute extreme values on a set D.

▶ EXAMPLE 4.18

Let $f(x, y) = 12 - x^2 - y^2$. Since $f(x, y) = 12 - (x^2 + y^2) \leq 12$ for all (x, y) in $\mathbb{R}^2$, it follows that $f(0, 0) = 12$ is the absolute maximum (and, of course, a local maximum) of f on $D = \mathbb{R}^2$. The given function has neither local nor absolute minimum.

The absolute maximum value of the function $h(x, y) = \sin x + \cos y$ on $\mathbb{R}^2$ is 2, and its absolute minimum on $\mathbb{R}^2$ is -2.

Consider the function $k(x, y) = x^2 + y^2 - 6x + 4y + 6$. Completing the square, we obtain $k(x, y) = (x - 3)^2 + (y + 2)^2 - 7$. Thus, k has absolute minimum $k(3, -2) = -7$. It has neither local nor absolute maximum values. ◀

We now turn to a study of local extreme values of a function $z = f(x, y)$ of two variables.

Assume that $\nabla f(\mathbf{x}_0) \neq \mathbf{0}$ at some point $\mathbf{x}_0$ in the domain of f. Then, in the direction of $\nabla f(\mathbf{x}_0)$, the values of f will increase. Thus, as long as the gradient at some point $\mathbf{x}_0$ is nonzero, a function can achieve larger values nearby, so it cannot have a local maximum at $\mathbf{x}_0$. Similarly, if $\nabla f(\mathbf{x}_0) \neq \mathbf{0}$, we can make the values of f smaller than $f(\mathbf{x}_0)$ by walking in the direction of $-\nabla f(\mathbf{x}_0)$. Thus, $\mathbf{x}_0$ is not a local minimum. This somewhat intuitive reasoning is proved correct by the following theorem.

THEOREM 4.9 Generalization of Fermat's Theorem

Let $f: U \subseteq \mathbb{R}^2 \to \mathbb{R}$ be a differentiable function defined on an open set U in $\mathbb{R}^2$. If $\mathbf{x}_0 = (x_0, y_0)$ is a local minimum or a local maximum, then $\nabla f(\mathbf{x}_0) = 0$. ◀

PROOF: Let $g(x) = f(x, y_0)$ and assume that f has a local maximum at $\mathbf{x}_0 = (x_0, y_0)$. In that case, g (which is a function of one variable) has a local maximum at x_0. By Fermat's Theorem (see Theorem 4.7), $g'(x_0) = 0$. But $g'(x_0) = (\partial f/\partial x)(x_0, y_0)$, and so $(\partial f/\partial x)(x_0, y_0) = 0$. Similarly, by considering the function $\bar{g}(y) = f(x_0, y)$, we conclude that $(\partial f/\partial y)(x_0, y_0) = 0$. The case of a local minimum is dealt with analogously. ◀

In Example 4.18 we showed that the function $f(x, y) = 12 - x^2 - y^2$ has a local maximum $f(0, 0) = 12$. Figure 4.9(a) illustrates level curves of f near $(0, 0)$. Since gradient vectors must be perpendicular to the level curves, they all must point toward the origin. Thus, at the origin, ∇f must be zero, or might not be defined. In Figure 4.9(b) we show gradient vectors of f near $(0, 0)$. (Vectors are not shown to scale.)

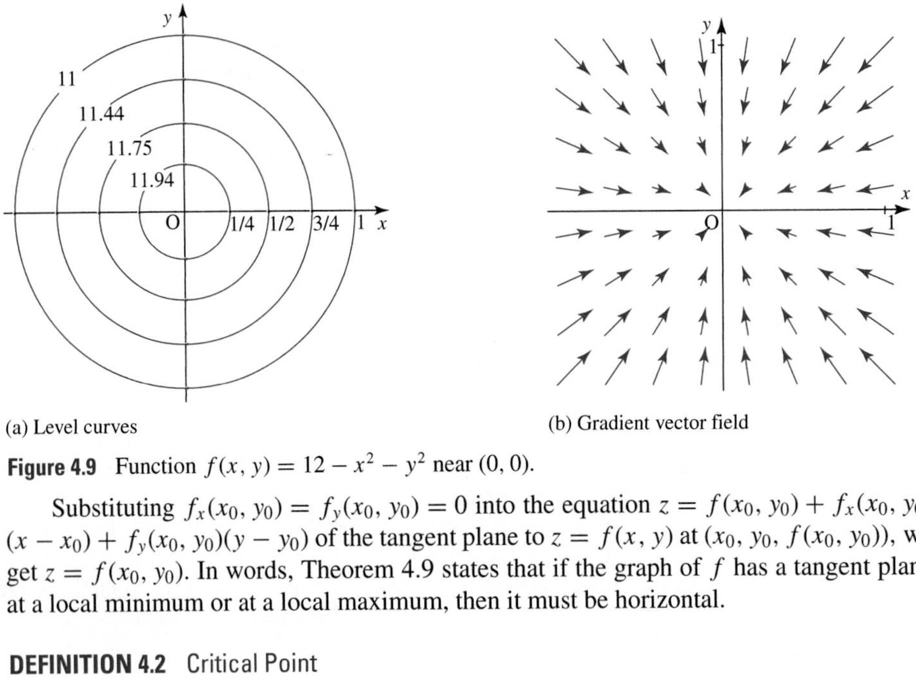

Figure 4.9 Function $f(x, y) = 12 - x^2 - y^2$ near $(0, 0)$.

Substituting $f_x(x_0, y_0) = f_y(x_0, y_0) = 0$ into the equation $z = f(x_0, y_0) + f_x(x_0, y_0)(x - x_0) + f_y(x_0, y_0)(y - y_0)$ of the tangent plane to $z = f(x, y)$ at $(x_0, y_0, f(x_0, y_0))$, we get $z = f(x_0, y_0)$. In words, Theorem 4.9 states that if the graph of f has a tangent plane at a local minimum or at a local maximum, then it must be horizontal.

DEFINITION 4.2 Critical Point

A point $\mathbf{x}_0 = (x_0, y_0)$ in the domain of a function $z = f(x, y)$ is called a *critical point* (or a *stationary point*) of f if $\nabla f(x_0, y_0) = \mathbf{0}$, or if at least one of the partial derivatives $f_x(x_0, y_0)$ or $f_y(x_0, y_0)$ does not exist. ◀

Theorem 4.9 claims that, in order to find either local or absolute extrema of a function on an open set, we have to identify all critical points. Note that, as in the case of a function of one variable, the theorem does *not* say that a critical point will actually yield a minimum or a maximum. Let us examine a few examples.

▶ EXAMPLE 4.19

Consider the function $f(x, y) = e^{-x^2-(y-1)^2}$. Find all critical points and, in each case, determine whether it is a local or global extreme value.

SOLUTION From $f_x = -2xe^{-x^2-(y-1)^2} = 0$ and $f_y = -2(y-1)e^{-x^2-(y-1)^2} = 0$, it follows that (0, 1) is the only critical point [note that both f_x and f_y are defined for all (x, y); thus, there are no critical points due to the nonexistence of first partials]. Since $-(x^2 + (y-1)^2) \le 0$ for all (x, y), we conclude that $f(x, y) = e^{-x^2-(y-1)^2} \le e^0 = 1$ for all (x, y). The value of f at (0, 1) is $f(0, 1) = 1$. Consequently, f has a local (and also global) maximum at (0, 1).

Note that the contour diagram of f consists of concentric circles centered at (0, 1). ◀

▶ EXAMPLE 4.20

Examine all critical points of the function $f(x, y) = 2x^2 - xy + y^2$.

SOLUTION From $f_x = 4x - y = 0$, we get $y = 4x$. Substituting this into $f_y = -x + 2y = 0$, we get $7x = 0$; that is, $x = 0$. Consequently, $y = 0$, and the only critical point is (0, 0).

Completing the square, we get

$$f(x, y) = 2x^2 - xy + y^2 = \left(\sqrt{2}x - \frac{1}{2\sqrt{2}}y\right)^2 + \left(\frac{7y}{8}\right)^2.$$

Thus, $f(x, y) \ge 0$ for all (x, y), and we conclude that $f(0, 0) = 0$ is a local (and also global) minimum. In Figure 4.10 we show the graph of the function $f(x, y)$ and its contour curves near the origin.

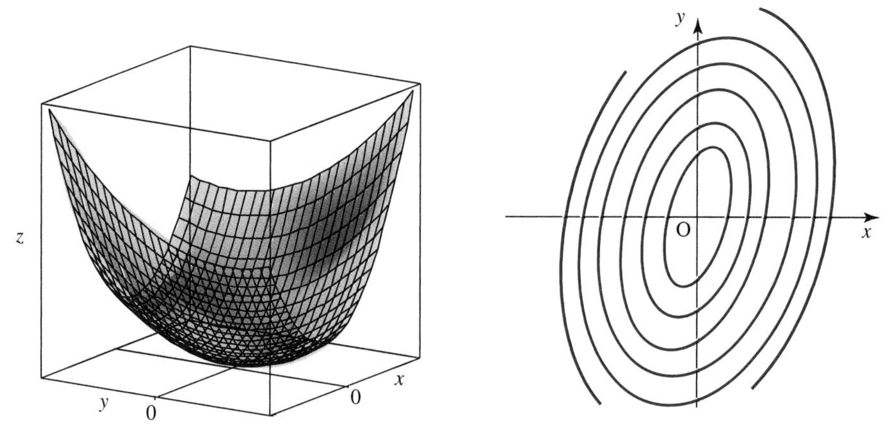

Figure 4.10 Graph of f in Example 4.20 and its contour diagram near (0, 0). ◀

▶ EXAMPLE 4.21

The graph of the function $f(x, y) = \sqrt{x^2 + y^2}$ is a cone whose vertex is at the origin and axis of rotation is the z-axis (see Figure 2.50 in Section 2.4). Examine all critical points of $f(x, y)$.

SOLUTION There are no points where both partial derivatives $f_x(x, y) = x/\sqrt{x^2 + y^2}$ and $f_y(x, y) = y/\sqrt{x^2 + y^2}$ are simultaneously zero. However, at (0, 0), f_x and f_y are not defined (see

Example 2.44). Thus, (0, 0) is a critical point. Since $f(x, y) = \sqrt{x^2 + y^2} \geq 0$ for all (x, y), it follows that (0, 0) is a local (and also global) minimum of f.

▶ EXAMPLE 4.22

Find all critical points of the function $f(x, y) = 2x^2 - y^2 + 3$.

SOLUTION

From $f_x = 0$ and $f_y = 0$, it follows that $4x = 0$ and $2y = 0$. Thus, (0, 0) is the only critical point of f.

Since $f(0, y) = -y^2 + 3 < 3$ (if $y \neq 0$), we conclude that $f(0, 0) = 3$ cannot be a local minimum (since f attains smaller values nearby on the y-axis). Similarly, $f(x, 0) = 2x^2 + 3 > 3$ (if $x \neq 0$) implies that $f(0, 0) = 3$ cannot be a local maximum. Thus, the critical point (0, 0) is neither a minimum point nor a maximum point of f.

DEFINITION 4.3 Saddle Point

A critical point that is neither a local minimum point nor a local maximum point is called a *saddle point*.

In other words, the critical point (0, 0) in Example 4.22 is a saddle point. Figure 4.11 shows the graph of the function and its level curves near (0, 0).

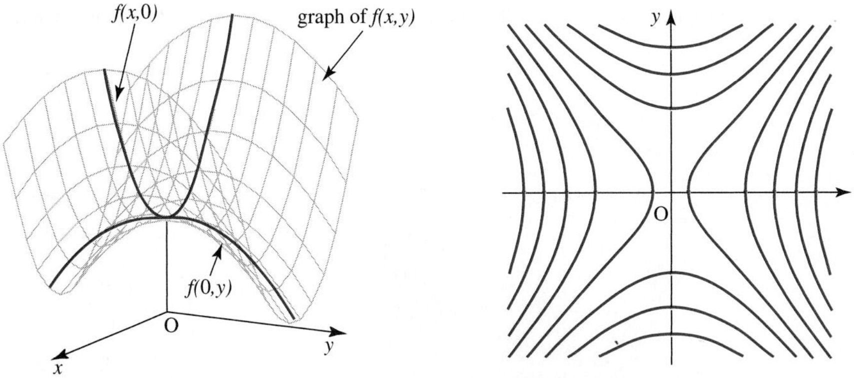

Figure 4.11 Function $f(x, y) = 2x^2 - y^2 + 3$ of Example 4.22.

▶ EXAMPLE 4.23

Examine the critical points of $z = x^{2/3}y^{2/3}$.

SOLUTION

The partials of z are $z_x = \frac{2}{3}x^{-1/3}y^{2/3}$ and $z_y = \frac{2}{3}x^{2/3}y^{-1/3}$. Now, $z_x = 0$ only when $y = 0$, but in this case z_y is not defined. Thus, there are no critical points coming from $z_x = z_y = 0$.

Since z_x does not exist when $x = 0$ and z_y does not exist when $y = 0$, it follows that all points with coordinates $(0, y)$ and $(x, 0)$, where $x, y \in \mathbb{R}$, are critical points. From $z = x^{2/3}y^{2/3} \geq 0$ for all x and y and $z(0, y) = z(x, 0) = 0$, we conclude that all critical points are local (and also global) minimum points. See Figure 4.12.

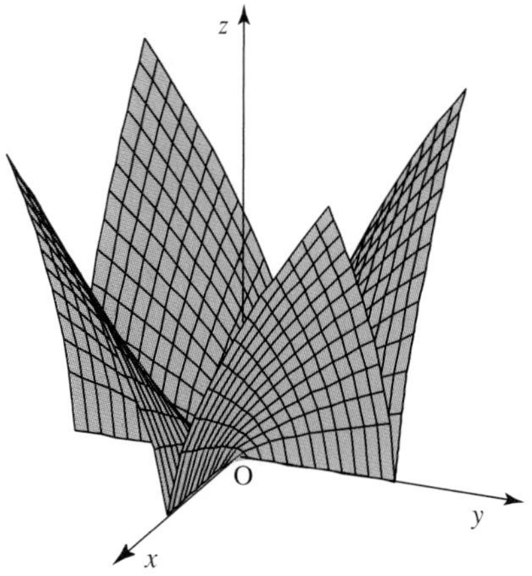

Figure 4.12 Graph of $z = x^{2/3}y^{2/3}$ for $-1 \leq x, y \leq 1$.

Examples 4.19 through 4.23 were carefully chosen: in each case, we were able to examine critical points by inspection (however, we cannot expect this to happen all the time).

Next, we derive a general test that will help us analyze critical points of C^2 functions of two variables. Recall that a function $f(x, y)$ is of class C^2 if its partial derivatives f_{xx}, f_{xy}, f_{yx}, and f_{yy} are continuous. In this case, $f_{xy} = f_{yx}$ (Theorem 4.1 in Section 4.1).

Let $\mathbf{x}_0 = (x_0, y_0)$ be a critical point of f. Since f is C^2, its first partials exist, and therefore, $f_x(x_0, y_0) = f_y(x_0, y_0) = 0$. Using the second-order Taylor formula from Theorem 4.3 and (4.20), we obtain

$$f(x, y) \approx T_2(x, y) = f(x_0, y_0) + \tfrac{1}{2}\big(f_{xx}(x_0, y_0)(x - x_0)^2 + 2f_{xy}(x_0, y_0)(x - x_0)(y - y_0) + f_{yy}(x_0, y_0)(y - y_0)^2\big),$$

and thus

$$f(x, y) - f(x_0, y_0) \approx \tfrac{1}{2}f_{xx}(x_0, y_0)(x - x_0)^2 + f_{xy}(x_0, y_0)(x - x_0)(y - y_0) + \tfrac{1}{2}f_{yy}(x_0, y_0)(y - y_0)^2. \tag{4.24}$$

To simplify calculations, we now assume that the critical point is located at (0, 0); that is, we write (4.24) as

$$f(x, y) - f(0, 0) \approx \tfrac{1}{2}f_{xx}(0, 0)\,x^2 + f_{xy}(0, 0)\,xy + \tfrac{1}{2}f_{yy}(0, 0)\,y^2. \tag{4.25}$$

Note that this is just a matter of translation. The right side of (4.25) is the polynomial

$$p(x, y) = ax^2 + bxy + cy^2, \tag{4.26}$$

where $a = f_{xx}(0, 0)/2$, $b = f_{xy}(0, 0)$, and $c = f_{yy}(0, 0)/2$. Note that (4.25) states that, near (0, 0), the graph of $f(x, y)$, up to a vertical shift, looks like the graph of $p(x, y)$. By analyzing $p(x, y)$, we will be able to determine the behavior of $f(x, y)$ at its critical point (0, 0).

▶ EXAMPLE 4.24 Properties of $p(x, y) = ax^2 + bxy + cy^2$

The first partials of p are $p_x = 2ax + by$ and $p_y = bx + 2cy$. Clearly, $p_x(0, 0) = p_y(0, 0) = 0$, so $(0, 0)$ is a critical point of $p(x, y)$. By completing the square, we get

$$p(x, y) = ax^2 + bxy + cy^2 = a\left(x^2 + \frac{bxy}{a} + \frac{cy^2}{a}\right)$$

$$= a\left[\left(x + \frac{b}{2a}y\right)^2 - \frac{b^2y^2}{4a^2} + \frac{cy^2}{a}\right] = a\left[\left(x + \frac{b}{2a}y\right)^2 + \frac{4ac - b^2}{4a^2}y^2\right]. \quad (4.27)$$

Let $D = 4ac - b^2$ (D is called the discriminant).

If $D > 0$, the expression in square brackets is positive or zero. If, moreover, $a > 0$, then $p(x, y) \geq 0$ for all (x, y) near $(0, 0)$. Since $p(0, 0) = 0$, it follows that $(0, 0)$ is a local minimum. If $D > 0$ and $a < 0$, then $p(x, y) \leq 0$ for all (x, y) near $(0, 0)$. Thus, $p(0, 0) = 0$ is a local maximum.

In the case $D < 0$, the expression in square brackets is a difference of two numbers—thus, it can be positive for some values (x, y) and negative for some other values (x, y) near $(0, 0)$. Therefore, $p(0, 0) = 0$ is neither minimum nor maximum. By Definition 4.3, it is a saddle point.

If $D = 0$, then $p(x, y) = a(x + by/2a)^2$. The point $(0, 0)$, and all other points on the line $x + by/2a = 0$ represent local minimum (if $a > 0$) or local maximum (if $a < 0$).

In our analysis, we assumed that $a \neq 0$. If $a = 0$ and $c \neq 0$, then we factor c out, complete the square, and continue as above. If $a = c = 0$, then $p(x, y) = bxy$. For (x, y) near $(0, 0)$, the expression bxy can be positive or negative. Thus, $(0, 0)$ is a saddle point, since it is neither minimum nor maximum. ◀

Now we go back to (4.24) and (4.25). To keep track of second partials, we use the *Hessian*

$$Hf(x, y) = \begin{bmatrix} f_{xx}(x, y) & f_{xy}(x, y) \\ f_{yx}(x, y) & f_{yy}(x, y) \end{bmatrix}.$$

Its determinant

$$D(x, y) = f_{xx}(x, y)f_{yy}(x, y) - (f_{xy}(x, y))^2$$

is the discriminant D appearing in Example 4.24. (When we want to emphasize that D is computed for a function f, we use D_f instead of D.)

By finishing Example 4.24, we completed the proof of the following theorem. We could generalize it to obtain the second derivative test for functions of any number of variables. An alternative proof is discussed in Exercise 2.

THEOREM 4.10 Second Derivatives Test

Assume that $z = f(x, y)$ is a C^2 function defined on an open set containing (x_0, y_0), and let (x_0, y_0) be a critical point of f [i.e., $f_x(x_0, y_0) = f_y(x_0, y_0) = 0$].

(a) If $D(x_0, y_0) > 0$ and $f_{xx}(x_0, y_0) < 0$, then $f(x_0, y_0)$ is a local maximum.

(b) If $D(x_0, y_0) > 0$ and $f_{xx}(x_0, y_0) > 0$, then $f(x_0, y_0)$ is a local minimum.

(c) If $D(x_0, y_0) < 0$, then $f(x_0, y_0)$ is a neither a local maximum nor a local minimum (i.e., it is a saddle point). ◀

If $f_{xx}(x_0, y_0) = 0$ and $f_{yy}(x_0, y_0) \neq 0$, then, in (a) and (b), we use $f_{yy}(x_0, y_0)$ instead of $f_{xx}(x_0, y_0)$. In the case $f_{xx}(x_0, y_0) = f_{yy}(x_0, y_0) = 0$ and $f_{xy}(x_0, y_0) \neq 0$, the discriminant $D(x_0, y_0)$ is negative, so case (c) applies. If $D(x_0, y_0) = 0$, then the Second Derivatives Test provides no answer: f could have a local extreme at (x_0, y_0), or (x_0, y_0) can be a saddle point. See Examples 4.27 and 4.28 below.

▶ EXAMPLE 4.25

Find and classify all critical points of the function $f(x, y) = x^3 - 3x^2 - 3y^2 + 3xy^2$.

SOLUTION

The first partials of f are computed to be $f_x = 3x^2 - 6x + 3y^2$ and $f_y = -6y + 6xy$. From $f_y = 6y(-1 + x) = 0$, it follows that either $y = 0$ or $x = 1$.

Substituting $y = 0$ into $f_x = 3x^2 - 6x + 3y^2 = 0$, we get $3x^2 - 6x = 3x(x - 2) = 0$; that is, $x = 0$ or $x = 2$. Substituting $x = 1$ into $f_x = 3x^2 - 6x + 3y^2 = 0$, we get $-3 + 3y^2 = 0$; that is, $y = \pm 1$. Thus, the critical points of f are $(0, 0)$, $(2, 0)$, $(1, 1)$, and $(1, -1)$.

We now apply the Second Derivatives Test. From $f_{xx} = 6x - 6$, $f_{yy} = -6 + 6x$ and $f_{xy} = f_{yx} = 6y$, we get

$$D(x, y) = (6x - 6)(-6 + 6x) - (6y)^2 = 36(x^2 - 2x + 1 - y^2).$$

Since $D(0, 0) = 36 > 0$ and $f_{xx}(0, 0) = -6 < 0$, we conclude that $f(0, 0) = 0$ is a local maximum. From $D(2, 0) = 36 > 0$ and $f_{xx}(2, 0) = 6 > 0$, it follows that $f(2, 0) = -4$ is a local minimum. The remaining two critical points, $(1, 1)$ and $(1, -1)$, are saddle points, since $D(1, \pm 1) = -36 < 0$. ◀

▶ EXAMPLE 4.26 Building an Optimal Box

Find the dimensions of the rectangular box (with a lid) with a volume of 100 cm^3 and the smallest possible surface area.

SOLUTION

Denote the length, width, and height of the box by x, y, and z, respectively ($x, y, z > 0$). It is given that $xyz = 100$. We have to make the surface area $\overline{S} = 2xy + 2xz + 2yz$ as small as possible.

We could build many boxes with a volume of 100 cm^3. For example, let $x = y = 1$ and $z = 100$; in that case, the surface area is $\overline{S} = 402$. If $x = 2$, $y = 5$, and $z = 10$, then $\overline{S} = 160$. If $x = y = 10$ and $z = 1$, then $\overline{S} = 240$, etc. We are asked to build the box (with a volume of 100 cm^3) for which the least amount of material is needed.

From $xyz = 100$, it follows that $z = 100/xy$. The surface area $\overline{S}$, expressed as a function of x and y only, is given by

$$S(x, y) = 2xy + 2x\frac{100}{xy} + 2y\frac{100}{xy} = 2xy + \frac{200}{y} + \frac{200}{x}.$$

(In other words, S is the restriction of the function $\overline{S}$ of three variables to the set of points that satisfy $xyz = 100$.)

Solving $S_x = 2y - 200/x^2 = 0$ and $S_y = 2x - 200/y^2 = 0$, we get $y = 100/x^2$ and $x = 100/y^2$. Combining the two equations implies that

$$x = \frac{100}{(100/x^2)^2} = \frac{x^4}{100}.$$

Thus, $x(1 - x^3/100) = 0$ and (since $x > 0$) $x = \sqrt[3]{100}$. It follows that $y = 100/(\sqrt[3]{100})^2 = \sqrt[3]{100}$. So, $(\sqrt[3]{100}, \sqrt[3]{100})$ is the only critical point. Note that $z = 100/xy = 100/(\sqrt[3]{100})^2 = \sqrt[3]{100}$.

Next, we use the Second Derivatives Test to show that $S(\sqrt[3]{100}, \sqrt[3]{100})$ is a minimum. From $S_{xx} = 400/x^3$, $S_{yy} = 400/y^3$, and $S_{xy} = S_{yx} = 2$, it follows that

$$D(x, y) = S_{xx}S_{yy} - (S_{xy})^2 = \frac{400^2}{x^3y^3} - 4.$$

Since $D(\sqrt[3]{100}, \sqrt[3]{100}) = 400^2/100^2 - 4 = 12 > 0$ and $S_{xx}(\sqrt[3]{100}, \sqrt[3]{100}) = 400/100 = 4 > 0$, we conclude that $S(\sqrt[3]{100}, \sqrt[3]{100})$ is a minimum. So the optimal box ("optimal" as defined in the text of this example) has the shape of a cube of side $\sqrt[3]{100}$. ◀

▶ EXAMPLE 4.27

Show that (0, 0) is the only critical point of the function $f(x, y) = x^3 - 3xy^2$. Determine whether it is a point where f attains its local extreme value or a saddle point.

SOLUTION From $f_x = 3x^2 - 3y^2 = 0$, it follows that $y^2 = x^2$ and $y = \pm x$. From $f_y = -6xy = 0$, it follows that $xy = 0$. The equation $xy = 0$ implies that either $x = 0$ (in which case, $y^2 = 0$ and $y = 0$) or $y = 0$ (in which case, $x^2 = 0$ and $x = 0$). Thus, (0, 0) is the only critical point.

The second partials are $f_{xx} = 6x$, $f_{yy} = -6x$, and $f_{xy} = f_{yx} = -6y$. Thus, $D(x, y) = -36x^2 - 36y^2 = -36(x^2 + y^2)$. Since $D(0, 0) = 0$, the Second Derivatives Test fails to provide an answer.

Note that $f(0, 0) = 0$. Consider points $(x, 0)$ on the x-axis; the corresponding values of f are $f(x, 0) = x^3$. This means that, near (0, 0), f attains both positive values (for $x > 0$) and negative values (for $x < 0$). Consequently, $f(0, 0) = 0$ cannot be a maximum or a minimum value. It follows that (0, 0) is a saddle point; see Figure 4.13. ◀

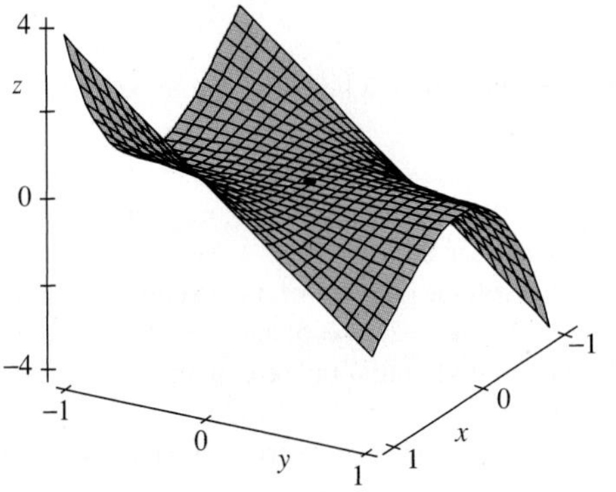

Figure 4.13 Function $f(x, y) = x^3 - 3xy^2$ has a saddle point at (0, 0).

▶ EXAMPLE 4.28

Consider the functions $f(x, y) = x^4 + y^4$, $g(x, y) = x^4 - y^4$ and $h(x, y) = -(x^4 + y^4)$. Show that, for each of the three functions, (0, 0) is the only critical point. Determine whether it is a point where a local extreme value occurs or a saddle point.

SOLUTION From $f_x = 4x^3 = 0$ and $f_y = 4y^3 = 0$, it follows that (0, 0) is the only critical point of f. The second partials are $f_{xx} = 12x^2$, $f_{yy} = 12y^2$, and $f_{xy} = f_{yx} = 0$; thus, $D_f(x, y) = 144x^2y^2$ and $D_f(0, 0) = 0$. This means that we cannot use the Second Derivatives Test. Since $f(x, y) \geq 0 = f(0, 0)$ for all (x, y), we conclude that $f(0, 0)$ is a local (also global) minimum.

From $g_x = 4x^3 = 0$ and $g_y = -4y^3 = 0$, we see that the only critical point of g is (0, 0). Next, $g_{xx} = 12x^2$, $g_{yy} = -12y^2$, and $g_{xy} = g_{yx} = 0$; thus, $D_g(x, y) = -144x^2y^2$ and $D_g(0, 0) = 0$. Consequently, we cannot use the Second Derivatives Test. The fact that $g(x, 0) = x^4 \geq 0$

implies that $g(0, 0) = 0$ cannot be a local maximum (since g attains larger values nearby). From $g(0, y) = -y^4 \leq 0$, we conclude that $g(0, 0) = 0$ cannot be a local minimum either, so $(0, 0)$ is a saddle point.

Finally, $h_x = -4x^3 = 0$ and $h_y = -4y^3 = 0$ imply that the only critical point of h is $(0, 0)$. Since $D_h = 144x^2y^2$ and $D_h(0, 0) = 0$, the Second Derivatives Test gives no information. However, from $h(x, y) = -(x^4 + y^4) \leq 0$ for all (x, y), it follows that $h(0, 0) = 0$ is a local (and also global) maximum. ◀

Absolute (Global) Extreme Values

The Extreme Value Theorem (see Theorem 4.6) claims that a continuous function $y = f(x)$ defined on a closed interval $[c, d]$ must have an absolute maximum and an absolute minimum. They occur either at the critical points inside $[c, d]$ or at the endpoints c and d. Let us consider an example.

▶ EXAMPLE 4.29

Find the absolute extreme values of the function $f(x) = x^4 - 2x^2 + 1$ on $[-1/2, 2]$.

SOLUTION

From $f'(x) = 4x^3 - 4x = 0$, we get that $x^3 - x = x(x^2 - 1) = 0$; thus, the critical points are $0, -1$, and 1. It remains to find the value of f at 0 and 1 (i.e., at the critical points inside $[-1/2, 2]$) and at $-1/2$ and 2 (i.e., at the endpoints of the given interval). From $f(0) = 1$, $f(1) = 0$, $f(-1/2) = 9/16$, and $f(2) = 9$, it follows that the absolute maximum is 9 (occurs at $x = 2$) and the absolute minimum is 0 (occurs at $x = 1$). ◀

We are now going to generalize the Extreme Value Theorem to functions of two variables. The assumption "closed interval $[c, d]$" in the statement of Theorem 4.6 will be replaced by "closed and bounded set in $\mathbb{R}^2$." We start by explaining what closed and bounded sets are.

DEFINITION 4.4 Closed Set in $\mathbb{R}^2$

A set $D \subseteq \mathbb{R}^2$ is called *closed* in $\mathbb{R}^2$ if it contains all its boundary points. ◀

Recall that an (open) ball of radius $r > 0$ centered at (a, b) is defined as the set $B(r) = \{(x, y) \mid \sqrt{(x - a)^2 + (y - b)^2} < r\}$. A point (a, b) in $D \subseteq \mathbb{R}^2$ is a boundary point of D if every open ball centered at (a, b) contains points that belong to D and points that do not belong to D.

The point (a, b) in Figure 4.14(a) is a boundary point of D. An open ball, centered at (a, b), no matter how small, contains both the points that belong to D and that do not belong to it. On the other hand, the point (a, b) in Figure 4.14(b) is not a boundary point. We can

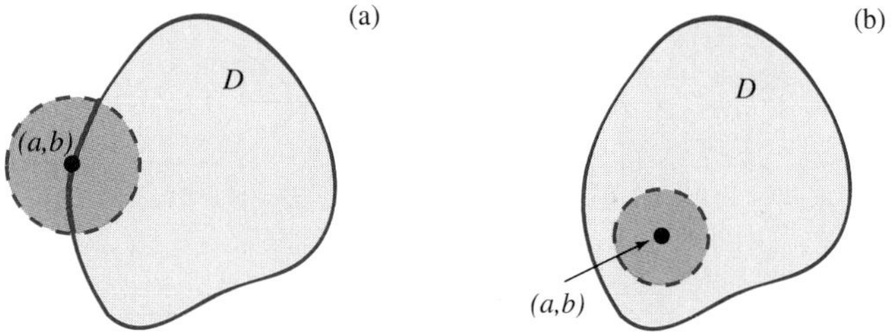

Figure 4.14 Set D with a boundary point (a) and an interior point (b).

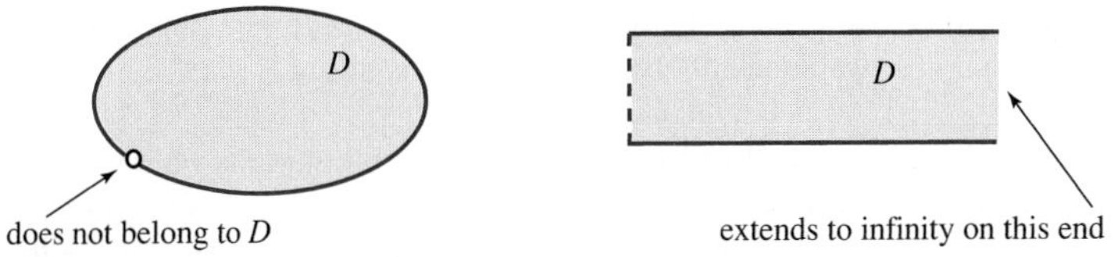

Figure 4.15 Sets that are not closed in $\mathbb{R}^2$.

find an open ball centered at (a, b) (such as the one shown in the picture) that contains only the points in D (and no points that do not belong to D). The point such as the point (a, b) in Figure 4.14(b) (which belongs to D but does not lie on its boundary) is called an *interior point*. By definition, if all boundary points are contained in D, then D is called a closed set.

The disk $D = \{(x, y) \mid x^2 + y^2 \leq 1\}$ contains the circle $\{(x, y) \mid x^2 + y^2 = 1\}$, which forms its boundary—thus, D is closed in $\mathbb{R}^2$. The first quadrant $\{(x, y) \mid x \geq 0, y \geq 0\}$ is closed in $\mathbb{R}^2$, since it contains all of its boundary points (namely, the positive x-axis, the positive y-axis, and the origin). The boundary of the right half-plane $D = \{(x, y) \mid x > 0\}$ is the y-axis. Since D does not contain the y-axis, it is not closed in $\mathbb{R}^2$.

The set $D = \{(x, y) \mid 0 \leq x \leq 2, 0 < y \leq 1\}$ is not closed in $\mathbb{R}^2$; for example, the point $(1, 0)$ belongs to the boundary of D but is not in D. Figure 4.15 shows two more sets that are not closed in $\mathbb{R}^2$. The set on the right is not closed because it does not contain its boundary (the segment on the left), and not because it extends to infinity.

A closed curve (such as a circle or an ellipse) is a closed set in $\mathbb{R}^2$.

DEFINITION 4.5 Bounded Set in $\mathbb{R}^2$

A set $D \subseteq \mathbb{R}^2$ is called *bounded* in $\mathbb{R}^2$ if it is contained inside some ball. ◀

For example, the set $D = \{(x, y) \mid -1 \leq x \leq 1, -1 \leq y \leq 1\}$ is bounded, since it is contained in the ball centered at the origin of radius 2 (actually, any ball centered at the origin of radius greater than $\sqrt{2}$ contains D). The sets shown in Figure 4.16 are bounded.

A set that extends to infinity cannot be bounded. For example, the horizontal strip $D = \{(x, y) \mid 0 \leq y \leq 1\}$ is not bounded. The region between the graph of $y = 1/x$ and the x-axis for $x > 0$ is not bounded. Figure 4.17 shows two sets that are not bounded in $\mathbb{R}^2$.

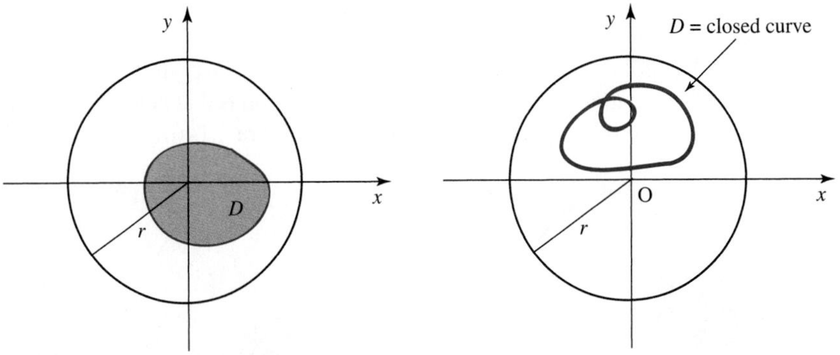

Figure 4.16 Bounded sets in $\mathbb{R}^2$.

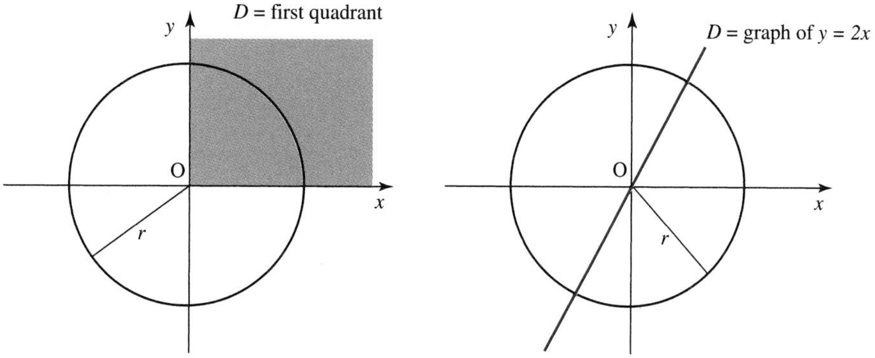

Figure 4.17 Sets that are not bounded in $\mathbb{R}^2$.

THEOREM 4.11 Extreme Value Theorem for Functions of Two Variables

Assume that a function $z = f(x, y)$ is continuous on a closed and bounded set $D \subseteq \mathbb{R}^2$. Then there exist points (a_1, b_1) and (a_2, b_2) in D such that $f(a_1, b_1)$ is an absolute minimum and $f(a_2, b_2)$ is an absolute maximum of f on D.

As in the case of a function of one variable, the theorem only guarantees the existence of absolute extrema. It does not tell us how to find them or at how many points in D they occur. Let us consider an example.

▶ EXAMPLE 4.30

Find the absolute minimum and absolute maximum of the function $f(x, y) = xy$ on the set $D = \{(x, y) \mid 0 \leq x \leq 1, 0 \leq y \leq 1\}$.

SOLUTION

Since $x \leq 1$ and $y \leq 1$, it follows that $f(x, y) = xy \leq 1$ for all $(x, y) \in D$. Thus, the absolute maximum is 1 and it occurs at the point $(1, 1)$ in D. The absolute minimum of f on D is zero, since $f(x, y) \geq 0$ whenever $x \geq 0$ and $y \geq 0$. It occurs at infinitely many points $(x, 0)$, $0 \leq x \leq 1$ and $(0, y)$, $0 \leq y \leq 1$.

The proof of Theorem 4.11 is beyond the scope of this book and we omit it here. However, to get a better feel of the theorem, we discuss two examples. They will show that if either assumption is removed, the theorem is no longer valid.

▶ EXAMPLE 4.31

Consider the function

$$f(x, y) = \begin{cases} x^2 + y^2 & \text{if } (x, y) \in D \text{ and } (x, y) \neq (0, 0) \\ 1 & \text{if } (x, y) = (0, 0) \end{cases}$$

where D is the disk $D = \{(x, y) \mid x^2 + y^2 \leq 1\}$.

The set D is closed and bounded, and, clearly, $f(x, y) > 0$ for all $(x, y) \in D$ [notice that the values $f(x, y)$ get arbitrarily close to 0]. However, f does not have an absolute minimum since there is no point (x, y) in D where $f(x, y) = 0$. The function f is not continuous at $(0, 0)$, since $0 = \lim_{(x,y)\to(0,0)} f(x, y) \neq f(0, 0) = 1$. Thus, Theorem 4.11 does not apply.

▶ EXAMPLE 4.32

Discuss the existence of absolute extrema of the function $f(x, y) = x + y$ on the sets $D_1 = \{(x, y) \mid 1 \leq x \leq 2, 1 < y < 2\}$ and $D_2 = \{(x, y) \mid 1 \leq x \leq 2, y \geq 0\}$.

SOLUTION

On the set D_1, $1 \leq x \leq 2$ and $1 < y < 2$. Thus, $x + y > 1 + 1 = 2$ and $x + y < 2 + 2 = 4$. By appropriately choosing a point (x, y) in D_1, $f(x, y)$ can be made arbitrarily close to 2 (or to 4), but never made equal to 2 (nor equal to 4). Consequently, f has neither an absolute minimum nor an absolute maximum. In this case, f is continuous, but D_1 is not closed (it is bounded); thus, Theorem 4.11 does not apply.

Now consider the set D_2. Since $f(x, y) = x + y$ approaches ∞ as y approaches ∞ (and $1 \leq x \leq 2$), it follows that f does not have an absolute maximum on D_2. This fact does not contradict Theorem 4.11 since D_2 is not bounded (for the record, it is closed).

We now describe a procedure for finding absolute extreme values.

Let $f(x, y)$ be a continuous function defined on a closed and bounded set D in $\mathbb{R}^2$. Remove the boundary points from D and denote the set thus obtained by U (hence, U is an open set in $\mathbb{R}^2$; it is called the *interior* of D); see Figure 4.18. Assume that the boundary ∂D (we denote a boundary of a set by ∂) of D is piecewise smooth: this means that ∂D consists of smooth curves whose endpoints are joined together to form a continuous curve; see Figure 4.18.

Theorem 4.9 states that if a differentiable function f attains its extreme value at a point $\mathbf{x}_0$ in (an open set) U, then $\mathbf{x}_0$ must be a critical point [i.e., $\nabla f(\mathbf{x}_0) = 0$]. In other words, if f has an extreme value in the *interior* of D, then it must occur at a critical point. This means that extreme values of f on D must occur either at critical points in the interior of D or at the points on the boundary of D.

Consequently, in order to find the absolute minimum and the absolute maximum of a differentiable function f on a closed and bounded set $D \subseteq \mathbb{R}^2$, we must:

(a) Compute the value of f at all critical points in the interior of D.

(b) Consider the restriction of f to the boundary of D and compute the value of that restriction at its critical points and at the endpoints of the boundary of D (if any).

(c) Choose the smallest and largest values among those from (a) and (b).

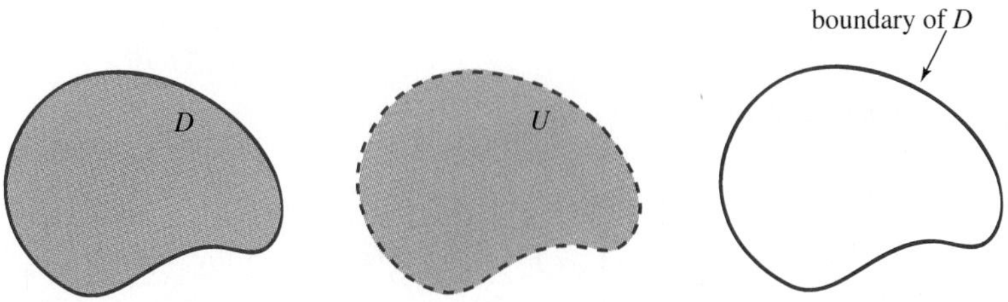

Figure 4.18 Set D, its interior U, and its piecewise smooth boundary ∂D.

▶ EXAMPLE 4.33

Find the absolute extreme values of the function $f(x, y) = 2x^2 + 2y^2 - x + y$ on the disk $D = \{(x, y) \mid x^2 + y^2 \leq 1\}$.

SOLUTION We first find critical points in the interior $U = \{(x, y) \mid x^2 + y^2 < 1\}$ of D. From $f_x = 4x - 1 = 0$ and $f_y = 4y + 1 = 0$, we get $(1/4, -1/4)$ to be the only critical point in U.

The boundary of D is the circle $x^2 + y^2 = 1$; parametrize it by $\mathbf{c}(t) = (\cos t, \sin t)$, $t \in [0, 2\pi]$ (see Section 2.5). Let g denote the function f, but viewed as a function on the boundary only; thus,

$$g(t) = f(\mathbf{c}(t)) = 2(\cos t)^2 + 2(\sin t)^2 - \cos t + \sin t = 2 - \cos t + \sin t,$$

where $t \in [0, 2\pi]$. Now apply the method for functions of one variable (see Example 4.29 and the text preceding it). From $g'(t) = \sin t + \cos t = 0$, it follows that $\sin t = -\cos t$ and $\tan t = -1$; thus, the critical points of g inside $[0, 2\pi]$ are $t = 3\pi/4$ and $t = 7\pi/4$. Consequently, the candidates for absolute extreme values (coming from the boundary) are $\mathbf{c}(3\pi/4) = (-\sqrt{2}/2, \sqrt{2}/2)$, $\mathbf{c}(7\pi/4) = (\sqrt{2}/2, -\sqrt{2}/2)$ (these are the critical points), and (the endpoints) $\mathbf{c}(0) = \mathbf{c}(2\pi) = (1, 0)$. Finally, from

$$\begin{aligned}
f(1/4, -1/4) &= \tfrac{1}{8} + \tfrac{1}{8} - \tfrac{1}{4} + \tfrac{1}{4} = \tfrac{1}{4}\\
f(-\sqrt{2}/2, \sqrt{2}/2) &= 1 + 1 + \tfrac{\sqrt{2}}{2} + \tfrac{\sqrt{2}}{2} = 2 + \sqrt{2}\\
f(\sqrt{2}/2, -\sqrt{2}/2) &= 1 + 1 - \tfrac{\sqrt{2}}{2} - \tfrac{\sqrt{2}}{2} = 2 - \sqrt{2}\\
f(1, 0) &= 2 + 0 - 1 + 0 = 1,
\end{aligned}$$

we conclude that $f(-\sqrt{2}/2, \sqrt{2}/2) = 2 + \sqrt{2}$ is the absolute maximum and $f(1/4, -1/4) = 1/4$ is the absolute minimum of f on D. ◀

▶ EXAMPLE 4.34

Find the absolute minimum and absolute maximum values of the function $f(x, y) = x^3 - 3xy + 3y^2$ on the rectangle $D = \{(x, y) \mid 0 \leq x \leq 1, 0 \leq y \leq 2\}$.

SOLUTION First, we must locate critical points of f. Combining $f_x = 3x^2 - 3y = 0$ and $f_y = -3x + 6y = 0$, we see that $y = x^2$ and $x = 2y$. Thus, $y = (2y)^2$, and $4y^2 - y = y(4y - 1) = 0$. So $y = 0$ [and hence $x = 0$] or $y = 1/4$ [and hence $x = 2(1/4) = 1/2$]. It follows that f has two critical points: $(0, 0)$ and $(1/2, 1/4)$. However, only $(1/2, 1/4)$ belongs to the interior of D. The corresponding value of f is $f(1/2, 1/4) = -1/16$.

The boundary of D consists of four line segments, $\mathbf{c}_1$, $\mathbf{c}_2$, $\mathbf{c}_3$, and $\mathbf{c}_4$; see Figure 4.19.

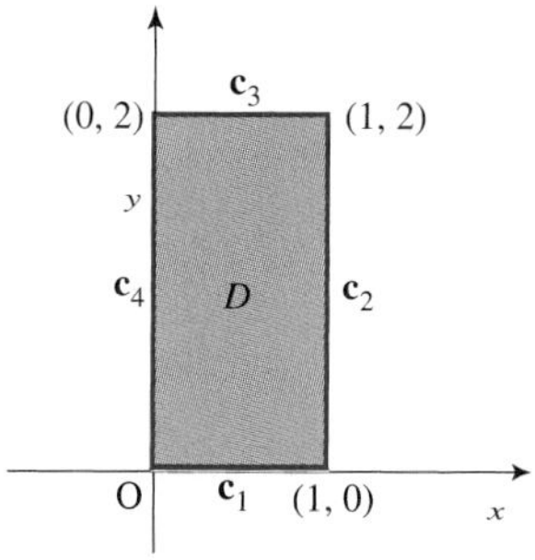

Figure 4.19 Domain D of the function $f(x, y)$ of Example 4.34.

Parametrize the segment $\mathbf{c}_1$ by $\mathbf{c}_1(t) = (t, 0)$, $0 \le t \le 1$. The value of f along $\mathbf{c}_1$ is $g_1(t) = f(\mathbf{c}_1(t)) = f(t, 0) = t^3$, $t \in [0, 1]$. Since t^3 is an increasing function, the candidates for extreme values are $g_1(0) = f(0, 0) = 0$ and $g_1(1) = f(1, 0) = 1$.

Parametrize the segment $\mathbf{c}_2$ by $\mathbf{c}_2(t) = (1, t)$, $0 \le t \le 2$. The value of f along $\mathbf{c}_2$ is $g_2(t) = f(\mathbf{c}_2(t)) = f(1, t) = 1 - 3t + 3t^2$, $t \in [0, 2]$. From $g_2'(t) = -3 + 6t = 0$, it follows that $t = 1/2$. Thus, the candidates for extreme values are $\mathbf{c}_2(1/2) = (1, 1/2)$ (critical point) and $(1, 0)$ and $(1, 2)$ (endpoints). The corresponding values of f are $f(1, 1/2) = 1/4$, $f(1, 0) = 1$ and $f(1, 2) = 7$.

To check the remaining two sides, we could imitate the above two computations; here is an alternative (not much different, though). On the line segment $\mathbf{c}_3$, the values of x range from 0 to 1, whereas the values of y are fixed at 2. Thus, the values of f along $\mathbf{c}_3$ are given by $g_3(x) = f(x, 2) = x^3 - 6x + 12$, $x \in [0, 1]$. From $g_3'(x) = 3x^2 - 6 = 0$, it follows that $x^2 = 2$ and $x = \pm\sqrt{2}$; thus, g_3 has no critical points inside $[0, 1]$. Consequently, the candidates for the extreme values are $f(0, 2) = 12$ and $f(1, 2) = 7$ (i.e., the values of f at endpoints).

Finally, along $\mathbf{c}_4$, $x = 0$ and $0 \le y \le 2$. Hence, $g_4(y) = f(0, y) = 3y^2$, $y \in [0, 2]$, represents the values of f along $\mathbf{c}_4$. Since $g_4(y)$ is increasing on $[0, 2]$, the candidates for extreme values are $g_4(0) = f(0, 0) = 0$ and $g_4(2) = f(0, 2) = 12$.

It follows that $f(0, 2) = 12$ is the absolute maximum and $f(1/2, 1/4) = -1/16$ is the absolute minimum of f on D. ◀

▶ EXAMPLE 4.35

Find the absolute minimum and absolute maximum values of the function $f(x, y) = e^{-x^2-y^2}$ on the set D shown in Figure 4.20. Assume that D contains its boundary.

SOLUTION

From $f_x = -2xe^{-x^2-y^2} = 0$ and $f_y = -2ye^{-x^2-y^2} = 0$, it follows that $(0, 0)$ is the only critical point inside D. The corresponding value is $f(0, 0) = 1$.

The boundary of D consists of the straight-line segment $\mathbf{c}_1$ joining $(1, 0)$ and $(0, 1)$ and the circular arc (denote it by $\mathbf{c}_2$) $x^2 + y^2 = 1$ connecting $(0, 1)$ and $(1, 0)$.

The segment $\mathbf{c}_1$ is given by $y = 1 - x$, where $0 \le x \le 1$. The values of f along $\mathbf{c}_1$ are given by

$$g_1(x) = f(x, 1 - x) = e^{-x^2-(1-x)^2} = e^{-2x^2+2x-1},$$

where $x \in [0, 1]$. From $g_1'(x) = (-4x + 2)e^{-2x^2+2x-1} = 0$, we get $x = 1/2$ (and so $y = 1 - 1/2 = 1/2$). Thus, the candidates for the absolute extrema are $g_1(1/2) = f(1/2, 1/2) = e^{-1/2}$ and the values at the endpoints, $g_1(0) = f(0, 1) = e^{-1}$ and $g_1(1) = f(1, 0) = e^{-1}$.

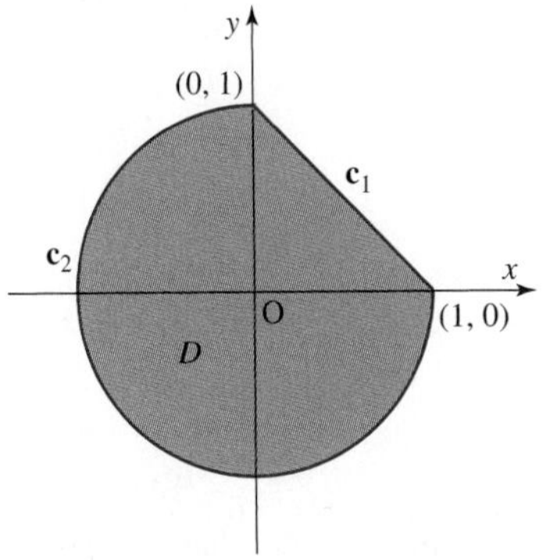

Figure 4.20 Set D of Example 4.35.

Parametrize $\mathbf{c}_2$ by $\mathbf{c}_2(t) = (\cos t, \sin t)$, $t \in [\pi/2, 2\pi]$. The values of f along $\mathbf{c}_2$ are given by $g_2(t) = e^{-(\cos t)^2-(\sin t)^2} = e^{-1}$. Thus, f is a constant function when viewed as a function on $\mathbf{c}_2$ only (we say that the restriction of f to $\mathbf{c}_2$ is a constant function).

Consequently, the absolute maximum of f is $f(0,0) = 1$ and the absolute minimum is e^{-1} (it occurs at all points on $\mathbf{c}_2$, including its endpoints).

Let us mention that results stated in this section generalize to functions of an arbitrary number of variables. The most important result is the generalization of the Extreme Value Theorem (see Theorem 4.11). It states that a continuous function f defined on a closed and bounded set $D \subseteq \mathbb{R}^m$, $m \geq 1$, attains its maximum and minimum values at some points $\mathbf{a}_1$ and $\mathbf{a}_2$ in D. ("Closed" is defined as in Definition 4.4, and "bounded" as in Definition 4.5, by replacing "open ball in $\mathbb{R}^2$" by "open ball in $\mathbb{R}^m$.")

There is an analogue of the Second Derivatives Test (with the same philosophy: second partials are used to determine what is happening at a critical point). Unfortunately, technical intricacies and difficulties increase proportionally to the number of variables.

EXERCISES 4.3

1. Show that if $f: \mathbb{R} \to \mathbb{R}$ is an even differentiable function, then $x = 0$ is a critical point of f. A function $f: \mathbb{R}^2 \to \mathbb{R}$ is called *even* if $f(-x, -y) = f(x, y)$ for all $(x, y) \in \mathbb{R}^2$. Assuming that f is differentiable, show that (0,0) is a critical point of f.

2. We discuss an alternative proof of the Second Derivatives Test, case (a).

(a) By Theorem 2.8 in Section 2.7, the directional derivative of f in the direction of a unit vector $\mathbf{u} = (u_1, u_2)$ is given by $D_\mathbf{u} f = \nabla f \cdot \mathbf{u} = f_x u_1 + f_y u_2$. By the same theorem, $D_\mathbf{u}(D_\mathbf{u} f) = \nabla(D_\mathbf{u} f) \cdot \mathbf{u} = \frac{\partial}{\partial x}(D_\mathbf{u} f)u_1 + \frac{\partial}{\partial y}(D_\mathbf{u} f)u_2$. Continue this calculation to show that $D_\mathbf{u}(D_\mathbf{u} f) = f_{xx}u_1^2 + 2f_{xy}u_1u_2 + f_{yy}u_2^2$.

(b) Complete the square to get $D_\mathbf{u}(D_\mathbf{u} f) = f_{xx}\left(u_1 + \frac{f_{xy}}{f_{xx}}u_2\right)^2 + \frac{u_2^2}{f_{xx}}\left(f_{xx}f_{yy} - f_{xy}^2\right)$.

(c) Explain why we can find an open ball B centered at (x_0, y_0) such that $f_{xx}(x, y) < 0$ and $D(x, y) > 0$ for all (x, y) in B. Show that $D_\mathbf{u}(D_\mathbf{u} f)(x, y) < 0$ for all (x, y) in B. Conclude that $f(x_0, y_0)$ is a local maximum.

Exercises 3 to 6: Looking at the gradient vector field of a differentiable function $f(x, y)$, identify points (or say there are none) where f has a local minimum, local maximum, or saddle point.

3.

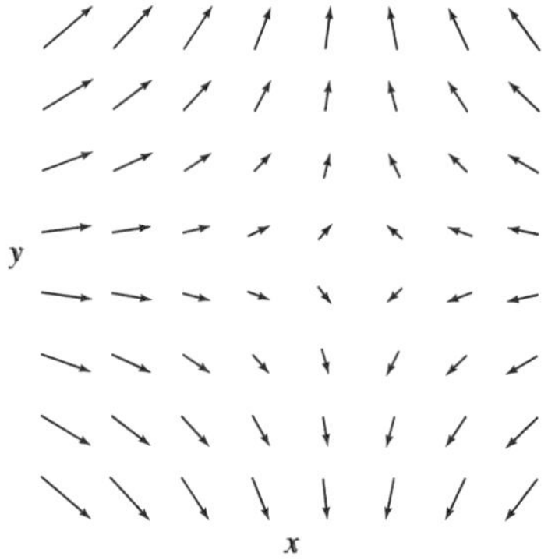

4.

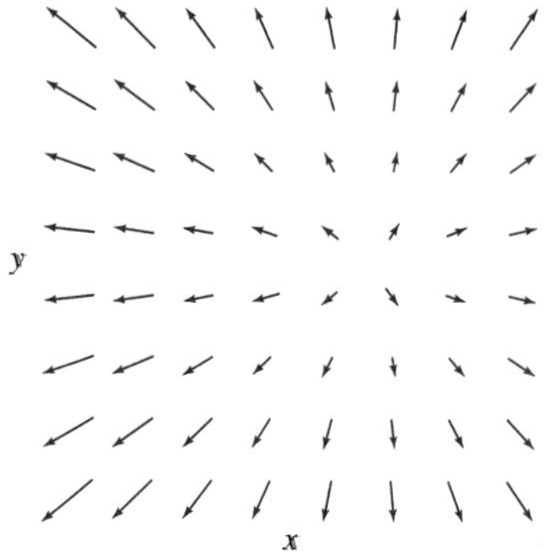

5.

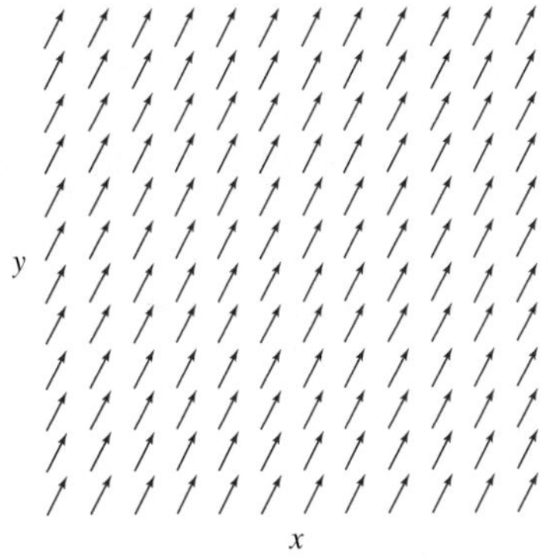

6.

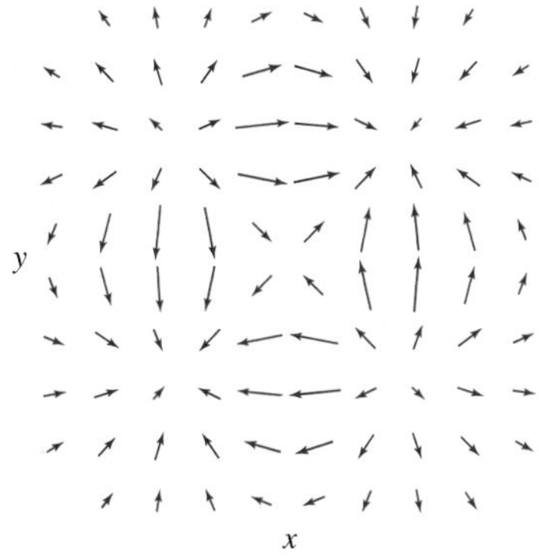

Exercises 7 to 16: Find all critical points (if any) of a given function $f(x, y)$ and determine whether they are local extreme points or saddle points.

7. $f(x, y) = x^2 + y^2 + xy^2$

8. $f(x, y) = x + y + \dfrac{2}{xy}$

9. $f(x, y) = xy + \dfrac{x+y}{xy}$

10. $f(x, y) = x^3 + y^3 + 3x^2y - 3y$

11. $f(x, y) = xye^{-x^2-y^2}$

12. $f(x, y) = e^x \sin y$

13. $f(x, y) = x \cos y$

14. $f(x, y) = \ln(x^2 + y^2 + 2)$

15. $f(x, y) = x \sin(x + y)$

16. $f(x, y) = (x + y)(xy - 1)$

17. Find the shortest distance from the point $(2, 0, 3)$ to the plane $x - y + z = 4$.

18. Find the shortest distance between the surface $z = 1/xy$ and the origin.

19. Find the dimensions of a closed, rectangular box of given volume $V > 0$ that has minimum surface area.

20. Find the point(s) on the surface $xyz + 1 = 0$ that are closest to the origin.

21. Find the volume of the largest (i.e., of maximum volume) rectangular box that can be inscribed into the sphere of radius $R > 0$.

22. Suppose that you have to build a rectangular box (with a lid) using $S > 0$ units^2 of material. Find the dimensions of the box that has the largest possible volume.

23. It was shown that the function $g(x, y) = x^4 - y^4$ of Example 4.28 has a saddle point at $(0, 0)$. Draw the contour curve that goes through $(0, 0)$. Add a few more level curves to your picture.

24. Find all points where the magnitude of the vector field $\mathbf{F} = (x - y)\mathbf{i} + (2x + y + 3)\mathbf{j}$ attains its local minimum.

25. A plane in a three-dimensional space, which is not parallel to any of the three coordinate planes, can be analytically described using the equation $x/a + y/b + z/c = 1$, where a, b, and c are its x-intercept, y-intercept, and z-intercept, respectively. Find the plane that passes through $(1, 1, 1)$ and is such that the solid in the first octant bounded by that plane has the smallest volume.

Exercises 26 to 29: Find the absolute minimum and absolute maximum of a given function $f(x, y)$ on a set D.

26. $f(x, y) = xy - 3x + y$; D is the triangular region with vertices $(0, 0)$, $(2, 0)$, and $(0, 2)$

27. $f(x, y) = \ln(x^2 + y + 1)$; D is the triangular region with vertices $(0, 0)$, $(1, 0)$, and $(1, 1)$

28. $f(x, y) = 2x^2 + y^2 + 2$; D is the disk $\{(x, y) \mid x^2 + y^2 \le 4\}$

29. $f(x, y) = \sin(xy)$; D is the rectangular region $\{(x, y) \mid 0 \le x \le \pi/2, \pi/2 \le y \le \pi\}$

30. The temperature at a point (x, y) on a metal plate D in the shape of the square $\{(x, y) \mid 0 \le x \le 1, 0 \le y \le 1\}$ is given by $T(x, y) = x^2 e^y$. Identify the warmest and coldest points in D.

31. The pressure at a point (x, y) on a membrane in the shape of the disk $\{(x, y) \mid x^2 + y^2 \le 1\}$ is given by $p(x, y) = e^{x^2+y^2+y} + 1$. Find the points on the membrane where the pressure is the strongest and where it is the weakest.

▶ 4.4 OPTIMIZATION WITH CONSTRAINTS AND LAGRANGE MULTIPLIERS

In this section, we study another common optimization problem: we are asked to find extreme values of a differentiable function f subject to certain conditions (or constraints) on its variables. Geometrically, we think of it in the following way: the conditions (one, or more) imposed on the variables define a subset D of the domain of f. We do not consider the values of f at all points in its domain, but restrict our attention to the values of f computed at the points that belong to D. (In other words, we are looking at the restriction of f to D.) Our task is to identify the minimum and maximum (if they exist) among these values of f. Let us consider two examples.

▶ EXAMPLE 4.36

Let $T(x, y) = x^2 - y + 200$ be the temperature at a point (x, y) on a thin metal plate in the shape of the disk $D = \{(x, y) \mid x^2 + y^2 \le 1\}$. Find the maximum temperature on the rim of D.

In other words, we have to find the maximum value of the function $T(x, y) = x^2 - y + 200$, but only among the values of T computed at the points (x, y) that satisfy $x^2 + y^2 = 1$. The equation $x^2 + y^2 = 1$ is called a constraint. ◀

▶ EXAMPLE 4.37 Building an Optimal Tank

Suppose that we have $S > 0$ units^2 of material available and have to build a cylindrical tank of maximum possible volume. Describe this problem as an optimization problem with a constraint.

SOLUTION

Denote the radius and height of the tank by r and h, respectively ($r > 0$, $h > 0$). We have to find the largest value of the volume $V(r, h) = \pi r^2 h$, but not among all possible values of r and h (which, by the way, would not make much sense, since then V could be made arbitrarily large). The values of r and h that we consider are related by the requirement that the surface area of the tank must be equal to S. Since the surface area of the tank is equal to $2\pi r^2 + 2\pi rh$, we get $2\pi r^2 + 2\pi rh = S$.

Thus, we are asked to find the maximum of the function $V(r, h) = \pi r^2 h$ subject to the constraint $2\pi r^2 + 2\pi rh = S$. ◀

To find extreme values of a function subject to a constraint, we apply the method of Lagrange multipliers. We will formulate the procedure for functions of two variables and will provide a geometric verification of it using the concept of a gradient vector. Then we will generalize the method to functions of three variables and to optimization problems involving two constraints. We start by considering functions of two variables.

Lagrange Multipliers for Functions of Two Variables

We are looking for the maximum value (minimum value) of a continuously differentiable (i.e., C^1) function $f(x, y)$ subject to a constraint $g(x, y) = k$. Assume that $g(x, y)$ is also a C^1 function. Let us think about it in geometric terms; see Figure 4.21. We use level curves to visualize the function $f(x, y)$, and the constraint $g(x, y) = k$ is shown as a curve in the xy-plane. As we walk along $g(x, y) = k$, we cross level curves of f. The problem of finding maximum (minimum) of f subject to $g(x, y) = k$ consists of finding the level curve of largest (smallest) value that we meet on our walk along the curve $g(x, y) = k$.

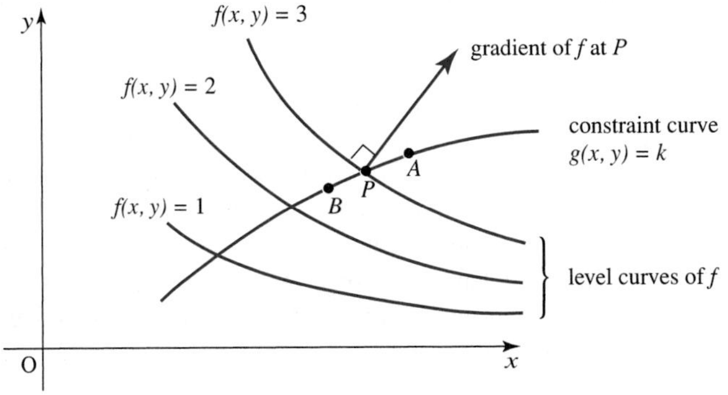

Figure 4.21 Geometric intuition about constrained maximum and minimum values.

Suppose that we are at the point P on $g(x, y) = k$ and that $\nabla f(P) \neq \mathbf{0}$. Recall that at P, f increases in *any* direction that makes an angle of less than $\pi/2$ with respect to $\nabla f(P)$ [recall that $D_{\mathbf{u}} f(P) = \|\nabla f(P)\| \cos\theta$, where θ is the angle between $\nabla f(P)$ and the unit vector $\mathbf{u}$].

Thus, if we walk along $g(x, y) = k$ from P toward A, f will increase (and thus P cannot be a maximum). Likewise, if we walk from P toward B, values of f will decrease (and so P cannot be a minimum). Consequently, as long as $\nabla f(P)$ is not perpendicular to $g(x, y) = k$, we can make f larger (or smaller) than $f(P)$. Thus, it seems that, when ($\nabla f(P) \neq \mathbf{0}$ and) $\nabla f(P)$ is perpendicular to $g(x, y) = k$, f could have an extreme value (maximum or minimum) subject to the given constraint at P.

Let us make this reasoning more formal. Suppose that $f(x, y)$ has a local maximum (or a local minimum) at (x_0, y_0) subject to $g(x, y) = k$. This means that $f(x_0, y_0) \geq f(x, y)$ [or $f(x_0, y_0) \leq f(x, y)$] for all (x, y) that are near (x_0, y_0) *and* that belong to the curve $g(x, y) = k$. Assume that $\nabla g(x_0, y_0) \neq \mathbf{0}$.

Recall that $g(x, y) = k$ can be viewed as a level curve of the surface $z = g(x, y)$ of value k. Thus, by Theorem 2.10 in Section 2.7, $\nabla g(x_0, y_0)$ is perpendicular to the curve $g(x, y) = k$ at (x_0, y_0); see Figure 4.22. Parametrize the curve $g(x, y) = k$ by $\mathbf{c}(t) = (x(t), y(t))$, $t \in [c, d]$, and assume that $\mathbf{c}(t_0) = (x_0, y_0)$ for some $t_0 \in (c, d)$. [The fact that $\nabla g(x_0, y_0)$ is perpendicular to $g(x, y) = k$ at (x_0, y_0) can now be written as $\nabla g(x_0, y_0) \cdot \mathbf{c}'(t_0) = 0$.]

Consider the function $\phi(t) = f(\mathbf{c}(t)) = f(x(t), y(t))$, $t \in [c, d]$. In words, $\phi(t)$ is the function f, but viewed as a function defined only at the points on the curve $g(x, y) = k$ [i.e., ϕ is the restriction of f to $g(x, y) = k$]. Note that $\phi(t_0) = f(\mathbf{c}(t_0)) = f(x_0, y_0)$. By assumption, $\phi(t)$ has a local maximum (local minimum) at $t = t_0$; thus, $\phi'(t_0) = 0$. Applying

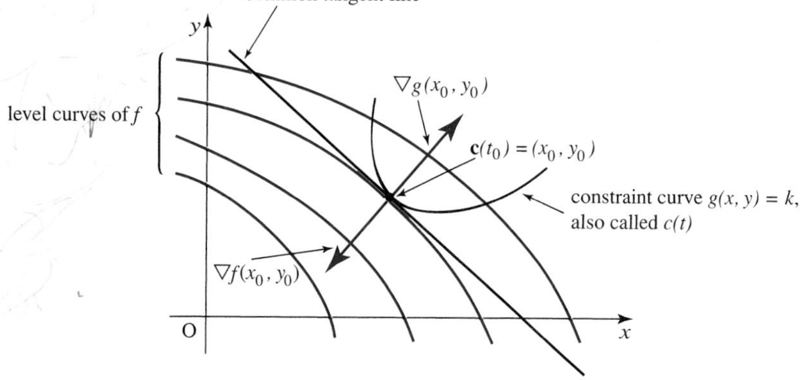

Figure 4.22 Geometric interpretation of the method of Lagrange multipliers.

the chain rule to $\phi(t) = f(x(t), y(t))$, we get

$$\begin{aligned}\phi'(t) &= \frac{\partial f}{\partial x}(x(t), y(t))x'(t) + \frac{\partial f}{\partial y}(x(t), y(t))y'(t)\\ &= \left(\frac{\partial f}{\partial x}(x(t), y(t)), \frac{\partial f}{\partial y}(x(t), y(t))\right) \cdot (x'(t), y'(t)) = \nabla f(\mathbf{c}(t)) \cdot \mathbf{c}'(t).\end{aligned}$$

Since $\phi'(t_0) = 0$, it follows that

$$0 = \phi'(t_0) = \nabla f(\mathbf{c}(t_0)) \cdot \mathbf{c}'(t_0) = \nabla f(x_0, y_0) \cdot \mathbf{c}'(t_0).$$

So, $\nabla f(x_0, y_0)$ (if nonzero) must also be perpendicular to $\mathbf{c}'(t)$, and thus to the curve $\mathbf{c}(t)$ at (x_0, y_0).

This geometric argument shows that both $\nabla f(x_0, y_0)$ and $\nabla g(x_0, y_0)$ are perpendicular to the curve $\mathbf{c}(t)$ at (x_0, y_0). Consequently (we are in two dimensions), they must be parallel; that is, there is a real number $\lambda \neq 0$ such that

$$\nabla f(x_0, y_0) = \lambda \nabla g(x_0, y_0).$$

If $\nabla f(x_0, y_0) = \mathbf{0}$, the above equation still holds (with $\lambda = 0$). Thus, we proved the following statement.

THEOREM 4.12 Lagrange Multipliers for Functions of Two Variables

Let $f, g: U \subseteq \mathbb{R}^2 \to \mathbb{R}^2$ be C^1 functions. If the function $f(x, y)$ has a local maximum or a local minimum subject to the constraint $g(x, y) = k$ at $\mathbf{x}_0 = (x_0, y_0)$, and if $\nabla g(\mathbf{x}_0) \neq \mathbf{0}$, then $\nabla f(x_0, y_0) = \lambda \nabla g(x_0, y_0)$, for some real number λ.

Geometrically, the points (x_0, y_0) where $\nabla f(x_0, y_0) = \lambda \nabla g(x_0, y_0)$ are the points where the level curves of f have the same tangent line as the curve $g(x, y) = k$; see Figure 4.22. This discussion leads to the following method (known as the *method of Lagrange multipliers*) for detecting extreme values subject to a constraint.

Assume that $\nabla g(x, y) \neq \mathbf{0}$ for all (x, y) that lie on the constraint curve $g(x, y) = k$. We first identify all points (x_0, y_0) that satisfy

$$\nabla f(x_0, y_0) = \lambda \nabla g(x_0, y_0) \qquad \text{and} \qquad g(x_0, y_0) = k, \tag{4.28}$$

where λ is a real number.

Suppose that the constraint curve $g(x, y) = k$ is a closed and bounded set in $\mathbb{R}^2$. By Theorem 4.11, f must have extreme values. Thus, the largest (smallest) of the values $f(x_0, y_0)$ for all points from (4.28) is the maximum (minimum) value of f subject to the constraint $g(x, y) = k$. If the curve $g(x, y) = k$ is not bounded or if it is not closed, additional arguments may be needed to determine whether (x_0, y_0) is a minimum point, a maximum point, or neither; see Examples 4.40 and 4.41 below.

Real number λ is called a *Lagrange multiplier*. When solving (4.28), it is not necessary to find the value(s) of λ. An alternative to (4.28) would be to consider the function

$$F(x, y, \lambda) = f(x, y) - \lambda(g(x, y) - k)$$

of *three* variables [$F(x, y, \lambda)$ is sometimes called the auxiliary function] and find its critical points. That this is so is easy to check: from $F_x = f_x - \lambda g_x = 0$, we get $f_x = \lambda g_x$; from $F_y = f_y - \lambda g_y = 0$, we get $f_y = \lambda g_y$ (these two, put together, give $\nabla f = \lambda \nabla g$); finally, from $F_\lambda = -(g(x, y) - k) = 0$, we get the constraint $g(x, y) = k$.

▶ EXAMPLE 4.38

Find the maximum and minimum values of the function $f(x, y) = 3x + y - 4$ on the curve $x^2 + 2y^2 = 38$.

SOLUTION

Let $g(x, y) = x^2 + 2y^2$. We are asked to find extreme values of $f(x, y) = 3x + y - 4$ subject to the constraint $g(x, y) = 38$. The constraint represents an ellipse, which is a closed and bounded set in $\mathbb{R}^2$. We compute the gradient vectors first: $\nabla f = (3, 1)$, $\nabla g = (2x, 4y)$. From $\nabla f = \lambda \nabla g$, it follows that $3 = 2x\lambda$ and $1 = 4y\lambda$. Eliminating λ from both equations, we get $\lambda = 3/2x$ and $\lambda = 1/4y$. Thus, $3/2x = 1/4y$; i. e., $2x = 12y$ and $x = 6y$. Substituting $x = 6y$ into the constraint $x^2 + 2y^2 = 38$, we get $38y^2 = 38$; thus, $y^2 = 1$ and $y = \pm 1$. From $x = 6y$, we get that $x = \pm 6$. Consequently, there are two candidates for extreme values, $(6, 1)$ and $(-6, -1)$.

From $\nabla g(x, y) = (2x, 4y) = \mathbf{0}$, it follows that $x = 0$ and $y = 0$. In other words, $\nabla g(x, y) \neq \mathbf{0}$ at all points on the constraint curve $x^2 + 2y^2 = 38$.

We compute $f(6, 1) = 15$ and $f(-6, -1) = -23$. Consequently, $f(6, 1) = 15$ is the maximum of f and $f(-6, -1) = -23$ is the minimum of f subject to the given constraint. Figure 4.23 shows the geometry of the Lagrange multiplier method: the points $(6, 1)$ and $(-6, -1)$ are the points where the level curves of f and the constraint curve $x^2 + 2y^2 = 38$ have the same tangent lines. Equivalently, at these points, the two vectors ∇f and ∇g are parallel. ◀

▶ EXAMPLE 4.39 Solution of Example 4.36

Find the maximum of $T(x, y) = x^2 - y + 200$ subject to the constraint $x^2 + y^2 = 1$.

SOLUTION

The constraint $g(x, y) = x^2 + y^2 = 1$ is a circle—thus, it is a closed and bounded set in $\mathbb{R}^2$. Substituting $\nabla T = (2x, -1)$ and $\nabla g = (2x, 2y)$ into $\nabla T = \lambda \nabla g$, we get that $2x = 2x\lambda$ and $-1 = 2y\lambda$. From the first equation, it follows that $2x(1 - \lambda) = 0$; so $x = 0$ or $\lambda = 1$. If $x = 0$, then the constraint

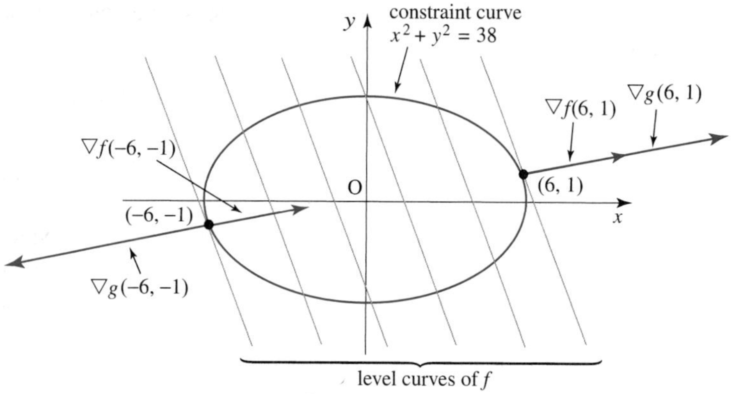

Figure 4.23 Geometric interpretation of Lagrange multiplier method.

$x^2 + y^2 = 1$ implies that $y^2 = 1$ and $y = \pm 1$. If $\lambda = 1$, then from $-1 = 2y\lambda$, we get $y = -1/2$. In that case (use the constraint $x^2 + y^2 = 1$ again), $x^2 = 3/4$ and $x = \pm\sqrt{3}/2$. Thus, there are four candidates for extreme values: $(0, 1)$, $(0, -1)$, $(\sqrt{3}/2, -1/2)$, and $(-\sqrt{3}/2, -1/2)$.

From $\nabla g(x, y) = (2x, 2y) = \mathbf{0}$, it follows that $x = 0$ and $y = 0$. Thus, $\nabla g(x, y) \neq \mathbf{0}$ for all (x, y) that satisfy $x^2 + y^2 = 1$.

From $T(0, 1) = 199$, $T(0, -1) = 201$, $T(\sqrt{3}/2, -1/2) = 3/4 + 1/2 + 200 = 201.25$, and $T(-\sqrt{3}/2, -1/2) = 3/4 + 1/2 + 200 = 201.25$, it follows that the maximum temperature of 201.25 (units) is reached at the points $(\sqrt{3}/2, -1/2)$ and $(-\sqrt{3}/2, -1/2)$ on the circle $x^2 + y^2 = 1$; see Figure 4.24.

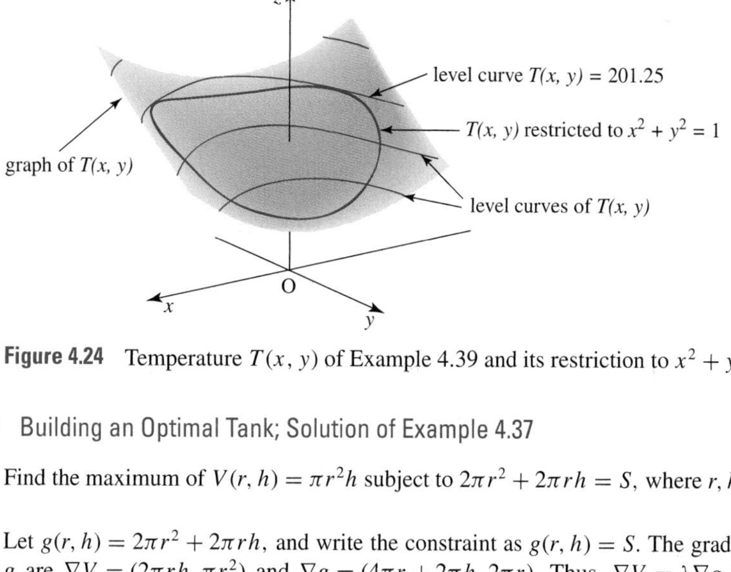

Figure 4.24 Temperature $T(x, y)$ of Example 4.39 and its restriction to $x^2 + y^2 = 1$. ◀

▶ EXAMPLE 4.40 Building an Optimal Tank; Solution of Example 4.37

Find the maximum of $V(r, h) = \pi r^2 h$ subject to $2\pi r^2 + 2\pi r h = S$, where $r, h > 0$.

SOLUTION Let $g(r, h) = 2\pi r^2 + 2\pi r h$, and write the constraint as $g(r, h) = S$. The gradient vectors of V and g are $\nabla V = (2\pi r h, \pi r^2)$ and $\nabla g = (4\pi r + 2\pi h, 2\pi r)$. Thus, $\nabla V = \lambda \nabla g$ implies that $2\pi r h = \lambda(4\pi r + 2\pi h)$ and $\pi r^2 = \lambda 2\pi r$; simplifying, we get the system

$$rh = \lambda(2r + h) \qquad \text{and} \qquad r = 2\lambda.$$

Substituting $\lambda = r/2$ into the first equation, we get $rh = r(2r + h)/2$; that is, $2h = 2r + h$ and $h = 2r$. Solving the constraint equation $2\pi r^2 + 2\pi rh = 2\pi r^2 + 2\pi r(2r) = S$ for r yields $6\pi r^2 = S$ and $r = \sqrt{S/6\pi}$ (since $r > 0$). It follows that $h = 2r = 2\sqrt{S/6\pi}$.

The equation $\nabla g = (4\pi r + 2\pi h, 2\pi r) = (0, 0)$ implies that $r = 0$ and $h = 0$. Thus, $\nabla g \neq \mathbf{0}$ for all r and h that satisfy the constraint.

From $2\pi r^2 + 2\pi rh = S$, we get $r^2 + rh = S/2\pi$, and, after completing the square,

$$\left(r + \frac{h}{2}\right)^2 - \frac{h^2}{4} = \frac{S}{2\pi}.$$

This equation represents a hyperbola (coordinate axes are r and h), which is an unbounded set (consequently, the Extreme Value Theorem does not apply).

So, there is only one candidate for extreme values, $(r = \sqrt{S/6\pi}, h = 2\sqrt{S/6\pi})$. The corresponding volume is

$$V(\sqrt{S/6\pi}, 2\sqrt{S/6\pi}) = \pi \left(\sqrt{\frac{S}{6\pi}}\right)^2 2\sqrt{\frac{S}{6\pi}} = \frac{S^{3/2}}{3\sqrt{6\pi}}.$$

We claim that this is the maximum. From $2\pi r^2 + 2\pi rh = S$, we get $h = S/2\pi r - r$ and $V = \pi r^2 h = Sr/2 - \pi r^3$. Thus, $V'' = -6\pi r < 0$ (for $r > 0$), and it follows that V, viewed as a function on the constraint only, is concave down. Since $V(0) = V(\sqrt{S/2\pi}) = 0$, V must have a positive maximum between 0 and $\sqrt{S/2\pi}$. ◀

This example should be viewed as an exercise in the use of the Lagrange multiplier method. Here is a faster alternative: the restriction of V to the constraint curve is $V(r) = Sr/2 - \pi r^3$. Now all we have to do is to find the absolute maximum of V (which is a function of one variable) on $[0, \sqrt{S/2\pi}]$.

Let us consider another example of a constraint that is not a bounded set.

▶ EXAMPLE 4.41

Find the maximum and minimum of the function $f(x, y) = x + y$ subject to the constraint $g(x, y) = xy = 1$.

SOLUTION

From $\nabla f = \lambda \nabla g$, we get $(1, 1) = \lambda(y, x)$. Thus, $1 = \lambda x$ and $1 = \lambda y$; eliminating λ, we get $\lambda = 1/x$ and $\lambda = 1/y$, and so $x = y$. From the constraint $xy = 1$, it follows that $x^2 = 1$ and $x = \pm 1$. Consequently, $y = \pm 1$.

From $\nabla g = (y, x) = \mathbf{0}$, it follows that $x = 0$ and $y = 0$; since $(0, 0)$ does not satisfy $xy = 1$, we conclude that $\nabla g(x, y) \neq \mathbf{0}$ for all points (x, y) on the constraint curve $xy = 1$. The constraint curve is a hyperbola $y = 1/x$, which is an unbounded set; thus, the Extreme Value Theorem cannot be applied.

The values of f at the two points that we identified are $f(1, 1) = 2$ and $f(-1, -1) = -2$. However, $f(1, 1) = 2$ is not a maximum. For example, if $x = 2$ and $y = 1/2$ (of course, they satisfy $xy = 1$), then $f(2, 1/2) = 5/2$. As a matter of fact, $x + y$ can be made larger than any positive number N: just take $x = N$ and $y = 1/N$; it follows that $f(N, 1/N) = N + 1/N > N$. Similar argument shows that $f(-1, -1) = -2$ is not a minimum. ◀

Although the Lagrange multiplier λ is present in calculations, its value is not needed to identify constrained extreme values. However, λ does have a meaning. Let us illustrate it in an example.

In Example 4.38, we showed that $f(6,1)=15$ was the maximum of the function $f(x,y)=3x+y-4$ subject to the constraint $x^2+2y^2=38$. The corresponding value $\lambda=1/4$ can be computed from $3=2x\lambda$ or from $1=4y\lambda$, by substituting $x=6$ or $y=1$.

Now, keeping $f(x,y)$ as is, we increase the constraint by one unit, so that it reads $x^2+2y^2=39$. Repeating the calculation with this new constraint, we see that the maximum occurs at $x=6\sqrt{39/38}$ and $y=\sqrt{39/38}$. The new maximum value of f is computed to be $f(6\sqrt{39/38},\sqrt{39/38})=15.24838$. We notice that the increase in maximum value, $15.24838-15=0.24838$, is approximately equal to the value $\lambda=1/4=0.25$!

This fact holds in general; that is, λ is approximately equal to the change in the maximum value of f as k in the constraint $g(x,y)=k$ increases by one unit. The proof of this fact is left as exercise (see Exercise 32).

We now state a general version of the method of Lagrange multipliers for functions of two variables. Keep in mind that the case $\nabla g(x_0,y_0)=\mathbf{0}$ [for a point (x_0,y_0) on the constraint curve] has not been included in our analysis [see (4.28)] and that extreme values can occur at endpoints.

To find maximum and minimum values of a C^1 function $f(x,y)$ subject to a constraint $g(x,y)=k$ [assume that $g(x,y)$ is also C^1], we identify all points (x_0,y_0) that satisfy any one of the following:

(a) $\nabla f(x_0,y_0)=\lambda\nabla g(x_0,y_0)$, $\lambda\in\mathbb{R}$, and $g(x_0,y_0)=k$.

(b) $\nabla g(x_0,y_0)=\mathbf{0}$ and $g(x_0,y_0)=k$.

(c) (x_0,y_0) is an endpoint of the curve $g(x,y)=k$.

If the constraint curve $g(x,y)=k$ is a closed and bounded set in $\mathbb{R}^2$, then the largest (smallest) of the values $f(x_0,y_0)$ for all points from (a)–(c) is the maximum (minimum) value of f subject to the constraint $g(x,y)=k$. If the curve $g(x,y)=k$ is not bounded or if it is not closed, additional arguments may be needed to determine whether (x_0,y_0) is a minimum point, a maximum point, or neither. Let us discuss an example.

▶ EXAMPLE 4.42

Find the point(s) on the straight-line segment joining $(-1,-3)$ and $(2,3)$ where the function $f(x,y)=x^2y+1$ attains its maximum value.

SOLUTION

The equation of the line through $(-1,-3)$ and $(2,3)$ is $y+3=2(x+1)$; that is, $y-2x+1=0$. We can rephrase the question as follows: maximize $f(x,y)=x^2y+1$ subject to the constraint $g(x,y)=y-2x+1=0$, where $-1\le x\le 2$ and $-3\le y\le 3$.

From $\nabla f=\lambda\nabla g$, it follows that $(2xy,x^2)=\lambda(-2,1)$; that is, $xy=-\lambda$ and $x^2=\lambda$. Combining the two equations, we get $xy=-x^2$ and $x(x+y)=0$. It follows that either $x=0$ (in which case, from $y-2x+1=0$, we get $y=-1$) or $x=-y$; in the latter case, $y-2x+1=0$ implies $3y+1=0$ and $y=-1/3$ (so that $x=1/3$). Thus, we found two points, $(0,-1)$ and $(1/3,-1/3)$, that are candidates for extreme values (note that both points belong to the given line segment).

Since $\nabla g=(-2,1)$, there are no points where $\nabla g=\mathbf{0}$. The points $(-1,-3)$ and $(2,3)$ are the endpoints of the constraint curve and must be included in the list of candidates for extreme values. From $f(0,-1)=1$, $f(1/3,-1/3)=26/27$, $f(-1,-3)=-2$, and $f(2,3)=13$, we conclude that $f(2,3)=13$ is the maximum of f on the given line segment. ◀

Generalizations of Lagrange Multipliers Method

We now generalize the method of Lagrange multipliers to functions of three variables. To recall: the problem is to find the maximum (minimum) value of a C^1 function $w = f(x, y, z)$ subject to a C^1 constraint $g(x, y, z) = k$. Assume that $\nabla g(x, y, z) \neq \mathbf{0}$ for all (x, y, z) that lie on the constraint surface $g(x, y, z) = k$. Using an argument similar to the one we used in the case of a function of two variables, we obtain the following procedure.

Find all points (x_0, y_0, z_0) that satisfy

$$\nabla f(x_0, y_0, z_0) = \lambda \nabla g(x_0, y_0, z_0) \quad \text{and} \quad g(x_0, y_0, z_0) = k, \tag{4.29}$$

where λ is a real number. If the constraint surface $g(x, y, z) = k$ is a closed and bounded set in $\mathbb{R}^3$ ("closed" and "bounded" are straightforward generalizations of definitions given in Section 4.3), then the largest (smallest) of the values $f(x_0, y_0, z_0)$ for all points from (4.29) is the maximum (minimum) value of f subject to the constraint $g(x, y, z) = k$. Otherwise, additional arguments may be needed to determine whether (x_0, y_0, z_0) is a minimum point, a maximum point, or neither.

▶ EXAMPLE 4.43

Find the points on the surface $z = 2xy + 4$ that are closest to the origin. You may assume that such points exist.

SOLUTION

The distance from a point (x, y, z) to the origin is given by the formula $d(x, y, z) = \sqrt{x^2 + y^2 + z^2}$. To simplify calculations, we consider its square; that is, we will minimize $f(x, y, z) = x^2 + y^2 + z^2$ subject to $g(x, y, z) = z - 2xy = 4$.

The gradient vectors are $\nabla f = (2x, 2y, 2z)$ and $\nabla g = (-2y, -2x, 1)$. From $\nabla f = \lambda \nabla g$, we get $2x = -2y\lambda$, $2y = -2x\lambda$, and $2z = \lambda$. So, we must solve the system

$$x = -y\lambda, \qquad y = -x\lambda, \qquad 2z = \lambda, \qquad \text{and} \qquad z = 2xy + 4$$

for x, y, and z. Substituting $x = -y\lambda$ into $y = -x\lambda$, we get $y = -(-y\lambda)\lambda = y\lambda^2$; that is, $y(1 - \lambda^2) = 0$. Thus, $y = 0$ or $\lambda = \pm 1$. If $y = 0$, then $x = 0$ and $z = 2xy + 4 = 4$. Thus, $(0, 0, 4)$ is a candidate for extreme values.

If $\lambda = 1$, then the above system implies that $x = -y$ and $z = 1/2$. From $z = 2xy + 4$, we get $1/2 = -2y^2 + 4$; thus, $y^2 = 7/4$ and $y = \pm\sqrt{7}/2$ [hence, $x = \mp\sqrt{7}/2$]. Consequently, we obtain two more candidates: $(-\sqrt{7}/2, \sqrt{7}/2, 1/2)$ and $(\sqrt{7}/2, -\sqrt{7}/2, 1/2)$. If $\lambda = -1$, then $x = y$ and $z = -1/2$, and [from $z = 2xy + 4$] we get $-1/2 = 2x^2 + 4$; thus, $x^2 = -9/4$ and there are no solutions for x.

Note that $\nabla g(x, y, z) = (-2y, -2x, 1) \neq (0, 0, 0)$ for all (x, y, z).

From $f(0, 0, 4) = 16$ and $f(-\sqrt{7}/2, \sqrt{7}/2, 1/2) = f(\sqrt{7}/2, -\sqrt{7}/2, 1/2) = 15/4$, it follows that the points $P_1(-\sqrt{7}/2, \sqrt{7}/2, 1/2)$ and $P_2(\sqrt{7}/2, -\sqrt{7}/2, 1/2)$ on the given surface are closest to the origin. Their distance from the origin is $\sqrt{15}/2$; see Figure 4.25. ◀

▶ EXAMPLE 4.44

A satellite in the shape of the sphere $x^2 + y^2 + z^2 = 1$ is located in a region in space where the temperature is given by $T(x, y, z) = 4x^2 + yz + 750$. Find the warmest point(s) on the surface of the satellite.

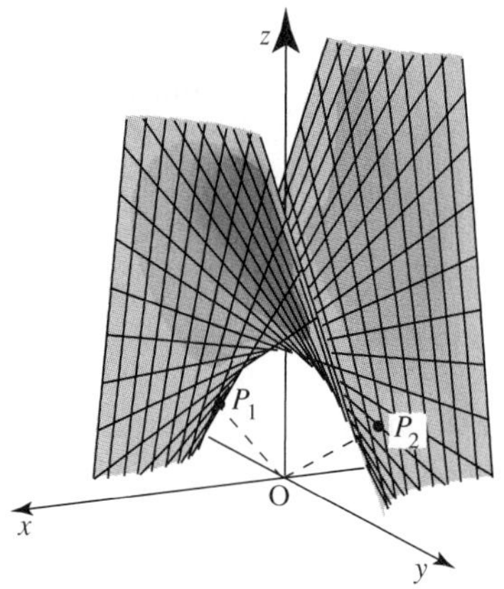

Figure 4.25 Points on the surface $z = 2xy + 4$ that are closest to the origin.

SOLUTION We are asked to find the maximum of $T(x, y, z) = 4x^2 + yz + 750$ subject to the constraint $g(x, y, z) = x^2 + y^2 + z^2 = 1$ (which is a sphere, and thus a closed and bounded set in $\mathbb{R}^3$). Substituting $\nabla T = (8x, z, y)$ and $\nabla g = (2x, 2y, 2z)$ into $\nabla T = \lambda \nabla g$, we get $8x = 2x\lambda$, $z = 2y\lambda$, and $y = 2z\lambda$. The first equation implies $x(4 - \lambda) = 0$; thus, $x = 0$ or $\lambda = 4$.

Consider the case $x = 0$ first. Combining the second and third equations, we get $y^2 = z^2$, and the constraint $x^2 + y^2 + z^2 = 1$ implies that $2z^2 = 1$ and $z = \pm 1/\sqrt{2}$. It follows that $y = \pm 1/\sqrt{2}$ for each of the two choices for z; so, we have found four points where extreme values could occur: $(0, 1/\sqrt{2}, 1/\sqrt{2})$, $(0, -1/\sqrt{2}, 1/\sqrt{2})$, $(0, 1/\sqrt{2}, -1/\sqrt{2})$, and $(0, -1/\sqrt{2}, -1/\sqrt{2})$.

Next, consider the case $\lambda = 4$. From $z = 8y$ and $y = 8z$, we get $y = z = 0$. Consequently, the constraint equation implies $x^2 = 1$ and $x = \pm 1$. So, there are two more candidates: $(1, 0, 0)$ and $(-1, 0, 0)$.

The only point where $\nabla g = (2x, 2y, 2z)$ is zero is $(0, 0, 0)$; however, it does not lie on the constraint surface.

From $f(0, 1/\sqrt{2}, 1/\sqrt{2}) = 750.5$, $f(0, -1/\sqrt{2}, 1/\sqrt{2}) = 749.5$, $f(0, 1/\sqrt{2}, -1/\sqrt{2}) = 749.5$, $f(0, -1/\sqrt{2}, -1/\sqrt{2}) = 750.5$, $f(1, 0, 0) = 754$, and $f(-1, 0, 0) = 754$, it follows that $(1, 0, 0)$ and $(-1, 0, 0)$ are the warmest points on the surface of the satellite. ◀

There is a more general version of the Lagrange multiplier method for functions of three variables. As in the case of two variables, it includes the points where $\nabla g(x_0, y_0, z_0) = \mathbf{0}$ [and $g(x_0, y_0, z_0) = k$], as well as the points on the boundary of the constraint surface $g(x, y, z) = k$.

Let us consider another generalization of the Lagrange multiplier method: suppose that we have to find extreme values of a C^1 function $f(x, y, z)$ subject to two constraints, $g_1(x, y, z) = k_1$ and $g_2(x, y, z) = k_2$ (assume that g_1 and g_2 are C^1 functions).

Using techniques similar to the one we used at the beginning of this section, we can prove the following: if f has an extreme value at a point (x_0, y_0, z_0) and if $\nabla g_1(x_0, y_0, z_0)$ and $\nabla g_2(x_0, y_0, z_0)$ are not parallel, then there exist real numbers λ_1 and λ_2 such that

$$\nabla f(x_0, y_0, z_0) = \lambda_1 \nabla g_1(x_0, y_0, z_0) + \lambda_2 \nabla g_2(x_0, y_0, z_0).$$

Thus, in order to find the maximum and minimum values of a C^1 function $f(x, y, z)$ subject to C^1 constraints $g_1(x, y, z) = k_1$ and $g_2(x, y, z) = k_2$, we must find all points (x_0, y_0, z_0)

that satisfy

$$\nabla f(x_0, y_0, z_0) = \lambda_1 \nabla g_1(x_0, y_0, z_0) + \lambda_2 \nabla g_2(x_0, y_0, z_0), \tag{4.30}$$

where λ_1 and λ_2 are real numbers and $g_1(x_0, y_0, z_0) = k_1$ and $g_2(x_0, y_0, z_0) = k_2$. It is assumed that $\nabla g_1(x_0, y_0, z_0)$ and $\nabla g_2(x_0, y_0, z_0)$ are not parallel vectors.

Geometrically, the two constraints $g_1(x, y, z) = k_1$ and $g_2(x, y, z) = k_2$ represent the intersection (call it D) of the surfaces $g_1(x, y, z) = k_1$ and $g_2(x, y, z) = k_2$ in space. If D is a closed and bounded set in $\mathbb{R}^3$, then f must have a maximum and minimum (subject to the given constraints). Otherwise, additional arguments may be needed to determine whether a point (x_0, y_0, z_0) is a minimum point, a maximum point, or neither.

We now illustrate this in an example.

▶ EXAMPLE 4.45

Find the maximum and minimum values of $f(x, y, z) = 2x + y - z$ subject to the constraints $2x + z = 2/\sqrt{5}$ and $y^2 + z^2 = 1$.

SOLUTION

Label the constraints as $g_1(x, y, z) = 2x + z = 2/\sqrt{5}$ and $g_2(x, y, z) = y^2 + z^2 = 1$. From $\nabla f = (2, 1, -1)$, $\nabla g_1 = (2, 0, 1)$, and $\nabla g_2 = (0, 2y, 2z)$ (notice that ∇g_1 and ∇g_2 are not parallel), using $\nabla f = \lambda_1 \nabla g_1 + \lambda_2 \nabla g_2$, we get $(2, 1, -1) = \lambda_1(2, 0, 1) + \lambda_2(0, 2y, 2z)$. Thus, we obtain the system

$$2 = 2\lambda_1, \qquad 1 = 2y\lambda_2, \qquad -1 = \lambda_1 + 2z\lambda_2,$$

which, combined with the two constraints $2x + z = 2/\sqrt{5}$ and $y^2 + z^2 = 1$, will give points where extreme values might occur. The constraint g_1 represents a plane and the constraint g_2 represents a cylinder. Their intersection is an ellipse, which is a closed and bounded set—thus, the minimum value and maximum value must exist.

From the first equation, $\lambda_1 = 1$. The second and the third equations imply that $\lambda_2 = 1/2y$ and $\lambda_2 = -1/z$; thus, $z = -2y$. Using the constraint $y^2 + z^2 = 1$, we get $5y^2 = 1$ and $y = \pm 1/\sqrt{5}$. It follows that $z = \mp 2/\sqrt{5}$. If $z = 2/\sqrt{5}$, then from $2x + z = 2/\sqrt{5}$, we get $x = 0$. Similarly, from $z = -2/\sqrt{5}$ (using $2x + z = 2/\sqrt{5}$), we obtain $2x = 4/\sqrt{5}$ and $x = 2/\sqrt{5}$. Thus, there are two candidates: $(2/\sqrt{5}, 1/\sqrt{5}, -2/\sqrt{5})$ and $(0, -1/\sqrt{5}, 2/\sqrt{5})$. It follows that the maximum of f is $f(2/\sqrt{5}, 1/\sqrt{5}, -2/\sqrt{5}) = 7/\sqrt{5}$ and the minimum of f is $f(0, -1/\sqrt{5}, 2/\sqrt{5}) = -3/\sqrt{5}$. ◀

It is possible to further generalize the method of Lagrange multipliers (so that it applies to functions of four, five, and more variables, and to more than two constraints).

▶ EXERCISES 4.4

1. Explain geometrically (i.e., by sketching level curves) why the function $f(x, y) = (x^2 + y^2)^{-1}$ cannot have a minimum or maximum subject to the constraint $x - y = 0$.

2. Minimize the function $f(x, y) = \sqrt{(x - 2)^2 + (y - 2)^2}$ subject to $x + y = 0$. Give a geometric interpretation of your answer.

Exercises 3 to 6: Shown is the gradient field of a C^1 function f. Find the approximate locations of the minimum and maximum of f subject to the given constraint curve.

3.

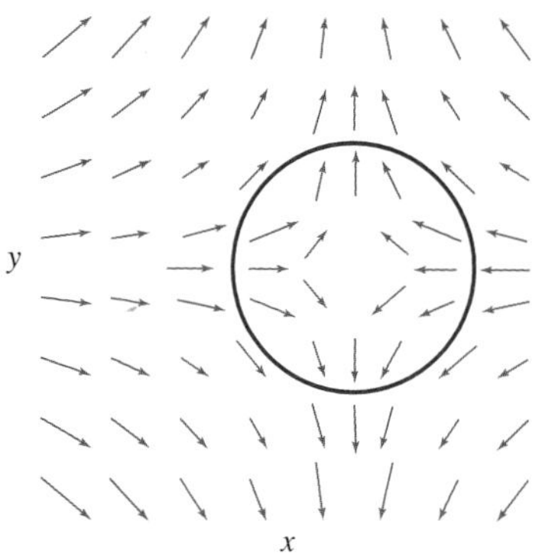

4.

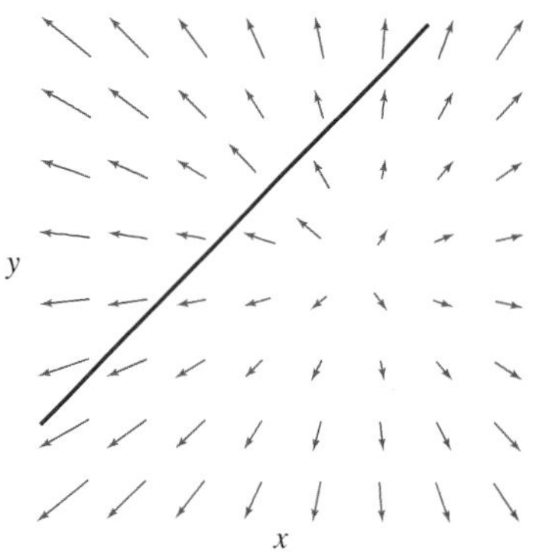

5.

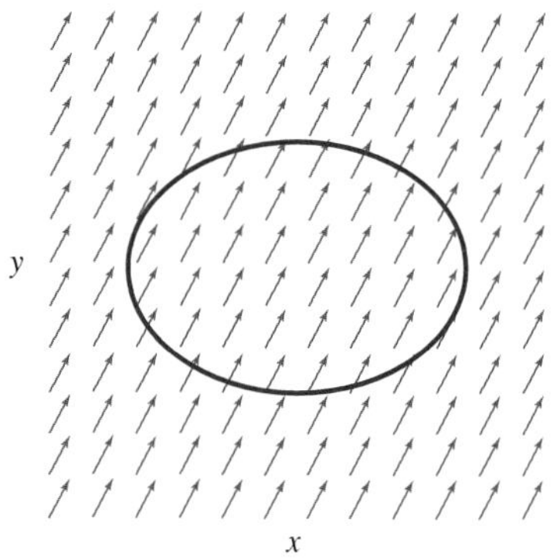

6.

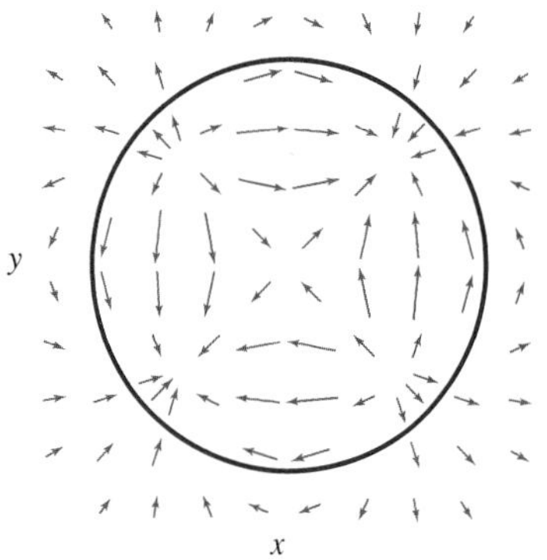

7. Sketch the level curves $f(x, y) = c$ of $f(x, y) = x^2 + y^2$ for $c = 1, 2, 4, 6, 9$, and 10. In the same coordinate system, sketch the graph of the constraint $(x - 1)^2 + y^2 = 4$.

(a) Looking at your picture, identify the minimum and maximum of f subject to the given constraint.

(b) Use the method of Lagrange multipliers to confirm your geometric reasoning.

8. Sketch the level curves $f(x, y) = c$ of $f(x, y) = 2x - y$ for $c = -3, -2, -1, 0, 1, 2$, and 3. In the same coordinate system, sketch the graph of the constraint $x^2 + y^2 = 1$.

(a) Looking at your picture, identify the minimum and maximum of f subject to the given constraint.

(b) Use the method of Lagrange multipliers to obtain the desired constrained minimum and maximum algebraically.

9. State the problem in Example 4.26 in Section 4.3 as a constrained optimization problem and solve it using Lagrange multipliers.

10. Explain why it does not make much sense to develop the Lagrange multipliers method to optimize a function $f(x, y)$ of two variables subject to two constraints $g_1(x, y) = k_1$ and $g_2(x, y) = k_2$.

Exercises 11 to 19: Find the extreme values (if any) of a function f subject to the given constraint.

11. $f(x, y) = 3xy$; $x^2 + y^2 = 4$

12. $f(x, y) = 4 - x^2 - y^2$; $y - 2x = 1$ (*Hint:* Find the maximum; argue that a minimum does not exist.)

13. $f(x, y) = 2x^2 - y^2$; $x^2 + y^2 = 1$

14. $f(x, y, z) = x - y + 4z$; $x^2 + y^2 + z^2 = 2$

15. $f(x, y) = x - y^2$; $x^3 - y^2 = 0$, $-1 \le y \le 1$

16. $f(x, y, z) = x^2 + 2y^2$; $x^2 + y^2 - z^2 = 1$ (*Hint:* Find the minimum; argue that a maximum does not exist.)

17. $f(x, y, z) = x + 2y - 4z$; $x^2 + y^2 + 2z^2 = 4$

18. $f(x, y, z) = xyz$; $x^2 + y^2 + z^2 = 9$

19. $f(x, y) = xy + 2y$; $x^2 + 2y^2 = 4$

20. Find all points on the curve $y^2x = 16$ that are closest to the origin.

21. Find the minimum distance from the surface $x^2 + y^2 - z^2 = 4$ to the origin.

22. The temperature at a point (x, y) on a metal plate in the shape of the disk $x^2 + y^2 \le 50$ is $T(x, y) = 2x^2 - xy + 2y^2 + 10$. Find the coldest point on the rim of the plate.

23. Find the point in the plane $x + y + 2z = 11$ that is closest to the point $(0, 1, 1)$.

24. Find the dimensions of a cylindrical can (with a lid) with a volume of 10 units^3 and minimum surface area.

25. Find the minimum of the function $f(x, y, z) = x^2 + y^2 + z^2$ subject to the constraints $2y + z = 6$ and $x - 2y = 4$.

26. Find the minimum and maximum of $f(x, y, z) = xy + z^2$ subject to the constraints $x + y = 0$ and $x^2 + y^2 + z^2 = 4$.

27. Find the point closest to the origin that belongs to the intersection of the planes $2z - y = 0$ and $x + y - z = 4$.

28. Find the extreme values of the function $f(x, y, z) = x + y + 4z$ along the ellipse that is the intersection of the cylinder $x^2 + y^2 = 82$ and the plane $z = 2x$.

29. Solve Exercise 21 in Section 4.3 using the method of Lagrange multipliers.

30. Solve Exercise 22 in Section 4.3 using the method of Lagrange multipliers.

31. Solve Exercise 25 in Section 4.3 using the method of Lagrange multipliers.

32. Let (x_0, y_0) be the point where a differentiable function $f(x, y)$ attains its maximum subject to the constraint $g(x, y) = k$.

(a) As k changes, so does the location of the constrained maximum; that is, x_0 and y_0 become functions of k. Consequently, $f(x_0, y_0)$ is a function of k. Show that, at the point (x_0, y_0), df/dk satisfies $\frac{df}{dk} = \frac{\partial f}{\partial x}\frac{dx_0}{dk} + \frac{\partial f}{\partial y}\frac{dy_0}{dk}$.

(b) Show that at (x_0, y_0), the right side is equal to $\lambda\left(\frac{\partial g}{\partial x}\frac{dx_0}{dk} + \frac{\partial g}{\partial y}\frac{dy_0}{dk}\right) = \lambda\frac{dg}{dk}$.

(c) Explain why $(dg/dk)(x_0, y_0) = 1$. Conclude that $\lambda = (df/dk)(x_0, y_0)$, and interpret the result.

▶ 4.5 FLOW LINES

Assume that the motion of a fluid is described by a vector field **F** (i.e., the value of **F** at a point gives the velocity of the fluid at that point). One way of visualizing **F** is to isolate a point in the fluid and follow its trajectory under the influence of the field. The path thus obtained is called a flow line.

A familiarity with basic concepts in the theory of ordinary differential equations is needed in this section.

DEFINITION 4.6 Flow Lines of a Vector Field

Let **F** be a continuous vector field defined on a subset U of $\mathbb{R}^2$ ($\mathbb{R}^3$). A *flow line* of **F** is a differentiable path $\mathbf{c}(t)$ in $\mathbb{R}^2$ ($\mathbb{R}^3$) that satisfies

$$\mathbf{c}'(t) = \mathbf{F}(\mathbf{c}(t)).$$

A flow line is also called an *integral curve,* a *streamline*, or a *line of force*. Let us emphasize that the definition requires that the vector field be independent of time. Of course, values of **F** may change from point to point, but the value of **F** at a particular point is the same for all times.

A vector field that describes the motion of the wind depends on time: at this moment, a vector **v** at a particular location P might point north, indicating a northerly wind. A moment later, the wind could change its intensity or direction at P, and can no longer be described by **v**. Flow lines of vector fields that depend on time are dealt with in more advanced texts.

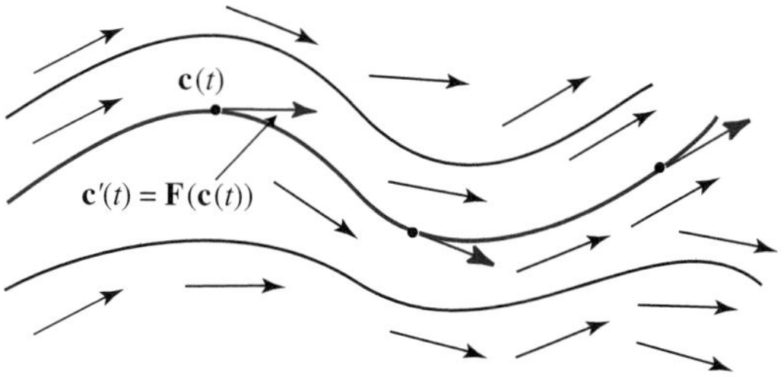

Figure 4.26 Flow lines of a vector field.

By definition, a path **c** is a flow line of **F** if its velocity vector $\mathbf{c}'(t)$ at $\mathbf{c}(t)$ coincides with the value $\mathbf{F}(\mathbf{c}(t))$ of the field **F** at the point $\mathbf{c}(t)$ on the curve; see Figure 4.26.

Pick a point P in the domain of a vector field **F**. Let us trace a flow line of **F** that goes through P. The only way to leave P is in the direction of the vector $\mathbf{F}(P)$ (that is the value of the field **F** at P); see Figure 4.27. Once we move to a nearby point, say, Q, we find another vector, $\mathbf{F}(Q)$, that tells us how to proceed. A moment later, arriving at a nearby point R, we proceed in the direction of $\mathbf{F}(R)$, etc. It follows that once we choose a starting point, our walk is completely determined by **F**.

This intuitive argument shows that a flow line is uniquely determined by a vector field and one point. In other words, distinct flow lines cannot intersect or touch each other. Here is the outline of a proof: suppose that two flow lines have a point P in common. By uniqueness, from P "onward" (i.e., in the direction of the vector field **F**) the two flow lines have to agree. Since the flow lines of $-\mathbf{F}$ are the flow lines of **F** with reversed orientation (see Exercise 13), by uniqueness again, the remaining parts of the flow lines have to overlap.

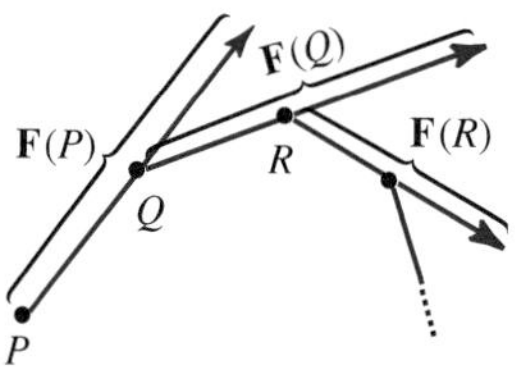

Figure 4.27 Tracing a flow line from P.

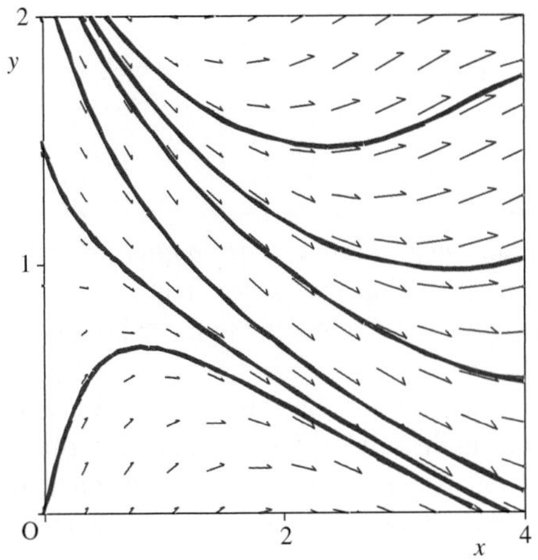

Figure 4.28 Flow lines of the vector field $\mathbf{F}(x, y) = (x + 0.5, \sin(x + 2y + 1))$.

Imagine that a vector field $\mathbf{F}$ describes the motion of a fluid. The flow line through a given point P describes the trajectory of a particle (located initially at P) participating in the motion of the fluid. Several flow lines of $\mathbf{F}(x, y) = (x + 0.5, \sin(x + 2y + 1))$ in $\mathbb{R}^2$ are shown in Figure 4.28.

Let $\mathbf{c}(t) = (x(t), y(t))$ be a differentiable curve and $\mathbf{F}(x, y) = (F_1(x, y), F_2(x, y))$ a vector field in $\mathbb{R}^2$. The equation $\mathbf{c}'(t) = \mathbf{F}(\mathbf{c}(t))$, or

$$(x'(t), y'(t)) = (F_1(x(t), y(t)), F_2(x(t), y(t))),$$

of the flow line $\mathbf{F}$ going through the point (x_0, y_0) can be written as the system of ordinary differential equations

$$\begin{aligned} x'(t) &= F_1(x(t), y(t)) \\ y'(t) &= F_2(x(t), y(t)) \end{aligned}$$

with initial condition $\mathbf{c}(0) = (x(0), y(0)) = (x_0, y_0)$.

From the theory of ordinary differential equations (more precisely, from the Existence and Uniqueness Theorem for systems of equations), it follows that continuous vector fields always have flow lines, and they are uniquely determined once a particular point has been specified (this argument formalizes our intuitive argument presented earlier; it generalizes, in a straightforward way, to vector fields and flow lines in $\mathbb{R}^3$).

▶ EXAMPLE 4.46

Consider the vector field $\mathbf{F}(x, y) = (-y, x)$. Verify that $\mathbf{c}_1(t) = (\cos t, \sin t)$ is the flow line of $\mathbf{F}$ going through $(1, 0)$, and $\mathbf{c}_2(t) = (3\cos t + \sin t, 3\sin t - \cos t)$ is its flow line containing $(3, -1)$.

SOLUTION Clearly, $\mathbf{c}_1(0) = (1, 0)$. Since $\mathbf{F}(\mathbf{c}_1(t)) = \mathbf{F}(\cos t, \sin t) = (-\sin t, \cos t)$ and $\mathbf{c}_1'(t)) = (-\sin t, \cos t)$, it follows that $\mathbf{c}_1'(t)) = \mathbf{F}(\mathbf{c}_1(t))$.

Likewise, from $\mathbf{c}_2'(t) = (-3\sin t + \cos t, 3\cos t + \sin t)$ and $\mathbf{F}(\mathbf{c}_2(t)) = \mathbf{F}(3\cos t + \sin t, 3\sin t - \cos t) = (-(3\sin t - \cos t), 3\cos t + \sin t)$, we conclude that $\mathbf{c}_2'(t)) = \mathbf{F}(\mathbf{c}_2(t))$. Since $\mathbf{c}_2(0) = (3, -1)$, it follows that $\mathbf{c}_2$ is the desired level curve of $\mathbf{F}$ containing $(3, -1)$. In Cartesian

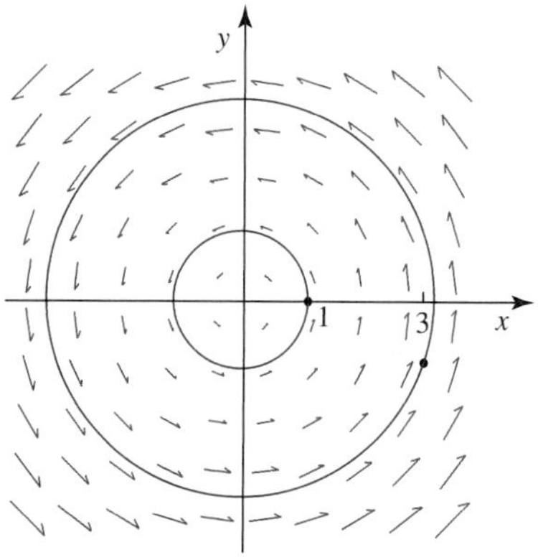

Figure 4.29 Flow lines of $\mathbf{F}(x, y) = (y, -x)$ from Example 4.46.

coordinates, $\mathbf{c}_2$ is given by $x^2 + y^2 = 10$ (thus, it is a circle of radius $\sqrt{10}$). The two flow lines are shown in Figure 4.29.

▶ EXAMPLE 4.47

Show that the curves

$$\mathbf{c}(t) = \left(\frac{C_1 \cos t + C_2 \sin t}{\sqrt{C_1^2 + C_2^2}}, \frac{-C_1 \sin t + C_2 \cos t}{\sqrt{C_1^2 + C_2^2}} \right)$$

(where C_1 and C_2 are constants and $C_1 \neq 0$ or $C_2 \neq 0$) are flow lines of the vector field

$$\mathbf{F} = \left(\frac{y}{\sqrt{x^2 + y^2}}, -\frac{x}{\sqrt{x^2 + y^2}} \right).$$

SOLUTION We have to show that $\mathbf{c}'(t) = \mathbf{F}(\mathbf{c}(t))$. Since

$$x^2 + y^2 = \frac{(C_1 \cos t + C_2 \sin t)^2}{C_1^2 + C_2^2} + \frac{(-C_1 \sin t + C_2 \cos t)^2}{C_1^2 + C_2^2} = 1,$$

it follows that

$$\mathbf{F}(\mathbf{c}(t)) = \left(\frac{-C_1 \sin t + C_2 \cos t}{\sqrt{C_1^2 + C_2^2}}, \frac{-C_1 \cos t - C_2 \sin t}{\sqrt{C_1^2 + C_2^2}} \right).$$

On the other hand, the tangent vector is computed to be

$$\mathbf{c}'(t) = \left(\frac{-C_1 \sin t + C_2 \cos t}{\sqrt{C_1^2 + C_2^2}}, \frac{-C_1 \cos t - C_2 \sin t}{\sqrt{C_1^2 + C_2^2}} \right).$$

Examples 4.46 and 4.47 show that different vector fields can have the same flow lines. In both cases, the flow lines are circles [with one difference: the flow line in Example 4.46 going through the origin consists of a single point (origin), but there is no flow line of the vector field in Example 4.47 that contains the origin].

▶ EXAMPLE 4.48 Flow Lines of an Electrostatic Field

Consider the electrostatic field $\mathbf{F}$ on $\mathbb{R}^2 - \{(0, 0)\}$ given by

$$\mathbf{F}(x, y) = \frac{1}{4\pi\epsilon_0}\frac{Qq}{\|\mathbf{r}\|^3}\mathbf{r} = \left(\frac{1}{4\pi\epsilon_0}\frac{Qq}{\|\mathbf{r}\|^3}x, \frac{1}{4\pi\epsilon_0}\frac{Qq}{\|\mathbf{r}\|^3}y\right)$$

(see Section 2.1), where $\mathbf{r} = x\mathbf{i} + y\mathbf{j}$. To find the flow lines of $\mathbf{F}$, we have to solve the system $(x', y') = \mathbf{F}(x, y)$, that is,

$$x' = \frac{K}{\|\mathbf{r}\|^3}x,$$
$$y' = \frac{K}{\|\mathbf{r}\|^3}y,$$

where $K = Qq/4\pi\epsilon_0$. Dividing the second equation by the first and simplifying, we get (keep in mind that $y' = dy/dt$ and $x' = dx/dt$)

$$\frac{y'}{y} = \frac{x'}{x},$$

and $(\ln y)' = (\ln x)'$. Integration gives $C_1 y = C_2 x$, where C_1 and C_2 are constants, which is the equation of a line through the origin (where the charge Q has been placed). Hence, the flow lines of this electrostatic field in the plane are straight lines through the origin. ◀

▶ EXAMPLE 4.49 Flow Lines of a Hurricane Vector Field

In Example 2.14 in Section 2.1, we introduced the vector field

$$\mathbf{H}(x, y) = \frac{(-x - y)\mathbf{i} + (x - y)\mathbf{j}}{x^2 + y^2}$$

as a model of the flow of air near and within a hurricane. We will now compute the flow lines of $\mathbf{H}$.

We express $\mathbf{H}$ in polar coordinates ($x = r\cos\theta$ and $y = r\sin\theta$):

$$\begin{aligned}\mathbf{H}(r, \theta) &= \frac{1}{r^2}(-(r\cos\theta + r\sin\theta)\mathbf{i} + (r\cos\theta - r\sin\theta)\mathbf{j}) \\ &= \frac{1}{r}(-\cos\theta\mathbf{i} - \sin\theta\mathbf{j} - \sin\theta\mathbf{i} + \cos\theta\mathbf{j}) = \frac{1}{r}(-\mathbf{e}_r + \mathbf{e}_\theta).\end{aligned}$$

In the last step, we used formula (2.32) from Section 2.8.

Let $\mathbf{c}$ be a flow line of $\mathbf{H}(r, \theta)$. Recall that in Section 2.8 [see formula (2.35) in Example 2.98], we computed the speed in polar coordinates, $\mathbf{c}'(t) = (dr/dt)\mathbf{e}_r + r(d\theta/dt)\mathbf{e}_\theta$. Therefore, from $\mathbf{H}(\mathbf{c}(t)) = \mathbf{c}'(t)$, we obtain

$$\frac{1}{r}(-\mathbf{e}_r + \mathbf{e}_\theta) = \frac{dr}{dt}\mathbf{e}_r + r\frac{d\theta}{dt}\mathbf{e}_\theta,$$

and thus,

$$\frac{dr}{dt} = -\frac{1}{r} \quad \text{and} \quad \frac{d\theta}{dt} = \frac{1}{r^2}.$$

Writing the first equation as $r(dr/dt) = -1$ and integrating with respect to t, we get $r^2 = -2t + C$, where C is a constant. So,

$$\frac{d\theta}{dt} = \frac{1}{-2t + C},$$

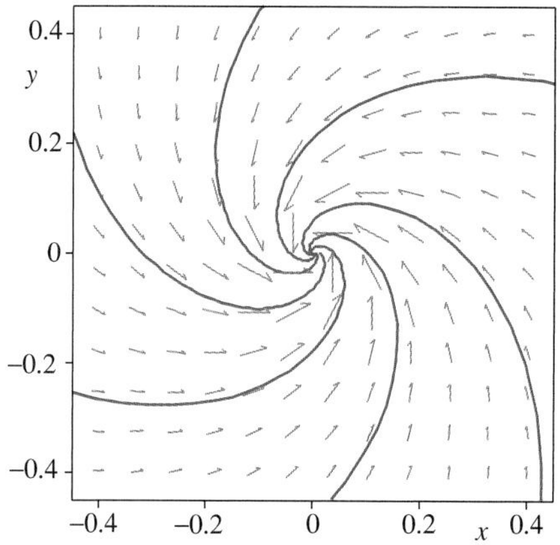

Figure 4.30 Flow lines $r = e^{-\theta+D}$ of the field **H** of Example 4.49.

and, after integration, $\theta = -\frac{1}{2}\ln(-2t + C) + D$, where D is another constant. Eliminating t, we get

$$\theta = -\frac{1}{2}\ln r^2 + D,$$

that is, $\theta = -\ln r + D$ and $\ln r = -\theta + D$. It follows that the flow lines of **H** are the spirals represented (in polar coordinates) by $r = e^{-\theta+D}$; see Figure 4.30.

▶ EXERCISES 4.5

1. Find the flow lines of the vector field $\mathbf{F}(x, y) = (x, 2y)$. Compare with the flow lines of the vector fields $\mathbf{F}_1(x, y) = (3x, 6y)$ and $\mathbf{F}_2(x, y) = (-2x, -4y)$.

2. Show that the curve $\mathbf{c}(t) = \left(\frac{3}{5}\cos t + \frac{4}{5}\sin t\right)\mathbf{i} + \left(-\frac{3}{5}\sin t + \frac{4}{5}\cos t\right)\mathbf{j}$ is the flow line of the vector field $\mathbf{F}(x, y) = y\mathbf{i}/\sqrt{x^2 + y^2} - x\mathbf{j}/\sqrt{x^2 + y^2}$ going through the point $(4/5, -3/5)$.

3. Find the flow line of the vector field $\mathbf{F}(x, y) = 2\|\mathbf{r}\|^{-1}\mathbf{r}$ (where $\mathbf{r} = x\mathbf{i} + y\mathbf{j}$) going through the point $(3, 2)$.

4. Find the flow line of the constant vector field $\mathbf{F}(x, y) = 3\mathbf{i} - 4\mathbf{j}$ that goes through the point $(-2, 1)$. (*Hint:* There is no need to solve differential equations.)

5. Find the flow line of the constant vector field $\mathbf{F}(x, y) = a\mathbf{i} + b\mathbf{j}$ (a, b are real numbers, $a \neq 0$, and/or $b \neq 0$) that goes through the origin. (*Hint:* There is no need to solve differential equations.)

Exercises 6 to 10: Sketch the vector field **F** and several of its flow lines.

6. $\mathbf{F}(x, y) = y\mathbf{i} - 2x\mathbf{j}$

7. $\mathbf{F}(x, y) = (x, x^2)$

8. $\mathbf{F}(x, y) = x\mathbf{i} + \mathbf{j}$

9. $\mathbf{F}(x, y) = (-2x, y)$

10. $\mathbf{F}(x, y) = \mathbf{i} + x\mathbf{j}$

11. Find a vector field for which the curve $\mathbf{c}(t) = (t^2, 2t, t)$, $t \in \mathbb{R}$ is a flow line.

12. Show that the curve $\mathbf{c}(t) = (e^t, 2\ln t, t^{-1})$, $t > 0$ is a flow line of the vector field $\mathbf{F}(x, y, z) = (x, 2z, -z^2)$.

13. Assume that $\mathbf{c}(t)$, $t \in [a, b]$, is a flow line of **F**. Show that $\boldsymbol{\gamma}(t) = \mathbf{c}(a + b - t)$, $t \in [a, b]$, is a flow line of $-\mathbf{F}$. Interpret this fact geometrically.

▶ 4.6 DIVERGENCE AND CURL OF A VECTOR FIELD

In this section, we introduce and discuss two operations involving partial derivatives, divergence and curl, and use them to investigate the rate of change of a vector field. We discuss examples that help us understand how divergence and curl relate to various physical concepts, such as fluid flow, rotation of a rigid body, or electrostatic and gravitational force fields.

Imagine that we look at diagrams of vector fields (for instance, Figures 4.29, 4.30, 4.33, 4.34, and 4.37) and try to identify the features that make them similar, or try to describe ways in which they differ from each other. Divergence and curl will help us answer these questions and, in general, will provide us with ways of describing various properties of vector fields.

Recall that a vector field $\mathbf{F}$ on $U \subseteq \mathbb{R}^3$ can be expressed as

$$\mathbf{F}(x, y, z) = (F_1(x, y, z), F_2(x, y, z), F_3(x, y, z)),$$

where the components F_1, F_2, and F_3 of $\mathbf{F}$ are real-valued functions. Sometimes, we express the variables x, y, and z as a vector $\mathbf{x} = (x, y, z)$ and write $\mathbf{F}(\mathbf{x}) = (F_1(\mathbf{x}), F_2(\mathbf{x}), F_3(\mathbf{x}))$. More often, we completely drop the variables and write $\mathbf{F} = (F_1, F_2, F_3)$.

DEFINITION 4.7 Divergence and Curl of a Vector Field

Let $\mathbf{F} = (F_1, F_2, F_3)\colon U \subseteq \mathbb{R}^3 \to \mathbb{R}^3$ be a differentiable vector field. The *divergence* of $\mathbf{F}$ is the real-valued function

$$div\,\mathbf{F} = \frac{\partial F_1}{\partial x} + \frac{\partial F_2}{\partial y} + \frac{\partial F_3}{\partial z}.$$

The *curl* of $\mathbf{F}$ is the vector field

$$curl\,\mathbf{F} = \left(\frac{\partial F_3}{\partial y} - \frac{\partial F_2}{\partial z}\right)\mathbf{i} + \left(\frac{\partial F_1}{\partial z} - \frac{\partial F_3}{\partial x}\right)\mathbf{j} + \left(\frac{\partial F_2}{\partial x} - \frac{\partial F_1}{\partial y}\right)\mathbf{k}.$$

The definition of the divergence generalizes in a straightforward way to a vector field $\mathbf{F}\colon U \subseteq \mathbb{R}^m \to \mathbb{R}^m$, $m \geq 2$. For example, $div\,\mathbf{F} = \partial F_1/\partial x + \partial F_2/\partial y$ if $\mathbf{F}(x, y) = (F_1(x, y), F_2(x, y))$ is a differentiable vector field on $U \subseteq \mathbb{R}^2$.

The curl is defined for vector fields in $\mathbb{R}^3$ only. To find the curl of a vector field $\mathbf{F}(x, y) = (F_1(x, y), F_2(x, y))$ in $\mathbb{R}^2$, we express it as the vector field $(F_1(x, y), F_2(x, y), 0)$ in $\mathbb{R}^3$ and then use Definition 4.7 to get

$$curl\,\mathbf{F} = \left(\frac{\partial F_2}{\partial x} - \frac{\partial F_1}{\partial y}\right)\mathbf{k}.$$

It follows that the curl (if nonzero) of a vector field in a plane always points in the direction perpendicular to the plane (see Examples 4.62 and 4.64).

DEFINITION 4.8 Scalar Curl

Let $\mathbf{F} = (F_1, F_2)\colon U \subseteq \mathbb{R}^2 \to \mathbb{R}^2$ be a differentiable vector field. The function

$$\frac{\partial F_2}{\partial x} - \frac{\partial F_1}{\partial y}$$

is called the *scalar curl* of $\mathbf{F}$.

So, the scalar curl of a vector field **F** in $\mathbb{R}^2$ is the **k** component of *curl* **F** (remember that **F** is viewed as a vector field in $\mathbb{R}^3$ whose **k** component is zero). We will use scalar curl later, in the contexts of conservative vector fields (Chapter 5) and Green's Theorem (Chapter 8).

▶ EXAMPLE 4.50

Let $\mathbf{F}(x, y, z) = (x^3, xy, e^{xyz})$. Compute $div\, \mathbf{F}$ and $curl\, \mathbf{F}$.

SOLUTION By definition,

$$div\, \mathbf{F} = \frac{\partial}{\partial x}(x^3) + \frac{\partial}{\partial y}(xy) + \frac{\partial}{\partial z}(e^{xyz}) = 3x^2 + x + xye^{xyz}$$

and

$$\begin{aligned} curl\, \mathbf{F} &= \left(\frac{\partial}{\partial y}(e^{xyz}) - \frac{\partial}{\partial z}(xy)\right)\mathbf{i} + \left(\frac{\partial}{\partial z}(x^3) - \frac{\partial}{\partial x}(e^{xyz})\right)\mathbf{j} + \left(\frac{\partial}{\partial x}(xy) - \frac{\partial}{\partial y}(x^3)\right)\mathbf{k} \\ &= xze^{xyz}\mathbf{i} - yze^{xyz}\mathbf{j} + y\mathbf{k}. \end{aligned}$$

◀

▶ EXAMPLE 4.51

The scalar curl of the vector field $\mathbf{F}(x, y) = (\sin(xy), \cos(xy))$ is the function

$$\frac{\partial}{\partial x}\Big(\cos(xy)\Big) - \frac{\partial}{\partial y}\Big(\sin(xy)\Big) = -y\sin(xy) - x\cos(xy).$$

◀

We are going to introduce a formalism that is often used in computations involving gradient, divergence, and curl.

In general, an *operator* acts on a function by assigning some other function to it. For example, the operator $A(y) = y''$ assigns the second derivative y'' to a twice differentiable function y; for example, $A(\sin x) = -\sin x$, $A(x^3 + 3) = 6x$, etc.

Define the operator ∇ (pronounced "del") by

$$\nabla = \frac{\partial}{\partial x}\mathbf{i} + \frac{\partial}{\partial y}\mathbf{j} + \frac{\partial}{\partial z}\mathbf{k} = \left(\frac{\partial}{\partial x}, \frac{\partial}{\partial y}, \frac{\partial}{\partial z}\right). \tag{4.31}$$

Its action on a differentiable real-valued function $f\colon \mathbb{R}^3 \to \mathbb{R}$ is defined by

$$\nabla f = \frac{\partial f}{\partial x}\mathbf{i} + \frac{\partial f}{\partial y}\mathbf{j} + \frac{\partial f}{\partial z}\mathbf{k} = \left(\frac{\partial f}{\partial x}, \frac{\partial f}{\partial y}, \frac{\partial f}{\partial z}\right), \tag{4.32}$$

that is, the ∇ operator assigns to f its gradient vector field $\nabla f = grad\, f$ (this justifies the notation ∇f that we have used already).

Now think of ∇ as a "vector" (it is not really a vector; for convenience, we will borrow a few words from the vector vocabulary) and let $\mathbf{F}\colon \mathbb{R}^3 \to \mathbb{R}^3$ be a vector field. Define the action of ∇ on **F** by the scalar product as follows:

$$\nabla \cdot \mathbf{F} = \left(\frac{\partial}{\partial x}, \frac{\partial}{\partial y}, \frac{\partial}{\partial z}\right) \cdot (F_1, F_2, F_3) = \frac{\partial F_1}{\partial x} + \frac{\partial F_2}{\partial y} + \frac{\partial F_3}{\partial z}.$$

The expression on the right side is the divergence of F, and hence,

$$div\, \mathbf{F} = \nabla \cdot \mathbf{F}. \tag{4.33}$$

If we consider the action of ∇ on $\mathbf{F}$ by the cross product,

$$\nabla \times \mathbf{F} = \begin{vmatrix} \mathbf{i} & \mathbf{j} & \mathbf{k} \\ \partial/\partial x & \partial/\partial y & \partial/\partial z \\ F_1 & F_2 & F_3 \end{vmatrix}$$
$$= \left(\frac{\partial F_3}{\partial y} - \frac{\partial F_2}{\partial z}\right)\mathbf{i} + \left(\frac{\partial F_1}{\partial z} - \frac{\partial F_3}{\partial x}\right)\mathbf{j} + \left(\frac{\partial F_2}{\partial x} - \frac{\partial F_1}{\partial y}\right)\mathbf{k},$$

we get that the resulting vector field is the curl of $\mathbf{F}$. Therefore,

$$curl\,\mathbf{F} = \nabla \times \mathbf{F} \tag{4.34}$$

(but we always keep in mind that this is not a real cross product: for instance, $\nabla \times \mathbf{F}$ is not, in general, perpendicular to $\mathbf{F}$).

▶ EXAMPLE 4.52 Divergence of the Gravitational Force Field

Compute the divergence of the gravitational force field $\mathbf{F} = -(GMm/\|\mathbf{r}\|^3)\,\mathbf{r}$ that we introduced in Example 2.10 in Section 2.1.

SOLUTION Writing $\mathbf{F}$ in components (with $\mathbf{r} = (x, y, z)$), we get

$$\mathbf{F} = -GMm\left(\frac{x}{(x^2+y^2+z^2)^{3/2}}, \frac{y}{(x^2+y^2+z^2)^{3/2}}, \frac{z}{(x^2+y^2+z^2)^{3/2}}\right).$$

It follows that

$$\frac{\partial F_1}{\partial x} = -GMm\frac{(x^2+y^2+z^2)^{3/2} - 3x^2(x^2+y^2+z^2)^{1/2}}{(x^2+y^2+z^2)^3} = -GMm\frac{y^2+z^2-2x^2}{(x^2+y^2+z^2)^{5/2}},$$

and similarly,

$$\frac{\partial F_2}{\partial y} = -GMm\frac{x^2+z^2-2y^2}{(x^2+y^2+z^2)^{5/2}} \quad \text{and} \quad \frac{\partial F_3}{\partial z} = -GMm\frac{x^2+y^2-2z^2}{(x^2+y^2+z^2)^{5/2}}.$$

Adding up the three partials, we get $div\,\mathbf{F} = 0$, in $\mathbb{R}^3 - \{(0,0,0)\}$. In words, the divergence of the gravitational force field is zero at all points where it is defined. ◀

▶ EXAMPLE 4.53 Divergence of an Electrostatic Field

Replacing constants G, m, and M in the previous example with $1/4\pi\epsilon_0$, q, and Q, we see that $div\,\mathbf{F} = 0$, where $\mathbf{F}$ now denotes an electrostatic force field (see Examples 2.11 and 2.12 in Section 2.1). Since $\mathbf{E} = \mathbf{F}/q$, it follows that $div\,\mathbf{E} = (div\,\mathbf{F})/q = 0$. Thus, the divergence of an electrostatic field vanishes in $\mathbb{R}^3 - \{(0,0,0)\}$. ◀

To arrive at a physical interpretation of the divergence, we study an example first.

Imagine that the vector field $\mathbf{F}(x, y) = (x^2, 0)$ represents the motion of a fluid, or a gas (of constant density, taken to be 1, for simplicity).

Place a small rectangle R with sides Δx and Δy into the flow, as shown in Figure 4.31. We are going to approximate the change in mass of fluid in R due to the flow. More precisely, we will estimate the *total outflow*

$$O = \text{mass outflow} - \text{mass inflow}.$$

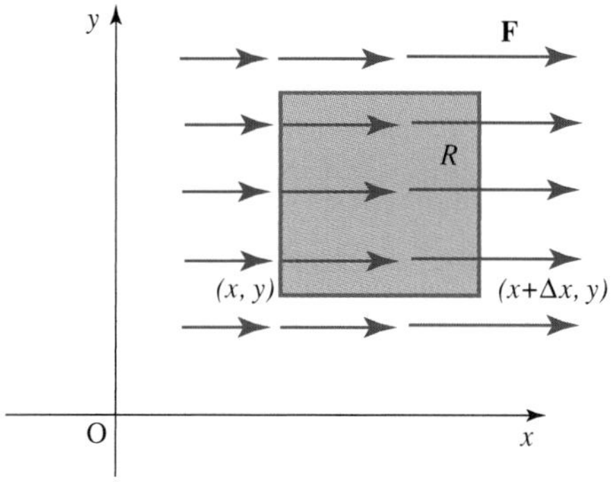

Figure 4.31 Small rectangle R placed in the flow given by $\mathbf{F}$.

The mass flows into R through its left side only. In time Δt, the mass inflow is equal to density (which is 1) times the area occupied by the particles that flow into R in time Δt. Thus,

$$\text{mass inflow} = x^2 \Delta t \Delta y$$

(particles enter R with speed x^2). The particles leave R along its right side, where $\mathbf{F}(x, y) = ((x + \Delta x)^2, 0)$. At that moment, their speed is $(x + \Delta x)^2$, and so

$$\text{mass outflow} = (x + \Delta x)^2 \Delta t \Delta y.$$

It follows that the total outflow is

$$O(\Delta t) = (x + \Delta x)^2 \Delta t \Delta y - x^2 \Delta t \Delta y = 2x \Delta x \Delta y \Delta t + (\Delta x)^2 \Delta y \Delta t.$$

Neglecting the small term $(\Delta x)^2 \Delta y \Delta t$, we get $O(\Delta t) = 2x \Delta x \Delta y \Delta t$, and

$$\frac{1}{\Delta x \Delta y} \frac{O(\Delta t)}{\Delta t} \approx 2x. \tag{4.35}$$

The expression $\Delta x \Delta y$ is the area of R. The left side in (4.35) represents the rate of change of the total outflow relative to the area of R. We note that the right side, $2x$, is the derivative of the x-component of the given field $\mathbf{F} = (x^2, 0)$.

Repeating this calculation for the vector field $\mathbf{F}(x, y) = (F_1(x, y), F_2(x, y)) = (f(x), 0)$, where $f(x)$ is a differentiable function of one variable, we will obtain (4.35) with $2x$ replaced by $f'(x)$ (see Exercise 17). Thus (at least) in the case that we studied,

$$\frac{1}{\Delta x \Delta y} \frac{O(\Delta t)}{\Delta t} \approx \mathit{div}\, \mathbf{F}(x, y).$$

In the limit, as $\Delta t \to 0$, we obtain

$$\frac{1}{A(R)} \frac{d}{dt} O(t) \bigg|_{t=0} \approx \mathit{div}\, \mathbf{F}(x, y),$$

where $A(R)$ is the area of R. Thus, we demonstrated that the divergence of $\mathbf{F}$ at a point (x, y) is approximately equal to the relative rate of change of the total outflow from R. The approximation improves as R gets smaller and smaller (i.e., as $\Delta x, \Delta y \to 0$).

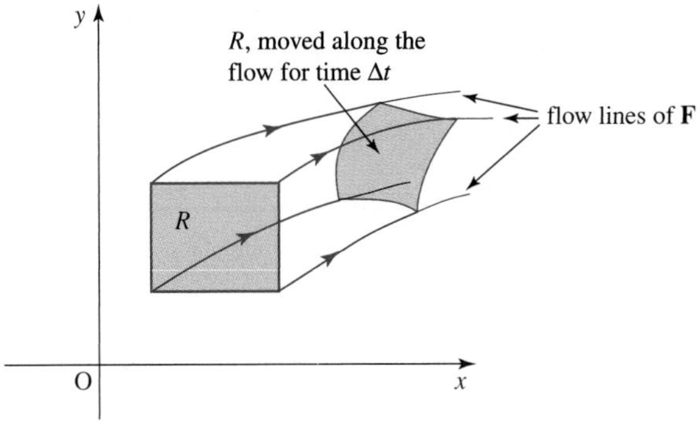

Figure 4.32 Small rectangle R subjected to the flow of $\mathbf{F}$.

In Section 8.6 we give a proof of this fact for a general vector field in $\mathbb{R}^2$ (including time-dependent vector fields). Interpretation for vector fields in $\mathbb{R}^3$ is the same—all we have to do is to replace area by volume.

Thus, if $\mathbf{F}$ represents the velocity vector field of a fluid (gas, vapor, etc.), then $div\,\mathbf{F}$ measures the rate of expansion per unit area (volume). In case $div\,\mathbf{F} < 0$, we say that the fluid (gas, vapor, etc.) is *compressing.* If the total outflow is zero, then the divergence of $\mathbf{F}$ is zero, and we say that $\mathbf{F}$ is *incompressible*.

Note that, since we took the density to be 1, the total outflow (as a number) is actually the change in the area of R as it is subjected to the flow for time Δt. More precisely, imagine that we take every point in R and move it along a flow line of $\mathbf{F}$ for time Δt.

The newly obtained region will, in general, have a different shape and size than the initial rectangle R; see Figure 4.32. The divergence of $\mathbf{F}$ tells us (approximately) at what relative rate (i.e., rate of change divided by the initial area) the area changes under the flow.

An interpretation of the divergence that involves integral theorems will be given in the last chapter (see Section 8.2).

In light of our discussion, we interpret intuitively the divergence of a vector field as a total outflow (or total outflux) per unit area (volume), or as a relative rate of change of area (volume). We now consider a few examples.

▶ EXAMPLE 4.54

Argue geometrically that the divergence of the vector field $\mathbf{F}(x, y) = x\mathbf{i}$ shown in Figure 4.33 is positive.

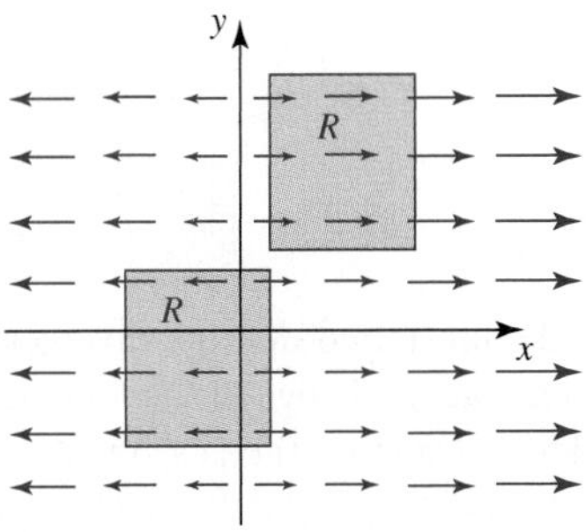

Figure 4.33 Vector field $\mathbf{F}(x, y) = x\mathbf{i}$.

SOLUTION In this case, there is no flow in the y-direction, and the flow in the x-direction increases as $|x|$ increases. The flow out of the rectangle R of small area is larger than the inflow (or there is no inflow at all!), and consequently, the divergence is positive. The definition yields $div\,\mathbf{F} = 1$, which confirms our conclusion.

▶ EXAMPLE 4.55

Determine the divergence of the vector field $\mathbf{F}$ in Figure 4.34.

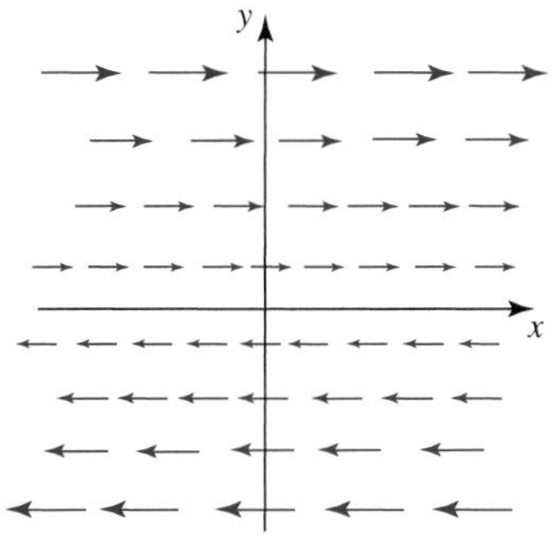

Figure 4.34 Field **F** of Example 4.55.

SOLUTION There is no flow in the direction of the y-axis, and the flow in the x-direction is constant along horizontal lines. Therefore, the inflow along any horizontal line is the same as the outflow, and the divergence is zero (at every point). One can use the definition $div\,\mathbf{F} = \partial F_1/\partial x + \partial F_2/\partial y$ and argue as follows: clearly, $F_2 = 0$, and hence, $\partial F_2/\partial y = 0$. The derivative $\partial F_1/\partial x$ represents the rate of change of the component F_1 in the x-direction, but F_1 does not change in the horizontal directions, and therefore, $\partial F_1/\partial x = 0$.

▶ EXAMPLE 4.56

Determine geometrically the sign of the divergence of the vector field $\mathbf{F}(x, y) = (x, y)$.

SOLUTION Consider a rectangle R placed within the flow of $\mathbf{F}$, as shown in Figure 4.35(a). Fluid flows into R along segments $\overline{AB}$ and $\overline{AD}$, and flows out along $\overline{BC}$ and $\overline{CD}$. Since $\|\mathbf{F}\|$ increases with distance,

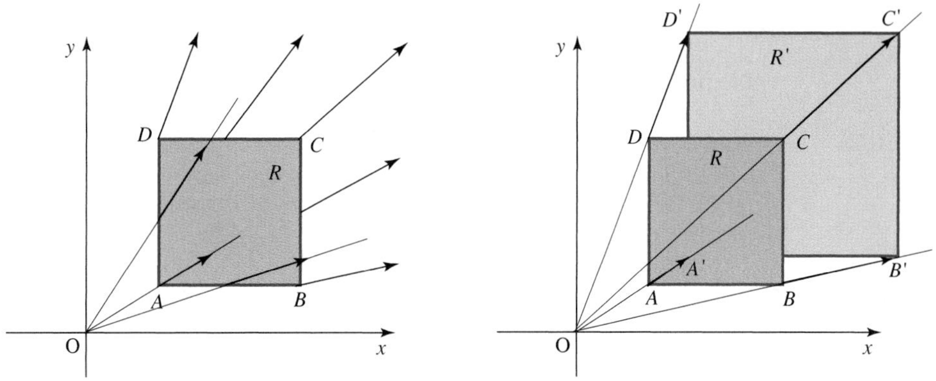

(a) Field **F** along sides of R. (b) R moved along the flow.

Figure 4.35 Geometric interpretation of divergence of **F**.

flow is stronger along $\overline{BC}$ and $\overline{CD}$ than along $\overline{AB}$ and $\overline{AD}$. Thus, the total outflux is positive, that is, $div\,\mathbf{F} > 0$.

If R contains the origin, the inflow is zero, so $div\,\mathbf{F} > 0$ in that case as well.

Alternatively, consider how the area of R changes along the flow. Note that the flow lines of $\mathbf{F}$ are straight lines through the origin (in the direction away from it). The flow will pull the points A, B, C, and D further from the origin, and further apart from each other, as shown in Figure 4.35(b). Thus, after some small time Δt, R is transformed to R' and (we use |..| to dente the length of a segment) $|\overline{A'B'}| > |\overline{AB}|$, $|\overline{B'C'}| > |\overline{BC}|$, $|\overline{C'D'}| > |\overline{CD}|$, and $|\overline{A'D'}| > |\overline{AD}|$. Thus, the area expands along the flow, and the divergence is positive.

A straightforward calculation shows that $div\,\mathbf{F} = 2 > 0$. ◀

▶ EXAMPLE 4.57

Explain why the divergence of the vector field $\mathbf{F}(x, y) = (-y, x)$ that describes rotational flow is equal to zero.

SOLUTION In Example 4.46 in Section 4.5, we showed that the flow lines of $\mathbf{F}(x, y) = (-y, x)$ are concentric circles, and the motion is counterclockwise. Place a small rectangle in the flow, as shown in Figure 4.36. Since the magnitude of the flow $\mathbf{F}(x, y) = (-y, x)$ is the same along all points on the same flow line, the inflow at A is the same as the outflow at A'. Since this is true for all points on the boundary of R, it follows that the divergence of $\mathbf{F}$ is zero. ◀

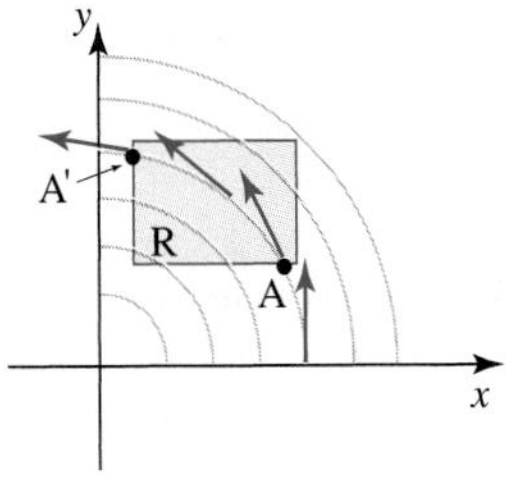

Figure 4.36 Field $\mathbf{F}(x, y) = (-y, x)$.

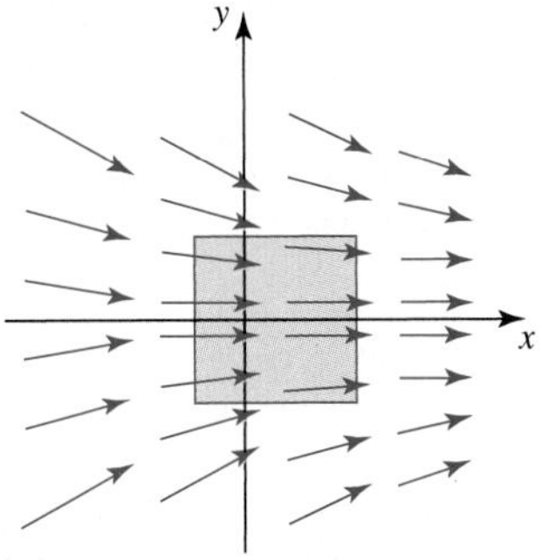

Figure 4.37 Field $\mathbf{F}$ of Example 4.58.

▶ EXAMPLE 4.58

Determine the sign of the divergence of the vector field $\mathbf{F}$ in Figure 4.37.

SOLUTION Place a rectangle of small area in the flow as shown: the mass (or fluid particles, or stream of sand particles in a sand storm, or electrons) flows in through three sides, and the inflow is stronger ($\|\mathbf{F}\|$ is larger) than the outflow through the right side. Consequently, the divergence of this field is negative. ◀

▶ EXAMPLE 4.59 Estimating Divergence of a Wind Field

Assume that a vector field $\mathbf{F} = (F_1(x, y), F_2(x, y))$ describes horizontal wind field over some region on Earth. Unlike the vector fields that we have studied so far, it is impossible to measure and to know the values of $\mathbf{F}(x, y)$ at *all* points that belong to a given region. Instead, measurements are made

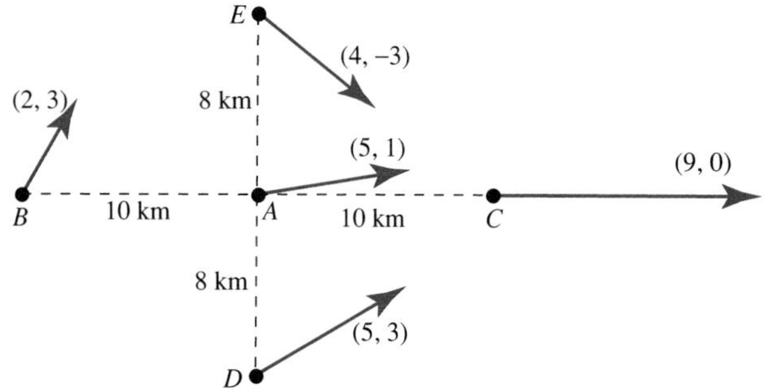

Figure 4.38 Wind velocity vectors, as measured by five weather stations.

at weather stations, located some distance apart from each other. Figure 4.38 shows wind velocity vectors at five locations [units are m/s; for instance, vector (2, 3) at B indicates that the x-component of the wind speed at B is 2 m/s, and the y-component is 3 m/s]. Vectors are not shown to scale with the distance between points.

We would like to estimate the divergence of the wind at A.

To get the divergence, we need to approximate the derivative of F_1 (x-component of $\mathbf{F}$) with respect to x and the derivative of F_2 (y-component of $\mathbf{F}$) with respect to y.

Going from B to A, we notice that the x-component changes by $5-2=3$ over a distance of 10 km; thus, $\partial F_1/\partial x \approx 3/10000 = 0.0003\,s^{-1}$. Between the stations A and C, we estimate $\partial F_1/\partial x \approx (9-5)/10000 = 0.0004\,s^{-1}$. Taking the average of the two estimates (common practice), we get $\partial F_1/\partial x \approx 0.00035\,s^{-1}$.

Going from D to A, we see that the y-component F_2 changes from 3 to 1; thus, $\partial F_2/\partial y \approx (1-3)/8000 = -0.00025\,s^{-1}$. From A to E, $\partial F_2/\partial y \approx (-3-1)/8000 = -0.0005\,s^{-1}$. We take the average of the two approximations $(-0.00025+(-0.0005))/2 = -0.000375$ as an estimate for $\partial F_2/\partial y$. Therefore,

$$div\,\mathbf{F} = \frac{\partial F_1}{\partial x} + \frac{\partial F_2}{\partial y} \approx 0.00035 - 0.000375 = -0.000025\,s^{-1}$$

is the desired estimate.

▶ EXAMPLE 4.60 Curl of an Electrostatic Field

Compute the curl of the electrostatic field $\mathbf{E}(\mathbf{r}) = Q\mathbf{r}/4\pi\epsilon_0\|\mathbf{r}\|^3$, where $\mathbf{r} = x\mathbf{i} + y\mathbf{j} + z\mathbf{k}$.

SOLUTION All terms in the expression for the curl involve derivatives, and therefore we can factor out the constant $Q/4\pi\epsilon_0$. We proceed as follows:

$$curl\,\mathbf{E} = \frac{Q}{4\pi\epsilon_0}\begin{vmatrix} \mathbf{i} & \mathbf{j} & \mathbf{k} \\ \partial/\partial x & \partial/\partial y & \partial/\partial z \\ x(x^2+y^2+z^2)^{-3/2} & y(x^2+y^2+z^2)^{-3/2} & z(x^2+y^2+z^2)^{-3/2} \end{vmatrix}$$

$$= \left(-\frac{3}{2}z\left(x^2+y^2+z^2\right)^{-5/2}2y + \frac{3}{2}y\left(x^2+y^2+z^2\right)^{-5/2}2z\right)\mathbf{i}$$

$$+ \left(-\frac{3}{2}z\left(x^2+y^2+z^2\right)^{-5/2}2x + \frac{3}{2}x\left(x^2+y^2+z^2\right)^{-5/2}2z\right)\mathbf{j}$$

$$+ \left(-\frac{3}{2}y\left(x^2+y^2+z^2\right)^{-5/2}2x + \frac{3}{2}x\left(x^2+y^2+z^2\right)^{-5/2}2y\right)\mathbf{k} = \mathbf{0},$$

in $\mathbb{R}^3 - \{(0,0,0)\}$.

▶ EXAMPLE 4.61

To illustrate the definition of the curl, let us compute the curl of the vector field **F** describing the counterclockwise rotational motion pictured in Figure 4.39 at the point A. Assume that the z-component F_3 of **F** is zero (therefore, $\partial F_3/\partial x = 0$ and $\partial F_3/\partial y = 0$), and that there is no variation of **F** in the z-direction, that is, $\partial F_1/\partial z = 0$ and $\partial F_2/\partial z = 0$.

Looking at points on the x-axis that are close to A (i.e., going from A' to A''), we see that the y-component F_2 increases, and hence $\partial F_2/\partial x > 0$ at A.

Let us determine the rate of change of F_1 with respect to y. Below the x-axis (say, at B'), F_1 is positive, is 0 at A, and negative above the x-axis (say, at B''). So F_1 changes from positive to negative values (i.e., it decreases) as we go through A in the y-direction, and consequently, $\partial F_1/\partial y < 0$ at A.

The derivatives $\partial F_2/\partial x$ and $\partial F_1/\partial y$ are the only nonzero terms in the expression for the curl (see Definition 4.7): it follows that, at A, $curl\,\mathbf{F} = C\mathbf{k}$, where C is positive. ◀

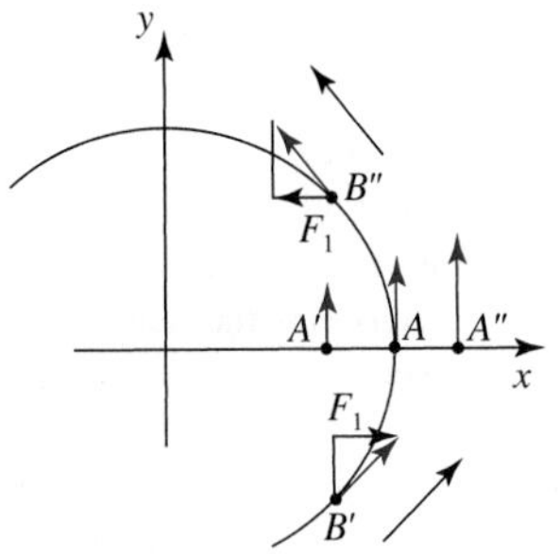

Figure 4.39 Field of Example 4.61.

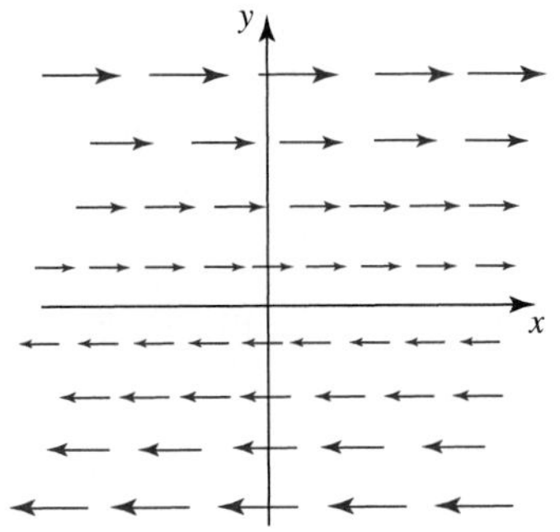

Figure 4.40 Field of Example 4.62.

▶ EXAMPLE 4.62

Determine the curl of the vector field in Figure 4.40. Assume that $\mathbf{F} = \mathbf{0}$ on the x-axis.

SOLUTION

Let us try to find the formula for **F**. Clearly, both **j** and **k** components are zero. The **i** component increases as we move up along the y-axis (increases from negative values below the x-axis toward positive values above it). Hence, $\mathbf{F} = g(y)\mathbf{i}$, where $g(0) = 0$ (so that $\mathbf{F} = \mathbf{0}$ on the x-axis) and g is increasing. By definition,

$$curl\,\mathbf{F} = \begin{vmatrix} \mathbf{i} & \mathbf{j} & \mathbf{k} \\ \partial/\partial x & \partial/\partial y & \partial/\partial z \\ g(y) & 0 & 0 \end{vmatrix} = -\frac{\partial g(y)}{\partial y}\mathbf{k},$$

with $\partial g(y)/\partial y > 0$ (since g is increasing). Therefore, $curl\,\mathbf{F}$ points into the page. ◀

▶ EXAMPLE 4.63 Rotation of a Rigid Body

The velocity vector of the rotation of a rigid body is given by $\mathbf{v} = \mathbf{w} \times \mathbf{r}$, where **r** denotes the position vector, $\mathbf{w} = (w_1, w_2, w_3)$ is the angular velocity vector (assumed to be constant), and **v** is the tangential velocity (see Example 1.36 in Section 1.5). Then

$$\mathbf{v} = \mathbf{w} \times \mathbf{r} = \begin{vmatrix} \mathbf{i} & \mathbf{j} & \mathbf{k} \\ w_1 & w_2 & w_3 \\ x & y & z \end{vmatrix} = (w_2 z - w_3 y)\mathbf{i} + (w_3 x - w_1 z)\mathbf{j} + (w_1 y - w_2 x)\mathbf{k}$$

and

$$curl\,\mathbf{v} = \begin{vmatrix} \mathbf{i} & \mathbf{j} & \mathbf{k} \\ \partial/\partial x & \partial/\partial y & \partial/\partial z \\ w_2 z - w_3 y & w_3 x - w_1 z & w_1 y - w_2 x \end{vmatrix} = 2(w_1\mathbf{i} + w_2\mathbf{j} + w_3\mathbf{k}),$$

that is, $curl\,\mathbf{v} = 2\mathbf{w}$. Therefore, the curl of the velocity vector of the rotation of a rigid body is parallel to the axis of rotation (given by the direction of $\mathbf{w}$) and its magnitude is twice the angular speed of rotation.

In the previous example, we gave an interpretation of the curl in the context of the rotation of a rigid body. Here is another interpretation connected with fluid flow. We present it now, although we will not be able to justify our statements until we introduce Stokes' Theorem in the last chapter (in Section 5.3, after introducing the circulation of a vector field, we give an intuitive explanation).

Suppose that a vector field $\mathbf{F}$ describes the flow of a fluid. If $curl\,\mathbf{F} = \mathbf{0}$ at some point P, then there are no rotations (whirlpools) in the flow at that point (the flow itself could be circular, but *within* the flow there are no rotations). More precisely, imagine that we place a coordinate system on a small floating device. "No rotations" means that the coordinate system does not rotate around its origin as our object moves along with the flow; see Figure 4.41(a).

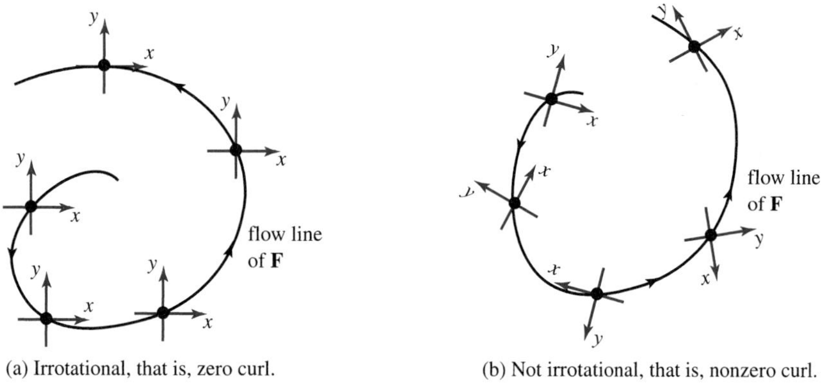

(a) Irrotational, that is, zero curl. (b) Not irrotational, that is, nonzero curl.

Figure 4.41 Vector fields in $\mathbb{R}^2$ (visualized through flow lines).

The condition $curl\,\mathbf{F} \neq \mathbf{0}$ allows rotations of a small coordinate system; see Figure 4.41(b). In other words, $curl\,\mathbf{F}$ is a measurement of the tendency of a fluid to swirl around an axis. We will explain this in more detail in Sections 5.3 and 8.3. A field $\mathbf{F}$ with $curl\,\mathbf{F} = \mathbf{0}$ at a point P is called *irrotational at P*.

EXAMPLE 4.64

Check that the vector field $\mathbf{F}_1 = y\mathbf{i}/(x^2 + y^2) - x\mathbf{j}/(x^2 + y^2)$ is irrotational whenever $(x, y) \neq (0, 0)$, but the vector field $\mathbf{F}_2 = y\mathbf{i} - x\mathbf{j}$ is not irrotational at any point.

SOLUTION By definition,

$$curl\,\mathbf{F}_1 = \begin{vmatrix} \mathbf{i} & \mathbf{j} & \mathbf{k} \\ \partial/\partial x & \partial/\partial y & \partial/\partial z \\ y/(x^2+y^2) & -x/(x^2+y^2) & 0 \end{vmatrix}$$

$$= 0\mathbf{i} + 0\mathbf{j} + \left(\frac{\partial}{\partial x}\left(\frac{-x}{x^2+y^2}\right) - \frac{\partial}{\partial y}\left(\frac{y}{x^2+y^2}\right)\right)\mathbf{k}$$

$$= \left(\frac{-(x^2+y^2)+x\cdot 2x}{(x^2+y^2)^2} - \frac{(x^2+y^2)-y\cdot 2y}{(x^2+y^2)^2}\right)\mathbf{k} = \mathbf{0}.$$

However,

$$curl\,\mathbf{F}_2 = \begin{vmatrix} \mathbf{i} & \mathbf{j} & \mathbf{k} \\ \partial/\partial x & \partial/\partial y & \partial/\partial z \\ y & -x & 0 \end{vmatrix} = \left(\frac{\partial}{\partial x}(-x) - \frac{\partial}{\partial y}(y)\right)\mathbf{k} = -2\mathbf{k}.$$

Note that $\mathbf{F}_2$ is defined for all $(x, y) \in \mathbb{R}^2$, whereas $\mathbf{F}_1$ is not defined at the origin. ◀

The vector fields $\mathbf{F}_1$ and $\mathbf{F}_2$ of the previous example have the same flow lines (they are concentric circles; that was shown in Examples 4.46 and 4.47 in Section 4.5). However, $\mathbf{F}_1$ is irrotational and $\mathbf{F}_2$ is not. This means that we cannot determine whether a vector field is irrotational or not just by looking at its flow lines.

Since gradient, divergence, and curl are built of (partial) derivatives, it is reasonable to expect that they satisfy properties of derivatives, namely linearity and the product rule—and indeed, they do. A precise formulation of these properties can be found in Section 4.8. In this section, we need only recognize the fact that a constant C can be factored out of divergence and curl; that is, $div\,(C\mathbf{F}) = C div\,\mathbf{F}$ and $curl\,(C\mathbf{F}) = C curl\,\mathbf{F}$ (the proofs are one-liners: just write out the left and right sides using the definitions and use the fact that the constant can be factored out of partial derivatives of real-valued functions).

THEOREM 4.13 Curl of Gradient and Divergence of Curl Are Zero

(a) Let f be a twice continuously differentiable real-valued function. Then

$$curl(grad\,f) = \mathbf{0}.$$

(b) Let $\mathbf{F}$ be a twice continuously differentiable vector field on $U \subseteq \mathbb{R}^3$. Then

$$div(curl\,\mathbf{F}) = 0.$$

◀

Using the ∇ formalism introduced in (4.31) through (4.34), identities (a) and (b) can be written as $\nabla \times \nabla f = \mathbf{0}$ and $\nabla \cdot (\nabla \times \mathbf{F}) = 0$. A twice continuously differentiable function (vector field) is also called a C^2 function (vector field).

PROOF: Both statements are proven by a straightforward computation with the use of appropriate definitions. To prove (a), we proceed as follows:

$$curl(grad\, f) = curl\left(\frac{\partial f}{\partial x}\mathbf{i} + \frac{\partial f}{\partial y}\mathbf{j} + \frac{\partial f}{\partial z}\mathbf{k}\right) = \begin{vmatrix} \mathbf{i} & \mathbf{j} & \mathbf{k} \\ \partial/\partial x & \partial/\partial y & \partial/\partial z \\ \partial f/\partial x & \partial f/\partial y & \partial f/\partial z \end{vmatrix}$$

$$= \left(\frac{\partial^2 f}{\partial y \partial z} - \frac{\partial^2 f}{\partial z \partial y}\right)\mathbf{i} + \left(\frac{\partial^2 f}{\partial z \partial x} - \frac{\partial^2 f}{\partial x \partial z}\right)\mathbf{j} + \left(\frac{\partial^2 f}{\partial x \partial y} - \frac{\partial^2 f}{\partial y \partial x}\right)\mathbf{k} = \mathbf{0},$$

due to the equality of mixed partial derivatives (see Theorem 4.1 in Section 4.1; this is where the assumption on f is needed). The statement (b) is proven analogously.

Part (a) of the theorem serves as a useful test for checking whether a given vector field $\mathbf{F}$ is conservative or not. If $\mathbf{F}$ is a conservative field, then $\mathbf{F} = -\nabla V$ for some real-valued function and $curl\,\mathbf{F} = -curl\,(\nabla f) = \mathbf{0}$.

Equivalently, a vector field $\mathbf{F}$ with $curl\,\mathbf{F} \neq \mathbf{0}$ *cannot* be conservative. In Section 5.4, we will show that under certain conditions on the domain U of $\mathbf{F}$ (i.e., that it be simply connected), the implication goes both ways: $\mathbf{F}$ is conservative in U if and only if $curl\,\mathbf{F} = \mathbf{0}$ in U. We will discuss the converse of (b) in Chapter 8.

DEFINITION 4.9 Laplace Operator

The action of the *Laplace operator* (also called the *Laplacian* or the *Laplacian operator*) Δ on a twice differentiable real-valued function f is defined by

$$\Delta f = div(grad\, f) = \nabla \cdot \nabla f.$$

Recall that we mentioned the Laplace operator in Example 4.10 in Section 4.1. The gradient of f is a vector field and therefore it makes sense to apply the divergence to it. Consequently, Δf is a real-valued function. The expression $\nabla \cdot \nabla f$ is sometimes abbreviated as $\nabla^2 f$. In Cartesian coordinates,

$$\Delta f = div\left(\frac{\partial f}{\partial x}\mathbf{i} + \frac{\partial f}{\partial y}\mathbf{j} + \frac{\partial f}{\partial z}\mathbf{k}\right) = \frac{\partial^2 f}{\partial x^2} + \frac{\partial^2 f}{\partial y^2} + \frac{\partial^2 f}{\partial z^2}.$$

The equation $\Delta f = 0$ is called *Laplace's equation*, and its solutions are called *harmonic functions*. The nonhomogeneous version of Laplace's equation $\Delta f = g$, where g is a continuous function, is (in some cases) called *Poisson's equation*.

Let $\mathbf{F} = (F_1, F_2, F_3)$ be a twice differentiable vector field. The action of the *Laplace operator* on $\mathbf{F}$ is defined by

$$\Delta \mathbf{F} = (\Delta F_1, \Delta F_2, \Delta F_3),$$

if the rectangular coordinates x, y, and z are used.

▶ **EXAMPLE 4.65** Gravitational Potential Satisfies Laplace's Equation

Show that the gravitational potential $V(x, y, z) = -GMm/\|\mathbf{r}\|$, $\mathbf{r} = x\mathbf{i} + y\mathbf{j} + z\mathbf{k}$ satisfies Laplace's equation.

SOLUTION In Example 2.40 in Section 2.4, it was shown that

$$\nabla V(x, y, z) = \frac{GMm}{\|\mathbf{r}\|^3}\mathbf{r} = -\mathbf{F},$$

where **F** is the gravitational force field. Therefore,

$$\Delta V = div(\nabla V) = -div\,\mathbf{F}.$$

The fact that $div\,\mathbf{F} = 0$ (see Example 4.52) completes the proof. ◀

▶ EXAMPLE 4.66 Laplace Operator Describes Diffusion

The concentration of a liquid changes ("diffuses") when some chemical is dissolved in it. The heat of a solid "diffuses," "flowing" from warmer regions toward cooler ones. Such processes of "transport" (or "transfer") are described by a *flux density vector field* **F**. In the case of heat transfer, $\mathbf{F} = -k\nabla T$, where T is the temperature (see Example 2.90). Heat transfer is a special case of *Fick's Law,* which states that the flux vector **F** is always parallel (and of the opposite direction) to the gradient of the "species" concentration:

$$\mathbf{F}(x, y, z) = -k\nabla f(x, y, z);$$

see Figure 4.42. For example, $f(x, y, z)$ could be the concentration of bacteria in air or the concentration of acid in a water solution. The symbol k ($k > 0$) denotes a constant, whose name (*conductivity, diffusivity*) depends on the process considered. The minus sign in Fick's Law indicates that the direction of the flow is always *away* from regions of higher concentration.

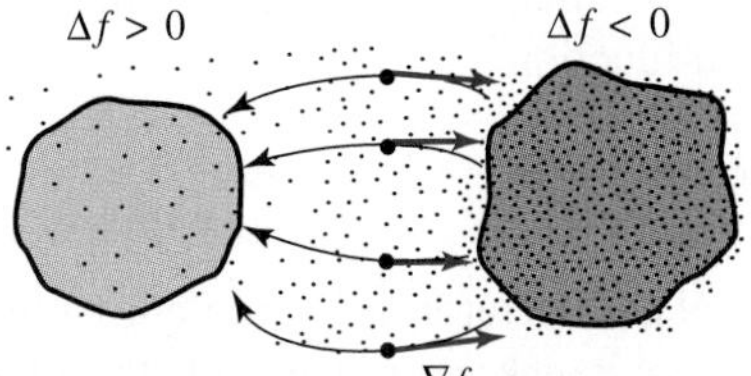

Figure 4.42 The Laplace operator describes a diffusion process.

The divergence of **F** is

$$div\,\mathbf{F}(x, y, z) = -k\,div\,(\nabla f(x, y, z)) = -k\Delta f(x, y, z).$$

We have seen that the divergence measures the net outflow of the "species" (i.e., "species that go out"—"species that go in"). At a point where the Laplacian Δ is negative, the outflow is positive, and the "species" must "go away" from that point; that is, the concentration decreases. Similarly, if $\Delta f(x, y, z) > 0$, then $div\,\mathbf{F} < 0$, and therefore, the inflow is larger than the outflow, and the concentration increases. (It is assumed, of course, that there are no outside "sources" or "sinks.") The equilibrium for a diffusion process is attained when the concentration "evens out" or "averages out"—in that case, the flow "stops" and the Laplacian of f is zero. ◀

▶ EXERCISES 4.6

Exercises 1 to 6: Let f be a scalar function and let **F** and **G** be vector fields in $\mathbb{R}^3$. State whether each expression is a scalar function, a vector field, or meaningless.

1. $grad\,(grad\,f)$
2. $curl\,(grad\,f) - \mathbf{G}$
3. $curl\,(\mathbf{F} - \mathbf{G}) \times grad\,(div\,\mathbf{F})$
4. $div\,(div\,\mathbf{F})$
5. $div\,(curl\,(grad\,f))$
6. $grad\,f^2 \times grad\,(\mathbf{F} \cdot \mathbf{G})$

Exercises 7 to 10: Find an example of a vector field (write down a formula, or make a sketch) that satisfies the following requirements.

7. $curl\, \mathbf{F} = \mathbf{0}$ and $div\, \mathbf{F} = 0$

8. $curl\, \mathbf{F} \neq \mathbf{0}$ and $div\, \mathbf{F} = 0$

9. $curl\, \mathbf{F} = \mathbf{0}$ and $div\, \mathbf{F} \neq 0$

10. $curl\, \mathbf{F} \neq \mathbf{0}$ and $div\, \mathbf{F} \neq 0$

11. Sketch a vector field in $\mathbb{R}^2$ whose divergence is positive at all points.

12. Sketch a vector field in $\mathbb{R}^2$ whose divergence is zero at all points.

Exercises 13 to 16: Find the curl and divergence of the vector field **F**.

13. $\mathbf{F}(x, y, z) = y^2 z\mathbf{i} - xz\mathbf{j} + xyz\mathbf{k}$

14. $\mathbf{F}(x, y, z) = (\ln z + xy)\mathbf{k}$

15. $\mathbf{F}(x, y, z) = (x^2 + y^2 + z^2)(3\mathbf{i} + \mathbf{j} - \mathbf{k})$

16. $\mathbf{F}(x, y, z) = e^{xy}\mathbf{i} + e^{yz}\mathbf{j} + e^{xz}\mathbf{k}$

17. Let $\mathbf{F}(x, y) = (f(x), 0)$, where $f(x)$ is a differentiable function of one variable. Show that the total outflow from a rectangle R with sides Δx and Δy placed in the flow (as in Figure 4.31) is given by $O(\Delta t) = (f(x + \Delta x) - f(x))\Delta t \Delta y$. Conclude that $\frac{1}{\Delta x \Delta y}\frac{O(\Delta t)}{\Delta t} \approx f'(x) = div\mathbf{F}$. Repeat the calculation with the vector field $\mathbf{F}(x, y) = (0, g(y))$ (g is a differentiable function of one variable) and show that the total outflow is again approximately equal to $div\mathbf{F}$.

18. What are the flow lines of the vector field $\mathbf{F}(x, y) = (-x, -y)$? Determine geometrically the sign of its divergence.

19. It can be easily checked that $curl\, \mathbf{r} = \mathbf{0}$, where $\mathbf{r} = x\mathbf{i} + y\mathbf{j} + z\mathbf{k}$. Interpret this result physically, by visualizing **r** as the velocity vector field of a fluid.

20. Consider the vector fields $\mathbf{F} = -y\mathbf{i} + x\mathbf{j}$, $\mathbf{G} = \mathbf{F}/\sqrt{x^2 + y^2}$, and $\mathbf{H} = \mathbf{F}/(x^2 + y^2)$. Compare their divergences and curls. Show that circles centered at the origin are the flow lines for all three vector fields. Describe their differences in physical terms.

Exercises 21 to 25: It will be shown in the next chapter that a vector field **F** defined on all of $\mathbb{R}^3$ (or all of $\mathbb{R}^2$) is conservative if and only if $curl\, \mathbf{F} = \mathbf{0}$. Determine whether the vector field **F** is conservative or not. If it is, find its potential function (i.e., find a real-valued function V such that $\mathbf{F} = -grad\, V$).

21. $\mathbf{F}(x, y, z) = \cos y\mathbf{i} + \sin x\mathbf{j} + \tan z\mathbf{k}$

22. $\mathbf{F}(x, y, z) = -y^2 z\mathbf{i} + (3y^2/2 - 2xyz)\mathbf{j} - xy^2\mathbf{k}$

23. $\mathbf{F}(x, y) = 3x^2 y\mathbf{i} + (x^3 + y^3)\mathbf{j}$

24. $\mathbf{F}(x, y, z) = x\mathbf{i} + y^2\mathbf{j} + z\mathbf{k}$

25. $\mathbf{F}(x, y, z) = -y\mathbf{i} - x\mathbf{j} - 3\mathbf{k}$

26. Check whether the vector field $\mathbf{F}(x, y) = \mathbf{i}/(x \ln xy) + \mathbf{j}/(y \ln xy)$ is conservative for $x, y > 0$, and if so, find all functions f such that $\mathbf{F} = grad\, f$.

27. Verify that $curl\,(grad\, f) = 0$ for the function $f(x, y, z) = (x^2 + y^2 + z^2)^{-1}$.

28. Verify that $\partial(curl\, \mathbf{F})_1/\partial x + \partial(curl\, \mathbf{F})_2/\partial y + \partial(curl\, \mathbf{F})_3/\partial z = 0$ for the vector field $\mathbf{F}(x, y, z) = 3x^3 y^2\mathbf{i} + yx^2\mathbf{j} - x^3 z^3\mathbf{k}$, where $(curl\, \mathbf{F})_1$, $(curl\, \mathbf{F})_2$, and $(curl\, \mathbf{F})_3$ are the components of $curl\, \mathbf{F}$.

29. Is there a C^2 vector field **F** such that $curl\, \mathbf{F} = xy^2\mathbf{i} + yz^2\mathbf{j} + zx^2\mathbf{k}$? Explain.

30. Is there a C^2 vector field **F** such that $curl\, \mathbf{F} = 2\mathbf{i} + \mathbf{j} + 3\mathbf{k}$? If so, find such a field.

31. A vector field **F** is irrotational if $curl\, \mathbf{F} = \mathbf{0}$. Show that any vector field of the form $\mathbf{F}(x, y, z) = f(x)\mathbf{i} + g(y)\mathbf{j} + h(z)\mathbf{k}$, where f, g, and h are differentiable real-valued functions of one variable, is irrotational.

32. A vector field **F** is incompressible if $div\, \mathbf{F} = 0$. Show that any vector field of the form $\mathbf{F}(x, y, z) = f(y, z)\mathbf{i} + g(x, z)\mathbf{j} + h(x, y)\mathbf{k}$, where f, g, and h are differentiable real-valued functions of two variables, is incompressible.

33. (For those familiar with complex numbers.) Show that the real and imaginary parts of the complex-valued function $z = (x - iy)^3$ (taken as the **i** and **j** components of a vector field whose **k** component is equal to 0) define an incompressible and irrotational vector field.

34. Find constants a, b, and c so that the vector field $\mathbf{F} = (3x - y + az)\mathbf{i} + (bx - z)\mathbf{j} + (4x + cy)\mathbf{k}$ is irrotational. Find the scalar function f so that $\mathbf{F} = \mathit{grad}\, f$.

35. Show that if the function f is harmonic (i.e., $\Delta f = 0$), then $\mathit{grad}\, f$ is not only an irrotational vector field but also an incompressible vector field.

36. Find the most general differentiable function $f(\|\mathbf{r}\|)$ defined on $\mathbb{R}^2$ such that the vector field $f(\|\mathbf{r}\|)\mathbf{r}$ is incompressible.

37. Show that the vector field $\mathbf{F} = (2x^2 + 8xy^2z)\mathbf{i} + (3x^3y - 3xy)\mathbf{j} - (4y^2z^2 + 2x^3z)\mathbf{k}$ is not incompressible but the vector field $\mathbf{G} = xyz^2\mathbf{F}$ is incompressible.

38. Prove that $\mathbf{F} \times \mathbf{G}$ is incompressible if the vector fields $\mathbf{F}$ and $\mathbf{G}$ are irrotational.

39. If f is a differentiable function of one variable, show that $f(\|\mathbf{r}\|)\mathbf{r}$ is an irrotational vector field.

Exercises 40 to 48: Prove the following identities, assuming that the functions and vector fields involved are differentiable as many times as needed. State those assumptions in each case. The vector $\mathbf{r}$ is the position vector $\mathbf{r} = x\mathbf{i} + y\mathbf{j} + z\mathbf{k}$.

40. $\mathit{div}\,(f\mathbf{F}) = f\, \mathit{div}\,\mathbf{F} + \mathbf{F} \cdot \mathit{grad}\, f$

41. $\mathit{curl}\,(f\mathbf{F}) = f\, \mathit{curl}\,\mathbf{F} + (\mathit{grad}\, f) \times \mathbf{F}$

42. $\mathit{curl}\,\mathbf{r} = \mathbf{0}$

43. $\mathit{div}\,\mathbf{r} = 3$

44. $\mathit{grad}\, \|\mathbf{r}\| = \dfrac{\mathbf{r}}{\|\mathbf{r}\|}$

45. $\mathit{div}\,(\|\mathbf{r}\|\mathbf{r}) = 4\|\mathbf{r}\|$

46. $\mathit{div}\,(\mathbf{F} \times \mathbf{G}) = \mathbf{G} \cdot \mathit{curl}\,\mathbf{F} - \mathbf{F} \cdot \mathit{curl}\,\mathbf{G}$

47. $\mathit{div}\,(\mathit{grad}\, f \times \mathit{grad}\, g) = 0$

48. $\Delta\|\mathbf{r}\|^3 = 12\|\mathbf{r}\|$

49. Evaluate the expression $\mathit{div}\,(\mathbf{F} \times \mathbf{r})$ if $\mathit{curl}\,\mathbf{F} = \mathbf{0}$, and $\mathbf{r} = x\mathbf{i} + y\mathbf{j} + z\mathbf{k}$.

50. Evaluate the expression $\mathit{curl}\,(f(\|\mathbf{r}\|)\mathbf{r})$, where f is a differentiable scalar function and $\mathbf{r} = x\mathbf{i} + y\mathbf{j} + z\mathbf{k}$.

▶ 4.7 IMPLICIT FUNCTION THEOREM

In this section, we state a general version of the Implicit Function Theorem. We have seen the importance of its special case in Section 3.1, where we studied curves defined by the equation $F(x, y) = 0$, for a C^1 function $F: \mathbb{R}^2 \to \mathbb{R}$.

We start by giving a straightforward generalization (without proof) of Theorem 3.1 from Section 3.1 to functions of many variables. We will denote points in $\mathbb{R}^{m+1}$ $(m \geq 1)$ by $(\mathbf{x}, z)$, where $\mathbf{x} \in \mathbb{R}^m$ and $z \in \mathbb{R}$.

THEOREM 4.14 Implicit Function Theorem, Special Case

Assume that a function $F: \mathbb{R}^{m+1} \to \mathbb{R}$ is of class C^1, $F(\mathbf{x}_0, z_0) = 0$, and $(\partial F/\partial z)(\mathbf{x}_0, z_0) \neq 0$ at a point $(\mathbf{x}_0, z_0)$ in its domain. Then:

(a) There exist an open ball $U \subseteq \mathbb{R}^m$ containing $\mathbf{x}_0$ and an open interval V containing z_0 such that there is a unique function $z = g(\mathbf{x})$, defined on U with values in V, satisfying

$$F(\mathbf{x}, g(\mathbf{x})) = 0$$

[i.e., $g(\mathbf{x})$ solves the equation $F(\mathbf{x}, z) = 0$ locally near $(\mathbf{x}_0, z_0)$].

(b) If $\mathbf{x}$ in U and z in V satisfy $F(\mathbf{x}, z) = 0$, then they are related by $z = g(\mathbf{x})$.

(c) The function $z = g(\mathbf{x})$ is of class C^1, and its partial derivatives are given by

$$\frac{\partial g}{\partial x_i}(\mathbf{x}) = -\frac{\dfrac{\partial F}{\partial x_i}(\mathbf{x}, z)}{\dfrac{\partial F}{\partial z}(\mathbf{x}, z)}, \tag{4.36}$$

for $i = 1, 2, \ldots, m$, provided that $(\partial F/\partial z)(\mathbf{x}, z) \neq 0$. ◀

Statements (a) and (b) establish the equality of the sets $\{(\mathbf{x}, z) \mid \mathbf{x} \in U, z \in V, \text{ and } z = g(\mathbf{x})\}$ and $\{(\mathbf{x}, z) \mid \mathbf{x} \in U, z \in V, \text{ and } F(\mathbf{x}, z) = 0\}$.

The following case will be of special interest.

DEFINITION 4.10 Implicitly Defined Surface

Let $F\colon \mathbb{R}^3 \to \mathbb{R}$ be a C^1 function. The set of points (x, y, z) in the domain of F where $F(x, y, z) = 0$ is called an *implicitly defined surface* in $\mathbb{R}^3$. ◀

Note that the above definition refers to the case where $m = 2$, and $\mathbf{x} = (x, y)$, so that $F(\mathbf{x}, z) = F(x, y, z)$. Theorem 4.14 states that, near the points where $(\partial F/\partial z)(x_0, y_0, z_0) \neq 0$, the equation $F(x, y, z) = 0$ can be solved (uniquely) in the form $z = g(x, y)$. In other words, near (x_0, y_0, z_0), the surface given implicitly by $F(x, y, z) = 0$ looks like the graph of a (unique) function $z = g(x, y)$.

The variables x, y, and z are interchangeable: for instance, if $(\partial F/\partial y)(x_0, y_0, z_0) \neq 0$, then, near (x_0, y_0, z_0), the graph of $F(x, y, z)$ looks like the graph of a surface $y = g_1(x, z)$. Likewise, whenever $(\partial F/\partial x)(x_0, y_0, z_0) \neq 0$, the equation $F(x, y, z) = 0$ has a local solution of the form $x = g_2(y, z)$.

For instance, the equation $F(x, y, z) = x^2 + y^2 + z^2 - 1 = 0$ represents the sphere of radius 1 centered at $(0, 0, 0)$. At all points where $\partial f/\partial z = 2z \neq 0$, we can solve locally for z as a function of x and y. More precisely, when $z > 0$, then $z = \sqrt{1 - x^2 - y^2}$, and when $z < 0$, then $z = -\sqrt{1 - x^2 - y^2}$. At the points $(x, y, 0)$ (they are all in the xy-plane) where we cannot represent z as a function of x and y, the tangent plane to the sphere is perpendicular to the xy-plane (note the similarity with the case discussed in Section 3.1).

Likewise, if $y \neq 0$, then $\partial F/\partial y = 2y \neq 0$, and we can solve the equation $F(x, y, z) = x^2 + y^2 + z^2 - 1 = 0$ for y; thus, when $y > 0$, then $y = \sqrt{1 - x^2 - z^2}$, and when $y < 0$, then $y = -\sqrt{1 - x^2 - z^2}$. At the points where $y = 0$, the tangent plane to the sphere is perpendicular to the xz-plane.

Assume that $z = g(x, y)$ is a local solution of $F(x, y, z) = 0$. Using the chain rule, we compute the partial derivative with repect to x:

$$D_1F(x, y, z)\frac{\partial x}{\partial x} + D_2F(x, y, z)\frac{\partial y}{\partial x} + D_3F(x, y, z)\frac{\partial z}{\partial x} = 0.$$

Since $\partial x/\partial x = 1$ and $\partial y/\partial x = 0$, we get $D_1F(x, y, z) + D_3F(x, y, z)(\partial z/\partial x) = 0$, and thus,

$$\frac{\partial g}{\partial x}(x, y) = \frac{\partial z}{\partial x}(x, y) = -\frac{D_1F(x, y, z)}{D_3F(x, y, z)},$$

which is the special case of (4.36). In the same way, starting with $F(\mathbf{x}, z) = 0$, we obtain the general formula (4.36).

▶ EXAMPLE 4.67

Consider the surface $F(x, y, z) = x + xe^y + y^2z - 1 = 0$. From $\partial F/\partial z = y^2 = 0$, we get $y = 0$. Thus, near a point (x_0, y_0, z_0), where $y_0 \neq 0$, the surface $F(x, y, z) = x + xe^y + y^2z - 1 = 0$ can be expressed uniquely as the graph of $z = g_1(x, y)$. In this case, we can solve explicitly:

$$z = g_1(x, y) = \frac{1 - x - xe^y}{y^2}.$$

Next, $\partial F/\partial y = xe^y + 2yz$. Thus, near a point (x_0, y_0, z_0), where $x_0e^{y_0} + 2y_0z_0 \neq 0$, the surface $F(x, y, z) = x + xe^y + y^2z - 1 = 0$ is the graph of a unique function $y = g_2(x, z)$. Note that we cannot solve for y explicitly.

Finally, $\partial F/\partial x = 1 + e^y \neq 0$ for all y. We conclude that near any point (x_0, y_0, z_0), the given surface can be represented as the graph of $x = g_3(y, z)$. In this case, we can compute the explicit form

$$x = g_3(y, z) = \frac{1 - y^2z}{1 + e^y}.$$

◀

We will return to the case of $F(x, y, z) = 0$ in Chapter 7, in our study of surfaces in $\mathbb{R}^3$.

Note that, in Theorem 3.1 in Section 3.1 and in Theorem 4.14 in this section, we were solving one equation for one variable. In general, we wish to solve a system of equations

$$F_1(\mathbf{x}, \mathbf{z}) = 0, \;\; F_2(\mathbf{x}, \mathbf{z}) = 0, \ldots, F_k(\mathbf{x}, \mathbf{z}) = 0, \tag{4.37}$$

where $\mathbf{x} = (x_1, x_2, \ldots, x_m)$ and $\mathbf{z} = (z_1, z_2, \ldots, z_k)$, and $F_1, F_2, \ldots, F_k$ are C^1 functions. The system (4.37) can be written as

$$\begin{aligned} &F_1(x_1, x_2, \ldots, x_m, z_1, z_2, \ldots, z_k) = 0 \\ &F_2(x_1, x_2, \ldots, x_m, z_1, z_2, \ldots, z_k) = 0 \\ &\vdots \\ &F_k(x_1, x_2, \ldots, x_m, z_1, z_2, \ldots, z_k) = 0. \end{aligned} \tag{4.38}$$

Let $\Delta(\mathbf{x}, \mathbf{z})$ denote the determinant of the matrix

$$\begin{bmatrix} \frac{\partial F_1}{\partial z_1}(\mathbf{x}, \mathbf{z}) & \frac{\partial F_1}{\partial z_2}(\mathbf{x}, \mathbf{z}) & \ldots & \frac{\partial F_1}{\partial z_k}(\mathbf{x}, \mathbf{z}) \\ \frac{\partial F_2}{\partial z_1}(\mathbf{x}, \mathbf{z}) & \frac{\partial F_2}{\partial z_2}(\mathbf{x}, \mathbf{z}) & \ldots & \frac{\partial F_2}{\partial z_k}(\mathbf{x}, \mathbf{z}) \\ \vdots & \vdots & & \vdots \\ \frac{\partial F_k}{\partial z_1}(\mathbf{x}, \mathbf{z}) & \frac{\partial F_k}{\partial z_2}(\mathbf{x}, \mathbf{z}) & \ldots & \frac{\partial F_k}{\partial z_k}(\mathbf{x}, \mathbf{z}) \end{bmatrix}. \tag{4.39}$$

The condition on partial derivatives in Theorem 4.14 will be replaced by the requirement that $\Delta(\mathbf{x}, \mathbf{z}) \neq 0$.

THEOREM 4.15 Implicit Function Theorem, General Case

Assume that $F_1, F_2, \ldots, F_k$ are real-valued, C^1 functions, defined on some subset of $\mathbb{R}^{m+k}$, where $m, k \geq 1$. If $F_1(\mathbf{x}_0, \mathbf{z}_0) = 0$, $F_2(\mathbf{x}_0, \mathbf{z}_0) = 0, \ldots, F_k(\mathbf{x}_0, \mathbf{z}_0) = 0$, and $\Delta(\mathbf{x}_0, \mathbf{z}_0) \neq 0$, then, near $(\mathbf{x}_0, \mathbf{z}_0)$, the system of equations (4.37) [or (4.38)] can be solved uniquely for $\mathbf{z} = (z_1, z_2, \ldots, z_k)$; that is, there exist unique C^1 functions $g_1, g_2, \ldots, g_k$, such that

$$\begin{aligned} z_1 &= g_1(x_1, x_2, \ldots, x_m) \\ z_2 &= g_2(x_1, x_2, \ldots, x_m) \\ &\vdots \\ z_k &= g_k(x_1, x_2, \ldots, x_m). \end{aligned} \tag{4.40}$$

The partial derivatives of g_i $(i = 1, 2, \ldots, k)$ are computed using implicit differentiation. ◀

In case of a small number of variables, instead of using subscripts as in (4.40), we use symbols such as x, y, z, u, v, etc.

We now explore several examples, to illustrate how Theorem 4.15 is used.

▶ EXAMPLE 4.68

Consider the equations

$$x + e^y uv = -2 \quad \text{and} \quad x^2 v + yu = 1.$$

We will show that, near the point $(\mathbf{x}, \mathbf{z}) = (x, y, u, v) = (1, 0, 3, -1)$, this system can be solved uniquely for u and v (as functions of x and y) and will compute their partial derivatives.

First, we rewrite the system as

$$\begin{aligned} F_1(x, y, u, v) &= x + e^y uv + 2 = 0 \\ F_2(x, y, u, v) &= x^2 v + yu - 1 = 0 \end{aligned} \tag{4.41}$$

(i.e., x and y are the x_i variables, and u and v are the z_i variables). The matrix in (4.39) is equal to

$$\begin{bmatrix} \partial F_1/\partial u & \partial F_1/\partial v \\ \partial F_2/\partial u & \partial F_2/\partial v \end{bmatrix} = \begin{bmatrix} e^y v & e^y u \\ y & x^2 \end{bmatrix}.$$

Thus,

$$\Delta(1, 0, 3, -1) = \det \begin{bmatrix} e^y v & e^y u \\ y & x^2 \end{bmatrix}_{\text{evaluated at } (1,0,3,-1)} = \det \begin{bmatrix} -1 & 3 \\ 0 & 1 \end{bmatrix} = -1 \neq 0.$$

The Implicit Function Theorem tells us that we can solve the system (4.41) for u and v as functions of x and y, near the given point.

Differentiating (4.41) with respect to x, we get

$$1 + e^y \frac{\partial u}{\partial x} v + e^y u \frac{\partial v}{\partial x} = 0 \quad \text{and} \quad 2xv + x^2 \frac{\partial v}{\partial x} + y \frac{\partial u}{\partial x} = 0.$$

At the point $(1, 0, 3, -1)$, the system reduces to

$$1 - \frac{\partial u}{\partial x} + 3 \frac{\partial v}{\partial x} = 0 \quad \text{and} \quad -2 + \frac{\partial v}{\partial x} = 0.$$

Solving this system, we get $(\partial u/\partial x)(1,0) = 7$ and $(\partial v/\partial x)(1,0) = 2$. The partial derivatives of v are computed analogously. ◀

▶ EXAMPLE 4.69 Polar Coordinates

Near which points is it possible to solve $x = r\cos\theta,\ y = r\sin\theta$ for r and θ? Compute $\partial r/\partial x$ and $\partial\theta/\partial x$.

SOLUTION We write the system as

$$\begin{aligned} F_1(x, y, r, \theta) &= x - r\cos\theta = 0 \\ F_2(x, y, r, \theta) &= y - r\sin\theta = 0, \end{aligned} \tag{4.42}$$

and compute

$$\Delta = \det\begin{bmatrix} \partial F_1/\partial r & \partial F_1/\partial\theta \\ \partial F_2/\partial r & \partial F_2/\partial\theta \end{bmatrix} = \det\begin{bmatrix} -\cos\theta & r\sin\theta \\ -\sin\theta & -r\cos\theta \end{bmatrix} = r.$$

Thus, near any point with $r \neq 0$, the given system can be solved uniquely for r and θ as functions of x and y [see (1.1) and (1.2) in Section 1.1 for expressions for r and θ].

Computing partial derivative of (4.42) with respect to x, we get

$$1 - \frac{\partial r}{\partial x}\cos\theta + r\sin\theta\frac{\partial\theta}{\partial x} = 0 \tag{4.43}$$

$$-\frac{\partial r}{\partial x}\sin\theta - r\cos\theta\frac{\partial\theta}{\partial x} = 0. \tag{4.44}$$

Multiplying the first equation by $\cos\theta$, the second equation by $\sin\theta$, and adding them up, we get $\cos\theta - \partial r/\partial x = 0$, i.e., $\partial r/\partial x = \cos\theta$. Similarly, we obtain $\partial\theta/\partial x = -\sin\theta/r$. ◀

▶ EXAMPLE 4.70 Spherical Coordinates

Near which points is it possible to solve the system

$$x = \rho\sin\phi\cos\theta, \quad y = \rho\sin\phi\sin\theta, \quad z = \rho\cos\phi$$

for ρ, θ, and ϕ?

SOLUTION Write

$$\begin{aligned} F_1(x, y, z, \rho, \theta, \phi) &= x - \rho\sin\phi\cos\theta = 0 \\ F_2(x, y, z, \rho, \theta, \phi) &= y - \rho\sin\phi\sin\theta = 0 \\ F_3(x, y, z, \rho, \theta, \phi) &= z - \rho\cos\phi = 0 \end{aligned}$$

and compute

$$\begin{aligned} \Delta &= \det\begin{bmatrix} \partial F_1/\partial\rho & \partial F_1/\partial\theta & \partial F_1/\partial\phi \\ \partial F_2/\partial\rho & \partial F_2/\partial\theta & \partial F_2/\partial\phi \\ \partial F_3/\partial\rho & \partial F_3/\partial\theta & \partial F_3/\partial\phi \end{bmatrix} \\ &= \det\begin{bmatrix} -\sin\phi\cos\theta & \rho\sin\phi\sin\theta & -\rho\cos\phi\cos\theta \\ -\sin\phi\sin\theta & -\rho\sin\phi\cos\theta & -\rho\cos\phi\sin\theta \\ -\cos\phi & 0 & \rho\sin\phi \end{bmatrix} = \rho^2\sin\phi. \end{aligned}$$

Thus, whenever $\rho^2\sin\phi \neq 0$, we can solve the above system for ρ, θ, and ϕ as functions of x, y, and z [recall that we actually solved the system in Section 2.8; see (2.37)]. ◀

As suggested, we will use the Implicit Function Theorem in computing the change of variables in multiple integration in Chapter 6 and in studying surfaces in Chapter 7.

▶ EXERCISES 4.7

1. Prove formula (4.36) in part (c) of Theorem 4.14.

2. For the function $v(x, y)$ defined implicitly by the system of equations given in Example 4.68, compute $(\partial u/\partial y)(1, 0)$ and $(\partial v/\partial y)(1, 0)$.

3. Provide details of the computations of the partial derivative $\partial\theta/\partial x$ in Example 4.69. Find $\partial r/\partial y$ and $\partial\theta/\partial y$.

4. Provide details of the computation of Δ in Example 4.70.

5. Using implicit differentiation, find $\partial\rho/\partial x$ and $\partial\phi/\partial x$ for the spherical coordinates in Example 4.70.

6. State the conditions under which the equation $F(x, y, z, w) = x^2y^2 - x^3w + ze^w = 0$ can be solved locally for w as a function of x, y, and z. Compute $\partial w/\partial y$ and $\partial w/\partial z$.

7. Determine whether the equation $F(x, y, z, w) = \sin(x + y + w) + z\cos(x + y + w) - 1 = 0$ can be solved for w as a function of x, y, and z, near the point $x = 0$, $y = 0$, $z = 1$, $w = 0$. If so, compute $\partial w/\partial x$ and $\partial w/\partial z$ at the given point.

8. Discuss the solvability of the system $u = x + 2y - z$, $v = x - y + 3z$, $w = 2x - z$ for x, y, and z in terms of u, v, and w.

9. Discuss the solvability of the system $u = x^2 + y^2$, $v = y^2 - z^2$, $w = xyz$ for x, y, and z as functions of u, v, and w.

10. Consider the system

$$F_1(x, y, u, v) = 2x - y + u - v = 0$$
$$F_2(x, y, u, v) = x + 4y + 4u + v = 0.$$

Discuss the solvability of the system for u and v as functions of x and y. Check your answer by explicitly calculating formulas for $u(x, y)$ and $v(x, y)$.

11. Consider the system

$$F_1(x, y, u, v) = ax + by - u = 0$$
$$F_2(x, y, u, v) = cx + dy - v = 0.$$

Discuss the solvability of the system for x and y as functions of u and v. Check your answer by explicitly calculating formulas for $x(u, v)$ and $y(u, v)$.

12. Determine whether the system $u = x^2 + yz$, $v = 2y - x - z^2$, $w = y^2 + xz$ can be solved, near $x = y = z = 0$, for x, y, and z as functions of u, v, and w.

13. Assume that the equation $F(x, y, z) = 0$, where F is a C^1 function, defines z implicitly as a function of x and y, that is, $z = g(x, y)$. Knowing that $F(3, 0, 1) = 0$ and $\nabla F(3, 0, 1) = (-4, -1, -6)$, compute the partial derivatives $g_x(3, 0)$ and $g_y(3, 0)$. Find the linear approximation of g at $(3, 0)$.

14. Determine whether the system $x + uyz + v = 1$, $y + uw + xz = 4$, $w + 2uy + z^2 = 0$ can be solved, near $x = 0$, $y = 0$, $z = 1$, for x, y, and z in terms of u, v, and w.

▶ 4.8 APPENDIX: SOME IDENTITIES OF VECTOR CALCULUS

In this section, we give some identities illustrating relationships among the differential operators *grad*, *div*, *curl*, and Δ and their properties with respect to algebraic operations on functions and vector fields, and with respect to vector operations.

Assume that f and g are differentiable (once or twice, as needed) real-valued functions and $\mathbf{F}$ and $\mathbf{G}$ are differentiable (once or twice, as needed) vector fields. If a statement involves a cross product, then f and g are assumed to be functions of three variables and $\mathbf{F}$ and $\mathbf{G}$ are assumed to be defined on a subset of $\mathbb{R}^3$. Let C denote a constant.

Linearity properties.

$$
\begin{aligned}
&grad\,(f+g) = grad\,f + grad\,g, && grad\,(Cf) = C\,grad\,f\\
&div\,(\mathbf{F}+\mathbf{G}) = div\,\mathbf{F} + div\,\mathbf{G}, && div\,(C\mathbf{F}) = C\,div\,\mathbf{F}\\
&curl\,(\mathbf{F}+\mathbf{G}) = curl\,\mathbf{F} + curl\,\mathbf{G}, && curl\,(C\mathbf{F}) = C\,curl\,\mathbf{F}\\
&\Delta(f+g) = \Delta f + \Delta g, && \Delta(Cf) = C\Delta f\\
&\Delta(\mathbf{F}+\mathbf{G}) = \Delta\mathbf{F} + \Delta\mathbf{G}, && \Delta(C\mathbf{F}) = C\Delta\mathbf{F}
\end{aligned}
$$

Recall that, in Cartesian coordinates, the action of the Laplace operator on a twice differentiable vector field $\mathbf{F}$ is defined componentwise; that is, $\Delta\mathbf{F} = (\Delta F_1, \Delta F_2, \Delta F_3)$.

Product rules.

$$
\begin{aligned}
&grad\,(fg)(\mathbf{x}) = g(\mathbf{x})\,grad\,f(\mathbf{x}) + f(\mathbf{x})\,grad\,g(\mathbf{x})\\
&grad\left(\frac{f}{g}\right)(\mathbf{x}) = \frac{g(\mathbf{x})\,grad\,f(\mathbf{x}) - f(\mathbf{x})\,grad\,g(\mathbf{x})}{g(\mathbf{x})^2}, \qquad \text{if } g(\mathbf{x}) \neq 0\\
&div\,(f\mathbf{F}) = f\,div\,\mathbf{F} + \mathbf{F}\cdot grad\,f\\
&div\,(\mathbf{F}\times\mathbf{G}) = \mathbf{G}\cdot curl\,\mathbf{F} - \mathbf{F}\cdot curl\,\mathbf{G} \qquad (4.44)\\
&curl\,(f\mathbf{F}) = f\,curl\,\mathbf{F} + grad\,f \times \mathbf{F}\\
&\Delta(fg) = g\Delta f + f\Delta g + 2\,grad\,f\cdot grad\,g
\end{aligned}
$$

Combinations of two operators.

$$div\,(grad\,f) = \Delta f \qquad (4.45)$$

$$div\,(curl\,\mathbf{F}) = 0 \qquad (4.46)$$

$$curl\,(grad\,f) = \mathbf{0} \qquad (4.47)$$

$$curl\,(curl\,\mathbf{F}) = grad\,(div\,\mathbf{F}) - \Delta\mathbf{F} \qquad (4.48)$$

$$
\begin{aligned}
&div\,(grad\,f \times grad\,g) = 0\\
&div\,(g\,grad\,f \times f\,grad\,g) = 0\\
&div\,(f\,grad\,g) = f\Delta g + grad\,f\cdot grad\,g\\
&div\,(f\,grad\,g - g\,grad\,f) = f\Delta g - g\Delta f
\end{aligned}
$$

Miscellanea. Let $\mathbf{r} = x\mathbf{i} + y\mathbf{j} + z\mathbf{k}$ and $\|\mathbf{r}\| = \sqrt{x^2+y^2+z^2}$. Assume that f is a differentiable real-valued function and let n be a real number. If there is a fraction involved (e.g., the exponent n could be negative), it is assumed that its denominator is not zero.

$$grad\left(\|\mathbf{r}\|^n\right) = n\|\mathbf{r}\|^{n-2}\,\mathbf{r} \tag{4.49}$$

$$grad\left(f^n\right) = nf^{n-1}\,grad\, f$$

$$grad\left(\frac{1}{\|\mathbf{r}\|}\right) = -\frac{\mathbf{r}}{\|\mathbf{r}\|^3} \tag{4.50}$$

$$div\,\frac{\mathbf{r}}{\|\mathbf{r}\|^3} = 0 \tag{4.51}$$

$$curl\,\mathbf{r} = \mathbf{0}$$

$$curl\,(\|\mathbf{r}\|^n\mathbf{r}) = \mathbf{0} \tag{4.52}$$

$$\Delta\left(\frac{1}{\|\mathbf{r}\|}\right) = 0 \tag{4.53}$$

$$\Delta\left(\|\mathbf{r}\|^n\right) = n(n+1)\|\mathbf{r}\|^{n-2}$$

Chain rule. If $f = f(u, v, w)$ and $u = u(x, y, z)$, $v = v(x, y, z)$, and $w = w(x, y, z)$, then

$$grad\, f(u, v, w) = \frac{\partial f}{\partial u}\,grad\, u + \frac{\partial f}{\partial v}\,grad\, v + \frac{\partial f}{\partial w}\,grad\, w. \tag{4.54}$$

Formula (4.45) is the definition of the Laplace operator, and (4.46) and (4.47) were proven in Theorem 4.13. Identity (4.50) was proven in Example 2.40 in Section 2.4 and (4.51) was proven in Example 4.65. Equation (4.53) follows from (4.45), (4.50), and (4.51), since

$$\Delta\left(\frac{1}{\|\mathbf{r}\|}\right) = div\left(grad\,\frac{1}{\|\mathbf{r}\|}\right) = -div\left(\frac{\mathbf{r}}{\|\mathbf{r}\|^3}\right) = 0$$

Next, we give the proofs of (4.44) and (4.54). Remaining proofs are left as exercises.

PROOF OF (4.44): The left side is computed to be

$$\begin{aligned}
div(\mathbf{F}\times\mathbf{G}) &= div\begin{vmatrix} \mathbf{i} & \mathbf{j} & \mathbf{k} \\ F_1 & F_2 & F_3 \\ G_1 & G_2 & G_3 \end{vmatrix} \\
&= div\left((F_2G_3 - F_3G_2)\mathbf{i} - (F_1G_3 - F_3G_1)\mathbf{j} + (F_1G_2 - F_2G_1)\mathbf{k}\right) \\
&= \frac{\partial}{\partial x}(F_2G_3 - F_3G_2) - \frac{\partial}{\partial y}(F_1G_3 - F_3G_1) + \frac{\partial}{\partial z}(F_1G_2 - F_2G_1) \\
&= \frac{\partial F_2}{\partial x}G_3 + F_2\frac{\partial G_3}{\partial x} - \frac{\partial G_2}{\partial x}F_3 - G_2\frac{\partial F_3}{\partial x} - \frac{\partial F_1}{\partial y}G_3 - F_1\frac{\partial G_3}{\partial y} \\
&\quad + \frac{\partial G_1}{\partial y}F_3 + G_1\frac{\partial F_3}{\partial y} + \frac{\partial F_1}{\partial z}G_2 + F_1\frac{\partial G_2}{\partial z} - \frac{\partial F_2}{\partial z}G_1 - F_2\frac{\partial G_1}{\partial z} \\
&= G_1\left(\frac{\partial F_3}{\partial y} - \frac{\partial F_2}{\partial z}\right) + G_2\left(\frac{\partial F_1}{\partial z} - \frac{\partial F_3}{\partial x}\right) + G_3\left(\frac{\partial F_2}{\partial x} - \frac{\partial F_1}{\partial y}\right) \\
&\quad - F_1\left(\frac{\partial G_3}{\partial y} - \frac{\partial G_2}{\partial z}\right) - F_2\left(\frac{\partial G_1}{\partial z} - \frac{\partial G_3}{\partial x}\right) - F_3\left(\frac{\partial G_2}{\partial x} - \frac{\partial G_1}{\partial y}\right).
\end{aligned}$$

On the other hand,

$$\mathbf{G}\cdot curl\,\mathbf{F}-\mathbf{F}\cdot curl\,\mathbf{G}=\mathbf{G}\cdot\begin{vmatrix}\mathbf{i}&\mathbf{j}&\mathbf{k}\\ \partial/\partial x&\partial/\partial y&\partial/\partial z\\ F_1&F_2&F_3\end{vmatrix}-\mathbf{F}\cdot\begin{vmatrix}\mathbf{i}&\mathbf{j}&\mathbf{k}\\ \partial/\partial x&\partial/\partial y&\partial/\partial z\\ G_1&G_2&G_3\end{vmatrix}$$

$$=(G_1,G_2,G_3)\cdot\left(\frac{\partial F_3}{\partial y}-\frac{\partial F_2}{\partial z},\frac{\partial F_1}{\partial z}-\frac{\partial F_3}{\partial x},\frac{\partial F_2}{\partial x}-\frac{\partial F_1}{\partial y}\right)$$

$$-(F_1,F_2,F_3)\cdot\left(\frac{\partial G_3}{\partial y}-\frac{\partial G_2}{\partial z},\frac{\partial G_1}{\partial z}-\frac{\partial G_3}{\partial x},\frac{\partial G_2}{\partial x}-\frac{\partial G_1}{\partial y}\right),$$

and we are done. ◀

PROOF OF (4.54): From the definition of the gradient and the chain rule, we get

$$grad\,f(u,v,w)=\frac{\partial}{\partial x}f(u,v,w)\mathbf{i}+\frac{\partial}{\partial y}f(u,v,w)\mathbf{j}+\frac{\partial}{\partial z}f(u,v,w)\mathbf{k}$$

$$=\left(\frac{\partial f}{\partial u}\frac{\partial u}{\partial x}+\frac{\partial f}{\partial v}\frac{\partial v}{\partial x}+\frac{\partial f}{\partial w}\frac{\partial w}{\partial x}\right)\mathbf{i}+\left(\frac{\partial f}{\partial u}\frac{\partial u}{\partial y}+\frac{\partial f}{\partial v}\frac{\partial v}{\partial y}+\frac{\partial f}{\partial w}\frac{\partial w}{\partial y}\right)\mathbf{j}$$

$$+\left(\frac{\partial f}{\partial u}\frac{\partial u}{\partial z}+\frac{\partial f}{\partial v}\frac{\partial v}{\partial z}+\frac{\partial f}{\partial w}\frac{\partial w}{\partial z}\right)\mathbf{k}.$$

To complete the proof, we rearrange terms by factoring out $\partial f/\partial u$, $\partial f/\partial v$, and $\partial f/\partial w$:

$$=\frac{\partial f}{\partial u}\left(\frac{\partial u}{\partial x}\mathbf{i}+\frac{\partial u}{\partial y}\mathbf{j}+\frac{\partial u}{\partial z}\mathbf{k}\right)+\frac{\partial f}{\partial v}\left(\frac{\partial v}{\partial x}\mathbf{i}+\frac{\partial v}{\partial y}\mathbf{j}+\frac{\partial v}{\partial z}\mathbf{k}\right)$$

$$+\frac{\partial f}{\partial w}\left(\frac{\partial w}{\partial x}\mathbf{i}+\frac{\partial w}{\partial y}\mathbf{j}+\frac{\partial w}{\partial z}\mathbf{k}\right)$$

$$=\frac{\partial f}{\partial u}grad\,u+\frac{\partial f}{\partial v}grad\,v+\frac{\partial f}{\partial w}grad\,w.$$ ◀

Differential operators. In cylindrical coordinates, with $\mathbf{F}=F_r\mathbf{e}_r+F_\theta\mathbf{e}_\theta+F_z\mathbf{e}_z$,

$$grad\,f=\frac{\partial f}{\partial r}\mathbf{e}_r+\frac{1}{r}\frac{\partial f}{\partial\theta}\mathbf{e}_\theta+\frac{\partial f}{\partial z}\mathbf{e}_z$$

$$div\,\mathbf{F}=\frac{1}{r}\frac{\partial(rF_r)}{\partial r}+\frac{1}{r}\frac{\partial F_\theta}{\partial\theta}+\frac{\partial F_z}{\partial z}$$

$$curl\,\mathbf{F}=\frac{1}{r}\begin{vmatrix}\mathbf{e}_r&r\mathbf{e}_\theta&\mathbf{e}_z\\ \frac{\partial}{\partial r}&\frac{\partial}{\partial\theta}&\frac{\partial}{\partial z}\\ F_r&rF_\theta&F_z\end{vmatrix}$$

$$\Delta f=\frac{1}{r}\left(\frac{\partial}{\partial r}\left(r\frac{\partial f}{\partial r}\right)+\frac{\partial}{\partial\theta}\left(\frac{1}{r}\frac{\partial f}{\partial\theta}\right)+\frac{\partial}{\partial z}\left(r\frac{\partial f}{\partial z}\right)\right)$$

$$=\frac{\partial^2 f}{\partial r^2}+\frac{1}{r}\frac{\partial f}{\partial r}+\frac{1}{r^2}\frac{\partial^2 f}{\partial\theta^2}+\frac{\partial^2 f}{\partial z^2}.$$

In spherical coordinates, with $\mathbf{F} = F_\rho \mathbf{e}_\rho + F_\theta \mathbf{e}_\theta + F_\phi \mathbf{e}_\phi$,

$$\begin{aligned}
grad\, f &= \frac{\partial f}{\partial \rho}\mathbf{e}_\rho + \frac{1}{\rho \sin\phi}\frac{\partial f}{\partial \theta}\mathbf{e}_\theta + \frac{1}{\rho}\frac{\partial f}{\partial \phi}\mathbf{e}_\phi \\
div\, \mathbf{F} &= \frac{1}{\rho^2 \sin\phi}\left(\frac{\partial}{\partial \rho}\left(\rho^2 \sin\phi F_\rho\right) + \frac{\partial}{\partial \theta}(\rho F_\theta) + \frac{\partial}{\partial \phi}\left(\rho \sin\phi F_\phi)\right)\right) \\
&= \frac{1}{\rho^2}\frac{\partial}{\partial \rho}\left(\rho^2 F_\rho\right) + \frac{1}{\rho \sin\phi}\frac{\partial}{\partial \theta}(F_\theta) + \frac{1}{\rho \sin\phi}\frac{\partial}{\partial \phi}\left(\sin\phi F_\phi\right) \\
curl\, \mathbf{F} &= \frac{1}{\rho^2 \sin\phi}\begin{vmatrix} \mathbf{e}_\rho & \rho\mathbf{e}_\phi & \rho \sin\phi \mathbf{e}_\theta \\ \frac{\partial}{\partial \rho} & \frac{\partial}{\partial \phi} & \frac{\partial}{\partial \theta} \\ F_\rho & \rho F_\phi & \rho \sin\phi F_\theta \end{vmatrix} \\
\Delta f &= \frac{1}{\rho^2 \sin\phi}\left(\frac{\partial}{\partial \rho}\left(\rho^2 \sin\phi \frac{\partial f}{\partial \rho}\right) + \frac{\partial}{\partial \theta}\left(\frac{1}{\sin\phi}\frac{\partial f}{\partial \theta}\right) + \frac{\partial}{\partial \phi}\left(\sin\phi \frac{\partial f}{\partial \phi}\right)\right) \\
&= \frac{\partial^2 f}{\partial \rho^2} + \frac{2}{\rho}\frac{\partial f}{\partial \rho} + \frac{1}{\rho^2 \sin^2\phi}\frac{\partial^2 f}{\partial \theta^2} + \frac{1}{\rho^2}\frac{\partial^2 f}{\partial \phi^2} + \frac{\cot\phi}{\rho^2}\frac{\partial f}{\partial \phi}.
\end{aligned}$$

▶ EXERCISES 4.8

Exercises 1 to 7: Prove the following identities.

1. $curl\,(f\mathbf{F}) = f\, curl\,\mathbf{F} + grad\, f \times \mathbf{F}$

2. $div\,(f\mathbf{F}) = f\, div\,\mathbf{F} + \mathbf{F} \cdot grad\, f$

3. $\Delta(fg) = g\Delta f + f\Delta g + 2\, grad\, f \cdot grad\, g$

4. $div\,(grad\, f \times grad\, g) = 0$

5. $div\,(g\, grad\, f \times f\, grad\, g) = 0$

6. $div\,(f\, grad\, g) = f\Delta g + grad\, f \cdot grad\, g$

7. $div\,(f\, grad\, g - g\, grad\, f) = f\Delta g - g\Delta f$

Exercises 8 to 10: Let $\mathbf{r} = x\mathbf{i} + y\mathbf{j} + z\mathbf{k}$. Prove the following identities and state the domain where each is valid.

8. $grad\, \|\mathbf{r}\|^{-1} = -\|\mathbf{r}\|^{-3}\mathbf{r}$

9. $div\, \|\mathbf{r}\|^{-3}\mathbf{r} = 0$

10. $\Delta\|\mathbf{r}\|^n = n(n+1)\|\mathbf{r}\|^{n-2}$

11. Find the gradient of f, if $f(x, y) = g(u(x, y), v(x, y), x, y)$, g is a differentiable function of four variables and u and v are differentiable functions of x and y.

12. Compute $grad\, f$, if $f(x, y) = g(x^2 y, x^3 - y, y^4)$ and g is differentiable.

Exercises 13 to 15: Let $\mathbf{r} = x\mathbf{i} + y\mathbf{j} + z\mathbf{k}$. Compute each expression.

13. $grad\,(\ln \|\mathbf{r}\|)$

14. $\Delta\left(div\,(\mathbf{r}/\|\mathbf{r}\|^2)\right)$

15. $\Delta(\ln \|\mathbf{r}\|)$

16. Let $f(r, \theta, z) = z \arctan\theta/r^2$ be the expression for the function f in cylindrical coordinates. Compute $grad\, f$ and Δf.

17. Let $\mathbf{F}(r, \theta, z) = r^2\mathbf{e}_r + rz\mathbf{e}_\theta + \sin\theta\mathbf{e}_z$ be the expression for the vector field $\mathbf{F}$ in cylindrical coordinates. Compute $div\,\mathbf{F}$ and $curl\,\mathbf{F}$.

18. Compute $grad\, f$ and Δf for the function $f(\rho, \theta, \phi) = \rho^2 \sin\phi$ in spherical coordinates.

19. Compute *div* $\mathbf{F}$ and *curl* $\mathbf{F}$ for the vector field $\mathbf{F}(\rho, \theta, \phi) = \theta\mathbf{e}_\rho + \cos\theta\cos\phi\mathbf{e}_\theta - \rho\mathbf{e}_\phi$ in spherical coordinates.

20. Express the heat equation $\partial U/\partial t = c\Delta U$ in spherical coordinates if $U = U(\rho, \theta, \phi, t)$ is independent of the following variables.

(a) θ (b) θ and ϕ (c) θ, ϕ and t (d) ρ and t

▶ CHAPTER REVIEW

CHAPTER SUMMARY

- **Higher-order partial derivatives.** Equality of mixed partial derivatives, wave equation, heat equation, Laplace's equation, first-order Taylor's formula, second-order Taylor's formula, linear and quadratic approximations, estimating the remainder in Taylor's formula.

- **Extreme values.** Fermat's Theorem and Second Derivative Test for functions of one variable, relative maximum and relative minimum, generalization of Fermat's Theorem to functions of several variables, critical point, saddle point, Second Derivatives Test for functions of two variables.

- **Optimization.** Absolute maximum and absolute minimum, closed and bounded sets, Extreme Value Theorem, finding absolute extreme values on closed and bounded sets, constrained optimization problem, Lagrange multipliers.

- **Vector fields.** Flow lines (integral curves), divergence, curl, scalar curl, physical interpretation of divergence, total outflow, incompressible vector field, curl and rotation of a rigid body, physical interpretation of curl, Laplace's operator and diffusion.

- **Implicit Function Theorem.** Implicitly defined surface, special and general cases of the Implicit Function Theorem, local solvability of a system of equations.

REVIEW

Discuss the following questions.

1. Give physical interpretations of the divergence and the curl. Sketch examples of vector fields $\mathbf{F}_1$, $\mathbf{F}_2$, and $\mathbf{F}_3$ such that *div* $\mathbf{F}_1 = 0$ and *curl* $\mathbf{F}_1 = \mathbf{0}$, *div* $\mathbf{F}_2 > 0$ and *curl* $\mathbf{F}_2 = \mathbf{0}$, *div* $\mathbf{F}_3 < 0$ and *curl* $\mathbf{F}_3 \neq \mathbf{0}$.

2. In Figure 4.43, several level curves of a differentiable function $f(x, y)$ are shown, together with a curve that represents the constraint function $g(x, y) = k$. Identify locations of points where

(a) $\nabla f = \lambda\nabla g$.

(b) f has a local minimum.

(c) f has a local minimum subject to $g(x, y) = k$.

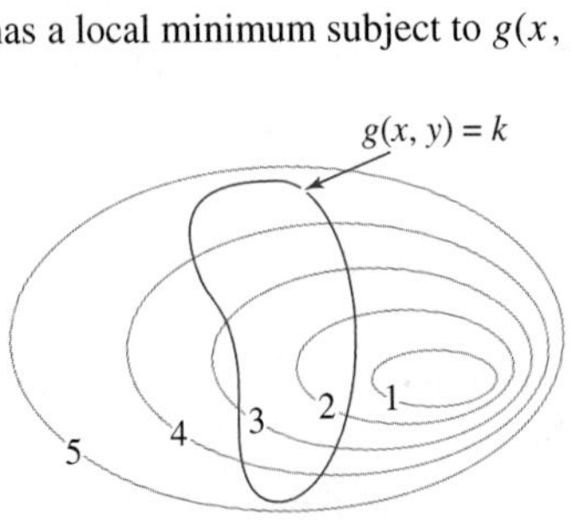

Figure 4.43 Level curves of Exercise 2.

3. Figure 4.44 shows several level curves of a function $f(x, y)$.

(a) Which of the four points A, B, C, D seem to be critical points? Determine whether each critical point you selected is a local maximum, minimum, or a saddle point.

(b) Show the direction of the gradient vector at several locations around each of the four points A, B, C, and D.

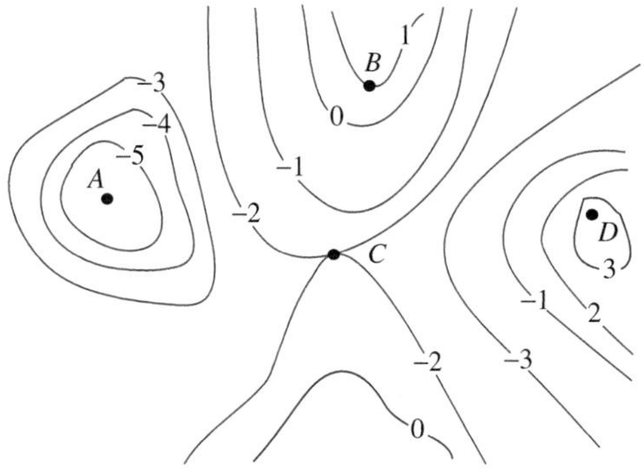

Figure 4.44 Level curves of Exercise 3.

4. Is there a C^2 vector field $\mathbf{F}$ such that $curl\,\mathbf{F} = xy^2\mathbf{i} + z\mathbf{k}$? Such that $curl\,\mathbf{F} = y\mathbf{i} + \mathbf{k}$? Explain your answers.

5. A function $f(x, y)$ satisfies the following conditions: $f(0, 0) = 4$, $f_x(0, 0) = f_y(0, 0) = 0$, $f_{xx}(0, 0) = 0$, $f_{yy}(0, 0) > 0$, and $f_{xy}(0, 0) > 0$. Determine whether f has a local minimum, local maximum, or a saddle point at $(0, 0)$. Sketch possible level curves of f near $(0, 0)$.

6. Sketch possible level curves near $(0, 0)$ for a function $f(x, y)$ that satisfies $f(0, 0) = 2$, $f_x(0, 0) = f_y(0, 0) = 0$, $f_{xx}(0, 0) > 0$, $f_{yy}(0, 0) > 0$, and $f_{xy}(0, 0) = 0$.

7. Explain what it means for a function $f(x, y)$ to have a saddle point at (x_0, y_0). Sketch the graph of a function $f(x, y)$ that has a saddle point; label the saddle point in your graph. Sketch the level curves of a function $f(x, y)$ that has a saddle point.

8. Assume that the vector $\mathbf{w}$ points in the direction of the largest increase of a function $f(x, y)$. Is it always true that f decreases most rapidly in the direction of $-\mathbf{w}$?

9. State the special case of the Implicit Function Theorem and explain how to use it to determine whether the equation $F(x, y, z) = x^3y - y^2 - \sin(x + yz) = 0$ can be solved near $(0, 0, 2)$ for z as a function of x and y.

10. What condition(s) must F_1 and F_2 satisfy so that the system $F_1(x, y, z, t, u, v) = 0$, $F_2(x, y, z, t, u, v) = 0$ can be solved locally for u and v?

11. Write down the first-order and second-order Taylor formulas for a function of one variable. What is a quadratic approximation? Give the formula for the nth-order Taylor polynomial.

12. Write down the second-order Taylor formula for a function $f(x, y)$ of two variables. Explain how to use it to determine the behavior of $f(x, y)$ at its critical points.

13. Give a geometric interpretation of a flow line. Explain how you can use the flow lines of a vector field to determine the sign of the divergence of the field.

TRUE/FALSE QUIZ

Determine whether the following statements are true or false. Give reasons for your answer.

1. The curve $\mathbf{c}(t) = (2\sin t, 2\cos t)$ is a flow line of the vector field $\mathbf{F}(x, y) = x\mathbf{i} + y\mathbf{j}$.

2. The equation $F(x, y, z, w) = \ln(x + y + w) + y^2zw - x^3 = 0$ can be solved for w near $(0, 1, 0, 0)$.

3. If $f_x(x_0, y_0) = 0$, then (x_0, y_0) is a critical point of $f(x, y)$.

4. The function $3xy + x$ is the second-order Taylor polynomial of $f(x, y) = x(y + 1)^3$.

5. If $f(x, y)$ has a local maximum at $(3, 1)$ subject to the constraint $g(x, y) = 2$, then $f(3, 1) = 2$.

6. The vector field $\mathbf{F}(x, y) = (x, -y)$ is both incompressible and irrotational.

7. The function $f(x, y) = |x|$ has a local minimum at $(0, 0)$.

8. If we remove the x-axis from the xy-plane, we obtain a closed set.

9. The set of all points whose distance from the point $(2, 3)$ is smaller than 10 is bounded.

10. Every function of two variables must have an absolute maximum on a closed and bounded set.

11. If $f_x(1, 1) = f_y(1, 1) = 0$ and $f_{xx}(1, 1) > 0$, then $f(x, y)$ has a local minimum at $(1, 1)$.

12. A vector field whose flow lines are circles can have a nonzero curl.

13. There is a C^2 function $f(x, y)$ such that $grad\, f = xe^y\mathbf{i} + e^y\mathbf{j}$.

14. $curl\, \mathbf{F} = \nabla \times \mathbf{F}$ is always perpendicular to $\mathbf{F}$.

REVIEW EXERCISES AND PROBLEMS

1. Show that $u(x, t) = e^{2k(x-4k^2t)}$ is a solution of the *linearized Korteweg–de Vries equation* $u_t + u_{xxx} = 0$.

2. Show that the vector field $\mathbf{F}(x, y) = e^x \cos y\,\mathbf{i} - e^x \sin y\,\mathbf{j}$ is incompressible and irrotational.

3. Let $\mathbf{c}(t)$ be a flow line of the vector field $\mathbf{F} = -grad\, f$ (i.e., $\mathbf{F}$ is a conservative field and f its potential function). Show that $f(\mathbf{c}(t))$ is a decreasing function (i.e., the potential decreases as one moves along its flow lines).

4. Real-valued functions $u(x, y)$ and $v(x, y)$ are said to satisfy the *Cauchy–Riemann equations* if $u_x = v_y$ and $u_y = -v_x$. Show that the following functions satisfy the Cauchy–Riemann equations:
(a) $u(x, y) = x^3 - 3xy^2$ and $v(x, y) = 3x^2y - y^3$.
(b) $u(x, y) = e^x \cos y$ and $v(x, y) = e^x \sin y$.

5. Show that if two functions $u(x, y)$ and $v(x, y)$ satisfy the Cauchy–Riemann equations (see Exercise 4) and u_{xy} and v_{xy} are continuous, then $u(x, y)$ and $v(x, y)$ satisfy Laplace's equation; that is, $u_{xx} + u_{yy} = 0$ and $v_{xx} + v_{yy} = 0$.

6. Find the second-order Taylor polynomial of $f(x, y) = \sin(x + y) + \cos(x - y)$ near $(0, 0)$.

7. Find the second-order Taylor polynomial of $f(x, y, z) = 1 + \sin(x + y + z^2) + 3\cos(2x - z)$ at $(0, 0, 0)$.

8. Discuss the solvability of the system $x + 3y^2 + z + w^2 + u^2 - v^2 = 0$, $x - y + 4z^2 + 2uv = 0$ for u and v in terms of remaining variables. Discuss the solvability of the same system for z and w as functions of x, y, u, and v.

9. Find an example (give a formula and make a sketch) of a vector field with constant negative divergence.

10. Find and classify all critical points of the function $f(x, y) = \sin x \cos y$.

11. Find and classify all critical points of the function $f(x, y) = x \sin y$.

12. Find the minimum and maximum of the function $f(x, y) = \sin^2 x + \sin^2 y$ subject to the constraint $x + y = 0$.

13. Using Lagrange multipliers, find the point in the plane $ax + by + cz + d = 0$ ($d \neq 0$) that is closest to the origin. You may assume that such point exists. Find the shortest distance between the given plane and the origin.

14. Show that, if $z = e^x + y$, and $x = r\cos\theta$ and $y = r\sin\theta$, then $z_{xx} + z_{yy} = z_{rr} + z_{\theta\theta}/r^2 + z_r/r$.

15. Find the second-order Taylor polynomial for the function $f(x, y) = \ln(x - 2y + 1)$ at $(2, 1)$.

16. Show that $\Delta(div(\|\mathbf{r}\|^{-2}\mathbf{r})) = 2\|\mathbf{r}\|^{-4}$, where $\mathbf{r} = x\mathbf{i} + y\mathbf{j} + z\mathbf{k} \neq \mathbf{0}$.

CHAPTER 5

Integration Along Paths

The next three chapters are devoted to the development of generalizations of the definite integral to various regions. We will learn how to integrate real-valued functions of several variables and vector fields along curves, over surfaces, and (in the case of real-valued functions only) over three-dimensional regions. In all cases, the constructions follow the same theme: an integral will be the limit of approximating sums that are called Riemann sums.

We start this chapter by studying paths (or parametrizations) as an analytic way of describing a curve in a plane or in space. Concepts relevant to integration are introduced and illustrated in a number of examples. We proceed by constructing path integrals of real-valued functions and vector fields (the latter are also known as line integrals). The work done by a force and the circulation of a vector field are presented as main applications of these concepts. As a consequence of the fact that the path integral of a real-valued function does not depend on the way a curve is traversed, we obtain the fact that the length of a curve does not depend on its parametrization.

A vector field is called a gradient vector field if it is the gradient of some real-valued function (probably the most important example is a conservative force field). It turns out that such vector fields possess remarkable properties when integrated along closed curves. A whole section is devoted to the investigation of gradient vector fields, their properties, and applications.

▶ 5.1 PATHS AND PARAMETRIZATIONS

In Chapters 2 and 3, we studied concepts and properties related to paths in a plane or in space. Now we turn our attention to those properties that are needed for the integration of functions and vector fields.

Recall that a path in $\mathbb{R}^2$ (or $\mathbb{R}^3$) is a function $\mathbf{c}(t)$: $[a, b] \to \mathbb{R}^2$ (or $\mathbb{R}^3$) defined on an interval $[a, b] \subseteq \mathbb{R}$, where a and b are real numbers (for purposes of integration we normally do not allow $a = -\infty$ or $b = \infty$; see Exercise 5 at the end of Section 5.2 for a case when $b = \infty$). The image of $\mathbf{c}(t)$ is called a curve and the function $\mathbf{c}(t)$ is said to parametrize the curve. We write $\mathbf{c}(t) = (x(t), y(t))$ (for a curve in $\mathbb{R}^2$) or $\mathbf{c}(t) = (x(t), y(t), z(t))$ (for a curve in $\mathbb{R}^3$), where the component functions $x(t)$, $y(t)$ and $z(t)$ are real-valued functions of one variable, defined on $[a, b]$. A path is continuous (differentiable) if and only if all of its component functions are continuous (differentiable). A curve is called continuous (differentiable) if it has a continuous (differentiable) parametrization; see Definition 2.18 in Section 2.5 and the comment following it.

Our next definition introduces paths that will be used in integration.

DEFINITION 5.1 C^1 Path and Piecewise C^1 Path

A path $\mathbf{c}(t)\colon [a,b] \to \mathbb{R}^2$ (or $\mathbb{R}^3$) is called a C^1 *path* if and only if its component functions have continuous derivatives (i.e., are C^1) on $[a,b]$. A path $\mathbf{c}$ is a *piecewise* C^1 path if its domain $[a,b]$ can be broken into subintervals $[a = t_1, t_2], [t_2, t_3], \ldots, [t_{n-1}, t_n = b]$ so that the restriction of $\mathbf{c}$ to each subinterval $[t_i, t_{i+1}]$, $i = 1, \ldots, n-1$, is a C^1 path.

In other words, a piecewise C^1 path is obtained by gluing together C^1 paths so that the terminal point of one path becomes the initial point of the following one. By definition, a C^1 path is also a piecewise C^1 path.

► EXAMPLE 5.1

Consider the path $\mathbf{c}(t) = (t, |t^2 - 1|)$, $t \in [-2, 3]$, shown in Figure 5.1. It is continuous, but not differentiable (and hence not C^1), since the component $y(t) = |t^2 - 1|$ does not have a derivative at $t = \pm 1$.

Now break up the interval $[-2, 3]$ into three subintervals $[-2, -1]$, $[-1, 1]$, and $[1, 3]$, and consider the restrictions

$$\mathbf{c}(t)|_{[-2,-1]} = (t, t^2 - 1)$$

[when $-2 \le t \le -1$, $t^2 - 1 \ge 0$, and therefore $|t^2 - 1| = t^2 - 1$],

$$\mathbf{c}(t)|_{[-1,1]} = (t, -t^2 + 1)$$

[when $-1 \le t \le 1$, $t^2 - 1 \le 0$, and therefore $|t^2 - 1| = -(t^2 - 1)$], and

$$\mathbf{c}(t)|_{[1,3]} = (t, t^2 - 1)$$

[when $t \ge 1$, $t^2 - 1 \ge 0$]. All components of all restrictions are polynomials, and hence continuously differentiable on the given intervals. It follows that the path $\mathbf{c}$ is built of three C^1 paths. We say that $\mathbf{c}$ is piecewise continuously differentiable, or piecewise C^1.

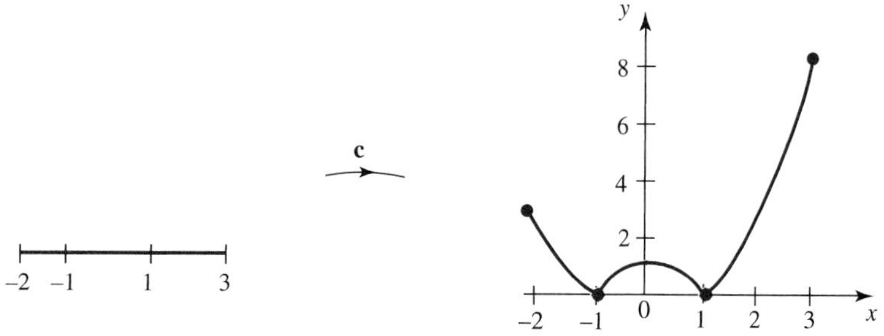

Figure 5.1 Piecewise C^1 path $\mathbf{c}(t) = (t, |t^2 - 1|)$, $t \in [-2, 3]$.

▶ EXAMPLE 5.2

The path $\mathbf{c}(t) = (\cos^3 t, \sin^3 t)$, $t \in [0, 2\pi]$, shown in Figure 5.2, is C^1, since both components of $\mathbf{c}'(t) = (-3\cos^2 t \sin t, 3\sin^2 t \cos t)$ are continuous on the interval $[0, 2\pi]$.

The path $\mathbf{c}(t) = (t^{1/3}, 2t^{1/3})$, $t \in [-1, 1]$, represents the straight-line segment joining $(-1, -2)$ and $(1, 2)$. It is not C^1 (not even piecewise C^1), since its derivative $\mathbf{c}'(t) = (t^{-2/3}/3, 2t^{-2/3}/3)$ does not exist when $t = 0$.

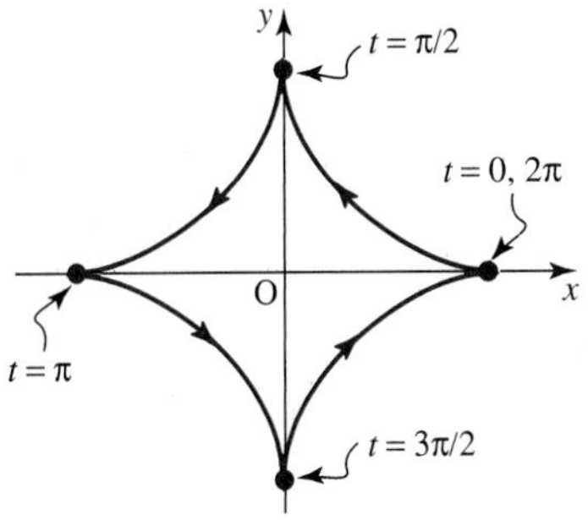

Figure 5.2 The path $\mathbf{c}(t) = (\cos^3 t, \sin^3 t)$, $t \in [0, 2\pi]$, is C^1. ◀

Examples 5.1 and 5.2 show that it is not possible to determine whether a path is C^1 or not just by looking at its image. In particular, the existence of a cusp does not mean that a path is not C^1. And a path that "looks like" it is C^1 (e.g., a line) is not necessarily C^1 (compare with the remark following Definition 2.18).

Recall that a function $\phi(t)$ is called *one-to-one* if $t_1 \neq t_2$ implies $\phi(t_1) \neq \phi(t_2)$ for all t_1, t_2 in its domain. In other words, a one-to-one function maps distinct points to distinct points. The equivalent statement, that $\phi(t)$ is one-to-one if and only if $\phi(t_1) = \phi(t_2)$ implies $t_1 = t_2$, provides a more "workable" version of the definition (it merely restates it in the following sense: if two points t_1 and t_2 get mapped by ϕ to the same point, then they must have been the same point to start with). A function $\phi(t)\colon [\alpha, \beta] \to [a, b]$, where $[\alpha, \beta]$ and $[a, b]$ are closed intervals in $\mathbb{R}$, is called *onto* if and only if the image of $\phi(t)$ is $[a, b]$; that is, for every c in $[a, b]$, there is a number t_0 in $[\alpha, \beta]$ such that $\phi(t_0) = c$.

DEFINITION 5.2 Bijective Function

A function $\phi\colon [\alpha, \beta] \to [a, b]$ is called *bijective* if and only if it is one-to-one and onto. ◀

Alternatively, ϕ is bijective if and only if it has an inverse $\phi^{-1}\colon [a, b] \to [\alpha, \beta]$.

The function $\phi\colon [0, 1] \to [1, 4]$ given by $\phi(t) = 1 + 3t$ is bijective [its inverse is $\phi^{-1}(t) = (t - 1)/3$], as is the map $\phi\colon [0, 1] \to [0, 1]$, $\phi(t) = t^3$ [the inverse map is $\phi^{-1}(t) = t^{1/3}$]. On the other hand, $\phi\colon [-1, 1] \to [0, 1]$ defined by $\phi(t) = t^2$ does not have an inverse [what would $\phi^{-1}(1)$ be, 1 or -1?], so it is not bijective.

Note that, in the case when ϕ is differentiable, the requirement that ϕ is one-to-one can be proven by showing that $\phi'(t) > 0$ for all $t \in (\alpha, \beta)$ [or $\phi'(t) < 0$ for all $t \in (\alpha, \beta)$]. This is intuitively clear: if a function strictly increases (strictly decreases) all the time, then it cannot have repeated values (see Exercise 9).

If $\phi\colon [\alpha, \beta] \to [a, b]$ is continuous and bijective, then it must map endpoints to endpoints (see Exercise 10). If, moreover, it is differentiable and $\phi' > 0$, then $\phi(\alpha) = a$ and $\phi(\beta) = b$; if $\phi' < 0$, then $\phi(\alpha) = b$ and $\phi(\beta) = a$.

In Section 2.5, we noticed that it is possible that infinitely many paths represent the same curve. However, they are not all unrelated to each other: among all paths that represent a curve we can always find so-called reparametrizations of a path. Let us make this concept precise.

DEFINITION 5.3 Reparametrization of a Path

Let $\mathbf{c}\colon [a, b] \to \mathbb{R}^2$ (or $\mathbb{R}^3$) be a C^1 path. The composition $\boldsymbol{\gamma} = \mathbf{c} \circ \phi\colon [\alpha, \beta] \to \mathbb{R}^2$ (or $\mathbb{R}^3$), where $\phi\colon [\alpha, \beta] \to [a, b]$ is C^1 and bijective, is called a *reparametrization* of $\mathbf{c}$.

The parametrization by arc-length that we studied in Section 3.3 is an example of a reparametrization (see Definition 3.6 and the text following it, and also Examples 3.29 and 3.30). For instance, $\mathbf{c}(s) = (3\cos(s/3), 3\sin(s/3))$ of Example 3.29 is a reparametrization of $\mathbf{c}(s) = (3\cos s, 3\sin s)$, where $\phi(s) = s/3$ (there was a reason why we used s, and not t, as the parameter).

EXAMPLE 5.3

Consider the helix $\mathbf{c}(t) = (\cos t, \sin t, t)$, $t \in [0, 2\pi]$, in $\mathbb{R}^3$. The initial point is $(1, 0, 0)$ and the terminal point is $(1, 0, 2\pi)$. The speed is computed to be $\|\mathbf{c}'(t)\| = \sqrt{2}$. Use $\phi_1\colon [0, \pi/2] \to [0, 2\pi]$, defined by $\phi_1(t) = 4t$, to reparametrize $\mathbf{c}(t)$ [ϕ_1 is clearly C^1 and a bijection: the inverse function is $\phi_1^{-1}(t) = t/4$]; set

$$\boldsymbol{\gamma}(t) = \mathbf{c}(\phi_1(t)) = \mathbf{c}(4t) = (\cos 4t, \sin 4t, 4t), \qquad t \in [0, \pi/2].$$

Curves $\boldsymbol{\gamma}$ and $\mathbf{c}$ have the same endpoints, but the speed of $\boldsymbol{\gamma}$ is $\|\boldsymbol{\gamma}'(t)\| = \sqrt{32} = 4\sqrt{2}$, that is, it is four times the speed of $\mathbf{c}$. Now define a new parametrization; let $\phi_2\colon [0, 2\pi] \to [0, 2\pi]$ be defined by $\phi_2(t) = 2\pi - t$ [ϕ_2 is bijective, since $\phi_2^{-1}(t) = 2\pi - t$]. Then

$$\boldsymbol{\gamma}(t) = \mathbf{c}(2\pi - t) = (\cos(2\pi - t), \sin(2\pi - t), 2\pi - t), \qquad t \in [0, 2\pi].$$

This time $\boldsymbol{\gamma}(0) = (1, 0, 2\pi)$ and $\boldsymbol{\gamma}(2\pi) = (1, 0, 0)$; that is, $\boldsymbol{\gamma}$ is traced in the direction opposite to $\mathbf{c}$. The speed of $\boldsymbol{\gamma}$ is $\sqrt{2}$.

The chain rule applied to the composition $\boldsymbol{\gamma}(t) = \mathbf{c}(\phi(t))$ implies that

$$\boldsymbol{\gamma}'(t) = \mathbf{c}'(\phi(t))\,\phi'(t).$$

In words, the velocity vector $\boldsymbol{\gamma}'(t)$ of the reparametrization $\boldsymbol{\gamma}$ at the point $\boldsymbol{\gamma}(t)$ is equal to the product of the scalar $\phi'(t)$ and the velocity vector $\mathbf{c}'(\phi(t))$ of $\mathbf{c}$ at the point $\mathbf{c}(\phi(t)) = \boldsymbol{\gamma}(t)$; that is, $\mathbf{c}'(\phi(t))$ and $\boldsymbol{\gamma}'(t)$ are parallel [of the same or opposite orientations, depending on the sign of $\phi'(t)$]. Moreover [using $\|\alpha\mathbf{v}\| = |\alpha| \cdot \|\mathbf{v}\|$, where α is a scalar and $\mathbf{v}$ a vector], we get the formula

$$\|\boldsymbol{\gamma}'(t)\| = |\phi'(t)|\,\|\mathbf{c}'(\phi(t))\|$$

stating the relationship between the speeds of $\boldsymbol{\gamma}$ and $\mathbf{c}$. In light of these observations, we interpret a reparametrization as a *change of speed and/or direction* (as we have done already in Example 5.3).

From Definition 5.3, it follows that a reparametrization $\phi: [\alpha, \beta] \to [a, b]$ maps the endpoints of $[\alpha, \beta]$ to the endpoints of $[a, b]$. There are two cases: either $\phi(\alpha) = a$ and $\phi(\beta) = b$, or $\phi(\alpha) = b$ and $\phi(\beta) = a$. In the former case, $\boldsymbol{\gamma}(\alpha) = \mathbf{c}(\phi(\alpha)) = \mathbf{c}(a)$ and $\boldsymbol{\gamma}(\beta) = \mathbf{c}(\phi(\beta)) = \mathbf{c}(b)$; that is, $\boldsymbol{\gamma}$ and $\mathbf{c}$ have the same initial point and the same terminal point; we say that ϕ is *orientation-preserving*. In the latter case, $\boldsymbol{\gamma}(\alpha) = \mathbf{c}(\phi(\alpha)) = \mathbf{c}(b)$ and $\boldsymbol{\gamma}(\beta) = \mathbf{c}(\phi(\beta)) = \mathbf{c}(a)$, and ϕ is *orientation-reversing*. This time, the initial point of $\mathbf{c}$ is the terminal point of $\boldsymbol{\gamma}$, and vice versa.

The reparametrization $\phi_1(t) = 4t$ in Example 5.3 is orientation-preserving and the reparametrization $\phi_2(t) = 2\pi - t$ in the same example is orientation-reversing.

After stating Definition 5.2, we argued that if $\phi' > 0$, then ϕ is orientation-preserving, and if $\phi' < 0$, then ϕ is orientation-reversing. The reverse implications are true, as the following theorem shows.

THEOREM 5.1 Orientation-Preserving and Orientation-Reversing Reparametrizations

A reparametrization ϕ is orientation-preserving if and only if $\phi' > 0$ (i.e., if and only if ϕ is a strictly increasing function), and orientation-reversing if and only if $\phi' < 0$ (i.e., if and only if ϕ is a strictly decreasing function). ◀

IDEA OF PROOF: Assume that ϕ is an orientation-preserving reparametrization and try to imagine what its graph would look like. By assumption, $\phi(\alpha) = a$ and $\phi(\beta) = b$, and consequently, the graph has to connect the points (α, a) and (β, b); see Figure 5.3. Since ϕ maps $[\alpha, \beta]$ into $[a, b]$, the graph has to be contained in the rectangle with vertices (α, a), (β, a), (β, b), and (α, b).

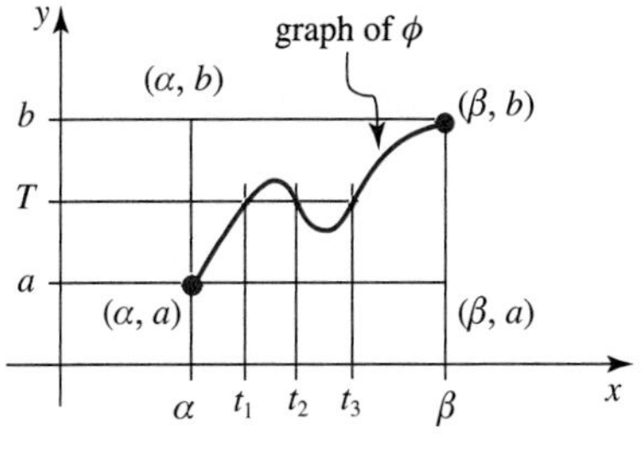

Figure 5.3 Graph of the function ϕ from the proof of Theorem 5.1.

Now suppose that the graph of ϕ starts decreasing at some point. Since it has to reach (β, b) it will have to start increasing again, so its shape will be something like the one shown in Figure 5.3. But then there are more points (in our case those are t_1, t_2, and t_3) that are mapped into the same point T, and this implies that ϕ is not one-to-one (and thus not bijective). Evidently, the assumption that ϕ decreases somewhere leads to a contradiction, so ϕ is always increasing.

A similar argument can be applied in the orientation-reversing case. ◀

▶ EXAMPLE 5.4

Let $\mathbf{c}: [a, b] \to \mathbb{R}^3$ be a curve in $\mathbb{R}^3$.

(a) Assume that the speed of $\mathbf{c}$ is constant, that is, $\|\mathbf{c}'(t)\| = C > 0$, for all t. Find an orientation-preserving reparametrization of $\mathbf{c}$ whose speed equals $S > 0$.

(b) Reparametrize **c** so that it takes T units of time to trace it.

(c) Reparametrize **c** so that it is traced in the opposite direction with the same speed C.

SOLUTION **(a)** The speed of $\mathbf{c}(t)$ and its reparametrization $\boldsymbol{\gamma}(t) = \mathbf{c}(\phi(t))$ are related by

$$\|\boldsymbol{\gamma}'(t)\| = |\phi'(t)|\,\|\mathbf{c}'(\phi(t))\|.$$

By assumption, the speed of **c** is equal to C everywhere; therefore, $\|\mathbf{c}'(\phi(t))\| = C$ and the above equation (together with the requirement on the speed of the reparametrization) implies that $|\phi'(t)| = S/C$; i.e., $\phi'(t) = \pm S/C$. We have to choose the plus sign in order to preserve orientation (see Theorem 5.1). Therefore, $\phi'(t) = S/C$ and $\phi(t) = St/C + D$, where D is a real number (as a matter of fact, we have found infinitely many desired reparametrizations). We need the domain $[\alpha, \beta]$ of ϕ. Since ϕ is orientation-preserving, it maps the initial (terminal) point to the initial (terminal) point. It follows that $\phi(\alpha) = S\alpha/C + D = a$ [and hence $\alpha = (a - D)C/S$] and $\phi(\beta) = S\beta/C + D = b$ [and hence $\beta = (b - D)C/S$]. We are done: the reparametrization of $\mathbf{c}(t)$, $t \in [a, b]$, is given by the composition

$$\boldsymbol{\gamma}(t) = \mathbf{c}(\phi(t)) = \mathbf{c}(St/C + D), \qquad t \in [(a - D)C/S, (b - D)C/S].$$

Let us check our result: the function ϕ is bijective, its inverse being $\phi^{-1}(t) = C(t - D)/S$. Since $\phi(t) = St/C + D$ is a polynomial (of degree 1), it is of class C^1, and therefore ϕ is a reparametrization. From

$$\boldsymbol{\gamma}\left(\frac{(a - D)C}{S}\right) = \mathbf{c}\left(\frac{S}{C}\,\frac{(a - D)C}{S} + D\right) = \mathbf{c}(a)$$

and

$$\boldsymbol{\gamma}\left(\frac{(b - D)C}{S}\right) = \mathbf{c}\left(\frac{S}{C}\,\frac{(b - D)C}{S} + D\right) = \mathbf{c}(b),$$

it follows that ϕ preserves orientation. Its speed is

$$\|\boldsymbol{\gamma}'(t)\| = \left\|\mathbf{c}'\left(\frac{St}{C} + D\right)\frac{S}{C}\right\| = \left\|\mathbf{c}'\left(\frac{St}{C} + D\right)\right\|\left|\frac{S}{C}\right| = C\,\frac{S}{C} = S,$$

since, by assumption, the speed of **c** is constant (equal to C).

(b) One needs $b - a$ units of time to trace **c**. We have to find a C^1, bijective map ϕ whose domain has length T; for example, $\phi\colon [0, T] \to [a, b]$. It follows that we need a function ϕ satisfying $\phi(0) = a$ and $\phi(T) = b$. One way to find it is to compute the equation of the line joining $(0, a)$ and (T, b) [its point-slope equation is $y - a = (b - a)x/T$] so that [set $x = t$, $y = \phi(t)$]

$$\phi(t) = \frac{b - a}{T}\,t + a, \qquad t \in [0, T].$$

ϕ is a polynomial (of degree 1) and hence continuously differentiable. Its inverse is given by $\phi^{-1}(t) = (t - a)T/(b - a)$. The composition

$$\boldsymbol{\gamma}(t) = \mathbf{c}(\phi(t)) = \mathbf{c}\left(\frac{b - a}{T}\,t + a\right), \qquad t \in [0, T],$$

is a desired parametrization.

(c) We need a map ϕ such that $\phi(a) = b$ and $\phi(b) = a$. Proceeding as in (b), we compute the equation of the line through (a, b) and (b, a): $y - b = (a - b)(x - a)/(b - a)$; that is, $y - b = -x + a$, and hence [replace y by $\phi(t)$ and x by t] $\phi(t) = a + b - t$. The function $\phi(t)$ is continuously differentiable and its inverse is $\phi^{-1}(t) = a + b - t$. It follows that the reparametrized curve

$$\boldsymbol{\gamma}(t) = \mathbf{c}(\phi(t)) = \mathbf{c}(a + b - t), \qquad t \in [a, b],$$

describes the curve $\mathbf{c}$ traced backward. Since $\phi'(t) = -1 < 0$, the reparametrization is orientation-reversing. [Alternatively, we could have computed the endpoints of $\boldsymbol{\gamma}$: $\boldsymbol{\gamma}(a) = \mathbf{c}(a - b - a) = \mathbf{c}(b)$ and $\boldsymbol{\gamma}(b) = \mathbf{c}(a - b - b) = \mathbf{c}(a)$.]

▶ EXAMPLE 5.5

Let $\mathbf{c}$ be the path in $\mathbb{R}^2$ given by $\mathbf{c}(t) = (t, t^2)$, $t \in [-1, 1]$. The image of $\mathbf{c}$ is the part of the parabola $y = x^2$ between the points $\mathbf{c}(-1) = (-1, 1)$ (the initial point of $\mathbf{c}$) and $\mathbf{c}(1) = (1, 1)$ (the terminal point of $\mathbf{c}$).

Define $\phi: [0, 1] \to [-1, 1]$ by $\phi(t) = -1 + 2t$. Clearly, $\phi(0) = -1$ and $\phi(1) = 1$ so ϕ is orientation-preserving [alternatively, we can apply Theorem 5.1 to $\phi'(t) = 2 > 0$]. The reparametrized curve $\mathbf{c}_1: [0, 1] \to \mathbb{R}^2$ has the equation

$$\mathbf{c}_1(t) = \mathbf{c}(\phi(t)) = (-1 + 2t, (-1 + 2t)^2) = (-1 + 2t, 1 - 4t + 4t^2).$$

The endpoints of the curve $\mathbf{c}_1$ are $\mathbf{c}_1(0) = (-1, 1)$ and $\mathbf{c}_1(1) = (1, 1)$, which confirms that the reparametrization is orientation-preserving.

On the other hand, the reparametrization $\phi: [-1, 1] \to [-1, 1]$ defined by $\phi(t) = -t$ gives the path

$$\mathbf{c}_2(t) = \mathbf{c}(\phi(t)) = \mathbf{c}(-t) = (-t, (-t)^2) = (-t, t^2),$$

which has an orientation opposite to $\mathbf{c}$. To verify this, all we have to do is to compute the endpoints: $\mathbf{c}_2(-1) = (1, 1)$ and $\mathbf{c}_2(1) = (-1, 1)$, or check the derivative: $\phi'(t) = -1 < 0$.

▶ EXAMPLE 5.6

Consider a reparametrization of the helix (see Example 5.3)

$$\mathbf{c}(t) = (\cos t, \sin t, t), \qquad t \in [0, 2\pi],$$

defined by $\phi:[0, \sqrt{2\pi}] \to [0, 2\pi]$, $\phi(t) = t^2$. (Notice that ϕ does have an inverse because t^2 is strictly increasing on $[0, \sqrt{2\pi}]$.) The curve

$$\mathbf{c}_1(t) = \mathbf{c}(\phi(t)) = (\cos t^2, \sin t^2, t^2), \qquad t \in [0, \sqrt{2\pi}],$$

has the speed (this time non-constant) $\|\mathbf{c}_1'(t)\| = \sqrt{8t^2} = 2\sqrt{2}t$. Curves $\mathbf{c}_1$ and $\mathbf{c}$ have the same orientation [since $\mathbf{c}(0) = \mathbf{c}_1(0) = (1, 0, 0)$ and $\mathbf{c}(2\pi) = \mathbf{c}_1(\sqrt{2\pi}) = (1, 0, 2\pi)$]. Therefore, ϕ is orientation-preserving.

Reparametrization can also be interpreted as a change in units. For instance, if the parameter t in $\mathbf{c}(t) = (\cos t, \sin t, t)$, $t \in [0, 2\pi]$, represents hours, then in the reparametrization $\mathbf{c}_1(t) = (\cos(t/60), \sin(t/60), t/60)$, $t \in [0, 120\pi]$, it represents minutes.

The definition of a one-to-one function that we gave earlier extends to paths. For example, the path $\mathbf{c}(t) = (2t, t^3 - t)$ is one-to-one, since $\mathbf{c}(t_1) = \mathbf{c}(t_2)$ implies

$(2t_1, t_1^3 - t_1) = (2t_2, t_2^3 - t_2)$ and therefore $t_1 = t_2$. However, $\mathbf{c}(t) = (t^2, \sin t)$, $t \in [-2\pi, 2\pi]$, is not one-to-one: both $t_1 = -\pi$ and $t_2 = \pi$ map to the same point, $\mathbf{c}(t_1) = \mathbf{c}(t_2) = (\pi^2, 0)$.

DEFINITION 5.4 Simple Curve

The image of a one-to-one, piecewise C^1 path $\mathbf{c}: [a, b] \to \mathbb{R}^2$ (or $\mathbf{c}: [a, b] \to \mathbb{R}^3$) is called a *simple curve.* ◀

In other words (since "one-to-one" means that no two points of the interval $[a, b]$ are mapped onto the same point on the curve), a simple curve cannot intersect itself; see Figures 5.4 and 5.5. The image of the path $\mathbf{c}(t)$ in Figure 5.1 or the parabola of Example 5.5 are simple curves.

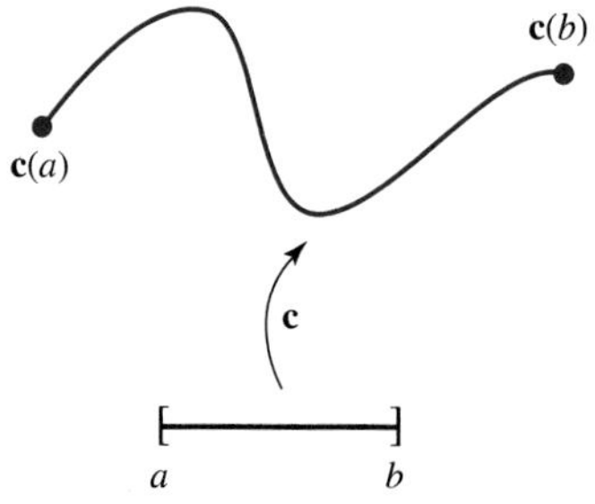

Figure 5.4 A simple curve.

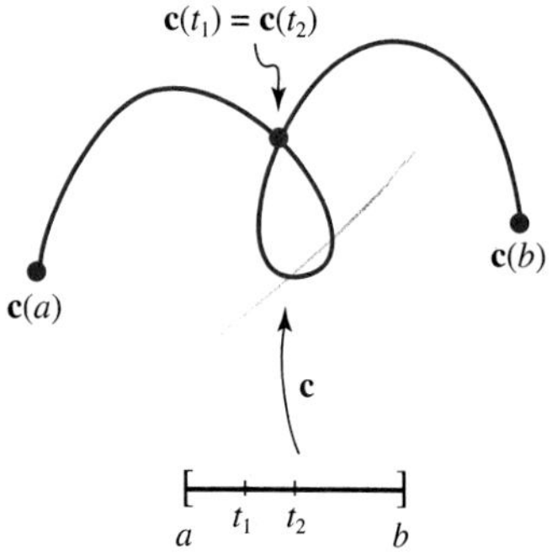

Figure 5.5 A curve that is not simple.

DEFINITION 5.5 Simple Closed Curve

The image of a piecewise C^1 path $\mathbf{c}: [a, b] \to \mathbb{R}^2$ (or $\mathbb{R}^3$) that is one-to-one on $[a, b)$ and is such that $\mathbf{c}(a) = \mathbf{c}(b)$ is called a *simple closed curve.* ◀

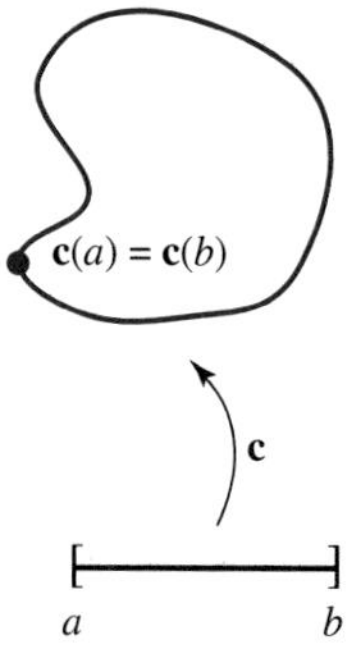

Figure 5.6 A simple closed curve.

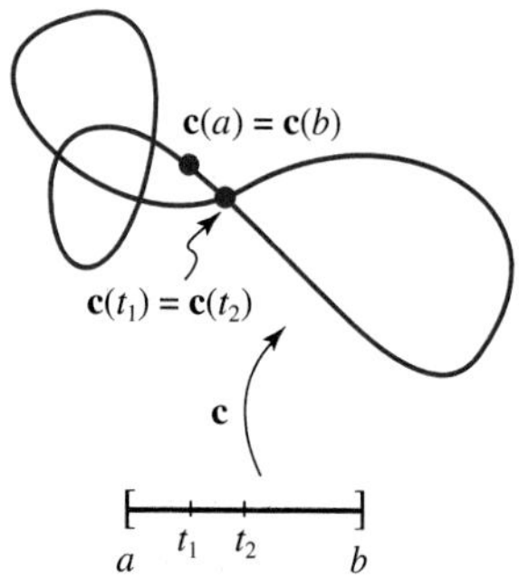

Figure 5.7 A closed curve that is not a simple closed curve.

In other words, no points other than the endpoints a and b of $[a, b]$ are mapped onto the same point; see Figures 5.6 and 5.7. It follows that a simple closed curve cannot intersect or retrace (partly or fully) itself. The image of the path $\mathbf{c}(t) = (\cos^3 t, \sin^3 t)$, $t \in [0, 2\pi]$, in Figure 5.2 is a simple closed curve.

▶ EXAMPLE 5.7

The circle $\mathbf{c}(t) = (\cos t, \sin t)$, $t \in [0, 2\pi]$, is a simple closed curve. The image of the map $\boldsymbol{\gamma}(t) = (\cos t, \sin t)$, $t \in [0, 6\pi]$, is the same circle, but $\boldsymbol{\gamma}$ is not simple: for example, $\boldsymbol{\gamma}(\pi) = \boldsymbol{\gamma}(3\pi) = \boldsymbol{\gamma}(5\pi) = (-1, 0)$. In fact, $\boldsymbol{\gamma}$ wraps around the circle $x^2 + y^2 = 1$ three times.

The image of the path $\mathbf{c}(t) = (\sin t, \sin 2t)$, $t \in [0, 2\pi]$, is a closed curve ($\mathbf{c}(0) = \mathbf{c}(2\pi) = (0, 0)$); however, it is not a simple closed curve, since it intersects itself: $\mathbf{c}(\pi) = (0, 0)$. It is an example of a so-called *figure-8 curve* and is shown in Figure 5.8. ◀

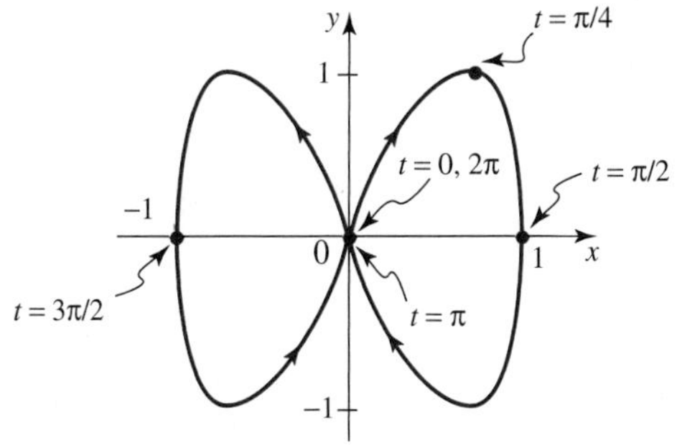

Figure 5.8 The image of $\mathbf{c}(t) = (\sin t, \sin 2t)$, $t \in [0, 2\pi]$, is a 'figure-8' curve.

DEFINITION 5.6 Orientation of a Curve

Let $\mathbf{c}$ be a simple curve and let A and B be its endpoints. There are two orientations associated with $\mathbf{c}$, defined by the choice of the initial and the terminal points (see Figure 5.9). If $\mathbf{c}$ is a simple closed curve, then the orientation is specified by one of the two possible ways of moving around $\mathbf{c}$. In $\mathbb{R}^2$, these two ways are usually referred to as the "clockwise" and the "counterclockwise" orientations; see Figure 5.10. A simple (closed) curve together with the choice of an orientation is called an *oriented simple (closed) curve.* ◀

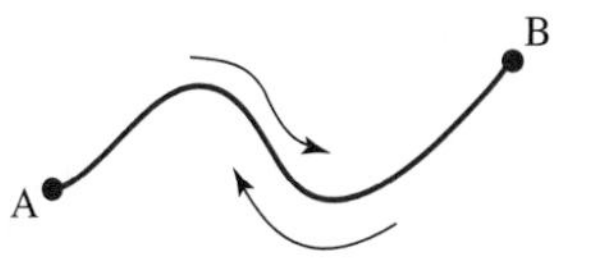

Figure 5.9 Possible orientations of a simple curve.

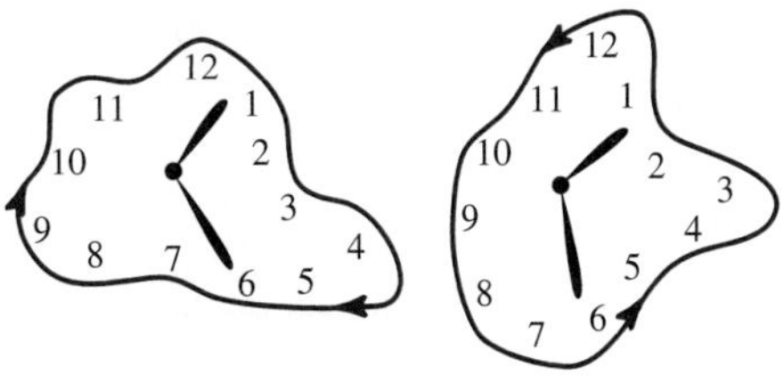

Figure 5.10 Clockwise and counterclockwise orientations in $\mathbb{R}^2$.

▶ EXERCISES 5.1

Exercises 1 to 7: State whether it is possible for the map ϕ to be a reparametrization of a path.

1. $\phi: [0, 1] \to [0, \ln 2]$, $\phi(t) = \ln(t + 1)$

2. $\phi: [-1, 1] \to [0, 1]$, $\phi(t) = t^2$

3. $\phi: [-8, 1] \to [-2, 1]$, $\phi(t) = t^{1/3}$

4. $\phi: [1, 2] \to [0, 3]$, $\phi(t) = t^2 - 1$

5. $\phi: [0, 1] \to [1, e]$, $\phi(t) = e^t$

6. $\phi: [-1, 1] \to [-\pi/4, \pi/4]$, $\phi(t) = \arctan t$

7. $\phi: [-2, 1] \to [0, 2]$, $\phi(t) = |t|$

8. Let $\mathbf{c}(t) = (t - 2, 3 - t - t^2)$, $t \in [0, 1]$. Is the reparametrization $\phi: [0, 3] \to [0, 1]$, given by $\phi(t) = 1 - t/3$, orientation-preserving or orientation-reversing?

9. Using the Mean Value Theorem, show that a differentiable function $\phi: [\alpha, \beta] \to [a, b]$ is one-to-one if $\phi'(t) > 0$ for all $t \in (\alpha, \beta)$ (or $\phi'(t) < 0$ for all $t \in (\alpha, \beta)$).

10. Explain why a continuous and bijective function $\phi: [\alpha, \beta] \to [a, b]$ must map endpoints to endpoints. Show that this statement is no longer true if ϕ is not continuous.

Exercises 11 to 16: Check whether the curve $\mathbf{c}(t)$ is simple or not, closed or not, simple closed or not.

11. $\mathbf{c}(t) = (\sin t, \cos t, (t - 2\pi)^2)$, $t \in [-2\pi, 6\pi]$

12. $\mathbf{c}(t) = (\sin t, \cos t, (t - 2\pi)^2)$, $t \in [-2\pi, 4\pi]$

13. $\mathbf{c}(t) = (t \sin t, t \cos t)$, $t \in [0, 2\pi]$

14. $\mathbf{c}(t) = (\sin 2t, t \cos t)$, $t \in [0, \pi/2]$

15. $\mathbf{c}(t) = (t - t^{-1}, t + t^{-1})$, $t \in [1, 2]$

16. $\mathbf{c}(t) = (t^2 - t, 3 - \sqrt{t^2 - t})$, $t \in [0, 1]$

17. Find a parametrization of the part of the curve $y = \sqrt{x^2 + 1}$ from $(-1, \sqrt{2})$ to $(1, \sqrt{2})$. Is your parametrization continuous? Differentiable? Piecewise C^1? C^1?

18. Find a parametrization of the curve $x^{2/3} + y^{2/3} = 1$. Is your parametrization continuous? Differentiable? Piecewise C^1? C^1?

19. Consider the following parametrizations of the straight-line segment from $(-1, 1)$ to $(1, 1)$. State which parametrizations are continuous, piecewise C^1 and C^1.

(a) $\mathbf{c}_1(t) = (t, 1)$, $-1 \le t \le 1$

(b) $\mathbf{c}_2(t) = \begin{cases} (-t^2, 1) & \text{if } -1 \le t \le 0 \\ (t^2, 1) & \text{if } 0 \le t \le 1 \end{cases}$

(c) $\mathbf{c}_3(t) = (t^{1/3}, 1)$, $-1 \le t \le 1$

(d) $\mathbf{c}_4(t) = (t^3, 1)$, $-1 \le t \le 1$

(e) $\mathbf{c}_5(t) = \begin{cases} (t, 1) & \text{if } -1 \le t \le 0 \\ (1 - t, 1) & \text{if } 0 \le t \le 1 \end{cases}$

20. Consider the curve $\mathbf{c}$ in $\mathbb{R}^2$ given by $\mathbf{c}(t) = (t, t^2)$, $t \in [-1, 2]$. State which of the following maps ϕ are reparametrizations of $\mathbf{c}$. Describe the curve $\mathbf{c}(\phi(t))$ for those ϕ that are reparametrizations:

(a) $\phi: [-1, \sqrt{3}] \to [-1, 2]$, $\phi(t) = t^2 - 1$

(b) $\phi: [-1/2, 1] \to [-1, 2]$, $\phi(t) = 2t$

(c) $\phi: [-1, 8] \to [-1, 2]$, $\phi(t) = t^{1/3}$

(d) $\phi: [-2/3, 1/3] \to [-1, 2]$, $\phi(t) = -3t$

21. Let $\mathbf{c}(t) = (t^2, 2 - t^2)$, $t \in [1, 3]$. Reparametrize $\mathbf{c}$ so that its speed is constant.

22. Let $\mathbf{c}(t) = (\cos 2\pi t, \sin 2\pi t, t)$, $0 \le t \le 1$.

(a) Reparametrize $\mathbf{c}$ so that its speed equals 1.

(b) Reparametrize $\mathbf{c}$ so that it takes 3 units of time to trace it.

(c) Reparametrize $\mathbf{c}$ so that it is traced in the opposite direction.

23. Let $\mathbf{c}$ be the circle $x^2 + y^2 = 1$, oriented clockwise. Find an orientation-preserving parametrization of $\mathbf{c}$ of constant speed S. Find an orientation-reversing parametrization of $\mathbf{c}$ of constant speed 1.

24. Assuming that the units are kilometers and hours, check that the speed of the path $\mathbf{c}(t) = (5\cos t, 5\sin t, 12t)$ is 13 km/h. Reparametrize $\mathbf{c}$ so that its speed is 13 mph.

▶ 5.2 PATH INTEGRALS OF REAL-VALUED FUNCTIONS

To motivate the definition of a path integral, let us first recall the construction of the definite integral of a function of one variable.

Definite Integral of a Real-Valued Function of One Variable

Assume that $y = f(x)$ is a continuous, positive function defined on an interval $[a, b]$. The graph of f, the vertical lines $x = a$ and $x = b$, and the x-axis define a region R in the xy-plane (called the *region below f over* $[a, b]$). We would like to find a way to compute the area of R.

Subdivide the interval $[a, b]$ into n subintervals $[a = t_1, t_2], [t_2, t_3], \ldots, [t_n, t_{n+1} = b]$ and construct rectangles $R_1, \ldots, R_n$ in the following way: the base of R_i, $i = 1, \ldots, n$, is the ith subinterval $[t_i, t_{i+1}]$ and its height is the value $f(t_i^*)$ of f at some point t_i^* in $[t_i, t_{i+1}]$; see Figure 5.11.

The area of R_i is $f(t_i^*)(t_{i+1} - t_i) = f(t_i^*)\Delta t_i$, where $\Delta t_i = t_{i+1} - t_i$. The rectangles $R_1, \ldots, R_n$ approximate the region R, and the sum of their areas

$$A_n = \sum_{i=1}^{n} f(t_i^*)\Delta t_i$$

approximates the area of R. It can be proven that the more rectangles we use, the better approximation we get; consequently, as $n \to \infty$, the sequence A_n of approximations of the area of R will approach the area of R; that is,

$$\text{area}(R) = \lim_{n\to\infty} A_n = \lim_{n\to\infty} \sum_{i=1}^{n} f(t_i^*)\Delta t_i.$$

We *define* the definite integral of f on $[a, b]$ as

$$\int_a^b f(x)\,dx = \text{area}\,(R) = \lim_{n\to\infty} \sum_{i=1}^{n} f(t_i^*)\Delta t_i,$$

provided that the limit exists.

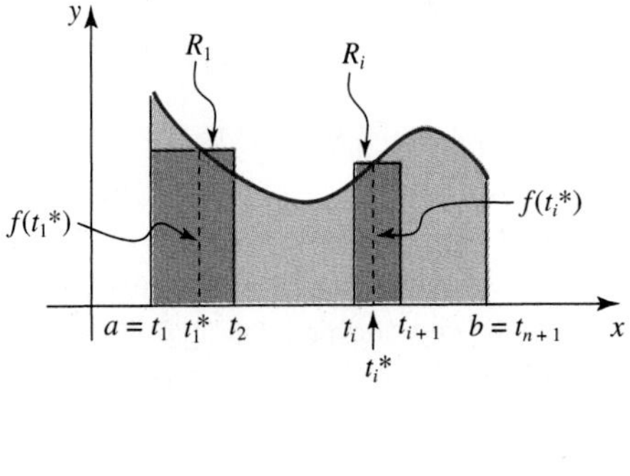

Figure 5.11 Approximating rectangles for the region R.

Now assume that f is not necessarily positive. It can be shown that the above construction works (i.e., the limit in question exists) whenever f is continuous, or, more generally, when it is piecewise continuous and bounded ("piecewise continuous" means that f is not continuous only at finitely many points; "bounded" means that there exists a number $M > 0$ such that $|f(x)| \leq M$ for all x in $[a, b]$). In other words, all discontinuities of f (if any) are "jumps" from one finite value to another finite value. Discontinuities like that of the function $y = 1/x$ at $x = 0$ are not allowed. For example, all continuous functions defined on a closed interval $[a, b]$ are bounded. Of course, in this general case, a definite integral does not necessarily represent an area.

It can be demonstrated that the limit in the definition is independent of the choices made in the construction (recall that we chose the subintervals and then selected a point in each).

DEFINITION 5.7 Definite Integral of a Function $y = f(x)$

The definite integral of a piecewise continuous and bounded function $y = f(x)$ defined on an interval $[a, b]$ is the real number

$$\int_a^b f(x)\,dx = \lim_{n\to\infty} \sum_{i=1}^{n} f(t_i^*)\Delta t_i.$$

The sum on the right side is called a *Riemann sum*. The definition clarifies the reference to the definite integral as a "limit of (Riemann) sums."

With the help of parametrizations, path integrals of scalar-valued functions and vector-valued functions will be reduced to definite integrals of real-valued functions of one variable.

Path Integral of a Real-Valued Function

So far, we have considered a function $y = f(x)$ defined on an interval $[a, b]$. To rephrase: we have considered a function defined at the points belonging to the straight-line segment from a to b on the x-axis. Now assume that a function f is defined at the points on a curve in a plane or in space (e.g., f could be the temperature or the density at points on a piece of metal wire). Is it possible to define (in a meaningful way) the definite integral of f along that curve?

The answer is yes—and all we have to do is to adjust the construction we described in the introduction.

Let $f\colon \mathbb{R}^2 \to \mathbb{R}$ be a continuous function and let $\mathbf{c}(t) = (x(t), y(t)) : [a, b] \to \mathbb{R}^2$ be a path in $\mathbb{R}^2$ (the construction for $\mathbb{R}^3$ and, in general, for $\mathbb{R}^n$, $n \geq 3$ is identical). The composition $f(\mathbf{c}(t))$ represents the values of f along the points on the curve $\mathbf{c}$ [e.g., if f is the electrostatic potential and $\mathbf{c}(t)$ is the trajectory of a charged particle, then $f(\mathbf{c}(t))$ describes the potential along the points on the trajectory].

Break up the interval $[a, b]$ into n subintervals $[a = t_1, t_2], [t_2, t_3], \ldots, [t_n, t_{n+1} = b]$ and approximate the curve $\mathbf{c}$ by the polygonal path p_n, whose vertices are $\mathbf{c}(a) = \mathbf{c}(t_1)$, $\mathbf{c}(t_2), \ldots, \mathbf{c}(t_n), \mathbf{c}(t_{n+1}) = \mathbf{c}(b)$ (this is the same type of polygonal path that we considered in deriving the formula for arc-length in Section 3.3). The length of the segment c_i connecting the points $\mathbf{c}(t_i)$ and $\mathbf{c}(t_{i+1})$ was approximated in (3.12) as $\ell(c_i) \approx \|\mathbf{c}'(t_i)\|\Delta t_i$, where

$\Delta t_i = t_{i+1} - t_i$. In the spirit of the construction in the introduction, we form the sums

$$A_n = \sum_{i=1}^{n} f(\mathbf{c}(t_i))\|\mathbf{c}'(t_i)\|\,\Delta t_i.$$

If f were positive, then A_n would represent the approximate area of the "fence" built along $\mathbf{c}$ from $\mathbf{c}(a)$ to $\mathbf{c}(b)$ whose height is determined by f; see Figures 5.12 and 5.13.

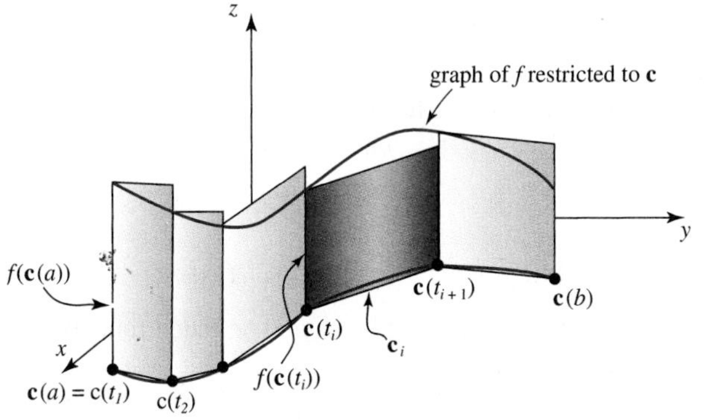

Figure 5.12 The sums A_n approximate the area of a "fence."

The integral of f along $\mathbf{c}$ is now defined as the limiting case of this construction as $n \to \infty$. With the understanding that A_n represents a Riemann sum of the *real-valued function* $f(\mathbf{c}(t))\|\mathbf{c}'(t)\|$ of one variable over $[a, b]$, the following definition becomes fully transparent.

DEFINITION 5.8 Path Integral of a Real-Valued Function

Let $\mathbf{c}\colon [a, b] \to \mathbb{R}^2$ be a C^1 path and let $f\colon \mathbb{R}^2 \to \mathbb{R}$ be a function such that the composition $f(\mathbf{c}(t))$ is continuous on $[a, b]$. The *path integral* $\int_{\mathbf{c}} f\,ds$ of f along $\mathbf{c}$ is given by

$$\int_{\mathbf{c}} f\,ds = \int_a^b f(\mathbf{c}(t))\|\mathbf{c}'(t)\|\,dt = \int_a^b f(x(t), y(t))\sqrt{(x'(t))^2 + (y'(t))^2}\,dt.$$

If $\mathbf{c}$ is a piecewise C^1 path consisting of C^1 paths $\mathbf{c}_j$, $j = 1, \ldots, m$, then

$$\int_{\mathbf{c}} f\,ds = \sum_{j=1}^{m} \int_{\mathbf{c}_j} f\,ds.$$ ◀

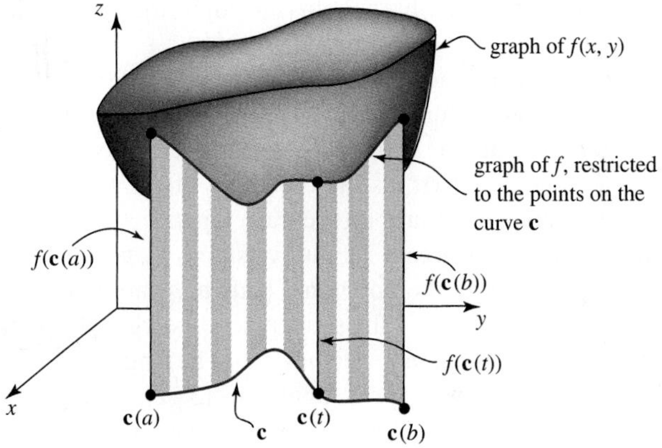

Figure 5.13 The integral $\int_{\mathbf{c}} f\,ds$ represents the area of the "fence" defined by f and $\mathbf{c}$.

The assumptions on f and $\mathbf{c}$ in the definition guarantee that the function $f(\mathbf{c}(t))\|\mathbf{c}'(t)\|$ is continuous, so that the definite integral $\int_a^b f(\mathbf{c}(t))\|\mathbf{c}'(t)\|dt$ makes sense. If $f\colon \mathbb{R}^3 \to \mathbb{R}$ and $\mathbf{c}\colon [a, b] \to \mathbb{R}^3$ is a C^1 path such that the composition $f(\mathbf{c}(t))$ is continuous on $[a, b]$, then

$$\int_{\mathbf{c}} f\,ds = \int_a^b f(\mathbf{c}(t))\|\mathbf{c}'(t)\|\,dt = \int_a^b f(x(t), y(t), z(t))\sqrt{(x'(t))^2 + (y'(t))^2 + (z'(t))^2}\,dt.$$

Sometimes, the notation $\oint_{\mathbf{c}} f\,ds$ is used for a path integral along closed paths.

Let us emphasize that, if $f(x, y) \geq 0$, then the path integral $\int_{\mathbf{c}} f\,ds$ along the curve $\mathbf{c}$ in the xy-plane represents the area of the region "along $\mathbf{c}$ and below the graph of f," as shown in Figure 5.13.

▶ EXAMPLE 5.8

Compute the path integral $\int_{\mathbf{c}} f\,ds$ of the function $f(x, y, z) = xyz$ along the path $\mathbf{c}(t) = (-\sin t, \sqrt{2}\cos t, \sin t)$, $t \in [0, \pi/2]$.

SOLUTION The values of $f(x, y, z) = xyz$ along the curve $\mathbf{c}(t)$ are

$$f(\mathbf{c}(t)) = f(-\sin t, \sqrt{2}\cos t, \sin t) = -\sqrt{2}\sin^2 t\cos t.$$

From $\mathbf{c}'(t) = (-\cos t, -\sqrt{2}\sin t, \cos t)$, it follows that $\|\mathbf{c}'(t)\| = \sqrt{2}$ and

$$\int_{\mathbf{c}} f\,ds = -\sqrt{2}\int_0^{\pi/2} (\sin^2 t\cos t)\sqrt{2}\,dt = -2\left.\frac{\sin^3 t}{3}\right|_0^{\pi/2} = -\frac{2}{3}$$

(the integral of $\sin^2 t\cos t$ was computed using the substitution $u = \sin t$). ◀

▶ EXAMPLE 5.9

Let $f(x, y) = 2x + y$. Consider the path integral $\int_{\mathbf{c}} f\,ds$ along the following paths in $\mathbb{R}^2$ joining the points $(1, 0)$ and $(0, 1)$ [see Figure 5.14]:

(a) Counterclockwise along the quarter circle $\mathbf{c}_1(t) = (\cos t, \sin t)$, $t \in [0, \pi/2]$.

(b) Along the straight-line segment $\mathbf{c}_2(t) = (1, 0) + t(-1, 1) = (1 - t, t)$, $t \in [0, 1]$, from $(1, 0)$ to $(0, 1)$.

(c) Along the piecewise C^1 path $\mathbf{c}_3(t)$ that consists of the path $\mathbf{c}_4(t) = (1 - t, 0)$, $t \in [0, 1]$ [from $(1, 0)$ to $(0, 0)$], followed by the path $\mathbf{c}_5(t) = (0, t)$, $t \in [0, 1]$ [from $(0, 0)$ to $(0, 1)$].

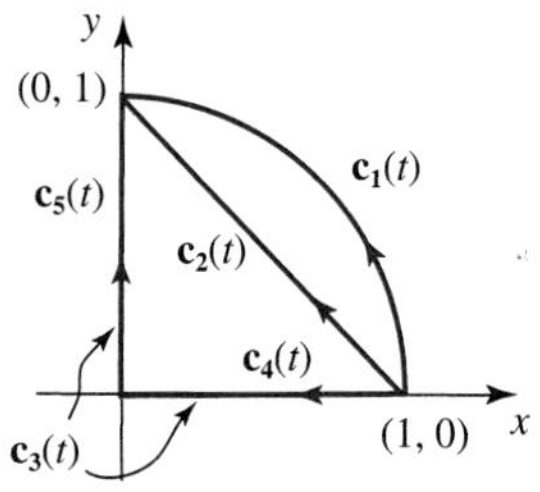

Figure 5.14 The paths of Example 5.9.

SOLUTION (a) The tangent vector is computed to be $\mathbf{c}_1'(t) = (-\sin t, \cos t)$, its norm is $\|\mathbf{c}_1'(t)\| = 1$, and hence,

$$\int_{\mathbf{c}_1} f\,ds = \int_0^{\pi/2} (2\cos t + \sin t)\,dt = (2\sin t - \cos t)\Big|_0^{\pi/2} = 3.$$

(b) This time, $\mathbf{c}_2'(t) = (-1, 1)$, $\|\mathbf{c}_2'(t)\| = \sqrt{2}$, and

$$\int_{\mathbf{c}_2} f\,ds = \int_0^1 (2-t)\sqrt{2}\,dt = \sqrt{2}\left(2t - \frac{t^2}{2}\right)\Big|_0^1 = \frac{3\sqrt{2}}{2}.$$

(c) Since $\|\mathbf{c}_4'(t)\| = 1$ and $\|\mathbf{c}_5'(t)\| = 1$, it follows that

$$\int_{\mathbf{c}_3} f\,ds = \int_{\mathbf{c}_4} f\,ds + \int_{\mathbf{c}_5} f\,ds = \int_0^1 (2-2t)\,dt + \int_0^1 t\,dt = \frac{3}{2}.$$

◀

▶ EXAMPLE 5.10

Compute $\int_{\mathbf{c}} e^{x+y}ds$, where $\mathbf{c}$ is the line segment from $(0, 0)$ to $(2, 1)$.

SOLUTION In order to compute the integral, we need a parametrization of the given line segment. Consider the path $\mathbf{c}_1(t) = (0, 0) + t(2, 1) = (2t, t)$, $t \in [0, 1]$. We compute $\|\mathbf{c}_1'(t)\| = \|(2, 1)\| = \sqrt{5}$ and

$$\int_{\mathbf{c}_1} e^{x+y}\,ds = \int_0^1 e^{3t}\sqrt{5}\,dt = \frac{\sqrt{5}}{3}e^{3t}\Big|_0^1 = \frac{\sqrt{5}}{3}(e^3 - 1).$$

What will happen if we try some other parametrization?

Consider the reparametrization [take $\phi(t) = 2t$] $\mathbf{c}_2(t) = (4t, 2t)$, $t \in [0, 1/2]$, of $\mathbf{c}_1$. Then $\mathbf{c}_2'(t) = (4, 2)$, $\|\mathbf{c}_2'(t)\| = \sqrt{20}$, and

$$\int_{\mathbf{c}_2} e^{x+y}\,ds = \int_0^{1/2} e^{6t}\sqrt{20}\,dt = \frac{\sqrt{20}}{6}e^{6t}\Big|_0^{1/2} = \frac{\sqrt{5}}{3}(e^3 - 1).$$

Now use $\phi(t) = t^2$ to reparametrize $\mathbf{c}_1$, thus obtaining $\mathbf{c}_3(t) = (2t^2, t^2)$, $t \in [0, 1]$. Then $\mathbf{c}_3'(t) = (4t, 2t)$, $\|\mathbf{c}_3'(t)\| = \sqrt{20t^2} = \sqrt{20}\,|t| = \sqrt{20}\,t$ (since $t \geq 0$), and

$$\int_{\mathbf{c}_3} e^{x+y}\,ds = \int_0^1 e^{3t^2}\sqrt{20}\,t\,dt = \sqrt{20}\frac{1}{6}e^{3t^2}\Big|_0^1 = \frac{\sqrt{5}}{3}(e^3 - 1).$$

It seems that, no matter what parametrization is used, we obtain the same result. ◀

▶ EXAMPLE 5.11

Consider the reparametrization

$$\mathbf{c}_1(t) = (-\sin 2t, \sqrt{2}\cos 2t, \sin 2t), \qquad t \in [0, \pi/4],$$

of the path $\mathbf{c}(t)$ in Example 5.8, and recompute the path integral of $f(x, y, z) = xyz$. It follows that $\mathbf{c}_1'(t) = (-2\cos 2t, -2\sqrt{2}\sin 2t, 2\cos 2t)$, $\|\mathbf{c}_1'(t)\| = \sqrt{8}$, and

$$\int_{\mathbf{c}_1} f\,ds = -\sqrt{2}\int_0^{\pi/4} (\sin^2 2t\cos 2t)\sqrt{8}\,dt = -4\frac{\sin^3 2t}{6}\Big|_0^{\pi/4} = -\frac{2}{3}.$$

Consider yet another parametrization

$$\mathbf{c}_2(t) = (-\cos(t/\sqrt{2}), \sqrt{2}\sin(t/\sqrt{2}), \cos(t/\sqrt{2})), \qquad t \in [0, \pi\sqrt{2}/2].$$

Then $\mathbf{c}_2'(t) = (\sin(t/\sqrt{2})/\sqrt{2}, \cos(t/\sqrt{2}), -\sin(t/\sqrt{2})/\sqrt{2})$, $\|\mathbf{c}_2'(t)\| = 1$, and

$$\int_{\mathbf{c}_2} f\,ds = -\sqrt{2}\int_0^{\pi\sqrt{2}/2} \sin\frac{t}{\sqrt{2}}\cos^2\frac{t}{\sqrt{2}}\,1\,dt = 2\,\frac{\cos^3(t/\sqrt{2})}{3}\bigg|_0^{\pi\sqrt{2}/2} = -\frac{2}{3}.$$ ◀

Examples 5.8 and 5.11 suggest that $\int_{\mathbf{c}} f\,ds$ might not depend on the parametrization: the given path was traversed first with a speed of $\sqrt{2}$, then twice that fast [that path was called $\mathbf{c}_1(t)$], and finally, in the opposite direction with unit speed [reparametrization $\mathbf{c}_2(t)$]. All integrations gave the same result: $-2/3$. We observed the same phenomenon occurring in Example 5.10. This is not a coincidence, but a consequence of the following theorem.

THEOREM 5.2 Independence of Path Integrals on Parametrization

Let $\mathbf{c}$ be a C^1 path in $\mathbb{R}^2$ (or $\mathbb{R}^3$), let f be a real-valued function continuous on the image of $\mathbf{c}$, and let $\boldsymbol{\gamma} = \mathbf{c} \circ \phi$ be a reparametrization of $\mathbf{c}$. Then

$$\int_{\mathbf{c}} f\,ds = \int_{\boldsymbol{\gamma}} f\,ds$$ ◀

PROOF: Let $\mathbf{c}(t)\colon [a,b] \to \mathbb{R}^2$ (or $\mathbb{R}^3$) be a parametrization of $\mathbf{c}$ and consider its reparametrization $\boldsymbol{\gamma}(t) = \mathbf{c}(\phi(t))\colon [\alpha,\beta] \to \mathbb{R}^2$ (or $\mathbb{R}^3$), where $\phi\colon [\alpha,\beta] \to [a,b]$. By definition,

$$\int_{\boldsymbol{\gamma}} f\,ds = \int_\alpha^\beta f(\boldsymbol{\gamma}(t))\,\|\boldsymbol{\gamma}'(t)\|\,dt.$$

Applying the chain rule $\boldsymbol{\gamma}'(t) = \mathbf{c}'(\phi(t))\,\phi'(t)$ and the identity $\|s\mathbf{v}\| = \|\mathbf{v}\|\,|s|$ (where $\mathbf{v}$ is a vector and s is a scalar), we get

$$\int_{\boldsymbol{\gamma}} f\,ds = \int_\alpha^\beta f(\mathbf{c}(\phi(t)))\,\|\mathbf{c}'(\phi(t))\,\phi'(t)\|\,dt = \int_\alpha^\beta f(\mathbf{c}(\phi(t)))\,\|\mathbf{c}'(\phi(t))\|\,|\phi'(t)|\,dt.$$

Removing the absolute value signs, we obtain

$$\int_{\boldsymbol{\gamma}} f\,ds = \begin{cases} \displaystyle\int_\alpha^\beta f(\mathbf{c}(\phi(t)))\,\|\mathbf{c}'(\phi(t))\|\,\phi'(t)\,dt & \text{if } \phi'(t) > 0 \\ \displaystyle -\int_\alpha^\beta f(\mathbf{c}(\phi(t)))\,\|\mathbf{c}'(\phi(t))\|\,\phi'(t)\,dt & \text{if } \phi'(t) < 0, \end{cases}$$

and continue by introducing a new variable $\tau = \phi(t)$, $d\tau = \phi'(t)\,dt$:

$$\int_{\boldsymbol{\gamma}} f\,ds = \begin{cases} \displaystyle\int_{\phi(\alpha)}^{\phi(\beta)} f(\mathbf{c}(\tau))\,\|\mathbf{c}'(\tau)\|d\tau & \text{if } \phi'(t) > 0 \\ \displaystyle -\int_{\phi(\alpha)}^{\phi(\beta)} f(\mathbf{c}(\tau))\,\|\mathbf{c}'(\tau)\|d\tau & \text{if } \phi'(t) < 0. \end{cases}$$

In the first integral, $\phi(\alpha) = a$ and $\phi(\beta) = b$, since $\phi' > 0$ implies that ϕ is an orientation-preserving reparametrization. In the latter case, $\phi(\alpha) = b$ and $\phi(\beta) = a$, since $\phi' < 0$ (so that ϕ is an orientation-reversing reparametrization). In any case, the integrals are equal to

$$\int_{\gamma} f\,ds = \int_a^b f(\mathbf{c}(\tau))\,\|\mathbf{c}'(\tau)\|d\tau = \int_{\mathbf{c}} f\,ds.$$ ◀

In Theorem 5.2, the assumption that f is continuous on the image of $\mathbf{c}$ means that the composition $f(\mathbf{c}(t))$ is continuous (which is needed to guarantee the existence of the path integral of f).

Substituting $f(x, y) = 1$ (or $f(x, y, z) = 1$) into the definition of the path integral, we get

$$\int_{\mathbf{c}} f\,ds = \int_a^b \|\mathbf{c}'(t)\|\,dt = \ell(\mathbf{c}).$$

In other words, the path integral of the constant function $f = 1$ gives the length of the path $\mathbf{c}$. Theorem 5.2 states that the above computation is independent of the paramaterization used. Therefore, in order to compute the length of a curve, we are free to choose any (C^1) parametrization we like (that is what we claimed in Section 3.3, but we did not give a justification).

Generalizing the definition of the average value of a function of one variable, we define the *average value of a function* f *along a curve* $\mathbf{c}$ ($\mathbf{c}$ is defined on an interval $[a, b]$) to be

$$\overline{f} = \frac{1}{\ell(\mathbf{c})}\int_{\mathbf{c}} f\,ds = \frac{1}{\ell(\mathbf{c})}\int_a^b f(\mathbf{c}(t))\,\|\mathbf{c}'(t)\|\,dt,$$

where $\ell(\mathbf{c})$ denotes the length of $\mathbf{c}$.

▶ EXAMPLE 5.12

Compute the average temperature of a wire in the shape of the helix

$$\mathbf{c}(t) = (\cos t, t/10, \sin t), \qquad t \in [0, 10\pi],$$

if the temperature at the point (x, y, z) in $\mathbb{R}^3$ is given by $T(x, y, z) = x^2 + y + z^2$.

SOLUTION

From $\mathbf{c}(t) = (\cos t, t/10, \sin t)$, we get $\mathbf{c}'(t) = (-\sin t, 1/10, \cos t)$ and

$$\|\mathbf{c}'(t)\| = \sqrt{(-\sin t)^2 + \left(\tfrac{1}{10}\right)^2 + (\cos t)^2} = \sqrt{1 + \tfrac{1}{100}} = \frac{\sqrt{101}}{10}.$$

The average temperature along the helix is

$$\overline{T} = \frac{1}{\ell(c)}\int_0^{10\pi}\left(\cos^2 t + \tfrac{1}{10}t + \sin^2 t\right)\frac{\sqrt{101}}{10}\,dt,$$

where

$$\ell(c) = \int_0^{10\pi} \|\mathbf{c}'(t)\|\,dt = \int_0^{10\pi} \frac{\sqrt{101}}{10}\,dt = \sqrt{101}\,\pi.$$

Hence,

$$\overline{T} = \frac{1}{\sqrt{101}\pi}\int_0^{10\pi}\left(1+\tfrac{1}{10}t\right)\frac{\sqrt{101}}{10}\,dt = \frac{1}{10\pi}\left(t+\tfrac{1}{20}t^2\right)\Big|_0^{10\pi} = 1+\frac{\pi}{2}.$$

It is worth repeating that in order to compute $\int_{\mathbf{c}} f\,ds$, it suffices to know the values of the function at the points on the curve only [that is the $f(\mathbf{c}(t))$ term in the path integral]. In light of this fact, we notice that Example 5.12 contains more data than needed—the temperature function was defined at all points in $\mathbb{R}^3$.

Further applications of path integrals are discussed in Section 7.5.

A curve in $\mathbb{R}^2$ (or $\mathbb{R}^3$) can be defined in various ways. For example, it can be described as the image of a map $\mathbf{c}: [a, b] \to \mathbb{R}^2$ (or $\mathbb{R}^3$), or as the graph of a function $f: \mathbb{R} \to \mathbb{R}$. Alternatively, we can use geometric terms, such as a "straight-line segment from A to B," or "circle of radius 4 centered at the origin," or the "intersection of the paraboloid $z = x^2 + 3y^2$ and the plane $-2x - y + 3z = 1$," etc.

Let $\mathbf{c}$ be a curve described in any of the ways given above, or in some other way. Assume that it is either a simple curve or a simple closed curve, endowed with an orientation (see Definitions 5.4, 5.5, and 5.6 at the end of Section 5.1). We would like to define an integral of a function along $\mathbf{c}$.

In order to compute a path integral, we need a parametrization. But how do we decide which one to use? The answer is—it does not matter! We define the integral of a real-valued function f along $\mathbf{c}$ as the path integral of f with respect to *any* smooth parametrization of $\mathbf{c}$. Here is why it works: it can be proved that any two one-to-one, C^1 maps (i.e., paths that parametrize a curve as a simple or a simple closed curve) that have the same image (i.e., represent the same curve) are reparametrizations of each other. And according to Theorem 5.2, the path integral does not depend on the parametrization used. Example 5.10 serves as an illustration of this fact.

A consequence of Theorem 5.2 states that when we integrate a *scalar* function along a curve, the orientation does not play any role. This sounds reasonable: for example, the average temperature of the wire should not depend on the way (i.e., on the direction in which) we measure the temperature at the points of the wire. The analogous statement does not hold for integrals of vector-valued functions, as we will witness in the next section.

However, the path integral *does* depend on the path used, as shown in Example 5.9. There is an important class of functions whose path integrals depend only on the endpoints, and not on the curve that joins them. Section 5.4 is devoted to a study of such functions.

▶ EXERCISES 5.2

1. Level curves of a linear function $f(x, y)$ are shown in Figure 5.15. Find the path integral of $f(x, y)$ along

(a) The line segment perpendicular to the level curves, from A to B

(b) The line segment that crosses all level curves at the angle of $\pi/4$, from C to D.

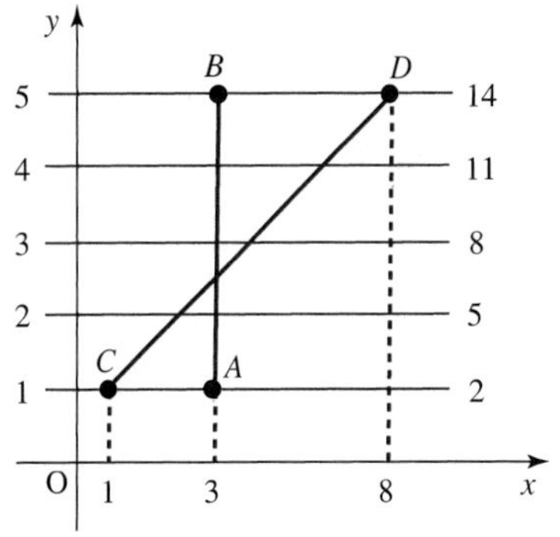

Figure 5.15 Level curves of Exercise 1.

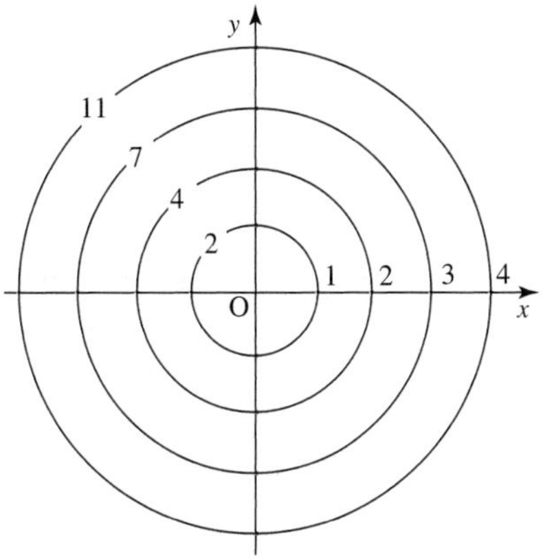

Figure 5.16 Level curves of Exercise 2.

2. The level curves of a function $f(x, y)$ are concentric circles centered at the origin; see Figure 5.16. Compute the path integral of $f(x, y)$ along

(a) The semicircle $x^2 + y^2 = 4$, $y \geq 0$

(b) Quarter-circle $x^2 + y^2 = 9$, $x \leq 0$, $y \leq 0$.

Exercises 3 to 11: Compute $\int_{\mathbf{c}} f\, ds$.

3. $f(x, y) = 2x - y$, $\mathbf{c}(t) = (e^t + 1, e^t - 2)$, $0 \leq t \leq \ln 2$

4. $f(x, y, z) = xy$, $\mathbf{c}(t) = (2\cos t, 3\sin t, 5t)$, $0 \leq t \leq \pi/2$

5. $f(x, y, z) = (x^2 + y^2 + z^2)^{-1}$, $\mathbf{c}(t) = (t, t, t)$, $1 \leq t < \infty$ (*Hint:* Take $1 \leq t \leq b$ and then compute the limit as b approaches ∞.)

6. $f(x, y) = x^3 + y^3$, $\mathbf{c}$ is the part of the curve $x^{2/3} + y^{2/3} = 1$ in the first quadrant

7. $f(x, y, z) = y - z^2$, $\mathbf{c}(t) = t^2\mathbf{i} + \ln t\mathbf{j} + 2t\mathbf{k}$, $1 \leq t \leq 4$

8. $f(x, y) = x^2 + 3y^2 - xy$, $\mathbf{c}$ is the circular arc of radius 3 in the xy-plane, from $(0, 3)$ to $(-3, 0)$

9. $f(x, y, z) = xyz$, $\mathbf{c}$ is the helix given by $\mathbf{c}(t) = (2\sin t, 4t, 2\cos t)$, $0 \leq t \leq 6\pi$

10. $f(x, y, z) = (x + y + z)/(x^2 + y^2 + z^2)$, $\mathbf{c}$ is the straight-line segment joining $(1, 1, 1)$ and (a, a, a), where $a \neq 1$

11. $f(x, y) = e^{x+3y}$, $\mathbf{c}$ is the line segment in $\mathbb{R}^2$ from $(0, 0)$ to $(3, -4)$

12. Compute $\int_{\mathbf{c}} f\, ds$, where $f(x, y, z) = x + 2y - z^2$, and $\mathbf{c}$ consists of the parabolic path $t\mathbf{i} + t^2\mathbf{j}$ from $(0, 0, 0)$ to $(1, 1, 0)$, followed by the straight line to $(1, -1, 1)$.

13. Compute $\int_{\mathbf{c}} f\, ds$, where $f(x, y, z) = x - 4y + z$, and $\mathbf{c}$ consists of the straight line from $(4, 2, 0)$ to $(0, 2, 0)$, followed by the circular path in the yz-plane (and above the xy-plane) with its center at the origin, from $(0, 2, 0)$ to $(0, -2, 0)$.

14. Let $f(x, y, z) = x - 3y^2 + z$ and let $\mathbf{c}$ be the straight-line segment from the origin to the point $(1, 1, 1)$. The four paths $\mathbf{c}_1(t) = (t, t, t)$, $t \in [0, 1]$, $\mathbf{c}_2(t) = (1 - t, 1 - t, 1 - t)$, $t \in [0, 1]$, $\mathbf{c}_3(t) = (e^t - 1, e^t - 1, e^t - 1)$, $t \in [0, \ln 2]$, and $\mathbf{c}_4(t) = (\ln t, \ln t, \ln t)$, $t \in [1, e]$, parametrize the given line segment.

(a) Describe their differences in terms of orientation and speed.

(b) Compute $\int_{\mathbf{c}_i} f\, ds$, $i = 1, \ldots, 4$.

15. Suppose that a continuous function f is integrated along two different paths joining the points $(1, 2)$ and $(3, -5)$, and two different answers are obtained. Is that possible, or has an error been made in the evaluation of integrals?

16. Compute the integral of $f(x, y) = xy - x - y + 1$ along the following curves connecting the points $(1, 0)$ and $(0, 1)$:

(a) $\mathbf{c}_1$: circular arc $\mathbf{c}_1(t) = (\cos t, \sin t)$, $0 \le t \le \pi/2$

(b) $\mathbf{c}_2$: straight-line segment $\mathbf{c}_2(t) = (1 - t, t)$, $0 \le t \le 1$

(c) $\mathbf{c}_3$: from $(1, 0)$ horizontally to the origin, then vertically to $(0, 1)$

(d) $\mathbf{c}_4$: from $(1, 0)$ vertically to $(1, 1)$, then horizontally to $(0, 1)$

(e) $\mathbf{c}_5$: circular arc $\mathbf{c}_5(t) = (\cos t, -\sin t)$, $0 \le t \le 3\pi/2$.

17. Compute the area of the part of the cylinder $x^2 + y^2 = 4$ between the xy-plane and the plane $z = y + 2$.

18. Compute the area of the part of the surface $y^2 = x$ defined by $0 \le x \le 2$, $0 \le z \le 2$.

19. Compute the area of the part of the surface $y = \sin x$, $0 \le x \le \pi/2$, above the xy-plane and below the surface $z = \sin x \cos x$.

20. Let $\mathbf{c}$ be the straight-line segment joining $(1, 0, 0)$ and $(0, 2, 0)$. Use a geometric argument (i.e., do not evaluate the integral) to find $\int_{\mathbf{c}} (x + 3y)\, ds$.

21. Use a geometric argument to find $\int_{\mathbf{c}} e^{x^2+y^2}\, ds$, where $\mathbf{c}$ is the circle centered at the origin of radius 4.

22. Argue geometrically that $\int_{\mathbf{c}} \sin(x^3)\, ds \ge 0$, where $\mathbf{c}$ is the graph of $y = \tan x$, $-\pi/4 \le x \le \pi/4$.

23. Is it possible that the average value of $f(x, y) = \sin x \cos y$ along some curve $\mathbf{c}$ is equal to 5?

24. Write down the version of the statement of Theorem 5.2 in the case where $\mathbf{c}$ is a piecewise C^1 path and prove it.

25. Find the average value of the function $f(x, y, z) = -\sqrt{x^2 + z^2}$ along the curve $\mathbf{c}(t) = (3\cos t)\mathbf{j} + (3\sin t)\mathbf{k}$, $0 \le t \le 2\pi$.

26. Find the average value $\overline{f}$ of the function $f(x, y, z) = 2x^2 - y^2$ along the unit circle in the xy-plane. Identify all points on $\mathbf{c}$ where the value of f is equal to $\overline{f}$.

27. Assume that $\mathbf{c}(t)\colon [a, b] \to \mathbb{R}^3$ represents a metal wire and that its density at a point (x, y, z) is given by the function $\rho(x, y, z)$. Explain how to use a path integral to compute the mass of the wire.

28. The density at a point (x, y) on a metal wire in the shape of a quarter-circle $x^2 + y^2 = 1$, $x, y \ge 0$, is given by $\rho(x, y) = 3 + 2xy$ g/cm (assume that the units along the coordinate axes are centimeters). Compute the mass of the wire.

29. Assume that a path $\mathbf{c}$ is given in polar coordinates by $r = r(\theta)$, $\theta_1 \le \theta \le \theta_2$. Show that $\int_{\mathbf{c}} f\, ds = \int_{\theta_1}^{\theta_2} f(r\cos\theta, r\sin\theta)\sqrt{r^2 + \left(\frac{dr}{d\theta}\right)^2}\, d\theta$.

30. Compute the path integral of the function $f(x, y) = x^2 + y^2$ along the curve $r = \sin\theta$, where $0 \le \theta \le \pi$.

▶ 5.3 PATH INTEGRALS OF VECTOR FIELDS

In this section, we are going to introduce one of the most important and useful concepts in vector calculus (and its applications), that of an integral of a vector field along a curve. It will be defined as the limiting case of a summation, in much the same way as the path

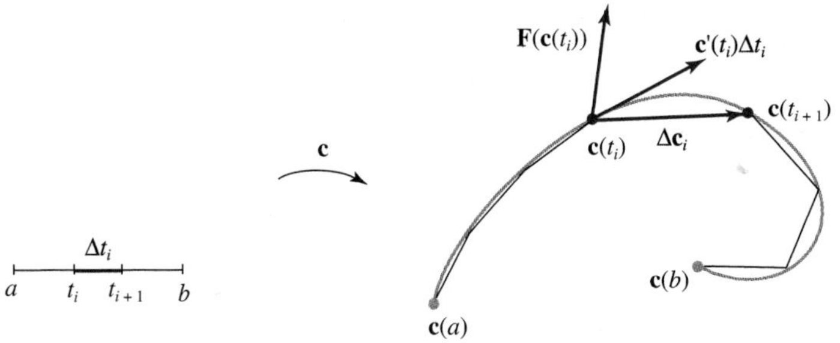

Figure 5.17 Construction of the path integral of **F**.

integral in the previous section (and as the double, triple, and surface integrals in the next two chapters). To lay the groundwork for the definition, let us consider the problem of computing the work done by a force.

The work W done by a constant force $\mathbf{F}$ on an object that moves from the position P to the position Q along a straight line is given by the dot product $W = \mathbf{F} \cdot \mathbf{d}$, where $\mathbf{d} = \overrightarrow{PQ}$ is the displacement vector. But what if the force is nonconstant and/or the path is not a straight line?

Assume that the force is given by a continuous vector field $\mathbf{F}$ (and thus could change from point to point) and the path (assumed to be C^1) is parametrized by $\mathbf{c}(t)\colon [a, b] \to \mathbb{R}^2$ (or $\mathbb{R}^3$).

We now present the idea, without giving a full proof. Subdivide the interval $[a, b]$ into n subintervals $[a = t_1, t_2], [t_2, t_3], \ldots, [t_n, t_{n+1} = b]$ and consider the displacement vectors $\Delta\mathbf{c}_i = \overrightarrow{\mathbf{c}(t_i)\mathbf{c}(t_{i+1})} = \mathbf{c}(t_{i+1}) - \mathbf{c}(t_i)$. Recall that, using the definition of the derivative, we showed that [see (3.11) in Section 3.3] $\Delta\mathbf{c}_i \approx \mathbf{c}'(t_i)\Delta t_i$, where $\mathbf{c}'(t_i)$ is the tangent vector at $\mathbf{c}(t_i)$ and $\Delta t_i = t_{i+1} - t_i$; see Figure 5.17.

For each i $(i = 1, \ldots, n)$, we form the dot product $\mathbf{F}(\mathbf{c}(t_i)) \cdot \Delta\mathbf{c}_i$, which is an approximation of the work done by $\mathbf{F}$ along the given curve from $\mathbf{c}(t_i)$ to $\mathbf{c}(t_{i+1})$. It is an approximation because instead of the (curved) path $\mathbf{c}$ from $\mathbf{c}(t_i)$ to $\mathbf{c}(t_{i+1})$, we took the displacement vector $\Delta\mathbf{c}_i$, and because we assumed that along the path from $\mathbf{c}(t_i)$ to $\mathbf{c}(t_{i+1})$ the force $\mathbf{F}$ is constant [and equal to $\mathbf{F}(\mathbf{c}(t_i))$]. Thus, the Riemann sum

$$W_n = \sum_{i=1}^{n} \mathbf{F}(\mathbf{c}(t_i)) \cdot \Delta\mathbf{c}_i \tag{5.1}$$

is an approximation of the work done by $\mathbf{F}$ along $\mathbf{c}$. Since $\Delta\mathbf{c}_i \approx \mathbf{c}'(t_i)\Delta t_i$, it follows that

$$W_n \approx \sum_{i=1}^{n} \mathbf{F}(\mathbf{c}(t_i)) \cdot \mathbf{c}'(t_i)\Delta t_i.$$

The exact *work* done is now defined as the limit of this construction as $n \to \infty$, that is,

$$W = \lim_{n\to\infty} W_n = \lim_{n\to\infty} \sum_{i=1}^{n} \mathbf{F}(\mathbf{c}(t_i)) \cdot \mathbf{c}'(t_i)\Delta t_i.$$

This limit is called the path integral (or the line integral) of $\mathbf{F}$ along $\mathbf{c}$, and is denoted by $\int_{\mathbf{c}} \mathbf{F} \cdot d\mathbf{s}$.

DEFINITION 5.9 Path Integral of a Vector Field. Work

Let $\mathbf{c}(t)\colon [a, b] \to \mathbb{R}^2$ (or $\mathbb{R}^3$) be a C^1 path, and let $\mathbf{F}\colon \mathbb{R}^2 \to \mathbb{R}^2$ (or $\mathbf{F}\colon \mathbb{R}^3 \to \mathbb{R}^3$) be a vector field such that the composition $\mathbf{F}(\mathbf{c}(t))$ is continuous on $[a, b]$. The *path integral* (or the *line integral*) $\int_\mathbf{c} \mathbf{F} \cdot d\mathbf{s}$ *of* $\mathbf{F}$ *along* $\mathbf{c}$ is given by

$$\int_\mathbf{c} \mathbf{F} \cdot d\mathbf{s} = \int_a^b \mathbf{F}(\mathbf{c}(t)) \cdot \mathbf{c}'(t)\, dt. \tag{5.2}$$

In the case where $\mathbf{F}$ represents a force, the path integral (5.2) gives the *work* done by $\mathbf{F}$ along $\mathbf{c}$. ◀

The term "line integral" refers to integrals of vector fields only, so there is no need to add "of a vector field." Thus, we speak of a "path integral of a vector field" or of a "line integral."

The integrand on the right side of (5.2) is the dot product of the value $\mathbf{F}(\mathbf{c}(t))$ of $\mathbf{F}$ at $\mathbf{c}(t)$ and the velocity vector $\mathbf{c}'(t)$ of $\mathbf{c}$ at $\mathbf{c}(t)$. Consequently, the path integral of a vector field reduces to the definite integral of a real-valued function of one variable.

Rewriting the definition of $\int_\mathbf{c} \mathbf{F} \cdot d\mathbf{s}$ as [assume that $\|\mathbf{c}'(t)\| \neq 0$]

$$\int_a^b \mathbf{F}(\mathbf{c}(t)) \cdot \mathbf{c}'(t)\, dt = \int_a^b \left(\mathbf{F}(\mathbf{c}(t)) \cdot \frac{\mathbf{c}'(t)}{\|\mathbf{c}'(t)\|} \right) \|\mathbf{c}'(t)\|\, dt,$$

we interpret the path integral of a *vector field* $\mathbf{F}$ as the path integral of the *scalar function* $\mathbf{F}(\mathbf{c}(t)) \cdot \mathbf{c}'(t)/\|\mathbf{c}'(t)\|$ [which is the component of $\mathbf{F}$ in the direction of the unit tangent vector to $\mathbf{c}$ at $\mathbf{c}(t)$] along the curve $\mathbf{c}$. Thus, only the tangential component of $\mathbf{F}$ contributes to the line integral $\int_\mathbf{c} \mathbf{F} \cdot d\mathbf{s}$. If $\mathbf{F}$ is perpendicular to $\mathbf{c}$, then the line integral $\int_\mathbf{c} \mathbf{F} \cdot d\mathbf{s}$ is zero. This means that the work done by a force acting perpendicular to the direction of the motion is zero.

It is worth repeating that $\int_\mathbf{c} \mathbf{F} \cdot d\mathbf{s}$ depends on the values of $\mathbf{F}$ along the curve and not at other points.

If $\mathbf{c}$ is a piecewise C^1 path, then

$$\int_\mathbf{c} \mathbf{F} \cdot d\mathbf{s} = \sum_{j=1}^{m} \int_{\mathbf{c}_j} \mathbf{F} \cdot d\mathbf{s},$$

where $\mathbf{c}_j$, $j = 1, \ldots, m$, are the pieces of $\mathbf{c}$ that are C^1.

▶ EXAMPLE 5.13

Compute the path integral $\int_\mathbf{c} \mathbf{F} \cdot d\mathbf{s}$ of the vector field $\mathbf{F}(x, y) = (-e^{x+y}, 3x)$ along the path $\mathbf{c}(t) = (t^2, 3 - 2t^2)$, $t \in [-1, 1]$.

SOLUTION From $x(t) = t^2$ and $y(t) = 3 - 2t^2$, we get the values

$$\mathbf{F}(\mathbf{c}(t)) = \mathbf{F}(t^2, 3 - 2t^2) = (-e^{3-t^2}, 3t^2)$$

of $\mathbf{F}$ along the curve. Since $\mathbf{c}'(t) = (2t, -4t)$, it follows that

$$\int_{\mathbf{c}} \mathbf{F} \cdot d\mathbf{s} = \int_{-1}^{1} \mathbf{F}(\mathbf{c}(t)) \cdot \mathbf{c}'(t)\, dt = \int_{-1}^{1} (-e^{3-t^2}, 3t^2) \cdot (2t, -4t)\, dt$$

$$= \int_{-1}^{1} (-2te^{3-t^2} - 12t^3)\, dt = (e^{3-t^2} - 3t^4)\Big|_{-1}^{1} = 0.$$

As a matter of fact, we did not have to evaluate the integral. The function $-2te^{3-t^2} - 12t^3$ is odd (i.e., symmetric with respect to the origin) and hence its integral over any interval $[-a, a]$, $a > 0$, is zero. ◀

▶ EXAMPLE 5.14 Work Done by a Force

Recall that the *work* done by a force $\mathbf{F}$ acting upon a particle that moves along the trajectory $\mathbf{c}(t)\colon [a, b] \to \mathbb{R}^3$ is given by the path integral

$$W = \int_a^b \mathbf{F}(\mathbf{c}(t)) \cdot \mathbf{c}'(t)\, dt.$$

Compute the work done by $\mathbf{F}(x, y, z) = (-y, x, 1)$ acting upon the particle that moves

(a) Radially away from the origin along $\mathbf{c}(t) = (t, t, t)$, $t \in [0, 1]$, and

(b) Along the helix $\mathbf{c}(t) = (\cos t, \sin t, t)$, $t \in [0, 2\pi]$.

SOLUTION

(a) In this case, $x(t) = y(t) = z(t) = t$, $\mathbf{F}(\mathbf{c}(t)) = (-t, t, 1)$, $\mathbf{c}'(t) = (1, 1, 1)$, and

$$W = \int_0^1 (-t, t, 1) \cdot (1, 1, 1)\, dt = \int_0^1 1\, dt = 1.$$

(b) This time, $\mathbf{F}(\mathbf{c}(t)) = (-\sin t, \cos t, 1)$, $\mathbf{c}'(t) = (-\sin t, \cos t, 1)$, and thus,

$$W = \int_0^{2\pi} (-\sin t, \cos t, 1) \cdot (-\sin t, \cos t, 1)\, dt = \int_0^{2\pi} 2\, dt = 4\pi.$$ ◀

To obtain a geometric interpretation of the path integral, we go back to the definition. If $\mathbf{F}$ is a continuous vector field and $\mathbf{c}$ is a C^1 path defined on $[a, b]$, then

$$\int_{\mathbf{c}} \mathbf{F} \cdot d\mathbf{s} = \int_a^b \mathbf{F}(\mathbf{c}(t)) \cdot \mathbf{c}'(t)\, dt = \int_a^b \|\mathbf{F}(\mathbf{c}(t))\|\, \|\mathbf{c}'(t)\| \cos\theta(t)\, dt,$$

where $\theta(t)$ is the angle between the vectors $\mathbf{F}(\mathbf{c}(t))$ and $\mathbf{c}'(t)$. The integrand is largest when $\theta(t) = 0$, equals zero if $\theta(t) = \pi/2$, and attains its minimum when $\theta(t) = \pi$. Consequently, the path integral is the largest for curves that are parallel to the vector field at all points (such curves are called flow lines and were discussed in Section 4.5) and remains large for curves whose direction does not differ much from that of $\mathbf{F}$ [i.e., $\theta(t)$ is small so that $\cos\theta(t)$ is close to 1]. The path integral is zero for curves running orthogonally to $\mathbf{F}$, and is smallest if the direction of the curve is opposite to $\mathbf{F}$. With this in mind, we interpret the path integral of a vector field as a measure of how well the curve "lines up" with the vector field (see Figure 5.18).

▶ EXAMPLE 5.15

Consider the vector field $\mathbf{F}(x, y) = \left(-y/\sqrt{x^2 + y^2}, x/\sqrt{x^2 + y^2}\right)$ in $\mathbb{R}^2$.

(a) Explain why $\int_{\mathbf{c}_1} \mathbf{F} \cdot d\mathbf{s} = 0$, if $\mathbf{c}_1(t) = (t, at)$, $t \in [1, 2]$, and a is a real number.

(b) Determine the sign of $\int_{\mathbf{c}_2} \mathbf{F} \cdot d\mathbf{s}$ if $\mathbf{c}_2(t) = (t, 1)$, $t \in [0, 4]$.

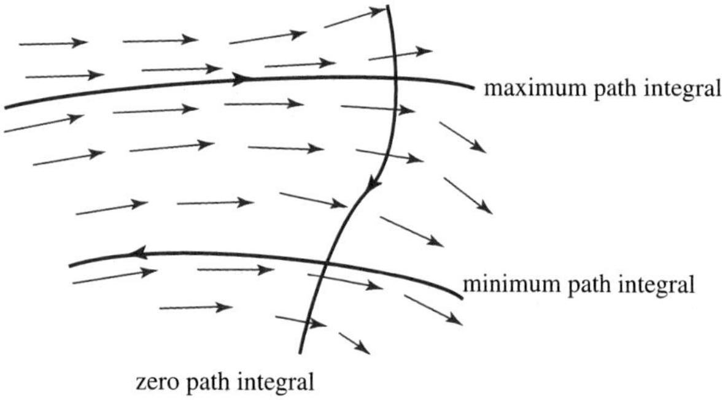

Figure 5.18 Path integral measures how well a curve "lines up" with a vector field.

SOLUTION The flow lines of $\mathbf{F}$ are circles [let us check: from $\mathbf{c}(t) = (\cos t, \sin t)$, we get $\mathbf{c}'(t) = (-\sin t, \cos t)$, and $\mathbf{F}(\mathbf{c}(t)) = (-\sin t/1, \cos t/1) = \mathbf{c}'(t)$].

(a) The path $\mathbf{c}_1(t)$ represents a part of the line through the origin of slope a. It intersects the flow lines of $\mathbf{F}$ orthogonally (this is just another way of saying that the directions of $\mathbf{c}_1'(t)$ and $\mathbf{F}$ [at a point $\mathbf{c}_1(t)$] are orthogonal) and therefore the path integral is zero.

(b) At all points on the given line segment, the angle $\theta(t)$ between $\mathbf{F}(\mathbf{c}_2(t))$ and $\mathbf{c}_2'(t)$ satisfies $\pi/2 < \theta(t) \leq \pi$; see Figure 5.19. Thus, all contributions $\|\mathbf{F}(\mathbf{c}_2(t))\| \, \|\mathbf{c}_2'(t)\| \cos\theta(t)$ to the integral are negative and it follows that the integral in question is negative.

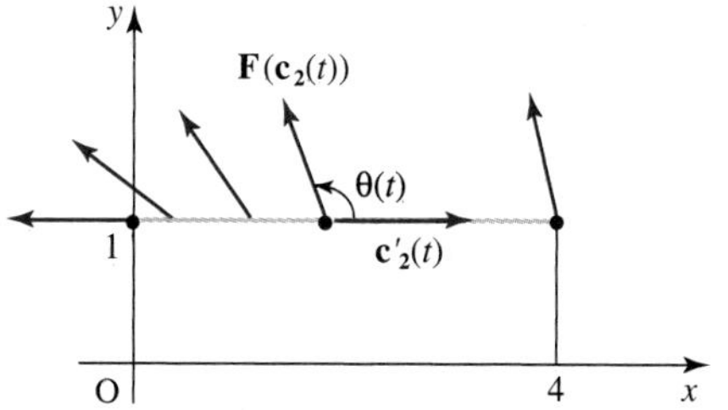

Figure 5.19 Vector field $\mathbf{F}$ along the segment from Example 5.15(b).

▶ EXAMPLE 5.16 Work Done by Gravity

Assume that the earth is a sphere. The work done by the force of gravity on a rocket launched vertically from the surface of the earth is negative. The work done by the force of gravity on a satellite orbiting at constant height is zero.

▶ EXAMPLE 5.17 Work Equals Gain in Kinetic Energy

Let $\mathbf{F}$ be a force acting on a particle of mass m moving along the trajectory $\mathbf{c}(t)$ from $\mathbf{c}(a)$ to $\mathbf{c}(b)$. By *Newton's Second Law,* $\mathbf{F}(\mathbf{c}(t)) = m\mathbf{a}(t) = m\mathbf{v}'(t)$, where $\mathbf{v}(t) = \mathbf{c}'(t)$ is the velocity of the particle. The work done by $\mathbf{F}$ is

$$W = \int_a^b \mathbf{F}(\mathbf{c}(t)) \cdot \mathbf{c}'(t)\, dt = \int_a^b m\mathbf{v}'(t) \cdot \mathbf{v}(t)\, dt$$

[use $(\|\mathbf{v}\|^2)' = (\mathbf{v} \cdot \mathbf{v})' = \mathbf{v}' \cdot \mathbf{v} + \mathbf{v} \cdot \mathbf{v}' = 2\mathbf{v} \cdot \mathbf{v}'$, to replace $\mathbf{v}' \cdot \mathbf{v}$ by $(\|\mathbf{v}\|^2)'/2$]

$$= m \int_a^b \frac{1}{2}\left(\|\mathbf{v}(t)\|^2\right)' dt$$

[by the Fundamental Theorem of Calculus]

$$= \tfrac{1}{2}m\|\mathbf{v}(t)\|^2\Big|_a^b = \tfrac{1}{2}m\|\mathbf{v}(b)\|^2 - \tfrac{1}{2}m\|\mathbf{v}(a)\|^2,$$

which is the difference of the final kinetic energy and the initial kinetic energy. ◀

In the last section, we proved that the path integral of a scalar function does not depend on the parametrization used. This is no longer true for integrals of vector fields: from the construction, we see that reversing the orientation of a path changes the sign of the path integral. Our next theorem states that it does not get any worse than that: the path integral actually depends on the orientation of a parametrization only (and not, for example, on its speed).

THEOREM 5.3 Path Integrals of Vector Fields and Parametrizations

Let $\mathbf{F}$ be a continuous vector field on $\mathbb{R}^2$ (or $\mathbb{R}^3$), let $\mathbf{c}\colon [a, b] \to \mathbb{R}^2$ (or $\mathbb{R}^3$) be a C^1 curve, and let $\boldsymbol{\gamma}(t) = \mathbf{c}(\phi(t))$ be a reparametrization of $\mathbf{c}$, where $\phi\colon [\alpha, \beta] \to [a, b]$. Then

$$\int_{\mathbf{c}} \mathbf{F} \cdot d\mathbf{s} = \begin{cases} \displaystyle\int_{\boldsymbol{\gamma}} \mathbf{F} \cdot d\mathbf{s} & \text{if } \phi \text{ is orientation-preserving} \\ \displaystyle -\int_{\boldsymbol{\gamma}} \mathbf{F} \cdot d\mathbf{s} & \text{if } \phi \text{ is orientation-reversing.} \end{cases}$$ ◀

PROOF: By definition of the path integral and the chain rule $\boldsymbol{\gamma}'(t) = \mathbf{c}'(\phi(t))\phi'(t)$, we obtain [keep in mind that $\boldsymbol{\gamma}\colon [\alpha, \beta] \to \mathbb{R}^2$ (or $\mathbb{R}^3$)]

$$\int_{\boldsymbol{\gamma}} \mathbf{F} \cdot d\mathbf{s} = \int_\alpha^\beta \mathbf{F}(\boldsymbol{\gamma}(t)) \cdot \boldsymbol{\gamma}'(t)\, dt = \int_\alpha^\beta \mathbf{F}(\mathbf{c}(\phi(t))) \cdot \mathbf{c}'(\phi(t))\, \phi'(t)\, dt.$$

Introduce the new variable $\tau = \phi(t)$; then $d\tau = \phi'(t)\, dt$ and

$$\int_{\boldsymbol{\gamma}} \mathbf{F} \cdot d\mathbf{s} = \int_{\phi(\alpha)}^{\phi(\beta)} \mathbf{F}(\mathbf{c}(\tau)) \cdot \mathbf{c}'(\tau)\, d\tau$$

[$\phi(\alpha) = a$ and $\phi(\beta) = b$ if ϕ is an orientation-preserving parametrization; $\phi(\alpha) = b$ and $\phi(\beta) = a$ if ϕ reverses the orientation]

$$= \begin{cases} \displaystyle\int_a^b \mathbf{F}(\mathbf{c}(\tau)) \cdot \mathbf{c}'(\tau) d\tau = \int_{\mathbf{c}} \mathbf{F} \cdot d\mathbf{s} & \text{if } \phi \text{ is orientation-preserving} \\ \displaystyle -\int_a^b \mathbf{F}(\mathbf{c}(\tau)) \cdot \mathbf{c}'(\tau) d\tau = -\int_{\mathbf{c}} \mathbf{F} \cdot d\mathbf{s} & \text{if } \phi \text{ is orientation-reversing.} \end{cases}$$ ◀

▶ EXAMPLE 5.18

Compute $\int_{\mathbf{c}} \mathbf{F} \cdot d\mathbf{s}$, where $\mathbf{F}(x, y, z) = xy\mathbf{i} + e^z\mathbf{j} + z\mathbf{k}$ and $\mathbf{c}$ is given by $\mathbf{c}(t) = (t^2, -t, t)$, $t \in [0, 1]$.

SOLUTION A straightforward computation gives

$$\int_{\mathbf{c}} \mathbf{F} \cdot d\mathbf{s} = \int_0^1 (-t^3, e^t, t) \cdot (2t, -1, 1)\, dt$$
$$= \int_0^1 (-2t^4 - e^t + t)\, dt = -\tfrac{2}{5}t^5 - e^t + \tfrac{1}{2}t^2 \Big|_0^1 = \tfrac{11}{10} - e.$$

Now reparametrize $\mathbf{c}$ by $\phi\colon [0, 1/2] \to [0, 1]$, $\phi(t) = 1 - 2t$. In other words, consider the curve $\mathbf{c}_1\colon [0, 1/2] \to \mathbb{R}^3$ given by

$$\mathbf{c}_1(t) = \mathbf{c}(\phi(t)) = \mathbf{c}(1 - 2t) = ((1 - 2t)^2, -(1 - 2t), 1 - 2t).$$

Since $\mathbf{c}_1(0) = (1, -1, 1) = \mathbf{c}(1)$ and $\mathbf{c}_1(1/2) = (0, 0, 0) = \mathbf{c}(0)$, it follows that $\mathbf{c}_1$ has an orientation opposite to that of $\mathbf{c}$. With the new parametrization,

$$\int_{\mathbf{c}_1} \mathbf{F} \cdot d\mathbf{s} = \int_0^{1/2} (-(1 - 2t)^3, e^{1-2t}, 1 - 2t) \cdot (-4(1 - 2t), 2, -2)\, dt$$
$$= \int_0^{1/2} \left(4(1 - 2t)^4 + 2e^{1-2t} - 2(1 - 2t)\right) dt$$
$$= \frac{-2(1 - 2t)^5}{5} - e^{1-2t} + \frac{(1 - 2t)^2}{2}\Bigg|_0^{1/2} = -\tfrac{11}{10} + e.$$

Hence, the integral of $\mathbf{F}$ along $\mathbf{c}$ is the negative of the integral along $\mathbf{c}_1$, as predicted by Theorem 5.3. ◀

Let $\mathbf{c}$ be a curve equipped with an orientation, assumed to be a simple curve or a simple closed curve. Theorem 5.3 states that in order to compute the integral of a vector field $\mathbf{F}$ along $\mathbf{c}$, we can use *any* orientation-preserving parametrization of $\mathbf{c}$. As a matter of fact, we could use an orientation-reversing parametrization as well, but must keep in mind that we have to change the sign of the result. Let us emphasize that, in contrast to the integral of a real-valued function, the integral of a vector field is *oriented*; that is, it depends on the direction in which the curve is traversed.

In our next example, we evaluate a line integral along a piecewise C^1 path.

▶ EXAMPLE 5.19

Compute the integral of $\mathbf{F}(x, y, z) = (y + z)\mathbf{i} + x\mathbf{j} + x\mathbf{k}$ along the following path: from $(1, 0, 0)$ counterclockwise along a circular path in the xy-plane to $(0, 1, 0)$, then along the straight line to $(0, 0, 1)$, and then along the straight line to $(1, 0, 1)$, as shown in Figure 5.20.

SOLUTION Parametrize the path $\mathbf{c}_1$ by $\mathbf{c}_1(t) = (\cos t, \sin t, 0)$, $t \in [0, \pi/2]$. Then

$$\int_{\mathbf{c}_1} \mathbf{F} \cdot d\mathbf{s} = \int_0^{\pi/2} (\sin t, \cos t, \cos t) \cdot (-\sin t, \cos t, 0)\, dt$$
$$= \int_0^{\pi/2} (\cos^2 t - \sin^2 t)\, dt = \int_0^{\pi/2} \cos 2t\, dt = \tfrac{1}{2} \sin 2t \Big|_0^{\pi/2} = 0.$$

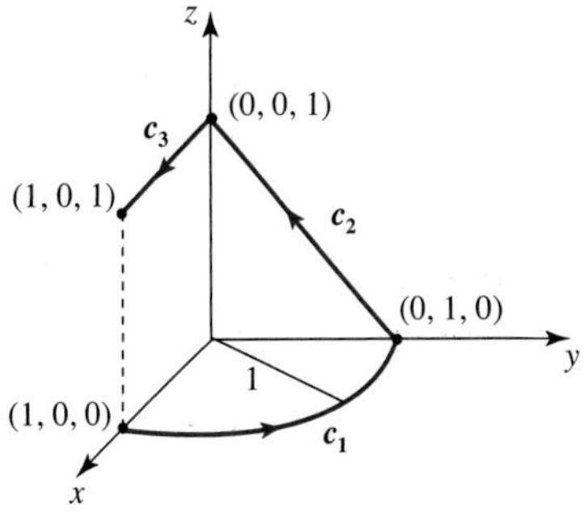

Figure 5.20 Path of Example 5.19.

Parametrize $\mathbf{c}_2$ by $\mathbf{c}_2(t) = (0, 1, 0) + t(0, -1, 1) = (0, 1 - t, t)$, $t \in [0, 1]$. It follows that

$$\int_{\mathbf{c}_2} \mathbf{F} \cdot d\mathbf{s} = \int_0^1 (1, 0, 0) \cdot (0, -1, 1)\, dt = \int_0^1 0\, dt = 0.$$

Finally, parametrize $\mathbf{c}_3$ by $\mathbf{c}_3(t) = (0, 0, 1) + t(1, 0, 0) = (t, 0, 1)$, $t \in [0, 1]$. Then

$$\int_{\mathbf{c}_3} \mathbf{F} \cdot d\mathbf{s} = \int_0^1 (1, t, t) \cdot (1, 0, 0)\, dt = \int_0^1 1\, dt = 1.$$

Thus,

$$\int_{\mathbf{c}} \mathbf{F} \cdot d\mathbf{s} = \int_{\mathbf{c}_1} \mathbf{F} \cdot d\mathbf{s} + \int_{\mathbf{c}_2} \mathbf{F} \cdot d\mathbf{s} + \int_{\mathbf{c}_3} \mathbf{F} \cdot d\mathbf{s} = 0 + 0 + 1 = 1.$$

The path integral of a vector field has several important physical interpretations. We discussed the work of a force in the introduction to this section. Now we turn to the circulation of a vector field, the interpretation coming from fluid mechanics, electromagnetism, and other disciplines.

DEFINITION 5.10 Circulation of a Vector Field

The line integral $\int_{\mathbf{c}} \mathbf{F} \cdot d\mathbf{s}$ of a continuous vector field $\mathbf{F}$ around an oriented simple closed curve $\mathbf{c}$ (assumed to be C^1) is called the *circulation of* $\mathbf{F}$ *around* $\mathbf{c}$.

In words, circulation describes the behavior of a vector field as "seen" from a closed curve. Here is an example.

▸ EXAMPLE 5.20

Let us compute the path integrals $\mathrm{circ}(\mathbf{F}_i) = \int_{\mathbf{c}} \mathbf{F}_i \cdot d\mathbf{s}$, $i = 1, 2, 3, 4$, around the unit circle (oriented counterclockwise) in $\mathbb{R}^2$ of the following vector fields: $\mathbf{F}_1(x, y) = (-y, x)$, $\mathbf{F}_2(x, y) = (-3y, 3x)$, $\mathbf{F}_3(x, y) = (x, y)$, and $\mathbf{F}_4(x, y) = (y, -x)$.

Parametrize the circle by $\mathbf{c}(t) = (\cos t, \sin t)$, $t \in [0, 2\pi]$. Then

$$\int_{\mathbf{c}} \mathbf{F}_1 \cdot d\mathbf{s} = \int_0^{2\pi} (-\sin t, \cos t) \cdot (-\sin t, \cos t)\, dt = \int_0^{2\pi} dt = 2\pi,$$

$$\int_{\mathbf{c}} \mathbf{F}_2 \cdot d\mathbf{s} = \int_0^{2\pi} (-3\sin t, 3\cos t) \cdot (-\sin t, \cos t)\, dt = \int_0^{2\pi} 3\, dt = 6\pi,$$

$$\int_{\mathbf{c}} \mathbf{F}_3 \cdot d\mathbf{s} = \int_0^{2\pi} (\cos t, \sin t) \cdot (-\sin t, \cos t)\, dt = \int_0^{2\pi} 0\, dt = 0$$

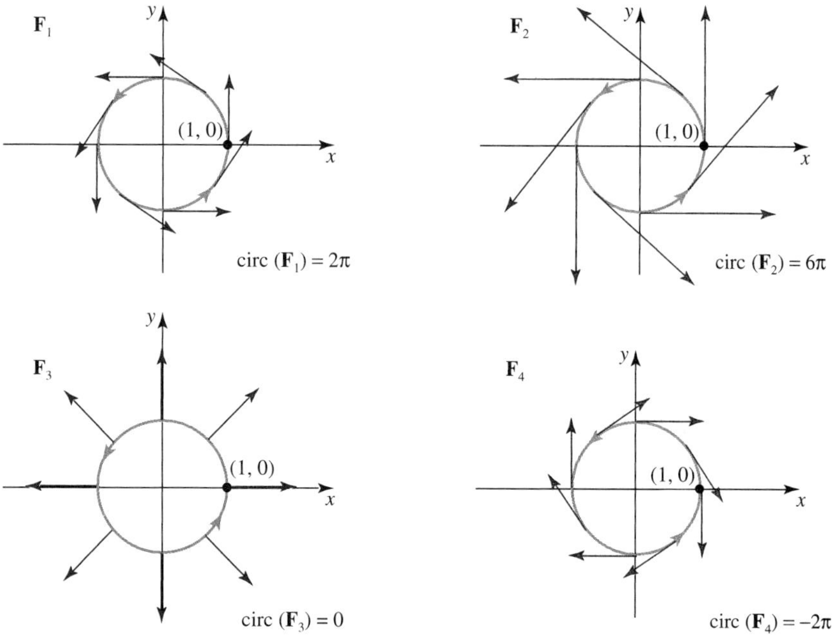

Figure 5.21 Vector fields of Example 5.20.

and

$$\int_{\mathbf{c}} \mathbf{F}_4 \cdot d\mathbf{s} = \int_0^{2\pi} (\sin t, -\cos t) \cdot (-\sin t, \cos t)\, dt = \int_0^{2\pi} (-1)\, dt = -2\pi.$$

Think of the given vector fields as velocity vector fields of a fluid. Suppose that a particle is placed at the point $(1, 0)$; see Figure 5.21. Subjected to $\mathbf{F}_1$, it will tend to turn counterclockwise, and subjected to $\mathbf{F}_2$, it will turn three times faster, since $\|\mathbf{F}_2\| = 3\|\mathbf{F}_1\|$; that is, $\mathbf{F}_2$ is three times "stronger" than $\mathbf{F}_1$. There will be no turning in the field $\mathbf{F}_3$; in fact, every particle just moves in a radial direction. Subjected to $\mathbf{F}_4$, a particle will tend to turn clockwise. Corresponding path integrals were positive in the first two cases, zero for $\mathbf{F}_3$ and negative for $\mathbf{F}_4$. Hence, loosely speaking, $\int_{\mathbf{c}} \mathbf{F} \cdot d\mathbf{s}$ measures the "turning of the fluid" in the counterclockwise sense. For example, a particle subjected to $\mathbf{F}_2$ will turn more (its integral was computed to be 6π) than if it were subjected to $\mathbf{F}_1$ (its integral was 2π). In the field $\mathbf{F}_4$, the "turning of the fluid" in the counterclockwise sense is negative, that is, the fluid turns clockwise. The comments we have just made justify calling the path integral $\int_{\mathbf{c}} \mathbf{F} \cdot d\mathbf{s}$ the *circulation of* $\mathbf{F}$ *around* $\mathbf{c}$.

Assume that $\mathbf{c}$ is a closed curve, oriented counterclockwise. We have already mentioned that since

$$\int_{\mathbf{c}} \mathbf{F} \cdot d\mathbf{s} = \int_a^b \mathbf{F}(\mathbf{c}(t)) \cdot \mathbf{c}'(t)\, dt = \int_{\mathbf{c}} \left(\mathbf{F}(\mathbf{c}(t)) \cdot \frac{\mathbf{c}'(t)}{\|\mathbf{c}'(t)\|} \right) ds,$$

the path integral of a vector field is actually the path integral of the tangential component of $\mathbf{F}$. Because an integral is a limit of sums, this means that $\int_{\mathbf{c}} \mathbf{F} \cdot d\mathbf{s}$ represents the *total tangential component of* $\mathbf{F}$ *around* $\mathbf{c}$. In other words, $\int_{\mathbf{c}} \mathbf{F} \cdot d\mathbf{s}$ is the total amount of counterclockwise turning of the fluid.

▶ EXAMPLE 5.21

Find the circulation of the constant vector field $\mathbf{F} = b\mathbf{i}$, $b > 0$ around a circle of radius a, oriented counterclockwise.

SOLUTION We will "add up" all tangential components of $\mathbf{F}$. At A and A' (see Figure 5.22), the vector field is orthogonal to the curve, so its tangential component is zero.

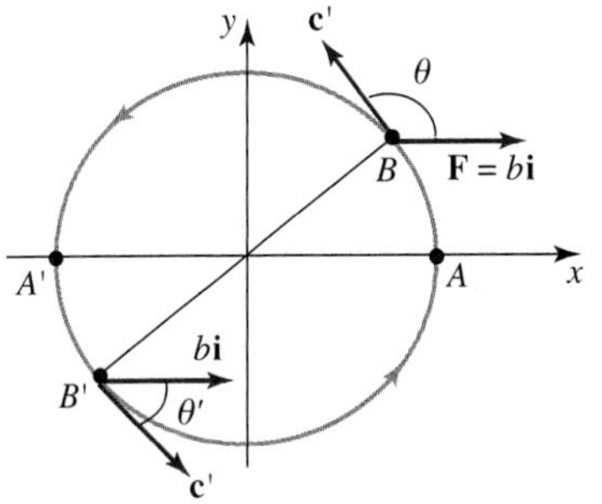

Figure 5.22 Computing circulation in Example 5.21.

Now consider a pair of diametrically opposite points B and B'. The contribution at B to the total tangential component is

$$\mathbf{F} \cdot \frac{\mathbf{c}'}{\|\mathbf{c}'\|}\bigg|_{\text{at } B} = \|\mathbf{F}\| \left\| \frac{\mathbf{c}'}{\|\mathbf{c}'\|} \right\| \cos\theta = b\cos\theta$$

and that of B' is

$$\mathbf{F} \cdot \frac{\mathbf{c}'}{\|\mathbf{c}'\|}\bigg|_{\text{at } B'} = \|\mathbf{F}\| \left\| \frac{\mathbf{c}'}{\|\mathbf{c}'\|} \right\| \cos\theta' = b\cos\theta'.$$

Since $\theta = \pi - \theta'$, it follows that $\cos\theta = -\cos\theta'$ and the two contributions cancel each other. Consequently, the circulation of $\mathbf{F}$ around the circle is zero. ◀

Now imagine walking around the circle and measuring the field $\mathbf{F}$ of the previous example. After completing one full revolution, we notice that there has been no change in the direction of $\mathbf{F}$. In general, "no circulation" means that, when we look under a microscope (i.e., for small circles around the point in question), we see no change in the vector field along the circle.

As an illustration, let us try to determine the circulation of the vector field in Figure 5.23 (pictured are the flow lines; i.e., the trajectories of particles subjected to the field).

To determine the circulation around Q, we choose a circle centered at Q (as shown), and, as we walk around in the counterclockwise direction, we record the vector defining the motion of the fluid corresponding to the point where we are. After completing our walk (i.e., we are back at the initial point of the walk), we realize that the corresponding vector did not turn/rotate, so there is no circulation around Q. Now take a circle around P and repeat the same experiment. This time, a vector will turn by 360° as we complete our counterclockwise walk. Therefore, there is a (positive) circulation at P.

In Section 4.6 we saw that the curl of a vector field at a given point (assuming that the vector field represents a fluid flow) is related to rotational (curling) aspects of the flow at that point. The circulation of a vector field is also related to rotational (turning) aspects of the flow—so, what is the connection between the curl and the circulation?

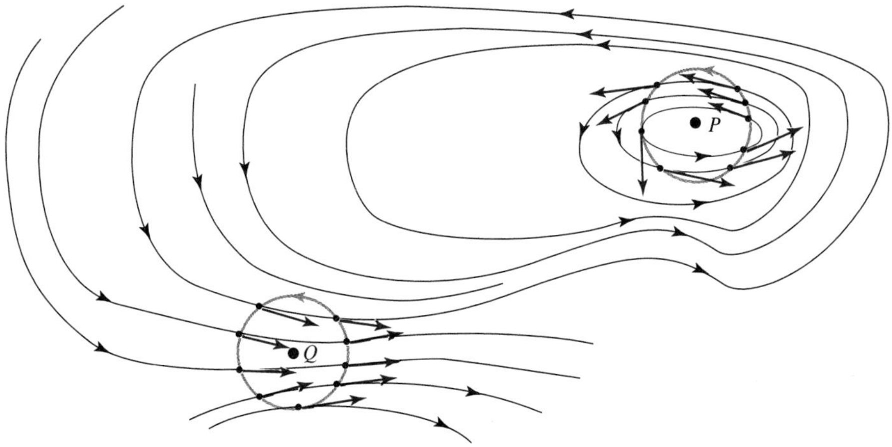

Figure 5.23 Determining the circulation of a vector field.

Here, we give an intuitive argument. We will be able to fully justify our answer in Section 8.3, after we introduce Stokes' Theorem.

Consider the flow given by a differentiable vector field $\mathbf{F}(x, y) = (F_1(x, y), F_2(x, y))$ in $\mathbb{R}^2$. We are interested in the behavior of the flow at some point P, so, for simplicity, we choose a coordinate system in which P is located at the origin. Take a small number $h > 0$. The image of the path $\mathbf{c}(t) = (h\cos t, h\sin t)$, $t \in [0, 2\pi]$, is a small circle (of raduis h) centered at the origin (i.e., at P) and oriented counterclockwise.

In order to find $\int_{\mathbf{c}} \mathbf{F} \cdot d\mathbf{s}$, we use the linear approximation for the components of $\mathbf{F}$:

$$\begin{aligned}\mathbf{F}(x, y) &= (F_1(x, y), F_2(x, y)) \\ &= \left(F_1(0,0) + \frac{\partial F_1}{\partial x}(0,0)\,x + \frac{\partial F_1}{\partial y}(0,0)\,y,\ F_2(0,0) + \frac{\partial F_2}{\partial x}(0,0)\,x + \frac{\partial F_2}{\partial y}(0,0)\,y\right).\end{aligned}$$

Thus,

$$\begin{aligned}\mathbf{F}(\mathbf{c}(t)) = \bigg(& F_1(0,0) + \frac{\partial F_1}{\partial x}(0,0)\,h\cos t + \frac{\partial F_1}{\partial y}(0,0)\,h\sin t, \\ & F_2(0,0) + \frac{\partial F_2}{\partial x}(0,0)\,h\cos t + \frac{\partial F_2}{\partial y}(0,0)\,h\sin t\bigg),\end{aligned}$$

and since $\mathbf{c}'(t) = (-h\sin t, h\cos t)$, we get

$$\begin{aligned}\int_{\mathbf{c}} \mathbf{F} \cdot d\mathbf{s} &= \int_0^{2\pi} \mathbf{F}(\mathbf{c}(t)) \cdot \mathbf{c}'(t)\,dt \\ &= \int_0^{2\pi} \bigg[h\left(F_2(0,0)\cos t - F_1(0,0)\sin t\right) \\ &\quad + h^2 \left(\frac{\partial F_2}{\partial x}(0,0)\cos^2 t - \frac{\partial F_1}{\partial y}(0,0)\sin^2 t \right. \\ &\quad + \left.\left(\frac{\partial F_2}{\partial y}(0,0) - \frac{\partial F_1}{\partial x}(0,0)\right)\sin t\cos t\right)\bigg]\,dt.\end{aligned}$$

Since $\int_0^{2\pi} \sin t\, dt = \int_0^{2\pi} \cos t\, dt = \int_0^{2\pi} \sin t \cos t\, dt = 0$ and $\int_0^{2\pi} \sin^2 t\, dt = \int_0^{2\pi} \cos^2 t\, dt = \pi$, it follows that

$$\int_{\mathbf{c}} \mathbf{F} \cdot d\mathbf{s} = h^2\pi \left(\frac{\partial F_2}{\partial x}(0,0) - \frac{\partial F_1}{\partial y}(0,0) \right) \tag{5.3}$$

and

$$\frac{1}{\pi h^2} \int_{\mathbf{c}} \mathbf{F} \cdot d\mathbf{s} = \frac{\partial F_2}{\partial x}(0,0) - \frac{\partial F_1}{\partial y}(0,0). \tag{5.4}$$

If we had used higher-order approximation (instead of linear), we would have obtained the terms involving h^3, h^4, h^5, etc. on the right side of (5.3). In that case, (5.4) would read

$$\frac{1}{\pi h^2} \int_{\mathbf{c}} \mathbf{F} \cdot d\mathbf{s} = \frac{\partial F_2}{\partial x}(0,0) - \frac{\partial F_1}{\partial y}(0,0) + \text{ terms involving } h, h^2, h^3, \text{etc.}$$

In any case,

$$\lim_{h \to 0} \frac{1}{\pi h^2} \int_{\mathbf{c}} \mathbf{F} \cdot d\mathbf{s} = \left(\frac{\partial F_2}{\partial x} - \frac{\partial F_1}{\partial y} \right)(0,0). \tag{5.5}$$

The expression

$$\frac{1}{\pi h^2} \int_{\mathbf{c}} \mathbf{F} \cdot d\mathbf{s}$$

is sometimes (for instance, in fluid dynamics) called the *vorticity* of $\mathbf{F}$ around the circle $\mathbf{c}$ of radius h centered at P. Briefly, the vorticity (of a vector field) is circulation per unit area.

Thus, the left side in (5.5) is infinitesimal vorticity of $\mathbf{F}$ at P. The right side is the scalar curl of $\mathbf{F}$ at P. In conclusion, the infinitesimal vorticity of a vector field $\mathbf{F}$ at a point P is equal to the scalar curl of $\mathbf{F}$ at P.

Quite often, in literature and in applications, we find the notation

$$\int_{\mathbf{c}} F_1 dx + F_2 dy + F_3 dz \tag{5.6}$$

for a line integral in $\mathbb{R}^3$ (or, $\int_{\mathbf{c}} F_1 dx + F_2 dy$ for a line integral in $\mathbb{R}^2$). By definition, (5.6) is equal to

$$\int_{\mathbf{c}} F_1 dx + F_2 dy + F_3 dz = \int_a^b \left(F_1 \frac{dx}{dt} + F_2 \frac{dy}{dt} + F_3 \frac{dz}{dt} \right) dt,$$

where $\mathbf{c}(t)\colon [a,b] \to \mathbb{R}^3$ is a C^1 path with components $\mathbf{c}(t) = (x(t), y(t), z(t))$. By removing terms that involve z, we obtain the definition of $\int_{\mathbf{c}} F_1 dx + F_2 dy$.

Let $\mathbf{F} = (F_1, F_2, F_3)$. Then (dropping independent variable from the notation)

$$\begin{aligned} \int_a^b \left(F_1 \frac{dx}{dt} + F_2 \frac{dy}{dt} + F_3 \frac{dz}{dt} \right) dt &= \int_a^b (F_1, F_2, F_3) \cdot (dx/dt, dy/dt, dz/dt)\, dt \\ &= \int_a^b \mathbf{F} \cdot \mathbf{c}'\, dt = \int_{\mathbf{c}} \mathbf{F} \cdot d\mathbf{s}. \end{aligned}$$

▶ EXAMPLE 5.22

Compute $\int_{\mathbf{c}} x^2dx + ydy + 2yzdz$ along the path $\mathbf{c}(t) = (1, t, -t^2)$, $t \in [0, 1]$.

SOLUTION By definition,

$$\int_{\mathbf{c}} x^2dx + ydy + 2yzdz = \int_0^1 \left(x^2\frac{dx}{dt} + y\frac{dy}{dt} + 2yz\frac{dz}{dt} \right) dt$$
$$= \int_0^1 \left(1 \cdot 0 + t \cdot 1 + 2t(-t^2)(-2t)\right) dt = \int_0^1 (t + 4t^4)\, dt = \frac{13}{10}. \quad ◀$$

We will further discuss the type of integration given by (5.6) in the context of differential forms in Chapter 8.

We will finish this section by introducing path integrals that appear in electromagnetism. A reader not familiar with the concepts of electric and magnetic fields may skip this part and move to the next section.

The *electric circulation* $\mathcal{E}$ of an electric field $\mathbf{E} = \mathbf{E}(x, y, z)$ (also called the *electromotive force*) is given by

$$\mathcal{E} = \int_{\mathbf{c}} \mathbf{E} \cdot d\mathbf{s},$$

where $\mathbf{c}$ is some closed curve in space. The *magnetic circulation* $\mathcal{B}$ is defined by

$$\mathcal{B} = \int_{\mathbf{c}} \mathbf{B} \cdot d\mathbf{s},$$

where $\mathbf{B}$ is the magnetic field and $\mathbf{c}$ denotes a closed curve in space.

▶ EXAMPLE 5.23

The magnetic field at a point (x, y, z) due to a single filament that carries a current I and whose direction is determined by the unit vector $\mathbf{u}$ is given by (see, e.g., J. D. Jackson, *Classical Electrodynamics*, 2nd ed. (Wiley: New York, 1975), pp. 169–170)

$$\mathbf{B}\,(x, y, z) = \frac{\mu_0 I}{2\pi} \frac{\mathbf{u} \times \mathbf{r}}{\|\mathbf{u} \times \mathbf{r}\|^2},$$

where $\mathbf{r} = x\mathbf{i} + y\mathbf{j} + z\mathbf{k}$, and μ_0 is a constant. Let us compute the magnetic circulation $\mathcal{B}$ along a circle $\mathbf{c}$ that lies in the yz-plane and encloses the filament. Place the filament so that $\mathbf{u} = \mathbf{i}$, as shown in Figure 5.24. Then

$$\mathbf{B} = \frac{\mu_0 I}{2\pi} \frac{\mathbf{i} \times (x\mathbf{i} + y\mathbf{j} + z\mathbf{k})}{\|\mathbf{i} \times (x\mathbf{i} + y\mathbf{j} + z\mathbf{k})\|^2} = \frac{\mu_0 I}{2\pi} \frac{-z\mathbf{j} + y\mathbf{k}}{y^2 + z^2} \tag{5.7}$$

and hence {$\mathbf{c}$ lies in the yz-plane, so we parametrize it by $\mathbf{c}(t) = (0, \cos t, \sin t)$, $t \in [0, 2\pi]$}, the magnetic circulation is computed to be

$$\mathcal{B} = \int_{\mathbf{c}} \mathbf{B} \cdot d\mathbf{s} = \int_0^{2\pi} \mathbf{B}(\mathbf{c}(t)) \cdot \mathbf{c}'(t)\, dt$$
$$= \frac{\mu_0 I}{2\pi} \int_0^{2\pi} \frac{-\sin t\mathbf{j} + \cos t\mathbf{k}}{1} \cdot (-\sin t\mathbf{j} + \cos t\mathbf{k})\, dt = \frac{\mu_0 I}{2\pi} \int_0^{2\pi} dt = \mu_0 I.$$

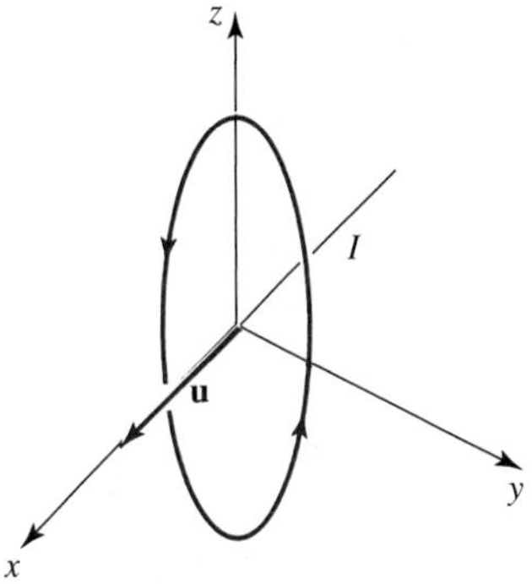

Figure 5.24 Filament of Example 5.23.

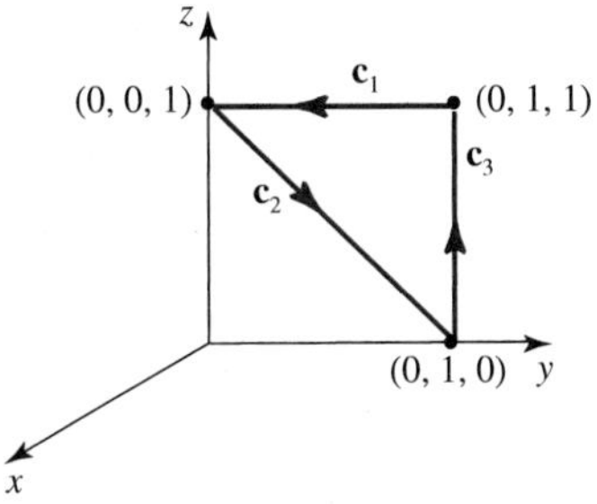

Figure 5.25 Piecewise C^1 curve.

Let us check, with an example, the physical fact that the magnetic circulation $\mathcal{B}$ vanishes along any closed curve that does not enclose the current filament. Consider the piecewise C^1 curve shown in Figure 5.25.

Parametrize the curve $\mathbf{c}_1$ by $\mathbf{c}_1(t) = (0,1,1) + t(0,-1,0) = (0, 1-t, 1), t \in [0,1]$. From (5.7), it follows that [use $\int 1/(x^2+1)\,dx = \arctan x$]

$$\begin{aligned}\int_{\mathbf{c}_1} \mathbf{B}\cdot d\mathbf{s} &= \frac{\mu_0 I}{2\pi}\int_0^1 \frac{(0,-1,1-t)}{(1-t)^2+1}\cdot(0,-1,0)\,dt \\ &= \frac{\mu_0 I}{2\pi}\int_0^1 \frac{1}{(1-t)^2+1}\,dt = \frac{\mu_0 I}{2\pi}\arctan(t-1)\Big|_0^1 = \frac{\mu_0 I}{2\pi}\frac{\pi}{4} = \frac{\mu_0 I}{8}.\end{aligned}$$

Parametrize $\mathbf{c}_2$ by $\mathbf{c}_2(t) = (0,0,1) + t(0,1,-1) = (0,t,1-t), t \in [0,1]$. Then

$$\begin{aligned}\int_{\mathbf{c}_2} \mathbf{B}\cdot d\mathbf{s} &= \frac{\mu_0 I}{2\pi}\int_0^1 \frac{(0,-1+t,t)}{(-1+t)^2+t^2}\cdot(0,1,-1)\,dt = \frac{\mu_0 I}{2\pi}\int_0^1 \frac{-1}{2t^2-2t+1}\,dt \\ &= \frac{\mu_0 I}{2\pi}\int_0^1 \frac{-1}{\frac{1}{2}\left((2t-1)^2+1\right)}\,dt = -\frac{\mu_0 I}{2\pi}\arctan(2t-1)\Big|_0^1 = -\frac{\mu_0 I}{2\pi}\frac{\pi}{2} = -\frac{\mu_0 I}{4}.\end{aligned}$$

Parametrize $\mathbf{c}_3$ by $\mathbf{c}_3(t) = (0,1,0) + t(0,0,1) = (0,1,t), t \in [0,1]$. Then

$$\begin{aligned}\int_{\mathbf{c}_3} \mathbf{B}\cdot d\mathbf{s} &= \frac{\mu_0 I}{2\pi}\int_0^1 \frac{(0,-t,1)}{1+t^2}\cdot(0,0,1)\,dt \\ &= \frac{\mu_0 I}{2\pi}\int_0^1 \frac{1}{1+t^2}\,dt = \frac{\mu_0 I}{2\pi}\arctan t\Big|_0^1 = \frac{\mu_0 I}{2\pi}\frac{\pi}{4} = \frac{\mu_0 I}{8}.\end{aligned}$$

Hence,

$$\mathcal{B} = \int_{\mathbf{c}_1} \mathbf{B}\cdot d\mathbf{s} + \int_{\mathbf{c}_2} \mathbf{B}\cdot d\mathbf{s} + \int_{\mathbf{c}_3} \mathbf{B}\cdot d\mathbf{s} = 0.$$ ◀

▶ EXERCISES 5.3

1. Consider the vector field $\mathbf{F}$ and the curves $\mathbf{c}_1$, $\mathbf{c}_2$, and $\mathbf{c}_3$ in Figure 5.26.

(a) Explain why $\int_{\mathbf{c}_1} \mathbf{F}\cdot d\mathbf{s} < 0$.

(b) Assume that $\mathbf{c}_2$ and $\mathbf{c}_3$ have the same speed. Which of the path integrals $\int_{\mathbf{c}_2} \mathbf{F}\cdot d\mathbf{s}$ or $\int_{\mathbf{c}_3} \mathbf{F}\cdot d\mathbf{s}$ is larger? Why?

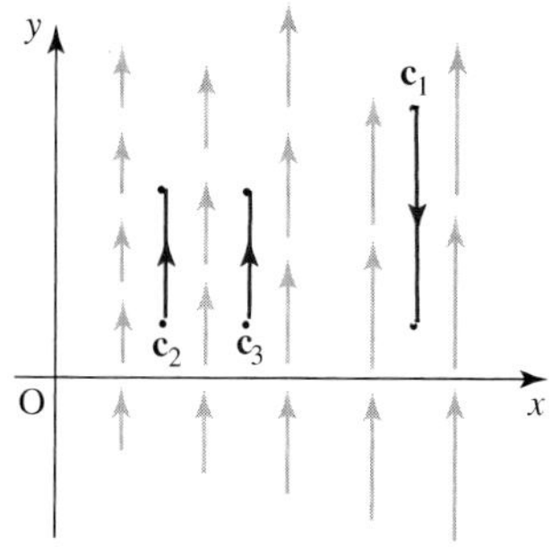

Figure 5.26 Diagram for Exercise 1.

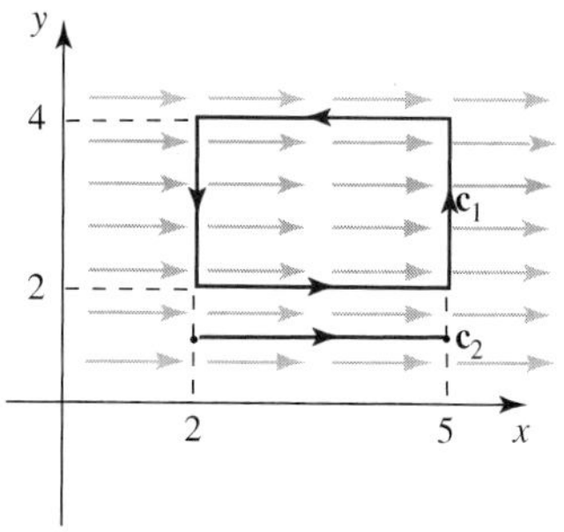

Figure 5.27 Diagram for Exercise 2.

2. Figure 5.27 shows a constant vector field $\mathbf{F}$.

(a) Compute $\int_{\mathbf{c}_1} \mathbf{F} \cdot d\mathbf{s}$ along the closed path $\mathbf{c}_1$.

(b) Assuming that $\|\mathbf{F}\| = 3/2$, compute $\int_{\mathbf{c}_2} \mathbf{F} \cdot d\mathbf{s}$

3. Let us compute the work W done by the force $\mathbf{F} = x\mathbf{i} + \mathbf{j}$ along the straight-line segment $\mathbf{c}(t) = (t, 1)$, $1 \le t \le 4$, using Riemann sums W_n defined in (5.1).

(a) Check that when the interval $[1, 4]$ is divided into n subintervals of equal length, the subdivision points are $\mathbf{c}(t_i) = (1 + 3(i-1)/n, 1)$, $i = 1, \ldots, n+1$.

(b) Show that $W_n = 3 + 9(n-1)/(2n)$.

(c) Conclude that $W = 15/2$. Check your answer by computing W using a path integral.

4. Using Riemann sums, as in Exercise 3, compute the work done by the force $\mathbf{F} = x\mathbf{i} + \mathbf{j}$ along the straight-line segment $\mathbf{c}(t) = (t, t)$, $1 \le t \le 2$. Verify your answer by computing the work using a path integral.

Exercises 5 to 12: Compute $\int_{\mathbf{c}} \mathbf{F} \cdot d\mathbf{s}$.

5. $\mathbf{F}(x, y) = y^2\mathbf{i} - x^2\mathbf{j}$, $\mathbf{c}$ is the part of the parabola $y = x^2$ from $(-1, 1)$ to $(1, 1)$

6. $\mathbf{F}(x, y) = x^2y\mathbf{i} + (y-1)\mathbf{j}$, $\mathbf{c}$ is the triangle with vertices $(0, 0)$, $(2, 0)$, and $(1, 1)$, oriented counterclockwise

7. $\mathbf{F}(x, y) = e^{x+y}\mathbf{i} - \mathbf{j}$, $\mathbf{c}$ is the boundary of the square with vertices $(0, 0)$, $(1, 0)$, $(1, 1)$, and $(0, 1)$, oriented clockwise

8. $\mathbf{F}(x, y, z) = (yz^2, xyz, 2x^2z)$, $\mathbf{c}$ consists of straight-line segments from $(-1, 2, -2)$ to $(-1, -2, -2)$, then to $(-1, -2, 0)$ and then to $(0, -2, 0)$

9. $\mathbf{F}(x, y, z) = (x^2, xy, 2z^2)$, $\mathbf{c}(t) = (\sin t, \cos t, t^2)$, $0 \le t \le \pi/2$

10. $\mathbf{F}(x, y) = e^{x+y}\mathbf{i} + e^{x-y}\mathbf{j}$, $\mathbf{c}$ is the triangle with vertices $(0, 0)$, $(0, 1)$, and $(1, 0)$, oriented counterclockwise

11. $\mathbf{F}(x, y) = 2xy\mathbf{i} + e^y\mathbf{j}$, $\mathbf{c}(t) = 4t^3\mathbf{i} + t^2\mathbf{j}$, $t \in [0, 1]$

12. $\mathbf{F}(x, y, z) = (xy, yz, xz)$, $\mathbf{c}$ consists of the straight-line segments from the origin to $(1, 0, 1)$, and then to $(1, 1, 0)$

13. Let $\mathbf{c}$ be an oriented C^1 path. A vector field $\mathbf{F}$ of constant magnitude $\|\mathbf{F}\| = k$ is tangent to $\mathbf{c}$ (at all points of $\mathbf{c}$) and points in the direction of $\mathbf{c}$. Find $\int_{\mathbf{c}} \mathbf{F} \cdot d\mathbf{s}$.

14. Let $\mathbf{F}$ be a continuous vector field defined on all of $\mathbb{R}^2$, and let $\mathbf{c}_1$ be a C^1 path from a point P to a point Q in $\mathbb{R}^2$. Define the piecewise C^1 path $\mathbf{c}_2$ as follows: $\mathbf{c}_2$ has the same image as $\mathbf{c}_1$; $\mathbf{c}_2$ starts at

P and stops at some point Q_1 before it reaches Q; then it moves back and stops at a point P_1 between P and Q_1. Finally, it moves from P_1 to Q. Explain why $\int_{\mathbf{c}_1} \mathbf{F}\,ds = \int_{\mathbf{c}_2} \mathbf{F}\,ds$.

15. Compute the work done when the force $\mathbf{F}(x, y) = x^3\mathbf{i} + (x + y)\mathbf{j}$ acts on a particle that moves from $(0, 0)$ to $(1, \pi^2/4)$ along the curve $\mathbf{c}(t) = \sin t\mathbf{i} + t^2\mathbf{j}$.

16. Assume that $\mathbf{F}$ is a constant force field acting in $\mathbb{R}^2$. Show that $\mathbf{F}$ does zero work on a particle that moves counterclockwise once around a circle in the xy-plane with constant speed.

17. Assume that the force $\mathbf{F} = C(x\mathbf{i} + y\mathbf{j})$ (C is a constant) acts on a particle moving in $\mathbb{R}^2$. Show that $\mathbf{F}$ does zero work if the particle moves counterclockwise once around a circle with constant speed.

18. The force between two positive electric charges [one, of charge ρ, is placed at the origin, and the other, of charge 1 C, is placed at (x, y)] is given by the formula $\mathbf{F} = \rho\mathbf{r}/\|\mathbf{r}\|^3$. How much work is needed in order to move the 1-C charge along the straight line from $(1, 0)$ to $(-1, 2)$ if the other charge remains at the origin?

19. Consider the force field $\mathbf{F}(x, y) = (y, 0)$. Compute the work done on a particle by the force $\mathbf{F}$ if the particle moves from $(0, 0)$ to $(1, 1)$ in each of the following ways:

(a) Along the x-axis to $(1, 0)$, then vertically up to $(1, 1)$

(b) Along the parabolic path $y = x^2$

(c) Along the path $y = x^4$

(d) Along the straight line

(e) Along the path $y = \sin(\pi x/2)$

(f) Along the y-axis to $(0, 1)$, then horizontally to $(1, 1)$

Interpret your results.

20. Compute $\int_{\mathbf{c}} 3(x + y)dx$ along the path $\mathbf{c}(t) = (e^t + 1, e^t - 2)$, $0 \le t \le 1$.

21. Compute $\int_{\mathbf{c}} (ydx + xdy)/(x^2 + y^2)$, where $\mathbf{c}$ is the circle centered at the origin of radius 2, oriented counterclockwise.

22. Compute $\int_{\mathbf{c}} xydx + ye^x dy$, where $\mathbf{c}$ is the rectangle with vertices $(0, 0)$, $(1, 0)$, $(1, 1)$, and $(0, 1)$, oriented counterclockwise.

23. Consider $\int_{\mathbf{c}_i} xydx + 2ydy$, where $\mathbf{c}_i$ is the straight-line segment joining the points $(0, 0)$ and $(1, 1)$ parametrized in the following ways:

(a) $\mathbf{c}_1(t) = (t, t)$, $t \in [0, 1]$

(b) $\mathbf{c}_2(t) = (\sin t, \sin t)$, $t \in [0, \pi/2]$

(c) $\mathbf{c}_3(t) = (\cos t, \cos t)$, $t \in [0, \pi/2]$

Are all the results the same? Explain why or why not.

24. Compute $\int_{\mathbf{c}} M(x, y, z)dx$, where M is a continuous function and $\mathbf{c}$ is any curve contained in a plane parallel to the yz-plane.

25. Show that the assumption "$\mathbf{c}$ is a C^1 curve" in Theorem 5.3 can be replaced by "$\mathbf{c}$ is a piecewise C^1 curve."

Exercises 26 to 29: In $\mathbb{R}^2$, the flux (flow) of a vector field $\mathbf{F}$ across a smooth closed curve $\mathbf{c}$ is defined as $\int_{\mathbf{c}} \mathbf{F} \cdot \mathbf{n}\,ds$, where $\mathbf{n}$ denotes the outward unit normal vector field along $\mathbf{c}$. The circulation of $\mathbf{F}$ is given by $\int_{\mathbf{c}} \mathbf{F} \cdot d\mathbf{s}$. Compute the flux and the circulation for the vector field $\mathbf{F}$ and the curve $\mathbf{c}$.

26. $\mathbf{F}(x, y) = 4x\mathbf{i} - 2y\mathbf{j}$, $\mathbf{c}$ is a circle of radius r, oriented clockwise

27. $\mathbf{F}(x, y) = x\mathbf{i} + y\mathbf{j}$, $\mathbf{c}$ is a circle of radius r, oriented counterclockwise

28. $\mathbf{F}(x, y) = x^2\mathbf{i} + y^2\mathbf{j}$, $\mathbf{c}$ is the semicircle of radius r from $(r, 0)$ to $(-r, 0)$, followed by the straight-line segment back to $(r, 0)$, oriented counterclockwise

29. $\mathbf{F}(x, y) = x\mathbf{i} + y\mathbf{j}$, $\mathbf{c}$ is the curve from Exercise 28

30. Let $\mathbf{F}(x, y) = P(x, y)\mathbf{i} + Q(x, y)\mathbf{j}$ be a continuous vector field. Show that its outward flux is given by $\int_{\mathbf{c}} \mathbf{F} \cdot \mathbf{n}\,ds = \int_{\mathbf{c}} P(x, y)dy - Q(x, y)dx$.

31. Assume that $\mathbf{F}$ is a differentiable vector field in $\mathbb{R}^3$. Using linear approximations of the components of $\mathbf{F}$ (as done in the text), prove the formula (5.3) for the circulation of $\mathbf{F}$ around $\mathbf{c}(t) = (h\cos t, h\sin t, 0)$, $t \in [0, 2\pi]$, for a small number $h > 0$. Then, in a similar way, compute the circulation of $\mathbf{F}$ around the small circle $\mathbf{c}(t) = (h\cos t, 0, h\sin t)$, $t \in [0, 2\pi]$, in the xz-plane. Finally, compute the circulation of $\mathbf{F}$ around the circle of radius h centered at the origin in the yz-plane. Interpret your answers.

▶ 5.4 PATH INTEGRALS INDEPENDENT OF PATH

In this section, we investigate gradient vector fields by studying their properties with respect to integration along curves. Probably the most famous examples of gradient vector fields are conservative force fields, two types of which (gravitational and electrostatic) have already been discussed in this book.

One of the most important theorems of vector calculus states that the path integral of a gradient vector field does not depend on the path, but only on its endpoints. We will prove this theorem in one special case and derive very useful consequences.

We start by defining a gradient vector field and addressing the first question that comes to mind: how does one identify a gradient vector field? Is there a test that can determine whether a given vector field is a gradient vector field?

DEFINITION 5.11 Gradient Vector Field

A vector field $\mathbf{F}: U \subseteq \mathbb{R}^2 \to \mathbb{R}^2$ (or $\mathbf{F}: U \subseteq \mathbb{R}^3 \to \mathbb{R}^3$) is called a *gradient vector field* if and only if there is a differentiable function $f: U \to \mathbb{R}$ such that $\mathbf{F} = \nabla f$. ◀

▶ EXAMPLE 5.24

The vector field $\mathbf{F}(x, y, z) = (2x + x^2 y)e^{xy}\mathbf{i} + x^3 e^{xy}\mathbf{j} + \mathbf{k}$ is a gradient vector field defined on $\mathbb{R}^3$, because if $f(x, y, z) = x^2 e^{xy} + z$, then

$$\nabla f(x, y, z) = (2xe^{xy} + x^2 y e^{xy})\mathbf{i} + x^3 e^{xy}\mathbf{j} + \mathbf{k} = \mathbf{F}(x, y, z).$$

The vector field $\mathbf{F}(x, y) = (x/\sqrt{x^2 + y^2})\mathbf{i} + (y/\sqrt{x^2 + y^2})\mathbf{j}$ is a gradient vector field defined on $\mathbb{R}^2 - \{(0, 0)\}$, because if $f(x, y) = \sqrt{x^2 + y^2}$, then $\nabla f(x, y) = \mathbf{F}(x, y)$. ◀

Recall that a vector field $\mathbf{F}$ is called conservative (see Example 2.89 in Section 2.7) if

$$\mathbf{F} = -\nabla V, \tag{5.8}$$

where V is a real-valued function, called the potential function.

For example, the gravitational field (see Example 2.40 in Section 2.4)

$$\mathbf{F}(x, y, z) = -GMm\mathbf{r}/\|\mathbf{r}\|^3,$$

where $\mathbf{r} = x\mathbf{i} + y\mathbf{j} + z\mathbf{k}$, is conservative; its potential function is $V = -GMm/\|\mathbf{r}\|$, defined on $\mathbb{R}^3 - \{(0, 0, 0)\}$. The electrostatic field [see Example 2.11 in Section 2.1; let ϵ denote the constant $(4\pi\epsilon_0)^{-1}$]

$$\mathbf{F}(x, y, z) = Qq\epsilon\mathbf{r}/\|\mathbf{r}\|^3$$

is also a conservative field; the function $V = Qq\epsilon/\|\mathbf{r}\|$ defined on $\mathbb{R}^3 - \{(0, 0, 0)\}$ is its potential function. (See the comment immediately following Example 2.93 in Section 2.7 for the use of the word "potential.")

Formally, there is no difference between a gradient vector field and a conservative vector field: if $\mathbf{F} = \nabla f$, then $\mathbf{F} = -\nabla(-f)$, so a gradient vector field is also a conservative field (with potential function $-f$). Likewise, if $\mathbf{F} = -\nabla V$, then $\mathbf{F} = \nabla(-V)$, and thus a conservative field is also a gradient field.

However, in applications (such as gravitational or electrostatic fields) we talk about conservative fields, and we use (5.8), since the signs of forces and potential functions have physical interpretation and meaning. In Example 2.91 in Section 2.7, we proved an important property of conservative vector fields (that explains why we call them "conservative"): the total energy of an object moving under the influence of such fields remains constant, e. g., is conserved. The minus sign in (5.8) was essential in the proof.

Let $\mathbf{F}$ be a gradient vector field; that is, $\mathbf{F} = \nabla f$ for some function f. Assuming that there is another function g such that $\mathbf{F} = \nabla g$, we would like to find the relation between f and g. In other words, we would like to identify all functions f that satisfy $\mathbf{F} = \nabla f$ for a given field $\mathbf{F}$. Since $\nabla f = \nabla g$, it follows that $\nabla(f - g) = \mathbf{0}$. Denoting $f - g$ by h, we get $\nabla h = \mathbf{0}$. Consequently, $\partial h/\partial x = 0$, $\partial h/\partial y = 0$, and $\partial h/\partial z = 0$; that is, h must be a constant function. Hence, $f - g = \text{constant}$, and

$$g = f + \text{constant}.$$

It follows that there are infinitely many functions f satisfying $\mathbf{F} = \nabla f$; however, they can differ from each other by, at most, an additive constant. Consequently, there are infinitely many potential functions for a given conservative field, and they all differ by a constant. With an extra condition (like prescribing the value of a potential function at a point, or the value of a limit, as in the next example), we can compute the constant and the potential function becomes uniquely determined.

▶ EXAMPLE 5.25

Any function of the form

$$V(x, y, z) = -\frac{GMm}{\|\mathbf{r}\|} + C$$

(C is a constant) is a potential function of the gravitational force field

$$\mathbf{F} = -\frac{GMm}{\|\mathbf{r}\|^3}\mathbf{r}$$

(see Example 2.40). The condition that the potential vanishes at infinity implies that

$$0 = \lim_{\|\mathbf{r}\|\to\infty} V = -\lim_{\|\mathbf{r}\|\to\infty} \frac{GMm}{\|\mathbf{r}\|} + C = C;$$

hence, $C = 0$. Thus,

$$V(x, y, z) = -\frac{GMm}{\|\mathbf{r}\|}$$

is the gravitational potential of $\mathbf{F}$, satisfying $\lim V = 0$ as $\|\mathbf{r}\| \to \infty$. ◀

Now we turn to the problem of determining whether a given vector field is a gradient vector field or not. First of all, we have to explain the assumption on the domain U of the vector field that will appear in the statement of the theorem.

DEFINITION 5.12 Connected Set

We say that a set $U \subseteq \mathbb{R}^2$ (or $\mathbb{R}^3$) is *connected* if any two points in U can be joined by a continuous curve that is completely contained in U. ◀

A set is connected if it is in "one piece." For example, $\mathbb{R}^2$, $\mathbb{R}^3$, or $\{(x, y)|y \geq x^2\}$ are connected. The set obtained from $\mathbb{R}^2$ by removing the x-axis is not connected. The set $\{(x, y)|xy > 0\}$ consists of the first and third quadrants in $\mathbb{R}^2$, without the coordinate axes. It is not connected.

DEFINITION 5.13 Simply-Connected Set

A set $U \subseteq \mathbb{R}^2$ (or $\mathbb{R}^3$) is *simply-connected* if it is connected and if every simple closed curve in U can be shrunk (without breaking) to a point without leaving U. ◀

Figure 5.28 shows examples of simply-connected sets: an arbitrary simple, closed curve is deformed, without breaking (this is sometimes called "continuously deformed") to a point in such a way that the deformations (from the initial curve to the point) do not leave the set. A plane in $\mathbb{R}^3$, the disk $\{(x, y)|x^2 + y^2 < 1\} \subseteq \mathbb{R}^2$, and $\mathbb{R}^3$ are examples of simply-connected sets.

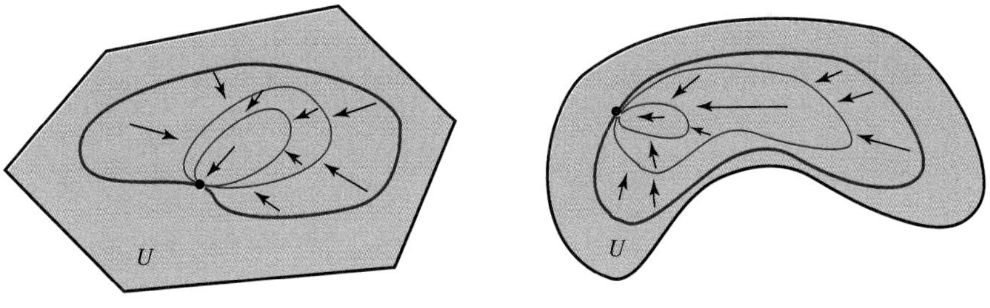

Figure 5.28 Simply-connected sets.

A plane with a hole [e.g., the set $\mathbb{R}^2 - \{(x, y)|x^2 + y^2 \leq 1\}$] or a "punctured" plane $\mathbb{R}^2 - \{(0, 0)\}$ are not simply-connected: it is impossible to deform a loop that goes around the hole (or around the deleted point) to a point without breaking the loop; see Figure 5.29.

The set U obtained from $\mathbb{R}^3$ by removing finitely many points is simply-connected. In the case of the "punctured" plane, we got stuck in deforming the curve when we encountered a hole. What we wanted to do was to "jump over" (or under) the hole to continue the deformation; since the curve was supposed to remain in the plane, that was not possible. In three-dimensional space, we can actually do that: since we have an extra dimension, we can easily avoid "holes" in deforming a curve to a point, as shown in Figure 5.30.

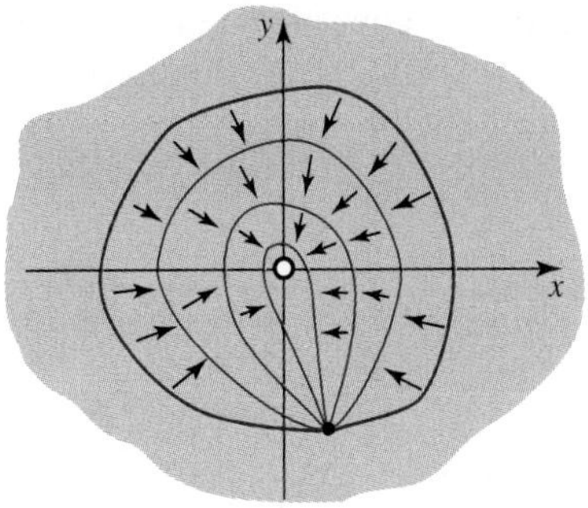

Figure 5.29 A "punctured" plane is not a simply-connected set.

Recall that, for a vector field $\mathbf{F} = (F_1, F_2, F_3)$ in $\mathbb{R}^3$,

$$curl\,\mathbf{F} = \left(\frac{\partial F_3}{\partial y} - \frac{\partial F_2}{\partial z}\right)\mathbf{i} + \left(\frac{\partial F_1}{\partial z} - \frac{\partial F_3}{\partial x}\right)\mathbf{j} + \left(\frac{\partial F_2}{\partial x} - \frac{\partial F_1}{\partial y}\right)\mathbf{k}.$$

The scalar curl of a vector field $\mathbf{F} = (F_1, F_2)$ in $\mathbb{R}^2$ is the function

$$\frac{\partial F_2}{\partial x} - \frac{\partial F_1}{\partial y}.$$

THEOREM 5.4 Necessary and Sufficient Conditions for a Gradient Vector Field

Let $\mathbf{F}$ be a C^1 vector field defined on an open, simply-connected set $U \subseteq \mathbb{R}^2$. Then $\mathbf{F}$ is a gradient vector field (i.e., $\mathbf{F} = \nabla f$ for some function f) if and only if the scalar curl of $\mathbf{F}$ is zero. A C^1 vector field $\mathbf{F}$ defined on an open, simply-connected set $U \subseteq \mathbb{R}^3$ is a gradient vector field if and only if $curl\,\mathbf{F} = \mathbf{0}$. ◀

Note that one implication is immediate: if $\mathbf{F}$ is in $\mathbb{R}^3$ and $\mathbf{F} = \nabla f$, then $curl\,\mathbf{F} = curl\,(\nabla f) = \mathbf{0}$ by Theorem 4.13 in Section 4.6. Likewise, if $\mathbf{F}$ is in $\mathbb{R}^2$ and $\mathbf{F} = \nabla f$, then the scalar curl is zero.

We are not going to present the proof of the other implication in its full generality, but in a special case where U is a star-shaped set; we will give another proof using Stokes'

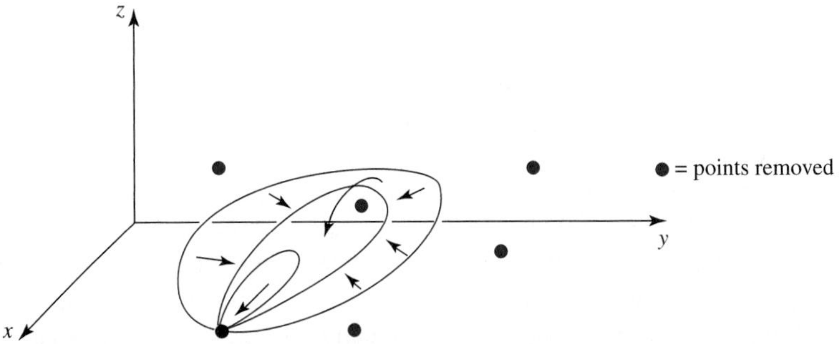

Figure 5.30 $\mathbb{R}^3$ with finitely many points removed is simply-connected.

Theorem in the last chapter. However, that proof will have a "weak" link: it will assume one fact that we will not be able to prove in this book.

DEFINITION 5.14 Star-Shaped Set

A set U is called *star-shaped* if there is a point A in U such that, for every point P in U, the entire line segment $\overline{AP}$ lies in U. ◀

A plane, all of $\mathbb{R}^3$, or the sets in Figure 5.31 are examples of star-shaped sets. (The set U on the right is a solid cylinder, that is, it contains the inside and the boundary of the cylinder.)

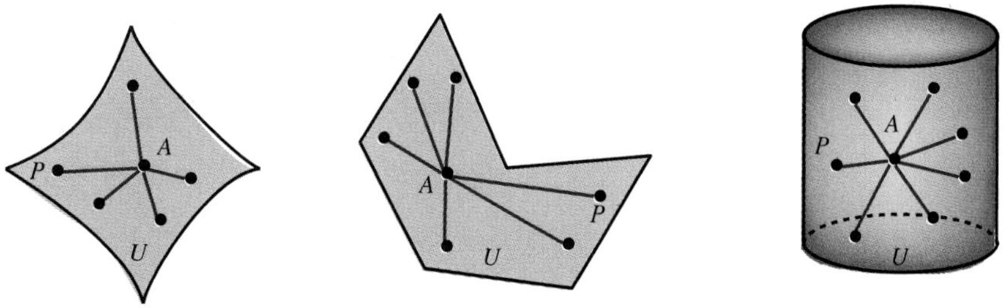

Figure 5.31 Star-shaped sets.

The set $U = \mathbb{R}^2 - \{(0, 0)\}$ is not star-shaped: no matter what point A is chosen, the line segment from A to the point P symmetric to A with respect to the origin (see Figure 5.32) has to go through the origin, and is not entirely in U.

A simple closed curve in a star-shaped set can always be continuously deformed to a point (this may seem reasonable, but a proof is beyond the scope of this book); therefore, a star-shaped set is always simply-connected. The converse is not true: $\mathbb{R}^3 - \{(0, 0, 0)\}$ is simply-connected, but not star-shaped.

In the proof of Theorem 5.4 (and later in this section), we will need the Fundamental Theorem of Calculus for functions of one variable, so we review it here.

Roughly speaking, the Fundamental Theorem of Calculus states that the derivative and integral are operations inverse to each other: when both are applied (in any order) to a

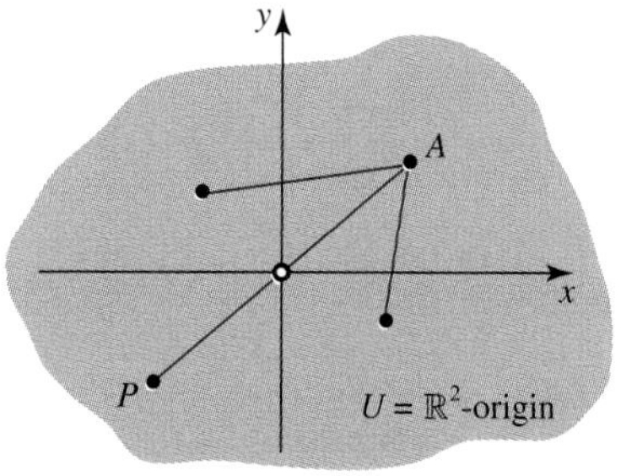

Figure 5.32 $\mathbb{R}^2 - \{(0, 0)\}$ is not a star-shaped set.

function, we get it back. More precisely, if f is a continuous function, then

$$\frac{d}{dx}\left(\int_a^x f(t)\,dt\right) = f(x), \tag{5.9}$$

where a is any constant. This formula is known as the Fundamental Theorem of Calculus, Part I. The second part of the theorem states that if f is continuously differentiable (i.e., if it is C^1), then

$$\int_a^b f'(t)\,dt = f(b) - f(a), \tag{5.10}$$

for any a and b in its domain.

We will also make use of the fact that it is possible to interchange partial derivatives and integrals of C^1 functions. For example,

$$\frac{\partial}{\partial x}\left(\int_a^b f(x, y, t)\,dt\right) = \int_a^b \frac{\partial f(x, y, t)}{\partial x}\,dt. \tag{5.11}$$

This formula is called *Leibniz's rule*. (For a proof, see W. Kaplan, *Advanced Calculus*, 4th ed. (Boston: Addison-Wesley, 1991), p. 266.)

PROOF OF THEOREM 5.4: Let $\mathbf{F} = (F_1, F_2)$ be a C^1 vector field defined on a star-shaped set U, such that its scalar curl is zero. Choose a coordinate system so that U is star-shaped with respect to the origin (i.e., the point A from the definition of a star-shaped set is the origin).

We define the real-valued function f in the following way: for a point (x, y) in U, let $f(x, y)$ be the value of the line integral of $\mathbf{F}$ along the straight-line segment from $(0, 0)$ to (x, y) (here we use the assumption that U is star-shaped!). Parametrizing the line segment by $\mathbf{c}(t) = (tx, ty)$, $t \in [0, 1]$, we get

$$\begin{aligned} f(x, y) &= \int_{\mathbf{c}} \mathbf{F} \cdot d\mathbf{s} = \int_0^1 \mathbf{F}(\mathbf{c}(t)) \cdot \mathbf{c}'(t)\,dt = \int_0^1 \mathbf{F}(tx, ty) \cdot (x, y)\,dt \\ &= \int_0^1 (F_1(tx, ty)x + F_2(tx, ty)y)\,dt. \end{aligned}$$

Thus, we define

$$f(x, y) = \int_0^1 (F_1(tx, ty)x + F_2(tx, ty)y)\,dt. \tag{5.12}$$

We will show that $\nabla f = \mathbf{F}$, and that will complete the proof.

First, we compute $\partial f/\partial x$. Interchanging the integral and partial derivatives [see (5.11); $\mathbf{F}$ is C^1, and so are its components], we get [in order to avoid writing partial derivatives as $\partial F_1(tx, ty)/\partial(tx)$, $\partial F_2(tx, ty)/\partial(tx)$, etc., we use $D_1F_1(tx, ty)$, $D_1F_2(tx, ty)$, etc. instead]

$$\begin{aligned} \frac{\partial f}{\partial x}(x, y) &= \int_0^1 \frac{\partial}{\partial x}(F_1(tx, ty)x + F_2(tx, ty)y)\,dt \\ &= \int_0^1 (D_1F_1(tx, ty)tx + F_1(tx, ty) + D_1F_2(tx, ty)ty)\,dt, \end{aligned} \tag{5.13}$$

where the first two terms in the integrand come from the product rule that we applied to compute the partial derivative of $F_1(tx, ty)\,x$ with respect to x. Since the scalar curl of $\mathbf{F}$ is zero, we get $D_1 F_2 = D_2 F_1$, and thus the integrand in (5.13) is equal to

$$F_1(tx, ty) + D_1 F_1(tx, ty)tx + D_2 F_1(tx, ty)ty.$$

Using the product and the chain rules, we check that this expression is the derivative of $tF_1(tx, ty)$ with respect to t:

$$\begin{aligned}\frac{\partial}{\partial t}(tF_1(tx, ty)) &= F_1(tx, ty) + t\frac{\partial}{\partial t}(F_1(tx, ty)) \\ &= F_1(tx, ty) + t\,(D_1 F_1(tx, ty)x + D_2 F_1(tx, ty)y). \qquad (5.14)\end{aligned}$$

Thus, from (5.13) and (5.14), we get

$$\frac{\partial f}{\partial x}(x, y) = \int_0^1 \frac{\partial}{\partial t}(tF_1(tx, ty))\,dt = tF_1(tx, ty)\Big|_0^1 = F_1(x, y).$$

In the above, we applied the Fundamental Theorem (5.10) to the function $tF_1(tx, ty)$.

The remaining identity $\partial f/\partial y = F_2$ is checked analogously. ◀

In the same way, we can prove the statement of the theorem for vector fields in $\mathbb{R}^3$ (see Exercise 12).

Theorem 5.4 states that a C^1 vector field $\mathbf{F} = (F_1, F_2)$ defined on an open, simply-connected set $U \subseteq \mathbb{R}^2$ is a gradient vector field (i.e., $\mathbf{F} = \nabla f$) if and only if its scalar curl is zero, that is, if and only if

$$\frac{\partial F_2}{\partial x} - \frac{\partial F_1}{\partial y} = 0.$$

Furthermore, a C^1 vector field $\mathbf{F} = (F_1, F_2, F_3)$ defined on an open, simply-connected set $U \subseteq \mathbb{R}^3$ is a gradient vector field if and only if *curl* $\mathbf{F} = \mathbf{0}$; that is, if and only if

$$\frac{\partial F_3}{\partial y} = \frac{\partial F_2}{\partial z}, \qquad \frac{\partial F_1}{\partial z} = \frac{\partial F_3}{\partial x}, \qquad \text{and} \qquad \frac{\partial F_2}{\partial x} = \frac{\partial F_1}{\partial y}.$$

The Fundamental Theorem of Calculus (5.10) implies that the definite integral of $f'(t)$ on $[a, b]$ depends on the value of f at the endpoints a and b *only*. Therefore, if f and g are *any* two C^1 functions that coincide at a and b [i.e., $f(a) = g(a)$ and $f(b) = g(b)$], then

$$\int_a^b f'(t)\,dt = \int_a^b g'(t)\,dt.$$

The generalization of this statement to functions of several variables is given in the following theorem.

THEOREM 5.5 Generalization of the Fundamental Theorem of Calculus

Let $f\colon \mathbb{R}^2$ (or $\mathbb{R}^3$) $\to \mathbb{R}$ be a C^1 function and let $\mathbf{c}\colon [a, b] \to \mathbb{R}^2$ (or $\mathbb{R}^3$) be a piecewise C^1 path. Then

$$\int_{\mathbf{c}} \nabla f \cdot d\mathbf{s} = f(\mathbf{c}(b)) - f(\mathbf{c}(a)).$$ ◀

PROOF: By definition of the path integral (assuming that $\mathbf{c}$ is C^1), we get

$$\int_{\mathbf{c}} \nabla f \cdot ds = \int_a^b \nabla f(\mathbf{c}(t)) \cdot \mathbf{c}'(t)\, dt.$$

The integrand is the product of the derivative of f and the derivative of $\mathbf{c}$, which "smells like" the chain rule. Indeed, as in Example 2.68 in Section 2.6, we obtain

$$(f(\mathbf{c}(t)))' = Df(\mathbf{c}(t)) \cdot D\mathbf{c}(t) = \nabla f(\mathbf{c}(t)) \cdot \mathbf{c}'(t)$$

and thus,

$$\int_{\mathbf{c}} \nabla f \cdot d\mathbf{s} = \int_a^b (f(\mathbf{c}(t)))'\, dt = f(\mathbf{c}(b)) - f(\mathbf{c}(a)),$$

by the Fundamental Theorem of Calculus (5.10) applied to the function $f(\mathbf{c}(t))$ of one variable.

Now assume that $[a, b]$ consists of two subintervals $[a = t_1, t_2]$ and $[t_2, t_3 = b]$ such that the restrictions $\mathbf{c_1} = \mathbf{c}|_{[t_1,t_2]}$ and $\mathbf{c_2} = \mathbf{c}|_{[t_2,t_3]}$ are C^1 (this is the proof in the piecewise C^1 case; for convenience, we assumed that $\mathbf{c}$ consists of two C^1 pieces; the general case of n C^1 pieces is approached in the same way). Then

$$\int_{\mathbf{c}} \nabla f \cdot d\mathbf{s} = \int_{\mathbf{c_1}} \nabla f \cdot d\mathbf{s} + \int_{\mathbf{c_2}} \nabla f \cdot d\mathbf{s}$$

($\mathbf{c_1}$ and $\mathbf{c_2}$ are C^1, and we proved the theorem in that case)

$$= f(\mathbf{c}(t_2)) - f(\mathbf{c}(t_1)) + f(\mathbf{c}(t_3)) - f(\mathbf{c}(t_2))$$
$$= f(\mathbf{c}(t_3)) - f(\mathbf{c}(t_1)) = f(\mathbf{c}(b)) - f(\mathbf{c}(a)).$$ ◀

▶ EXAMPLE 5.26

Let $f(x, y, z) = e^{x^2+y^2+z^2}$ and let $\mathbf{c}(t) = (\cos t, t, \sin t)$, $t \in [0, 2\pi]$. Then

$$\int_{\mathbf{c}} \nabla f \cdot ds = f(\mathbf{c}(2\pi)) - f(\mathbf{c}(0)) = f(1, 2\pi, 0) - f(1, 0, 0) = e^{1+4\pi^2} - e.$$ ◀

Theorem 5.5 provides a powerful tool for computing path integrals: if we know that we are integrating the gradient of a function, we do not need to parametrize the path(s) in question; all we have to do is to evaluate the function at the terminal point and at the initial point and subtract. The next three examples will serve as illustrations of this principle.

▶ EXAMPLE 5.27

Let us go back to Example 5.19 of the previous section. The vector field $\mathbf{F} = (y + z)\mathbf{i} + x\mathbf{j} + x\mathbf{k}$ is a gradient vector field since

$$\nabla(xy + xz) = (y + z)\mathbf{i} + x\mathbf{j} + x\mathbf{k}.$$

The application of Theorem 5.5 greatly simplifies the computation of the path integral:

$$\int_{\mathbf{c}} \mathbf{F} \cdot d\mathbf{s} = \int_{\mathbf{c}} \nabla(xy + xz) \cdot d\mathbf{s} = (xy + xz)\Big|_{(1,0,0)}^{(1,0,1)} = 1.$$ ◀

▶ EXAMPLE 5.28

Compute the work of the electrostatic force $\mathbf{F}(\mathbf{r}) = (Qq/4\pi\epsilon_0)\mathbf{r}/\|\mathbf{r}\|^3$ acting on a charge q that moves from the point A to the point B along a curve $\mathbf{c}$.

SOLUTION

A straightforward computation gives the relation $\mathbf{F}(\mathbf{r}) = -\nabla V(\mathbf{r})$, where $V(\mathbf{r}) = (Qq/4\pi\epsilon_0)\|\mathbf{r}\|^{-1}$ is the electrostatic potential and $\mathbf{r} = x\mathbf{i} + y\mathbf{j} + z\mathbf{k}$.

Let $\mathbf{r} = \mathbf{r}(t)$: $[a, b] \to \mathbb{R}^3$ be a parametrization of $\mathbf{c}$, $\mathbf{r}(a) = A$ and $\mathbf{r}(b) = B$. The work of $\mathbf{F}$ is computed to be

$$W = \int_{\mathbf{c}} \mathbf{F} \cdot d\mathbf{s} = -\int_{\mathbf{c}} \nabla V(\mathbf{r}) \cdot d\mathbf{s}$$

$$= -V(\mathbf{r})\Big|_{\mathbf{r}(a)}^{\mathbf{r}(b)} = V(\mathbf{r}(a)) - V(\mathbf{r}(b)) = \frac{Qq}{4\pi\epsilon_0}\left(\frac{1}{\|\mathbf{r}(a)\|} - \frac{1}{\|\mathbf{r}(b)\|}\right).$$

▶ EXAMPLE 5.29

Compute $\int_{\mathbf{c}} \mathbf{F} \cdot d\mathbf{s}$, where $\mathbf{F} = 2x\mathbf{i} + 2e^z\mathbf{j} + 2ye^z\mathbf{k}$, and $\mathbf{c}$ is the piecewise C^1 curve shown in Figure 5.33.

SOLUTION

A straightforward computation shows that $curl\,\mathbf{F} = \mathbf{0}$. Since $\mathbf{F}$ is C^1 and defined on $\mathbb{R}^3$ (which is open and simply-connected), it follows (by Theorem 5.4) that $\mathbf{F} = \nabla f$ for some scalar function f, and thus,

$$\int_{\mathbf{c}} \mathbf{F} \cdot d\mathbf{s} = \int_{\mathbf{c}} \nabla f d\mathbf{s} = f(1, 1, 0) - f(0, 0, 0), \tag{5.15}$$

by Theorem 5.5. So all we have to do is to find f (the method was explained and illustrated in Example 2.94 in Section 2.7). The equation $\mathbf{F} = \nabla f$ implies that

$$\frac{\partial f}{\partial x} = 2x, \qquad \frac{\partial f}{\partial y} = 2e^z, \qquad \text{and} \qquad \frac{\partial f}{\partial z} = 2ye^z.$$

Integrating the first equation with respect to x, we get

$$f(x, y, z) = x^2 + C(y, z), \tag{5.16}$$

where $C(y, z)$ denotes a function of y and z only. Differentiating (5.16) with respect to y and substituting into $\partial f/\partial y = 2e^z$ yield $\partial C(y, z)/\partial y = 2e^z$. Thus, $C(y, z) = 2ye^z + C(z)$, after integration with respect to y.

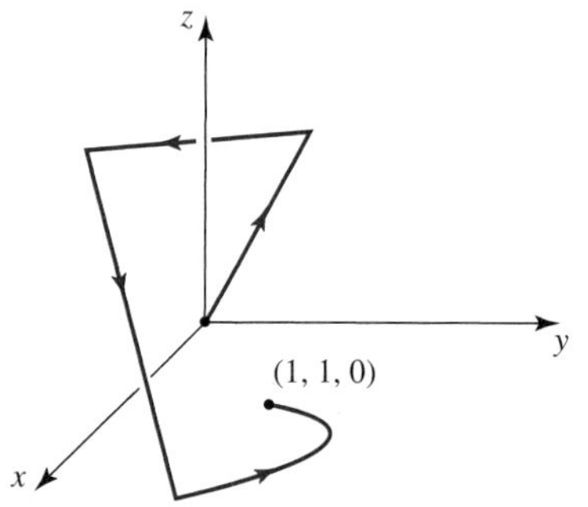

Figure 5.33 The curve of Example 5.29.

We have recovered some parts of f—so far, $f(x, y, z) = x^2 + 2ye^z + C(z)$, and the function $C(z)$ is yet to be found. We now differentiate this expression with respect to z and combine with the equation for $\partial f/\partial z$, thus obtaining

$$\frac{\partial f}{\partial z} = 2ye^z + \frac{dC(z)}{dz} = 2ye^z;$$

hence, $dC(z)/dz = 0$ and $C(z) = C$ (where C is a constant).

Therefore, $f(x, y, z) = x^2 + 2ye^z + C$, and (5.15) implies that

$$\int_{\mathbf{c}} \mathbf{F} \cdot d\mathbf{s} = f(1, 1, 0) - f(0, 0, 0) = 3.$$

◀

Assume that $\mathbf{F}$ is a C^1 gradient vector field, so that $\mathbf{F} = \nabla f$ for some function f. An immediate consequence of the Fundamental Theorem 5.5 is the fact that if $\mathbf{c}\colon [a, b] \to \mathbb{R}^2$ (or $\mathbb{R}^3$) is an oriented, simple *closed* curve [i.e., $\mathbf{c}(a) = \mathbf{c}(b)$], then

$$\int_{\mathbf{c}} \mathbf{F} \cdot d\mathbf{s} = \int_{\mathbf{c}} \nabla f \cdot d\mathbf{s} = f(\mathbf{c}(b)) - f(\mathbf{c}(a)) = 0. \tag{5.17}$$

Now take a curve $\mathbf{c}$ in the domain U (assumed open and connected) of $\mathbf{F}$ whose initial point is A and terminal point is B. Let $\overline{\mathbf{c}}$ be any curve in U with the same initial and terminal points as $\mathbf{c}$, see Figure 5.34.

Consider the following curve (call it Γ): along $\mathbf{c}$ from A to B and then along $\overline{\mathbf{c}}$ in the opposite direction back to A. Assume that Γ does not intersect itself (i.e., that it is a simple closed curve). Then (5.17) implies that

$$\int_{\Gamma} \mathbf{F} \cdot d\mathbf{s} = 0.$$

On the other hand,

$$0 = \int_{\Gamma} \mathbf{F} \cdot d\mathbf{s} = \int_{\mathbf{c}} \mathbf{F} \cdot d\mathbf{s} + \int_{\overline{\mathbf{c}}^{opp}} \mathbf{F} \cdot d\mathbf{s} = \int_{\mathbf{c}} \mathbf{F} \cdot d\mathbf{s} - \int_{\overline{\mathbf{c}}} \mathbf{F} \cdot d\mathbf{s},$$

by Theorem 5.3 in Section 5.3, where $\overline{\mathbf{c}}^{opp}$ denotes the curve $\overline{\mathbf{c}}$ traversed in the opposite direction. Hence,

$$\int_{\mathbf{c}} \mathbf{F} \cdot d\mathbf{s} = \int_{\overline{\mathbf{c}}} \mathbf{F} \cdot d\mathbf{s};$$

that is, if $\mathbf{F}$ is a gradient vector field, then the path integral of $\mathbf{F}$ is the same for any curve joining A and B.

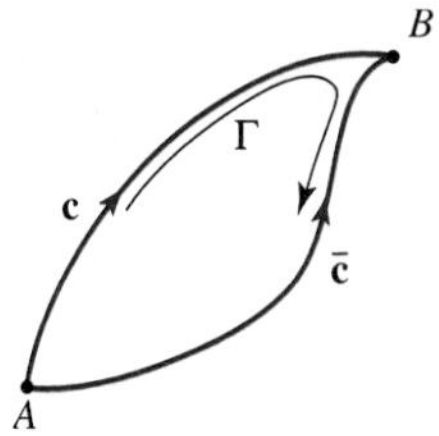

Figure 5.34 Curve Γ made of $\mathbf{c}$ and $\overline{\mathbf{c}}$.

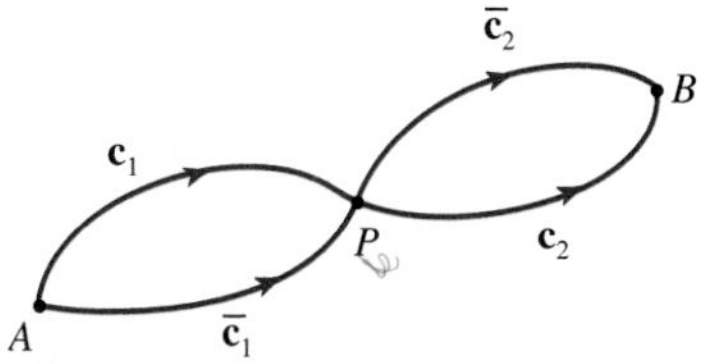

Figure 5.35 Curve Γ intersects itself.

It remains to discuss the case where $\mathbf{c}$ and $\overline{\mathbf{c}}$ intersect so that Γ is not simple (assume that they intersect at one point, P; the same argument will prove the case for any number of intersections). Let $\mathbf{c}_1$ be the part of $\mathbf{c}$ from A to P and let $\mathbf{c}_2$ be the part from P to B. Similarly, $\overline{\mathbf{c}}_1$ is the part of $\overline{\mathbf{c}}$ from A to P and $\overline{\mathbf{c}}_2$ is the part from P to B; see Figure 5.35. Then, since $\mathbf{c}_1$ and $\overline{\mathbf{c}}_1$ and $\mathbf{c}_2$ and $\overline{\mathbf{c}}_2$ do not intersect,

$$\int_{\mathbf{c}_1} \mathbf{F} \cdot d\mathbf{s} = \int_{\overline{\mathbf{c}}_1} \mathbf{F} \cdot d\mathbf{s} \quad \text{and} \quad \int_{\mathbf{c}_2} \mathbf{F} \cdot d\mathbf{s} = \int_{\overline{\mathbf{c}}_2} \mathbf{F} \cdot d\mathbf{s}$$

and therefore,

$$\int_{\mathbf{c}} \mathbf{F} \cdot d\mathbf{s} = \int_{\mathbf{c}_1} \mathbf{F} \cdot d\mathbf{s} + \int_{\mathbf{c}_2} \mathbf{F} \cdot d\mathbf{s} = \int_{\overline{\mathbf{c}}_1} \mathbf{F} \cdot d\mathbf{s} + \int_{\overline{\mathbf{c}}_2} \mathbf{F} \cdot d\mathbf{s} = \int_{\overline{\mathbf{c}}} \mathbf{F} \cdot d\mathbf{s}.$$

The conclusions of our argument are summarized in the next theorem.

THEOREM 5.6 Path-Independence of Integrals of Vector Fields

Let $\mathbf{F}$ be a C^1 gradient vector field defined on an open and connected set U. If $\mathbf{c}$ is an oriented, simple closed curve in U, then

$$\int_{\mathbf{c}} \mathbf{F} \cdot d\mathbf{s} = 0.$$

If $\mathbf{c}$ and $\overline{\mathbf{c}}$ are two oriented, simple curves in U with the same initial and terminal points, then

$$\int_{\mathbf{c}} \mathbf{F} \cdot d\mathbf{s} = \int_{\overline{\mathbf{c}}} \mathbf{F} \cdot d\mathbf{s}.$$ ◀

In Examples 5.27 and 5.29, we computed the path integral of a gradient vector field $\mathbf{F} = \nabla f$ by finding the function f and then using the Fundamental Theorem 5.5. Theorem 5.6 suggests an alternative that does not require that we compute the function f. The next two examples illustrate how this is done.

▶ EXAMPLE 5.30

Let us revisit Example 5.19 of the previous section once again (we have also discussed it in Example 5.27). Since

$$curl\,\mathbf{F} = \begin{vmatrix} \mathbf{i} & \mathbf{j} & \mathbf{k} \\ \partial/\partial x & \partial/\partial y & \partial/\partial z \\ y+z & x & x \end{vmatrix} = -(1-1)\mathbf{j} + (1-1)\mathbf{k} = \mathbf{0},$$

$\mathbf{F}$ is a gradient vector field, and by Theorem 5.6, the integral $\int_{\mathbf{c}} \mathbf{F} \cdot d\mathbf{s}$ does not depend on the curve, just on the endpoints. So take $\mathbf{c}$ to be the simplest possible curve: the straight-line segment from $(1, 0, 0)$ to $(1, 0, 1)$. Parametrize it by

$$\mathbf{c}(t) = (1, 0, 0) + t(0, 0, 1) = (1, 0, t), \qquad t \in [0, 1].$$

Then

$$\int_{\mathbf{c}} \mathbf{F} \cdot d\mathbf{s} = \int_0^1 (t, 1, 1) \cdot (0, 0, 1)\, dt = \int_0^1 1\, dt = 1.$$ ◀

▶ **EXAMPLE 5.31**

We will recompute the path integral of Example 5.29.

Since $curl\,(2x\mathbf{i}+2e^z\mathbf{j}+2ye^z\mathbf{k})=\mathbf{0}$, the integral $\int_{\mathbf{c}}\mathbf{F}\cdot d\mathbf{s}$ is independent of path. So take the straight-line segment $c(t)=(t,t,0)$, $t\in[0,1]$, joining $(0,0,0)$ and $(1,1,0)$. Then

$$\int_{\mathbf{c}}\mathbf{F}\cdot d\mathbf{s}=\int_0^1(2t,2,2t)\cdot(1,1,0)\,dt=\int_0^1(2t+2)\,dt=3.$$

Theorem 5.6 states that a gradient vector field is path-independent. Next, we prove the converse statement. ◀

THEOREM 5.7 Independence of $\int_{\mathbf{c}}\mathbf{F}\cdot d\mathbf{s}$ on the Path Implies That **F** Is a Gradient Vector Field

Let $\mathbf{F}$ be a C^1 vector field defined on an open, connected set $U\subseteq\mathbb{R}^2$ (or $\mathbb{R}^3$). The independence of the integral $\int_{\mathbf{c}}\mathbf{F}\cdot d\mathbf{s}$ on the path implies that the vector field $\mathbf{F}$ is a gradient vector field. ◀

PROOF: For simplicity, we will consider the case when $\mathbf{F}$ is defined on $U\subseteq\mathbb{R}^2$ (the case $U\subseteq\mathbb{R}^3$ is discussed analogously). Assume that $\int_{\mathbf{c}}\mathbf{F}\cdot d\mathbf{s}$ does not depend on the path $\mathbf{c}$. We have to find a scalar function $f\colon U\subseteq\mathbb{R}^2\to\mathbb{R}$ such that $\nabla f=\mathbf{F}=(F_1,F_2)$.

Define

$$f(x,y)=\int_{\mathbf{c}}\mathbf{F}\cdot d\mathbf{s},$$

where $\mathbf{c}$ is *any* curve joining $(0,0)$ and the point (x,y) in U. If $(0,0)$ does not belong to U, choose any point in U as the initial point of integration. By assumption, $\int_{\mathbf{c}}\mathbf{F}\cdot d\mathbf{s}$ depends only on the endpoints, and not on the curve joining them. Therefore, take $\mathbf{c}$ to be the following piecewise C^1 curve: along the x-axis from $(0,0)$ to $(x,0)$ and then along the straight line to (x,y); see Figure 5.36(a).

Parametrize $\mathbf{c}_1$ by $\mathbf{c}_1(t)=(t,0)$, $t\in[0,x]$. Then $\mathbf{c}_1'(t)=(1,0)$ and

$$\begin{aligned}\int_{\mathbf{c}_1}F\cdot ds&=\int_0^x F(\mathbf{c}_1(t))\cdot\mathbf{c}_1'(t)\,dt\\&=\int_0^x(F_1(t,0),F_2(t,0))\cdot(1,0)\,dt=\int_0^x F_1(t,0)\,dt.\end{aligned}$$

Parametrize $\mathbf{c}_2$ by $\mathbf{c}_2(t)=(x,t)$, $t\in[0,y]$. Then $\mathbf{c}_2'(t)=(0,1)$ and

$$\begin{aligned}\int_{\mathbf{c}_2}\mathbf{F}\cdot d\mathbf{s}&=\int_0^y\mathbf{F}(\mathbf{c}_2(t))\cdot\mathbf{c}_2'(t)\,dt\\&=\int_0^y\big(F_1(x,t),F_2(x,t)\big)\cdot(0,1)\,dt=\int_0^y F_2(x,t)\,dt.\end{aligned}$$

Hence,

$$f(x,y)=\int_0^x F_1(t,0)\,dt+\int_0^y F_2(x,t)\,dt.$$

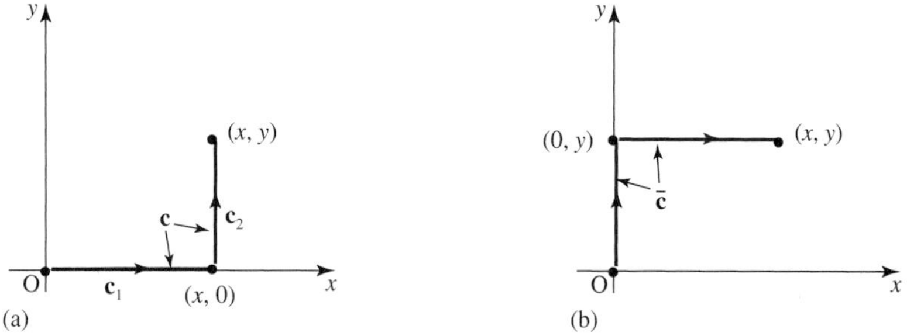

Figure 5.36 Paths of integration in the proof of Theorem 5.7.

We have to check that $\nabla f = \mathbf{F}$. The first integral depends on x only, and the second one depends on x and y. Therefore,

$$\frac{\partial f}{\partial y}(x, y) = 0 + \frac{\partial}{\partial y}\left(\int_0^y F_2(x, t)\,dt\right) = F_2(x, y),$$

where the right side was computed using the Fundamental Theorem of Calculus (5.9) with $f(t) = F_2(x, t)$.

Choosing a path $\bar{\mathbf{c}}$ joining (0, 0) and (x, y) shown in Figure 5.36(b), we get

$$f(x, y) = \int_0^y F_2(0, t)\,dt + \int_0^x F_1(t, y)\,dt. \tag{5.18}$$

Differentiating (5.18) with respect to x, we get $\partial f/\partial x(x, y) = F_1(x, y)$. Thus,we conclude that $\nabla f(x, y) = \mathbf{F}(x, y)$.

We omitted one technical point in the proof: what if the paths that we considered do not belong to U?

Figure 5.37(a) shows one such case: neither of the paths (call them $\mathbf{c}$ and $\bar{\mathbf{c}}$) belongs to U. In the proof, we learned how to express f in terms of integrals with respect to horizontal and vertical line segments. In the general case, all we have to do is to consider polygonal paths joining (0, 0) and (x, y) with the property that all of their segments are either horizontal or vertical, as in Figure 5.37(b).

We have proven a number of useful results in this section. Let us put them all together.

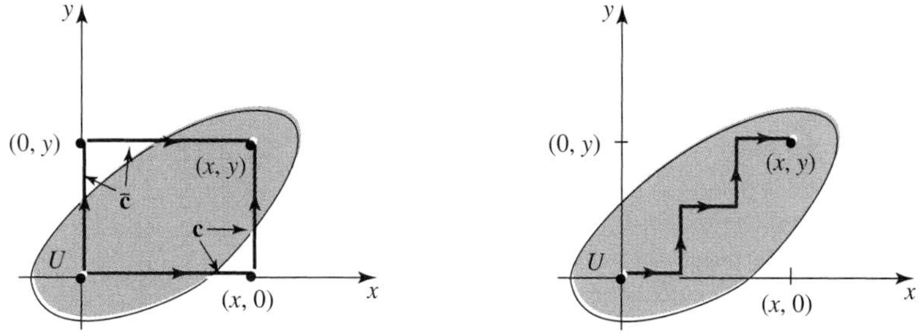

(a) Paths $\mathbf{c}$ and $\bar{\mathbf{c}}$ (b) Polygonal path from (0, 0) to (x, y)

Figure 5.37 A case not considered in the proof of Theorem 5.7.

THEOREM 5.8 Properties of a Gradient Vector Field

Let $\mathbf{F}$ be a C^1 vector field defined on an open, connected set $U \subseteq \mathbb{R}^2$ (or $\mathbb{R}^3$). The following statements are equivalent:

(a) $\mathbf{F}$ is a gradient vector field; in other words, there exists a differentiable function $f: U \subseteq \mathbb{R}^2$ (or $\mathbb{R}^3$) $\to \mathbb{R}$ such that $\mathbf{F} = \nabla f$.

(b) For any oriented, simple closed curve $\mathbf{c}$,

$$\int_{\mathbf{c}} \mathbf{F} \cdot d\mathbf{s} = 0.$$

(c) $\mathbf{F}$ is path-independent: for any two oriented, simple curves $\mathbf{c}_1$ and $\mathbf{c}_2$ having the same initial and terminal points,

$$\int_{\mathbf{c}_1} \mathbf{F} \cdot d\mathbf{s} = \int_{\mathbf{c}_2} \mathbf{F} \cdot d\mathbf{s}.$$

Any of the three statements (a), (b), or (c) implies the following:

(d) Scalar curl of $\mathbf{F}$ is zero (if $\mathbf{F}: U \subseteq \mathbb{R}^2 \to \mathbb{R}^2$), or *curl* $\mathbf{F} = \mathbf{0}$ (if $\mathbf{F}: U \subseteq \mathbb{R}^3 \to \mathbb{R}^3$).

If, in addition, U is simply-connected, then (d) implies (a), (b), and (c) as well. ◀

PROOF: The implications (a) $\Rightarrow$ (b) and (b) $\Rightarrow$ (c) are contained in Theorem 5.6. The statement (c) $\Rightarrow$ (a) is proven in Theorem 5.7.

The implication (a) $\Rightarrow$ (d) is the trivial part of Theorem 5.4 (which does not require the assumption of simple-connectedness). The fact that the reverse implication (d) $\Rightarrow$ (a) is true was proven in Theorem 5.4. ◀

▶ EXAMPLE 5.32

Compute $\int_{\mathbf{c}} \mathbf{F} \cdot d\mathbf{s}$ if $\mathbf{F} = (-y/(x^2+y^2), x/(x^2+y^2))$ and $\mathbf{c}$ is the unit circle $\mathbf{c}(t) = (\cos t, \sin t)$, $t \in [0, 2\pi]$.

SOLUTION

Since $\mathbf{F}(\mathbf{c}(t)) = (-\sin t, \cos t)$ and $\mathbf{c}'(t) = (-\sin t, \cos t)$, it follows that

$$\int_{\mathbf{c}} \mathbf{F} \cdot d\mathbf{s} = \int_0^{2\pi} \mathbf{F}(\mathbf{c}(t)) \cdot \mathbf{c}'(t)\, dt = \int_0^{2\pi} 1\, dt = 2\pi.$$

Since the integral is not zero, we conclude that $\mathbf{F}$ is not path-independent. According to Theorem 5.8 [$\mathbf{F}$ is defined on $U = \mathbb{R}^2 - \{(0,0)\}$, which is open and connected], $\mathbf{F}$ cannot be a gradient vector field. But notice that

$$\frac{\partial F_2}{\partial x} = \frac{y^2 - x^2}{(x^2+y^2)^2} = \frac{\partial F_1}{\partial y},$$

that is, the scalar curl of $\mathbf{F}$ is zero!

However, this is not a contradiction: the vector field $\mathbf{F}$ is C^1 on U, but U is not *simply-connected*—and therefore, the equivalence of statements (b) and (d) in Theorem 5.8 cannot be applied.

Note that if $f(x, y) = \arctan(y/x)$, then

$$\nabla f = \nabla \arctan(y/x) = \frac{1}{1+(y/x)^2}\left(-\frac{y}{x^2}, \frac{1}{x}\right) = \left(-\frac{y}{x^2+y^2}, \frac{x}{x^2+y^2}\right).$$

This fact does not violate our conclusion that **F** is not a gradient vector field. For **F** to be a gradient vector field, we need $\mathbf{F} = \nabla f$ on $U = \mathbb{R}^2 - \{(0, 0)\}$. However, f is defined on the set $\{(x, y)|x \neq 0\}$, which is a proper subset of U. ◀

▶ EXAMPLE 5.33

Compute $\int_{\mathbf{c}} \mathbf{F} \cdot d\mathbf{s}$ if $\mathbf{F} = (x/(x^2 + y^2), y/(x^2 + y^2))$ and **c** is the unit circle $\mathbf{c}(t) = (\cos t, \sin t)$, $t \in [0, 2\pi]$.

SOLUTION Since $\mathbf{F}(\mathbf{c}(t)) = (\cos t, \sin t)$, we get

$$\int_{\mathbf{c}} \mathbf{F} \cdot d\mathbf{s} = \int_0^{2\pi} \mathbf{F}(\mathbf{c}(t)) \cdot \mathbf{c}'(t)\, dt = \int_0^{2\pi} (\cos t, \sin t) \cdot (-\sin t, \cos t)\, dt = \int_0^{2\pi} 0\, dt = 0.$$

Note that we did not have to calculate the path integral. Since **F** points radially (away from the origin), it is perpendicular to **c** at all points on **c**. Thus, $\int_{\mathbf{c}} \mathbf{F} \cdot d\mathbf{s} = 0$.

The above calculation does not suffice to prove the path-independence of **F**. To prove it, we would have to show that $\int_{\mathbf{c}} \mathbf{F} \cdot d\mathbf{s} = 0$ for *every* oriented, simple closed curve.

We can easily check that the scalar curl of **F** is zero. However, this does not imply that **F** is path-independent [implication (d) ⇒ (a) in Theorem 5.8] because the domain of **F** is $\mathbb{R}^2 - \{(0, 0)\}$, which is not a simply-connected set.

A straightforward calculation shows that $\mathbf{F} = \nabla \left(\frac{1}{2} \ln (x^2 + y^2)\right)$, that is, **F** is a gradient vector field on $U = \mathbb{R}^2 - \{(0, 0)\}$. Thus, it is path-independent by implication (a) ⇒ (c) in Theorem 5.8. (Note that U is open and connected, so Theorem 5.8 applies.) ◀

▶ EXAMPLE 5.34

The curl of the vector field $\mathbf{F} = (x, y, z)/(x^2 + y^2 + z^2)^{3/2}$ is zero (see Exercise 26).

The domain of **F** is the "punctured" space $U = \mathbb{R}^3 - \{(0, 0, 0)\}$. It is a simply-connected space and thus [using the implication (d) ⇒ (a) in Theorem 5.8] we conclude that **F** is a gradient vector field. In fact, $\mathbf{F} = \nabla f$, where $f(x, y, z) = -(x^2 + y^2 + z^2)^{-1/2}$ (see Exercise 26). Since U is also connected, it follows that [by the implication (a) ⇒ (c) in Theorem 5.8] **F** is a path-independent vector field. ◀

▶ EXAMPLE 5.35

Consider the vector field $\mathbf{F}(x, y, z) = (-y/(x^2 + y^2), x/(x^2 + y^2), z)$. The domain of **F** is the set $U = \{(x, y, z)|x \neq 0 \text{ and/or } y \neq 0\} = \mathbb{R}^3 - \{z\text{-axis}\}$.

It can be easily checked that *curl* $\mathbf{F} = \mathbf{0}$. This fact does not imply that **F** is a gradient vector field, since the domain of **F** is not a simply-connected set. The path integral of **F** around the circular path $\mathbf{c}(t) = (\cos t, \sin t, 0)$, $t \in [0, 2\pi]$, is

$$\begin{aligned}\int_{\mathbf{c}} \mathbf{F} \cdot d\mathbf{s} &= \int_0^{2\pi} \mathbf{F}(\mathbf{c}(t)) \cdot \mathbf{c}'(t)\, dt \\ &= \int_0^{2\pi} (-\sin t, \cos t, 0) \cdot (-\sin t, \cos t, 0)\, dt = \int_0^{2\pi} 1\, dt = 2\pi.\end{aligned}$$

Thus, **F** is not path-independent, and using Theorem 5.8 we conclude that it is not a gradient vector field.

Note that the relation $\nabla(\arctan (y/x) + z^2/2) = \mathbf{F}$ holds for all (x, y, z) such that $x \neq 0$ (i.e., it holds on $\mathbb{R}^3$ with the yz-plane removed). Since this set is not the domain of **F**, the above relation does not contradict the fact that **F** is not a gradient vector field. ◀

Next, we restate Theorem 5.8 in the context of conservative force fields. Recall that a force field is called conservative if $\mathbf{F} = -\nabla V$ (V is a potential function), and keep in mind that the path integral of $\mathbf{F}$ represents the work done by $\mathbf{F}$.

THEOREM 5.9 Conservative Force Fields

Let $\mathbf{F}$ be a C^1 force field defined on an open, connected set $U \subseteq \mathbb{R}^2$ (or $\mathbb{R}^3$). The following statements are equivalent:

(a) $\mathbf{F}$ is conservative.

(b) The work of the force $\mathbf{F}$ along any oriented, simple closed curve is zero (i.e., mechanical energy is conserved).

(c) The work of $\mathbf{F}$ does not depend on the path, but only on its endpoints.

Any of the three statements (a), (b), or (c) implies the following:

(d) $\mathbf{F}$ is irrotational; that is, the scalar curl of $\mathbf{F}$ is zero (if $\mathbf{F}\colon U \subseteq \mathbb{R}^2 \to \mathbb{R}^2$), or $\mathit{curl}\,\mathbf{F} = \mathbf{0}$ (if $\mathbf{F}\colon U \subseteq \mathbb{R}^3 \to \mathbb{R}^3$).

If, in addition, U is simply-connected, then (d) implies (a), (b), and (c) as well. ◀

Recall that the total energy of a particle moving under the influence of a conservative vector field is conserved (i.e., remains constant). (Most important examples of conservative vector fields are gravitational and electrostatoic fields.) Part (c) of Theorem 5.9 states that conservative vector fields are independent of path. For instance, the work done by gravity is the same for all paths that lead from the same point at the foot of the hill to its top.

Note that both gravitational and electrostatic fields have the same domain, $U = \mathbb{R}^3 - \{(0, 0, 0)\}$. Since U is a simply-connected set, all statements (a)–(d) of Theorem 5.9 hold for these fields.

▶ EXAMPLE 5.36

A particle of mass m moves in the xy-plane subject to the force field $\mathbf{F} = -C(x(1 + y^2)\mathbf{i} + x^2y\mathbf{j})$, where C is a constant.

(a) Show that $\mathbf{F}$ is conservative.

(b) Find the potential energy $V(x, 0)$ if $V(0, 0) = 0$.

(c) Find the potential energy $V(x, y)$ at a general position (x, y).

SOLUTION The domain $U = \mathbb{R}^2$ of $\mathbf{F}$ is simply-connected.

(a) A vector field $\mathbf{F} = F_1\mathbf{i} + F_2\mathbf{j}$ is conservative if and only if $\mathit{curl}\,\mathbf{F} = \mathbf{0}$, if and only if $\partial F_2/\partial x = \partial F_1/\partial y$. In our case, $F_1 = -Cx(1 + y^2)$ and $F_2 = -Cx^2y$, and hence,

$$\frac{\partial F_2}{\partial x} = -Cy \cdot 2x, \qquad \frac{\partial F_1}{\partial y} = -Cx \cdot 2y.$$

(b) Part (a) implies that there exists a potential function (i.e., a scalar function V such that $\mathbf{F} = -\nabla V$) and that the integral along a curve depends only on its endpoints. Let us compute $\int_{\mathbf{c}} \mathbf{F} \cdot d\mathbf{s}$ along the

straight-line segment (call it $\mathbf{c}$) from $(0,0)$ to $(x,0)$. Parametrize it by $\mathbf{c}(t)=(t,0)$, $t\in[0,x]$. Then

$$\int_{\mathbf{c}} \mathbf{F}\cdot d\mathbf{s} = \int_0^x \mathbf{F}(\mathbf{c}(t))\cdot\mathbf{c}'(t)\,dt = \int_0^x (-Ct)\mathbf{i}\cdot\mathbf{i}\;dt = -C\frac{t^2}{2}\bigg|_0^x = -C\frac{x^2}{2}.$$

The Fundamental Theorem of Calculus (Theorem 5.5) implies that

$$\int_{\mathbf{c}} \mathbf{F}\cdot d\mathbf{s} = -\int_c \nabla V(x,y)\cdot d\mathbf{s} = -V(x,y)\bigg|_{(0,0)}^{(x,0)} = -V(x,0)+V(0,0) = -V(x,0),$$

since $V(0,0)=0$ by assumption. Combining the two expressions, we get $V(x,0)=Cx^2/2$.

(c) This time we have to reach a general position (x,y) from the origin. As in (b), take a straight-line segment joining $(0,0)$ and (x,y), that is, $\mathbf{c}(t)=(tx,ty)$, $t\in[0,1]$. Then

$$\begin{aligned}
\int_{\mathbf{c}} \mathbf{F}\cdot d\mathbf{s} &= \int_0^1 \mathbf{F}(\mathbf{c}(t))\cdot\mathbf{c}'(t)\,dt \\
&= \int_0^1 (-Cxt(1+y^2t^2)\mathbf{i} - Cx^2t^2yt\mathbf{j})\cdot(x\mathbf{i}+y\mathbf{j})\,dt \\
&= \int_0^1 (-Cx^2t - Cx^2y^2t^3 - Cx^2y^2t^3)\,dt \\
&= -Cx^2\frac{t^2}{2} - 2Cx^2y^2\frac{t^4}{4}\bigg|_0^1 = -\frac{Cx^2}{2} - \frac{Cx^2y^2}{2} = -\frac{C}{2}(x^2+x^2y^2).
\end{aligned}$$

By the Fundamental Theorem of Calculus,

$$\int_{\mathbf{c}} \mathbf{F}\cdot d\mathbf{s} = -\int_c \nabla V(x,y)\cdot d\mathbf{s} = -V(x,y)\bigg|_{(0,0)}^{(x,y)} = -V(x,y)+V(0,0) = -V(x,y),$$

[since $V(0,0)=0$], and hence,

$$V(x,y) = \frac{C}{2}(x^2+x^2y^2).$$

Here is another way to compute (c). In part (b), we came to the point $(x,0)$, so let us continue from there to (x,y), that is, walk along the curve $\mathbf{c}(t)=(x,t)$, $t\in[0,y]$. Then

$$\begin{aligned}
\int_{\mathbf{c}} \mathbf{F}\cdot d\mathbf{s} &= \int_0^y \mathbf{F}(\mathbf{c}(t))\cdot\mathbf{c}'(t)\,dt \\
&= \int_0^y \left(-Cx(1+t^2)\mathbf{i} - Cx^2t\mathbf{j}\right)\cdot\mathbf{j}\;dt = -C\int_0^y x^2t\,dt = -C\frac{x^2y^2}{2}.
\end{aligned}$$

As before, combine this result with the Fundamental Theorem of Calculus:

$$\int_{\mathbf{c}} \mathbf{F}\cdot d\mathbf{s} = -\int_{\mathbf{c}} \nabla V(x,y)\cdot d\mathbf{s} = -V(x,y)\bigg|_{(x,0)}^{(x,y)} = -V(x,y)+V(x,0)$$

to get

$$V(x,y) = C\frac{x^2y^2}{2} + V(x,0) = C\frac{x^2y^2}{2} + C\frac{x^2}{2} = \frac{C}{2}(x^2y^2+x^2).$$

Another way to solve parts (b) and (c) would be to compute V from the system of differential equations

$$\frac{\partial V}{\partial x} = Cx(1+y^2) \qquad \text{and} \qquad \frac{\partial V}{\partial y} = Cx^2 y$$

with the initial condition $V(0,0)=0$. Integrating the first equation with respect to x, we get $V(x,y) = Cx^2(1+y^2)/2 + D(y)$, where $D(y)$ depends (possibly) on y. Computing the derivative of $V(x,y)$ with respect to y and combining with $\partial V/\partial y = Cx^2 y$ yield

$$C\frac{x^2}{2}2y + \frac{dD(y)}{dy} = Cx^2 y;$$

hence, $dD(y)/dy = 0$ and $D(y) = D$ (D is a real number). Therefore, $V = Cx^2(1+y^2)/2 + D$. The condition $V(0,0)=0$ implies that $D=0$, and thus, $V = Cx^2(1+y^2)/2$. ◀

▶ EXERCISES 5.4

Exercises 1 to 10: State which of the following sets are connected, simply-connected, and/or star-shaped:

1. $\mathbb{R}^2$, with the circle $x^2+y^2=1$ removed

2. The set $\{(x,y)|y<|x|\}$ in $\mathbb{R}^2$

3. $\mathbb{R}^3$, with the circle $x^2+y^2=1$, $z=1$ removed

4. The set $\{(x,y,z)|x^2+y^2+z^2 \le 1 \text{ and } z \neq 1\}$ in $\mathbb{R}^3$

5. $\mathbb{R}^3$, with the sphere $x^2+y^2+z^2=1$ removed

6. $\mathbb{R}^3$, with the ball $x^2+y^2+z^2 \le 1$ removed

7. $\mathbb{R}^3$, with the helix $\mathbf{c}(t) = (\cos t, \sin t, t)$, $t \in [0,\pi]$ removed

8. The set $\{(x,y)|x^2+y^2<1 \text{ or } x^2+y^2>2\}$ in $\mathbb{R}^2$

9. The region inside the polygonal line joining the points (0, 5), (2, 0), (2, 3), (4, 3), (4, 5), and (0, 5) (in that order)

10. The set $\{(x,y)|x^2-y^2<0\}$ in $\mathbb{R}^2$

11. Explain why the sphere $\{(x,y,z)|x^2+y^2+z^2=1\}$ and the sphere without the "North Pole" $\{(x,y,z)|x^2+y^2+z^2=1 \text{ and } z<1\}$ are simply-connected sets.

12. Let $\mathbf{F} = (F_1, F_2, F_3)$ be a C^1 vector field defined on a star-shaped set U, such that $curl\,\mathbf{F} = \mathbf{0}$. Define the function $f(x,y,z)$ as in the proof of Theorem 5.4.
(a) Show that $f(x,y,z) = \int_0^1 (F_1(tx,ty,tz)x + F_2(tx,ty,tz)y + F_3(tx,ty,tz)z)\,dt$.
(b) Derive the formula $\partial f/\partial x = \int_0^1 A\,dt$, where

$$A = F_1(tx,ty,tz) + D_1F_1(tx,ty,tz)tx + D_2F_1(tx,ty,tz)ty + D_3F_1(tx,ty,tz)tz.$$

(c) Show that the integrand in (b) is equal to $(\partial/\partial t)(tF_1(tx,ty,tz))$ and conclude that $\partial f/\partial x = F_1$.

13. Compute $\int_{\mathbf{c}} \mathbf{F}\cdot d\mathbf{s}$, where $\mathbf{F} = y^2\cos x\,\mathbf{i} + 2y\sin x\,\mathbf{j}$, and $\mathbf{c}$ is any path starting at (1, 1) and ending at (1, 3).

14. Compute the path integral $\int_{\mathbf{c}} \mathbf{F}\cdot d\mathbf{s}$, where $\mathbf{F} = (\cos(xy) - xy\sin(xy))\mathbf{i} - x^2\sin(xy)\mathbf{j}$, and $\mathbf{c}(t) = (e^t\cos t, e^t\sin t)$, $0 \le t \le \pi$.

15. Check that the vector field $\mathbf{F} = (-y/(x^2+y^2), x/(x^2+y^2), 1)$ satisfies $curl\,\mathbf{F} = \mathbf{0}$ in $\mathbb{R}^3 - \{(0,0,0)\}$, but is not conservative in $\mathbb{R}^3$. In order to show that $\mathbf{F}$ is not conservative, compute path integrals of $\mathbf{F}$ along the curves $\mathbf{c}_1(t) = (\cos t, \sin t, 0)$, $0 \le t \le \pi$, and $\mathbf{c}_2(t) = (\cos t, -\sin t, 0)$, $0 \le t \le \pi$, joining $(1, 0)$ and $(-1, 0)$. Explain why this does not contradict Theorem 5.8.

16. Let $\mathbf{F}(x, y) = y\mathbf{i} + x\mathbf{j}$.

(a) Compute $\int_{\mathbf{c}} \mathbf{F} \cdot d\mathbf{s}$ along the circular path from $(1, 0)$ counterclockwise to $(0, -1)$, then along the y-axis from $(0, -1)$ to $(0, 2)$, and then along the straight line from $(0, 2)$ to $(1, 0)$.

(b) Show that $\mathbf{F}(x, y)$ is a gradient vector field and use this fact to check your answer in (a).

Exercises 17 to 21: Determine whether $\mathbf{F}$ is a gradient vector field, and if so, specify its domain U and find all functions f such that $\mathbf{F} = \nabla f$.

17. $\mathbf{F} = (4x^2 - 4y^2 + x)\mathbf{i} + (7xy + \ln y)\mathbf{j}$

18. $\mathbf{F} = (3x^2 \ln x + x^2)\mathbf{i} + x^3 y^{-1}\mathbf{j}$

19. $\mathbf{F} = 2x \ln y\mathbf{i} + (2y + x^2/y)\mathbf{j}$

20. $\mathbf{F} = (yz + e^x \sin z)\mathbf{i} + (xz + y^2 - e^y)\mathbf{j} + (xy + e^x \cos z)\mathbf{k}$

21. $\mathbf{F} = y \cos(xy)\mathbf{i} + (x \cos(xy) - z \sin y)\mathbf{j} + \cos y\mathbf{k}$

Exercises 22 to 24: Evaluate the following integrals:

22. $\displaystyle\int_{(0,1,0)}^{(3,3,1)} (4xy - 2xy^2z^2)\,dx + (2x^2 - 2x^2yz^2)\,dy - 2x^2y^2z\,dz$

23. $\displaystyle\int_{(0,0,0)}^{(\pi,\pi/2,\pi/3)} \cos x \tan z\,dx + dy + \sin x \sec^2 z\,dz$

24. $\displaystyle\int_{(1,2,1)}^{(2,2,2)} x^{-2}dx + z^{-1}dy + yz^{-2}\,dz$

25. Provide omitted detail in the proof of Theorem 5.7: parametrize the path $\bar{\mathbf{c}}$ in Figure 5.36(b) to obtain the formula (5.18). Then show that $\partial f/\partial x(x, y) = F_1(x, y)$.

26. Let $\mathbf{F}(x, y, z) = (x, y, z)/(x^2 + y^2 + z^2)^{3/2}$. Show that $curl\,\mathbf{F} = \mathbf{0}$. Show that $\nabla(x^2 + y^2 + z^2)^{-1/2} = -\mathbf{F}$ (see Example 2.40 in Section 2.4).

27. Consider the vector field $\mathbf{F}(x, y) = -(1 + x)ye^x\mathbf{i} - xe^x\mathbf{j}$.

(a) Show that $\mathbf{F}(x, y)$ is conservative.

(b) Using the Fundamental Theorem of Calculus, find the potential energy $V(x, 0)$ along the x-axis if $V(0, 0) = 0$.

(c) Using the Fundamental Theorem of Calculus, find the potential energy $V(0, y)$ along the y-axis if $V(0, 0) = 0$.

28. Let $\mathbf{F}(x, y, z) = x^3y\mathbf{i} + z^2\mathbf{k}$. Does there exist a function f such that $\mathbf{F} = \nabla f$?

29. Find $\int_{\mathbf{c}} \mathbf{F} \cdot d\mathbf{s}$, where $\mathbf{F}(x, y) = 2xye^y\mathbf{i} + x^2e^y(1 + y)\mathbf{j}$ and $\mathbf{c}$ is the straight-line segment from $(0, 0)$ to $(3, -2)$,

(a) Using a parametrization for $\mathbf{c}$.

(b) Using the fact that $curl\,\mathbf{F} = \mathbf{0}$.

30. Check that $\mathbf{F}(x, y, z) = 2xy^2\mathbf{i} + (2x^2y + e^z)\mathbf{j} + ye^z\mathbf{k}$ is a conservative force field in $\mathbb{R}^3$. Find the work done by $\mathbf{F}$ on an object that moves from $(0, 2, -1)$ to $(3, 2, 0)$.

31. Show that the vector field $\mathbf{F}(\mathbf{r}) = \|\mathbf{r}\|^2\mathbf{r}$ is a gradient vector field ($\mathbf{r} = x\mathbf{i} + y\mathbf{j} + z\mathbf{k}$), and find a function f such that $\nabla f = \mathbf{F}$.

32. Show that the vector field $\mathbf{F}(\mathbf{r}) = \|\mathbf{r}\|\mathbf{r}$ is a gradient vector field ($\mathbf{r} = x\mathbf{i} + y\mathbf{j} + z\mathbf{k}$), and find a function f such that $\nabla f = \mathbf{F}$.

33. Compute $\int_{\mathbf{c}} \mathbf{F} \cdot d\mathbf{s}$, where $\mathbf{F} = (\ln(x + y^2) + x/(x + y^2))\mathbf{i} + (2xy/(x + y^2))\mathbf{j}$, and $\mathbf{c}$ is the part of the curve $y = x^3$ from $(1, 1)$ to $(2, 8)$. Use the fact (check it!) that $\mathbf{F}$ is a gradient vector field.

▶ CHAPTER REVIEW

CHAPTER SUMMARY

- **Paths and curves.** C^1 path and piecewise C^1 path, reparametrization, orientation-preserving and orientation-reversing reparametrization, simple curve, simple closed curve, orientation of a curve.
- **Path integral of a real-valued function.** Definition, independence on parametrization, computing path integrals.
- **Path integral of a vector field.** Definition, work due to a force, dependence on parametrization, techniques of computation of path integrals, circulation of a vector field and its relation to curl.
- **Independence on path.** Gradient vector field, path-independent vector fields, generalization of the Fundamental Theorem of Calculus, properties of gradient vector fields and conservative fields.

REVIEW

Discuss the following questions.

1. Define continuous, C^1, and piecewise C^1 paths. Is it possible to parametrize the same curve using a continuous path, a C^1 path, and a piecewise C^1 path? If possible, find an example.

2. Define a reparametrization of a path and describe the ways of checking whether it is orientation-preserving or orientation-reversing. Explain to what extent the path integral of a vector field depends on reparametrization.

3. Define the circulation of a vector field and explain its physical significance. Draw an example of a vector field that has nonzero circulation at more than one point.

4. Let $\mathbf{F}$ be an inverse-square law force field and let $\mathbf{c}$ be any curve lying on the sphere $x^2 + y^2 + z^2 = a^2$. Explain why $\mathbf{F}$ does zero work in moving the particle along $\mathbf{c}$.

5. Suppose that the path integral of a C^1 vector field $\mathbf{F}$ is zero for *all* paths connecting two points A and B in space. Is $\mathbf{F}$ a gradient vector field?

6. Define and give examples of connected, simply-connected, and star-shaped sets in $\mathbb{R}^2$ and $\mathbb{R}^3$. Find an example of a set in $\mathbb{R}^3$ that is not star-shaped. Is every star-shaped set simply-connected?

7. Give a definition of a conservative vector field. If $\mathbf{F}_1$ and $\mathbf{F}_2$ are conservative vector fields, is their sum a conservative vector field? If it is, find its potential function. Discuss the same question for the cross product $\mathbf{F}_1 \times \mathbf{F}_2$.

8. List the equivalent properties of a gradient vector field. State the assumptions on the domain of the field needed for the statements to hold.

TRUE/FALSE QUIZ

Determine whether the following statements are true or false. Give reasons for your answer.

1. If the gradient of a function f is perpendicular (at all points) to a curve joining the points P and Q, then $f(P) = f(Q)$.

2. A constant vector field in $\mathbb{R}^2$ [i.e., $\mathbf{F}(\mathbf{x}) = \mathbf{a}$, for all $\mathbf{x}$ in $\mathbb{R}^2$] is conservative.

3. A vector field of constant magnitude is path-independent.

4. $\phi(t) = 3 - e^t$ in an orientation-reversing reparametrization.

5. The circulation of a conservative vector field in $\mathbb{R}^2$ is zero.

6. If $\mathbf{c}$ is the circle $x^2 + y^2 = 1$, then $\int_{\mathbf{c}} 2\, ds = 2\pi$.

7. The work done by the force $\mathbf{F}(x, y) = y\mathbf{i}$ along any vertical path in $\mathbb{R}^2$ is zero.

8. If $\int_{\mathbf{c}} \mathbf{F} \cdot d\mathbf{s} = 0$ for all circles centered at the origin, then $\mathbf{F}$ is path-independent.

9. If two vector fields $\mathbf{F}$ and $\mathbf{G}$ have the same circulation along a given closed curve, then $\mathbf{F} = \mathbf{G}$.

10. $\mathbf{F}(x, y) = 2xy\mathbf{i} + (x^2 - y^2)\mathbf{j}$ is a gradient vector field.

11. A gradient vector field defined on all of $\mathbb{R}^2$ is path-independent.

12. If vector fields $\mathbf{F}$ and $\mathbf{G}$ are path-independent, then the vector field $\mathbf{H} = 3\mathbf{F} - 2\mathbf{G}$ is also path-independent.

REVIEW EXERCISES AND PROBLEMS

1. Let $\mathbf{c}(t) = (t^{3/2}, t)$, $t \in [0, 1]$. Find an orientation-preserving reparametrization ϕ of $\mathbf{c}$ such that $\mathbf{c}(\phi(t))$ has constant speed.

2. In this exercise, we investigate an estimate for a path integral.

(a) Assume that $|f(x, y)| \leq M$ for all points on the curve $\mathbf{c}(t)$ in $\mathbb{R}^2$, $t \in [a, b]$. Show that $\left|\int_{\mathbf{c}} f\, ds\right| \leq M\ell$, where ℓ is the length of $\mathbf{c}$.

(b) Consider $f(x, y) = x^2 + y^2$ and $\mathbf{c}(t) = (\cos t, \sin t)$, $t \in [0, 4\pi]$. Show that $\int_{\mathbf{c}} f\, ds = 4\pi$. Clearly, $|f(x, y)| = |x^2 + y^2| \leq 1 = M$ for all points on $\mathbf{c}$; furthermore, $\ell = 2\pi$, being the length of the circle. Hence, $M\ell = 2\pi$, and the inequality in (a) seems to be false. What is wrong with this argument?

3. What is the value of $\int_{\mathbf{c}} \mathbf{T} \cdot d\mathbf{s}$, where $\mathbf{T}$ is the unit tangent vector field to $\mathbf{c}$? What is the value of $\int_{\mathbf{c}} \nabla f \cdot d\mathbf{s}$ if $\mathbf{c}$ is a contour curve of f?

4. Consider the vector field $\mathbf{F}(x, y) = y\mathbf{i}$. Find paths $\mathbf{c}_1$, $\mathbf{c}_2$, and $\mathbf{c}_3$ for which $\int_{\mathbf{c}_1} \mathbf{F} \cdot d\mathbf{s} > 0$, $\int_{\mathbf{c}_2} \mathbf{F} \cdot d\mathbf{s} = 0$, and $\int_{\mathbf{c}_3} \mathbf{F} \cdot d\mathbf{s} < 0$. Answer the same question for the vector field $\mathbf{F}(x, y) = x\mathbf{i} + y\mathbf{j}$.

5. Show that the vector field $\mathbf{F}(x, y) = (3x^2 + 8xy)\mathbf{i} + 4x^2\mathbf{j}$ is a gradient vector field. Find f such that $\nabla f = \mathbf{F}$.

6. Find the line integral of $\mathbf{F}(x, y) = (x^2 + y)\mathbf{i} - 3xy^2\mathbf{j}$ counterclockwise around the circumference of the square with vertices $(0, 0)$, $(2, 0)$, $(2, 2)$, and $(0, 2)$.

7. Find the change in f as (x, y) moves from $(1, 2)$ to $(-2, 1)$, if $\nabla f(x, y) = x^2\mathbf{i} - y^3\mathbf{j}$.

8. Let f be a C^1 function and let $\mathbf{c}$ be a smooth curve with endpoints (x_0, y_0) and (x_1, y_1). Show that $\int_{\mathbf{c}}(f_x dx + f_y dy) = f(x_1, y_1) - f(x_0, y_0)$.

9. A radial force acting on a particle is given by $\mathbf{F} = a\,\|\mathbf{r}\|\,\mathbf{r}$, where a is a constant and $\mathbf{r} = x\mathbf{i} + y\mathbf{j} + z\mathbf{k}$. The potential of a conservative force is defined to be the work done by the force on a particle as it moves from $(0, 0, 0)$ to (x, y, z). Assuming that its value at the origin is zero, find the potential of the radial force $\mathbf{F}$.

10. Compute $\int_{\mathbf{c}} f\, ds$ if $f(x, y, z) = (x + yz)/(x^2 + y^2 + z^2)$ and $\mathbf{c}$ is the straight-line segment joining $(1, 1, 1)$ and (a, a, a), where $a \neq 1$. Describe what happens as $a \to 0$ and as $a \to \infty$.

11. Let $\mathbf{F} = -\nabla f$ (i.e., $\mathbf{F}$ is a conservative force field) and assume that a particle of mass m moves along the curve $\mathbf{c}(t)$ in this field. Show that the quantity $f(\mathbf{c}(t)) + m\|\mathbf{c}'(t)\|^2/2$ is constant in time and give a physical interpretation.

12. Consider the gravitational force field $\mathbf{F}(\mathbf{r}) = -\mathbf{r}/\|\mathbf{r}\|^3$ (where $\mathbf{r} = x\mathbf{i} + y\mathbf{j} + z\mathbf{k}$) defined on $\mathbb{R}^3 - \{(0, 0, 0)\}$. Find the work W done by $\mathbf{F}$ as a particle moves from $A(x_A, y_A, z_A)$ to $B(x_B, y_B, z_B)$.

Identify all points in $\mathbb{R}^3 - \{(0, 0, 0)\}$ to which the particle can move from A so that the work done by $\mathbf{F}$ is W.

13. A satellite orbits the earth at height h_1. Find the work done by the gravity in moving the satellite to a new orbit of height $h_2 > h_1$.

14. Assume that a vector field $\mathbf{F}$ is path-independent. Determine whether the vector field $\mathbf{H} = f\mathbf{F}$, where f is a differentiable function, is also path-independent.

CHAPTER 6

Double and Triple Integrals

In order to compute an integral along a curve **c** in $\mathbb{R}^2$ or $\mathbb{R}^3$, we have to find a suitable parametrization of **c** first. Then, with the help of that analytic description of the curve, we reduce the integral along **c** to the definite integral over an interval of real numbers.

Now we move one dimension higher: our goal is to define integrals over surfaces in space. The construction proceeds in the same way as for curves: a surface will be described in analytic terms (i.e., will be "parametrized"). A surface integral will then be defined in terms of a double integral over a region in $\mathbb{R}^2$. That will be done in the next chapter.

In this chapter, we define and study double integrals over various regions in $\mathbb{R}^2$. We start by considering the simplest possible regions in the plane—rectangles. Our approach is certainly not new: the double integral is constructed as the limit of approximating Riemann sums. Next, the definition is extended to more general regions in $\mathbb{R}^2$ (called elementary regions) and the properties of double integrals are studied and illustrated by examples. Two sections are devoted to the evaluation of double integrals, with special attention given to the Change of Variables Theorem.

Taking advantage of the construction of the double integral, we define, in analogous terms, the triple integral over a region in space. Triple integrals will appear in various applications in Section 7.5 and in the Divergence Theorem in Chapter 8.

6.1 DOUBLE INTEGRALS: DEFINITION AND PROPERTIES

In this section, we define the double integral in order to generalize the concept of the definite integral to functions of two variables (the extension to functions of three variables will be discussed in Section 6.5).

The definite integral of a real-valued functon of one variable is defined as a limit of approximating sums, called Riemann sums. For a non-negative function f (f is non-negative if $f \geq 0$), a Riemann sum represents the sum of areas of rectangles that approximate a region under the graph of f. In this case, a definite integral can be interpreted as the area of a region in $\mathbb{R}^2$. As a matter of fact, one could *define* the area of a region as a limit of corresponding Riemann sums; that is, as a definite integral.

One of many reasons why definite integrals are so important is that they allow us to do more complicated integrations: in particular, integrals of real-valued functions of several variables and vector fields along curves in $\mathbb{R}^2$ and $\mathbb{R}^3$ are reduced to definite integrals over intervals on the x-axis.

By analogy with this situation, the double integral of a non-negative function of two variables over a region in $\mathbb{R}^2$ can be interpreted as volume. Moreover, integrals over surfaces in space will be defined in terms of integration in $\mathbb{R}^2$. In the first stage of our construction, we will define double integrals over special regions in $\mathbb{R}^2$ that we now introduce.

DEFINITION 6.1 Rectangle

A *rectangle* (or a *closed rectangle*) $R = [a, b] \times [c, d]$ is the set of all points (x, y) in $\mathbb{R}^2$ such that $a \le x \le b$ and $c \le y \le d$. ◀

Figure 6.1 shows two rectangles, $R_1 = [2, 4] \times [1, 4]$ and $R_2 = [-3, 0] \times [-2, 2]$.

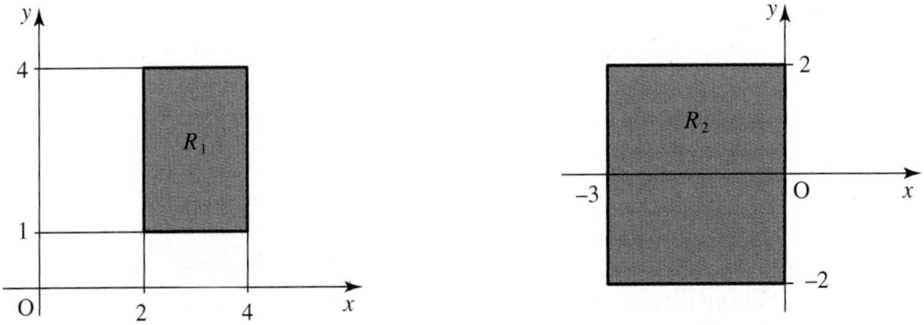

Figure 6.1 Rectangles $R_1 = [2, 4] \times [1, 4]$ and $R_2 = [-3, 0] \times [-2, 2]$ in $\mathbb{R}^2$.

Let $z = f(x, y)$: $U \subseteq \mathbb{R}^2 \to \mathbb{R}$ be a continuous function and let $R = [a, b] \times [c, d]$ be a rectangle contained in its domain U. Assume that $f(x, y) \ge 0$ for all $(x, y) \in R$. The graph of $z = f(x, y)$, the xy-plane, and the four vertical planes $x = a$, $x = b$, $y = c$, and $y = d$ (see Figure 6.2) define a three-dimensional solid W, called the "solid under the surface $z = f(x, y)$ above R." Our aim is to find the volume of W.

Once again, we use the method of constructing Riemann sums, also called the *method of exhaustion*. Formalized by Bernhard Riemann in the 19th century, it has been known and used for well over 2000 years.

Divide the intervals $[a, b]$ and $[c, d]$ into n subintervals

$$[a = x_1, x_2], [x_2, x_3], \dots, [x_n, x_{n+1} = b]$$

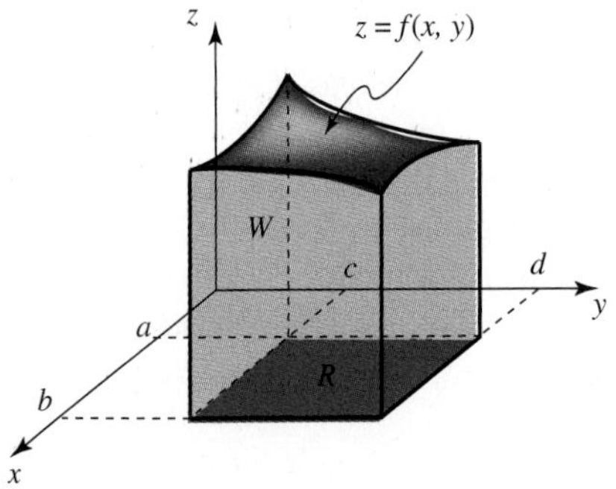

Figure 6.2 Solid under the surface $z = f(x, y)$ above the region R.

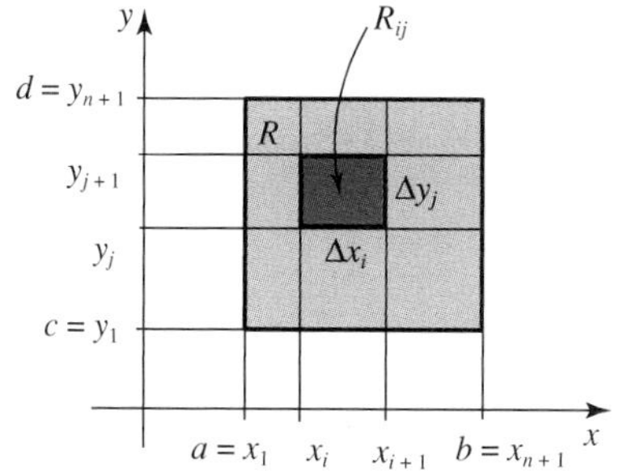

Figure 6.3 Subdivision of R into n^2 rectangles.

and

$$[c = y_1, y_2], [y_2, y_3], \ldots, [y_n, y_{n+1} = d]$$

and form the rectangles $R_{ij} = [x_i, x_{i+1}] \times [y_j, y_{j+1}]$, $i, j = 1, \ldots, n$; see Figure 6.3. We have thus obtained a division of R into n^2 subrectangles R_{ij}. The sides of R_{ij} are $\Delta x_i = x_{i+1} - x_i$ and $\Delta y_j = y_{j+1} - y_j$, and its area is $\Delta A_{ij} = \Delta x_i \Delta y_j$.

Choose a point (x_i^*, y_j^*) in each R_{ij} and build a parallelepiped (rectangular box) over R_{ij} whose height is equal to the value $f(x_i^*, y_j^*)$ of f at (x_i^*, y_j^*). The volume of the parallelepiped, $f(x_i^*, y_j^*)\Delta A_{ij}$, approximates the volume of the three-dimensional region under the surface $z = f(x, y)$ and above R_{ij}; see Figure 6.4. The sum of the volumes of all n^2 parallelepipeds thus obtained [also called the *(double) Riemann sum*]:

$$\mathcal{R}_n = \sum_{i=1}^{n} \sum_{j=1}^{n} f(x_i^*, y_j^*)\Delta A_{ij}$$

approximates the volume of W.

It can be shown that this sequence converges as $n \to \infty$, irrespective of the choices for x_i, y_j and (x_i^*, y_j^*) made in the construction. This is intuitively clear, since as $n \to \infty$, each rectangle shrinks to a point, and there is less and less freedom to choose a point (x_i^*, y_j^*) inside it. The rigorous proof of this fact will not be given here. We have assumed that f is continuous; more general conditions will be spelled out in a moment.

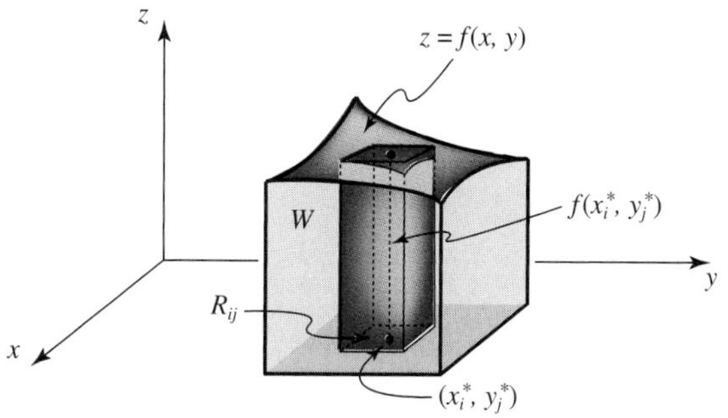

Figure 6.4 Approximating parallelepiped defined by the subrectangle R_{ij}.

We define the double integral of f (recall that $f \geq 0$) over the rectangle R to be

$$\iint_R f\,dA = \lim_{n\to\infty} \mathcal{R}_n,$$

that is, to be equal to the volume $v(W)$ of the solid W.

▶ EXAMPLE 6.1

Approximate the value of $\iint_R (x^2+y^2)\,dA$, where $R = [0,1] \times [-1,2]$, using the double Riemann sum $\mathcal{R}_3$ [take (x_i^*, y_j^*) to be the center of each subrectangle].

SOLUTION Let $f(x,y) = x^2 + y^2$. Form the partition of R into 9 rectangles $R_{11}, R_{12}, \ldots, R_{33}$ using vertical lines $x = 0$, $x = 1/3$, $x = 2/3$, and $x = 1$ and horizontal lines $y = -1$, $y = 0$, $y = 1$, and $y = 2$ as shown in Figure 6.5. The sum $\mathcal{R}_3$ has nine terms, each in the form (value of f at the center of the subrectangle) · (area of the subrectangle). For example, the contribution of $R_{32} = [2/3, 1] \times [0, 1]$ to $\mathcal{R}_3$ is

$$f\left(\frac{5}{6}, \frac{1}{2}\right) \cdot \left(\frac{1}{3}\right) \cdot 1 = \left(\frac{34}{36}\right) \cdot \left(\frac{1}{3}\right) = \frac{34}{108}.$$

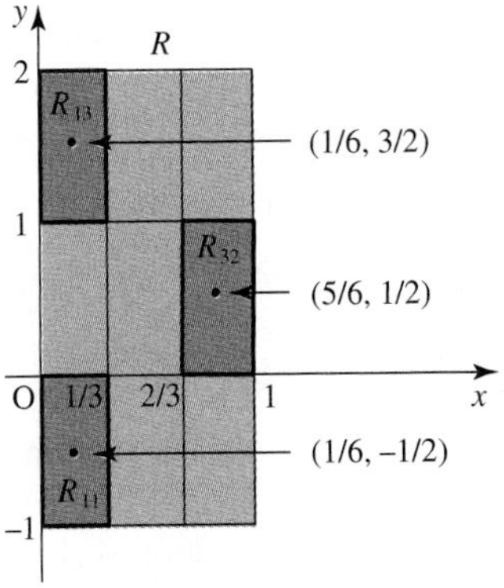

Figure 6.5 Subdivision of R.

Therefore [$\Delta A_{ij} = 1/3$, for all i and j, is the area of R_{ij}],

$$\begin{aligned}\mathcal{R}_3 = {} & f\left(\frac{1}{6}, -\frac{1}{2}\right)\Delta A_{11} + f\left(\frac{1}{6}, \frac{1}{2}\right)\Delta A_{12} + f\left(\frac{1}{6}, \frac{3}{2}\right)\Delta A_{13} \\ & + f\left(\frac{3}{6}, -\frac{1}{2}\right)\Delta A_{21} + f\left(\frac{3}{6}, \frac{1}{2}\right)\Delta A_{22} + f\left(\frac{3}{6}, \frac{3}{2}\right)\Delta A_{23} \\ & + f\left(\frac{5}{6}, -\frac{1}{2}\right)\Delta A_{31} + f\left(\frac{5}{6}, \frac{1}{2}\right)\Delta A_{32} + f\left(\frac{5}{6}, \frac{3}{2}\right)\Delta A_{33}.\end{aligned}$$

It follows that (factor out $\Delta A_{ij} = 1/3$)

$$\begin{aligned}\mathcal{R}_3 = {} & \frac{1}{3}\left(\frac{1}{36} + \frac{1}{4} + \frac{1}{36} + \frac{1}{4} + \frac{1}{36} + \frac{9}{4} + \frac{9}{36} + \frac{1}{4} + \frac{9}{36} + \frac{1}{4} + \frac{9}{36} + \frac{9}{4}\right. \\ & \left. + \frac{25}{36} + \frac{1}{4} + \frac{25}{36} + \frac{1}{4} + \frac{25}{36} + \frac{9}{4}\right) = \frac{1}{3}\left(\frac{105}{36} + \frac{33}{4}\right) = \frac{402}{108} \approx 3.722.\end{aligned}$$

The value 3.722 for $\mathcal{R}_3$ is an approximation of the volume of the solid region W shown in Figure 6.6.

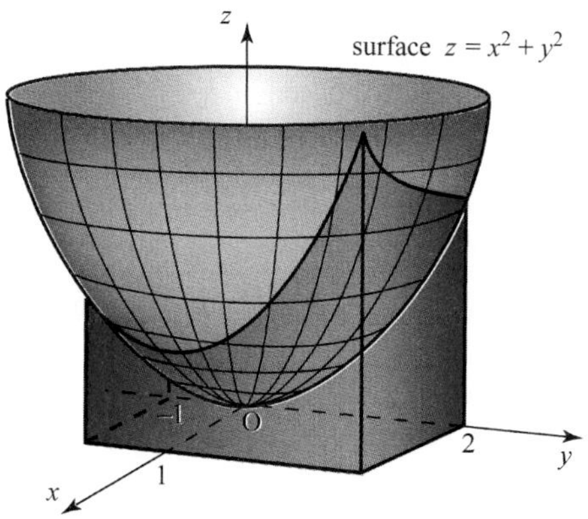

Figure 6.6 Solid W of Example 6.1.

To obtain a better approximation of the integral in Example 6.1, we have to increase the number of subrectangles and repeat the computation (e.g., $\mathcal{R}_4 = 3.844$, $\mathcal{R}_5 = 3.900$, $\mathcal{R}_{10} = 3.975$, $\mathcal{R}_{20} = 3.994$, etc.). The exact value of the volume ($\iint_R (x^2 + y^2)\, dA = 4$) will be computed in Example 6.10 in Section 6.2. Indeed, the sequence of Riemann sums $\mathcal{R}_3, \mathcal{R}_4, \mathcal{R}_5, \mathcal{R}_{10}, \mathcal{R}_{20}$ seems to tend to 4.

In Example 6.1, values for f were calculated at the centers of subrectangles. From the comment preceding this example, we notice that we could have chosen (x_i^*, y_j^*) any way we liked—we would have obtained different approximations, but their limits as $n \to \infty$ are the same.

The construction of the double integral that we have described and illustrated in this section can be carried out in general, for functions that also assume negative values; in such cases, the double integral does not necessarily represent volume.

Recall that a function $f(x, y)$ is *bounded on a set* $R \subseteq \mathbb{R}^2$ if there is a constant $M \geq 0$ such that $|f(x, y)| \leq M$ for all $(x, y) \in R$.

DEFINITION 6.2 Integrable Function and Double Integral

A function $f(x, y)$ defined on a rectangle $R \subseteq \mathbb{R}^2$ is called *integrable on* R if the limit of the sequence of Riemann sums

$$\mathcal{R}_n = \sum_{i=1}^{n} \sum_{j=1}^{n} f(x_i^*, y_j^*) \Delta A_{ij},$$

as $n \to \infty$, exists and does not depend on the way the points (x_i^*, y_j^*) are chosen in each subrectangle R_{ij}.

If f is integrable, then the *double integral* $\iint_R f\, dA$ *of* f *over the rectangle* R is given by

$$\iint_R f\, dA = \lim_{n\to\infty} \mathcal{R}_n = \lim_{n\to\infty} \sum_{i=1}^{n} \sum_{j=1}^{n} f(x_i^*, y_j^*) \Delta A_{ij}.$$

The following theorem probably answers the first question that comes to mind: we know what integrable functions are, but how do we find or identify them?

THEOREM 6.1 Integrable Functions

Let $f\colon R \subseteq \mathbb{R}^2 \to \mathbb{R}$ be a bounded function defined on a rectangle R in $\mathbb{R}^2$. Assume that the set of points where f is not continuous consists of a finite number of continuous curves and/or a finite number of points. Then f is integrable over R. ◀

The definition of a curve actually includes the case of a point [e.g., the image of the path $\mathbf{c}(t) = (a_1, a_2)$, $t \in [0, 1]$, is the curve that consists of the single point (a_1, a_2)], so it was not necessary to mention "points" separately in the theorem.

It can be proved (see an advanced calculus or analysis text) that all continuous functions defined on a closed rectangle R are bounded. Consequently, all continuous functions are integrable over rectangles in $\mathbb{R}^2$. Figure 6.7 shows a discontinuous function that is still integrable on R (the function is not continuous at points that belong to curves $\mathbf{c}$ and $\mathbf{l}$).

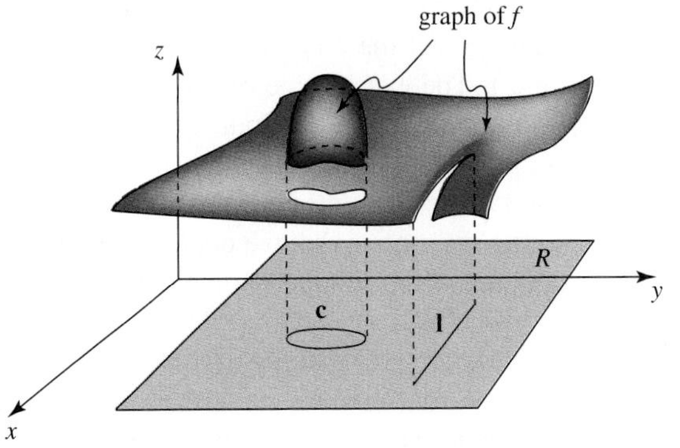

Figure 6.7 Example of an integrable function over $R \subseteq \mathbb{R}^2$.

The proof of Theorem 6.1 is quite long and technically involved and will not be presented here.

Let us repeat that if $f \geq 0$, then the double integral $\iint_R f\,dA$ represents volume (and hence $\iint_R f\,dA \geq 0$). A number of symbols, such as $\int_R f$, $\int_R f\,dA$, $\iint_R f\,dA$, $\iint_R f\,dxdy$, and some others, are used to denote double integrals.

▶ EXAMPLE 6.2

Find $\iint_R 5\,dA$, where $R = [-2, 2] \times [0, 3]$.

SOLUTION

The graph of $f(x, y) = 5$ is the plane parallel to the xy-plane, five units above it. Hence, $\iint_R 5\,dA$ can be interpreted as the volume of the rectangular box (parallelepiped) of height 5 built over the rectangle $[-2, 2] \times [0, 3]$. Therefore, $\iint_R 5\,dA = 4 \cdot 3 \cdot 5 = 60$. ◀

▶ EXAMPLE 6.3

Explain why $\iint_R y^3\,dA = 0$, where $R = [0, 1] \times [-1, 1]$.

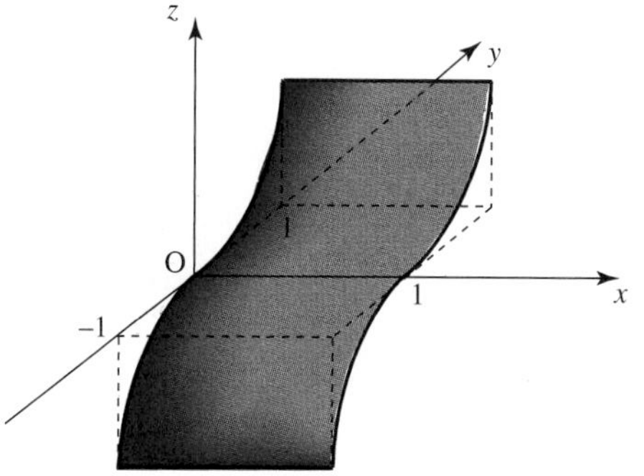

Figure 6.8 The surface $z = y^3$, $0 \le x \le 1$, $-1 \le y \le 1$.

SOLUTION

The graph of $z = y^3$ is a surface obtained by sliding the graph of $z = y^3$ in the yz-plane along the x-axis from 0 to 1; see Figure 6.8. For $y > 0$, it lies above the xy-plane, and for $y < 0$, it lies below the xy-plane, and the two parts are symmetric with respect to the x-axis.

Let us compare the contributions to the Riemann sum $\mathcal{R}_n$ that come from a subrectangle R_{ij} and its symmetric (with respect to the x-axis) counterpart. We know that we are free to choose a point in each subrectangle any way we like—all sequences of Riemann sums will give the same limit, since $z = f(x, y) = y^3$ is continuous and hence integrable. Figure 6.9(a) shows a partition into subrectangles for $\mathcal{R}_5$, and Figure 6.9(b) for $\mathcal{R}_6$.

The contribution of R_{ij} to the Riemann sum is $f(x_i^*, y_j^*)\Delta A_{ij} = (y_j^*)^3 \Delta A_{ij}$, where ΔA_{ij} is the area of R_{ij} and (x_i^*, y_j^*) is a point in R_{ij}. If R_{ij} is halved by the x-axis [that will happen if n is odd, as in Figure 6.9(a)], then take $(x_i^*, y_j^*) = (x_i^*, 0)$, in which case the contribution to the Riemann sum will be $f(x_i^*, 0)\Delta A_{ij} = (0)^3 \Delta A_{ij} = 0$. Therefore, we can drop all such subrectangles from our consideration. To compute the contribution of the symmetric counterpart R'_{ij} of R_{ij} [which is not halved by the x-axis], select the (symmetric) point $(x_i^*, -y_j^*)$ in R'_{ij}; thus, its contribution $f(x_i^*, -y_j^*)\Delta A_{ij} = -(y_j^*)^3 \Delta A_{ij}$ cancels the contribution of R_{ij} [all subrectangles have the same area]. It follows that every Riemann sum thus constructed is zero, and therefore,

$$\iint_R f\,dA = \lim_{n\to\infty} \mathcal{R}_n = \lim_{n\to\infty} 0 = 0.$$

◀

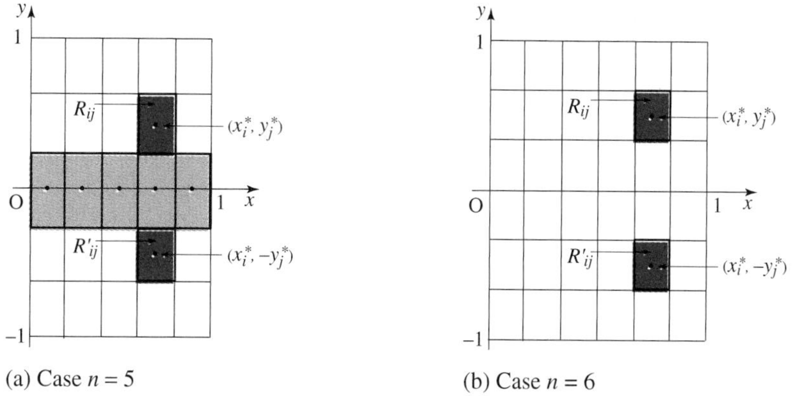

Figure 6.9 Subrectangles for Riemann sums in Example 6.3.

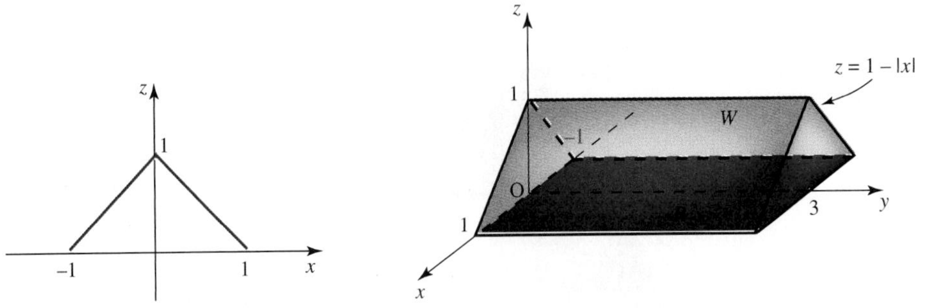

(a) Graph of $z = 1 - |x|, -1 \leq x \leq 1$ (b) Surface $z = 1 - |x|$, over $[-1, 1] \times [0, 3]$

Figure 6.10 Solid W of Example 6.4.

▶ EXAMPLE 6.4

Find $\iint_R f\, dA$, if $f(x, y) = 1 - |x|$ and $R = [-1, 1] \times [0, 3]$.

SOLUTION

Since $z = f(x, y) = 1 - |x| \geq 0$ for $-1 \leq x \leq 1$, the double integral in question represents the volume of the solid W defined over the rectangle $[-1, 1] \times [0, 3]$. W is the prism (triangular solid) obtained by sliding the graph of $z = 1 - |x|$ [see Figure 6.10(a)] along the y-axis from 0 to 3; see Figure 6.10(b). Its volume is $(1/2) \cdot 2 \cdot 1 \cdot 3 = 3$ and hence $\iint_R (1 - |x|)\, dA = 3$. ◀

Assume that $f(x, y) = 1$ at all points that belong to a rectangle $R \subseteq \mathbb{R}^2$. For every n, the Riemann sum

$$\mathcal{R}_n = \sum_{i=1}^{n} \sum_{j=1}^{n} f(x_i^*, y_j^*) \Delta A_{ij} = \sum_{i=1}^{n} \sum_{j=1}^{n} \Delta A_{ij}$$

is equal to the area of R. Taking the limit as $n \to \infty$, we conclude that $\iint_R 1\, dA$ (also denoted by $\iint_R dA$) is equal to the area of R.

THEOREM 6.2 Properties of Double Integrals

Let f and g be integrable functions defined on a rectangle R and let C be a constant. Then

(a) The function $f + g$ is integrable, and

$$\iint_R (f + g)\, dA = \iint_R f\, dA + \iint_R g\, dA.$$

(b) The function Cf is integrable, and

$$\iint_R Cf\, dA = C \iint_R f\, dA.$$

(c) If R is divided into n rectangles R_i $(i = 1, \ldots, n)$ that are mutually disjoint (i.e., their intersection may contain only the bounding line segments), then f is integrable over each R_i and

$$\iint_R f\,dA = \sum_{i=1}^{n} \iint_{R_i} f\,dA.$$

(d) If $f(x, y) \le g(x, y)$ on R, then

$$\iint_R f\,dA \le \iint_R g\,dA.$$

(e) The absolute value of the double integral satisfies

$$\left|\iint_R f\,dA\right| \le \iint_R |f|\,dA.$$

PROOF: The properties (a)–(d) can be understood by visualizing a double integral as the limit of Riemann sums. The Riemann sums for $f + g$ in (a) can be written as

$$\sum_{i=1}^{n}\sum_{j=1}^{n}(f+g)(x_i^*, y_j^*)\Delta A_{ij} = \sum_{i=1}^{n}\sum_{j=1}^{n} f(x_i^*, y_j^*)\Delta A_{ij} + \sum_{i=1}^{n}\sum_{j=1}^{n} g(x_i^*, y_j^*)\Delta A_{ij}.$$

Now take the limit of this equation as $n \to \infty$. The limits on the right side exist (since, by assumption, f and g are integrable), and therefore, the right side is equal to $\iint_R f\,dA + \iint_R g\,dA$. It follows that the limit on the left side exists as well, and is (by definition) equal to the double integral $\iint_R (f + g)\,dA$.

Statement (b) is proven analogously. To prove (d), consider the Riemann sums

$$\sum_{i=1}^{n}\sum_{j=1}^{n} f(x_i^*, y_j^*)\Delta A_{ij} \quad \text{and} \quad \sum_{i=1}^{n}\sum_{j=1}^{n} g(x_i^*, y_j^*)\Delta A_{ij},$$

where the same points $(x_i^*, y_j^*) \in R_{ij}$ are chosen for both sums. Since $f(x_i^*, y_j^*) \le g(x_i^*, y_j^*)$, it follows that

$$\sum_{i=1}^{n}\sum_{j=1}^{n} f(x_i^*, y_j^*)\Delta A_{ij} \le \sum_{i=1}^{n}\sum_{j=1}^{n} g(x_i^*, y_j^*)\Delta A_{ij}$$

holds for all Riemann sums thus constructed. The claim of (d) is now obtained by computing the limit of both sides as $n \to \infty$.

Clearly, we can extend property (d) to the situation where $f \le g \le h$, in which case

$$\iint_R f\,dA \le \iint_R g\,dA \le \iint_R h\,dA.$$

To prove (e), we first recall two statements about absolute values. First, for any real number a, $-|a| \le a \le |a|$. If a and A are two real numbers, $A \ge 0$ and $-A \le a \le A$, then $|a| \le A$. Applying (d) to $-|f(x, y)| \le f(x, y) \le |f(x, y)|$ (that is the first statement with $a = f(x, y)$), we get

$$-\iint_R |f(x, y)|\,dA \le \iint_R f(x, y)\,dA \le \iint_R |f(x, y)|\,dA.$$

Now apply the second statement with $a = \iint_R f(x, y)\,dA$ and $A = \iint_R |f(x, y)|\,dA$ to get

$$\left|\iint_R f(x, y)\,dA\right| \le \iint_R |f(x, y)|\,dA.$$

Although statement (c) may be intuitively clear, its proof is quite involved and will not be discussed here. ◀

▶ EXAMPLE 6.5

Show that $\iint_R x \sin(x + y)\,dA \le 8\pi^3$, where $R = [0, 2\pi] \times [-\pi, \pi]$.

SOLUTION

Since $\sin(x + y) \le 1$ and $0 \le x \le 2\pi$, it follows that $x \sin(x + y) \le 2\pi$. Applying Theorem 6.2(d), we get

$$\iint_R x \sin(x + y)\,dA \le \iint_R 2\pi\,dA.$$

The integral on the right side is the double integral of the constant positive function $f(x, y) = 2\pi$, and represents the volume of the rectangular box of height 2π whose base is the rectangle $[0, 2\pi] \times [-\pi, \pi]$. Hence, $\iint_R 2\pi\,dA = 8\pi^3$, and we are done. ◀

▶ EXAMPLE 6.6

Let

$$f(x, y) = \begin{cases} 2x^2 + 1 & \text{if } x^2 + y^2 < 1 \\ 0 & \text{if } x^2 + y^2 \ge 1 \end{cases}.$$

Is f integrable on the rectangle $R_1 = [0, 1/2] \times [0, 1/2]$? If it is, find an upper bound for the double integral $\iint_{R_1} f(x, y)\,dA$ [i.e., find a number M such that $\iint_{R_1} f(x, y)\,dA \le M$]. Is f integrable on $R_2 = [-1/2, 1/2] \times [-2, 2]$?

SOLUTION

The given function f is equal to 0, except inside the circle $x^2 + y^2 < 1$, where $f(x, y) = 2x^2 + 1$. Since

$$|f(x, y)| = 2x^2 + 1 \le 2(x^2 + y^2) + 1 < 3$$

for (x, y) inside the circle $x^2 + y^2 < 1$, it follows that $|f(x, y)| < 3$ for *all* (x, y); that is, the function f is bounded. The limit of $f(x, y)$ as (x, y) approaches any point on the circle $x^2 + y^2 = 1$ does not exist: an approach from inside the circle gives a number greater than or equal to 1 (since $2x^2 + 1 \ge 1$), but an approach from outside gives 0. It follows that f is not continuous at the points on the circle $x^2 + y^2 = 1$.

Since R_1 is inside the disk $\{x^2 + y^2 < 1\}$, $f(x, y) = 2x^2 + 1$ on R_1. Consequently, f is continuous on R_1 and therefore integrable. From $f(x, y) < 3$, it follows that

$$\iint_{R_1} f(x, y)\,dA \le \iint_{R_1} 3\,dA = 3A(R_1) = \tfrac{3}{4},$$

where $A(R_1) = 1/4$ is the area of R_1. Notice that we can easily get a better estimate: if $(x, y) \in R_1$, then $0 \le x \le 1/2$, and the largest value of $f(x, y) = 2x^2 + 1$ is $2(1/2)^2 + 1 = 3/2$. Therefore,

$$\iint_{R_1} f(x, y)\,dA \le \iint_{R_1} \tfrac{3}{2}\,dA = \tfrac{3}{2}A(R_1) = \tfrac{3}{8}.$$

The function f is continuous on R_2 except along the two curves that are the parts of the circle $x^2 + y^2 = 1$ inside R_2. By Theorem 6.1, f is integrable over R_2.

In the next section, we will define double integrals over more general regions and will extend Theorem 6.2 to such integrals. We will use that opportunity to discuss more examples and illustrate various properties of double integrals.

EXERCISES 6.1

1. Figure 6.11 shows the level curves of a continuous function $f(x, y)$. Using the double Riemann sum $\mathcal{R}_4$ find an estimate for $\iint_R f\,dA$, where $R = [1, 5] \times [0, 2]$. (There are many possible values for $\mathcal{R}_4$.)

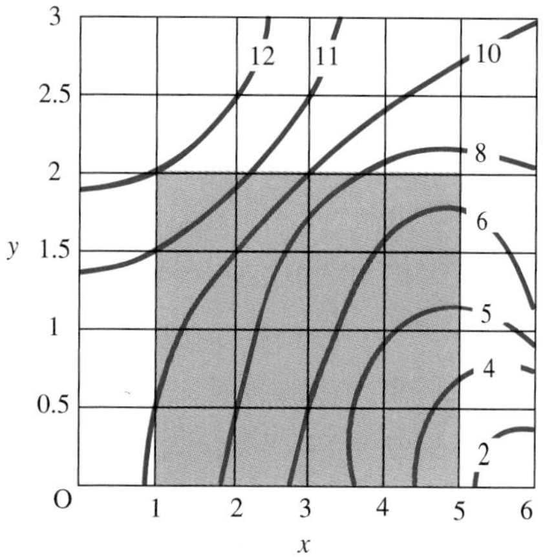

Figure 6.11 Diagram for Exercise 1.

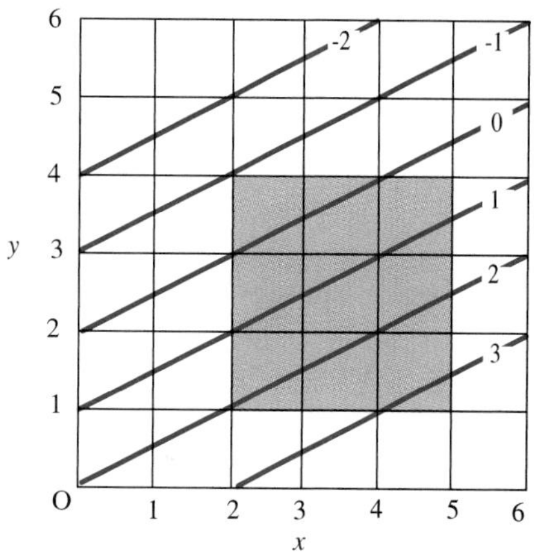

Figure 6.12 Diagram for Exercise 2.

2. Level curves of a linear function $f(x, y)$ are shown in Figure 6.12. Estimate $\iint_R f\,dA$, where $R = [2, 5] \times [1, 4]$, using the double Riemann sum $\mathcal{R}_3$. (There are many possible values for $\mathcal{R}_3$.)

3. Compute the (double) Riemann sum $\mathcal{R}_3$ for the function $f(x, y) = x + 4y^2 - 1$ defined on the rectangle $R = [0, 3] \times [0, 2]$. Take (x_i^*, y_j^*) to be the upper left corner of each subrectangle R_{ij}.

4. Compute the (double) Riemann sum $\mathcal{R}_4$ for the function $f(x, y) = 2xy + 1$ defined on the rectangle $R = [0, 4] \times [0, 1]$. Take (x_i^*, y_j^*) to be the lower left corner of each subrectangle R_{ij}.

5. Compute the Riemann sum $\mathcal{R}_2$ for $f(x, y) = x^2 - y^2$ defined on the rectangle $R = [0, 5] \times [0, 3]$. Take (x_i^*, y_j^*) to be the upper left corner of each subrectangle R_{ij}.

6. Compute the Riemann sum $\mathcal{R}_3$ for $f(x, y) = xy - y^2 - x$ defined on the rectangle $R = [0, 5] \times [0, 4]$. Take (x_i^*, y_j^*) to be the center of each subrectangle R_{ij}.

7. Compute the Riemann sum $\mathcal{R}_2$ for $f(x, y) = (x - y)e^{xy}$ defined on the rectangle $R = [-1, 1] \times [0, 1]$. Take (x_i^*, y_j^*) to be the lower right corner of each subrectangle R_{ij}.

8. Let R be the square with vertices $(0, 0)$, $(0, 4)$, $(4, 4)$, and $(4, 0)$ and let $f(x, y)$ be the distance from the point (x, y) to the x-axis.

(a) Estimate $\iint_R f\,dA$ by partitioning the square R into four squares and using the midpoint of each square.

(b) Show that $0 \le \iint_R f\,dA \le 64$.

9. Determine whether the integral $\iint_R xy^2\,dA$, where $R = [-2, 1] \times [-1, 1]$, is positive or negative.

10. In Theorem 6.2 we listed some properties of double integrals.

(a) Provide details of the proof of part (b).

(b) If $h \ge 0$ then $\iint_R h\,dA \ge 0$ (since it represents the volume of a region). Use this fact to give another proof of part (d).

(c) Find an example of f and R such that $\left|\iint_R f\,dA\right| < \iint_R |f|\,dA$.

11. Evaluate the integral $\iint_R (3 - x)\,dA$, where $R = [0, 1] \times [0, 2]$, by identifying it as the volume of a solid.

12. Show that $\iint_R K\,dA = K(b - a)(d - c)$ (K is a constant), where $R = [a, b] \times [c, d]$.

13. Show that $4 \le \iint_R e^{x^2+y^2}\,dA \le 4e^2$, where $R = [-1, 1] \times [-1, 1]$.

14. Find an upper bound and a lower bound for the integral $\iint_R e^{-x^2-y^2}\,dA$, where R is the rectangle with vertices $(0, 0)$, $(1, 0)$, $(1, 1)$, and $(0, 1)$ (if $m \le \iint_R f\,dA \le M$, then m is a lower bound, and M is an upper bound for the integral of f over R).

15. Find an upper bound for the integral $\iint_R \sin(2x)\cos(5y)\,dA$, where R is the rectangle with vertices $(0, 0)$, $(2\pi, 0)$, $(2\pi, \pi)$, and $(0, \pi)$.

16. Find an example of a function such that $\left|\iint_R f\,dA\right| < \iint_R |f|\,dA$, where $R = [0, 1] \times [0, 1]$.

17. Let

$$f(x, y) = \begin{cases} 3 & \text{if } x \ge 0 \\ x + 3 & \text{if } x < 0 \end{cases}$$

Is f integrable over $R = [-1, 1] \times [0, 2]$? If so, find $\iint_R f\,dA$.

18. Let

$$f(x, y) = \begin{cases} \sqrt{4 - x^2} & \text{if } 0 \le x \le 2 \\ 2 & \text{otherwise} \end{cases}$$

Is f continuous on $R = [-2, 2] \times [0, 2]$? Integrable over R? If f is integrable, find $\iint_R f\,dA$ by interpreting it as a volume.

19. Is it true that $\iint_R (x^2 + y^2)\,dA \le \iint_R (x^3 + y^3)\,dA$ if $R = [0, 1] \times [0, 1]$? If $R = [1, 2] \times [2, 3]$?

20. Is the function $f(x, y) = 2x(x^2 + y^2 - 10)^{-1}$ integrable over $[0, 1] \times [0, 1]$? Over $[-3, 3] \times [-3, 3]$? Over $[-10, 10] \times [-10, 10]$?

21. The function $f(x, y) = e^{1/(x^2+y^2)}$ is not continuous at $(0, 0)$. Is it integrable on $[-1, 1] \times [-1, 1]$?

22. Is the function

$$f(x, y) = \begin{cases} 3 & \text{if } (x, y) = (0, 0) \\ x^2 + y^2 & \text{otherwise} \end{cases}$$

integrable on any rectangle that contains the origin?

6.2 DOUBLE INTEGRALS OVER GENERAL REGIONS

Guided by the geometric concept of the volume of a solid, we constructed the double integral of a real-valued function of two variables as a limit of approximating (Riemann) sums. In this section, we will start developing methods of evaluating double integrals, not only over rectangles but also over more general regions in $\mathbb{R}^2$.

Let us once again recall the definition of the definite integral of a function f over $[a, b]$:

$$\int_a^b f(x)\,dx = \lim_{n\to\infty} \sum_{i=1}^{n} f(x_i^*)\Delta x_i$$

(points $a = x_1, x_2, \ldots, x_{n+1} = b$ define the subdivision of the interval $[a, b]$, $\Delta x_i = x_{i+1} - x_i$, and x_i^* is any point in $[x_i, x_{i+1}]$).

Assume that $f(x) \geq 0$ [so that $\int_a^b f(x)dx$ represents the area of the region D below f; see Figure 6.13], and rephrase the definition as follows: consider the cross sections of D with respect to lines parallel to the y-axis. At a location x_i^* on the x-axis, the cross section is the line segment of length $f(x_i^*)$. "Thicken it" a bit, to obtain a thin rectangle with base length Δx_i. The Riemann sum that approximates the area of D can be viewed as the sum of all such "thickened" cross sections of D, from a to b.

Now we move one dimension higher: Let W be a solid in $\mathbb{R}^3$ placed so that the points a and b on the x-axis represent its minimum and maximum distances from the reference plane (in our case it is the yz-plane), as shown in Figure 6.14.

Consider the cross sections of W defined by planes parallel to the yz-plane: let $A(x_i^*)$ be the area of the cross section defined by the plane x_i^* units away from the yz-plane. The volume of a thin disk defined by that cross section is $A(x_i^*)\Delta x_i$, and the sum $\sum_{i=1}^{n} A(x_i^*)\Delta x_i$ of all thin disks thus obtained approximates the volume $v(W)$ of W. Since

$$v(W) = \lim_{n\to\infty} \sum_{i=1}^{n} A(x_i^*)\Delta x_i,$$

the definition of the definite integral implies that

$$v(W) = \int_a^b A(x)dx. \tag{6.1}$$

Thus, the volume of W is obtained by integrating (i.e., by adding up) its cross sections with respect to planes parallel to a reference plane. Formula (6.1) is called *Cavalieri's principle.* A consequence of this principle is the fact that two solids W_1 and W_2

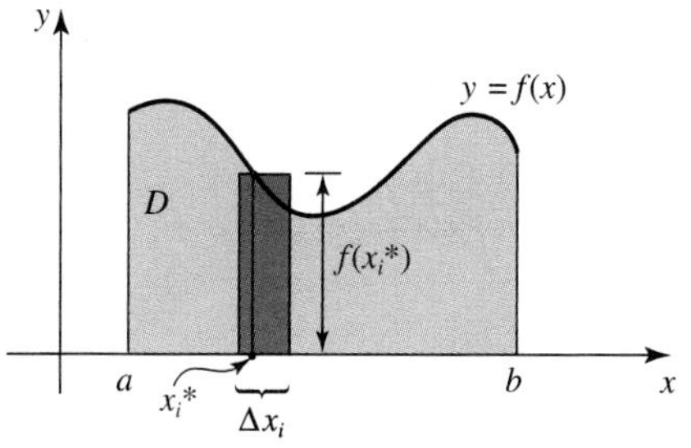

Figure 6.13 A "thickened" cross section has area $f(x_i^*)\Delta x_i$.

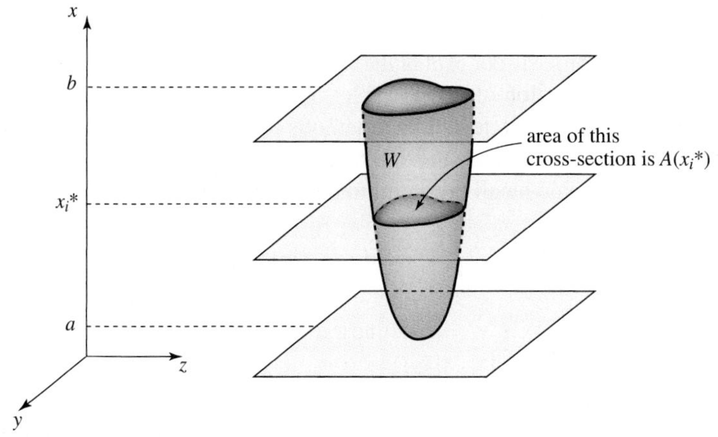

Figure 6.14 Volume of W by cross sections.

with the same cross-sectional areas $A_1(x) = A_2(x)$ for every x have the same volume; see Figure 6.15.

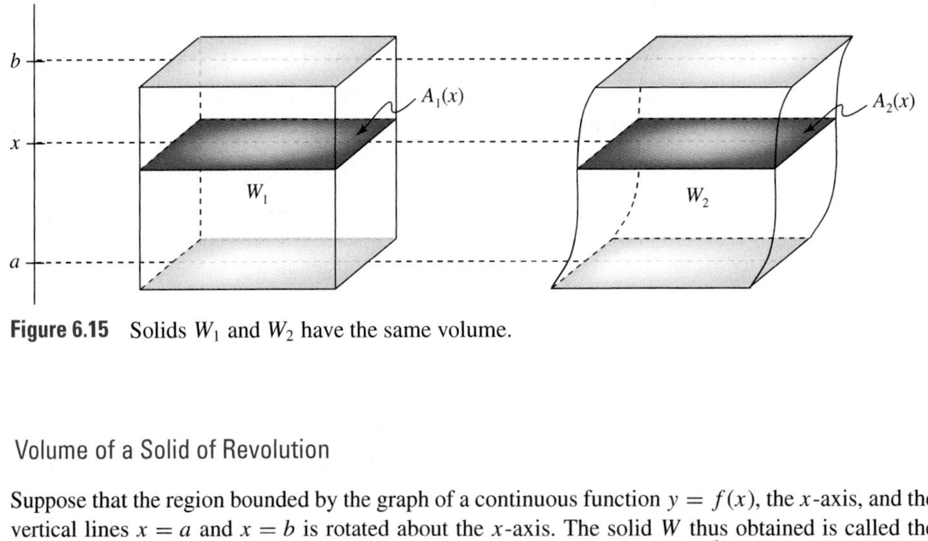

Figure 6.15 Solids W_1 and W_2 have the same volume.

▶ EXAMPLE 6.7 Volume of a Solid of Revolution

Suppose that the region bounded by the graph of a continuous function $y = f(x)$, the x-axis, and the vertical lines $x = a$ and $x = b$ is rotated about the x-axis. The solid W thus obtained is called the *solid of revolution*. Using Cavalieri's principle, we get the formula $v(W) = \int_a^b A(x)dx$ for its volume, where $A(x)$ denotes the cross-sectional area. The cross section at x is the disk of radius $f(x)$; see Figure 6.16. Consequently, $A(x) = \pi(f(x))^2$, and

$$v(W) = \pi \int_a^b (f(x))^2 \, dx.$$ ◀

As an illustration, consider the graph of the semicircle $y = \sqrt{1 - x^2}$, $x \in [-1, 1]$. The solid of revolution is a sphere of radius 1 (see Figure 6.17), and its volume is

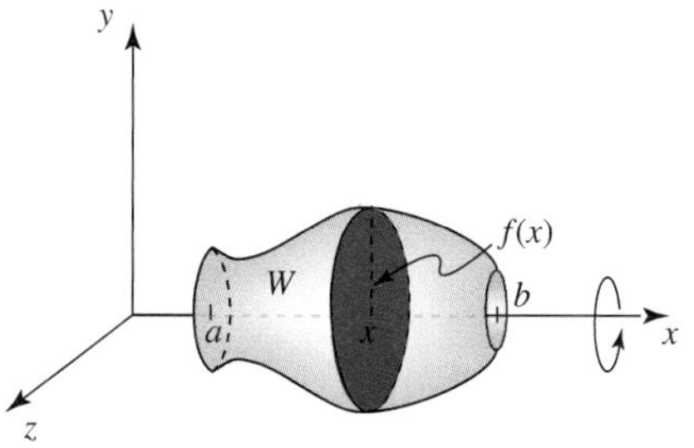

Figure 6.16 Solid of revolution.

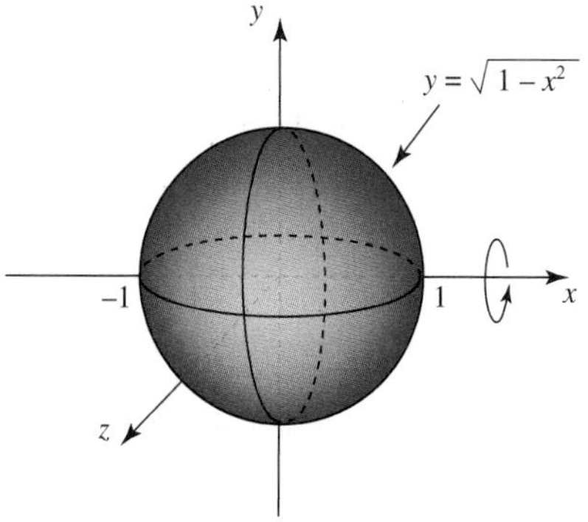

Figure 6.17 Sphere is a solid of revolution.

computed to be

$$v = \pi \int_{-1}^{1} (1 - x^2)\, dx = \pi \left(x - \frac{x^3}{3} \right) \Bigg|_{-1}^{1} = \frac{4\pi}{3}.$$

Suppose that $z = f(x, y)$ is a continuous and non-negative function defined on a rectangle $R = [a, b] \times [c, d]$ in $\mathbb{R}^2$ ["non-negative" means that $f(x, y) \geq 0$ for all (x, y) in R]. We are going to compute the volume of the "solid W below f" [i.e., the solid bounded by the surface $z = f(x, y)$, the xy-plane, and the four planes $x = a$, $x = b$, $y = c$, and $y = d$] using Cavalieri's principle.

Consider the cross sections of W by planes parallel to the yz-plane. Fix a value of x, $a \leq x \leq b$. The cross section at x is the region $D(x)$ under $f(x, y)$ defined on the interval $[c, d]$; see Figure 6.18(a). (The function f is now viewed as a function of one variable, y, since x is held fixed.) The area of $D(x)$ is computed using the definite integral

$$A(x) = \int_{c}^{d} f(x, y)\, dy.$$

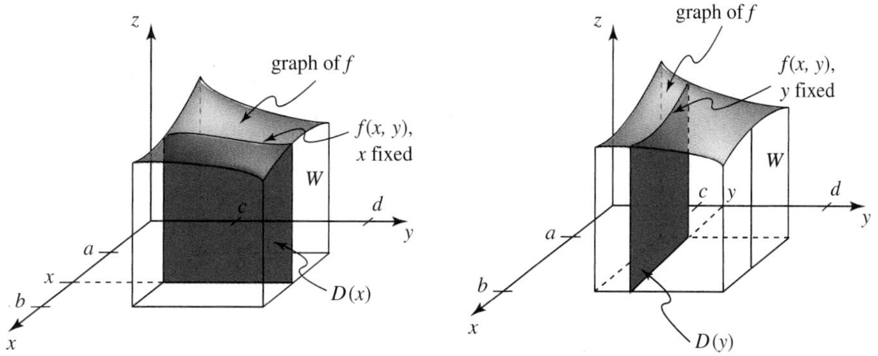

(a) Parallel to the yz-plane, at x

(b) Parallel to the xz-plane, at y

Figure 6.18 Cross sections of W.

Since y is the variable of integration, it will vanish when the integral is evaluated; what will remain is an expression involving x. By Cavalieri's principle, the volume of W is

$$v(W) = \int_a^b \big(\text{cross-sectional area at } x\big)\, dx = \int_a^b A(x)\, dx = \int_a^b \left(\int_c^d f(x, y) dy \right) dx.$$

Now fix a value for y, $c \le y \le d$, and repeat this process. The cross section (this time parallel to the xz-plane) is the region $D(y)$ under $f(x, y)$ (f is now viewed as a function of x only) defined on $a \le x \le b$; see Figure 6.18(b). The area of $D(y)$ is $A(y) = \int_a^b f(x, y)\, dx$, and by Cavalieri's principle,

$$v(W) = \int_c^d \big(\text{cross-sectional area at } y\big)\, dy = \int_c^d A(y)\, dy = \int_c^d \left(\int_a^b f(x, y) dx \right) dy.$$

Our somewhat intuitive argument in the previous section leads to the interpretation of the double integral as volume. And we have just computed the volume of W in two different ways. Therefore,

$$\iint_R f\, dA = \int_a^b \left(\int_c^d f(x, y) dy \right) dx = \int_c^d \left(\int_a^b f(x, y) dx \right) dy,$$

where $R = [a, b] \times [c, d]$. This formula holds in general and not only for non-negative functions.

THEOREM 6.3 Fubini's Theorem

If $f = f(x, y)$ is a continuous function defined on a rectangle $R = [a, b] \times [c, d]$ in $\mathbb{R}^2$, then

$$\iint_R f\, dA = \int_a^b \left(\int_c^d f(x, y) dy \right) dx = \int_c^d \left(\int_a^b f(x, y) dx \right) dy. \quad ◀$$

The theorem states that the double integral $\iint_R f\, dA$ over a rectangle is computed as an *iterated integral*: we integrate with respect to y first, and then integrate the result with respect to x, or alternatively, we do the two integrals in the reversed order. The statement of the theorem remains true in the more general case when f is bounded on R, discontinuous along finitely many curves, and when the iterated integrals exist.

The proof of the theorem is omitted here.

▶ EXAMPLE 6.8

Evaluate $\iint_R 6x^2 y\, dA$, where $R = [-1, 1] \times [0, 4]$.

SOLUTION The function $f(x, y) = 6x^2 y$ is continuous, and by Fubini's Theorem,

$$\iint_R 6x^2 y\, dA = \int_{-1}^1 \left(\int_0^4 6x^2 y dy \right) dx = \int_0^4 \left(\int_{-1}^1 6x^2 y dx \right) dy.$$

Let us evaluate the first iterated integral

$$\int_{-1}^{1}\left(\int_{0}^{4} 6x^2 y dy\right) dx.$$

To find $\int_0^4 6x^2 y\, dy$, we view x as a constant and evaluate it as the definite integral of a function of one variable:

$$\int_0^4 6x^2 y\, dy = 6x^2 \int_0^4 y\, dy = 6x^2\left(\frac{y^2}{2}\bigg|_0^4\right) = 6x^2(8-0) = 48x^2.$$

Therefore,

$$\int_{-1}^{1}\left(\int_{0}^{4} 6x^2 y dy\right) dx = \int_{-1}^{1} 48x^2\, dx = 16x^3\bigg|_{-1}^{1} = 32.$$

The second iterated integral is computed analogously: fixing y and integrating with respect to x, we get

$$\int_{-1}^{1} 6x^2 y\, dx = 6y\int_{-1}^{1} x^2\, dx = 6y\left(\frac{x^3}{3}\bigg|_{-1}^{1}\right) = 4y.$$

Therefore,

$$\int_{0}^{4}\left(\int_{-1}^{1} 6x^2 y dx\right) dy = \int_0^4 4y\, dy = 2y^2\bigg|_0^4 = 32.$$ ◀

▶ EXAMPLE 6.9

Evaluate $\iint_R x^2 e^y\, dA$, where $R = [0,1]\times[0,2]$.

SOLUTION By Fubini's Theorem,

$$\iint_R x^2 e^y\, dA = \int_0^1\left(\int_0^2 x^2 e^y dy\right) dx = \int_0^1 x^2\left(e^y\bigg|_0^2\right) dx$$

$$= \int_0^1 (e^2-1)x^2\, dx = (e^2-1)\frac{x^3}{3}\bigg|_0^1 = \frac{e^2-1}{3}.$$

Reversing the order, we get

$$\iint_R x^2 e^y\, dA = \int_0^2\left(\int_0^1 x^2 e^y dx\right) dy = \int_0^2 e^y\left(\frac{x^3}{3}\bigg|_0^1\right) dy$$

$$= \frac{1}{3}\int_0^2 e^y\, dy = \frac{1}{3}e^y\bigg|_0^2 = \frac{e^2-1}{3}.$$ ◀

We will see in the next section that a particular order of iterated integration could lead to either significantly easier or harder computations. It gets worse: sometimes, only one (or possibly neither) of the iterated integrals can be evaluated exactly as compact formulas [i.e., using exact methods of integration, unlike power series methods or approximate (numerical) integration].

▶ EXAMPLE 6.10

In Example 6.1, we approximated the value of $\iint_R (x^2 + y^2)\,dA$, where $R = [0, 1] \times [-1, 2]$, using Riemann sums. Let us now compute the exact value of the integral using Fubini's Theorem:

$$\iint_R (x^2 + y^2)\,dA = \int_0^1 \left(\int_{-1}^2 (x^2 + y^2) dy \right) dx = \int_0^1 \left(x^2 y + \frac{y^3}{3} \Big|_{-1}^2 \right) dx$$

$$= \int_0^1 (3x^2 + 3)\,dx = (x^3 + 3x) \Big|_0^1 = 4.$$

◀

Recall that the double integral $\iint_R f\,dA$, for $f \geq 0$, can be interpreted as the volume of the solid under f and above R.

▶ EXAMPLE 6.11

Find the volume of the solid W bounded by the surface $z = 5 - 2x - y^2$, the three coordinate planes, and the planes $x = 1$ and $y = 1$.

SOLUTION The solid W is the solid that lies above the rectangle $R = [0, 1] \times [0, 1]$ in the xy-plane and is bounded from above by the graph of $z = 5 - 2x - y^2$ (notice that $z > 0$ on R). The volume of W is

$$v(W) = \iint_R (5 - 2x - y^2)\,dA,$$

and by Fubini's Theorem,

$$v(W) = \int_0^1 \left(\int_0^1 (5 - 2x - y^2) dx \right) dy$$

$$= \int_0^1 \left(5x - x^2 - xy^2 \Big|_0^1 \right) dy = \int_0^1 (4 - y^2)\,dy = \left(4y - \frac{y^3}{3} \right) \Big|_0^1 = \frac{11}{3}.$$

◀

Having learned how to integrate over rectangles, we now move one step further, to integration over more general regions in $\mathbb{R}^2$.

Recall that a region (i.e., a subset) D of $\mathbb{R}^2$ is called *bounded* if it is contained in some ball (see Definition 4.5 in Section 4.3). In the context of integration, we prefer to use rectangles, and say that the region is *bounded* if it can be enclosed in a rectangle $R = [a, b] \times [c, d] \subseteq \mathbb{R}^2$, see Figure 6.19. (The two definitions are equivalent.)

The collection of boundary points (see Section 2.3 for the definition) of D forms the *boundary* ∂D of D (see Figure 6.20). A region is called *closed* if it contains all of its boundary.

The boundaries of regions that we will meet in this book consist of at most finitely many curves (and/or points; recall that a point is a special case of a curve). For example, the boundary of the disk $D = \{(x, y) \,|\, x^2 + y^2 \leq 1\}$ is the circle $\partial D = \{(x, y) \,|\, x^2 + y^2 = 1\}$.

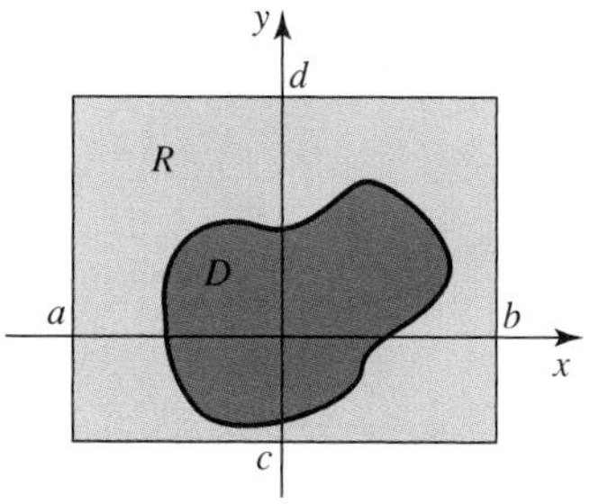

Figure 6.19 Bounded region in $\mathbb{R}^2$.

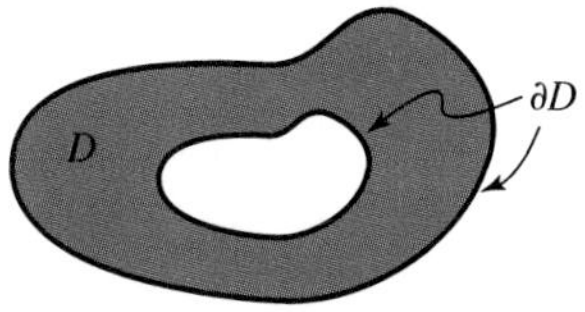

Figure 6.20 Boundary of a region D.

Since ∂D is contained in D, the disk D is closed. The boundary ∂D of the "punctured disk" $D = \{(x, y) \mid x^2 + y^2 \leq 1 \text{ and } (x, y) \neq (0, 0)\}$ consists of the circle $\{(x, y) \mid x^2 + y^2 = 1\}$ and the point $(0, 0)$. The "punctured disk" is not closed, since it does not contain $(0, 0)$. The boundary of the annulus $\{(x, y) \mid 1 \leq x^2 + y^2 \leq 4\}$ consists of two circles

$$\{(x, y) \mid x^2 + y^2 = 1\} \quad \text{and} \quad \{(x, y) \mid x^2 + y^2 = 4\}.$$

The boundary of the rectangle $R = [a, b] \times [c, d] = \{(x, y) \mid a \leq x \leq b, c \leq y \leq d\}$ consists of four straight-line segments: $\{(x, c) \mid a \leq x \leq b\}$, $\{(x, d) \mid a \leq x \leq b\}$, $\{(a, y) \mid c \leq y \leq d\}$, and $\{(b, y) \mid c \leq y \leq d\}$. Clearly, $\partial R \subseteq R$, and therefore R is closed (which justifies the term "closed rectangle" that we have used already).

The boundary ∂D of the region $D = \{(x, y) \mid x^2 + y^2 \leq 1 \text{ and } y > 0\}$ consists of the semicircle $x^2 + y^2 = 1$, $y \geq 0$ and the line segment from $(-1, 0)$ to $(1, 0)$. D contains only a part of its boundary and is therefore not closed.

Assume that $z = f(x, y)$ is a continuous function defined on a closed and bounded region D in $\mathbb{R}^2$. Since D is bounded, it is possible to choose a rectangle R such that $D \subseteq R$. Consider the function $F(x, y)$ defined on R by

$$F(x, y) = \begin{cases} f(x, y) & \text{if } (x, y) \in D \\ 0 & \text{if } (x, y) \notin D, \text{ but } (x, y) \in R. \end{cases}$$

In words, $F(x, y)$ equals $f(x, y)$ on D, and is zero for points in R that do not belong to D; see Figure 6.21. We say that F is an *extension by zero* (a *trivial extension*) of f to R.

It can be proved that the function F thus defined is continuous, except possibly along the curve(s) that form the boundary of D. Since f is bounded (as a continuous function defined on a closed and bounded set; see Appendix A for the precise statement), it follows that F is bounded, and Theorem 6.1 implies that F is integrable over R.

DEFINITION 6.3 Double Integral over a Closed and Bounded Region

Let $f = f(x, y)$ be a continuous function defined on a closed and bounded region $D \subseteq \mathbb{R}^2$. The double integral $\iint_D f \, dA$ of f over D is given by

$$\iint_D f \, dA = \iint_R F \, dA.$$

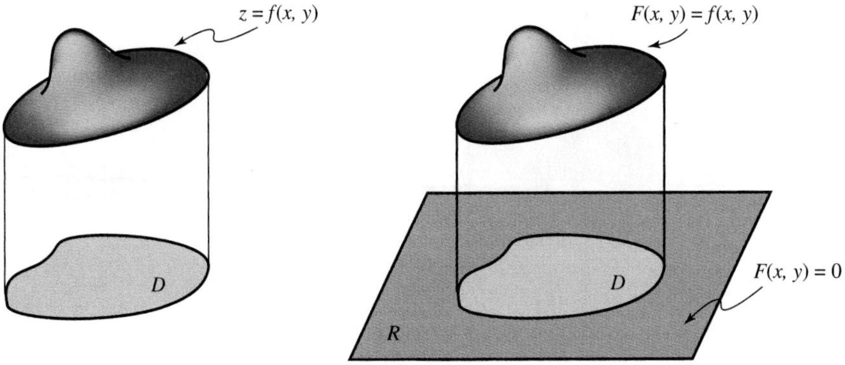

Figure 6.21 Extension by zero of f on D to F on R.

In other words, the double integral of f over a closed and bounded region D is defined as the double integral of the extension F of f over a rectangle R that contains D. The extension is defined so that it agrees with f on D and is zero otherwise. Consequently, the points outside of D do not contribute anything to the integral $\iint_D f\,dA$. This means that it does not matter what rectangle R is taken in Definition 6.3, as long as it encloses D.

From the construction, it follows that, for $f \geq 0$, $\iint_D f\,dA$ represents the volume of the solid under the graph of f and above the region D in the xy-plane.

Although we gave the definition of the double integral for any closed and bounded region, we will have to restrict our study to some special regions. They are introduced in our next definition.

DEFINITION 6.4 Regions of Type 1, 2, and 3. Elementary Regions

A *region of type 1* is a subset D of $\mathbb{R}^2$ of the form

$$D = \{(x, y) \mid a \leq x \leq b, \phi(x) \leq y \leq \psi(x)\},$$

where $\phi(x)$ and $\psi(x)$ are continuous functions defined on $[a, b]$ satisfying $\phi(x) \leq \psi(x)$. A *region of type 2* is defined by

$$D = \{(x, y) \mid c \leq y \leq d, \phi(y) \leq x \leq \psi(y)\},$$

where $\phi(y)$ and $\psi(y)$ are continuous functions defined on $[c, d]$ satisfying $\phi(y) \leq \psi(y)$.

We say that D is a *region of type 3* if it is of both type 1 and type 2. A region of type 1, 2, or 3 is called an *elementary region*. ◀

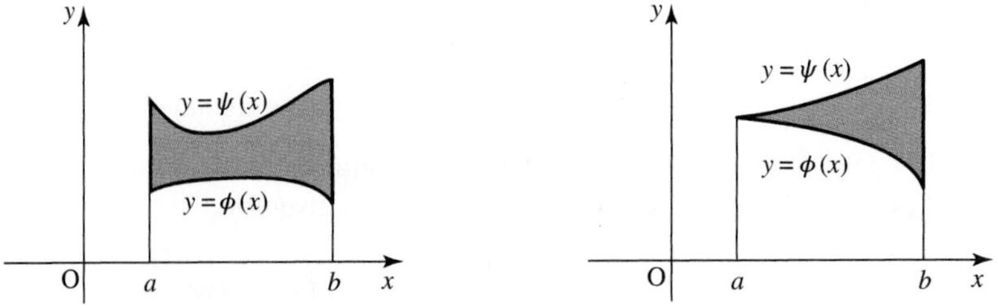

Figure 6.22 Regions of type 1.

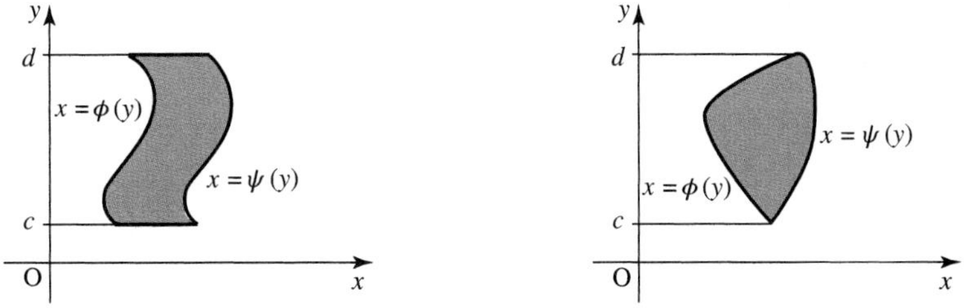

Figure 6.23 Regions of type 2.

By definition, all elementary regions are closed and bounded. A region of type 1 is also called *x-simple* or *vertically simple*, and a region of type 2 is referred to as *y-simple* or *horizontally simple*. A region of type 1 is bounded from above and below by graphs of continuous functions; see Figure 6.22 (i.e., the "top" and "bottom" are curves, whereas the "left" and "right" sides are either vertical lines or points).

Figure 6.23 shows two regions of type 2 (the "sides" are curves, whereas the "top" and "bottom" are horizontal lines or points).

Examples of regions of type 3 include all rectangles in $\mathbb{R}^2$, the disk $\{(x, y) \mid x^2 + y^2 \leq 1\}$, and the set $D = \{(x, y) \mid x^3 \leq y \leq \sqrt{x}, 0 \leq x \leq 1\}$ shown in Figure 6.24.

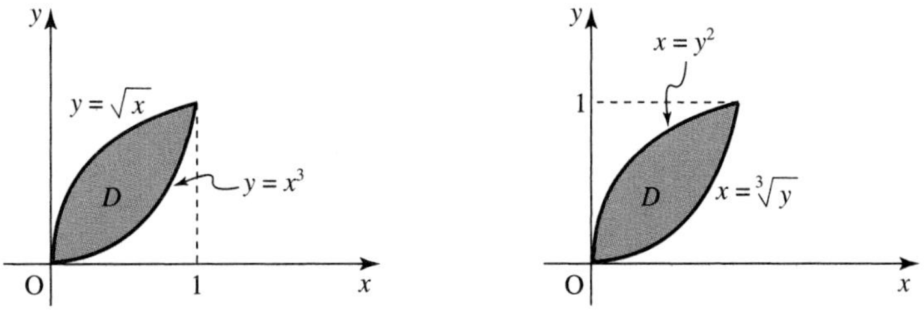

(a) As a type-1 region (b) As a type-2 region

Figure 6.24 The region $D = \{(x, y) \mid x^3 \leq y \leq \sqrt{x}, 0 \leq x \leq 1\}$.

Now let us evaluate $\iint_D f\, dA$ if D is a region of type 1. Choose a rectangle $R = [a, b] \times [c, d]$ as shown in Figure 6.25. By Definition 6.3,

$$\iint_D f\, dA = \iint_R F\, dA,$$

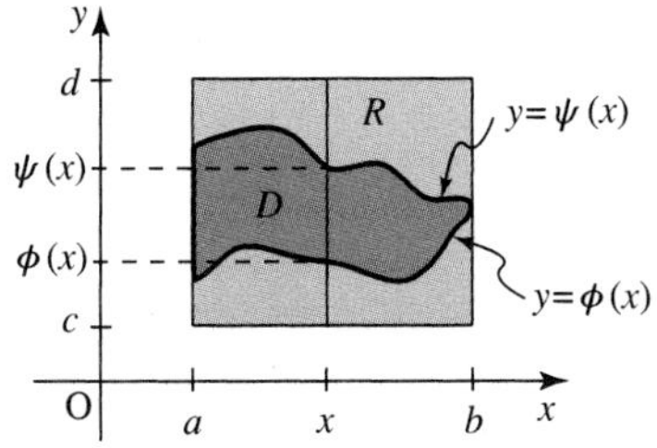

Figure 6.25 Integral over a region of type 1.

where F is the extension by zero of f on R. From Fubini's Theorem, we get

$$\iint_R F\,dA = \int_a^b \left(\int_c^d F(x, y)\,dy \right) dx.$$

By construction, $F(x, y) = 0$ if $(x, y) \notin D$; that is, when $y < \phi(x)$ and $y > \psi(x)$. Therefore, (in this computation x is fixed; we are doing the inner integration)

$$\begin{aligned} \int_c^d F(x, y)\,dy &= \int_c^{\phi(x)} F(x, y)\,dy + \int_{\phi(x)}^{\psi(x)} F(x, y)\,dy + \int_{\psi(x)}^{d} F(x, y)\,dy \\ &= \int_{\phi(x)}^{\psi(x)} F(x, y)\,dy, \end{aligned}$$

since the first and third integrals in the sum are zero. Since $F = f$ on D, we get

$$\int_{\phi(x)}^{\psi(x)} F(x, y)\,dy = \int_{\phi(x)}^{\psi(x)} f(x, y)\,dy,$$

and therefore,

$$\iint_R F\,dA = \int_a^b \left(\int_{\phi(x)}^{\psi(x)} f(x, y)dy \right) dx.$$

The double integral over a region of type 2 is computed analogously. We formulate these results in the statement of the next theorem.

THEOREM 6.4 Iterated Integrals over Elementary Regions

Assume that $f\colon D \subseteq \mathbb{R}^2 \to \mathbb{R}$ is a continuous function.

(a) If D is a region of type 1, then

$$\iint_D f\,dA = \int_a^b \left(\int_{\phi(x)}^{\psi(x)} f(x, y)\,dy \right) dx.$$

(b) If D is a region of type 2, then

$$\iint_D f\,dA = \int_c^d \left(\int_{\phi(y)}^{\psi(y)} f(x, y)\,dx \right) dy.$$

(c) If D is a region of type 3, then either (a) or (b) can be used to evaluate the double integral over D. ◀

Let us emphasize that, in (a), the limits of the inner integration are functions of x [this means that the boundary curves have to be expressed in the form $y = \phi(x)$ and $y = \psi(x)$]. In (b), the limits of the inner integration are functions of y [consequently, the boundary curves have to be expressed in the form $x = \phi(y)$ and $x = \psi(y)$].

▶ EXAMPLE 6.12

Evaluate $\iint_D e^{2x+y}\,dA$, where D is the region bounded by the lines $y = 2x$, $y = x$, $x = 1$, and $x = 2$.

SOLUTION The region D is of type 3; see Figure 6.26(a). For convenience, we choose to view D as a type-1 region. It follows that

$$\iint_D e^{2x+y}\,dA = \int_1^2 \left(\int_x^{2x} e^{2x+y}\,dy \right) dx.$$

Before evaluating the integral, let us explain how the limits of integration were set. The outer integration is with respect to x, so its limits are defined by the interval on the x-axis that contains the values of x (in this case, it is $[1, 2]$). The inner integration (i.e., the first to be evaluated) is with respect to y. Think of arrows "traveling" parallel to the y-axis, starting below D and finishing above D, as shown in Figure 6.26(b). They enter D along the line $y = x$ (that is the lower limit of integration) and leave D along $y = 2x$ (that is the upper limit).

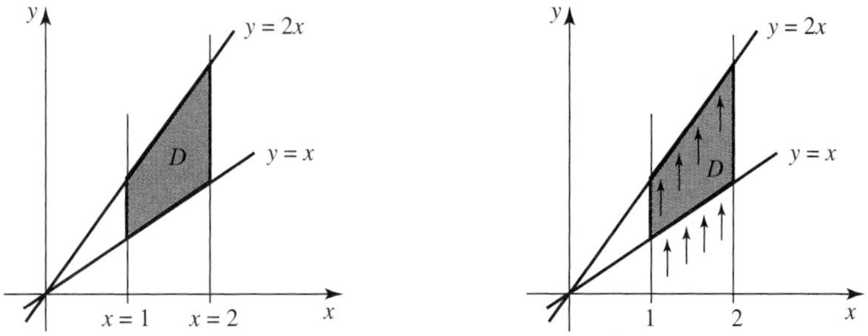

(a) D is a type-3 region (b) "Arrows argument" for inner integration

Figure 6.26 Region of Example 6.12.

We evaluate the integral as an iterated integral

$$\iint_D e^{2x+y}\,dA = \int_1^2 \left(\int_x^{2x} e^{2x+y}\,dy \right) dx = \int_1^2 \left(e^{2x+y}\Big|_x^{2x} \right) dx$$
$$= \int_1^2 (e^{4x} - e^{3x})\,dx = \left(\frac{e^{4x}}{4} - \frac{e^{3x}}{3} \right)\Bigg|_1^2 = \tfrac{1}{4}(e^8 - e^4) - \tfrac{1}{3}(e^6 - e^3).$$

In Exercise 10, we set up and evaluate the same integral, using the fact that D is also a region of type 2. ◀

▶ EXAMPLE 6.13

Evaluate $\iint_D 2y\,dA$, where D is the region in the xy-plane bounded by $y = x - 6$ and $y^2 = x$.

SOLUTION To make an accurate sketch and to set up the limits of integration, we need to find the points of intersection of the two curves. Substituting $x = y^2$ into $y = x - 6$, we get $y^2 - y - 6 = 0$ and $y = -2$

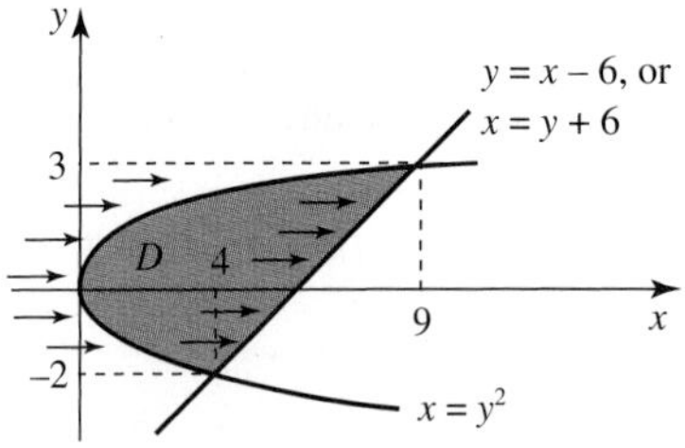

Figure 6.27 Region D of Example 6.13.

or $y = 3$. From either of the two equations, we compute the corresponding x-coordinates, thus getting $(4, -2)$ and $(9, 3)$ as the points of intersection.

The region D can be viewed as a type-2 region (Figure 6.27): y-coordinates of all points in D belong to the interval $[-2, 3]$, and arrows "traveling" parallel to the x-axis enter D along $x = y^2$ and leave along $y = x - 6$ (we need it in the form $x = y + 6$). Therefore,

$$\iint_D 2y\,dA = \int_{-2}^{3}\left(\int_{y^2}^{y+6} 2y\,dx\right)dy = \int_{-2}^{3}\left(2yx\,\Big|_{y^2}^{y+6}\right)dy$$
$$= \int_{-2}^{3} 2y(y+6-y^2)\,dy = \left(\frac{2y^3}{3} + 6y^2 - \frac{y^4}{2}\right)\Big|_{-2}^{3} = \frac{125}{6}.$$

In Exercise 11, we evaluate the same integral, with D viewed as a region of type 1. ◀

It follows from the definition given at the beginning of this section that, in the general case of a closed and bounded region D, the double integral $\iint_D f\,dA$ over D of a non-negative function f represents the volume (of the solid above D and under the graph of f).

Substituting $f(x, y) = 1$ into Theorem 6.4(a), we get

$$\iint_D 1\,dA = \int_a^b \left(\int_{\phi(x)}^{\psi(x)} 1\,dy\right)dx = \int_a^b (\psi(x) - \phi(x))dx,$$

which is the formula for the area of the region D (of type 1) between two curves given in standard calculus texts. Similarly, from Theorem 6.4(b) it follows that

$$\iint_D 1\,dA = \int_c^d \left(\int_{\phi(y)}^{\psi(y)} 1\,dx\right)dy = \int_c^d (\psi(y) - \phi(y))dy,$$

if D is a region of type 2. In any case,

$$\iint_D dA = \text{area of } D,$$

where D is an elementary region. As a matter of fact, this formula works for *any* region. In the following example, we use Riemann sums to verify this.

▶ EXAMPLE 6.14

Using Riemann sums, compute $\iint_D 1\,dA$, where D is any closed and bounded region.

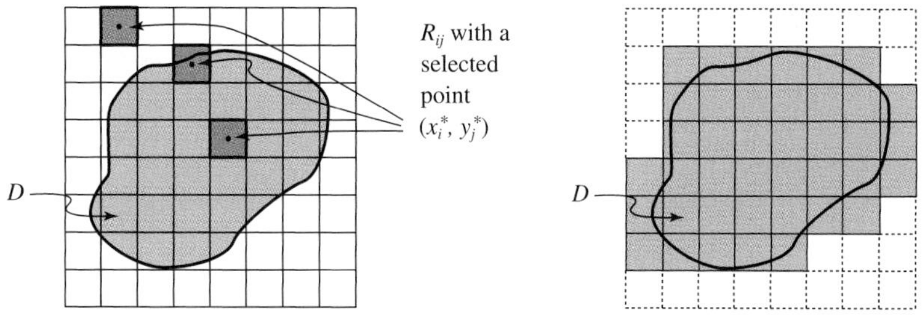

(a) Possible positions of R_{ij} in D (b) Rectangles R_{ij} "circumscribe" D

Figure 6.28 Region D of Example 6.14.

SOLUTION First of all, enclose D in a rectangle R and extend the function $f(x, y) = 1$ defined on D by zero to the function $F(x, y)$ on R; that is, $F(x, y) = f(x, y) = 1$ for all points in D, and $F(x, y) = 0$ for points in R that are not in D. Divide R into n^2 subrectangles R_{ij} and form the Riemann sum

$$\mathcal{R}_n = \sum_{i=1}^{n} \sum_{j=1}^{n} F(x_i^*, y_j^*) \Delta A_{ij},$$

where ΔA_{ij} is the area of R_{ij}. If the subrectangle R_{ij} is inside D, take (x_i^*, y_j^*) to be any point in it; then $F(x_i^*, y_j^*) = 1$. If R_{ij} has no points in common with D, then $F(x_i^*, y_j^*) = 0$, no matter what point (x_i^*, y_j^*) is selected. Finally, if R_{ij} intersects D, take (x_i^*, y_j^*) to be any point in the intersection of D and R_{ij}; see Figure 6.28(a).

It follows that

$$\mathcal{R}_n = \sum^{D} F(x_i^*, y_j^*) \Delta A_{ij} + \sum^{D'} F(x_i^*, y_j^*) \Delta A_{ij}$$

where Σ^D represents the sum over all rectangles that intersect D (including, of course, those that are completely contained in D) and $\Sigma^{D'}$ is the sum over those R_{ij} that have no points in common with D. By construction, $\Sigma^{D'} F(x_i^*, y_j^*) \Delta A_{ij} = 0$, and therefore,

$$\mathcal{R}_n = \sum^{D} F(x_i^*, y_j^*) \Delta A_{ij} = \sum^{D} \Delta A_{ij}$$

[since $F(x_i^*, y_j^*) = 1$ for all R_{ij} included in Σ^D]. In words, $\mathcal{R}_n$ is the sum of areas of those R_{ij} that "circumscribe" D, as shown in Figure 6.28(b). As $n \to \infty$, the rectangles that contribute to $\mathcal{R}_n$ will "circumscribe" D closer and closer, and in the limit,

$$\lim_{n \to \infty} \mathcal{R}_n = \iint_D 1 \, dA = \text{area of } D.$$

The properties (a), (b), (d), and (e) of Theorem 6.2 continue to hold for double integrals over general closed and bounded regions. Now we add a few more properties to this list.

THEOREM 6.5 Properties of Double Integrals

Let D be a closed and bounded region in $\mathbb{R}^2$.

(a) Assume that f is an integrable function defined on D. If D is divided into n mutually disjoint elementary regions D_i, then f is integrable over each D_i and

$$\iint_D f \, dA = \sum_{i=1}^{n} \iint_{D_i} f \, dA$$

("mutually disjoint" means that the intersections of any two, three, etc., elementary regions are either empty or contain curves and/or points only).

(b) Assume that f is a continuous function defined on D. There exist real numbers m and M such that

$$mA(D) \leq \iint_D f\,dA \leq MA(D),$$

where $A(D)$ is the area of D.

(c) Assume that f is a continuous function defined on D. There exists a point (x_0, y_0) in D such that

$$\iint_D f\,dA = f(x_0, y_0)A(D),$$

where $A(D)$ is the area of D. ◀

Part (c) of the theorem is called the *Mean Value Theorem for Integrals*. (In Exercise 39 we review the Mean Value Theorem for a definite integral of a function of one variable.) We can rephrase it as follows: there exists a point (x_0, y_0) in D such that

$$f(x_0, y_0) = \frac{1}{A(D)} \iint_D f\,dA. \tag{6.2}$$

The right side is the "total value of f over D" divided by the area of D—that is, the *average value of f over D*. Formula (6.2) states that, if f is continuous, then there must be a point in D where the average value is attained (we will soon illustrate this point with an example).

The real numbers $mA(D)$ and $MA(D)$ in (b) are called a lower bound and an upper bound for the double integral $\iint_D f\,dA$.

PROOF: Part (a) of the theorem is intuitively clear (its special case appeared in the previous section). However, the proof is technically involved and will not be presented here.

A continuous function defined on a closed and bounded set D has a minimum (call it m) and a maximum (call it M); see Appendix A. Hence, $m \leq f(x, y) \leq M$, and therefore,

$$\iint_D m\,dA \leq \iint_D f\,dA \leq \iint_D M\,dA$$

[by the extension of Theorem 6.2(d) to D]. The conclusion of part (b) now follows from $\iint_D m\,dA = m\iint_D dA = mA(D)$ and $\iint_D M\,dA = MA(D)$.

Divide the formula in (b) by $A(D)$ to get

$$m \leq \frac{1}{A(D)} \iint_D f\,dA \leq M.$$

In words, the average value of f is a number between the minimum m and the maximum M of f. By the Intermediate Value Theorem (see Appendix A for the precise statement), a continuous function assumes every value between its minimum and maximum. Therefore, there must be a point (x_0, y_0) in D where

$$f(x_0, y_0) = \frac{1}{A(D)} \iint_D f\,dA.$$ ◀

▶ EXAMPLE 6.15

Let $T(x, y) = 120 + 30x^2 - 18y$ be the temperature (in °C) at a point (x, y) on a metal plate D in the shape of a triangle with vertices $(0, 0)$, $(1, 0)$, and $(1, 1)$. Compute the average temperature $\overline{T}$ on the plate.

SOLUTION The region D is of both type 1 and type 2. As a type-1 region, it is bounded by the curves $y = 0$ and $y = x$ over the interval $0 \le x \le 1$. It follows that (the area of D is $1/2$—no integration needed!)

$$\begin{aligned}\overline{T} &= \frac{1}{A(D)} \iint_D (120 + 30x^2 - 18y)\, dA \\ &= 2\int_0^1 \left(\int_0^x (120 + 30x^2 - 18y)\, dy \right) dx = 2\int_0^1 \left(120y + 30x^2 y - 9y^2 \Big|_0^x \right) dx \\ &= 2\int_0^1 (120x + 30x^3 - 9x^2)\, dx = 2(60x^2 + 7.5x^4 - 3x^3)\Big|_0^1 = 129°\text{C}.\end{aligned}$$

The Mean Value Theorem claims that there must be a point (or points) in D where the temperature is exactly 129°C. Let us identify such point(s). From $T = 120 + 30x^2 - 18y = 129$, we get $18y = 30x^2 - 9$, that is, $y = 5x^2/3 - 1/2$. It follows that the temperature at all points on the parabola $y = 5x^2/3 - 1/2$ which belong to D (see Figure 6.29) is equal to the average value of 129 °C.

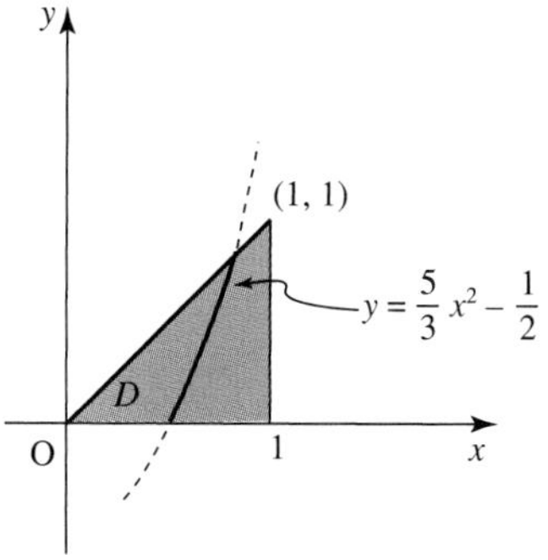

Figure 6.29 Points in D where the temperature equals its average value. ◀

▶ EXAMPLE 6.16

Find an upper bound and a lower bound for $\iint_D 3\sin^2(3x - y^4)\, dA$, where D is the circle $\{(x, y) | x^2 + y^2 \le 9\}$.

SOLUTION Since $-1 \le \sin a \le 1$ for any real number a, it follows that $0 \le \sin^2(3x - y^4) \le 1$ and $0 \le 3\sin^2(3x - y^4) \le 3$; that is, $m = 0$ and $M = 3$. The area of D is 9π, and therefore,

$$0 = \iint_D 0\, dA \le \iint_D 3\sin^2(3x - y^4)\, dA \le \iint_D 3\, dA = 27\pi.$$ ◀

▶ EXAMPLE 6.17

Find the volume of the solid W below the graph of $z = 12 + 2xy^2$ and above the triangle in the xy-plane with vertices $(-1, 0, 0)$, $(1, 0, 0)$, and $(0, 1, 0)$.

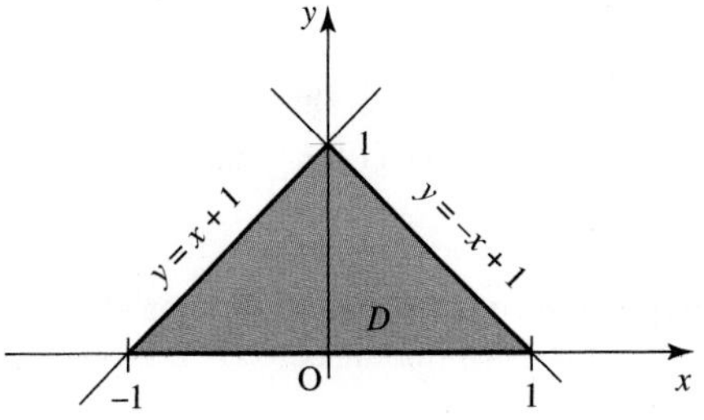

Figure 6.30 Triangle D of Example 6.17.

SOLUTION

The volume of W is given by the double integral $v(W) = \iint_D (12 + 2xy^2)\,dA$, where D is the given triangle. D is bounded by the x-axis and the lines $y = x + 1$ and $y = -x + 1$; see Figure 6.30. Using the fact that D is of type 2, we get

$$\begin{aligned} v(W) &= \iint_D (12 + 2xy^2)\,dA \\ &= \int_0^1 \left(\int_{y-1}^{-y+1} (12 + 2xy^2)\,dx \right) dy = \int_0^1 \left(12x + x^2y^2 \Big|_{y-1}^{-y+1} \right) dy \\ &= \int_0^1 \left(12(-y+1) + (-y+1)^2y^2 - 12(y-1) - (y-1)^2y^2 \right) dy \\ &= 24 \int_0^1 (1-y)\,dy = 24 \left(y - \frac{y^2}{2} \right) \Big|_0^1 = 12. \end{aligned}$$

◀

Now assume that $f(x, y) \geq g(x, y) \geq 0$ for all (x, y) in a region $D \subseteq \mathbb{R}^2$, and consider the solid W over D bounded from above by the graph of $z = f(x, y)$ and from below by the graph of $z = g(x, y)$; see Figure 6.31.

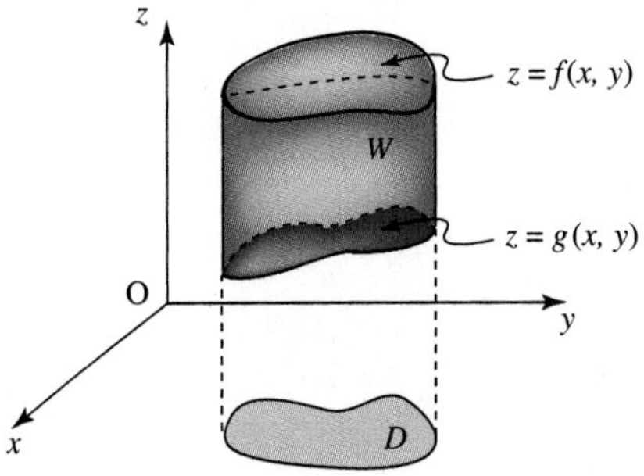

Figure 6.31 Solid W bounded by the graphs of $z = f(x, y)$ and $z = g(x, y)$.

The double integral $\iint_D f\,dA$ represents the volume of the solid under $z = f(x, y)$ over the region D in the xy-plane. Similarly, $\iint_D g\,dA$ is the volume of the solid under $z = g(x, y)$ over D. The volume of W is their difference

$$v(W) = \iint_D f\,dA - \iint_D g\,dA = \iint_D (f - g)\,dA.$$

▶ **EXAMPLE 6.18**

Find the volume of the solid W bounded from above by the surface $z = 4e^{-x}$, from below by the plane $z = 1$ and from the sides by vertical planes $y = 1$, $y = 2$, and $x = 0$; see Figure 6.32.

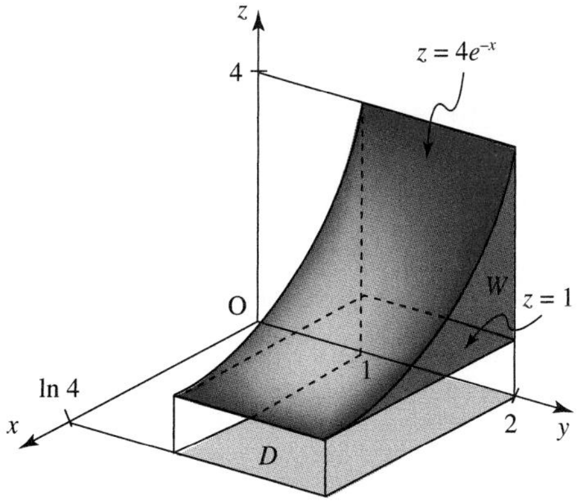

Figure 6.32 Solid W of Example 6.18.

SOLUTION The surface $z = 4e^{-x}$ and the plane $z = 1$ intersect when $4e^{-x} = 1$, that is, when $x = -\ln(1/4) = \ln 4$ (in words, they intersect along the line $x = \ln 4$ in the plane $z = 1$).

The region D of integration is the rectangle $0 \le x \le \ln 4$, $1 \le y \le 2$, and the volume of W is given by $v(W) = \iint_D (4e^{-x} - 1)\, dA$, since, on D, $4e^{-x} \ge 1$. It follows that

$$\begin{aligned} v(W) &= \iint_D (4e^{-x} - 1)\, dA = \int_1^2 \left(\int_0^{\ln 4} (4e^{-x} - 1)\, dx \right) dy \\ &= \int_1^2 \left(-4e^{-x} - x \Big|_0^{\ln 4} \right) dy = \int_1^2 (3 - \ln 4)\, dy \\ &= 3 - \ln 4 \approx 1.61. \end{aligned}$$

Additional examples and some techniques for computing double integrals will be presented in the following two sections.

▶ EXERCISES 6.2

Exercises 1 to 4: Identify the regions below as type 1, type 2, or type 3 [for a region of type 1, state explicitly the functions $y = \phi(x)$ and $y = \psi(x)$ and the values of a and b; likewise, give all necessary detail for a region of type 2].

1. Disk $D = \{(x, y) \mid x^2 + y^2 \le 1\}$

2.

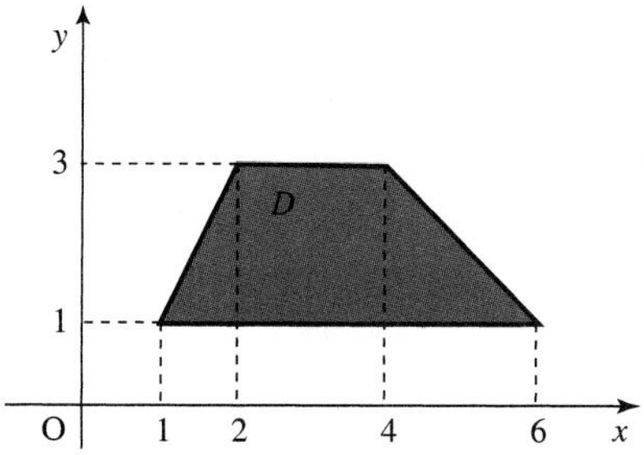

3.

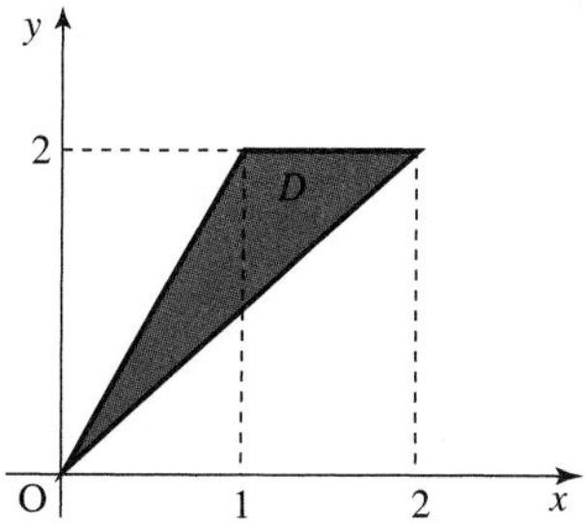

4. The region D bounded by the curves $y = 1 - x^2$ and $y = -3$

5. Suppose that $f(x, y) \leq 0$ for all $(x, y) \in D \subseteq \mathbb{R}^2$. What is a geometric meaning of $\iint_D f\,dA$? If $f(x, y) \geq g(x, y)$ for all $(x, y) \in D$, what geometric interpretation could be given to $\iint_D (f - g)\,dA$? (Note that we do not assume that f and g are positive.)

6. Using Cavalieri's principle, find the volume of the solid W in Figure 6.33.

7. Using Cavalieri's principle, find the volume of a cone of radius r and height h.

8. Find the volume of the solid obtained by rotating the graph of $y = \ln x$, $1 \leq x \leq 2$, about the x-axis. Now imagine that the same graph is rotated about the y-axis. Find the volume of the solid thus obtained.

9. Using Cavalieri's principle, find the volume of the solid W in Figure 6.34.

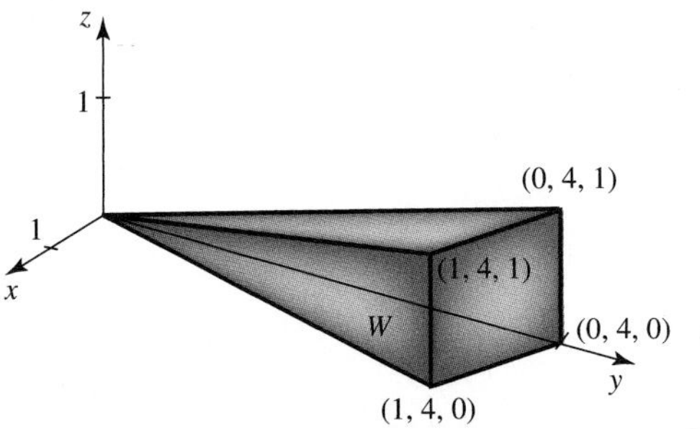

Figure 6.33 Solid W of Exercise 6.

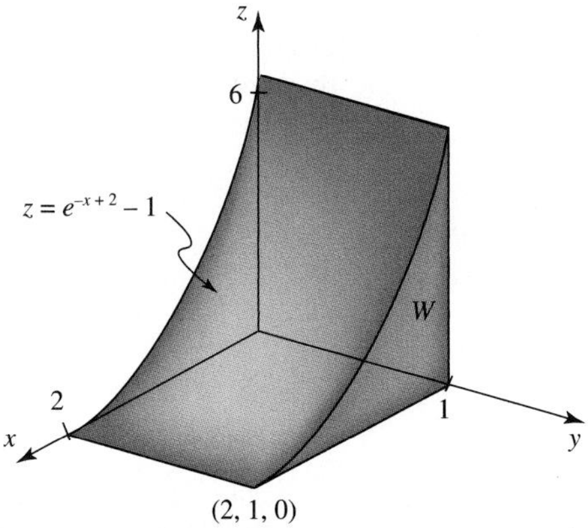

Figure 6.34 Solid W of Exercise 9.

10. Consider the double integral of Example 6.12. Show that, when D is viewed as a type-2 region, $\iint_D e^{2x+y}\,dA = \int_1^2 \left(\int_1^y e^{2x+y}\,dx\right) dy + \int_2^4 \left(\int_{y/2}^2 e^{2x+y}\,dx\right) dy$. Evaluate this integral, thus checking the result of Example 6.12.

11. Consider the double integral of Example 6.13. Show that, when D is viewed as a type-1 region, $\iint_D 2y\,dA = \int_0^4 \left(\int_{-\sqrt{x}}^{\sqrt{x}} 2y\,dy\right) dx + \int_4^9 \left(\int_{x-6}^{\sqrt{x}} 2y\,dy\right) dx$. By evaluating this integral, check the result of Example 6.13.

Exercises 12 to 17: Evaluate the following iterated integrals:

12. $\displaystyle\int_0^1 \left(\int_0^x \cos(x^2)\,dy\right) dx$

13. $\displaystyle\int_0^2 \left(\int_{2-x}^{x+1} (xe^y - 2y - 1)\,dy\right) dx$

14. $\displaystyle\int_{-1}^1 \left(\int_0^{3x} e^{x+3y}\,dy\right) dx$

15. $\displaystyle\int_0^\pi \left(\int_0^{\cos\theta} \rho^2 \sin\theta\,d\rho\right) d\theta$

16. $\displaystyle\int_1^2 \left(\int_0^{y/2} x\sqrt{x^2 + y^2}\,dx\right) dy$

17. $\displaystyle\int_0^{\pi/2} \left(\int_0^{\sin y} x\cos y\,dx\right) dy$

Exercises 18 to 24: Evaluate $\iint_D f\,dA$ for the function f and the region $D \subseteq \mathbb{R}^2$.

18. $f(x, y) = e^{-x-3y}$, $D = [0, \ln 2] \times [0, \ln 3]$

19. $f(x, y) = xy^{-1} - x^2y^2$, $D = [0, 2] \times [3, 4]$

20. $f(x, y) = xye^{x^2}$, $D = [-1, 1] \times [0, 1]$

21. $f(x, y) = 2xy - y$, $D = \{(x, y) \mid 0 \le y \le 1, -y \le x \le 1 + y\}$

22. $f(x, y) = e^x$, $D = \{(x, y) \mid 0 \le x \le 3, x \le y \le 2x^2\}$

23. $f(x, y) = x^{-2/3}$, D is the region in the first quadrant bounded by the parabolas $y = x^2$ and $y = 4 - x^2$

24. $f(x, y) = \ln(xy)$, D is the triangular region bounded by the lines $y = 1$, $y = x$, and $x = 0$

25. Find an upper bound and a lower bound for $\iint_D e^{-x-y}\,dA$, where $D = [-1, 1] \times [0, 2]$.

26. Find an upper bound and a lower bound for the double integral $\iint_D x^2 \sin(x^2 - y)\,dA$, where D is the disk $x^2 + y^2 \le 1$.

Exercises 27 to 30: Find the volume of the solid in $\mathbb{R}^3$.

27. The solid under the plane $x + y/2 + z = 6$ and above the rectangle $[-1, 1] \times [0, 2]$

28. The solid in the first octant bounded by the cylindrical sheet $z = -y^2 + 9$ and the plane $x = 2$

29. The solid between the planes $x + y + z = 1$ and $x + y + 2z = 1$ in the first octant

30. The solid below $z = 9 - x^2 - y^2$ and above the triangle in the xy-plane with the vertices $(0, 0, 0)$, $(1, 0, 0)$, and $(0, 2, 0)$

31. Let $D \subseteq \mathbb{R}^2$ be an elementary region and let f and g be continuous, real-valued functions on D. Show that if $\iint_D f\,dA = \iint_D g\,dA$ then there exists a point (x_0, y_0) in D such that $f(x_0, y_0) = g(x_0, y_0)$.

Exercises 32 to 36: Find the area of the region $D \subseteq \mathbb{R}^2$.

32. Bounded by $y = 2x$, $y = 5x$, and $x^2 + y^2 = 1$

33. The ellipse with the semiaxes $a > 0$ (in the x-direction) and $b > 0$ (in the y-direction)

34. Below $y = x^{-1}$, between $x = a$ and $x = b$, where $a, b > 0$

35. Between $y = x^2$ and $y = 4 - x^2$, to the right of the y-axis

36. Inside the disk $x^2 + y^2 \le 2$ and outside the square $[-1, 1] \times [-1, 1]$

37. Let $f(x, y) = k(x^2 + y^2)$ describe the temperature ($k > 0$ is a constant) at points on a rectangular metal plate $R = [0, 1] \times [0, 2]$. Find all points (x_0, y_0) in R that satisfy the conclusion of the Mean Value Theorem.

38. Find the point (x_0, y_0) from the Mean Value Theorem if $f(x, y) = x^2$ and D is the triangle defined by the coordinate axes and the line $x + y = 1$.

39. Assume that a function $y = f(x)$ is continuous on an interval $[a, b]$.

(a) Explain why there exist real numbers m and M such that $m \le f(x) \le M$ for all x in $[a, b]$.

(b) By integrating the inequality in (a), prove that $m \le \frac{1}{b - a} \int_a^b f(x)\,dx \le M$.

(c) Explain why there exists a number x_0 in $[a, b]$ satisfying $f(x_0) = \frac{1}{b - a} \int_a^b f(x)\,dx$.

(d) The expression on the right side of the equation in (c) is called the *average value* of f on $[a, b]$. Explain in words the meaning of the formula in (c).

▶ 6.3 EXAMPLES AND TECHNIQUES OF EVALUATION OF DOUBLE INTEGRALS

In this section, we give examples of computations of double integrals and illustrate two elementary techniques. Due to its importance, the change of variables technique will be presented separately in the following section.

▶ EXAMPLE 6.19

Evaluate $\iint_R y\sin(xy)\,dA$, where R is the rectangle $R = [1,3]\times[0,\pi/2]$.

SOLUTION Using Fubini's Theorem, we get

$$\iint_R y\sin(xy)\,dA = \int_0^{\pi/2}\left(\int_1^3 y\sin(xy)\,dx\right)dy = \int_0^{\pi/2}\left(-\cos(xy)\Big|_{x=1}^{x=3}\right)dy$$

$$= \int_0^{\pi/2}(\cos y - \cos(3y))\,dy = \left(\sin y - \frac{1}{3}\sin(3y)\right)\Big|_0^{\pi/2} = \frac{4}{3}.$$

Now let us reverse the order of integration:

$$\iint_R y\sin(xy)\,dA = \int_1^3\left(\int_0^{\pi/2} y\sin(xy)\,dy\right)dx$$

[use integration by parts with $u = y$, $du = dy$, $dv = \sin(xy)\,dy$, and $v = -\cos(xy)/x$]

$$= \int_1^3\left(-\frac{y}{x}\cos(xy)\Big|_0^{\pi/2} + \int_0^{\pi/2}\frac{1}{x}\cos(xy)\,dy\right)dx$$

[use substitution $u = xy$, $du = xdy$]

$$= \int_1^3\left(-\frac{\pi}{2x}\cos\left(\frac{\pi}{2}x\right) + \frac{1}{x^2}\sin(xy)\Big|_0^{\pi/2}\right)dx$$

$$= \int_1^3\left(-\frac{\pi}{2x}\cos\left(\frac{\pi}{2}x\right) + \frac{1}{x^2}\sin\left(\frac{\pi}{2}x\right)\right)dx$$

[apply integration by parts (with $u = \pi/(2x)$, $du = -\pi dx/(2x^2)$, $dv = \cos(\pi x/2)dx$, and $v = (2/\pi)\sin(\pi x/2)$) to the first integrand, keep the second one]

$$= -\left(\frac{1}{x}\sin\left(\frac{\pi}{2}x\right)\Big|_1^3 + \int_1^3\frac{1}{x^2}\sin\left(\frac{\pi}{2}x\right)dx\right) + \int_1^3\frac{1}{x^2}\sin\left(\frac{\pi}{2}x\right)dx$$

$$= -\frac{1}{x}\sin\left(\frac{\pi}{2}x\right)\Big|_1^3 = \frac{4}{3},$$

since two integrals cancel each other. ◀

This example shows that a specific order of integration could lead to significantly easier (or harder) computation. Sometimes, it may even be impossible to evaluate an iterated integral using methods that give exact solutions in compact form (unlike, for example, numerical or power series methods). Our next example illustrates this point.

▶ EXAMPLE 6.20

Evaluate $\iint_D e^{x/y}\,dA$, where D is the region between the curves $y = x$ and $x = y^3$ in the first quadrant.

SOLUTION

Combining $y = x$ and $x = y^3$, we get $y^3 - y = 0$, and therefore $y = -1$, 0, or 1. The points of intersection are $(-1, -1)$, $(0, 0)$, and $(1, 1)$. The region D is shown in Figure 6.35. Using the fact that it is a type-2 region, we get

$$\iint_D e^{x/y}\,dA = \int_0^1 \left(\int_{y^3}^{y} e^{x/y}\,dx \right) dy$$

$$= \int_0^1 \left(ye^{x/y} \Big|_{x=y^3}^{x=y} \right) dy = \int_0^1 \left(ye - ye^{y^2} \right) dy$$

(substitute $u = y^2$, $du = 2ydy$ in the second integral)

$$= \left(\frac{y^2}{2} e - \frac{1}{2} e^{y^2} \right) \Big|_0^1 = \frac{1}{2}.$$

If we view D as a type-1 region, we end up with the iterated integral

$$\iint_R e^{x/y}\,dA = \int_0^1 \left(\int_x^{\sqrt[3]{x}} e^{x/y}\,dy \right) dx$$

that cannot be solved by exact means.

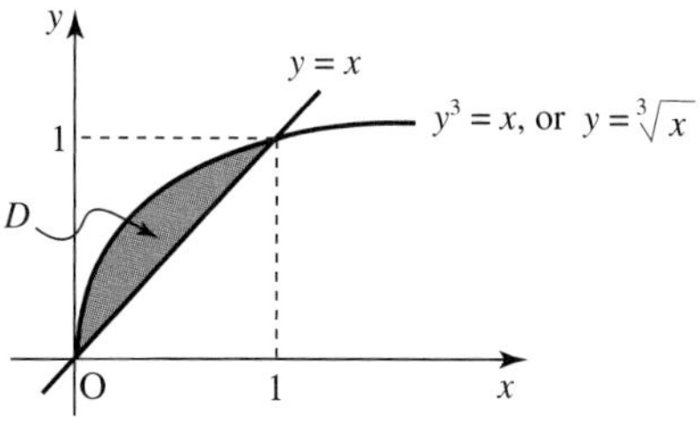

Figure 6.35 Region of Example 6.20.

▶ EXAMPLE 6.21

Find the volume of the solid W in the first octant bounded by the parabolic cylinder $x^2 + z = 9$ and the vertical plane $x + 3y = 3$.

SOLUTION

To obtain the graph of $x^2 + z = 9$, draw the parabola $z = 9 - x^2$ in the xz-plane first. Since there is no mention of y in the equation (and hence no restriction on its values), the same graph must be repeated for every y; that is, for every plane parallel to the xz-plane. Mechanically, the graph is obtained by moving the xz-plane that contains the parabola along the y-axis (keeping the origin on the y-axis, the plane has to remain perpendicular to the xy-plane and parallel to its initial position); see Figure 6.36(a).

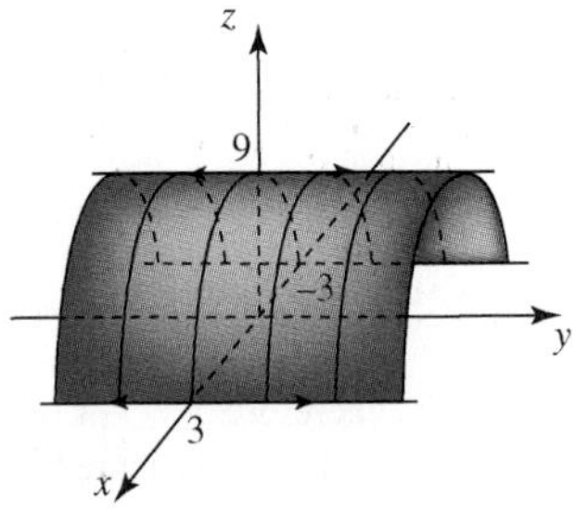

(a) Parabolic sheet $z = 9 - x^2$

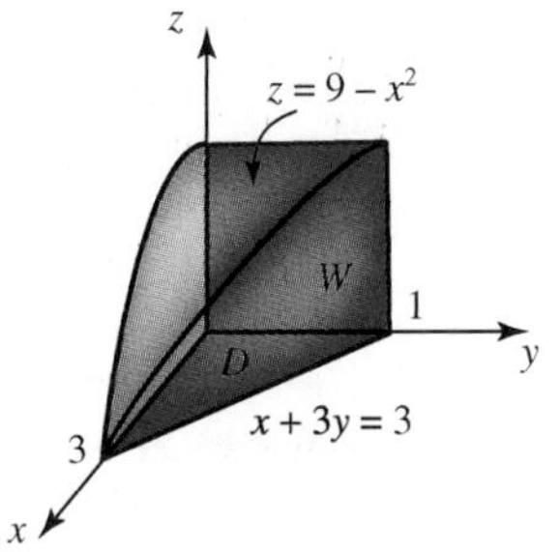

(b) Solid W above D and under $z = 9 - x^2$

Figure 6.36 Region of Example 6.21.

The region D in the xy-plane is determined from $x^2 = 9$ (substitute $z = 0$ into $x^2 + z = 9$), $x + 3y = 3$, and $x, y \geq 0$ (W is in the first octant). The solid W is that part of the three-dimensional region under the graph of $z = 9 - x^2$ that lies above D; see Figure 6.36(b). Its volume is (view D as a type-2 region)

$$v(W) = \iint_D (9 - x^2)\, dA = \int_0^1 \left(\int_0^{3-3y} (9 - x^2)\, dx \right) dy.$$

The y-coordinates of all points in D are between 0 and 1. Arrows "traveling" parallel to the x-axis enter D along the y-axis (hence, $x = 0$) and exit D along the line $x + 3y = 3$ (hence, $x = 3 - 3y$). The iterated integral is equal to

$$\int_0^1 \left(9x - \frac{x^3}{3} \Big|_0^{3-3y} \right) dy = \int_0^1 (18 - 27y^2 + 9y^3)\, dy = \frac{45}{4}.$$ ◀

▶ EXAMPLE 6.22

Find the volume of the region W in the first octant bounded by the planes $x + y + z = 1$ and $x + y + z/2 = 1$.

SOLUTION A good way to visualize a plane is to compute its x-intercept, y-intercept, and z-intercept and join them with line segments (thus representing the part of the plane that belongs to a particular octant). This is how the sketch in Figure 6.37 was obtained. [Set $y = z = 0$; both equations give $x = 1$, and hence $(1, 0, 0)$ is the x-intercept for both planes. Now set $x = z = 0$ to get the y-intercept, etc.] The region D in the xy-plane is determined by $x \geq 0$, $y \geq 0$ and $x + y = 1$ (substitute $z = 0$ into both equations).

The volume of W is the difference of the volumes of the solid under $z = 2 - 2x - 2y$ and the solid under $z = 1 - x - y$ (and above D). Hence (view D as a region of type 1),

$$\begin{aligned} v(W) &= \iint_D (2 - 2x - 2y)\, dA - \iint_D (1 - x - y)\, dA \\ &= \iint_D (1 - x - y)\, dA = \int_0^1 \left(\int_0^{1-x} (1 - x - y)\, dy \right) dx \\ &= \int_0^1 \left(y - xy - \tfrac{1}{2}y^2 \Big|_0^{1-x} \right) dx = \int_0^1 \left(\tfrac{1}{2}y - x + \tfrac{1}{2}x^2 \right) dx = \tfrac{1}{6}. \end{aligned}$$

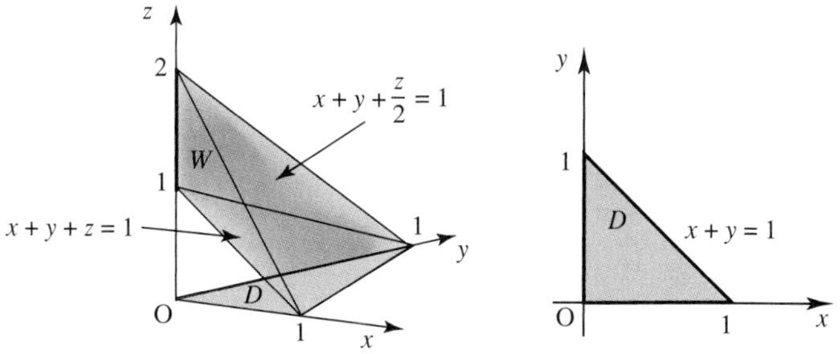

Figure 6.37 Solid W and region D of Example 6.22.

▶ EXAMPLE 6.23

Set up iterated integrals for $\iint_D x^3y\,dA$, where D is the region in the xy-plane bounded by the parabola $y = x^2$ and the line $y = x + 2$.

SOLUTION

Combining $y = x^2$ and $y = x + 2$, we get $x^2 - x - 2 = 0$ and $x = -1, 2$. The points of intersection are $(-1, 1)$ and $(2, 4)$. D can be viewed as a type-1 region (defined for $-1 \le x \le 2$, between $y = x^2$ and $y = x + 2$); see Figure 6.38(a). Therefore,

$$\iint_D f\,dA = \int_{-1}^{2} \left(\int_{x^2}^{x+2} x^3y\,dy \right) dx.$$

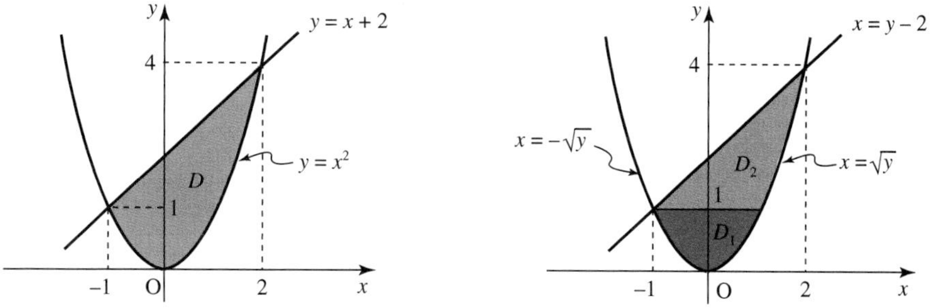

(a) As a type-1 region

(b) As a type-2 region

Figure 6.38 Region D of Example 6.23.

Thinking of the reversed order, consider the inner integration (that will be with respect to x). Arrows that "travel" parallel to the x-axis do not enter D along the same curve: some enter along the parabola (if $0 \le y \le 1$), and some along the line (if $1 \le y \le 4$). We break up D into two type-2 regions accordingly: D_1 is defined by $0 \le y \le 1$ and D_2 by $1 \le y \le 4$, see Figure 6.38(b). There is one more issue: we need to express x in terms of y; from $y = x^2$ we get $x = \pm\sqrt{y}$. The equation $x = \sqrt{y}$ represents the part of the parabola in the first quadrant, and $x = -\sqrt{y}$ is its symmetric part in the second quadrant. By Theorem 6.5(a),

$$\iint_D x^3y\,dA = \iint_{D_1} x^3y\,dA + \iint_{D_2} x^3y\,dA,$$

where

$$\iint_{D_1} x^3 y \, dA = \int_0^1 \left(\int_{-\sqrt{y}}^{\sqrt{y}} x^3 y \, dx \right) dy$$

and

$$\iint_{D_2} x^3 y \, dA = \int_1^4 \left(\int_{y-2}^{\sqrt{y}} x^3 y \, dx \right) dy.$$

▶ EXAMPLE 6.24 Separation of Variables

We are going to evaluate the integral $\iint_R f \, dA$, where $R = [a, b] \times [c, d]$ is a rectangle in $\mathbb{R}^2$ and f is a function of the form $f(x, y) = g(x)h(y)$ [i.e., the variables are separated, as in $f(x, y) = x^2 \cos y$ or in $f(x, y) = e^{x+y} = e^x e^y$; on the other hand, the variables in $f(x, y) = (x + y)^2$ cannot be separated].

By Fubini's Theorem,

$$\iint_R f \, dA = \int_c^d \left(\int_a^b f(x, y) \, dx \right) dy = \int_c^d \left(\int_a^b g(x)h(y) \, dx \right) dy$$

[$h(y)$ is constant for inner integration, so factor it out]

$$= \int_c^d h(y) \left(\int_a^b g(x) \, dx \right) dy = \left(\int_a^b g(x) \, dx \right) \left(\int_c^d h(y) \, dy \right).$$

The definite integral $\int_a^b g(x) \, dx$ is a real number, so we factored it out of the integration with respect to y in the last step.

▶ EXAMPLE 6.25

Compute $\iint_D e^{x+y} \, dA$, where $D = [0, 1] \times [0, 1]$, using separation of variables.

SOLUTION Since $e^{x+y} = e^x e^y$, it follows that

$$\iint_{[0,1]\times[0,1]} e^{x+y} \, dA = \int_0^1 \left(\int_0^1 e^x e^y \, dx \right) dy = \left(\int_0^1 e^x \, dx \right) \left(\int_0^1 e^y \, dy \right)$$

$$= \left(\int_0^1 e^x \, dx \right)^2 = \left(e^x \Big|_0^1 \right)^2 = (e - 1)^2.$$

In our computation, we used the fact that $\int_0^1 e^x dx = \int_0^1 e^y dy$.

The formulas in Theorem 6.4 provide two different ways of computing double integrals over regions of type 3. This fact can be used to simplify the computation of iterated integrals. We have already witnessed that the order of integration chosen in the first part of Example 6.19 was significantly easier to handle than the reversed order. Sometimes, it is even impossible to compute the iterated integral (as a compact formula) without reversing the order of integration, as Example 6.20 and couple of examples that follow illustrate.

The idea of this technique is simple: given an iterated integral, all we have to do is to reconstruct the double integral (i.e., find the region D) that corresponds to that iterated integral. Then we try to evaluate the double integral as an iterated integral in the reversed order of integration.

▶ EXAMPLE 6.26

Evaluate the iterated integral

$$\int_0^1 \left(\int_{\sqrt{y}}^1 (x^2 + 2y)\, dx \right) dy$$

by reversing the order of integration.

SOLUTION The given integral is an iterated integral of the double integral $\iint_D (x^2 + 2y)\, dA$, where D is the region (see Figure 6.39) described by $0 \le y \le 1$ and $\sqrt{y} \le x \le 1$. It is the region bounded by the graph of $y = x^2$, the x-axis, and the vertical line $x = 1$. Therefore,

$$\int_D (x^2 + 2y)\, dA = \int_0^1 \left(\int_0^{x^2} (x^2 + 2y)\, dy \right) dx$$

$$= \int_0^1 \left((x^2 y + y^2) \Big|_{y=0}^{y=x^2} \right) dx = \int_0^1 (x^4 + x^4)\, dx = \frac{2x^5}{5}\Big|_0^1 = \frac{2}{5}.$$ ◀

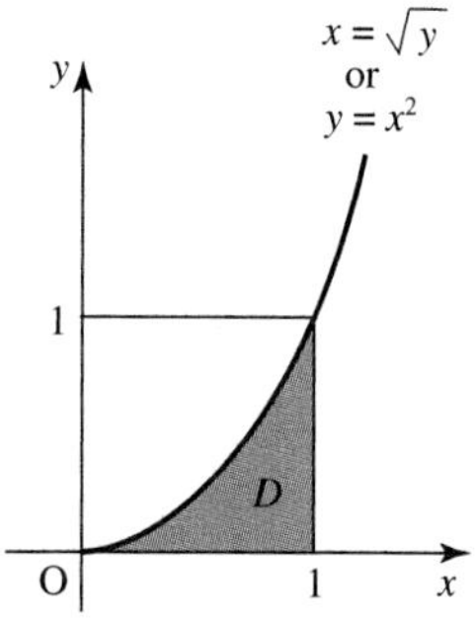

Figure 6.39 Region D of Example 6.26.

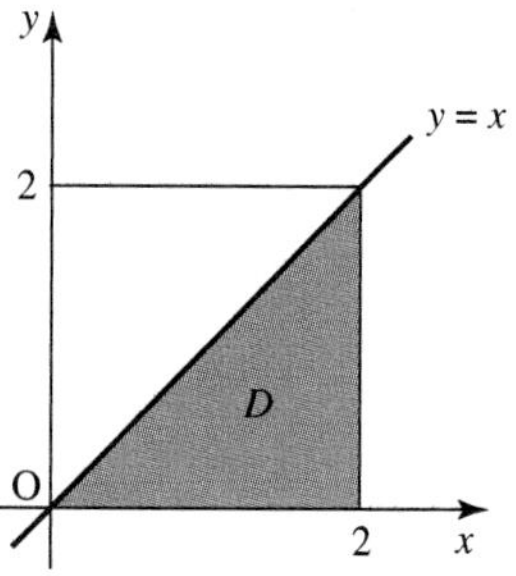

Figure 6.40 Region D of Example 6.27.

▶ EXAMPLE 6.27

Evaluate the iterated integral

$$\int_0^2 \left(\int_y^2 e^{x^2}\, dx \right) dy.$$

SOLUTION Since it is impossible to find a formula for $\int e^{x^2} dx$, we have to reverse the order of integration, hoping that we will get an integrand we can handle. The given integral is equal to $\iint_D e^{x^2}\, dA$, where D is the region described by $0 \le y \le 2$ and $y \le x \le 2$. In other words, D is the triangular region bounded by the graph of $y = x$, the x-axis, and the vertical line $x = 2$ (see Figure 6.40). Hence,

$$\int_0^2 \left(\int_y^2 e^{x^2}\, dx \right) dy = \int_0^2 \left(\int_0^x e^{x^2}\, dy \right) dx$$

$$= \int_0^2 \left(y e^{x^2} \Big|_{y=0}^{y=x} \right) dx = \int_0^2 x e^{x^2}\, dx = \frac{1}{2} e^{x^2}\Big|_0^2 = \frac{1}{2} e^4 - \frac{1}{2},$$

where the substitution $u = x^2$, $du = 2x dx$ was used to compute the integral of $x e^{x^2}$. ◀

▶ EXAMPLE 6.28

Compute

$$\int_0^1 \left(\int_x^1 \frac{\cos y}{y}\, dy \right) dx.$$

SOLUTION The integral of $\cos y/y$ cannot be computed (exactly, as a compact formula), so we reverse the order of integration. The region of integration, given by the inequalities $0 \le x \le 1$ and $x \le y \le 1$, is the triangle with sides $y = x$, $y = 1$, and the y-axis shown in Figure 6.41. It follows that

$$\int_0^1 \left(\int_x^1 \frac{\cos y}{y}\, dy \right) dx = \int_0^1 \left(\int_0^y \frac{\cos y}{y}\, dx \right) dy$$

$$= \int_0^1 \left(\frac{\cos y}{y} x \Big|_{x=0}^{x=y} \right) dy = \int_0^1 \cos y\, dy = \sin y \Big|_0^1 = \sin 1.$$

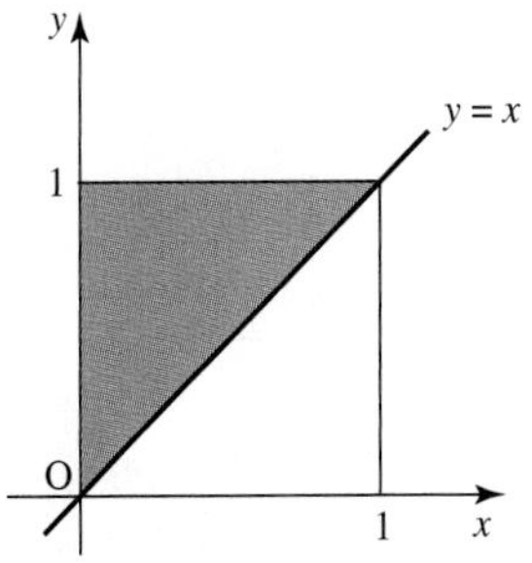

Figure 6.41 Region of Example 6.28.

▶ EXERCISES 6.3

Exercises 1 to 4: Evaluate $\iint_D f\, dA$ for the function f and the region $D \subseteq \mathbb{R}^2$.

1. $f(x, y) = ye^x$, D is the triangular region bounded by the lines $y = 1$, $y = 2x$, and $x = 0$

2. $f(x, y) = y(x^2 + y^2)^{3/4}$, D is the disk $x^2 + y^2 \le 9$

3. $f(x, y) = y^2$, D is the triangular region bounded by the lines $y = 2x$, $y = 5x$, and $x = 2$

4. $f(x, y) = (2x - x^2)^{-1/2}$, D is the triangular region bounded by the lines $y = -x + 1$, $y = 0$, and $x = 0$

Exercises 5 to 9: Find the volume of the solid in $\mathbb{R}^3$.

5. The solid bounded by the cylinder $y^2 + z^2 = 4$ and the plane $x + y = 2$ in the first octant

6. The solid under the paraboloid $z = x^2 + y^2$ and above the region in the xy-plane bounded by the parabola $y = x^2$ and the line $y = x$

7. The solid bounded by the planes $y = 3x$, $y = 0$, $z = 0$, and $x + y + z = 4$

8. The solid bounded by the cylinders $x^2 + y^2 = 1$ and $x^2 + z^2 = 1$

9. The solid under the surface $z = xy$ and above the triangle in the xy-plane with vertices $(0, 1)$, $(1, 1)$, and $(1, 2)$

10. Compute $\int_0^{\pi/3} \left(\int_0^{\pi/2} \cos(x + y)\, dx \right) dy$ using $\cos(x + y) = \cos x \cos y - \sin x \sin y$ and separation of variables. Check your result by direct evaluation.

11. Compute $\int_0^1 \left(\int_0^2 (1-x-y+xy)\,dx\right) dy$ using separation of variables. Check your result by direct evaluation.

Exercises 12 to 15: Sketch the region of integration and reverse the order of integration. Do not solve the integrals.

12. $\displaystyle\int_0^{\pi} \left(\int_0^{\sin(x/2)} x^3y^2\,dy\right) dx$

13. $\displaystyle\int_0^1 \left(\int_{\arctan x}^{\pi/4} (y^2-x)\,dy\right) dx$

14. $\displaystyle\int_0^1 \left(\int_{x/2}^{x} x^2y^2\,dy\right) dx + \int_1^2 \left(\int_{x/2}^{1} x^2y^2\,dy\right) dx$

15. $\displaystyle\int_1^2 \left(\int_1^{2y} \frac{\ln x}{x}\,dx\right) dy$

Exercises 16 to 22: Evaluate the following integrals by reversing the order of integration.

16. $\displaystyle\int_0^1 \left(\int_y^1 e^{x^2}\,dx\right) dy$

17. $\displaystyle\int_0^1 \left(\int_0^{\arccos y} x\,dx\right) dy$

18. $\displaystyle\int_0^1 \left(\int_{y^{1/3}}^1 e^{x^4}\,dx\right) dy$

19. $\displaystyle\int_0^3 \left(\int_{x^2}^9 x\cos(2y^2)\,dy\right) dx$

20. $\displaystyle\int_{1/2}^1 \left(\int_1^{2y} \frac{\ln x}{x}\,dx\right) dy + \int_1^2 \left(\int_y^2 \frac{\ln x}{x}\,dx\right) dy$

21. $\displaystyle\int_1^2 \left(\int_1^{\sqrt{y}} 5\,dx\right) dy + \int_2^4 \left(\int_{y/2}^{\sqrt{y}} 5\,dx\right) dy$

22. $\displaystyle\int_0^1 \left(\int_0^{\arcsin x} y^2\,dy\right) dx$

23. Compute $\iint_D e^{x^3}\,dA$ over the region bounded by $y=x^2$, $x=1$, and $y=0$.

24. Compute the area of the region $x^2+y^2 \le 9$ that lies to the left of the line $x=1/5$.

▶ 6.4 CHANGE OF VARIABLES IN A DOUBLE INTEGRAL

Sometimes, the evaluation of a double integral $\iint_D f\,dA$ is difficult because either the region D is geometrically complicated, or the function f and/or D give rise to an integrand that is hard to handle. One possible way to solve this problem is to use the change of variables technique.

Let us recall how change of variables (also known as the substitution rule) works for functions of one variable. Consider the definite integral $\int_1^2 e^{5x}\,dx$. Let $u=5x$, so that $x=u/5$; this means that x is now viewed as a function of u, $x(u)=u/5$. Then $dx = x'(u)\,du = (1/5)\,du$ and

$$\int_1^2 e^{5x}\,dx = \int_5^{10} e^u \tfrac{1}{5}\,du,$$

where the limits of integration have been changed accordingly (when $x=1$, $u=5$; when $x=2$, $u=10$). One more example: consider the integral $\int_1^2 e^{-5x}\,dx$. Using $u=-5x$, so that $x = -u/5$, we get $dx = (-1/5)du$ and

$$\int_1^2 e^{-5x}dx = \int_{-5}^{-10} e^u\left(-\tfrac{1}{5}\right)du = \int_{-10}^{-5} e^u\tfrac{1}{5}\,du,$$

where the minus sign was used to switch the limits of integration. In general, to compute $\int_a^b f(x)dx$, we set $x = x(u)$. Then $dx = x'(u)du = (dx/du)du$ and

$$\int_I f(x)dx = \int_{a^*}^{b^*} f(x(u))\frac{dx}{du}du = \int_{I^*} f(x(u))\left|\frac{dx}{du}\right|du,$$

where $I = [a, b]$, $x(a^*) = a$, $x(b^*) = b$, and $I^* = [\min\{a^*, b^*\}, \max\{a^*, b^*\}]$. Whenever we use interval notation, like $[\alpha, \beta]$, we must have $\alpha \le \beta$. This is what the fuss involving min and max and the absolute value is all about (recall the example of $\int_1^2 e^{-5x}dx$ above).

It follows that the change of variables for functions of one variable can be written as

$$\int_I f(x)dx = \int_{I^*} f(x(u))\left|\frac{dx}{du}\right|du.$$

And this is precisely the formula that we will be able to generalize to functions of two (and later, three) variables. The ingredients in the integral on the right side are the composition of f and x (remember that x is now a function of u) and the transformed interval I^* (the function x maps I^* to I).

Assume that $f = f(x, y)$ is a function of two variables. A *change of variables* is defined by setting

$$x = x(u, v), \qquad y = y(u, v),$$

where x and y are differentiable functions of u and v. In terms of functions, the change of variables can be described as a differentiable map $T\colon D^* \to D$, where D^* and D are regions in $\mathbb{R}^2$ and

$$T(u, v) = (x(u, v), y(u, v))$$

(instead of the usual notation T_1, T_2 for the components of T, we use x and y, to emphasize that we are thinking of them as forming a coordinate system in $\mathbb{R}^2$).

A successful change of variables (i.e., a good choice of T) will reduce the integration over a region D to an integration over a simpler region D^*. Before stating the theorem, we have to understand how T works. So, first of all, we study the properties of maps $T\colon \mathbb{R}^2 \to \mathbb{R}^2$.

A good way to understand functions is to draw their graphs. However, drawing the graph of T would require four dimensions: two for the independent variables and two for the range (a different view of T, that of a vector field in $\mathbb{R}^2$, has been studied before—unfortunately, it is of no help to us now). To overcome this problem, we proceed as follows: we use one coordinate system as the domain of T (and call it the *uv-plane* or the *uv-coordinate system*) and another as the range (and call it the *xy-plane* or the *xy-coordinate system*). We visualize T by investigating its effect on different objects (points, lines, rectangles, regions, etc.) in the domain.

▶ EXAMPLE 6.29

Consider the map $T\colon \mathbb{R}^2 \to \mathbb{R}^2$ given by $T(u, v) = (u + 2v, 3u - v)$. Find the image $T(D^*)$ of the rectangle $D^* = [0, 1] \times [0, 2]$ under T.

SOLUTION

Let us first compute the image of a horizontal line $v = k$ (k is a constant) in the uv-plane. Since $T(u, k) = (u + 2k, 3u - k)$, it follows that in the xy-plane, $x = u + 2k$ and $y = 3u - k$. Eliminating u (e.g., compute u from the first equation and substitute into the second), we get $y = 3x - 7k$. Therefore, horizontal lines $v = k$ map to parallel lines (all with slope 3) $y = 3x - 7k$.

Similarly, the image of a vertical line $u = k$ is $T(k, v) = (k + 2v, 3k - v)$. Hence, $x = k + 2v$, $y = 3k - v$, and (after eliminating v) $y = -x/2 + 7k/2$. In words, vertical lines $u = k$ map to parallel lines $y = -x/2 + 7k/2$.

The rectangle D^* is bounded by horizontal lines $v = 0$ and $v = 2$ and vertical lines $u = 0$ and $u = 1$. We can now find its image. T maps the line $v = 0$ to the line $y = 3x$ and the line $v = 2$ to the line $y = 3x - 14$. Moreover, any horizontal line $v = k$, $0 < k < 2$ (i.e., between $v = 0$ and $v = 2$) is mapped to the line $y = 3x - 7k$ (where $0 < 7k < 14$) that lies between the images $y = 3x$ and $y = 3x - 14$ of $v = 0$ and $v = 2$. Therefore, T maps the horizontal strip $0 \le v \le 2$ to the (slanted) strip bounded by $y = 3x$ and $y = 3x - 14$; see Figure 6.42.

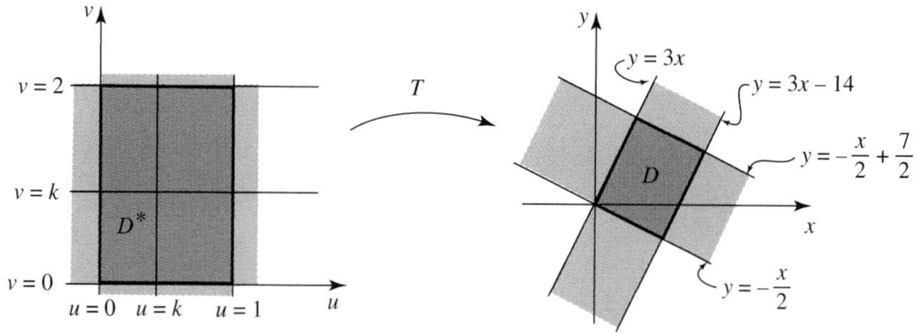

Figure 6.42 Map T of Example 6.29.

Similarly, T maps $u = 0$ to $y = -x/2$, $u = 1$ to $y = -x/2 + 7/2$, and every vertical line $u = k$, $0 < k < 1$, between $u = 0$ and $u = 1$ to the line $y = -x/2 + 7k/2$ between the images $y = -x/2$ and $y = -x/2 + 7/2$ of $u = 0$ and $u = 1$ in the xy-plane. Consequently, T maps the vertical strip $0 \le u \le 1$ to the (slanted) strip bounded by $y = -x/2$ and $y = -x/2 + 7/2$.

Therefore, the image D of the rectangle D^* is the parallelogram bounded by $y = 3x$, $y = 3x - 14$, $y = -x/2$, and $y = -x/2 + 7/2$. ◀

Sometimes, we use a table to keep track of how boundary curves map under T. Shown below is the table for the mapping T from the previous example.

D^* (uv-plane)	D (xy-plane)
$v = 0$	$y = 3x$
$v = 2$	$y = 3x - 14$
$u = 0$	$y = -x/2$
$u = 1$	$y = -x/2 + 7/2$

▶ EXAMPLE 6.30 Polar Coordinates

The change from Cartesian to polar coordinates can be described as $x = r\cos\theta$, $y = r\sin\theta$, or as the mapping

$$T(r, \theta) = (r\cos\theta, r\sin\theta)$$

(following tradition, we use r and θ instead of u and v). Compute the image of the rectangle $[1, \sqrt{2}] \times [0, \pi]$ in the $r\theta$-plane under the map T.

SOLUTION

We proceed as in the previous example: the image of a horizontal line $\theta = k$ (k is a constant) is $T(r, k) = (r\cos k, r\sin k)$; then $x = r\cos k$, $y = r\sin k$, and (divide the two equations) $y/x = \tan k$; that is, $y = (\tan k)x$.

It follows that the horizontal lines $\theta = k$ map to the lines through the origin with slope $\tan k$ (if $k \neq \pi/2$). If $k = \pi/2$, then $x = 0$, $y = r$ and the image is the y-axis (as suspected); see Figure 6.43.

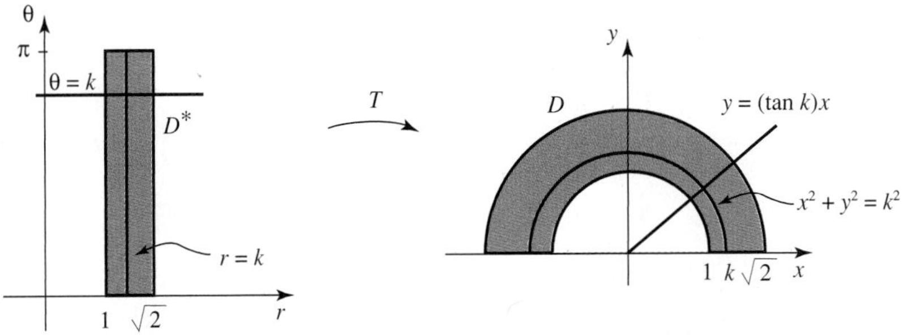

Figure 6.43 Polar mapping of Example 6.30.

The images of vertical lines $r = k \geq 0$ are $T(k, \theta) = (k\cos\theta, k\sin\theta)$; that is, $x = k\cos\theta$, $y = k\sin\theta$, and $x^2 + y^2 = k^2$. In words, vertical lines $r = k$ map to circles of radius k, if $k > 0$. If $k = 0$, the whole line maps to the origin. Therefore,

D^* ($r\theta$-plane)	D (xy-plane)
$\theta = 0$	$y = 0$
$\theta = \pi$	$y = 0$
$r = 1$	circle $x^2 + y^2 = 1$
$r = \sqrt{2}$	circle $x^2 + y^2 = 2$

The rectangle D^* is bounded by $r = 1$, $r = \sqrt{2}$, $\theta = 0$, and $\theta = \pi$. Its image $D = T(D^*)$ is bounded by the image curves: circles of radii 1 and $\sqrt{2}$ and the line $y = 0$. Since $r\sin\theta \geq 0$ for $0 \leq \theta \leq \pi$, D is the upper half of the annulus of radii 1 and $\sqrt{2}$. ◀

▶ EXAMPLE 6.31

Find the image of the rectangle $D^* = [1, 2] \times [0, 4]$ under $T(u, v) = (2u, 1)$.

SOLUTION The image of a horizontal line $v = k$ is $T(u, k) = (2u, 1)$; that is, $x = 2u$ (u runs over all real numbers) and $y = 1$. Therefore, every horizontal line maps to the same horizontal line, $y = 1$. The vertical line $u = k$, $1 \le k \le 2$, maps to $T(k, v) = (2k, 1)$ (k is fixed here); that is, to the point $(2k, 1)$ on the line $y = 1$. It follows that T "squishes" the rectangle D^* onto the line segment D on the line $y = 1$, from $x = 2$ to $x = 4$. ◀

Not every function $T: \mathbb{R}^2 \to \mathbb{R}^2$ can represent a change of variables (the map of Example 6.31 is certainly not likely to be such a map).

Although we defined one-to-one and onto functions earlier, we repeat the definition here, in the context of mappings from $\mathbb{R}^2$ to $\mathbb{R}^2$.

DEFINITION 6.5 One-to-One and Onto Function

A function $T: D^* \subseteq \mathbb{R}^2 \to \mathbb{R}^2$ is called *one-to-one* if for each (u, v) and (u', v') in D^*, $T(u, v) = T(u', v')$ implies that $u = u'$ and $v = v'$.

A function $T: D^* \subseteq \mathbb{R}^2 \to \mathbb{R}^2$ is called *onto* D if for every point $(x, y) \in D$ there is a point (u, v) in D^* such that $T(u, v) = (x, y)$. ◀

According to Definition 6.5, a function or a map (they are synonyms) is one-to-one if whenever two points have the same image, they are actually the same point. Equivalently, a one-to-one function maps different points to different points. The function T of Example 6.31 is not one-to-one, since, for example, both $(1, 0)$ and $(1, 1)$ (actually all points on $u = 1$) map onto the point $(2, 1)$. The function T is onto D if every point in D is "hit" by one (or more) points from D^*.

▶ EXAMPLE 6.32

Show that the polar map of Example 6.30 is onto the semiannulus

$$D = \{(x, y) \mid 1 \le x^2 + y^2 \le 2,\ y \ge 0\}.$$

SOLUTION Pick any point $(x_0, y_0) \in D$. We have to find a point (or points) in D^* that is (are) mapped to (x_0, y_0). The point (x_0, y_0) lies on the circle $x^2 + y^2 = x_0^2 + y_0^2$ of radius $\sqrt{x_0^2 + y_0^2}$ and on the line $y = \arctan(y_0/x_0)x$ through the origin; see Figure 6.44. In Example 6.30, we showed that horizontal lines map to lines through the origin, and vertical lines map to circles. In particular, the vertical line $r = \sqrt{x_0^2 + y_0^2}$ will map to the circle of radius $\sqrt{x_0^2 + y_0^2}$ and the horizontal line $\theta = \arctan(y_0/x_0)$ will map to the line of slope $\arctan(y_0/x_0)$ through the origin. The point A that is the intersection of the two lines in the $r\theta$-plane is mapped to (x_0, y_0). ◀

We have also noticed that the function T distorts the region it is applied to. We now introduce the function that will measure that distortion. Its precise meaning will be discussed later in this section.

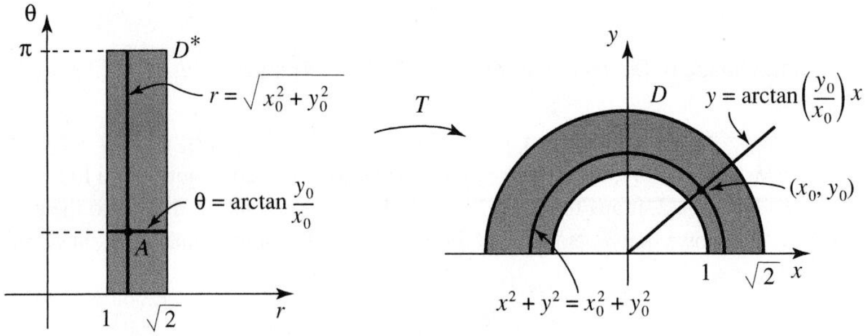

Figure 6.44 The polar map is onto the semiannulus.

DEFINITION 6.6 Jacobian of a Function $T: \mathbb{R}^2 \to \mathbb{R}^2$

Let $T(u, v) = (x(u, v), y(u, v))$ be a C^1 function defined on a region $D^* \subseteq \mathbb{R}^2$. The determinant of the derivative DT of T is called the *Jacobian* of T and is denoted by $J(x, y; u, v)$ or $\partial(x, y)/\partial(u, v)$. Hence,

$$\frac{\partial(x, y)}{\partial(u, v)} = \begin{vmatrix} \partial x/\partial u & \partial x/\partial v \\ \partial y/\partial u & \partial y/\partial v \end{vmatrix}.$$

◀

▶ EXAMPLE 6.33

Consider the mapping (of Example 6.30) $T(r, \theta) = (r\cos\theta, r\sin\theta)$ that defines polar coordinates in the plane. Show that it is one-to-one on $D^* = [1, \sqrt{2}] \times [0, \pi]$ and compute its Jacobian.

SOLUTION To prove that T is one-to-one, we have to show that $T(r, \theta) = T(r', \theta')$ implies that $r = r'$ and $\theta = \theta'$ [i.e., if two points (r, θ) and (r', θ') were mapped by T into the same point, then (r, θ) and (r', θ') must have been the same point to begin with]. Now $T(r, \theta) = T(r', \theta')$ implies that $(r\cos\theta, r\sin\theta) = (r'\cos\theta', r'\sin\theta')$, and hence,

$$r\cos\theta = r'\cos\theta' \tag{6.3}$$

and

$$r\sin\theta = r'\sin\theta', \tag{6.4}$$

for $1 \le r, r' \le \sqrt{2}$ and $0 \le \theta, \theta' \le \pi$.

If $\theta = \pi/2$, then (6.3) reads $r'\cos\theta' = 0$; hence, $\cos\theta' = 0$ (since $1 \le r' \le \sqrt{2}$, so r' cannot be 0) and therefore $\theta' = \pi/2$ (which is the only solution such that $0 \le \theta' \le \pi$). Now (6.4) with $\theta = \theta' = \pi/2$ gives $r = r'$. The case $\theta' = \pi/2$ is dealt with analogously.

If $\theta, \theta' \neq \pi/2$, then we can divide (6.4) by (6.3), thus getting $\tan\theta = \tan\theta'$. Since $0 \le \theta, \theta' \le \pi$, we conclude that either $\theta = 0$ and $\theta' = \pi$, or $\theta = \pi$ and $\theta' = 0$, or $\theta = \theta'$. If $\theta = 0, \theta' = \pi$, then (6.3) gives $r = -r'$, which is impossible since $1 \le r, r' \le \sqrt{2}$ (i.e., both r and r' are positive). The second case is ruled out in the same way. In the third case, either (6.3) or (6.4) imply that $r = r'$.

Therefore, $(r, \theta) = (r', \theta')$, so T is a one-to-one mapping on $D^* = [1, \sqrt{2}] \times [0, \pi]$. Moreover, T is C^1 (the component functions are C^1), and the Jacobian is computed to be

$$\frac{\partial(x, y)}{\partial(r, \theta)} = \begin{vmatrix} \cos\theta & -r\sin\theta \\ \sin\theta & r\cos\theta \end{vmatrix} = r. \tag{6.5}$$

Formula (6.5) is worth remembering, as it will be used quite often. ◀

EXAMPLE 6.34

Show that the map $T(u, v) = (u + 2v, 3u - v)$ of Example 6.29 is one-to-one and onto D and compute its Jacobian.

SOLUTION

From $T(u, v) = T(u', v')$, it follows that $(u + 2v, 3u - v) = (u' + 2v', 3u' - v')$; that is, $u + 2v = u' + 2v'$ and $3u - v = 3u' - v'$. Multiply the second equation by 2 and add the two equations, thus obtaining $7u = 7u'$; that is, $u = u'$. Either of the two equations now yields $v = v'$. Therefore, T is one-to-one.

To show that T is onto, pick any point (x_0, y_0) in D. We have to find a point (u_0, v_0) in D^* such that $T(u_0, v_0) = (x_0, y_0)$. We could argue geometrically as in the case of the polar coordinates mapping: (x_0, y_0) belongs to the intersection of the two lines that are parallel to the sides of the parallelogram. All we have to do is to identify the two lines in D^* that map to those lines and find their intersection. Here is an alternative way: from $T(u_0, v_0) = (u_0 + 2v_0, 3u_0 - v_0) = (x_0, y_0)$, it follows that $u_0 + 2v_0 = x_0$ and $3u_0 - v_0 = y_0$. Solving for u_0 and v_0, we get $u_0 = (x_0 + 2y_0)/7$ and $v_0 = (3x_0 - y_0)/7$. To check that, we compute

$$
\begin{aligned}
T(u_0, v_0) &= T\left(\frac{x_0 + 2y_0}{7}, \frac{3x_0 - y_0}{7}\right) \\
&= \left(\frac{x_0 + 2y_0}{7} + 2\frac{3x_0 - y_0}{7}, 3\frac{x_0 + 2y_0}{7} - \frac{3x_0 - y_0}{7}\right) = (x_0, y_0).
\end{aligned}
$$

Clearly, T is a C^1 function. Its Jacobian T is

$$
\frac{\partial(x, y)}{\partial(u, v)} = \begin{vmatrix} 1 & 2 \\ 3 & -1 \end{vmatrix} = -7.
$$

Our next definition introduces a special class of maps, called affine (and in a special case, linear) maps.

Identify a point (u, v) in $\mathbb{R}^2$ with the tip of the vector $\mathbf{u} = (u, v)$, that is represented as a 2×1 matrix $\begin{bmatrix} u \\ v \end{bmatrix}$. Let A be a 2×2 matrix $A = \begin{bmatrix} a_{11} & a_{12} \\ a_{21} & a_{22} \end{bmatrix}$ and let $\mathbf{b} = \begin{bmatrix} b_1 \\ b_2 \end{bmatrix}$.

DEFINITION 6.7 Linear and Affine Maps

A map $T\colon \mathbb{R}^2 \to \mathbb{R}^2$ defined by

$$
T\begin{bmatrix} u \\ v \end{bmatrix} = \begin{bmatrix} a_{11} & a_{12} \\ a_{21} & a_{22} \end{bmatrix}\begin{bmatrix} u \\ v \end{bmatrix} + \begin{bmatrix} b_1 \\ b_2 \end{bmatrix}
$$

is called an *affine map*. If $b = \begin{bmatrix} 0 \\ 0 \end{bmatrix}$, then T it is called a *linear map*.

With the identifications announced before the definition, we write $T(\mathbf{u}) = A\mathbf{u} + \mathbf{b}$, where the operations on the right side are matrix operations, and the resultant 2×1 matrix is interpreted as a point in $\mathbb{R}^2$. The formula for T can be expanded as

$$
T(u, v) = (a_{11}u + a_{12}v + b_1, a_{21}u + a_{22}v + b_2).
$$

We have already met a linear map in Example 6.29 (and also in Section 1.4, where we called it a linear vector field) and an affine map in Example 6.31. Our next theorem will simplify computations with affine (and therefore also linear) maps.

THEOREM 6.6 Properties of Affine Maps

Let $T(\mathbf{u}) = A\mathbf{u} + \mathbf{b}$ be an affine map and assume that $\det(A) \neq 0$. Then

(a) T is one-to-one.

(b) T maps lines to lines, parallel lines to parallel lines, and the intersection of two lines to the intersection of their images.

(c) If D^* is a parallelogram, then $T(D^*)$ is a parallelogram.

(d) If D is a parallelogram, and D^* is a region such that $T(D^*) = D$, then D^* is a parallelogram. ◀

The proofs are technical and are left as exercises (see Exercise 12).

When we use the change of variables theorem, we will have to find the region D^* from $T(D^*) = D$; that is, the setup will be such that the given region of integration will be the range of T, and the new region of integration that is obtained from the change of variables will be the domain of T.

In the case where D is a rectangle (or polygon, in general), Theorem 6.6 will help us find D^*: all we need is to identify the vertices of D^* and connect them with straight lines.

▶ EXAMPLE 6.35

The change of variables $x = u + 2v$, $y = u - v$ can be represented as a map $T\colon \mathbb{R}^2 \to \mathbb{R}^2$ given by

$$T(u, v) = (u + 2v, u - v).$$

Let D be the square $[0, 1] \times [0, 1]$. Find the region D^* that maps to D and compute the Jacobian of T.

SOLUTION

The map T is linear, and therefore in order to find D^*, we have to find the points that map to the vertices $(0, 0)$, $(1, 0)$, $(0, 1)$, and $(1, 1)$ of D, and then join them with straight lines.

From $u + 2v = 0$ and $u - v = 0$, it follows that $(u = 0, v = 0)$ maps to $(x = 0, y = 0)$. From $u + 2v = 1$ and $u - v = 0$, we see that the point $(u = 1/3, v = 1/3)$ maps to $(x = 1, y = 0)$. Similarly, we check that $(2/3, -1/3)$ maps to $(0, 1)$ and that $(1, 0)$ maps to $(1, 1)$; therefore, D^* is the parallelogram with vertices $(0, 0)$, $(1/3, 1/3)$, $(2/3, -1/3)$, and $(1, 0)$.

The fact that T is one-to-one follows from

$$\det(A) = \det \begin{bmatrix} 1 & 2 \\ 1 & -1 \end{bmatrix} = -3 \neq 0$$

and Theorem 6.6(a). The map T is clearly C^1 ($u + 2v$ and $u - v$ are C^1 functions of u and v), and its Jacobian is

$$\frac{\partial(x, y)}{\partial(u, v)} = \begin{vmatrix} 1 & 2 \\ 1 & -1 \end{vmatrix} = -3.$$

◀

The fact that $\det(A)$ is equal to the Jacobian of T is not a coincidence, but holds for all affine maps (the proof is straightforward, see Exercise 11).

Recall that a change of variables $x = x(u, v)$, $y = y(u, v)$, where x and y are C^1 functions of u and v, can be described as a C^1, one-to-one mapping $T(u, v) = (x(u, v), y(u, v))$.

THEOREM 6.7 Change of Variables Formula

Let D and D^* be elementary regions in $\mathbb{R}^2$ and let $T\colon D^* \to D$ be a C^1, one-to-one map such that $T(D^*) = D$. For any integrable function $f\colon D \to \mathbb{R}$,

$$\iint_D f(x, y)\, dA = \iint_{D^*} f(x(u, v), y(u, v)) \left| \frac{\partial(x, y)}{\partial(u, v)} \right| dA^*, \tag{6.6}$$

where $\dfrac{\partial(x, y)}{\partial(u, v)}$ is the Jacobian of T.

In words, the integral of f over D equals the integral over D^* of the composition $f \circ T$, multiplied by the absolute value of the Jacobian of T. Instead of giving the (technical and somewhat long) proof of the theorem, we will explain the geometric meaning of (6.6).

Subdivide the region D^* into small rectangles in the usual way: enclose D^* into a rectangle, form a division of that rectangle into n^2 (n is a large integer) subrectangles, and consider only those subrectangles whose intersection with D^* is non-empty (see Figure 6.45). Choose one of those subrectangles and name it R^*. Label its sides by Δu and Δv and assume that one of its vertices is located at (u_0, v_0). Label the diametrically opposite vertex by $(u, v) = (u_0 + \Delta u, v_0 + \Delta v)$.

The area of R^* is $\Delta u \Delta v$, and the sum $\sum \Delta u \Delta v$ over all subrectangles that intersect D^* (the others were thrown away) approximates the area of D^*.

The image $R = T(R^*)$ of the rectangle R^* is probably no longer a rectangle or a parallelogram. It is the region in $\mathbb{R}^2$ bounded by the images by T of the horizontal lines ($v = v_0$ and $v = v_0 + \Delta v$) and vertical lines ($u = u_0$ and $u = u_0 + \Delta u$) that form the boundary of R^*. We will now approximate the area of R. Recall that a function $T\colon \mathbb{R}^2 \to \mathbb{R}^2$, $T(u, v) = (x(u, v), y(u, v))$, maps points to points, and its derivative $DT\colon \mathbb{R}^2 \to \mathbb{R}^2$,

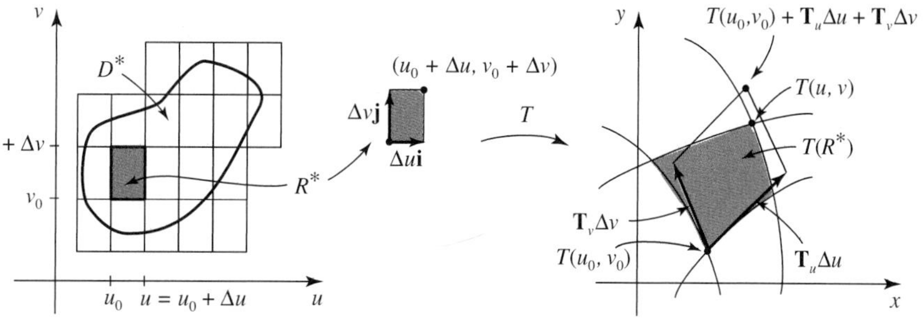

Figure 6.45 Small rectangle R^* and its image $R = T(R^*)$.

given by

$$DT(u,v) = \begin{bmatrix} \frac{\partial x}{\partial u} & \frac{\partial x}{\partial v} \\ \frac{\partial y}{\partial u} & \frac{\partial y}{\partial v} \end{bmatrix},$$

maps tangent vectors to tangent vectors. The vector $\Delta u\mathbf{i}$ is tangent to the line $v = v_0$ at (u_0, v_0). Therefore, its image

$$\begin{aligned} DT(u_0,v_0)(\Delta u\mathbf{i}) &= \begin{bmatrix} \frac{\partial x}{\partial u}(u_0,v_0) & \frac{\partial x}{\partial v}(u_0,v_0) \\ \frac{\partial y}{\partial u}(u_0,v_0) & \frac{\partial y}{\partial v}(u_0,v_0) \end{bmatrix} \cdot \begin{bmatrix} \Delta u \\ 0 \end{bmatrix} \\ &= \frac{\partial x}{\partial u}\bigg|_{(u_0,v_0)} \Delta u\mathbf{i} + \frac{\partial y}{\partial u}\bigg|_{(u_0,v_0)} \Delta u\mathbf{j} = \mathbf{T}_u(u_0,v_0)\Delta u, \end{aligned}$$

where $\mathbf{T}_u(u_0, v_0) = (\partial x/\partial u)(u_0, v_0)\mathbf{i} + (\partial y/\partial u)(u_0, v_0)\mathbf{j}$, is the vector tangent to the image of the line $v = v_0$ at $T(u_0, v_0)$. All functions involved (T, $\partial x/\partial u$, $\partial x/\partial v$, etc.) are evaluated at (u_0, v_0). We will keep this in mind but, for simplicity, will drop (u_0, v_0) from the notation. Similarly, at (u_0, v_0), the image of $\Delta v\mathbf{j}$ under DT is computed to be

$$DT(\Delta v\mathbf{j}) = \begin{bmatrix} \frac{\partial x}{\partial u} & \frac{\partial x}{\partial v} \\ \frac{\partial y}{\partial u} & \frac{\partial y}{\partial v} \end{bmatrix} \cdot \begin{bmatrix} 0 \\ \Delta v \end{bmatrix} = \frac{\partial x}{\partial v}\Delta v\mathbf{i} + \frac{\partial y}{\partial v}\Delta v\mathbf{j} = \mathbf{T}_v\Delta v,$$

where $\mathbf{T}_v = (\partial x/\partial v)\mathbf{i} + (\partial y/\partial v)\mathbf{j}$.

On the other hand,

$$T(u,v) \approx T(u_0,v_0) + DT(u_0,v_0)\begin{bmatrix} u - u_0 \\ v - v_0 \end{bmatrix},$$

where $u = u_0 + \Delta u$, $v = v_0 + \Delta v$ [read formula (2.20) in Section 2.4 with $\mathbf{F} = T$, $\mathbf{a} = (u_0, v_0)$ and $\mathbf{x} = (u, v)$]. Therefore,

$$\begin{aligned} T(u,v) \approx T(u_0,v_0) + \begin{bmatrix} \frac{\partial x}{\partial u} & \frac{\partial x}{\partial v} \\ \frac{\partial y}{\partial u} & \frac{\partial y}{\partial v} \end{bmatrix}\begin{bmatrix} \Delta u \\ \Delta v \end{bmatrix} &= T(u_0,v_0) + \begin{bmatrix} \frac{\partial x}{\partial u}\Delta u + \frac{\partial x}{\partial v}\Delta v \\ \frac{\partial y}{\partial u}\Delta u + \frac{\partial y}{\partial v}\Delta v \end{bmatrix} \\ = T(u_0,v_0) + \left(\frac{\partial x}{\partial u}\Delta u + \frac{\partial x}{\partial v}\Delta v\right)\mathbf{i} + \left(\frac{\partial y}{\partial u}\Delta u + \frac{\partial y}{\partial v}\Delta v\right)\mathbf{j} \\ = T(u_0,v_0) + \mathbf{T}_u\Delta u + \mathbf{T}_v\Delta v. \end{aligned}$$

It follows that the area $A(R)$ of R is approximated by the area of the parallelogram spanned by $\mathbf{T}_u \Delta u$ and $\mathbf{T}_v \Delta v$, which is equal to $\|\mathbf{T}_u \Delta u \times \mathbf{T}_v \Delta v\|$. But

$$\mathbf{T}_u \Delta u \times \mathbf{T}_v \Delta v = \begin{vmatrix} \mathbf{i} & \mathbf{j} & \mathbf{k} \\ \frac{\partial x}{\partial u}\Delta u & \frac{\partial y}{\partial u}\Delta u & 0 \\ \frac{\partial x}{\partial v}\Delta v & \frac{\partial y}{\partial v}\Delta v & 0 \end{vmatrix} = \Delta u \Delta v \left(\frac{\partial x}{\partial u}\frac{\partial y}{\partial v} - \frac{\partial x}{\partial v}\frac{\partial y}{\partial u} \right) \mathbf{k}$$

$$= \begin{vmatrix} \frac{\partial x}{\partial u} & \frac{\partial x}{\partial v} \\ \frac{\partial y}{\partial u} & \frac{\partial y}{\partial v} \end{vmatrix} \Delta u \Delta v \, \mathbf{k} = \frac{\partial(x, y)}{\partial(u, v)} \Delta u \Delta v \, \mathbf{k},$$

and hence (Δu and Δv are positive),

$$A(R) = \|\mathbf{T}_u \Delta u \times \mathbf{T}_v \Delta v\| = \left| \frac{\partial(x, y)}{\partial(u, v)} \right| \Delta u \Delta v;$$

that is, $A(R)$ is approximately equal to the absolute value of the Jacobian multiplied by $\Delta u \Delta v$. Thus, the area of the image R under a map T of a small rectangle R^* is approximately equal to the absolute value of the Jacobian of T multiplied by the area of R^*. The sum

$$\sum_R \left| \frac{\partial(x, y)}{\partial(u, v)} \right| \Delta u \Delta v$$

(over all rectangles R) approximates the area of $T(D^*) = D$. Taking the limit as the rectangles become smaller and smaller, we determine that

$$\int_D \left| \frac{\partial(x, y)}{\partial(u, v)} \right| du dv \tag{6.7}$$

gives the area of D. Hence, the (integral of the absolute value of the) Jacobian describes how area changes under the map T.

▶ EXAMPLE 6.36

Approximate the area of the image R of a small rectangle R^* with sides $\Delta u = 0.1$, $\Delta v = 0.2$ and one vertex at $(u_0, v_0) = (2, 3)$ under the map $T(u, v) = (uv^2, u + v)$.

SOLUTION The area of R^* is 0.02. The area of R is approximately equal to the product of the absolute value of the Jacobian $\det(DT)$ at (2, 3) and the area of R^*. Therefore,

$$A(R) \approx \left| \det \begin{bmatrix} v^2 & 2uv \\ 1 & 1 \end{bmatrix}_{at\ (2,3)} \right| A(R^*) = |v^2 - 2uv|\Big|_{(2,3)} 0.02 = 0.06.$$

Let us look a bit closer into the way T distorts the area. The approximate area of the image R_1 of the rectangle with sides $\Delta u = 0.1$ and $\Delta v = 0.2$ (same sides as the rectangle R) and one vertex at $(1, 4)$ is

$$A(R_1) \approx \left| \det \begin{bmatrix} v^2 & 2uv \\ 1 & 1 \end{bmatrix}_{at\ (1,4)} \right| \Delta u \Delta v = |v^2 - 2uv| \Big|_{(1,4)} 0.02 = 0.16.$$

The approximate area of image R_2 of the rectangle with sides $\Delta u = 0.1$ and $\Delta v = 0.2$ and one vertex at $(0, 10)$ is

$$A(R_2) \approx |v^2 - 2uv| \Big|_{(0,10)} 0.02 = 2.$$

So, not only does T distort the area, but at different points the distortions are different. ◀

▶ EXAMPLE 6.37 Jacobian for Affine Maps

If T is an affine map (and, in particular, a linear map), then the above approximate computation becomes exact, since T maps parallelograms to parallelograms (so that there are no curves to be approximated by straight-line segments). Let us look at several maps and compute their Jacobians.

(a) Let $T\colon \mathbb{R}^2 \to \mathbb{R}^2$ be the translation $T(u, v) = (u + 1, v)$ in the horizontal direction. The Jacobian is

$$\frac{\partial(x, y)}{\partial(u, v)} = \begin{vmatrix} 1 & 0 \\ 0 & 1 \end{vmatrix} = 1,$$

and hence there is no change in the area; see Figure 6.46.

(b) Let $T(u, v) = (au, bv)$, where $a, b > 0$. In words, T is an expansion in the x-direction if $a > 1$ and an expansion in the y-direction if $b > 1$. If $a < 1$, then T is a contraction in the x-direction, and if $b < 1$, a contraction in the y-direction. The Jacobian of T is

$$\frac{\partial(x, y)}{\partial(u, v)} = \begin{vmatrix} a & 0 \\ 0 & b \end{vmatrix} = ab,$$

and the area changes by the factor ab; see Figure 6.47.

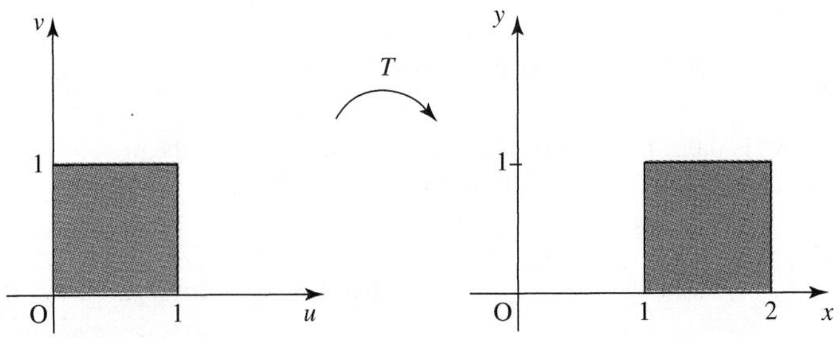

Figure 6.46 The map $T(u, v) = (u + 1, v)$ preserves area.

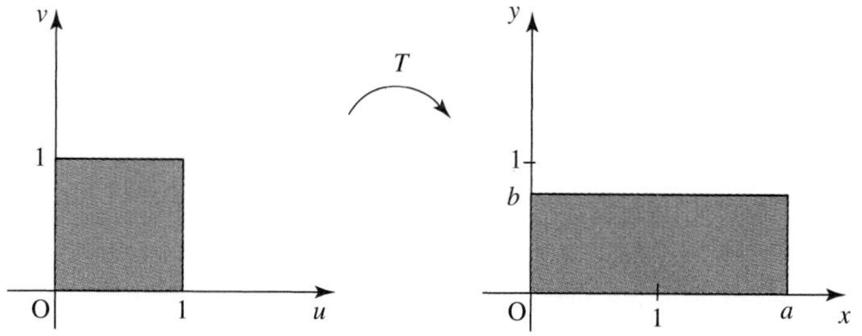

Figure 6.47 The map $T(u, v) = (au, bv)$ changes the area by the factor ab (pictured is a case for which $a > 1$ and $b < 1$).

(c) Let $T(u, v) = (u \cos\phi - v \sin\phi, u \sin\phi + v \cos\phi)$; that is, T is a rotation about the origin through the angle ϕ in the counterclockwise direction; see Figure 6.48. Then

$$\frac{\partial(x, y)}{\partial(u, v)} = \begin{vmatrix} \cos\phi & -\sin\phi \\ \sin\phi & \cos\phi \end{vmatrix} = 1,$$

so there is no change in the area.

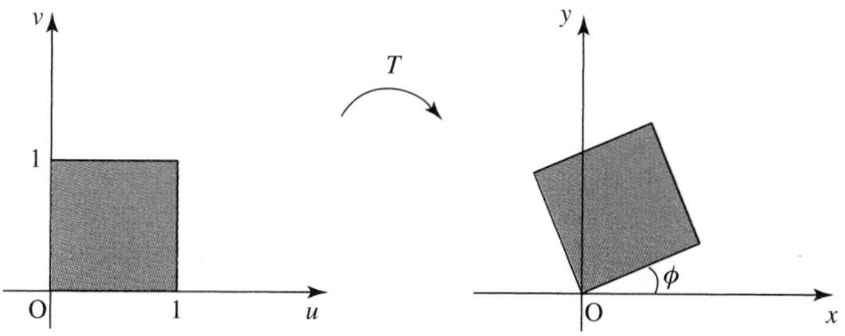

Figure 6.48 The map $T(u, v) = (u \cos\phi - v \sin\phi, u \sin\phi + v \cos\phi)$ preserves area. ◀

▶ EXAMPLE 6.38

Evaluate $\iint_D e^{x^2+y^2}\, dA$, where $D = \{(x, y) \mid 1 \le x^2 + y^2 \le 2, y \ge 0\}$ is the region of Example 6.30.

SOLUTION

Let $x = r\cos\theta$, $y = r\sin\theta$, with $1 \le r \le \sqrt{2}$ and $0 \le \theta \le \pi$. Then

$$\left|\frac{\partial(x, y)}{\partial(r, \theta)}\right| = |r| = r,$$

as in (6.5), and by Theorem 6.7,

$$\iint_D e^{x^2+y^2}\, dA = \iint_{D^*} e^{r^2} r\, dA^*,$$

where D^* is the rectangle $[1, \sqrt{2}] \times [0, \pi]$ in the $r\theta$-plane (the calculations were done in Example 6.30). Therefore,

$$\iint_D e^{x^2+y^2}\, dA = \int_0^{\pi} \left(\int_1^{\sqrt{2}} e^{r^2} r\, dr \right) d\theta$$

(substitute $u = r^2$, $du = 2rdr$)

$$= \int_0^{\pi} \left(\frac{1}{2} e^{r^2} \Big|_1^{\sqrt{2}} \right) d\theta = \left(\frac{1}{2}e^2 - \frac{1}{2}e \right) \int_0^{\pi} d\theta = \frac{1}{2}(e^2 - e)\pi.$$

▶ EXAMPLE 6.39

Compute $\iint_D (x + y)\, dA$, where D is the region $0 \le x \le 1, 0 \le y \le x$

(a) By using the change of variables $x = u + v$, $y = u - v$, and

(b) By direct computation.

SOLUTION Notice that $T(u, v) = (x, y) = (u + v, u - v)$ is a linear map and $A = \begin{bmatrix} 1 & 1 \\ 1 & -1 \end{bmatrix}$ its corresponding matrix.

(a) The Jacobian of the change of variables is

$$\frac{\partial(x, y)}{\partial(u, v)} = \begin{vmatrix} 1 & 1 \\ 1 & -1 \end{vmatrix} = -2,$$

and therefore,

$$\iint_D (x + y)\, dA = \iint_{D^*} 2u\, |-2|\, dA^*,$$

where D^* is the region that maps to D under the map $T(u, v) = (u + v, u - v)$.

The region D is the triangular region with vertices $(0, 0)$, $(1, 0)$, and $(1, 1)$. Now $x = u + v = 0$ and $y = u - v = 0$ imply that $u = v = 0$; that is, the point $(0, 0)$ in the uv-plane maps to the point $(0, 0)$ in the xy-plane. Analogously, $x = u + v = 1$ and $y = u - v = 0$ imply that $u = v = 1/2$; that is, the point $(1/2, 1/2)$ in the uv-plane maps to the point $(1, 0)$ in the xy-plane. Similarly, the point $(1, 0)$ in the uv-plane maps to $(1, 1)$ in the xy-plane; see Figure 6.49. The fact that triangles map to triangles follows from Theorem 6.6(b).

It follows that D^* is the triangle with vertices $(0, 0)$, $(1/2, 1/2)$, and $(1, 0)$. The fact that T is one-to-one follows from the fact that T is linear, since $\det(A) \neq 0$ (see Theorem 6.6(a)). Hence,

$$\iint_D (x + y)\, dA = 4 \int_0^{1/2} \left(\int_v^{1-v} u du \right) dv = 2 \int_0^{1/2} \left(u^2 \Big|_{u=v}^{u=1-v} \right) dv$$

$$= 2 \int_0^{1/2} (1 - 2v)\, dv = 2(v - v^2) \Big|_0^{1/2} = \frac{1}{2}.$$

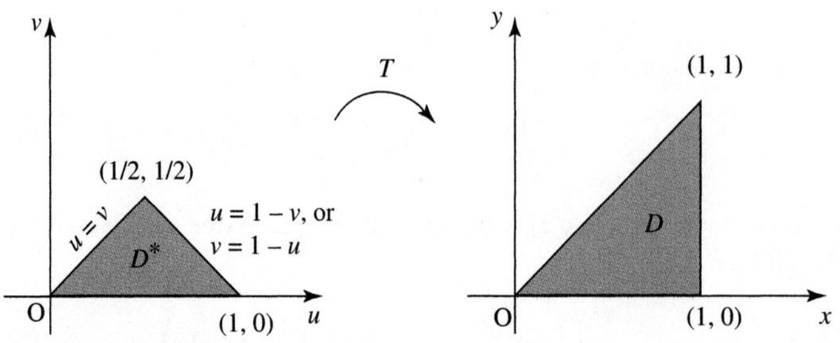

Figure 6.49 The map $T(u, v) = (u + v, u - v)$.

(b) Direct computation gives

$$\iint_D (x+y)\,dx\,dy = \int_0^1 \left(\int_0^x (x+y)\,dy \right) dx = \int_0^1 \left(xy + \frac{y^2}{2}\Big|_{y=0}^{y=x} \right) dx$$

$$= \frac{3}{2} \int_0^1 x^2\,dx = \frac{3}{2} \frac{x^3}{3}\Big|_0^1 = \frac{1}{2}.$$

▶ EXAMPLE 6.40

Using the change of variables $x = v$ and $y = u/v$, transform the integral $\iint_D x^2y^2\,dA$, where D is the region in the first quadrant bounded by the parabolas $y = x^2$ and $y = 2x^2$ and the hyperbolas $xy = 1$ and $xy = 2$.

SOLUTION The change of variables function T is defined by $T(u, v) = (v, u/v)$. First of all, we have to find the region D^* such that $T(D^*) = D$. From $y = x^2$, we get $u/v = v^2$ and $v = u^{1/3}$. From $y = 2x^2$, we get $u/v = 2v^2$ and $v = (u/2)^{1/3}$. Similarly, $xy = 1$ implies $u = 1$ and $xy = 2$ implies $u = 2$. It follows that D^* is the region of type 1 in the uv-plane, defined by $1 \le u \le 2$ and $(u/2)^{1/3} \le v \le u^{1/3}$; see Figure 6.50.

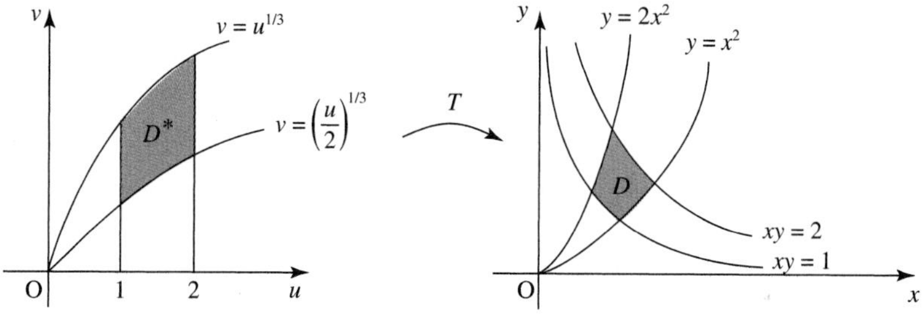

Figure 6.50 The function T maps the region D^* to D.

The function T is C^1 except at $v = 0$ (but that is irrelevant, since D^* is away from the u-axis), and its Jacobian is

$$\frac{\partial(x, y)}{\partial(u, v)} = \begin{vmatrix} 0 & 1 \\ 1/v & -u/v^2 \end{vmatrix} = \frac{1}{v}.$$

Therefore,

$$\iint_D x^2y^2\,dA = \iint_{D^*} v^2 \left(\frac{u}{v}\right)^2 \left|\frac{1}{v}\right| dA^* = \int_1^2 \left(\int_{(u/2)^{1/3}}^{u^{1/3}} \frac{u^2}{v}\,dv \right) du.$$

Sometimes, it is more or less obvious what T, that is, what change of variables (change of coordinates), to choose; such cases include polar coordinates (as in Example 6.38), or cylindrical and spherical coordinates (see Section 6.5). In general, however, identifying a change of variables that makes a given integration easier could be quite difficult.

▶ EXERCISES 6.4

Exercises 1 to 5: Evaluate the given double integral by converting to polar coordinates.

1. $\int_0^2 \left(\int_0^{\sqrt{4-x^2}} e^{x^2+y^2}\,dy \right) dx$

2. $\iint_D xy\, dA$, where D is the region in the first quadrant bounded by $x^2 + y^2 = 4$, $y = x/\sqrt{2}$, and $y = 0$

3. $\int_{-1}^{1} \left(\int_0^{\sqrt{1-x^2}} \arctan(y/x)\, dy \right) dx$

4. $\iint_D \sqrt{x^2 + y^2}\, dA$, where D is the semicircle $(x-1)^2 + y^2 = 1$, $y \geq 0$

5. $\iint_D \sqrt{2x^2 + 2y^2 + 3}\; dA$, where D is the disk $x^2 + y^2 \leq 1$

6. Convert the double integral $\int_0^2 \left(\int_0^{\sqrt{4x-x^2}} (x^2 + y^2)^{3/4}\, dy \right) dx$ to polar coordinates. Do not evaluate it.

7. Find the volume of the solid under the paraboloid $z = x^2 + y^2$, inside the cylinder $x^2 + y^2 = 5$, and above the xy-plane.

8. Find the volume of the solid between the paraboloids $z = 3x^2 + 3y^2$ and $z = 12 - 3x^2 - 3y^2$.

9. Find the volume of the solid inside the cylinder $x^2 + y^2 = 4$, inside the ellipsoid $4x^2 + 4y^2 + z^2 = 64$, and above the xy-plane.

10. Using a double integral, compute the volume of a sphere of radius $a > 0$.

11. Show that if $T(\mathbf{u}) = A\mathbf{u} + \mathbf{b}$ is an affine map, then $\det(A)$ is equal to the Jacobian of T.

12. Consider the map $T(\mathbf{u}) = A\mathbf{u} + \mathbf{b}$, where $A = \begin{bmatrix} a_{11} & a_{12} \\ a_{21} & a_{22} \end{bmatrix}$, $\mathbf{b} = \begin{bmatrix} b_1 \\ b_2 \end{bmatrix}$, and assume that $\det(A) \neq 0$.

(a) Show that T is one-to-one. [*Hint*: You will need the fact that if $\det(A) \neq 0$, then the 2×2 system $a_{11}X + a_{12}Y = 0$ and $a_{21}X + a_{22}Y = 0$ has a unique solution $X = Y = 0$.]

(b) Compute the image of the line $\mathbf{l}_1(t) = (w_1 + tv_1, w_2 + tv_2)$, $t \in \mathbb{R}$, under T. Next, compute the image of the line $\mathbf{l}_2(t) = (\overline{w_1} + tv_1, \overline{w_2} + tv_2)$, $t \in \mathbb{R}$, parallel to $\mathbf{l}_1(t)$. Conclude that parallel lines map to parallel lines.

(c) Consider two lines $\mathbf{l}_1(t) = (tv_1, tv_2)$, $t \in \mathbb{R}$, and $\mathbf{l}_2(t) = (tw_1, tw_2)$, $t \in \mathbb{R}$, that intersect at the origin. The origin is mapped to (b_1, b_2) under T. Compute the images of $\mathbf{l}_1(t)$ and $\mathbf{l}_2(t)$ and find their point of intersection.

(d) Show that $S(\mathbf{x}) = A^{-1}\mathbf{x} - A^{-1}\mathbf{b}$ is the inverse map of T, where A^{-1} is the inverse matrix of A [i.e., show that $S \circ T(u, v) = (u, v)$ and $T \circ S(x, y) = (x, y)$].

(e) Using (d), prove statement (d) of Theorem 6.6.

13. Approximate the area of the image of a small rectangle with sides $\Delta u = 0.1$ and $\Delta v = 0.05$ and one vertex located at (2, 4), under the mapping T defined by $T(u, v) = (\sqrt{u^2 + v^2}, uv)$.

14. Approximate the area of the image of a small rectangle with sides $\Delta u = 0.03$ and $\Delta v = 0.1$ and one vertex located at $(-2, 1)$, under the mapping T defined by $T(u, v) = (u \sin v, u \cos v)$.

15. Describe in words the map $T\colon \mathbb{R}^2 \to \mathbb{R}^2$, $T(u, v) = (au, v + b)$, where $a > 1$ and $b > 0$. What is the relation between the area of a region D and the area of its image $T(D)$?

16. Consider the map $T\colon \mathbb{R}^2 \to \mathbb{R}^2$ given by $T(u, v) = (u + v, v)$. Describe the region to which T maps a square whose sides are parallel to the coordinate axes. Compute the Jacobian of T and interpret the result geometrically.

17. Evaluate $\iint_D (5x + y^2 + x^2)\, dA$, where D is the part of the annulus $1 \leq x^2 + y^2 \leq 4$ in the upper half-plane.

18. Using a double integral, find the area of the region enclosed by one loop of the curve $r = \cos 2\theta$. [*Hint*: Sketch the curve.]

19. Find the area of the region inside the cardioid $r = 1 - \sin\theta, 0 \le \theta \le 2\pi$.

20. Express the volume of the right circular cone of radius r and height h as a double integral in polar coordinates.

21. Find the area of the region in the first quadrant bounded by the curves $r = \theta$ and $r = 2\theta$.

22. Compute the integral $\iint_D (4x + 6y)\, dA$, where D is the region bounded by the lines $4y = x - 3$, $4y = x + 2$, $2x + 3y = 6$, and $2x + 3y = 17$. [*Hint*: Use the change of variables $x = 4u - 3v$, $y = u + 2v$.]

23. Compute the integral $\iint_D (x^2 - y^2)\, dA$, where D is the region bounded by the curves $xy = 1$, $y = x - 1$, and $y = x + 1$, [*Hint*: Use the change of variables $x = u + v$, $y = -u + v$.]

24. Compute the integral $\iint_D (2x - y)\, dA$, where D is the region bounded by the curves $y = 2x$, $x = 2y$, and $x + y = 6$. [*Hint*: Use the change of variables $x = u - v$, $y = u + v$.]

25. Compute the volume of the wedge cut from the cylinder $x^2 + y^2 = 9$ by the planes $z = 0$ and $z = y + 3$.

26. Compute the volume of the solid below the plane $z = y + 4$ and above the disk $x^2 + y^2 \le 1$.

27. Compute the integral $\iint_D xy^3\, dA$, where D is the region in the first quadrant bounded by the lines $x = 1$ and $x = 2$ and the hyperbolas $xy = 1$ and $xy = 3$. [*Hint*: Use the change of variables $x = v$, $y = u/v$.]

28. Compute the integral $\iint_D 5\, dA$, where D is the region inside the ellipse $4x^2 + 2y^2 = 1$. [*Hint*: Define a change of variables so that the region of integration becomes a circle.]

29. Evaluate $\iint_D 5(x + y)\, dA$, where D is the region bounded by the lines $3x - 2y = 5$, $3x - 2y = -2$, $x + y = -2$, and $x + y = 1$ using a suitable change of variables.

30. Evaluate $\iint_D x^2\, dA$, where D is the region $0 \le \frac{1}{9}x^2 + y^2 \le 1$ using a suitable change of variables.

31. Evaluate $\iint_D e^x\, dA$, where D is the region defined by $x + y = 0$, $x + y = 2$, $y = x$, and $y = 2x$.

32. Evaluate $\iint_D (x^2 - y^2)\, dA$, where D is the region in the first quadrant bounded by the curves $x^2 - y^2 = 1$, $x^2 - y^2 = 2$, $y = 0$, and $y = x/2$, using the change of variables $x = u \cosh v$ and $y = u \sinh v$.

33. Evaluate $\iint_D \sin \frac{x+y}{x-y}\, dA$, where D is the region bounded by the lines $x - y = 1$, $x - y = 5$, and the coordinate axes.

▶ 6.5 TRIPLE INTEGRALS

The definition, properties, and methods of evaluation of triple integrals are analogous to those of double integrals. Nevertheless, for the sake of completeness, we will briefly go through the relevant concepts.

Assume that $f = f(x, y, z)$ is a *bounded* function defined on a closed and bounded solid W in $\mathbb{R}^3$. Recall that a function $f\colon W \to \mathbb{R}$ is bounded if there exists $M \ge 0$ such that $|f(x, y, z)| \le M$ for all $(x, y, z) \in W$. The fact that W is closed means that it contains the surface(s) that constitute(s) its boundary.

Enclose W into a big rectangular box (this is possible due to the assumption that W is bounded) and divide it into n^3 subboxes W_{ijk}, $i, j, k = 1, \ldots, n$, with faces parallel to

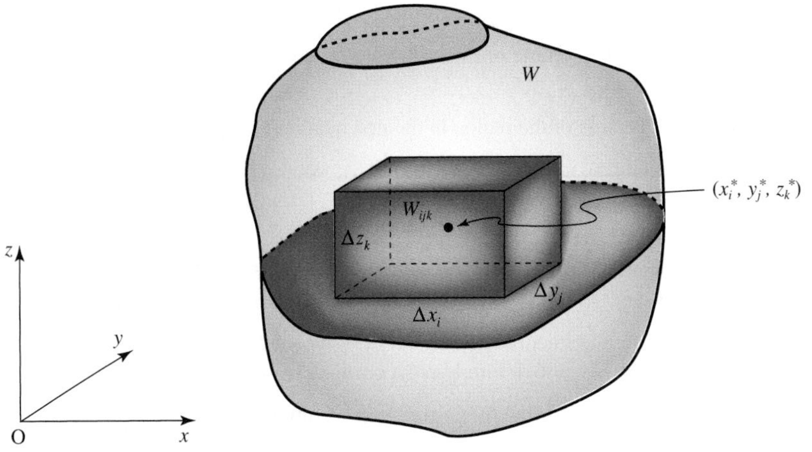

Figure 6.51 A subbox W_{ijk} from the definition of the Riemann sum of f.

coordinate planes; see Figure 6.51. Form the *(triple) Riemann sum*

$$\mathcal{R}_n = \sum_i \sum_j \sum_k f(x_i^*, y_j^*, z_k^*)\Delta x_i \Delta y_j \Delta z_k,$$

where (x_i^*, y_j^*, z_k^*) is any point in W_{ijk}, the product $\Delta x_i \Delta y_j \Delta z_k$ is the volume of W_{ijk}, and the sums run over those W_{ijk} that have a non-empty intersection with W.

DEFINITION 6.8 Triple Integral

The *triple integral* $\iiint_W f\, dV$ *of* f *over* W is defined by

$$\iiint_W f\, dV = \lim_{n\to\infty} \mathcal{R}_n,$$

whenever the limit on the right side exists. ◀

It can be proved that the definition does not depend on the way W is divided into subboxes W_{ijk} or on the choice of the point (x_i^*, y_j^*, z_k^*) in each W_{ijk}.

A function f for which the above triple integral is defined is called *integrable*. Continuous functions, as well as some bounded discontinuous functions (those whose points of discontinuity lie on the graph of a continuous function of two variables), are integrable (compare with the integrability of functions of two variables over regions in $\mathbb{R}^2$ discussed in Section 6.2). Continuous functions defined on a closed and bounded set are bounded (see Appendix A), so there was no need to say "continuous bounded functions" in the previous sentence.

From the construction, it follows that the integral $\iiint_W 1\, dV$ represents the *volume* of the solid W. Various symbols, such as $\int_W f dV$, $\int_W f dxdydz$, $\iiint_W f\, dxdydz$, and some others, are used to denote triple integrals. As in the two-dimensional case, the triple integral $\iiint_W f\, dV$ over the rectangular parallelepiped (i.e., the "rectangular box")

$$W = [a_1, b_1] \times [a_2, b_2] \times [a_3, b_3] = \{(x, y, z) \mid a_1 \le x \le b_1, a_2 \le y \le b_2, a_3 \le z \le b_3\}$$

can be computed as an iterated integral. This fact is contained in the following theorem that we present without proof.

THEOREM 6.8 Fubini's Theorem

Assume that $f = f(x, y, z)$ is a continuous function defined on a rectangular box $W = [a_1, b_1] \times [a_2, b_2] \times [a_3, b_3]$ in $\mathbb{R}^3$. Then

$$\iiint_W f\,dV = \int_{a_3}^{b_3} \left(\int_{a_2}^{b_2} \left(\int_{a_1}^{b_1} f(x, y, z) dx \right) dy \right) dz$$

$$= \int_{a_1}^{b_1} \left(\int_{a_3}^{b_3} \left(\int_{a_2}^{b_2} f(x, y, z) dy \right) dz \right) dx$$

$$= \int_{a_3}^{b_3} \left(\int_{a_1}^{b_1} \left(\int_{a_2}^{b_2} f(x, y, z) dy \right) dx \right) dz,$$

etc. There are six iterated integrals altogether.

In Section 6.2 , we defined elementary regions in $\mathbb{R}^2$ as regions of type 1, 2, or 3. To distinguish between those and their analogues in three dimensions, we will add (2D) or (3D) to their name, thus indicating their dimension. A *region of type 1(3D)* is the set of all points (x, y, z) such that:

(a) Points (x, y) belong to an elementary (2D) region D in $\mathbb{R}^2$, and

(b) $\kappa_1(x, y) \le z \le \kappa_2(x, y)$, where κ_1 and κ_2 denote continuous functions such that if $\kappa_1(x, y) = \kappa_2(x, y)$, then the point (x, y) belongs to the boundary of D.
In other words, if surfaces that bound a type-1(3D) region intersect, then they intersect at points on the boundary of D, and not in the inside of D; see Figure 6.52.

A *region of type 2(3D)* is defined similarly, with x and z interchanged. This means that the back and front of a type-2(3D) region are the surfaces $x = \kappa_1(y, z)$ and $x = \kappa_2(y, z)$, and the points (y, z) belong to an elementary (2D) region in the yz-plane. The surfaces $\kappa_1(y, z)$ and $\kappa_2(y, z)$ can only intersect along the boundary of that region; see Figure 6.53.

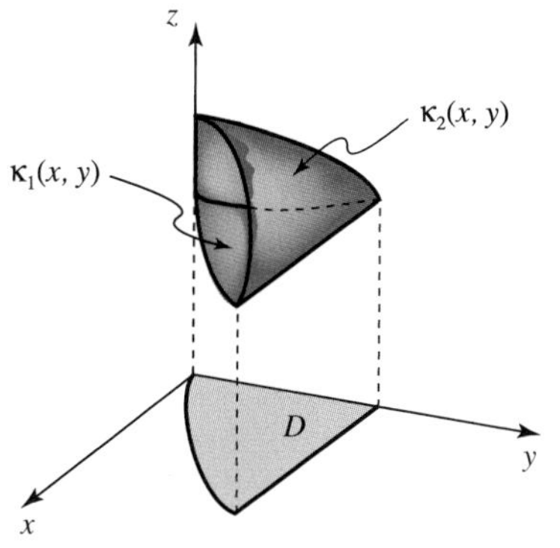

Figure 6.52 A region of type 1(3D).

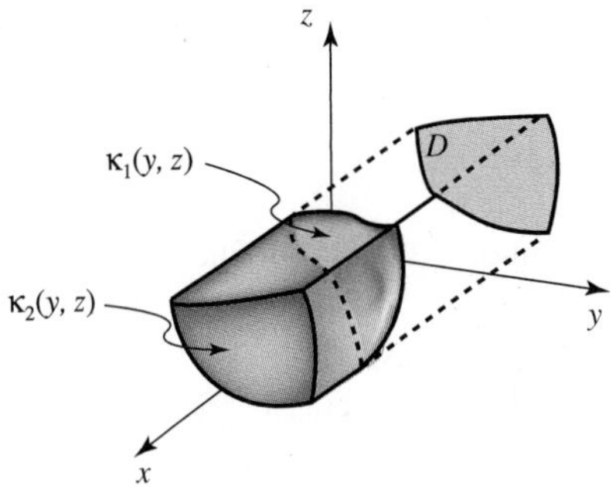

Figure 6.53 A region of type 2(3D).

A *region of type 3(3D)* is defined by interchanging y and z in the definition of a region of type 1(3D). A region is of *type 4(3D)* if it is of types 1(3D), 2(3D), and 3(3D). A three-dimensional region is called an *elementary 3D region* if it belongs to one (or more) of the four types defined above.

Now assume that W is of type 1(3D). Then either

$$\begin{cases} a \leq x \leq b, & \phi(x) \leq y \leq \psi(x) \\ \kappa_1(x, y) \leq z \leq \kappa_2(x, y) \end{cases}$$

if $D \subseteq \mathbb{R}^2$ is of type 1(2D), or

$$\begin{cases} c \leq y \leq d, & \phi(y) \leq x \leq \psi(y) \\ \kappa_1(x, y) \leq z \leq \kappa_2(x, y) \end{cases}$$

if $D \subseteq \mathbb{R}^2$ is of type 2(2D). Consequently,

$$\begin{aligned} \iiint_W f\, dV &= \iint_D \left(\int_{\kappa_1(x,y)}^{\kappa_2(x,y)} f(x, y, z) dz \right) dA \\ &= \int_a^b \left(\int_{\phi(x)}^{\psi(x)} \left(\int_{\kappa_1(x,y)}^{\kappa_2(x,y)} f(x, y, z) dz \right) dy \right) dx \end{aligned}$$

or

$$\begin{aligned} \iiint_W f\, dV &= \iint_D \left(\int_{\kappa_1(x,y)}^{\kappa_2(x,y)} f(x, y, z) dz \right) dA \\ &= \int_c^d \left(\int_{\phi(y)}^{\psi(y)} \left(\int_{\kappa_1(x,y)}^{\kappa_2(x,y)} f(x, y, z) dz \right) dx \right) dy. \end{aligned}$$

THEOREM 6.9 Change of Variables in a Triple Integral

Let W and W^* be elementary (3D) regions in $\mathbb{R}^3$, and let

$$T = T(u, v, w) = (x(u, v, w), y(u, v, w), z(u, v, w)) \colon W^* \to W$$

be a C^1, one-to-one function such that $T(W^*) = W$. For an integrable function $f: W \to \mathbb{R}$,

$$\iiint_W f\,dV = \iiint_{W^*} f(x(u,v,w), y(u,v,w), z(u,v,w)) \left| \frac{\partial(x,y,z)}{\partial(u,v,w)} \right| dV^*.$$

The integrand on the right side is the composition $f \circ T$ (i.e., f expressed in terms of "new" variables u, v, and w) multiplied by the absolute value of the Jacobian

$$\frac{\partial(x,y,z)}{\partial(u,v,w)} = det(DT) = \begin{vmatrix} \partial x/\partial u & \partial x/\partial v & \partial x/\partial w \\ \partial y/\partial u & \partial y/\partial v & \partial y/\partial w \\ \partial z/\partial u & \partial z/\partial v & \partial z/\partial w \end{vmatrix}.$$

This is a straightforward generalization of the Change of Variables Theorem for double integrals (see Section 6.4). Also, note that (as before) the new region of integration W^* is the domain of the change of variables T, whereas the initial region of integration W is the range of T.

► EXAMPLE 6.41

Evaluate $\iiint_W f\,dV$, where $f(x,y,z) = ze^{x+y}$ and W is the parallelepiped $0 \le x \le 1$, $0 \le y \le 2$, and $0 \le z \le 3$.

SOLUTION By Fubini's Theorem for triple integrals (f is continuous), we can compute this integral in any of the six iterated versions. For example,

$$\iiint_W f\,dV = \int_0^1 \left(\int_0^3 \left(\int_0^2 ze^{x+y}\,dy \right) dz \right) dx.$$

Since $f(x,y,z) = ze^{x+y} = ze^x e^y$, we can separate the integrations and proceed as follows:

$$\iiint_W f\,dV = \left(\int_0^1 e^x\,dx \right) \left(\int_0^3 z\,dz \right) \left(\int_0^2 e^y\,dy \right)$$

$$= \left(e^x \Big|_0^1 \right) \left(\frac{z^2}{2} \Big|_0^3 \right) \left(e^y \Big|_0^2 \right) = \frac{9}{2}(e-1)(e^2-1).$$

► EXAMPLE 6.42

Evaluate $\iiint_W 2y\,dV$, where W is the solid in the first octant bounded by the plane $x + 2y + z = 6$; see Figure 6.54. Express the given integral as an iterated integral in various orders of integration.

SOLUTION The solid W is a type-1(3D) region: the corresponding elementary (2D) region D is the triangle bounded by the x-axis, the y-axis, and the line $x + 2y = 6$ (which is the intersection of the plane $x + 2y + z = 6$ and the xy-plane). The bottom and top surfaces are $z = \kappa_1(x,y) = 0$ and

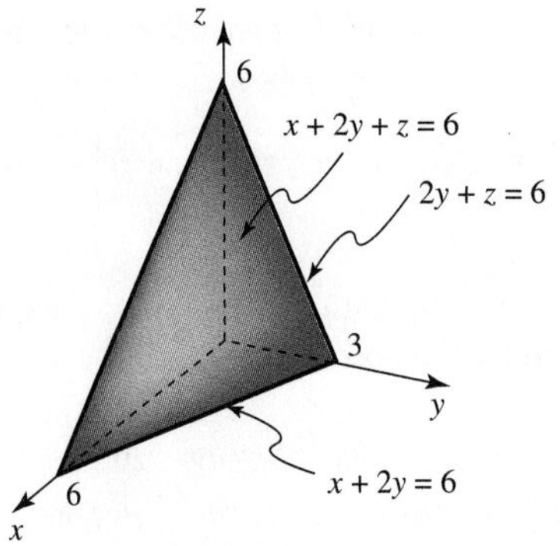

Figure 6.54 Region of Example 6.42.

$z = \kappa_2(x, y) = 6 - x - 2y$. It follows that

$$\iiint_W 2y\,dV = \iint_D \left(\int_0^{6-x-2y} 2y dz \right) dA$$

$$= \int_0^3 \left(\int_0^{6-2y} \left(\int_0^{6-x-2y} 2y dz \right) dx \right) dy$$

$$= \int_0^3 \left(\int_0^{6-2y} \left(12y - 2xy - 4y^2 \right) dx \right) dy$$

$$= \int_0^3 \left(12xy - x^2y - 4y^2x \Big|_{x=0}^{x=6-2y} \right) dy$$

$$= \int_0^3 (36y - 24y^2 + 4y^3)\,dy = (18y^2 - 8y^3 + y^4)\Big|_0^3 = 27.$$

There are other ways to write $\iiint_W 2y\,dV$ as an iterated integral. For instance,

$$\iiint_W 2y\,dV = \iint_D \left(\int_0^{6-x-2y} 2y dz \right) dA$$

$$= \int_0^6 \left(\int_0^{3-x/2} \left(\int_0^{6-x-2y} 2y dz \right) dy \right) dx.$$

The solid W is also a type-2(3D) region: points (y, z) belong to the triangle in the yz-plane bounded by the lines $y = 0$ (z-axis), $z = 0$ (y-axis), and $2y + z = 6$. Furthermore, $0 \le x \le 6 - 2y - z$; that is, the sides of W are the surfaces $x = 0$ and $x = 6 - 2y - z$. Hence,

$$\iiint_W 2y\,dV = \int_0^6 \left(\int_0^{3-z/2} \left(\int_0^{6-2y-z} 2y dx \right) dy \right) dz$$

$$= \int_0^3 \left(\int_0^{6-2y} \left(\int_0^{6-2y-z} 2y dx \right) dz \right) dy.$$

Two more equivalent integrals can be obtained by using the fact that W is also a type-3(3D) region. ◀

▶ EXAMPLE 6.43

Compute $\iiint_W (x^2 + y^2)\,dV$, where W is the solid bounded by the cylinder $x^2 + y^2 \le 4$ and the planes $z = 0$ and $z = 2$.

SOLUTION The solid W can be viewed as a type-1(3D) region: the corresponding elementary (2D) region is the disk $D = \{(x, y)|x^2 + y^2 \le 4\}$ in the xy-plane, and W is bounded from below and above by the planes $z = \kappa_1(x, y) = 0$ and $z = \kappa_2(x, y) = 2$. Since D is a disk, we pass to cylindrical coordinates $x = r\cos\theta$, $y = r\sin\theta$, $z = z$. The Jacobian is computed to be

$$\frac{\partial(x, y, z)}{\partial(r, \theta, z)} = \begin{vmatrix} \cos\theta & -r\sin\theta & 0 \\ \sin\theta & r\cos\theta & 0 \\ 0 & 0 & 1 \end{vmatrix} = r,$$

and therefore,

$$\iiint_W (x^2 + y^2)\, dV = \iiint_{W^*} r^2\,|r|\,dV^*,$$

where W^* is the rectangular box $[0, 2] \times [0, 2\pi] \times [0, 2]$ (the first interval represents r, the second θ, and the last one z). It follows that

$$\iiint_{W^*} r^2\,|r|\,dV^* = \iint_{[0,2]\times[0,2\pi]} \left(\int_0^2 r^3 dz\right) dA$$

$$= \int_0^{2\pi} \left(\int_0^2 \left(\int_0^2 r^3 dz\right) dr\right) d\theta = \int_0^{2\pi} \left(\int_0^2 2r^3\, dr\right) d\theta = 2\pi\, \frac{r^4}{2}\bigg|_0^2 = 16\pi. \quad ◀$$

► **EXAMPLE 6.44**

Express the integral $\int_{-1}^{1}(\int_0^1(\int_{-\sqrt{x}}^{\sqrt{x}} 2xydz)dx)dy$ as an iterated integral in some other order of integration (other than just switching dx and dy).

SOLUTION The given integral is equal to the triple integral $\iiint_W 2xy\,dV$, where W is determined by $-1 \le y \le 1$, $0 \le x \le 1$, and $-\sqrt{x} \le z \le \sqrt{x}$. It follows that W is a type-1(3D) region defined by $D = [0, 1] \times [-1, 1]$ and $-\sqrt{x} = \kappa_1(x, y) \le z \le \kappa_2(x, y) = \sqrt{x}$; see Figure 6.55.

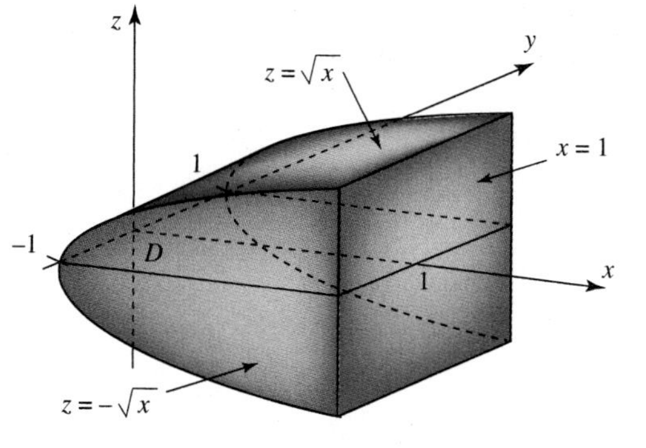

Figure 6.55 The three-dimensional region of Example 6.44.

Think of W as a type-2(3D) region, where D is the rectangle $D = [-1, 1] \times [-1, 1]$ in the yz-plane [since $-\sqrt{x} \le z \le \sqrt{x}$ and $0 \le x \le 1$, it follows that $-1 \le z \le 1$; that is how the second interval in D was computed]. W is bounded on the sides by $x = z^2$ and $x = 1$. Therefore,

$$\iiint_W 2xy\,dV = \iint_D \left(\int_{z^2}^1 2xydx\right) dA = \int_{-1}^1 \left(\int_{-1}^1 \left(\int_{z^2}^1 2xy\,dx\right) dy\right) dz. \quad ◀$$

► **EXAMPLE 6.45**

Set up an iterated integral (in Cartesian coordinates) for the volume of the solid ball W in $\mathbb{R}^3$ given by $x^2 + y^2 + z^2 \le 1$.

SOLUTION Think of the ball W as a type-1(3D) region: the corresponding elementary (2D) region is the disk $D = \{(x, y) | x^2 + y^2 \leq 1\}$ in the xy-plane, and W is determined by the inequalities $-\sqrt{1 - x^2 - y^2} \leq z \leq \sqrt{1 - x^2 - y^2}$. Thus, the volume of W is

$$\iiint_W 1\, dV = \iint_D \left(\int_{-\sqrt{1-x^2-y^2}}^{\sqrt{1-x^2-y^2}} dz \right) dA = \int_{-1}^{1} \left(\int_{-\sqrt{1-x^2}}^{\sqrt{1-x^2}} \left(\int_{-\sqrt{1-x^2-y^2}}^{\sqrt{1-x^2-y^2}} dz \right) dy \right) dx.$$

By switching the outer two integrations, and using the fact that W is also a type-2(3D) and type-3(3D) region, we get five more iterated integrals for the required volume. ◀

The iterated integral in the previous example is not easy to evaluate. By switching to spherical coordinates $x = \rho \sin\phi \cos\theta$, $y = \rho \sin\phi \sin\theta$, and $z = \rho \cos\phi$, the calculations will be simplified; the reason is that, because

$$\begin{aligned} x^2 + y^2 + z^2 &= \rho^2 \sin^2\phi \cos^2\theta + \rho^2 \sin^2\phi \sin^2\theta + \rho^2 \cos^2\phi \\ &= \rho^2(\sin^2\phi(\cos^2\theta + \sin^2\theta) + \cos^2\phi) = \rho^2, \end{aligned}$$

the (Cartesian) equation of the sphere $x^2 + y^2 + z^2 = 1$ becomes $\rho = 1$ in spherical coordinates. We use this fact in the following example.

▶ EXAMPLE 6.46

Compute the volume of the solid W that lies inside the sphere $x^2 + y^2 + z^2 = a^2$ and outside the sphere $x^2 + y^2 + z^2 = b^2$, $b < a$.

SOLUTION To simplify calculations, we use spherical coordinates. The Jacobian of the change of variables is computed to be

$$\begin{aligned} \frac{\partial(x, y, z)}{\partial(\rho, \phi, \theta)} &= \begin{vmatrix} \sin\phi\cos\theta & \rho\cos\phi\cos\theta & -\rho\sin\phi\sin\theta \\ \sin\phi\sin\theta & \rho\cos\phi\sin\theta & \rho\sin\phi\cos\theta \\ \cos\phi & -\rho\sin\phi & 0 \end{vmatrix} \\ &= \sin\phi\cos\theta(\rho^2\sin^2\phi\cos\theta) - \rho\cos\phi\cos\theta(-\rho\sin\phi\cos\phi\cos\theta) \\ &\quad -\rho\sin\phi\sin\theta(-\rho\sin^2\phi\sin\theta - \rho\cos^2\phi\sin\theta) \\ &= \rho^2\sin^3\phi\cos^2\theta + \rho^2\cos^2\phi\cos^2\theta\sin\phi + \rho^2\sin^3\phi\sin^2\theta + \rho^2\sin\phi\cos^2\phi\sin^2\theta \end{aligned}$$

(combine the first and third, and the second and fourth terms)

$$\begin{aligned} &= \rho^2\sin^3\phi + \rho^2\sin\phi\cos^2\phi = \rho^2\sin\phi(\sin^2\phi + \cos^2\phi) \\ &= \rho^2\sin\phi. \end{aligned} \tag{6.8}$$

The spheres $x^2 + y^2 + z^2 = a^2$ and $x^2 + y^2 + z^2 = b^2$ have equations $\rho = a$ and $\rho = b$ in spherical coordinates. It follows that

$$\begin{aligned} \text{volume}(W) &= \int_0^{2\pi} \left(\int_0^{\pi} \left(\int_b^a 1 \cdot \rho^2 \sin\phi\, d\rho \right) d\phi \right) d\theta \\ &= \int_0^{2\pi} \left(\int_0^{\pi} \sin\phi\, d\phi \right) \left(\int_b^a \rho^2\, d\rho \right) d\theta \\ &= \frac{4\pi}{3}(a^3 - b^3). \end{aligned}$$

There is a much faster way to compute the volume in question: subtract the volume of the smaller ball ($4\pi b^3/3$) from the volume of the larger ball ($4\pi a^3/3$). However, the point of this example was to illustrate the use of spherical coordinates in integration.

▶ EXAMPLE 6.47

Find the volume of the solid region ("ice-cream cone with a scoop of ice-cream") that lies inside the sphere $x^2 + y^2 + z^2 = z$ and above the cone $z^2 = x^2 + y^2$, $z \geq 0$, shown in Figure 6.56.

SOLUTION

As in the previous example, we use spherical coordinates. In order to compute the limits of integration (and thus obtain the solid W^* that is mapped onto W, see Theorem 6.9), we transform the given equations. The equation of the sphere $x^2 + y^2 + z^2 = z$ transforms to $\rho^2 = \rho \cos\phi$ (since $x^2 + y^2 + z^2 = \rho^2$) or $\rho = \cos\phi$. Hence, $0 \leq \rho \leq \cos\phi$ are the limits for ρ. The equation of the cone $z^2 = x^2 + y^2$ transforms to

$$\rho^2 \cos^2\phi = \rho^2 \sin^2\phi \cos^2\theta + \rho^2 \sin^2\phi \sin^2\theta = \rho^2 \sin^2\phi,$$

or $\rho\cos\phi = \pm\rho\sin\phi$. Hence (divide by $\rho\cos\phi$), $\tan\phi = \pm 1$ or $\phi = \pm\pi/4$. Since $\phi \geq 0$ (by the definition of spherical coordinates), it follows that $0 \leq \phi \leq \pi/4$ are the limits of integration for ϕ. Finally, $0 \leq \theta \leq 2\pi$ and

$$\begin{aligned}
\text{volume}(W) &= \int_0^{2\pi} \left(\int_0^{\pi/4} \left(\int_0^{\cos\phi} \rho^2 \sin\phi d\rho \right) d\phi \right) d\theta \\
&= \int_0^{2\pi} \left(\int_0^{\pi/4} \sin\phi \left.\frac{\rho^3}{3}\right|_0^{\cos\phi} d\phi \right) d\theta \\
&= \frac{1}{3} \int_0^{2\pi} \left(\int_0^{\pi/4} \sin\phi \cos^3\phi \, d\phi \right) d\theta = -\frac{1}{3} \int_0^{2\pi} \left(\frac{1}{4} \cos^4\phi \Big|_0^{\pi/4} \right) d\theta \\
&= -\frac{1}{3}\frac{1}{4} \left(\left(\frac{\sqrt{2}}{2} \right)^4 - 1 \right) \int_0^{2\pi} d\theta = \frac{\pi}{8}.
\end{aligned}$$

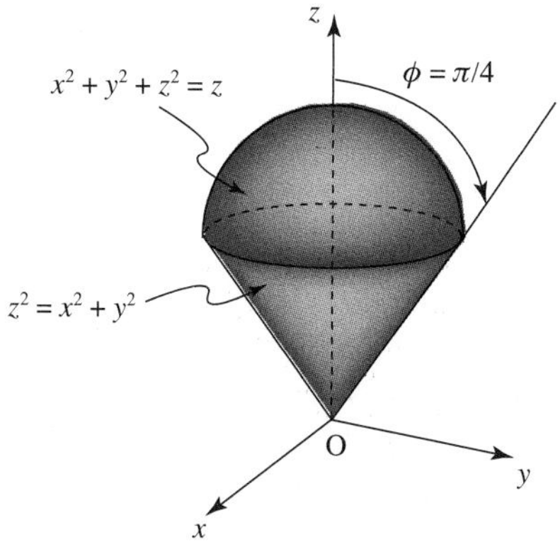

Figure 6.56 "Ice cream cone" of Example 6.47.

▶ EXERCISES 6.5

Exercises 1 to 9: Compute the value of the triple integral $\iiint_W f\, dV$.

1. $f(x, y, z) = 2x - y - z$, W is the parallelepiped defined by the planes $x = 3$, $y = 2$, and $z = 2$ in the first octant

2. $f(x, y, z) = xye^{y+z}$, W is the rectangular box $0 \le x \le 2, 0 \le y \le 1, 0 \le z \le \ln 2$

3. $f(x, y, z) = y^2$, W is the tetrahedron in the first octant bounded by $x + y + z = 1$

4. $f(x, y, z) = yz$, W is the three-dimensional solid that lies under the parabolic sheet $z = 4 - y^2$ and above the rectangle $[0, 1] \times [0, 2]$ in the xy-plane

5. $f(x, y, z) = 2z - 5$, W is the three-dimensional solid between the surfaces $z = 2y^2$ and $z = 8 - 2y^2$, for $0 \le x \le 1$

6. $f(x, y, z) = x + y$, W is the three-dimensional solid below the paraboloid $z = 1 - x^2 - y^2$, inside the cylinder $x^2 + y^2 = 1$, and above the xy-plane

7. $f(x, y, z) = 4y$, W is the three-dimensional solid bounded by the paraboloids $z = 12 - x^2 - y^2$ and $z = 2x^2 + 2y^2$

8. $f(x, y, z) = 3 + 2x$, W is the three-dimensional solid inside the cone $y^2 = x^2 + z^2$, between the planes $y = 2$ and $y = 5$

9. $f(x, y, z) = xyz$, W is the three-dimensional solid inside the sphere $x^2 + y^2 + z^2 = 2$ and inside the cylinder $x^2 + y^2 = 1$

Exercises 10 to 16: Evaluate the iterated integral and describe the region of integration (parentheses have been dropped from the notation).

10. $\int_0^2 \int_0^x \int_0^3 xy\, dz dy dx$

11. $\int_0^1 \int_0^{2-2y} \int_0^{4-2x-4y} 3\, dz dx dy$

12. $\int_0^1 \int_0^y \int_{-1}^{5y} x\, dz dx dy$

13. $\int_0^{2\pi} \int_0^2 \int_0^{r^2} r^2 z\, dz dr d\theta$

14. $\int_{-1}^1 \int_0^{1-x^2} \int_0^{\sqrt{y}} x^2 y^2 z^2\, dz dy dx$

15. $\int_{-\sqrt{8}}^{\sqrt{8}} \int_{-\sqrt{8-x^2}}^{\sqrt{8-x^2}} \int_{-3}^{8-x^2-y^2} 2\, dz dy dx$

16. $\int_0^1 \int_0^{\sqrt{1-y^2}} \int_0^{\sqrt{1-x^2-y^2}} (2x - y)\, dz dx dy$

Exercises 17 to 25: Find the volume of the solid in $\mathbb{R}^3$.

17. The tetrahedron with vertices $(0, 0, 0)$, $(1, 0, 0)$, $(0, 1, 0)$, and $(0, 0, 1)$

18. Inside $x^2 + y^2 + z^2 = a^2$ and outside $x^2 + y^2 + z^2 = b^2$, $b < a$

19. Inside the sphere $x^2 + y^2 + z^2 = 4$ and outside the ellipsoid $4x^2 + 4y^2 + z^2 = 4$

20. Above the cone $z^2 = x^2 + y^2$, inside the sphere $x^2 + y^2 + z^2 = 1$, and above the xy-plane

21. Inside the cylinder $(x - 1)^2 + y^2 = 1$, outside the cone $z^2 = x^2 + y^2$, and above the xy-plane

22. The part of the cone $2z^2 = x^2 + y^2$ between the horizontal planes $z = 1$ and $z = 2$

23. The smaller of the two solids above the xy-plane, bounded by the paraboloid $z = 16 - x^2 - y^2$ and by the vertical plane $x + y = 4$; do not evaluate the integral—just set it up in Cartesian and cylindrical coordinates.

24. Above the plane $z = a$ and inside the sphere $x^2 + y^2 + z^2 = b^2$, $b > a$

25. Inside the sphere $x^2 + y^2 + z^2 = a^2$ and inside the cylinder $x^2 + y^2 = b^2$, $b \leq a$

26. Express the integral $\int_0^1 \int_0^y \int_0^x x^2yz\,dzdxdy$ in the remaining five orders of integration.

27. Express the integral $\int_{-\sqrt{2}}^{\sqrt{2}} \int_{-\sqrt{2-x^2}}^{\sqrt{2-x^2}} \int_{x^2+y^2-2}^{2-x^2-y^2} (x^2 + y^2 - 2)\,dzdydx$ in cylindrical coordinates.

28. Express the integral $\int_0^6 \int_0^{1-x/6} \int_0^{2-x/3-2y} xyz\,dzdydx$ in the remaining five orders of integration. What is the region of integration?

29. Compute the volume of the solid bounded by the following pairs of parallel planes: $x + y = 1$, $x + y = 3$, $y + 2z = -2$, $y + 2z = 4$, $2x + y + z = -1$, and $2x + y + z = 1$.

30. Find the volume of the three-dimensional solid below $z = x^2 + y^2$ and above the region $xy = 1$, $xy = 3$, $x^2 - y^2 = 1$, $x^2 - y^2 = 4$ in the first quadrant.

31. Consider the triple integral $\iiint_W 2x\,dV$, where W is the solid three-dimensional region bounded by the surfaces $z = x^2 + y^2$, $z = 2(x^2 + y^2)$, and $z = 1$. Express it as an iterated integral in cylindrical coordinates. Do not evaluate it.

▶ CHAPTER REVIEW

CHAPTER SUMMARY

- **Double integrals.** Definition using Riemann sums, integrable functions, properties of double integrals, Cavalieri's principle, Fubini's Theorem, elementary regions, iterated integrals over elementary regions, techniques of evaluation of double integrals.

- **Change of variables.** Linear and affine maps, Jacobian, Change of Variables Formula for double and triple integrals, Jacobian for polar, cylindrical, and spherical coordinates.

- **Triple integrals.** Definition and properties, evaluation of triple integrals as iterated integrals, Fubini's Theorem.

REVIEW

Discuss the following questions.

1. Explain how to approximate the volume of a solid under the graph of a positive function $z = f(x, y)$ and above a rectangle R in the xy-plane using double Riemann sums. Suppose that you have used a double Riemann sum to approximate the volume of the solid under the graph of $z = e^{-x}$ and above $[0, 1] \times [-1, 1]$, using the upper left corner of each subrectangle. Is your estimate an overestimate or underestimate?

2. One of the properties of double integrals is that $\left|\iint_R f\,dA\right| \leq \iint_R |f|\,dA$, where R is a rectangle. Find examples of functions (and rectangles) such that $\left|\iint_R f\,dA\right| = \iint_R |f|\,dA$ and $\left|\iint_R f\,dA\right| < \iint_R |f|\,dA$.

3. State Cavalieri's principle (for the computation of volume). Formulate its analog in the case of area and use it to prove that any two triangles with the same base and same height have equal areas. Explain how to obtain a formula for the volume of a solid of revolution using Cavalieri's principle.

4. State Fubini's Theorem and explain its meaning.

5. Find examples of type-2(2D) regions. Describe how to extend the definition of the double integral from a rectangle to a region of type 2. Express a double integral over a region of type 2 as an iterated integral.

6. Without integration, find the value of $\iint_D f\,dA$, where $f(x, y) = \sqrt{1 - x^2 - y^2}$ and D is the unit disk $x^2 + y^2 \leq 1$. Again, without integration, find $\iint_R \sin^3 x\,dA$, where $R = [-\pi, \pi] \times [0, 1]$.

7. Is it possible that the average value of $f(x, y) = e^{x+y} \sin(3x - y^4)$ over the square $D = [-1, 1] \times [-1, 1]$ can be 9? Explain your answer.

8. Define the Jacobian of the function $T\colon \mathbb{R}^2 \to \mathbb{R}^2$ and explain its geometric significance. If T is defined by $T(\mathbf{x}) = A\mathbf{x}$, where A is a 2×2 matrix, what is its Jacobian?

9. State the Change of Variables Theorems for double and triple integrals.

TRUE/FALSE QUIZ

Determine whether the following statements are true or false. Give reasons for your answer.

1. The Jacobian $\partial(x, y)/\partial(u, v)$ is a positive number.

2. If $D = [1, 2] \times [-2, -1]$, then $\iint_D dA = 1$.

3. The region of integration in $\int_0^2 \left(\int_{-\sqrt{x}}^{\sqrt{x}} f(x, y)\,dy\right) dx$ is a circle of radius 2.

4. An affine map maps parallel lines to parallel lines.

5. $\int_{-2}^{2} \left(\int_{-1}^{x} 2\,dy\right) dx = \int_{-1}^{x} \left(\int_{-2}^{2} 2\,dx\right) dy$.

6. $D = \{(x, y) | 1 \leq x^2 + y^2 \leq 2\}$ is a region of type 2(2D).

7. $D = \{(x, y) | 1 \leq x^2 + y^2 \leq 2,\ x \geq 0\}$ is a region of type 2(2D).

8. If $T(\mathbf{u}) = A\mathbf{u} + \mathbf{b}$ is an affine map, then $\det(A)$ is constant.

9. $\int_0^1 \left(\int_0^x \sin(x^2 + y^2)\,dy\right) dx < 1$.

10. The map $T(u, v) = (2u, v/2)$ does not change the area.

11. The map $T(u, v) = (2u, v/2)$ maps squares into squares.

12. $\left|\iint_R f\,dA\right| < \iint_R |f|\,dA$.

REVIEW EXERCISES AND PROBLEMS

1. Evaluate the iterated integral $\int_0^{2\pi} \int_0^{2+\cos\theta} \int_0^{\pi/4} \rho^2 \sin\phi\,d\phi d\rho d\theta$ and describe the region of integration.

2. The amount of radioactivity at a point (x, y) near a nuclear power plant is given by the function $f(x, y)$. Let D denote some geographical region. What is the meaning of $\iint_D f\,dA$? What is the meaning of $\iint_D f\,dA / \iint_D dA$?

3. Find the volume of the wedge cut out from the cylinder $x^2 + y^2 = a^2$, $a > 0$, by the planes $z = 0$ and $z = my + am$, $m > 0$.

4. Compute the volume of the region outside the paraboloid $z = x^2 + y^2$, inside the cylinder $x^2 + 4y^2 = 4$, and above the xy-plane using the change of variables $x = 2r\cos\theta$, $y = r\sin\theta$, $z = z$.

5. Let $f(x, y)$ be a continuous function on a rectangle $R \subseteq \mathbb{R}^2$ and assume that $f \geq 0$ on R. Prove that if $\iint_R f\,dA = 0$, then $f(x, y) = 0$ for all (x, y) in R.

Exercises 6 to 9: Express $\iint_D f\,dA$ as an iterated integral in polar coordinates, where D is the given region in $\mathbb{R}^2$.

6. $D = \{(x, y) | 1 \leq x^2 + y^2 \leq 3,\ x \leq 0\}$

7.

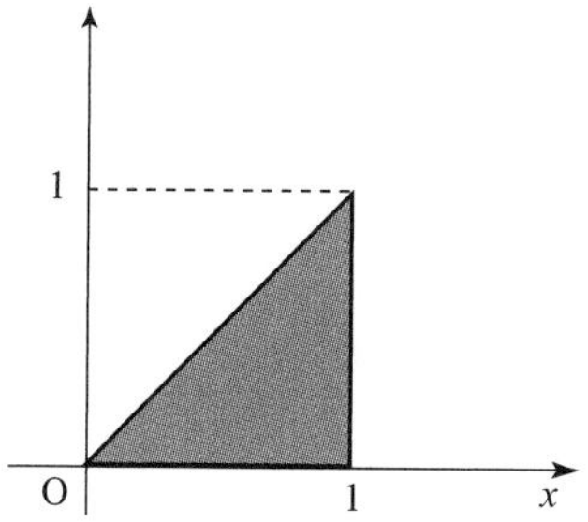

8.

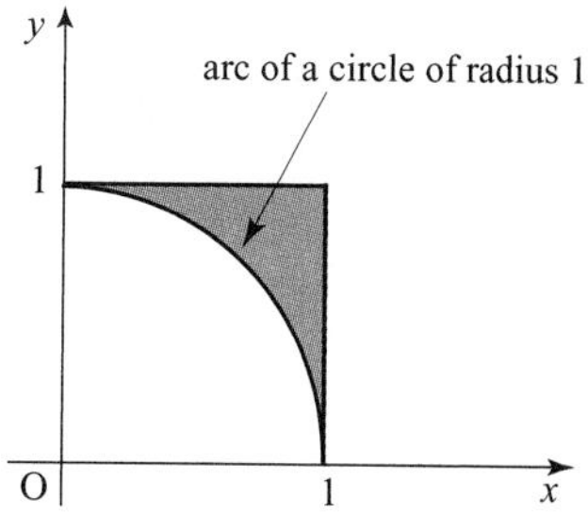

9. $D = \{(x, y)|x^2 + y^2 \le 2,\ x \le 0,\ y \le 0\}$

10. Figure 6.57 shows several level curves of a function $f(x, y)$. Using Riemann sums $\mathcal{R}_2$ and $\mathcal{R}_4$, estimate $\iint_D f\,dA$, where D is the rectangle $[0, 4] \times [0, 2]$. (There are many correct answers.)

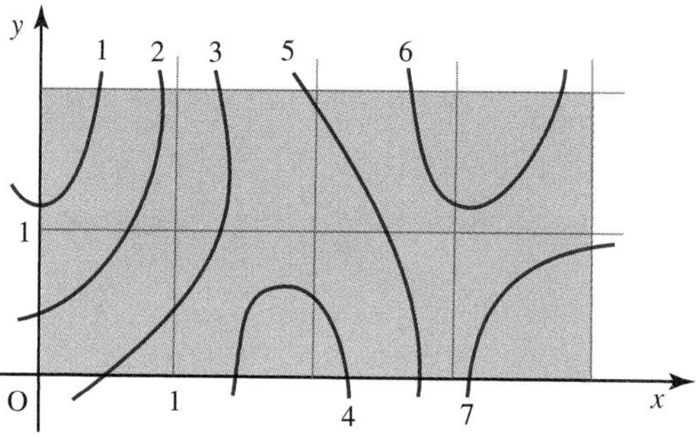

Figure 6.57 Level curves of Exercise 10.

11. Find the area of the region inside both circles $r = 4\cos\theta$ and $r = 2$.

12. Find the volume of the three-dimensional solid in the first octant bounded by the surfaces $xy = 1$, $xy = 2$, $xz = 1$, $xz = 3$, $y + z = 0$, and $y + z = 2$ (use a suitable change of variables).

13. Let D be a type-1 region defined by $a \le x \le b$ and $-\psi(x) \le y \le \psi(x)$, where $\psi(x) \ge 0$ is continuous on $[a, b]$. Let $f(x, y)$ be a continuous function on D such that $f(x, -y) = -f(x, y)$ for all $(x, y) \in D$. Using a Riemann sum argument, prove that $\iint_D f\,dA = 0$.

14. Express the iterated integral $\int_0^1 \left(\int_0^{\sqrt{a^2-x^2}} \left(\int_0^{\sqrt{a^2-x^2-y^2}} xyz\,dz \right) dy \right) dx$ as an iterated integral in two other orders of itegration. What solid is involved?

15. Compute the volume of the tetrahedron with vertices $(0, 0, 0)$, $(2, 0, 0)$, $(1, 4, 0)$, and $(0, 0, 10)$ using a triple integral.

16. Evaluate the integral $\int_0^1 \left(\int_{-y}^{y} y\,dx \right) dy$ by converting to polar coordinates. Check your answer by direct calculation.

17. In this exercise, we evaluate the integral $\int_{-\infty}^{\infty} e^{-x^2}\,dx$, which is one of the most important integrals in the theory of probability.

(a) Explain why $\int_{-\infty}^{\infty} \left(\int_{-\infty}^{\infty} e^{-x^2-y^2}\,dy \right) dx = \left(\int_{-\infty}^{\infty} e^{-x^2}\,dx \right)^2$.

(b) Compute the iterated integral in (a) using polar coordinates.

(c) Conclude that $\int_{-\infty}^{\infty} e^{-x^2}\,dx = \sqrt{\pi}$.

18. Find the area of the ellipse $x^2/a^2 + y^2/b^2 = 1$ by using a double integral and the change of variables $x = au$, $y = bv$.

19. Find the volume of the ellipsoid $x^2/a^2 + y^2/b^2 + z^2/c^2 = 1$ by using a triple integral and the change of variables $x = au$, $y = bv$, $z = cw$.

20. What is the relation between the Jacobians $\partial(x, y)/\partial(u, v)$ and $\partial(u, v)/\partial(x, y)$ in the case $u = 2x - y$, $v = 4x + 3y$? In the general case? Prove it.

21. Let $D = [0, 1] \times [0, 1]$. Express $\iint_D f\, dA$ as an iterated integral in polar coordinates.

CHAPTER 7

Integration Over Surfaces, Properties, and Applications of Integrals

By analogy with the description of a curve using a function defined on an interval of real numbers, we define a surface as the image of a function (that is also called a parametrization) defined on a subset of $\mathbb{R}^2$. To gain a better feel for parametrizations, we examine in detail many cases, such as a sphere, cylinder, cone, torus, and surface that is the graph of a function of two variables. In a separate section, we study a number of surfaces (some coming from applications). We proceed by discussing smoothness, orientation, and other concepts relevant for integration.

The surface integrals of real-valued functions and vector fields are defined in terms of integration over elementary regions in $\mathbb{R}^2$. The sections devoted to these topics provide a rich source of worked examples and discuss important applications, such as surface area and the flux of a vector field. A way of constructing surface integrals by projection (needed in the case where a parametrization is not available or not convenient) is also discussed.

In the last section, we unify various types of integration defined in the last three chapters into a single concept. Further examples, properties, and some physical applications are discussed as well.

7.1 PARAMETRIZED SURFACES

So far we have investigated (double) integrals over various regions in the plane $\mathbb{R}^2$. Our next goal is to define integration over more general regions in space (such as the surface of a sphere, cylinder, torus, etc.). Since this generalization is very similar in spirit to the definition of the path integral, we start by recalling the main idea.

A curve in $\mathbb{R}^3$ is represented by the equation $\mathbf{c}(t) = (x(t), y(t), z(t))$, $t \in [a, b]$, called the parametric representation. The path integral of a real-valued function of several variables or of a vector field is defined as the definite integral over the interval $[a, b] \subseteq \mathbb{R}$ of a function of (one variable) t—hence, integration over a curve is reduced to integration along an interval on the x-axis. In this section, we start developing analogous concepts, by first defining a surface and its parametric representation. In Sections 7.3 and 7.4, surface integrals of

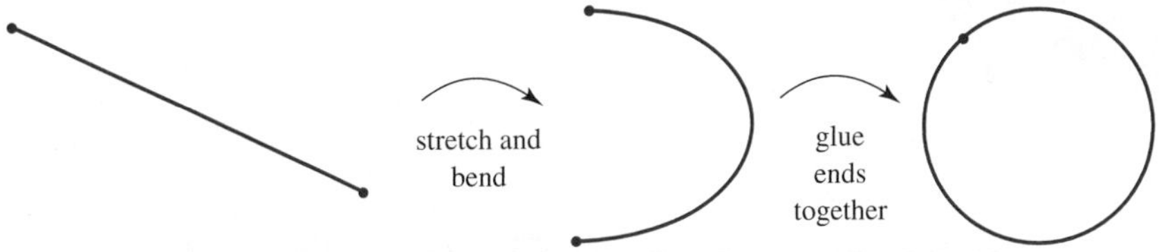

Figure 7.1 Deforming an interval produces a curve.

real-valued functions of several variables and surface integrals of vector fields will be defined in terms of integration over elementary regions in $\mathbb{R}^2$.

Before giving the definition of a surface, let us examine what objects we would like to call surfaces.

We can think of the image of a curve $\mathbf{c}: [a, b] \to \mathbb{R}^3$ in a mechanical way: assume that the interval $[a, b]$ is made of some material that can be deformed. By bending, stretching, and pasting, we can deform $[a, b]$ to "fit over" the image of $\mathbf{c}$. Hence, we think of $\mathbf{c}$ as being made from the interval $[a, b]$ by "deforming" it (without breaking), as shown in Figure 7.1.

Similarly, a surface is a geometric object that can be obtained by deforming (i.e., stretching, bending, twisting, pasting, but not breaking) a region in the plane $\mathbb{R}^2$. Figure 7.2 demonstrates the deformation of a rectangle into a surface called a helicoid.

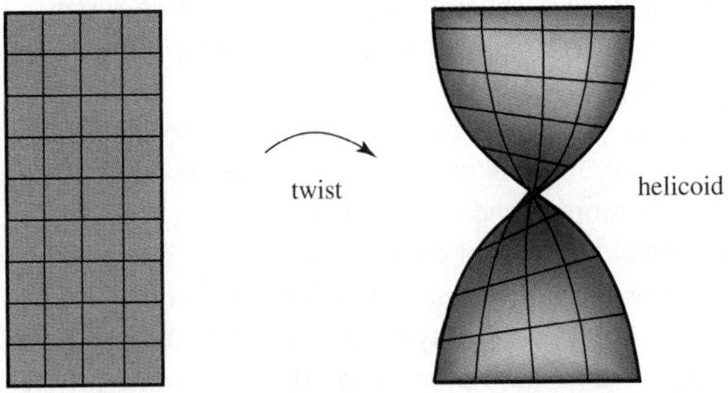

Figure 7.2 Deformation of a rectangle into a helicoid.

In Figure 7.3, a rectangle is first deformed into a cylinder, and then into a torus. (All surfaces shown here will be studied in detail in this chapter.) Parallel to the distinction between a path and a curve, we make a distinction between a parametrized surface (analytic description) and a surface (geometric object).

DEFINITION 7.1 Parametrized Surface and Surface

A *parametrized surface,* or a *parametrization of a surface,* is a map $\mathbf{r}: D \to \mathbb{R}^3$, where D is a region in (i.e., a subset of) $\mathbb{R}^2$ and $\mathbf{r}$ is one-to-one except possibly on the boundary of D. The image $S = \mathbf{r}(D)$ is called a *surface.* ◀

Figure 7.4 illustrates the distinction made in the definition.

The map $\mathbf{r}$ is the map (described in the introduction) that "deforms" the region D into the surface S. The assumption that $\mathbf{r}$ is one-to-one guarantees that the surface does not intersect itself [a similar requirement (simple curve) was needed for integration along

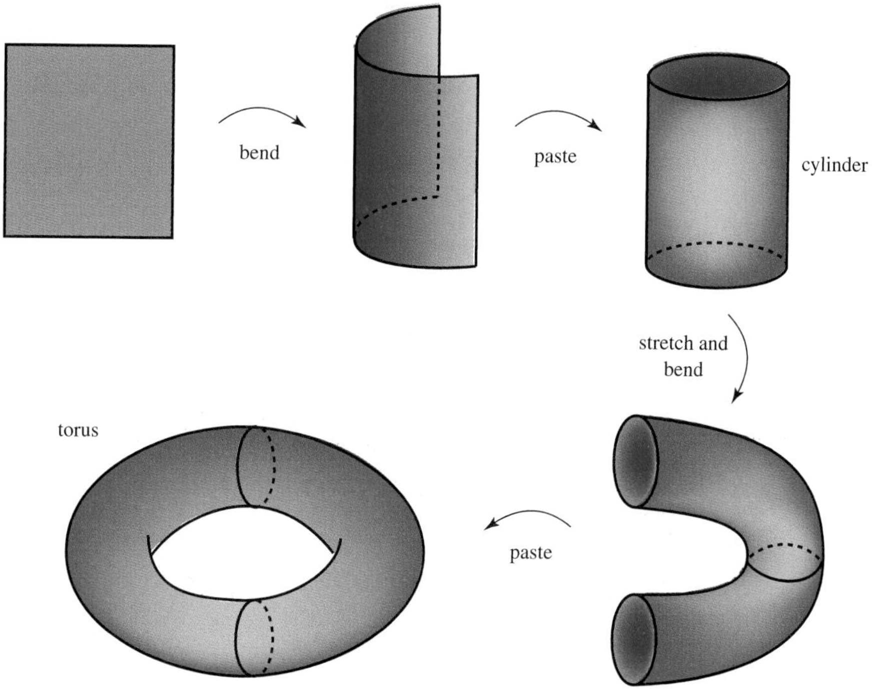

Figure 7.3 Deformation of a rectangle into a cylinder and torus.

curves]. The fact that on the boundary of D, $\mathbf{r}$ does not have to be one-to-one means that we are allowed to "glue together" points on the boundary of D to form a surface (think of making a cylinder from a sheet of paper).

In components, a parametrization $\mathbf{r}$ can be expressed as

$$\mathbf{r}(u, v) = (x(u, v), y(u, v), z(u, v)), \quad (u, v) \in D.$$

Recall that $\mathbf{r}(u,v)$ is continuous (respectively, differentiable, C^1) if and only if its components $x(u, v), y(u, v), z(u, v)\colon D \to \mathbb{R}$ are continuous (respectively, differentiable, C^1).

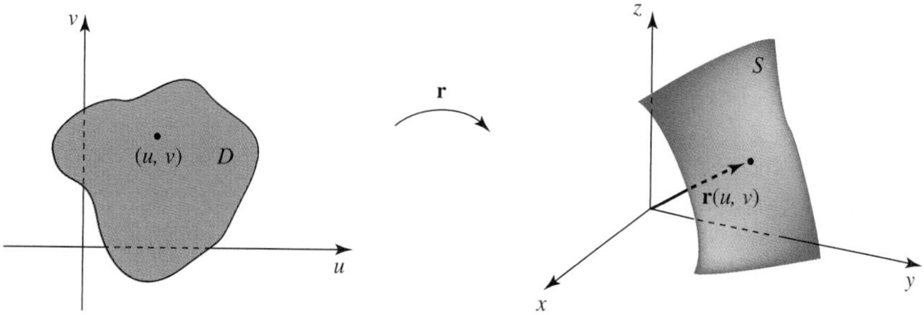

Figure 7.4 S is a surface, and $\mathbf{r}$ is its parametrization (or parametrized surface).

DEFINITION 7.2 Continuous, Differentiable, and C^1 Surface

A surface S is called *continuous* (respectively, *differentiable,* C^1) if and only if it has a parametrization $\mathbf{r}(u, v)\colon D \to \mathbb{R}^3$ that is continuous (respectively, differentiable, C^1). ◀

Now we examine several parametrized surfaces and surfaces.

▶ EXAMPLE 7.1 Plane

A plane in $\mathbb{R}^3$ can be represented in the form $ax + by + cz + d = 0$ [see Example 1.26 in Section 1.3 and (1.13) in the text following it]. If $c \neq 0$, then $z = (-ax - by - d)/c$, and thus (taking $u = x$ and $v = y$ for the parameters), we obtain

$$\mathbf{r}(u, v) = (u, v, mu + nv + p), \tag{7.1}$$

where $m = -a/c$, $n = -b/c$, $p = -d/c$, and $u, v \in \mathbb{R}$ (hence, $D = \mathbb{R}^2$). For instance, the plane $3x + y - 2z - 1 = 0$ can be represented parametrically as $\mathbf{r}(u, v) = (u, v, 3u/2 + v/2 - 1/2)$, where $u, v \in \mathbb{R}$. The xy-plane $z = 0$ has a parametric equation $\mathbf{r}(u, v) = (u, v, 0)$, $u, v \in \mathbb{R}$.

To understand the parametrization (7.1) better, we investigate how $\mathbf{r}$ maps horizontal and vertical lines. The image of a horizontal line $v = v_0 =$ constant is given by

$$\mathbf{r}(u, v_0) = (u, v_0, mu + nv_0 + p) = (0, v_0, nv_0 + p) + u(1, 0, m);$$

that is, it is a line going through $(0, v_0, nv_0 + p)$ whose direction vector is $(1, 0, m)$. Likewise, the image by $\mathbf{r}$ of a vertical line $u = u_0 =$ constant is

$$\mathbf{r}(u_0, v) = (u_0, v, mu_0 + nv + p) = (u_0, 0, mu_0 + p) + v(0, 1, n).$$

Consequently, vertical lines in the domain of $\mathbf{r}$ are mapped to parallel lines with the direction vector $(0, 1, n)$. Note that the dot product of $(1, 0, m)$ and $(0, 1, n)$ is mn. Thus, when $mn \neq 0$, $\mathbf{r}$ does not preserve angles; it maps horizontal and vertical lines in its domain into two families of parallel lines that cross each other at the angle $\theta = \arccos\left(mn(1 + m^2)^{-1/2}(1 + n^2)^{-1/2}\right)$ [recall Theorem 1.5 in Section 1.3]; see Figure 7.5.

The above discussion covers the case of the plane $ax + by + cz + d = 0$ when $c \neq 0$. In the case $a \neq 0$, we obtain the parametrization $\mathbf{r}(u, v) = (mu + nv + p, u, v)$, for some constants m, n, and p, and $u, v \in \mathbb{R}$. In a similar way, we obtain a parametrization when $b \neq 0$. For instance, the plane $x = 1$ can be written as $\mathbf{r}(u, v) = (1, u, v)$, $u, v \in \mathbb{R}$. The plane $3x + y - 2z - 1 = 0$ can be

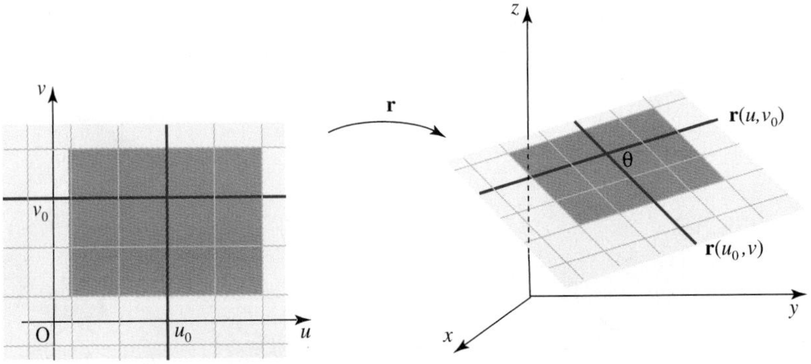

Figure 7.5 Parametric representation of a plane.

represented as $\mathbf{r}(u, v) = (-u/3, +2v/3 + 1/3, u, v)$, $u, v \in \mathbb{R}$, or as $\mathbf{r}(u, v) = (u, -3u + 2v + 1, v)$, $u, v \in \mathbb{R}$.

▶ EXAMPLE 7.2 Cylinder

The cylinder $x^2 + y^2 = a^2$, $0 \le z \le b$, of radius a and height b can be represented by

$$\mathbf{r}(u, v) = (a\cos u, a\sin u, v), \quad 0 \le v \le b, \quad 0 \le u \le 2\pi.$$

[Hence, the domain D of $\mathbf{r}$ is the rectangle $D = [0, 2\pi] \times [0, b]$ in the uv-plane.] The curves $u = u_0 =$ constant are vertical line segments (v is the parameter)

$$\mathbf{r}(u_0, v) = (a\cos u_0, a\sin u_0, v) = (a\cos u_0, a\sin u_0, 0) + v(0, 0, 1), \qquad 0 \le v \le b;$$

see Figure 7.6. For $v = v_0 =$ constant, we get

$$\mathbf{r}(u, v_0) = (a\cos u, a\sin u, v_0), \qquad 0 \le u \le 2\pi;$$

that is, the circle that is the intersection of the cylinder and the plane $z = v_0$.

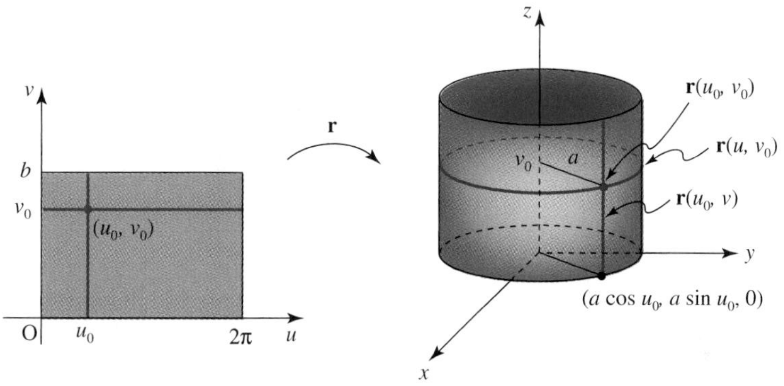

Figure 7.6 Parametric representation of a cylinder.

In words, $\mathbf{r}$ takes the rectangle $[0, 2\pi] \times [0, b]$, bends it (see Figure 7.3), and then glues its left side $\{(0, v) | 0 \le v \le b\}$ to its right side $\{(2\pi, v) | 0 \le v \le b\}$. Thus, $\mathbf{r}$ is not one-to-one on the left and the right sides of D. (It is one-to-one everywhere else.)

In Figure 7.7, we show how the cylinder is built: vertical line segments (left), which are the images by $\mathbf{r}$ of vertical line segments in D, and circles in horizontal planes (right), which are the images by $\mathbf{r}$ of horizontal line segments in D.

We could have described the same cylinder as

$$\mathbf{r}(u, v) = (a\cos mu, a\sin mu, nv), \qquad 0 \le v \le b/n, \quad 0 \le u \le 2\pi/m,$$

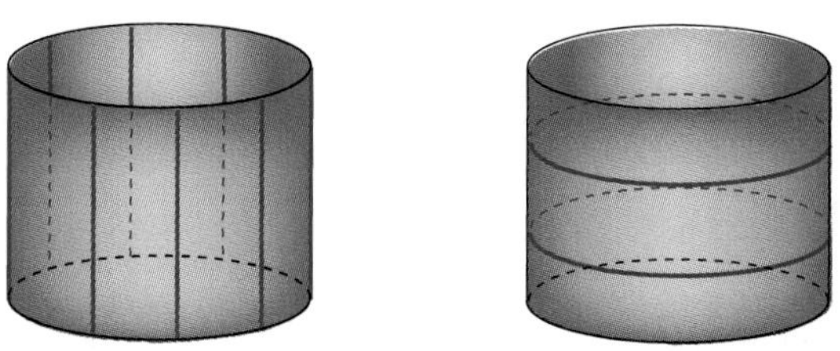

Figure 7.7 Anatomy of the parametrized cylinder in Example 7.2.

where $m, n > 0$. This shows that a surface (like a curve) can have infinitely many parametric representations.

▶ EXAMPLE 7.3 Cone

The parametric representation of the cone $x^2 + y^2 = z^2$, $0 \le z \le b$, of height b is given by

$$\mathbf{r}(u, v) = (v \cos u, v \sin u, v), \qquad (u, v) \in D = [0, 2\pi] \times [0, b].$$

The curve corresponding to $v = v_0$,

$$\mathbf{r}(u, v_0) = (v_0 \cos u, v_0 \sin u, v_0), \qquad 0 \le u \le 2\pi,$$

is the circle of radius v_0 in the horizontal plane $z = v_0$, if $v_0 > 0$ [when $v_0 = 0$ the circle collapses to the point $(0, 0, 0)$]; see Figure 7.8. The curve corresponding to $u = u_0$,

$$\mathbf{r}(u_0, v) = (v \cos u_0, v \sin u_0, v) = v(\cos u_0, \sin u_0, 1), \qquad 0 \le v \le b,$$

is a straight-line segment from the origin (when $v = 0$) to the point $(b \cos u_0, b \sin u_0, b)$, when $v = b$.

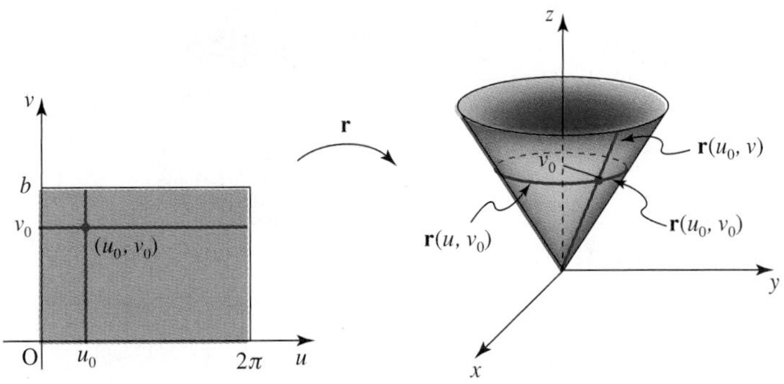

Figure 7.8 Parametric representation of a cone. ◀

In terms of a deformation, the cone of the previous example is obtained in the following way: the rectangle $[0, 2\pi] \times [0, b]$ is first deformed into a triangle. The bottom horizontal line segment $v = 0$ is compressed to a point, and the top segment $v = b$ is expanded or compressed (depending on b) to the length $2\pi b$ (which is the length of the circle that forms the "rim" of the cone); see Figure 7.9. Horizontal line segment $v = v_0$, $0 \le v_0 \le b$ is

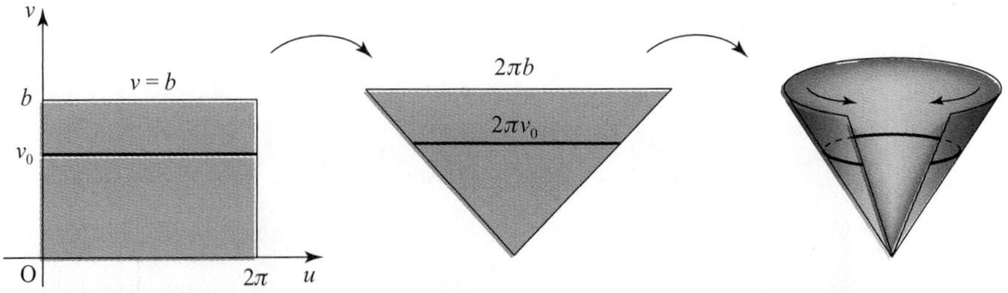

Figure 7.9 Deformation of a rectangle into a cone.

deformed into a line segment of length $2\pi v_0$. The cone is now obtained by gluing together the two (slanted) sides of the triangle.

The map $\mathbf{r}$ thus described is not one-to-one on the left, right, and bottom sides of the rectangle. It is one-to-one on the top side, except at its ends.

EXAMPLE 7.4 Sphere

A parametric representation of the sphere $x^2 + y^2 + z^2 = a^2$ of radius a centered at the origin is given by (see Figure 7.10)

$$\mathbf{r}(u, v) = (a \cos v \cos u, a \cos v \sin u, a \sin v), \quad 0 \le u \le 2\pi, \quad -\pi/2 \le v \le \pi/2. \tag{7.2}$$

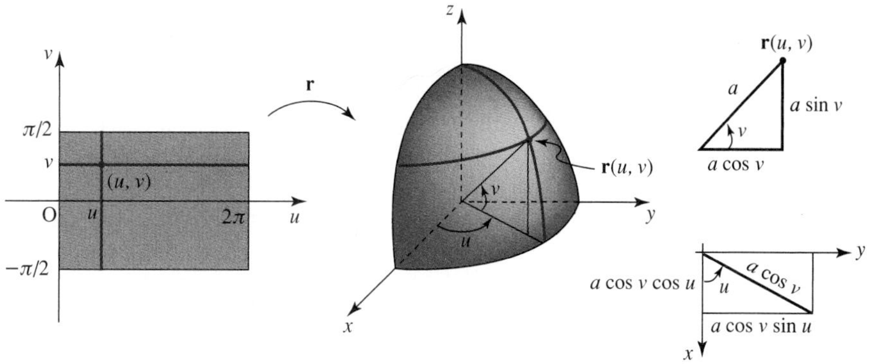

Figure 7.10 Geometric meaning of parameters u and v for the sphere.

Since the coordinates are $x = a \cos v \cos u$, $y = a \cos v \sin u$, and $z = a \sin v$, it follows that $x^2 + y^2 + z^2 = a^2 \cos^2 v + a^2 \sin^2 v = a^2$. Inequalities $0 \le u \le 2\pi$ and $-\pi/2 \le v \le \pi/2$ state that the domain D is the rectangle $[0, 2\pi] \times [-\pi/2, \pi/2]$ in the uv-plane.

Let us examine the map $\mathbf{r}$. The image of the line segment $\{(u, 0) | 0 \le u \le 2\pi\}$ is

$$\mathbf{r}(u, 0) = (a \cos u, a \sin u, 0), \quad 0 \le u \le 2\pi,$$

which is the circle of radius a in the xy-plane centered at $(0, 0)$ and represents the "equator" of the sphere. The image of the segment $\{(u, v_0) | 0 \le u \le 2\pi\}$ (v_0 is fixed) parallel to the u-axis is

$$\mathbf{r}(u, v_0) = (a \cos v_0 \cos u, a \cos v_0 \sin u, a \sin v_0), \quad 0 \le u \le 2\pi,$$

which is the circle of radius $a \cos v_0$, obtained as the intersection of the plane $z = a \sin v_0$ (parallel to the xy-plane) and the sphere. In other words, it is a "parallel" (of latitude v_0) of the sphere S; see Figure 7.11.

The image of the vertical segment $\{(0, v) | -\pi/2 \le v \le \pi/2\}$ is given by

$$\mathbf{r}(0, v) = (a \cos v, 0, a \sin v), \quad -\pi/2 \le v \le \pi/2,$$

which is the semicircle in the xz-plane, a "meridian" of S. In general, the image of the vertical segment $\{(u_0, v) | -\pi/2 \le v \le \pi/2\}$ (u_0 is fixed) is the curve

$$\mathbf{r}(u_0, v) = (a \cos u_0 \cos v, a \sin u_0 \cos v, a \sin v), \quad -\pi/2 \le v \le \pi/2,$$

which is the "meridian" corresponding to the angle u_0 measured from the positive direction of the x-axis (think of longitudinal lines on the earth).

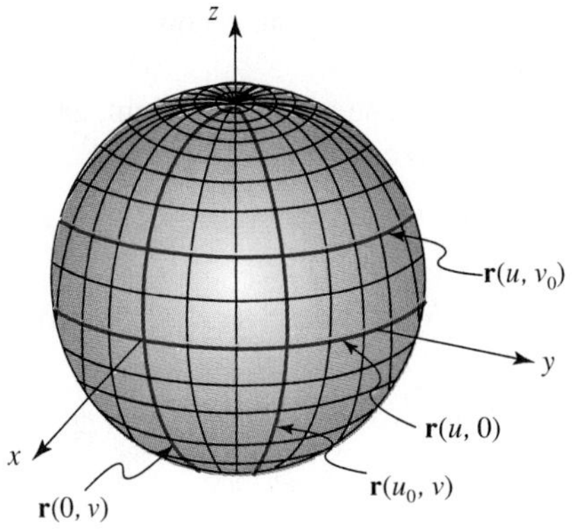

Figure 7.11 Parametric representation of the sphere.

By adjusting parameters u and v, we obtain various parts of the sphere. For instance, the parametrization (7.2) with $0 \le u \le 2\pi$ and $0 \le v \le \pi/2$ describes the upper hemisphere. If we restrict (7.2) to $0 \le u \le \pi/2$ and $0 \le v \le \pi/2$, we obtain the part of the sphere that lies in the first octant.

There are other representations of the sphere. For example, letting u and v represent spherical coordinates ($u = \theta$, $v = \phi$; see Definition 2.24 in Section 2.8), we get

$$\mathbf{r}(u, v) = (a \cos u \sin v, a \sin u \sin v, a \cos v), \quad 0 \le u \le 2\pi, \quad 0 \le v \le \pi.$$

In this case, the parameters u and v have interpretations different from those given in Example 7.4.

DEFINITION 7.3 Tangent Vector to a Surface

We say that a vector $\mathbf{v}$ is *tangent* to a surface S at a point P if $\mathbf{v}$ is a tangent vector, at P, to some curve that is contained in S. ◀

Assume that a surface S is represented by $\mathbf{r}(u, v) = (x(u, v), y(u, v), z(u, v))$, where $(u, v) \in D \subseteq \mathbb{R}^2$. This parametrization not only describes how the surface is constructed, but also provides us with two special tangent vectors that we now construct.

Pick a point $\mathbf{r}(u_0, v_0)$ on S. The restriction of $\mathbf{r}$ to the curve $v = v_0$ in $D \subseteq uv$-plane is the curve

$$\mathbf{c}_{v_0}(u) = \mathbf{r}(u, v_0) = (x(u, v_0), y(u, v_0), z(u, v_0)), \qquad (u, v_0) \in D,$$

on S going through the point $\mathbf{r}(u_0, v_0)$; see Figure 7.12 (the parameter of the curve is called u, rather than t; v_0 is fixed). The tangent vector $\mathbf{T}_u$ to $\mathbf{c}_{v_0}$ at $\mathbf{r}(u_0, v_0)$ is given by

$$\mathbf{T}_u(u_0, v_0) = \mathbf{c}'_{v_0}(u_0) = \left(\left.\frac{\partial x}{\partial u}\right|_{(u_0,v_0)}, \left.\frac{\partial y}{\partial u}\right|_{(u_0,v_0)}, \left.\frac{\partial z}{\partial u}\right|_{(u_0,v_0)} \right).$$

Similarly, the tangent to the curve (here u_0 is fixed and v is the parameter)

$$\mathbf{c}_{u_0}(v) = \mathbf{r}(u_0, v) = (x(u_0, v), y(u_0, v), z(u_0, v)), \qquad (u_0, v) \in D,$$

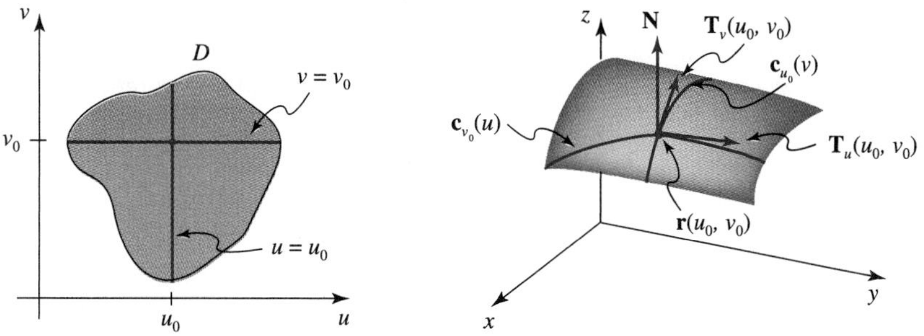

Figure 7.12 Tangent and normal vectors to a surface.

at the point $\mathbf{r}(u_0, v_0)$ is given by

$$\mathbf{T}_v(u_0, v_0) = \mathbf{c}'_{u_0}(v_0) = \left(\frac{\partial x}{\partial v}\bigg|_{(u_0,v_0)}, \frac{\partial y}{\partial v}\bigg|_{(u_0,v_0)}, \frac{\partial z}{\partial v}\bigg|_{(u_0,v_0)} \right).$$

The components of the tangent vector $\mathbf{T}_u(u_0, v_0)$ are the partial derivatives of the components $x(u, v)$, $y(u, v)$, and $z(u, v)$ with respect to u, evaluated at (u_0, v_0). Using the notation introduced in Section 2.4, we write $\mathbf{T}_u(u_0, v_0) = (\partial\mathbf{r}/\partial u)(u_0, v_0)$. Similarly, $\mathbf{T}_v(u_0, v_0) = (\partial\mathbf{r}/\partial v)(u_0, v_0)$.

DEFINITION 7.4 Normal Vector to a Surface

The vector $\mathbf{N}(u_0, v_0) = \mathbf{T}_u(u_0, v_0) \times \mathbf{T}_v(u_0, v_0)$, perpendicular to both $\mathbf{T}_u(u_0, v_0)$ and $\mathbf{T}_v(u_0, v_0)$ [and therefore perpendicular to the plane tangent to the surface S at $\mathbf{r}(u_0, v_0)$], is called a *normal vector* (or the *surface normal*) to S at $\mathbf{r}(u_0, v_0)$.

The coordinate expression for $\mathbf{N}$ is given by [drop (u_0, v_0) to keep the notation simpler]

$$\mathbf{N} = \mathbf{T}_u \times \mathbf{T}_v = \begin{vmatrix} \mathbf{i} & \mathbf{j} & \mathbf{k} \\ \partial x/\partial u & \partial y/\partial u & \partial z/\partial u \\ \partial x/\partial v & \partial y/\partial v & \partial z/\partial v \end{vmatrix} \tag{7.3}$$

$$= \left(\frac{\partial y}{\partial u}\frac{\partial z}{\partial v} - \frac{\partial y}{\partial v}\frac{\partial z}{\partial u}, \frac{\partial z}{\partial u}\frac{\partial x}{\partial v} - \frac{\partial z}{\partial v}\frac{\partial x}{\partial u}, \frac{\partial x}{\partial u}\frac{\partial y}{\partial v} - \frac{\partial x}{\partial v}\frac{\partial y}{\partial u} \right), \tag{7.4}$$

or, using the Jacobian (see Definition 6.6),

$$\mathbf{N} = \left(\frac{\partial(y, z)}{\partial(u, v)}, \frac{\partial(z, x)}{\partial(u, v)}, \frac{\partial(x, y)}{\partial(u, v)} \right). \tag{7.5}$$

DEFINITION 7.5 Smooth Parametrization and Smooth Surface

A parametrization $\mathbf{r} = \mathbf{r}(u, v)\colon D \subseteq \mathbb{R}^2 \to \mathbb{R}^3$ is said to be *smooth at* (u_0, v_0) [or *smooth at a point* $\mathbf{r}(u_0, v_0)$] if $\mathbf{r}$ is differentiable at (u_0, v_0) and if $\mathbf{N}(u_0, v_0) \neq \mathbf{0}$. It is *smooth* if it is smooth at all points in D.

A surface is called *smooth* if it has a smooth parametrization.

If $\mathbf{N}(u_0, v_0) = \mathbf{0}$ or if $\mathbf{N}(u_0, v_0)$ does not exist, we say that the parametrization $\mathbf{r}$ is not smooth at (u_0, v_0).

We have already seen that there could be an infinite number of parametrizations with the same image; that is, representing the same surface. A surface is called smooth if *one* of them turns out to be smooth. (Recall that we defined a smooth curve in the same way.)

▶ EXAMPLE 7.5

Consider the cone $\mathbf{r}(u, v) = v\cos u\mathbf{i} + v\sin u\mathbf{j} + v\mathbf{k}$, $0 \le u \le 2\pi$, $0 \le v \le b$, of Example 7.3. The tangents are $\mathbf{T}_u = -v\sin u\mathbf{i} + v\cos u\mathbf{j}$ and $\mathbf{T}_v = \cos u\mathbf{i} + \sin u\mathbf{j} + \mathbf{k}$, and the surface normal $\mathbf{N}$ is computed to be

$$\mathbf{N} = \mathbf{T}_u \times \mathbf{T}_v = \begin{vmatrix} \mathbf{i} & \mathbf{j} & \mathbf{k} \\ -v\sin u & v\cos u & 0 \\ \cos u & \sin u & 1 \end{vmatrix} = v\cos u\mathbf{i} + v\sin u\mathbf{j} - v\mathbf{k}.$$

At the vertex of the cone (which corresponds to $v = 0$ and $0 \le u \le 2\pi$), $\mathbf{N}(u, 0) = (0, 0, 0)$; therefore, this parametrization is not smooth at the vertex. It is smooth at any other point, since in that case $v > 0$ and hence $\mathbf{N}(u, v) \neq \mathbf{0}$.

Except at the vertex, the $\mathbf{k}$ component of $\mathbf{N}$ is negative. In other words, the normal vector $\mathbf{N}$ to the cone (parametrized as above, not in every case!) points downward; that is, away from the cone. ◀

Note that we cannot conclude from Example 7.5 that a cone is not a smooth surface (to prove that, we would have to examine all parametrizations and show that none is smooth).

▶ EXAMPLE 7.6

Compute the normal vector $\mathbf{N} = \mathbf{T}_u \times \mathbf{T}_v$ to the surface $\mathbf{r}(u, v) = (u, u, v^3)$, $u, v \in \mathbb{R}$.

SOLUTION By definition,

$$\mathbf{N} = \mathbf{T}_u \times \mathbf{T}_v = \begin{vmatrix} \mathbf{i} & \mathbf{j} & \mathbf{k} \\ 1 & 1 & 0 \\ 0 & 0 & 3v^2 \end{vmatrix} = 3v^2\mathbf{i} - 3v^2\mathbf{j},$$

and it follows that the parametrization is smooth at all points $\mathbf{r}(u, v)$ for which $v \neq 0$. The same surface (it is the vertical plane that crosses the xy-plane along the line segment $y = x$) can be parametrized by $\overline{\mathbf{r}}(u, v) = (u, u, v)$, $u, v \in \mathbb{R}$. In this case,

$$\overline{\mathbf{N}} = \overline{\mathbf{T}}_u \times \overline{\mathbf{T}}_v = \begin{vmatrix} \mathbf{i} & \mathbf{j} & \mathbf{k} \\ 1 & 1 & 0 \\ 0 & 0 & 1 \end{vmatrix} = \mathbf{i} - \mathbf{j},$$

which is always a nonzero vector. ◀

This example shows that a plane can have a parametrization that is not smooth. Since the parametrization $\overline{\mathbf{r}}$ is smooth, we conclude that the plane in question is a smooth surface.

▶ EXAMPLE 7.7

Consider the parametrization of the sphere

$$\mathbf{r}(u, v) = (a\cos v\cos u, a\cos v\sin u, a\sin v), \quad 0 \le u \le 2\pi, \quad -\pi/2 \le v \le \pi/2$$

$(a > 0)$ that we studied in Example 7.4. Compute the normal vector $\mathbf{N} = \mathbf{T}_u \times \mathbf{T}_v$.

SOLUTION The tangent vectors are given by $\mathbf{T}_u(u, v) = (-a\cos v\sin u, a\cos v\cos u, 0)$ and $\mathbf{T}_v(u, v) = (-a\sin v\cos u, -a\sin v\sin u, a\cos v)$, and thus,

$$\mathbf{N}(u, v) = \mathbf{T}_u(u, v) \times \mathbf{T}_v(u, v) = \begin{vmatrix} \mathbf{i} & \mathbf{j} & \mathbf{k} \\ -a\cos v\sin u & a\cos v\cos u & 0 \\ -a\sin v\cos u & -a\sin v\sin u & a\cos v \end{vmatrix}$$

$$= (a^2\cos^2 v\cos u, a^2\cos^2 v\sin u, a^2\cos v\sin v\sin^2 u + a^2\sin v\cos v\cos^2 u)$$

$$= a\cos v(a\cos v\cos u, a\cos v\sin u, a\sin v) = a\cos v\,\mathbf{r}(u, v).$$

It follows that the vector normal to the sphere is parallel to the position vector and

$$\|\mathbf{N}(u, v)\| = |a\cos v|\|\mathbf{r}(u, v)\| = a^2\cos v,$$

since $\|\mathbf{r}(u, v)\| = a$ and $\cos v \ge 0$ for $-\pi/2 \le v \le \pi/2$. So, this particular parametrization is not smooth when $v = \pm\pi/2$; that is, at the North and South Poles $(0, 0, \pm a)$. ◀

▶ EXAMPLE 7.8

Compute the surface normal $\mathbf{N}$ to the surface S given by $\mathbf{r}(u, v) = (u, v, |u|), -1 \le u \le 1, 0 \le v \le 1$.

SOLUTION Clearly, $\mathbf{T}_v = (0, 1, 0)$. If $u > 0$, then $\mathbf{r}(u, v) = (u, v, u)$ and $\mathbf{T}_u = (1, 0, 1)$, and if $u < 0$, then $\mathbf{r}(u, v) = (u, v, -u)$ and $\mathbf{T}_u = (1, 0, -1)$. It follows that

$$\mathbf{N}(u, v) = (\mathbf{T}_u \times \mathbf{T}_v)(u, v) = (\mathbf{i} + \mathbf{k}) \times \mathbf{j} = \mathbf{k} - \mathbf{i}$$

at points $\mathbf{r}(u, v)$ where $u > 0$, and

$$\mathbf{N}(u, v) = (\mathbf{T}_u \times \mathbf{T}_v)(u, v) = (\mathbf{i} - \mathbf{k}) \times \mathbf{j} = \mathbf{k} + \mathbf{i}$$

at points $\mathbf{r}(u, v)$ where $u < 0$.

Since the function $|u|$ is not differentiable at $u = 0$, it follows that the tangent vector $\mathbf{T}_u$ is not defined at points $\mathbf{r}(0, v)$ for $0 \le v \le 1$, and so the normal vector $\mathbf{N}(0, v)$, for $0 \le v \le 1$, does not exist. Consequently, $\mathbf{r}$ is not a smooth parametrization at points along the edge $\{(u, v) | u = 0, 0 \le v \le 1\}$; see Figure 7.13. ◀

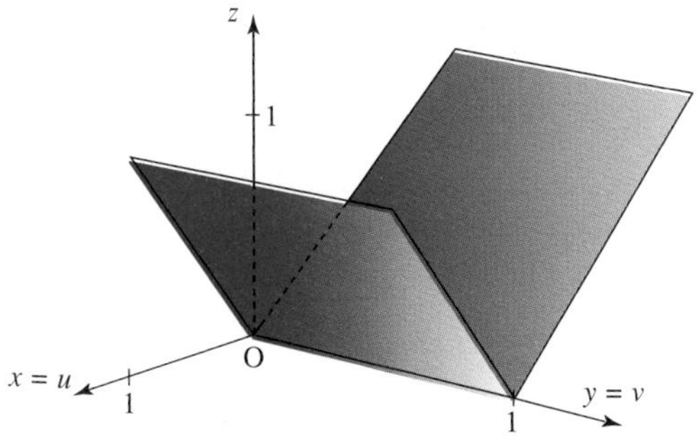

Figure 7.13 Parametrized surface $\mathbf{r}(u, v) = (u, v, |u|)$ is not smooth along the edge.

▶ EXAMPLE 7.9 Graph of a Differentiable Function $z = f(x, y)$ Is a Smooth Surface in $\mathbb{R}^3$

The graph of a differentiable function $f(x, y): D \subseteq \mathbb{R}^2 \to \mathbb{R}$ is a surface in $\mathbb{R}^3$ that can be parametrized by

$$\mathbf{r}(u, v) = (u, v, f(u, v)), \quad (u, v) \in D.$$

Instead of drawing two separate coordinate systems, one for the domain D (the "uv-coordinate system") and one for the surface S (the "xyz-coordinate system") as done in the previous examples, we merge together the two by placing the uv-plane over the xy-plane. Figure 7.14 shows the graph of $z = x^2y - y^3$ in this newly constructed coordinate system: both the surface and its domain are shown. Keeping the identification $u = x$ and $v = y$ in mind, we sometimes write the parametrization as $\mathbf{r}(x, y) = (x, y, f(x, y))$, $(x, y) \in D$ [compare with Example 2.52 in Section 2.5, where the graph of $y = f(x)$ was viewed as a parametrized curve].

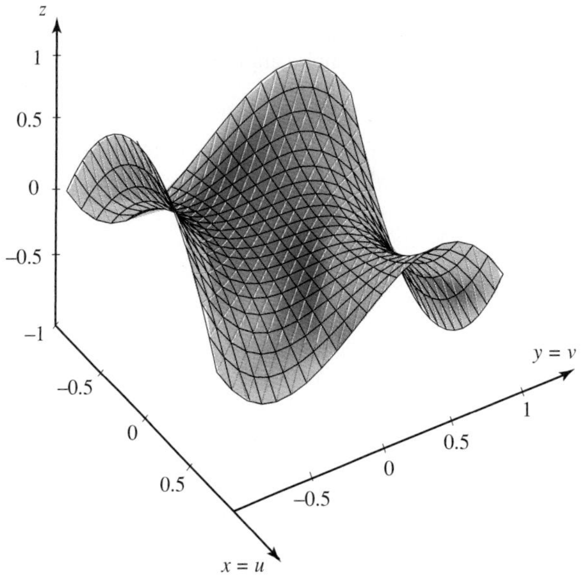

Figure 7.14 Graph of $z = f(x, y) = x^2y - y^3$ is a smooth surface.

The tangent vectors are given by $\mathbf{T}_u = (1, 0, \partial f/\partial u)$ and $\mathbf{T}_v = (0, 1, \partial f/\partial v)$, and therefore,

$$\mathbf{N} = \mathbf{T}_u \times \mathbf{T}_v = \left(1, 0, \frac{\partial f}{\partial u}\right) \times \left(0, 1, \frac{\partial f}{\partial v}\right) = \left(-\frac{\partial f}{\partial u}, -\frac{\partial f}{\partial v}, 1\right).$$

Since $\mathbf{N}(u, v) \neq \mathbf{0}$ for all (u, v), it follows that the graph of a differentiable function is a smooth surface. The normal vector $\mathbf{N}$ points upward, since its $\mathbf{k}$ component is positive. ◀

DEFINITION 7.6 Tangent Plane to a Surface

Let $\mathbf{r} = \mathbf{r}(u, v)$ be a smooth parametrization of a surface S, and let $\mathbf{r}(u_0, v_0)$ be a point on S. The *tangent plane* to S at $\mathbf{r}(u_0, v_0)$ is the plane that contains $\mathbf{r}(u_0, v_0)$, and is spanned by the vectors $\mathbf{T}_u(u_0, v_0)$ and $\mathbf{T}_v(u_0, v_0)$. ◀

The equation of the tangent plane is

$$a(x - x_0) + b(y - y_0) + c(z - z_0) = 0,$$

where $\mathbf{N}(u_0, v_0) = (a, b, c)$ is the surface normal $\mathbf{N}(u_0, v_0) = \mathbf{T}_u(u_0, v_0) \times \mathbf{T}_v(u_0, v_0)$, and $\mathbf{r}(u_0, v_0) = (x_0, y_0, z_0)$ is the point of tangency with the surface. A parametrized surface that is not smooth at (u_0, v_0) does not have a tangent plane at that point.

▶ EXAMPLE 7.10

Let S be the surface parametrized by $\mathbf{r}(u, v) = (u \cos v, u \sin v, v)$, where $u > 0$ and $v \geq 0$. Compute the equation of the plane tangent to S at $\mathbf{r}(1, \pi) = (-1, 0, \pi)$.

SOLUTION The tangent vectors to S at $\mathbf{r}(1, \pi)$ are

$$\mathbf{T}_u(1, \pi) = \frac{\partial \mathbf{r}}{\partial u}\bigg|_{(1,\pi)} = (\cos v, \sin v, 0)\bigg|_{(1,\pi)} = (-1, 0, 0), \quad \text{and}$$

$$\mathbf{T}_v(1, \pi) = \frac{\partial \mathbf{r}}{\partial v}\bigg|_{(1,\pi)} = (-u \sin v, u \cos v, 1)\bigg|_{(1,\pi)} = (0, -1, 1),$$

and therefore, $\mathbf{N}(1, \pi) = (\mathbf{T}_u \times \mathbf{T}_v)(1, \pi) = (0, 1, 1)$. The equation of the tangent plane is given by $0(x + 1) + 1(y - 0) + 1(z - \pi) = 0$, that is, $y + z - \pi = 0$. ◀

▶ EXAMPLE 7.11

Compute the equation of the plane tangent to the graph of a differentiable function $z = f(x, y)$ at (x_0, y_0, z_0).

SOLUTION The normal vector $\mathbf{N}$ is (see Example 7.9; replace u by x and v by y)

$$\mathbf{N} = \left(-\frac{\partial f}{\partial x}(x_0, y_0), -\frac{\partial f}{\partial y}(x_0, y_0), 1\right). \tag{7.6}$$

Consequently, the tangent plane has the equation (after multiplying by -1)

$$\frac{\partial f}{\partial x}(x_0, y_0)(x - x_0) + \frac{\partial f}{\partial y}(x_0, y_0)(y - y_0) - (z - z_0) = 0.$$ ◀

A surface may be *one-sided* or *two-sided.* A two-sided surface is a surface whose sides are separated in the sense that a person walking along one side of the surface can never reach the other side of it without crossing an edge or breaking through the surface. The sphere, the cone, the plane, the torus, the graph of $z = f(x, y)$, the surface in Figure 7.15—actually all the surfaces we have encountered so far—are two-sided.

Probably the most famous one-sided surface is the *Möbius strip*; see Figure 7.16. It can be made from a rectangular piece of paper, with one side longer than the other: just twist one smaller side by 180° and glue it to the other smaller side.

This time, a person walking along "one side" will suddenly find herself or himself on "the other side" of it. Replacing a walking person by a vector normal to the surface, we see that the normal, starting at the point P, returns to P after a continuous motion, but points in the opposite direction. This can only happen on one-sided surfaces.

On a two-sided surface, there are two normal directions given by the unit normal vectors $\mathbf{n}_1$ and $\mathbf{n}_2$ such that $\mathbf{n}_2 = -\mathbf{n}_1$; see Figure 7.17.

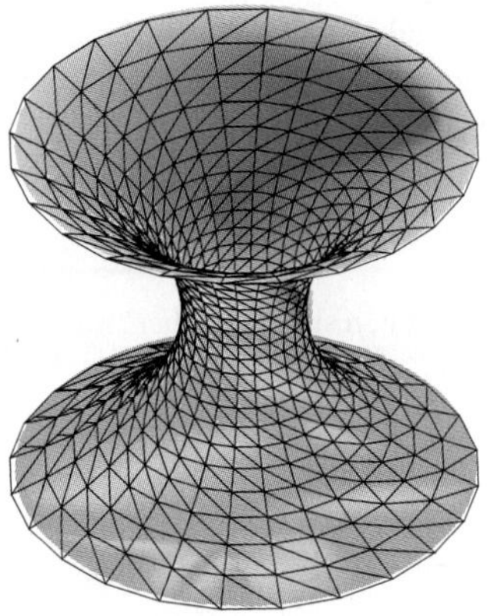

Figure 7.15 Two-sided surface.

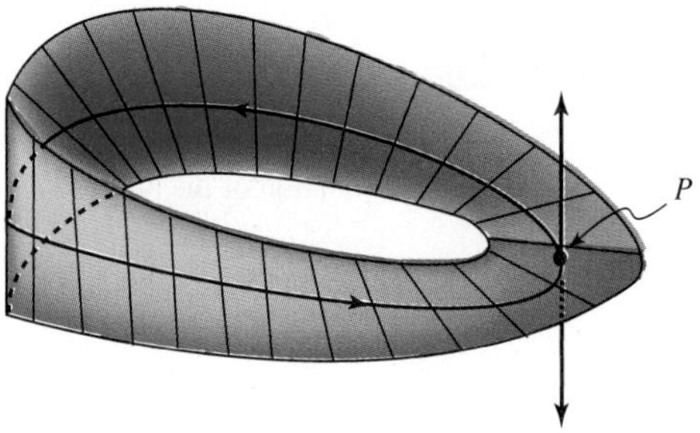

Figure 7.16 Möbius strip has only one side.

DEFINITION 7.7 Orientation of a Surface. Outside and Inside of a Surface

The *orientation* of a two-sided surface is a continuous choice of unit normal vectors. That choice is called the *positive orientation*. It determines the side of the surface (the one unit normal vectors point away from), called the *outside* or *positive side* of the surface. The other side is called the *inside* or *negative side.* ◀

▶ **EXAMPLE 7.12**

The plane given by the equation $x + 2y + z = 3$ can be parametrized as $\mathbf{r}(u, v) = (u, v, 3 - u - 2v)$, where $u, v \in \mathbb{R}$. From $\mathbf{T}_u = (1, 0, -1)$ and $\mathbf{T}_v = (0, 1, -2)$, we get $\mathbf{N} = \mathbf{T}_u \times \mathbf{T}_v = (1, 2, 1)$, so, the unit normal vectors are $\pm(1, 2, 1)/\sqrt{6}$.

By choosing $\mathbf{n}$ to be $(1, 2, 1)/\sqrt{6}$, we defined the positive orientation, and the outside of the plane (or the positive side) is the side we see if we stand, say, at the point A in the first octant; see Figure 7.18. We can see the negative side (or the inside) of the plane by standing, for example, at the origin. [Of course, we could have picked $-(1, 2, 1)/\sqrt{6}$ to determine the positive orientation.] ◀

▶ **EXAMPLE 7.13**

A normal vector to the sphere S of radius 1 in $\mathbb{R}^3$ [think of the sphere as the level surface of $f(x, y, z) = x^2 + y^2 + z^2$ of value 1] is given by the gradient, $\nabla f = (2x, 2y, 2z)$. Let us choose a

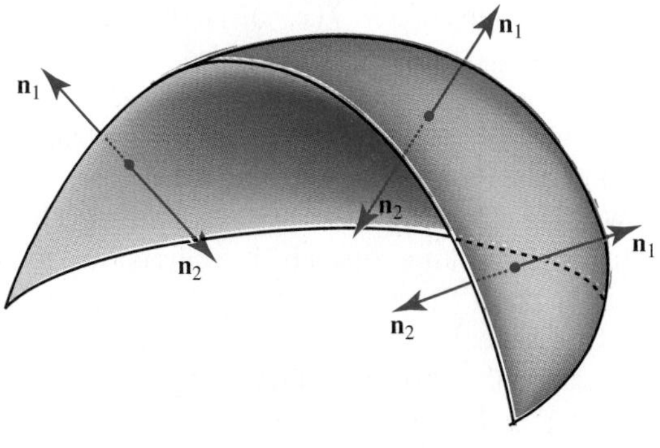

Figure 7.17 Normal directions on a two-sided surface.

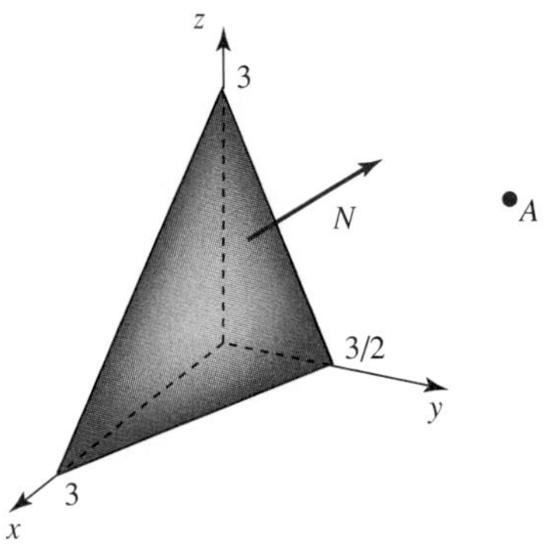

Figure 7.18 From the point A, we see the outside of the plane $\mathbf{r}(u, v) = (u, v, 3 - u - 2v)$.

unit normal to be

$$\mathbf{n} = \frac{\nabla f}{\|\nabla f\|} = \frac{1}{2\sqrt{x^2 + y^2 + z^2}}(2x, 2y, 2z) = (x, y, z)$$

(clearly, the other choice for a unit normal would be the negative of $\mathbf{n}$).

The vector $\mathbf{n}$ points away from the outside of the sphere, which agrees with our intuition. Vector $-\mathbf{n}$ points into the inside of S. In other words, the side of the sphere that we see if we stand, say, at the point $(10, 10, 0)$ is the positive orientation of S. The side we see standing at the origin is the inside, or the negative orientation.

DEFINITION 7.8 Orientation-Preserving and Orientation-Reversing Parametrization

Assume that S is an oriented surface, oriented by the choice of a unit normal vector field $\mathbf{n}$. Let $\mathbf{r}(u, v)$ be some parametrization of S, and let $\mathbf{N}$ be the corresponding surface normal vector field to S. The unit normal $\mathbf{N}/\|\mathbf{N}\| = \mathbf{T}_u \times \mathbf{T}_v/\|\mathbf{T}_u \times \mathbf{T}_v\|$ is either equal to $\mathbf{n}$ or to $-\mathbf{n}$. If $\mathbf{N}/\|\mathbf{N}\| = \mathbf{n}$, we say that the parametrization $\mathbf{r}$ is *orientation-preserving,* and if $\mathbf{N}/\|\mathbf{N}\| = -\mathbf{n}$, then $\mathbf{r}$ is *orientation-reversing*.

EXAMPLE 7.14

Let S be the sphere $x^2 + y^2 + z^2 = 1$, oriented by the unit vector $\mathbf{n}$ that we computed in Example 7.13. Consider the following parametrization of S:

$$\mathbf{r}(u, v) = (\cos u \sin v, \sin u \sin v, \cos v), \qquad 0 \le u \le 2\pi, \quad 0 \le v \le \pi.$$

[Note that $\|\mathbf{r}(u, v)\| = 1$.] The surface normal vector $\mathbf{N}$ is given by

$$\begin{aligned} \mathbf{N} = \mathbf{T}_u \times \mathbf{T}_v &= \begin{vmatrix} \mathbf{i} & \mathbf{j} & \mathbf{k} \\ -\sin u \sin v & \cos u \sin v & 0 \\ \cos u \cos v & \sin u \cos v & -\sin v \end{vmatrix} \\ &= (-\cos u \sin^2 v, -\sin u \sin^2 v, -\sin v \cos v) \\ &= -\sin v(\cos u \sin v, \sin u \sin v, \cos v) = -\sin v \cdot \mathbf{r}(u, v). \end{aligned}$$

It follows that the unit normal (whenever $v \neq 0, \pi$)

$$\frac{\mathbf{N}}{\|\mathbf{N}\|} = \frac{-\sin v\ \mathbf{r}(u,v)}{|-\sin v| \cdot \|\mathbf{r}(u,v)\|} = \frac{-\sin v}{|-\sin v|}\mathbf{r}(u,v) = -\mathbf{r}(u,v)$$

points in the direction opposite to $\mathbf{r}(u,v)$ [since $0 < v < \pi$, it follows that $\sin v > 0$ and therefore $-\sin v/|-\sin v| = -\sin v/\sin v = -1$; this explains the appearance of the minus sign in front of $\mathbf{r}(u,v)$]. Therefore, $\mathbf{N}/\|\mathbf{N}\| = -\mathbf{n}$; that is, the above parametrization is orientation-reversing.

On the other hand, the parametrization

$$\mathbf{r}(u,v) = (\cos v \cos u, \cos v \sin u, \sin v), \quad 0 \le u \le 2\pi, \quad -\pi/2 \le v \le \pi/2,$$

of S given in Examples 7.4 and 7.7 (with $a = 1$) yields the unit normal vector (whenever $v \neq -\pi/2, \pi/2$)

$$\frac{\mathbf{N}}{\|\mathbf{N}\|} = \frac{\cos v\ \mathbf{r}(u,v)}{\|\cos v\ \mathbf{r}(u,v)\|} = \frac{\cos v}{|\cos v|}\mathbf{r}(u,v) = \mathbf{r}(u,v).$$

This time, $\mathbf{N}/\|\mathbf{N}\|$ has the same direction as $\mathbf{r}(u,v)$, since $\cos v > 0$ for $-\pi/2 < v < \pi/2$. So, $\mathbf{N}/\|\mathbf{N}\| = \mathbf{n}$, and it follows that this is an orientation-preserving parametrization. ◂

DEFINITION 7.9 Orientation of the Graph of $z = f(x, y)$

The orientation of the graph S of a differentiable function $z = f(x, y)$ is defined as follows: the outside of S is the side away from which the unit normal vector

$$\mathbf{n} = \frac{\mathbf{N}}{\|\mathbf{N}\|} = \frac{\left(-\frac{\partial f}{\partial x}, -\frac{\partial f}{\partial y}, 1\right)}{\sqrt{\left(\frac{\partial f}{\partial x}\right)^2 + \left(\frac{\partial f}{\partial y}\right)^2 + 1}}$$

points. ◂

In other words, the positive orientation is determined by the choice of the normal with **k** component equal to $+1$, and the negative orientation corresponds to the normal whose **k** component is -1 (the normal was computed in Example 7.9).

▸ EXAMPLE 7.15

The normal **N** to the graph S of the function $f(x, y) = x^2y - y^3$ (shown in Figure 7.14) is given by $\mathbf{N} = (-\partial f/\partial x, -\partial f/\partial y, 1) = (-2xy, -x^2 + 3y^2, 1)$; see (7.6). The corresponding unit normal field is $\mathbf{n} = \mathbf{N}/\|\mathbf{N}\| = (-2xy, -x^2 + 3y^2, 1)/\|\mathbf{N}\|$.

Thus, $\mathbf{n}(0, 0, 0) = (0, 0, 1) = \mathbf{k}$, and it follows that the outside of S is the side that we see when we look at S from high up on the z-axis. [Note that, in order to determine the orientation (of an orientable surface), we need to find the normal **n** at only *one* point on the surface.] ◂

Note that we have used the word "surface" in different contexts: level surface, the graph of a function of two variables, and parametrized surface. In Exercise 6 in the chapter review section (see Review Exercises and Problems), we show that these three concepts coincide.

▸ EXERCISES 7.1

1. Consider the parametric representations of the cylinder and the sphere given in Examples 7.2 and 7.4. Describe how **r** maps the boundary of the rectangle D in each case.

2. Consider the parametrization $\mathbf{r}(u,v) = (a\cos u \sin v, a \sin u \sin v, a\cos v)$, $0 \le u \le 2\pi$, $0 \le v \le \pi$ of the sphere of radius a. Give a geometric interpretation of the parameters u and v.

Exercises 3 to 12: Find a parametrization of each surface in $\mathbb{R}^3$.

3. Upper hemisphere $x^2 + y^2 + z^2 = a^2$, $z \ge 0$

4. Quarter-sphere $x^2 + y^2 + z^2 = a^2$, $z \ge 0$, $x \ge 0$

5. Sphere of radius 2, centered at $(-2, 3, 7)$

6. The part of the upper hemisphere $x^2 + y^2 + z^2 = a^2$, $z \ge 0$, cut out by the cone $z^2 = x^2 + y^2$

7. The part of the plane $z - 3y + x = 2$ inside the cylinder $x^2 + y^2 = 4$

8. The graph of $x^2 + y^2 - z^2 = 1$

9. The part of the plane $x + 2y + z = 6$ in the first octant

10. The part of the cone $x^2 + y^2 = z^2$ in the first octant

11. The part of the paraboloid $z = x^2 + y^2$ in the first octant

12. The surface obtained by rotating the circle $(y-3)^2 + z^2 = 1$, $x = 0$ about the z-axis

Exercises 13 to 20: For each parametrized surface $\mathbf{r}(u,v)$ in $\mathbb{R}^3$,

(a) Find the tangent vectors $\mathbf{T}_u$ and $\mathbf{T}_v$ and the normal $\mathbf{N}$, and

(b) Find all points where $\mathbf{r}(u,v)$ is smooth.

13. $\mathbf{r}(u,v) = (2u, u^2 + v, v^2)$, $u, v \ge 0$

14. $\mathbf{r}(u,v) = (u, e^u \sin v, e^u \cos v)$, $0 \le v \le 2\pi$, $u \in \mathbb{R}$

15. $\mathbf{r}(u,v) = (\sin u \cos v, \sin u \sin v, 2\cos u)$, $0 \le u, v \le 2\pi$

16. $\mathbf{r}(u,v) = (u^2 + v^2, u^2 - v^2, 2uv)$, $0 \le u, v \le 1$

17. $\mathbf{r}(u,v) = ((1 + \cos v)\cos u, (1 + \cos v)\sin u, \sin v)$, $0 \le u, v \le 2\pi$

18. $\mathbf{r}(u,v) = (u, \cos v, \sin v)$, $0 \le u \le 1$, $0 \le v \le \pi$

19. $\mathbf{r}(u,v) = (u, v, 1 - (u^2 + v^2))$, $u, v \ge 0$

20. $\mathbf{r}(u,v) = (u, |u|, v)$, $-1 \le u \le 1$, $0 \le v \le 2$

21. Find an equation of the plane tangent to $\mathbf{r}(u,v) = (e^u, e^v, uv)$ at $(1, 1, 0)$.

22. Let S be the surface $z = 10 - x^2 - 2y^2$. Compute the equation of the plane tangent to it at the point $(1, 2, 1)$ in three different ways:

(a) By using the parametrization $\mathbf{r}(u,v) = (u, v, 10 - u^2 - 2v^2)$

(b) By viewing S as the graph of the function $f(x,y) = 10 - x^2 - 2y^2$

(c) By viewing S as the level surface of $f(x,y,z) = z + x^2 + 2y^2$.

23. Find an equation of the plane tangent to the graph of $y = x^2 + 2xz$ at $(1, 3, 1)$.

24. Find an equation of the plane tangent to the graph of $y = f(x,z)$ at (x_0, y_0, z_0).

25. Show that the plane tangent to the cone $z^2 = x^2 + y^2$ (at any point where it exists) goes through the origin.

26. Consider the following parametrizations:

(a) $\mathbf{r}_1(u,v) = (u, v, 1)$, $-1 \le u, v \le 1$

(b) $\mathbf{r}_2(u,v) = (2u, 3v, 1)$, $-1/2 \le u \le 1/2$, $-1/3 \le v \le 1/3$

(c) $\mathbf{r}_3(u,v) = (u^3, v^3, 1)$, $-1 \le u, v \le 1$

(d) $\mathbf{r}_4(u,v) = (\sin u, \sin v, 1)$, $0 \le u, v \le 2\pi$

Check that the images of $\mathbf{r}_1, \ldots, \mathbf{r}_4$ represent the same set. State which parametrizations are continuous, differentiable, C^1. State which parametrizations are smooth at (0, 0, 1). Compute the tangent plane at (0, 0, 1) for those parametrizations.

27. Find a parametrization of the ellipsoid $x^2/a^2 + y^2/b^2 + z^2/c^2 = 1$.

28. Let S be the surface $\mathbf{r}(u, v) = (u^2, 2uv, 0)$, $-\infty < u, v < \infty$. Find an orientation-reversing parametrization of S.

29. Find the points (if any) on the surface $\mathbf{r}(u, v) = (u^2 v, uv^2, 1)$ where the tangent plane is parallel to the plane $z = x - y$.

30. Find all points (if any) (x, y, z) on the paraboloid $z = 2 - x^2 - y^2$ where the normal vector is parallel to the vector joining the origin and the point (x, y, z).

31. Consider a differentiable parametrized surface $\mathbf{r}(u, v)\colon D \subseteq \mathbb{R}^2 \to \mathbb{R}^3$, and pick a point (u_0, v_0) in the domain of $\mathbf{r}$ where $\mathbf{T}_u(u_0, v_0) \times \mathbf{T}_v(u_0, v_0) \neq \mathbf{0}$.

(a) Recall that the derivative $D\mathbf{r}(u_0, v_0)$ is a linear map from $\mathbb{R}^2$ to $\mathbb{R}^3$. Find its matrix representation. Show that the range of $D\mathbf{r}(u_0, v_0)$ is the plane spanned by $\mathbf{T}_u(u_0, v_0)$ and $\mathbf{T}_v(u_0, v_0)$.

(b) Show that the plane tangent to the image of $\mathbf{r}$ at $\mathbf{r}(u_0, v_0)$ can be represented as $\mathbf{p}(u, v) = \mathbf{r}(u_0, v_0) + D\mathbf{r}(u_0, v_0)\begin{bmatrix} u - u_0 \\ v - v_0 \end{bmatrix}$. Thus (as expected), the derivative $D\mathbf{r}$ enters into the equation of the tangent plane (thought of as the linear approximation).

▶ 7.2 WORLD OF SURFACES

In this section, we study various surfaces, to understand better how parametrizations work, to provide more examples of implicitly defined surfaces and the Implicit Function Theorem, to hint at some (of many) applications of surfaces, and because we will need these surfaces in this chapter and also in Chapter 8. Differential geometry is one of several mathematical disciplines that are dedicated to exploring geometric objects such as surfaces.

▶ EXAMPLE 7.16 Surface of Revolution

As the graph of a differentiable function $y = f(x)$, $x \in [a, b]$, is rotated about the x-axis, it generates a *surface of revolution* S; see Figure 7.19. (Recall Example 6.7 in Section 6.2, where we studied solids of revolution.)

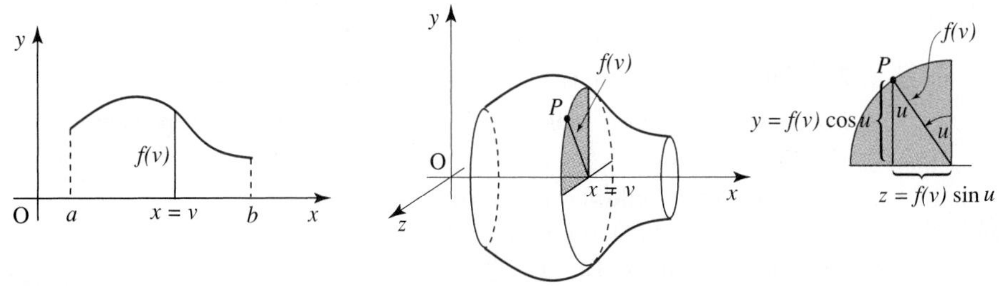

Figure 7.19 Surface of revolution obtained by rotating $y = f(x)$ about the x-axis.

Let $v = x$, and denote by u the angle of rotation (measured counterclockwise from the xy-plane). The surface S can be parametrized by

$$\mathbf{r}(u, v) = (v,\ f(v)\cos u,\ f(v)\sin u),$$

where $a \le v \le b$ and $0 \le u \le 2\pi$. In Exercise 1 we investigate the map $\mathbf{r}$, by determining how it maps horizontal and vertical lines in its domain.

From $\mathbf{T}_u = (0, -f(v)\sin u,\ f(v)\cos u)$ and $\mathbf{T}_v = (1,\ f'(v)\cos u,\ f'(v)\sin u)$, we compute

$$\mathbf{N}(u, v) = \mathbf{T}_u(u, v) \times \mathbf{T}_v(u, v) = (-f(v)f'(v),\ f(v)\cos u,\ f(v)\sin u),$$

and $\|\mathbf{N}(u, v)\|^2 = f^2(v)((f'(v))^2 + 1)$. Thus, whenever $f(v) \neq 0$, the normal $\mathbf{N}(u, v)$ is nonzero, and the parametrized surface S is smooth.

Figure 7.20 shows the surface of revolution generated by revolving the graph of $y = f(x) = \ln x$, $0.5 \le x \le 2$, about the x-axis. The surface shrinks to a point when $x = 1$ (i.e., when $f(1) = \ln 1 = 0$), and it appears to be non-smooth there. [Of course, the above calculation does not prove that S is not smooth].

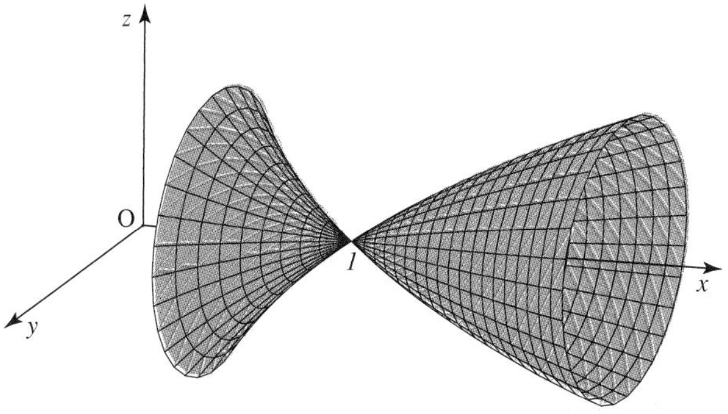

Figure 7.20 Parametrized non- smooth surface of revolution.

In the above, we considered x-axis as the axis of revolution. In a similar way, we obtain parametrizations for rotations about other coordinate axes. For instance, if $x = f(z)$, $z \in [c, d]$, is rotated about the z-axis, the resulting surface of revolution can be parametrized as $\mathbf{r}(u, v) = (f(v)\cos u,\ f(v)\sin u,\ v)$, where $c \le v \le d$ and $0 \le u \le 2\pi$.

To obtain the parametrization of the cylinder in Example 7.2, we take $f(z) = a$, $0 \le z \le b$, thus getting $\mathbf{r}(u, v) = (a\cos u, a\sin u, v)$. For the cone in Example 7.3, we take $f(z) = z$, $0 \le z \le b$, and so $\mathbf{r}(u, v) = (v\cos u, v\sin u, v)$.

In Exercise 2 we show that when the image of a differentiable path $\mathbf{c}(t) = (f(t), g(t))$, $t \in [a, b]$, in the xz-plane (where $f(t) \ge 0$ for $t \in [a, b]$) is rotated about the z-axis, the resulting surface of revolution can be parametrized as

$$\mathbf{r} = (f(v)\cos u,\ f(v)\sin u,\ g(v)), \tag{7.7}$$

with $a \le v \le b$ and $0 \le u \le 2\pi$. The parametrization of the sphere in Example 7.4 could have been obtained by taking $f(v) = a\cos v$ and $g(v) = a\sin v$.

▶ EXAMPLE 7.17 Flow of a Fluid in a Tube

A fluid flowing in a straight, horizontal tube of constant radius R does not flow at the same rate: it flows faster near the center of the tube, whereas near the walls of the tube, the flow is much slower (it is zero at the walls).

Let r denote the distance from the center of the tube, $0 \leq r \leq R$. It can be demonstrated that the dependence of the velocity on the location r within a tube is given by $v(r) = v_m(1 - r^2/R^2)$, where v_m is the maximum velocity (i.e., the velocity at the center of the tube).

The diagram in Figure 7.21(a) is sometimes called the *velocity profile*. When this graph is rotated about the axis that goes through the center of the tube, we obtain the three-dimensional velocity profile that shows how fluid moves through the tube; see Figure 7.21(b). Besides applications in engineering and physics, this profile is sometimes used in medicine: it represents a rough approximation of the blood flow in arteries. ◀

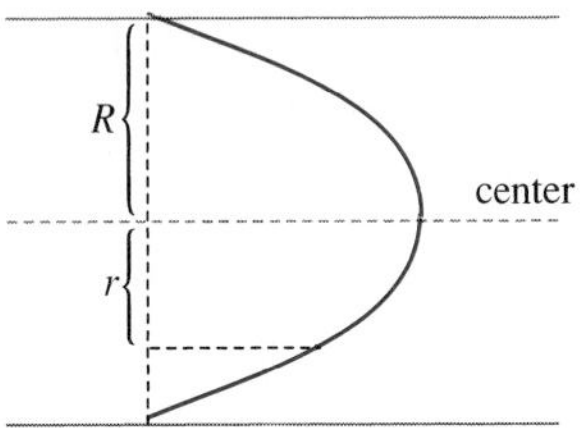

(a) (b)

Figure 7.21 Two-dimensional and three-dimensional velocity profiles.

▶ EXAMPLE 7.18 Torus

Place a circle of radius ρ so that the distance between its center C and the z-axis is R, where $R > \rho$. Assume that C lies on the x-axis, as shown in Figure 7.22. The surface of revolution obtained by rotating the circle about the z-axis is called a *torus;* see Figure 7.23.

To find a parametric representation of the torus, we select the parameters in the following way (see Figure 7.24): u is the angle of rotation measured counterclockwise from the x-axis, and v is the angle that will be used to determine the location of a point on the circle.

Pick a point P on the torus; we need to express its x, y, and z coordinates in terms of u and v. Let P' be the orthogonal projection of P onto the line through O and C, see Figures 7.24 and 7.25. Then $\overline{OP'} = R + \rho \cos v$, and thus, from the triangle $OP'P''$, we compute $x = \overline{OP'} \cos u = (R + \rho \cos v) \cos u$ and $y = \overline{OP'} \sin u = (R + \rho \cos v) \sin u$. Since $z = \rho \sin v$, we

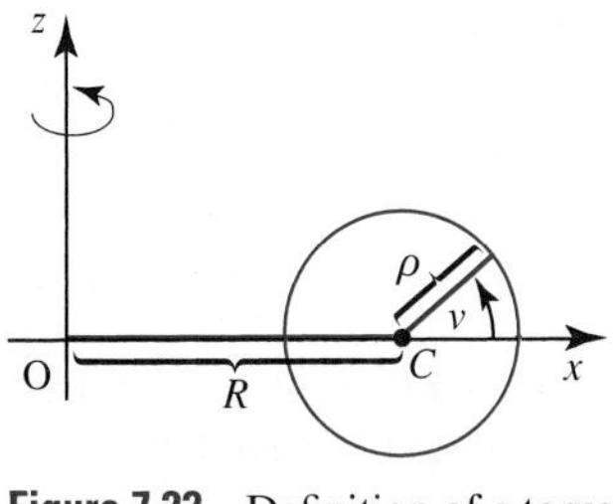

Figure 7.22 Definition of a torus.

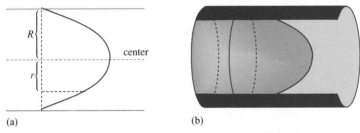

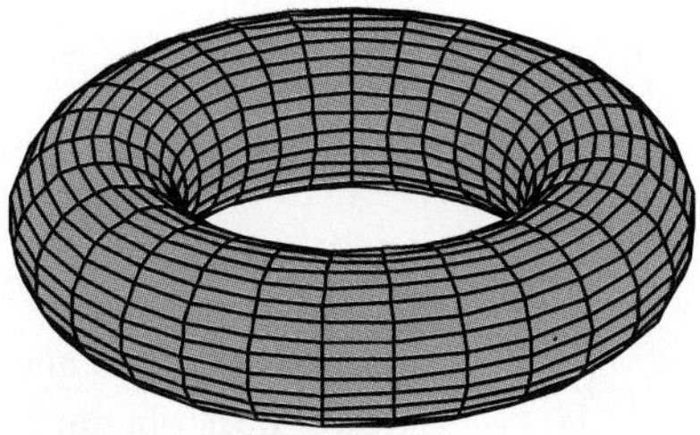

Figure 7.23 Torus.

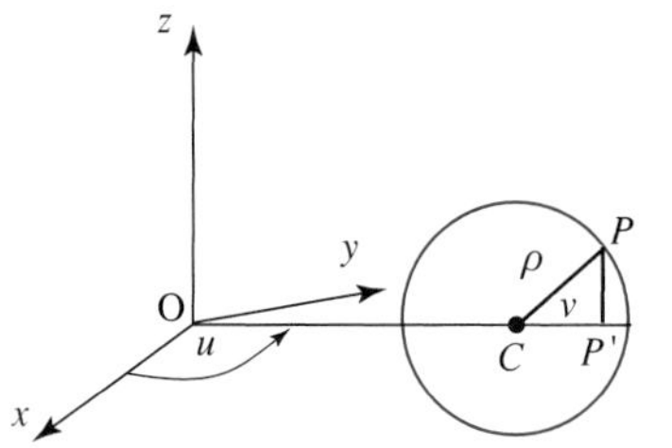

Figure 7.24 The parameters u and v.

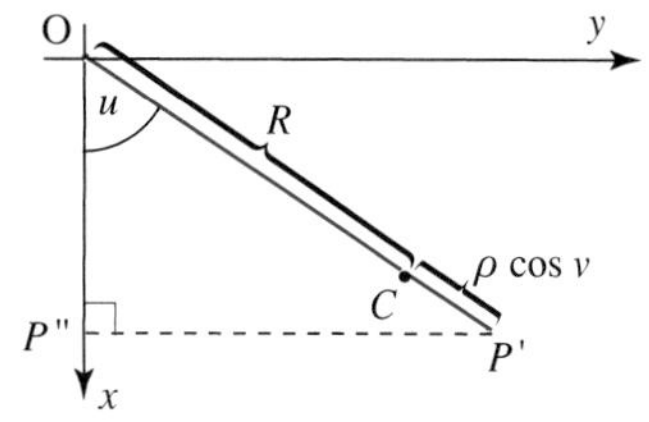

Figure 7.25 Computing parametrization.

obtain the parametrization

$$\mathbf{r}(u, v) = ((R + \rho \cos v) \cos u, (R + \rho \cos v) \sin u, \rho \sin v), \quad 0 \le u, v \le 2\pi. \tag{7.8}$$

Note that we could have obtained the parametrization (7.8) from (7.7), by taking $f(v) = R + \rho \cos v$ and $g(v) = \rho \sin v$. The domain D of $\mathbf{r}$ is the rectangle $[0, 2\pi] \times [0, 2\pi]$. How does $\mathbf{r}$ deform D to produce the torus?

The image of a horizontal segment $v = v_0$ is

$$\mathbf{r}(u, v_0) = ((R + \rho \cos v_0) \cos u, (R + \rho \cos v_0) \sin u, \rho \sin v_0), \quad 0 \le u \le 2\pi,$$

which is the circle of radius $R + \rho \cos v_0$ obtained by intersecting the torus with the horizontal plane $z = \rho \sin v_0$. Thus, the two points $A(0, v_0)$ and $A'(2\pi, v_0)$ are mapped into the same point (see Figure 7.26), and we conclude that the left side and the right side of D become glued together by $\mathbf{r}$. The image of the vertical segment $u = u_0$ is

$$\mathbf{r}(u_0, v) = ((R + \rho \cos v) \cos u_0, (R + \rho \cos v) \sin u_0, \rho \sin v), \quad 0 \le v \le 2\pi.$$

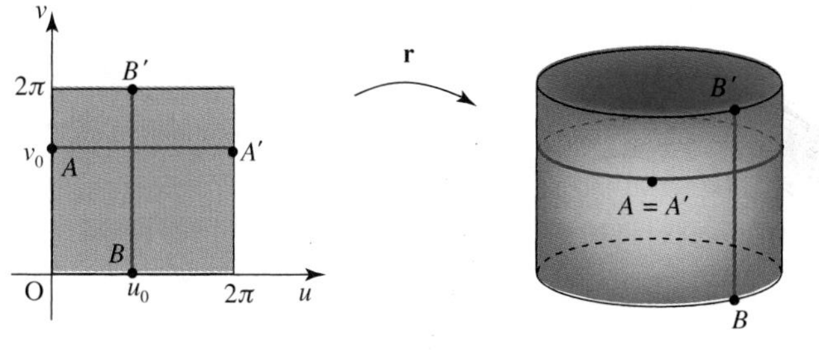

Figure 7.26 Deforming D to obtain the torus.

It is the circle that, by rotation, generates the torus, at the moment when the angle of rotation is u_0. Since $\mathbf{r}(u_0, 0) = \mathbf{r}(u_0, 2\pi)$, the points B and B' (Figure 7.26) are mapped into the same point. So, $\mathbf{r}$ glues together the top and the bottom circles, bending the cylinder so that the vertical segment BB' becomes a circle (see Figure 7.3 in Section 7.1). ◀

▶ EXAMPLE 7.19 Möbius Strip

Recall that we can make a *Möbius strip* from a rectangular piece of paper, with one side longer than the other: twist one smaller side by 180° and glue it to the other smaller side. Let us use this "mechanical" description to find a parametrization of the Möbius strip.

Start with the circle $\mathbf{c}(t) = (\cos t, \sin t, 0)$, $t \in [0, 2\pi]$, and imagine a line segment perpendicular to the xy-plane whose midpoint is at a point A on that circle.

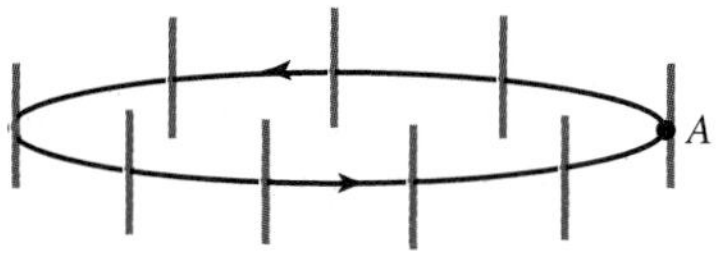

Figure 7.27 Trying to get a Möbius strip.

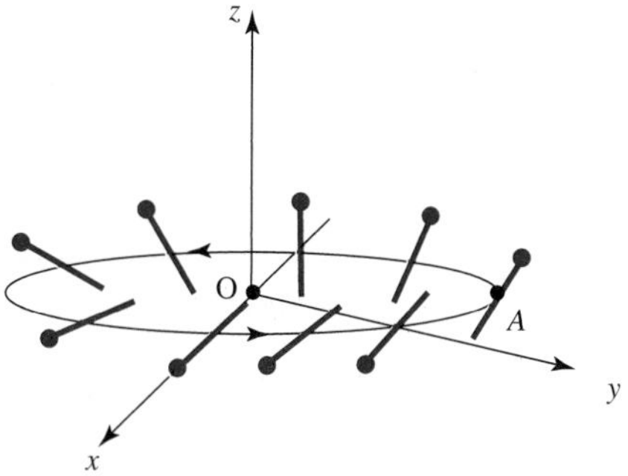

Figure 7.28 Computing parametrization.

As A completes one full revolution around the circle, the corresponding line segments describe a cylinder of height equal to the length of that segment; see Figure 7.27 (that's not exactly what we want, but . . .).

Now imagine that, as A moves around the circle, the segment moves with A in such a way that its angle with respect to the xy-plane changes—as A completes one full revolution, the line segment is rotated through 180°; that is, ends up upside-down; see Figure 7.28. This rotation of the segment creates the Möbius strip. To describe it, we use $\theta = t/2$ (as t goes from 0 to 2π, θ goes from 0 to π—just what we need). Let u be the distance from A to a point P on the segment, if P is above the xy-plane (if it lies below the xy-plane, we take $-u$).

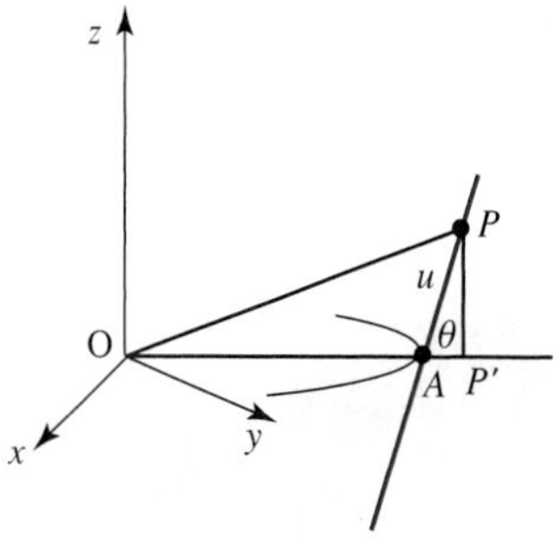

Figure 7.29 Computing parametrization.

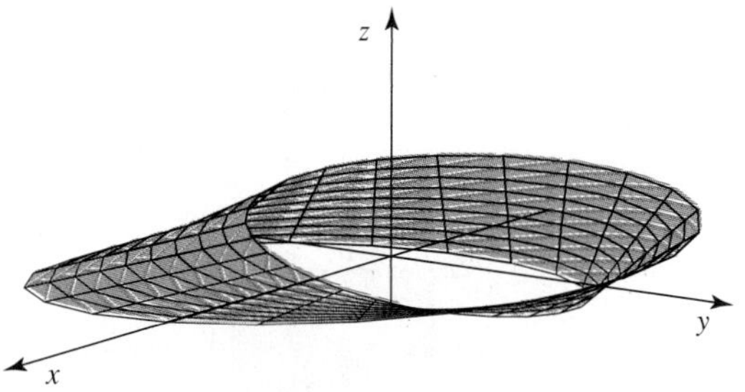

Figure 7.30 A Möbius strip.

The point A is on $\mathbf{c}$, and thus, $\overrightarrow{OA} = (\cos t, \sin t, 0)$, for some fixed value of t. Now $\overrightarrow{AP} = \overrightarrow{AP'} + \overrightarrow{P'P}$; see Figure 7.29. Since $\overrightarrow{AP'}$ is parallel to $\overrightarrow{OA}$ and $\|\overrightarrow{AP'}\| = u|\cos\theta|$, it follows that

$$\overrightarrow{AP'} = u\cos\theta \frac{\overrightarrow{OA}}{\|\overrightarrow{OA}\|} = u\cos(t/2)(\cos t, \sin t, 0).$$

Since $\overrightarrow{P'P}$ is perpendicular to the xy-plane and $\|\overrightarrow{P'P}\| = u|\sin\theta|$, it follows that $\overrightarrow{P'P} = u\sin\theta\,\mathbf{k} = (0, 0, u\sin(t/2))$. Hence,

$$\begin{aligned}\overrightarrow{OP} &= \overrightarrow{OA} + \overrightarrow{AP'} + \overrightarrow{P'P}\\ &= (\cos t, \sin t, 0) + u\cos(t/2)(\cos t, \sin t, 0) + (0, 0, u\sin(t/2)),\end{aligned}$$

and so the Möbius strip can be parametrized as

$$\mathbf{r}(t, u) = (\cos t\,(1 + u\cos(t/2)), \sin t\,(1 + u\cos(t/2)), u\sin(t/2)),$$

where $0 \le t \le 2\pi$ and $-1/2 \le u \le 1/2$ (we assume that the generating segment is of length 1); see Figure 7.30. ◀

▶ EXAMPLE 7.20 Whitney Umbrella

Consider the parametrization $\mathbf{r}(u, v) = (uv, u, v^2)$, where $-1 \le u \le 1$ and $-1 \le v \le 1$. Note that $\mathbf{r}(0, 1/2) = \mathbf{r}(0, -1/2) = (0, 0, 1/4)$, and because the points $(0, 1/2)$ and $(0, -1/2)$ are not the boundary points of the domain $D = [-1, 1] \times [-1, 1]$ of $\mathbf{r}$, the function $\mathbf{r}$ *does not* represent a parametrized surface (see Definition 7.1 in Section 7.1). As a matter of fact, $\mathbf{r}$ is not one-to-one at all points $\{(0, v) | -1 \le v \le 1\}$ (see Exercise 5).

The graph of $\mathbf{r}$ is called a *Whitney umbrella*. It is an example of a self-intersecting surface (and, as we showed, it is not a parametrized surface). Two views of the surface are shown in Figure 7.31. ◀

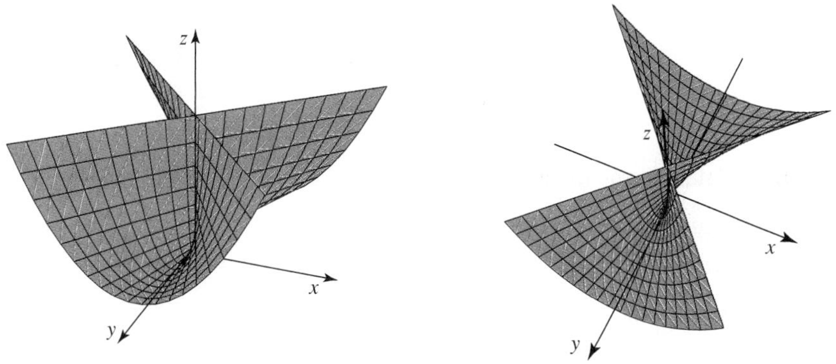

Figure 7.31 Two views of a Whitney umbrella.

▶ EXAMPLE 7.21 Helicoid

The image of a parametrized surface $\mathbf{r}(u, v) = (u \cos v, u \sin v, v)$, $0 \le u \le 1$, $0 \le v \le 2\pi$, is called a *helicoid.*

For a fixed value $u = u_0$,

$$\mathbf{r}(u_0, v) = (u_0 \cos v, u_0 \sin v, v), \qquad 0 \le v \le 2\pi,$$

is a helix of pitch 2π (see Example 3.8 in Section 3.1). For a fixed value $v = v_0$,

$$\mathbf{r}(u, v_0) = (u \cos v_0, u \sin v_0, v_0), \qquad 0 \le u \le 1,$$

is a straight-line segment in the plane $z = v_0$, whose vertical projection onto the xy-plane makes an angle of v_0 rad, measured counterclockwise from the positive x-axis.

Figure 7.32 shows how helices of various sizes are put together to form the helicoid. Mechanically, we obtain the helicoid in the following way: place a segment of length 1 along the x-axis, one end at the origin, and the other at $(1, 0, 0)$. As we rotate the line segment about the z-axis, we lift it (see Figure 7.32) at the rate at which the angle of rotation is changing (e.g., when the segment is rotated through the angle of $\pi/4$ rad, it is lifted $\pi/4$ units above the xy-plane).

The shape of a helicoid has been known for a long time. For instance, Greek mathematician Archimedes constructed a device (known as *Archimedes' screw*) that has been used to raise water (or other liquids) to a higher elevation. A woodcut in Figure 7.33 shows Archimedes' screw pumping water from a water source into a bucket (from Virtruvius's *De Architectura,* 1522 edition). The helicoid is a generic shape of a spiral staircase. In various modifications, we can identify it in a variety of screws, instruments, machine parts, as a design element in architecture, etc. ◀

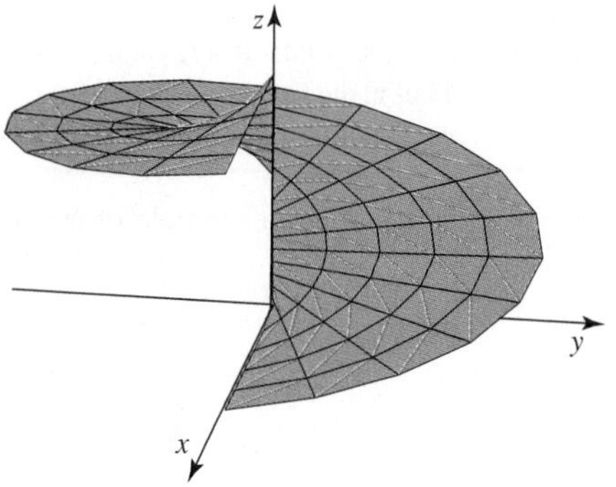

Figure 7.32 Helicoid.

Figure 7.33 Archimedes' screw.

▶ EXAMPLE 7.22 Catenoid and Minimal Surfaces

Rotation of the catenary curve (see Example 3.5 in Section 3.1) generates a surface of revolution, called a *catenoid.* In Exercise 16, we show that the catenoid can be parametrized by $r(u, v) = (\cosh v \cos u, \cosh v \sin u, v)$, where $0 \leq u \leq 2\pi$, $-1 \leq v \leq 1$. The image of $\mathbf{r}$ is shown in Figure 7.34.

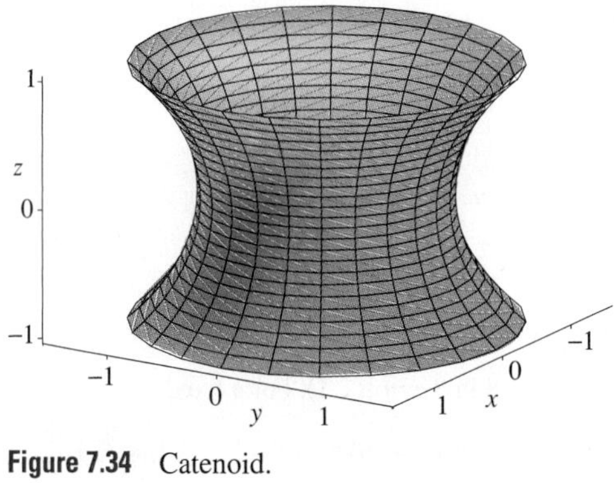

Figure 7.34 Catenoid.

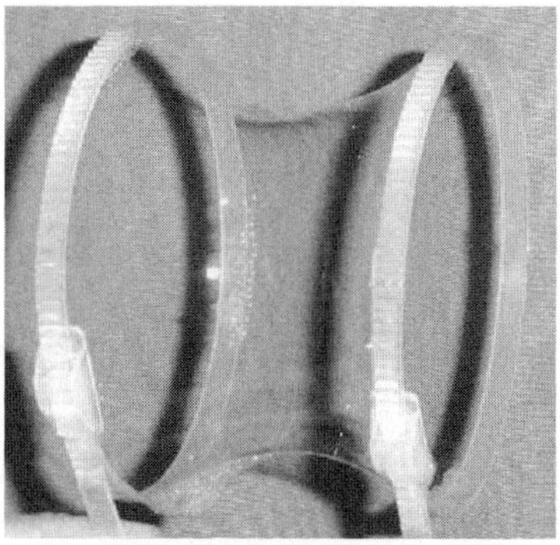

Figure 7.35 Soap film.

If we shape a wire [say, as a circle, ellipse, helix, or some (not necessarily) closed curve in three dimensions, etc.], dip it into a soap solution, and pull it out carefully, a thin soap film surface will be formed. That surface is an example of an important family of surfaces, called *minimal surfaces.* Given the shape of the wire [i.e., given the curve(s) that will form the boundary], the soap film surface is the surface with the smallest area whose boundary is the given curve (that is why it is called minimal).

If we dip two rings into a soap solution and keep them parallel as we pull them out, we will obtain a catenoid; see Figure 7.35. Thus, the catenoid is a minimal surface. The plane and helicoid are also minimal surfaces. (We do not give proofs of these facts here.)

Minimal surfaces have been studied in mathematics for several centuries. They have been used in engineering and architecture (lightweight structures), and have been identified in nature (for instance, in the shape of cell membranes). ◀

Figure 7.36 Two views of the Bilbao Guggenheim Museum.

▶ EXAMPLE 7.23 Surfaces in Architecture

Helicoids and minimal surfaces are part of a large spectrum of surfaces that have been used in modern (and not only modern) architecture. For instance, the Bilbao Guggenheim Museum in Bilbao, Spain (designed by Frank O, Gehry, and completed in 1997), is built of a number of surfaces of varying shape and curvature (see Figure 7.36).

The "sea shells" of the Sydney Opera House in Sydney, Australia (designed by Jorn Utzon and built in three stages, from 1957 to 1973; Figures 7.37 and 7.38) were constructed using spheres and ellipsoids.

The shape of the office tower at 30 St. Mary Axe in London, England (designed by Sir Norman Foster and Ken Shuttleworth, and completed in 2004; see Figure 7.39), is a circular ellipsoid (or ellipsoid of rotation; for somewhat obvious reasons, the tower has been nicknamed "The Gherkin"). The particular shape of the tower allowed for many innovations, including obtaining the optimal flow of winds around the building, a reduction in the amount of energy used, and an increase in the amount of natural light available in offices within the building. ◀

▶ EXAMPLE 7.24 Hyperbolic Paraboloid

The image of the parametrized surface $\mathbf{r}(u, v) = (u, v, uv)$, $-1 \le u, v \le 1$, is called a *hyperbolic paraboloid*. In Exercise 15 we show that, although the surface is curved (looks like a saddle), it contains two families of straight-line segments; see Figure 7.40. In other words: for any point on the hyperbolic paraboloid there are two line segments that go through it and are contained in the surface (such surfaces are called *doubly ruled surfaces*).

We recognize the shape of the hyperbolic paraboloid in the roof of the Catalano House (designed by Eduardo Catalano, and built in 1954) in Raleigh, North Carolina, shown in Figure 7.41. ◀

Figure 7.37 Sydney Opera House.

To make a hyperbolic paraboloid, we stretch a canvas over a rectangular frame, which is built so that it is flexible at all vertices. All we have to do is to twist the opposite sides in opposite directions, as illustrated in Figure 7.42.

▶ EXAMPLE 7.25 Sea Shells and Horns

We have already constructed a surface based on the helix—helicoid in Example 7.21. We can reformulate the construction of the helicoid in the following way: as we walk along the helix, we keep drawing horizontal line segments from the point where we are standing to the z-axis (think of a spiral staircase!).

Figure 7.38 Sydney Opera House, detail.

Figure 7.39 "The Gherkin", London.

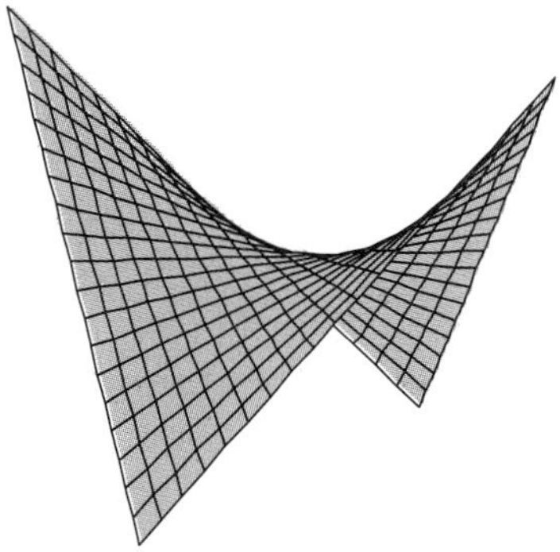

Figure 7.40 Hyperbolic paraboloid.

Figure 7.41 Roof of the Catalano House.

If, instead of drawing horizontal line segments, we draw circles of increasing radii (centered at the point on the helix where we are standing), we will obtain surfaces such as those shown in Figure 7.43.

In Exercise 24, we show that such surfaces can be represented in parametric form as $\mathbf{r}(u, v) = (f(u, v), g(u, v), h(u, v))$, where

$$f(u, v) = a \cos u + ru \left(-\cos v \cos u + b \sin v \sin u/\sqrt{a^2 + b^2}\right),$$

$$g(u, v) = a \sin u + ru \left(-\cos v \sin u - b \sin v \cos u/\sqrt{a^2 + b^2}\right),$$

$$h(u, v) = bu + aru \sin v/\sqrt{a^2 + b^2},$$

and $a, b, r > 0$, $0 \leq v \leq 2\pi$, $0 \leq u \leq 2\pi$ (or another value for u, depending on the height we need). The surfaces in Figure 7.43 approximate the shapes of sea shells or animals' horns and could be used to study their growth. ◀

Implicitly Defined Surfaces

Recall that the set S of points (x, y, z) in the domain of a C^1 function $F\colon \mathbb{R}^3 \to \mathbb{R}$ where $F(x, y, z) = 0$ is called an implicitly defined surface (see Definition 4.10 in Section 4.7). The Implicit Function Theorem (see Theorem 4.14 and the text following Definition 4.10) states that if $\nabla F(x_0, y_0, z_0) \neq \mathbf{0}$, then, near (x_0, y_0, z_0), the surface S looks like the graph of a real-valued function of two variables (for an illustration, see Example 4.67).

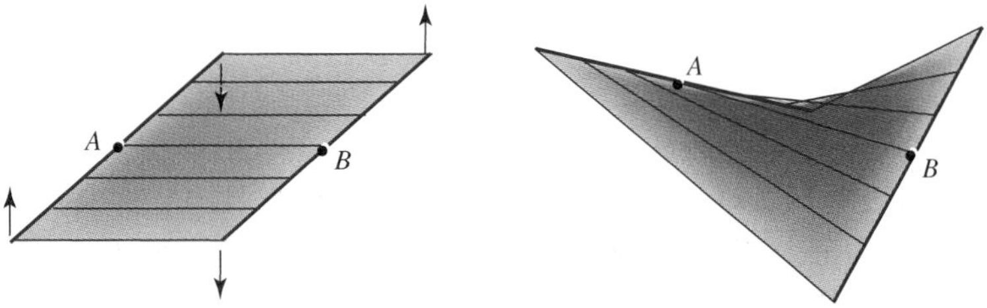

Figure 7.42 How to make a hyperbolic paraboloid.

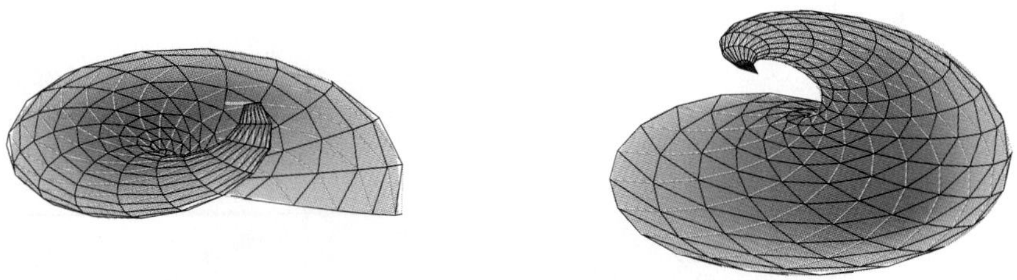

Figure 7.43 Two views of the parametrized surface in Example 7.25.

Note that S can also be viewed as a level surface of F of value zero. Thus, by Theorem 2.11 in Section 2.7, the vector $\nabla F(x, y, z)$ is normal to S. We illustrate this observation in our next example, where we compute an equation of a tangent plane to a surface.

▶ EXAMPLE 7.26

In Example 4.67 in Section 4.7, we studied the surface S, defined implicitly by the equation $F(x, y, z) = x + xe^y + y^2z - 1 = 0$.

The vector $\mathbf{N}(x, y, z) = \nabla F(x, y, z) = (1 + e^y, xe^y + 2yz, y^2)$ is normal to S at (x, y, z). Since $1 + e^y \neq 0$ for all y, we conculde that $\mathbf{N} \neq \mathbf{0}$ for all points on S (thus, S is a smooth surface). Its tangent plane at (x_0, y_0, z_0) is given by

$$(1 + e^{y_0})(x - x_0) + (x_0e^{y_0} + 2y_0z_0)(y - y_0) + y_0^2(z - z_0) = 0.$$

Pick a point on S, say, $(0, 2, 1/4)$. The tangent plane to S at $(0, 2, 1/4)$ is computed to be $(1 + e^2)x + (y - 2) + 4(z - 1/4) = 0$, that is, $(1 + e^2)x + y + 4z - 3 = 0$.

Let us look at another way of calculating a normal vector to S. Because the partial derivative $(\partial f/\partial z)(0, 2, 1/4) = 4$ is not zero, the Implicit Function Theorem implies that, near $(0, 2, 1/4)$, the surface S is the graph of a (unique) function $z = g(x, y)$. Recall that the normal to S at $(0, 2, 1/4)$ is given by $\overline{\mathbf{N}} = (-\partial g/\partial x, -\partial g/\partial y, 1)$, evaluated at $(0, 2, 1/4)$; see (7.6) in Example 7.11. Using formula (4.36) in part (c) of the Implicit Function Theorem, we get

$$\frac{\partial g}{\partial x} = -\frac{\partial F/\partial x}{\partial F/\partial z} = -\frac{1 + e^y}{y^2},$$

and thus, $(\partial g/\partial x)(0, 2) = -(1 + e^2)/4$. Similarly,

$$\frac{\partial g}{\partial y} = -\frac{\partial F/\partial y}{\partial F/\partial z} = -\frac{xe^y + 2yz}{y^2},$$

and $(\partial g/\partial y)(0, 2) = -1/4$. Thus, the normal is $\overline{\mathbf{N}} = ((1 + e^2)/4, 1/4, 1)$. Of course, the two normals $\mathbf{N}$ and $\overline{\mathbf{N}}$ are parallel.

Note that in this case, we can solve for g explicitly $[z = g(x, y) = (1 - x - xe^y)/y^2]$ and compute the partial derivatives directly. ◀

▶ EXAMPLE 7.27 Quadric (Quadratic) Surfaces

A *quadric surface* (also known as a *quadratic surface*) is the set of points (x, y, z) in $\mathbb{R}^3$ where

$$F(x, y, z) = Ax^2 + By^2 + Cz^2 + Dxy + Exz + Fyz + Gx + Hy + Iz + J = 0 \qquad (7.9)$$

$(A, B, \ldots, J$ are constants). In this example, we briefly discuss several quadric surfaces.

The equation $x^2/a^2 + y^2/b^2 + z^2/c^2 = 1$ represents an *ellipsoid* with semiaxes $a > 0$, $b > 0$, and $c > 0$; see Figure 7.44. If $a = b$, then it is an ellipsoid of revolution [obtained by revolving the

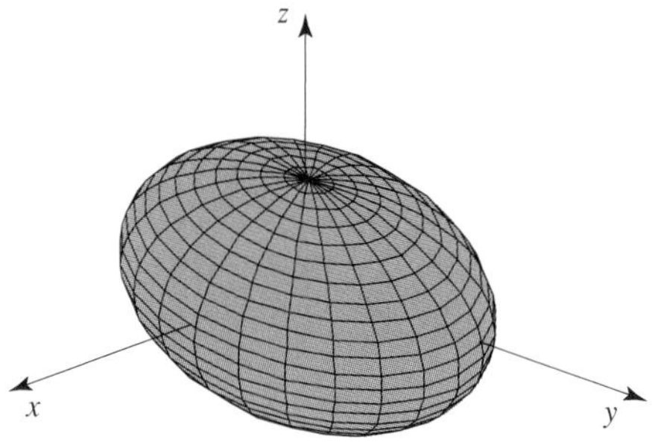

Figure 7.44 Ellipsoid.

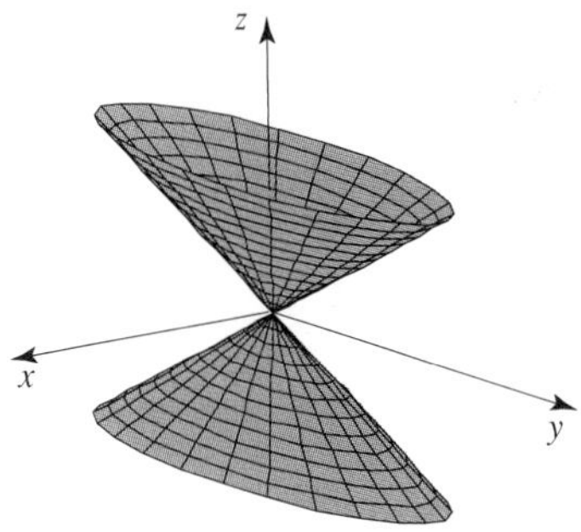

Figure 7.45 Cone.

ellipse $x^2/a^2 + z^2/c^2 = 1$ (in the xz-plane) about the z-axis]. Likewise, if $b = c$ or $a = c$, we obtain ellipsoids of revolution. If $a = b = c$, the above equation represents the sphere of radius a.

The equation $x^2/a^2 + y^2/b^2 - z^2/c^2 = 0$ represents a *cone* (in this general case, it is not a surface of revolution!); see Figure 7.45. The level curve of value $z = z_0$ is the ellipse $x^2/a^2 + y^2/b^2 = z_0^2/c^2$. If $a = b$, the above equation represents the cone that is obtained by rotating one of the lines $z = \pm cx/a$ about the z-axis.

A *hyperboloid of one sheet* is represented by the equation $x^2/a^2 + y^2/b^2 - z^2/c^2 = 1$; see Figure 7.46. Level curves $z = z_0$ are ellipses $x^2/a^2 + y^2/b^2 = 1 + z_0^2/c^2$ that increase in size as the distance z_0 from the origin increases. The intersections of the hyperboloid with the xz-plane and yz-plane are hyperbolas.

The surface defined by $x^2/a^2 + y^2/b^2 - z^2/c^2 = -1$ is a *hyperboloid of two sheets*; see Figure 7.47. The level curve of value $z = z_0$ is the ellipse $x^2/a^2 + y^2/b^2 = z_0^2/c^2 - 1$, unless $z_0 = c$ [in which case, it is a point $(0, 0, z_0)$] or $z_0 < c$ [then it is an empty set].

Other quadric surfaces include the *elliptic cylinder* $x^2/a^2 + y^2/b^2 = 1$, then the *hyperbolic cylinder* $x^2/a^2 - y^2/b^2 = 1$ and *parabolic cylinder* $x^2/a^2 + y = 0$. Futhermore, there are the *elliptic paraboloid* $x^2/a^2 + y^2/b^2 - z/c = 0$ and the *hyperbolic paraboloid* $x^2/a^2 - y^2/b^2 - z/c = 0$. Some properties of these cylinders and paraboloids are discussed in the exercises.

In some cases, (7.9) does not represent a set that (in a meaningful way) can be called a surface. For instance, the set defined by $F(x, y, z) = x^2 + y^2 + z^2 + 1 = 0$ is empty. The equation $F(x, y, z) = x^2 + y^2 + z^2 = 0$ gives a single point, whereas $F(x, y, z) = x^2 + y^2 = 0$ represents a line. The quadratic equation $F(x, y) = x^2 = 0$ describes a plane (we would like planes to represent

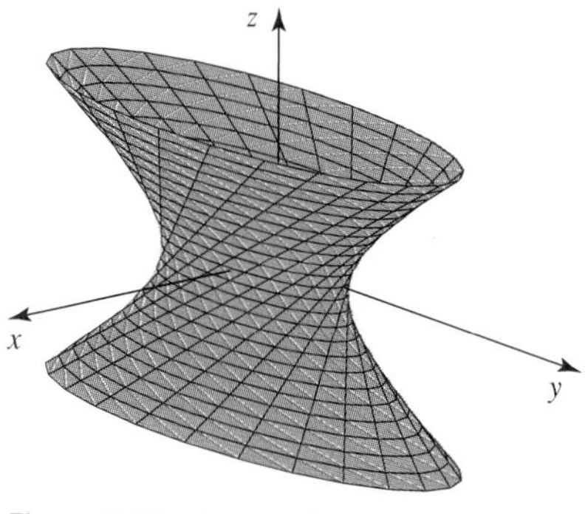

Figure 7.46 Hyperboloid of one sheet.

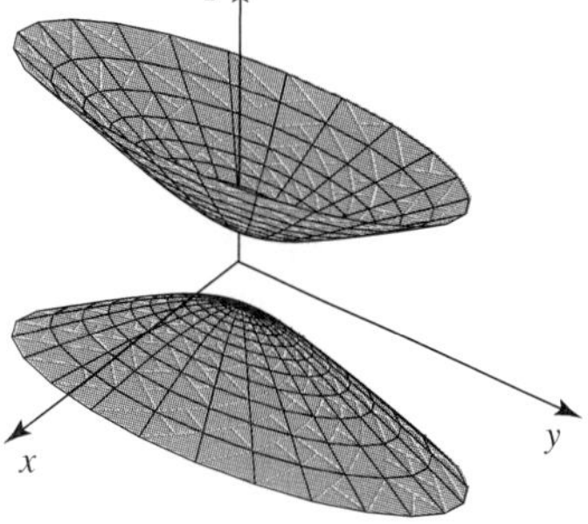

Figure 7.47 Hyperboloid of two sheets.

linear functions!). In order to exclude the sets like the above, we need to impose certain conditions on F. (We will not do it here.) ◀

Clearly, the equation $F(x, y, z) = 0$ can represent sets with a wide variety of properties. Two more examples: the set defined implicitly by $\sin(x^2 + y^2 + z^2 - 1) = 0$ consists of (infinitely many) spheres centered at the origin, of radii $\sqrt{k\pi + 1}$, $k \geq 0$. The set $\sin(x - y) = 0$ consists of infinitely many parallel planes.

▶ EXERCISES 7.2

1. Consider the parametrization of the surface of revolution given in Example 7.16. Determine how $\mathbf{r}$ maps horizontal lines ($v = v_0$) and vertical lines ($u = u_0$) in its domain. Are the images of the two sets of lines perpendicular to each other?

2. Denote by $\mathbf{c}$ the image of a differentiable path $\mathbf{c}(t) = (f(t), g(t))$, $t \in [a, b]$, in the xz-plane, where $f(t) \geq 0$ for $t \in [a, b]$. Show that $\mathbf{r} = (f(v)\cos u,\ f(v)\sin u,\ g(v))$ is a parametrization of a surface of revolution obtained by rotating the curve $\mathbf{c}$ about the z-axis. Determine the domain of $\mathbf{r}$. Is $\mathbf{r}$ necessarily a parametrized surface in the sense of our Definition 7.1?

3. Find a parametrization of the surface of revolution obtained when the curve (in the xy-plane) $\mathbf{c}(t) = (t^2, t^3)$, $1 \leq t \leq 2$, is rotated about the y-axis.

4. In the rv-coordinate system, sketch the two-dimensional profile $v(r) = v_m(1 - r^2/R^2)$ from Example 7.17. Give a parametric representation of the three-dimensional profile [i.e, parametric representation of the surface obtained by revolving the graph of $v(r)$ about the v-axis].

5. Prove that the function $\mathbf{r}(u, v) = (uv, u, v^2)$ defined in Example 7.20 is one-to-one on $D = [-1, 1] \times [-1, 1]$, except at the points (u, v) where $u = 0$.

6. What is the image S of the parametrization $\mathbf{r}(u, v) = (u, v, u^2 + v^2)$, $u, v \in \mathbb{R}$? Using cylindrical coordinates, give another parametrization of S. Is S orientable?

7. Prove that the function $\mathbf{r}(u, v) = (u\cos v, u\sin v, v)$ in Example 7.21 is indeed the parametrization of a surface (see Definition 7.1).

8. Convert parametrizations $\mathbf{r}(u, v) = (a(\cos u \mp v\sin u), b(\sin u \pm v\cos u), \pm cv)$, where $0 \leq u \leq 2\pi$, $v \in \mathbb{R}$, and $a, b, c > 0$, into implicit equations and identify the two surfaces.

9. The circle $(x - 3)^2 + (y - 4)^2 = 4$ is rotated about the x-axis, producing a torus. Find its parametric representation. Find a parametric representation of the torus obtained by rotating the given circle about the y-axis.

10. The curve $(x - 4)^2 + y^2 = 1$, $y \geq 0$, is rotated about the y-axis. Find a parametric representation of the surface of revolution thus obtained.

11. Given in Figure 7.48 is the domain D of the torus from Example 7.18. We showed that the line segment AA' is mapped by $\mathbf{r}$ into a circle (see Figure 7.26).

(a) Explain why all four points B, B', B'', and B''' in Figure 7.48(a) are mapped by $\mathbf{r}$ into the same point.

(b) Assume that A is half-way between B and B''' and A' is half-way between B' and B''; see Figure 7.48(a). Explain why the line segments BA' and AB'' are mapped by $\mathbf{r}$ into a continuous curve. Why is that curve closed?

(c) Describe in words how $\mathbf{r}$ maps the four line segments in Figure 7.48(b). [See Example 3.9 and Figure 3.10(b) in Section 3.1.]

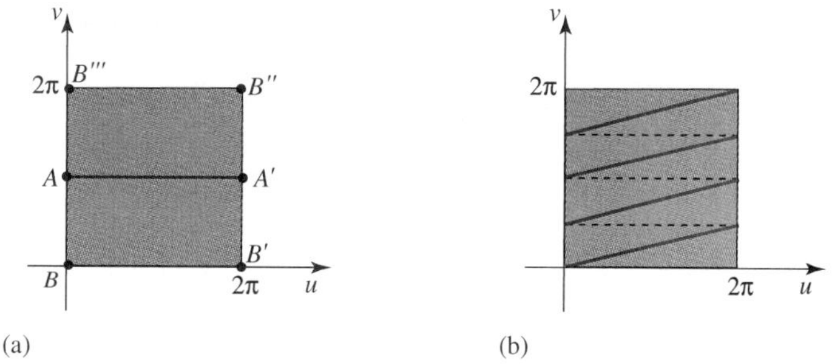

Figure 7.48 Domain D of the torus of Exercise 11.

12. Compute the normal vector to the Möbius strip of Example 7.19 and relate your answer to the fact that the Möbius strip is not an orientable surface.

13. What do we obtain if we allow the line segment in Example 7.19 to rotate through the angle of 360°, rather than through 180°? Is the resulting surface one-sided or two-sided?

14. Consider the parametrization $\mathbf{r}(t, u)$ of the Möbius strip of Example 7.19. For a fixed $t = t_0$, describe the curve $\mathbf{r}(t_0, u)$. What are the curves $\mathbf{r}(t, 0)$, $0 \le t \le 2\pi$, and $\mathbf{r}(t, 1/2)$, $0 \le t \le 4\pi$?

15. Consider the hyperbolic paraboloid $\mathbf{r}(u, v) = (u, v, uv)$, $-1 \le u, v \le 1$, of Example 7.24.

(a) Explain why the name of the surface includes the words "parabola" and "hyperbola." [*Hint*: Convert the given parametrization into the form $z = g(x, y)$ and analyze level curves $z = c$ and cross-sections with planes perpendicular to the xy-plane.]

(b) Show that, for a fixed value $u = u_0$, $r(u_0, v)$ is a line segment. Describe how this line segment changes as u_0 changes from -1 to 1.

(c) Show that, for a fixed value $v = v_0$, $r(u, v_0)$ is a line segment. Describe how this line segment changes as v_0 changes from -1 to 1.

16. Using the way we parametrize surfaces of revolution, show that the catenoid can be parametrized by $r(u, v) = (\cosh v \cos u, \cosh v \sin u, v)$. Find a possible domain of $\mathbf{r}$ and explain what the parameters u and v represent.

17. Check that $\mathbf{r}(u, v) = (a \cos v \cos u, b \cos v \sin u, c \sin v)$ is a parametrization of the ellipsoid $x^2/a^2 + y^2/b^2 + z^2/c^2 = 1$. Find the domain of $\mathbf{r}$ so that its image is the given ellipsoid.

18. What ellipse needs to be revolved (and about what axis) so that the resulting surface of revolution is the ellipsoid $x^2/a^2 + y^2/b^2 + z^2/b^2 = 1$? Using this fact, obtain a parametric representation of this ellipsoid.

19. Find a parametrization of the cone $x^2/a^2 + y^2/b^2 - z^2/c^2 = 0$. (*Hint*: Look at Example 7.3.) Find the surface normal vector $\mathbf{N}$. What can you say about the smoothness of the cone, based on your parametrization?

20. Show that $\mathbf{r}_1(u, v) = (a \cosh u \cos v, b \cosh u \sin v, c \sinh u)$ is a parametrization of the hyperboloid of one sheet. Find the domain of $\mathbf{r}_1$. Prove that the hyperboloid of two sheets can be represented parametrically as $\mathbf{r}_2(u, v) = (a \sinh u \cos v, b \sinh u \sin v, \pm c \cosh u)$; state the domain of $\mathbf{r}_2$.

21. Consider the elliptic cylinder $x^2/a^2 + y^2/b^2 = 1$ of Example 7.27. Compute the level curves $z = z_0$ and identify the curves that are intersections of the cylinder with the xz-plane and yz-plane. Find a parametrization of the cylinder that makes it a smooth surface.

22. Repeat Exercise 21 for the hyperbolic cylinder $x^2/a^2 - y^2/b^2 = 1$ and the parabolic cylinder $x^2/a^2 + y = 0$.

23. Find a smooth parametric representation of the elliptic paraboloid $x^2/a^2 + y^2/b^2 - z/c = 0$. Why is the surface called paraboloid?

24. In this exercise, we derive the parametrization given in Example 7.25.

(a) Write down a parametrization of the helix (see Example 3.8 in Section 3.1).

(b) Write down a parametrization of the circle of radius r in the plane spanned by the normal and binormal vectors to the helix (see Example 3.43 in Section 3.5).

(c) Combine (a) and (b), replacing r by ru (so that the radius increases as u increases) to obtain the desired parametrization.

▶ 7.3 SURFACE INTEGRALS OF REAL-VALUED FUNCTIONS

The path integral of a real-valued function provides a way of investigating the values of a function along a given curve. It is defined as a limit of approximating sums, each summand being of the form (value of f at a point on the curve) · (length of an approximation of a small part of the curve near that point by a straight-line segment). If $f \equiv 1$, then the path integral represents the length of a curve.

The surface integral is a higher-dimensional analogue of the path integral. This time, what interests us are the values of a function f at points that belong to a surface S in $\mathbb{R}^3$. We will build approximating sums in the form (value of f at a point on the surface) · (area of an approximation of a small part of the surface near that point by a parallelogram), and define the surface integral as the limit of such sums as the number of approximating parallelograms approaches infinity. In the special case when $f \equiv 1$, we will obtain a formula for surface area.

Let $\mathbf{r}: D \to \mathbb{R}^3$ be a differentiable parametrization of a surface S in $\mathbb{R}^3$ and let D be an elementary region in $\mathbb{R}^2$. Assume, for simplicity, that D is a rectangle. If D is not a rectangle, enclose it with a rectangle (this can be done since D is an elementary region, and therefore bounded), subdivide that rectangle into small rectangles, and consider only those that have a nonempty intersection with D. We have already used this approach in Section 6.2, when we defined double integrals over general regions; see Definition 6.3 and the text preceding and following it; see also Example 6.14.

Divide D into n^2 rectangles, choose one of them, name it R (we drop the subscripts for simplicity, and instead of using R_{ij}, we use R to denote a generic rectangle that belongs to a subdivision) and label its sides Δu and Δv. For small Δu and Δv, using a linear approximation (see Exercise 1), we get that

$$\mathbf{r}(u + \Delta u, v) - \mathbf{r}(u, v) \approx \frac{\partial \mathbf{r}}{\partial u} \Delta u$$

and

$$\mathbf{r}(u, v + \Delta v) - \mathbf{r}(u, v) \approx \frac{\partial \mathbf{r}}{\partial v} \Delta v$$

[we keep in mind that the partial derivatives $\partial \mathbf{r}/\partial u$ and $\partial \mathbf{r}/\partial v$ are computed at (u, v), but we do not indicate it in the notation, to make it simpler].

It follows that the image of the lower side Δu of R (see Figure 7.49) is a curve $\mathbf{r}(\Delta u)$ whose length (assuming that R is small) can be approximated as

$$\ell(\mathbf{r}(\Delta u)) \approx \|\mathbf{r}(u + \Delta u, v) - \mathbf{r}(u, v)\| \approx \left\| \frac{\partial \mathbf{r}}{\partial u} \Delta u \right\|.$$

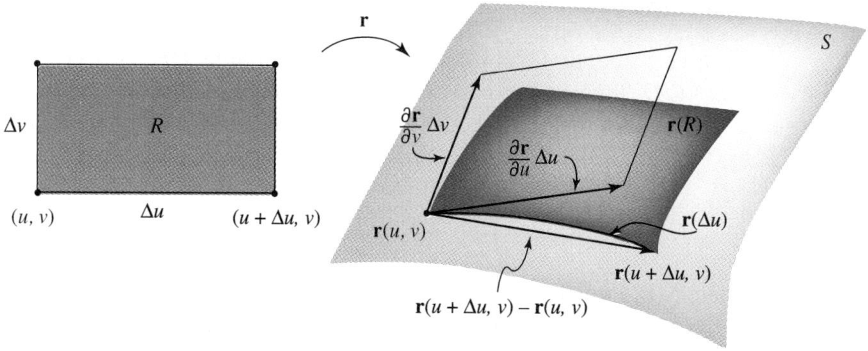

Figure 7.49 Approximating the patch $\mathbf{r}(R)$ by a parallelogram.

Similarly, the image $\mathbf{r}(\Delta v)$ of the left side of the rectangle (labeled as Δv) has length

$$\ell(\mathbf{r}(\Delta v)) \approx \|\mathbf{r}(u, v + \Delta v) - \mathbf{r}(u, v)\| \approx \left\| \frac{\partial \mathbf{r}}{\partial v} \Delta v \right\|.$$

Now consider the image $\mathbf{r}(R)$ of R [$\mathbf{r}(R)$ is sometimes called a *patch*]. For small Δu and Δv, $\mathbf{r}(R)$ can be approximated by the parallelogram spanned by

$$\frac{\partial \mathbf{r}}{\partial u} \Delta u = \mathbf{T}_u \Delta u \qquad \text{and} \qquad \frac{\partial \mathbf{r}}{\partial v} \Delta v = \mathbf{T}_v \Delta v.$$

The area of the patch $\mathbf{r}(R)$ on S is approximately equal to

$$\|\mathbf{T}_u \Delta u \times \mathbf{T}_v \Delta v\| = \|\mathbf{T}_u \times \mathbf{T}_v\| \Delta u \Delta v$$

(recall that the area of a parallelogram spanned by two vectors equals the magnitude of their cross product). Form the sum $\mathcal{R}_n = \sum_R \|\mathbf{T}_u \times \mathbf{T}_v\| \Delta u \Delta v$ of areas of patches $\mathbf{r}(R)$ for all n^2 rectangles R that form a subdivision of D. As $n \to \infty$, the sums $\mathcal{R}_n$ approach the double integral

$$\iint_D \|\mathbf{T}_u \times \mathbf{T}_v\| \, dA.$$

On the other hand, the patches $\mathbf{r}(R)$ approximate the surface S better and better as n keeps increasing. Therefore, we define the *surface area* $A(S)$ of S to be

$$A(S) = \iint_D \|\mathbf{T}_u \times \mathbf{T}_v\| \, dA.$$

Now take any continuous function f defined on S, and form the Riemann sums in the usual way:

$$\mathcal{R}_n = \sum_R f(\mathbf{r}(u, v)) \|\mathbf{T}_u \times \mathbf{T}_v\| \Delta u \Delta v.$$

The summation goes over n^2 rectangles R that form a subdivision of D. The surface integral of f over S will be defined (as usual) as the limit of $\mathcal{R}_n$ as $n \to \infty$. The assumption on the continuity of f guarantees that this limit exists (we will not prove that). Recall that the cross product $\mathbf{T}_u \times \mathbf{T}_v$ of tangent vectors is the surface normal vector $\mathbf{N}$.

DEFINITION 7.10 Surface Integral of a Real-Valued Function

Let S be a smooth, C^1 surface in $\mathbb{R}^3$ parametrized by $\mathbf{r}(u, v) = (x(u, v), y(u, v), z(u, v))$, $(u, v) \in D$ (thus, $\mathbf{r}$ is of class C^1), where D is an elementary region in $\mathbb{R}^2$. Assume that $f: S \to \mathbb{R}$ is a real-valued continuous function. The *surface integral of f over S*, denoted by $\iint_S f\, dS$, is defined by the formula

$$\iint_S f\, dS = \iint_D f(\mathbf{r}(u, v))\|\mathbf{N}(u, v)\|\, dA,$$

where $\mathbf{N}(u, v) = \mathbf{T}_u(u, v) \times \mathbf{T}_v(u, v)$. ◀

Let us emphasize that only the values of f at points on the surface S are relevant in the computation of the surface integral $\iint_S f\, dS$.

If $f \equiv 1$, the integral $\iint_S dS$ gives the surface area of S.

▶ EXAMPLE 7.28

Compute the surface integral $\iint_S xy\, dS$, where S denotes the surface of the cylinder $x^2 + y^2 = 4$, with $-1 \le z \le 1$.

SOLUTION Parametrize S by (see Example 7.2 in Section 7.1) $\mathbf{r}(u, v) = (2\cos u, 2\sin u, v)$, $(u, v) \in D$, where D is the rectangle $[0, 2\pi] \times [-1, 1]$ in the uv-plane. (Clearly, $\mathbf{r}$ is of class C^1.) The tangent vectors are $\mathbf{T}_u = (-2\sin u, 2\cos u, 0)$ and $\mathbf{T}_v = (0, 0, 1)$, and thus,

$$\mathbf{N}(u, v) = \mathbf{T}_u(u, v) \times \mathbf{T}_v(u, v) = (2\cos u, 2\sin u, 0).$$

It follows that $\|\mathbf{N}(u, v)\| = 2$ and

$$\iint_S xy\, dS = \iint_D (2\cos u)(2\sin u)2\, dA$$

(use the double-angle formula $\sin 2u = 2\sin u\, \cos u$)

$$= 4\int_{-1}^{1}\left(\int_0^{2\pi} \sin 2u\, du\right) dv = 4\int_{-1}^{1}\left(-\frac{1}{2}\cos 2u\Big|_0^{2\pi}\right) dv = 0.$$ ◀

▶ EXAMPLE 7.29

Find the surface integral $\iint_S \sqrt{x^2 + y^2 + 1}\, dS$, where S is the helicoid $\mathbf{r}(u, v) = (u\cos v, u\sin v, v)$, $0 \le u \le 1$, $0 \le v \le 2\pi$; see Example 7.21 in Section 7.2.

SOLUTION The tangent vectors $\mathbf{T}_u$ and $\mathbf{T}_v$ to S are

$$\mathbf{T}_u = (\cos v, \sin v, 0) \qquad \text{and} \qquad \mathbf{T}_v = (-u\sin v, u\cos v, 1).$$

Thus, $\mathbf{N} = \mathbf{T}_u \times \mathbf{T}_v = (\sin v, -\cos v, u)$, $\|\mathbf{N}\| = \sqrt{1 + u^2}$, and

$$\iint_S \sqrt{x^2 + y^2 + 1}\, dS = \iint_{[0,1]\times[0,2\pi]} \sqrt{u^2 + 1}\sqrt{u^2 + 1}\, dA$$

$$= \int_0^{2\pi}\left(\int_0^1 (u^2 + 1)\, du\right) dv = \int_0^{2\pi}\left(\frac{u^3}{3} + u\Big|_0^1\right) dv = \frac{8\pi}{3}.$$ ◀

THEOREM 7.1 Surface Integrals Are Independent of Parametrization

Let f be a real-valued continuous function. The surface integral $\iint_S f\,dS$ does not depend on the parametrization of S, provided it is C^1 and smooth. ◀

The theorem states that if S is paramterized by smooth C^1 parametrizations $\mathbf{r}(u, v)\colon D \to \mathbb{R}^3$ and $\mathbf{r}^*(u^*, v^*)\colon D^* \to \mathbb{R}^3$, then

$$\iint_S f\,dS = \iint_D f(\mathbf{r}(u, v))\|\mathbf{T}_u \times \mathbf{T}_v\|\,dA = \iint_{D^*} f(\mathbf{r}^*(u^*, v^*))\|\mathbf{T}_{u^*} \times \mathbf{T}_{v^*}\|\,dA^*.$$

Consequently, the notation $\iint_S f\,dS$ for the surface integral is justified—there is no need to mention the parametrization that is used. In particular, if $f \equiv 1$, then

$$A(S) = \iint_D \|\mathbf{T}_u \times \mathbf{T}_v\|\,dA = \iint_{D^*} \|\mathbf{T}_{u^*} \times \mathbf{T}_{v^*}\|\,dA^*,$$

that is, the surface area $A(S)$ does not depend on a parametrization of the surface (as expected).

The proof of this theorem is similar to that of the theorem stating that the path integrals of scalar functions are independent of parametrization and will be omitted.

▶ **EXAMPLE 7.30** Surface Integral over the Graph of $z = f(x, y)$

Find a formula for the surface integral $\iint_S g\,dS$, where $g(x, y, z)\colon S \subseteq \mathbb{R}^3 \to \mathbb{R}$ is a continuous function, and S is the graph of a C^1 function $z = f(x, y)$, defined on an elementary region $D \subseteq \mathbb{R}^2$.

SOLUTION Let $x = u$ and $y = v$, so that $z = f(u, v)$. The map $\mathbf{r}(u, v) = (u, v, f(u, v))$, $(u, v) \in D$, parametrizes the surface S (see Example 7.9). It follows that $\mathbf{T}_u = (1, 0, \partial f/\partial u)$, $\mathbf{T}_v = (0, 1, \partial f/\partial v)$, and $\mathbf{N}(u, v) = \mathbf{T}_u \times \mathbf{T}_v = (-\partial f/\partial u, -\partial f/\partial v, 1)$. Hence,

$$\iint_S g\,dS = \iint_D g(u, v, f(u, v))\sqrt{1 + \left(\frac{\partial f}{\partial u}\right)^2 + \left(\frac{\partial f}{\partial v}\right)^2}\,dA. \quad ◀$$

▶ **EXAMPLE 7.31**

Compute the integral of $g(x, y, z) = \arctan(y/x)$ over the surface S that consists of the part of the graph of $z = x^2 + y^2$ between $z = 1$ and $z = 2$.

SOLUTION To parametrize S, we take $x = u$ and $y = v$, so that $\mathbf{r}(u, v) = (u, v, u^2 + v^2)$, where $1 \le u^2 + v^2 \le 2$. It follows that $\mathbf{T}_u = (1, 0, 2u)$, $\mathbf{T}_v = (0, 1, 2v)$, $\mathbf{N} = \mathbf{T}_u \times \mathbf{T}_v = (-2u, -2v, 1)$, and

$$\iint_S \arctan\left(\frac{y}{x}\right)\,dS = \iint_D \arctan\left(\frac{v}{u}\right)\sqrt{4u^2 + 4v^2 + 1}\,dA,$$

where D is the annulus $1 \le u^2 + v^2 \le 2$ in the uv-plane (obtained by combining $z = x^2 + y^2$ and $1 \le z \le 2$). Changing to polar coordinates, we get [recall that $\arctan(y/x) = \theta$]

$$\iint_S \arctan\left(\frac{y}{x}\right)\,dS = \int_0^{2\pi}\left(\int_1^{\sqrt{2}} \theta\sqrt{1 + 4r^2}\,r\,dr\right)d\theta$$

$$= \left(\int_0^{2\pi} \theta\,d\theta\right)\left(\frac{1}{12}\left(1 + 4r^2\right)^{3/2}\Big|_1^{\sqrt{2}}\right) = \frac{\pi^2}{6}\left(9^{3/2} - 5^{3/2}\right). \quad ◀$$

▶ EXAMPLE 7.32

Compute the surface area $A(S)$ of the part S of the plane $2x + y + z = 4$ in the first octant.

SOLUTION Parametrize the plane by $x = u$, $y = v$, and $z = 4 - 2u - v$, where (x, y) belongs to the triangular region D in the xy-plane bounded by the x-axis, y-axis, and line $2x + y = 4$; see Figure 7.50. In the spirit of the definition of a parametrization, we have to describe D as the triangular region bounded by the u-axis, v-axis, and line $2u + v = 4$ (however, it might be more convenient to imagine D as a subset of the xy-plane). The area of S is given by

$$A(S) = \iint_D \|\mathbf{T}_u \times \mathbf{T}_v\| \, dA.$$

From $\mathbf{r}(u, v) = (u, v, 4 - 2u - v)$, it follows that $\mathbf{T}_u = (1, 0, -2)$, $\mathbf{T}_v = (0, 1, -1)$, and $\|\mathbf{T}_u \times \mathbf{T}_v\| = \|(2, 1, 1)\| = \sqrt{6}$. Thus,

$$A(S) = \int_0^2 \left(\int_0^{4-2u} \sqrt{6}\, dv \right) du = \sqrt{6} \int_0^2 (4 - 2u)\, du = \sqrt{6}\,(4u - u^2)\Big|_0^2 = 4\sqrt{6}.$$

Two alternative ways of calculating the area are suggested in Exercise 4. ◀

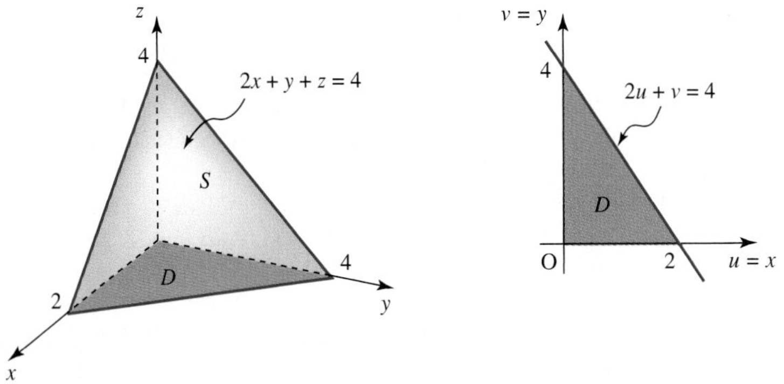

Figure 7.50 Surface S of the plane in Example 7.32.

▶ EXAMPLE 7.33

Compute the surface area of the torus

$$\mathbf{r}(u, v) = ((R + \rho\cos v)\cos u, (R + \rho\cos v)\sin u, \rho\sin v), \qquad 0 \le u, v \le 2\pi,$$

that we discussed in Example 7.18 in Section 7.2.

SOLUTION The surface normal vector $\mathbf{N} = \mathbf{T}_u \times \mathbf{T}_v$ is computed to be

$$\begin{aligned}\mathbf{N}(u, v) &= \begin{vmatrix} \mathbf{i} & \mathbf{j} & \mathbf{k} \\ -(R + \rho\cos v)\sin u & (R + \rho\cos v)\cos u & 0 \\ -\rho\sin v\cos u & -\rho\sin v\sin u & \rho\cos v \end{vmatrix} \\ &= ((R + \rho\cos v)\rho\cos u\cos v, (R + \rho\cos v)\rho\sin u\cos v, (R + \rho\cos v)\rho\sin v) \\ &= (R + \rho\cos v)\rho(\cos u\cos v, \sin u\cos v, \sin v),\end{aligned}$$

and its norm is given by

$$\|\mathbf{N}(u,v)\|^2 = (R+\rho\cos v)^2\rho^2\|(\cos u\cos v, \sin u\cos v, \sin v)\|^2 = (R+\rho\cos v)^2\rho^2.$$

Consequently, the surface area of the torus is computed to be

$$A(S) = \iint_{[0,2\pi]\times[0,2\pi]} (R+\rho\cos v)\rho\, dA = \rho\int_0^{2\pi}\left(\int_0^{2\pi}(R+\rho\cos v)dv\right)du$$
$$= \rho\int_0^{2\pi}\left(Rv+\rho\sin v\Big|_0^{2\pi}\right)du = \rho\int_0^{2\pi} 2\pi R\, du = 4\pi^2 R\rho.$$

▶ EXAMPLE 7.34

The electrostatic potential at $(0, 0, -a)$ induced by a charge of constant charge density σ on the hemisphere S defined by $x^2+y^2+z^2=a^2$, $z\geq 0$, is given by

$$U = c\iint_S \frac{\sigma}{\sqrt{x^2+y^2+(z+a)^2}}\,dS,$$

where the value of the constant c depends on the units that are used. Show that $U = 2\pi\sigma ac\left(2-\sqrt{2}\right)$.

SOLUTION

Parametrize the hemisphere by $\mathbf{r}(u,v) = (a\cos u\sin v, a\sin u\sin v, a\cos v)$, where $0\leq u\leq 2\pi$ and $0\leq v\leq \pi/2$. Then

$$x^2+y^2+(z+a)^2 = a^2\cos^2 u\sin^2 v + a^2\sin^2 u\sin^2 v + a^2\cos^2 v + 2a^2\cos v + a^2$$
$$= a^2\sin^2 v + a^2\cos^2 v + 2a^2\cos v + a^2 = 2a^2(1+\cos v)$$

and

$$\mathbf{N} = \mathbf{T}_u\times\mathbf{T}_v = \begin{vmatrix} \mathbf{i} & \mathbf{j} & \mathbf{k} \\ -a\sin u\sin v & a\cos u\sin v & 0 \\ a\cos u\cos v & a\sin u\cos v & -a\sin v \end{vmatrix}$$
$$= -a\sin v(a\sin v\cos u, a\sin v\sin u, a\cos v) = -a\sin v\,\mathbf{r}(u,v).$$

It follows that $\|\mathbf{N}\| = |-a\sin v|\,\|\mathbf{r}(u,v)\| = a^2\sin v$ [since $\|\mathbf{r}(u,v)\| = a$], and hence,

$$U = c\iint_S \frac{\sigma}{\sqrt{x^2+y^2+(z+a)^2}}\,dS = c\int_0^{2\pi}\left(\int_0^{\pi/2}\frac{\sigma}{a\sqrt{2}\sqrt{1+\cos v}}a^2\sin v\,dv\right)du$$

[using the substitution $t = 1+\cos v$]

$$= c\int_0^{2\pi}\left(\frac{a\sigma}{\sqrt{2}}\left(-2(1+\cos v)^{1/2}\right)\Big|_{v=0}^{v=\pi/2}\right)dv$$
$$= c\int_0^{2\pi}\frac{a\sigma}{\sqrt{2}}\left(-2+2\sqrt{2}\right)dv = 2\pi\frac{ac\sigma}{\sqrt{2}}\sqrt{2}\left(2-\sqrt{2}\right) = 2\pi a\sigma c\left(2-\sqrt{2}\right).$$

It is possible to define the surface integral over a more general class of surfaces, as will now be introduced.

DEFINITION 7.11 Piecewise C^1 Smooth Surface

A surface S is called a *piecewise C^1 smooth surface* if it is a disjoint union of surfaces S_i (that means that the most two S_i's can have in common are points and/or curves), parametrized by $\mathbf{r}_i\colon D_i \to \mathbb{R}^3$, $i = 1, \ldots, n$, such that

(a) D_i is an elemetary region in $\mathbb{R}^2$,

(b) $\mathbf{r}_i$ is one-to-one, except possibly on the boundary of D_i,

(c) $\mathbf{r}_i$ is of class C^1, except possibly on the boundary of D_i, and

(d) $S_i = \mathbf{r}_i(D_i)$ is a smooth surface, except possibly at a finite number of points. ◀

For a continuous function f and a piecewise C^1 smooth surface S, we define

$$\iint_S f\,dS = \sum_{i=1}^{n} \iint_{S_i} f\,dS;$$

that is, the integral of f over S is the sum of integrals of f over S_i. Here is an example.

▶ EXAMPLE 7.35

Compute $\iint_S yz\,dS$, where S consists of the hemisphere $x^2+y^2+z^2=1$, $z \geq 0$, and the disk $0 \leq x^2+y^2 \leq 1$, in the xy-plane.

SOLUTION

The surface S is piecewise C^1 smooth (see the parametrizations below), and

$$\iint_S yz\,dS = \iint_{S_1} yz\,dS + \iint_{S_2} yz\,dS,$$

where S_1 is the surface of the hemisphere and S_2 is the disk. The hemisphere can be parametrized by $\mathbf{r}_1(u, v) = (\cos u \sin v, \sin u \sin v, \cos v)$, where $0 \leq u \leq 2\pi$, $0 \leq v \leq \pi/2$. Repeating the computation of Example 7.34 with $a = 1$, we get $\|\mathbf{N}_1\| = \sin v$, and hence,

$$\begin{aligned}\iint_{S_1} yz\,dS &= \iint_{[0,2\pi]\times[0,\pi/2]} \sin u \sin v \cos v \; \sin v\,dA \\ &= \int_0^{2\pi}\left(\int_0^{\pi/2} \sin u \sin^2 v \cos v dv\right) du \\ &= \left(\int_0^{2\pi} \sin u\,du\right)\left(\int_0^{\pi/2} \sin^2 v \cos v\,dv\right) = 0,\end{aligned}$$

by separation of variables, using the fact that $\int_0^{2\pi} \sin u\,du = 0$.

Paramterize S_2 by $\mathbf{r}_2(u, v) = (u, v, 0)$, $(u, v) \in D$, where $D = \{(u, v) \mid u^2 + v^2 \leq 1\}$. It follows that (since $\mathbf{N} = \mathbf{k}$)

$$\iint_{S_2} yz\,dS = \iint_D (v)(0)(1)\,dA = 0.$$

Therefore, $\iint_S f\,dS = 0$. (Note that $\mathbf{r}_1$ and $\mathbf{r}_2$ satisfy Definition 7.11, i.e., S is indeed a piecewise C^1 smooth surface.) ◀

So far, we have learned how to set up a surface integral if a parametrization of a surface is known explicitly (or if it can be constructed). We have also derived a formula in the case

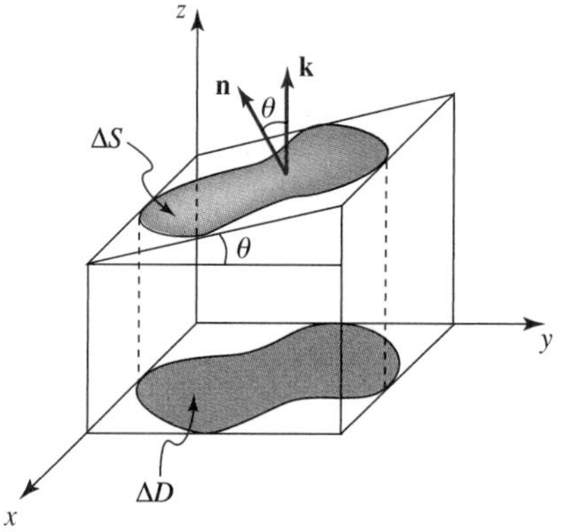

Figure 7.51 Area Cosine Principle.

where a surface is given as the graph of a function $z = f(x, y)$. But how do we set up a surface integral if there is no (explicit) parametrization? To be more specific: suppose that a surface is given in the form $F(x, y, z) = C$, where F is a C^1 function and $C \in \mathbb{R}$ (i.e., it is given as the level surface of a C^1 function). We need to find a way to compute its (surface) area and then (as in the beginning of this section) generalize the construction to obtain the surface integral of a continuous function $f: \mathbb{R}^3 \to \mathbb{R}$.

One thing is immediate: since the gradient vector is perpendicular to a level surface, it follows that either $\mathbf{n} = \nabla F/\|\nabla F\|$ or $\mathbf{n} = -\nabla F/\|\nabla F\|$ is the unit normal vector (provided that $\nabla F \neq \mathbf{0}$) that determines the orientation of the surface.

In order to derive a formula for the area, we take a more intuitive approach, based on the following principle (which is sometimes called *the Area Cosine Principle*). Consider a plane region ΔS that makes an angle θ $(0 \leq \theta \leq \pi/2)$ with respect to the xy-plane; see Figure 7.51. Let ΔD be the region in the xy-plane that is the orthogonal projection of ΔS. Let us compare the areas of ΔS and ΔD.

The distances in the x-direction remain the same for both ΔS and its projection. However, the distances in the y-direction do not: the distance in the projection ΔD is distorted (shortened) by the factor of $\cos\theta$. Therefore,

$$\text{area}(\Delta D) = \text{area}\,(\Delta S)\,\cos\theta$$

(see Exercise 19), where $0 \leq \theta \leq \pi/2$. Notice that θ is also the angle between the upward unit normal $\mathbf{n}$ to ΔS and the vector $\mathbf{k}$ (unit normal to ΔD), so that $\cos\theta = \mathbf{n} \cdot \mathbf{k}/(\|\mathbf{n}\|\,\|\mathbf{k}\|) = \mathbf{n} \cdot \mathbf{k}$, and thus,

$$\text{area}(\Delta S) = \frac{\text{area}\,(\Delta D)}{\mathbf{n} \cdot \mathbf{k}}.$$

Technical issue: assume that ΔS is oriented by the downward-pointing normal $-\mathbf{n}$. In that case, the angle between the normal $-\mathbf{n}$ and $\mathbf{k}$ is $\pi - \theta$, and their dot product is negative. Therefore, in order to include both orientations in the formula, we need the absolute value, so that

$$\text{area}(\Delta S) = \frac{\text{area}(\Delta D)}{|\mathbf{n} \cdot \mathbf{k}|}.$$

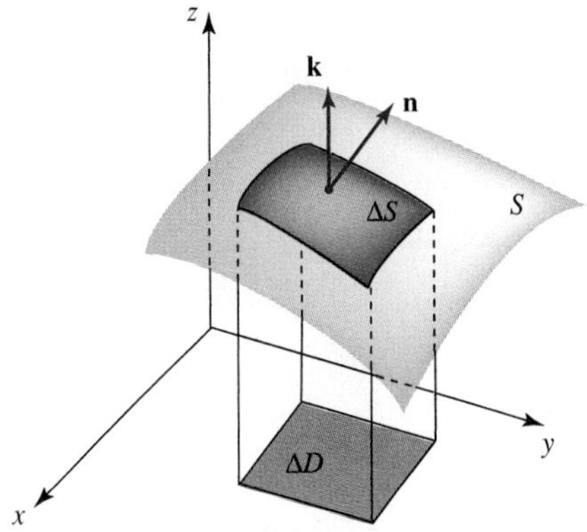

Figure 7.52 area $(\Delta S) \approx$ area $(\Delta D)/|\mathbf{n} \cdot \mathbf{k}|$.

Now let S be a surface oriented by the unit normal $\mathbf{n}$, and assume that every line perpendicular to the xy-plane intersects S in at most one point (so that we can project S onto the xy-plane). Take a small patch ΔS on S (which is so small that it can be assumed flat; i.e., can be approximated by a plane region) and consider its projection ΔD onto the xy-plane; see Figure 7.52. By the Area Cosine Principle, the area of ΔS is approximately equal to area$(\Delta D)/|\mathbf{n} \cdot \mathbf{k}|$, and

$$\text{area}(S) \approx \sum \frac{\text{area}(\Delta D)}{|\mathbf{n} \cdot \mathbf{k}|},$$

where the summation goes over all small patches ΔS. In the limit, as the size ΔS approaches 0, we see that

$$\text{area}(S) = \iint_D \frac{dA}{|\mathbf{n} \cdot \mathbf{k}|}, \tag{7.10}$$

where D is the projection of S onto the xy-plane, and dA refers to integration with respect to x and y.

▶ EXAMPLE 7.36

Find the surface area of the part S of the paraboloid $z = 2 - x^2 - y^2$ that lies above the xy-plane.

SOLUTION

View S as the level surface $F(x, y, z) = x^2 + y^2 + z = 2$. Its normal is $\nabla F = (2x, 2y, 1)$, and hence, the (upward) unit normal to S is

$$\mathbf{n} = \left(2x/\sqrt{4x^2+4y^2+1}, 2y\Big/\sqrt{4x^2+4y^2+1}, 1/\sqrt{4x^2+4y^2+1}\right).$$

From $z = 2 - x^2 - y^2 \geq 0$, it follows that $x^2 + y^2 \leq 2$ [i.e., the projection D of S onto the xy-plane is the disk $x^2 + y^2 \leq 2$]. Since $|\mathbf{n} \cdot \mathbf{k}| = 1/\sqrt{4x^2+4y^2+1}$, it follows that the area of S is

$$\text{area}(S) = \iint_D \sqrt{4x^2+4y^2+1}\, dA.$$

Passing to polar coordinates, we get (recall that $dA = r dr d\theta$)

$$\text{area}(S) = \int_0^{2\pi} \left(\int_0^{\sqrt{2}} \sqrt{4r^2+1}\, r dr \right) d\theta$$

$$= \int_0^{2\pi} \left(\frac{1}{12}(4r^2+1)^{3/2} \Big|_0^{\sqrt{2}} \right) d\theta = \frac{2}{12}\pi(9^{3/2}-1) = \frac{13}{3}\pi.$$ ◀

Repeating the above construction for projections with respect to the remaining two coordinate planes, we determine that $\text{area}(S) = \iint_D dA/|\mathbf{n}\cdot\mathbf{i}|$, where D is the projection of S onto the yz-plane and dA refers to integration with respect to y and z, and $\text{area}(S) = \iint_D dA/|\mathbf{n}\cdot\mathbf{j}|$, where D is the projection of S onto the xz-plane and dA refers to integration with respect to x and z. Of course, in every case we have to make sure that it is possible to project S onto the required plane.

In general, if S is a C^1 surface in $\mathbb{R}^3$ oriented by the unit normal $\mathbf{n}$ and $f(x, y, z)$ is a continuous real-valued function on S, then

$$\iint_S f\, dS = \iint_D f(x, y, z) \frac{dA}{|\mathbf{n}\cdot\mathbf{k}|}, \tag{7.11}$$

where D is the projection of S onto the xy-plane. As usual, dA refers to the corresponding integration (in this case, with respect to x and y). By considering the remaining two projections, we obtain two more formulas for $\iint_S f\, dS$.

▶ EXAMPLE 7.37

Let $f(x, y, z) = 4xz$ and assume that S is the surface $x^2 + y + 2z = 4$ in the first octant oriented by the upward-pointing normal; see Figure 7.53. Set up (do not evaluate) double integrals that would evaluate the surface integral $\iint_S f\, dS$ using projections.

SOLUTION The surface S can be viewed as the level surface $F(x, y, z) = x^2 + y + 2z = 4$. From $\nabla F = (2x, 1, 2)$, it follows that the (upward) unit normal (the z-component has to be positive) is $\mathbf{n} = \nabla F/\|\nabla F\| = (2x, 1, 2)/\sqrt{4x^2+5}$.

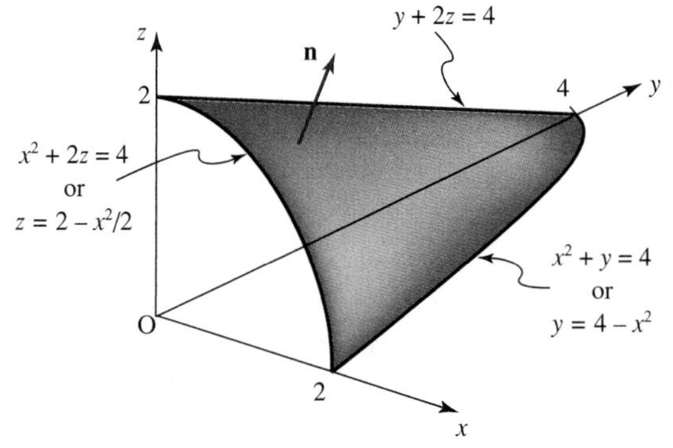

Figure 7.53 Surface $x^2 + y + 2z = 4$ of Example 7.37.

The projection D_1 of S onto the xz-plane is the region bounded by $x = 0$, $z = 0$, and (substitute $y = 0$ into the equation for S) $z = 2 - x^2/2$. Since $|\mathbf{n} \cdot \mathbf{j}| = 1/\sqrt{4x^2 + 5}$, we get

$$\iint_S f\,dS = \iint_{D_1} f(x, y, z)\frac{dA}{|\mathbf{n} \cdot \mathbf{j}|} = \iint_{D_1} 4xz\frac{dA}{1/\sqrt{4x^2+5}},$$

where dA refers to integration with respect to x and z. Hence,

$$\iint_S f\,dS = \int_0^2 \left(\int_0^{2-x^2/2} 4xz\sqrt{4x^2+5}\,dz \right) dx.$$

The projection D_2 of S onto the yz-plane is the triangle bounded by $y = 0$, $z = 0$, and (substitute $x = 0$ into the equation for S) $y + 2z = 4$. It follows that

$$\iint_S f\,dS = \iint_{D_2} f(x, y, z)\frac{dA}{|\mathbf{n} \cdot \mathbf{i}|},$$

where dA refers to integration with respect to y and z, and $|\mathbf{n} \cdot \mathbf{i}| = |2x/\sqrt{4x^2+5}| = 2x/\sqrt{4x^2+5}$, since $x \geq 0$. Thus,

$$\iint_S f\,dS = \iint_{D_2} 4xz\frac{dA}{2x/\sqrt{4x^2+5}} = \iint_{D_2} 2z\sqrt{4x^2+5}\,dA.$$

Since the integration is with respect to y and z, we have to eliminate x. From $x^2 + y + 2z = 4$, we get $x^2 = 4 - y - 2z$, and therefore,

$$\iint_S f\,dS = \iint_{D_2} 2z\sqrt{21 - 4y - 8z}\,dA = \int_0^4 \left(\int_0^{2-y/2} 2z\sqrt{21 - 4y - 8z}\,dz \right) dy.$$

The projection D_3 of S onto the xy-plane is the region bounded by $x = 0$, $y = 0$, and $y = 4 - x^2$. Since $|\mathbf{n} \cdot \mathbf{k}| = 2/\sqrt{4x^2 + 5}$, it follows that

$$\iint_S f\,dS = \iint_{D_3} f(x, y, z)\frac{dA}{|\mathbf{n} \cdot \mathbf{k}|} = \iint_{D_3} 4xz\frac{dA}{2/\sqrt{4x^2+5}},$$

where dA refers to integration with respect to x and y. Eliminating z and setting up the limits of integration, we get

$$\iint_S f\,dS = \iint_{D_3} 2x\left(2 - \frac{x^2}{2} - \frac{y}{2}\right)\sqrt{4x^2+5}\,dA$$

$$= \int_0^2 \left(\int_0^{4-x^2} x(4 - x^2 - y)\sqrt{4x^2+5}\,dy \right) dx.$$

◀

▶ EXERCISES 7.3

1. Let $\mathbf{r}: D \to \mathbb{R}^3$ be a differentiable parametrization of a surface in $\mathbb{R}^3$; in components, $\mathbf{r}(u, v) = (x(u, v), y(u, v), z(u, v))$, for $(u, v) \in D \subseteq \mathbb{R}^2$.

(a) Explain why $x(u + \Delta u, v) - x(u, v) \approx (\partial x/\partial u)(u, v)\,\Delta u$.

(b) Using similar approximations for the y and z components of $\mathbf{r}$, show that $\mathbf{r}(u + \Delta u, v) - \mathbf{r}(u, v) \approx (\partial \mathbf{r}/\partial u)(u, v)\,\Delta u$.

2. Consider the surface of revolution S obtained when the graph of a C^1 function $y = f(x)$, $x \in [a, b]$, is rotated about the x-axis (see Example 7.16 in Section 7.2).

(a) Show that the surface area is given by $A(S) = 2\pi \int_a^b |f(x)|\sqrt{1 + (f'(x))^2}\,dx$.

(b) Show that $A(S)$ from (a) is equal to the path integral $\int_{\mathbf{c}} 2\pi|f(x)|\,ds$, where $\mathbf{c}(t) = (t, f(t))$, $t \in [a, b]$. Explain in words how to compute the surface area of a surface of rotation using path integrals.

3. Show that $A(S) = 2\pi \int_a^b |x|\sqrt{1 + (f'(x))^2}\,dx$ gives the surface area of the surface of revolution S obtained when the graph of a C^1 function $y = f(x)$, $x \in [a, b]$, is rotated about the y-axis.

4. The surface S in Example 7.32 is a triangle whose vertices lie on the coordinate axes, as shown in Figure 7.50.

(a) Compute the area of S using Heron's formula $A = \sqrt{s(s-a)(s-b)(s-c)}$ for the area of the triangle with sides a, b, and c, where $s = (a + b + c)/2$.

(b) Compute the area of S using vector product.

Exercises 5 to 12: Compute $\iint_S f\,dS$ in each case.

5. $f(x, y, z) = xy$, S is the part of the paraboloid $z = x^2 + y^2$ that lies inside the cylinder of radius 2 whose axis of rotation is the z-axis

6. $f(x, y, z) = 2z(x^2 + y^2)$, S is the surface parametrized by $\mathbf{r}(u, v) = (\cos u, \sin u, v)$, $0 \le u \le \pi$, $0 \le v \le 2$

7. $f(x, y, z) = y + x$, S is the tetrahedron with vertices $(0, 0, 0)$, $(2, 0, 0)$, $(0, 2, 0)$, and $(0, 0, 2)$

8. $f(x, y, z) = x^2 + y^2$, S is the part of the cone $z^2 = x^2 + y^2$ between $z = 1$ and $z = 4$

9. $f(x, y, z) = 2y - x$, S is the part of the cone $x^2 = y^2 + z^2$, $x \le 1$, in the first octant

10. $f(x, y, z) = 8y$, S is the parabolic sheet $z = 1 - y^2$, $0 \le x \le 2$, $0 \le y \le 1$

11. $f(x, y, z) = (4x^2 + 4y^2 + 1)^{-1/2}$, S is the part of the paraboloid $z = 4 - x^2 - y^2$ above the xy-plane

12. $f(x, y, z) = \sqrt{x^2 + y^2}$, S is the helicoidal surface $\mathbf{r}(u, v) = (u\cos v, u\sin v, v)$, $0 \le u \le 1$, $0 \le v \le 4\pi$

13. Compute the surface area of the part of the surface $\mathbf{r}(u, v) = (2u\cos v, 2u\sin v, v)$, where $0 \le u \le 2$, $0 \le v \le \pi$.

14. Compute the surface area of the part of the cylinder $x^2 + z^2 = 1$, $z \ge 0$, between the planes $y = 0$ and $z = y + 1$.

15. Compute the surface area of a cone of radius r and height h, using surface integrals.

16. Find the area of the triangle with vertices $(1, 2, 0)$, $(3, 0, 7)$, and $(-1, 0, 0)$ using a surface integral. Check your answer using the cross product.

17. Let S be the sphere $x^2 + y^2 + z^2 = a^2$. Find $\iint_S x\,dS$, $\iint_S x^2\,dS$, and $\iint_S x^3\,dS$ without evaluating surface integrals using a parametrization.

18. Compute the surface area of the part of the plane $z = 0$ defined by $-1 \le x \le 1$, $-1 \le y \le 1$ using the following parametrizations:

(a) $\mathbf{r}(u, v) = (u, v, 0)$, $-1 \le u, v \le 1$

(b) $\mathbf{r}(u, v) = (u^3, v, 0)$, $-1 \le u, v \le 1$

(c) $\mathbf{r}(u, v) = (u^{1/3}, v^{1/3}, 0)$, $-1 \le u, v \le 1$

(d) $\mathbf{r}(u, v) = (\sin u, \sin v, 0)$, $0 \le u, v \le 2\pi$

The results in (a), (b), and (c) are the same. Why is the result in (d) different?

19. Let S be the rectangle in the plane $z = my$, $m > 0$, lying directly above the rectangle $R = [0, a] \times [0, b]$, $a, b > 0$, in the xy-plane. Show that (area of S) $= \sqrt{m^2 + 1} \cdot$ (area of R). Let α be the angle between $\mathbf{k}$ and the upward normal to S. Conclude that (area of S) $= \sec\alpha \cdot$ (area of R).

20. Consider the integral $\iint_S f(x, y, z)\,dS$, where S is a surface symmetric with respect to the xz-plane. If $f(x, -y, z) = -f(x, y, z)$, what is the value of $\iint_S f\,dS$? Using your result, recompute the surface integral in Example 7.35.

21. Evaluate $\iint_S f\,dS$, where $f(x, y, z) = 4xy$ and S is the parabolic sheet $z = 1 - y^2$ in the first octant, bounded by the plane $x = 2$.

22. Using the Area Cosine Principle, find the formula for the area of an ellipse with semi-axes a and b.

23. Compute $\iint_S y\,dS$, where S is the part of the surface $x + y^2 + z = 4$ in the first octant, using a projection of S onto one of the coordinate planes.

24. Let S be the part of the plane $x + y + 2z = 4$ in the first octant, oriented by the upward-pointing normal. Compute $\iint_S (xy^2 + z^2)\,dS$

(a) Using a parametrization of S,

(b) By viewing S as the graph of the function $z = 2 - x/2 - y/2$ and using the formula of Example 7.30, and

(c) By using any of the three projections of S onto the coordinate planes.

25. Find the area of the hemisphere S defined by $x^2 + y^2 + z^2 = a^2, a > 0, y \geq 0$, using a projection of S onto a coordinate plane.

26. Find the surface area of the strip on the sphere $x^2 + y^2 + z^2 = a^2$ $(a > 0)$, defined by the angles ϕ_1 and ϕ_2, where $\phi_1 < \phi_2$ (ϕ_1 and ϕ_2 are defined in the same way as the angle ϕ in spherical coordinates).

▶ 7.4 SURFACE INTEGRALS OF VECTOR FIELDS

The aim of this section is to give a generalization of integrals of scalar functions to integrals of vector functions over surfaces in $\mathbb{R}^3$. An important application of this concept is the flux of a vector field.

DEFINITION 7.12 Surface Integral of a Vector Field

Let S be a smooth surface in $\mathbb{R}^3$ parametrized by a C^1 map $\mathbf{r} = \mathbf{r}(u, v)\colon D \to \mathbb{R}^3$ (where D is an elementary region in $\mathbb{R}^2$) and let $\mathbf{F}\colon S \subseteq \mathbb{R}^3 \to \mathbb{R}^3$ be a continuous vector field on S. The *surface integral* $\iint_S \mathbf{F} \cdot d\mathbf{S}$ *of* $\mathbf{F}$ *over* S is defined by

$$\iint_S \mathbf{F} \cdot d\mathbf{S} = \iint_D \mathbf{F}(\mathbf{r}(u, v)) \cdot \mathbf{N}(u, v)\,dA,$$

where $\mathbf{N}(u, v) = \mathbf{T}_u(u, v) \times \mathbf{T}_v(u, v)$. ◀

The surface integral of a vector field depends only on the values of the vector field at points on the surface. According to the definition, it is reduced to a double integral of the *real-valued* function $\mathbf{F}(\mathbf{r}(u, v)) \cdot \mathbf{N}(u, v)$ over an elementary region D. (Note the analogy with the path integrals of vector fields that we studied in Section 5.3.)

Insight: assume that $\mathbf{F}$ represents the velocity of a fluid (and that it is constant, which implies that $\|\mathbf{F}\| = C$). Take a surface S to be a subset of the xy-plane (thus, its normal is $\mathbf{N} = \mathbf{k}$), as shown in Figure 7.54.

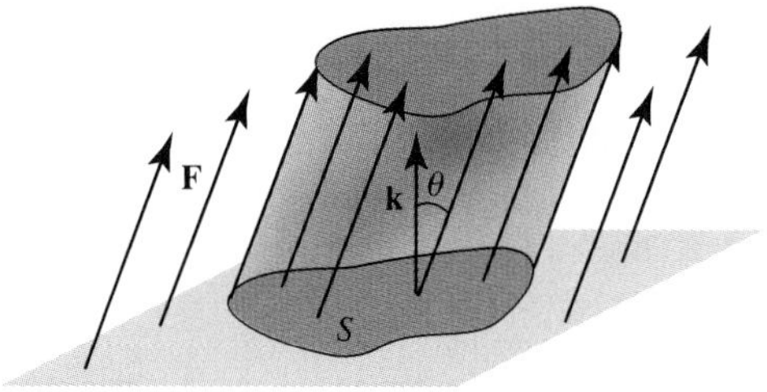

Figure 7.54 Surface integral is related to fluid flow.

In this case,

$$\iint_S \mathbf{F} \cdot d\mathbf{S} = \iint_S \mathbf{F}(\mathbf{r}(u, v)) \cdot \mathbf{N}(u, v)\, dA,$$

$$= \iint_S \|\mathbf{F}(\mathbf{r}(u, v))\| \cdot \|\mathbf{N}(u, v)\| \cos\theta\, dA = C\cos\theta \iint_S dA$$

[since $\|\mathbf{F}(\mathbf{r}(u, v))\| = C$ and $\|\mathbf{N}(u, v)\| = 1$; θ is the angle between **F** and **k**]

$$= A(S)\, C\cos\theta,$$

where $A(S)$ is the area of S. The expression $A(S)\, C\cos\theta$ is the volume [$A(S)$ is the base area and $C\cos\theta$ is the height] that flows through the surface S in a unit of time. Thus (at least in this special case), the surface integral is related to the rate of fluid flow (later in this section we will show that this is true in general).

▶ EXAMPLE 7.38

Compute $\iint_S \mathbf{F} \cdot d\mathbf{S}$, where $\mathbf{F} = y\mathbf{i} - x\mathbf{j} + z^2\mathbf{k}$ and S is the helicoid given by $\mathbf{r}(u, v) = (u\cos v, u\sin v, v)$, where $0 \le u \le 1$ and $0 \le v \le \pi/2$.

SOLUTION The value of **F** at the points on the helicoid S is $\mathbf{F}(\mathbf{r}(u, v)) = u\sin v\mathbf{i} - u\cos v\mathbf{j} + v^2\mathbf{k}$. The vector **N** normal to S is given by the cross product

$$\mathbf{N} = \mathbf{T}_u \times \mathbf{T}_v = \begin{vmatrix} \mathbf{i} & \mathbf{j} & \mathbf{k} \\ \cos v & \sin v & 0 \\ -u\sin v & u\cos v & 1 \end{vmatrix} = \sin v\mathbf{i} - \cos v\mathbf{j} + u\mathbf{k},$$

where $\mathbf{T}_u = \partial\mathbf{r}/\partial u$ and $\mathbf{T}_v = \partial\mathbf{r}/\partial v$ are tangent vectors. By definition,

$$\iint_S \mathbf{F} \cdot d\mathbf{S} = \iint_{[0,1]\times[0,\pi/2]} (u\sin v\mathbf{i} - u\cos v\mathbf{j} + v^2\mathbf{k}) \cdot (\sin v\mathbf{i} - \cos v\mathbf{j} + u\mathbf{k})\, dA$$

$$= \int_0^{\pi/2} \left(\int_0^1 (u + uv^2) du \right) dv = \int_0^{\pi/2} \left(\frac{u^2}{2} + \frac{u^2v^2}{2} \Big|_{u=0}^{u=1} \right) dv$$

$$= \int_0^{\pi/2} \left(\frac{1}{2} + \frac{v^2}{2} \right) dv = \left(\frac{v}{2} + \frac{v^3}{6} \right) \Big|_0^{\pi/2} = \frac{\pi}{4} + \frac{\pi^3}{48}.$$

◀

▶ EXAMPLE 7.39

Compute the surface integral $\iint_S (y^3\mathbf{i} + x^3\mathbf{j} + 3z^2\mathbf{k}) \cdot d\mathbf{S}$ over the surface S parametrized by $\mathbf{r}: D \to \mathbb{R}^3$, where $\mathbf{r}(u, v) = (u, v, u^2 + v^2)$ and $D = \{(u, v) | u^2 + v^2 \le 4\}$.

SOLUTION

The tangent vectors are $\mathbf{T}_u = (1, 0, 2u)$ and $\mathbf{T}_v = (0, 1, 2v)$, and the surface normal is $\mathbf{N} = \mathbf{T}_u \times \mathbf{T}_v = (-2u, -2v, 1)$. It follows that $\mathbf{F}(\mathbf{r}(u, v)) = (v^3, u^3, 3(u^2 + v^2)^2)$, and therefore,

$$\begin{aligned}\mathbf{F} \cdot \mathbf{N} &= (v^3, u^3, 3(u^2 + v^2)^2) \cdot (-2u, -2v, 1) = -2uv^3 - 2u^3v + 3(u^2 + v^2)^2 \\ &= -2uv(u^2 + v^2) + 3(u^2 + v^2)^2 = (u^2 + v^2)(3(u^2 + v^2) - 2uv).\end{aligned}$$

The surface integral of $\mathbf{F}$ over S is

$$\begin{aligned}\iint_S \mathbf{F} \cdot d\mathbf{S} &= \iint_{\{0 \le u^2+v^2 \le 4\}} \mathbf{F} \cdot \mathbf{N}\, dA \\ &= \iint_{\{0 \le u^2+v^2 \le 4\}} (u^2 + v^2)(3(u^2 + v^2) - 2uv)\, dA\end{aligned}$$

[passing to polar coordinates $u = r\cos\theta$, $v = r\sin\theta$]

$$= \int_0^{2\pi} \left(\int_0^2 r^2 \left(3r^2 - 2r^2 \sin\theta \cos\theta\right) r dr \right) d\theta$$

[using $2\cos\theta \sin\theta = \sin 2\theta$]

$$\begin{aligned}&= \int_0^{2\pi} \left(\int_0^2 \left(3r^5 - r^5 \sin 2\theta\right) dr \right) d\theta = \int_0^{2\pi} \left(\frac{r^6}{2} - \frac{r^6}{6} \sin 2\theta \Big|_0^2 \right) d\theta \\ &= \int_0^{2\pi} \left(32 - \frac{32}{3} \sin 2\theta\right) d\theta = \left(32\theta + \frac{32}{6} \cos 2\theta\right) \Big|_0^{2\pi} = 64\pi.\end{aligned}$$

◀

Assume that a surface S is smooth; that is, $\|\mathbf{N}(u, v)\| \neq 0$ for all (u, v). From the definition of the surface integral, it follows that

$$\iint_S \mathbf{F} \cdot d\mathbf{S} = \iint_D \mathbf{F}(\mathbf{r}(u, v)) \cdot \mathbf{N}(u, v)\, dA = \iint_D \mathbf{F}(\mathbf{r}(u, v)) \cdot \frac{\mathbf{N}(u, v)}{\|\mathbf{N}(u, v)\|} \|\mathbf{N}(u, v)\|\, dA.$$

This means that the surface integral of a *vector* function $\mathbf{F}$ reduces to the surface integral of the *scalar* function $\mathbf{F}(\mathbf{r}(u, v)) \cdot \mathbf{N}(u, v)/\|\mathbf{N}(u, v)\|$, which is the component of $\mathbf{F}$ in the normal direction. (Recall that we had a similar situation in Section 5.3, where we reduced the path integral of a vector field to the integral of its tangential component.)

Denoting $\mathbf{N}(u, v)/\|\mathbf{N}(u, v)\|$ by $\mathbf{n}(u, v)$ (so $\mathbf{n}$ is a unit normal vector), we write

$$\iint_S \mathbf{F} \cdot d\mathbf{S} = \iint_D \mathbf{F}(\mathbf{r}(u, v)) \cdot \mathbf{n}(u, v)\ \|\mathbf{N}(u, v)\|\, dA = \iint_S \mathbf{F} \cdot \mathbf{n}\, dS.$$

This formula can sometimes simplify the computation of a surface integral, as in the case presented in the following example.

▶ EXAMPLE 7.40

Compute $\iint_S (x\mathbf{i} + y\mathbf{j} + z\mathbf{k})\,dS$, where S is the surface of the sphere $x^2 + y^2 + z^2 = 1$ oriented by the outward-pointing normal.

SOLUTION

We can solve this problem without the use of a parametrization. Notice that the direction of a normal vector to the sphere is radial; that is, it has the direction of the line joining the point on the sphere and its center. An outward-pointing normal is given by $\mathbf{n} = x\mathbf{i} + y\mathbf{j} + z\mathbf{k}$. Since $\sqrt{x^2 + y^2 + z^2} = 1$, it follows that $\mathbf{n}$ is actually a unit normal. Therefore,

$$\mathbf{F} \cdot \mathbf{n} = (x\mathbf{i} + y\mathbf{j} + z\mathbf{k}) \cdot (x\mathbf{i} + y\mathbf{j} + z\mathbf{k}) = x^2 + y^2 + z^2 = 1,$$

and

$$\iint_S \mathbf{F} \cdot d\mathbf{S} = \iint_S \mathbf{F} \cdot \mathbf{n}\,dS = \iint_S 1\,dS = 4\pi,$$

since the surface area of a sphere of radius 1 is 4π. ◀

In order to compute the integral of a vector field along a curve, we have to specify the orientation on the curve and then choose a parametrization that preserves this orientation. Similarly, the surface integral $\iint_S \mathbf{F} \cdot d\mathbf{S}$ of a vector field $\mathbf{F}$ is an oriented integral: once the orientation of S has been specified (by prescribing the normal direction or by declaring one of the sides of S to be the outside), we have to choose a parametrization that respects our choice of orientation. In Examples 7.38 and 7.39, the surface was defined using a parametrization, and that implicitly contains the information on the orientation.

▶ EXAMPLE 7.41

Compute $\iint_S \mathbf{F} \cdot d\mathbf{S}$ if $\mathbf{F} = xyz\mathbf{i}$ and S is the part of the surface of the sphere $x^2 + y^2 + z^2 = 4$ in the first octant, oriented by the outward-pointing normal (i.e., the one pointing away from the origin).

SOLUTION

Parametrize S by $\mathbf{r}(u, v) = (2\cos v \cos u, 2\cos v \sin u, 2\sin v)$, where the parameters satisfy $0 \le u \le \pi/2$ and $0 \le v \le \pi/2$. The surface normal $\mathbf{N} = 2\cos v\, \mathbf{r}(u, v)$ was computed in Example 7.7. It points away from the origin, as required. Thus,

$$\begin{aligned}\mathbf{F} \cdot \mathbf{N} &= (8\cos^2 v \sin v \cos u \sin u, 0, 0) \cdot 2\cos v(2\cos v \cos u, 2\cos v \sin u, 2\sin v)\\ &= 32\cos^4 v \sin v \cos^2 u \sin u,\end{aligned}$$

and therefore,

$$\begin{aligned}\iint_S \mathbf{F} \cdot d\mathbf{S} &= \iint_{[0,\pi/2]\times[0,\pi/2]} \mathbf{F} \cdot \mathbf{N}\,dA\\ &= \iint_{[0,\pi/2]\times[0,\pi/2]} 32\cos^4 v \sin v \cos^2 u \sin u\,dA\\ &= 32\left(\int_0^{\pi/2} \cos^4 v \sin v\,dv\right)\left(\int_0^{\pi/2} \cos^2 u \sin u\,du\right)\\ &= 32\left(-\frac{\cos^5 v}{5}\Big|_0^{\pi/2}\right)\left(-\frac{\cos^3 v}{3}\Big|_0^{\pi/2}\right) = \frac{32}{15}.\end{aligned}$$

◀

▶ EXAMPLE 7.42

Find the integral $\iint_S ((z-x)\mathbf{i} - y\mathbf{j} - 2y^2\mathbf{k}) \cdot d\mathbf{S}$, where S is the graph of the function $z = x + 2y^2 - 3$, $0 \le x \le 1$, $-1 \le y \le 1$, oriented by the upward-pointing normal.

SOLUTION Choose the parametrization $\mathbf{r}(u,v) = u\mathbf{i} + v\mathbf{j} + (u + 2v^2 - 3)\mathbf{k}$, $0 \le u \le 1$, $-1 \le v \le 1$. Then $\mathbf{T}_u = \mathbf{i} + \mathbf{k}$, $\mathbf{T}_v = \mathbf{j} + 4v\mathbf{k}$, and $\mathbf{N} = -\mathbf{i} - 4v\mathbf{j} + \mathbf{k}$. Since the $\mathbf{k}$ component of $\mathbf{N}$ is positive, $\mathbf{N}$ points upward, and $\mathbf{r}(u,v)$ is an orientation-preserving parametrization of S. By definition,

$$\begin{aligned}\iint_S ((z-x)\mathbf{i} - y\mathbf{j} - 2y^2\mathbf{k})\, dS &= \iint_{[0,1]\times[-1,1]} \mathbf{F}(\mathbf{r}(u,v)) \cdot \mathbf{N}(u,v) dA \\ &= \iint_{[0,1]\times[-1,1]} ((2v^2 - 3)\mathbf{i} - v\mathbf{j} - 2v^2\mathbf{k}) \cdot (-\mathbf{i} - 4v\mathbf{j} + \mathbf{k})\, dA \\ &= \iint_{[0,1]\times[-1,1]} 3\, dA = 6,\end{aligned}$$

since the last double integral equals three times the area of the rectangle $[0, 1] \times [-1, 1]$. ◀

THEOREM 7.2 Dependence of Surface Integral on Parametrization

Let S be an oriented surface, and $\mathbf{r}: D \to \mathbb{R}^3$ and $\mathbf{r}^*: D^* \to \mathbb{R}^3$ two C^1 parametrizations of S with corresponding normal vectors $\mathbf{N} = \mathbf{T}_u \times \mathbf{T}_v$ and $\mathbf{N}^* = \mathbf{T}_u^* \times \mathbf{T}_v^*$. Assume that $\mathbf{F}$ is a continuous vector field on S. Then,

$$\iint_D \mathbf{F}(\mathbf{r}(u,v)) \cdot \mathbf{N}(u,v)\, dA = \iint_{D^*} \mathbf{F}(\mathbf{r}^*(u,v)) \cdot \mathbf{N}^*(u,v)\, dA^*$$

if $\mathbf{r}$ and $\mathbf{r}^*$ have the same orientation, and

$$\iint_D \mathbf{F}(\mathbf{r}(u,v)) \cdot \mathbf{N}(u,v)\, dA = -\iint_{D^*} \mathbf{F}(\mathbf{r}^*(u,v)) \cdot \mathbf{N}^*(u,v)\, dA^*$$

if $\mathbf{r}$ and $\mathbf{r}^*$ have opposite orientations. ◀

Theorem 7.2 says that we do not have to worry too much when evaluating the surface integrals of vector fields. Given an oriented surface, we can take any smooth parametrization and evaluate $\iint_S \mathbf{F} \cdot d\mathbf{S}$. If that parametrization is orientation-preserving, we have the result. If it is orientation-reversing, the result is the negative of what we obtained.

The proof of the theorem is similar to the proof of the corresponding theorem for path integrals. Its crucial ingredient is the change of variables formula for double integrals.

▶ EXAMPLE 7.43

Let us compute $\iint_S \mathbf{F} \cdot d\mathbf{S}$ with $\mathbf{F}$ and S as in Example 7.41, but this time with the parametrization $\mathbf{r}(u,v) = (2\cos u \sin v, 2\sin u \sin v, 2\cos v)$, $0 \le u, v \le \pi/2$.

SOLUTION The surface normal $\mathbf{N}$ is $\mathbf{N} = -2\sin v\, \mathbf{r}(u,v)$; see Example 7.34. Its direction is opposite to that of the radius vector $\mathbf{r}(u,v)$; that is, $\mathbf{N}$ points toward the origin (this follows from the fact that $\sin v \ge 0$ for $0 \le v \le \pi/2$). Hence, the parametrization given in this example is orientation-reversing, and according to Theorem 7.2, the result should be the negative of the result obtained in Example 7.41.

Indeed,

$$\mathbf{F} \cdot \mathbf{N} = 8 \cos u \sin u \sin^2 v \cos v \cdot (-2 \sin v) \cdot 2 \cos u \sin v$$

and

$$\begin{aligned}\iint_S \mathbf{F} \cdot d\mathbf{S} &= \iint_{[0,\pi/2]\times[0,\pi/2]} \mathbf{F} \cdot \mathbf{N}\, dA \\ &= \iint_{[0,\pi/2]\times[0,\pi/2]} (-32) \sin^4 v \cos v \cos^2 u \sin u\, dA \\ &= -32 \left(\int_0^{\pi/2} \sin^4 v \cos v\, dv \right) \left(\int_0^{\pi/2} \cos^2 u \sin u\, du \right) \\ &= -32 \left(\frac{\sin^5 v}{5} \bigg|_0^{\pi/2} \right) \left(-\frac{\cos^3 u}{3} \bigg|_0^{\pi/2} \right) = -\frac{32}{15}.\end{aligned}$$

EXAMPLE 7.44

Consider the surface integral $\iint_S (4x^2 + 4y^2 + 1)\mathbf{k} \cdot d\mathbf{S}$, where S is the part of the surface $z = 1 - (x^2 + y^2)$ that lies above the xy-plane, oriented by an upward-pointing normal. Use the projection of S onto the xy-plane (see the end of Section 7.3) to evaluate the integral.

SOLUTION

Viewing S as the level surface $F(x, y, z) = x^2 + y^2 + z = 1$, we compute the unit normal vector to S to be

$$\nabla F / \|\nabla F\| = \pm (2x, 2y, 1)/\sqrt{4x^2 + 4y^2 + 1}.$$

We need the z-component of $\mathbf{n}$ to be positive; hence, $\mathbf{n} = (2x, 2y, 1)/\sqrt{4x^2 + 4y^2 + 1}$ and

$$\iint_S (4x^2 + 4y^2 + 1)\mathbf{k} \cdot d\mathbf{S} = \iint_S (4x^2 + 4y^2 + 1)\mathbf{k} \cdot \mathbf{n}\, dS.$$

The projection D of S onto the xy-plane is the disk $x^2 + y^2 \le 1$. Therefore [notice that $\mathbf{n} \cdot \mathbf{k} = 1/\sqrt{4x^2 + 4y^2 + 1} > 0$],

$$\begin{aligned}\iint_S (4x^2 + 4y^2 + 1)\mathbf{k} \cdot \mathbf{n}\, dS &= \iint_D (4x^2 + 4y^2 + 1)\mathbf{k} \cdot \mathbf{n} \frac{dA}{|\mathbf{n} \cdot \mathbf{k}|} \\ &= \iint_D (4x^2 + 4y^2 + 1)\, dA,\end{aligned}$$

where dA refers to integration with respect to x and y. Passing to polar coordinates, we get

$$\begin{aligned}\iint_D (4x^2 + 4y^2 + 1)\, dA &= \int_0^{2\pi} \left(\int_0^1 (4r^2 + 1)\, r dr \right) d\theta \\ &= \int_0^{2\pi} \left(\left(r^4 + \tfrac{1}{2}r^2\right) \bigg|_0^1 \right) d\theta = \tfrac{3}{2} 2\pi = 3\pi.\end{aligned}$$

Let us now discuss the geometric meaning of the surface integral $\iint_S \mathbf{F} \cdot d\mathbf{S}$ of a vector field $\mathbf{F}$ over a surface S. Suppose that $\mathbf{r}\colon D \to \mathbb{R}^3$ is a C^1 parametrization of S and $\mathbf{F}\colon S \subseteq \mathbb{R}^3 \to \mathbb{R}^3$ is a continuous vector field. As in the previous section, assume that D is a rectangle and subdivide it into n^2 rectangles. Select one rectangle (name it R) with one vertex at (u, v) and sides Δu and Δv, as in Figure 7.55. The surface integral $\iint_S \mathbf{F} \cdot d\mathbf{S}$ can

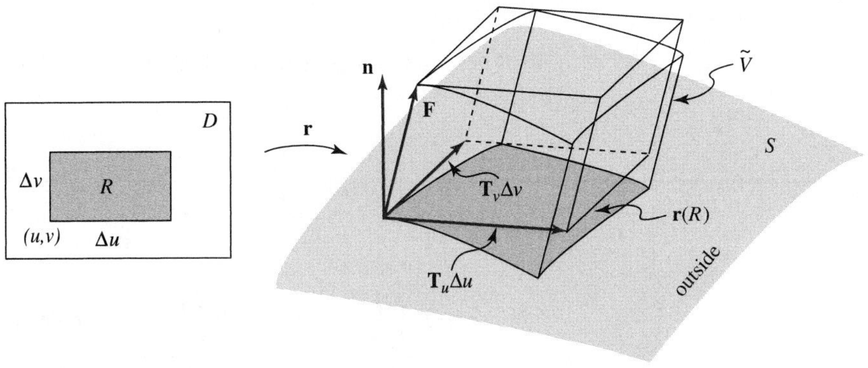

Figure 7.55 Geometric meaning of the surface integral of a vector field.

be written as

$$\iint_S \mathbf{F} \cdot d\mathbf{S} = \iint_D \mathbf{F} \cdot (\mathbf{T}_u \times \mathbf{T}_v)\, dA = \lim_{\text{size}(R)\to 0} \sum_R \mathbf{F} \cdot (\mathbf{T}_u \times \mathbf{T}_v)\Delta u \Delta v,$$

where $\mathbf{T}_u$ and $\mathbf{T}_v$ are the tangent vectors and the summation goes over all rectangles that form the subdivision of D.

Each summand $\mathbf{F} \cdot (\mathbf{T}_u \times \mathbf{T}_v)\Delta u \Delta v = \mathbf{F} \cdot (\mathbf{T}_u \Delta u \times \mathbf{T}_v \Delta v)$ is the scalar triple product of $\mathbf{F}$, $\mathbf{T}_u \Delta u$, and $\mathbf{T}_v \Delta v$. Consequently, the absolute value

$$|\mathbf{F} \cdot (\mathbf{T}_u \Delta u \times \mathbf{T}_v \Delta v)|$$

represents the volume of the parallelepiped with sides $\mathbf{F}$, $\mathbf{T}_u \Delta u$, and $\mathbf{T}_v \Delta v$.

Suppose that the orientation of S is given by $\mathbf{n} = +\mathbf{T}_u \times \mathbf{T}_v / \|\mathbf{T}_u \times \mathbf{T}_v\|$, as shown in Figure 7.55. The dot product $\mathbf{F} \cdot (\mathbf{T}_u \times \mathbf{T}_v)\Delta u \Delta v$ is positive if $\mathbf{F}$ points away from the outside of S (since the angle between $\mathbf{F}$ and the normal $\mathbf{n}$ is smaller than $\pi/2$). If $\mathbf{F}$ points away from the inside, then $\mathbf{F} \cdot (\mathbf{T}_u \times \mathbf{T}_v)\Delta u \Delta v$ is negative. Therefore,

$$\iint_S \mathbf{F} \cdot d\mathbf{S} = \lim_{\text{size}(R)\to 0} \Big(\sum\nolimits^{+} \mathbf{F} \cdot (\mathbf{T}_u \Delta u \times \mathbf{T}_v \Delta v) + \sum\nolimits^{-} \mathbf{F} \cdot (\mathbf{T}_u \Delta u \times \mathbf{T}_v \Delta v) \Big),$$

where $\sum^{+}$ denotes the sum over all R's for which $\mathbf{F} \cdot (\mathbf{T}_u \Delta u \times \mathbf{T}_v \Delta v)$ is positive, and $\sum^{-}$ is the sum over all R's for which $\mathbf{F} \cdot (\mathbf{T}_u \Delta u \times \mathbf{T}_v \Delta v)$ is negative.

Think of $\mathbf{F}$ as the velocity vector of a fluid and consider $\sum^{+}$ first. The patch $\mathbf{r}(R)$ on S is approximated by the parallelogram spanned by $\mathbf{T}_u \Delta u$ and $\mathbf{T}_v \Delta v$. Assume that R [and hence $\mathbf{r}(R)$] are so small that $\mathbf{F}$ is (approximately) constant on $\mathbf{r}(R)$. The amount of fluid that flows across the patch $\mathbf{r}(R)$ is equal to the volume of the "twisted parallelepiped" $\tilde{V}$ (see Figure 7.55), and can be approximated by the volume of the parallelepiped spanned by $\mathbf{F}$, $\mathbf{T}_u \Delta u$, and $\mathbf{T}_v \Delta v$. Therefore, $\sum^{+}$ approximates the net volume of fluid that flows outward across S per unit time. Similarly, $\sum^{-}$ gives an approximation for the net volume of fluid that flows inward across S per unit time.

Consequently, $\iint_S \mathbf{F} \cdot d\mathbf{S}$ is the net volume of fluid that flows across S per unit time, which is the rate of fluid flow. That is the reason why the integral $\iint_S \mathbf{F} \cdot d\mathbf{S} = \iint_S \mathbf{F} \cdot \mathbf{n}\, dS$ is called the *(outward) flux* of $\mathbf{F}$ across the surface S.

▶ EXAMPLE 7.45

Compute the flux of the vector field $\mathbf{F} = (\sin x, z, y)$ across the part of the cylinder $y^2 + z^2 = 4$ bounded by $-1/2 \le x \le 1/2$, $y \ge 0$, and $z \ge 0$.

SOLUTION Parametrize the surface of the cylinder (call it S) by $\mathbf{r}(u, v) = (u, 2\cos v, 2\sin v)$, where $-1/2 \le u \le 1/2$ and $0 \le v \le \pi/2$. Then $\mathbf{T}_u = (1, 0, 0)$, $\mathbf{T}_v = (0, -2\sin v, 2\cos v)$, and the normal is $\mathbf{N} = (0, -2\cos v, -2\sin v)$. The flux of $\mathbf{F}$ is computed to be

$$\begin{aligned}\iint_S \mathbf{F} \cdot d\mathbf{S} &= \iint_{[-1/2,1/2]\times[0,\pi/2]} (\sin u, 2\sin v, 2\cos v) \cdot (0, -2\cos v, -2\sin v)\, dA \\ &= \int_{-1/2}^{1/2} \left(\int_0^{\pi/2} (-8\sin v \cos v)\, dv \right) du \\ &= -4 \int_{-1/2}^{1/2} \left(\int_0^{\pi/2} \sin 2v dv \right) du = -4 \int_{-1/2}^{1/2} 1\, du = -4.\end{aligned}$$

So the rate of fluid flow across S in the direction $\mathbf{N} = (0, -2\cos v, -2\sin v)$ is -4. The normal points downward, toward the x-axis (since the z-component $z = -2\sin v$ is negative); the fluid flows in the direction opposite to that of the normal. ◀

As in the case of real-valued functions, the surface integral of a vector field can be defined over a piecewise smooth surface (as the sum of the surface integrals over the disjoint pieces of S that are smooth). Here is an example.

▶ EXAMPLE 7.46

Compute $\iint_S \mathbf{F} \cdot d\mathbf{S}$, where $\mathbf{F} = x\mathbf{i} + z\mathbf{k}$ and S is the surface of the cube bounded by the six planes $x = 0$, $y = 0$, $z = 0$, $x = 1$, $y = 1$, and $z = 1$ and oriented by an outward normal.

SOLUTION The surface of a cube is piecewise smooth: it consists of six smooth surfaces that are the faces of the cube (the most that two faces have in common is a common edge, so they are disjoint). We have to integrate $\mathbf{F}$ over all six surfaces and add up the results.

Parametrize the bottom surface by $\mathbf{r}_1(u, v) = (u, v, 0)$, $0 \le u, v \le 1$. Then $\mathbf{n}_1 = -\mathbf{k}$ (see Figure 7.56; no computation for $\mathbf{n}_1$ is needed!) and

$$\iint_{S_1} \mathbf{F} \cdot d\mathbf{S} = \iint_{S_1} \mathbf{F} \cdot \mathbf{n}_1\, dS = \iint_{[0,1]\times[0,1]} (u\mathbf{i}) \cdot (-\mathbf{k})\, dA = 0.$$

Parametrize the top surface by $\mathbf{r}_2(u, v) = (u, v, 1)$, $0 \le u, v \le 1$. Then $\mathbf{n}_2 = \mathbf{k}$ and

$$\iint_{S_2} \mathbf{F} \cdot d\mathbf{S} = \iint_{S_2} \mathbf{F} \cdot \mathbf{n}_2\, dS = \iint_{[0,1]\times[0,1]} (u\mathbf{i} + \mathbf{k}) \cdot \mathbf{k}\, dA = \iint_{[0,1]\times[0,1]} dA = 1$$

(no integration was done here; the last integral was interpreted as area).

Similarly, $\mathbf{r}_3(u, v) = (u, 0, v)$, and $\mathbf{r}_4(u, v) = (u, 1, v)$, $0 \le u, v \le 1$ parametrize the left and right sides of S. The unit normals are $\mathbf{n}_3 = -\mathbf{j}$ and $\mathbf{n}_4 = \mathbf{j}$, and since $\mathbf{F} \cdot \mathbf{n}_3 = (u\mathbf{i} + v\mathbf{k}) \cdot \mathbf{j} = 0$ and $\mathbf{F} \cdot \mathbf{n}_4 = (u\mathbf{i} + v\mathbf{k}) \cdot (-\mathbf{j}) = 0$, both integrals $\iint_{S_3} \mathbf{F} \cdot d\mathbf{S}$ and $\iint_{S_4} \mathbf{F} \cdot d\mathbf{S}$ are zero.

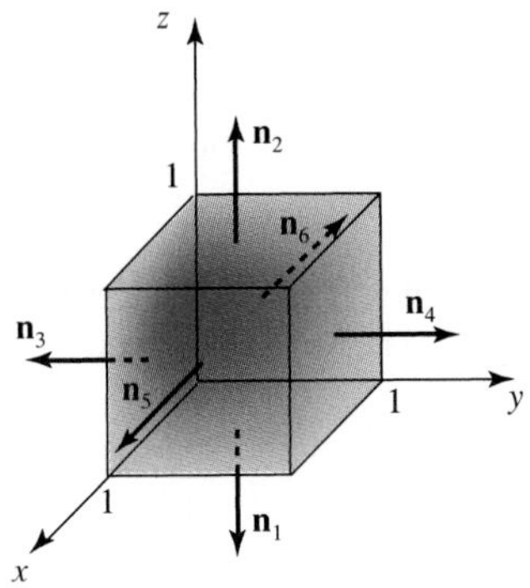

Figure 7.56 Surface of the cube oriented by outward-pointing normals.

Finally, parametrize the front and back sides of S by $\mathbf{r}_5(u, v) = (1, u, v)$, and $\mathbf{r}_6(u, v) = (0, u, v)$, $0 \le u, v \le 1$. Then $\mathbf{n}_5 = \mathbf{i}$ and $\mathbf{n}_6 = -\mathbf{i}$, and

$$\iint_{S_5} \mathbf{F} \cdot d\mathbf{S} = \iint_{S_5} \mathbf{F} \cdot \mathbf{n}_5 \, dS = \iint_{[0,1]\times[0,1]} (\mathbf{i} + v\mathbf{k}) \cdot \mathbf{i} \, dA = 1$$

and

$$\iint_{S_6} \mathbf{F} \cdot d\mathbf{S} = \iint_{S_6} \mathbf{F} \cdot \mathbf{n}_6 \, dS = \iint_{[0,1]\times[0,1]} (v\mathbf{k}) \cdot (-\mathbf{i}) \, dA = 0.$$

Therefore, $\iint_S \mathbf{F} \cdot d\mathbf{S} = 0 + 1 + 0 + 0 + 1 + 0 = 2$ is the flux out across S. ◀

▶ EXERCISES 7.4

1. The surface S is defined by $x^2 + y^2 = 1$, $1 \le z \le 2$; assume that it is oriented by the outward-pointing normal vector. Among the vector fields $\mathbf{F}_1(x, y, z) = \mathbf{i}$, $\mathbf{F}_2(x, y, z) = x\mathbf{i} + y\mathbf{j}$, and $\mathbf{F}_3(x, y, z) = \mathbf{k}$, identify those whose (outward) flux is zero.

2. Consider the graph of the function $g(x, y) = 1 - x^2 - y^2$, $-1 \le x, y \le 1$, oriented by the upward-pointing normal. Determine the sign of the outward flux of the following fields: $\mathbf{F}_1(x, y, z) = \mathbf{k}$, $\mathbf{F}_2(x, y, z) = x\mathbf{i} + y\mathbf{j}$, and $\mathbf{F}_3(x, y, z) = -x\mathbf{i} - y\mathbf{j}$.

Exercises 3 to 6: Compute the flux of the vector field $\mathbf{F} = \mathbf{i} - 2\mathbf{j} + 4\mathbf{k}$ across the given region (assumed subset of a plane), with the indicated orientation.

3.

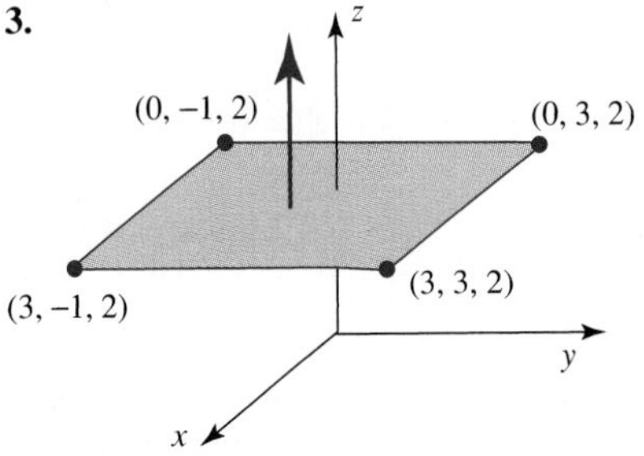

4.

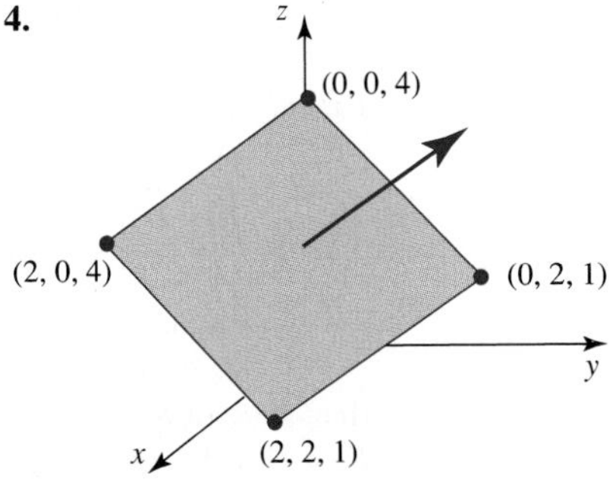

5.

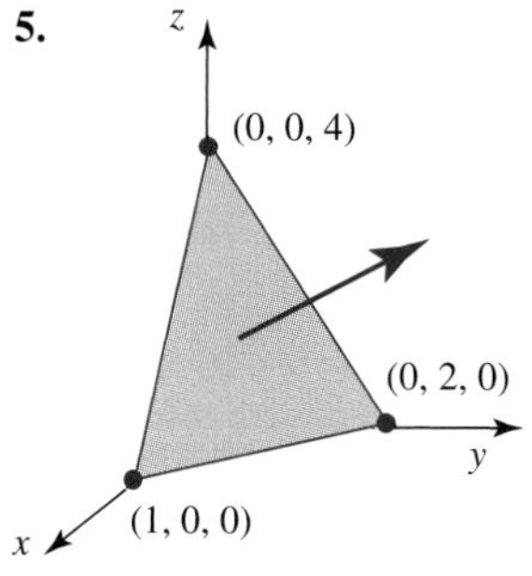

6.

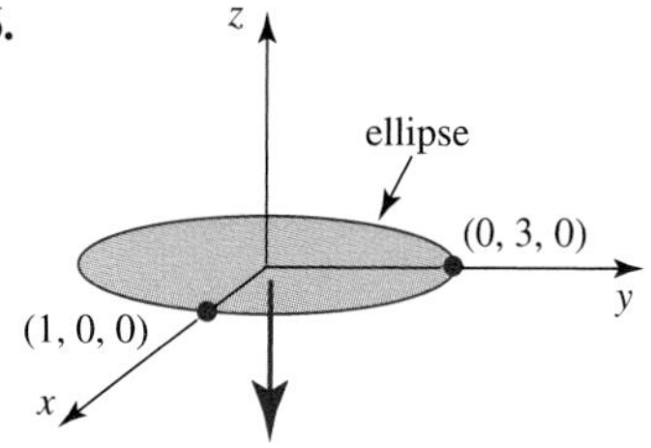

Exercises 7 to 14: Compute $\iint_S \mathbf{F} \cdot d\mathbf{S}$.

7. $\mathbf{F} = 4y\mathbf{i} + (3x - 1)\mathbf{j} + z\mathbf{k}$, S is the part of the plane $3x + y - z = 1$ (with the upward normal) inside the vertical cylinder of radius 2 whose axis of symmetry is the z-axis

8. $\mathbf{F} = x\mathbf{i} + y\mathbf{j}$, S is the part of the cone $z = \sqrt{x^2 + y^2}$ (oriented with the inward normal) inside the vertical cylinder $x^2 + y^2 = 9$

9. $\mathbf{F} = x^2\mathbf{i} + 2z\mathbf{k}$, S is the hemisphere $x^2 + y^2 + z^2 = 9$, $z \geq 0$, oriented with an outward normal

10. $\mathbf{F} = y\mathbf{i} - x\mathbf{j} + \mathbf{k}$, S is the surface parametrized by $\mathbf{r}(u, v) = (u \cos v, u \sin v, v)$, $0 \leq u \leq 1$, $0 \leq v \leq 4\pi$, and oriented with an upward-pointing normal

11. $\mathbf{F} = z\mathbf{k}$, S is the paraboloid $z = x^2 + y^2$ (oriented with the normal pointing away from it) between the planes $z = 1$ and $z = 2$

12. $\mathbf{F} = 2\mathbf{i} - xy\mathbf{j}$, S is the graph of the function $z = f(x, y) = x^2y^3 - 1$, where $0 \leq x, y \leq 1$, oriented by the upward-pointing normal

13. $\mathbf{F} = x\mathbf{i} + y\mathbf{j} + z\mathbf{k}$, S is the surface parametrized by $\mathbf{r}(u, v) = (e^u \cos v, e^u \sin v, v)$, $0 \leq u \leq \ln 2$, $0 \leq v \leq \pi$, and oriented with an upward-pointing normal

14. $\mathbf{F} = x^2y\mathbf{i} - (y + x)\mathbf{j} - z^2x\mathbf{k}$, S is the part of the plane $x + 2y + 8z = 8$ in the first octant with the normal vector pointing upward

15. Let $T(x, y, z) = x^2 + y^2 + 3z^2$ be the temperature at a point (x, y, z) in $\mathbb{R}^3$. Compute the heat flux outward across the surface $x^2 + y^2 = 1$, $-1 \leq z \leq 1$.

16. Let $T(x, y, z) = e^{-x^2-y^2-z^2}$ be the temperature at a point (x, y, z) in $\mathbb{R}^3$. Compute the heat flux outward across the sphere $x^2 + y^2 + z^2 = 1$.

17. Consider the vector field $\mathbf{F} = c\mathbf{k}$, where c is a constant.

(a) Compute the flux $\iint_S \mathbf{F} \cdot \mathbf{n}\, dS$, where S is the hemisphere $x^2 + y^2 + z^2 = a^2$, $z \geq 0$, with the outward-pointing normal.

(b) Compute the flux of $\mathbf{F}$ across the disk $x^2 + y^2 \leq a^2$ in the xy-plane, with the upward-pointing normal.

(c) Why are the answers in (a) and (b) the same?

Exercises 18 to 23: Find the flux of $\mathbf{F}$ across the surface S.

18. $\mathbf{F} = x\mathbf{i} + y\mathbf{j} + \mathbf{k}$, out of the closed region bounded by the paraboloid $z = 2x^2 + 2y^2$ and the plane $z = 4$

19. $\mathbf{F} = x\mathbf{i}$, out of the closed region bounded by the paraboloids $z = x^2 + y^2$ and $z = 12 - x^2 - y^2$

20. $\mathbf{F} = x\mathbf{i}$, out of the closed region bounded by the spheres $x^2 + y^2 + z^2 = a^2$ and $x^2 + y^2 + z^2 = b^2$, $a > b$

21. $\mathbf{F} = y^3(\mathbf{j} - \mathbf{k})$, across the part of the plane $2x + y + z = 16$ in the first octant in the direction away from the origin

22. $\mathbf{F} = \mathbf{i} + xy\mathbf{j}$, across the closed cylinder (and in the direction away from it) $x^2 + y^2 = 1$, with the top disk at $z = 2$ and the bottom disk in the xy-plane

23. $\mathbf{F} = x^2y\mathbf{i} + xy^3\mathbf{j} + 2xyz\mathbf{k}$, upward across the surface $z = 2x^2y$, $0 \le x \le 1$, $0 \le y \le 2$

24. Compute the flux through the surface of the plane $z = 2$, $0 \le x, y \le a$ $(a > 0)$ of the constant unit vector field $\mathbf{F}$ that makes an angle of α rad $(0 \le \alpha \le \pi/2)$ with respect to the plane.

25. Let $\mathbf{F} = F_\rho \mathbf{e}_\rho + F_\theta \mathbf{e}_\theta + F_\phi \mathbf{e}_\phi$ be the representation of the vector field $\mathbf{F}$ in spherical coordinates. Show that the flux of $\mathbf{F}$ out of the sphere $x^2 + y^2 + z^2 = a^2$, $a > 0$, satisfies $\iint_S \mathbf{F} \cdot \mathbf{S} = a^2 \int_0^{2\pi} \left(\int_0^\pi F_\rho \sin\phi d\phi \right) d\theta$.

26. Let S be the (closed) surface consisting of the part of the cone $z^2 = x^2 + y^2$, $1 \le z \le 2$ together with the top and bottom disks (in the planes $z = 2$ and $z = 1$). Show that the vector fields $\mathbf{F}_1 = x\mathbf{i} + 2y\mathbf{j} + 3z\mathbf{k}$ and $\mathbf{F}_2 = (y^2 + z)\mathbf{i} + (6y + x)\mathbf{j}$ have the same outward flux.

27. Let $\mathbf{F} = (x + y)\mathbf{i} + \mathbf{j} + z\mathbf{k}$ and assume that S is the part of the plane $x + 2y + 8z = 8$ in the first octant, oriented by the downward-pointing normal. Compute the surface integral $\iint_S \mathbf{F} \cdot d\mathbf{S}$ using a projection of S onto a coordinate plane.

28. Let S be the part of the plane $z = 2$ defined by $0 \le x, y \le a$, $a > 0$. Let c be a positive constant.

(a) Compute the flux of a vertical field $\mathbf{F} = c\mathbf{k}$ across S.

(b) Compute the flux of $\mathbf{F} = cz\mathbf{k}$ across S.

(c) Compute the flux of $\mathbf{F} = cz^2\mathbf{k}$ across S.

(d) Compute the flux of $\mathbf{F} = c(\mathbf{j} + \mathbf{k})/\sqrt{2}$ across S.

(e) Compute the flux of $\mathbf{F} = f(x, y, z)\mathbf{i} + g(x, y, z)\mathbf{j}$ across S.

(f) Interpret the results of (a)–(e).

29. Compute $\iint_S \mathbf{F} \cdot d\mathbf{S}$, where $\mathbf{F}$ and S are as in Exercise 9, using the projection of S onto the xy-plane.

30. Let S be the graph of a C^1 function $z = f(x, y)$, $(x, y) \in D$, oriented as required by Definition 7.9. Show that, for a continuous vector field $\mathbf{F} = (F_1, F_2, F_3)$, $\iint_S \mathbf{F} \cdot d\mathbf{S} = \iint_D (-F_1(\partial f/\partial x) - F_2(\partial f/\partial y) + F_3)\, dA$.

▶ 7.5 INTEGRALS: PROPERTIES AND APPLICATIONS

Although we have defined all kinds of integrals in the last three chapters, we have to admit that we were in a way repeating the same things over and over again (such as, for example, the construction of Riemann sums). As a conclusion, we will present a unified view of the integrations we have discussed, list their properties, and show several applications.

Notation

Throughout this section, we will use $\mathcal{M}$ to denote either of the following:

(a) A curve $\mathbf{c}$ (that will be viewed in applications as an approximation of a thin wire);

(b) A plane region D (that will represent a thin flat plate in $\mathbb{R}^2$);

(c) A surface S (that will represent a thin sheet (possibly curved) in $\mathbb{R}^3$); or

(d) A solid W (that will represent a three-dimensional solid in $\mathbb{R}^3$).

We will use $\mathbf{x}$ to denote x, (x, y) or (x, y, z), depending on the context. A common name for geometric objects described in (a)–(d) is a *manifold (with a boundary)*. The integral $\int_{\mathcal{M}} f\,d\mu$ is interpreted accordingly, as follows:

(a) The path integral $\int_{\mathbf{c}} f ds$ of a real-valued function f. The path integral $\int_{\mathbf{c}} \mathbf{F} \cdot d\mathbf{s}$ of a vector field can be reduced to the path integral of a real-valued function (namely of its tangential component). In the special case when $\mathbf{c}$ is an interval $[a, b]$ on the x-axis, the path integral is the definite integral $\int_a^b f(x)dx$ of a real-valued function of one variable.

(b) The double integral $\iint_D f dA$ of a real-valued function f.

(c) The surface integral $\iint_S f\,dS$ of a real-valued function f. The surface integral $\iint_S \mathbf{F} \cdot d\mathbf{S}$ of a vector field can be reduced to the surface integral of a real-valued function (normal component of the vector field).

(d) The triple integral $\iiint_W f\,dV$ of a real-valued function f.

Fundamental Theorem of Calculus

Recall that, if the function $f\colon [a, b] \to \mathbb{R}$ is continuously differentiable (i.e., if it is C^1), then

$$\int_a^b f'(x)dx = f(b) - f(a).$$

In Section 5.4, we proved the generalization

$$\int_{\mathbf{c}} \nabla f \cdot d\mathbf{s} = f(\mathbf{c}(b)) - f(\mathbf{c}(a))$$

for any C^1 function f and a piecewise C^1 path $\mathbf{c}\colon [a, b] \to \mathbb{R}^2$ (or $\mathbb{R}^3$). What is common to both is that the integral of the *derivative* of a function is computed by evaluating the function at the endpoints (of an interval, or of a curve). With the proper interpretation of the derivative, in Chapter 8, we will be able to extend this important fact to more general integrations.

Length, Area, Volume

We have already seen that the integral $\mu(\mathcal{M}) = \int_{\mathcal{M}} d\mu$ of the constant function $f(x, y, z) = 1$ (or $f(x, y) = 1$) has a geometric meaning.

(a) If $\mathcal{M} = \mathbf{c}$ is a curve parametrized by $\mathbf{c}(t)$, $t \in [a, b]$, then $\int_{\mathbf{c}} ds = \int_a^b \|\mathbf{c}'(t)\|\,dt$ gives its length $\ell(\mathbf{c})$.

(b) If $\mathcal{M} = D$ is a region in the xy-plane, then $\iint_D dA$ gives the area $A(D)$ of D.

(c) Let $\mathcal{M} = S$ be a surface parametrized by $\mathbf{r}(u, v)$, $(u, v) \in D$. Then $\iint_S dS = \iint_D \|\mathbf{N}\|\,dA$ computes the surface area $A(S)$ of S.

(d) Taking $\mathcal{M}$ to be a three-dimensional solid W, we get the formula $\iiint_W dV$ for its volume $v(W)$.

Average Value of a Function

The *average value* of a real-valued function $f = f(x, y, z)$ [or $f = f(x, y)$] is given by

$$\overline{f} = \frac{1}{\mu(\mathcal{M})} \int_{\mathcal{M}} f d\mu,$$

where $\mu(\mathcal{M})$ denotes the length, area, or volume, depending on $\mathcal{M}$.

▶ EXAMPLE 7.47

Compute the average temperature of the cylindrical sheet S described by $x^2 + y^2 = 4$, $0 \le z \le 1$, if the temperature distribution function is $T(x, y, z) = z^2$.

SOLUTION Parametrize the surface of the cylindrical sheet by $\mathbf{r}(u, v) = (2\cos u, 2\sin u, v)$, where $0 \le u \le 2\pi$ and $0 \le v \le 1$. Then

$$\mathbf{N} = \mathbf{T}_u \times \mathbf{T}_v = (-2\sin u, 2\cos u, 0) \times (0, 0, 1) = (2\cos u, 2\sin u, 0)$$

($\|\mathbf{N}\| = 2$), and hence, the average temperature is

$$\overline{T} = \frac{1}{A(S)} \iint_S z^2 \, dS = \frac{1}{A(S)} \iint_{[0,2\pi]\times[0,1]} v^2 \, 2 \, dA,$$

where $A(S)$ is the surface area of S. We do not need to integrate to compute it: cutting the cylinder vertically and unwrapping it, we get a rectangle of base length (equal to the circumference of the circle) 4π and height 1. Therefore, $A(S) = 4\pi$ and

$$\overline{T} = \frac{1}{4\pi} \int_0^{2\pi} \left(\int_0^1 2v^2 dv \right) du = \frac{1}{4\pi} \int_0^{2\pi} \frac{2}{3} du = \frac{1}{3}.$$

◀

Properties of Integrals

Assume that functions f and g are integrable, so that $\int_{\mathcal{M}} f d\mu$ and $\int_{\mathcal{M}} g d\mu$ are defined. Let $\mathcal{M}$ denote a piecewise smooth C^1 curve, a piecewise smooth C^1 surface (in particular, a plane region), or a solid three-dimensional region. Then

(a) Integrals are *linear* (C denotes a constant):

$$\int_{\mathcal{M}} (f + g) d\mu = \int_{\mathcal{M}} f d\mu + \int_{\mathcal{M}} g d\mu,$$

$$\int_{\mathcal{M}} C f d\mu = C \int_{\mathcal{M}} f d\mu.$$

(b) Integrals are *additive*: if $\mathcal{M}$ is divided into n (elementary) regions $\mathcal{M}_1, \ldots, \mathcal{M}_n$ that are mutually disjoint, then

$$\int_{\mathcal{M}} f d\mu = \int_{\mathcal{M}_1} f d\mu + \cdots + \int_{\mathcal{M}_n} f d\mu.$$

(c) If $f(\mathbf{x}) \le g(\mathbf{x})$ for all $\mathbf{x} \in \mathcal{M}$, then

$$\int_{\mathcal{M}} f d\mu \le \int_{\mathcal{M}} g d\mu.$$

If $m \le f(\mathbf{x}) \le M$ for all $\mathbf{x} \in \mathcal{M}$ (m and M are constants), then

$$m\,\mu(\mathcal{M}) \le \int_{\mathcal{M}} f\,d\mu \le M\,\mu(\mathcal{M}),$$

where $\mu(\mathcal{M})$ represents length, area, or volume, depending on $\mathcal{M}$.

(d) The inequality for the absolute value:

$$\left|\int_{\mathcal{M}} f\,d\mu\right| \le \int_{\mathcal{M}} |f|\,d\mu.$$

(e) Mean Value Theorem for integrals: If f is a continuous function, then there exists a point $\mathbf{x}_0$ in $\mathcal{M}$ such that

$$\int_{\mathcal{M}} f\,d\mu = f(\mathbf{x}_0)\,\mu(\mathcal{M}). \tag{7.12}$$

(f) Let $\overline{\mathcal{M}}$ be a collection of all piecewise smooth curves, or all piecewise smooth surfaces, or all solid three-dimensional regions (for which it is possible to define the triple integral). Suppose that

$$\int_{\mathcal{M}} f\,d\mu = 0$$

for *all* $\mathcal{M}$ in $\overline{\mathcal{M}}$ that belong to the domain U of f, and assume that f is continuous on U. Then f is identically zero [i.e., $f(\mathbf{x}) = 0$ for all $\mathbf{x} \in U$].

(g) Assume that $f = f(\mathbf{x}, t)$ is differentiable. Then

$$\frac{\partial}{\partial t}\left(\int_{\mathcal{M}} f\,d\mu\right) = \int_{\mathcal{M}} \frac{\partial f}{\partial t}\,d\mu. \tag{7.13}$$

The proofs of all statements except (e) and (f) are based on considering the appropriate Riemann sums (and we have done so in some cases in the last three chapters). Property (e) was proven in the special case of double integrals in Section 6.2. Rather that (re)proving these statements, we are going to make a few comments.

We can rewrite (e) as

$$f(\mathbf{x}_0) = \frac{1}{\mu(\mathcal{M})}\int_{\mathcal{M}} f\,d\mu, \tag{7.14}$$

noticing that the right side is the average value of f. Therefore, the Mean Value Theorem for Integrals implies that there must be a point in $\mathcal{M}$ at which a continuous function f attains its average value.

We will use property (f) later, in the following context: assume that f and g are continuous on a set U and

$$\int_{\mathcal{M}} f\,d\mu = \int_{\mathcal{M}} g\,d\mu$$

holds for *every* $\mathcal{M} \subseteq U$ in $\overline{\mathcal{M}}$. Then $\int_{\mathcal{M}}(f - g)d\mu = 0$, for all $\mathcal{M}$, and it follows by (f) that $f = g$ on U. In words, if the integrals of two functions agree for *all* manifolds (i.e., for all curves or for all surfaces, etc.), then the two functions are equal. To repeat, (f) has to hold for all $\mathcal{M}$ in $\overline{\mathcal{M}}$: it is easy to cook up examples where $\int_{\mathcal{M}} f\,d\mu = \int_{\mathcal{M}} g\,d\mu$ for some

$\mathcal{M}$, but $f \neq g$. For example, $\iint_{[0,1]\times[0,1]} 2x\,dA = \iint_{[0,1]\times[0,1]} 1\,dA$ (both are equal to 1), but $2x \neq 1$.

Property (g) will be used in the next chapter in computations dealing with time-dependent electric and magnetic fields; it will allow us to change the order of the time-derivative and integral over a surface or over a solid three-dimensional region.

Total Mass and Total Charge

Let $\rho = \rho(\mathbf{x})$ denote the *mass density* (i.e., mass per unit length, area, or volume, depending on the context) of an object represented as a manifold $\mathcal{M}$. The *total mass* of $\mathcal{M}$ is given by the integral

$$m = \int_{\mathcal{M}} \rho\, d\mu.$$

In the case when $\rho(\mathbf{x})$ represents the *charge density* (i.e., charge per unit length, area, or volume, depending on the context), the above formula gives the *total charge* contained in $\mathcal{M}$. If ρ is constant, we say that $\mathcal{M}$ is *homogeneous* (if ρ denotes the mass density) or that it is *uniformly charged* (if ρ denotes the charge density). Then

$$m = \rho \int_{\mathcal{M}} d\mu = \rho\mu(\mathcal{M}),$$

where the meaning of $\mu(\mathcal{M})$ depends on the context.

▶ EXAMPLE 7.48

Find the mass of a thin-walled conical funnel $z = \sqrt{x^2 + y^2}$, $1 \le z \le 2$ (see Figure 7.57) whose density is $\rho(x, y, z) = 2 + z$.

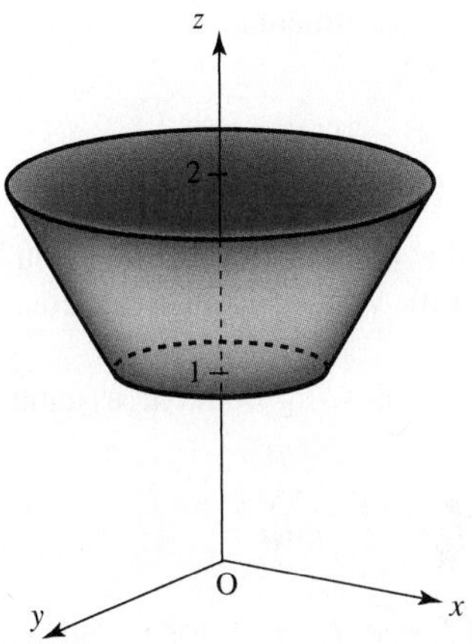

Figure 7.57 Conical funnel of Example 7.48.

SOLUTION The mass of the funnel is given by $m = \int_{\mathcal{M}} \rho\, d\mu$, where $\mathcal{M}$ denotes the surface S of the funnel and $\int_{\mathcal{M}}$ represents the surface integral (so that $d\mu = dS$).

Thus, $m = \iint_S \rho\, dS = \iint_S (2+z)\, dS$, where $\mathbf{r}(u,v) = (u\cos v, u\sin v, u)$, $1 \le u \le 2$, $0 \le v \le 2\pi$ parametrizes S. We compute $\mathbf{T}_u = (\cos v, \sin v, 1)$, $\mathbf{T}_v = (-u\sin v, u\cos v, 0)$, and $\mathbf{N} = \mathbf{T}_u \times \mathbf{T}_v = (-u\cos v, -u\sin v, u)$, $\|\mathbf{N}\| = u\sqrt{2}$. It follows that

$$\begin{aligned} m &= \iint_{[1,2]\times[0,2\pi]} (2+u)\, \|\mathbf{N}\| dA \\ &= \int_0^{2\pi} \left(\int_1^2 (2+u)\sqrt{2}\, u\, du \right) dv = \sqrt{2} \int_0^{2\pi} \left(u^2 + \frac{1}{3}u^3 \right)\Bigg|_1^2 dv = \frac{32}{3}\pi\sqrt{2}. \end{aligned}$$

Center of Mass

Intuitively speaking, the *center of mass* (or the *center of gravity*) of $\mathcal{M}$ is the point $C\mathcal{M}$ on which $\mathcal{M}$ balances when supported at that point. However, the center of mass might not be a point of $\mathcal{M}$: the ring $\{1 \le x^2 + y^2 \le 2\}$ of uniform density does not contain the origin, which is its center of mass. In some situations, the object $\mathcal{M}$ behaves as if all of its mass were concentrated at its center of mass (e.g., in case of motion by translation).

The coordinates of the center of mass of an object $\mathcal{M}$ are

$$\overline{x} = \frac{1}{m} \int_{\mathcal{M}} x\rho\, d\mu, \qquad \overline{y} = \frac{1}{m} \int_{\mathcal{M}} y\rho\, d\mu, \qquad \overline{z} = \frac{1}{m} \int_{\mathcal{M}} z\rho\, d\mu,$$

where m denotes the total mass and ρ is the mass density (if $\mathcal{M}$ represents an object in $\mathbb{R}^2$, then $\overline{z} = 0$, and if it represents an object in $\mathbb{R}$, then $\overline{y} = 0$ and $\overline{z} = 0$). The integral expressions in the above formulas are called *moments*, and are denoted as follows: if $\mathcal{M} \subseteq \mathbb{R}^3$, then

$M_{yz} = \int_{\mathcal{M}} x\rho(x,y,z) d\mu$ is the moment with respect to the yz-plane;

$M_{xz} = \int_{\mathcal{M}} y\rho(x,y,z) d\mu$ is the moment with respect to the xz-plane; and

$M_{xy} = \int_{\mathcal{M}} z\rho(x,y,z) d\mu$ is the moment with respect to the xy-plane.

If $\mathcal{M} \subseteq \mathbb{R}^2$, then

$M_x = \int_{\mathcal{M}} y\rho(x,y) d\mu$ is the moment about the x-axis; and

$M_y = \int_{\mathcal{M}} x\rho(x,y) d\mu$ is the moment about the y-axis.

The formulas given here represent a generalization of the center of mass formulas (discussed in Section 1.2) for a finite number of masses to a continuous mass distribution given by the density function. The case of finitely many (say, n) masses can be recovered by replacing the integral with the sum and density by the masses involved. For example, the expression $\int_{\mathcal{M}} \rho\, d\mu$ becomes $\sum_{i=1}^n m_i$, and $\int_{\mathcal{M}} x\rho\, d\mu$ becomes $\sum_{i=1}^n x_i m_i$.

▶ EXAMPLE 7.49

Find the center of mass of a thin wire in the shape of the helix $\mathbf{c}(t) = (\cos t, \sin t, t+1)$, $t \in [0, 6\pi]$, if the density function is constant and equal to ρ.

SOLUTION

From the parametrization for $\mathbf{c}$, we get $\mathbf{c}'(t) = (-\sin t, \cos t, 1)$, and hence, $\|\mathbf{c}'(t)\| = \sqrt{2}$. The mass of the wire is computed to be

$$m = \int_{\mathbf{c}} \rho ds = \rho \int_0^{6\pi} \|\mathbf{c}'(t)\|\, dt = \rho \int_0^{6\pi} \sqrt{2}\, dt = 6\sqrt{2}\pi\rho.$$

The moments are computed similarly:

$$M_{yz} = \int_{\mathbf{c}} x\rho ds = \rho \int_0^{6\pi} \cos t \|\mathbf{c}'(t)\|\, dt = \rho\sqrt{2} \int_0^{6\pi} \cos t\, dt = 0,$$

$$M_{xz} = \int_{\mathbf{c}} y\rho ds = \rho \int_0^{6\pi} \sin t \|\mathbf{c}'(t)\|\, dt = \rho\sqrt{2} \int_0^{6\pi} \sin t\, dt = 0,$$

and

$$M_{xy} = \int_{\mathbf{c}} z\rho ds = \rho \int_0^{6\pi} (t+1)\|\mathbf{c}'(t)\|\, dt = \rho\sqrt{2} \int_0^{6\pi} (t+1)\, dt = 6\sqrt{2}\pi\rho(3\pi+1).$$

The coordinates of the center of mass are

$$\overline{x} = \frac{0}{m} = 0, \quad \overline{y} = \frac{0}{m} = 0, \quad \text{and} \quad \overline{z} = \frac{6\sqrt{2}\pi\rho(3\pi+1)}{6\sqrt{2}\pi\rho} = 3\pi + 1.$$ ◀

▶ EXAMPLE 7.50

Find the mass and center of mass of the thin plate

$$D = \{(x, y) \;\big|\; 0 \le x \le 1, x^2 \le y \le \sqrt{x}\}$$

in the xy-plane, if the density function is $\rho(x, y) = ax$, $a > 0$.

SOLUTION

The mass is

$$m = \iint_D ax\, dA = \int_0^1 \left(\int_{x^2}^{\sqrt{x}} axdy \right) dx = a \int_0^1 \left(x^{3/2} - x^3\right) dx = \frac{3a}{20}.$$

The moment M_x about the x-axis is

$$M_x = \iint_D axy\, dA = \int_0^1 \left(\int_{x^2}^{\sqrt{x}} axydy \right) dx = \frac{a}{2} \int_0^1 \left(x^2 - x^5\right) dx = \frac{a}{12}.$$

Similarly, the moment M_y about the y-axis is computed to be

$$M_y = \iint_D ax^2\, dA = \int_0^1 \left(\int_{x^2}^{\sqrt{x}} ax^2dy \right) dx = a \int_0^1 \left(x^{5/2} - x^4\right) dx = \frac{3a}{35}.$$

Hence, the coordinates of the center of mass are

$$\overline{x} = \frac{M_y}{m} = \frac{3a/35}{3a/20} = \frac{4}{7} \quad \text{and} \quad \overline{y} = \frac{M_x}{m} = \frac{a/12}{3a/20} = \frac{5}{9}.$$ ◀

▶ EXAMPLE 7.51

Find the center of mass of the solid region W of constant density a that is inside the sphere $x^2 + y^2 + z^2 = 1$ and above the cone $z^2 = x^2 + y^2$, $z \ge 0$.

SOLUTION The fact that the solid W is homogeneous (i.e., has constant density) and symmetric with respect to the z-axis implies that the center of mass lies on the z-axis. Consequently, all we have to compute are the mass of W and the moment M_{xy}. Using spherical coordinates

$$x = \rho \sin\phi \cos\theta, \qquad y = \rho \sin\phi \sin\theta, \qquad z = \rho \cos\phi$$

with $dV = \rho^2 \sin\phi \, d\rho d\phi d\theta$ [see (6.8) in Section 6.5], we compute the mass of W to be

$$m = \iiint_W a\, dV = a \int_0^{2\pi} \left(\int_0^{\pi/4} \left(\int_0^1 \rho^2 \sin\phi d\rho \right) d\phi \right) d\theta.$$

The equation of the sphere $x^2 + y^2 + z^2 = 1$ transforms into $\rho^2 = 1$, that is, $\rho = 1$; and the limits of integration of ρ are $0 \le \rho \le 1$. The equation of the cone $z^2 = x^2 + y^2$ transforms to $\rho^2 \cos^2\phi = \rho^2 \sin^2\phi$, or $\rho\cos\phi = \pm\rho\sin\phi$ and $\tan\phi = \pm 1$ (after dividing by $\rho\cos\phi$). Therefore, $\phi = \pi/4$ (by definition, $\phi \ge 0$) and the limits are $0 \le \phi \le \pi/4$. Continuing the computation, we get

$$m = a \int_0^{2\pi} \left(\int_0^{\pi/4} \frac{1}{3} \sin\phi d\phi \right) d\theta = \frac{a}{3} 2\pi \left(1 - \frac{\sqrt{2}}{2} \right) = \frac{a\pi(2 - \sqrt{2})}{3}.$$

Similarly,

$$\begin{aligned} M_{xy} &= \iiint_W az\, dV = a \int_0^{2\pi} \left(\int_0^{\pi/4} \left(\int_0^1 \rho^3 \sin\phi \cos\phi d\rho \right) d\phi \right) d\theta \\ &= a \int_0^{2\pi} \left(\int_0^{\pi/4} \frac{1}{4} \sin\phi \cos\phi d\phi \right) d\theta = \frac{a}{4} \left(\int_0^{2\pi} d\theta \right) \left(\int_0^{\pi/4} \sin\phi \cos\phi \, d\theta \right) \\ &= \frac{a}{4} 2\pi \left. \frac{\sin^2\phi}{2} \right|_0^{\pi/4} = \frac{a\pi}{8}. \end{aligned}$$

It follows that the coordinates of the center of mass are

$$\overline{x} = 0, \quad \overline{y} = 0, \quad \overline{z} = \frac{M_{xy}}{m} = \frac{a\pi/8}{a\pi(2 - \sqrt{2})/3} = \frac{3}{8(2 - \sqrt{2})}.$$ ◀

Moments of Inertia

The *moment of inertia* is a measure of the rotational inertia of a body. In other words, it is the "opposition" we feel when we try to change the speed of rotation about an axis. The formulas given below extend the definition of the moments of inertia of a single particle (which is the product of the mass and the square of the distance to the axis) to the moments of inertia of a mass distribution (density) ρ over a manifold (that could be a curve, a plane region, a surface in $\mathbb{R}^3$, or a three-dimensional solid region).

Assume that $\mathcal{M} \subseteq \mathbb{R}^2$. The *moment of inertia about the x-axis* is defined by

$$I_x = \int_{\mathcal{M}} y^2 \rho(x, y) d\mu,$$

and the *moment of inertia about the y-axis* is

$$I_y = \int_{\mathcal{M}} x^2 \rho(x, y) d\mu.$$

The *moment of inertia about the origin* (or the *polar moment of inertia*) is given by

$$I_o = \int_{\mathcal{M}} (x^2 + y^2) \rho(x, y) d\mu.$$

If $\mathcal{M} \subseteq \mathbb{R}^3$, then the *moments of inertia about the coordinate axes* are given by

$$I_x = \int_{\mathcal{M}} (y^2 + z^2)\rho(x, y, z)d\mu,$$

$$I_y = \int_{\mathcal{M}} (x^2 + z^2)\rho(x, y, z)d\mu, \quad \text{and}$$

$$I_z = \int_{\mathcal{M}} (x^2 + y^2)\rho(x, y, z)d\mu.$$

▶ EXAMPLE 7.52

Compute the moment of inertia of the homogeneous thin-walled torus S given by (assume that the density is $\rho = 1$)

$$\mathbf{r}(u, v) = ((R + \cos v)\cos u, (R + \cos v)\sin u, \sin v), \qquad 0 \le u, v \le 2\pi,$$

with respect to the z-axis.

SOLUTION By definition,

$$I_z = \iint_S 1 \cdot (x^2 + y^2)\,dS = \iint_{[0,2\pi]\times[0,2\pi]} (x^2 + y^2)\|\mathbf{N}\|\,dA$$

[the surface normal $\mathbf{N}$ was computed in Example 7.33 (with the inner radius $\rho = 1$) to be $\|\mathbf{N}\| = R + \cos v$; ρ in that example denotes the inner radius of the torus and has nothing to do with the density ρ used in this section]. Continuing the computation, we get

$$I_z = \int_0^{2\pi} \left(\int_0^{2\pi} (R + \cos v)^3 du \right) dv$$

$$= \left(\int_0^{2\pi} du \right) \left(\int_0^{2\pi} (R^3 + 3R^2 \cos v + 3R\cos^2 v + \cos^3 v)\,dv \right)$$

[use $\cos^2 v = 1/2 + (\cos 2v)/2$ for the third integral and $\int \cos^3 v dv = \int \cos^2 v \cos v dv = \int (1 - \sin^2 v)\cos v dv = \int \cos v dv - \int \sin^2 v \cos v dv = \sin v - \sin^3 v/3$ for the fourth integral]

$$= 2\pi \left(R^3 v + 3R^2 \sin v + \frac{3R}{2}\left(v + \frac{1}{2}\sin 2v \right) + \sin v - \frac{\sin^3 v}{3} \right)\Bigg|_0^{2\pi}$$

$$= 2\pi \left(R^3 2\pi + \frac{3R}{2} 2\pi \right) = 2\pi^2 R(2R^2 + 3).$$

Therefore, the moment of inertia of the thin-walled homogeneous torus S about the z-axis is $2\pi^2 R(2R^2 + 3)$. ◀

▶ EXAMPLE 7.53

Compute the moment of inertia about the y-axis of the solid cylinder $0 \le x^2 + y^2 \le 4$, $0 \le z \le 1$ of constant density $a = 1$.

SOLUTION The moment of inertia about the y-axis is $I_y = \iiint_W (x^2+z^2)a\,dV$. Using cylindrical coordinates $x = r\cos\theta$, $y = r\sin\theta$, $z = z$, with $dV = r dz dr d\theta$ (see Example 6.43), we get

$$\begin{aligned} I_y &= \int_0^{2\pi}\left(\int_0^2\left(\int_0^1 (r^2\cos^2\theta + z^2) r dz\right) dr\right) d\theta \\ &= \int_0^{2\pi}\left(\int_0^2 \left(r^3 z\cos^2\theta + \tfrac{1}{3} r z^3\right)\Big|_0^1 dr\right) d\theta = \int_0^{2\pi}\left(\int_0^2 \left(r^3\cos^2\theta + \tfrac{1}{3}r\right) dr\right) d\theta \\ &= \int_0^{2\pi} \left(\tfrac{1}{4}r^4\cos^2\theta + \tfrac{1}{6}r^2\right)\Big|_0^2 d\theta = \int_0^{2\pi}\left(4\cos^2\theta + \tfrac{2}{3}\right) d\theta \\ &= \left(4\left(\tfrac{1}{2}\theta + \tfrac{1}{4}\sin 2\theta\right) + \tfrac{2}{3}\theta\right)\Big|_0^{2\pi} = \tfrac{16}{3}\pi. \end{aligned}$$

The desired moment of inertia is $I_y = 16\pi/3 \approx 16.76$. ◀

▶ EXAMPLE 7.54 Steiner's Theorem

The moment of inertia of an object $\mathcal{M}$ with respect to an axis ℓ is defined by the integral $I_\ell = \int_{\mathcal{M}} \rho d^2 d\mu$, where d denotes the (perpendicular) distance from a point on $\mathcal{M}$ to the axis ℓ and ρ is the density of $\mathcal{M}$. Let $I_{C\mathcal{M}}$ be the moment of inertia of $\mathcal{M}$ with respect to an axis $\ell_{C\mathcal{M}}$ parallel to ℓ that goes through its center of mass. Prove *Steiner's Theorem,* which states that the moment of inertia with respect to the line ℓ parallel to $\ell_{C\mathcal{M}}$ is $I_\ell = k^2 m + I_{C\mathcal{M}}$, where k is the distance between ℓ and $\ell_{C\mathcal{M}}$, and m is the mass of $\mathcal{M}$; see Figure 7.58.

SOLUTION Assume, for simplicity, that the mass m is contained in $\mathbb{R}^2$ and that the y-axis coincides with the axis $\ell_{C\mathcal{M}}$. Then

$$\begin{aligned} I_\ell &= \iint_D \rho d^2\, dA = \iint_D \rho (k-x)^2\, dA \\ &= \iint_D \left(\rho k^2 - 2\rho k x + \rho x^2\right) dA \\ &= k^2 \iint_D \rho\, dA - 2k\iint_D \rho x\, dA + \iint_D \rho x^2\, dA. \end{aligned}$$

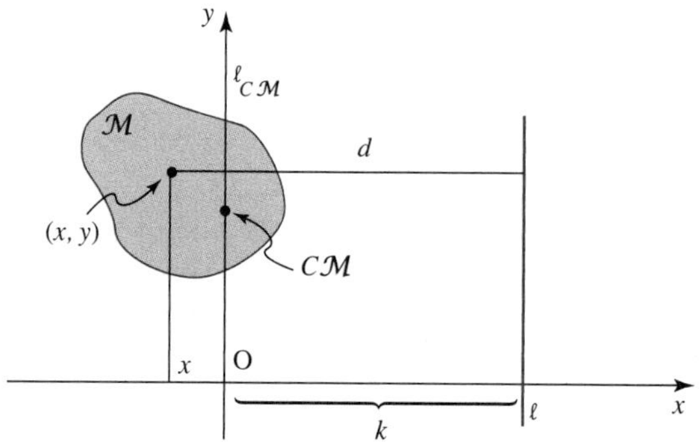

Figure 7.58 Moment of inertia with respect to the line ℓ.

The first integral is the product of k^2 and the integral of the density (which is the mass), that is, $k^2 m$. The third one is the moment of inertia with respect to the y-axis, or $I_y = I_{\mathcal{CM}}$ in our notation (the y-axis is the same as $\ell_{\mathcal{CM}}$). The second integral is equal to

$$-2k \iint_D \rho x \, dA = -2km \frac{1}{m} \iint_D \rho x \, dA = -2km\overline{x},$$

where $\overline{x}$ denotes the x-coordinate of the center of mass. By our choice, the center of mass lies on the y-axis and hence $\overline{x} = 0$. Therefore, the second integral vanishes, and

$$I_\ell = k^2 m + I_{\mathcal{CM}}.$$

▶ EXERCISES 7.5

1. Compute the average pressure on the hemispherical surface $x^2 + y^2 + z^2 = a^2$, $a \le 1$, $z \ge 0$, if the pressure is given by $p(x, y, z) = 1 - z^2$.

2. Let D be a disk of radius r and let $f(x, y)$ be the distance function from the point (x, y) to the origin. Find the average value of f over D.

3. Let S be the part of the cone $z^2 = x^2 + y^2$ such that $1 \le z \le 4$. Compute the average distance from a point on the cone to the z-axis.

4. Find the mass of a thin wire in the form of the helix $\mathbf{c}(t) = \sin t\mathbf{i} - \cos t\mathbf{j} + 4t\mathbf{k}$, $\pi \le t \le 6\pi$, made of a material whose density per unit length is given by $\rho(x, y, z) = 3(y^2 + z^2)$.

5. Find the total charge contained in the tetrahedron with vertices $(0, 0, 0)$, $(1, 0, 0)$, $(0, 3, 0)$, and $(0, 0, 4)$, if the charge density is $\rho(x, y, z) = 2x^2 + y$.

6. Find the mass of a metal sheet in the form of the paraboloid $z = x^2 + y^2$, $2 \le z \le 3$, if the density is given by $\rho(x, y, z) = 3z$.

7. Find the center of mass of a uniform wire of density ρ, in the form of the semicircle $x^2 + y^2 = a^2$, $y \ge 0$.

8. Find the mass and center of mass of a uniform wire of density ρ in the form of the helix $\mathbf{c}(t) = (2\cos t, 2\sin t, 3t)$, $t \in [0, 6\pi]$.

9. Find the mass and center of mass of a wire shaped as a quarter circle $x^2 + y^2 = a^2$, $a > 0$, $x, y \ge 0$, with the density $\rho = bx$ (b is a constant).

10. Find the moment of inertia about the z-axis of the homogeneous thin-walled torus of density ρ generated by revolving the circle centered at $(c, 0, 0)$ of radius a ($a < c$) in the xz-plane around the z-axis.

11. Find the center of mass and moment of inertia about the z-axis of the surface of the paraboloid $z = 2(x^2 + y^2)$, $0 \le z \le 8$ of constant density ρ.

12. Find the polar moment of inertia of the spherical surface $x^2 + y^2 + z^2 = a^2$ of constant density ρ.

13. Find the coordinates of the center of mass of the solid upper hemisphere V given by $x^2 + y^2 + z^2 \le a^2$, $z \ge 0$, of constant density $\rho(x, y, z) = b$.

14. Find the mass of a thin-walled metal funnel in the shape of the cone $z^2 = 4(x^2 + y^2)$, $1 \le z \le 4$, with the density $\rho(x, y, z) = 5 - z$.

15. Assume that V is a solid region filled with a liquid and S is a surface in contact with the liquid. The magnitude of the hydrostatic force exerted by the liquid is given by the formula $F = \iint_S \mu(z_o - z)\, dS$, where μ is the specific weight of the liquid (the units are weight/volume) and z_o is the z-coordinate of the highest point in V. Assume that a tank in the shape of a hemisphere of radius a with a flat bottom is filled with a liquid of constant specific weight μ. Compute the magnitude of the hydrostatic force on the tank.

16. Assume that the tank from Exercise 15 is turned upside down. Compute the magnitude of the hydrostatic force on the tank.

17. Compute the magnitude of the hydrostatic force (see Exercise 15) on a cylindrical tank with a flat top and bottom of radius 2 m and height 10 m. Assume that the liquid has a constant specific weight μ.

18. Compute the magnitude of the hydrostatic force (see Exercise 15) on a cylindrical tank with a hemispherical top and flat bottom, if the cylindrical part of it has radius 2 m and height 10 m. Assume that the liquid has a constant specific weight μ.

19. Find a function f and curve $\mathbf{c}$ in $\mathbb{R}^3$ so that $\left|\int_{\mathbf{c}} f\,ds\right| < \int_{\mathbf{c}} |f|$.

20. Using $\ln(1+u) \le u$ (which holds for $u \ge 0$), estimate $\iint_D \ln\left(1 + \frac{x}{2} + \frac{y}{3}\right) dA$, where D is the triangle in the first quadrant bounded by the line $y = -x + 3$.

21. Estimate $\iint_S (z^2 + 1)\,dS$ from above and below, where S is the upper hemisphere $x^2 + y^2 + z^2 = 1$, $z \ge 0$.

22. Estimate $\iiint_V (x^2 + y^2 + z^2)^{-1}\,dV$ from above and below, where V is the rectangular box $0 \le x, y \le 3$, $2 \le z \le 4$.

23. Estimate $\int_{\mathbf{c}} e^{\sin(x+y)}\,ds$, where $\mathbf{c}$ is the helix $\mathbf{c}(t) = (2\cos t, 2\sin t, t+2)$, $\pi \le t \le 4\pi$.

24. Find the center of mass of the rectangle $R = [0,1] \times [-1,1]$ in $\mathbb{R}^3$ whose density is given by $\rho(x,y) = (1 + y^2)$. Using Steiner's Theorem, find the moment of inertia of R about the line $y = 2$.

▶ CHAPTER REVIEW

CHAPTER SUMMARY

- **Parametrized surfaces and surfaces.** Continuous, differentiable, and C^1 surface, piecewise differentiable, piecewise C^1 surface, smooth parametrization and smooth surface, tangent and normal vectors, tangent plane, orientation of a surface.
- **Surfaces.** Graph of a function of two variables, parametrized surface, surface defined implicitly, various examples, such as plane, cylinder, cone, sphere, surfaces of revolution, torus, Möbius band, helicoid, quadratic (quadric) surfaces.
- **Surface integral of a real-valued function.** Definition, surface area, independence on parametrization, computing surface integrals, area cosine principle.
- **Surface integral of a vector field.** Definition, outward flux, dependence on parametrization, computing surface integrals.
- **Properties and applications.** Fundamental Theorem of Calculus, length, area, and volume, average value, properties of integrals, total mass and total charge, center of mass, moments of inertia, Steiner's Theorem.

REVIEW

Discuss the following questions.

1. Describe how to parametrize the curve $\mathbf{c}$ that is the graph of $y = f(x)$, $x \in [a,b]$, and the surface S that is the graph of $z = f(x,y)$, $(x,y) \in D \subseteq \mathbb{R}^2$. Define the positive orientation of S.

2. If you cut a Möbius strip along its middle circle, how many surfaces do you get, and are they one-sided or two-sided?

3. Define the surface integral of a real-valued function f. Write down a parametrization of a surface S that is the graph of a continuous function $z = g(x, y)$ and an expression for the surface integral of f over S.

4. Explain the Area Cosine Principle. Suppose that you are sitting high on the z-axis and looking at a plane region directly below you. If you see a square of side 1, and the actual plane region is a rectangle with sides 1 and 2, under what angle with respect to the xy-plane was the rectangle placed?

5. Is it true that the surface integral of a constant vector field $\mathbf{F}$ over a sphere is zero? Explain your answer.

6. Suppose that a plate in the shape of the disk $\{(x, y)|x^2 + y^2 \leq 2\}$ is made of several materials with different densities. Is it still possible for its center $(0, 0)$ to be the center of mass?

7. What is the flux of $\mathbf{F} = \mathbf{i} + \mathbf{j}$ across the surface of the cube in the first octant bounded by $x = 1$, $y = 1$, and $z = 1$?

8. Describe the ellipsoid $x^2 + y^2/2 + z^2/2 = 1$ as a surface of revolution, and write down its parametrization. Set up a formula for its surface area. Set up a formula for its volume (it is a solid of revolution!).

9. What is the flux of $\mathbf{F} = \mathbf{r}/\|\mathbf{r}\|$ out of the sphere $x^2 + y^2 + z^2 = 1$? Determine the sign of the flux of $\mathbf{F} = \mathbf{r}/\|\mathbf{r}\|$ through the sphere $x^2 + y^2 + (z - 2)^2 = 1$, oriented by the outward normal.

TRUE/FALSE QUIZ

Determine whether the following statements are true or false. Give reasons for your answer.

1. The image of $\mathbf{r}(u, v) = (u, u, v)$, $u, v \in \mathbb{R}$, is a plane perpendicular to the xy-plane.

2. $\mathbf{r}(u, v) = (\cos u, v, \sin u)$, $0 \leq u \leq 2\pi$, $v \in \mathbb{R}$, parametrizes the cylinder $x^2 + z^2 = 1$.

3. Surface normal vectors to the cylinder $x^2 + z^2 = 1$ are parallel to the y-axis.

4. The plane tangent to the surface $\mathbf{r}(u, v) = (u^2 + v^2, u^2 - 2v^2, v)$ at $(1, 1, 0)$ is parallel to the xy-plane.

5. If S is a sphere of radius 1, then $\iint_S dS = 4\pi$.

6. $\iint_S dS = \pi$, where S is the part of the plane $z = 0$ inside the ellipse $x^2 + y^2/4 = 1$.

7. If $\iint_S d\mathbf{S} = 2$, then $\iint_{S^*} d\mathbf{S}^* = -2$, where S and S^* are the same surface, except that the orientation of S^* is opposite to that of S.

8. The flux of $\mathbf{F}(x, y, z) = \mathbf{i} + \mathbf{j}$ through the disk $x^2 + y^2 \leq 4$, $z = 4$, is zero.

9. The flux of $\mathbf{F}(x, y, z) = x^2\mathbf{i}$ through any surface is positive.

10. The surface integral of a vector field is a vector field.

11. Assume that the rectangles S_1 and S_2 have the same area. Then $\iint_{S_1} \mathbf{F} \cdot d\mathbf{S} = \iint_{S_2} \mathbf{F} \cdot d\mathbf{S}$ for the vector field $\mathbf{F}(x, y, z) = \mathbf{i}$.

12. Let $S = \{(x, y, z)|x^2 + y^2 \leq 1, z = 1\}$. If $\mathbf{F}$ is parallel to the xy-plane, then $\iint_S \mathbf{F} \cdot d\mathbf{S} = 0$.

13. Let $S = \{(x, y, z)|x^2 + y^2 \leq 1, z = 1\}$. If $\iint_S \mathbf{F} \cdot d\mathbf{S} = 0$, then $\mathbf{F}$ must be parallel to the xy-plane.

14. If $f(x, y, z) > 0$ for all points on a surface S, then $\iint_S f\, dS > 0$.

15. If W is the ball $x^2 + y^2 + z^2 \le 4$, then $\iiint_W dV = 16\pi$.

16. If $f(x) > 0$ for all x in $[a, b]$, then $|\int_a^b f(x)dx| = \int_a^b |f(x)|dx$.

REVIEW EXERCISES AND PROBLEMS

1. Describe the ellipsoid $x^2/a^2 + y^2/a^2 + z^2/c^2 = 1$ as a surface of revolution and write down its parametric representation.

2. Using the fact that $\cosh^2 t - \sinh^2 t = 1$ for all t, write down the parametrization of the hyperboloid $x^2/a^2 + y^2/b^2 - z^2/c^2 = 1$.

3. Find a function f and surface S in $\mathbb{R}^3$ so that $\left|\iint_S f\,dS\right| < \iint_S |f|\,dS$.

4. Let f be a continuous function defined on a region D that is divided into n mutually disjoint regions $D_1, \ldots, D_n$. Find the average of f over D, if the areas of $D_1, \ldots, D_n$ and the averages of f over each D_i, $i = 1, \ldots, n$ are known.

5. The formula $\mathbf{E}(\mathbf{r}) = \frac{Q}{4\pi\epsilon_0}\frac{\mathbf{r}}{\|\mathbf{r}\|^3}$ describes the electrostatic field $\mathbf{E}(\mathbf{r})$ at the point $\mathbf{r} \neq \mathbf{0}$ due to a charge Q placed at the origin (see Example 2.12 in Section 2.1).

(a) Compute the flux of $\mathbf{E}$ through the surface of the cylinder $x^2 + y^2 = R^2$, $-H \le z \le H$, oriented by the outward-pointing normal.

(b) Show that the flux of $\mathbf{E}$ through the closed cylinder [i.e., it consists of the surface from (a), together with the two disks that close it from above and below], oriented by the outward-pointing normal, is equal to Q/ϵ_0. Compare with Gauss' Law (Example 8.12 in Section 8.2).

6. Using the Implicit Function Theorem, we showed that, near a point (x_0, y_0, z_0), the level surface $F(x, y, z) = 0$ of a C^1 function F is the graph of a C^1 function of two variables, provided that $\nabla F(x_0, y_0, z_0) \neq \mathbf{0}$.

(a) Show that a C^1 surface $\mathbf{r}(u, v)$ has the same property: namely, near (u_0, v_0), it is the graph of a C^1 function of two variables, if $\mathbf{T}_u(u_0, v_0) \times \mathbf{T}_v(u_0, v_0) \neq \mathbf{0}$. (Thus, our use of the word "surface" in these different contexts has been justified.)

(b) From (a) we conlcude that if the z-component of $\mathbf{T}_u(u_0, v_0) \times \mathbf{T}_v(u_0, v_0)$ is nonzero, then the image of $\mathbf{r}$ near $\mathbf{r}(u_0, v_0)$ is the graph of some function $z = g(x, y)$. Let (x_0, y_0) be the point in the domain of g such that $(x_0, y_0, g(x_0, y_0)) = \mathbf{r}(u_0, v_0)$. Show that the tangent plane at $\mathbf{r}(u_0, v_0)$ [as defined in this chapter, that is, the plane spanned by $\mathbf{T}_u(u_0, v_0)$ and $\mathbf{T}_v(u_0, v_0)$] is the same as the tangent to the graph of $z = g(x, y)$ at $(x_0, y_0, g(x_0, y_0))$.

7. Consider the hyperbolic paraboloid $x^2/a^2 - y^2/b^2 - z/c = 0$ of Example 7.27 in Section 7.2. Compute the level curves $z = z_0$, and compute the curves that are intersections of the paraboloid with the xz-plane and yz-plane. Using a graphing device, sketch the graph of the paraboloid and identify the curves that you obtained. Find a parametrization of the paraboloid that makes it a smooth surface.

8. Compute the surface area of the helicoid of Example 7.21 in Section 7.2.

9. The velocity vector field of a fluid is given by $\mathbf{F}(x, y, z) = 2x\mathbf{i} + y\mathbf{k}$ m/s. Find the volume of fluid (in cubic meters per second) that flows through the surface $x + 2y + 3z = 6$ in the first octant, oriented by the upward-pointing normal.

10. Find the total flux of $\mathbf{F}(x, y, z) = x\mathbf{i} + y\mathbf{j} + z\mathbf{k}$ out of the closed cylinder of radius 4, whose axis of symmetry is the x-axis, where $0 \le x \le 10$.

11. Let S be a level surface of a function $f\colon \mathbb{R}^3 \to \mathbb{R}$ and D its projection onto the xy-plane. Find an expression for the surface area of S. Then use it to find a formula for the surface area of the graph of $z = g(x, y)$, $(x, y) \in D$.

12. Find an example of a non-constant function defined on the homogeneous disk $D = \{(x, y) | x^2 + y^2 \leq 1\}$ with the property that its average value is attained at the center of mass of D.

13. Let S be a surface in $\mathbb{R}^3$ parametrized by $\mathbf{r}(u, v)\colon D \subseteq \mathbb{R}^2 \to \mathbb{R}^3$, and let $E = \|\partial\mathbf{r}/\partial u\|^2$, $F = (\partial\mathbf{r}/\partial u) \cdot (\partial\mathbf{r}/\partial v)$, and $G = \|\partial\mathbf{r}/\partial v\|^2$. Prove that $\iint_S dS = \iint_D (EG - F^2)\, dA$. The quantities E, F, and G are said to form the *Second Fundamental Form* of S and are used to investigate geometric properties of S.

14. Estimate $\iiint_V (\sin(zx) + e^{x^2+y})\, dV$ from above and below, where V is the rectangular box $-1 \leq x, y \leq 1, 0 \leq z \leq 4\pi$.

15. Find the moments of inertia about the coordinate axes of a wire in the shape of the curve $\mathbf{c}(t) = (t^2, 1, -t^2)$, $t \in [0, 1]$, with density $\rho(x, y, z) = y + 1$.

CHAPTER 8

Classical Integration Theorems of Vector Calculus

The Fundamental Theorem of Calculus states that the definite integral of the derivative of a function depends not on the values of the function on the whole interval of integration but only on the values at the endpoints. We generalized this important result to the integral of a gradient vector field along any curve in a plane or in space.

In this chapter, we continue our investigation of the relation between the concepts of integration and differentiation. The results, contained in the theorems of Green, Gauss, and Stokes (the so-called Classical Integration Theorems of Vector Calculus), are all variations on the same theme applied to different types of integration. Green's Theorem relates the path integral of a vector field along an oriented, simple closed curve in the xy-plane to the double integral of its derivative (to be precise, the *curl*) over the region enclosed by that curve. Gauss' Divergence Theorem extends this result to closed surfaces and Stokes' Theorem generalizes it to simple closed curves in space. Several versions of these theorems are presented, together with a number of worked examples and applications. In particular, we obtain physical interpretations of divergence and curl.

Next, we introduce differential forms, investigate basic properties, and study the integration of forms. Besides various uses in applications, we show how the three classical integration theorems become special cases of an integration theorem for differential forms.

The last two sections are devoted to applications of vector calculus in electromagnetism and fluid flow. The emphasis is placed not on explaining the details of the theory but on identifying physical quantities involved as mathematical objects and showing how to use calculus in manipulating them to obtain meaningful results. The guiding idea is to explain, line by line, all the details and intricacies of mathematical arguments, so that, when studying a text in electromagnetism or fluid flow, the reader will be prepared to smoothly cover the mathematical side and concentrate on understanding the physics of it.

▶ 8.1 GREEN'S THEOREM

One of the theorems that establishes a relationship between the concepts of the derivative and the integral is Green's Theorem. It states that the integral of a vector field along a closed curve in a plane is equal to the integral of the derivative of that vector field over the two-dimensional region enclosed by the curve.

Before stating the theorem, we will identify what curves and regions are involved and then consider some special cases.

ASSUMPTION 8.1 Regions Involved in Green's Theorem

Assume that a region $D \subseteq \mathbb{R}^2$ is either an elementary region (those were defined in Section 6.2) or the union $D = D_1 \cup D_2 \cup \cdots \cup D_n$ of elementary regions D_i, $i = 1, \ldots, n$, such that the most any two of them can have in common is their (common) boundary curve (or boundary curves).

The boundary ∂D of D consists of a finite number of simple closed curves that are oriented by the following rule: if we walk along the boundary curve in the positive direction (i.e., we follow the positive orientation), then the region is on our left. ◀

Let us consider a few examples. The disk $D = \{(x, y) | x^2 + y^2 \leq 1\}$ is an elementary region (to be precise: a region of type 3), whose boundary is the circle $\partial D = \{(x, y) | x^2 + y^2 = 1\}$ oriented counterclockwise (that is the positive orientation).

The annulus $D = \{(x, y) | 1 \leq x^2 + y^2 \leq 4\}$ is not an elementary region, but can be divided into four pieces, each of which is an elementary region; see Figure 8.1(a).

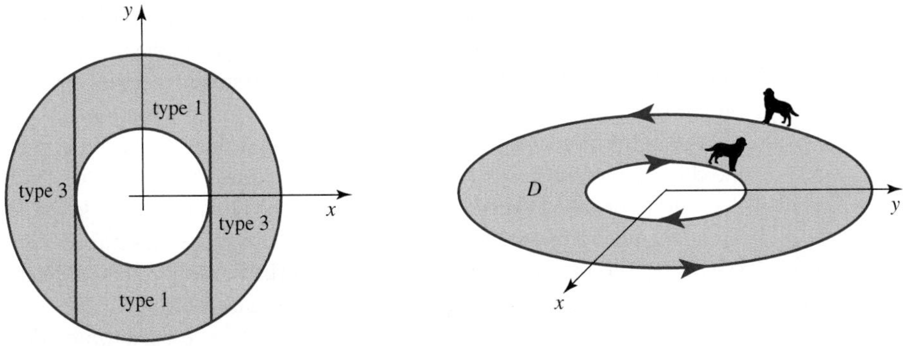

(a) D is a union of elementary regions. (b) Positive orientation of the boundary ∂D.

Figure 8.1 Annulus $D = \{(x, y) | 1 \leq x^2 + y^2 \leq 4\}$.

The boundary ∂D of D consists of two circles (of radii 1 and 2), and the positive orientation is given by the counterclockwise orientation of the outer circle and the clockwise orientation of the inner one, as shown in Figure 8.1(b).

Next, we consider several special cases.

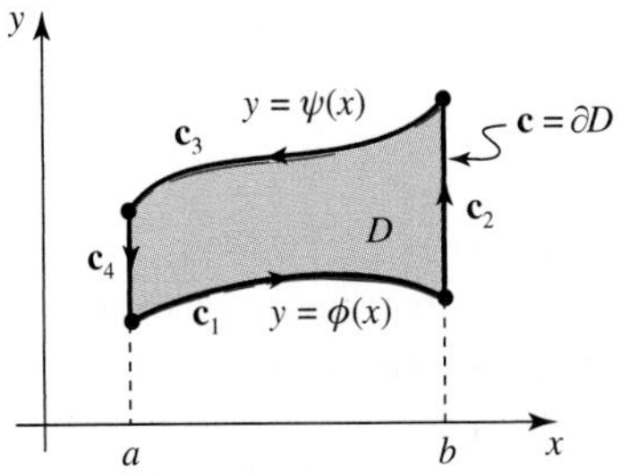

Figure 8.2 Type-1 region D.

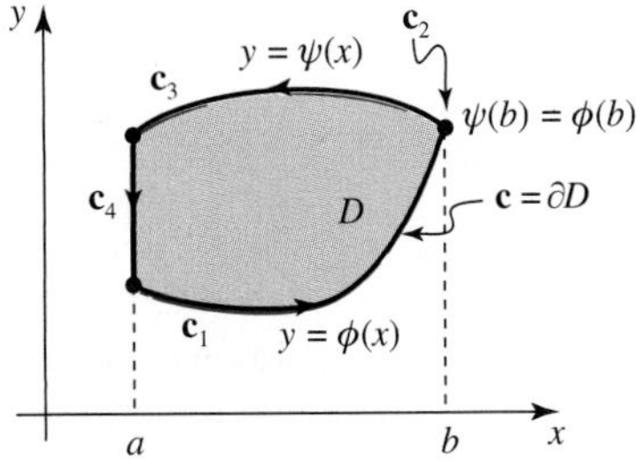

Figure 8.3 Type-1 region D.

Let D be a region of type 1 and let $\mathbf{c} = \partial D$ be its positively oriented boundary; see Figures 8.2 and 8.3. The curve $\mathbf{c}$ is piecewise smooth. It consists of the graphs $\mathbf{c}_1$ and $\mathbf{c}_3$ of continuous functions $y = \phi(x)$ and $y = \psi(x)$ defined on $[a, b]$ and at most two vertical segments: along $x = a$ from $\psi(a)$ to $\phi(a)$ [call it $\mathbf{c}_4$; if $\psi(a) = \phi(a)$, then $\mathbf{c}_4$ collapses to a point] and along $x = b$ from $\phi(b)$ to $\psi(b)$ [call it $\mathbf{c}_2$; if $\psi(b) = \phi(b)$, then $\mathbf{c}_2$ becomes a point; see Figure 8.3].

Take the vector field $\mathbf{F}_1(x, y) = (P(x, y), 0)$, where $P(x, y)$ is a C^1 function, and compute its integral along the boundary curve $\mathbf{c}$:

$$\int_{\mathbf{c}} \mathbf{F}_1 \cdot d\mathbf{s} = \int_{\mathbf{c}_1} \mathbf{F}_1 \cdot d\mathbf{s} + \int_{\mathbf{c}_2} \mathbf{F}_1 \cdot d\mathbf{s} + \int_{\mathbf{c}_3} \mathbf{F}_1 \cdot d\mathbf{s} + \int_{\mathbf{c}_4} \mathbf{F}_1 \cdot d\mathbf{s}.$$

Parametrize $\mathbf{c}_1$ by $\mathbf{c}_1(t) = (t, \phi(t))$, $t \in [a, b]$. Then $\mathbf{c}_1'(t) = (1, \phi'(t))$ and

$$\begin{aligned} \int_{\mathbf{c}_1} \mathbf{F}_1 \cdot d\mathbf{s} &= \int_a^b \mathbf{F}_1(\mathbf{c}_1(t)) \cdot \mathbf{c}_1'(t)\, dt \\ &= \int_a^b (P(t, \phi(t)), 0) \cdot (1, \phi'(t))\, dt = \int_a^b P(t, \phi(t))\, dt. \end{aligned}$$

Parametrize the line segment (if not a point) $\mathbf{c}_2$ by $\mathbf{c}_2(t) = (b, t)$, $t \in [\phi(b), \psi(b)]$. Then $\mathbf{c}_2'(t) = (0, 1)$ and

$$\begin{aligned} \int_{\mathbf{c}_2} \mathbf{F}_1 \cdot d\mathbf{s} &= \int_{\phi(b)}^{\psi(b)} \mathbf{F}_1(\mathbf{c}_2(t)) \cdot \mathbf{c}_2'(t)\, dt \\ &= \int_{\phi(b)}^{\psi(b)} (P(b, t), 0) \cdot (0, 1)\, dt = \int_{\phi(b)}^{\psi(b)} 0\, dt = 0. \end{aligned}$$

An analogous computation would show that $\int_{\mathbf{c}_4} \mathbf{F}_1 \cdot d\mathbf{s} = 0$ (clearly, it does not matter whether $\mathbf{c}_2$ and/or $\mathbf{c}_4$ are segments or points—in any case, the integrals are zero).

Now consider the parametrization $\overline{\mathbf{c}}_3(t) = (t, \psi(t))$, $t \in [a, b]$. The orientation of $\overline{\mathbf{c}}_3$ is not what we need [it goes from $(a, \psi(a))$ to $(b, \psi(b))$], but we will fix it. Replacing $\phi(t)$ by $\psi(t)$ in the evaluation of $\int_{\mathbf{c}_1} \mathbf{F}_1 \cdot d\mathbf{s}$, we get

$$\int_{\overline{\mathbf{c}}_3} \mathbf{F}_1 \cdot d\mathbf{s} = \int_a^b P(t, \psi(t))\, dt.$$

Therefore,

$$\int_{\mathbf{c}_3} \mathbf{F}_1 \cdot d\mathbf{s} = -\int_{\overline{\mathbf{c}}_3} \mathbf{F}_1 \cdot d\mathbf{s} = -\int_a^b P(t, \psi(t))\, dt,$$

and the integral of $\mathbf{F}_1$ along $\mathbf{c}$ is

$$\int_{\mathbf{c}} \mathbf{F}_1 \cdot d\mathbf{s} = \int_{\mathbf{c}_1} \mathbf{F}_1 \cdot d\mathbf{s} + \int_{\mathbf{c}_3} \mathbf{F}_1 \cdot d\mathbf{s} = \int_a^b \big(P(t, \phi(t)) - P(t, \psi(t))\big)\, dt.$$

Rewrite the above integral using x instead of t:

$$\int_{\mathbf{c}} \mathbf{F}_1 \cdot d\mathbf{s} = \int_a^b (P(x, \phi(x)) - P(x, \psi(x)))\, dx, \tag{8.1}$$

and consider the difference $P(x, \phi(x)) - P(x, \psi(x))$ in the integrand. The Fundamental Theorem of Calculus $\int_a^b f'(y)dy = f(b) - f(a)$, applied to the function $f(y) = P(x, y)$, reads

$$\int_a^b \frac{\partial}{\partial y} P(x, y)\, dy = P(x, b) - P(x, a).$$

Replacing a by $\psi(x)$ and b by $\phi(x)$ in the above formula and reading it from right to left, we get

$$P(x, \phi(x)) - P(x, \psi(x)) = \int_{\psi(x)}^{\phi(x)} \frac{\partial}{\partial y} P(x, y)\, dy,$$

and that is exactly what we need! Continuing from (8.1),we get

$$\begin{aligned}\int_{\mathbf{c}} \mathbf{F}_1 \cdot d\mathbf{s} &= \int_a^b \big(P(x, \phi(x)) - P(x, \psi(x))\big)\, dx \\ &= \int_a^b \left(\int_{\psi(x)}^{\phi(x)} \frac{\partial P(x, y)}{\partial y} dy \right) dx = -\int_a^b \left(\int_{\phi(x)}^{\psi(x)} \frac{\partial P(x, y)}{\partial y} dy \right) dx.\end{aligned}$$

The expression we obtained is the iterated integral of $\partial P/\partial y$ over the region D. Hence (dropping variables x and y),

$$\int_{\mathbf{c}} \mathbf{F}_1 \cdot d\mathbf{s} = -\iint_D \frac{\partial P}{\partial y}\, dA, \tag{8.2}$$

where $\mathbf{F}_1(x, y) = (P(x, y), 0)$.

Next, consider a type-2 region D with positively oriented boundary $\mathbf{c} = \partial D$ and integrate the vector field $\mathbf{F}_2(x, y) = (0, Q(x, y))$ along $\mathbf{c}$ [here, $Q(x, y)$ is a C^1 function of x and y]. Proceeding as in the previous case, we obtain

$$\int_{\mathbf{c}} \mathbf{F}_2 \cdot d\mathbf{s} = \iint_D \frac{\partial Q}{\partial x}\, dA. \tag{8.3}$$

The details of the computation are left to the reader (see Exercise 8).

Finally, consider a type-3 region D, its positively oriented boundary curve $\mathbf{c}$, and a C^1 vector field $\mathbf{F}(x, y) = \mathbf{F}_1(x, y) + \mathbf{F}_2(x, y) = (P(x, y), Q(x, y))$. Adding up the formulas (8.2) and (8.3) for the path integrals of $\mathbf{F}_1$ and $\mathbf{F}_2$, we get

$$\int_{\mathbf{c}} \mathbf{F} \cdot d\mathbf{s} = \int_{\mathbf{c}} \mathbf{F}_1 \cdot d\mathbf{s} + \int_{\mathbf{c}} \mathbf{F}_2 \cdot d\mathbf{s} = \iint_D \left(\frac{\partial Q}{\partial x} - \frac{\partial P}{\partial y} \right) dA.$$

THEOREM 8.1 Green's Theorem

Let D be a region in $\mathbb{R}^2$ that satisfies Assumption 8.1, and let $\mathbf{c} = \partial D$ be its positively oriented boundary. If $\mathbf{F}(x, y) = (P(x, y), Q(x, y))$ is a C^1 vector field on D, then

$$\int_{\mathbf{c}} \mathbf{F} \cdot d\mathbf{s} = \iint_D \left(\frac{\partial Q(x, y)}{\partial x} - \frac{\partial P(x, y)}{\partial y} \right) dA.$$

Note that the integrand on the right side is the scalar curl of $\mathbf{F}$.

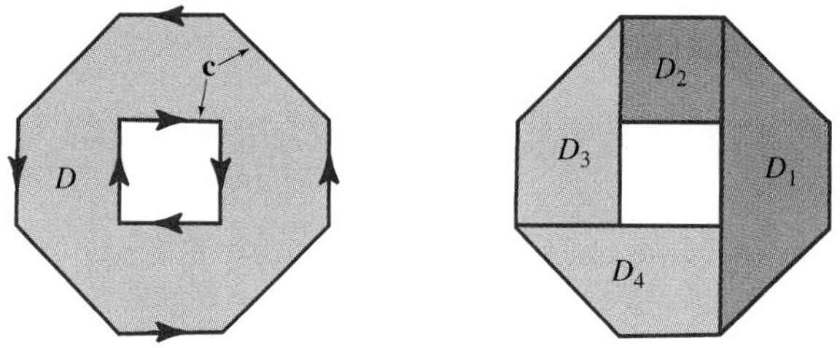

Figure 8.4 Region D divided into four type-3 regions.

In numerous situations (not just in applications), we use the notation $\int_{\mathbf{c}} Pdx + Qdy$ for the path integral $\int_{\mathbf{c}} \mathbf{F} \cdot d\mathbf{s}$ of the vector field $\mathbf{F} = (P, Q)$ along a curve $\mathbf{c}$ [see formula (5.6) in Section 5.3 and the text following it]. With this in mind, we write Green's Theorem in a (probably) more familiar form as

$$\int_{\mathbf{c}} Pdx + Qdy = \iint_D \left(\frac{\partial Q}{\partial x} - \frac{\partial P}{\partial y} \right) dA. \tag{8.4}$$

PROOF OF THEOREM 8.1: We are not going to prove the theorem in its full generality, that is, for an arbitrary region D satisfying Assumption 8.1. The discussion preceding the statement of the theorem actually serves as a proof in the case where D is a region of type 3.

To give the flavor of the general case, consider a region D that is not an elementary region, but can be broken into a union of type-3 regions that satisfy Assumption 8.1; see Figure 8.4. (*Note:* There are many ways to divide a given region into type-3 regions.) Let $\mathbf{c} = \partial D$ be the positively oriented boundary of D.

Consider the regions D_1 and D_2. Orienting the boundary ∂D_1 by our convention, as shown in Figure 8.5(a), and applying Green's Theorem for type-3 regions, we get

$$\int_{\partial D_1} \mathbf{F} \cdot d\mathbf{s} = \iint_{D_1} \left(\frac{\partial Q}{\partial x} - \frac{\partial P}{\partial y} \right) dA,$$

that is, (by $\boldsymbol{\gamma}$ we denote the curve that is the common boundary of D_1 and D_2),

$$\int_{\mathbf{c}_1} \mathbf{F} \cdot d\mathbf{s} + \int_{\boldsymbol{\gamma}} \mathbf{F} \cdot d\mathbf{s} = \iint_{D_1} \left(\frac{\partial Q}{\partial x} - \frac{\partial P}{\partial y} \right) dA. \tag{8.5}$$

The orientation convention for ∂D_2 requires that we orient the common boundary of D_1 and D_2 in the direction opposite to that of $\boldsymbol{\gamma}$; we denote it by $\overline{\boldsymbol{\gamma}}$, see Figure 8.5(b).

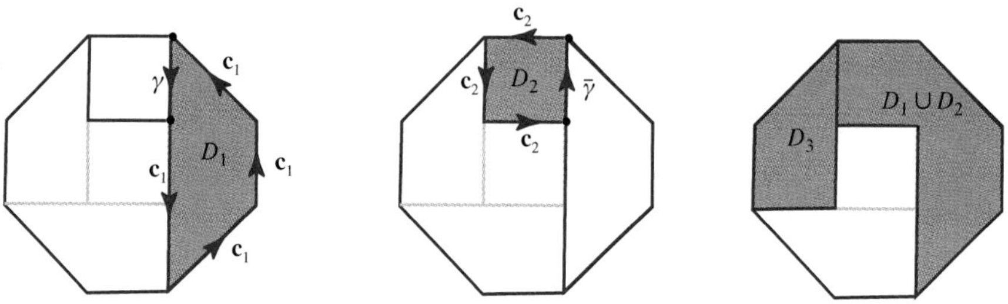

(a) (b) (c)

Figure 8.5 Applying Green's Theorem to type-3 regions.

Applying Green's Theorem to D_2, we get

$$\int_{\mathbf{c}_2} \mathbf{F}\cdot d\mathbf{s} + \int_{\overline{\gamma}} \mathbf{F}\cdot d\mathbf{s} = \iint_{D_2} \left(\frac{\partial Q}{\partial x} - \frac{\partial P}{\partial y}\right) dA. \tag{8.6}$$

Adding up (8.5) and (8.6), we obtain (keep in mind that $\int_{\overline{\gamma}} \mathbf{F}\cdot d\mathbf{s} = -\int_{\gamma} \mathbf{F}\cdot d\mathbf{s}$, since γ and $\overline{\gamma}$ are of opposite orientations)

$$\int_{\mathbf{c}_1} \mathbf{F}\cdot d\mathbf{s} + \int_{\mathbf{c}_2} \mathbf{F}\cdot d\mathbf{s} = \iint_{D_1} \left(\frac{\partial Q}{\partial x} - \frac{\partial P}{\partial y}\right) dA + \iint_{D_2} \left(\frac{\partial Q}{\partial x} - \frac{\partial P}{\partial y}\right) dA,$$

and therefore,

$$\int_{\partial(D_1\cup D_2)} \mathbf{F}\cdot d\mathbf{s} = \iint_{D_1\cup D_2} \left(\frac{\partial Q}{\partial x} - \frac{\partial P}{\partial y}\right) dA.$$

Thus, we proved Green's Theorem for the region $D_1 \cup D_2$. Now, we take the regions $D_1 \cup D_2$ and D_3 [see Figure 8.5(c)], and, imitating the above calculations, prove that Green's Theorem holds for $D_1 \cup D_2 \cup D_3$. Continuing in the same way, we complete the proof of the theorem. ◀

Green's Theorem gives us options: we can either compute a path integral directly, using a parametrization, or convert it to a double integral and then evaluate (of course, the curve and region involved have to satisfty the assumptions of the theorem).

▶ EXAMPLE 8.1

Evaluate $\int_{\mathbf{c}} \mathbf{F}\cdot d\mathbf{s}$, where $\mathbf{F}(x, y) = e^x\mathbf{i} + 2x\mathbf{j}$ and $\mathbf{c}$ is the boundary of the region D shown in Figure 8.6, oriented counterclockwise.

SOLUTION

The region D is of type 1, the vector field $\mathbf{F}$ is C^1, and the orientation convention has been respected—therefore, we can apply Green's Theorem. It follows that

$$\int_{\mathbf{c}} \mathbf{F}\cdot d\mathbf{s} = \iint_D \left(\frac{\partial}{\partial x}(2x) - \frac{\partial}{\partial y}(e^x)\right) dA = \iint_D 2\,dA.$$

Using iterated integrals, we get

$$\iint_D 2\,dA = \int_0^{\pi} \left(\int_{\sin x}^{2+\cos x} 2\,dy\right) dx$$

$$= 2\int_0^{\pi} (\cos x - \sin x + 2)\,dx = 2\,(\sin x + \cos x + 2x)\Big|_0^{\pi} = 4\pi - 4. \quad ◀$$

▶ EXAMPLE 8.2

Evaluate the path integral $\int_{\mathbf{c}} (y - \sin x)\,dx + \cos x\,dy$, where $\mathbf{c}$ is the triangle with vertices $(0, 0)$, $(\pi/2, 0)$, and $(\pi/2, 1)$, both directly and using Green's theorem.

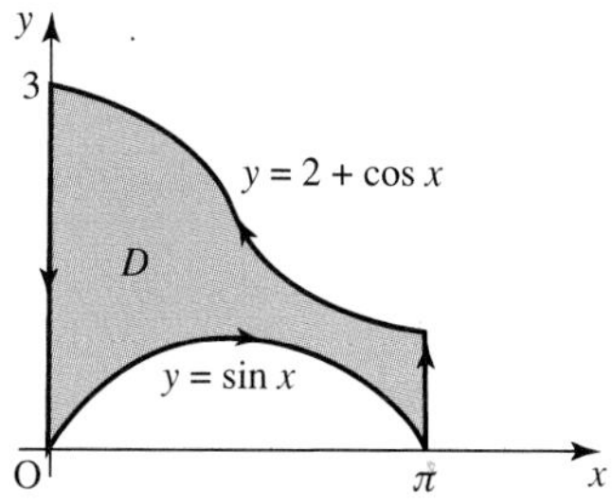

Figure 8.6 Region D of Example 8.1.

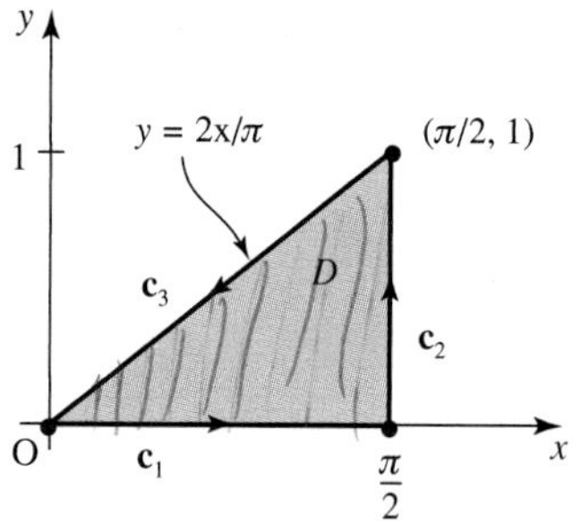

Figure 8.7 Triangle of Example 8.2.

SOLUTION Orient the curve $\mathbf{c}$ counterclockwise, so that the region lies to the left of it; see Figure 8.7. Let $\mathbf{c}_1(t) = (t, 0)$, $t \in [0, \pi/2]$. Then

$$\int_{\mathbf{c}_1} (y - \sin x)\,dx + \cos x\,dy = \int_0^{\pi/2} ((y - \sin x)x' + \cos x\, y')\,dt$$
$$= -\int_0^{\pi/2} \sin t\,dt = \cos t\Big|_0^{\pi/2} = -1.$$

Parametrize $\mathbf{c}_2$ by $\mathbf{c}_2(t) = (\pi/2, t)$, $t \in [0, 1]$. Then

$$\int_{\mathbf{c}_2} (y - \sin x)\,dx + \cos x\,dy = \int_0^1 ((t-1)\cdot 0 + 0 \cdot 1)\,dt = 0.$$

Parametrize $\mathbf{c}_3$ by $\mathbf{c}_3(t) = (\pi/2, 1) + t(-\pi/2, -1) = (\pi/2 - t\pi/2, 1 - t)$, $t \in [0, 1]$. Then

$$\int_{\mathbf{c}_3} (y - \sin x)\,dx + \cos x\,dy$$
$$= \int_0^1 \left[\left(1 - t - \sin\left(\tfrac{\pi}{2} - t\tfrac{\pi}{2}\right)\right)\left(-\tfrac{\pi}{2}\right) + \cos\left(\tfrac{\pi}{2} - t\tfrac{\pi}{2}\right)(-1)\right] dt$$
$$= -\frac{\pi}{2}\left(t - \frac{t^2}{2} - \frac{2}{\pi}\cos\left(\tfrac{\pi}{2} - t\tfrac{\pi}{2}\right)\right)\Big|_0^1 + \frac{2}{\pi}\sin\left(\tfrac{\pi}{2} - t\tfrac{\pi}{2}\right)\Big|_0^1 = -\frac{\pi}{4} + 1 - \frac{2}{\pi}.$$

Adding the three integrals, we get

$$\int_{\mathbf{c}} (y - \sin x)\,dx + \cos x\,dy = -\frac{\pi}{4} - \frac{2}{\pi}.$$

On the other hand, the application of Green's Theorem gives (the equation of $\mathbf{c}_3$ is $y = 2x/\pi$)

$$\int_{\mathbf{c}} (y - \sin x)\,dx + \cos x\,dy = \iint_D (-\sin x - 1)\,dA$$
$$= \int_0^1 \left(\int_{\pi y/2}^{\pi/2} (-\sin x - 1)\,dx\right) dy = \int_0^1 (\cos x - x)\Big|_{\pi y/2}^{\pi/2} dy$$
$$= \int_0^1 \left(-\frac{\pi}{2} - \cos\left(\frac{\pi}{2}y\right) + \frac{\pi}{2}y\right) dy$$
$$= \left(-\frac{\pi}{2}y - \frac{2}{\pi}\sin\left(\frac{\pi}{2}y\right) + \frac{\pi}{2}\frac{y^2}{2}\right)\Big|_0^1$$
$$= -\frac{\pi}{2} - \frac{2}{\pi} + \frac{\pi}{4} = -\frac{\pi}{4} - \frac{2}{\pi}.$$

Next, we give a vector-field interpretation of the double integral appearing in Green's Theorem. If we write the vector field $\mathbf{F}$ as $\mathbf{F}(x, y) = P(x, y)\mathbf{i} + Q(x, y)\mathbf{j} + 0\mathbf{k}$, we find its curl to be (P and Q do not depend on z)

$$curl\,\mathbf{F} = \begin{vmatrix} \mathbf{i} & \mathbf{j} & \mathbf{k} \\ \partial/\partial x & \partial/\partial y & \partial/\partial z \\ P & Q & 0 \end{vmatrix} = \left(\frac{\partial Q}{\partial x} - \frac{\partial P}{\partial y}\right)\mathbf{k}.$$

Computing the dot product of both sides by $\mathbf{k}$ yields

$$curl\,\mathbf{F}\cdot\mathbf{k} = \left(\frac{\partial Q}{\partial x} - \frac{\partial P}{\partial y}\right),$$

which is precisely the integrand from the theorem. Hence, the right side in Theorem 8.1 is the double integral of the scalar function $curl\,\mathbf{F}\cdot\mathbf{k}$ (i.e., the scalar curl of $\mathbf{F}$) over the region D.

Putting together the above remarks, we obtain the vector form of Green's theorem:

$$\int_{\mathbf{c}=\partial D} \mathbf{F}\cdot d\mathbf{s} = \iint_D curl\,\mathbf{F}\cdot\mathbf{k}\; dA.$$

Remember that, in the above statement, D satisfies Assumption 8.1, $\mathbf{c}$ is the positively oriented boundary of D, and $\mathbf{F}$ is a C^1 vector field defined on D.

A direct consequence of Green's Theorem is that if a simple closed curve $\mathbf{c}$ encloses a region D and f is twice continuously differentiable, then, with $\mathbf{F} = \nabla f$,

$$\int_{\mathbf{c}} \nabla f\cdot d\mathbf{s} = \iint_D curl\,(\nabla f)\cdot\mathbf{k}\; dA = 0,$$

since $curl\,(\nabla f) = \mathbf{0}$. Thus, we reestablished the fact that the path integral of a gradient vector field around a simple closed curve is zero (see Theorem 5.8 in Section 5.4).

▶ EXAMPLE 8.3

Compute $\int_{\mathbf{c}} \mathbf{F}\cdot d\mathbf{s}$, where $\mathbf{F} = e^y\mathbf{i} + \sin x\,\mathbf{j}$ and $\mathbf{c}$ is the boundary of the rectangle $[0, \pi] \times [0, 1]$.

SOLUTION

Rather than breaking $\mathbf{c}$ into four smooth curves and evaluating $\int_{\mathbf{c}} \mathbf{F}\cdot d\mathbf{s}$ over each curve, we use Green's Theorem. Orienting $\mathbf{c}$ counterclockwise (so that the rectangle is on its left) and computing $curl\,\mathbf{F} = (0, 0, \cos x - e^y)$, we proceed as follows:

$$\begin{aligned}\int_{\mathbf{c}} \mathbf{F}\cdot d\mathbf{s} &= \iint_{[0,\pi]\times[0,1]} curl\,\mathbf{F}\cdot\mathbf{k}\; dA = \iint_{[0,\pi]\times[0,1]} (\cos x - e^y)\, dA \\ &= \int_0^1 \left(\int_0^\pi (\cos x - e^y)\, dx\right) dy = \int_0^1 (\sin x - xe^y)\Big|_0^\pi\, dy \\ &= -\int_0^1 \pi e^y\, dy = \pi(1 - e).\end{aligned}$$

◀

▶ EXAMPLE 8.4

Compute $\int_{\mathbf{c}} \mathbf{F}\cdot d\mathbf{s}$, if $\mathbf{F} = xy^2\mathbf{i} - x^2y\mathbf{j}$ and $\mathbf{c}$ is the boundary of the region $x \geq 0$, $0 \leq y \leq 1 - x^2$, shown in Figure 8.8.

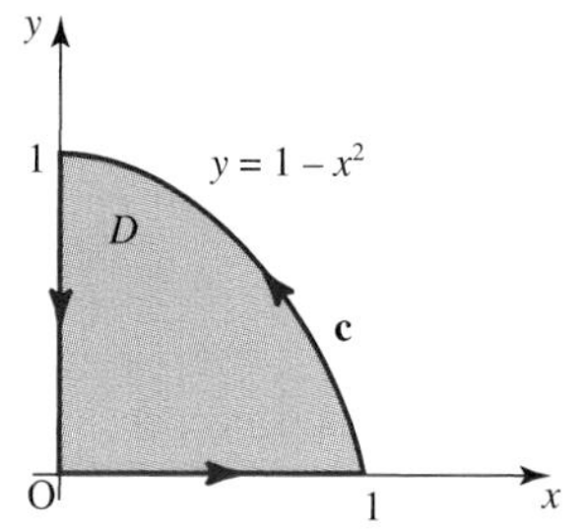

Figure 8.8 Region D and its boundary **c** of Example 8.4.

SOLUTION

The *curl* of **F** is computed to be $(0, 0, -4xy)$ and, hence, by Green's Theorem (view D as a type-1 region)

$$\begin{aligned}\int_{\mathbf{c}} \mathbf{F} \cdot d\mathbf{s} &= \iint_D (-4xy\mathbf{k}) \cdot \mathbf{k}\, dA \\ &= \int_0^1 \left(\int_0^{1-x^2} -4xy dy \right) dx = \int_0^1 \left(-2xy^2 \Big|_0^{1-x^2} \right) dx \\ &= \int_0^1 -2x(1-x^2)^2 dx = \frac{(1-x^2)^3}{3} \Big|_0^1 = -\frac{1}{3}.\end{aligned}$$

◀

Green's Theorem gives new formulas for computing the area of a region in the plane. Take a region D that satisfies Assumption 8.1 and let $\mathbf{c} = \partial D$ be its positively oriented boundary. Substituting $P(x, y) = -y$, $Q(x, y) = 0$ in (8.4), we get

$$\int_{\mathbf{c}} (-y)\, dx = \iint_D 1\, dA = A(D),$$

where $A(D)$ is the area of D. Similarly, with $P(x, y) = 0$, $Q(x, y) = x$, we get

$$\int_{\mathbf{c}} x\, dy = \iint_D 1\, dA = A(D).$$

Adding up the two expressions, we obtain (after dividing by 2)

$$\frac{1}{2} \int_{\mathbf{c}} x\, dy - y\, dx = A(D).$$

These formulas provide three different ways of computing the area of a region as a path integral along its (closed) boundary.

▶ EXAMPLE 8.5

Using a path integral, find the area of the region D in the first quadrant bounded by $y = x$ and $y = x^3$.

SOLUTION

We will use the first of the three formulas derived above. The boundary $\mathbf{c} = \partial D$ of D consists of the part of the cubic parabola $y = x^3$ from $(0, 0)$ to $(1, 1)$ (call it $\mathbf{c}_1$) and of the straight-line segment from $(1, 1)$ to $(0, 0)$ (call it $\mathbf{c}_2$); see Figure 8.9.

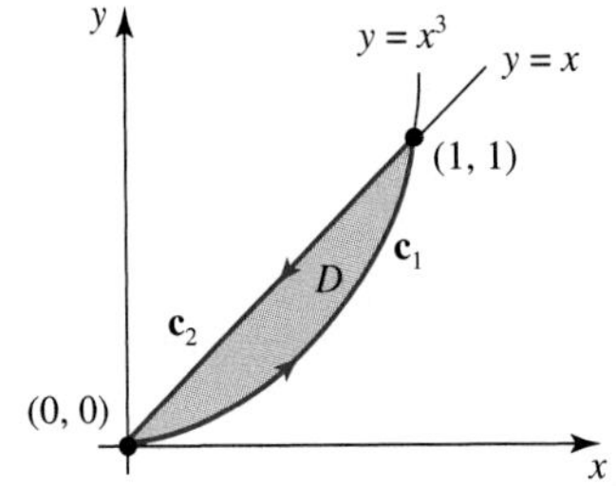

Figure 8.9 Region D of Example 8.5.

The curve $\mathbf{c}_1$ is given by $\mathbf{c}_1(t) = (t, t^3)$, $t \in [0, 1]$, and

$$\int_{\mathbf{c}_1} (-y)\,dx = \int_0^1 \left(-y\frac{dx}{dt}\right) dt = -\int_0^1 t^3\,dt = -\frac{1}{4}.$$

The curve $\mathbf{c}_2$ is given by $\mathbf{c}_2(t) = (1-t, 1-t)$, $t \in [0, 1]$, and hence,

$$\int_{\mathbf{c}_2} (-y)\,dx = \int_0^1 -(1-t)(-1)\,dt = \left(t - \frac{1}{2}t^2\right)\bigg|_0^1 = \frac{1}{2}.$$

The area of D is thus

$$A(D) = \int_{\mathbf{c}} (-y)\,dx = \int_{\mathbf{c}_1} (-y)\,dx + \int_{\mathbf{c}_2} (-y)\,dx = \frac{1}{4}.$$

To check our answer, we use the double integral:

$$A(D) = \int_0^1 \left(\int_{x^3}^{x} 1\,dy\right) dx = \int_0^1 \left(y\,\bigg|_{x^3}^{x}\right) dx = \int_0^1 (x - x^3)\,dx = \frac{1}{4}.$$

◀

▶ EXAMPLE 8.6 Double Integral of the Laplacian

Let $D \subseteq \mathbb{R}^2$ be a region that satisfies Assumption 8.1 and let $\mathbf{c} = \partial D$ be its C^1, positively oriented, smooth boundary curve [recall that "smooth" implies that $\|\mathbf{c}'(t)\| \neq 0$]. Assume that the function $f\colon D \to \mathbb{R}$ is twice continuously differentiable (it is also called a C^2 function). By definition, the Laplace operator of f is equal to

$$\Delta f = \frac{\partial^2 f}{\partial x^2} + \frac{\partial^2 f}{\partial y^2},$$

and therefore,

$$\iint_D \Delta f\,dA = \iint_D \left(\frac{\partial^2 f}{\partial x^2} + \frac{\partial^2 f}{\partial y^2}\right) dA = \iint_D \left(\frac{\partial}{\partial x}\left(\frac{\partial f}{\partial x}\right) + \frac{\partial}{\partial y}\left(\frac{\partial f}{\partial y}\right)\right) dA$$

using Green's Theorem (8.4) with $Q = \partial f/\partial x$ and $P = -\partial f/\partial y$,

$$= \int_{\mathbf{c}=\partial D} \left(-\frac{\partial f}{\partial y}\mathbf{i} + \frac{\partial f}{\partial x}\mathbf{j}\right) \cdot d\mathbf{s} = \int_a^b \left(-\frac{\partial f}{\partial y}x' + \frac{\partial f}{\partial x}y'\right) dt,$$

where $\mathbf{c}(t) = (x(t), y(t))$, $t \in [a, b]$, is a parametrization of $\mathbf{c}$.

Let us interpret the integrand in the above expression: it is the dot product of the vectors $(\partial f/\partial x, \partial f/\partial y)$ and $(y'(t), -x'(t))$. The first factor is the gradient of f. The dot product of the second factor $\mathbf{N} = (y'(t), -x'(t))$ with the tangent $\mathbf{c}'(t) = (x'(t), y'(t))$ is zero. Consequently, $\mathbf{N}$ is a vector normal to $\mathbf{c}$. Notice that $\|\mathbf{N}\| = \|\mathbf{c}'(t)\|$. Moreover, we know that $\mathbf{N}$ is an outward normal, since its

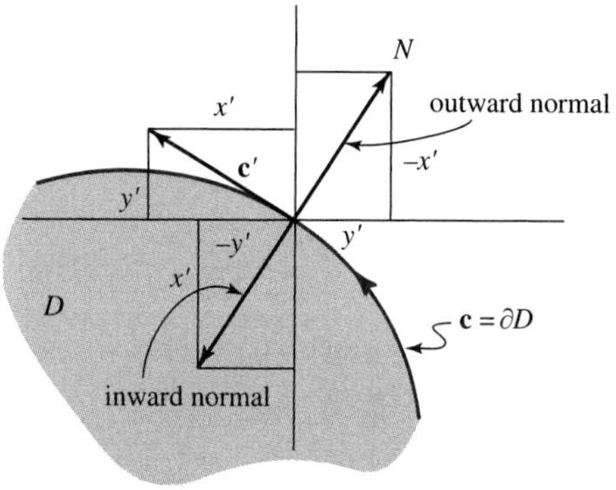

Figure 8.10 Inward and outward normal vectors to a curve $\mathbf{c} = \partial D$.

x-component coincides with the y-component of the tangent (the x-component of the inward normal equals the negative of the y-component of the tangent); see Figure 8.10.

It follows that

$$\iint_D \Delta f\, dA = \int_a^b \nabla f \cdot \mathbf{N}\, dt = \int_a^b \left(\nabla f \cdot \frac{\mathbf{N}}{\|\mathbf{N}\|}\right) \|\mathbf{N}\|\, dt$$

[since $\mathbf{c}$ is smooth and $\|\mathbf{N}\| = \|\mathbf{c}'(t)\|$, it follows that $\|\mathbf{N}\| \neq 0$]

$$= \int_a^b \left(\nabla f \cdot \frac{\mathbf{N}}{\|\mathbf{N}\|}\right) \|\mathbf{c}'\|\, dt = \int_{\mathbf{c}} \left(\nabla f \cdot \frac{\mathbf{N}}{\|\mathbf{N}\|}\right) ds$$

[by definition of the path integral $\int_{\mathbf{c}} g ds = \int_a^b g\|\mathbf{c}'\|dt$, with $g = \nabla f \cdot \mathbf{N}/\|\mathbf{N}\|$, read backward]

$$= \int_{\mathbf{c}} \nabla f \cdot \mathbf{n}\, ds = \int_{\mathbf{c}} D_{\mathbf{n}} f\, ds$$

(here $\mathbf{n} = \mathbf{N}/\|\mathbf{N}\|$ is the unit outward normal), by definition of the directional derivative, $D_{\mathbf{n}} f = \nabla f \cdot \mathbf{n}$. The directional derivative $D_{\mathbf{n}} f$ is called the *normal derivative of* f and is denoted by $\partial f/\partial n$. Hence,

$$\iint_D \Delta f\, dA = \int_{\mathbf{c}=\partial D} \frac{\partial f}{\partial n}\, ds.$$

◀

▶ EXAMPLE 8.7

Compute $\int_{\mathbf{c}}(\partial f/\partial n)ds$, where $f(x, y) = x^2 + y^2$ and $\mathbf{c}$ is the circle $x^2 + y^2 = 4$, both directly and using the formula derived in the previous example.

SOLUTION

We compute the path integral first. Parametrize $\mathbf{c}$ by $\mathbf{c}(t) = (2\cos t, 2\sin t)$, $t \in [0, 2\pi]$. Then $\mathbf{c}'(t) = (-2\sin t, 2\cos t)$ and thus $\mathbf{N} = (2\cos t, 2\sin t)$ is an outward normal. The normal derivative is computed to be

$$\frac{\partial f}{\partial n} = \nabla f \cdot \mathbf{n} = (2x, 2y) \cdot \frac{\mathbf{N}}{\|\mathbf{N}\|} = (4\cos t, 4\sin t) \cdot \frac{(2\cos t, 2\sin t)}{2} = 4,$$

and hence [$\ell(\mathbf{c})$ denotes the length of $\mathbf{c}$],

$$\int_{\mathbf{c}} \frac{\partial f}{\partial n}\, ds = \int_{\mathbf{c}} 4\, ds = 4\ell(\mathbf{c}) = 16\pi.$$

Alternatively (the Laplacian is computed to be $\Delta f = 2 + 2 = 4$),

$$\int_{\mathbf{c}} \frac{\partial f}{\partial n} ds = \iint_{\{x^2+y^2\le 4\}} \Delta f \, dA = 4 \text{ area}(\{x^2 + y^2 \le 4\}) = 16\pi.$$

▶ EXERCISES 8.1

1. Consider a constant vector field $\mathbf{F}(x, y) = a\mathbf{i} + b\mathbf{j}$ in $\mathbb{R}^2$ $(a, b \in \mathbb{R})$, and let $\mathbf{c}$ be any simple closed curve in $\mathbb{R}^2$.

(a) Without using Green's Theorem, find the circulation $\int_{\mathbf{c}} \mathbf{F} \cdot d\mathbf{s}$ of $\mathbf{F}$ around $\mathbf{c}$.

(b) Use Green's Theorem to confirm your answer to (a).

2. Let $\mathbf{F}$ be a C^1 vector field in $\mathbb{R}^2$ whose scalar curl at $(3, -2)$ is equal to 7. Approximate the counterclockwise circulation $\int_{\mathbf{c}} \mathbf{F} \cdot d\mathbf{s}$ of $\mathbf{F}$ around the circle $\mathbf{c}$ of radius 0.02 centered at $(3, -2)$.

Exercises 3 to 7: Compute $\int_{\mathbf{c}} \mathbf{F} \cdot d\mathbf{s}$ using Green's Theorem.

3. $\mathbf{F} = -2y\mathbf{i} + x\mathbf{j}$, $\mathbf{c}(t) = (2\cos t, \sin t)$, $t \in [0, 2\pi]$

4. $\mathbf{F} = (x^2 + 1)^{-1}\mathbf{j}$, $\mathbf{c}$ is the boundary of the rectangle $[0, 2] \times [0, 3]$, oriented counterclockwise

5. $\mathbf{F} = e^{x+y}\mathbf{j} - e^{x-y}\mathbf{i}$, $\mathbf{c}$ is the boundary of the triangle defined by the lines $y = 0$, $x = 1$, and $y = x$, oriented counterclockwise

6. $\mathbf{F} = (2 - y^3)\mathbf{i} + (y + x^3 + 2)\mathbf{j}$, $\mathbf{c}$ is the circle of radius 5 centered at the origin and oriented counterclockwise

7. $\mathbf{F} = 2x^2y^2\mathbf{i} - x\mathbf{j}$, $\mathbf{c}$ consists of the curve $y = 2x^3$ from $(0, 0)$ to $(1, 2)$ followed by the straight-line segment from $(1, 2)$ back to $(0, 0)$, oriented counterclockwise

8. Let D be a type-2 region given by $c \le y \le d$ and $\phi(y) \le x \le \psi(y)$, let $\mathbf{c}$ be its positively oriented boundary, and let $\mathbf{F}_2(x, y) = (0, Q(x, y))$.

(a) Show that the integral of $\mathbf{F}_2$ along the line segments $y = c$ and $y = d$ is zero.

(b) Prove that $\int_{\mathbf{c}} \mathbf{F}_2 \cdot d\mathbf{s} = \int_c^d Q(\psi(y), y) - Q(\phi(y), y)\, dy$.

(c) Show that $Q(\psi(y), y) - Q(\phi(y), y) = \int_{\phi(y)}^{\psi(y)} (\partial Q(x, y)/\partial x)\, dx$.

(d) Conclude that $\int_{\mathbf{c}} \mathbf{F}_2 \cdot d\mathbf{s} = \iint_D (\partial Q/\partial x)\, dA$.

Exercises 9 to 13: Compute $\int_{\mathbf{c}} \mathbf{F} \cdot d\mathbf{s}$ directly, or using Green's Theorem.

9. $\mathbf{F} = x^2y^2\mathbf{i} + y^4\mathbf{j}$, $\mathbf{c}$ is the curve $x^2 + y^2 = 1$, oriented counterclockwise

10. $\mathbf{F} = (2x + 3y + 2)\mathbf{i} - (x - 4y + 3)\mathbf{j}$, $\mathbf{c}$ is the ellipse $x^2 + 4y^2 = 4$, oriented clockwise

11. $\mathbf{F} = \cosh y\mathbf{i} + x \sinh y\mathbf{j}$, $\mathbf{c}$ is the boundary of the triangle defined by the lines $y = 4x$, $y = 2x$, and $x = 1$, oriented counterclockwise [recall that $\cosh y = (e^y + e^{-y})/2$ and $\sinh y = (e^y - e^{-y})/2$]

12. $\mathbf{F} = e^x(\mathbf{i} + \mathbf{j})$, $\mathbf{c}$ is the boundary of the triangle with vertices $(0, 0)$, $(1, 2)$, and $(0, 2)$, oriented counterclockwise

13. $\mathbf{F} = \arctan(y/x)\mathbf{i} + \arctan(x/y)\mathbf{j}$, $\mathbf{c}$ is the circle $x^2 + y^2 = 2$, oriented counterclockwise

14. Assume that the curves involved are oriented counterclockwise.

(a) Compute $\displaystyle\int_{\mathbf{c}} \frac{xdy - ydx}{x^2 + y^2}$, where $\mathbf{c}$ is the circle $x^2 + y^2 = 1$.

(b) Compute $\displaystyle\int_{\mathbf{c}} \frac{xdy - ydx}{x^2 + y^2}$, where $\mathbf{c}$ is the circle $(x - 1)^2 + (y - 1)^2 = 1$.

15. Using a path integral, compute the area of the region D bounded by the curves $y = 2x^2$ and $y = 4x$.

16. Using a path integral, compute the area of the region D bounded by the curves $x = y^2$, $x = 2$, and $x = 3$.

17. Using a path integral, compute the area of the region D in the first quadrant bounded by the astroid $x^{2/3} + y^{2/3} = 1$.

18. Using a path integral, compute the area of the region bounded by the x-axis and the cycloid $\mathbf{c}(t) = (t - \sin t, 1 - \cos t)$, where $0 \le t \le 2\pi$.

19. Compute the work of the force $\mathbf{F} = x\mathbf{i} + (x^2 + 3y^2)\mathbf{j}$ done on a particle that moves along the straight-line segments from $(0, 0)$ to $(3, 0)$, then from $(3, 0)$ to $(1, 2)$, and then from $(1, 2)$ back to $(0, 0)$.

20. Let D be a region that satisfies Assumption 8.1, with a positively oriented boundary $\partial D = \mathbf{c}$. Assume that D is of constant density ρ. Express its mass m and moments M_x and M_y (with respect to the y-axis and x-axis) in terms of path integrals.

21. Let D be a region that satisfies Assumption 8.1, with a positively oriented boundary $\partial D = \mathbf{c}$. Assume that D is of constant density ρ. Express its moments of inertia about the x-axis and y-axis in terms of path integrals.

22. Let D be the disk $x^2 + y^2 \le 1$, let $\mathbf{c}$ be its positively oriented boundary, and let $f(x, y) = x^2 + 3y^2$. By computing both sides, check that $\iint_D \Delta f \, dA = \int_{\mathbf{c}} D_{\mathbf{n}} f \, ds$, where $\mathbf{n}$ is the outward normal to $\mathbf{c}$ and $D_{\mathbf{n}} f$ is the directional derivative in the direction of the normal. Δf denotes the Laplacian of f, defined by $\Delta f = f_{xx} + f_{yy}$.

23. Check that (see Exercise 22 for the notation) $\iint_D \Delta f \, dA = \int_{\mathbf{c}} D_{\mathbf{n}} f \, ds$ for the function $f(x, y) = e^x \cos y$, where D is the rectangle $[0, 1] \times [0, 2]$ and $\mathbf{c}$ is its positively oriented boundary.

24. Check that (see Exercise 22 for the notation) $\iint_D \Delta f \, dA = \int_{\mathbf{c}} D_{\mathbf{n}} f \, ds$ for the function $f(x, y) = e^{x+y}$, where D is the rectangle $[0, 1] \times [0, 1]$ and $\mathbf{c}$ is its positively oriented boundary.

8.2 THE DIVERGENCE THEOREM

The Divergence Theorem (or Gauss' Divergence Theorem) is similar to Green's Theorem: it relates an integral over a closed geometric object (a closed surface) to an integral over the region (in this case, a three-dimensional solid region) enclosed by it.

Elementary regions in $\mathbb{R}^3$ are regions in $\mathbb{R}^3$ bounded by surfaces that are graphs of real-valued functions of two variables. Depending on which of the variables are involved, the regions are called type 1, type 2, or type 3. A region is of type 1 if its "bottom" and "top" sides are graphs of continuous functions $\kappa_1(x, y)$ and $\kappa_2(x, y)$. A region is of type 2 if its "back" and "front" sides are graphs of continuous functions $\kappa_1(y, z)$ and $\kappa_2(y, z)$, and of type 3 if its "left" and "right" sides are graphs of continuous functions $\kappa_1(x, z)$ and $\kappa_2(x, z)$ [of course, the names "top," "bottom," "left," etc., for sides depend on the point from which we look at the xyz-coordinate system; see Section 6.5 for precise definitions; here, we drop the notation (3D) since the context will clearly distinguish between two-dimensional and three-dimensional elementary regions].

A region is of type 4 if it is of type 1, type 2, and type 3. For example, a rectangular box whose sides are parallel to coordinate axes is of type 4. The ball $\{(x, y, z) | x^2 + y^2 + z^2 \le 1\}$ and upper half-ball $\{(x, y, z) | x^2 + y^2 + z^2 \le 1, z \ge 0\}$ are of type 4.

Before giving the statement of the theorem, we describe the regions that will be involved.

ASSUMPTION 8.2 Regions Involved in Gauss' Divergence Theorem

Assume that $W \subseteq \mathbb{R}^3$ is either a region of type 4, or can be broken into pieces, each of which is a region of type 4. The most any two pieces are allowed to have in common is their (common) boundary surface (or boundary surfaces).

The boundary ∂W of W is a closed surface or union of closed surfaces. It can be oriented in two ways: either by choosing an outward normal (i.e., a normal that points away from the solid region W into space) or an inward normal (that points into the region W). We define the *positive* orientation as the choice of an outward normal. ◀

For example, the solid region W between two cubes, one placed inside the other one, satisfies Assumption 8.2: it can be broken into six (or more) parallelepipeds, see Figure 8.11.

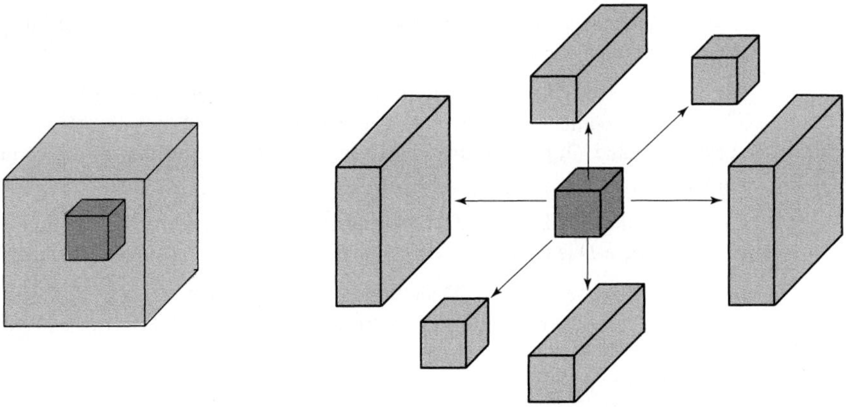

Figure 8.11 The solid region between two cubes satisfies Assumption 8.2.

The boundary of W consists of 12 squares, six of which are the faces of the larger cube, and the other six are the faces of the smaller cube. The positive orientation of ∂W is shown in Figure 8.12. The normals to the faces of the larger cube point outward (i.e., away from the cubes). The normals to the six smaller faces point into the smaller cube.

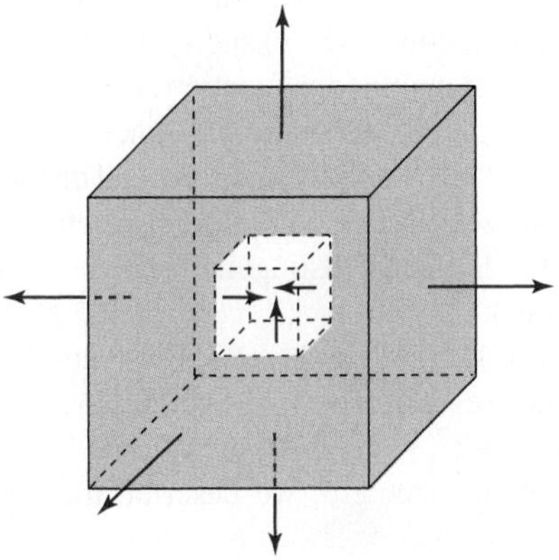

Figure 8.12 Positive orientation of the boundary ∂W of W (not shown on all faces).

If $\mathbf{F}$ is the velocity of a fluid, then the surface integral (see Section 7.4) $\iint_S \mathbf{F} \cdot d\mathbf{S} = \iint_S \mathbf{F} \cdot \mathbf{n}\, dS$ measures [for S closed and oriented by the (unit) outward normal $\mathbf{n}$] the total volume of fluid leaving the three-dimensional region enclosed by the surface S per unit time. On the other hand, the three-dimensional analogue of our interpretation of the divergence given in Section 4.6 relates the total fluid outflow (from the solid region) to the divergence of the vector field that represents the fluid velocity. This observation is formalized in the statement of our next theorem.

THEOREM 8.2 Divergence Theorem of Gauss

Let W be the region in $\mathbb{R}^3$ that satisfies Assumption 8.2 and let ∂W be its positively oriented boundary. Then

$$\iint_{S=\partial W} \mathbf{F} \cdot d\mathbf{S} = \iiint_W \mathit{div}\, \mathbf{F}\, dV,$$

for a C^1 vector field $\mathbf{F}$ defined on W.

PROOF: We will not be able to prove the theorem for a general region that satisfies Assumption 8.2; instead, we concentrate on a special case of a region of type 4. As we will see soon, the proof is very similar to the proof of special cases of Green's Theorem.

Consider a type-1 region W defined by

$$W = \{(x, y, z) | \kappa_1(x, y) \le z \le \kappa_2(x, y), (x, y) \in D\},$$

where D is an elementary region in $\mathbb{R}^2$. The boundary $S = \partial W$ is, in general, a piecewise smooth surface that consists of the "bottom" and "top" surfaces [that are the graphs of $z = \kappa_1(x, y)$ and $z = \kappa_2(x, y)$] and (at most) four vertical sides. Figure 8.13 shows regions W with four and two vertical sides. Orient S by the outward-pointing normal vector field $\mathbf{N}$ (i.e., by our convention, positive orientation).

We will compute $\iint_S \mathbf{F} \cdot d\mathbf{S}$, where $\mathbf{F} = R(x, y, z)\mathbf{k}$ is a C^1 vector field defined on W. On any of the vertical sides, $\mathbf{N}$ is parallel to the xy-plane, and therefore,

$$\mathbf{F} \cdot \mathbf{N} = R(x, y, z)\,\mathbf{k} \cdot \mathbf{N} = 0.$$

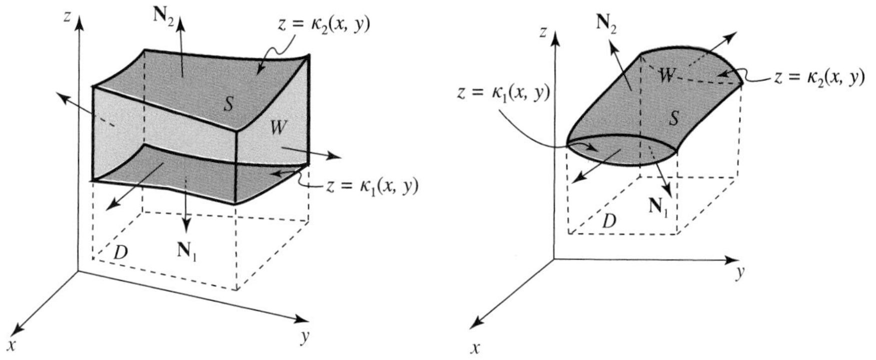

Figure 8.13 Two type-1 regions W with positively oriented boundary surfaces S.

Consequently, vertical sides do not contribute anything to the integral $\iint_S \mathbf{F} \cdot d\mathbf{S}$ (which is certainly not a surprise: **F** represents a flow in the vertical direction only, and so there is no flux through the sides that are parallel to the flow).

Parametrize the bottom surface S_1 using $\mathbf{r}_1(u, v) = (u, v, \kappa_1(u, v))$, where $(u, v) \in D$. The outward normal to S_1 is $\mathbf{N}_1 = -(-\partial\kappa_1/\partial u, -\partial\kappa_1/\partial v, 1)$. Since the **k**-component of $\mathbf{N}_1$ is -1, $\mathbf{N}_1$ points downward (as it should; in Section 7.1, this choice of the normal was called the negative orientation of the surface). From

$$\mathbf{F}(\mathbf{r}_1(u, v)) \cdot \mathbf{N}_1(u, v) = R(u, v, \kappa_1(u, v))\,\mathbf{k} \cdot \left(\frac{\partial\kappa_1}{\partial u}\mathbf{i} + \frac{\partial\kappa_1}{\partial v}\mathbf{j} - \mathbf{k}\right) = -R(u, v, \kappa_1(u, v)),$$

it follows that

$$\iint_{S_1} \mathbf{F} \cdot d\mathbf{S} = -\iint_D R(u, v, \kappa_1(u, v))\, dA = -\iint_D R(x, y, \kappa_1(x, y))\, dA,$$

after replacing u by x and v by y.

The integral over S_2 is computed analogously, the only difference being the choice of a normal. Parametrize S_2 by $\mathbf{r}_2(u, v) = (u, v, \kappa_2(u, v))$, where $(u, v) \in D$ and choose $\mathbf{N}_2 = +(-\partial\kappa_2/\partial u, -\partial\kappa_2/\partial v, 1)$. Proceeding as above, we obtain

$$\iint_{S_2} \mathbf{F} \cdot d\mathbf{S} = \iint_D R(x, y, \kappa_2(x, y))\, dA,$$

and therefore,

$$\iint_S \mathbf{F} \cdot d\mathbf{S} = \iint_D \left(R(x, y, \kappa_2(x, y)) - R(x, y, \kappa_1(x, y))\right) dA. \tag{8.7}$$

The Fundamental Theorem of Calculus $\int_a^b f'(z)\, dz = f(b) - f(a)$, applied to the function $f(z) = R(x, y, z)$, gives

$$\int_a^b \frac{\partial}{\partial z} R(x, y, z)\, dz = R(x, y, b) - R(x, y, a).$$

Replacing a by $\kappa_1(x, y)$ and b by $\kappa_2(x, y)$, we get

$$\int_{\kappa_1(x,y)}^{\kappa_2(x,y)} \frac{\partial}{\partial z} R(x, y, z)\, dz = R(x, y, \kappa_2(x, y)) - R(x, y, \kappa_1(x, y)), \tag{8.8}$$

and that is the integrand we are looking for! That is, combining (8.7) and (8.8), we obtain

$$\iint_S \mathbf{F} \cdot d\mathbf{S} = \iint_D \left(\int_{\kappa_1(x,y)}^{\kappa_2(x,y)} \frac{\partial}{\partial z} R(x, y, z)\, dz\right) dA.$$

The integral on the right is the iterated integral of the triple integral of $\partial R/\partial z$ over W; hence,

$$\iint_S \mathbf{F} \cdot d\mathbf{S} = \iint_S R\mathbf{k} \cdot d\mathbf{S} = \iiint_W \frac{\partial R}{\partial z}\, dV.$$

Similar computations will prove that

$$\iint_S P\mathbf{i} \cdot d\mathbf{S} = \iiint_W \frac{\partial P}{\partial x}\, dV$$

for a type-2 region W and

$$\iint_S Q\mathbf{j}\cdot d\mathbf{S} = \iiint_W \frac{\partial Q}{\partial y}\,dV$$

for a type-3 region W. Adding up the three expressions, we see that

$$\iint_S (P\mathbf{i}+Q\mathbf{j}+R\mathbf{k})\cdot d\mathbf{S} = \iiint_W \left(\frac{\partial P}{\partial x}+\frac{\partial Q}{\partial y}+\frac{\partial R}{\partial z}\right)dV,$$

that is,

$$\iint_S \mathbf{F}\cdot d\mathbf{S} = \iiint_W \mathit{div}\,\mathbf{F}\,dV,$$

where $\mathbf{F} = P\mathbf{i}+Q\mathbf{j}+R\mathbf{k}$ and W is a type-4 region.

EXAMPLE 8.8

Verify the Divergence Theorem for the vector field $\mathbf{F} = y^2\mathbf{i}+x^2\mathbf{j}+z^2\mathbf{k}$, where S is the surface of the cylinder $x^2+y^2=4$, $0\le z\le 5$, together with the top disk $\{(x,y)|x^2+y^2\le 4, z=5\}$ and the bottom disk $\{(x,y)|x^2+y^2\le 4, z=0\}$, oriented by the outward normal.

SOLUTION Let us first compute $\iint_S \mathbf{F}\cdot d\mathbf{S}$ directly. The top disk S_1 can be parametrized using $\mathbf{r}_1(u,v)=(u,v,5)$, where $u^2+v^2\le 4$. Then $\mathbf{T}_u^1=(1,0,0)$, $\mathbf{T}_v^1=(0,1,0)$, the normal is $\mathbf{N}_1=\mathbf{T}_u^1\times\mathbf{T}_v^1=(0,0,1)$, and

$$\begin{aligned}\iint_{S_1}\mathbf{F}\cdot d\mathbf{S} &= \iint_{\{u^2+v^2\le 4\}}(v^2,u^2,25)\cdot(0,0,1)\,dA\\ &= \iint_{\{u^2+v^2\le 4\}} 25\,dA = 25\ \text{area}(\{u^2+v^2\le 4\}) = 100\pi.\end{aligned}$$

In words, the total flux out of the cylinder through its top side equals 100π. Similarly, the bottom disk S_2 can be parametrized by $\mathbf{r}_2(u,v)=(u,v,0)$, $u^2+v^2\le 4$. The tangents and normal are as for the top side. However, to comply with the outward orientation requirement for the surface, we take $\mathbf{N}_2=(0,0,-1)$ as the normal; see Figure 8.14. Hence,

$$\iint_{S_2}\mathbf{F}\cdot d\mathbf{S} = \iint_{\{u^2+v^2\le 4\}}(v^2,u^2,0)\cdot(0,0,-1)\,dA = 0.$$

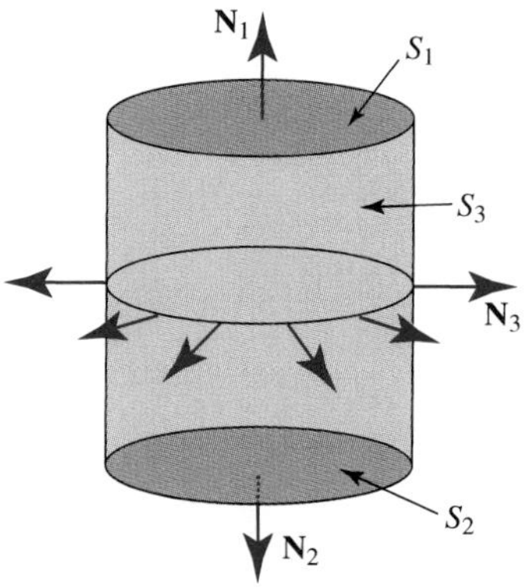

Figure 8.14 Surface of the cylinder oriented by the outward normal.

Finally, parametrize S_3 (the curved surface of the cylinder) by

$$\mathbf{r}_3(u, v) = (2\cos u, 2\sin u, v), \qquad 0 \le u \le 2\pi, \quad 0 \le v \le 5.$$

Then $\mathbf{T}_u^3 = (-2\sin u, 2\cos u, 0)$, $\mathbf{T}_v^3 = (0, 0, 1)$, $\mathbf{N}_3 = (2\cos u, 2\sin u, 0)$, and ($\mathbf{N}_3$ points outward: it is the position vector of a point on the circle in the xy-plane)

$$\begin{aligned}\iint_{S_3} \mathbf{F}\cdot d\mathbf{S} &= \iint_{[0,2\pi]\times[0,5]} (4\sin^2 u, 4\cos^2 u, v^2)\cdot(2\cos u, 2\sin u, 0)\,dA \\ &= \int_0^5 \left(\int_0^{2\pi} (8\sin^2 u\cos u + 8\cos^2 u\sin u)\,du\right) dv \\ &= 8\int_0^5 \left(\tfrac{1}{3}\sin^3 u - \tfrac{1}{3}\cos^3 u\right)\Big|_0^{2\pi} dv = 8\int_0^5 0\,dv = 0.\end{aligned}$$

Adding up the three integrals, we get $\iint_S \mathbf{F}\cdot d\mathbf{S} = 100\pi$.

The Divergence Theorem claims that the same result will be obtained by computing the integral $\iiint_W div\,\mathbf{F}\,dV$, where W denotes the *solid* cylinder $\{(x, y, z)|x^2 + y^2 \le 4, 0 \le z \le 5\}$. Indeed ($div\,\mathbf{F} = 2z$),

$$\begin{aligned}\iiint_W div\,\mathbf{F}\,dV &= \iint_{\{x^2+y^2\le 4\}} \left(\int_0^5 2z\,dz\right) dA \\ &= 25\iint_{\{x^2+y^2\le 4\}} dA = 25\text{area}(\{(x, y)|x^2 + y^2 \le 4\}) = 100\,\pi.\end{aligned}$$

◀

▶ EXAMPLE 8.9

Compute the (outward) flux of $\mathbf{F} = xy^2\mathbf{i} + y^3\mathbf{j} + 4x^2z\mathbf{k}$ through the surface of the (solid) cylinder W given by $x^2 + y^2 \le 4$, $0 \le z \le 5$.

SOLUTION The Divergence Theorem applied to the closed surface S enclosing W ($\partial W = S$) implies that

$$\begin{aligned}\iint_S (xy^2\mathbf{i} + y^3\mathbf{j} + 4x^2z\mathbf{k})\cdot d\mathbf{S} &= \iiint_W (y^2 + 3y^2 + 4x^2)\,dV \\ &= \iint_{\{x^2+y^2\le 4\}} \left(\int_0^5 (4x^2 + 4y^2)dz\right) dA \\ &= 20\iint_{\{x^2+y^2\le 4\}} (x^2 + y^2)\,dA\end{aligned}$$

[using polar coordinates $x = r\cos\theta$, $y = r\sin\theta$; hence, $dA = r dr d\theta$]

$$= 20\int_0^{2\pi}\left(\int_0^2 r^2\,r dr\right) d\theta = 20\cdot 2\pi\cdot 4 = 160\,\pi.$$

◀

▶ EXAMPLE 8.10

Compute $\iint_S (x^2\mathbf{i} - (2x - 1)y\mathbf{j} + 4z\mathbf{k})\cdot d\mathbf{S}$, where S is the surface of the cone $x^2 + y^2 = z^2$, $0 \le z \le 2$, oriented by the outward normal.

SOLUTION S is a closed surface enclosing a solid three-dimensional region W. By the Divergence Theorem,

$$\begin{aligned}\iint_S (x^2\mathbf{i}-(2x-1)y\mathbf{j}+4z\mathbf{k})\cdot d\mathbf{S} &= \iiint_W (2x-(2x-1)+4)\,dV\\ &= 5\iint_{\{x^2+y^2\le 4\}}\left(\int_{\sqrt{x^2+y^2}}^{2} dz\right)dA\\ &= 5\iint_{\{x^2+y^2\le 4\}}\left(2-\sqrt{x^2+y^2}\right)dA = 5\int_0^{2\pi}\left(\int_0^2 (2-r)r\,dr\right)d\theta\\ &= 5\left(\int_0^{2\pi} d\theta\right)\left(\int_0^2 (2r-r^2)\,dr\right) = 5\cdot 2\pi\left(r^2-\frac{1}{3}r^3\right)\Big|_0^2 = \frac{40\pi}{3}.\end{aligned}$$

There is a faster way to compute this integral. From the second line in the above computation, it follows that the result is 5 volume(W), where W is the cone with base radius 2 and height 2. Therefore, the answer is $5(1/3)8\pi = 40\pi/3$. ◀

Sometimes, in calculating $\iint_S \mathbf{F}\cdot d\mathbf{S}$, we end up with an integral that cannot be solved in a compact form (we need to use power series, for instance). If S happens to be a closed surface, and the solid bounded by it satisfies Assumption 8.2, then we can use the Divergence Theorem to try to solve the given integral (see Exercise 29).

The Divergence Theorem of Gauss states that the integral $\iint_S \mathbf{F}\cdot d\mathbf{S} = \iint_S \mathbf{F}\cdot\mathbf{n}\,dS$ of the normal component $\mathbf{F}\cdot\mathbf{n}$ of a vector field $\mathbf{F}$ over a surface equals the integral of the divergence of $\mathbf{F}$ over the three-dimensional region enclosed by that surface.

Keeping this fact in mind, we now discuss the analogue of Gauss' Theorem in two dimensions. Let $D \subseteq \mathbb{R}^2$ be a region to which Green's Theorem applies (see Assumption 8.1 in Section 8.1). Parametrize its boundary curve by $\mathbf{c}(t) = (x(t), y(t))$, $t \in [a,b]$, so that $\mathbf{c}$ is positively oriented (i.e., D is on its left). In Example 8.6, we showed that $\mathbf{N} = (y'(t), -x'(t))$ is the outward normail to $\mathbf{c}$. The unit outward normal is given by $\mathbf{n} = \mathbf{N}/\sqrt{(x'(t))^2+(y'(t))^2}$.

THEOREM 8.3 Divergence Theorem in the Plane

Let $\mathbf{F} = P(x,y)\mathbf{i} + Q(x,y)\mathbf{j}$ be a C^1 vector field on $D \subseteq \mathbb{R}^2$, where D is a region that satisfies Assumption 8.1. By $\mathbf{c}$ we denote its positively oriented boundary. Then

$$\int_{\mathbf{c}=\partial D} \mathbf{F}\cdot\mathbf{n}\,ds = \iint_D div\,\mathbf{F}\,dA,$$

where $\mathbf{n}$ is the outward unit normal vector to $\mathbf{c}$. ◀

PROOF: Parametrize $\mathbf{c}$ by $\mathbf{c}(t) = (x(t), y(t))$, $t \in [a,b]$. By the definition of the path integral (drop t to keep notation simple),

$$\begin{aligned}\int_{\mathbf{c}=\partial D} \mathbf{F}\cdot\mathbf{n}\,ds &= \int_a^b \mathbf{F}(\mathbf{c})\cdot\mathbf{n}\,\|\mathbf{c}'\|\,dt\\ &= \int_a^b (P(x,y)\mathbf{i}+Q(x,y)\mathbf{j})\cdot\frac{y'\mathbf{i}-x'\mathbf{j}}{\sqrt{(x')^2+(y')^2}}\cdot\sqrt{(x')^2+(y')^2}\,dt\\ &= \int_a^b (P(x,y)\,y' - Q(x,y)\,x')\,dt\end{aligned}$$

[see formula (5.6) in Section 5.3 and the text following it]

$$= \int_{\mathbf{c}} P(x, y)\, dy - Q(x, y)\, dx$$

[using Green's Theorem, see formula (8.4)]

$$= \iint_D \left(\frac{\partial P}{\partial x} + \frac{\partial Q}{\partial y} \right) dA = \iint_D \operatorname{div} \mathbf{F}\, dA.$$

The integral $\int_{\mathbf{c}} \mathbf{F} \cdot \mathbf{n}\, ds$, where $\mathbf{c}$ is a simple closed curve and $\mathbf{n}$ is the outward unit normal to $\mathbf{c}$, is called the *(outward) flux of* $\mathbf{F}$ *across* $\mathbf{c}$ or the *(outward) flux of* $\mathbf{F}$ *across the region enclosed by* $\mathbf{c}$.

▶ EXAMPLE 8.11

Compute the integral of the normal component of $\mathbf{F}(x, y) = x^2\mathbf{i} + xy\mathbf{j}$ along the unit circle, oriented counterclockwise.

SOLUTION Let us compute $\int_{\mathbf{c}} \mathbf{F} \cdot \mathbf{n}\, ds$ as a line integral first. Parametrize $\mathbf{c}$ by $\mathbf{c}(t) = (\cos t, \sin t)$, $t \in [0, 2\pi]$. Then $\mathbf{c}'(t) = (-\sin t, \cos t)$ is the tangent field and $\mathbf{n}(t) = (\cos t, \sin t)$ is the outward unit normal field ($\mathbf{n}$ is clearly an outward normal, since it is equal to the position vector of a point on the circle). It follows that $[\|\mathbf{c}'(t)\| = 1]$

$$\int_{\mathbf{c}} \mathbf{F} \cdot \mathbf{n}\, ds = \int_0^{2\pi} \mathbf{F}(\mathbf{c}(t)) \cdot \mathbf{n}(\mathbf{c}(t)) \|\mathbf{c}'(t)\|\, dt$$

$$= \int_0^{2\pi} (\cos^2 t, \sin t \cos t) \cdot (\cos t, \sin t)\, dt = \int_0^{2\pi} (\cos^3 t + \sin^2 t \cos t)\, dt$$

$$= \int_0^{2\pi} (\cos t(1 - \sin^2 t) + \sin^2 t \cos t)\, dt = \int_0^{2\pi} \cos t\, dt = 0.$$

By the Divergence Theorem in the plane,

$$\int_{\mathbf{c}} \mathbf{F} \cdot \mathbf{n}\, ds = \iint_D \operatorname{div} \mathbf{F}\, dA = \iint_{\{x^2+y^2 \le 1\}} 3x\, dA$$

[passing to polar coordinates]

$$= 3 \int_0^{2\pi} \left(\int_0^1 r^2 \cos\theta\, dr \right) d\theta = 3 \left(\int_0^{2\pi} \cos\theta\, d\theta \right) \left(\int_0^1 r^2\, dr \right) = 0.$$

▶ EXAMPLE 8.12 Gauss' Law for Electrostatic Fields

Consider the electrostatic field

$$\mathbf{E}(\mathbf{r}) = \frac{1}{4\pi\epsilon_0} \frac{Q}{\|\mathbf{r}\|^3} \mathbf{r},$$

in $\mathbb{R}^3 - \{(0, 0, 0)\}$ (where $\mathbf{r} = x\mathbf{i} + y\mathbf{j} + z\mathbf{k}$) due to a charge Q located at the origin.

Take a small sphere S_1 (of radius a) centered at the origin and let S_2 be any closed surface containing S_1. Denote by $\mathbf{n_1}$ and $\mathbf{n_2}$ the corresponding unit outward-pointing normal vectors; see Figure 8.15.

Consider the solid three-dimensional region W between S_1 and S_2: its boundary ∂W consists of S_1 and S_2 and is oriented by $\mathbf{n_2}$ on S_2 and by $-\mathbf{n_1}$ on S_1 (the normal $\mathbf{n_1}$ points into the solid W and is

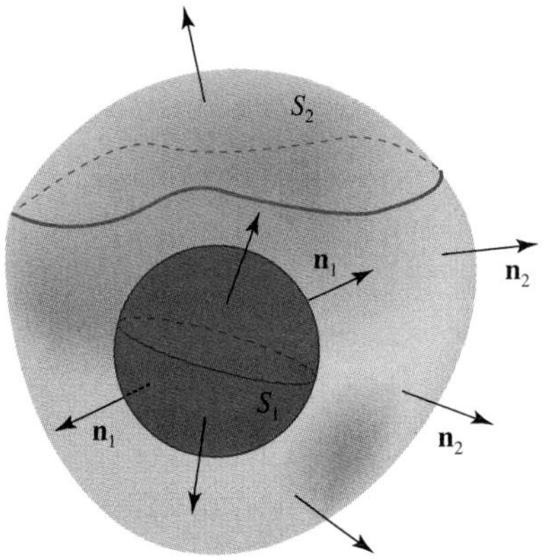

Figure 8.15 Surfaces S_1 and S_2 oriented by outward-pointing normal vectors.

the inward normal for W). By the Divergence Theorem,

$$\iiint_W div\,\mathbf{E}\,dV = \iint_{\partial W} \mathbf{E}\cdot d\mathbf{S} = \iint_{\partial W} \mathbf{E}\cdot\mathbf{n}\,dS$$

(∂W consists of S_1 and S_2)

$$= \iint_{S_1} \mathbf{E}\cdot(-\mathbf{n}_1)\,dS + \iint_{S_2} \mathbf{E}\cdot\mathbf{n}_2\,dS.$$

But $div\,\mathbf{E} = 0$ in W (that was computed in Section 4.6) and the above computation implies that

$$\iint_{S_2} \mathbf{E}\cdot\mathbf{n}_2\,dS = \iint_{S_1} \mathbf{E}\cdot\mathbf{n}_1\,dS$$

for *any* closed surface S_2 that contains S_1.

Let us compute the integral $\iint_{S_1} \mathbf{E}\cdot\mathbf{n}_1\,dS$. Instead of writing down a parametrization, we argue as follows: S_1 is a sphere, so a normal at a point (x, y, z) on S_1 has the same direction as the position vector $\mathbf{r} = x\mathbf{i} + y\mathbf{j} + z\mathbf{k}$ of that point. Hence, the unit outward normal is $\mathbf{n}_1 = \mathbf{r}/\|\mathbf{r}\|$. Consequently,

$$\iint_{S_1} \mathbf{E}\cdot\mathbf{n}_1\,dS = \iint_{S_1} \frac{1}{4\pi\epsilon_0}\frac{Q}{\|\mathbf{r}\|^3}\mathbf{r}\cdot\frac{\mathbf{r}}{\|\mathbf{r}\|}\,dS = \frac{Q}{4\pi\epsilon_0}\iint_{S_1}\frac{1}{\|\mathbf{r}\|^2}\,dS$$

($\|\mathbf{r}\| = a$, since S_1 is a sphere of radius a)

$$= \frac{Q}{4\pi\epsilon_0}\frac{1}{a^2}\iint_{S_1} dS = \frac{Q}{4\pi\epsilon_0}\frac{1}{a^2}4\pi a^2 = \frac{Q}{\epsilon_0}$$

[recall that $\iint_{S_1} dS =$ (surface) area of $S = 4\pi a^2$]. Therefore,

$$\iint_{S_2} \mathbf{E}\cdot\mathbf{n}_2\,dS = \frac{Q}{\epsilon_0},$$

that is, the electric flux of $\mathbf{E}$ through *any* closed surface that contains the charge Q (located at the origin) is Q/ϵ_0 (consequently, it does not depend on that surface).

This is a special case (a single charge Q is involved) of *Gauss' Law*, which states that the net charge enclosed by a (closed) surface S is

$$Q = \epsilon_0 \iint_S \mathbf{E}\cdot d\mathbf{S},$$

where $\mathbf{E}$ is the electrostatic field due to the charge and ϵ_0 is a constant. From the Divergence Theorem, it follows that

$$Q = \epsilon_0 \iint_S \mathbf{E} \cdot d\mathbf{S} = \epsilon_0 \iiint_W div\, \mathbf{E}\, dV,$$

where W is the three-dimensional solid enclosed by S (i.e., $\partial W = S$). On the other hand, $Q = \iiint_W \rho\, dV$ (where ρ is the charge density; see Section 7.5), and therefore,

$$\epsilon_0 \iiint_W div\, \mathbf{E}\, dV = \iiint_W \rho\, dV$$

and

$$\iiint_W \left(\frac{\rho}{\epsilon_0} - div\, \mathbf{E} \right) dV = 0.$$

Since W is an arbitrary three-dimensional region, we conclude that [see Section 7.5, part (f) in the Properties of Integrals]

$$div\, \mathbf{E} - \frac{\rho}{\epsilon_0} = 0.$$

This equation relates the charge distribution and the resulting electric field; it is called *Maxwell's first equation* (see Section 8.5). ◀

Fluid Flow and Heat Flow

With the help of the Divergence Theorem of Gauss, we can give an interpretation of divergence related to fluid flow.

Let $\mathbf{F}(x, y, z)$ be the velocity vector field of a fluid. Assume that the flow is *steady,* so that $\mathbf{F}$ does not depend on time. Denote by $\rho(x, y, z)$ the mass density of the fluid. Let P be a point inside a small closed surface S placed in the flow, as in Figure 8.16.

The outward flux across S is given by the integral $\iint_S \rho\, \mathbf{F} \cdot d\mathbf{S}$, where $\rho\mathbf{F}$ describes the rate of flow of mass per unit area. The Divergence Theorem implies that

$$\iint_S \rho\, \mathbf{F} \cdot d\mathbf{S} = \iiint_W div\, (\rho\, \mathbf{F})\, dV, \tag{8.9}$$

where W is a three-dimensional solid enclosed by S.

By the Mean Value Theorem for integrals [use (7.12) or (7.14) in Section 7.5 with $f = div\, (\rho\, \mathbf{F})$, $\mathcal{M} = W$, and $d\mu = dV$], it follows that there is a point Q in W (i.e., inside S)

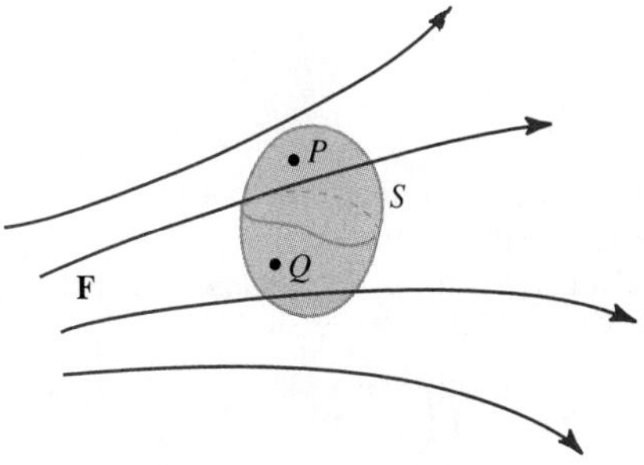

Figure 8.16 Closed surface in a fluid flow.

such that $[v(W)$ denotes the volume of $W]$

$$div\,(\rho\,\mathbf{F})(Q) = \frac{1}{v(W)} \iiint_W div\,(\rho\,\mathbf{F})\,dV.$$

Combining the above, we get

$$div\,(\rho\,\mathbf{F})(Q) = \frac{1}{v(W)} \iint_S \rho\,\mathbf{F} \cdot d\mathbf{S},$$

and hence,

$$div\,(\rho\,\mathbf{F})(P) = \lim_{W \to P} \frac{1}{v(W)} \iint_S \rho\,\mathbf{F} \cdot d\mathbf{S}. \tag{8.10}$$

The limit is taken as W shrinks to a point, and therefore, Q approaches P (since both Q and P are inside S). The left side of (8.10) is called the *source intensity at* P. Equation (8.10) states that the divergence at P is the net rate of outward flux (at P) per unit volume.

A point P is called a *source* if $div\,(\rho\mathbf{F})$ at P is positive (then the net flow is outward, hence the name). If $div\,(\rho\mathbf{F})(P)$ is negative, the net flow is inward and P is called a *sink.*

Next, we discuss an application of the Divergence Theorem related to *heat flow*. Denote by $u(x, y, z, t)$ the temperature at a point (x, y, z) in a solid W, measured at time t. Assume that the function u is twice continuously differentiable. If the temperature inside W is not constant, heat transfer will occur from regions with a higher temperature toward regions with a lower temperature. Heat transfers are described by the *heat flow equation* (see Example 2.90 in Section 2.7 and Example 4.66 in Section 4.6)

$$\mathbf{F} = -k\,\nabla u, \tag{8.11}$$

where k is a positive constant (called the *heat conductivity*), and ∇u is computed by keeping t fixed.

The gradient of u points in the direction of the largest increase in temperature, and heat flows in the opposite direction, that is, in the direction of the largest decrease in u; see Figure 8.17. This explains the appearance of the minus sign in formula (8.11). Since the

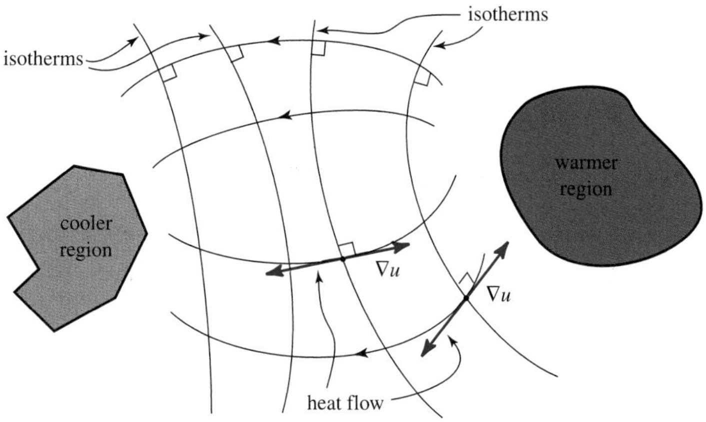

Figure 8.17 Heat flow.

gradient vector field is always perpendicular to level curves (in this case, of the temperature function), heat flow occurs in directions perpendicular to isotherms.

Consider a three-dimensional region W with boundary S inside the flow. The rate of heat flow across S out of W is given by

$$\iint_S \mathbf{F} \cdot d\mathbf{S} = \iint_S \mathbf{F} \cdot \mathbf{n}\, dS,$$

where $\mathbf{n}$ is the unit outward normal. From (8.11) and the Divergence Theorem, we obtain

$$\begin{aligned} \iint_S \mathbf{F} \cdot d\mathbf{S} &= \iint_S -k\,\nabla u \cdot d\mathbf{S} \\ &= -k \iiint_W div\,(\nabla u)\, dV = -k \iiint_W \Delta u\, dV, \end{aligned} \tag{8.12}$$

where $\Delta u = div\,(\nabla u)$ denotes the Laplacian of u.

The net amount of heat inside S is given by $\mathcal{H} = \iiint_W \sigma\rho\, u\, dV$, where $\sigma =$ constant is the *specific heat* of the material and ρ denotes its mass density. We assume that $\partial\rho/\partial t = 0$; that is, that the density remains constant with respect to time. The time rate of change of $\mathcal{H}$ is [the function $\sigma\rho\, u$ is differentiable, and by (7.13) in Section 7.5 we are allowed to switch the time derivative and the triple integral]

$$\frac{\partial\mathcal{H}}{\partial t} = \frac{\partial}{\partial t}\left(\iiint_W \sigma\rho\, u\, dV\right) = \iiint_W \frac{\partial}{\partial t}(\sigma\rho\, u)\, dV = \iiint_W \sigma\rho\frac{\partial u}{\partial t}\, dV, \tag{8.13}$$

by the product rule $\partial(\sigma\rho u)/\partial t = (\partial\sigma/\partial t)\rho u + (\partial\rho/\partial t)\sigma u + (\partial u/\partial t)\sigma\rho = (\partial u/\partial t)\sigma\rho$, since ρ and σ are constant with respect to time.

Since the time rate of change of the heat inside S (which is the boundary of W) equals the rate at which heat "flows" from the outside *into* W, it follows that

$$\frac{\partial\mathcal{H}}{\partial t} = \iint_S \mathbf{F} \cdot (-\mathbf{n})\, dS = -\iint_S \mathbf{F} \cdot d\mathbf{S},$$

where $-\mathbf{n}$ is the inward normal. Using (8.12) and (8.13), we conclude that

$$\iiint_W \sigma\rho\frac{\partial u}{\partial t}\, dV = k \iiint_W \Delta u\, dV,$$

that is,

$$\iiint_W \left(\sigma\rho\frac{\partial u}{\partial t} - k\Delta u\right) dV = 0,$$

where W is *any* solid three-dimensional region. Hence [see Properties of Integrals, part (f) in Section 7.5], we get $\sigma\rho\,\partial u/\partial t = k\Delta u$, or, by taking $c^2 = k\sigma^{-1}\rho^{-1}$,

$$\frac{\partial u}{\partial t} = c^2\Delta u,$$

which is the *heat equation*.

▶ EXERCISES 8.2

1. Consider the vector field $\mathbf{F} = \mathbf{r}/\|\mathbf{r}\|^3$, where $\mathbf{r} \neq \mathbf{0}$.

(a) Show that $div\,\mathbf{F} = 0$.

(b) Find $\iint_{S_1} \mathbf{F} \cdot d\mathbf{S}$, where S_1 is the sphere of radius 1 centered at the origin, oriented by the outward normal. Can the Divergence Theorem be used to compute this integral?

(c) If possible, use the Divergence Theorem to compute $\iint_{S_2} \mathbf{F} \cdot d\mathbf{S}$, where S_2 is the sphere of radius 1 centered at the point $(0, 0, 2)$, oriented by the outward normal.

2. Let $\mathbf{F}$ be the velocity vector of a fluid, and assume that the only information known about it is that $div\,\mathbf{F}(3, 0, -1) = 4$. Approximate the flux out of a sphere of radius 0.1 centered at $(3, 0, -1)$. Give a reason why your answer is an approximation and not the actual value of the flux.

3. Assume that $\mathbf{F}$ is a vector field such that $div\,\mathbf{F}(x, y, z) = 3$ for all $(x, y, z) \in \mathbb{R}^3$. Find the flux of $\mathbf{F}$ out of the parallelepiped with sides 3, 2, and 5.

4. Find the flux of the vector field $\mathbf{F} = \mathbf{c} \times \mathbf{r}$, where $\mathbf{c}$ is a constant vector and $\mathbf{r} = x\mathbf{i} + y\mathbf{j} + z\mathbf{k}$, out of any sphere of radius 1.

Exercises 5 to 13: Evaluate the surface integral $\iint_S \mathbf{F} \cdot d\mathbf{S}$, where S is a closed surface oriented by an outward normal.

5. $\mathbf{F}(x, y, z) = (y^2 + \sin z)\mathbf{i} + (e^{\sin z} + 2)\mathbf{j} + (xy + \ln x)\mathbf{k}$, S is the surface of the cube $0 \leq x, y, z \leq 1$

6. $\mathbf{F}(x, y, z) = (x^2 + z^2)\mathbf{i} + (y^2 + z^2)\mathbf{k}$, S is the surface of the parallelepiped $0 \leq x, y \leq 2$, $0 \leq z \leq 4$

7. $\mathbf{F}(x, y, z) = (x + y^2 + 1)\mathbf{i} + (y + xz)\mathbf{j}$, S consists of the part of the cone $z^2 = x^2 + y^2$ bounded by the disks $0 \leq x^2 + y^2 \leq 1$, $z = 1$, and $0 \leq x^2 + y^2 \leq 4$, $z = 2$

8. $\mathbf{F}(x, y, z) = (2x + 3y)\mathbf{i} - (4y + 3z)\mathbf{j} + 4z\mathbf{k}$, S consists of the paraboloid $z = x^2 + y^2$, $0 \leq z \leq 1$, and the disk $0 \leq x^2 + y^2 \leq 1$, $z = 1$

9. $\mathbf{F}(x, y, z) = -e^x \cos y\mathbf{i} + e^x \sin y\mathbf{j} + \mathbf{k}$, S is the surface of the sphere $x^2 + y^2 + z^2 = 1$

10. $\mathbf{F}(x, y, z) = x^{-1}\mathbf{i} + z^{-1}\mathbf{j} - yz^{-1}\mathbf{k}$, S is the surface of the parallelepiped $1 \leq x \leq 2$, $2 \leq y, z \leq 4$

11. $\mathbf{F}(x, y, z) = x^2\mathbf{i} + xy\mathbf{j} + xz\mathbf{k}$, S consists of the upper hemisphere $x^2 + y^2 + z^2 = 1$, $z \geq 0$, and the disk $0 \leq x^2 + y^2 \leq 1$ in the xy-plane

12. $\mathbf{F}(x, y, z) = 2x\mathbf{i} + xy^2\mathbf{j} + xyz\mathbf{k}$, S is the boundary of the three-dimensional solid inside $x^2 + y^2 = 2$, outside $x^2 + y^2 = 1$, and between the planes $z = 0$ and $z = 4$

13. $\mathbf{F}(x, y, z) = ye^z\mathbf{i} + yz\mathbf{k}$, S is the surface of the tetrahedron in the first octant, bounded by the plane $x + 2y + z = 4$

14. Let $\mathbf{r} = x\mathbf{i} + y\mathbf{j} + z\mathbf{k}$. Prove that $\iint_S \mathbf{r} \cdot d\mathbf{S} = 3v(W)$, where $v(W)$ is the volume of the three-dimensional region W, bounded by S (i.e., $\partial W = S$).

15. Let W be a solid three-dimensional region that satisfies Assumption 8.2, and denote by S its positively oriented boundary. Prove that $\iiint_W \|\mathbf{r}\|^{-2}\, dV = \iint_S \|\mathbf{r}\|^{-2}\mathbf{r} \cdot d\mathbf{S}$.

16. Use the Divergence Theorem to compute $\iint_S (x + y)\, dS$, where S consists of the upper hemisphere $x^2 + y^2 + z^2 = 1$, $z \geq 0$ (oriented positively), and the disk $0 \leq x^2 + y^2 \leq 1$ in the xy-plane.

17. Use the Divergence Theorem to compute the surface integral $\iint_S xyz\, dS$, where S is the sphere $x^2 + y^2 + z^2 = 1$, oriented by the outward-pointing normal.

18. Compute $\iint_S (x^2 + y^2)\, dS$, where S consists of the part of the paraboloid $z = 2(x^2 + y^2)$ between $z = 0$ and $z = 4$, together with the top disk $0 \leq x^2 + y^2 \leq 2$, $z = 4$, oriented by the outward normal.

19. Let W be a solid three-dimensional region that satisfies Assumption 8.2, bounded by a closed, positively oriented surface S. Show that, for C^2 functions f and g, $\iint_S f\nabla g\, d\mathbf{S} = \iiint_W (f\Delta g + \nabla f \cdot \nabla g)\, dV$.

20. Compute $\iint_S \mathbf{c} \cdot d\mathbf{S}$, if $\mathbf{c}$ is a constant vector field and S is a closed surface.

21. Let W be a solid three-dimensional region that satisfies Assumption 8.2, bounded by a closed, positively oriented surface S. Show that $\iint_S D_{\mathbf{n}} f\, dS = \iiint_W \Delta f\, dV$, where f is of class C^2 and $D_{\mathbf{n}} f$ denotes the directional derivative of f in the direction of the outward unit normal to S.

22. Let $\mathbf{F}$ be a C^1 vector field in $\mathbb{R}^2$. Assume that $\int_{\mathbf{c}} \mathbf{F} \cdot \mathbf{n}\, ds = 0$ for any closed curve $\mathbf{c}$ in $\mathbb{R}^2$ (with the outward normal $\mathbf{n}$). What (if anything) can be said about the divergence of $\mathbf{F}$?

23. Let $\mathbf{F} = y\mathbf{i}/(x^2 + y^2) - x\mathbf{j}/(x^2 + y^2)$. Compute the outward flux $\int_{\mathbf{c}} \mathbf{F} \cdot \mathbf{n}\, ds$ of $\mathbf{F}$ across the rectangle $R = [-1, 1] \times [-1, 2]$.

Exercises 24 to 28: Find the outward flux $\int_{\mathbf{c}} \mathbf{F} \cdot \mathbf{n}\, ds$ and the counterclockwise circulation $\int_{\mathbf{c}} \mathbf{F} \cdot d\mathbf{s}$ of the vector field $\mathbf{F}$ along the curve $\mathbf{c}$.

24. $\mathbf{F}(x, y) = (2x - 1 + y)\mathbf{i} - (x - 3y)\mathbf{j}$, $\mathbf{c}$ is the square with the vertices $(-1, -1)$, $(1, -1)$, $(1, 1)$, and $(-1, 1)$

25. $\mathbf{F}(x, y) = (x^2 + y^2)\mathbf{i} - xy\mathbf{j}$, $\mathbf{c}$ is the triangle defined by $y = x$, $y = 2x$, and $x = 1$

26. $\mathbf{F}(x, y) = 2xy\mathbf{i} + 3y^2\mathbf{j}$, $\mathbf{c}$ is the boundary of the region in the first quadrant defined by $y = x^2$ and $y = 1$

27. $\mathbf{F}(x, y) = e^x e^y \mathbf{i} + 2e^y \mathbf{j}$, $\mathbf{c}$ is the boundary of the rectangle $R = [0, 3] \times [0, 4]$ oriented counterclockwise

28. $\mathbf{F}(x, y) = e^x \cos y\mathbf{i} + (xy + e^x \sin y)\mathbf{j}$, $\mathbf{c}$ is the boundary of the region defined by the curves $y = \ln x$, $y = 0$, and $x = e$

29. Consider the vector field $\mathbf{F}(x, y, z) = e^{-x^2}\mathbf{k}$, and let W be the cube $[0, 1] \times [0, 1] \times [0, 1]$. The boundary $\partial W = S$ consists of six squares.

(a) Try to evaluate $\iint_S \mathbf{F} \cdot d\mathbf{S}$ directly, that is, by computing surface integrals.

(b) Use the Divergence Theorem to compute $\iint_S \mathbf{F} \cdot d\mathbf{S}$.

▶ 8.3 STOKES' THEOREM

Stokes' Theorem is similar in spirit to Green's Theorem: it relates the path integral of a vector field around a closed curve $\mathbf{c}$ in $\mathbb{R}^3$ to an integral over a surface S whose boundary is $\mathbf{c}$. As usual, we have to make precise the assumptions on the curves and surfaces involved. We will do it in two stages: first for a surface that is the graph of a function $z = f(x, y)$ and then for a general parametrized surface.

Let S be a surface defined as the graph of a function $z = f(x, y)$, where $(x, y) \in D$. Assume that the domain $D \subseteq \mathbb{R}^2$ is a region to which Green's Theorem applies (see Assumption 8.1 in Section 8.1). The boundary ∂D of D is a simple closed curve (i.e., a closed curve that does not intersect itself), or several such curves, oriented positively (as we walk along the boundary, the region D is on our left). Parametrize S by (for convenience, we depart from using the standard parameters u and v and use x and y instead)

$$\mathbf{r}(x, y) = (x, y, f(x, y)), \qquad (x, y) \in D,$$

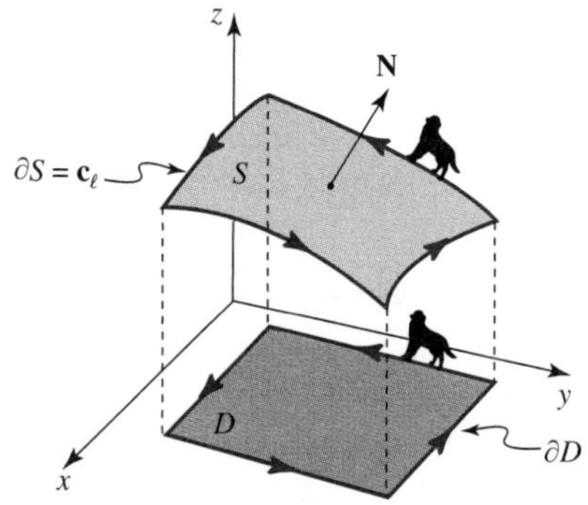

Figure 8.18 "Lifting" the positive orientation to the boundary of the surface.

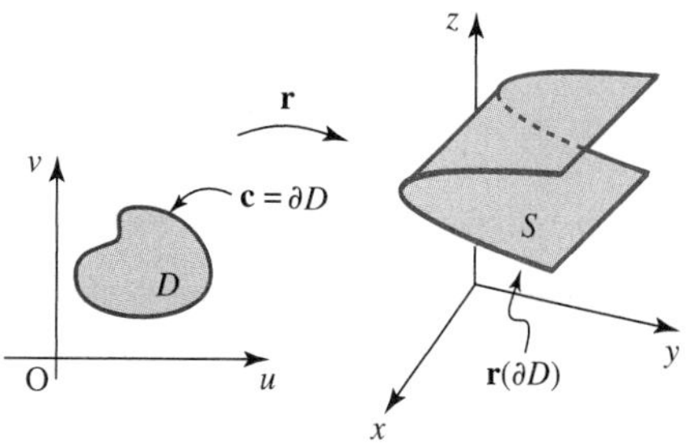

Figure 8.19 An attempt to define the boundary of a parametrized surface.

and choose the upward normal $\mathbf{N} = (-\partial f(x, y)/\partial x, -\partial f(x, y)/\partial y, 1)$ as the orientation of S. The positive orientation of the boundary ∂S of S is defined by "lifting" the positive orientation of ∂D, as shown in Figure 8.18.

In other words, ∂S is oriented as follows: as we walk on the positive side of S (the outward normal points away from it) along the boundary, the surface S is on our left.

To be precise, let $\mathbf{c}(t) = (x(t), y(t))$, $t \in [a, b]$, be an orientation-preserving parametrization of the boundary of D. The boundary curve $\mathbf{c}_\ell = \partial S$ can be parametrized as

$$\mathbf{c}_\ell(t) = (x(t), y(t), f(x(t), y(t))), \qquad t \in [a, b].$$

The orientation coming from this parametrization (i.e., the direction of increasing values for t) defines the positive orientation of the boundary ∂S of S.

Now let us look at the case of a parametrized surface S given by $\mathbf{r} = \mathbf{r}(u, v)$: $D \to \mathbb{R}^3$, where D is a region in $\mathbb{R}^2$ to which Green's Theorem applies. Consider the boundary curve $\mathbf{c} = \partial D$ of D. We are tempted to define the boundary of S as the image of $\mathbf{c}$ under $\mathbf{r}$; see Figure 8.19. However, that will not work.

Take, for example, the parametrization of the sphere (discussed in Example 7.4 of Section 7.1) $\mathbf{r}(u, v) = (\cos v \cos u, \cos v \sin u, \sin v)$, $(u, v) \in D$, where D is the rectangle $[0, 2\pi] \times [-\pi/2, \pi/2]$. Horizontal segments $v = -\pi/2$ and $v = \pi/2$ that form part of the boundary ∂D map to the North and South Poles, respectively, and the vertical segments $u = 0$ and $u = 2\pi$ are both mapped to the same semicircle in the xz-plane. In other words, the boundary of D is mapped into the half-meridian of the sphere; see Figure 8.20 (this is certainly not what we would consider the boundary of a sphere).

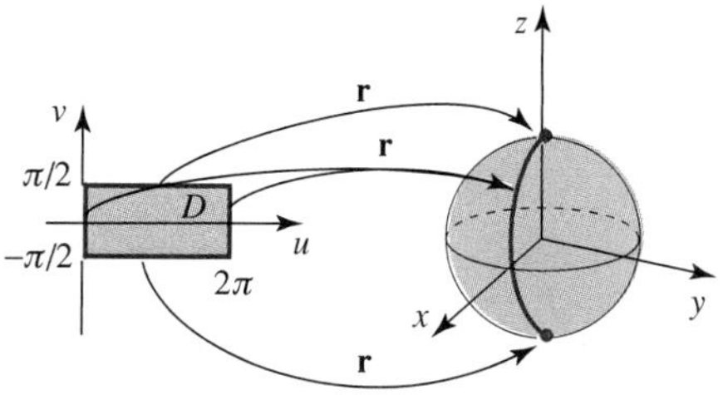

Figure 8.20 A half-meridian on the sphere is not its boundary.

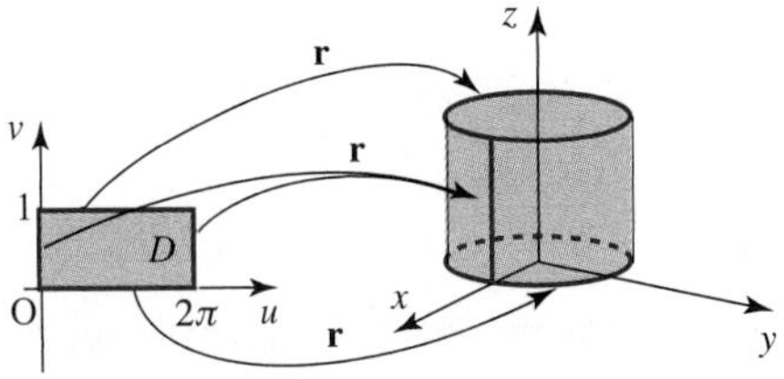

Figure 8.21 "Weird boundary" of the cylinder.

Parametrize the cylinder by $\mathbf{r}(u, v) = (\cos u, \sin u, v)$, $(u, v) \in D = [0, 2\pi] \times [0, 1]$ as in Example 7.2 in Section 7.1. The image of ∂D under $\mathbf{r}$ consists of the top and bottom circles joined by a vertical segment, as shown in Figure 8.21. This curve (it is not simple!) does not represent our common-sense notion of the boundary of a cylinder.

Analyzing these examples, we notice that problems arise when different points on the boundary of D map to the same point. Therefore, to get a meaningful concept of the boundary ∂S of a parametrized surface, we will have to assume that $\mathbf{r}$ is one-to-one on D. In that case, ∂S is defined as the image $\mathbf{r}(\partial D)$ of the boundary ∂D of D under $\mathbf{r}$. If ∂D is parametrized by the map $\mathbf{c}(t) = (x(t), y(t))$, $t \in [a, b]$ (that is orientation-preserving in the sense that, for motion along $\mathbf{c}$, the region D is on the left), then the boundary ∂S has the "lifted" parametrization $\mathbf{c}_\ell = \mathbf{r}(\mathbf{c}(t))$ that determines its orientation.

In other words, the situation for one-to-one parametrizations is the same as for surfaces defined as graphs of functions.

Some surfaces might not have a one-to-one parametrization (or if there is one, it is too complicated to work with). In these cases, we try representing the surface as the graph of a function of two variables (or break it up into pieces and represent each piece as the graph).

We are now ready to give the statement of the theorem.

THEOREM 8.4 Stokes' Theorem

Let S be an oriented surface parametrized by a one-to-one parametrization $\mathbf{r}: D \subseteq \mathbb{R}^2 \to \mathbb{R}^3$, where D is a region to which Green's Theorem applies (i.e., D satisfies Assumption 8.1). Let ∂S be the positively oriented piecewise smooth boundary of S. Then

$$\int_{\partial S} \mathbf{F} \cdot d\mathbf{s} = \iint_S \mathit{curl}\, \mathbf{F} \cdot d\mathbf{S},$$

for a C^1 vector field $\mathbf{F}$ on S. ◀

Note that the parametrization $\mathbf{r}(u, v) = (u, v, f(u, v))$ of the graph of the function $f(x, y)$ is always one-to-one.

If a surface S has no boundary (for instance, S could be a sphere or an ellipsoid), then $\iint_S \mathit{curl}\, \mathbf{F} \cdot d\mathbf{S} = 0$ (see Exercise 2 for an intuitive argument).

PROOF OF THEOREM 8.4: We will give a proof of the theorem in the case where S is the graph of a C^2 function $z = f(x, y)$, $(x, y) \in D$ and D is an elementary region in $\mathbb{R}^2$.

Let $\mathbf{F}(x, y, z) = (P(x, y, z), Q(x, y, z), R(x, y, z))$ be a C^1 vector field defined on S. Parametrize the boundary of S (as explained in the introduction) by

$$\mathbf{c}_\ell(t) = (x(t), y(t), f(x(t), y(t))), \qquad t \in [a, b],$$

where $\mathbf{c}(t) = (x(t), y(t))$ represents the boundary of D. For simplicity, we drop independent variables from the notation. The path integral of $\mathbf{F}$ along the boundary of S is

$$\int_{\mathbf{c}_\ell} \mathbf{F} \cdot d\mathbf{s} = \int_a^b \mathbf{F} \cdot \mathbf{c}_\ell' \, dt = \int_a^b (P, Q, R) \cdot \left(\frac{dx}{dt}, \frac{dy}{dt}, \frac{dz}{dt}\right) dt$$
$$= \int_a^b \left(P\frac{dx}{dt} + Q\frac{dy}{dt} + R\frac{dz}{dt}\right) dt.$$

By the chain rule [$z = f(x, y)$, and $x = x(t)$, $y = y(t)$], with the usual convenient sloppiness in notation (dz/dx instead of df/dx, etc.), we get

$$\frac{dz}{dt} = \frac{\partial z}{\partial x}\frac{dx}{dt} + \frac{\partial z}{\partial y}\frac{dy}{dt},$$

and therefore,

$$\int_{\mathbf{c}_\ell} \mathbf{F} \cdot d\mathbf{s} = \int_a^b \left(P\frac{dx}{dt} + Q\frac{dy}{dt} + R\left(\frac{\partial z}{\partial x}\frac{dx}{dt} + \frac{\partial z}{\partial y}\frac{dy}{dt}\right)\right) dt$$
$$= \int_a^b \left(\left(P + R\frac{\partial z}{\partial x}\right)\frac{dx}{dt} + \left(Q + R\frac{\partial z}{\partial y}\right)\frac{dy}{dt}\right) dt$$
$$= \int_{\mathbf{c}} \left(P + R\frac{\partial z}{\partial x}\right) dx + \left(Q + R\frac{\partial z}{\partial y}\right) dy$$

[for the last step, see formula (5.6) in Section 5.3 and the text following it]

$$= \iint_D \left(\frac{\partial}{\partial x}\left(Q + R\frac{\partial z}{\partial y}\right) - \frac{\partial}{\partial y}\left(P + R\frac{\partial z}{\partial x}\right)\right) dA,$$

by Green's Theorem; see (8.4) in Section 8.1. Consider the integrand in this double integral. Since $Q = Q(x, y, z) = Q(x, y, f(x, y))$, the chain rule implies that

$$\frac{\partial}{\partial x}(Q) = D_1 Q \frac{\partial x}{\partial x} + D_2 Q \frac{\partial y}{\partial x} + D_3 Q \frac{\partial f(x, y)}{\partial x},$$

where $D_i Q$ represents the derivative of Q with respect to its ith variable ($i = 1, 2, 3$). Writing $D_1 Q$ as $\partial Q/\partial x$, $D_3 Q$ as $\partial Q/\partial z$, and $\partial f/\partial x$ as $\partial z/\partial x$, we get

$$\frac{\partial}{\partial x}(Q) = \frac{\partial Q}{\partial x} + \frac{\partial Q}{\partial z}\frac{\partial z}{\partial x}.$$

Other partial derivatives are computed similarly. It follows that, by the product and chain rules,

$$\frac{\partial}{\partial x}\left(Q + R\frac{\partial z}{\partial y}\right) - \frac{\partial}{\partial y}\left(P + R\frac{\partial z}{\partial x}\right)$$
$$= \frac{\partial Q}{\partial x} + \frac{\partial Q}{\partial z}\frac{\partial z}{\partial x} + \left(\frac{\partial R}{\partial x} + \frac{\partial R}{\partial z}\frac{\partial z}{\partial x}\right)\frac{\partial z}{\partial y} + R\frac{\partial^2 z}{\partial x \partial y}$$
$$- \left(\frac{\partial P}{\partial y} + \frac{\partial P}{\partial z}\frac{\partial z}{\partial y} + \left(\frac{\partial R}{\partial y} + \frac{\partial R}{\partial z}\frac{\partial z}{\partial y}\right)\frac{\partial z}{\partial x} + R\frac{\partial^2 z}{\partial y \partial x}\right)$$
$$= \frac{\partial Q}{\partial x} - \frac{\partial P}{\partial y} + \frac{\partial z}{\partial x}\left(\frac{\partial Q}{\partial z} - \frac{\partial R}{\partial y}\right) + \frac{\partial z}{\partial y}\left(\frac{\partial R}{\partial x} - \frac{\partial P}{\partial z}\right),$$

and consequently,

$$\int_{\mathbf{c}_\ell} \mathbf{F}\cdot d\mathbf{s} = \iint_D \left(\frac{\partial Q}{\partial x} - \frac{\partial P}{\partial y} + \frac{\partial z}{\partial x}\left(\frac{\partial Q}{\partial z} - \frac{\partial R}{\partial y}\right) + \frac{\partial z}{\partial y}\left(\frac{\partial R}{\partial x} - \frac{\partial P}{\partial z}\right)\right) dA.$$

Notice that we used the equality of mixed partials $\partial^2 z/\partial x\,\partial y = \partial^2 z/\partial y\,\partial x$ to cancel two terms in the above computation [this is where we need $z = f(x, y)$ to be of class C^2].

On the other hand, since

$$curl\,\mathbf{F} = \left(\frac{\partial R}{\partial y} - \frac{\partial Q}{\partial z}, \frac{\partial P}{\partial z} - \frac{\partial R}{\partial x}, \frac{\partial Q}{\partial x} - \frac{\partial P}{\partial y}\right)$$

and (use z instead of f) $\mathbf{N} = (-\partial z/\partial x, -\partial z/\partial y, 1)$, it follows that

$$\begin{aligned}\iint_S curl\,\mathbf{F}\cdot d\mathbf{S} &= \iint_D curl\,\mathbf{F}\cdot\mathbf{N}\,dA \\ &= \iint_D \left(\frac{\partial R}{\partial y} - \frac{\partial Q}{\partial z}, \frac{\partial P}{\partial z} - \frac{\partial R}{\partial x}, \frac{\partial Q}{\partial x} - \frac{\partial P}{\partial y}\right)\cdot\left(-\frac{\partial z}{\partial x}, -\frac{\partial z}{\partial y}, 1\right) dA \\ &= \iint_D \left(-\frac{\partial z}{\partial x}\left(\frac{\partial R}{\partial y} - \frac{\partial Q}{\partial z}\right) - \frac{\partial z}{\partial y}\left(\frac{\partial P}{\partial z} - \frac{\partial R}{\partial x}\right) + \frac{\partial Q}{\partial x} - \frac{\partial P}{\partial y}\right) dA.\end{aligned}$$

We are done! Both $\int_{\mathbf{c}_\ell} \mathbf{F}\cdot d\mathbf{s}$ and $\iint_S curl\,\mathbf{F}\cdot d\mathbf{S}$ have been shown to be equal to the same integral, and are therefore equal to each other. ◀

▶ EXAMPLE 8.13

Evaluate the path integral $\int_{\mathbf{c}} \mathbf{F}\cdot d\mathbf{s}$, where $\mathbf{F} = 4z\mathbf{i} - 2x\mathbf{j} + 2x\mathbf{k}$, and $\mathbf{c}$ is the intersection of the cylinder $x^2 + y^2 = 1$ and the plane $z = y + 1$, oriented counterclockwise as seen by a person standing on the plane; see Figure 8.22.

SOLUTION First, we compute $\int_{\mathbf{c}} \mathbf{F}\cdot d\mathbf{s}$ directly. Parametrize $\mathbf{c}$ by $\mathbf{c}(t) = (\cos t, \sin t, \sin t + 1)$, $t \in [0, 2\pi]$. Then $\mathbf{c}'(t) = (-\sin t, \cos t, \cos t)$ and

$$\begin{aligned}\int_{\mathbf{c}} \mathbf{F}\cdot d\mathbf{s} &= \int_0^{2\pi} (4\sin t + 4, -2\cos t, 2\cos t)\cdot(-\sin t, \cos t, \cos t)\,dt \\ &= -4\int_0^{2\pi} \sin^2 t\,dt - 4\int_0^{2\pi} \sin t\,dt \\ &= -4\left(\frac{1}{2}t - \frac{1}{4}\sin 2t\right)\Big|_0^{2\pi} + 4\cos t\Big|_0^{2\pi} = -4\pi.\end{aligned}$$

Since

$$curl\,\mathbf{F} = \begin{vmatrix} \mathbf{i} & \mathbf{j} & \mathbf{k} \\ \partial/\partial x & \partial/\partial y & \partial/\partial z \\ 4z & -2x & 2x \end{vmatrix} = 2\mathbf{j} - 2\mathbf{k},$$

Stokes' Theorem implies that (the assumption on the orientation is satisfied!)

$$\int_c (4z\mathbf{i} - 2x\mathbf{j} + 2x\mathbf{k})\cdot d\mathbf{s} = \iint_S (2\mathbf{j} - 2\mathbf{k})\cdot d\mathbf{S},$$

where S is the part of the surface of the plane $z = y + 1$ cut out by the cylinder $x^2 + y^2 = 1$.

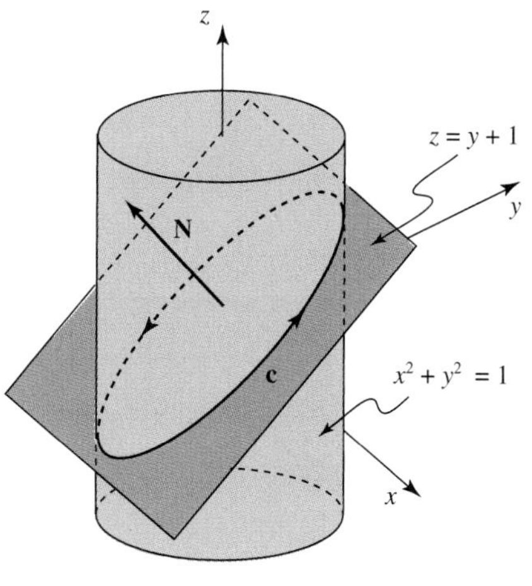

Figure 8.22 The cylinder $x^2 + y^2 = 1$ and the plane $z = y + 1$ of Example 8.13.

Parametrize S by $\mathbf{r}(u, v) = (u, v, v + 1)$, $u^2 + v^2 \leq 1$. Then $\mathbf{T}_u = (1, 0, 0)$, $\mathbf{T}_v = (0, 1, 1)$, $\mathbf{N} = (0, -1, 1)$, and

$$\iint_S (2\mathbf{j} - 2\mathbf{k}) \cdot d\mathbf{S} = \iint_{\{u^2+v^2 \leq 1\}} (0, 2, -2) \cdot (0, -1, 1)\, dA$$

$$= \iint_{\{u^2+v^2 \leq 1\}} (-4)\, dA = -4 \operatorname{area}(\{u^2 + v^2 \leq 1\}) = -4\pi.$$

▶ EXAMPLE 8.14

Evaluate $\int_{\mathbf{c}} \mathbf{F} \cdot d\mathbf{s}$, where $\mathbf{F} = 2yz\mathbf{i} + xz\mathbf{j} + xy\mathbf{k}$, and $\mathbf{c}$ is the intersection of the cylinder $x^2 + y^2 = 1$ and the parabolic sheet $z = y^2$, with the orientation indicated in Figure 8.23.

SOLUTION By Stokes' Theorem

$$\int_{\mathbf{c}} \mathbf{F} \cdot d\mathbf{s} = \iint_S \mathit{curl}\, \mathbf{F} \cdot d\mathbf{S},$$

where S is the surface $z = y^2$, with $x^2 + y^2 \leq 1$. Parametrize S by $\mathbf{r}(u, v) = (u, v, v^2)$, $u^2 + v^2 \leq 1$. Then $\mathbf{T}_u = (1, 0, 0)$, $\mathbf{T}_v = (0, 1, 2v)$, and $\mathbf{N} = (0, -2v, 1)$. The normal $\mathbf{N}$ points upward, and with

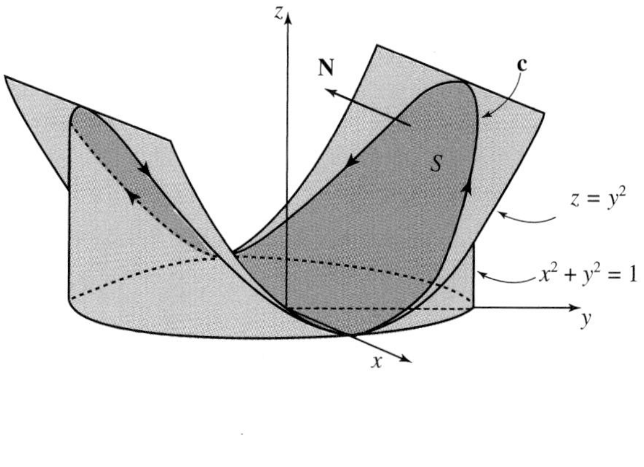

Figure 8.23 Intersection of the cylinder $x^2 + y^2 = 1$ and the parabolic sheet $z = y^2$.

the given orientation of **c**, the surface S lies to its left. Now $curl\,\mathbf{F} = (0, y, -z)$, and therefore,

$$\int_{\mathbf{c}} \mathbf{F} \cdot d\mathbf{s} = \iint_S curl\,\mathbf{F} \cdot d\mathbf{S} = \iint_S curl\,\mathbf{F} \cdot \mathbf{N}\, dS$$

$$= \iint_{\{u^2+v^2\le 1\}} (0, v, -v^2) \cdot (0, -2v, 1)\, dA = \iint_{\{u^2+v^2\le 1\}} (-3v^2)\, dA$$

(passing to polar coordinates $u = r\cos\theta$, $v = r\sin\theta$, $dA = r dr d\theta$)

$$= -3\int_0^{2\pi}\left(\int_0^1 r^2 \sin^2\theta\, r dr\right) d\theta = -3\left(\int_0^{2\pi} \sin^2\theta\, d\theta\right)\left(\int_0^1 r^3\, dr\right)$$

$$= -3\left(\frac{1}{2}\theta - \frac{1}{4}\sin 2\theta\right)\Big|_0^{2\pi}\left(\frac{1}{4}r^4\right)\Big|_0^1 = -\frac{3\pi}{4}.$$

◀

▶ EXAMPLE 8.15

Evaluate the path integral of the vector field $\mathbf{F} = z^2\mathbf{i} + x^2\mathbf{j} + y^2\mathbf{k}$ along the boundary **c** of the square $0 \le x \le 1, 0 \le y \le 1$ in the plane $z = 1$, oriented clockwise as seen from the origin; see Figure 8.24.

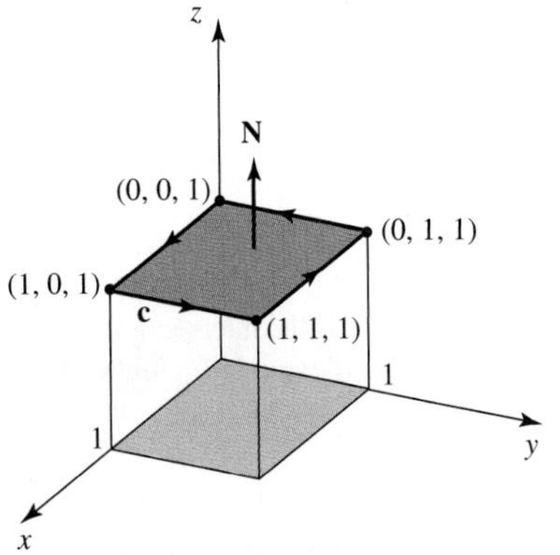

Figure 8.24 Square in Example 8.15.

SOLUTION Since $curl\,\mathbf{F} = 2y\mathbf{i} + 2z\mathbf{j} + 2x\mathbf{k}$, Stokes' Theorem gives

$$\int_{\mathbf{c}} (z^2\mathbf{i} + x^2\mathbf{j} + y^2\mathbf{k}) \cdot d\mathbf{s} = \iint_S (2y\mathbf{i} + 2z\mathbf{j} + 2x\mathbf{k}) \cdot d\mathbf{S},$$

where S is the square $0 \le x \le 1$, $0 \le y \le 1$, $z = 1$.

Parametrizing S by $\mathbf{r}(u, v) = (u, v, 1)$, $0 \le u, v \le 1$, we get $\mathbf{N} = \mathbf{T}_u \times \mathbf{T}_v = (0, 0, 1) = \mathbf{k}$. Next, we check the orientation. The normal $\mathbf{N} = \mathbf{k}$ defines the outside to be the side of the surface that can be seen from a point high on the z-axis, say, $(0, 0, 10)$. Alternatively, it is the side away from which **N** points. The positive orientation of **c** is therefore the counterclockwise orientation, as seen from the point $(0, 0, 10)$; observe Figure 8.24. Seen from the origin, this orientation is clockwise, as required. Hence,

$$\iint_S (2y\mathbf{i} + 2z\mathbf{j} + 2x\mathbf{k}) \cdot d\mathbf{S} = \iint_{[0,1]\times[0,1]} (2v\mathbf{i} + 2\mathbf{j} + 2u\mathbf{k}) \cdot \mathbf{k}\, dA$$

$$= \int_0^1 \left(\int_0^1 2u du\right) dv = 1.$$

◀

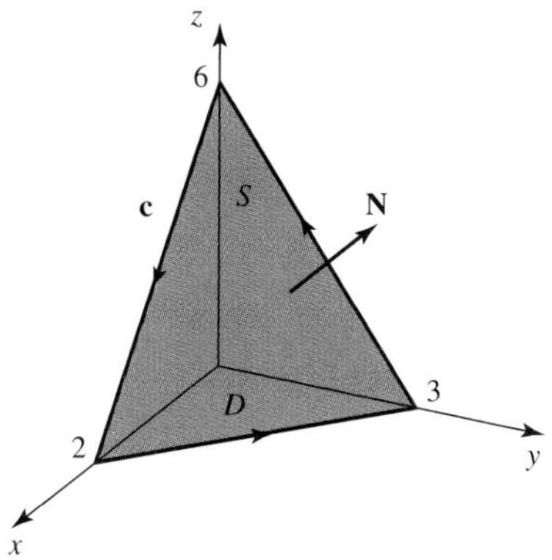

Figure 8.25 Triangle in Example 8.16.

▶ EXAMPLE 8.16

Evaluate $\int_{\mathbf{c}} \mathbf{F} \cdot d\mathbf{s}$, where $\mathbf{F} = (x + y)\mathbf{i} + (2x - z)\mathbf{j} + y\mathbf{k}$ and $\mathbf{c}$ is the boundary of the triangle with vertices (2, 0, 0), (0, 3, 0), and (0, 0, 6), oriented as shown in Figure 8.25.

SOLUTION

Let us take the part of the plane bounded by $\mathbf{c}$ as the surface of integration. Its equation is $x/2 + y/3 + z/6 = 1$, or $z = 6 - 3x - 2y$, and can be represented parametrically as $\mathbf{r}(u, v) = (u, v, -3u - 2v + 6)$, $\mathbf{u} \in D$, where D represents the triangle in the xy-plane defined by $x \geq 0$, $y \geq 0$ and (substitute $z = 0$ in $z = 6 - 3x - 2y$ to get) $3x + 2y = 6$.

We compute $\mathbf{T}_u = (1, 0, -3)$, $\mathbf{T}_v = (0, 1, -2)$, $\mathbf{N} = (3, 2, 1)$ (the orientation convention works!) and $curl\,\mathbf{F} = (2, 0, 1)$. Therefore,

$$\int_{\mathbf{c}} \mathbf{F} \cdot d\mathbf{s} = \iint_S curl\,\mathbf{F} \cdot d\mathbf{S}$$

$$= \iint_D (2\mathbf{i} + \mathbf{k}) \cdot (3\mathbf{i} + 2\mathbf{j} + \mathbf{k})\, dA = 7 \iint_D dA = 7\,\text{area}\,(D) = 21.$$ ◀

Stokes' Theorem states that, in order to compute the surface integral $\iint_S curl\,\mathbf{F} \cdot d\mathbf{S}$, all we really need are the values of $\mathbf{F}$ on the boundary of S, and nowhere else! Therefore, as long as two surfaces S_1 and S_2 have the same boundary $\partial S_1 = \partial S_2$ (with the orientation requirement fulfilled), $\iint_{S_1} curl\,\mathbf{F} \cdot d\mathbf{S}_1 = \iint_{S_2} curl\,\mathbf{F} \cdot d\mathbf{S}_2$, for a C^1 vector field $\mathbf{F}$. Consequently, in computing the path integral around a simple closed curve using Stokes' Theorem, we are free to choose any surface that is bounded by the given curve (with the proper orientation).

We illustrate this observation in our next example.

▶ EXAMPLE 8.17

Evaluate $\int_{\mathbf{c}} (2\mathbf{i} + x\mathbf{j} + y^2\mathbf{k}) \cdot d\mathbf{s}$ as a path integral, where $\mathbf{c}$ is the circle $x^2 + y^2 = 1$, $z = 1$ oriented counterclockwise as seen from a point $(0, 0, z)$ (with $z > 1$) on the z-axis. Check the result by applying Stokes' Theorem.

SOLUTION

Parametrize $\mathbf{c}$ by $\mathbf{c}(t) = (\cos t, \sin t, 1)$, $t \in [0, 2\pi]$. Then

$$\int_c (2\mathbf{i} + x\mathbf{j} + y^2\mathbf{k}) \cdot d\mathbf{s} = \int_0^{2\pi} (2, \cos t, \sin^2 t) \cdot (-\sin t, \cos t, 0)\, dt$$

$$= \int_0^{2\pi} (-2\sin t + \cos^2 t)\, dt = \left(2\cos t + \frac{1}{2}t + \frac{1}{4}\sin 2t\right)\Bigg|_0^{2\pi} = \pi.$$

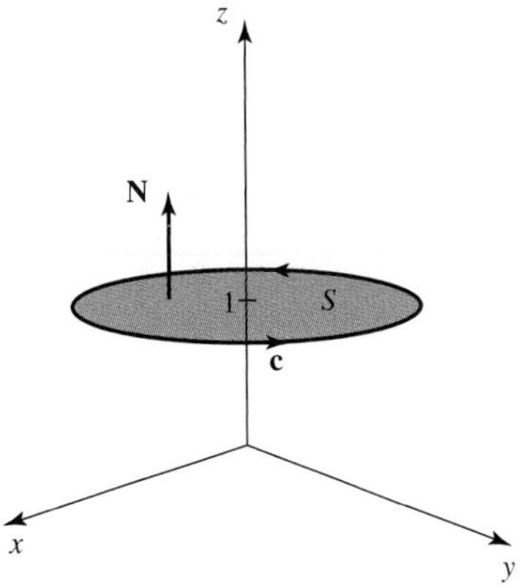

Figure 8.26 The disk in Example 8.17.

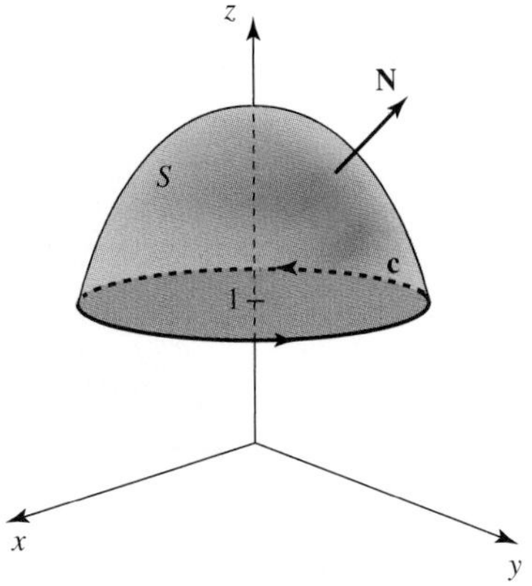

Figure 8.27 Paraboloid $z = 2 - x^2 - y^2$.

In order to use Stokes' Theorem, we need a surface S whose boundary is $\mathbf{c}$ with the correct orientation. Let us take the surface S to be the disk $\{x^2 + y^2 \le 1, z = 1\}$, and parametrize it by $\mathbf{r}(u, v) = (u, v, 1)$, $u^2 + v^2 \le 1$. Then $\mathbf{T}_u = (1, 0, 0)$, $\mathbf{T}_v = (0, 1, 0)$, and $\mathbf{N} = (0, 0, 1) = \mathbf{k}$. By our conventions, $\mathbf{N}$ is an outward normal (z-component is $+1$) and the outside of S is the top side of the disk. The corresponding orientation of the boundary is the counterclockwise orientation, as needed; see Figure 8.26. It follows that ($curl\,\mathbf{F} = 2y\mathbf{i} + \mathbf{k}$)

$$\int_{\mathbf{c}} \mathbf{F} \cdot d\mathbf{s} = \iint_S curl\,\mathbf{F} \cdot d\mathbf{S} = \iint_{\{u^2+v^2\le 1\}} (2y\mathbf{i} + \mathbf{k}) \cdot \mathbf{k}\, dA = \iint_{\{u^2+v^2\le 1\}} 1\, dA = \pi.$$

Although it is simpler to do so, we don't have to use the disk as the surface for integration in Stokes' Theorem. Let S be the surface of the paraboloid $z = 2 - x^2 - y^2$ between $z = 1$ and $z = 2$, see Figure 8.27. The boundary of S is the intersection of $z = 2 - x^2 - y^2$ and $z = 1$; that is, the circle $x^2 + y^2 = 1$ in the plane $z = 1$. Parametrize S by $\mathbf{r}(u, v) = (u, v, 2 - u^2 - v^2)$, where $u^2 + v^2 \le 1$; the condition $u^2 + v^2 \le 1$ was obtained by combining $z = 2 - x^2 - y^2$ and $1 \le z \le 2$.

We compute $\mathbf{T}_u = (1, 0, -2u)$, $\mathbf{T}_v = (0, 1, -2v)$, and $\mathbf{N} = (2u, 2v, 1)$. With this choice of $\mathbf{N}$, the induced orientation on the boundary of S is the counterclockwise orientation, exactly as needed. Hence,

$$\int_{\mathbf{c}} \mathbf{F} \cdot d\mathbf{s} = \iint_S curl\,\mathbf{F} \cdot d\mathbf{S} = \iint_{\{u^2+v^2\le 1\}} (2y\mathbf{i} + \mathbf{k}) \cdot \mathbf{N}\, dA$$

$$= \iint_{\{u^2+v^2\le 1\}} (2v, 0, 1) \cdot (2u, 2v, 1)\, dA = \iint_{\{u^2+v^2\le 1\}} (4uv + 1)\, dA$$

(passing to polar coordinates $u = r\cos\theta$, $v = r\sin\theta$, $dA = r dr d\theta$)

$$= \int_0^{2\pi} \left(\int_0^1 (4r^2 \cos\theta \sin\theta + 1) r dr \right) d\theta$$

$$= \int_0^{2\pi} \left(r^4 \cos\theta \sin\theta + \frac{1}{2} r^2 \right) \Big|_0^1 d\theta = \int_0^{2\pi} \left(\cos\theta \sin\theta + \frac{1}{2} \right) d\theta$$

$$= \left(-\frac{1}{4} \cos 2\theta + \frac{1}{2}\theta \right) \Big|_0^{2\pi} = \pi.$$

Consider yet another surface, the paraboloid S given by $z = x^2 + y^2$, $0 \le z \le 1$; see Figure 8.28. Its boundary is the circle $x^2 + y^2 = 1$, $z = 1$, as required. Parametrize S by $\mathbf{r}(u, v) = (u, v, u^2 + v^2)$,

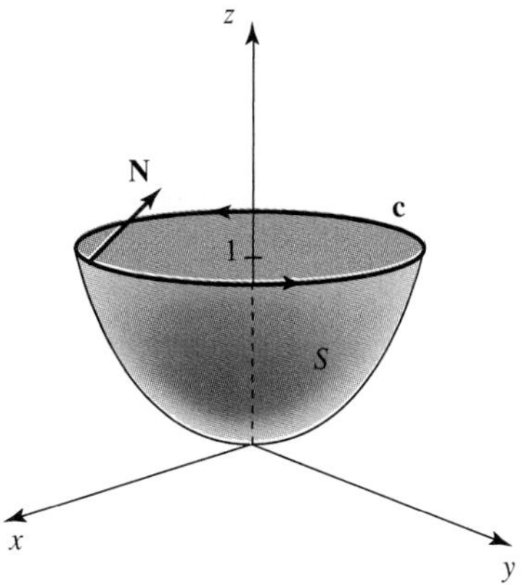

Figure 8.28 The paraboloid $z = x^2 + y^2$ in Example 8.17.

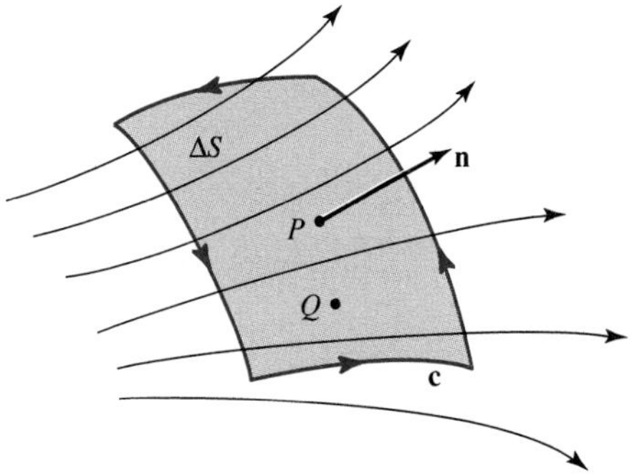

Figure 8.29 Surface ΔS placed in the flow of $\mathbf{F}$.

$u^2 + v^2 \leq 1$. The surface normal $\mathbf{N}$ is $\mathbf{N} = \mathbf{T}_u \times \mathbf{T}_v = (1, 0, 2u) \times (0, 1, 2v) = (-2u, -2v, 1)$. It points inward, into the paraboloid. Once again, it gives the right orientation of the boundary circle ∂S. Hence,

$$\int_{\mathbf{c}} \mathbf{F} \cdot d\mathbf{s} = \iint_S curl\, \mathbf{F} \cdot d\mathbf{S} = \iint_{\{u^2+v^2 \leq 1\}} (2y\mathbf{i} + \mathbf{k}) \cdot \mathbf{N}\, dA$$

$$= \iint_{\{u^2+v^2 \leq 1\}} (2v, 0, 1) \cdot (-2u, -2v, 1)\, dA = \iint_{\{u^2+v^2 \leq 1\}} (-4uv + 1)\, dA$$

(passing to polar coordinates and continuing as in the previous case)

$$= \left(\frac{1}{4} \cos 2\theta + \frac{1}{2}\theta \right) \Big|_0^{2\pi} = \pi.$$

◀

We will now use Stokes' Theorem to give a physical interpretation of the curl. Let $\mathbf{F}$ be a velocity vector field of a fluid (assume that $\mathbf{F}$ is C^1). The path integral $\int_{\mathbf{c}} \mathbf{F} \cdot d\mathbf{s}$ around a simple closed curve $\mathbf{c}$ describes the circulation of $\mathbf{F}$ (see Section 5.3); that is, it measures the "turning of the fluid." In other words, the integral $\int_{\mathbf{c}} \mathbf{F} \cdot d\mathbf{s}$ represents the "total velocity" of the fluid around $\mathbf{c}$. We are going to use this description to give an interpretation of $curl\, \mathbf{F}$.

Take a small surface ΔS whose boundary is a positively oriented, simple closed curve $\mathbf{c}$ and place it in the fluid, as shown in Figure 8.29. Choose a point P in ΔS that does not lie on $\mathbf{c}$. The surface integral of *curl* $\mathbf{F}$ over ΔS is equal to

$$\iint_{\Delta S} curl\, \mathbf{F} \cdot d\mathbf{S} = \iint_{\Delta S} curl\, \mathbf{F} \cdot \mathbf{n}\, dS,$$

where $\mathbf{n}$ is the unit normal to ΔS that satisfies the orientation convention. The average value $\overline{curl\, \mathbf{F} \cdot \mathbf{n}}$ of the *scalar* function *curl* $\mathbf{F} \cdot \mathbf{n}$ over ΔS is given by

$$\overline{curl\, \mathbf{F} \cdot \mathbf{n}} = \frac{1}{A(\Delta S)} \iint_{\Delta S} curl\, \mathbf{F} \cdot \mathbf{n}\, dS,$$

where $A(\Delta S)$ denotes the area of ΔS. The Mean Value Theorem for integrals [see (7.14) in Section 7.5] states that the average value of a continuous function must be attained somewhere; that is, there is a point Q in ΔS such that

$$\overline{curl\ \mathbf{F}\cdot\mathbf{n}} = curl\ \mathbf{F}(Q)\cdot\mathbf{n}(Q).$$

From Stokes' Theorem, we get that

$$\frac{1}{A(\Delta S)}\iint_{\Delta S} curl\ \mathbf{F}\cdot\mathbf{n}\,dS = \frac{1}{A(\Delta S)}\int_{\partial(\Delta S)=\mathbf{c}} \mathbf{F}\cdot d\mathbf{s}.$$

The integral on the right is called the *circulation density*. Combining the above yields

$$curl\ \mathbf{F}(Q)\cdot\mathbf{n}(Q) = \overline{curl\ \mathbf{F}\cdot\mathbf{n}} = \frac{1}{A(\Delta S)}\int_{\mathbf{c}} \mathbf{F}\cdot d\mathbf{s},$$

and, by taking the limit, we get

$$curl\ \mathbf{F}(P)\cdot\mathbf{n}(P) = \lim_{\Delta S\to P}\frac{1}{A(\Delta S)}\int_{c} \mathbf{F}\cdot d\mathbf{s}. \tag{8.14}$$

As ΔS shrinks to the point P, the point Q, being in ΔS, must approach P; this explains the appearance of P on the left side. Equation (8.14) states that the normal component of the curl at P (left side) is the limit of the circulation per unit area (right side). Recall that, in the context of fluid dynamics, the limit of circulation per unit area is called infinitesimal vorticity [see Section 5.3, formula (5.5), and the text following it].

As an illustration, consider the vector field $\mathbf{F}_1$ of Example 5.20 in Section 5.3. The counterclockwise circulation around the unit circle was computed to be 2π; dividing by the area (of the disk) enclosed by $\mathbf{c}$, we get 2. On the other hand, $curl\,\mathbf{F}_1 = 2\mathbf{k}$, and (since $\mathbf{c}$ is in the xy-plane) $\mathbf{n} = \mathbf{k}$, so that $curl\,\mathbf{F}_1\cdot\mathbf{n} = 2$ as well.

Since the circulation $\int_{\mathbf{c}}\mathbf{F}\cdot d\mathbf{s}$ of $\mathbf{F}$ gives the total velocity of the fluid measured around a small closed curve $\mathbf{c}$, the normal component $curl\,\mathbf{F}\cdot\mathbf{n}$ can be interpreted as measuring the "total turning" of the fluid around an axis parallel to $\mathbf{n}$; this formalizes our somewhat intuitive reasoning in Section 5.3.

▶ EXAMPLE 8.18 Green's Theorem Is a Special Case of Stokes' Theorem

Assume that $\mathbf{F} = P(x,y)\mathbf{i} + Q(x,y)\mathbf{j}$ is a C^1 vector field in $\mathbb{R}^2$ and $\mathbf{c}$ a simple closed curve that bounds a region $D\subseteq\mathbb{R}^2$ to which Green's Theorem applies (see Assumption 8.1). Parametrize D by $\mathbf{r}(u,v) = (u,v,0)$, $(u,v)\in D$. Then $\mathbf{N} = \mathbf{T}_u\times\mathbf{T}_v = (1,0,0)\times(0,1,0) = (0,0,1) = \mathbf{k}$, and it follows that

$$\int_{\mathbf{c}}\mathbf{F}\cdot d\mathbf{s} = \iint_D curl\,\mathbf{F}\cdot d\mathbf{S} = \iint_D curl\,\mathbf{F}\cdot\mathbf{k}\,dA,$$

by Stokes' Theorem and the definition of the surface integral of the function $curl\,\mathbf{F}$. The left side is equal to

$$\int_{\mathbf{c}}\mathbf{F}\cdot d\mathbf{s} = \int_a^b (P\mathbf{i}+Q\mathbf{j})\cdot(x'\mathbf{i}+y'\mathbf{j})\,dt = \int_{\mathbf{c}} P dx + Q dy,$$

where $\mathbf{c}(t) = (x(t), y(t))$, $t\in[a,b]$, parametrizes the curve $\mathbf{c}$. The right side is computed to be

$$\iint_D curl\,\mathbf{F}\cdot\mathbf{k}\,dA = \iint_D\left(\frac{\partial Q}{\partial x}-\frac{\partial P}{\partial y}\right)\mathbf{k}\cdot\mathbf{k}\,dA = \iint_D\left(\frac{\partial Q}{\partial x}-\frac{\partial P}{\partial y}\right)dA.$$

Combining the above, we get

$$\int_{\mathbf{c}} Pdx + Qdy = \iint_D \left(\frac{\partial Q}{\partial x} - \frac{\partial P}{\partial y} \right) dA,$$

and that is the statement of Green's Theorem; see formula (8.4).

When talking about gradient vector fields in Section 5.4, we stated the fact that $\int_{\mathbf{c}} \mathbf{F} \cdot d\mathbf{s} = 0$ for any oriented, simple closed curve $\mathbf{c}$ is equivalent to $curl\,\mathbf{F} = \mathbf{0}$ if the domain U of $\mathbf{F}$ is simply-connected [read Theorem 5.8, following the equivalences (b) $\Leftrightarrow$ (a) $\Leftrightarrow$ (d)]. However, we gave the proof only in the case where U is a star-shaped set. We will now outline the proof in a general case.

Let $\mathbf{F}$ be a vector field that is defined and is C^1 on a simply-connected set $U \subseteq \mathbb{R}^3$, and assume that $curl\,\mathbf{F} = \mathbf{0}$. Find a surface S that does not go through the points where $\mathbf{F}$ is not defined or not C^1 and whose boundary is a closed curve $\mathbf{c}$ (this can always be done; however, the proof is beyond the scope of this book). By Stokes' Theorem,

$$\int_{\mathbf{c}} \mathbf{F} \cdot d\mathbf{s} = \iint_S curl\,\mathbf{F} \cdot d\mathbf{S} = 0,$$

and we are done.

▶ EXERCISES 8.3

1. Let $\mathbf{F}(x, y, z) = -y\mathbf{i} + x\mathbf{j}$. Use Stokes' Theorem to find the path integral $\int_{\mathbf{c}} \mathbf{F} \cdot d\mathbf{s}$, where:

(a) $\mathbf{c}$ is the boundary of the square with vertices $(1, 0, 1)$, $(1, 1, 1)$, $(0, 1, 1)$, and $(0, 0, 1)$, oriented counterclockwise, as seen from above.

(b) $\mathbf{c}$ is the boundary of the square with vertices $(1, 1, 0)$, $(0, 1, 0)$, $(0, 1, 1)$, and $(1, 1, 1)$, oriented by the normal $\mathbf{n} = \mathbf{j}$.

2. Take a closed surface S oriented by the outward normal, and break it into two parts S_1 and S_2 that share a boundary curve $\mathbf{c}$. Assume that S_1 and S_2 are oriented by the same normal as S, and that they satisfy the assumptions of Stokes' Theorem. Show that $\iint_{S_1} curl\,\mathbf{F} \cdot d\mathbf{S} = -\iint_{S_2} curl\,\mathbf{F} \cdot d\mathbf{S}$, for a C^1 vector field $\mathbf{F}$ defined on S. Conclude that $\iint_S curl\,\mathbf{F} \cdot d\mathbf{S} = 0$.

3. Compute $\int_{\mathbf{c}} \mathbf{F} \cdot d\mathbf{s}$ directly, and then use Stokes' Theorem: let $\mathbf{F} = (x+1)^2\mathbf{i} - x^2\mathbf{k}$, and let $\mathbf{c}$ be the intersection of the cylinder $x^2 + 2x + y^2 = 3$ and the plane $z = x$, oriented counterclockwise, as seen from above.

Exercises 4 to 10: Find the circulation $\int_{\mathbf{c}} \mathbf{F} \cdot d\mathbf{s}$ of the vector field $\mathbf{F}$ along the curve $\mathbf{c}$ in the given direction.

4. $\mathbf{F}(x, y, z) = y^2\mathbf{i} - x\mathbf{j} + z^2\mathbf{k}$, $\mathbf{c}$ is the ellipse $x^2 + 4y^2 = 4$, $z = 0$, oriented counterclockwise

5. $\mathbf{F}(x, y, z) = (2x + y)\mathbf{i} - (3x - y - z)\mathbf{k}$, $\mathbf{c}$ is the boundary of the triangle cut out from the plane $x + 4y + 3z = 1$ by the first octant, oriented clockwise as seen from the origin

6. $\mathbf{F}(x, y, z) = x^2\mathbf{i} + y^2\mathbf{j} + z^2\mathbf{k}$, $\mathbf{c}$ is the boundary of the circle $x^2 + y^2 = 4$ in the plane $z = 4$, oriented counterclockwise as seen from the origin

7. $\mathbf{F}(x, y, z) = (x^2 + z^2)\mathbf{i} + y^2z^2\mathbf{j}$, $\mathbf{c}$ is the boundary of the rectangle cut out from the plane $y = z$ by the planes $x = 1$, $x = 2$, $y = 0$, and $y = 4$, oriented counterclockwise as seen from above

8. $\mathbf{F}(x, y, z) = -2y\mathbf{i} + z\mathbf{j} - z\mathbf{k}$, $\mathbf{c}$ is the intersection of the cylinder $z^2 + x^2 = 1$ and the plane $y = x + 1$, oriented counterclockwise as seen from the origin

9. $\mathbf{F}(x, y, z) = y^2(\mathbf{i} + \mathbf{j} + \mathbf{k})$, $\mathbf{c}$ is the circle on the sphere $x^2 + y^2 + z^2 = 1$ defined by $z = 1/2$, oriented clockwise as seen from the origin

10. $\mathbf{F}(x, y, z) = 2x\mathbf{i} + y^2\mathbf{k}$, $\mathbf{c}$ is the boundary of the paraboloid $z = 4 - x^2 - y^2$ in the first octant, oriented clockwise as seen from the origin

11. Let $\mathbf{F}$ be a constant vector field. A surface S in $\mathbb{R}^3$ and its boundary curve $\mathbf{c}$ are assumed to satisfy the assumptions of Stokes' Theorem. Show that $\iint_S \mathbf{F} \cdot d\mathbf{S} = \frac{1}{2}\int_{\mathbf{c}} (\mathbf{F} \times \mathbf{r}) \cdot d\mathbf{s}$, where $\mathbf{r} = x\mathbf{i} + y\mathbf{j} + z\mathbf{k}$.

12. Compute $\int_{\mathbf{c}} (2\mathbf{i} + x\mathbf{j} + y^2\mathbf{k}) \cdot d\mathbf{s}$, where $\mathbf{c}$ is the circle $x^2 + y^2 = 1$, $z = 1$, oriented counterclockwise (as seen from above), by using the fact that $\mathbf{c}$ is the boundary of the cone $z^2 = x^2 + y^2$, $z = 1$.

13. Consider the vector field $\mathbf{F} = -2y\mathbf{i}/(x^2 + y^2) + 2x\mathbf{j}/(x^2 + y^2)$. Compute the counterclockwise circulation of $\mathbf{F}$ along the circle $x^2 + y^2 = 1$, $z = 0$, directly. Can you compute $\iint_S \mathit{curl}\,\mathbf{F} \cdot d\mathbf{S}$, over the disk S in the xy-plane enclosed by $\mathbf{c}$? Explain why your answers do not violate Stokes' Theorem.

Exercises 14 to 21: Compute the circulation $\int_{\mathbf{c}} \mathbf{F} \cdot d\mathbf{s}$ of the vector field $\mathbf{F}$ along the curve $\mathbf{c}$ by direct computation, using the Fundamental Theorem of Calculus or using Stokes' Theorem.

14. $\mathbf{F}(x, y) = 3xe^{-y}\mathbf{i}$, $\mathbf{c}$ consists of the path $y = x^2$ from $(0, 0)$ to $(2, 4)$, followed by the straight line from $(2, 4)$ back to $(0, 0)$

15. $\mathbf{F}(x, y) = 2x\mathbf{i}/(x^2 + y) + \mathbf{j}/(x^2 + y)$, $\mathbf{c}$ is the boundary of the rectangle $[1, 2] \times [0, 1]$, oriented counterclockwise

16. $\mathbf{F}(x, y) = x \sin y\mathbf{i} + y \sin x\mathbf{j}$, $\mathbf{c}$ is the boundary of the triangle defined by the lines $y = x$, $y = y = \pi x/2$, and $x = 1$, oriented counterclockwise

17. $\mathbf{F}(x, y, z) = y\mathbf{i} + 2z\mathbf{j} + 3x\mathbf{k}$, $\mathbf{c}$ is the intersection of the cylinder $x^2 + y^2 = 1$ and the plane $z = y$, oriented counterclockwise as seen from above

18. $\mathbf{F}(x, y) = (2xy\mathbf{i} + \mathbf{j})e^{x^2}$, $\mathbf{c}$ consists of the straight-line segments from $(0, 0)$ to $(1, 1)$, then from $(1, 1)$ to $(0, 2)$, and then from $(0, 2)$ back to $(0, 0)$

19. $\mathbf{F}(x, y, z) = x\mathbf{i} - yz\mathbf{j} + \mathbf{k}$, $\mathbf{c}$ is the intersection of the paraboloid $z = x^2 + y^2$ and the plane $z = 2y$, oriented counterclockwise as seen from above

20. $\mathbf{F}(x, y, z) = 5\mathbf{i} + 2\mathbf{j} + z\mathbf{k}$, $\mathbf{c}$ is the ellipse $y^2 + 4z^2 = 4$ in the plane $x = 2$, oriented clockwise as seen from the origin

21. $\mathbf{F}(x, y, z) = (2x + y)\mathbf{i} + (2y - x)\mathbf{j}$, $\mathbf{c}$ is the helix $\mathbf{c}(t) = (\cos t, \sin t, t)$, $t \in [0, 3\pi]$, followed by the line segment from $(-1, 0, 3\pi)$ back to $(1, 0, 0)$

22. Show that if the curve $\mathbf{c} = \partial S$ and the surface S satisfy the assumptions of Stokes' Theorem, then $\int_{\mathbf{c}} f \nabla g \cdot d\mathbf{s} = \iint_S (\nabla f \times \nabla g) \cdot d\mathbf{S}$.

23. Show that if the curve $\mathbf{c} = \partial S$ and the surface S satisfy the assumptions of Stokes' Theorem, then $\int_{\mathbf{c}} f \nabla f \cdot d\mathbf{s} = 0$

24. Set up the integral for the counterclockwise circulation of the vector field $\mathbf{F} = e^x\mathbf{i}/(x^2 + 1)$ around the unit circle in the xy-plane. Then evaluate it using Stokes' Theorem.

▶ 8.4 DIFFERENTIAL FORMS AND CLASSICAL INTEGRATION THEOREMS

There are several reasons why we introduce and study differential forms. They provide a useful way of formalizing certain concepts, and also, they appear often in applications of vector calculus (see, for instance, Section 8.5). Moreover, we will be able to show that the three classical integration theorems that we discussed in Sections 8.1-8.3 have exactly the same form, that is, they are all special cases of a single theorem about differential forms.

Differential Forms

Throughout this section, U will denote an open set in $\mathbb{R}^3$. All functions are assumed to be differentiable as many times as needed (e.g., $\partial^2 f/\partial x^2$ means that f is assumed to be twice differentiable).

DEFINITION 8.3 0-Forms

A (differential) 0-form on U [or a *(differential) form of degree 0*] is a real-valued (differentiable) function $f: U \subseteq \mathbb{R}^3 \to \mathbb{R}$. Two 0-forms f_1 and f_2 on U can be added or multiplied, thus giving 0-forms $f_1 + f_2$ and $f_1 \cdot f_2$ (these operations are, of course, just the usual addition and multiplication of real-valued functions). A *zero 0-form* is a constant function $f(x, y, z) = 0$. ◀

DEFINITION 8.4 1-Forms

Formal expressions dx, dy, and dz are called *basic 1-forms*. A *(differential) 1-form* on a set U [or a *(differential) form of degree 1*] is a combination

$$\alpha = f(x, y, z)\,dx + g(x, y, z)\,dy + h(x, y, z)\,dz,$$

where the components (or coefficients) $f(x, y, z)$, $g(x, y, z)$, and $h(x, y, z)$ are (differentiable) real-valued functions defined on U. ◀

For the moment, we do not attach any meaning to dx, dy, and dz. We can think of them as forming a basis of some (three-dimensional) vector space, where the rôle of the scalars is played by the real-valued functions. Such a structure, strictly speaking, is not a vector space (it is called a vector bundle). Nevertheless, it helps to use vector space terminology. For example, we say that a 1-form is a linear combination of basic 1-forms dx, dy, and dz. The order of dx, dy, and dz in a 1-form is not relevant. The *zero 1-form* is the 1-form $0 = 0dx + 0dy + 0dz$, with components $f(x, y, z) = 0$, $g(x, y, z) = 0$, and $h(x, y, z) = 0$.

▶ EXAMPLE 8.19

The form $\alpha = (2x^2 + y^2)dx + e^{xy}dy + 3yz^3dz$ is a 1-form whose components are $f(x, y, z) = 2x^2 + y^2$, $g(x, y, z) = e^{xy}$, and $h(x, y, z) = 3yz^3$. The form $dx + 2xdy$ has components $f(x, y, z) = 1$, $g(x, y, z) = 2x$, and $h(x, y, z) = 0$. The components of the 1-form dz are $f(x, y, z) = 0$, $g(x, y, z) = 0$, and $h(x, y, z) = 1$. ◀

As in the case of a vector space, we can define operations of addition and multiplication by scalars (which are here replaced by functions). Let $\alpha = f_1dx + g_1dy + h_1dz$ and $\beta = f_2dx + g_2dy + h_2dz$ be 1-forms. Then $\alpha + \beta = (f_1 + f_2)dx + (g_1 + g_2)dy + (h_1 + h_2)dz$ and, if p is a function $p: U \to \mathbb{R}$, then $p\alpha = pf_1dx + pg_1dy + ph_1dz$. In words, 1-forms are added by adding up their respective components. Multiplying a 1-form by a function amounts to multiplying each component by that function.

DEFINITION 8.5 2-Forms

Expressions $dxdy$, $dydz$, and $dzdx$ are called *basic 2-forms*. A *(differential) 2-form* on a set U [or a *(differential) form of degree* 2] is a combination

$$\alpha = f(x, y, z)dxdy + g(x, y, z)dydz + h(x, y, z)dydx,$$

where the components $f(x, y, z)$, $g(x, y, z)$, and $h(x, y, z)$ are (differentiable) real-valued functions on U. ◀

Basic two-forms are "built" of 1-forms (we will clarify this later) that appear in cyclic order (dx, dy, dz, dx, etc.). Think of basic 2-forms as the basis of some three-dimensional vector space. A 2-form is then a linear combination of basic 2-forms. The order of $dxdy$, $dydz$, and $dzdx$ in a 2-form does not play any role. The addition of 2-forms and the product of a 2-form and a function are defined "componentwise"; as for 1-forms, let $\alpha = f_1dxdy + g_1dydz + h_1dzdx$ and $\beta = f_2dxdy + g_2dydz + h_2dzdx$ be 2-forms and let p be a function $p: U \to \mathbb{R}$. Then $\alpha + \beta = (f_1 + f_2)dxdy + (g_1 + g_2)dydz + (h_1 + h_2)dzdx$ and $p\alpha = pf_1dxdy + pg_1dydz + ph_1dzdx$.

▶ **EXAMPLE 8.20**

Zero 2-form is the 2-form $0 = 0dxdy + 0dydz + 0dzdx$. The expressions $\alpha = 3yzdxdy + (x^2 + y^2)dydz + dzdx$ and $\beta = z^3dxdy - x^2dydz$ are 2-forms defined on $U = \mathbb{R}^3$. Their sum is $\alpha + \beta = (3yz + z^3)dxdy + y^2dydz + dzdx$ and the product of α and e^{xy} is $e^{xy}\alpha = 3yze^{xy}dxdy + (x^2 + y^2)e^{xy}dydz + e^{xy}dzdx$. ◀

There will be no need to add forms of different degrees, such as 1-forms and 2-forms.

DEFINITION 8.6 3-Forms

The *basic 3-form* is the expression $dxdydz$. A *(differential) 3-form* on U [or a *(differential) form of degree 3*] is an expression $\alpha = f(x, y, z)\,dxdydz$, where $f(x, y, z)$ is a real-valued function on U. ◀

This time, $dxdydz$ is the only "basis" element, so the vector space considered here is one-dimensional. If $\alpha = f_1dxdydz$ and $\beta = f_2dxdydz$ are 3-forms and p is a function, then the sum $\alpha + \beta$ is a 3-form $\alpha + \beta = (f_1 + f_2)dxdydz$, and the product of p and α is a 3-form $p\alpha = pf_1dxdydz$. The *zero 3-form* is the 3-form $0 = 0dxdydz$.

DEFINITION 8.7 Wedge Product

The *wedge product* is an operation on forms that satisfies the following properties:

(a) If α is a k-form and β is an l-form, $0 \le k, l \le 3$, then their wedge product is a $(k + l)$-form $\alpha \wedge \beta$.

(b) Anticommutativity: $\alpha \wedge \beta = (-1)^{kl} \beta \wedge \alpha$, where α is a k-form and β is an l-form.

(c) Associativity: $(\alpha \wedge \beta) \wedge \gamma = \alpha \wedge (\beta \wedge \gamma)$, for any forms α, β, and γ.

(d) If 0 is a zero form of any degree, then $\alpha \wedge 0 = 0$ for any α.

(e) If f is a real-valued function (also called a 0-form), then the wedge product $f \wedge \alpha$ is the product $f \cdot \alpha$ of a form and a function; that is, $f \wedge \alpha = f \cdot \alpha$.

(f) Distributivity: if α and β are of the same degree, and γ is any form, then $(\alpha + \beta) \wedge \gamma = \alpha \wedge \gamma + \beta \wedge \gamma$.

(g) Homogeneity with respect to functions: if f is a real-valued function, then $f(\alpha \wedge \beta) = (f\alpha) \wedge \beta = \alpha \wedge (f\beta)$.

(h) The wedge products of basic 1-forms are basic 2-forms: $dx \wedge dy = dxdy$, $dy \wedge dz = dydz$, and $dz \wedge dx = dzdx$. Moreover, $dy \wedge dx = dydx$, $dz \wedge dy = dzdy$, and $dx \wedge dz = dxdz$. The basic 3-form is the wedge product of basic 1-forms in their cyclic order, that is,

$$dx \wedge dy \wedge dz = (dx \wedge dy) \wedge dz = dx \wedge (dy \wedge dz) = dxdydz.$$

From anticommutativity of forms and (h), it follows that

$$dydx \overset{\text{(h)}}{=} dy \wedge dx \overset{\text{(b)}}{=} -dx \wedge dy \overset{\text{(h)}}{=} -dxdy;$$

similarly, we show that $dzdy = dz \wedge dy = -dy \wedge dz = -dydz$, and $dzdx = dx \wedge dz = -dz \wedge dx = -dzdx$. The equality $dx \wedge dx = -dx \wedge dx$, which was obtained by switching dx and dx using (b) with $k = l = 1$, implies that $dx \wedge dx = 0$ (thus, $dy \wedge dy = 0$ and $dz \wedge dz = 0$).

A triple product in which a basic 1-form occurs more than once is zero. For example,

$$dxdy \wedge dx \overset{\text{(h)}}{=} (dx \wedge dy) \wedge dx \overset{\text{(b)}}{=} (-dy \wedge dx) \wedge dx \overset{\text{(c)}}{=} -dy \wedge (dx \wedge dx) = -dy \wedge 0 \overset{\text{(d)}}{=} 0,$$

since $dx \wedge dx = 0$. Similar computations lead to the following conclusion: since there are only three basic 1-forms in $\mathbb{R}^3$, every 4-form (written as the wedge product of 1-forms) has at least one of dx, dy, or dz repeated—and therefore is equal to zero. Thus, all 4-forms, 5-forms, and, in general, all forms of degree greater than 3 in $\mathbb{R}^3$ are zero. That is the reason why we say that there are no (nontrivial) forms of degree greater than 3 in $\mathbb{R}^3$.

The basic 3-form $dxdydz$ can be written in different ways as a product of 1-forms and 2-forms; for example,

$$\begin{aligned} dxdydz &\overset{\text{(h)}}{=} dx \wedge (dy \wedge dz) \overset{\text{(b)}}{=} dx \wedge (-dz \wedge dy) \overset{\text{(g)}}{=} -dx \wedge (dz \wedge dy) \\ &\overset{\text{(h)}}{=} -dxdzdy \overset{\text{(h)}}{=} (-dx \wedge dz) \wedge dy \overset{\text{(h)}}{=} -dxdz \wedge dy = \text{etc.} \end{aligned}$$

► EXAMPLE 8.21

Let $\alpha = x^2dx + ydy$. Compute $\alpha \wedge \alpha$.

SOLUTION By definition,

$$\begin{aligned} \alpha \wedge \alpha &= (x^2dx + ydy) \wedge (x^2dx + ydy) \\ &\overset{\text{(f)}}{=} x^2dx \wedge x^2dx + ydy \wedge x^2dx + x^2dx \wedge ydy + ydy \wedge ydy \\ &\overset{\text{(g)}}{=} x^4dx \wedge dx + x^2ydy \wedge dx + x^2ydx \wedge dy + y^2dy \wedge dy = 0. \end{aligned}$$

The first and last terms are zero since $dx \wedge dx = 0$ and $dy \wedge dy = 0$. The middle two terms cancel due to the anticommutativity (b).

Examine the computation in the previous example. As an alternative, by imitating the way we proved that $dx \wedge dx = 0$, we realize that $\alpha \wedge \alpha = 0$ for any 1-form α. This identity also holds for 2-forms and 3-forms in $\mathbb{R}^3$ [that could be checked by direct computation; actually, it is enough to notice that the wedge product of a 2-form (or 3-form) with itself is a 4-form (or 6-form), and all such forms are zero in $\mathbb{R}^3$]. However, $\alpha \wedge \alpha \neq 0$, in general, for forms in $\mathbb{R}^m$.

▶ EXAMPLE 8.22

Let $\alpha = ydxdy + xzdydz$ and let $\beta = dx + ydy + z^2dz$. Compute $\alpha \wedge \beta$.

SOLUTION Using distributivity property (f) and homogeneity (g), we get

$$\begin{aligned}\alpha \wedge \beta &= (ydxdy + xzdydz) \wedge (dx + ydy + z^2dz)\\ &= ydxdy \wedge dx + xzdydz \wedge dx + y^2dxdy \wedge dy\\ &\quad + xyzdydz \wedge dy + z^2ydxdy \wedge dz + xz^3dydz \wedge dz.\end{aligned}$$

All triple products with a repeated basic 1-form are zero. Hence,

$$\alpha \wedge \beta = xzdydz \wedge dx + z^2ydxdy \wedge dz,$$

and since (no parentheses needed, due to associativity)

$$dydz \wedge dx = dy \wedge dz \wedge dx = -dy \wedge dx \wedge dz = dx \wedge dy \wedge dz = dxdydz$$

(anticommutativity has been used twice), we get $\alpha \wedge \beta = (xz + z^2y)dxdydz$.

DEFINITION 8.7 Differential of a Form

The *differential* is an operation that assigns a $(k+1)$-form $d\alpha$ to a k-form α $(0 \leq k \leq 3)$ according to the following rules:

(a) If $f: U \to \mathbb{R}$ is a 0-form, then df is the 1-form

$$df = \frac{\partial f}{\partial x}dx + \frac{\partial f}{\partial y}dy + \frac{\partial f}{\partial z}dz.$$

(b) If $\alpha = fdx + gdy + hdz$ is a 1-form, then $d\alpha$ is the 2-form

$$d\alpha = df \wedge dx + dg \wedge dy + dh \wedge dz,$$

where df, dg, and dh are computed by (a), since f, g, and h are functions (or 0-forms).

(c) If $\alpha = fdxdy + gdydz + hdzdx$ is a 2-form, then $d\alpha$ is the 3-form

$$d\alpha = df \wedge dxdy + dg \wedge dydz + dh \wedge dzdx,$$

where df, dg, and dh are computed by (a).

(d) If $\alpha = fdxdydz$ is a 3-form, then $d\alpha = 0$.

▶ EXAMPLE 8.23

Compute the differential $d\alpha$ of the following forms:

(a) The 0-form $\alpha = xy + e^{yz}$ **(b)** The 1-form $\alpha = ydx$

(c) The 2-form $\alpha = x^3 dxdy - x\cos z dydz$ **(d)** The 2-form $\alpha = dzdx$

SOLUTION

(a) Using part (a) of Definition 8.8, we get

$$\begin{aligned} d\alpha &= \frac{\partial}{\partial x}(xy + e^{yz})\,dx + \frac{\partial}{\partial y}(xy + e^{yz})\,dy + \frac{\partial}{\partial z}(xy + e^{yz})\,dz \\ &= ydx + (x + ze^{yz})dy + ye^{yz}dz. \end{aligned}$$

(b) The form α has only one nonzero component. By (b) of Definition 8.8,

$$\begin{aligned} d\alpha &= d(y) \wedge dx = \left(\frac{\partial}{\partial x}(y)\,dx + \frac{\partial}{\partial y}(y)\,dy + \frac{\partial}{\partial z}(y)\,dz\right) \wedge dx \\ &= dy \wedge dx = -dx \wedge dy = -dxdy. \end{aligned}$$

(c) By definition,

$$\begin{aligned} d\alpha &= d(x^3) \wedge dxdy + d(-x\cos z) \wedge dydz \\ &= \left(\frac{\partial}{\partial x}(x^3)\,dx + \frac{\partial}{\partial y}(x^3)\,dy + \frac{\partial}{\partial z}(x^3)\,dz\right) \wedge dxdy \\ &\quad - \left(\frac{\partial}{\partial x}(x\cos z)\,dx + \frac{\partial}{\partial y}(x\cos z)\,dy + \frac{\partial}{\partial z}(x\cos z)\,dz\right) \wedge dydz \\ &= 3x^2 dx \wedge dxdy - (\cos z dx - x\sin z dz) \wedge dydz = -\cos z dxdydz. \end{aligned}$$

(d) $\alpha = 1 \cdot dzdx$ and, hence,

$$d\alpha = \left(\frac{\partial}{\partial x}(1)\,dx + \frac{\partial}{\partial y}(1)\,dy + \frac{\partial}{\partial z}(1)\,dz\right) \wedge dzdx = 0 \wedge dzdx = 0.$$

◀

Gradient, Divergence, and Curl as Differentials on Forms

We will now see how the three vector differential operators can be interpreted as differentials on forms. By definition, the differential of the 0-form f (which is a function) is given by

$$df = \frac{\partial f}{\partial x}dx + \frac{\partial f}{\partial y}dy + \frac{\partial f}{\partial z}dz = \nabla f \cdot d\mathbf{s},$$

where $d\mathbf{s} = dx\mathbf{i} + dy\mathbf{j} + dz\mathbf{k}$ is a formal expression called the *line element*, and ∇f is the gradient of f. Take a 1-form $\alpha = F_1 dx + F_2 dy + F_3 dz$ and a vector field $\mathbf{F} = (F_1, F_2, F_3)$. The differential of α is computed to be

$$\begin{aligned} d\alpha &= dF_1 \wedge dx + dF_2 \wedge dy + dF_3 \wedge dz \\ &= \left(\frac{\partial F_1}{\partial x}dx + \frac{\partial F_1}{\partial y}dy + \frac{\partial F_1}{\partial z}dz\right) \wedge dx + \left(\frac{\partial F_2}{\partial x}dx + \frac{\partial F_2}{\partial y}dy + \frac{\partial F_2}{\partial z}dz\right) \wedge dy \\ &\quad + \left(\frac{\partial F_3}{\partial x}dx + \frac{\partial F_3}{\partial y}dy + \frac{\partial F_3}{\partial z}dz\right) \wedge dz \\ &= \left(\frac{\partial F_2}{\partial x} - \frac{\partial F_1}{\partial y}\right) dx \wedge dy + \left(\frac{\partial F_3}{\partial y} - \frac{\partial F_2}{\partial z}\right) dy \wedge dz + \left(\frac{\partial F_1}{\partial z} - \frac{\partial F_3}{\partial x}\right) dz \wedge dx. \end{aligned}$$

The coefficients of $d\alpha$ coincide with the components of $curl\,\mathbf{F}$ (see Definition 4.7): the coefficient of $dy \wedge dz$ equals the $\mathbf{i}$ component of $curl\,\mathbf{F}$, the coefficient of $dz \wedge dx$ equals the $\mathbf{j}$ component of $curl\,\mathbf{F}$, and the coefficient of $dx \wedge dy$ equals the $\mathbf{k}$ component of $curl\,\mathbf{F}$. Thus, $d\alpha = 0$ if and only if $curl\,\mathbf{F} = \mathbf{0}$.

Finally, let $\alpha = F_1 dydz + F_2 dzdx + F_3 dxdy$ be a 2-form. Then

$$\begin{aligned} d\alpha &= dF_1 \wedge dydz + dF_2 \wedge dzdx + dF_3 \wedge dxdy \\ &= \left(\frac{\partial F_1}{\partial x}dx + \frac{\partial F_1}{\partial y}dy + \frac{\partial F_1}{\partial z}dz\right) \wedge dydz + \left(\frac{\partial F_2}{\partial x}dx + \frac{\partial F_2}{\partial y}dy + \frac{\partial F_2}{\partial z}dz\right) \wedge dzdx \\ &+ \left(\frac{\partial F_3}{\partial x}dx + \frac{\partial F_3}{\partial y}dy + \frac{\partial F_3}{\partial z}dz\right) \wedge dxdy \\ &= \left(\frac{\partial F_1}{\partial x} + \frac{\partial F_2}{\partial y} + \frac{\partial F_3}{\partial z}\right) dxdydz. \end{aligned}$$

The coefficient of $dxdydz$ is the divergence of the vector field $\mathbf{F} = (F_1, F_2, F_3)$.

To summarize: the differential of a 0-form corresponds to the gradient, the differential of a 1-form corresponds to the curl, and the differential of a 2-form corresponds to the divergence. In other words, the "classical" vector calculus operations can be considered as special cases of the differential d acting on differential forms.

THEOREM 8.5 Properties of the Differential

For k-forms α_1, α_2, and α and l-form β, the following identities hold:

(a) $d(\alpha_1 + \alpha_2) = d\alpha_1 + d\alpha_2$

(b) $d(d\alpha) = 0$

(c) $d(\alpha \wedge \beta) = d\alpha \wedge \beta + (-1)^k \alpha \wedge d\beta$.

PROOF:

(a) If α_1 and α_2 are 3-forms, then $d\alpha_1 = 0$ and $d\alpha_2 = 0$ by definition. Their sum $\alpha_1 + \alpha_2$ is again a 3-form, and hence, $d(\alpha_1 + \alpha_2) = 0$. Therefore, (a) holds for 3-forms. Now let $\alpha_1 = f_1$ and $\alpha_2 = f_2$ be 0-forms. Then

$$\begin{aligned} d(f_1 + f_2) &= \frac{\partial}{\partial x}(f_1 + f_2)dx + \frac{\partial}{\partial y}(f_1 + f_2)dy + \frac{\partial}{\partial z}(f_1 + f_2)dz \\ &= \frac{\partial f_1}{\partial x}dx + \frac{\partial f_1}{\partial y}dy + \frac{\partial f_1}{\partial z}dz + \frac{\partial f_2}{\partial x}dx + \frac{\partial f_2}{\partial y}dy + \frac{\partial f_2}{\partial z}dz \\ &= df_1 + df_2, \end{aligned}$$

so (a) is true for 0-forms. This identity is now used to prove the statement for 1-forms and 2-forms. Let $\alpha_1 = f_1 dx + g_1 dy + h_1 dz$ and $\alpha_2 = f_2 dx + g_2 dy +$

$h_2 dz$ be 1-forms. By definition of the differential and the distributivity of the wedge product,

$$\begin{aligned}
d(\alpha_1+\alpha_2) &= d(f_1dx+g_1dy+h_1dz+f_2dx+g_2dy+h_2dz)\\
&= d\left((f_1+f_2)dx+(g_1+g_2)dy+(h_1+h_2)dz\right)\\
&= d(f_1+f_2)\wedge dx+d(g_1+g_2)\wedge dy+d(h_1+h_2)\wedge dz\\
&= (df_1+df_2)\wedge dx+(dg_1+dg_2)\wedge dy+(dh_1+dh_2)\wedge dz\\
&= df_1\wedge dx+df_2\wedge dx+dg_1\wedge dy+dg_2\wedge dy+dh_1\wedge dz+dh_2\wedge dz\\
&= (df_1\wedge dx+dg_1\wedge dy+dh_1\wedge dz)+(df_2\wedge dx+dg_2\wedge dy+dh_2\wedge dz)\\
&= d\alpha_1+d\alpha_2.
\end{aligned}$$

An analogous computation gives the proof of (a) for 2-forms.

(b) There are two trivial cases: if α is a 3-form, then $d\alpha=0$, and so $d(d\alpha)=d(0)=0$. If α is a 2-form, then $d\alpha$ is a 3-form, and $d(d\alpha)=0$ (we have used the fact that, by definition, the differential of any 3-form is zero). We have actually proved the remaining parts of (b) before: if $\alpha=f$ is a 0-form, then $d(d\alpha)=d(df)$ can be interpreted as *curl*(*grad* f), which was proven to be zero in Theorem 4.13. And if α is a 1-form, then $d(d\alpha)$ corresponds to the *div* of the *curl*, which is again zero, by the same theorem.

Comment: we can use (a) to simplify the computations in the proof. More precisely, if statements (b) or (c) hold for expressions of the form (function · basic form), such as f_1dx, f_2dy or f_3dxdy, then, by (a), they must hold for the sum (e.g., for $f_1dx+g_1dy+h_1dz$, which is a general 1-form).

Another proof of (b): in light of the previous comment, take $\alpha=fdx$. Then (whenever differentiating, we immediately drop terms where some basic form appears more than once)

$$\begin{aligned}
d\left(d(fdx)\right) &= d(df\wedge dx)=d\left(\left(\frac{\partial f}{\partial x}dx+\frac{\partial f}{\partial y}dy+\frac{\partial f}{\partial z}dz\right)\wedge dx\right)\\
&= d\left(\frac{\partial f}{\partial y}dy\wedge dx+\frac{\partial f}{\partial z}dz\wedge dx\right)\\
&= \frac{\partial^2 f}{\partial z\partial y}dz\wedge dy\wedge dx+\frac{\partial^2 f}{\partial y\partial z}dy\wedge dz\wedge dx\\
&= \left(-\frac{\partial^2 f}{\partial z\partial y}+\frac{\partial^2 f}{\partial y\partial z}\right)dxdydz=0,
\end{aligned}$$

by the equality of mixed second partial derivatives. The remaining cases in (b) are verified analogously.

(c) This statement is proved in a similar way, by considering forms of different degrees. To illustrate the proof, take, for example, 1-forms $\alpha=fdx$ and $\beta=gdy$. Then

$$\begin{aligned}
d(\alpha\wedge\beta) &= d(fdx\wedge gdy)=d(fg)\wedge dx\wedge dy=\left(\frac{\partial}{\partial z}(fg)dz\right)\wedge dx\wedge dy\\
&= \left(\frac{\partial f}{\partial z}g+f\frac{\partial g}{\partial z}\right)dz\wedge dx\wedge dy=\left(\frac{\partial f}{\partial z}g+f\frac{\partial g}{\partial z}\right)dx\wedge dy\wedge dz,
\end{aligned}$$

by the product rule. The right side of (c) is

$$\begin{aligned}
d\alpha \wedge \beta + (-1)^1 \alpha \wedge d\beta &= d(f dx) \wedge g dy - (f dx) \wedge d(g dy) \\
&= \left(\frac{\partial f}{\partial y} dy \wedge dx + \frac{\partial f}{\partial z} dz \wedge dx \right) \wedge g dy - f dx \wedge \left(\frac{\partial g}{\partial x} dx \wedge dy + \frac{\partial g}{\partial z} dz \wedge dy \right) \\
&= \frac{\partial f}{\partial z} g dz \wedge dx \wedge dy - f \frac{\partial g}{\partial z} dx \wedge dz \wedge dy = \left(\frac{\partial f}{\partial z} g + f \frac{\partial g}{\partial z} \right) dx dy dz.
\end{aligned}$$

◀

Integration of Differential Forms

In this subsection, we define the path and surface integrals of differential forms and relate them to path and surface integrals of vector fields.

DEFINITION 8.9 Path Integral of a Differential Form

Let $\alpha = F_1 dx + F_2 dy + F_3 dz$ be a differentiable 1-form and let $\mathbf{c}: [a, b] \to \mathbb{R}^3$ be a C^1 path. The *path integral of* α *along* $\mathbf{c}$ is defined by

$$\int_{\mathbf{c}} \alpha = \int_{\mathbf{c}} F_1 dx + F_2 dy + F_3 dz = \int_a^b \left(F_1 \frac{dx}{dt} + F_2 \frac{dy}{dt} + F_3 \frac{dz}{dt} \right) dt.$$

◀

▶ EXAMPLE 8.24

Compute the integral of the 1-form $\alpha = x^2 dx + y dy + 2yz dz$ along the curve $\mathbf{c}(t) = (1, t, -t^2)$, $t \in [0, 1]$.

SOLUTION By definition (of course, $dx/dt = x'$, $dy/dt = y'$, and $dz/dt = z'$),

$$\begin{aligned}
\int_{\mathbf{c}} x^2 dx + y dy + 2yz dz &= \int_0^1 \left(x^2 \frac{dx}{dt} + y \frac{dy}{dt} + 2yz \frac{dz}{dt} \right) dt \\
&= \int_0^1 \left(1 \cdot 0 + t \cdot 1 + 2t(-t^2)(-2t) \right) dt = \int_0^1 (t + 4t^4) dt = \frac{13}{10}.
\end{aligned}$$

◀

Let $\alpha = F_1 dx + F_2 dy + F_3 dz$ be a 1-form and let $\mathbf{F} = (F_1, F_2, F_3)$ be a vector field with the same component functions. Take a C^1 curve $\mathbf{c} = (x(t), y(t), z(t))$ defined on an interval $[a, b]$. Then (dropping the independent variable from the notation)

$$\begin{aligned}
\int_{\mathbf{c}} \mathbf{F} \cdot d\mathbf{s} &= \int_a^b \mathbf{F}(\mathbf{c}(t)) \cdot \mathbf{c}'(t) dt \\
&= \int_a^b (F_1, F_2, F_3) \cdot (x', y', z') dt = \int_a^b (F_1 x' + F_2 y' + F_3 z') dt;
\end{aligned}$$

that is, by Definition 8.9,

$$\int_{\mathbf{c}} \mathbf{F} \cdot d\mathbf{s} = \int_{\mathbf{c}} \alpha. \tag{8.15}$$

In words, the path integral of a vector field can be interpreted as the path integral of the corresponding 1-form. In particular, if $\mathbf{F} = (F_1, F_2)$, then $\int_{\mathbf{c}} \mathbf{F} \cdot d\mathbf{s} = \int_{\mathbf{c}} \alpha$, where

$\alpha = F_1dx + F_2dy$. This is not only a useful formalism—its importance can best be understood and appreciated in the context of classical integration theorems presented later in this section.

In the last subsection, we discovered a close relationship between functions and their derivatives and differential forms. We now continue to explore this relationship.

DEFINITION 8.10 Closed Differential Form

A differential form α defined on $U \subseteq \mathbb{R}^3$ is called *closed* if $d\alpha = 0$.

A constant real-valued function is a closed 0-form. The form $\alpha = y^2dx + 2xydy + dz$ is closed since $d\alpha = d(y^2dx + 2xydy + dz) = 2ydy \wedge dx + 2ydx \wedge dy = 0$. The differential of the 2-form $\beta = x^2dxdy + ye^zdydz$ is zero, and hence β is closed. The definition of the differential implies that every 3-form defined on $U \subseteq \mathbb{R}^3$ is closed.

DEFINITION 8.11 Exact Differential Form

A k-form α $(1 \le k \le 3)$ defined on $U \subseteq \mathbb{R}^3$ is *exact* if there exists a $(k-1)$-form β such that $d\beta = \alpha$.

For example, the 3-form $\alpha = 2xydxdydz$ is exact since $d(x^2ydydz) = 2xydxdydz$.

The form β in Definition 8.11 is not uniquely determined by α; as a matter of fact, there are infinitely many choices. For example, $d(x^2ydydz + g(x, y)dxdy + h(x, z)dzdx) = 2xydxdydz$, where g and h are any differentiable functions (of the variables indicated). The fact that $d(e^xdy + ydz) = e^xdxdy + dydz$ proves that the 2-form $e^xdxdy + dydz$ is exact.

Let α be an exact form; that is, $\alpha = d\beta$ for some form β. Then $d\alpha = d(d\beta) = 0$ by Theorem 8.5, which proves that α is closed. In other words, every exact form is closed. The converse of this statement does not hold in general (i.e., for any form defined on any subset $U \subseteq \mathbb{R}^3$).

Our next theorem states the conditions under which the converse is true (recall that we defined simply-connected and star-shaped sets in Section 5.4).

THEOREM 8.6 Closed Forms Are Exact

(a) Assume that α is a closed 1-form defined on an open, simply-connected set $U \subseteq \mathbb{R}^3$. Then α is exact.

(b) Assume that α is a closed 2-form defined on an open, star-shaped set $U \subseteq \mathbb{R}^3$. Then α is exact.

Since star-shaped sets are simply-connected, the condition on U in (b) is stronger than the condition in (a). The proof of part (b) is similar to the proof of the statement that $curl\,\mathbf{F} = \mathbf{0}$ implies $\mathbf{F} = \nabla f$ for star-shaped sets (see Theorem 5.4 in Section 5.4). Part (a) is a consequence of Stokes' Theorem and another theorem that we will have to take for granted (although its statement is intuitively clear, its proof is beyond the scope of this book), and will be discussed at the end of this section.

Let us now translate the statements from the language of differential forms to the language of vector differential operators. Assume that U is a simply-connected set and let $\alpha = F_1 dx + F_2 dy + F_3 dz$ be a 1-form and $\mathbf{F} = F_1\mathbf{i} + F_2\mathbf{j} + F_3\mathbf{k}$ the corresponding vector field defined on U. Theorem 8.6(a) and the remark preceding it state that the 1-form α is exact if and only if it is closed; that is,

$$\alpha = df \text{ for some differentiable function } f \text{ if and only if } d\alpha = 0.$$

This equivalence can be interpreted as

$$\mathbf{F} = \nabla f \text{ for some differentiable function } f \text{ if and only if } \mathit{curl}\, \mathbf{F} = \mathbf{0},$$

which is precisely the statement of Theorem 5.4.

Interpreting the second part of Theorem 8.6 in terms of vector differential operators actually produces a new statement. Assume that U is a star-shaped set and consider $\mathbf{F} = F_1\mathbf{i} + F_2\mathbf{j} + F_3\mathbf{k}$ and $\beta = F_1 dydz + F_2 dxdz + F_3 dxdy$. The equivalence

$$\beta = d\gamma \text{ (i.e., } \beta \text{ is exact) if and only if } d\beta = 0 \text{ (i.e., } \beta \text{ is closed)}$$

means that

$$\mathbf{F} = \mathit{curl}\, \mathbf{H} \text{ if and only if } \mathit{div}\, \mathbf{F} = 0. \tag{8.16}$$

DEFINITION 8.12 Surface Integral of a Differential Form

Let $\alpha = F_1(x, y, z)dydz + F_2(x, y, z)dzdx + F_3(x, y, z)dxdy$ be a 2-form defined on an open subset $U \subseteq \mathbb{R}^3$ and let $\mathbf{r}(u, v) = (x(u, v), y(u, v), z(u, v))\colon D \to \mathbb{R}^3$, $(u, v) \in D$, be a C^1 parametric representation of a smooth surface $S \subseteq U$ (D is an elementary region in $\mathbb{R}^2$). The *(surface) integral of α over S* is defined by

$$\begin{aligned}\int_S \alpha &= \int_S F_1 dydz + F_2 dzdx + F_3 dxdy \\ &= \iint_D \Big(F_1(x(u, v), y(u, v), z(u, v)) \frac{\partial(y, z)}{\partial(u, v)} + F_2(x(u, v), y(u, v), z(u, v)) \frac{\partial(z, x)}{\partial(u, v)} \\ &\quad + F_3(x(u, v), y(u, v), z(u, v)) \frac{\partial(x, y)}{\partial(u, v)} \Big) dA,\end{aligned}$$

where $\partial(x, y)/\partial(u, v)$, $\partial(y, z)/\partial(u, v)$, and $\partial(z, x)/\partial(u, v)$ denote the Jacobian determinants

$$\frac{\partial(x, y)}{\partial(u, v)} = \begin{vmatrix} \frac{\partial x}{\partial u} & \frac{\partial x}{\partial v} \\ \frac{\partial y}{\partial u} & \frac{\partial y}{\partial v} \end{vmatrix}, \quad \frac{\partial(y, z)}{\partial(u, v)} = \begin{vmatrix} \frac{\partial y}{\partial u} & \frac{\partial y}{\partial v} \\ \frac{\partial z}{\partial u} & \frac{\partial z}{\partial v} \end{vmatrix} \quad \text{and} \quad \frac{\partial(z, x)}{\partial(u, v)} = \begin{vmatrix} \frac{\partial z}{\partial u} & \frac{\partial z}{\partial v} \\ \frac{\partial x}{\partial u} & \frac{\partial x}{\partial v} \end{vmatrix}.$$ ◀

Thus, $\int_S \alpha$ is evaluated (once an orientation-preserving parametrization of S is chosen) as the double integral of a real-valued function over an elementary region in $\mathbb{R}^2$. That real-valued function is the dot product of the vector field $\mathbf{F} = (F_1, F_2, F_3)$ and the surface normal

$$\mathbf{N}(u, v) = \mathbf{T}_u(u, v) \times \mathbf{T}_v(u, v) = \left(\frac{\partial(y, z)}{\partial(u, v)}, \frac{\partial(z, x)}{\partial(u, v)}, \frac{\partial(x, y)}{\partial(u, v)} \right),$$

[see (7.5)] and is therefore equal to

$$\int_S \alpha = \iint_D \mathbf{F}(\mathbf{r}(u,v)) \cdot \mathbf{N}(u,v)\, dudv = \iint_S \mathbf{F} \cdot d\mathbf{S}.$$

It follows that the integral of a 2-form $\alpha = F_1 dydz + F_2 dzdx + F_3 dxdy$ over a surface S is just the surface integral of the corresponding vector field $\mathbf{F} = (F_1, F_2, F_3)$.

▶ EXAMPLE 8.25

Compute $\int_S \alpha$ if $\alpha = 2ydxdy - xzdydz$ and S is parametrized by $\mathbf{r}(u,v) = (u+2v, u^2, uv)$, where $0 \le u, v \le 1$.

SOLUTION

By definition,

$$\int_S \alpha = \int_S 2ydxdy - xzdydz = \iint_{[0,1]\times[0,1]} \left(2y \frac{\partial(x,y)}{\partial(u,v)} - xz \frac{\partial(y,z)}{\partial(u,v)} \right) dA,$$

where

$$\frac{\partial(x,y)}{\partial(u,v)} = \begin{vmatrix} 1 & 2 \\ 2u & 0 \end{vmatrix} = -4u \quad \text{and} \quad \frac{\partial(y,z)}{\partial(u,v)} = \begin{vmatrix} 2u & 0 \\ v & u \end{vmatrix} = 2u^2.$$

Hence,

$$\begin{aligned} \int_S \alpha &= \iint_{[0,1]\times[0,1]} \left(2u^2(-4u) - (u+2v)uv\,2u^2\right) dA, \\ &= -\int_0^1 \left(\int_0^1 \left(8u^3 + 2u^4 v + 4u^3 v^2\right) du \right) dv \\ &= -\int_0^1 \left(\left(2u^4 + \frac{2u^5 v}{5} + u^4 v^2\right)\Bigg|_{u=0}^{u=1} \right) dv = -\int_0^1 \left(2 + \frac{2v}{5} + v^2\right) dv = -\frac{38}{15}. \end{aligned}$$

◀

Classical Integration Theorems Revisited

Recall that the path integral $\int_{\mathbf{c}} \mathbf{F} \cdot d\mathbf{s}$ of the vector field $\mathbf{F} = (P, Q)$ is equal to the path integral $\int_{\mathbf{c}} Pdx + Qdy$ of the corresponding 1-form $\alpha = Pdx + Qdy$ [see the previous subsection and also (5.6) in Section 5.3]. Using the definition, we compute the differential of α to be

$$\begin{aligned} d\alpha &= dP \wedge dx + dQ \wedge dy \\ &= \left(\frac{\partial P}{\partial x} dx + \frac{\partial P}{\partial y} dy \right) \wedge dx + \left(\frac{\partial Q}{\partial x} dx + \frac{\partial Q}{\partial y} dy \right) \wedge dy = \left(\frac{\partial Q}{\partial x} - \frac{\partial P}{\partial y} \right) dxdy, \end{aligned}$$

hence obtaining yet another form of Green's Theorem.

THEOREM 8.7 Green's Theorem for Differential Forms

Let $\alpha = P(x,y)dx + Q(x,y)dy$ be a C^1 1-form, defined on a region $D \subseteq \mathbb{R}^2$ that satisfies Assumption 8.1 in Section 8.1, and let $\mathbf{c} = \partial D$ be its positively oriented boundary. Then

$$\int_{\partial D} \alpha = \int_D d\alpha.$$

◀

The integral on the right side is the integral of a 2-form $d\alpha$. In words, the integral of a 1-form α over a boundary of D equals the integral of its differential $d\alpha$ over D. We will soon realize that this formulation is a common theme of all integration theorems.

Let us clarify a technical issue: the right side of the formula in Theorem 8.7 is the integral $\int_D f(x, y)dxdy$ of the 2-form $f(x, y)dxdy = (\partial Q/\partial x - \partial P/\partial y)dxdy$ over a region D in the xy-plane. In general, to evaluate such integrals [for any $f(x, y)$], we parametrize D by $\mathbf{r}(u, v) = (u, v, 0)$, where $(u, v) \in D$; then (see Definition 8.12)

$$\int_D f(x, y)dxdy = \iint_D f(u, v)\frac{\partial(x, y)}{\partial(u, v)}\,dA = \iint_D f(u, v)\,dA,$$

since

$$\frac{\partial(x, y)}{\partial(u, v)} = \begin{vmatrix} 1 & 0 \\ 0 & 1 \end{vmatrix} = 1.$$

Now $\iint_D f(u, v)\,dA$ is a double integral over a region in $\mathbb{R}^2$, and is computed as an iterated integral. Replacing u and v by x and y, we get $\iint_D f(u, v)\,dA = \iint_D f(x, y)\,dA$, and hence,

$$\int_D f(x, y)\,dxdy = \iint_D f(x, y)\,dA.$$

This means that an integral of a 2-form $f(x, y)\,dxdy$ over a region D in the xy-plane can be interpreted as a double integral of $f(x, y)$ over D (i.e., as an iterated integral).

▶ EXAMPLE 8.26

Let $\alpha = xydx - x^2ydy$ be a 1-form and let $\mathbf{c}$ be the circle $\mathbf{c}(t) = (\cos t, \sin t)$, $t \in [0, 2\pi]$. Compute $\int_{\mathbf{c}} \alpha$, first directly, and then using Green's theorem.

SOLUTION By definition,

$$\int_{\mathbf{c}} \alpha = \int_c xydx - x^2ydy = \int_0^{2\pi} \left(xy\frac{dx}{dt} - x^2y\frac{dy}{dt}\right) dt$$

$$= \int_0^{2\pi} \left(-\cos t\sin^2 t - \sin t\cos^3 t\right) dt = \left(-\tfrac{1}{3}\sin^3 t + \tfrac{1}{4}\cos^4 t\right)\Big|_0^{2\pi} = 0.$$

Since $d\alpha = (ydx + xdy) \wedge dx - (2xydx + x^2dy) \wedge dy = (-x - 2xy)dxdy$, we get

$$\int_{\mathbf{c}} \alpha = \int_D d\alpha = -\iint_{\{0 \le x^2+y^2 \le 1\}} (x + 2xy)\,dA,$$

using Theorem 8.7. Passing to polar coordinates,

$$\int_{\mathbf{c}} \alpha = -\int_0^{2\pi} \left(\int_0^1 \left(r\cos\theta + 2r^2\cos\theta\sin\theta\right) rdr\right) d\theta$$

$$= -\int_0^{2\pi} \left(\tfrac{1}{3}r^3\cos\theta + \tfrac{1}{2}r^4\cos\theta\sin\theta\right)\Big|_0^1 \,d\theta$$

$$= -\int_0^{2\pi} \left(\tfrac{1}{3}\cos\theta + \tfrac{1}{2}\cos\theta\sin\theta\right) d\theta = -\left(\tfrac{1}{3}\sin\theta + \tfrac{1}{4}\sin^2\theta\right)\Big|_0^{2\pi} = 0.$$ ◀

THEOREM 8.8 Gauss' Divergence Theorem for Differential Forms

Let W be region in $\mathbb{R}^3$ that satisfies Assumption 8.2 of Section 8.2, and let ∂W be its positively oriented boundary. Assume that α is a C^1 2-form defined on an open set U containing W. Then

$$\int_{\partial W} \alpha = \int_W d\alpha.$$

PROOF: This theorem is a straightforward translation of the Divergence Theorem into the language of differential forms. Let $\alpha = F_1 dydz + F_2 dzdx + F_3 dxdy$, where $F_1 = F_1(x, y, z)$, $F_2 = F_2(x, y, z)$, and $F_3 = F_3(x, y, z)$ are C^1 real-valued functions defined on U. The surface integral in Gauss' Theorem,

$$\iint_{S=\partial W} \mathbf{F} \cdot d\mathbf{S} = \iiint_W div\, \mathbf{F}\, dV,$$

where $\mathbf{F} = F_1\mathbf{i} + F_2\mathbf{j} + F_3\mathbf{k}$, is equal to $\int_{\partial W} \alpha$. Furthermore (we immediately drop terms that are zero),

$$\begin{aligned} d\alpha &= \frac{\partial F_1}{\partial x} dx \wedge dydz + \frac{\partial F_2}{\partial y} dy \wedge dzdx + \frac{\partial F_3}{\partial z} dz \wedge dxdy \\ &= \left(\frac{\partial F_1}{\partial x} + \frac{\partial F_2}{\partial y} + \frac{\partial F_3}{\partial z} \right) dxdydz. \end{aligned}$$

By definition, the integral $\int_W f dxdydz$ of the 3-form $f dxdydz$ [$f = f(x, y, z)$ is a real-valued function] is equal to $\iiint_W f\, dV$. Therefore,

$$\int_W d\alpha = \int_W \left(\frac{\partial F_1}{\partial x} + \frac{\partial F_2}{\partial y} + \frac{\partial F_3}{\partial z} \right) dxdydz = \iiint_W div\, \mathbf{F}\, dV,$$

and the statement of the theorem is established. ◀

▶ EXAMPLE 8.27

Let W be a region in $\mathbb{R}^3$ to which Gauss' Divergence Theorem applies. Prove that $\int_{\partial W} xdydz + ydzdx + zdxdy = 3v(W)$, where $v(W)$ denotes the volume of W (compare with area formulas obtained from Green's Theorem).

SOLUTION By Theorem 8.8, we obtain

$$\begin{aligned} \int_{\partial W} xdydz + ydzdx + zdxdy &= \int_W d\,(xdydz + ydzdx + zdxdy) \\ &= \int_W dx \wedge dydz + dy \wedge dzdx + dz \wedge dxdy \\ &= 3\int_W dxdydz = 3\int_W dV = 3v(W). \end{aligned}$$ ◀

▶ EXAMPLE 8.28

Using Gauss' Divergence Theorem, evaluate $\int_S \alpha$, where $\alpha = xdydz + ydzdx$ and S is the closed upper hemisphere of radius 1 (i.e., it consists of the upper hemisphere $\{(x, y, z) | x^2 + y^2 + z^2 = 1,$

$z \geq 0\}$ together with the disk $\{(x, y, z) | x^2 + y^2 \leq 1,\ z = 0\}$ in the xy-plane). Check the result by direct computation.

SOLUTION Since $d\alpha = dx \wedge dydz + dy \wedge dzdx = 2dxdydz$, Theorem 8.8 implies that

$$\int_{S=\partial W} xdydz + ydzdx = \int_W 2dxdydz,$$

where W denotes the upper-half ball of radius 1. Therefore,

$$\int_S xdydz + ydzdx = 2\int_W dxdydz = 2v(W) = \frac{4\pi}{3}.$$

Now we compute the same integral by computing the surface integral of α. Parametrize the upper hemisphere S_1 (see Section 7.1) by $\mathbf{r}_1(u, v) = (\cos v \cos u, \cos v \sin u, \sin v)$, where $0 \leq u \leq 2\pi$, $0 \leq v \leq \pi/2$. By the definition of the integral of a 2-form,

$$\int_{S_1} \alpha = \iint_{[0,2\pi]\times[0,\pi/2]} \left(\cos v \cos u \frac{\partial(y, z)}{\partial(u, v)} + \cos v \sin u \frac{\partial(z, x)}{\partial(u, v)} \right) dA,$$

where

$$\frac{\partial(y, z)}{\partial(u, v)} = \begin{vmatrix} \cos v \cos u & -\sin v \sin u \\ 0 & \cos v \end{vmatrix} = \cos^2 v \cos u$$

and

$$\frac{\partial(z, x)}{\partial(u, v)} = \begin{vmatrix} 0 & \cos v \\ -\cos v \sin u & -\sin v \cos u \end{vmatrix} = \cos^2 v \sin u.$$

Hence,

$$\int_{S_1} \alpha = \iint_{[0,2\pi]\times[0,\pi/2]} (\cos^3 v \cos^2 u + \cos^3 v \sin^2 u) dA$$

$$= \int_0^{\pi/2} \left(\int_0^{2\pi} \cos^3 v\ du \right) dv = 2\pi \int_0^{\pi/2} \cos^3 v\ dv$$

[$\int \cos^3 v dv = \int \cos v(1 - \sin^2 v)dv = \int \cos v dv - \int \cos v \sin^2 v dv = \sin v - (\sin^3 v)/3$]

$$= 2\pi \left(\sin v - \frac{1}{3} \sin^3 v \right) \Big|_0^{\pi/2} = \frac{4\pi}{3}.$$

Parametrize the lower side S_2 by $\mathbf{r}_2(u, v) = (u, v, 0)$, $u^2 + v^2 \leq 1$. Then

$$\int_{S_2} \alpha = \iint_{\{u^2+v^2\leq 1\}} \left(u \frac{\partial(y, z)}{\partial(u, v)} + v \frac{\partial(z, x)}{\partial(u, v)} \right) dA,$$

where

$$\frac{\partial(y, z)}{\partial(u, v)} = \begin{vmatrix} 0 & 1 \\ 0 & 0 \end{vmatrix} = 0 \quad \text{and} \quad \frac{\partial(z, x)}{\partial(u, v)} = \begin{vmatrix} 0 & 0 \\ 1 & 0 \end{vmatrix} = 0.$$

Hence, $\int_{S_2} \alpha = 0$ and

$$\int_S \alpha = \int_{S_1} \alpha + \int_{S_2} \alpha = \frac{4\pi}{3} + 0 = \frac{4\pi}{3},$$

as expected. ◀

THEOREM 8.9 Stokes' Theorem for Differential Forms

Let $\alpha = Pdx + Qdy + Rdz$ be a C^1 1-form defined on an open set U in $\mathbb{R}^3$. With the assumptions on $S \subseteq U$ and ∂S as in Theorem 8.4,

$$\int_{\partial S} \alpha = \int_S d\alpha.$$

PROOF: The statement of this theorem is a rewrite of Stokes' Theorem. To show that, take a vector field $\mathbf{F} = P\mathbf{i} + Q\mathbf{j} + R\mathbf{k}$. The equality $\iint_{\partial S} \mathbf{F} \cdot d\mathbf{s} = \int_{\partial S} \alpha$ holds by definition of the path integral of a 1-form. We have shown earlier that $d\alpha$ corresponds to *curl* $\mathbf{F}$, and hence, $\int_S d\alpha = \iint_S curl\,\mathbf{F} \cdot d\mathbf{S}$. ◄

After reading the statements of Green's and Gauss' Theorems for differential forms, the statement of Theorem 8.9 comes as no surprise. They are all the same! Using the notation and terminology introduced in Section 7.5, we write all three theorems as

$$\int_{\partial \mathcal{M}} \alpha = \int_{\mathcal{M}} d\alpha, \tag{8.17}$$

where α is a differential form and $\mathcal{M}$ is a manifold (i.e., a curve, a region in a plane, a surface, or a three-dimensional solid). And that is all we need! With the correct interpretation, we can recover all three integration theorems from (8.17). For example, if α is a 1-form and $\partial\mathcal{M}$ is a simple, closed curve in a plane, then $\int_{\partial\mathcal{M}} \alpha$ represents the path integral of a vector field along the curve. The right side is the integral over the region (in a plane) enclosed by $\partial\mathcal{M}$ (i.e., it is the double integral over $\mathcal{M}$) of the differential $d\alpha$; that is, of the curl of the vector field. So, in this case we get Green's Theorem.

Now let $\alpha = f$ be a 0-form (i.e., a real-valued function) and let $\mathcal{M} = \mathbf{c}$ be a curve with the initial point A and the terminal point B. In this case, the differential $d\alpha$ is the gradient of f, and $\partial\mathcal{M}$ consists of the two points, A and B. The left side of (8.17) is the "integral" of f over two points A and B, and is interpreted as the real number $f(B) - f(A)$. The right side is the path integral of the gradient, and hence (8.17) reads $\int_{\mathbf{c}} \nabla f \cdot d\mathbf{s} = f(B) - f(A)$. We recognize this equation as the statement of the generalization of the Fundamental Theorem of Calculus discussed in Section 5.4.

The underlying theme (and this is really important) in (8.17) is that the integral of the derivative of a function over a manifold $\mathcal{M}$ does not depend on all of $\mathcal{M}$, but only on its boundary (e.g., there are ways of finding the temperature of the core of the earth that do not require we actually dig a hole to the core).

Comparing differential-form versions of Green's and Stokes' Theorems, we notice that Green's Theorem is a special case of Stokes' Theorem.

Translated into the language of differential forms, we have shown that if a 1-form α, defined on a simply-connected set $U \subseteq \mathbb{R}^3$, is closed (i.e., $d\alpha = 0$; recall that d for 1-forms is *curl* for vector fields), then its integral $\int_{\mathbf{c}} \alpha$ along any oriented, simple, closed curve is zero. Interpreting the equivalence (b) $\Leftrightarrow$ (a) of Theorem 5.8, we conclude that there must be a 0-form (i.e., a function) f such that $\alpha = df$. In other words, α is exact, as claimed in Theorem 8.6(a).

▶ EXERCISES 8.4

Exercises 1 to 9: Let $f = e^{x+y}$, $\alpha = 3xdx + yzdy$, $\beta = (x^2 + y^2)dz$, $\gamma = 2dxdy - x\cos y\, dydz$, $\mu = e^{-x}dxdz + e^{-y}dzdy$, and $\nu = \sin x\, dxdydz$ be differential forms defined on $\mathbb{R}^3$. Determine whether the following expressions are defined and, if so, evaluate them.

1. $f\beta$ **2.** $f\mu - \alpha \wedge \alpha$ **3.** $\alpha \wedge \beta$

4. $x\alpha + 2yz\beta$ **5.** $\beta - \mu$ **6.** $\nu - (\alpha \wedge \nu)$

7. $(\beta \wedge \gamma) - \nu$ **8.** $-\mu + (\alpha \wedge \beta)$ **9.** $\nu \wedge \alpha$

10. Let $\alpha = x^2ydx - y^3dy + y^2dz$ and let $\beta = ydxdy + zdydz + xdzdx$. Verify each identity.

(a) $\alpha \wedge \beta = \beta \wedge \alpha$ (b) $\alpha \wedge \alpha = 0$ (c) $\beta \wedge \beta = 0$

Exercises 11 to 18: Find $d\alpha$ for the form α.

11. $\alpha = e^{xyz}$, on $\mathbb{R}^3$ **12.** $\alpha = (x^2 + y^2)\, dxdz$, on $\mathbb{R}^3$

13. $\alpha = x^2(dxdy + dydz)$, on $\mathbb{R}^3$ **14.** $\alpha = \sin x \cos y\, dxdydz$, on $\mathbb{R}^3$

15. $\alpha = \arctan x$, on $\mathbb{R}^2$ **16.** $\alpha = xdy$, on $\mathbb{R}^2$

17. $\alpha = xdx/(x^2 + y^2) - ydy/(x^2 + y^2)$, on $\mathbb{R}^2 - \{(0, 0)\}$

18. $\alpha = dydzdx$, on $\mathbb{R}^3$

19. Let $\alpha = x^3dy - 2ydx + zdz$. Check that $d(d\alpha) = 0$.

20. Let $\alpha = xydz - 2dx + ydy$ and $\beta = dxdz$. Verify each identity.

(a) $d(\alpha \wedge \beta) = d\alpha \wedge \beta - \alpha \wedge d\beta$ (b) $d(\beta \wedge \alpha) = d\beta \wedge \alpha + \beta \wedge d\alpha$

21. Consider the 0-form (i.e., a real-valued function) $f = \ln(x^2 + y^2 + z^2 + 1)$. Show that the components of df coincide with the components of *grad* f. What vector identity is represented by $d(df) = 0$?

Exercises 22 to 24: Compute $\int_{\mathbf{c}} \alpha$ for the 1-form α along the curve $\mathbf{c}$.

22. $\alpha = 2x^2y^2dx - y^3xdy$, $\mathbf{c}$ is the line segment from $(-1, 0)$ to $(1, 1)$, followed by the segment from $(1, 1)$ to $(0, 0)$

23. $\alpha = ydy - xyzdz$, $\mathbf{c}(t) = (\sqrt{t}, t^2, t^3)$, $0 \le t \le 1$

24. $\alpha = (x^3 + xy^2)\, dx - (x^2 + y^2)\, dy + 2zdz$, $\mathbf{c}(t) = (3\sin t, 3\cos t, 2t)$, $0 \le t \le \pi/2$

25. State which of the following forms are closed, and/or exact, or neither.

(a) $\alpha = ye^{xy}dx + xe^{xy}dy$ (b) $\alpha = xdx + ydy + zdz$

(c) $\alpha = -ydx + xdy + zdz$ (d) $\alpha = -2ydxdy - ydzdy + zdxdz$

(e) $\alpha = \sin x\, dxdy + (\sin y + \cos z)\, dydz$ (f) $\alpha = \sin(xyz^2)\, dxdydz$

26. Let α be an exact form (i.e., $\alpha = df$ for some differentiable function f) defined on $\mathbb{R}^3$ and let $\mathbf{c}(t)$, $t \in [a, b]$, be a smooth differentiable curve. Show that $\int_{\mathbf{c}} \alpha = f(\mathbf{c}(b)) - f(\mathbf{c}(a))$.

27. Show that the form $\alpha = ydx + xdy + 4dz$ is exact and compute $\int_{\mathbf{c}} \alpha$ along any path from $(0, 0, 0)$ to $(0, 4, 1)$.

Exercises 28 to 31: Compute $\int_S \alpha$.

28. $\alpha = dxdy$, S is the surface of the sphere $x^2 + y^2 + z^2 = a^2$

29. $\alpha = xdxdy + ydydz + zdzdx$, S is the surface of the cylinder $x^2 + y^2 = 1$, $0 \le z \le 3$

30. $\alpha = xydydz - x^2yzdzdx$, S is the surface $\mathbf{r}(u, v) = (u^2 - v^2, uv, 1 - v)$, $0 \le u, v \le 1$

31. $\alpha = dxdy + zdzdx$, S is the part of the paraboloid $z = x^2 + y^2$ below $z = 4$

Exercises 32 to 35: Compute $\int_{\mathbf{c}} \alpha$ using Green's Theorem.

32. $\alpha = y^2dx + xydy$, $\mathbf{c}$ is the boundary of the triangle determined by the equations $y = x$, $x = 0$, and $y = 1$, with counterclockwise orientation

33. $\alpha = xdx + xdy$, $\mathbf{c}$ consists of the semicircle $x^2 + y^2 = 1$, $y \ge 0$, followed by the straight-line segment from $(-1, 0)$ to $(1, 0)$, with counterclockwise orientation

34. $\alpha = xdx + (x^2 + y^2)dy$, $\mathbf{c}$ is the boundary of the annulus $1 \le x^2 + y^2 \le 2$; the outer circle is oriented counterclockwise, and the inner one has clockwise orientation

35. $\alpha = (2x + 2y^2)dx - 3xdy$, $\mathbf{c}$ is the boundary of the rectangle $[0, 1] \times [-2, 2]$, with counterclockwise orientation

36. Let f be a C^1 function and let $\mathbf{c}$ be a smooth curve with endpoints (x_0, y_0) and (x_1, y_1). Show that $\int_{\mathbf{c}}(f_xdx + f_ydy) = f(x_1, y_1) - f(x_0, y_0)$.

Exercises 37 to 40: Use Gauss' Divergence Theorem to compute $\int_S \alpha$.

37. $\alpha = ydydz + xzdzdx + zdxdy$, S is the boundary of the region inside the cylinder $x^2 + y^2 = 1$, above the xy-plane, and below the paraboloid $z = x^2 + y^2$

38. $\alpha = zdxdy$, S is the boundary of the ellipsoid $x^2/a^2 + y^2/b^2 + z^2/c^2 = 1$

39. $\alpha = xdydz + ydzdx + zdxdy$, S is the boundary of the region $1 \le x^2 + y^2 + z^2 \le 4$

40. $\alpha = \ln(x^2 + y^2)dydz$, S is the boundary of the cylinder $x^2 + y^2 = 4$, $1 \le z \le 2$

8.5 VECTOR CALCULUS IN ELECTROMAGNETISM

The purpose of this section is to explain and illustrate the use of concepts and tools of vector calculus in electromagnetism. We will not attempt to give a presentation covering fully the background needed for the formulas and laws that will be discussed. Instead, we will identify physical quantities as mathematical objects and show how to manipulate them to obtain meaningful physical quantities.

Formulas from electromagnetism appear in various references in different forms (that differ at most by constants), due to different choices of physical units. The constants that we use are the permittivity of vacuum ϵ_0 (it has appeared already in formulas for the electrostatic field and electrostatic potential) and the *permeability of vacuum* μ_0 (which can be, for example, determined from $\mu_0\epsilon_0 = c^{-2}$, where c is the speed of light in vacuum).

Point Charges

Recall that the electrostatic field $\mathbf{E}(x_0, y_0, z_0)$ at the point $P(x_0, y_0, z_0)$ due to a single charge Q located at $\mathbf{r}_Q = (x_Q, y_Q, z_Q)$ is defined as force per unit charge; that is,

$$\mathbf{E}(x_0, y_0, z_0) = \frac{1}{4\pi\epsilon_0} \frac{Q}{\|\mathbf{r}_0 - \mathbf{r}_Q\|^2}\mathbf{u},$$

where $\mathbf{r}_0 = (x_0, y_0, z_0)$ and $\mathbf{u} = (\mathbf{r}_0 - \mathbf{r}_Q)/\|\mathbf{r}_0 - \mathbf{r}_Q\|$ is the unit vector in the direction from the source Q toward the point P.

The electrostatic field at $P(x_0, y_0, z_0)$ due to n charges $Q_1, \ldots, Q_n$ located at $\mathbf{r}_1, \ldots, \mathbf{r}_n$ is the vector sum

$$\mathbf{E}(x_0, y_0, z_0) = \sum_{i=1}^{n} \frac{1}{4\pi\epsilon_0} \frac{Q_i}{\|\mathbf{r}_0 - \mathbf{r}_i\|^2} \mathbf{u}_i,$$

($\mathbf{u}_i = (\mathbf{r}_0 - \mathbf{r}_i)/\|\mathbf{r}_0 - \mathbf{r}_i\|$) of individual electrostatic fields.

Charge Density Function

The above approach, although convenient, is not always satisfactory. There are situations when one has to consider electric charge as being "spread" over some region, rather than concentrated at particular point(s). Such a distribution of charge is described by the *charge density function* $\rho(x, y, z)$. It is defined in the following way: take a region ΔW that contains a point (x, y, z) and assume that it encloses the total charge of ΔQ. The average charge in ΔW is $\Delta Q/\Delta W$ and the limit of these averages as ΔW shrinks to the "point" (x, y, z) is the charge density [by "point" (x, y, z) we actually mean a very small region that contains the point (x, y, z) and is still large enough to contain many charged particles].

Given the charge density function, the total charge contained in a solid region W is given by (this circularity—defining ρ using Q and then defining Q using ρ seems to be unavoidable)

$$Q = \iiint_W \rho\, dV, \tag{8.18}$$

and the electrostatic field is obtained as the integral version of the above summation formula:

$$\mathbf{E}(x_0, y_0, z_0) = \frac{1}{4\pi\epsilon_0} \iiint_W \frac{\rho\mathbf{u}}{\|\mathbf{r}_0 - \mathbf{r}\|^2}\, dV,$$

where $\mathbf{r} = (x, y, z)$ and $\mathbf{u} = (\mathbf{r}_0 - \mathbf{r})/\|\mathbf{r}_0 - \mathbf{r}\|$ (the triple integral of a vector is computed as the triple integral of its components).

Current Density Vector Field

One way to describe *current* is to use the *current density vector field* $\mathbf{J}(x, y, z)$. It is defined as the vector whose magnitude is the current per unit area and whose direction is the direction of the current flow. To be precise, place a small surface ΔS (with a unit normal $\mathbf{n}$) containing the point (x, y, z) in the current. If the total current flowing through ΔS is ΔI, then $\Delta I = \mathbf{J} \cdot \mathbf{n}\, A(\Delta S)$, where $A(\Delta S)$ is the area of ΔS. The quantity $\mathbf{J} \cdot \mathbf{n}(x, y, z)$ is now computed as the limit as ΔS collapses to the "point" (x, y, z). If charges have well-ordered motion given by a velocity vector field $\mathbf{v}$, then $\mathbf{J} = \rho\mathbf{v}$. The total current flowing through the surface S placed in the flow is

$$I = \iint_S \mathbf{J} \cdot \mathbf{n}\, dS = \iint_S \mathbf{J} \cdot d\mathbf{S}.$$

Gauss' Theorem

Gauss' Theorem (see Example 8.12 in Section 8.2) states that the net charge Q enclosed by a closed surface S is $Q = \epsilon_0 \iint_S \mathbf{E}(x, y, z) \cdot d\mathbf{S}$, where $\mathbf{E}(x, y, z)$ is the electrostatic field.

▶ EXAMPLE 8.29

Find the charge contained in the solid upper hemisphere W of radius 1 if the electric field is given by $\mathbf{E} = x\mathbf{i} + y\mathbf{j} + z\mathbf{k}$.

SOLUTION By Gauss' Law, the total charge in W is $Q = \epsilon_0 \iint_S \mathbf{E}(x, y, z)\cdot d\mathbf{S}$, where $S = \partial W$. Using the Divergence Theorem of Gauss, we get $Q = \epsilon_0 \iiint_W div\, \mathbf{E}(x, y, z)\, dV$.

Notice that this is the formula for the total charge (8.18), where the charge density is given by the scalar function $\epsilon_0 div\, \mathbf{E}$. In our case [$v(W)$ denotes the volume of W],

$$Q = \epsilon_0 \iiint_W div\, \mathbf{E}(x, y, z)\, dV = \epsilon_0 \iiint_W 3\, dV = 3\epsilon_0 v(W) = 2\pi\epsilon_0.$$ ◀

Ampère's Law

Let $\mathbf{B}(x, y, z)$ denote a magnetic field at a point (x, y, z) in space. The *magnetic circulation* is defined as the path integral $\mathcal{B} = \int_{\mathbf{c}} \mathbf{B}(x, y, z) \cdot d\mathbf{s}$, where $\mathbf{c}$ is a closed contour in space. From standard physical arguments (such as Biot–Savart's Law), it follows that (here presented without proof) $\mathcal{B} = \mu_0 I$, where I is the current $I = \iint_S \mathbf{J}(x, y, z) \cdot d\mathbf{S}$. The vector field $\mathbf{J}$ is the current density and S is any surface bounded by the contour $\mathbf{c}$. Stokes' Theorem

$$\mathcal{B} = \int_{\mathbf{c}} \mathbf{B}(x, y, z) \cdot d\mathbf{s} = \iint_S curl\, \mathbf{B}(x, y, z) \cdot d\mathbf{S},$$

combined with the above expression for $\mathcal{B}$ yields

$$\iint_S (curl\, \mathbf{B}(x, y, z) - \mu_0 \mathbf{J}(x, y, z)) \cdot d\mathbf{S} = 0$$

for any S such that $\partial S = \mathbf{c}$. It follows that the integrand is zero; that is,

$$curl\, \mathbf{B}(x, y, z) = \mu_0 \mathbf{J}(x, y, z),$$

which is known as *Ampère's Law*.

Time-Changing Electric and Magnetic Fields

From this moment on, we assume that the electric field $\mathbf{E}(x, y, z, t)$, the magnetic field $\mathbf{B}(x, y, z, t)$, the charge density $\rho(x, y, z, t)$, and current density $\mathbf{J}(x, y, z, t)$ all change with time. It is assumed that they are continuously differentiable (C^1) functions of the arguments listed. Differential operators *grad, div,* and *curl* are computed by keeping t fixed (i.e., the partial derivatives are taken with respect to "space" variables x, y, and z only).

Let $\mathbf{c}$ be a simple closed curve. The *circulation* $\mathcal{E}(t)$ *of the electric field* $\mathbf{E}(x, y, z, t)$ is given by

$$\mathcal{E}(t) = \int_{\mathbf{c}} \mathbf{E}(x, y, z, t) \cdot d\mathbf{s},$$

and the *magnetic circulation* $\mathcal{B}(t)$ is the path integral

$$\mathcal{B}(t) = \int_{\mathbf{c}} \mathbf{B}(x, y, z, t) \cdot d\mathbf{s}.$$

Let S be a surface in $\mathbb{R}^3$ that satisfies the assumptions of Stokes' Theorem and let ∂S be its positively oriented boundary. The *magnetic flux* $\Phi(t)$ is defined as

$$\Phi(t) = \iint_S \mathbf{B}(x, y, z, t) \cdot d\mathbf{S},$$

and the *flux* $\Psi(t)$ *of the electric field* is defined as

$$\Psi(t) = \iint_S \mathbf{E}(x, y, z, t) \cdot d\mathbf{S}.$$

Conservation of Charge, Continuity Equation

The conservation of charge principle states that if an amount of charge leaves a solid three-dimensional region W enclosed by the surface S (i.e., $S = \partial W$), then the charge inside V must decrease accordingly.

Let W contain a charge density $\rho(x, y, z, t)$, with a current density $\mathbf{J}(x, y, z, t)$ on the boundary surface S. Assume that both ρ and $\mathbf{J}$ are continuously differentiable functions.

The flux integral

$$\iint_S \mathbf{J}(x, y, z, t) \cdot d\mathbf{S} = \iint_S \mathbf{J}(x, y, z, t) \cdot \mathbf{n}\, dS,$$

with the unit normal $\mathbf{n}$ oriented outward, represents the charge per unit time leaving W through S. The total charge inside W at any time t is $Q(t) = \iiint_W \rho(x, y, z, t)\, dV$, and is changing at the rate

$$\frac{\partial Q(t)}{\partial t} = \frac{\partial}{\partial t}\left(\iiint_W \rho(x, y, z, t)\, dV\right) = \iiint_W \frac{\partial \rho(x, y, z, t)}{\partial t}\, dV \tag{8.19}$$

(the function ρ is differentiable, and therefore we are allowed to switch the integration and the time-derivative). By the conservation law,

$$\iint_S \mathbf{J}(x, y, z, t) \cdot d\mathbf{S} = -\frac{\partial Q(t)}{\partial t}. \tag{8.20}$$

If a charge leaves W, then the flux integral on the left is positive; at the same time, the charge inside is decreasing. Hence, $\partial Q/\partial t < 0$ and $-\partial Q/\partial t$ is positive. Similarly, if a charge enters W, the flux integral is negative (i.e., the outward flux is negative); the charge inside is increasing. Hence, $\partial Q/\partial t > 0$ and $-\partial Q/\partial t < 0$. This explains the appearance of the minus sign in (8.20).

We express the left side in (8.20) as a volume integral using the Divergence Theorem:

$$\iint_S \mathbf{J}(x, y, z, t) \cdot d\mathbf{S} = \iiint_W div\, \mathbf{J}(x, y, z, t)\, dV. \tag{8.21}$$

Substituting this equation and (8.19) into (8.20), we get

$$\iiint_W div\, \mathbf{J}(x, y, z, t)\, dV = -\iiint_W \frac{\partial \rho(x, y, z, t)}{\partial t}\, dV$$

and therefore,

$$\iiint_W \left(div\, \mathbf{J}(x, y, z, t) + \frac{\partial \rho(x, y, z, t)}{\partial t} \right) dV = 0.$$

Since this equation holds for any region V, it follows that

$$div\, \mathbf{J}(x, y, z, t) + \frac{\partial \rho(x, y, z, t)}{\partial t} = 0. \tag{8.22}$$

Equation (8.22) is called the *continuity equation* and is a basic equation of electromagnetism (similar equations appear in other applications). If $\partial\rho/\partial t = 0$ at all points, the continuity equation gives $div\, \mathbf{J}(x, y, z, t) = 0$, which is the condition for *steady currents.* In that case,

$$\iint_S \mathbf{J}(x, y, z, t) \cdot d\mathbf{S} = \iiint_W div\, \mathbf{J}(x, y, z, t)\, dV = 0;$$

that is, the total current leaving any closed surface S is zero. In other words, charge does not "accumulate" or "disappear" at some point.

Faraday's Law

Faraday's Law states that the circulation of $\mathbf{E}$ around a simple closed curve $\mathbf{c}$ equals negative rate of change of magnetic flux through a surface S bounded by $\mathbf{c}$,

$$\mathcal{E}(t) = -\frac{\partial \Phi(t)}{\partial t}.$$

The circulation $\mathcal{E}(t)$ is computed, by Stokes' Theorem, to be

$$\mathcal{E}(t) = \int_{\mathbf{c}=\partial S} \mathbf{E}(x, y, z, t) \cdot d\mathbf{s} = \iint_S curl\, \mathbf{E}(x, y, z, t) \cdot d\mathbf{S}.$$

When we substitute this and

$$\frac{\partial \Phi(t)}{\partial t} = \frac{\partial}{\partial t}\left(\iint_S \mathbf{B}(x, y, z, t) \cdot d\mathbf{S} \right) = \iint_S \frac{\partial \mathbf{B}(x, y, z, t)}{\partial t} \cdot d\mathbf{S}$$

[we are allowed to switch the integral and partial derivative since $\mathbf{B}$ is assumed differentiable] into Faraday's Law, we obtain

$$\iint_S curl\, \mathbf{E}(x, y, z, t) \cdot d\mathbf{S} = -\iint_S \frac{\partial \mathbf{B}(x, y, z, t)}{\partial t} \cdot d\mathbf{S},$$

that is,

$$\iint_S \left(curl\, \mathbf{E}(x, y, z, t) + \frac{\partial \mathbf{B}(x, y, z, t)}{\partial t} \right) \cdot d\mathbf{S} = 0.$$

Since this equation holds for any surface S, it follows that

$$curl\, \mathbf{E}(x, y, z, t) = -\frac{\partial \mathbf{B}(x, y, z, t)}{\partial t},$$

which is known as Maxwell's second equation.

It follows that Faraday's Law implies Maxwell's second equation. As a matter of fact, the two are equivalent. To demonstrate it, we start with Maxwell's second equation and will derive Faraday's Law. By Stokes' Theorem,

$$\mathcal{E}(t) = \int_{\partial S} \mathbf{E}(x, y, z, t) \cdot d\mathbf{s} = \iint_S curl\, \mathbf{E}(x, y, z, t) \cdot d\mathbf{S}$$

(using Maxwell's equation and switching the time derivative and the integral)

$$= \iint_S -\frac{\partial \mathbf{B}(x, y, z, t)}{\partial t} \cdot d\mathbf{S} = -\frac{\partial}{\partial t}\left(\iint_S \mathbf{B}(x, y, z, t) \cdot d\mathbf{S} \right) = -\frac{\partial \Phi(t)}{\partial t},$$

where

$$\Phi(t) = \iint_S \mathbf{B}(x, y, z, t) \cdot d\mathbf{S}$$

is the magnetic flux.

Let us point out that a number of laws that we have encountered so far (and there will be more in the remaining part of this section) come in two forms. They either claim something about an integral of a field (circulation or flux), or (expressed in a differential form) about the properties of a field at a point. For example, the (integral) law $\mathcal{B}(t) = \int_{\mathbf{c}} \mathbf{B}(x, y, z, t) \cdot d\mathbf{s} = \mu_0 I(t)$ has its (differential) counterpart $curl\, \mathbf{B}(x, y, z, t) = \mu_0 \mathbf{J}(x, y, z, t)$, known as Ampère's Law. We showed in Section 8.2 that the integral statement of Gauss' Law $\iint_S \mathbf{E}(x, y, z, t) \cdot d\mathbf{S} = Q/\epsilon_0$ can be written as the formula $div\, \mathbf{E}(x, y, z, t) = \rho(x, y, z, t)/\epsilon_0$ giving the value of $div\, \mathbf{E}$ at a point (it is known as Maxwell's first equation). Furthermore, Maxwell's second equation $curl\, \mathbf{E}(x, y, z, t) = -\partial \mathbf{B}(x, y, z, t)/\partial t$ and Faraday's Law $\int_{\mathbf{c}} \mathbf{E}(x, y, z, t) \cdot d\mathbf{s} = -\partial \Phi(t)/\partial t$ represent two different viewpoints of the same physical fact.

Therefore, there are not that many formulas after all. As a matter of fact, a complete set of laws that relate electric and magnetic fields to each other and to the charges and currents that produce them consists of four equations, known as *Maxwell's equations*.

Maxwell's Equations

We have already discussed two equations: the first one is Gauss' Law

$$div\, \mathbf{E}(x, y, z, t) = \frac{\rho(x, y, z, t)}{\epsilon_0}, \tag{8.23}$$

written in differential form (actually all four equations can be written in differential form). The second equation is Faraday's Law

$$curl\, \mathbf{E}(x, y, z, t) = -\frac{\partial \mathbf{B}(x, y, z, t)}{\partial t}. \tag{8.24}$$

The generalized form of Ampère's Law

$$curl\, \mathbf{B}(x, y, z, t) = \mu_0 \left(\mathbf{J}(x, y, z, t) + \epsilon_0 \frac{\partial \mathbf{E}(x, y, z, t)}{\partial t} \right), \tag{8.25}$$

[earlier in the section, we assumed that $\partial \mathbf{E}(x, y, z, t)/\partial t = 0$] and the requirement that there be no magnetic sources present:

$$div\,\mathbf{B}(x, y, z, t) = 0. \tag{8.26}$$

complete the list of Maxwell's equations.

In the absence of charges [i.e., $\rho(x, y, z, t) = 0$; e.g., in the case of electromagnetic waves propagating in a vacuum] and currents [i.e., $\mathbf{J}(x, y, z, t) = \mathbf{0}$], Maxwell's equations read [from now on, we drop the list of variables (x, y, z, t) from the notation]

$$div\,\mathbf{E} = 0 \tag{8.27}$$

$$curl\,\mathbf{E} = -\frac{\partial \mathbf{B}}{\partial t} \tag{8.28}$$

$$curl\,\mathbf{B} = \frac{1}{c^2}\frac{\partial \mathbf{E}}{\partial t} \tag{8.29}$$

$$div\,\mathbf{B} = 0. \tag{8.30}$$

Maxwell's equations are not symmetric in $\mathbf{B}$ and $\mathbf{E}$ (i.e., interchanging $\mathbf{B}$ and $\mathbf{E}$ does not yield the same equations). Nevertheless, we are going to show that $\mathbf{B}$ and $\mathbf{E}$ satisfy the same differential equation (in this special case when $\rho = 0$ and $\mathbf{J} = \mathbf{0}$). We will accomplish that by computing the Laplacian of $\mathbf{B}$ and $\mathbf{E}$ [recall that the Laplacian of the vector field $\mathbf{F} = (F_1, F_2, F_3)$ is given by $\Delta\mathbf{F} = (\Delta F_1, \Delta F_2, \Delta F_3)$] using the transformation formula

$$curl\,(curl\,\mathbf{F}) = grad\,(div\,\mathbf{F}) - \Delta\mathbf{F}; \tag{8.31}$$

see Section 4.8. We first use (8.28)

$$curl\,(curl\,\mathbf{E}) = curl\left(-\frac{\partial \mathbf{B}}{\partial t}\right) = -curl\left(\frac{\partial \mathbf{B}}{\partial t}\right),$$

then interchange the derivatives $curl$ and $\partial/\partial t$ (this can be done whenever $\mathbf{B}$ is a differentiable vector field), and use (8.29)

$$= -\frac{\partial}{\partial t}(curl\,\mathbf{B}) = -\frac{\partial}{\partial t}\left(\frac{1}{c^2}\frac{\partial \mathbf{E}}{\partial t}\right) = -\frac{1}{c^2}\frac{\partial^2 \mathbf{E}}{\partial t^2}.$$

Substituting this into (8.31) and using (8.27), we get

$$\Delta\mathbf{E} = grad\,(div\,\mathbf{E}) - curl\,(curl\,\mathbf{E}) = \frac{1}{c^2}\frac{\partial^2 \mathbf{E}}{\partial t^2}.$$

Similarly, by (8.29) and (8.28),

$$curl\,(curl\,\mathbf{B}) = curl\left(\frac{1}{c^2}\frac{\partial \mathbf{E}}{\partial t}\right) = \frac{1}{c^2}curl\left(\frac{\partial \mathbf{E}}{\partial t}\right) = \frac{1}{c^2}\frac{\partial}{\partial t}(curl\,\mathbf{E}) = -\frac{1}{c^2}\frac{\partial^2 \mathbf{B}}{\partial t^2},$$

and

$$\Delta\mathbf{B} = grad\,(div\,\mathbf{B}) - curl\,(curl\,\mathbf{B}) = \frac{1}{c^2}\frac{\partial^2 \mathbf{B}}{\partial t^2},$$

since $div\,\mathbf{B} = 0$ by (8.30). Therefore,

$$\Delta\mathbf{E} = \frac{1}{c^2}\frac{\partial^2 \mathbf{E}}{\partial t^2} \quad \text{and} \quad \Delta\mathbf{B} = \frac{1}{c^2}\frac{\partial^2 \mathbf{B}}{\partial t^2};$$

that is, both **B** and **E** satisfy the same *(higher-dimensional) wave equation* in the special case when $\rho = 0$ and $\mathbf{J} = \mathbf{0}$.

Poynting Vector

Let us for the moment stay with the special case of Maxwell's equations (8.27)–(8.30). Computing the dot product of (8.29) (with $1/c^2$ replaced by $\mu_0\epsilon_0$)

$$curl\,\mathbf{B} = \mu_0\epsilon_0\frac{\partial \mathbf{E}}{\partial t}$$

with $\mathbf{E}$, and noticing that $(\|\mathbf{E}\|^2)' = (\mathbf{E}\cdot\mathbf{E})' = \mathbf{E}'\cdot\mathbf{E} + \mathbf{E}\cdot\mathbf{E}' = 2\mathbf{E}\cdot\mathbf{E}'$, we obtain

$$\mathbf{E}\cdot curl\,\mathbf{B} = \mu_0\epsilon_0\,\mathbf{E}\cdot\frac{\partial \mathbf{E}}{\partial t} = \frac{1}{2}\mu_0\epsilon_0\frac{\partial}{\partial t}\left(\|\mathbf{E}\|^2\right). \tag{8.32}$$

Similarly, the dot product of $curl\,\mathbf{E} = -\partial\mathbf{B}/\partial t$ [this is Eq. (8.28)] with $\mathbf{B}$ gives

$$\mathbf{B}\cdot curl\,\mathbf{E} = -\mathbf{B}\cdot\frac{\partial \mathbf{B}}{\partial t} = -\frac{1}{2}\frac{\partial}{\partial t}\left(\|\mathbf{B}\|^2\right). \tag{8.33}$$

Now subtract (8.32) from (8.33):

$$\mathbf{B}\cdot curl\,\mathbf{E} - \mathbf{E}\cdot curl\,\mathbf{B} = -\frac{\partial}{\partial t}\left(\frac{1}{2}\|\mathbf{B}\|^2 + \frac{1}{2}\mu_0\epsilon_0\|\mathbf{E}\|^2\right)$$

and use the formula $\mathbf{B}\cdot curl\,\mathbf{E} - \mathbf{E}\cdot curl\,\mathbf{B} = div\,(\mathbf{E}\times\mathbf{B})$ from Section 4.8 to simplify the left side and obtain

$$div\,(\mathbf{E}\times\mathbf{B}) = -\mu_0\frac{\partial}{\partial t}\left(\frac{1}{2\mu_0}\|\mathbf{B}\|^2 + \frac{1}{2}\epsilon_0\|\mathbf{E}\|^2\right). \tag{8.34}$$

The expression in parentheses on the right side of (8.34) is called the *total energy density* Ω in an electromagnetic field: it is equal to the sum $\Omega = \Omega_e + \Omega_m$ of the *energy density of the electric field* $\Omega_e = \frac{1}{2}\epsilon_0\|\mathbf{E}\|^2$ and the *energy density of the magnetic field* $\Omega_m = \frac{1}{2\mu_0}\|\mathbf{B}\|^2$. We can rewrite (8.34) as

$$div\,\mathbf{P} = -\frac{\partial \Omega}{\partial t},$$

where $\mathbf{P} = \mu_0^{-1}\mathbf{E}\times\mathbf{B}$. The vector $\mathbf{P}$ is called the *Poynting vector* and indicates the magnitude and direction of the energy flow (time rate of change of the total energy density) in an electromagnetic field.

Vector and Scalar Potentials

Now we go back to the general form of Maxwell's equations (8.23)–(8.26). Consider the electrostatic field in $\mathbb{R}^3$, given by $\mathbf{E} = \frac{1}{4\pi\epsilon_0}\frac{Q}{\|\mathbf{r}\|^3}\mathbf{r}$, where $\mathbf{r} \neq \mathbf{0}$. Since $curl\,\mathbf{E} = \mathbf{0}$ (we have shown that in Section 4.6), it follows by Theorem 5.8 in Section 5.4 that there is a *scalar potential* ϕ such that

$$\mathbf{E} = -grad\,\phi. \tag{8.35}$$

According to Maxwell's equation (8.26), the divergence of the magnetic field $\mathbf{B}$ vanishes, and consequently [see (8.16)], there exists a *vector potential* $\mathbf{A}$ such that

$$\mathbf{B} = curl\,\mathbf{A}. \tag{8.36}$$

Now assume that both $\mathbf{E}$ and $\mathbf{B}$ vary with time. Then, in the absence of magnetic charge, (8.36) still holds, but (8.35) is no longer true, since $curl\,\mathbf{E} \neq \mathbf{0}$. Therefore, we have to define a new scalar potential for this (new, time-dependent) situation. By (8.24) and (8.36),

$$curl\,\mathbf{E} = -\frac{\partial \mathbf{B}}{\partial t} = -\frac{\partial}{\partial t}(curl\,\mathbf{A}) = -curl\left(\frac{\partial \mathbf{A}}{\partial t}\right),$$

and thus $curl\,(\mathbf{E} + \partial\mathbf{A}/\partial t) = \mathbf{0}$. This identity implies that the vector field $\mathbf{E} + \partial\mathbf{A}/\partial t$ has a potential function. So we *define* the *scalar potential* ϕ by

$$grad\,\phi = -\left(\mathbf{E} + \frac{\partial \mathbf{A}}{\partial t}\right). \tag{8.37}$$

Substituting (8.37) into (8.23), we get

$$\begin{aligned}\frac{\rho}{\epsilon_0} &= div\,\mathbf{E} = div\left(-grad\,\phi - \frac{\partial \mathbf{A}}{\partial t}\right)\\ &= -div\,(grad\,\phi) - div\left(\frac{\partial \mathbf{A}}{\partial t}\right) = -\Delta\phi - \frac{\partial}{\partial t}(div\,\mathbf{A}).\end{aligned}$$

In this computation, we used the definition of the Laplace operator $\Delta\phi = div\,(grad\,\phi)$ and the fact that for a differentiable function the derivatives $\partial/\partial t$ and div can be interchanged. Rewrite the above as

$$\Delta\phi + \frac{\partial}{\partial t}(div\,\mathbf{A}) = -\frac{\rho}{\epsilon_0}. \tag{8.38}$$

We will now compute both sides in Maxwell's equation (8.25), which states that

$$curl\,\mathbf{B} = \mu_0\mathbf{J} + \mu_0\epsilon_0\frac{\partial \mathbf{E}}{\partial t} = \mu_0\mathbf{J} + \frac{1}{c^2}\frac{\partial \mathbf{E}}{\partial t}. \tag{8.39}$$

The left side can be expressed using (8.31) as

$$curl\,\mathbf{B} = curl\,(curl\,\mathbf{A}) = grad\,(div\,\mathbf{A}) - \Delta\mathbf{A}. \tag{8.40}$$

Substituting (8.37) into the right side of (8.39), we obtain

$$\begin{aligned}\mu_0\mathbf{J} + \frac{1}{c^2}\frac{\partial \mathbf{E}}{\partial t} &= \mu_0\mathbf{J} + \frac{1}{c^2}\frac{\partial}{\partial t}\left(-grad\,\phi - \frac{\partial \mathbf{A}}{\partial t}\right)\\ &= \mu_0\mathbf{J} - \frac{1}{c^2}\frac{\partial}{\partial t}(grad\,\phi) - \frac{1}{c^2}\frac{\partial^2 \mathbf{A}}{\partial t^2} = \mu_0\mathbf{J} - \frac{1}{c^2}grad\,\frac{\partial\phi}{\partial t} - \frac{1}{c^2}\frac{\partial^2 \mathbf{A}}{\partial t^2}.\end{aligned} \tag{8.41}$$

Identity (8.39), together with (8.40) and (8.41), gives

$$grad\,(div\,\mathbf{A}) - \Delta\mathbf{A} = \mu_0\mathbf{J} - \frac{1}{c^2}grad\,\frac{\partial\phi}{\partial t} - \frac{1}{c^2}\frac{\partial^2 \mathbf{A}}{\partial t^2},$$

or, after we rearrange terms,

$$\Delta\mathbf{A} - \frac{1}{c^2}\frac{\partial^2\mathbf{A}}{\partial t^2} - grad\left(div\,\mathbf{A} + \frac{1}{c^2}\frac{\partial\phi}{\partial t}\right) = -\mu_0\mathbf{J}. \tag{8.42}$$

Equations (8.38) and (8.42) give the relations between the scalar potential ϕ and the vector potential $\mathbf{A}$.

Lorentz Gauge

Let us first examine how much freedom we have in choosing $\mathbf{A}$ such that $\mathbf{B} = curl\,\mathbf{A}$. In other words: suppose that $\mathbf{B} = curl\,\mathbf{A}_0$ for some $\mathbf{A}_0$—what is the relation between $\mathbf{A}$ and $\mathbf{A}_0$? Do they have to be equal?

Since $\mathbf{B} = curl\,\mathbf{A}$ and $\mathbf{B} = curl\,\mathbf{A}_0$, it follows that $curl\,\mathbf{A} = curl\,\mathbf{A}_0$ and $curl\,(\mathbf{A} - \mathbf{A}_0) = \mathbf{0}$. A vector field whose $curl$ is zero has a potential function. Hence, $\mathbf{A} - \mathbf{A}_0 = grad\,f$ for some scalar function f, and the answer to our question is the following: instead of taking $\mathbf{A}$, we could take $\mathbf{A} - grad\,f$ and still keep the equality $\mathbf{B} = curl\,\mathbf{A}$.

Now suppose that we took

$$\mathbf{A}_0 = \mathbf{A} - grad\,f \tag{8.43}$$

instead of $\mathbf{A}$ in (8.36). The only change we have to make is to define a new scalar potential ϕ_0; that is, the one that corresponds to $\mathbf{A}_0$ by means of (8.37). Hence, by (8.37) and (8.43),

$$\begin{aligned} grad\,\phi_0 &= -\mathbf{E} - \frac{\partial\mathbf{A}_0}{\partial t} = -\mathbf{E} - \frac{\partial}{\partial t}(\mathbf{A} - grad\,f) \\ &= -\mathbf{E} - \frac{\partial\mathbf{A}}{\partial t} + \frac{\partial}{\partial t}(grad\,f) = grad\,\phi + grad\left(\frac{\partial f}{\partial t}\right); \end{aligned}$$

that is, $grad\,\phi_0 = grad\,(\phi + \partial f/\partial t)$. Therefore,

$$\phi_0 = \phi + \frac{\partial f}{\partial t} + \text{constant} = \phi + \frac{\partial f}{\partial t} \tag{8.44}$$

(take the constant to be 0, for simplicity).

Therefore, instead of choosing $\mathbf{A}$ and ϕ, we could choose $\mathbf{A}_0$ and ϕ_0, given by (8.43) and (8.44), with any scalar function f. We are going to make use of this freedom in choosing f to simplify equations (8.38) and (8.42). To be precise, we will try to get rid of the term

$$div\,\mathbf{A} + \frac{1}{c^2}\frac{\partial\phi}{\partial t}$$

in (8.42). Using (8.43) and (8.44), we get

$$\begin{aligned} div\,\mathbf{A}_0 + \frac{1}{c^2}\frac{\partial\phi_0}{\partial t} &= div\,(\mathbf{A} - grad\,f) + \frac{1}{c^2}\frac{\partial}{\partial t}\left(\phi + \frac{\partial f}{\partial t}\right) \\ &= div\,\mathbf{A} + \frac{1}{c^2}\frac{\partial\phi}{\partial t} - \left(div\,(grad\,f) - \frac{1}{c^2}\frac{\partial^2 f}{\partial t^2}\right), \end{aligned}$$

or

$$div\,\mathbf{A} + \frac{1}{c^2}\frac{\partial\phi}{\partial t} = div\,\mathbf{A}_0 + \frac{1}{c^2}\frac{\partial\phi_0}{\partial t} + \left(\Delta f - \frac{1}{c^2}\frac{\partial^2 f}{\partial t^2}\right). \tag{8.45}$$

From the theory of partial differential equations (nonhomogeneous wave equations), it follows that (under general conditions, which are fulfilled in our case) there exists a scalar function f such that

$$\Delta f - \frac{1}{c^2}\frac{\partial^2 f}{\partial t^2} = -\left(div\,\mathbf{A}_0 + \frac{1}{c^2}\frac{\partial \phi_0}{\partial t}\right);$$

that is, we can find f such that the right side of (8.45) is zero. Consequently, this choice for f implies, by (8.45),

$$div\,\mathbf{A} + \frac{1}{c^2}\frac{\partial \phi}{\partial t} = 0. \tag{8.46}$$

The scalar potential ϕ and vector potential $\mathbf{A}$ satisfying (8.46) are said to satisfy the *Lorentz gauge*. Substituting the Lorentz gauge condition (8.46) into (8.38) and (8.42) gives

$$\Delta\phi + \frac{\partial}{\partial t}\left(-\frac{1}{c^2}\frac{\partial \phi}{\partial t}\right) = \Delta\phi - \frac{1}{c^2}\frac{\partial^2 \phi}{\partial t^2} = -\frac{\rho}{\epsilon_0} \tag{8.47}$$

and

$$\Delta\mathbf{A} - \frac{1}{c^2}\frac{\partial^2 \mathbf{A}}{\partial t^2} = -\mu_0\mathbf{J}. \tag{8.48}$$

Equations (8.47) and (8.48) are a decoupled [and thus simpler; ϕ and $\mathbf{A}$ are separated, which was not the case in (8.38) and (8.42)] pair of wave equations for ϕ and $\mathbf{A}$.

Electromagnetic Potential as a Differential Form

Recall that the vector potential $\mathbf{A} = (A_x, A_y, A_z)$ is defined by

$$\mathbf{B} = curl\,\mathbf{A}, \tag{8.49}$$

where $\mathbf{B}$ is a time-changing magnetic field. Until the end of this section, we will use subscripts to denote the components of a vector and "∂" notation for partial derivatives.

The scalar potential ϕ is determined from [see (8.37)]

$$grad\,\phi = -\mathbf{E} - \frac{\partial \mathbf{A}}{\partial t}, \tag{8.50}$$

where $\mathbf{E}$ denotes a time-changing electric field.

We have to slightly generalize the definition of a differential form. Recall that forms were built of basic forms corresponding to the coordinate functions x, y, and z. Our generalization consists of including the time t as a coordinate, so that the basic 1-forms are dt, dx, dy, and dz. A (differential) 1-form is an expression

$$\alpha = f(t,x,y,z)dt + g(t,x,y,z)dx + h(t,x,y,z)dy + k(t,x,y,z)dz,$$

where f, g, h, and k are real-valued functions. This time there are more forms: there are six basic 2-forms (*dtdx, dtdy, dtdz, dxdy, dydz,* and *dzdx*), four basic 3-forms (*dtdxdy, dtdydz, dtdzdx,* and *dxdydz*), and (a new one!) a basic 4-form *dtdxdydz* (we decided to put *dt* first in order to obtain correct signs in the formulas that we will derive).

The wedge product and differential are defined as in Section 8.4, keeping in mind that the degrees go up to four and that the differential of a 4-form is zero. For example, the differential of the 1-form α given above is $d\alpha = df \wedge dt + dg \wedge dx + dh \wedge dy + dk \wedge dz$, where

$$df = \frac{\partial f}{\partial t}dt + \frac{\partial f}{\partial x}dx + \frac{\partial f}{\partial y}dy + \frac{\partial f}{\partial z}dz,$$

with similar expressions for dg, dh, and dk.

Define the *electromagnetic potential* to be the 1-form

$$\mathcal{A} = -\phi dt + A_x dx + A_y dy + A_z dz,$$

where A_x, A_y, A_z, and ϕ are defined in (8.49) and (8.50). The differential of $\mathcal{A}$ is

$$\begin{aligned} d\mathcal{A} &= -d\phi \wedge dt + dA_x \wedge dx + dA_y \wedge dy + dA_z \wedge dz \\ &= -\frac{\partial \phi}{\partial x}dxdt - \frac{\partial \phi}{\partial y}dydt - \frac{\partial \phi}{\partial z}dzdt + \frac{\partial A_x}{\partial t}dtdx + \frac{\partial A_x}{\partial y}dydx + \frac{\partial A_x}{\partial z}dzdx \\ &\quad + \frac{\partial A_y}{\partial t}dtdy + \frac{\partial A_y}{\partial x}dxdy + \frac{\partial A_y}{\partial z}dzdy + \frac{\partial A_z}{\partial t}dtdz + \frac{\partial A_z}{\partial x}dxdz + \frac{\partial A_z}{\partial y}dydz \end{aligned}$$

[combine terms together using $dxdt = -dtdx$, $dydx = -dxdy$, etc.]

$$\begin{aligned} &= \left(\frac{\partial A_x}{\partial t} + \frac{\partial \phi}{\partial x}\right)dtdx + \left(\frac{\partial A_y}{\partial t} + \frac{\partial \phi}{\partial y}\right)dtdy + \left(\frac{\partial A_z}{\partial t} + \frac{\partial \phi}{\partial z}\right)dtdz \\ &\quad + \left(\frac{\partial A_y}{\partial x} - \frac{\partial A_x}{\partial y}\right)dxdy + \left(\frac{\partial A_z}{\partial y} - \frac{\partial A_y}{\partial z}\right)dydz + \left(\frac{\partial A_x}{\partial z} - \frac{\partial A_z}{\partial x}\right)dzdx. \end{aligned}$$

Let us identify the expression we have obtained. Rewriting (8.50) in components,

$$\left(\frac{\partial \phi}{\partial x}, \frac{\partial \phi}{\partial y}, \frac{\partial \phi}{\partial z}\right) = -\left(E_x, E_y, E_z\right) - \left(\frac{\partial A_x}{\partial t}, \frac{\partial A_y}{\partial t}, \frac{\partial A_z}{\partial t}\right),$$

we realize that the terms in parentheses appearing in the first three summands are the components of $-\mathbf{E}$. Since

$$\begin{aligned} curl\,\mathbf{A} &= \begin{vmatrix} \mathbf{i} & \mathbf{j} & \mathbf{k} \\ \partial/\partial x & \partial/\partial y & \partial/\partial z \\ A_x & A_y & A_z \end{vmatrix} \\ &= \left(\frac{\partial A_z}{\partial y} - \frac{\partial A_y}{\partial z}, \frac{\partial A_x}{\partial z} - \frac{\partial A_z}{\partial x}, \frac{\partial A_y}{\partial x} - \frac{\partial A_x}{\partial y}\right), \end{aligned} \tag{8.51}$$

the remaining three summands contain components $(curl\,\mathbf{A})_z$, $(curl\,\mathbf{A})_x$, and $(curl\,\mathbf{A})_y$ of $curl\,\mathbf{A}$. But $\mathbf{B} = curl\,\mathbf{A}$ by (8.49), so these components are just $\mathbf{B}_z$, $\mathbf{B}_x$, and $\mathbf{B}_y$. Therefore,

$$d\mathcal{A} = -E_x dtdx - E_y dtdy - E_z dtdz + B_z dxdy + B_x dydz + B_y dzdx. \tag{8.52}$$

The differential form $d\mathcal{A}$ carries information on both the electric and magnetic field, and is called the *electromagnetic tensor.*

Theorem 8.5 states that the differential applied twice to any differential form gives zero. So let us expand $d(d\mathcal{A}) = 0$ to see what will come out of it. Starting from (8.52), we get

$$
\begin{aligned}
0 = d(d\mathcal{A}) &= -dE_x \wedge dtdx - dE_y \wedge dtdy - dE_z \wedge dtdz \\
&\quad + dB_z \wedge dxdy + dB_y \wedge dzdx + dB_x \wedge dydz \\
&= -\frac{\partial E_x}{\partial y} dydtdx - \frac{\partial E_x}{\partial z} dzdtdx - \frac{\partial E_y}{\partial x} dxdtdy - \frac{\partial E_y}{\partial z} dzdtdy \\
&\quad - \frac{\partial E_z}{\partial x} dxdtdz - \frac{\partial E_z}{\partial y} dydtdz + \frac{\partial B_z}{\partial t} dtdxdy + \frac{\partial B_z}{\partial z} dzdxdy \\
&\quad + \frac{\partial B_y}{\partial t} dtdzdx + \frac{\partial B_y}{\partial y} dydzdx + \frac{\partial B_x}{\partial t} dtdydz + \frac{\partial B_x}{\partial x} dxdydz
\end{aligned}
$$

[use anticommutativity $dydxdt = -dydtdx = dtdydx = -dtdxdy$, etc.]

$$
\begin{aligned}
&= \left(\frac{\partial E_y}{\partial x} - \frac{\partial E_x}{\partial y} + \frac{\partial B_z}{\partial t} \right) dtdxdy + \left(\frac{\partial E_z}{\partial y} - \frac{\partial E_y}{\partial z} + \frac{\partial B_x}{\partial t} \right) dtdydz \\
&\quad + \left(\frac{\partial E_x}{\partial z} - \frac{\partial E_z}{\partial x} + \frac{\partial B_y}{\partial t} \right) dtdzdx + \left(\frac{\partial B_x}{\partial x} + \frac{\partial B_y}{\partial y} + \frac{\partial B_z}{\partial z} \right) dxdydz.
\end{aligned}
$$

The 3-form above is zero, and that implies that all four of its components have to be zero. The last one reads $div\,\mathbf{B} = 0$, which is Maxwell's equation (8.26)! Computing $curl\,\mathbf{E}$ exactly as in (8.51), we get that (recall that we use subscripts to denote the components of a vector)

$$
\begin{aligned}
\frac{\partial E_y}{\partial x} - \frac{\partial E_x}{\partial y} + \frac{\partial B_z}{\partial t} = 0 \quad &\text{implies} \quad (curl\,\mathbf{E})_z + \left(\frac{\partial \mathbf{B}}{\partial t} \right)_z = 0, \\
\frac{\partial E_z}{\partial y} - \frac{\partial E_y}{\partial z} + \frac{\partial B_x}{\partial t} = 0 \quad &\text{implies} \quad (curl\,\mathbf{E})_x + \left(\frac{\partial \mathbf{B}}{\partial t} \right)_x = 0, \quad \text{and} \\
\frac{\partial E_x}{\partial z} - \frac{\partial E_z}{\partial x} + \frac{\partial B_y}{\partial t} = 0 \quad &\text{implies} \quad (curl\,\mathbf{E})_y + \left(\frac{\partial \mathbf{B}}{\partial t} \right)_y = 0.
\end{aligned}
$$

In vector notation, $curl\,\mathbf{E} = -\partial\mathbf{B}/\partial t$, which is Maxwell's second equation (8.24).

▶ EXERCISES 8.5

1. Let $\mathbf{A} = 2xz^2\mathbf{i} + xy\mathbf{j} + yz\mathbf{k}$. Find a vector field $\mathbf{A}_0$ defined on $\mathbb{R}^3$ (that differs nontrivially from $\mathbf{A}$; i.e., differs by more than just a constant) such that $curl\,\mathbf{A}_0 = curl\,\mathbf{A}$. Describe all such vector fields.

2. Maxwell's equations for a steady-state charge distribution (i.e., $d\rho/dt = 0$) and for a divergence-free current distribution (i.e., $div\,\mathbf{J} = 0$) are $curl\,\mathbf{E} = \mathbf{0}$, $div\,\mathbf{E} = \rho/\epsilon_0$, $div\,\mathbf{B} = 0$, and $curl\,\mathbf{B} = \mu_0\mathbf{J}$.
(a) Show that $\mathbf{E} = (\rho/\epsilon_0)x\mathbf{i} + z\mathbf{j} + y\mathbf{k}$ and $\mathbf{B} = -xy\mathbf{i} + x\mathbf{j} + yz\mathbf{k}$ are examples of electric and magnetic fields that satisfy Maxwell's equations and find $\mathbf{J}$.
(b) Compute the Poynting vector $\mathbf{P} = {\mu_0}^{-1}\mathbf{E} \times \mathbf{B}$.

3. Verify that Maxwell's equations (8.23)–(8.26) imply the equation of continuity for $\mathbf{J}$ and ρ; that is, show that $div\,\mathbf{J} + \partial\rho/\partial t = 0$.

4. Show that if a vector field $\mathbf{A}$ and scalar function ϕ satisfy $div\,\mathbf{A} + \frac{\partial\phi}{\partial t} = 0$, $\Delta\phi - \frac{\partial^2\phi}{\partial t^2} = -\rho\mu_0$, and $\Delta\mathbf{A} - \frac{\partial^2\mathbf{A}}{\partial t^2} = -\mu_0\mathbf{J}$, then $\mathbf{E} = c^2\left(-grad\,\phi - \frac{\partial\mathbf{A}}{\partial t}\right)$ and $\mathbf{B} = curl\,\mathbf{A}$ satisfy Maxwell's equations (8.23)–(8.26).

5. Show that $\mathbf{E} = e^t\left((x+y)\mathbf{i} + (y+z)\mathbf{j} + z\mathbf{k}\right)$ and $\mathbf{B} = e^t\mathbf{i} + (x^2 - z^2)\mathbf{j} + (e^t + 1)\mathbf{k}$ satisfy Maxwell's equations [see (8.23)–(8.26)] with $\rho = 3\epsilon_0 e^t$ and $\mathbf{J} = \left(\frac{2z}{\mu_0} - \epsilon_0 e^t(x+y)\right)\mathbf{i} - \epsilon_0 e^t(y+z)\mathbf{j} + \left(\frac{2x}{\mu_0} - \epsilon_0 e^t z\right)\mathbf{k}$.

6. Let $\mathbf{E}(x, y, z, t) = e^t\left(x^2\mathbf{i} + y^2\mathbf{j} + z^2\mathbf{k}\right)$ be a time-dependent electric field, $t \geq 0$.

(a) Find the total charge density $\rho(x, y, z, t)$ from the first Maxwell's equation $div\,\mathbf{E} = \rho/\epsilon_0$.

(b) Find the scalar potential ϕ (recall that $\mathbf{E} = -grad\,\phi$).

(c) Check that the vector field $\mathbf{B} = y\mathbf{i} + xz\mathbf{j} + xy\mathbf{k}$ satisfies Maxwell's equations $curl\,\mathbf{E} = -\partial\mathbf{B}/\partial t$ and $div\,\mathbf{B} = 0$.

(d) Find the vector potential $\mathbf{A}$ (recall that $\mathbf{A}$ is determined from $\mathbf{B} = -curl\,\mathbf{A}$).

(e) Check that the scalar and vector potentials satisfy $\Delta\phi + (\partial/\partial t)div\,\mathbf{A} = -\rho/\epsilon_0$.

7. Express Maxwell's equation $curl\,\mathbf{E} = -\partial\mathbf{B}/\partial t$ in cylindrical coordinates.

8. Express Maxwell's equation $curl\,\mathbf{B} = \mu_0\left(\mathbf{J} + \epsilon_0\partial\mathbf{E}/\partial t\right)$ in cylindrical coordinates.

9. Show that $\mathbf{E} = \sin x \sin t\mathbf{j}$ and $\mathbf{B} = \sin x \cos t\mathbf{k}$ satisfy some (but not all) of Maxwell's equations (8.27)–(8.30) if we take $c = 1$. (This means that $\mathbf{E}$ and $\mathbf{B}$ are not realistic electric and magnetic fields.)

▶ 8.6 VECTOR CALCULUS IN FLUID FLOW

In this section, we examine several applications of vector calculus concepts in fluid flow. Although we do attempt to provide physical context, our primary focus is on identifying mathematical objects involved and understanding how to work with them.

Physical Interpretation of Divergence

Consider the motion of a compressible fluid (such as a vapor or a gas) in $\mathbb{R}^2$, with no sources or sinks (i.e., there are no points where extra amounts of fluid are produced or where fluid disappears). We choose $\mathbb{R}^2$ rather than $\mathbb{R}^3$ to simplify calculations.

Let $\mathbf{v}(x, y, t)$ be a vector field describing the flow of the fluid and let $\rho(x, y, t)$ be its density. Assume that both $\mathbf{v}$ and ρ are differentiable functions of their variables [the variable t denotes time and the variables (x, y) give the location of a point in the flow]. Consider a small rectangle R with sides Δx and Δy; see Figure 8.30. We are going to approximate the flux across the boundary of R; that is, the change of mass of fluid in R due to the flow.

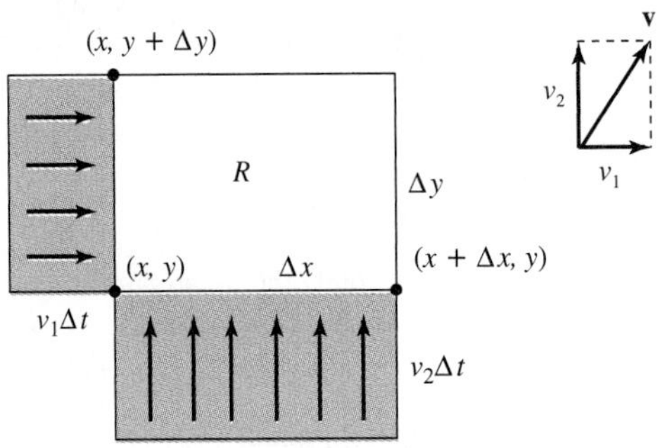

Figure 8.30 Mass inflow into R.

By the conservation of mass principle,

$$\text{loss of mass in } R = \text{outflowing mass} - \text{inflowing mass}. \tag{8.53}$$

Furthermore,

$$\text{loss of mass in } R = \text{time rate change of the mass} \cdot \text{time}. \tag{8.54}$$

Since mass = density · area, and the area of R is fixed,

$$\text{loss of mass in } R = \text{time rate change of the density} \cdot \text{area} \cdot \text{time} = -\frac{\partial \rho}{\partial t}\Delta x \Delta y \Delta t.$$

Let us clarify the appearance of the minus sign in the formula. If the derivative of the density $\partial\rho/\partial t$ is positive at some point in R, then the density increases at that point; that is, more mass is coming in than is flowing out. Hence, the *gain* in mass is $(\partial\rho/\partial t)\Delta x \Delta y \Delta t$ (which is positive), and the *loss* is $-(\partial\rho/\partial t)\Delta x \Delta y \Delta t$. If $\partial\rho/\partial t < 0$, then the density is decreasing, and more mass is flowing out than is coming in. Hence, the *gain* in the mass is $(\partial\rho/\partial t)\Delta x \Delta y \Delta t$ (which is now negative); that is, the *loss* is $-(\partial\rho/\partial t)\Delta x \Delta y \Delta t$. (A little dictionary for translating into everyday language: positive gain means gain, negative gain means loss, positive loss means loss, and negative loss means gain.)

Now let us compute the inflowing mass. Only the x-component v_1 of $\mathbf{v}$ contributes to the flow through the left side of R, and hence (approximately),

$$\text{mass inflow (left side)} = \text{density} \cdot \text{"area"}$$

(where "area" = area occupied by particles that flow into R in time Δt)

$$= \text{density} \cdot \text{base} \cdot \text{height} = \rho(x, y, t)v_1(x, y, t)\Delta t \Delta y$$

(since base = velocity · time). We assumed that the left side (of length Δy) is so small that ρ and v_1 have approximately the same values [equal to $\rho(x, y, t)$ and $v_1(x, y, t)$, respectively] at points near that side. With a similar assumption for the lower side

$$\text{mass inflow (lower side)} = \rho(x, y, t)v_2(x, y, t)\Delta t \Delta x,$$

since only the y-component v_2 of $\mathbf{v}$ contributes to the inflow. The total inflow into R is

$$\begin{aligned}
&\rho(x, y, t)v_1(x, y, t)\Delta t \Delta y + \rho(x, y, t)v_2(x, y, t)\Delta t \Delta x \\
&\qquad = \Delta t\big(\rho(x, y, t)v_1(x, y, t)\Delta y + \rho(x, y, t)v_2(x, y, t)\Delta x\big) \\
&\qquad = \Delta t \Delta x \Delta y \left(\rho(x, y, t)\frac{v_1(x, y, t)}{\Delta x} + \rho(x, y, t)\frac{v_2(x, y, t)}{\Delta y}\right).
\end{aligned}$$

Analogous computations give (see Figure 8.31)

$$\text{mass outflow (right side)} = \rho(x + \Delta x, y, t)v_1(x + \Delta x, y, t)\Delta t \Delta y$$

[we assume that Δy is so small that the density ρ and component v_1 have approximately the same values (equal to $\rho(x + \Delta x, y, t)$ and $v_1(x + \Delta x, y, t)$, respectively) at points near the right edge of R]. With a similar assumption for the upper side,

$$\text{mass outflow (upper side)} = \rho(x, y + \Delta y, t)v_2(x, y + \Delta y, t)\Delta t \Delta x.$$

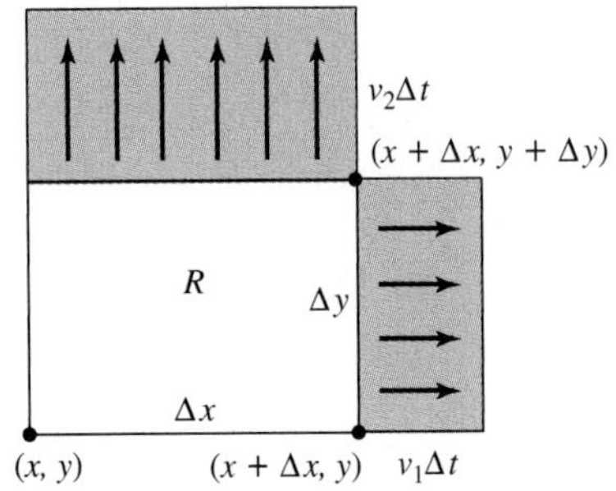

Figure 8.31 Mass outflow from R.

The total outflow from R equals

$$\begin{aligned}
&\rho(x+\Delta x, y, t)v_1(x+\Delta x, y, t)\Delta t\Delta y + \rho(x, y+\Delta y, t)v_2(x, y+\Delta y, t)\Delta t\Delta x \\
&\quad = \Delta t\big(\rho(x+\Delta x, y, t)v_1(x+\Delta x, y, t)\Delta y + \rho(x, y+\Delta y, t)v_2(x, y+\Delta y, t)\Delta x\big) \\
&\quad = \Delta t\Delta x\Delta y\left(\rho(x+\Delta x, y, t)\frac{v_1(x+\Delta x, y, t)}{\Delta x} + \rho(x, y+\Delta y, t)\frac{v_2(x, y+\Delta y, t)}{\Delta y}\right).
\end{aligned}$$

Now substitute the expressions for the loss of mass, total inflow, and total outflow into the conservation of mass formula (8.53):

$$\begin{aligned}
-\frac{\partial\rho}{\partial t}\Delta x\Delta y\Delta t = \Delta x\Delta y\Delta t&\left(\frac{\rho(x+\Delta x, y, t)v_1(x+\Delta x, y, t) - \rho(x, y, t)v_1(x, y, t)}{\Delta x}\right. \\
&\left.+\frac{\rho(x, y+\Delta y, t)v_2(x, y+\Delta y, t) - \rho(x, y, t)v_2(x, y, t)}{\Delta y}\right),
\end{aligned}$$

divide by $\Delta x\Delta y\Delta t$, and let $\Delta x \to 0$ and $\Delta y \to 0$, thus getting

$$-\frac{\partial\rho}{\partial t} = \frac{\partial}{\partial x}\big(\rho(x, y, t)v_1(x, y, t)\big) + \frac{\partial}{\partial y}\big(\rho(x, y, t)v_2(x, y, t)\big)$$

(by using the definition of the partial derivatives of ρv_1 with respect to x and ρv_2 with respect to y), that is,

$$\frac{\partial\rho}{\partial t} + div\,(\rho\mathbf{v}) = 0. \tag{8.55}$$

This equation is called the *continuity equation* for compressible fluid flow.

In light of our discussion, we interpret intuitively the divergence of a vector field as a total outflow (or total outflux) per unit area (recall the interpretation of the divergence given in Section 8.2.) Recall that we also discussed the continuity equation in the context of electromagnetism [see (8.22) in Section 8.5].

Fluid Dynamics

Let $\mathbf{v}(\mathbf{x}, t) = \mathbf{v}(x, y, z, t)$ be the velocity of a fluid (such as air or water) at a location (x, y, z) in $\mathbb{R}^3$ at time t. By $p(\mathbf{x}, t)$ we denote the pressure, and by $\rho(\mathbf{x}, t)$ the density of the fluid [both $p(\mathbf{x}, t)$ and $\rho(\mathbf{x}, t)$ are scalar quantities]. Repeating the derivation of (8.55), we arrive at the continuity equation

$$\frac{\partial\rho(x,y,z,t)}{\partial t} + div\,(\rho(x,y,z,t)\mathbf{v}(x,y,z,t)) = 0 \tag{8.56}$$

for a *three-dimensional* fluid flow. An alternative version of the continuity equation (drop variables from the notation)

$$\frac{\partial\rho}{\partial t} + \nabla\rho\cdot\mathbf{v} + \rho\,div\,\mathbf{v} = 0 \tag{8.57}$$

is deduced in Exercise 1.

In the special case where the density is constant in time (i.e., when $\partial\rho/\partial t = 0$), the continuity equation implies that $div\,(\rho\mathbf{v}) = 0$. Moreover, if the density is constant in space as well (i.e., $\partial\rho/\partial x = \partial\rho/\partial y = \partial\rho/\partial z = 0$), then $div\,(\rho\mathbf{v}) = \rho\ div\,\mathbf{v} = 0$ implies that $div\,\mathbf{v} = 0$. In this case, the total outflow from a region equals the total inflow into it. Whenever $div\,\mathbf{v} = 0$, we say that the fluid, or the vector field $\mathbf{v}$, are *incompressible*.

Let W be a solid three-dimensional region to which the Divergence Theorem applies, and denote by $S = \partial W$ its outward-oriented boundary surface. The surface integral

$$\iint_S \rho\mathbf{v}\cdot d\mathbf{S}$$

gives the total mass of fluid that flows out across the surface S per unit time (S is oriented by its outward normal). By the Divergence Theorem,

$$\iint_S \rho\mathbf{v}\cdot d\mathbf{S} = \iiint_W div\,(\rho\mathbf{v})\,dV. \tag{8.58}$$

▶ EXAMPLE 8.30 Divergence Theorem Implies Continuity Equation

The mass contained inside a closed surface S (which is the boundary of a solid region W in space) is given by $m(t) = \iiint_W \rho\,dV$ (see Section 7.5; in general, ρ is a function of time, and thus mass depends on t). The quantity

$$\frac{dm(t)}{dt} = \frac{d}{dt}\left(\iiint_W \rho\,dV\right) = \iiint_W \frac{\partial\rho}{\partial t}\,dV$$

represents the rate of change of mass inside S per unit time. Combining the above, we conclude that

$$\iint_S \rho\mathbf{v}\cdot d\mathbf{S} = -\iiint_W \frac{\partial\rho}{\partial t}\,dV \tag{8.59}$$

(see Exercise 2). Using the Divergence Theorem [see formula (8.58)],

$$\iiint_W div\,(\rho\mathbf{v})\,dV = \iint_S \rho\mathbf{v}\cdot d\mathbf{S} = -\iiint_W \frac{\partial\rho}{\partial t}\,dV.$$

Since this equality holds true for any region W, we conclude that $div\,(\rho\mathbf{v}) = -\partial\rho/\partial t$ [see (f) in the subsection entitled "Properties of Integrals" in Section 7.5]. ◀

Euler's Equation

The continuity equation is not the only equation that describes the motion of a fluid. We now derive another important equation, in the special case of a perfect fluid.

A fluid is called *perfect* if for any solid three-dimensional region W contained in it, the forces of pressure act in the direction of the normal to the boundary of W. We assume that the region W satisfies the assumptions of the Divergence Theorem, and denote by $S = \partial W$ its boundary, oriented by the outward-pointing unit normal $\mathbf{n}$. The fact that the fluid is perfect can be expressed by saying that the total force due to the pressure p is given by

$$\mathbf{F}_S = -\iint_S p\,\mathbf{n}\,dS \tag{8.60}$$

(since $\mathbf{F}_S$ acts in the direction of $-\mathbf{n}$). The quantities on both sides of (8.60) are vectors. If $\mathbf{n} = n_1\mathbf{i} + n_2\mathbf{j} + n_3\mathbf{k}$, then

$$\iint_S p\,\mathbf{n}\,dS = \mathbf{i}\iint_S pn_1\,dS + \mathbf{j}\iint_S pn_2\,dS + \mathbf{k}\iint_S pn_3\,dS,$$

where the integrals on the right side are surface integrals of scalar functions.

For an arbitrary, fixed vector $\mathbf{a}$ in $\mathbb{R}^3$, we get

$$\mathbf{F}_S \cdot \mathbf{a} = -\iint_S p\mathbf{n}\cdot\mathbf{a}\,dS = -\iint_S p\mathbf{a}\cdot\mathbf{n}\,dS = -\iiint_W div\,(p\mathbf{a})\,dV$$

by the Divergence Theorem. Proceed by using the product rule (see Section 4.8)

$$div\,(p\mathbf{a}) = \nabla p\cdot\mathbf{a} + p\,(div\,\mathbf{a}) = \nabla p\cdot\mathbf{a}$$

($\mathbf{a}$ is constant and thus $div\,\mathbf{a} = 0$) to get

$$\mathbf{F}_S\cdot\mathbf{a} = -\iiint_W \nabla p\cdot\mathbf{a}\,dV.$$

Thus (see Exercise 3),

$$\mathbf{F}_S = -\iiint_W \nabla p\,dV. \tag{8.61}$$

Assume that $\mathbf{F}_e$ is an external force per unit mass, acting on the fluid. The total resultant external force acting on the region W is given by

$$\mathbf{F}_E = \iiint_W \rho\mathbf{F}_e\,dV. \tag{8.62}$$

The integral on the right side of (8.62) is a vector, whose **i**th (**j**th, **k**th) component is the triple integral over W of the scalar function ρ multiplied by the **i**th (**j**th, **k**th) component of $\mathbf{F}_e$). Using Newton's Second Law, we obtain

$$\frac{d}{dt}\iiint_W \rho\mathbf{v}\,dV = \mathbf{F}_S + \mathbf{F}_E = -\iiint_W \nabla p\,dV + \iiint_W \rho\mathbf{F}_e\,dV. \tag{8.63}$$

The left side in (8.63) is the time rate change of the total momentum of the fluid contained in W. It can be shown that, when $div\,\mathbf{v} = 0$,

$$\frac{d}{dt}\iiint_W \rho\mathbf{v}\,dV = \iiint_W \rho\frac{d\mathbf{v}}{dt}\,dV \tag{8.64}$$

(this is a consequence of the so-called transport theorem in fluid dynamics, see Example 8.33 at the end of this section). Thus,

$$\iiint_W \rho \frac{d\mathbf{v}}{dt}\, dV = -\iiint_W \nabla p\, dV + \iiint_W \rho \mathbf{F}_e\, dV,$$

and since W is an arbitrary region, we conclude that $\rho(d\mathbf{v}/dt) = -\nabla p + \rho F_e$, or

$$\rho \left(\frac{d\mathbf{v}}{dt} - \mathbf{F}_e \right) = -\nabla p. \tag{8.65}$$

Equation (8.65) is called *Euler's equation for a perfect fluid.* In Exercise 4, we show that $d\mathbf{v}/dt = \mathbf{v} \cdot \nabla \mathbf{v} + \partial \mathbf{v}/\partial t$, where $\mathbf{v} \cdot \nabla \mathbf{v}$ [sometimes denoted by $(\mathbf{v} \cdot \nabla)\mathbf{v}$] is the vector field whose **i**th (**j**th, **k**th) component is the dot product of **v** with the gradient of the **i**th (**j**th, **k**th) component of **v**. Thus, we can write (8.65) in the form

$$\rho \left(\frac{\partial \mathbf{v}}{\partial t} + \mathbf{v} \cdot \nabla \mathbf{v} - \mathbf{F}_e \right) = -\nabla p. \tag{8.66}$$

In Exercise 7, we derive an expression for $\mathbf{v} \cdot \nabla \mathbf{v}$ that involves the cross product. Using it, we write (8.66) as

$$\rho \left(\frac{\partial \mathbf{v}}{\partial t} + \frac{1}{2} \nabla \left(\|\mathbf{v}\|^2 \right) - \mathbf{v} \times curl\, \mathbf{v} - \mathbf{F}_e \right) = -\nabla p.$$

▶ EXAMPLE 8.31 Fluid at Rest

We consider the case where fluid is at rest, that is, when $\mathbf{v} = \mathbf{0}$. The equation (8.66) implies that $\rho(-\mathbf{F}_e) = -\nabla p$, and thus,

$$\frac{1}{\rho} \nabla p = \mathbf{F}_e. \tag{8.67}$$

Since ∇p is perpendicular to the level surfaces of p, equation (8.67) implies that level surfaces of p (i.e., surfaces of constant pressure) are always perpendicular to the external force.

Consider a mass of air at rest, subject to the gravitational force of the earth. In that case, $\mathbf{F}_e = -g\mathbf{k}$, and since (8.67) reads

$$\frac{1}{\rho} \left(\frac{\partial p}{\partial x}\mathbf{i} + \frac{\partial p}{\partial y}\mathbf{j} + \frac{\partial p}{\partial z}\mathbf{k} \right) = -g\mathbf{k},$$

we get that $\partial p/\partial x = \partial p/\partial y = 0$ and $\partial p/\partial z = -\rho g$. After integration, $p(x, y, z) = -\rho g z + c$, where c is a constant. Thus, the pressure decreases linearly with height. ◂

Transport Theorems

Sometimes, to study the properties of a fluid flow, we focus on a specific portion of the fluid and study how it changes as it is moved by the flow. For instance, we might want to compute the time derivative of the volume integral of a density function. In general, we are interested in computing quantities related to a surface or solid in motion (such as the time rate of change of a surface or volume integral). Here we explain how to approach these types of calculations.

Consider a vector field **F** that changes with time, that is, $\mathbf{F} = \mathbf{F}(\mathbf{x}, t)$, where $\mathbf{x} = (x, y, z)$ in $\mathbb{R}^3$. Written in components, $\mathbf{F}(\mathbf{x}, t) = (F_1(\mathbf{x}, t), F_2(\mathbf{x}, t), F_3(\mathbf{x}, t))$, where F_1,

F_2, and F_3 are real-valued functions. Differential operators are computed with respect to space variables only; for instance, $div\,\mathbf{F} = \partial F_1/\partial x + \partial F_2/\partial y + \partial F_3/\partial z$, or $grad\,F_1 = (\partial F_1/\partial x)\mathbf{i} + (\partial F_1/\partial y)\mathbf{j} + (\partial F_1/\partial z)\mathbf{k}$, etc.

Imagine a surface moving through $\mathbb{R}^3$. We use S_t to describe what the surface looks like at time $t \geq 0$ (so S_0 represents the "initial" or "starting" surface). Let ∂S_t denote the positively oriented boundary of S_t. The flux of $\mathbf{F}$ through S_t, at time t, is given by the surface integral

$$\Phi(t) = \iint_{S_t} \mathbf{F}(\mathbf{x}, t) \cdot d\mathbf{S}. \tag{8.68}$$

The function $\Phi(t)$ depends on t because the vector field $\mathbf{F}$ is changing with time and also because the surface S_t moves thorugh space. We would like to compute $d\Phi/dt$.

Parametrize the surface S_t by

$$\mathbf{r}(u, v, t) = (x(u, v, t), y(u, v, t), z(u, v, t)),$$

where $(u, v) \in D$, and D is an elementary region in $\mathbb{R}^2$. By ∂D we denote the positively oriented boundary (curve) of D. Note that $\mathbf{r}(u, v, 0)$ is the parametrization of the "initial" surface S_0. For a fixed value of t, the function $\mathbf{r}(u, v, t)$ parametrizes the surface S_t. We assume that all parametrizations $\mathbf{r}(u, v, t)$ are defined over the same region D.

If we fix (u, v), we obtain the function $\mathbf{r}(u, v, t)$ of one variable (so its image is a curve), that describes how the point $\mathbf{r}(u, v, t = 0)$, initially on S_0, travels, carried by the flow. Thus, the velocity of a point $\mathbf{r}(u, v, t)$ on S_t, at time t, is given by $\mathbf{v}(u, v, t) = \partial\mathbf{r}(u, v, t)/\partial t$. Using the parametrization $\mathbf{r}(u, v, t)$, we write the flow $\Phi(t)$ as

$$\Phi(t) = \iint_{S_t} \mathbf{F} \cdot d\mathbf{S} = \iint_D \mathbf{F}(\mathbf{r}(u, v, t)) \cdot \left(\frac{\partial \mathbf{r}}{\partial u} \times \frac{\partial \mathbf{r}}{\partial v}\right) du dv$$

[recall that the normal to S_t is given by $(\partial\mathbf{r}/\partial u) \times (\partial\mathbf{r}/\partial v)$]. Thus,

$$\begin{aligned} \frac{d\Phi(t)}{dt} &= \iint_D \left(\frac{d}{dt}\mathbf{F}(\mathbf{r}(u, v, t))\right) \cdot \left(\frac{\partial \mathbf{r}}{\partial u} \times \frac{\partial \mathbf{r}}{\partial v}\right) du dv \\ &\quad + \iint_D \mathbf{F}(\mathbf{r}(u, v, t)) \cdot \frac{\partial}{\partial t}\left(\frac{\partial \mathbf{r}}{\partial u} \times \frac{\partial \mathbf{r}}{\partial v}\right) du dv. \end{aligned} \tag{8.69}$$

In Exercise 4, we show that $d\mathbf{F}/dt = \mathbf{v} \cdot \nabla\mathbf{F} + \partial\mathbf{F}/\partial t$, where $\mathbf{v} \cdot \nabla\mathbf{F}$ is a vector function, defined by $\mathbf{v} \cdot \nabla\mathbf{F} = (\mathbf{v} \cdot \nabla F_1, \mathbf{v} \cdot \nabla F_2, \mathbf{v} \cdot \nabla F_3)$, where ∇F_i, $i = 1, 2, 3$, is the gradient of the component F_i of $\mathbf{F}$. Thus, we write the first integral in (8.69) as

$$\iint_D \frac{d\mathbf{F}}{dt} \cdot \left(\frac{\partial \mathbf{r}}{\partial u} \times \frac{\partial \mathbf{r}}{\partial v}\right) du dv = \iint_D \left(\mathbf{v} \cdot \nabla\mathbf{F} + \frac{\partial \mathbf{F}}{\partial t}\right) \cdot \left(\frac{\partial \mathbf{r}}{\partial u} \times \frac{\partial \mathbf{r}}{\partial v}\right) du dv. \tag{8.70}$$

Now we work on the second integral in (8.69). By the product rule,

$$\mathbf{F} \cdot \frac{\partial}{\partial t}\left(\frac{\partial \mathbf{r}}{\partial u} \times \frac{\partial \mathbf{r}}{\partial v}\right) = \mathbf{F} \cdot \left(\frac{\partial^2 \mathbf{r}}{\partial t\, \partial u} \times \frac{\partial \mathbf{r}}{\partial v} + \frac{\partial \mathbf{r}}{\partial u} \times \frac{\partial^2 \mathbf{r}}{\partial t\, \partial v}\right)$$

[assuming that **r** is C^2, we are allowed to interchange second partial derivatives]

$$= \mathbf{F}\cdot\left(\frac{\partial}{\partial u}\left(\frac{\partial \mathbf{r}}{\partial t}\times\frac{\partial \mathbf{r}}{\partial v}\right)-\frac{\partial \mathbf{r}}{\partial t}\times\frac{\partial^2 \mathbf{r}}{\partial u\,\partial v}+\frac{\partial}{\partial v}\left(\frac{\partial \mathbf{r}}{\partial u}\times\frac{\partial \mathbf{r}}{\partial t}\right)-\frac{\partial^2 \mathbf{r}}{\partial v\,\partial u}\times\frac{\partial \mathbf{r}}{\partial t}\right)$$
$$= \mathbf{F}\cdot\left(\frac{\partial}{\partial u}\left(\mathbf{v}\times\frac{\partial \mathbf{r}}{\partial v}\right)-\frac{\partial}{\partial v}\left(\mathbf{v}\times\frac{\partial \mathbf{r}}{\partial u}\right)\right),$$

since $\mathbf{v} = \partial\mathbf{r}/\partial t$. Using the product rule again, we get

$$\mathbf{F}\cdot\frac{\partial}{\partial t}\left(\frac{\partial \mathbf{r}}{\partial u}\times\frac{\partial \mathbf{r}}{\partial v}\right)=\frac{\partial}{\partial u}\left(\mathbf{F}\cdot\left(\mathbf{v}\times\frac{\partial \mathbf{r}}{\partial v}\right)\right)-\frac{\partial \mathbf{F}}{\partial u}\cdot\left(\mathbf{v}\times\frac{\partial \mathbf{r}}{\partial v}\right)$$
$$-\frac{\partial}{\partial v}\left(\mathbf{F}\cdot\left(\mathbf{v}\times\frac{\partial \mathbf{r}}{\partial u}\right)\right)+\frac{\partial \mathbf{F}}{\partial v}\cdot\left(\mathbf{v}\times\frac{\partial \mathbf{r}}{\partial u}\right).$$

In Exercise 8, we show that

$$\frac{\partial \mathbf{F}}{\partial v}\cdot\left(\mathbf{v}\times\frac{\partial \mathbf{r}}{\partial u}\right)-\frac{\partial \mathbf{F}}{\partial u}\cdot\left(\mathbf{v}\times\frac{\partial \mathbf{r}}{\partial v}\right)=\left((\mathit{div}\,\mathbf{F})\cdot\mathbf{v}-\mathbf{v}\cdot\nabla\mathbf{F}\right)\cdot\left(\frac{\partial \mathbf{r}}{\partial u}\times\frac{\partial \mathbf{r}}{\partial v}\right). \tag{8.71}$$

Combining (8.70) and (8.71), we get

$$\frac{d\Phi(t)}{dt}=\iint_D\left(\mathbf{v}\cdot\nabla\mathbf{F}+\frac{\partial \mathbf{F}}{\partial t}+(\mathit{div}\,\mathbf{F})\cdot\mathbf{v}-\mathbf{v}\cdot\nabla\mathbf{F}\right)\cdot\left(\frac{\partial \mathbf{r}}{\partial u}\times\frac{\partial \mathbf{r}}{\partial v}\right)dudv$$
$$+\iint_D\left[\frac{\partial}{\partial u}\left(\mathbf{F}\cdot\left(\mathbf{v}\times\frac{\partial \mathbf{r}}{\partial v}\right)\right)-\frac{\partial}{\partial v}\left(\mathbf{F}\cdot\left(\mathbf{v}\times\frac{\partial \mathbf{r}}{\partial u}\right)\right)\right]dudv$$

[now apply Green's Theorem (see (8.4)) to the second integral]

$$=\iint_D\left(\frac{\partial \mathbf{F}}{\partial t}+(\mathit{div}\,\mathbf{F})\cdot\mathbf{v}\right)\cdot\left(\frac{\partial \mathbf{r}}{\partial u}\times\frac{\partial \mathbf{r}}{\partial v}\right)dudv$$
$$+\int_{\partial D}\mathbf{F}\cdot\left(\mathbf{v}\times\frac{\partial \mathbf{r}}{\partial u}\right)du+\mathbf{F}\cdot\left(\mathbf{v}\times\frac{\partial \mathbf{r}}{\partial v}\right)dv. \tag{8.72}$$

Using the properties of the scalar triple product (see Exercise 10 in Section 1.5), we get

$$\mathbf{F}\cdot\left(\mathbf{v}\times\frac{\partial \mathbf{r}}{\partial u}\right)=\frac{\partial \mathbf{r}}{\partial u}\cdot(\mathbf{F}\times\mathbf{v})=(\mathbf{F}\times\mathbf{v})\cdot\frac{\partial \mathbf{r}}{\partial u}.$$

Therefore, the path integral in (8.72) is equal to

$$\int_{\partial D}(\mathbf{F}\times\mathbf{v})\cdot\frac{\partial \mathbf{r}}{\partial u}du+(\mathbf{F}\times\mathbf{v})\cdot\frac{\partial \mathbf{r}}{\partial v}dv=\int_{\partial D}(\mathbf{F}\times\mathbf{v})\cdot d\mathbf{s},$$

and, finally from (8.72),

$$\frac{d\Phi(t)}{dt}=\iint_{S_t}\left(\frac{\partial \mathbf{F}}{\partial t}+(\mathit{div}\,\mathbf{F})\cdot\mathbf{v}\right)\cdot d\mathbf{S}+\int_{\partial D}(\mathbf{F}\times\mathbf{v})\cdot d\mathbf{s}. \tag{8.73}$$

The formula (8.73) is called the *flux transport theorem.*

Note that, in the case where we can extend **v** to a C^1 vector field in a region in $\mathbb{R}^3$ that contains all surfaces S_t, the path integral in (8.73) may be transformed into a surface integral using Stokes' Theorem:

$$\int_{\partial D} (\mathbf{F} \times \mathbf{v}) \cdot d\mathbf{s} = \iint_{S_t} curl\,(\mathbf{F} \times \mathbf{v}) \cdot d\mathbf{S}.$$

Thus, in some cases (for instance, if a surface is being transported inside a fluid), we get

$$\frac{d\Phi(t)}{dt} = \iint_{S_t} \left(\frac{\partial \mathbf{F}}{\partial t} + (div\,\mathbf{F}) \cdot \mathbf{v} + curl\,(\mathbf{F} \times \mathbf{v}) \right) \cdot d\mathbf{S}. \tag{8.74}$$

Next, we derive a version of the transport theorem for volumes. Consider a function $\rho = \rho(x, y, z, t)$, where $(x, y, z) \in \mathbb{R}^3$ and t denotes time (for instance, ρ could be the density of a fluid). Define

$$\Psi(t) = \iiint_{W_t} \rho(x, y, z, t)\, dV,$$

where W_t represents a region in $\mathbb{R}^3$ at time t (imagine a portion of a fluid transported by the flow). We would like to compute $d\Psi(t)/dt$.

To start, we use the fact (we will not prove it here) that a differentiable, scalar function ρ (defined on a bounded, connected subset of $\mathbb{R}^3$) can be written as a divergence of some C^2 vector field $\mathbf{F}$; that is, $\rho = div\,\mathbf{F}$. Thus, using the Divergence Theorem, we get

$$\iiint_{W_t} \rho\, dV = \iiint_{W_t} div\,\mathbf{F}\, dV = \iint_{S_t} \mathbf{F} \cdot d\mathbf{S},$$

where S_t is the positively oriented boundary of W_t. Using the transport theorem (8.74),

$$\frac{d\Psi(t)}{dt} = \frac{d}{dt}\left(\iint_{S_t} \mathbf{F} \cdot d\mathbf{S} \right) = \iint_{S_t} \left(\frac{\partial \mathbf{F}}{\partial t} + (div\,\mathbf{F}) \cdot \mathbf{v} + curl\,(\mathbf{F} \times \mathbf{v}) \right) \cdot d\mathbf{S}. \tag{8.75}$$

Since $div\,(curl\,(\mathbf{F} \times \mathbf{v})) = 0$, the Divergence Theorem implies that

$$\iint_{S_t} curl\,(\mathbf{F} \times \mathbf{v}) \cdot d\mathbf{S} = 0.$$

Again, by the Divergence Theorem,

$$\iint_{S_t} \frac{\partial \mathbf{F}}{\partial t} \cdot d\mathbf{S} = \iiint_{W_t} div \left(\frac{\partial \mathbf{F}}{\partial t} \right) dV = \iiint_{W_t} \frac{\partial}{\partial t}(div\,\mathbf{F})\, dV$$

[we were allowed to interchange second partial derivatives since $\mathbf{F}$ is C^2]

$$= \iiint_{W_t} \frac{\partial \rho}{\partial t}\, dV,$$

because $\rho = div\,\mathbf{F}$. Thus, (8.75) implies that

$$\frac{d\Psi(t)}{dt} = \iiint_{W_t} \frac{\partial \rho}{\partial t}\, dV + \iint_{S_t} \rho \mathbf{v} \cdot d\mathbf{S}. \tag{8.76}$$

The formula (8.76) is known as *Reynold's transport theorem.*

EXAMPLE 8.32 Vector Form of Transport Theorem.

Show that

$$\frac{d}{dt}\iiint_{W_t} \rho\,\mathbf{v}\,dV = \iiint_{W_t}\left(\frac{\partial}{\partial t}(\rho\mathbf{v}) + \mathbf{v}\cdot\nabla(\rho\mathbf{v}) + (\rho\mathbf{v})\,div\,\mathbf{v}\right)dV. \tag{8.77}$$

SOLUTION By definition,

$$\iiint_{W_t} \rho\,\mathbf{v}\,dV = \mathbf{i}\iiint_{W_t} \rho v_1\,dV + \mathbf{j}\iiint_{W_t} \rho v_2\,dV + \mathbf{k}\iiint_{W_t} \rho v_3\,dV,$$

where $\mathbf{v} = (v_1, v_2, v_3)$. The formula (8.76) and the Divergence Theorem imply that

$$\frac{d}{dt}\iiint_{W_t} \rho v_1\,dV = \iiint_{W_t} \frac{\partial(\rho v_1)}{\partial t}\,dV + \iint_{S_t} \rho v_1 \mathbf{v}\cdot d\mathbf{S}$$
$$= \iiint_{W_t}\left(\frac{\partial(\rho v_1)}{\partial t} + div\,(\rho v_1\mathbf{v})\right)dV.$$

By the product rule, $div\,(\rho v_1\mathbf{v}) = \nabla(\rho v_1)\cdot\mathbf{v} + \rho v_1 div\,\mathbf{v}$, and consequently,

$$\frac{d}{dt}\iiint_{W_t} \rho v_1\,dV = \iiint_{W_t}\left(\frac{\partial(\rho v_1)}{\partial t} + \mathbf{v}\cdot\nabla(\rho v_1) + (\rho v_1)div\,\mathbf{v}\right)dV.$$

Thus, we have established the equality of **i** components of (8.77). In the same way, we prove the equality of **j** and **k** components.

EXAMPLE 8.33

Let us establish formula (8.64) that we used earlier, in the derivation of Euler's equation.

Assuming $div\,\mathbf{v} = 0$, we obtain, using (8.77),

$$\frac{d}{dt}\iiint_{W_t} \rho\,\mathbf{v}\,dV = \iiint_{W_t}\left(\rho\frac{\partial\mathbf{v}}{\partial t} + \frac{\partial\rho}{\partial t}\mathbf{v} + \mathbf{v}\cdot\nabla(\rho\mathbf{v})\right)dV$$

[now use the fact that $d\mathbf{v}/dt = \mathbf{v}\cdot\nabla\mathbf{v} + \partial\mathbf{v}/\partial t$; see Exercise 4]

$$= \iiint_{W_t}\left(\rho\frac{d\mathbf{v}}{dt} - \rho\mathbf{v}\cdot\nabla\mathbf{v} + \frac{\partial\rho}{\partial t}\mathbf{v} + \mathbf{v}\cdot\nabla(\rho\mathbf{v})\right)dV$$
$$= \iiint_{W_t}\left(\rho\frac{d\mathbf{v}}{dt} - \rho\mathbf{v}\cdot\nabla\mathbf{v} - div\,(\rho\mathbf{v})\mathbf{v} + \mathbf{v}\cdot\nabla(\rho\mathbf{v})\right)dV,$$

by the continuity equation (8.55). Since $\mathbf{v}\cdot\nabla(\rho\mathbf{v}) = \rho\mathbf{v}\cdot\nabla\mathbf{v} + div\,(\rho\mathbf{v})\mathbf{v}$ (see Exercise 9), it follows that

$$\frac{d}{dt}\iiint_{W_t} \rho\,\mathbf{v}\,dV = \iiint_{W_t} \rho\frac{d\mathbf{v}}{dt}\,dV,$$

as claimed in (8.64).

EXERCISES 8.6

1. Using a product rule formula for the divergence, show that (8.56) implies (8.57).

2. Justify the apearance of the minus sign in the formula (8.59). (*Hint:* Consider cases $\partial\rho/\partial t > 0$ and $\partial\rho/\partial t < 0$.)

3. Show that if **v** and **w** are vectors in $\mathbb{R}^3$ such that $\mathbf{v}\cdot\mathbf{a} = \mathbf{w}\cdot\mathbf{a}$ for all vectors **a** in $\mathbb{R}^3$, then $\mathbf{v} = \mathbf{w}$. Show that if **v** and **w** are vector fields in $\mathbb{R}^3$ satisfying $\mathbf{v}\cdot\mathbf{a} = \iiint_W \mathbf{w}\cdot\mathbf{a}\,dV$ for any constant vector **a**, then $\mathbf{v} = \iiint_W \mathbf{w}\,dV$.

4. Let $\mathbf{v} = \mathbf{v}(\mathbf{x}, t) = (v_1(\mathbf{x}, t), v_2(\mathbf{x}, t), v_3(\mathbf{x}, t))$. Show that $d\mathbf{v} = (d\mathbf{r} \cdot \nabla v_1)\mathbf{i} + (d\mathbf{r} \cdot \nabla v_2)\mathbf{j} + (d\mathbf{r} \cdot \nabla v_3)\mathbf{k} + (\partial\mathbf{v}/\partial t)dt$, where ∇v_i represents the gradient of the component v_i of $\mathbf{v}$, $i = 1, 2, 3$. Conclude that $d\mathbf{v}/dt = \mathbf{v} \cdot \nabla\mathbf{v} + \partial\mathbf{v}/\partial t$. Show that, if $\mathbf{F} = \mathbf{F}(\mathbf{x}, t)$, then $d\mathbf{F}/dt = \mathbf{v} \cdot \nabla\mathbf{F} + \partial\mathbf{F}/\partial t$.

5. Let $\mathbf{v}(\mathbf{x}, t) = (x + 3t, 2y + 2t, 3z + t)$. Compute $\mathbf{v} \cdot \nabla\mathbf{v}$.

6. Compute $\mathbf{v} \cdot \nabla\mathbf{v}$ if $\mathbf{v}(\mathbf{x}, t) = (yz, xzt, xyt^2)$.

7. Show that $\nabla(\mathbf{v}\cdot \mathbf{w}) = \mathbf{v} \cdot \nabla\mathbf{w} + \mathbf{w} \cdot \nabla\mathbf{v} + \mathbf{v} \times curl\,\mathbf{w} + \mathbf{w} \times curl\,\mathbf{v}$, and then let $\mathbf{w} = \mathbf{v}$ to derive the formula $\mathbf{v} \cdot \nabla\mathbf{v} = \frac{1}{2}\nabla\left(\|\mathbf{v}\|^2\right) - \mathbf{v} \times curl\,\mathbf{v}$.

8. Let $\mathbf{F}(x, y, z, t) = (F_1(x, y, z, t), F_2(x, y, z, t), F_3(x, y, z, t))$ be a time-changing vector field in $\mathbb{R}^3$, and let $\mathbf{r}(u, v, t) = (x(u, v, t), y(u, v, t), z(u, v, t))$. Prove that the formula (8.71) holds for $\mathbf{v} = \mathbf{i}$, $\mathbf{v} = \mathbf{j}$, and $\mathbf{v} = \mathbf{k}$. Explain why this suffices to prove that (8.71) holds for any vector $\mathbf{v}$ in $\mathbb{R}^3$.

9. Prove that $\mathbf{v} \cdot \nabla(\rho\mathbf{v}) = \rho\mathbf{v} \cdot \nabla\mathbf{v} + div\,(\rho\mathbf{v})\mathbf{v}$, for a vector function $\mathbf{v}$ in $\mathbb{R}^3$ satisfying $div\ \mathbf{v} = 0$, and a scalar function ρ.

▶ CHAPTER REVIEW

CHAPTER SUMMARY

- **Integration theorems.** Green's Theorem, computation of path integrals around closed curves, area as path integral, Divergence Theorem, interpretation of the divergence, Stokes' Theorem.
- **Differential forms.** Definition of a form, wedge product and differential, vector differential operators as differentials on forms, closed and exact forms, integration of forms, unified version of classical integration theorems.
- **Applications in electromagnetism and fluid flow.** Gauss' Theorem, conservation of charge, continuity equation, Faraday's law, Maxwell's equations, vector and scalar potentials, interpretation of divergence, Euler's equation, transport theorems.

REVIEW

Discuss the following questions.

1. Write down different versions of Green's Theorem and list the assumptions needed for the results to hold. What are advantages and disadvantages of using Green's Theorem (as compared to computing path integrals)?

2. Let $\mathbf{c}_1$ and $\mathbf{c}_2$ be simple closed curves in $\mathbb{R}^2$, both oriented counterclockwise and such that $\mathbf{c}_1$ is completely contained in the region enclosed by $\mathbf{c}_2$. Express the double integral of $curl\,\mathbf{F}$ over the region between $\mathbf{c}_1$ and $\mathbf{c}_2$ in terms of path integrals.

3. The integral of $\mathbf{F} = -y\mathbf{i}/(x^2 + y^2) + x\mathbf{j}/(x^2 + y^2)$ along the circle $x^2 + y^2 = 1$ is 2π. The double integral of $curl\,\mathbf{F}$ over the disk $\{(x, y)|x^2 + y^2 \leq 1\}$ seems to be zero. Is that a contradiction to Green's Theorem?

4. Define the normal derivative $D_{\mathbf{n}}f$ of a real-valued function f. Explain how to evaluate the path integral of $D_{\mathbf{n}}f$ in terms of a double integral.

5. Give the statement of the Divergence Theorem of Gauss. Explain why we say that it is a generalization of Green's Theorem to one dimension higher.

6. Explain how to recognize/define the outward normal to a closed plane curve. State the two-dimensional version of the Divergence Theorem.

7. Define the boundary of a parametrized surface. Explain how to define the orientation of its boundary curve(s).

8. Write down the differential-form version of the classical integration theorems. Explain how to obtain particular versions (Green's, Gauss' and Stokes') and the Fundamental Theorem of Calculus and its generalization (see Section 5.4) from it.

9. Let $\mathbf{F} = (\tan z, x, yz)$. Is it true that the path integral of $\mathbf{F}$ along the unit circle $\mathbf{c}$ in the xy-plane must be equal to the surface integral of *curl* $\mathbf{F}$ along *any* surface whose boundary is $\mathbf{c}$?

10. In defining differential forms in $\mathbb{R}^3$, we defined forms of degree 0, 1, 2, and 3. Why didn't we define 4-forms, 5-forms, etc.?

11. Define the differential of a form and explain how gradient, divergence, and curl can be interpreted as special cases. Suppose that there are four basic 1-forms, say, dx, dy, dz, and dt. How many basic 2-forms and 3-forms are there? Are there any nontrivial 4-forms, 5-forms, etc.?

12. Consider Maxwell's equations (8.27)–(8.30). Write down the corresponding integral versions. Which version would you call "global" and which would you call "local"?

TRUE/FALSE QUIZ

Determine whether the following statements are true or false. Give reasons for your answer.

1. A constant vector field in $\mathbb{R}^2$ is path-independent.

2. It is possible to use Green's Theorem to calculate the path integral $\int_{\mathbf{c}} (\mathbf{i}/x + \mathbf{j}) \cdot d\mathbf{s}$, where $\mathbf{c}$ is the circle $x^2 + y^2 = 1$, oriented positively.

3. If $\partial Q/\partial x = \partial P/\partial y$ for a C^1 vector field $\mathbf{F} = (P, Q)$, then $\mathbf{F}$ is a gradient vector field.

4. If $\mathbf{F} = x\mathbf{i} + y\mathbf{j}$ and C is a simple closed curve, then $\int_{\mathbf{c}} \mathbf{F} \cdot d\mathbf{s} = 0$ by Green's Theorem.

5. It is known that $div\, \mathbf{F}(x, y, z) = 0$ (for all (x, y, z)) for a C^1 vector field $\mathbf{F}$, defined on a region W that satisfies Assumption 8.2. Then $\iint_S \mathbf{F} \cdot d\mathbf{S} = 0$, where $S = \partial W$.

6. Let S_a and S_b be spheres centered at the origin of radii a and b, and let $\mathbf{F} = \mathbf{r}/\|\mathbf{r}\|^3$. If $a > b$, then the flux of $\mathbf{F}$ through S_a is larger than the flux of $\mathbf{F}$ through S_b.

7. For any closed, positively oriented surface S that bounds a three-dimensional region W satisfying Assumption 8.2, and for any C^2 vector field $\mathbf{F}$ on W, $\iint_S curl\, \mathbf{F} \cdot d\mathbf{S} = 0$.

8. If $\mathbf{F} = curl\, \mathbf{H}$, then $div\, \mathbf{F} = 0$.

9. The differential of the form $dxdy$ in $\mathbb{R}^3$ is zero.

10. A closed 1-form defined on the set $U = \{(x, y) | 1 < x^2 + y^2 < 2\} \subseteq \mathbb{R}^2$ must be exact.

11. The continuity equation for compressible fluid flow states that $d\rho/dt = -div\,(\rho\mathbf{v})$.

12. If $div\, \mathbf{v} = 0$ and ρ is a function, then $div\,(\rho\mathbf{v}) = 0$.

13. If $div\, \mathbf{F} = 0$, then there is a vector field $\mathbf{H}$ such that $\mathbf{F} = curl\, \mathbf{H}$.

REVIEW EXERCISES AND PROBLEMS

1. Show that the flux $\iint_S \mathbf{F} \cdot d\mathbf{S}$ of the gravitational field $\mathbf{F} = -GM\mathbf{r}/\|\mathbf{r}\|^3$ through a sphere centered at the origin does not depend on its radius.

2. Assume that f is a C^2, harmonic function (i.e., $f_{xx} + f_{yy} = 0$) on a set that contains a region D which satisfies Assumption 8.1. Show that $\int_{\partial D}(f_x dy - f_y dx) = 0$, where ∂D is the boundary of D oriented counterclockwise.

3. Find the area of the region bounded by the ellipse $x^2/a^2 + y^2/b^2 = 1$.

4. Compute $\int_{\mathbf{c}} y\,dx$, where $\mathbf{c}$ consists of the circles $x^2+y^2=16$ (oriented counterclockwise), $(x-2)^2+y^2=1$ (oriented clockwise), and $(x+2)^2+y^2=1$ (oriented clockwise).

5. Assume that f is a harmonic function (i.e., $f_{xx}+f_{yy}=0$) on a set that contains a region D which satisfies Assumption 8.1. Let ϕ be any differentiable function. Show that $\int_{\partial D}\phi(f_x dy - f_y dx) = \int_D(\phi_x f_x + \phi_y f_y)\,dxdy$, where ∂D is the boundary of D oriented counterclockwise.

6. Let $\alpha = -y(1-(x^2+y^2))^{-1}dx + x(1-(x^2+y^2))^{-1}dy$. Evaluate $\int_{\mathbf{c}_\epsilon}\alpha$, where $\mathbf{c}_\epsilon$ is the circle $\mathbf{c}_\epsilon(t)=(\epsilon\cos t, \epsilon\sin t)$, $0\le t\le 2\pi$, and $\epsilon>0$, $\epsilon\ne 1$. Compute $\lim\int_{\mathbf{c}_\epsilon}\alpha$ as $\epsilon\to 0$, $\epsilon\to 1$, and $\epsilon\to\infty$.

7. Assume that $\alpha = x^2dx + xydy$ and $\mathbf{c}$ is the triangle with vertices $(0,0)$, $(2,0)$, and $(0,1)$, oriented counterclockwise. Evaluate $\int_{\mathbf{c}}\alpha$ directly and using Green's Theorem.

8. Evaluate the surface integral $\iint_S \mathbf{F}\cdot d\mathbf{S}$, where $\mathbf{F}=(x^3+3yz)\mathbf{i}+3z^2y\mathbf{j}+3y^2z\mathbf{k}$, and S is the boundary of the three-dimensional region between the spheres $x^2+y^2+z^2=1$ and $x^2+y^2+z^2=9$ oriented by the outward-pointing normal.

9. Let W be a solid three-dimensional region (that satisfies the assumptions of the Divergence Theorem) bounded by a closed, positively oriented surface S. Show that $\iint_S(f\nabla g - g\nabla f)\cdot d\mathbf{S} = \iint_W(f\Delta g - g\Delta f)\,dV$, where f and g are of class C^2.

10. Consider the vector field $\mathbf{F}=x(x^2+y^2)^{-1}\mathbf{i}+y(x^2+y^2)^{-1}\mathbf{j}$ in $\mathbb{R}^2$.

(a) Check that the divergence of $\mathbf{F}$ is zero at all points except at the origin.

(b) Compute the path integral of $\mathbf{F}\cdot\mathbf{n}$ along the circle $\mathbf{c}_\epsilon$ of radius ϵ, $\epsilon>0$ ($\mathbf{n}$ is the outward unit normal).

(c) By (a), $\iint_{D_\epsilon} \mathit{div}\,\mathbf{F}\,dA = 0$, where D_ϵ is the disk $0<x^2+y^2\le\epsilon$. Explain why the fact that the result in (b) is not zero does not violate the Divergence Theorem in the plane.

11. Use Gauss' Theorem to compute $\int_S\alpha$, where $\alpha=(x^2+2xy)dydz+(2y+x^2z)dzdx+4x^2y^3dxdy$ and S is the boundary of the region in the first octant cut out by the cylinder $x^2+y^2=9$ and the plane $z=1$.

12. Find the circulation of the vector field $\mathbf{F}=e^xyz\mathbf{i}-e^y\mathbf{j}+e^z\mathbf{k}$ along the boundary $\mathbf{c}$ of the triangle with vertices $(0,0,1)$, $(0,-2,0)$, and $(1,0,0)$, oriented clockwise as seen from the origin.

13. Let S be a surface and $\mathbf{c}=\partial S$ its boundary curve. Show that if a vector field $\mathbf{F}$ is perpendicular to $\mathbf{c}$, then $\iint_S \mathit{curl}\,\mathbf{F}\cdot d\mathbf{S}=0$.

14. Consider the integral $\iint_D e^{x^2}\,dA$, where D is the triangle defined by the lines $y=x$, $x=1$, and $y=0$. Follow the steps to evaluate it using Stokes' Theorem.

(a) Write e^{x^2} as the dot product $e^{x^2}\mathbf{k}\cdot\mathbf{k}$ of the vector field $\mathbf{F}=e^{x^2}\mathbf{k}$ and the unit normal $\mathbf{N}=\mathbf{k}$ to D. Check that the triangle D can be parametrized so that the normal is $\mathbf{k}$. Check that $\mathit{curl}\,(-ye^{x^2}\mathbf{i})=\mathbf{F}$.

(b) From (a), it follows that $\iint_D e^{x^2}\,dA = \iint_D \mathit{curl}\,(-ye^{x^2}\mathbf{i})\cdot\mathbf{N}\,dA$. Now use Stokes' Theorem to evaluate the integral on the right side.

(c) Check your result in (b) by evaluating the double integral using an appropriate order of integration.

15. Let $\mathcal{B}=B_xdtdx+B_ydtdy+B_zdtdz+c^{-2}(E_xdydz+E_ydzdx+E_zdxdy)$ and let $\mathcal{D}=J_xdtdydz+J_ydtdzdx+J_zdtdxdy-\rho dxdydz$, where B_x, B_y, B_z and E_x, E_y, E_z are the components of time-changing magnetic and electric fields; J_x, J_y, and J_z are the components of the total current density, and ρ is the total charge density.

(a) Show that Maxwell's equations (8.23)–(8.26) imply $d\mathcal{B}+\mu_0\mathcal{D}=0$.

(b) Conclude from (a) that $d\mathcal{D}=0$.

(c) Interpret $d\mathcal{D}=0$ in physical terms.

(d) Find α such that $d\alpha=\mathcal{D}$. [*Hint:* Use (a)].

Exercise 16 and 17: Consider differential forms corresponding to the coordinates t, x, y, and z. There are six basic 2-forms: *dtdx, dtdy, dtdz, dxdy, dydz*, and *dzdx*. We define the operation $*$ (called the Hodge star operator, or just star) that maps 2-forms to 2-forms, in the following way: first we define $*$ on basic 2-forms and then extend it (using linearity) to all 2-forms:

(i) $*dxdy = dtdz$, $*dydz = dtdx$, $*dzdx = dtdy$, $*dtdx = -c^{-2}dydz$, $*dtdy = -c^{-2}dzdx$, and $*dtdz = -c^{-2}dxdy$

(ii) Let $\alpha = fdtdx + gdtdy + hdtdz + kdxdy + ldydz + mdzdx$ be a 2-form, where f, g, h, k, l, and m are differentiable functions of x, y, z, and t. Then $*\alpha = -c^{-2}fdydz - c^{-2}gdzdx - c^{-2}hdxdy + kdtdz + ldtdx + mdtdy$.

16. Let $\alpha = x^2tdxdy + zt^2dtdz - xy^2ztdtdy$.

(a) Compute $*\alpha$.

(b) Show that $*(*\alpha) = -c^{-2}\alpha$.

(c) Compute $\alpha \wedge (*\alpha)$.

17. Recall that the electromagnetic tensor $\mathcal{E}$ is given by $\mathcal{E} = -E_x dtdx - E_y dtdy - E_z dtdz + B_z dxdy + B_x dydz + B_y dzdx$. We showed that $\mathcal{E} = d\mathcal{A}$, where $\mathcal{A}$ is the 1-form $\mathcal{A} = -\phi dt + A_x dx + A_y dy + A_z dz$, called the electromagnetic potential.

(a) Since $\mathcal{E} = d\mathcal{A}$, it follows that $d\mathcal{E} = 0$ (why?). Interpret this differential-form equation by analyzing its components.

(b) Compute $*\mathcal{E}$.

(c) Interpret the equation $d(*\mathcal{E}) = 0$.

18. Let S be a surface in $\mathbb{R}^3$ and let $\mathbf{c}$ be its boundary curve. Assuming that S and $\mathbf{c}$ satisfy the assumptions of Stokes' Theorem, show that $\int_{\mathbf{c}} (f\nabla g + g\nabla f) \cdot d\mathbf{s} = 0$ for C^2 functions f and g.

19. Let $\alpha = f(x, y)dx + g(y, z)dy + h(x, z)dz$ be a 1-form in $\mathbb{R}^3$, and let f, g, and h be differentiable functions of the variables indicated. Compute $d\alpha$ and check that $d(d\alpha) = 0$. What assumption on the functions f, g, and h is needed for this identity to hold?

APPENDIX

A

Various Results Used in This Book and Proofs of Differentiation Theorems

In this appendix, we give (mostly) technical proofs of theorems about properties of derivatives that were stated in Chapter 2. We start by quoting a few results that have been used in various situations in this book. Some are needed again in the proofs presented here.

THEOREM 1 Intermediate Value Theorem

Assume that $g\colon [a, b] \to \mathbb{R}$ is a continuous function defined on the closed interval $[a, b]$ and let N be any number between $g(a)$ and $g(b)$. Then there exists a number $c \in [a, b]$ such that $g(c) = N$. ◀

THEOREM 2 Mean Value Theorem

Assume that $g\colon [a, b] \to \mathbb{R}$ is a function continuous on the closed interval $[a, b]$ and differentiable on the open interval (a, b). Then there is a number $c \in (a, b)$ such that $g(b) - g(a) = g'(c)(b - a)$. ◀

THEOREM 3 Extreme Value Theorem

Let $f\colon V \subseteq \mathbb{R}^m \to \mathbb{R}$ be a continuous function defined on a closed and bounded set $V \subseteq \mathbb{R}^m$, $m \geq 1$. There is a point $\mathbf{a}_1$ in V such that $f(\mathbf{a}_1) \geq f(\mathbf{x})$ for all $\mathbf{x} \in V$, and there is a point $\mathbf{a}_2$ in V such that $f(\mathbf{a}_2) \leq f(\mathbf{x})$ for all $\mathbf{x} \in V$. ◀

In words, a continuous function defined on a closed and bounded set has a minimum and a maximum. It follows that a continuous function defined on a closed and bounded set is bounded.

The following two inequalities will be used repeatedly in the proofs in this appendix.

THEOREM 4 Triangle Inequality

Let $\mathbf{v}$ and $\mathbf{w}$ be vectors in $\mathbb{R}^m$, $m \geq 1$. Then $\|\mathbf{v} + \mathbf{w}\| \leq \|\mathbf{v}\| + \|\mathbf{w}\|$. ◀

THEOREM 5 Cauchy–Schwarz Inequality

Let **v** and **w** be vectors in $\mathbb{R}^m$, $m \geq 1$. Then $|\mathbf{v} \cdot \mathbf{w}| \leq \|\mathbf{v}\| \, \|\mathbf{w}\|$. ◀

The Cauchy–Schwarz Inequality implies that $|\mathbf{v} \cdot \mathbf{w}|^2 \leq \|\mathbf{v}\|^2 \, \|\mathbf{w}\|^2$ for $\mathbf{v}, \mathbf{w} \in \mathbb{R}^m$. Consequently, if $\mathbf{v} = (v_1, \ldots, v_m)$ and $\mathbf{w} = (w_1, \ldots, w_m)$, then

$$(v_1 w_1 + \cdots + v_m w_m)^2 \leq (v_1 + \cdots + v_m)^2 (w_1 + \cdots + w_m)^2. \tag{1}$$

See Exercise 10 in Section 1.3 for the versions of this inequality in $\mathbb{R}^2$ and in $\mathbb{R}^3$.

Let A be an $n \times m$ matrix and let $\mathbf{x}$ be a vector in $\mathbb{R}^m$. By $A \cdot \mathbf{x}$, we mean a matrix product of A and $\mathbf{x}$, where $\mathbf{x}$ is thought of as an $m \times 1$ matrix. In that case, $A \cdot \mathbf{x}$ is an $n \times 1$ matrix, or a vector in $\mathbb{R}^n$; we use $\|A \cdot \mathbf{x}\|$ to denote its norm.

The norm of an $n \times m$ matrix $A = [a_{ij}]$ is defined by

$$\|A\| = \left(\sum_{i=1}^{n} \sum_{j=1}^{m} a_{ij}^2 \right)^{1/2}.$$

THEOREM 6 Inequality for the Norm of a Matrix

Let A be an $n \times m$ matrix and let $\mathbf{x} \in \mathbb{R}^m$ be any vector. Then $\|A \cdot \mathbf{x}\| \leq \|A\| \, \|\mathbf{x}\|$. ◀

PROOF: Let $\mathbf{x} = (x_1, \ldots, x_m)$; the product $A \cdot \mathbf{x}$ is a vector whose components are

$$A \cdot \mathbf{x} = \begin{bmatrix} a_{11} & a_{12} & \ldots & a_{1m} \\ a_{21} & a_{22} & \ldots & a_{2m} \\ \vdots & \vdots & & \vdots \\ a_{n1} & a_{n2} & \ldots & a_{nm} \end{bmatrix} \cdot \begin{bmatrix} x_1 \\ x_2 \\ \vdots \\ x_m \end{bmatrix} = \begin{bmatrix} a_{11}x_1 + a_{12}x_2 + \cdots + a_{1m}x_m \\ a_{21}x_1 + a_{22}x_2 + \cdots + a_{2m}x_m \\ \vdots \\ a_{n1}x_1 + a_{n2}x_2 + \cdots + a_{nm}x_m \end{bmatrix}.$$

The square of its magnitude is

$$\begin{aligned} \|A \cdot \mathbf{x}\|^2 &= (a_{11}x_1 + a_{12}x_2 + \cdots + a_{1m}x_m)^2 + (a_{21}x_1 + a_{22}x_2 + \cdots + a_{2m}x_m)^2 \\ &\quad + \cdots + (a_{n1}x_1 + a_{n2}x_2 + \cdots + a_{nm}x_m)^2. \end{aligned}$$

Using (1) with $v_1 = a_{11}, v_2 = a_{12}, \ldots, v_m = a_{1m}$ and $w_1 = x_1, w_2 = x_2, \ldots, w_m = x_m$ we get

$$(a_{11}x_1 + a_{12}x_2 + \cdots + a_{1m}x_m)^2 \leq \left(a_{11}^2 + a_{12}^2 + \cdots + a_{1m}^2\right)\left(x_1^2 + x_2^2 + \cdots + x_m^2\right).$$

In a similar way, we estimate the remaining terms in the expression for $\|A \cdot \mathbf{x}\|^2$, thus obtaining

$$\begin{aligned} \|A \cdot \mathbf{x}\|^2 &\leq \left(a_{11}^2 + a_{12}^2 + \cdots + a_{1m}^2\right)\left(x_1^2 + x_2^2 + \cdots + x_m^2\right) \\ &\quad + \left(a_{21}^2 + a_{22}^2 + \cdots + a_{2m}^2\right)\left(x_1^2 + x_2^2 + \ldots + x_m^2\right) \\ &\quad + \cdots + \left(a_{n1}^2 + a_{n2}^2 + \cdots + a_{nm}^2\right)\left(x_1^2 + x_2^2 + \cdots + x_m^2\right) \\ &= \left(a_{11}^2 + a_{12}^2 + \cdots + a_{nm}^2\right)\left(x_1^2 + x_2^2 + \cdots + x_m^2\right) \\ &= \left(\sum_{i=1}^{n} \sum_{j=1}^{m} a_{ij}^2 \right) \|\mathbf{x}\|^2. \end{aligned}$$

Consequently,

$$\|A \cdot \mathbf{x}\| \leq \left(\sum_{i=1}^{n}\sum_{j=1}^{m} a_{ij}^2\right)^{1/2} \|\mathbf{x}\| = \|A\|\,\|\mathbf{x}\|.$$

THEOREM 7 An Estimate for $\|\mathbf{F}(\mathbf{x}) - \mathbf{F}(\mathbf{a})\|$

Let $\mathbf{F}\colon U \subseteq \mathbb{R}^m \to \mathbb{R}^n$ be differentiable at $\mathbf{a} \in \mathbb{R}^m$. Then

$$\|\mathbf{F}(\mathbf{x}) - \mathbf{F}(\mathbf{a})\| \leq (1 + \|D\mathbf{F}(\mathbf{a})\|)\,\|\mathbf{x} - \mathbf{a}\|,$$

for $\mathbf{x}$ near $\mathbf{a}$, such that $\mathbf{x} \neq \mathbf{a}$. $D\mathbf{F}(\mathbf{a})$ denotes the derivative of $\mathbf{F}$ evaluated at $\mathbf{a}$.

PROOF: By Definition 2.13, $\mathbf{F}$ is differentiable at $\mathbf{a}$ if

$$\lim_{\mathbf{x}\to\mathbf{a}} \frac{\|\mathbf{F}(\mathbf{x}) - \mathbf{F}(\mathbf{a}) - D\mathbf{F}(\mathbf{a})(\mathbf{x} - \mathbf{a})\|}{\|\mathbf{x} - \mathbf{a}\|} = 0.$$

Interpret this definition in terms of ϵ and δ: choose $\epsilon = 1$; then there is a $\delta > 0$ such that $0 < \|\mathbf{x} - \mathbf{a}\| < \delta$ implies

$$\frac{\|\mathbf{F}(\mathbf{x}) - \mathbf{F}(\mathbf{a}) - D\mathbf{F}(\mathbf{a})(\mathbf{x} - \mathbf{a})\|}{\|\mathbf{x} - \mathbf{a}\|} < \epsilon = 1,$$

that is,

$$\|\mathbf{F}(\mathbf{x}) - \mathbf{F}(\mathbf{a}) - D\mathbf{F}(\mathbf{a})(\mathbf{x} - \mathbf{a})\| < \|\mathbf{x} - \mathbf{a}\|. \tag{2}$$

By the Triangle Inequality and (2),

$$\begin{aligned}
\|\mathbf{F}(\mathbf{x}) - \mathbf{F}(\mathbf{a})\| &= \|\mathbf{F}(\mathbf{x}) - \mathbf{F}(\mathbf{a}) - D\mathbf{F}(\mathbf{a})(\mathbf{x} - \mathbf{a}) + D\mathbf{F}(\mathbf{a})(\mathbf{x} - \mathbf{a})\| \\
&\leq \|\mathbf{F}(\mathbf{x}) - \mathbf{F}(\mathbf{a}) - D\mathbf{F}(\mathbf{a})(\mathbf{x} - \mathbf{a})\| + \|D\mathbf{F}(\mathbf{a})(\mathbf{x} - \mathbf{a})\| \\
&\leq \|\mathbf{x} - \mathbf{a}\| + \|D\mathbf{F}(\mathbf{a})\|\,\|\mathbf{x} - \mathbf{a}\|,
\end{aligned}$$

since, by Theorem 6,

$$\|D\mathbf{F}(\mathbf{a})(\mathbf{x} - \mathbf{a})\| \leq \|D\mathbf{F}(\mathbf{a})\|\,\|\mathbf{x} - \mathbf{a}\|.$$

It follows that

$$\|\mathbf{F}(\mathbf{x}) - \mathbf{F}(\mathbf{a})\| \leq \big(1 + \|D\mathbf{F}(\mathbf{a})\|\big)\|\mathbf{x} - \mathbf{a}\|,$$

for $0 < \|\mathbf{x} - \mathbf{a}\| < \delta$.

Next, we turn to the proofs of the theorems in Sections 2.4, 2.6, and 4.1. For convenience, we recall their statements.

THEOREM 2.4 Differentiable Functions Are Continuous

Let $\mathbf{F}\colon U \subseteq \mathbb{R}^m \to \mathbb{R}^n$ be a vector-valued function and let $\mathbf{a} \in U$. If $\mathbf{F}$ is differentiable at $\mathbf{a}$, then it is continuous at $\mathbf{a}$.

PROOF: We have to show that $\lim_{\mathbf{x}\to\mathbf{a}} \mathbf{F}(\mathbf{x}) = \mathbf{F}(\mathbf{a})$, or, equivalently, $\lim_{\mathbf{x}\to\mathbf{a}} \|\mathbf{F}(\mathbf{x}) - \mathbf{F}(\mathbf{a})\| = 0$. Since $\mathbf{F}$ is differentiable at $\mathbf{a}$, it follows from Theorem 7 that there exists $\delta > 0$ such that

$$\|\mathbf{F}(\mathbf{x}) - \mathbf{F}(\mathbf{a})\| \leq \big(1 + \|D\mathbf{F}(\mathbf{a})\|\big)\,\|\mathbf{x} - \mathbf{a}\|, \tag{3}$$

where $0 < \|\mathbf{x} - \mathbf{a}\| < \delta$. We will now use the definition of a limit: choose *any* $\overline{\epsilon} > 0$ and let $\overline{\delta} < \min\{\delta, \overline{\epsilon}(1 + \|D\mathbf{F}(\mathbf{a})\|)^{-1}\}$. Because $\overline{\delta} < \delta$, (3) holds for $0 < \|\mathbf{x} - \mathbf{a}\| < \overline{\delta}$, so

$$\|\mathbf{F}(\mathbf{x}) - \mathbf{F}(\mathbf{a})\| < (1 + \|D\mathbf{F}(\mathbf{a})\|)\,\overline{\delta} < (1 + \|D\mathbf{F}(\mathbf{a})\|)\frac{\overline{\epsilon}}{1 + \|D\mathbf{F}(\mathbf{a})\|} = \overline{\epsilon},$$

since $\overline{\delta} < \overline{\epsilon}(1 + \|D\mathbf{F}(\mathbf{a})\|)^{-1}$.

THEOREM 2.5 Continuity of Partial Derivatives Implies Differentiability

Let $\mathbf{F}: U \subseteq \mathbb{R}^m \to \mathbb{R}^n$ be a vector-valued function with components $F_1, \dots, F_n: U \subseteq \mathbb{R}^m \to \mathbb{R}$. If all partial derivatives $\partial F_i/\partial x_j$ $(i = 1, \dots, n,\ j = 1 \dots, m)$ are continuous at $\mathbf{a}$, then $\mathbf{F}$ is differentiable at $\mathbf{a}$.

PROOF: Consider the case $m = 2$ and $n = 1$; that is, let $f(x, y): U \subseteq \mathbb{R}^2 \to \mathbb{R}$ and assume that $\partial f/\partial x$ and $\partial f/\partial y$ are continuous at $\mathbf{a} = (a, b) \in U$. According to Definition 2.13, we have to show that

$$\lim_{\mathbf{x}\to\mathbf{a}} \frac{|f(x, y) - f(a, b) - Df(a, b)(x - a, y - b)|}{\|(x - a, y - b)\|} = 0,$$

where $Df(a, b) = [\,\partial f/\partial x(a, b) \quad \partial f/\partial y(a, b)\,]$. Write $f(x, y) - f(a, b)$ as $f(x, y) - f(a, b) = f(x, y) - f(a, y) + f(a, y) - f(a, b)$. By the Mean Value Theorem (see Theorem 2) applied to $g(x) = f(x, y)$ (y is kept fixed), it follows that there is a number c_1 between x and a such that

$$f(x, y) - f(a, y) = \frac{\partial f}{\partial x}(c_1, y)(x - a).$$

Similarly, there is a number c_2 between y and b such that

$$f(a, y) - f(a, b) = \frac{\partial f}{\partial y}(a, c_2)(y - b).$$

It follows that

$$f(x, y) - f(a, b) = \frac{\partial f}{\partial x}(c_1, y)(x - a) + \frac{\partial f}{\partial y}(a, c_2)(y - b)$$

and

$$\begin{aligned}
&|f(x, y) - f(a, b) - Df(a, b)(x - a, y - b)| \\
&\quad = \left|\frac{\partial f}{\partial x}(c_1, y)(x - a) + \frac{\partial f}{\partial y}(a, c_2)(y - b) - \frac{\partial f}{\partial x}(a, b)(x - a) - \frac{\partial f}{\partial y}(a, b)(y - b)\right| \\
&\quad = \left|\left(\frac{\partial f}{\partial x}(c_1, y) - \frac{\partial f}{\partial x}(a, b)\right)(x - a) + \left(\frac{\partial f}{\partial y}(a, c_2) - \frac{\partial f}{\partial y}(a, b)\right)(y - b)\right|
\end{aligned}$$

(continue using the Triangle Inequality and the inequalities $|x-a| \le \sqrt{(x-a)^2+(y-b)^2} = \|(x-a, y-b)\|$ and $|y-b| \le \sqrt{(x-a)^2+(y-b)^2} = \|(x-a, y-b)\|$)

$$\le \left|\frac{\partial f}{\partial x}(c_1, y) - \frac{\partial f}{\partial x}(a, b)\right| |x-a| + \left|\frac{\partial f}{\partial y}(a, c_2) - \frac{\partial f}{\partial y}(a, b)\right| |y-b|$$

$$\le \left(\left|\frac{\partial f}{\partial x}(c_1, y) - \frac{\partial f}{\partial x}(a, b)\right| + \left|\frac{\partial f}{\partial y}(a, c_2) - \frac{\partial f}{\partial y}(a, b)\right|\right) \|(x-a, y-b)\|.$$

Finally,

$$\frac{|f(x, y) - f(a, b) - Df(a, b)(x-a, y-b)|}{\|(x-a, y-b)\|}$$
$$\le \left|\frac{\partial f}{\partial x}(c_1, y) - \frac{\partial f}{\partial x}(a, b)\right| + \left|\frac{\partial f}{\partial y}(a, c_2) - \frac{\partial f}{\partial y}(a, b)\right|.$$

As $x \to a$ (hence, $c_1 \to a$) and $y \to b$,

$$\frac{\partial f}{\partial x}(c_1, y) - \frac{\partial f}{\partial x}(a, b) \to 0,$$

by the continuity of $\partial f/\partial x$ at (a, b). Similarly, as $x \to a$ and $y \to b$,

$$\frac{\partial f}{\partial y}(a, c_2) - \frac{\partial f}{\partial y}(a, b) \to 0,$$

and we are done.

The technique of this proof can be applied to any function $f: \mathbb{R}^m \to \mathbb{R}$ with $m \ge 2$. The general case of a vector-valued function $\mathbf{F}: \mathbb{R}^m \to \mathbb{R}^n$ is dealt with componentwise.

THEOREM 2.6 Properties of Derivatives

(a) Assume that the functions $\mathbf{F}, \mathbf{G}: U \subseteq \mathbb{R}^m \to \mathbb{R}^n$ are differentiable at $\mathbf{a} \in U$. Then the sum $\mathbf{F} + \mathbf{G}$ and the difference $\mathbf{F} - \mathbf{G}$ are differentiable at $\mathbf{a}$ and

$$D(\mathbf{F}+\mathbf{G})(\mathbf{a}) = D\mathbf{F}(\mathbf{a}) + D\mathbf{G}(\mathbf{a}) \quad \text{and} \quad D(\mathbf{F}-\mathbf{G})(\mathbf{a}) = D\mathbf{F}(\mathbf{a}) - D\mathbf{G}(\mathbf{a}).$$

(b) If the function $\mathbf{F}: U \subseteq \mathbb{R}^m \to \mathbb{R}^n$ is differentiable at $\mathbf{a} \in U$ and $c \in \mathbb{R}$ is a constant, then the product $c\mathbf{F}$ is differentiable at $\mathbf{a}$ and

$$D(c\mathbf{F})(\mathbf{a}) = c\, D\mathbf{F}(\mathbf{a}).$$

(c) If the real-valued functions $f, g: U \subseteq \mathbb{R}^m \to \mathbb{R}$ are differentiable at $\mathbf{a} \in U$, then their product fg is differentiable at $\mathbf{a}$ and

$$D(fg)(\mathbf{a}) = g(\mathbf{a})Df(\mathbf{a}) + f(\mathbf{a})Dg(\mathbf{a}).$$

(d) If the real-valued functions $f, g: U \subseteq \mathbb{R}^m \to \mathbb{R}$ are differentiable at $\mathbf{a} \in U$, then their quotient f/g is differentiable at $\mathbf{a}$ if $g(\mathbf{a}) \ne 0$, and

$$D\left(\frac{f}{g}\right)(\mathbf{a}) = \frac{g(\mathbf{a})Df(\mathbf{a}) - f(\mathbf{a})Dg(\mathbf{a})}{g(\mathbf{a})^2},$$

if $g(\mathbf{a}) \ne 0$.

(e) If the vector-valued functions $\mathbf{v}, \mathbf{w}: U \subseteq \mathbb{R} \to \mathbb{R}^n$ are differentiable at $a \in U$, then their dot (scalar) product $\mathbf{v} \cdot \mathbf{w}$ is differentiable at a and

$$(\mathbf{v} \cdot \mathbf{w})'(a) = \mathbf{v}'(a) \cdot \mathbf{w}(a) + \mathbf{v}(a) \cdot \mathbf{w}'(a).$$

(f) If the vector-valued functions $\mathbf{v}, \mathbf{w}: U \subseteq \mathbb{R} \to \mathbb{R}^3$ are differentiable at $a \in U$, their cross (vector) product $\mathbf{v} \times \mathbf{w}$ is differentiable at a and

$$(\mathbf{v} \times \mathbf{w})'(a) = \mathbf{v}'(a) \times \mathbf{w}(a) + \mathbf{v}(a) \times \mathbf{w}'(a).$$

PROOF:

(a) Let $\mathbf{H} = \mathbf{F} + \mathbf{G}$; we have to show that $D\mathbf{H}(\mathbf{a}) = D\mathbf{F}(\mathbf{a}) + D\mathbf{G}(\mathbf{a})$, that is,

$$\lim_{\mathbf{x}\to\mathbf{a}} \frac{\|\mathbf{H}(\mathbf{x}) - \mathbf{H}(\mathbf{a}) - (D\mathbf{F}(\mathbf{a}) + D\mathbf{G}(\mathbf{a}))(\mathbf{x}-\mathbf{a})\|}{\|\mathbf{x}-\mathbf{a}\|} = 0.$$

Using the Triangle Inequality, we get

$$\begin{aligned} &\frac{\|\mathbf{H}(\mathbf{x}) - \mathbf{H}(\mathbf{a}) - (D\mathbf{F}(\mathbf{a}) + D\mathbf{G}(\mathbf{a}))(\mathbf{x}-\mathbf{a})\|}{\|\mathbf{x}-\mathbf{a}\|} \\ &\quad = \frac{\|\mathbf{F}(\mathbf{x}) - \mathbf{F}(\mathbf{a}) - D\mathbf{F}(\mathbf{a})(\mathbf{x}-\mathbf{a}) + \mathbf{G}(\mathbf{x}) - \mathbf{G}(\mathbf{a}) - D\mathbf{G}(\mathbf{a})(\mathbf{x}-\mathbf{a})\|}{\|\mathbf{x}-\mathbf{a}\|} \\ &\quad \le \frac{\|\mathbf{F}(\mathbf{x}) - \mathbf{F}(\mathbf{a}) - D\mathbf{F}(\mathbf{a})(\mathbf{x}-\mathbf{a})\|}{\|\mathbf{x}-\mathbf{a}\|} + \frac{\|\mathbf{G}(\mathbf{x}) - \mathbf{G}(\mathbf{a}) - D\mathbf{G}(\mathbf{a})(\mathbf{x}-\mathbf{a})\|}{\|\mathbf{x}-\mathbf{a}\|}. \end{aligned}$$

By assumption, both terms approach 0 as $\mathbf{x} \to \mathbf{a}$, and we are done. The proof of the formula for the derivative of the difference is analogous.

(b) This statement follows immediately from

$$\frac{\|c\mathbf{F}(\mathbf{x}) - c\mathbf{F}(\mathbf{a}) - cD\mathbf{F}(\mathbf{a})(\mathbf{x}-\mathbf{a})\|}{\|\mathbf{x}-\mathbf{a}\|} = |c|\frac{\|\mathbf{F}(\mathbf{x}) - \mathbf{F}(\mathbf{a}) - D\mathbf{F}(\mathbf{a})(\mathbf{x}-\mathbf{a})\|}{\|\mathbf{x}-\mathbf{a}\|}$$

and the fact that $\mathbf{F}$ is differentiable.

(c) We have to prove that

$$\lim_{\mathbf{x}\to\mathbf{a}} \frac{|(fg)(\mathbf{x}) - (fg)(\mathbf{a}) - (g(\mathbf{a})Df(\mathbf{a}) + f(\mathbf{a})Dg(\mathbf{a}))(\mathbf{x}-\mathbf{a})|}{\|\mathbf{x}-\mathbf{a}\|} = 0.$$

The numerator can be written as

$$|(fg)(\mathbf{x}) - (fg)(\mathbf{a}) - (g(\mathbf{a})Df(\mathbf{a}) + f(\mathbf{a})Dg(\mathbf{a}))(\mathbf{x}-\mathbf{a})|$$

[add and subtract $f(\mathbf{x})g(\mathbf{a})$]

$$\begin{aligned} &= |f(\mathbf{x})g(\mathbf{x}) - f(\mathbf{a})g(\mathbf{a}) - g(\mathbf{a})Df(\mathbf{a})(\mathbf{x}-\mathbf{a}) - f(\mathbf{a})Dg(\mathbf{a})(\mathbf{x}-\mathbf{a}) \\ &\quad + f(\mathbf{x})g(\mathbf{a}) - f(\mathbf{x})g(\mathbf{a})| \end{aligned}$$

[add and subtract $f(\mathbf{x})Dg(\mathbf{a})(\mathbf{x}-\mathbf{a})$]

$$\begin{aligned} &= |g(\mathbf{a})(f(\mathbf{x}) - f(\mathbf{a}) - Df(\mathbf{a})(\mathbf{x}-\mathbf{a})) + f(\mathbf{x})(g(\mathbf{x}) - g(\mathbf{a}) - Dg(\mathbf{a})(\mathbf{x}-\mathbf{a})) \\ &\quad + (f(\mathbf{x}) - f(\mathbf{a}))Dg(\mathbf{a})(\mathbf{x}-\mathbf{a})| \end{aligned}$$

[use the Triangle Inequality]

$$\leq |g(\mathbf{a})|\,|f(\mathbf{x}) - f(\mathbf{a}) - Df(\mathbf{a})(\mathbf{x}-\mathbf{a})| + |f(\mathbf{x})|\,|g(\mathbf{x}) - g(\mathbf{a}) - Dg(\mathbf{a})(\mathbf{x}-\mathbf{a})|$$
$$+|f(\mathbf{x}) - f(\mathbf{a})|\,|Dg(\mathbf{a})(\mathbf{x}-\mathbf{a})|. \tag{4}$$

By assumption, f and g are differentiable, and therefore,

$$\frac{|f(\mathbf{x}) - f(\mathbf{a}) - Df(\mathbf{a})(\mathbf{x}-\mathbf{a})|}{\|\mathbf{x}-\mathbf{a}\|} \to 0 \quad \text{and} \quad \frac{|g(\mathbf{x}) - g(\mathbf{a}) - Dg(\mathbf{a})(\mathbf{x}-\mathbf{a})|}{\|\mathbf{x}-\mathbf{a}\|} \to 0$$

as $\mathbf{x} \to \mathbf{a}$. Finally, from Theorem 6, it follows that

$$\frac{|f(\mathbf{x}) - f(\mathbf{a})|\,|Dg(\mathbf{a})(\mathbf{x}-\mathbf{a})|}{\|\mathbf{x}-\mathbf{a}\|} \leq \frac{|f(\mathbf{x}) - f(\mathbf{a})|\,\|Dg(\mathbf{a})\|\,\|\mathbf{x}-\mathbf{a}\|}{\|\mathbf{x}-\mathbf{a}\|}$$
$$= |f(\mathbf{x}) - f(\mathbf{a})|\,\|Dg(\mathbf{a})\|.$$

Since f is differentiable, it is continuous, and therefore, $|f(\mathbf{x}) - f(\mathbf{a})| \to 0$ as $\mathbf{x} \to \mathbf{a}$. So all three terms in (4) divided by $\|\mathbf{x}-\mathbf{a}\|$ approach 0 as $\mathbf{x} \to \mathbf{a}$.

(d) The proof is similar to (c).

(e) If $\mathbf{v}, \mathbf{w}: U \subseteq \mathbb{R} \to \mathbb{R}^n$, then their dot product $\mathbf{v} \cdot \mathbf{w}$ is a function from U into $\mathbb{R}$; therefore, "to be differentiable" means "to have a derivative (as a single-variable function)." Using the definition, we get

$$(\mathbf{v}\cdot\mathbf{w})'(a) = \lim_{x\to a} \frac{(\mathbf{v}\cdot\mathbf{w})(x) - (\mathbf{v}\cdot\mathbf{w})(a)}{x-a}$$

[add and subtract $\mathbf{v}(x)\cdot\mathbf{w}(a)$]

$$= \lim_{x\to a} \frac{\mathbf{v}(x)\cdot\mathbf{w}(x) - \mathbf{v}(a)\cdot\mathbf{w}(a) + \mathbf{v}(x)\cdot\mathbf{w}(a) - \mathbf{v}(x)\cdot\mathbf{w}(a)}{x-a}$$
$$= \lim_{x\to a} \left(\frac{\mathbf{v}(x)\cdot(\mathbf{w}(x) - \mathbf{w}(a))}{x-a} + \frac{\mathbf{w}(a)\cdot(\mathbf{v}(x) - \mathbf{v}(a))}{x-a} \right)$$
$$= \mathbf{v}(a)\cdot\mathbf{w}'(a) + \mathbf{w}(a)\cdot\mathbf{v}'(a).$$

(f) Proceed as in (e).

THEOREM 2.7 Chain Rule

Suppose that $\mathbf{F}: U \subseteq \mathbb{R}^m \to \mathbb{R}^n$ is differentiable at $\mathbf{a} \in U$, U is open in $\mathbb{R}^m$, $\mathbf{G}: V \subseteq \mathbb{R}^n \to \mathbb{R}^p$ is differentiable at $\mathbf{F}(\mathbf{a}) \in V$, V is open in $\mathbb{R}^n$, and $\mathbf{F}(U) \subseteq V$ (so that the composition $\mathbf{G} \circ \mathbf{F}$ is defined). Then $\mathbf{G} \circ \mathbf{F}$ is differentiable at $\mathbf{a}$ and

$$D(\mathbf{G}\circ\mathbf{F})(\mathbf{a}) = D\mathbf{G}(\mathbf{F}(\mathbf{a}))\cdot D\mathbf{F}(\mathbf{a}),$$

where $\cdot$ denotes matrix multiplication.

PROOF: Denote the composition $\mathbf{G} \circ \mathbf{F}$ by $\mathbf{H}$. We have to prove that

$$\lim_{\mathbf{x}\to\mathbf{a}} \frac{\|\mathbf{H}(\mathbf{x}) - \mathbf{H}(\mathbf{a}) - D\mathbf{G}(\mathbf{F}(\mathbf{a}))\cdot D\mathbf{F}(\mathbf{a})(\mathbf{x}-\mathbf{a})\|}{\|\mathbf{x}-\mathbf{a}\|} = 0.$$

Write the numerator as

$$\|\mathbf{H}(\mathbf{x}) - \mathbf{H}(\mathbf{a}) - D\mathbf{G}(\mathbf{F}(\mathbf{a}))(\mathbf{F}(\mathbf{x}) - \mathbf{F}(\mathbf{a})) \\ + D\mathbf{G}(\mathbf{F}(\mathbf{a}))(\mathbf{F}(\mathbf{x}) - \mathbf{F}(\mathbf{a}) - D\mathbf{F}(\mathbf{a})(\mathbf{x} - \mathbf{a}))\|$$

(use the Triangle Inequality)

$$\leq \|\mathbf{H}(\mathbf{x}) - \mathbf{H}(\mathbf{a}) - D\mathbf{G}(\mathbf{F}(\mathbf{a}))(\mathbf{F}(\mathbf{x}) - \mathbf{F}(\mathbf{a}))\| \\ + \|D\mathbf{G}(\mathbf{F}(\mathbf{a}))(\mathbf{F}(\mathbf{x}) - \mathbf{F}(\mathbf{a}) - D\mathbf{F}(\mathbf{a})(\mathbf{x} - \mathbf{a}))\|. \tag{5}$$

We analyze each term separately. Since $\mathbf{F}$ is differentiable, Theorem 7 implies that for some $\delta_1 > 0$,

$$\|\mathbf{F}(\mathbf{x}) - \mathbf{F}(\mathbf{a})\| \leq M\,\|\mathbf{x} - \mathbf{a}\|, \tag{6}$$

where $0 < \|\mathbf{x} - \mathbf{a}\| < \delta_1$; the constant $(1 + \|D\mathbf{F}(\mathbf{a})\|)$ is denoted by M. Hence,

$$\|\mathbf{F}(\mathbf{x}) - \mathbf{F}(\mathbf{a})\| < M\delta_1. \tag{7}$$

Let $\epsilon > 0$. Since $\mathbf{G}$ is differentiable at $\mathbf{F}(\mathbf{a})$, there exists $\delta_2 > 0$ such that

$$\frac{\|\mathbf{G}(\mathbf{y}) - \mathbf{G}(\mathbf{F}(\mathbf{a})) - D\mathbf{G}(\mathbf{F}(\mathbf{a}))(\mathbf{y} - \mathbf{F}(\mathbf{a}))\|}{\|\mathbf{y} - \mathbf{F}(\mathbf{a})\|} < \frac{\epsilon}{2} \tag{8}$$

whenever $0 < \|\mathbf{y} - \mathbf{F}(\mathbf{a})\| < \delta_2$. We can assume that $\delta_2 < 1$. (We took $\epsilon/2$ intead of ϵ for strategic reasons.)

If $M\delta_1 < \delta_2$, then from (7) we get $\|\mathbf{F}(\mathbf{x}) - \mathbf{F}(\mathbf{a})\| < M\delta_1 < \delta_2$. If $M\delta_1 \geq \delta_2$, then instead of taking $\mathbf{x}$ such that $0 < \|\mathbf{x} - \mathbf{a}\| < \delta_1$, take $\mathbf{x}$ that satisfies $0 < \|\mathbf{x} - \mathbf{a}\| < \delta_2/M \leq \delta_1$. In that case, (6) implies that $\|\mathbf{F}(\mathbf{x}) - \mathbf{F}(\mathbf{a})\| \leq M\|\mathbf{x} - \mathbf{a}\| < \delta_2$. Therefore, if $0 < \|\mathbf{x} - \mathbf{a}\| < \min\{\delta_2/M, \delta_2\}$, then $\|\mathbf{F}(\mathbf{x}) - \mathbf{F}(\mathbf{a})\| < \delta_2 < 1$. But in that case [read (8) with $\mathbf{y} = \mathbf{F}(\mathbf{x})$],

$$\frac{\|\mathbf{G}(\mathbf{F}(\mathbf{x})) - \mathbf{G}(\mathbf{F}(\mathbf{a})) - D\mathbf{G}(\mathbf{F}(\mathbf{a}))(\mathbf{F}(\mathbf{x}) - \mathbf{F}(\mathbf{a}))\|}{\|\mathbf{F}(\mathbf{x}) - \mathbf{F}(\mathbf{a})\|} < \frac{\epsilon}{2},$$

or

$$\|\mathbf{H}(\mathbf{x}) - \mathbf{H}(\mathbf{a}) - D\mathbf{G}(\mathbf{F}(\mathbf{a}))(\mathbf{F}(\mathbf{x}) - \mathbf{F}(\mathbf{a}))\| < \frac{\epsilon}{2}\|\mathbf{F}(\mathbf{x}) - \mathbf{F}(\mathbf{a})\| < \frac{\epsilon}{2}.$$

Now look at the last term in (5). By Theorem 7 [with $A = D\mathbf{G}(\mathbf{F}(\mathbf{a}))$], it follows that

$$\|D\mathbf{G}(\mathbf{F}(\mathbf{a}))(\mathbf{F}(\mathbf{x}) - \mathbf{F}(\mathbf{a}) - D\mathbf{F}(\mathbf{a})(\mathbf{x} - \mathbf{a}))\| \\ \leq \|D\mathbf{G}(\mathbf{F}(\mathbf{a}))\|\,\|\mathbf{F}(\mathbf{x}) - \mathbf{F}(\mathbf{a}) - D\mathbf{F}(\mathbf{a})(\mathbf{x} - \mathbf{a})\|.$$

Since $\mathbf{F}$ is differentiable, there exists $\delta_3 > 0$ such that (instead of ϵ, take $\epsilon/[2\|D\mathbf{G}(\mathbf{F}(\mathbf{a}))\|]$)

$$\|\mathbf{F}(\mathbf{x}) - \mathbf{F}(\mathbf{a}) - D\mathbf{F}(\mathbf{a})(\mathbf{x} - \mathbf{a}))\| < \frac{\epsilon}{2\|D\mathbf{G}(\mathbf{F}(\mathbf{a}))\|},$$

whenever $0 < \|\mathbf{x} - \mathbf{a}\| < \delta_3$.

Let $\delta = \min\{\delta_2/M, \delta_2, \delta_3\}$. Then for all $\mathbf{x}$, $0 < \|\mathbf{x} - \mathbf{a}\| < \delta$, (5) is less than

$$\frac{\epsilon}{2} + \|D\mathbf{G}(\mathbf{F}(\mathbf{a}))\|\,\frac{\epsilon}{2\|D\mathbf{G}(\mathbf{F}(\mathbf{a}))\|} = \epsilon,$$

and we are done. ◀

THEOREM 4.1 Equality of Mixed Partial Derivatives

Let f be a real-valued function of m variables $x_1, \ldots, x_m$ with continuous second-order partial derivatives (i.e., of class C^2). Then

$$\frac{\partial^2 f}{\partial x_i \partial x_j} = \frac{\partial^2 f}{\partial x_j \partial x_i},$$

for all $i, j = 1, \ldots, m$.

PROOF: Assume that $f = f(x, y)$; the general case is proven analogously. Consider the expression

$$G(h, k) = f(x + h, y + k) - f(x, y + k) - f(x + h, y) - f(x, y) = \overline{f}(x + h) - \overline{f}(x),$$

where $\overline{f}(x) = f(x, y + k) - f(x, y)$. The function $\overline{f}$ is a function of one variable; by the Mean Value Theorem (see Theorem 2), exists a number c_x between $x + h$ and x such that

$$\overline{f}(x + h) - \overline{f}(x) = \overline{f}'(c_x)h = \left(\frac{\partial f}{\partial x}(c_x, y + k) - \frac{\partial f}{\partial x}(c_x, y)\right)h.$$

Now view $\partial f/\partial x$ as a function of its second variable and use the Mean Value Theorem again; so there exists a number c_y between $y + k$ and y such that

$$\frac{\partial f}{\partial x}(c_x, y + k) - \frac{\partial f}{\partial x}(c_x, y) = \frac{\partial}{\partial y}\left(\frac{\partial f}{\partial x}\right)(c_x, c_y)k.$$

It follows that

$$G(h, k) = \frac{\partial^2 f}{\partial y \partial x}(c_x, c_y)hk \quad \text{and} \quad \frac{\partial^2 f}{\partial y \partial x}(c_x, c_y) = \frac{G(h, k)}{hk}.$$

As $h \to 0$ and $k \to 0$, $c_x \to x$ and $c_y \to y$, and therefore,

$$\frac{\partial^2 f}{\partial y \partial x}(x, y) = \lim_{k,h \to 0} \frac{G(h, k)}{hk}.$$

Starting with $G(h, k) = \overline{\overline{f}}(y + k) - \overline{\overline{f}}(y)$, where $\overline{\overline{f}}(y) = f(x + h, y) - f(x, y)$ and proceeding as before, we arrive at

$$\frac{\partial^2 f}{\partial x \partial y}(x, y) = \lim_{k,h \to 0} \frac{G(h, k)}{hk}.$$

It follows that

$$\frac{\partial^2 f}{\partial y \partial x}(x, y) = \frac{\partial^2 f}{\partial x \partial y}(x, y).$$

APPENDIX B

Answers to Odd-Numbered Exercises

CHAPTER 1

Section 1.1

1. For example, $\mathbf{v} = \mathbf{i}$ and $\mathbf{w} = 4\mathbf{i}$ satisfy $\|\mathbf{v} + \mathbf{w}\| = \|\mathbf{v}\| + \|\mathbf{w}\|$; if $\mathbf{v} = \mathbf{i}$ and $\mathbf{w} = \mathbf{i} + \mathbf{j}$, then $\|\mathbf{v} + \mathbf{w}\| = \sqrt{5} < \|\mathbf{v}\| + \|\mathbf{w}\| = 1 + \sqrt{2}$; vectors $\mathbf{v} = -\mathbf{i} + \mathbf{j}$ and $\mathbf{w} = \mathbf{i}$ are such that $\|\mathbf{v} + \mathbf{w}\| = 1 < (\|\mathbf{v}\| + \|\mathbf{w}\|)/2 = (\sqrt{2} + 1)/2$.
3. Take $\mathbf{v} = (v_1, v_2) \in \mathbb{R}^2$ (same proof works in $\mathbb{R}^n$, $n \geq 2$); if $\|\mathbf{v}\| = 0$, then $\sqrt{v_1^2 + v_2^2} = 0$ implies $v_1 = v_2 = 0$. If $\mathbf{v} = (0, 0)$, then $\|\mathbf{v}\| = 0$.
5. $(\sqrt{8}, \pi/4)$, $(\sqrt{8}, 3\pi/4)$, $(\sqrt{8}, 7\pi/4)$, $(\sqrt{8}, 5\pi/4)$.
7. The difference of lengths of two sides in a triangle is smaller than or equal to the length of the third side; equality holds in the case where the two vectors are parallel and point in the same direction.
9. The smallest value of $\|\mathbf{v} + \mathbf{w}\|$ is 0, and its largest value is 2.
11. $\sqrt{5}$.
13. 1.
15. Let $A(a_1, a_2)$, $B(b_1, b_2)$ be points in $\mathbb{R}^2$. Construct the right triangle whose hypotenuse is $\overline{AB}$ and the sides are parallel to the coordinate axes and use the Pythagorean Theorem (consider the cases when A and B lie on the same horizontal or on the same vertical line separately). Use a similar approach for a proof in $\mathbb{R}^3$.
17. Representatives of $\mathbf{v}$ are $\overrightarrow{A_1B_1}$, $\overrightarrow{A_2B_2}$, $\overrightarrow{A_3B_3}$, and $\overrightarrow{A_4B_4}$, where $A_1(0, 1, 1)$, $B_1(0, 3, 0)$, $A_2(0, 3, 0)$, $B_2(0, 5, -1)$, $A_3(8, 9, -4)$, $B_3(8, 11, -5)$, $A_4(10, -1, 4)$, and $B_4(10, 1, 3)$.
19. $\mathbf{a} - 2\mathbf{b} = 8\mathbf{i} - \mathbf{j} - \mathbf{k}$, $\mathbf{a} - \mathbf{c}/\|\mathbf{c}\| = \mathbf{i} - \mathbf{j} + \mathbf{k}$, and $3\mathbf{a} + \mathbf{c} - \mathbf{j} + \mathbf{k} = 8\mathbf{i} - 4\mathbf{j} + 4\mathbf{k}$; the unit vector in the direction of $\mathbf{b} + 2\mathbf{a}$ is $(\mathbf{i} - 2\mathbf{j} + 3\mathbf{k})/\sqrt{14}$.
21. $-3\mathbf{i} = 3(\cos\pi\,\mathbf{i} + \sin\pi\,\mathbf{j})$, $\mathbf{i}/2 - \mathbf{j} \approx 1.11803(\cos 5.17604\,\mathbf{i} + \sin 5.17604\,\mathbf{j})$, and $\mathbf{i} - 4\mathbf{j} \approx 4.12311(\cos 4.95737\,\mathbf{i} + \sin 4.95737\,\mathbf{j})$.
23. Let $\mathbf{v} = (v_1, v_2, v_3)$ and compute both sides using definitions of vector operations. The fact that they are equal follows from properties of real numbers.

Section 1.2

1. $\mathbf{l}(t) = (1 - 3t, 3 - 15t)$, $t \in \mathbb{R}$; to get infinitely many parametrizations, replace $3\mathbf{v}$ by $m\mathbf{v}$, $m \neq 0$.
3. In all three cases, $\mathbf{l}(t) = (1 - 3t, 1 + 3t)$. For the line, $t \in \mathbb{R}$; for the half-line, $t \geq 0$; for the line segment, $0 \leq t \leq 1$.
5. $\mathbf{l}(t) = (2, 1) + t(-3, 4)$, $t \in \mathbb{R}$; $\mathbf{l}'(t) = (2, 1) + t(-3, 4)$, $t \geq 0$.
7. Let $\mathbf{p}$ be the position vector of a point in the rectangle or on its boundary; then $\mathbf{p} = (1 + 3\alpha)\mathbf{i} + (1 + \beta)\mathbf{j}$, where $0 \leq \alpha, \beta \leq 1$.
9. Relative position is $\pm(\mathbf{i} + 2\mathbf{j} - 3\mathbf{k})$; distance is $\sqrt{14}$.
11. The particle will cross the xy-plane at $(15, 2, 0)$, when $t = 4$.
13. $\mathbf{F} = 5\sqrt{3}\mathbf{i} + 5\mathbf{j}$.
15. $(1, -1/2)$.
17. Mass of the third stone is 4 kg.
19. The center of mass is located on the axis of symmetry of the triangle; that is, on the line through one of its vertices perpendicular to the opposite side, $\sqrt{3}/3$ units away from that side.
21. Approximately $5.861\mathbf{i} + 5.439\mathbf{j}$.

Section 1.3

1. Write $\mathbf{v} = (v_1, \ldots, v_n)$ and $\mathbf{w} = (w_1, \ldots, w_n)$ and use the definition of the dot product and properties of real numbers.
3. Let $\mathbf{u} = (u_1, u_2, u_3)$, $\mathbf{v} = (v_1, v_2, v_3)$, and $\mathbf{w} = (w_1, w_2, w_3)$, and compute both sides of each formula using the definition of the dot product. The identities follow from the properties of real numbers.
5. Write $\mathbf{a} = a_\mathbf{u}\mathbf{u} + a_\mathbf{v}\mathbf{v} + a_\mathbf{w}\mathbf{w}$, where $a_\mathbf{u}$, $a_\mathbf{v}$, and $a_\mathbf{w}$ are real numbers. To get $a_\mathbf{u}$, compute the dot product of $\mathbf{a}$ with $\mathbf{u}$ and use the assumptions. The remaining two coefficients are obtained analogously.
7. Approximately 1.6515, 1.1867, and 0.3034 rad.
9. The dot product of the two vectors is zero.
11. All three angles are the same, approximately equal to 0.9553 rad.
13. Use the formula for the angle between two vectors; the assumption guarantees that the two angles are equal.
15. $\mathbf{i} + 2\mathbf{j} = \frac{3}{2}(\mathbf{i} + \mathbf{j}) - \frac{1}{2}(\mathbf{i} - \mathbf{j})$.
17. $3x - z + 2 = 0$.
19. $3x - y + 7 = 0$.
21. $3x + 2y - 8z = 0$.
23. Orient the sides $\mathbf{p}_1, \ldots, \mathbf{p}_n$ of a polygon so that the terminal point of one side is the initial point of the neighboring one. The work of $\mathbf{F}$ is $W = \mathbf{F} \cdot \mathbf{p}_1 + \cdots + \mathbf{F} \cdot \mathbf{p}_n$. Since $\mathbf{p}_1 + \cdots + \mathbf{p}_n = \mathbf{0}$, it follows that $W = 0$.
25. $\mathbf{i} = (1/6)\mathbf{u} + (1/6)\mathbf{w}$.
27. Let $\mathbf{a} \in \mathbb{R}^3$ be orthogonal to $\mathbf{u}$, $\mathbf{v}$, and $\mathbf{w}$. Using $\mathbf{a} = a_\mathbf{u}\mathbf{u} + a_\mathbf{v}\mathbf{v} + a_\mathbf{w}\mathbf{w}$ and the assumptions, show that $\mathbf{a} = \mathbf{0}$.
29. $-x + 3y + 3z + 5 = 0$.
31. If $C \neq 0$, then $\mathbf{p} = \mathbf{a} + t\mathbf{v} + s\mathbf{w}$, where $\mathbf{a} = (0, 0, -D/C)$, $\mathbf{v} = (1, 0, -A/C)$, and $\mathbf{w} = (0, 1, -B/C)$. Similar parametric equations are obtained when $A \neq 0$ or $B \neq 0$.

Section 1.4

1. $\begin{bmatrix} -16 & 10 \\ 8 & -16 \end{bmatrix}$.
3. Not defined.
5. $\begin{bmatrix} 1 & -3 \\ 7 & -13 \end{bmatrix}$.
7. $\begin{bmatrix} 0 & 25 & -20 \\ 16 & -33 & 28 \\ -8 & -71 & 56 \end{bmatrix}$.
9. $\begin{bmatrix} 59 & 42 \\ -98 & 255 \end{bmatrix}$.
11. $\begin{bmatrix} -5/3 & 7/3 \\ -8/3 & 1/3 \end{bmatrix}$.
13. $\begin{bmatrix} 0 & 0 \\ -10 & -1 \end{bmatrix}$.
15. $\mathbf{F}_C$ assigns to a vector its symmetric image with respect to the line $y = x$; $\mathbf{F}_{I_2}$ maps a vector into itself.
17. -7.
19. r
21. -20.
23. Let $A = [a_{ij}]$, $B = [b_{ij}]$, $(i, j = 1, 2)$, and show that $\det(AB)$ and $\det(A)\det(B)$ are equal to the same number. Then $\det(AB) = \det(A)\det(B) = \det(B)\det(A) = \det(BA)$.
25. Let $A = \begin{bmatrix} a_{11} & a_{12} \\ a_{21} & a_{22} \end{bmatrix}$; then $B = \begin{bmatrix} a_{11} & a_{12} \\ \alpha a_{11} & \alpha a_{12} \end{bmatrix}$ and $\det(B) = 0$.
27. Every time we interchange two rows or two columns in a matrix, its determinant changes sign; consequently, $\det(B) = \det(A)$.
29. The expression for $\det(A)$ consists of six terms. Each term contains *exactly one* entry from each row of A. If we multiply all elements in one row of A by α, each of the six terms will contain the factor α, and therefore, $\det(B) = \alpha\det(A)$.
31. Let $A = \begin{bmatrix} 1 & 2 \\ 0 & 1 \end{bmatrix}$ and $B = \begin{bmatrix} 1 & 0 \\ 1 & 2 \end{bmatrix}$. Then $(AB)^t = \begin{bmatrix} 3 & 1 \\ 4 & 2 \end{bmatrix}$ and $A^t B^t = \begin{bmatrix} 1 & 1 \\ 2 & 4 \end{bmatrix}$. To prove $(AB)^t = B^t A^t$, write $A = \begin{bmatrix} a_{11} & a_{12} \\ a_{21} & a_{22} \end{bmatrix}$ and $B = \begin{bmatrix} b_{11} & b_{12} \\ b_{21} & b_{22} \end{bmatrix}$, and verify that the two sides are equal.
33. $\mathbf{F}_A(x\mathbf{i} + y\mathbf{j}) = y\mathbf{j}$; $\mathbf{F}_A$ is the orthogonal projection onto the y-axis. $\mathbf{F}_B(x\mathbf{i} + y\mathbf{j}) = 2(x\mathbf{i} + y\mathbf{j})$; $\mathbf{F}_B$ magnifies the vector by the factor of 2, that is, it assigns the vector of the same direction and orientation, and twice the magnitude of the original vector [and $\mathbf{F}_B(\mathbf{0}) = \mathbf{0}$]. $\mathbf{F}_C(x\mathbf{i} + y\mathbf{j}) = x\mathbf{i} - y\mathbf{j}$; $\mathbf{F}_C$ assigns to each vector its mirror image with respect to the x-axis.
35. $\mathbf{F}_{A+B} = \mathbf{F}_A + \mathbf{F}_B$; $\mathbf{F}_{2AB} = 2(\mathbf{F}_A \circ \mathbf{F}_B)$ ($\circ$ denotes composition); $\mathbf{F}_{7B} = 7\mathbf{F}_B$.

Section 1.5

1. $\mathbf{v} \times \mathbf{w} = 12\mathbf{i} - 5\mathbf{j} + 8\mathbf{k}$, $(\mathbf{v} + \mathbf{w}) \times \mathbf{k} = 4\mathbf{i} - 3\mathbf{j}$, $(2\mathbf{v} - \mathbf{w}) \times (\mathbf{v} + \mathbf{w}) = 36\mathbf{i} - 15\mathbf{j} + 24\mathbf{k}$, and $\mathbf{i} \times (\mathbf{v} \times \mathbf{w}) = -8\mathbf{j} - 5\mathbf{k}$.
3. Expand both sides using the definition of the cross product.
5. $\pm(\mathbf{i} - \mathbf{j})/\sqrt{2}$.
7. For example, $\mathbf{u} = \mathbf{i}$, $\mathbf{v} = \mathbf{i}$, and $\mathbf{w} = \mathbf{j}$.
9. 1.
11. Prove the identity first for $\mathbf{a} = \mathbf{i}$, $\mathbf{a} = \mathbf{j}$, and $\mathbf{a} = \mathbf{k}$. Then use the properties of the dot and cross products (Theorems 1.2 and 1.9) to prove it in the general case.
13. Let $\mathbf{v}$ be the vector from $(0, 2)$ to $(3, 2)$ and let $\mathbf{w}$ be the vector from $(0, 2)$ to $(1, 1)$. The area of the parallelogram spanned by $\mathbf{v}$ and $\mathbf{w}$ is $\|\mathbf{v} \times \mathbf{w}\| = 3$.
15. The four points do not belong to the same plane.
17. Let $\mathbf{v}$ be the vector from $(1, 0, 0)$ to $(-3, 2, 2)$, and let $\mathbf{w}$ be the vector from $(1, 0, 0)$ to $(3, -1, -1)$; show that $\mathbf{v} \times \mathbf{w} = \mathbf{0}$.
19. $y = 1$.
21. $-x + z + 2 = 0$.
23. $\sqrt{8/3}$.
25. The magnitude of the torque is the largest on the equator, and the smallest on the North Pole and on the South Pole.
27. The magnitude of $\mathbf{v}$ is the largest on the edge of the disk and the smallest at its center.

Chapter Review

True/False Quiz

1. True.
3. False.
5. False.
7. False.
9. False.
11. True.
13. True.

Review Exercises and Problems

1. The sum of the squares of lengths of the diagonals in a parallelogram is equal to the sum of the squares of lengths of its four sides.
3. $\cos\alpha = \mathbf{v} \cdot \mathbf{i}/\|\mathbf{v}\|$, $\cos\beta = \mathbf{v} \cdot \mathbf{j}/\|\mathbf{v}\|$, and $\cos\gamma = \mathbf{v} \cdot \mathbf{k}/\|\mathbf{v}\|$; $\mathbf{v} = \|\mathbf{v}\|(\cos\alpha\mathbf{i} + \cos\beta\mathbf{j} + \cos\gamma\mathbf{k})$.
5. $11/\sqrt{6} \approx 4.491$.
7. If $A = \begin{bmatrix} 0 & 1 \\ 0 & 0 \end{bmatrix}$, then $A^2 = \begin{bmatrix} 0 & 0 \\ 0 & 0 \end{bmatrix}$. Assume that $A = \begin{bmatrix} a & b \\ c & d \end{bmatrix}$, and show that $AA^t = 0$ implies that $a^2 + b^2 = 0$ and $c^2 + d^2 = 0$.
9. Equation of the plane Π is $8x - 10y - 7z + 12 = 0$. Planes perpendicular to Π that contain the given point are represented by $(t - 2s)x - (2t + 3s)y + (4t + 2s)z + 3t + 8s = 0$, where $t, s \in \mathbb{R}$ and at least one of t or s is not zero.
11. Both have the volume of 1. The tips of both $\mathbf{c}$ and $\mathbf{a} + \mathbf{c}$ are at the same distance from the plane spanned by $\mathbf{a}$ and $\mathbf{b}$. Thus, the two parallelepipeds have the same height.
13. Interpret the determinant as a scalar triple product.
15. (a) For example, $A = \begin{bmatrix} 2 & -1 \\ 4 & -2 \end{bmatrix}$ and $\mathbf{v} = \begin{bmatrix} 3 \\ 6 \end{bmatrix}$. (b) $\mathbf{F}_A(\mathbf{v}) = \mathbf{0}$, written in components, gives a system of equations for the components of $\mathbf{v}$. Use the fact that $\det(A) \neq 0$ to show that the only solution is $\mathbf{v} = 0$.
17. (a) Car A: $(10 - 50t)\mathbf{i}$; car B: $80t\mathbf{j}$. (b) $d(t) = 10\sqrt{89t^2 - 10t + 1}$. (d) Approximately 8.47998 km, reached when $t \approx 0.05618$ h. (e) After approximately 2.17427 h.

CHAPTER 2

Section 2.1

1. Components of $\mathbf{G}(\mathbf{r})$ are $2x\sqrt{x^2 + y^2 + z^2}$, $2y\sqrt{x^2 + y^2 + z^2}$, $2z\sqrt{x^2 + y^2 + z^2}$. Component functions of $\mathbf{F}(\mathbf{r})$ are $-GMmx(x^2 + y^2 + z^2)^{-3/2}$, $-GMmy(x^2 + y^2 + z^2)^{-3/2}$, and $-GMmz(x^2 + y^2 + z^2)^{-3/2}$.
3. Domain is $\mathbb{R}^2$ without the circle of radius 1 centered at the origin.
5. Defined whenever $y \neq 0$.
7. The range of $f(x, y) = 3x + y - 7$ is $\mathbb{R}$. If $a \neq 0$ and/or $b \neq 0$, the range of $f(x, y) = ax + by + c$ is

$\mathbb{R}$. If $a=b=0$, the range consists of the single number, i. e., it is equal to $\{c\}$.

9. Domain is $\mathbb{R}^3$; range is $(0, 1]$.

11. Domain is the set $\mathbb{R}^2 - \{(0, 0)\}$; range is $[0, 3]$.

13. Domain is the set $\{(x, y) \mid x > 0 \text{ and } y > 0\}$; range is $\{(c, d) \mid c \in \mathbb{R} \text{ and } d \geq 0\}$.

15. Domain is $\mathbb{R}^2$; range is $[0, \infty)$.

17. Domain is the set $\{(x, y) \mid x \neq 0,\ y \in \mathbb{R}\}$; range is $\mathbb{R}$.

19. (a) $W(10, 0) = 19.335$, $W(-10, 0) = 6.905$, $W(-20, 0) = 0.69$; these values would imply that, when there is no wind, we would feel warmer than it actually is. (b) For instance, we might require that $v \geq 5$, and consider a reasonable interval for T, say, $-70 \leq T \leq 10$. (c) At constant temperature, the stronger the wind, the colder we will feel; thus, the wind chill decreases. (d) With an increase in temperature (assuming constant wind speed), we will feel warmer, that is, the wind chill is increasing.

21. Domain is $\mathbb{R}^2$, range is $[-1, 1] \times [-1, 1]$.

23. $((x-1)\mathbf{i} + (y-2)\mathbf{j} + (z+2)\mathbf{k}) / \sqrt{(x-1)^2 + (y-2)^2 + (z+2)^2}$.

25. $\mathbf{F}$ is the unit vector field in the radial direction, pointing away from the origin.

27. $\mathbf{F}(\mathbf{x})$ is the vector field of Exercise 25 moved so that the origin is located at $(1, 2)$.

29. Since $\mathbf{F}(x, y)$ does not depend on y, its value at all points on a vertical line is the same (vectors not drawn to scale).

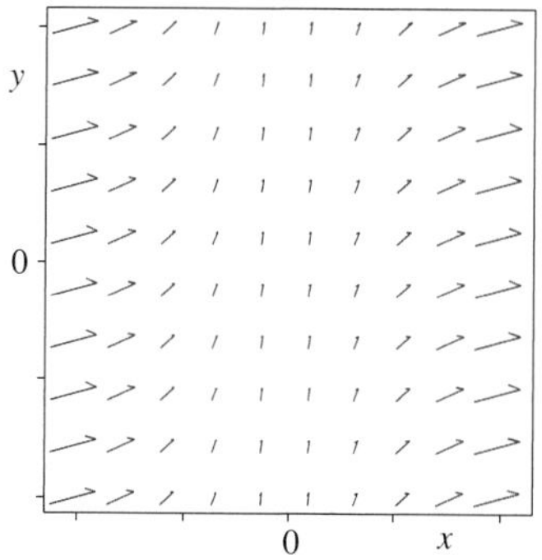

31. $\mathbf{F}$ assigns to a vector its mirror image with respect to the x-axis (vectors not drawn to scale).

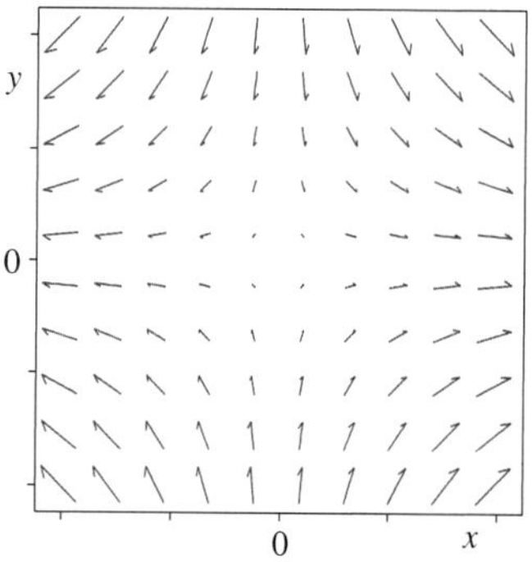

33. (a) Show that $\mathbf{F}_1 \cdot \mathbf{r} = 0$, where $\mathbf{r} = x\mathbf{i} + y\mathbf{j}$. (b) $\|\mathbf{F}_1\| = 1/\sqrt{x^2+y^2}$. (c) See the figure below (vectors not drawn to scale).

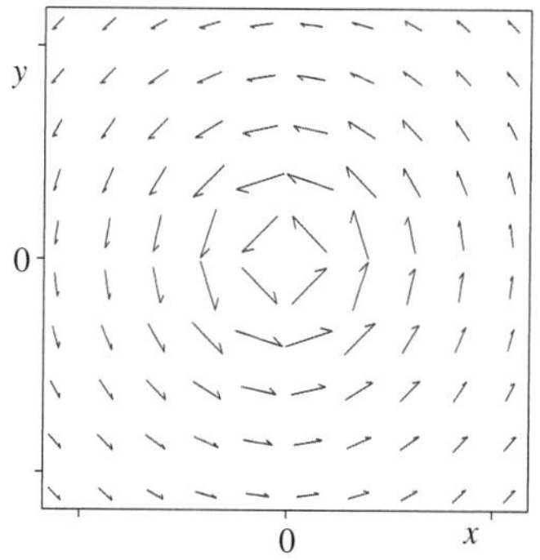

Section 2.2

1. There are no level curves of value $c \neq 1$; the "level curve" (or, better yet, the level set) of value $c = 1$ is the xy-plane.

3. There are no level curves of value $c > 3$; the "level curve" of value $c = 3$ is the origin; if $c < 3$, the level curve is a circle centered at the origin, of radius $\sqrt{3-c}$.

5. There are no level curves of value $c \leq 0$; the level curve of value $c > 0$ is the hyperbola $y = (\ln c)/x$ if $c \neq 1$ and the x-axis if $c = 1$.

7. The level curve of value $c \neq 0$ is the line $y = x/c$ with the origin removed. If $c = 0$, the level curve is the y-axis without the origin.

9. The level curve of value c is the graph of $y = \sin x$ moved up c units if $c \geq 0$ or down $|c|$ units if $c < 0$.

11. The level curve of value $c > 0$ is the hyperbola with asymptotes $y = \pm x$ and x-intercepts at $(\pm\sqrt{c}, 0)$; the level curve of value $c < 0$ is the hyperbola with asymptotes $y = \pm x$ and y-intercepts at $(0, \pm\sqrt{-c})$; the level curve of value $c = 0$ consists of two lines, $y = x$ and $y = -x$, that intersect at the origin.

13. (a) $f(x, y) = -2x - 2y - 8$. (b) $f(x, y) = x - y/2 + 2$.

15. The level curve $c = 0$ is the set $\{(x, y) \mid x^2 + y^2 \geq 1\}$. For $0 < c < 1$, the level curve is the circle of radius $\sqrt{1-c}$ centered at the origin. When $c = 1$, the level curve consists of the single point $(0, 0)$. If $c < 0$ or $c > 1$, the level set is empty.

17. There are no level curves if $c < 0$ or if $c > 1$. If $c = 1$, the level curve is the point $(0, 0)$. For $0 \leq c < 1$, we get an ellipse centered at the origin with semiaxes $\sqrt{1-c^2}/3$ and $\sqrt{1-c^2}/2$.

19. The level surface of value c is the parabolic sheet $y = x^2 + c$.

21. There are no level surfaces for $c < 1$; the "level surface" of value $c = 1$ is the origin; if $c > 1$, the level surface is the ellipsoid with semiaxes $\sqrt{c-1}/2$ (in the x-direction), $\sqrt{c-1}/2$ (in the y-direction), and $\sqrt{c-1}$ (in the z-direction), symmetric with respect to the origin.

23. The level surface of value c is the plane parallel to the yz-plane that crosses the x-axis at $((c-4)/3, 0, 0)$.

25. (a) In $\mathbb{R}^2$, $x = a$ is the line, parallel to the y-axis, going through $(a, 0)$. In $\mathbb{R}^3$, $x = a$ is the plane, parallel to the yz-plane, going through $(a, 0, 0)$. (b) In $\mathbb{R}^2$, $y = b$ is the line, parallel to the x-axis, going through $(0, b)$. In $\mathbb{R}^3$, $y = b$ is the plane, parallel to the xz-plane, going through $(0, b, 0)$.

27. (a) Sphere, radius a, center at origin. (b) Sphere, radius a, center at (m, n, p).

29. Upper hemisphere (i.e., the part of the sphere centered at the origin of radius $\sqrt{2}$, above and in the xy-plane).

31. Cylinder of radius 3 whose axis of (rotational) symmetry is the z-axis.

33. Plane, parallel to the xy-plane, 4 units above it.

35. Paraboloid, symmetric with respect to the z-axis, turned upside-down, with the vertex at $(0, 0, 2)$.

37. Inverted cone, symmetric with respect to the z-axis, whose vertex is at the origin.

39. Level curves are part of the line that goes through the origin of slope Rn/c in the third quadrant, if $c < 0$; part of the line that goes through the origin of slope Rn/c in the first quadrant, if $c > 0$. If $c = 0$, the level curve is the (positive) V-axis.

41. (a) y-axis. (b) If $|c| < 1/4$, the level curve is a circle; when $c = \pm 1/4$, the level curve collapses to $(\pm 2, 0)$.

43. $f(x, y) = 3y$.

45. $f(x, y, z) = 2x + 3y - 4z + d$, where d is any real number.

Section 2.3

1. (a) $f(x)$ is not defined at a. (b) $f(x)$ is defined at a, and $f(a) \neq L$. (c) $f(x)$ is defined at a, and $f(a) = L$.

3. The radius has to be smaller than $\sqrt{-\ln 0.99}$.

5. See the figure below.

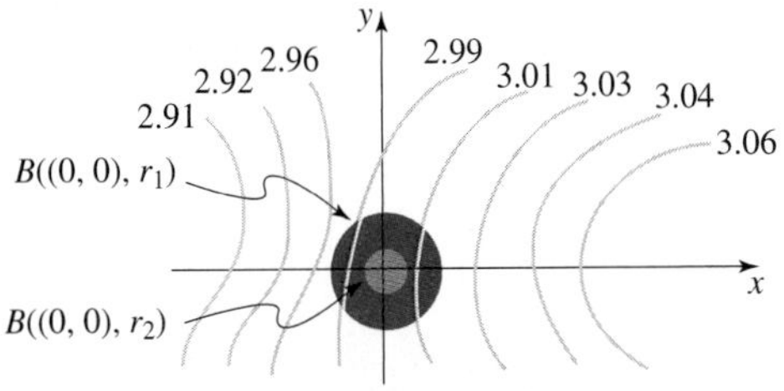

7. Choose approaches $x = 0$ and $y = 0.2x$.

9. 0.

11. Does not exist.

13. 0.

15. Does not exist.

17. Does not exist.

19. Does not exist.

21. Does not exist.

23. $\mathbf{F}$ is not continuous at $(0, 0, 0)$.

25. $f(x, y)$ is continuous for all $(x, y) \in \mathbb{R}^2$.

27. f is not continuous at $\mathbf{x} = \mathbf{x}_0$.

29. It is impossible to define $f(0, 0)$ so as to make f continuous at $(0, 0)$.

31. All points in U are interior points; the boundary of U consists of the x-axis and the y-axis.

33. Let $\mathbf{x} = (x_1, \ldots, x_m)$ and $\mathbf{a} = (a_1, \ldots, a_m)$; then $f(\mathbf{x}) = x_1a_1 + \cdots + x_ma_m$ is a first degree polynomial and therefore continuous.

35. Defining $\mathbf{F}(0, 2) = (2, 2)$ will make $\mathbf{F}$ continuous at $(0, 2)$.

37. Defining $\mathbf{F}(1, 0) = (\sin 1, 0, 1)$ will make $\mathbf{F}$ continuous at $(1, 0)$.

39. The limit is 0. Define $f(-1, -2) = 0$.

Section 2.4

1. U is open in $\mathbb{R}^2$.
3. U is not open in $\mathbb{R}^2$.
5. U is open in $\mathbb{R}^3$.
7. (a) $(\partial f/\partial x)(5, 3) < 0$. (b) $(\partial f/\partial x)(10, 5) > (\partial f/\partial x)(10, 3)$.
9. $f_x = yx^{y-1} + y/x$; $f_y = x^y \ln x + \ln x$.
11. $f_x = 1/(x+y+z^2)$; $f_z = 2z/(x+y+z^2)$.
13. $f_x = e^{xy} \sin y(y \cos x - \sin x)$; $f_y = e^{xy} \cos x(x \sin y + \cos y)$.
15. $\partial f/\partial x_i = x_i/\sqrt{x_1^2 + \cdots + x_m^2}$.
17. $f_x = xe^{-x^2}$; $f_y = 0$.
19. $z_x = f'(x)$; $z_y = g'(y)$.
21. $z_x = f'(x)/g(y)$; $z_y = -f(x)g'(y)/g(y)^2$.
23. In the northern direction.
25. (a) $f_y(2, 3) = 24e^{-22}$. (b) The slope is $24e^{-22}$. (c) $f_y(2, 3)$ is the slope of the tangent line (at the point where $y = 3$) to the curve $z = f(2, y)$ that is the intersection of the graph of $z = f(x, y)$ and the vertical plane $x = 2$.
27. $\begin{bmatrix} 0 & 1 \\ 1 & 0 \\ 0 & 0 \end{bmatrix}$.
29. $\begin{bmatrix} 1 & 1 & 0 \\ 2 & 2 & 1 \end{bmatrix}$.
31. $[2a_1 \quad 2a_2 \quad 2a_3]$.
33. It suffices to show that $-\nabla(1/\|\mathbf{r}\|) = \mathbf{r}/\|\mathbf{r}\|^3$. To prove this, write $\mathbf{r} = x\mathbf{i} + y\mathbf{j} + z\mathbf{k}$ and verify that the two sides are equal.
35. Interpreted as a vector, $\nabla f(\mathbf{x}) = \mathbf{x}/\|\mathbf{x}\|$; its domain is $\mathbb{R}^3 - \{(0, 0, 0)\}$.
37. $L_{(2,-1)}(x, y) = \ln 4 + \frac{3}{4}(x-2) + \frac{1}{2}(y+1)$.
39. $L_{(3,2)}(x, y) = 4x - y - 2$.
41. $L_{(0,1,1)}(x, y, z) = (y+z)/\sqrt{2}$.
43. Use the fact that, for (x, y) near $(2, 3)$, $f(x, y) \approx L_{(2,3)}(x, y)$.
45. (a) Compute the limit using the approach $y = b$, $x \to a$, and use the definition of the partial derivative $\partial f/\partial x$. (b) Use the approach $x = a$, $y \to b$.
47. Use the fact that $f(x, y) = \sqrt{x^3 + y^3} \approx L_{(1,2)}(x, y)$, where $L_{(1,2)}(x, y)$ is the linear approximation of f at $(1, 2)$; it follows that $f(0.99, 2.02) \approx 3.035$; the calculator value is 3.0352441 (accuracy depends on the calculator used).
49. $7.95 \ln 1.02 \approx 0.16$; the calculator value is 0.1574309.
51. Approximate $f(x, y) = \int_x^y e^{-t^2} dt$ using the linear approximation $L_{(1,1)}(x, y) = e^{-1}(y - x)$; it follows that $f(0.995, 1.02) \approx L_{(1,1)}(0.995, 1.02) = 0.025e^{-1}$.
53. $\Delta f \approx 0.2950836$; $df \approx 0.2456344$.
55. $\Delta f \approx -0.080384$. $df = -0.08$.
57. $dP = (Rn/V)dT - (RnT/V^2)dV$. The coefficient Rn/V of dT is positive, since the increase in temperature (keeping the volume V fixed) will increase the pressure. The coefficient $-RnT/V^2$ of dV is negative, since the increase in volume (keeping the temperature T fixed) will decrease the pressure.
59. The error in computing the volume V is approximately 4.5% (i.e., $\Delta V \approx 0.045V$).
61. Approximately 5%.
63. $2x + 4y + z - 11 = 0$.
65. f_x is given by $f_x(x, y) = 2xy/(x^2 + y^2)$ if $(x, y) \neq (0, 0)$ and $f_x(0, 0) = 0$. Since the limit of f_x as $(x, y) \to (0, 0)$ does not exist, f_x is not continuous at $(0, 0)$.
67. All partial derivatives of the components of $\mathbf{F}$ exist at $(0, 0)$, and $D\mathbf{F} = \begin{bmatrix} 1 & 0 \\ 0 & 0 \end{bmatrix}$. Prove $\lim_{(x,y)\to(0,0)} \|\mathbf{F}(x, y) - \mathbf{F}(0, 0) - D\mathbf{F}(0, 0) \cdot (x, y)\| / \|(x, y)\| = 0$ using polar coordinates.
69. (a) No. (b) No. (c) No.

Section 2.5

1. $\mathbf{c}(t) = (3 - 3t, 1 + 4t, -2 + 2t)$, $t \in \mathbb{R}$.
3. $\mathbf{c}(t) = (\sqrt{5} \cos t, \sqrt{5} \sin t)$, $t \in [0, 2\pi]$.
5. $\mathbf{c}(t) = (-2 + 3\cos t, -1 + \sin t)$, $t \in [0, 2\pi]$.
7. $\mathbf{c}(t) = (1 + t^2, t)$, $0 \le t \le 1$.
9. $\mathbf{c}(t) = (\cos^3 t, \sin^3 t)$, $t \in [0, 2\pi]$.
11. $\mathbf{c}(t)$ is the part of the parabola $y = (x+3)^2$ between $(-3, 0)$ and $(-1, 4)$, oriented from $(-3, 0)$ to $(-1, 4)$.
13. $\mathbf{c}(t)$ is the right branch of the hyperbola $x^2 - y^2 = 4$; as $t \to \infty$, both components of $\mathbf{c}(t)$ approach $+\infty$; see the figure.

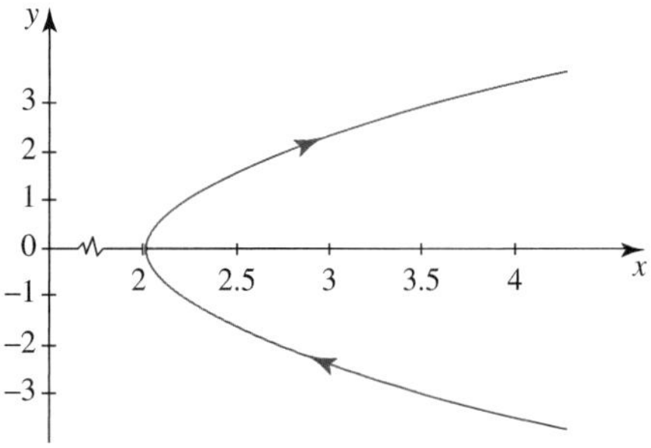

15. $\mathbf{c}(t)$ is the part of the graph of $y = e^{3x}$, oriented from (0, 1) to (ln 2, 8).
17. $\mathbf{c}(t)$ is the circle of radius 1 in the plane $z = 3$ with its center at (0, 0, 3). The initial and terminal points are at (1, 0, 3), and the orientation is counterclockwise, as seen from high up the z-axis [say, from the point (0, 0, 10)].
19. $\mathbf{c}(t)$ represents a circular helix with the initial point (1, 0, 0) and the terminal point $(-1, 0, \pi^3)$.
21. $\mathbf{c}(t)$ is the part of the graph of $y = \arctan x$ from $(-1, -\pi/4)$ to $(1, \pi/4)$.
23. $\mathbf{c}(t)$ is the part of the right branch of the hyperbola $x^2 - y^2 = 4$, from (2, 0) to (5/2, 3/2).
25. $\mathbf{c}(t)$ is a clockwise spiral—it starts at (0, 1) and ends at $(0, e^{\pi/2})$; see the figure.

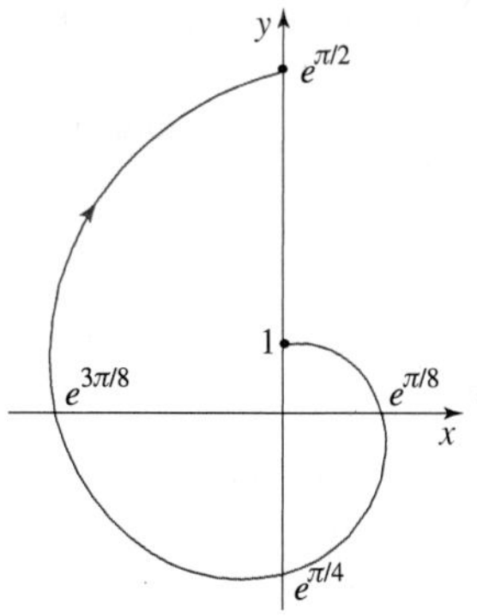

27. In all four cases, $x^2 + y^2 = 4$; $\mathbf{c}_1$, $\mathbf{c}_3$, and $\mathbf{c}_4$ are oriented clockwise, and $\mathbf{c}_2$ is oriented counterclockwise; $\mathbf{c}_1$, $\mathbf{c}_2$, and $\mathbf{c}_4$ wind around the circle once, while $\mathbf{c}_3$ winds around it three times. To get new parametrizations, adjust the existing ones, like $\mathbf{c}_5(t) = (2\cos 5t, 2\sin 5t)$, $t \in [0, 4\pi]$, or invent more "exotic" ones, like $\mathbf{c}_6(t) = (\sqrt{2}(\sin t + \cos t), \sqrt{2}(\sin t - \cos t))$, $t \in [0, 2\pi]$.
29. For example, $\mathbf{c}(t) = ((t+1)^{2/3}, 2(t+1)^{2/3})$ is not differentiable at $t = -1$.
31. $\mathbf{c}(t)$ is continuous, not differentiable, and not C^1.
33. $\mathbf{c}(t)$ is C^1, hence also continuous and differentiable.
35. $\mathbf{c}'(t) = \begin{bmatrix} (t+1)e^t \\ -te^t \\ e^t \end{bmatrix}$, $\mathbf{c}''(t) = \begin{bmatrix} (t+2)e^t \\ -(t+1)e^t \\ e^t \end{bmatrix}$.
37. $\mathbf{v}(t) = (3t^2, -t^{-2}, 0)$; speed $= \sqrt{9t^4 + t^{-4}}$; $\mathbf{a}(t) = (6t, 2t^{-3}, 0)$.
39. $\mathbf{v}(t) = 2e^{2t}(\sin(2t) + \cos(2t))\mathbf{i} + 2e^{2t}(\cos(2t) - \sin(2t))\mathbf{j}$; speed $= 2\sqrt{2}e^{2t}$; the acceleration is $\mathbf{a}(t) = 8e^{2t}\cos(2t)\mathbf{i} - 8e^{2t}\sin(2t)\mathbf{j}$.
41. The velocity is $(t^{-1/2}/2, 1, 3t^{1/2}/2)$; speed $= (4 + 9t + t^{-1})^{1/2}/2$; the acceleration is $(-t^{-3/2}/4, 0, 3t^{-1/2}/4)$.

Section 2.6

1. Show that $\partial f/\partial y = -2yD_1g + 2yD_2g$ and $\partial f/\partial x = 2xD_1g - 2xD_2g$.
3. $g'(t) = (\sin t + t\cos t)D_1f + (\cos t - t\sin t)D_2f + D_3f$ [here we use shorter notation D_if instead of $D_if(t\sin t, t\cos t, t)$, $i = 1, 2, 3$].
5. $(f \circ \mathbf{c})'(t) = 2\sin t\cos^2 t - \sin^3 t$.
7. $(f \circ \mathbf{c})'(t) = 2t\sin t + t^2\cos t - 2t\sin t^2$.
9. Using D_ig for the derivative of g with respect to its ith variable, we get $f_x = 2xyD_1g + 2D_2g + D_3g$ and $f_y = x^2D_1g + 5D_2g + D_4g$, where D_ig, $i = 1, 2, 3, 4$, are evaluated at $(x^2y, 2x + 5y, x, y)$.
11. $F_x(x, y) = (D_1f)h'(x) + (D_3f)k_x(x, y)$, $F_y(x, y) = (D_2f)g'(y) + (D_3f)k_y(x, y)$, where D_1f, D_2f, and D_3f are computed at $(h(x), g(y), k(x, y))$.
13. $\nabla(fg)(x, y) = [(2x + y)\ln(xy) + x + y \quad x\ln(xy) + (x^2 + xy)/y]$; the derivative of f/g at (2, 2) is $D(f/g)(2, 2) \approx [2.24672 \quad -0.63867]$.
15. $D(f/x)(1, \pi, -1) = [4 \quad 1 \quad -\pi]$; $D(x^2yf)(2, 0, 1) = [0 \quad 4 \quad 0]$.
17. $\partial w/\partial\rho = f_x\sin\phi\cos\theta + f_y\sin\phi\sin\theta + f_z\cos\phi$; $\partial w/\partial\theta = -\rho\sin\phi(f_x\sin\theta - f_y\cos\theta)$; $\partial w/\partial\phi = f_x\rho\cos\phi\cos\theta + f_y\rho\cos\phi\sin\theta - f_z\rho\sin\phi$.
19. $(\mathbf{v} \times \mathbf{w})'(t) = 2te^t(2 + t)\mathbf{i} - 8t^3\mathbf{k}$.
21. $D(g \circ \mathbf{F})(0, 0) = [1 \quad 1]$.
23. $\partial w/\partial x = D_1f + g_xD_2f$; $\partial w/\partial z = g_zD_2f + D_3f$; all partial derivatives D_if, $i = 1, 2, 3$ are computed at $(x, g(x, z), z)$.
25. $D\mathbf{F}(\mathbf{x}) = A$; to prove differentiability, use Definition 2.13 or Theorem 2.5.
27. $(df/dt)(0) = 8$.
29. $Df(\mathbf{x}) = -2\sin\left(\|\mathbf{F}(\mathbf{x})\|^2\right)\mathbf{F}(\mathbf{x}) \cdot D\mathbf{F}(\mathbf{x})$ (the dot represents matrix multiplication).

Section 2.7

1. If we use $f(2, 1) = 16$ and $f(2, 1.5) = 18$, the estimate is 4.
3. If we use $f(3, 2) = 18$ and $f(2, 2.5) = 20$, the estimate is $4/\sqrt{5}$.
5. If we use $f(4, 1) = 12$ and $f(3, 1) = 14$, the estimate is 2.
7. $-3/\sqrt{5}$.
9. $\nabla f(0, -1, 2) = 2e^{-5}\mathbf{j} - 4e^{-5}\mathbf{k}$; the directional derivative is $-2e^{-5}/\sqrt{3}$.
11. $(3\ln 4 + 12)/5$.
13. 0.
15. $D_{\mathbf{u}} f(0, 0)$ does not exist in any direction.
17. Maximum rate of change is $\sqrt{10}$; the direction is $(\sqrt{2}, 2\sqrt{2})$.
19. Maximum rate of change equals $\sqrt{77}/4$; the direction is $3\mathbf{i}/2 - 5\mathbf{j}/4 - \mathbf{k}$.
21. Maximum rate of change equals $\sqrt{13}$; the direction is $(2, -3)$.
23. (a) $-400e^{-2}/\sqrt{2}$. (b) In the direction of $\nabla P(0, 1) = (0, -400e^{-2})$; it decreases most rapidly in the opposite direction. (c) $400e^{-2}$. (d) $\pm\mathbf{i}$.
25. In the direction of the vector $(3/5, 4/5)$ and in the direction of the vector $(1, 0)$.
27. Directions that make an angle θ, $0 \le \theta \le \arccos(0.8)$, with respect to the direction of the gradient.
29. Approximately 10.61 m per 100 m.
31. $24/\sqrt{10}$.
33. Use the definition of the gradient.
35. Use the definition of the gradient and the quotient rule.
37. Approximately 1.35748 rad.
39. $\pi y + 3z - 2\pi = 0$.
41. $2x - 3y - 4\sqrt{2}\,z + 6 = 0$.
43. Normal line: $\mathbf{l}_N(t) = (2, 4 + 2t, \sqrt{3} + 2t\sqrt{3})$, $t \in \mathbb{R}$; tangent plane: $y + \sqrt{3}z - 7 = 0$.
45. Normal line: $\mathbf{l}_N(t) = (0, e + t, 1)$, $t \in \mathbb{R}$; tangent plane: $y = 1$.
47. Normal line: $\mathbf{l}_N(t) = (3t, -2 - 2t)$, $t \in \mathbb{R}$; tangent line: $\mathbf{l}_T(t) = (2t, -2 + 3t)$, $t \in \mathbb{R}$.
49. Normal line: $\mathbf{l}_N = (2t + \ln 2, \pi/2)$, $t \in \mathbb{R}$; tangent line: $\mathbf{l}_T(t) = (\ln 2, t + \pi/2)$, $t \in \mathbb{R}$.
51. $(1/2, 3/2, -5/2)$.
53. Unit normal vectors are $\pm(\pi, 2, 2)/\sqrt{\pi^2 + 8}$.
55. Using the gradient, find a parametrization of a line normal to the sphere. Show that that line contains the center of the sphere.
57. Show that the normal vectors of the two families of curves are perpendicular.
59. Use the fact that the gradient vector is perpendicular to a level surface.
61. $V(x, y) = -xe^{xy} + C$, where C is a constant.
63. $V(x, y, z) = x^2y^2 + x^3z - yz^3 + C$, where C is a constant.
65. $V(x, y) = -x^2y^2/2 - x^3y + C$, where C is a constant.

Section 2.8

1. $(4, \pi, 0)$; $(0, \theta, 3)$ for $0 \le \theta < 2\pi$; $(2, \pi/2, 4)$; $(3.60555, 5.30039, -1)$.
3. T maps the given cube onto a rectangular box whose sides are $2a$ and the height is a and then rotates it for π rad about the z-axis.
5. Paraboloid: $z = 4 - r^2$ (in cylindrical coordinates), $\rho\cos\phi = 4 - \rho^2\sin^2\phi$ (in spherical coordinates); plane: $r\cos\theta + 2r\sin\theta - z = 0$ (in cylindrical coordinates), $\sin\phi\cos\theta + 2\sin\phi\sin\theta - \cos\phi = 0$ (in spherical coordinates).
7. $r = C$ is the cylinder of radius C (if $C > 0$) whose axis of rotation is the z-axis; $\theta = C$ is the plane perpendicular to the xy-plane that contains the z-axis and whose intersection with the xy-plane makes the angle θ with respect to the positive x-axis. $z = C$ is the plane parallel to the xy-plane. The coordinate curve $r = C_1$, $\theta = C_2$ is the line perpendicular to the xy-plane, crossing it at the point with polar coordinates (C_1, C_2); the coordinate curve $r = C_1$, $z = C_2$ is the circle of radius C_1 in the plane $z = C_2$, centered at $(0, 0, C_2)$; the coordinate curve $\theta = C_1$, $z = C_2$ is the line in the plane $z = C_2$ crossing the z-axis at $(0, 0, C_2)$; when that line is projected onto the xy-plane, it makes the angle θ with respect to the positive x-axis.
9. $\mathbf{i} = \sin\phi\cos\theta\mathbf{e}_\rho - \sin\theta\mathbf{e}_\theta + \cos\phi\cos\theta\mathbf{e}_\phi$, $\mathbf{j} = \sin\phi\sin\theta\mathbf{e}_\rho + \cos\theta\mathbf{e}_\theta + \cos\phi\sin\theta\mathbf{e}_\phi$, $\mathbf{k} = \cos\phi\mathbf{e}_\rho - \sin\phi\mathbf{e}_\phi$.
11. $\mathbf{F}(r, \theta, z) = (\cos\theta + \sin\theta)\mathbf{e}_r + (\cos\theta - \sin\theta)\mathbf{e}_\theta$, in cylindrical coordinates. In spherical coordinates, $\mathbf{F}(\rho, \theta, \phi) = \sin\phi(\cos\theta + \sin\theta)\mathbf{e}_\rho + (\cos\theta - \sin\theta)\mathbf{e}_\theta + \cos\phi(\cos\theta + \sin\theta)\mathbf{e}_\phi$.
13. $\mathbf{F}(r, \theta, z) = -r\mathbf{e}_\theta$ (in cylindrical coordinates); $\mathbf{F}(\rho, \theta, \phi) = -\rho\sin\phi\mathbf{e}_\theta$ (in spherical coordinates).

15. $d\mathbf{e}_\theta/dt = -\sin\phi(d\theta/dt)\mathbf{e}_\rho - \cos\phi(d\theta/dt)\mathbf{e}_\phi$; $d\mathbf{e}_\phi/dt = -(d\phi/dt)\mathbf{e}_\rho + \cos\phi(d\theta/dt)\mathbf{e}_\theta$.

17. In cylindrical coordinates, $ds^2 = dr^2 + r^2 d\theta^2 + dz^2$; in spherical coordinates, $ds^2 = d\rho^2 + \rho^2\sin^2\phi d\theta^2 + \rho^2 d\phi^2$.

Chapter Review

True/False Quiz

1. False.
3. True.
5. False.
7. True.
9. True.
11. True.
13. False.

Review Exercises and Problems

1. For instance, $\mathbf{G}(\mathbf{r}) = (1, 1)/\|\mathbf{r} - (2, 3)\|$.

3. Keeping L fixed, at 1.3, in order to increase the production from 1.2 to 1.4, to 1.6, etc. (i.e., at a constant rate), we need to increase K so that the rate of increase of K increases as well. The same conclusion stands when K is kept fixed.

5. There are no level curves if $c > 1$; when $c \le 1$, a level curve consists of the circle $x^2 + y^2 = 9$ (if $c = 1$); the circles $x^2 + y^2 = 9 \pm \sqrt{1-c}$ (if $-80 < c < 1$); the circle $x^2 + y^2 = 18$ and the point $(0, 0)$ (if $c = -80$); the circle $x^2 + y^2 = 9 + \sqrt{1-c}$ (if $c < -80$).

7. The limit is zero; define $f(0, 0) = 0$.

9. (a) $(\partial T/\partial x)(3/2, 2) \approx -0.6103$; the cross-section is a sine curve of period 4 and amplitude $30e^{-4}$; as we move along the rod from $x = 3/2$ forward, we will experience a decrease in temperature of approximately 0.6103 degrees/unit distance.
(b) $(\partial T/\partial t)(3/2, 2) \approx -0.7771$; the cross-section is the curve $y(t) = 15\sqrt{2}e^{-2t}$; at time $t = 2$, the temperature decreases at a rate of approximately 0.7771 degrees/unit of time.

11. Using the assumption, show that $f(0, 0) = 0$ and $f_x(0, 0) = f_y(0, 0) = 0$. To verify part (b) of Definition 2.13, use the assumption once again.

13. $\partial F/\partial t = gR^2(R+r)^{-2}(\partial m/\partial t) - 2mgR^2(R+r)^{-3}(\partial r/\partial t)$.

15. Use the "ϵ-δ definition" of continuity.

17. $\frac{dy}{dx} = \frac{dy/dt}{dx/dt}$ and $\frac{d^2y}{dx^2} = \frac{(d^2y/dt^2)(dx/dt) - (dy/dt)(d^2x/dt^2)}{(dx/dt)^3}$

19. Compute the derivative of $f(t\mathbf{x}) = t^p f(\mathbf{x})$ with respect to t and evaluate for $t = 1$.

21. $(\partial f/\partial x)(\mathbf{a}) = 7\sqrt{2}/2$ and $(\partial f/\partial y)(\mathbf{a}) = -3\sqrt{2}/2$.

CHAPTER 3

Section 3.1

1. $\mathbf{c}'(t) = (2 - 2t^2, -4t)/(1+t^2)^2$. The speed $\|\mathbf{c}'(t)\| = 2/(1+t^2)$ is largest (equal to 2) when $t = 0$.

3. For instance, $\mathbf{c}(t) = (\cos t^3, \sin t^3)$, where $t \in [0, \sqrt[3]{2\pi}]$. Actually, we can use any parametrization of the form $\mathbf{c}(t) = (\cos f(t), \sin f(t))$, where the range of $f(t)$ contains an interval of length 2π, $f(t)$ is differentiable, and $f'(t) \ne 0$ for some t in its domain (there are many possible parametrizations).

5. Straightforward calculation: use the product and chain rules to compute $x'(t)$ and $y'(t)$ and substitute $t = 0$ and $t = 1$.

7. (a) Since $-1 \le \cos 2t, \sin 3t \le 1$ for all t, it follows that the image of $\mathbf{c}(t)$ is contained in the square $-1 \le x, y \le 1$. The curve touches $x = 1$ at $\mathbf{c}(0) = \mathbf{c}(\pi) = \mathbf{c}(2\pi) = (1, 0)$; touches $x = -1$ at $\mathbf{c}(\pi/2) = (-1, -1)$ and $\mathbf{c}(3\pi/2) = (-1, 1)$; touches $y = 1$ at $\mathbf{c}(\pi/6) = (1/2, 1)$ and $\mathbf{c}(3\pi/2) = (-1, 1)$; touches $y = -1$ at $\mathbf{c}(\pi/2) = (-1, -1)$ and $\mathbf{c}(11\pi/6) = (1/2, -1)$. (b) $\mathbf{c}(0) = \mathbf{c}(2\pi) = (1, 0)$ is the initial and terminal point; see the figure below.

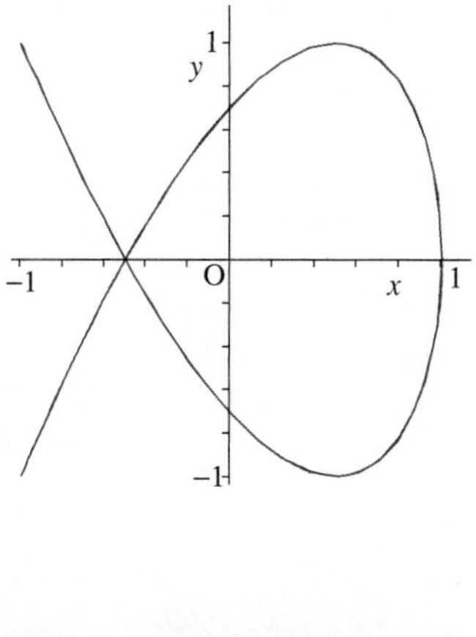

9. (a) $F(x, y) = y^2 - 4x^2(1 - x^2) = 0$, $-1 \le x, y \le 1$. Symmetry follows from the fact that replacing x by $-x$ or y by $-y$ does not change $F(x, y)$. (b) $y = g(x) = 2x\sqrt{1 - x^2}$. (c) The local solution from (b) is the part of the curve shown below in the first quadrant.

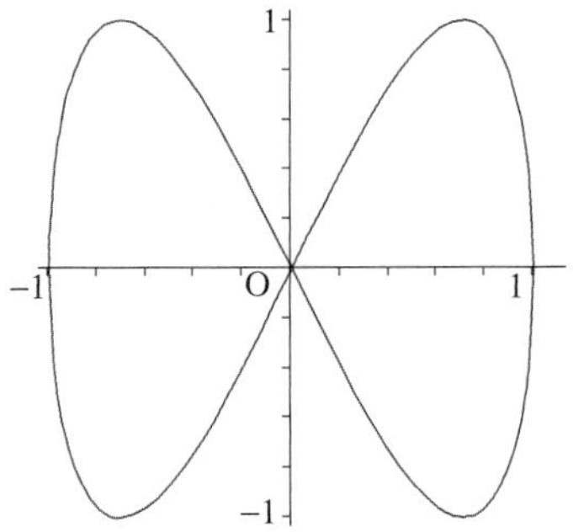

11. One possible parametrization is $\mathbf{c}(t) = (t, -\frac{5}{4} + \frac{3}{4}t, -\frac{13}{4} + \frac{7}{4}t)$, $t \in \mathbb{R}$.

13. $\mathbf{c}(t) = (-2 + 2\cos t, 3, 2 + 2\sin t)$, $t \in [0, 2\pi]$.

15. $\mathbf{c}(t) = (\sqrt{2}\sin t, \pm\sqrt{2}\sin t, \sqrt{2}\cos t)$, $t \in [0, 2\pi]$.

17. Line $\mathbf{l}$ is perpendicular to normal vectors $\mathbf{N}_1 = (1, 2, -1)$ and $\mathbf{N}_2 = (2, -1, -1)$ to the two planes. (b) For instance, $(2, 1, 0)$. (c) $\mathbf{l}(t) = (2, 1, 0) + t(3, 1, 5)$, $t \in \mathbb{R}$ (there are many possible parametrizations).

19. $L_1(x) = 0.3989x + 1.1077$.

21. $g'(2) = 1$. The linear approximation is $L_2(x) = x + 2$.

23. There are many possible answers. (a) $F(x, y) = (x - 1)^2 + (y + 4)^2$. (b) $F(x, y) = (x + y)^2 - 1$. (c) $F(x, y) = (y - x)y$. (d) $F(x, y) = e^{x+y}$.

25. Straightforward calculation.

27. The angle θ satisfies $\cos\theta = a/\sqrt{a^2 + 1}$. Thus, the angles that the spiral makes with rays emanating from the origin are all equal.

29. $f(t) = t$: spirals outward, windings equally spaced, oriented counterclockwise; $f(t) = t^2$: spirals outward, distance between windings increases as t increases, oriented counterclockwise; $f(t) = \ln t$: spirals outward, distance between windings decreases as t increases, oriented counterclockwise; $f(t) = 1/t$: spirals inward, curls around the origin, distance between windings decreases as t increases, oriented counterclockwise.

31. The position is $\mathbf{c}(12) = (-0.8160, 0.5781, 12)$, where $\mathbf{c}(t) = (\cos\sqrt{8}t, \sin\sqrt{8}t, t)$, $t \ge 0$, is the trajectory of the object.

33. $\mathbf{c}(t) = (t, 10\cosh 0.031336t)$, $t \in [-50, 50]$.

Section 3.2

1. Velocity is $\mathbf{c}'(\theta) = (1 - \cos\theta, \sin\theta)$; speed is $\sqrt{2(1 - \cos\theta)}$. Speed is largest (equal to 2) when $\theta = \pi + 2\pi k$, where k is an integer.

3. $\|\mathbf{c}'(t)\| = \sqrt{a^2 + b^2}$. Since $\mathbf{c}'(t) \cdot \mathbf{c}''(t) = 0$, it follows that at every point on the helix, the velocity vector is perpendicular to the acceleration.

5. $\mathbf{v}(t) = (-t + 1, t + 2, 0)$; $\mathbf{c}(t) = (t - t^2/2, 2 + 2t + t^2/2, 0)$.

7. $\mathbf{v}(t) = (t^2/2, t + 1, t)$; $\mathbf{c}(t) = (t^3/6 - 2, t^2/2 + t, t^2/2 + 3)$.

9. $\mathbf{v}(t) = (t^2/2)\mathbf{i} + (2 + t^3/3)\mathbf{j} + (-3 + t^2/2)\mathbf{k}$; $\mathbf{c}(t) = (4 + t^3/6)\mathbf{i} + (2 + 2t + t^4/12)\mathbf{j} + (-6 - 3t + t^3/6)\mathbf{k}$.

11. The particle reaches its maximum speed of approximately 8.0002 (units) at $t = 4$.

13. The particle reaches its highest position when $t = 12$.

15. The range of the projectile is approximately 43,302 m; maximum height ($\approx$18,750 m) is reached when $t \approx 61.86$ s; the speed at the time of impact is 700 m/s.

17. $\mathbf{l}(t) = (\sqrt{3}, t)$, $t \in \mathbb{R}$.

19. $\mathbf{l}(t) = (-3t, 3, 2\pi + 4t)$, $t \in \mathbb{R}$.

21. $\mathbf{l}(t) = (-1, 1 + t)$, $t \in \mathbb{R}$.

23. Use the product rule and chain rule.

25. $\mathbf{v}(t) = -e^{-t}(\cos t + \sin t, \sin t - \cos t)$; it is horizontal at $\mathbf{c}(\pi/4)$, $\mathbf{c}(5\pi/4)$, and $\mathbf{c}(9\pi/4)$ and vertical at $\mathbf{c}(3\pi/4)$, $\mathbf{c}(7\pi/4)$, and $\mathbf{c}(11\pi/4)$.

27. The vector $-\mathbf{j}$ is tangent to the image of $\mathbf{c}(t)$ under $\mathbf{F}$ at $t = 0$.

29. The required vector is $(a_{11}c_1 + a_{12}c_2)\mathbf{i} + (a_{21}c_1 + a_{22}c_2)\mathbf{j}$.

Section 3.3

1. $\ell(p_5) \approx 1.17488$.

3. (a) Use the formula for the distance between two points. (b) Recall the Mean Value Theorem: if $f(t)$ is differentiable on (a, b) and continuous on $[a, b]$, then there exists $t_0 \in (a, b)$ such that $(f(b) - f(a))/(b - a) = f'(t_0)$. (c) The length of the polygonal path $\ell(p_n) = \sum_{i=1}^{n} \sqrt{\left(x'(t_i^*)\right)^2 + \left(y'(t_i^{**})\right)^2}\,\Delta t$ is a Riemann sum corresponding to the integral $\int_a^b \|\mathbf{c}'(t)\|\,dt$.

5. (a) Use the definition to compute $y'(0)$. (b) Functions $\cos^2 t$ and $\cos^3 t$ are differentiable for all t. (c) $\mathbf{c}'(0)=\mathbf{0}$ (note: $\mathbf{c}'=\mathbf{0}$ at some other values of t as well).
7. π.
9. $\sqrt{2}(e^{\pi}-1)$.
11. Approximately 2.08581.
13. 12.
15. $\ell(\mathbf{c})=-\sqrt{a^2+1}/a$.
17. It is not possible to evaluate the integrals for the lengths of the given curves; the idea is to compare the integrands.
19. $a\pi/4$.
21. $8^{3/2}-5^{3/2}$.
23. 4; integration is a bit tricky. *Hint:* Multiply and divide by $\sqrt{1-\sin\theta}$.
25. $s(t)=t^2+t,\ 0\le t\le 2\pi$.
27. $s(t)=13t,\ t\in[0,\pi/4]$; $\mathbf{c}(s)=5\cos(s/13)\mathbf{i}+5\sin(s/13)\mathbf{j}+(12s/13)\mathbf{k},\ s\in[0,13\pi/4]$.
29. $s(t)=\sqrt{2}(e^t-1),\ 0\le t\le 1$; $\mathbf{c}(s)=(1+s/\sqrt{2})[\cos(\ln(1+s/\sqrt{2}))\,\mathbf{i}+\sin(\ln(1+s/\sqrt{2}))\,\mathbf{j}]$, $s\in[0,\sqrt{2}(e-1)]$.
31. Since $\mathbf{c}_2(t)=\mathbf{c}_1(t/2)$ and $\mathbf{c}_3(t)=\mathbf{c}_1(-t)$, it follows that the paths $\mathbf{c}_1$, $\mathbf{c}_2$, and $\mathbf{c}_3$ are reparametrizations of each other, and so represent the same curve; its length is $\pi\sqrt{5}$.
33. $\mathbf{T}(1)=-\mathbf{i}$.
35. No.

Section 3.4

1. $\mathbf{a}_T=(16t^2,8t,16t^2)/(8t^2+1)$; $\mathbf{a}_N=(2,-8t,2)/(8t^2+1)$.
3. $\mathbf{a}_T=\mathbf{0}$; $\mathbf{a}_N=\mathbf{a}=-12\sin t\,\mathbf{j}-12\cos t\,\mathbf{k}$.
5. $\mathbf{a}_T=(\sin t(1-\cos t)\mathbf{i}+\sin^2 t\mathbf{j})/(2-2\cos t)$; and $\mathbf{a}_N=(\sin t(1-\cos t)\mathbf{i}+(2\cos t-2\cos^2 t-\sin^2 t)\mathbf{j})/(2-2\cos t)$.
7. $\|\mathbf{c}'\|=\sqrt{(x')^2+(y')^2}$; $\mathbf{T}'=((x')^2+(y')^2)^{-3/2}\left(x''(y')^2-x'y'y'',\ y''(x')^2-x'y'x''\right)$; $\|\mathbf{T}'\|=((x')^2+(y')^2)^{-1}|x''y'-x'y''|$; so $\kappa=\|\mathbf{T}'\|/\|\mathbf{c}'\|=|x''y'-x'y''|/((x')^2+(y')^2)^{3/2}$
9. $\mathbf{c}(t)$ is a line; its curvature is zero.
11. $\kappa(t)=2(4t^2+1)^{-3/2}$; the maximum is 2, occurs at $\mathbf{c}(0)=(0,3)$; as $t\to\infty$, $\kappa(t)\to 0$.
13. $\mathbf{T}(t)=(\sin t+\cos t,0,\cos t-\sin t)/\sqrt{2}$; $\mathbf{N}(t)=(\cos t-\sin t,0,-\sin t-\cos t)/\sqrt{2}$; curvature is $\kappa(t)=\sqrt{2}e^{-t}/2$; $\mathbf{a}_N=e^t(\cos t-\sin t,0,-\sin t-\cos t)$.
15. $\mathbf{T}(t)=(1,-\sin t,-\cos t)/\sqrt{2}$; $\mathbf{N}(t)=(0,-\cos t,\sin t)$; the curvature is $\kappa(t)=1/2$; $\mathbf{a}_N=(0,-\cos t,\sin t)$.
17. $\mathbf{T}(t)=(-\cos t-\sin t,-\sin t+\cos t,-1)/\sqrt{3}$; $\mathbf{N}(t)=(\sin t-\cos t,-\sin t-\cos t,0)/\sqrt{2}$; $\kappa(t)=e^t\sqrt{2}/3$; $\mathbf{a}_N=e^{-t}(\sin t-\cos t,-\sin t-\cos t,0)$.
19. $(x-241/12)^2+(y+143)^2=145^3/144$.
21. $x+2z-\pi=0$.
23. $\kappa(x_0)=2(1+4x_0^2)^{-3/2}$.
25. Maximum curvature occurs at $x=1/\sqrt{2}$; its value is $\kappa(1/\sqrt{2})\approx 0.3849$.
27. $2y+z-2=0$.
29. $(x-\pi/2)^2+y^2=1$.

Section 3.5

1. $d(\mathbf{T}(s)\cdot\mathbf{T}(s))/ds=0$; $(d\mathbf{c}(t)/dt)\cdot\mathbf{T}(t)=\|\mathbf{c}'(t)\|$; $d\mathbf{N}(s)/ds\cdot\mathbf{B}(s)=\tau(s)$.
3. Yes.
5. $\mathbf{T}(\pi/2)=(-\mathbf{i}+\mathbf{k})/\sqrt{2}$; $\mathbf{N}(\pi/2)=(-\mathbf{i}-\mathbf{k})/\sqrt{2}$; $\mathbf{B}(\pi/2)=-\mathbf{j}$.
7. $\mathbf{T}(t)=(\cos t/\sqrt{2},\cos t/\sqrt{2},-\sin t)$; $\mathbf{N}(t)=(-\sin t/\sqrt{2},-\sin t/\sqrt{2},-\cos t)$; the binormal is $\mathbf{B}(t)=\mathbf{T}(t)\times\mathbf{N}(t)=(-1,1,0)/\sqrt{2}$.
9. Since $d\mathbf{N}/ds+\kappa\mathbf{T}$ is perpendicular to both $\mathbf{T}$ and $\mathbf{N}$, it must be parallel to $\mathbf{B}$. We can define τ so that $d\mathbf{N}/ds+\kappa\mathbf{T}=\tau\mathbf{B}$.
11. Differentiate $\mathbf{c}'(t)=(ds/dt)\mathbf{T}$ with respect to t and use formula (3.19).
13. Use $\mathbf{c}'(t)=\|\mathbf{c}'(t)\|\mathbf{T}(t)=(ds/dt)\mathbf{T}(t)$ and the formula of Exercise 11.
15. $b/(a^2+b^2)$.
17. The torsion is zero; the curvature $\kappa(t)=1/(2a(t^2+1)^{3/2})$ is largest when $t=0$.
19. $\kappa(t)=(3+2\sin t\cos t-2\sin t-2\cos t)^{1/2}(4-2\sin t-2\cos t)^{-3/2}$; $\tau(t)=-(3+2\sin t\cos t-2\sin t-2\cos t)^{-1}$.

Chapter Review

True/False Quiz

1. False.
3. True.
5. True.
7. False.

Review Exercises and Problems

1. Initial point is $(0, 2)$, terminal point is $(3, 2)$. From $t=0$ to $t=1$: constant speed (equal to 1), $\mathbf{c}(1)=(1, 2)$; from $t=1$ to $t=2$: constant speed (equal to $\sqrt{2}$), $\mathbf{c}(2)=(2, 1)$; from $t=2$ to $t=3$: constant speed (equal to 1), $\mathbf{c}(3)=(3, 1)$; from $t=3$ to $t=4$: object rests at $(3, 1)$; from $t=4$ to $t=5$: constant speed (equal to 1), $\mathbf{c}(5)=(3, 2)$.
3. $(4, 0, 4)$ and $(12/5, 16/5, 4)$.
5. $\mathbf{W}=\tau\mathbf{T}+\kappa\mathbf{B}$.
7. Recall that $\kappa(s)=\|d\mathbf{T}(s)/ds\|=\lim_{\Delta s\to 0}\|\mathbf{T}(s+\Delta s)-\mathbf{T}(s)\|/|\Delta s|$. To compute the limit, use $\|\mathbf{T}(s+\Delta s)-\mathbf{T}(s)\|=2\sin(|\theta(s+\Delta s)-\theta(s)|/2)$, where θ is the angle between the tangent and the positive x-axis, and the approximation $\sin a\approx a$ for a close to 0.
9. Use Newton's Second Law and Exercise 23 in Section 3.2 to show that the derivative of $r^2(d\theta/dt)$ with respect to t is zero.
11. The speed is $e^t\sqrt{3}$; the magnitude of the acceleration is $e^t\sqrt{5}$.
13. (a) Compute $(A(t)-A(t_0))/(t-t_0)$ and then let $t\to t_0$. The number $A(t)-A(t_0)$ is the area of the elliptic sector between $\mathbf{r}(t)$ and $\mathbf{r}(t_0)$; approximate it by a circular sector. (b) Let $\mathbf{r}=(\|\mathbf{r}(t)\|\cos\theta(t), \|\mathbf{r}(t)\|\sin\theta(t))$; compute $\mathbf{v}$ and then substitute into $\mathbf{d}=\mathbf{r}\times\mathbf{v}$. (c) The rate at which the area is swept by $\mathbf{r}(t)$ is constant.
15. $0.1499\cdot 10^{12}$ m.
17. The lengths are $\ell(\mathbf{c}_1)=\sqrt{2}\approx 1.41421$; $\ell(\mathbf{c}_2)\approx 1.53775$; $\ell(\mathbf{c}_3)=\pi/2\approx 1.57080$; and $\ell(\mathbf{c}_4)\approx 1.46370$; corresponding times are $T_1=2/\sqrt{g}$; $T_2\approx 1.84030/\sqrt{g}$; $T_3\approx 1.85407/\sqrt{g}$; and $T_4\approx 1.88734/\sqrt{g}$; $\mathbf{c}_2$ is the "fastest" curve.

CHAPTER 4

Section 4.1

1. $f_x(P)<0$, $f_y(P)=0$, $f_{xx}(P)<0$, $f_{xy}(P)=0$, and $f_{yy}(P)=0$.
3. $f_x(P)>0$, $f_y(P)>0$, $f_{xx}(P)=0$, $f_{xy}(P)=0$, and $f_{yy}(P)=0$.
5. $z_{xx}=y^2e^{xy}-2/x^2$; $z_{yy}=x^2e^{xy}-3/y^2$; $z_{xy}=z_{yx}=e^{xy}(1+xy)$.
7. $z_{xx}=5(x^2+y^2)^{1/2}(4x^2+y^2)$; $z_{yy}=5(x^2+y^2)^{1/2}(x^2+4y^2)$; $z_{xy}=z_{yx}=15xy(x^2+y^2)^{1/2}$.
9. $z_{xx}=z_{xy}=z_{yx}=z_{yy}=2\cos(2x+2y)$.
11. $z_{xx}=a^2f''(ax+by)+a^2y^{-2}g''(ax/y)$; $z_{xy}=abf''(ax+by)-a^2xy^{-3}g''(ax/y)-ay^{-2}g'(ax/y)$; $z_{yx}=z_{xy}$; $z_{yy}=b^2f''(ax+by)+a^2x^2y^{-4}g''(ax/y)+2axy^{-3}g'(ax/y)$.
13. $w_{xyzx}=48xyz^3$.
15. Both sides are equal to $z=xe^y+ye^x$.
17. A C^2 function of m variables has $m(m+1)/2$ different second-order partial derivatives.
19. Use f', f'', g', and g'' to denote the derivatives of f and g; substituting $u_{xx}=2f'+xf''+yg''$, $u_{xy}=f'+xf''+g'+yg''$, and $u_{yy}=2g'+xf''+yg''$ into $u_{xx}-2u_{xy}+u_{yy}$, we get 0.
21. Straightforward calculation.
23. Show that $f_{xx}=-f_{yy}=2xy/(x^2+y^2)^2$.
25. $V_{xx}=GMm(y^2+z^2-2x^2)(x^2+y^2+z^2)^{-5/2}$, $V_{yy}=GMm(x^2+z^2-2y^2)(x^2+y^2+z^2)^{-5/2}$, and $V_{zz}=GMm(x^2+y^2-2z^2)(x^2+y^2+z^2)^{-5/2}$.
27. (a) Initial temperature at ends is $T(0, 0)=T(\pi, 0)=1$. Warmest point is $x=\pi/2$, $T(\pi/2, 0)=2$. (b) Straightforward calculation. (c) As $t\to\infty$, $T(x, t)\to 1$ for all x; see the figure below.

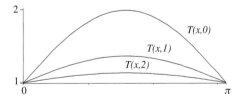

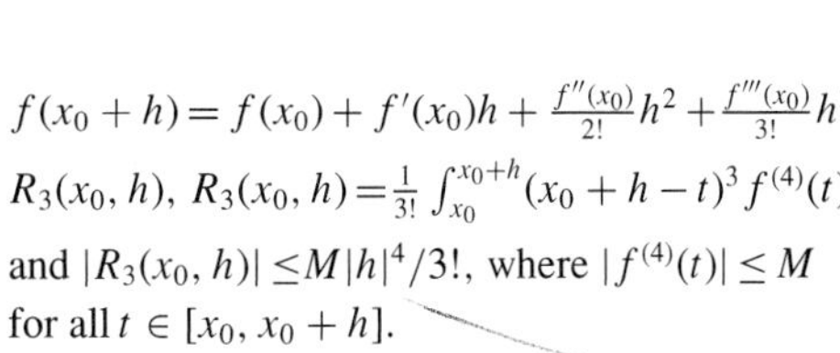

Section 4.2

1. If $h<0$, then $\int_{x_0}^{x_0+h}|f(t)|\,dt<0$ and the estimate we used no longer works. The mentioned step is correct because $|\int_a^b f(t)\,dt|=|-\int_b^a f(t)\,dt|=|\int_b^a f(t)\,dt|$ for all a and b.
3. $f(x_0+h)=f(x_0)+f'(x_0)h+\frac{f''(x_0)}{2!}h^2+\frac{f'''(x_0)}{3!}h^3+R_3(x_0, h)$, $R_3(x_0, h)=\frac{1}{3!}\int_{x_0}^{x_0+h}(x_0+h-t)^3f^{(4)}(t)\,dt$ and $|R_3(x_0, h)|\le M|h|^4/3!$, where $|f^{(4)}(t)|\le M$ for all $t\in[x_0, x_0+h]$.

5. $T_2(3,h)=\sqrt{3}+\sqrt{3}\,h/6-\sqrt{3}\,h^2/72$; error: $|R_2(3,h)|\le 3/(64\sqrt{2})$.

7. $T_2(x)=\frac{\sqrt{2}}{2}+\frac{\sqrt{2}}{2}(x-\frac{\pi}{4})-\frac{\sqrt{2}}{4}(x-\frac{\pi}{4})^2$; $R_2(x)=-\frac{1}{2}\int_{\pi/4}^{x}(x-t)^2\cos t\,dt$.

9. $T_2(x)=\ln 4+\frac{1}{4}(x-4)-\frac{1}{32}(x-4)^2$; $R_2(x)=\int_4^x (x-t)^2 t^{-3}\,dt$.

11. $T_2(x)=1+x-x^2/2$; $T_2(0.1)=1.0950000$. $T_3(x)=1+x-x^2/2-x^3/6$; $T_3(0.1)=1.0948333$. Exact value: $\sin 0.1+\cos 0.1=1.0948376$.

13. $(\partial F/\partial y)(0,2)=8\ne 0$, so the Implicit Function Theorem applies. $T_2(x)=2-x^2/8$.

15. $T_2(x)=x-x^2$; $T_2(0.2)=0.16$. Error: $|R_2(0,0.2)|\le 0.012$.

17. $T_2(\mathbf{x}_0,\mathbf{h})=f+f_xh_1+f_yh_2+f_zh_3+\frac{1}{2}(f_{xx}h_1^2+2f_{xy}h_1h_2+2f_{xz}h_1h_3+f_{yy}h_2^2+2f_{yz}h_2h_3+f_{zz}h_3^2)$, where f and all its partial derivatives are computed at $\mathbf{x}_0=(x_0,y_0,z_0)$, and $\mathbf{h}=(h_1,h_2,h_3)$. In case of a function of m variables,

$T_2(\mathbf{x}_0,\mathbf{h})=f(\mathbf{x}_0)+\sum_{i=1}^{m} f_{x_i}(\mathbf{x}_0)\,h_i+\frac{1}{2}\left(\sum_{i=1}^{m}\sum_{j=1}^{m} f_{x_ix_j}(\mathbf{x}_0)\,h_ih_j\right)$.

19. (a) Start with $F(t_0+h)=F(t_0)+F'(t_0)h+R_1(t_0,h)$ and substitute $t_0=0$ and $h=1$. For $G(t)$, use the calculation done before Theorem 4.3. (b) Apply the Mean Value Theorem with $h(t)=1-t$ and with each of $g(t)=f_{xx}(\mathbf{x}_0+t\mathbf{h})$, $g(t)=f_{xy}(\mathbf{x}_0+t\mathbf{h})$, and $g(t)=f_{yy}(\mathbf{x}_0+t\mathbf{h})$. (c) $R_1(\mathbf{x}_0,\mathbf{h})=\sum_{i=1}^{m}\sum_{j=1}^{m} f_{x_ix_j}(\mathbf{c}_{ij})\,h_ih_j$, where $\mathbf{c}_{ij}$, $i,j=1,\ldots,m$ lie on the line joining $\mathbf{x}_0$ and $\mathbf{x}_0+\mathbf{h}$.

21. $f(x_0+h_1,y_0+h_2)=f+f_xh_1+f_yh_2+\frac{1}{2!}(f_{xx}h_1^2+2f_{xy}h_1h_2+f_{yy}h_2^2)+\frac{1}{3!}(f_{xxx}h_1^3+3f_{xxy}h_1^2h_2+3f_{xyy}h_1h_2^2+f_{yyy}h_2^3)+R_3(0,1)$; all terms on the right side are evaluated at (x_0,y_0). $R_3(0,1)=\frac{1}{3!}\int_0^1(1-t)^3G(t)\,dt$, and $G(t)=f_{xxxx}h_1^4+4f_{xxxy}h_1^3h_2+6f_{xxyy}h_1^2h_2^2+4f_{xyyy}h_1h_2^3+f_{yyyy}h_2^4$; all derivatives are evaluated at (x_0+th_1,y_0+th_2).

23. $T_2(x,y)=1+(x-1)^2-2(x-1)(y-1)+(y-1)^2=f(x,y)$.

25. $T_2(x,y)=x-2y+\pi$.

27. $T_2(x,y)=\frac{3}{2}-\frac{1}{2}x-\frac{1}{4}y+\frac{1}{2}(x-1)^2+\frac{1}{4}(x-1)(y-2)+\frac{1}{8}(y-2)^2$.

29. $T_1(x,y)=\frac{15}{8}+\frac{1}{8}x+\frac{1}{2}y$, $T_2(x,y)=\frac{15}{8}+\frac{1}{8}x+\frac{1}{2}y-\frac{1}{512}(x-5)^2-\frac{1}{64}(x-5)(y-3)-\frac{1}{32}(y-3)^2$. $T_1(4.9,3.1)=4.0375000$, $T_2(4.9,3.1)=4.0373242$; $f(4.9,3.1)=4.0373258$.

31. $T_2(x,y)=xy$. The contour diagram consists of hyperbolas $y=c/x$ (if $c\ne 0$) and of the lines $x=0$ and $y=0$ (if $c=0$).

33. $f(x,y)=e^{x^2-y^2}$, $T_2(x,y)=e^{-1}-2e^{-1}(y-1)+e^{-1}x^2+e^{-1}(y-1)^2$; $T_2(0.03,0.95)=0.405918$.

35. $f(x,y)=x\arctan y$, $T_2(x,y)=xy$; $T_2(3.98,0.02)=0.0796$.

Section 4.3

1. Differentiate $f(-x)=f(x)$ and substitute $x=0$. Compute partial derivatives of $f(-x,-y)=f(x,y)$ and substitute $x=y=0$.

3. $\nabla f=\mathbf{0}$ at the point P, see the figure below; P is a saddle point.

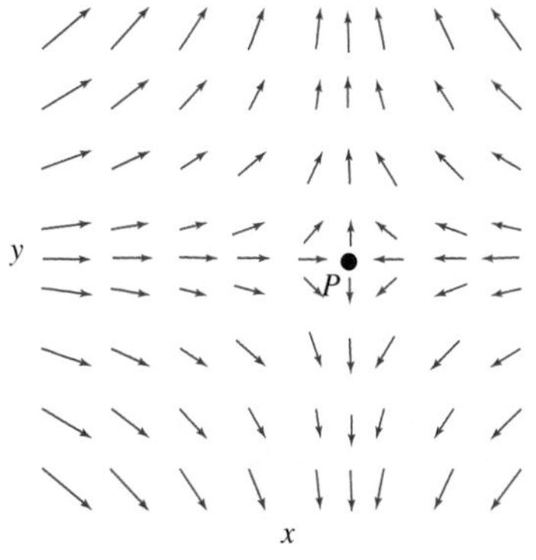

5. $\nabla f\ne\mathbf{0}$ at all points, so f has no critical points.

7. Local minimum $f(0,0)=0$; saddle points at $(-1,\pm\sqrt{2})$.

9. Local minimum $f(1,1)=3$.

11. Saddle point at (0, 0); local minimum $f(-1/\sqrt{2},1/\sqrt{2})=f(1/\sqrt{2},-1/\sqrt{2})=-e^{-1}/2$; local maximum $f(1/\sqrt{2},1/\sqrt{2})=f(-1/\sqrt{2},-1/\sqrt{2})=e^{-1}/2$.

13. Saddle points at $(0,\pi/2+k\pi)$ (k is an integer).

15. Saddle points at $(0,k\pi)$ (k is an integer).

17. $1/\sqrt{3}$.

19. Cube of side $\sqrt[3]{V}$.

21. Volume is $8R^3/3^{3/2}$; the box is a cube of side $2R/\sqrt{3}$.

23. The contour curve through (0, 0) consists of lines $y = \pm x$. See the figure below.

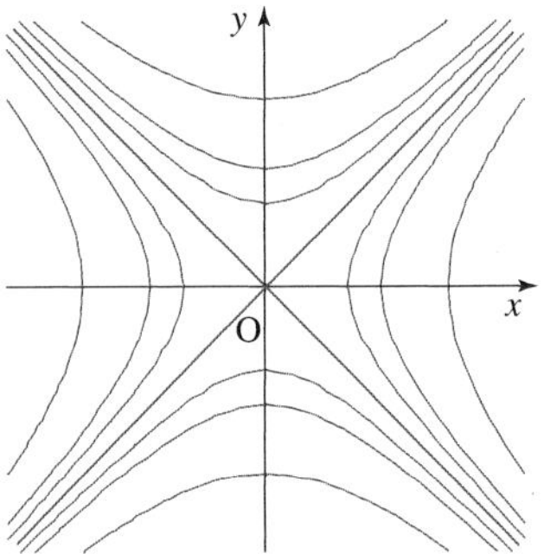

25. $x/3 + y/3 + z/3 = 1$.

27. Absolute minimum $f(0, 0) = 0$; absolute maximum $f(1, 1) = \ln 3$.

29. Absolute maximum is 1; it occurs at all points (x, y) inside D such that $xy = \pi/2$, and also at $(1, \pi/2)$ and $(1/2, \pi)$. Absolute minimum is -1; it occurs at $(3/2, \pi)$ and $(\pi/2, 3)$.

31. $p(0, 1) = e^2 + 1$ is the absolute maximum and $p(0, -1/2) = e^{-1/4} + 1$ is the absolute minimum of p on the given disk.

Section 4.4

1. ∇f is parallel to the constraint curve.

3. Identify points where the gradient of f is perpendicular to the constraint curve; minimum at P_1 and P_3, maximum at P_2 and P_4.

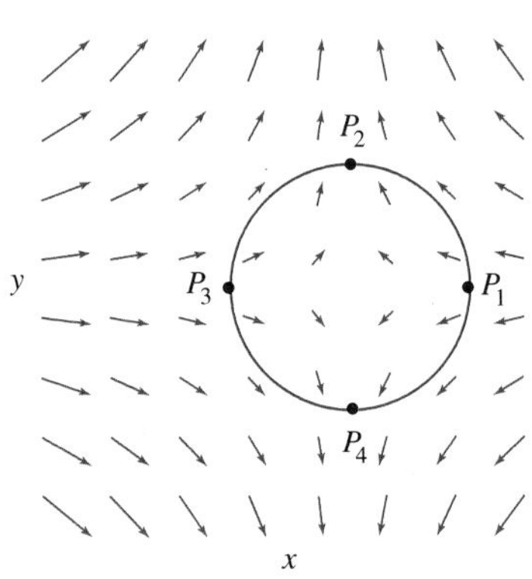

5. f has a maximum at P_1 and minimum at P_2.

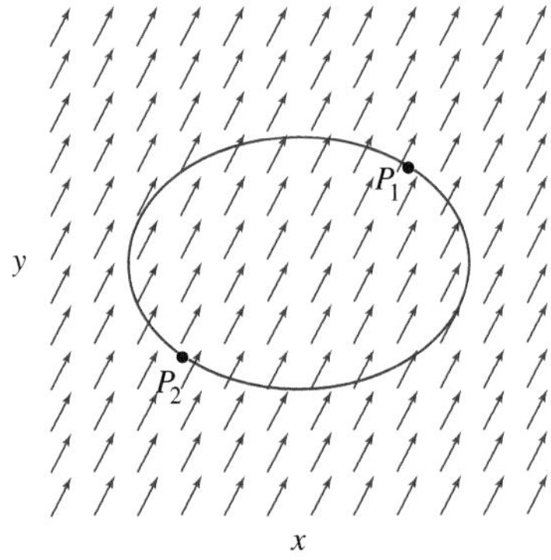

7. (a), (b) $f(3, 0) = 9$ is maximum, and $f(-1, 0) = 1$ is minimum.

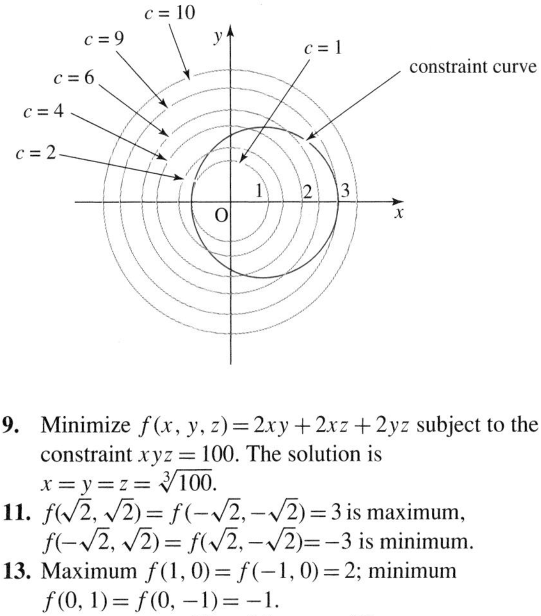

9. Minimize $f(x, y, z) = 2xy + 2xz + 2yz$ subject to the constraint $xyz = 100$. The solution is $x = y = z = \sqrt[3]{100}$.

11. $f(\sqrt{2}, \sqrt{2}) = f(-\sqrt{2}, -\sqrt{2}) = 3$ is maximum, $f(-\sqrt{2}, \sqrt{2}) = f(\sqrt{2}, -\sqrt{2}) = -3$ is minimum.

13. Maximum $f(1, 0) = f(-1, 0) = 2$; minimum $f(0, 1) = f(0, -1) = -1$.

15. Maximum $f(3^{-1/2}, 3^{-3/4}) = 2(3^{-3/2}) \approx 0.3849$; minimum $f(-1, 1) = -2$.

17. $f(2/\sqrt{13}, 4/\sqrt{13}, -4/\sqrt{13}) = -6/\sqrt{13}$ is maximum subject to the given constraint. Minimum value is $f(-2/\sqrt{13}, -4/\sqrt{13}, 4/\sqrt{13}) = -26/\sqrt{13}$.

19. Maximum $f(1, \sqrt{3/2}) = 3\sqrt{3/2}$; minimum $f(1, -\sqrt{3/2}) = -\sqrt{3/2}$.

21. Minimum distance is 2.

23. $(4/3, 7/3, 11/3)$; the distance is $\sqrt{32/3}$.

25. $f(44/9, 4/9, 56/9) = 5088/81$.

27. $(14/3, -4/3, -2/3)$; the distance is $\sqrt{24}$.

29. Volume is $8R^3/3^{3/2}$; the box is a cube of side $2R/\sqrt{3}$.

31. $a = b = c = 3$; volume is $9/2$.

Section 4.5

1. The flow lines of **F**, $\mathbf{F}_1$, and $\mathbf{F}_2$ are parabolas $y = Cx^2$, $C =$ constant. $\mathbf{F}_1$ has the "fastest" flow lines, while those of **F** have the smallest speed; the flow lines of $\mathbf{F}_1$ "flow" in the same direction as those of **F**, while the flow lines of $\mathbf{F}_2$ "flow" in the opposite direction.

3. $\mathbf{c}(t) = t(6/\sqrt{13}, 4/\sqrt{13})$, $t \geq 0$.

5. Line $\mathbf{c}(t) = (at, bt)$, $t \in \mathbb{R}$.

7. See the figure below.

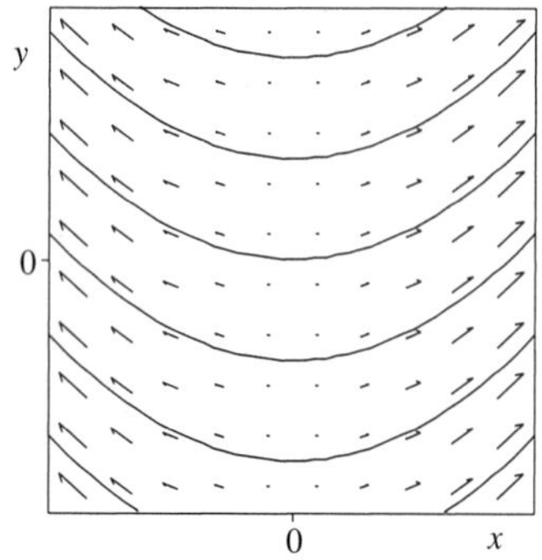

9. See the figure below.

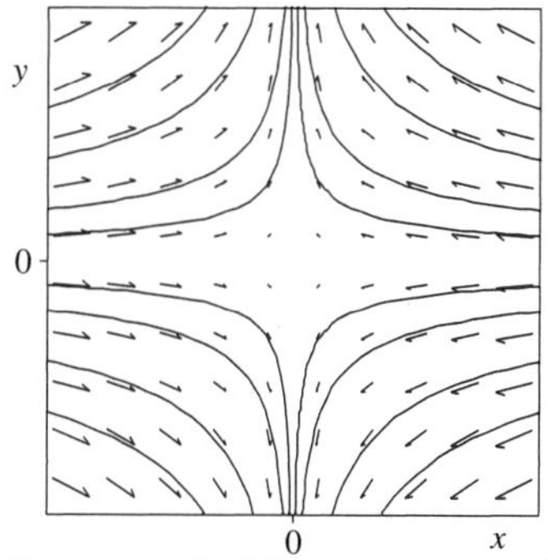

11. There are infinitely many answers. For example, $\mathbf{F}(x, y, z) = (y, y/z, 2z/y)$.

13. To get the flow lines of $-\mathbf{F}$, take the flow lines of **F** and reverse their orientation.

Section 4.6

1. The expression is meaningless.

3. Vector field.

5. Scalar function.

7. For instance, **F** is a constant vector field [$\mathbf{F}(\mathbf{x}) = \mathbf{a}$ for all **x**].

9. For instance, $\mathbf{F}(x, y, z) = f(x)\mathbf{i}$, where f is a differentiable function of one variable such that $f'(x) \neq 0$.

11. For instance, $\mathbf{F}(x, y, z) = f(x)\mathbf{i}$, where f is a positive and increasing function for all x [for instance, $f(x) = e^x$].

13. $\mathit{curl}\,\mathbf{F} = (xz + x)\mathbf{i} - (yz - y^2)\mathbf{j} - (z + 2yz)\mathbf{k}$; $\mathit{div}\,\mathbf{F} = xy$.

15. $\mathit{curl}\,\mathbf{F} = -2(y + z)\mathbf{i} + 2(x + 3z)\mathbf{j} + 2(x - 3y)\mathbf{k}$; $\mathit{div}\,\mathbf{F} = 6x + 2y - 2z$.

17. Imitate the calculation done after Example 4.53: replace $\mathbf{F}(x, y) = (x^2, 0)$ by $\mathbf{F}(x, y) = (f(x), 0)$ and argue similarly in the case $\mathbf{F}(x, y) = (0, g(y))$.

19. **r** is a radial vector field; since there are no rotations within the flow, $\mathit{curl}\,\mathbf{r} = \mathbf{0}$.

21. **F** is not conservative.

23. $\mathbf{F}(x, y)$ is conservative; $V(x, y) = -x^3y - y^4/4 + C$, where C is a constant.

25. $\mathbf{F}(x, y, z)$ is conservative; $V(x, y, z) = xy + 3z + C$, where C is a constant.

27. Use definitions and the chain rule.

29. No, since $\mathit{div}\,(\mathit{curl}\,\mathbf{F}) \neq 0$.

31. Show that $\mathit{curl}\,\mathbf{F} = \mathbf{0}$.

33. $\mathbf{F}(x, y) = (x^3 - 3xy^2, -3x^2y + y^3, 0)$; check that $\mathit{div}\,\mathbf{F} = 0$ and $\mathit{curl}\,\mathbf{F} = \mathbf{0}$.

35. Use $\mathit{curl}\,(\mathit{grad} f) = \mathbf{0}$ and the definition of Δf.

37. $\mathit{div}\,\mathbf{F} = 4x + 8y^2z + 3x^3 - 3x - 8y^2z + 2x^3 \neq 0$; $\mathit{div}\,(xyz^2\mathbf{F}) = 0$.

39. Use the chain rule to show that $\mathit{curl}\,\mathbf{F} = \mathbf{0}$.

41. Use definitions to show that both sides are equal to the same vector; the identity holds if both f and **F** are differentiable.

43. $\mathit{div}\,(x\mathbf{i} + y\mathbf{j} + z\mathbf{k}) = 3$.

45. Write $\|\mathbf{r}\|\mathbf{r} = \sqrt{x^2 + y^2 + z^2}\,(x\mathbf{i} + y\mathbf{j} + z\mathbf{k})$ and compute the divergence; alternatively, use Exercise 40 and then Exercises 43 and 44.

47. Use definitions; f and g must be of class C^2.

49. $\mathit{div}\,(\mathbf{F} \times \mathbf{r}) = 0$.

Section 4.7

1. Differentiate $F(x_1, \ldots, x_m, g(x_1, \ldots, x_m)) = 0$ with respect to x_i and use the chain rule.
3. $\partial r/\partial y = \sin\theta$, $\partial r/\partial\theta = \cos\theta/r$.
5. $\partial\rho/\partial x = \sin\phi\cos\theta$, $\partial\phi/\partial x = \cos\theta\cos\phi/\rho$.
7. Yes, it can be solved. $(\partial w/\partial x)(0, 0, 1) = -1$, $(\partial w/\partial z)(0, 0, 1) = -1$.
9. The given system can be solved uniquely for x, y, z as functions of u, v, w near all points (x, y, z, u, v, w) that satisfy $x^2y^2 + x^2z^2 - y^2z^2 \neq 0$.
11. The given system can be solved uniquely for x and y as functions of u and v near all points, whenever $ad - bc \neq 0$; $x = (du - bv)/(ad - bc)$ and $y = (-cu + av)/(ad - bc)$.
13. $g_x(3, 0) = 2/3$, $g_y(3, 0) = 1/6$. The linear approximation is $L_{(3,0)}(x, y) = 2x/3 + y/6 - 1$.

Section 4.8

1. Straightforward calculation.
3. $\Delta(fg) = f_{xx}g + 2f_xg_x + fg_{xx} + f_{yy}g + 2f_yg_y + fg_{yy}$; $g\Delta f = g(f_{xx} + f_{yy})$; $f\Delta g = f(g_{xx} + g_{yy})$; $grad f \cdot grad\, g = f_xg_x + f_yg_y$.
5. An alternative to a somewhat long computation is to simplify the expression using product rules of the form $div\,(\mathbf{F} \times \mathbf{G})$ and $curl\,(f\mathbf{F})$, together with the fact that the scalar triple product of vectors with two equal factors is zero.
7. $div\,(f\, grad\, g) = f\, div\,(grad\, g) + (grad\, g) \cdot (grad f)$; interchange f and g to obtain the expression for $div\,(g\, grad f)$ and subtract.
9. Use the product rule for $div\,(f\mathbf{F})$ with $f = \|\mathbf{r}\|^{-3}$ and $\mathbf{F} = \mathbf{r}$ and the fact that (check it!) $grad\,\|\mathbf{r}\|^{-3} = -3\|\mathbf{r}\|^{-5}\mathbf{r}$.
11. $grad f = (D_1g\,u_x + D_2g\,v_x + D_3g, D_1g\,u_y + D_2g\,v_y + D_4g)$.
13. $\|\mathbf{r}\|^{-2}\mathbf{r}$.
15. $\|\mathbf{r}\|^{-2}$.
17. $div\,\mathbf{F} = 3r$; $curl\,\mathbf{F} = \frac{1}{r}(\cos\theta - r^2)\mathbf{e}_r + 2z\mathbf{e}_z$.
19. $div\,\mathbf{F} = 2\theta/\rho - (\sin\theta\cot\phi)/\rho - \cot\phi$; the curl is computed to be $curl\,\mathbf{F} = \cos\theta\cos(2\phi)\mathbf{e}_\rho/(\rho\sin\phi) - (\frac{1}{2}\cos\theta\sin(2\phi) - 1)\mathbf{e}_\phi/(\rho\sin\phi) - 2\mathbf{e}_\theta$.

Chapter Review

True/False Quiz

1. False.
3. False.
5. True.
7. True.
9. True.
11. False.
13. False.

Review Exercises and Problems

1. Show that $u_t = -u_{xxx} = -8k^3e^{2k(x-4k^2t)}$.
3. Show that $(f(\mathbf{c}(t)))' = -\|\mathbf{c}'(t)\|^2 \leq 0$.
5. Straightforward calculation.
7. $T_2(x, y, z) = 4 + x + y - 6x^2 - z^2/2 + 6xz$.
9. There are many correct answers. For instance, $\mathbf{F}(x, y) = (-2x, -3y)$.
11. Critical points: $(0, k\pi)$, where k is an integer; all are saddle points.
13. The point $(-ad/(a^2 + b^2 + c^2), -bd/(a^2 + b^2 + c^2), -cd/(a^2 + b^2 + c^2))$ is closest to the origin; the distance is $|d|/\sqrt{a^2 + b^2 + c^2}$.
15. $T_2(x, y) = x - 2y - x^2/2 + xy - 2y^2$.

CHAPTER 5

Section 5.1

1. Yes.
3. No, since ϕ is not differentiable at 0.
5. Yes.
7. No, since ϕ is not bijective.
9. Use proof by contradiction.
11. Not simple; closed; not simple closed.
13. Simple; not closed; not simple closed.
15. Simple; not closed; not simple closed.
17. $\mathbf{c}(t) = (t, \sqrt{t^2 + 1})$, $t \in [-1, 1]$, is C^1 (hence differentiable and continuous).

19. $\mathbf{c}_1$, $\mathbf{c}_2$, and $\mathbf{c}_4$ are C^1 paths; $\mathbf{c}_3$ is continuous, but not piecewise C^1 and not C^1; $\mathbf{c}_5$ is piecewise C^1, but not continuous, and not C^1.

21. $\phi(t) = \pm\sqrt{Ct + D}$, $t \in [1, 3]$; the constants C and D are chosen so that ϕ is defined.

23. $(\cos(St), -\sin(St))$, $t \in [0, 2\pi/S]$, is orientation-preserving of speed S; $(\cos t, \sin t)$, $t \in [0, 2\pi]$, is orientation-reversing of speed 1.

Section 5.2

1. (a) 32. (b) $8\sqrt{65}$.

3. $11\sqrt{2}/2$.

5. $\sqrt{3}/3$.

7. $16 \ln 4 + (\ln 4)^2/2 - 547.5$.

9. $-48\pi\sqrt{5}$.

11. $5(1 - e^{-9})/9 \approx 0.5556$.

13. -16.

15. Yes, that is possible.

17. 8π.

19. $(2^{3/2} - 1)/3 \approx 0.6095$.

21. $8\pi e^{16}$.

23. No.

25. $-6/\pi$.

27. Mass $= \int_{\mathbf{c}} \rho\, ds$.

29. Parametrize $\mathbf{c}$ by $\mathbf{c}(\theta) = (r\cos\theta, r\sin\theta)$, $\theta_1 \le \theta \le \theta_2$, and use Definition 5.8.

Section 5.3

1. (a) $\mathbf{F}$ and $\mathbf{c}_1'$ are of opposite orientations.
(b) $\int_{\mathbf{c}_3} \mathbf{F}\cdot d\mathbf{s} > \int_{\mathbf{c}_2} \mathbf{F}\cdot d\mathbf{s}$.

3. (b) $\Delta\mathbf{c}_i = \frac{3}{n}\mathbf{i}$, $\mathbf{F}(\mathbf{c}(t_i)) = \left(1 + \frac{3}{n}(i-1)\right)\mathbf{i} + \mathbf{j}$, and $W_n = \sum_{i=1}^{n} \frac{3}{n}\left(1 + \frac{3}{n}(i-1)\right)$. (c) Compute $\lim_{n\to\infty} W_n$.

5. $2/5$.

7. $e^2 - 2e + 1$.

9. $(2/3)(\pi/2)^6 \approx 10.0145$.

11. $e + 11$.

13. $\int_{\mathbf{c}} \mathbf{F}\cdot d\mathbf{s} = k\ell(\mathbf{c})$, where $\ell(\mathbf{c})$ is the length of $\mathbf{c}$.

15. Approximately 5.2940.

17. The direction of $\mathbf{F}$ is orthogonal to all circles centered at the origin.

19. (a) 0. (b) $1/3$. (c) $1/5$. (d) $1/2$.
(e) $2/\pi \approx 0.6366$. (f) 1.

21. 0.

23. (a) $4/3$. (b) $4/3$. (c) $-4/3$; path integral depends on the orientation.

25. Express the path integral along a curve as the sum along its pieces that are smooth and apply Theorem 5.3 to each piece.

27. The circulation is 0; the flux is $2\pi r$.

29. The circulation is 0; the flux is πr.

31. If $\mathbf{c}(t) = (h\cos t, h\sin t, 0)$, $t \in [0, 2\pi]$, then $\int_{\mathbf{c}} \mathbf{F}\cdot d\mathbf{s} = h^2\pi\cdot$($\mathbf{k}$ component of $curl\,\mathbf{F}$); if $\mathbf{c}(t) = (h\cos t, 0, h\sin t)$, $t \in [0, 2\pi]$, then $\int_{\mathbf{c}} \mathbf{F}\cdot d\mathbf{s} = -h^2\pi\cdot$($\mathbf{j}$ component of $curl\,\mathbf{F}$); if $\mathbf{c}(t) = (0, h\cos t, h\sin t)$, $t \in [0, 2\pi]$, then $\int_{\mathbf{c}} \mathbf{F}\cdot d\mathbf{s} = h^2\pi\cdot$($\mathbf{i}$ component of $curl\,\mathbf{F}$).

Section 5.4

1. Not connected; not simply-connected; not star-shaped.

3. Connected; not simply-connected; not star-shaped.

5. Not connected; not simply-connected; not star-shaped.

7. Connected; simply-connected; not star-shaped.

9. Connected; simply-connected; star-shaped.

11. Draw a sphere and a simple closed curve on it. If a curve encloses the "North Pole," then push it around the other side.

13. $8\sin 1$.

15. Integral along $\mathbf{c}_1$ is π, whereas the integral along $\mathbf{c}_2$ is $-\pi$; the domain of $\mathbf{F}$ is not simply-connected, and Theorem 5.8 does not apply.

17. $\mathbf{F}$ is not a gradient vector field.

19. $\mathbf{F}$ is a gradient field on $U = \{(x, y) | y > 0\}$; $f = x^2 \ln y + y^2 + C$, C is a constant.

21. $\mathbf{F}$ is a gradient vector field on $\mathbb{R}^3$; $f = \sin(xy) + z\cos y + C$, where C is a constant.

23. $\pi/2$.

25. Let $\overline{\mathbf{c}}_1(t) = (0, t)$, $t \in [0, y]$; then $\int_{\overline{\mathbf{c}}_1} \mathbf{F}\cdot d\mathbf{s} = \int_0^y F_2(0, t)\,dt$. Let $\overline{\mathbf{c}}_2(t) = (t, y)$, $t \in [0, x]$; then $\int_{\overline{\mathbf{c}}_2} \mathbf{F}\cdot d\mathbf{s} = \int_0^x F_1(t, y)\,dt$. To compute the derivative of $f(x, y) = \int_0^y F_2(0, t)\,dt + \int_0^x F_1(t, y)\,dt$, use the Fundamental Theorem of Calculus.

27. (a) $curl\,\mathbf{F} = \mathbf{0}$. (b) $V(x, 0) = 0$. (c) $V(0, y) = 0$.

29. (a), (b) $-18e^{-2}$.

31. $f = \|\mathbf{r}\|^4/4 + C$, where C is a constant.

33. $\int_{\mathbf{c}} \mathbf{F}\cdot d\mathbf{s} = \ln 2178$.

Chapter Review

True/False Quiz

1. True.
3. False.
5. True.
7. True.
9. False.
11. True.

Review Exercises and Problems

1. $\phi(t) = (Ct + D)^{2/3} - 4/9$, where $C > 0$ and D are constants.
3. $\int_{\mathbf{c}} \mathbf{T} \cdot d\mathbf{s} = \ell(\mathbf{c})$, where $\ell(\mathbf{c})$ is the length of $\mathbf{c}$. $\int_{\mathbf{c}} \nabla f \cdot d\mathbf{s} = 0$.
5. Scalar curl of $\mathbf{F}$ is zero; $f(x, y) = 4x^2y + x^3 + C$.
7. $3/4$.
9. $-a\|\mathbf{r}\|^3/3$.
11. Using Newton's Second Law, show that $dE(t)/dt = 0$. Conservation of Energy Law.
13. $W = -mg(h_2 - h_1)$.

CHAPTER 6

Section 6.1

1. $\mathcal{R}_4 = 56.5$.
3. $448/9$.
5. $-75/2$.
7. $(1 + \sqrt{e})/4$.
9. Negative.
11. 5.
13. Conclude that $1 \le e^{x^2+y^2} \le e^2$ and integrate over R.
15. $2\pi^2$.
17. f is integrable; $\iint_R f\, dA = 11$.
19. Not true if $R = [0, 1] \times [0, 1]$; true if $R = [1, 2] \times [2, 3]$.
21. No.

Section 6.2

1. D is of type 1: $\phi(x) = -\sqrt{1-x^2}$, $\psi(x) = \sqrt{1-x^2}$, $a = -1$, and $b = 1$; D is of type 2: $\phi(y) = -\sqrt{1-y^2}$, $\psi(y) = \sqrt{1-y^2}$, $c = -1$, and $d = 1$. Thus, D is of type 3.
3. D is of type 1: $\phi(x) = x$, $\psi(x) = 2x$ (if $0 \le x \le 1$) and $\psi(x) = 2$ (if $1 \le x \le 2$), $a = 0$, and $b = 2$; D is of type 2: $\phi(y) = y/2$, $\psi(y) = 2$, $c = 0$, and $d = 2$. Thus, D is of type 3.
5. $\iint_D f\, dA$ is the negative of the volume of the solid region below D and above f. $\iint_D (f - g)\, dA$ is the volume of the solid region between f and g and above $D \subseteq \mathbb{R}^2$.
7. $r^2h\pi/3$.
9. $e^2 - 3$.
11. Break D into regions D_1 and D_2; D_1 is bounded by $x = 0$, $x = 4$, $y = -\sqrt{x}$, and $y = \sqrt{x}$; D_2 is bounded by $x = 4$, $x = 9$, $y = -x + 6$, and $y = \sqrt{x}$.
13. $e^3 - e^2 + e - 5$.
15. 0.
17. $1/6$.
19. Approximately -32.3135.
21. 0.
23. Approximately 23.0906.
25. $4e^{-3} \le \iint_D e^{-x-y}\, dA \le 4e$.
27. 22.
29. $1/12$.
31. Use the remark immediately following the statement of Theorem 7.
33. $ab\pi$.
35. $8\sqrt{2}/3$.
37. Points in R that lie on the circle centered at $(0, 0)$ of radius $\sqrt{5/3}$.
39. (a) Use the Extreme Value Theorem. (c) Use the Intermediate Value Theorem. (d) There is a point in the interval $[a, b]$ where f attains its average value.

Section 6.3

1. $-2e^{1/2} + 7/2$.
3. 156.
5. $2\pi - 8/3$.
7. $8/3$.
9. $11/24$.
11. 0.
13. $\int_0^{\pi/4} \left(\int_0^{\tan y} (y^2 - x)dx \right) dy$.

15. $\int_1^2 \left(\int_1^2 \frac{\ln x}{x} dy\right) dx + \int_2^4 \left(\int_{x/2}^2 \frac{\ln x}{x} dy\right) dx$.
17. $-1+\pi/2$.
19. $(\sin 162)/8$.
21. $10/3$.
23. $(e-1)/3$.

Section 6.4

1. $\pi(e^4-1)/4$.
3. $\pi^2/4$.
5. $\pi(5^{3/2}-3^{3/2})/3$.
7. $25\pi/2$.
9. Approximately 93.9578.
11. Follows from $DT = A$.
13. Approximately 0.01342.
15. T is an expansion by the factor a in the u-direction and a translation by b in the v-direction; the area of the image of D is equal to $a \cdot \text{area}(D)$.
17. $15\pi/4$.
19. $3\pi/2$.
21. $\pi^3/2^4$.
23. 0.
25. 27π.
27. $15/2$.
29. $-21/2$.
31. $1+2e-3e^{2/3}$.
33. 0; use the change of variables $x=\frac{1}{2}(u+v)$, $y=\frac{1}{2}(u-v)$.

Section 6.5

1. 12.
3. $1/60$.
5. $32\sqrt{2}$.
7. 0.
9. 0.
11. 4; three-dimensional solid in the first octant bounded by the plane $2x+4y+z=4$.
13. $128\pi/7$; three-dimensional solid inside the cylinder $x^2+y^2=4$, above the xy-plane, and below the paraboloid $z=x^2+y^2$.
15. 112π; three-dimensional solid inside the cylinder $x^2+y^2=8$, bounded from below by the plane $z=-3$ and from above by the paraboloid $z=8-x^2-y^2$.
17. $1/6$.
19. 8π.
21. $32/9$.
23. $\int_0^4 \left(\int_{4-x}^{\sqrt{16-x^2}} (16-x^2-y^2)\, dy\right) dx$;

    $\int_0^{\pi/2} \left(\int_{4/(\sin\theta+\cos\theta)}^4 (16-r^2) r\, dr\right) d\theta$.
25. $4\pi(a^3-(a^2-b^2)^{3/2})/3$.
27. $\int_0^{2\pi} \left(\int_0^{\sqrt{2}} \left(\int_{r^2-2}^{2-r^2} (r^2-2) r\, dz\right) dr\right) d\theta$.
29. 8.
31. $\int_0^{2\pi} \left(\int_0^{\sqrt{1/2}} \left(\int_{r^2}^{2r^2} 2r^2\cos\theta\, dz\right) dr\right) d\theta +$
    $\int_0^{2\pi} \left(\int_{\sqrt{1/2}}^{1} \left(\int_{r^2}^{1} 2r^2\cos\theta\, dz\right) dr\right) d\theta$.

Chapter Review

True/False Quiz

1. False.
3. False.
5. False.
7. True.
9. True.
11. False.

Review Exercises and Problems

1. $11\pi(2-\sqrt{2})/3$. The region of integration is the solid inside the inverted cone [with the vertex at (0, 0, 0) and angle $\pi/4$ with respect to the positive z-axis with the z-axis as the axis of rotation] bounded from above by the graph of $\rho = 2+\cos\theta$.
3. $a^3 m\pi$.
5. Let $\iint_R f\, dA = 0$ and assume that $f(x_0, y_0) > 0$ for some (x_0, y_0). Use the fact that a continuous function with $f(x_0, y_0) > 0$ must be positive on a ball centered at (x_0, y_0) to prove that $\iint_R f\, dA > 0$, thus getting a contradiction.
7. $\iint_D f\, dA = \int_0^{\pi/4} \left(\int_0^{\sec\theta} f(r\cos\theta, r\sin\theta) r\, dr\right) d\theta$. In the reversed order, $\iint_D f\, dA =$
   $\int_0^1 \left(\int_0^{\pi/4} f(r\cos\theta, r\sin\theta) r\, d\theta\right) dr +$
   $\int_1^{\sqrt{2}} \left(\int_{\arccos(1/r)}^{\pi/4} f(r\cos\theta, r\sin\theta) r\, d\theta\right) dr$.
9. $\iint_D f\, dA = \int_\pi^{3\pi/2} \left(\int_0^1 f(r\cos\theta, r\sin\theta) r\, dr\right) d\theta$.

11. $(8\pi/3) - 2\sqrt{3}$.
13. Use the idea of Example 6.3 in Section 6.1.
15. $40/3$.
17. (a) Use separation of variables; see Example 6.24 in Section 6.3. (b) The iterated integral is $\int_0^{2\pi}\left(\int_0^{\infty} e^{-r^2} r\,dr\right)d\theta = \pi$.
19. $4abc\pi/3$.
21. $\int_0^1\left(\int_0^{\pi/2} f(r\cos\theta, r\sin\theta)r\,d\theta\right)dr + \int_1^{\sqrt{2}}\left(\int_{\arccos(1/r)}^{\arcsin(1/r)} f(r\cos\theta, r\sin\theta)r\,d\theta\right)dr$; reversing the order, $\int_0^{\pi/4}\left(\int_0^{\sec\theta} f(r\cos\theta, r\sin\theta)r\,dr\right)d\theta + \int_{\pi/4}^{\pi/2}\left(\int_0^{\csc\theta} f(r\cos\theta, r\sin\theta)r\,dr\right)d\theta$.

CHAPTER 7

Section 7.1

1. Sphere: the top and bottom line segments are mapped to the North and South Poles, respectively, while the vertical boundary line segments are mapped to the meridian that is the intersection of the sphere and the xz-plane with $x \geq 0$. Cylinder: the vertical boundary line segments are mapped to the line that is the intersection of the cylinder and the xz-plane with $x \geq 0$. Horizontal lines $v = 0$ and $v = b$ are mapped to the bottom and top boundary circles.
3. $\mathbf{r}(u, v) = (a\cos v\cos u, a\cos v\sin u, a\sin v)$, $0 \leq u \leq 2\pi$, $0 \leq v \leq \pi/2$.
5. $\mathbf{r}(u, v) = (2\cos v\cos u - 2, 2\cos v\sin u + 3, 2\sin v + 7)$, $0 \leq u \leq 2\pi$, $-\pi/2 \leq v \leq \pi/2$; or, adjust any other parametrization of the sphere.
7. $\mathbf{r}(u, v) = (u, v, 2 + 3v - u)$, $(u, v) \in D$, where D is the disk $u^2 + v^2 \leq 4$.
9. $\mathbf{r}(u, v) = (u, v, 6 - u - 2v)$, $(u, v) \in D$, where D is the triangular region bounded by the coordinate lines and the line $u + 2v = 6$.
11. $\mathbf{r}(u, v) = (u, v, u^2 + v^2)$, $u, v \geq 0$.
13. (a) $\mathbf{T}_u = (2, 2u, 0)$, $\mathbf{T}_v = (0, 1, 2v)$; $\mathbf{N} = (4uv, -4v, 2)$. (b) S is smooth for all $u, v \geq 0$.
15. (a) $\mathbf{T}_u = (\cos u\cos v, \cos u\sin v, -2\sin u)$, $\mathbf{T}_v = (-\sin u\sin v, \sin u\cos v, 0)$; the surface normal is $\mathbf{N} = \sin u(2\sin u\cos v, 2\sin u\sin v, \cos u)$. (b) S is not smooth at $(0, 0, \pm 2)$.
17. (a) $\mathbf{T}_u = (1 + \cos v)(-\sin u, \cos u, 0)$, $\mathbf{T}_v = (-\sin v\cos u, -\sin v\sin u, \cos v)$; $\mathbf{N} = (1 + \cos v)(\cos u\cos v, \sin u\cos v, \sin v)$. (b) S is not smooth at $(0, 0, 0)$.
19. (a) $\mathbf{T}_u = (1, 0, -2u)$, $\mathbf{T}_v = (0, 1, -2v)$; $\mathbf{N} = (2u, 2v, 1)$. (b) S is smooth for all $u, v \geq 0$.
21. $z = 0$.
23. $4x - y + 2z - 3 = 0$.
25. The cone is not smooth only at $\mathbf{r}(u, 0) = (0, 0, 0)$. At any other point [call it $\mathbf{r}(u_0, v_0)$], the tangent plane has the equation $(v_0\cos u_0)x + (v_0\sin u_0)y - v_0 z = 0$.
27. $\mathbf{r}(u, v) = (a\cos v\cos u, b\cos v\sin u, c\sin v)$, $0 \leq u \leq 2\pi$, $-\pi/2 \leq v \leq \pi/2$ (of course, this is not the only possible parametrization).
29. There are no such points.
31. (a) $D\mathbf{r}(u_0, v_0)\begin{bmatrix} u \\ v \end{bmatrix} = A\begin{bmatrix} u \\ v \end{bmatrix}$,

$$A = \begin{bmatrix} \partial x/\partial u(u_0, v_0) & \partial x/\partial v(u_0, v_0) \\ \partial y/\partial u(u_0, v_0) & \partial y/\partial v(u_0, v_0) \\ \partial z/\partial u(u_0, v_0) & \partial z/\partial v(u_0, v_0) \end{bmatrix}.$$

(b) Use (a) and the fact that $D\mathbf{r}(u_0, v_0)$ maps tangent vectors based at (u_0, v_0) to tangent vectors based at $\mathbf{r}(u_0, v_0)$.

Section 7.2

1. $\mathbf{r}$ maps a horizontal line $v = v_0$ onto a circle in the plane $x = v_0$, centered at $(v_0, 0, 0)$ of radius $|f(v_0)|$. $\mathbf{r}$ maps a vertical line $u = u_0$ onto a curve that is obtained by rotating the graph of $y = f(x)$ through the angle of u_0 rad. The images of two sets of lines are perpendicular to each other.
3. $\mathbf{r}(u, v) = (v^2\cos u, v^3, v^2\sin u)$, where $1 \leq v \leq 2$ and $0 \leq u \leq 2\pi$.
5. Start with $\mathbf{r}(u_1, v_1) = \mathbf{r}(u_2, v_2)$ and prove that if $u_1, u_2 \neq 0$, then $u_1 = u_2$ and $v_1 = v_2$. Then show that $\mathbf{r}(0, v) = (0, 0, v^2)$ is not one-to-one.
7. Prove that $\mathbf{r}(u, v)$ is one-to-one for all $(u, v) \in [0, 1] \times [0, 2\pi]$.
9. About x-axis: $\mathbf{r}(t, u) = (2\cos t + 3, (2\sin t + 4)\cos u, (2\sin t + 4)\sin u)$, $0 \leq u, t \leq 2\pi$. About y-axis: $\mathbf{r}(t, u) = ((2\cos t + 3)\cos u, 2\sin t + 4, (2\cos t + 3)\sin u)$, $0 \leq u, t \leq 2\pi$.
11. (a) Determine what happens to the points B, B', B'', and B''' as D gets deformed into a torus. (b) A and A' are the same point on the cylinder. The curve is closed because, by (a), B and B'' end up at the same

point. (c) $\mathbf{r}$ maps the segments into a closed, continuous curve that wraps around the torus four times.

13. We get a doubly twisted strip. To obtain its parametrization, replace $t/2$ by t in the parametrization given in Example 7.19. It is a two-sided surface (thus orientable).

15. (a) The level curves $z=c$ are hyperbolas $y=c/x$, when $c \neq 0$. The intersection of $z=xy$ with the plane $y=mx$ is the parabola $z=mx^2$. (b) The change in u_0 produces a twist in a pair of parallel sides. (c) The change in v_0 produces a twist in the other pair of parallel sides.

17. The domain could be $0 \le u \le 2\pi$, $-\pi/2 \le v \le \pi/2$.

19. $\mathbf{r}(u, v) = (av\cos u, bv\sin u, cv)$, where $0 \le u \le 2\pi$ and $v \in \mathbb{R}$. The surface normal is $(bcv\cos u, acv\sin u, -abv)$. Not smooth at $(0, 0, 0)$, smooth everywhere else.

21. The level curve of value z_0 is the ellipse with semiaxes a and b and the center at $(z_0, 0, 0)$. The cylinder intersects the xz-plane along $x = \pm a$, and the yz-plane along $y = \pm b$. $\mathbf{r}(u, v) = (a\cos u, b\sin u, v)$, where $0 \le u \le 2\pi$, $v \in \mathbb{R}$, is its smooth parametrization.

23. $\mathbf{r}(u, v) = (u, v, c(u^2/a^2 + v^2/b^2))$, $u, v \in \mathbb{R}$. The surface intersects any vertical plane along a parabola.

Section 7.3

1. (a) Use the definition of the partial derivative of x with respect to u.

3. Use a path integral, or parametrize S by $\mathbf{r}(u, v) = (u\cos v, f(u), u\sin v)$, $a \le u \le b$, $0 \le v \le 2\pi$, and use a surface integral.

5. 0.

7. $(16 + 8\sqrt{3})/3$.

9. $(4 - \pi)\sqrt{2}/6$.

11. 4π.

13. Approximately 29.1966.

15. $\pi r\sqrt{h^2 + r^2}$.

17. $\iint_S x\,dS = \iint_S x^3\,dS = 0$; $\iint_S x^2\,dS = 4\pi a^4/3$.

19. The sides of S are a and $b\sqrt{1 + m^2}$.

21. $4(5^{3/2} - 1)/3$.

23. Approximately 10.4652.

25. $2\pi a^2$.

Section 7.4

1. $\mathbf{F}_1$ and $\mathbf{F}_2$.

3. 48.

5. 4.

7. 0.

9. 36π.

11. $-3\pi/2$.

13. $3\pi^2/4$.

15. -8π.

17. (a) $\pi a^2 c$. (b) $\pi a^2 c$. (c) The flow is vertical and the projections of both surfaces onto the xy-plane are the same.

19. 36π.

21. 0.

23. -2.

25. Use $\mathbf{r}(\theta, \phi) = a\cos\theta\sin\phi\mathbf{i} + a\sin\theta\sin\phi\mathbf{j} + a\cos\phi\mathbf{k}$, where $0 \le \theta \le 2\pi$ and $0 \le \phi \le \pi$; the surface normal is $a\sin\phi\,\mathbf{r}(\theta, \phi) = a^2\sin\phi\mathbf{e}_\rho$, see (2.38) in Section 2.8; now use the definition to compute the given surface integral.

27. $-52/3$.

29. 36π.

Section 7.5

1. $1 - a^2/3$.

3. $42/15$.

5. $19/10$.

7. $(0, 2a/\pi)$.

9. Mass is $a^2 b$; the center of mass is at $(a\pi/4, a/2)$.

11. The center of mass is at $(0, 0, 4.6695)$; the moment of inertia about the z-axis is approximately 84.4635ρ.

13. $(0, 0, 3a/8)$.

15. $a^3\mu\pi$.

17. $100\mu\pi$.

19. For example, $f(x, y, z) = z$ and $\mathbf{c}(t) = (0, 0, t)$, $-1 \le t \le 1$.

21. $2\pi \le \iint_S (z^2 + 1)\,dS \le 4\pi$.

23. $3\pi\sqrt{5}/e \le \iint_{\mathbf{c}} e^{\sin(x+y)} ds \le 3\pi e\sqrt{5}$.

Chapter Review

True/False Quiz

1. True.
3. False.
5. True.
7. False.
9. False.
11. False.
13. False.
15. False.

Review Exercises and Problems

1. $\mathbf{r}(u, t) = (a \cos t \cos u, a \cos t \sin u, c \sin t)$, $0 \le t, u \le 2\pi$.
3. Let $f(x, y, z) = z$; take S to be the part of the plane $z = x$, $-1 \le x \le 1$, $0 \le y \le 1$.
5. (a) $(QH/\epsilon_0)(R^2 + H^2)^{-1/2}$.
(b) $(QH/2\epsilon_0)(-(R^2 + H^2)^{-1/2} + 1/H)$ is the flux through each of the top and bottom disks.
7. Level curve of value z_0 is the hyperbola [centered at $(0, 0)$ with asymptotes $y = \pm bx/a$] if $z_0 \ne 0$ or a pair of lines if $z_0 = 0$. $\mathbf{r}(u, v) = (u, v, cu^2/a^2 - cv^2/b^2)$, $u, v \in \mathbb{R}$ is a smooth parametrization of the surface. See figure below for the graph of the paraboloid.

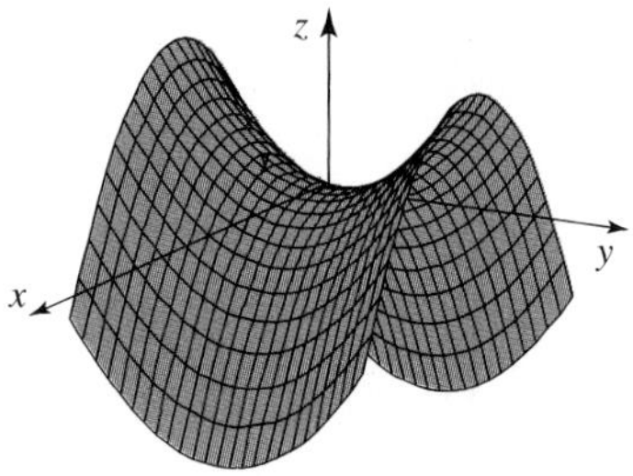

9. 21 m^3/s.
11. $A(S) = \iint_D |f_z|^{-1}(f_x^2 + f_y^2 + f_z^2)^{1/2}\, dA$;
$A(S) = \iint_D (g_x^2 + g_y^2 + 1)^{1/2}\, dA$.
13. Use formula (7.4) in Section 7.1 to show that $\|\mathbf{N}\| = EG - F^2$.
15. $I_x = I_z = 8\sqrt{2}/3$, $I_y = 4\sqrt{2}/3$.

CHAPTER 8

Section 8.1

1. (a), (b) 0.
3. 6π.
5. $-2e + (e^2 + 5)/2$.
7. $-109/90$.
9. 0.
11. 0.
13. 0.
15. $8/3$.
17. $3\pi/32$.
19. 8.
21. $I_x = \rho \int_{\mathbf{c}} (C_1(x)\mathbf{i} + (xy^2 + C_2(y))\mathbf{j}) \cdot d\mathbf{s}$,
$I_y = \rho \int_{\mathbf{c}} (D_1(x)\mathbf{i} + (\frac{1}{3}x^3 + D_2(y))\mathbf{j}) \cdot d\mathbf{s}$, where $C_1(x)$, $C_2(y)$, $D_1(x)$, and $D_2(y)$ are functions of the variables indicated (there are infinitely many answers).
23. Both sides are equal to 0.

Section 8.2

1. (b) 4π. The Divergence Theorem cannot be used since $\mathbf{F}$ is not defined at the origin. (c) $\iint_{S_2} \mathbf{F} \cdot d\mathbf{S} = 0$.
3. 90.
5. 0.
7. $14\pi/3$.
9. 0.
11. 0.
13. $8/3$.
15. Show that $div\,(\|\mathbf{r}\|^{-2}\mathbf{r}) = \|\mathbf{r}\|^{-2}$.
17. 0.
19. Using a product rule from Section 4.8, show that $div\,(f\nabla g) = f\Delta g + \nabla f \cdot \nabla g$.
21. Start with $D_{\mathbf{n}} f = \nabla f \cdot \mathbf{n}$, integrate over S, and use the Divergence Theorem.
23. 0.
25. Flux: $1/3$; circulation: $-3/2$.
27. Flux: $(e^4 - 1)(e^3 + 5)$; circulation: $-(e^3 - 1)(e^4 - 1)$.
29. (a) We end up with $\int e^{-x^2} dx$. (b) $\iint_S \mathbf{F} \cdot d\mathbf{S} = 0$.

Section 8.3

1. (a) 2. (b) 0.
3. 0.
5. $5/12$.
7. -16.
9. 0.
11. Show that *curl* $(\mathbf{F} \times \mathbf{r}) = 2\mathbf{F}$ and use Stokes' Theorem.
13. The circulation is 4π; Stokes' Theorem does not apply.
15. 0.
17. 2π
19. 0.
21. -3π.
23. Show that *curl* $(f\nabla f) = \mathbf{0}$ and use Stokes' Theorem.

Section 8.4

1. $e^{x+y}(x^2 + y^2)dz$.
3. $3x(x^2 + y^2)dxdz + yz(x^2 + y^2)dydz$.
5. Not defined.
7. $(2x^2 + 2y^2 - \sin x)dxdydz$.
9. 0 (zero 4-form).
11. $e^{xyz}(yzdx + xzdy + xydz)$.
13. $2xdxdydz$.
15. $(1 + x^2)^{-1}dx$.
17. $d\alpha = 4xy(x^2 + y^2)^{-2}dxdy$.
19. $d\alpha = (3x^2 + 2)dxdy$; $d(d\alpha) = 0$.
21. $d(df) = 0$ represents *curl* $\nabla f = \mathbf{0}$.
23. $5/34$.
25. (a) Closed and exact. (b) Closed and exact. (c) Not closed (thus not exact). (d) Closed and exact. (e) Closed and exact. (f) Exact.
27. α is exact because $d(xy + 4z + C) = \alpha$; $\int_{\mathbf{c}} \alpha = 4$.
29. 0.
31. 4π.
33. $\pi/2$.
35. -12.
37. $\pi/2$.
39. 28π.

Section 8.5

1. For example, $\mathbf{A}_0 = (2xz^2 + f(x))\mathbf{i} + (xy + g(y))\mathbf{j} + (yz + h(z))\mathbf{k}$, where f, g, and h are differentiable functions; $\mathbf{A}_0 = \mathbf{A} + grad f$, where $f(x, y, z)$ is a differentiable function.
3. Compute *div* $\mathbf{J}$ from (8.25) and then use (8.23).
5. Use definitions of *div* and *curl*.
7. $\frac{1}{r}\left(\frac{\partial E_z}{\partial \theta} - r\frac{\partial E_\theta}{\partial z}\right) = -\frac{\partial B_r}{\partial t}$;
$\frac{1}{r}\left(-\frac{\partial E_z}{\partial r} + \frac{\partial E_r}{\partial z}\right) = -\frac{\partial B_\theta}{\partial t}$;
$\frac{1}{r}\left(E_\theta + r\frac{\partial E_\theta}{\partial r} - \frac{\partial E_r}{\partial \theta}\right) = -\frac{\partial B_z}{\partial t}$.
9. Equation (8.29) does not hold.

Section 8.6

1. Straightforward (keep in mind that derivatives are taken with respect to x, y, and z).
3. Substitute $\mathbf{a} = \mathbf{i}$, and then $\mathbf{a} = \mathbf{j}$ and $\mathbf{a} = \mathbf{k}$ to show that $\mathbf{v} = \mathbf{w}$. Use the same strategy to prove the other statement.
5. $\mathbf{v} \cdot \nabla\mathbf{v} = (x + 3t, 4y + 4t, 9z + 3t)$.
7. Compare the $\mathbf{i}$ components (then also $\mathbf{j}$ and $\mathbf{k}$ components) of vectors on both sides.
9. Straightforward calculation; use definitions.

Chapter Review

True/False Quiz

1. True.
3. False.
5. True.
7. True.
9. True.
11. True.
13. True.

Review Exercises and Problems

1. $\iint_S \mathbf{F}\cdot d\mathbf{S} = -4\pi GM$.
3. $ab\pi$.
5. Use Green's Theorem.
7. $\int_c \alpha = 1/3$.
9. Use the fact that $div\,(f\nabla g - g\nabla f) = f\Delta g - g\Delta f$ (see Section 4.8).
11. $\int_S \alpha = 36 + 9\pi/2$.
13. Use Stokes' Theorem, or the definition of the path integral.
15. (a) Straightforward calculation with differentials. (b) Differential applied twice is zero. (c) Continuity equation for $\mathbf{J}$ and ρ. (d) $\alpha = -\mathcal{B}/\mu_0$.
17. (a) We get $curl\,\mathbf{E} = -\partial\mathbf{B}/\partial t$ and $div\,\mathbf{B} = 0$; see the end of Section 8.5. (b) $*\mathcal{E} = \mathcal{B}$. (c) $d(*\mathcal{E}) = 0$ represents $div\,\mathbf{E} = 0$ and $curl\,\mathbf{B} = \mathbf{0}$.
19. $d\alpha = -(f_y(x, y)dxdy + g_z(y, z)dydz + h_x(x, z)dzdx)$; f, g, and h must be C^2.

Index

absolute maximum, 244
absolute minimum, 244
acceleration, 120
  centripetal, 184, 188
  Coriolis, 188
  in cylindrical coordinates, 154
  normal component of, 202
  tangential component of, 202
addition of vectors
  parallelogram law, 7
  triangle law, 7
affine map, 407
  Jacobian for, 412
Ampère's Law, 555
angle between vectors, 22, 24
angular momentum, 189
angular velocity vector, 44
anticommutativity
  of cross product, 40
  of wedge product, 539
antiderivative, 182
arc-length, 193
arc-length function, 196
  parametrization by, 196
Archimedes's screw, 453
area
  below the graph of $y = f(x)$, 316
  integral formula for, 386
  of a "fence"along a curve, 318
  of a plane region, 386, 485
  surface, 485
  using Green's Theorem, 507
Area Cosine Principle, 469
associativity
  of addition of matrices, 31
  of addition of vectors, 7
  of multiplication of matrices, 35
  of wedge product, 539
average value of a function, 322, 393, 486

barometric formula, 106
basic 1-form, 537
basic 2-form, 538
basic 3-form, 538
basic 4-form, 563
Bézier curves, 167
bijective function (map), 308
Bilbao Guggenheim Museum, 455
binormal vector, 210
"bound"vector, 5
boundary, 380
  curve, orientation of, 525
  of a parametrized surface, 525
  of a region, orientation, 500
  point, 81, 253
  surface, positive orientation of, 512
bounded function, 317, 367, 417
bounded region, 380
bounded set, 254, 380
brachistochrone, 166, 218

C^1 curve, 118
C^1 function, 108
C^1 path, 118, 307
C^1 surface, 434
C^2 function, 220
Cartesian coordinate system, 1–2
Catalano House, 455, 457
catenary curve, 168
catenoid, 454
Cauchy–Riemann equations, 304
Cauchy–Schwarz inequality, 28, 50, 582
Cavalieri's principle, 375
center of gravity, *see* center of mass
center of mass, 18, 489
centripetal acceleration, 184, 188
centripetal force, 184
centroid, *see* center of mass
chain rule, 126, 299, 587
change of variables, 130
  notational convention, 130
  in definite integral, 401–402
  in double integral, 409
  in triple Integral, 420
charge density function, 554
circle
  curvature of, 204
  equation, 4
  osculating, 207
  parametrization of, 114
circle of curvature, 207
circulation
  electric, 337, 555
  magnetic, 337, 555
  of a vector field, 332–333
circulation density, 534
clockwise orientation, 314
closed differential form, 545
closed set, 253–254, 380
Cobb-Douglas production function, 55
commutativity
  of addition of matrices, 31
  of addition of vectors, 7
  of dot product, 21
component(s) of a function, 85
component(s) of a vector, 4
composition of functions
  chain rule, 126, 299, 587
  continuity of, 89
cone, 459
  parametrization, 436
connected set, 343
conservation of charge, 556
Conservation of Energy Law, 146
conservative force (field), 100, 145, 341
  equivalent properties, 356
Continuity Equation, 557, 568
continuous curve, 118
continuous function(s), 87–88, 118
  composition, 89
  list of, 88, 89
  on a set, 88
  on an interval, 87, 118
  properties of, 88–89
continuous path, 118
continuous surface, 434
continuously differentiable function, 108
contour curve, *see* level curve
contour diagram, 65
contour surface, *see* level surface
coordinate axis (axes), 1–2
coordinate system
  Cartesian (rectangular), 1–2
  cylindrical, 152
  polar, 3
  spherical, 155–156
coordinates of a vector, 4
Coriolis acceleration, 188
Coulomb's Law, 58
counterclockwise orientation, 314
critical point, 246

cross (vector) product, 39
  anticommutativity of, 40
  distributivity of, 40
  geometric properties of, 39
  right-hand rule, 43
cross-section, 65
curl, 278, 506
  as differential on 1-forms, 542
  as operator on vector fields, 280
  divergence of curl is zero, 288
  in context of fluid flow, 287
  in context of rotation of a rigid body, 286
  in cylindrical coordinates, 300
  in spherical coordinates, 301
  of a vector field, 278
  of an electrostatic field, 285
  of gradient is zero, 288
  physical meaning, 287, 335–336, 533–534
  properties of, 298–300
  scalar, of a vector field, 278, 335–336, 502
current, 554
current density vector field, 554
curvature, 204
  circle, 204
  formula(s) for, 208–209
  helix, 205
  line, 204
  parabola, 205
curve, 112, *see also* path
  C^1, 118
  contour, 63–65
  continuous, 118
  curvature, 204
  differentiable, 118
  endpoints of, 113
  equipotential, 69
  implicitly defined, 173
  initial point of, 113
  length of, 193
  level, 63–65
  "mechanical definition," 432
  negative orientation of, 114
  orientation-preserving parametrization of, 310
  orientation-reversing parametrization of, 310
  oriented simple, 314
  parametric equations of, 112
  parametrization of, 112
  positive orientation of, 114
  simple closed, 313
  simple, 313
  smooth, 193
  tangent vector to, 120
  terminal point of, 113
  torsion, 212
  unit tangent vector to, 197
cycloid, 165
cylinder
  parametrization, 435
cylindrical coordinates, 152
  acceleration in, 154
  Jacobian, 406
  orthonormal basis, 153
  unit tangent vectors in, 153
  velocity in, 154

decreasing returns to scale, 161
definite integral of a function, 316
derivative
  directional, 136, 138
  geometric interpretation of, 187
  matrix, of a function, 98, 101
  mixed partial derivatives, 221, 589
  partial, 95, 221, 589
  partial, on a surface, 131
  properties of, 123, 585–586
determinant of a matrix, 36
differentiable
  curve, 118
  function, 101
  path, 118
  surface, 434
differential (of a form), 540
  properties of, 542
  relation to *grad*, *curl*, *div*, 541–542
differential (of a function), 105
differential form(s), *see also* 0-form, 1-form, 2-form, 3-form, 4-form
  addition of, 537
  basic 1-form, 537
  basic 2-form, 538
  basic 3-form, 538
  basic 4-form, 563
  closed, 545
  components of, 537–538
  differential of, 540
  exact, 545
  Gauss' Divergence Theorem for, 549
  Green's Theorem for, 547
  multiplication by a function, 537
  of degree 0, 537
  of degree 1, 537
  of degree 2, 538
  of degree 3, 538
  Stokes' Theorem for, 551
  wedge product of, 538
  zero 0-form, 537
diffusion equation, 229
directed line segment, 5
direction cosines, 50
directional derivative, 136–137
  as slope of tangent, 136
  coordinate description of, 138
  of gravitational potential, 140
displacement vector, 14
distance function, 53
distributivity
  of cross product, 40
  of dot product, 21
  of multiplication of matrices, 35
  of multiplication of matrix by scalar, 31–32
  of multiplication of vector by scalar, 7
  of wedge product, 539
divergence, 278
  as differential on 2-forms, 542
  as total outflow, 280–282
  geometric interpretation, 283
  in cylindrical coordinates, 300
  in spherical coordinates, 301
  of a vector field, 278
  of an electrostatic field, 280
  of curl is zero, 288
  of gravitational force field, 280
  physical interpretation, 281–282, 566
  properties of, 298–299
Divergence Theorem of Gauss, 513
  for differential forms, 549
  in the plane, 517
dot product, 21
  commutativity of, 21
  distributivity of, 21
  geometric version of, 22
double integral, 367, 381
  as area, 370
  as volume, 366, 368, 378
  change of variables in, 409
  iterated, 378
  iterated, over elementary regions, 384
  notation for, 368
  over a closed and bounded region, 381
  over a rectangle, 366
  properties of, 387–388
  reversing the order of (iterated) integration, 394–397
  separation of variables, 398
  techniques of evaluation, 394–400
doubly ruled surfaces, 455

electric circulation of an electric field, 337
electric field
circulation of, 337, 555
energy density, 560
flux of, 556
time-changing, 555
electromagnetic
field, 560
force, 59
potential, 564
tensor, 564
electromotive force, 337
electrostatic field, 58–59
curl of, 285
divergence of, 280
flow lines of, 276
Gauss' Law for, 518
electrostatic potential, 58
electrostatic potential energy, 59
elementary region, 382
elementary 3D region, 420
ellipse
parametrization of, 114
ellipsoid, 458
elliptic cylinder, 459
elliptic paraboloid, 459
endpoint(s) of a curve, 113
energy density
of electric field, 560
of magnetic field, 560
total, 560
equation(s)
Cauchy–Riemann, 304
Continuity, 557, 568
diffusion, 229
Euler's, 571
heat, 225, 521–522
Korteweg-de Vries, 227, 304
Laplace's, 226–227, 289
Maxwell's, 520, 557–559
parametric, *see* parametric equation(s)
Poisson's, 227, 289
soliton, 227
wave, 223
equilibrium (of forces), 17
equipotential curve, 64
of electric dipole, 69
equipotential surface
of gravitational potential, 70–71, 140–141
Euler's equation, 571
exact differential form, 545
exponential atmosphere, 106
Extreme Value Theorem, 243, 255, 581
extreme values, 244

Faraday's Law, 557
Fermat's Theorem, 244, 246
Fick's Law, 290
"figure 8"curve, 314
flow
description using Laplace's operator, 290
heat 225–226
incompressible, 282, 569
irrotational, 287
line, 273
uniform sink, 62
vortex, 62
flow line, 273
of an electrostatic field, 276
fluid
incompressible, 569
is compressing, 282
perfect, 570
flux
across a region, 518
integral, 474
magnetic, 556
net outward, 474, 480
of an electric field, 556, 519
outward, of a vector field, 474, 480
flux density vector field, 290
force(s)
centripetal, 184
conservative, 100
gravitational, 57–58
electromagnetic, 59, 126
electrostatic, 58
in equilibrium, 17
Lorentz, 59
magnetic, 59
resultant, 17
torque of, 44, 189
work of, 27, 326–328
force field, *see* force(s)
form, *see* differential form
4-form
basic, 563
differential of, 563
"free"vector, 5
Frenet frame, 211
Fubini's Theorem, 378, 419
function, *see also* map
affine, 407
arc-length, 196
average value of, 322, 486
bijective function, 308
bounded, 317, 367, 417
C^1, 108
C^2, 220
Cobb-Douglas production, 55
component(s), 53, 85
continuous, 87, 88
continuously differentiable, 108
derivative (matrix) of, 98, 101
differentiable, 101
differential of, 105
directional derivative of, 136
distance, 53
good approximation of, 102–104
gradient of, 99, 135
graph of, 62–63
harmonic, 227, 289
integrable, 367, 418
limit of, 77–78, 80
linear, 35, 53
linear approximation of, 101, 104
maximum rate of change of, 140
one-to-one, 308, 405
onto, 308, 405
partial derivative of, 95
path, 100, 112
potential, 100, 145
projection, 53, 89
real-valued, of m variables, 53
scalar, of m variables, 53
scalar-valued, of m variables, 53
Taylor formula for, 234
twice continuously differentiable, 220
vector, of m variables, 53
vector-valued, of m variables, 53
Fundamental Theorem of Calculus, 346
generalization of, 347, 551

Gauss' Divergence Theorem 513, 517, 549
Gauss' Law for electrostatic fields, 518
Gauss' Theorem, 555
global maximum, 244
global minimum, 244
good approximation of a function, 102–103
gradient, 99, 135, *see also* gradient vector field
as differential on 0-forms, 541–542
as maximum rate of change, 140
in cylindrical coordinates, 300
in spherical coordinates, 301
is perpendicular to level curves, 141
is perpendicular to level surfaces, 142
of a distance function, 141
of a function, 99, 135

gradient—(*continued*)
of the gravitational potential, 100
properties of, 298–300
gradient vector field, 341
properties of, 354
graph of a function, 62, 63, 442
curvature of, 209
Implicit Function Theorem, 176, 292
is a smooth surface, 442
length of, 195
normal vector to, 443
orientation of $z = f(x, y)$, 446
parametrization of $y = f(x)$, 115
parametrization of $z = f(x, y)$, 442
tangent plane to, 443
gravitational constant, 58
gravitational force (field), 75, 58
divergence of, 280
gravitational potential, 58, 70, 100, 140
satisfies Laplace's equation, 289
gravitational potential energy, 59
Green's Theorem, 502, 534
for differential forms, 547
formulas for area, 507
regions involved in, 500
special case of Stokes' Theorem, 534
vector form, 506

harmonic function, 227, 289
heat conductivity, 521
heat equation, 225, 521, 522
heat flow, 225–226
heat flux vector field, 146
helicoid, 432, 453, 455
helix, 170
curvature of, 205, 213
length of, 194
parametrization of, 117–118
TNB frame for, 211
torsion of, 213
Hessian matrix, 239, 240, 250
Hodge star operator, 579
homogeneous object, 488
horizontally simple region, 383
hyperbolic cylinder, 459
hyperbolic paraboloid, 455, 459
hyperboloid
of one sheet, 459
of two sheets, 459

identity matrix, 31
Implicit Function Theorem
for curves, 176
general case, 295
special case, 292
implicitly defined curve, 173
implicitly defined surface, 293, 457
incompressible fluid, 569
incompressible vector field, 282
increasing returns to scale, 161
indefinite integral of a vector-valued function, 182
inequality
Cauchy–Schwarz, 28, 50
Triangle, 8, 50
initial point of a curve, 113
inner product, *see* dot product
inside of a surface, 444
instantaneous velocity vector, 120
integrable function, 367, 418
integral(s), *see also* double integral, path integral, surface integral, triple integral
additivity of, 486
definite, 316–317
double, 367, 381
double, of the Laplacian, 508
flux, 340, 481
indefinite, of a vector-valued function, 182
inequality for absolute value, 487
iterated, 378
line, 327
linearity of, 486
Mean Value Theorem for, 487
oriented, 330–331, 478
path, of a 1-form, 544
path, of a real-valued function, 318
path, of a vector field, 327
surface, of a real-valued function, 464
surface, of a 2-form, 546
surface, of a vector field, 474
triple, 418
integral curve, *see* flow line
interior of a set, 256
interior point, 81, 254
Intermediate Value Theorem, 581
irrotational vector field, 287
isobar, 64
isomer, 64
isotherm, 64
iterated integral, 378
over elementary regions, 384
iterated partial derivative, 220

Jacobian, 406, 411
components of normal vector, 439
for affine maps, 412
geometric meaning of, 410–411
in change of variables formula, 409, 421
in cylindrical coordinates, 423
in spherical coordinates, 424
in polar coordinates, 406

k-cross operator, 47
Kepler's Laws, 185, 217
kinetic energy, 126, 146, 234
Korteweg-de Vries equation, 227, 304

Lagrange multiplier, 264
Laplace operator, 226–227, 289
acting on vector fields, 289, 298
describes diffusion, 290
double integral of, 508
in cylindrical coordinates, 300
in spherical coordinates, 301
properties of, 298–299
Laplace equation, 226–227, 289
Laplacian, *see* Laplace's operator, 227
Leibniz's rule, 346
lemniscate, 173
length
arc-length, 193
in polar coordinates, 200
of a curve, 193
of a path, 192
of a polygonal path, 192
of a vector, 6
of the graph of $y = f(x)$, 195
level curve, 63, 65
gradient is perpendicular to, 141
level set, 65
level surface, 64, 65
gradient is perpendicular to, 142
limit of a function, 77–80
line
curvature of, 204
parametrization of, 11, 114
tangent, 119–120
line element, 541
line integral, *see* path integral
line of force, *see* flow line
line segment
directed, 5
parametrization of, 114
linear approximation, 101, 103, 104
linear function, 35, 53, 407
level curves of, 74
linear map, 407
linear combination (basic forms), 537–538
linear vector field, 35

linearization, *see* linear approximation
Lissajous curve, 169
local linearity, 102
local maximum, 244
local minimum, 244
logarithmic spiral, 167
Lorentz force, 59
Lorentz gauge, 563

magnetic circulation, 337
magnetic field, 59
  circulation, 555
  energy density of, 560
  flux, 556
  time-changing, 555
magnitude of a vector, 6
manifold with a boundary, 485
map, *see also* function
  affine, 407
  affine, properties of, 408
  bijective, 308
  inverse, 308
  Jacobian for, 412–413
  linear, 407
  one-to-one, 405
  onto, 405
  polar 404, 406
matrix (matrices)
  column of, 30
  determinant of, 36
  diagonal of, 31
  difference of, 31
  elementary operations, 31
  equal, 31
  Hessian, 239, 240, 250
  i-th column of, 30
  i-th row of, 30
  identity, 31
  main diagonal of, 31
  multiplication by a scalar, 31
  of order $m \times n$, 30
  of type $m \times n$, 30
  off-diagonal elements of, 31
  product of, 33
  row of, 30
  square, of order n, 30, 31
  sum of, 31
  transpose of, 37
  zero, 31
Maxwell's equations, 520, 557–559
Mean Value Theorem, 581
  for integrals, 388
Method of Lagrange multipliers, 263
minimal surface, 454
Möbius strip, 443, 451
moment (center of mass)
  about x-axis, 489
  about xy-plane, 18, 489
  about xz-plane, 18, 489
  about y-axis, 489
  about yz-plane, 18, 489
moment of inertia, 491
  about coordinate axes in $\mathbb{R}^3$, 492
  about origin, 491
  about x-axis, 491
  about y-axis, 491
  polar, 491
  Steiner's Theorem, 493

n-dimensional vector, 4
negative orientation of a curve, 114
negative side of a surface, 444
Newton's Law of Gravitation, 57, 185
Newton's Law of Cooling, 226
Newton's Second Law, 126, 146, 185, 223, 329
norm of a vector, 6
normal (component of) acceleration, 202
normal derivative, 509
normal unit vector, 443
normal vector
  inward, to a curve, 508–509
  orientation of a surface, 444, 445
  outward, to a curve, 508–509
  principal unit, 206
  to a plane, 28
  to a sphere, 441, 444
  to a surface, 439, 442–443
  to the graph of $z = f(x, y)$, 443
  unit, 206, 446
normalizing a vector, 8
1-form, 537, 563
  basic, 537
  differential of, 540
  electromagnetic potential, 564
  path integral of, 544
  wedge product of, 539
  zero, 537

one-to-one function, 405
onto function, 405
open ball, 79
open set, 93
operator, *see also* curl, divergence, gradient, Laplace operator
  differential, 298–301, 541–542
  k-cross, 47
  Hodge star, 579
orientation
  of $\mathbb{R}^2$, 8
  of $\mathbb{R}^3$, 8
orientation convention for the boundary of a plane region, 500
orientation of a path/curve, 113–114
  as boundary of a plane region, 500
  clockwise, 314
  counterclockwise, 314
  graph of a function, 62, 63
  negative, 114
  positive, 114
orientation of a surface, 444, *see also* surface
  boundary of, positive, 512
  graph of $z = f(x, y)$, 446
  of a boundary curve, 525
  positive, 444
orientation-preserving parametrization
  of a curve, 310
  of a surface, 445
orientation-reversing parametrization,
  of a curve, 310
  of a surface, 445
oriented simple curve, 314
origin, 1
orthogonal (vector) projection, 25
orthogonal trajectories, 151
orthogonal vectors, 23
  test for orthogonality, 23
orthonormal (set of) vectors, 23
osculating circle, 207
osculating plane, 207
outside of a surface, 444
outward flux, 518

parabola
  curvature, 205
parabolic cylinder, 459
parallel vectors, 8, 45
parallelepiped
  spanned by three vectors, 46–47
  volume, 46–47
parallelogram
  area, 46
  spanned by two vectors, 45–46
Parallelogram Law, 7
parameter(s), 11–12, 13–14
parametric equation, *see* parametrization
parametric representation, *see* parametrization
parametrization (of a curve), 112
  by arc-length, 196
  circle, 114
  cycloid, 166

parametrization—(*continued*)
ellipse, 114
graph of $y = f(x)$, 115
helix, 117–118
line, 11–12, 114
line segment, 114
smooth, 193
spiral, 167
parametrization (of a surface), 432
catenoid, 454
cone, 436
cylinder, 435
helicoid, 453
hyperbolic paraboloid, 455
Möbius strip, 451
orientation-preserving, 445
orientation-reversing, 445
plane, 13–14
smooth, 439
sphere, 437
surface of revolution, 448
torus, 450
Whitney umbrella, 453
parametrized surface, 432
see also parametrization (of a surface)
partial derivative, 95
iterated, 220
mixed, 221
notation for, 95–96
on a surface, 131
second-order, 220
patch, 463
path, 100, 112, 181, *see also* curve
C^1, 118, 307
continuous, 118
differentiable, 118
length of, 192
piecewise C^1, 307
polygonal, 191
reparametrization of, 309
smooth, 193
path integral
dependence on parametrization, 330
independence of parametrization, 321
of a real-valued function, 318
of a vector field, 327
of a 1-form, 544
path independence of, 354
perfect fluid, 570
permittivity of vacuum, 18, 58, 553
perpendicular vectors, *see* orthogonal vectors
piecewise C^1 path, 307
piecewise C^1 smooth surface, 468
plane
equation in space, 27
normal vector to, 28
osculating, 207
parametrization of, 14
spanned by two vectors, 13
tangent, to a surface, 104, 143, 440, 442
point
boundary, 81, 253
critical, 246
interior, 81, 254
saddle, 248
stationary, 246
point charges, 553
Poisson's equation, 227, 289
polar axis, 3
polar coordinates, 3, 132
polar form of a vector, 9
polar moment of inertia, 491
pole, 3
position vector, 5
positive orientation of a curve, 114
positive side of a surface, 444
potential, 145, 147
due to a point charge, 147
electromagnetic, 564
electrostatic, 58
function, 100, 145
function, finding it, 147
gravitational, 58, 70, 100, 140
of a dipole, 69
scalar, 560, 561
vector, 561
potential energy, 59, 146, 147
potential function, *see* potential
Poynting vector, 560
pressure in an ideal gas, 55
principal unit normal vector, 206
Principle of Superposition, 18
product
cross, of vectors, 39
dot, of vectors, 21, *see also* dot product
inner, of vectors, *see* dot product
of matrices, 33
scalar, of vectors, *see* dot product
scalar triple, of vectors, 42
projection
orthogonal, 25
scalar projection of a vector, 25
projection function, 53, 89
properties of
addition of matrices, 31
addition of vectors, 7
affine maps, 408
continuous functions, 88–89
cross (vector) product of vectors, 40
derivatives, 123, 585
determinant of a matrix, 37
differential of a form, 542
differential operators, 298–301, 541–542
dot (scalar) product of vectors, 21
double integrals, 370, 387–388
gradient, 143–144
gradient vector field, 354
multiplication of matrices by scalars, 31
multiplication of vectors by scalars, 7
wedge product, 539

quadratic surface, 458
quadric surface, 458

$\mathbb{R}^2$ (two-dimensional space), 2
$\mathbb{R}^3$ (three-dimensional space), 2
$\mathbb{R}^n$ (n-dimensional space), 2
real-valued function, 53
rectangle, 364
region
below f over $[a, b]$, 316
boundary of, 380–381
bounded, 380–381
closed, 380
elementary, 382
elementary 3D, 420
horizontally simple, 383
of type 1, 382
of type 1(3D), 419
of type 2, 382
of type 2(3D), 419
of type 3, 382
of type 3(3D), 420
of type 4(3D), 420
vertically simple, 383
x-simple, 383
y-simple, 383
relative maximum, 244
relative minimum, 244
relative position, 14
relative speed, 16
relative velocity, 16
reparametrization of a path, 309
as change of speed and/or direction, 309
orientation-preserving, 310
orientation-reversing, 310
representative of a vector, 5
resultant force, 17
Reynold's Transport Theorem, 574
Riemann sum, 317, 318

Riemann sum—(*continued*)
double 264–265, 267
triple, 418
rotation of a rigid body, 286

saddle point, 248
scalar (quantity), 1
scalar curl of a vector field, 278
scalar field, 57
scalar potential, 560
scalar product, *see* dot product
scalar projection, 25
scalar triple product, 42
Second Derivative Test, 244
Second Derivatives Test, 250
Second Fundamental Form, 498
Serret–Frenet formulas, 212
set
boundary of, 380
bounded, 254, 380
closed, 254, 380
connected, 343
interior of, 256
open, 93
simply connected, 343
star-shaped, 345
simple closed curve, 313
simple curve, 313
simply connected set, 343
sink, 521
smooth
curve, 193
parametrization, 193
path, 193
surface, 439
solid between f and g, 390
volume of, 390
solid in $\mathbb{R}^3$
Cavalieri's principle for, 375
volume of, 364–365, 377, 382, 418
solid of revolution, 376
volume of, 376
solid under $z = f(x, y)$, 364–365, 377, 382, 386
volume, 364–365, 377, 382, 386
soliton equation, 227
soliton wave, 227–228
source, 521
source intensity, 521
specific heat, 522
speed, 15, 120
relative, 16
sphere
parametrization of, 437
spherical coordinates, 155–156
Jacobian, 424
orthonormal basis, 157
unit tangent vectors in, 157
spiral, 167
standard basis of $\mathbb{R}^3$, 8
standard unit vectors, 8
star-shaped set, 345
stationary point, 246
steady current, 557
Steiner's Theorem, 493
Stokes' Theorem, 526, 534
for differential forms, 551
streamline, *see* flow line
surface, 63, 432
area, 463
area, integral formula, 464
boundary, positive orientation, 512
C^1, 434
continuous, 434
contour, 64, 65
differentiable, 434
doubly ruled, 455
equipotential,70–71, 140–141
graph of $z = f(x, y)$,
implicitly defined, 293, 457
inside, 444
integral, of scalar function, 464
integral, of vector field, 474
level surface, 64, 65
"mechanical"definition, 432–433
minimal, 454
negative side, 444
normal unit vector, 443
normal vector to, 439
of revolution, 448–449
one-sided, 443
orientation of, 444
orientation-preserving parametrization of, 445
orientation-reversing parametrization of, 445
oriented, 445
outside, 444
parametrization of, 432
parametrized, 432
partial derivatives on, 131
piecewise C^1 smooth, 468
positive orientation, 444
positive side, 444
quadratic, 458
quadric, 458
smooth, 439
smooth parametrization of, 439
tangent plane to, 442
tangent vector, 438
two-sided, 443
surface area, 463
integral formula, 464
surface integral
dependence on parametrization, 478
independence of parametrization, 465
of scalar function, 464
of vector field, 474
over the graph of $z = f(x, y)$, 465
surface normal vector, 439
to a cone, 440
to a sphere, 441
to a torus, 466
to the graph of $z = f(x, y)$, 442
surface of revolution, surface of revolution, 448
Sydney Opera House, 455, 456

tangent line, 119–120
tangent plane, 103, 442
tangent vector, 100, 119–120
tangential (component of) acceleration, 202
tangential velocity vector, 44
tautochrone, 166, 218
Taylor formula
first-order, 231, 235, 239
for functions of one variable, 234
nth-order, 234
second-order, 233, 236, 240
Taylor polynomial
first-order, 231
first-order, remainder, 231
nth-order, 234
nth-order, remainder, 234
second-order, 233, 236–237
second-order, remainder, 233, 236–237
terminal point of a curve, 113
theorem
Cauchy–Schwarz Inequality, 582
Chain Rule, 126, 587
Change of Variables, in Double Integral, 409
Change of Variables, in Triple Integral, 420
Divergence Theorem in the Plane, 517
Divergence Theorem of Gauss, 513
Divergence Theorem of Gauss for Differential Forms, 549
Equality of Mixed Partial Derivatives, 221, 589
Extreme Value Theorem, 243, 255, 581
Fermat's Theorem, 244, 246
Flux Transport, 573

theorem—(*continued*)
- Fubini's Theorem, 378, 419
- Fundamental Theorem of Calculus, 346, 347, 551
- Gauss' Theorem, 555
- Green's Theorem, 502, 534
- Green's Theorem for Differential Forms, 547
- Implicit Function Theorem for Curves, 176
- Implicit Function Theorem, 292, 295
- Intermediate Value Theorem, 581
- Lagrange Multipliers, 263
- Mean Value Theorem, 581
- Mean Value Theorem for Integrals, 388, 487
- Reynold's Transport Theorem, 574
- Second Derivative Test, 244
- Second Derivatives Test, 250
- Steiner's Theorem, 493
- Stokes' Theorem, 526, 534
- Stokes' Theorem for Differential Forms, 551
- Taylor's Formula, 231, 234–235, 239–240
- Triangle Inequality, 581

three-dimensional vector, 4
3-form, 538
- as wedge product, 539
- basic, 538, 563
- differential of, 540
- zero, 538

TNB frame, 211
tolerance, 79
torque, 44, 189
torsion, 212
- formula for, 214
- of a plane curve, 212
- of a helix, 213

torus, 450
- parametrization of, 450
- surface area of, 466

torus knot, 170
total energy, 146
total energy density, 560
total outflow, 280
transport theorem
- flux, 573
- Reynold's, 574
- vector form of, 575

transpose of a matrix, 37
Triangle inequality, 8, 50, 581
Triangle Law, 7
triple integral, 418
twice continuously differentiable function, 220
two-dimensional vector, 4
2-form, 538
- as wedge product, 539
- basic, 538, 563
- differential of, 540
- electromagnetic tensor, 564
- surface integral of, 546
- zero, 538

uniform sink flow, 62
uniformly charged object, 488
unit normal vector, 206
unit tangent vector, 197
unit vector, 8
- standard, 8

uv-coordinate system, 402
uv-plane, 402

vector (quantity), 1
vector(s), 1, 4
- acceleration, 120
- addition of, 6–7
- angle between, 22, 24
- angular velocity vector, 44
- "barb", 19
- binormal, 210
- "bound", 5
- column vector, 30
- components of, 4
- components of, in cylindrical coordinates, 153–154
- components of, in spherical coordinates, 157
- coordinates of, 4
- cross product, 39
- direction angles, 50
- direction cosines, 50
- displacement, 14
- dot (scalar) product of, 21
- "free", 5
- gradient, *see* gradient, gradient vector field
- "hunting arrow", 19
- in terms of orthogonal vectors, 24
- instantaneous velocity, 120
- inward normal to a curve, 509
- length of, 6
- magnitude of, 6
- multiplication by scalars, 6–7
- *n*-dimensional vector, 4
- norm of, 6
- normal, to a plane, 28
- normal, to a surface, 439
- normal unit, to a curve, 206
- normal unit, to a surface, 443
- of opposite directions, 8
- of same direction, 8
- on meteorological maps, 19
- orthogonal, 23
- orthogonal (vector) projection of, 25
- orthonormal, 23
- orthonormal, in cylindrical coordinates, 153
- orthonormal, in spherical coordinates, 157
- outward normal to a curve, 509
- parallel, 8
- perpendicular, 23
- polar form of, 9
- position, 5
- Poynting, 560
- principal unit normal, 206
- representative of, 5
- row, 30
- scalar projection of, 25
- standard unit, 8
- tangent, 120
- tangent, to a curve, 100
- tangent, to a surface, 438
- tangential velocity, 44
- three-dimensional, 4
- two-dimensional, 4
- unit, 8
- unit normal, 206
- unit tangent, 197
- velocity, 100, 120
- velocity, of a rotating body, 44
- "whisker", 19
- zero, 6

vector field, 56, *see also* vector(s)
- circulation around a curve, 332–333
- components of, 53, 153–154, 157
- conservative, 100, 145, 341
- curl of, 278, 280
- current density, 554
- divergence of, 278
- electric, 556, 519
- electromagnetic, 560
- electrostatic, 58, 59
- flow line of, 273
- flux (outward) of, 474, 480, 518
- flux density, 290
- gradient, 341
- gravitational, 58
- heat flux, 146
- in space, 56
- in the plane, 56

incompressible, 282, 569
irrotational, 287
Laplace, acting on, 289, 298
linear, 35
magnetic, 59
outward flux of, 480, 518
path integral of, 327
scalar curl of, 278, 335–336, 502
surface integral of, 474
velocity, 15, 120
vector potential, 561
vector product *see* cross product
vector-valued function, 53, *see also* vector field(s)
antiderivative of, 182
continuity, 88
derivative, 98, 101
differentiability, 101
limit of, 85
velocity, 15, 120
angular, 44
in cylindrical coordinates, 154–155
of rotation of a rigid body, 286
relative, 16
vector, 100
vertically simple region, 383
volume, 368
Cavalieri's principle for, 375
of a solid, 364–365, 377, 382, 418
of a solid between f and g, 390
of a solid of revolution, 376–377
of a solid under $z = f(x, y)$, 364–365, 377, 382, 386
vortex flow, 62
vorticity, 336

wave equation, 223
wedge product, 538–539
anticommutativity of, 539
associativity of, 539
distributivity of, 539
homogeneity of, 539
of basic forms, 539
Whitney umbrella, 453
wind chill index, 55, 68
work, 27, 326–328

x-axis, 1, 2
x-coordinate, 1, 2
x-simple region, 383
xy-coordinate system, 2, 402
xy-plane, 2, 402
xz-plane, 2

y-axis, 1, 2
y-coordinate, 1, 2
y-simple region, 383
yz-plane, 2

z-axis, 1, 2
z-coordinate, 2
0-form, 537
zero, 537
differential of, 540
zero 0-form, 537
zero 1-form, 537
zero 2-form, 538
zero 3-form, 538
zero matrix, 31
zero vector, 6

eighth edition

Calculus

SINGLE VARIABLE

HOWARD ANTON

Drexel University

IRL BIVENS

Davidson College

STEPHEN DAVIS

Davidson College

WILEY

JOHN WILEY & SONS, INC.

Associate Publisher: Laurie Rosatone
Freelance Developmental Editor: Anne Scanlan-Rohrer
Senior Marketing Manager: Angela Battle
Associate Editor: Jennifer Battista
Editorial Assistants: Danielle Amico/Kelly Boyle
Senior Production Editor: Ken Santor
Senior Designer: Karin Kincheloe
Cover Design: David Levy
Cover Photo: © Arthur Tilley/Taxi Getty Images
Text Design: Nancy Field
Photo Editor: Hilary Newman/Ellinor Wagoner
Illustration Editor: Sigmund Malinowski
Illustration Studio: Techsetters, Inc.

This book was set in Times Roman by Techsetters, Inc., and printed and bound by Von Hoffmann Press. The cover was printed by Von Hoffmann Press.

This book is printed on acid-free paper. ♾

The paper in this book was manufactured by a mill whose forest management programs include sustained yield harvesting of its timberlands. Sustained yield harvesting principles ensure that the numbers of trees cut each year does not exceed the amount of new growth.

ISBN 0-471-48273-0

Printed in the United States of America

10 9 8 7 6 5 4 3 2 1

c h a p t e r   t e n

INFINITE SERIES

Great fleas have little fleas upon their backs to bite 'em, And little fleas have lesser fleas, and so ad infinitum. And the great fleas themselves, in turn, have greater fleas to go on; While these again have greater still, and greater still, and so on.

—Augustus De Morgan
Mathematician

In this chapter we will be concerned with infinite series, which are sums that involve infinitely many terms. Infinite series play a fundamental role in both mathematics and science—they are used, for example, to approximate trigonometric functions and logarithms, to solve differential equations, to evaluate difficult integrals, to create new functions, and to construct mathematical models of physical laws. Since it is impossible to add up infinitely many numbers directly, one goal will be to define exactly what we mean by the sum of an infinite series. However, unlike finite sums, it turns out that not all infinite series actually have a sum, so we will need to develop tools for determining which infinite series have sums and which do not. Once the basic ideas have been developed we will begin to apply our work; we will show how infinite series are used to evaluate such quantities as ln 2, e, sin 3°, and π, how they are used to create functions, and finally, how they are used to model physical laws.

Photo: *Perspective creates the illusion that the sequence of telephone poles continues indefinitely but converges toward a single point infinitely far away.*

10.1 SEQUENCES

In everyday language, the term "sequence" means a succession of things in a definite order—chronological order, size order, or logical order, for example. In mathematics, the term "sequence" is commonly used to denote a succession of numbers whose order is determined by a rule or a function. In this section, we will develop some of the basic ideas concerning sequences of numbers.

DEFINITION OF A SEQUENCE

Stated informally, an ***infinite sequence***, or more simply a ***sequence***, is an unending succession of numbers, called ***terms***. It is understood that the terms have a definite order; that is, there is a first term a_1, a second term a_2, a third term a_3, a fourth term a_4, and so forth. Such a sequence would typically be written as

$$a_1, a_2, a_3, a_4, \ldots$$

where the dots are used to indicate that the sequence continues indefinitely. Some specific examples are

$$1, 2, 3, 4, \ldots, \qquad 1, \tfrac{1}{2}, \tfrac{1}{3}, \tfrac{1}{4}, \ldots,$$
$$2, 4, 6, 8, \ldots, \qquad 1, -1, 1, -1, \ldots$$

Each of these sequences has a definite pattern that makes it easy to generate additional terms if we assume that those terms follow the same pattern as the displayed terms. However, such patterns can be deceiving, so it is better to have a rule or formula for generating the terms. One way of doing this is to look for a function that relates each term in the sequence to its term number. For example, in the sequence

$$2, 4, 6, 8, \ldots$$

each term is twice the term number; that is, the nth term in the sequence is given by the formula $2n$. We denote this by writing the sequence as

$$2, 4, 6, 8, \ldots, 2n, \ldots$$

We call the function $f(n) = 2n$ the *general term* of this sequence. Now, if we want to know a specific term in the sequence, we need only substitute its term number in the formula for the general term. For example, the 37th term in the sequence is $2 \cdot 37 = 74$.

▶ **Example 1** In each part, find the general term of the sequence.

(a) $\frac{1}{2}, \frac{2}{3}, \frac{3}{4}, \frac{4}{5}, \ldots$ (b) $\frac{1}{2}, \frac{1}{4}, \frac{1}{8}, \frac{1}{16}, \ldots$

(c) $\frac{1}{2}, -\frac{2}{3}, \frac{3}{4}, -\frac{4}{5}, \ldots$ (d) $1, 3, 5, 7, \ldots$

Table 10.1.1

TERM NUMBER	1	2	3	4	⋯	n	⋯
TERM	$\frac{1}{2}$	$\frac{2}{3}$	$\frac{3}{4}$	$\frac{4}{5}$	⋯	$\frac{n}{n+1}$	⋯

Solution (a). In Table 10.1.1, the four known terms have been placed below their term numbers, from which we see that the numerator is the same as the term number and the denominator is one greater than the term number. This suggests that the nth term has numerator n and denominator $n + 1$, as indicated in the table. Thus, the sequence can be expressed as

$$\frac{1}{2}, \frac{2}{3}, \frac{3}{4}, \frac{4}{5}, \ldots, \frac{n}{n+1}, \ldots$$

Table 10.1.2

TERM NUMBER	1	2	3	4	⋯	n	⋯
TERM	$\frac{1}{2}$	$\frac{1}{2^2}$	$\frac{1}{2^3}$	$\frac{1}{2^4}$	⋯	$\frac{1}{2^n}$	⋯

Solution (b). In Table 10.1.2, the denominators of the four known terms have been expressed as powers of 2 and the first four terms have been placed below their term numbers, from which we see that the exponent in the denominator is the same as the term number. This suggests that the denominator of the nth term is 2^n, as indicated in the table. Thus, the sequence can be expressed as

$$\frac{1}{2}, \frac{1}{4}, \frac{1}{8}, \frac{1}{16}, \ldots, \frac{1}{2^n}, \ldots$$

Solution (c). This sequence is identical to that in part (a), except for the alternating signs. Thus, the nth term in the sequence can be obtained by multiplying the nth term in part (a) by $(-1)^{n+1}$. This factor produces the correct alternating signs, since its successive values, starting with $n = 1$, are $1, -1, 1, -1, \ldots$. Thus, the sequence can be written as

$$\frac{1}{2}, -\frac{2}{3}, \frac{3}{4}, -\frac{4}{5}, \ldots, (-1)^{n+1}\frac{n}{n+1}, \ldots$$

Table 10.1.3

TERM NUMBER	1	2	3	4	⋯	n	⋯
TERM	1	3	5	7	⋯	$2n-1$	⋯

Solution (d). In Table 10.1.3, the four known terms have been placed below their term numbers, from which we see that each term is one less than twice its term number. This suggests that the nth term in the sequence is $2n - 1$, as indicated in the table. Thus, the sequence can be expressed as

$$1, 3, 5, 7, \ldots, 2n - 1, \ldots \blacktriangleleft$$

When the general term of a sequence

$$a_1, a_2, a_3, \ldots, a_n, \ldots \tag{1}$$

is known, there is no need to write out the initial terms, and it is common to write only the general term enclosed in braces. Thus, (1) might be written as

$$\{a_n\}_{n=1}^{+\infty} \quad \text{or as} \quad \{a_n\}_{n=1}^{\infty}$$

For example, here are the four sequences in Example 1 expressed in brace notation.

Consider the sequence whose general term is

$$f(n) = \tfrac{1}{3}(3 - 5n + 6n^2 - n^3)$$

Calculate the first three terms, and make a conjecture about the fourth term. Check your conjecture by calculating the fourth term. What message does this convey?

SEQUENCE	BRACE NOTATION
$\frac{1}{2}, \frac{2}{3}, \frac{3}{4}, \frac{4}{5}, \ldots, \frac{n}{n+1}, \ldots$	$\left\{\frac{n}{n+1}\right\}_{n=1}^{+\infty}$
$\frac{1}{2}, \frac{1}{4}, \frac{1}{8}, \frac{1}{16}, \ldots, \frac{1}{2^n}, \ldots$	$\left\{\frac{1}{2^n}\right\}_{n=1}^{+\infty}$
$\frac{1}{2}, -\frac{2}{3}, \frac{3}{4}, -\frac{4}{5}, \ldots, (-1)^{n+1}\frac{n}{n+1}, \ldots$	$\left\{(-1)^{n+1}\frac{n}{n+1}\right\}_{n=1}^{+\infty}$
$1, 3, 5, 7, \ldots, 2n-1, \ldots$	$\{2n-1\}_{n=1}^{+\infty}$

The letter n in (1) is called the ***index*** for the sequence. It is not essential to use n for the index; any letter not reserved for another purpose can be used. For example, we might view the general term of the sequence $a_1, a_2, a_3, \ldots$ to be the kth term, in which case we would denote this sequence as $\{a_k\}_{k=1}^{+\infty}$. Moreover, it is not essential to start the index at 1; sometimes it is more convenient to start it at 0 (or some other integer). For example, consider the sequence

$$1, \frac{1}{2}, \frac{1}{2^2}, \frac{1}{2^3}, \ldots$$

One way to write this sequence is

$$\left\{\frac{1}{2^{n-1}}\right\}_{n=1}^{+\infty}$$

However, the general term will be simpler if we think of the initial term in the sequence as the zeroth term, in which case we can write the sequence as

$$\left\{\frac{1}{2^n}\right\}_{n=0}^{+\infty}$$

We began this section by describing a sequence as an unending succession of numbers. Although this conveys the general idea, it is not a satisfactory mathematical definition because it relies on the term "succession," which is itself an undefined term. To motivate a precise definition, consider the sequence

$$2, 4, 6, 8, \ldots, 2n, \ldots$$

If we denote the general term by $f(n) = 2n$, then we can write this sequence as

$$f(1), f(2), f(3), \ldots, f(n), \ldots$$

which is a "list" of values of the function

$$f(n) = 2n, \quad n = 1, 2, 3, \ldots$$

whose domain is the set of positive integers. This suggests the following definition.

10.1.1 DEFINITION. A ***sequence*** is a function whose domain is a set of integers. Specifically, we will regard the expression $\{a_n\}_{n=1}^{+\infty}$ to be an alternative notation for the function $f(n) = a_n$, $n = 1, 2, 3, \ldots$.

When the starting value for the index of a sequence is not relevant to the discussion, it is common to use a notation such as $\{a_n\}$ in which there is no reference to the starting value of n. We can distinguish between different sequences by using different letters for their general terms; thus, $\{a_n\}$, $\{b_n\}$, and $\{c_n\}$ denote three different sequences.

GRAPHS OF SEQUENCES

Since sequences are functions, it makes sense to talk about the graph of a sequence. For example, the graph of the sequence $\{1/n\}_{n=1}^{+\infty}$ is the graph of the equation

$$y = \frac{1}{n}, \quad n = 1, 2, 3, \ldots$$

Because the right side of this equation is defined only for positive integer values of n, the graph consists of a succession of isolated points (Figure 10.1.1a). This is in distinction to the graph of

$$y = \frac{1}{x}, \quad x \geq 1$$

which is a continuous curve (Figure 10.1.1b).

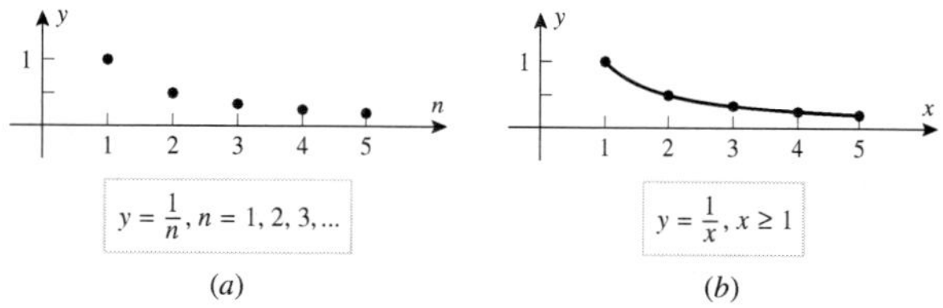

Figure 10.1.1

LIMIT OF A SEQUENCE

Since sequences are functions, we can inquire about their limits. However, because a sequence $\{a_n\}$ is only defined for integer values of n, the only limit that makes sense is the limit of a_n as $n \to +\infty$. In Figure 10.1.2 we have shown the graphs of four sequences, each of which behaves differently as $n \to +\infty$:

- The terms in the sequence $\{n+1\}$ increase without bound.
- The terms in the sequence $\{(-1)^{n+1}\}$ oscillate between -1 and 1.
- The terms in the sequence $\{n/(n+1)\}$ increase toward a "limiting value" of 1.
- The terms in the sequence $\left\{1 + \left(-\frac{1}{2}\right)^n\right\}$ also tend toward a "limiting value" of 1, but do so in an oscillatory fashion.

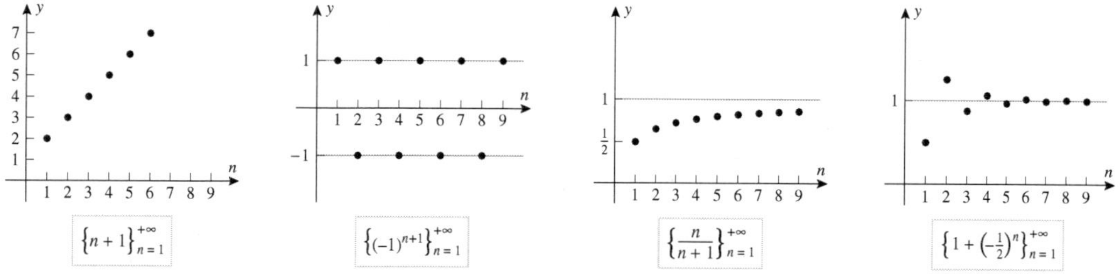

Figure 10.1.2

Informally speaking, the limit of a sequence $\{a_n\}$ is intended to describe how a_n behaves as $n \to +\infty$. To be more specific, we will say that *a sequence* $\{a_n\}$ *approaches a limit* L *if the terms in the sequence eventually become arbitrarily close to* L. Geometrically, this means that for any positive number ϵ there is a point in the sequence after which all terms lie between the lines $y = L - \epsilon$ and $y = L + \epsilon$ (Figure 10.1.3).

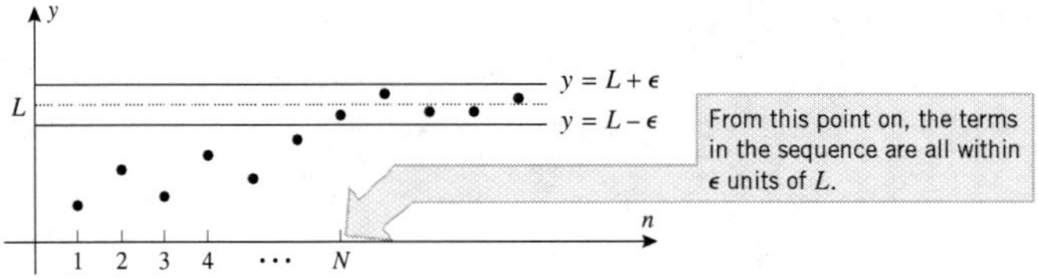

Figure 10.1.3

The following definition makes these ideas precise.

How would you define

$$\lim_{n \to +\infty} a_n = +\infty$$

and

$$\lim_{n \to +\infty} a_n = -\infty?$$

10.1.2 DEFINITION. A sequence $\{a_n\}$ is said to ***converge*** to the ***limit*** L if given any $\epsilon > 0$, there is a positive integer N such that $|a_n - L| < \epsilon$ for $n \geq N$. In this case we write

$$\lim_{n \to +\infty} a_n = L$$

A sequence that does not converge to some finite limit is said to ***diverge***.

► **Example 2** The first two sequences in Figure 10.1.2 diverge, and the second two converge to 1; that is,

$$\lim_{n \to +\infty} \frac{n}{n+1} = 1 \quad \text{and} \quad \lim_{n \to +\infty} \left[1 + \left(-\tfrac{1}{2}\right)^n\right] = 1 \blacktriangleleft$$

The following theorem, which we state without proof, shows that the familiar properties of limits apply to sequences. This theorem ensures that the algebraic techniques used to find limits of the form $\lim_{x \to +\infty}$ can also be used for limits of the form $\lim_{n \to +\infty}$.

10.1.3 THEOREM. *Suppose that the sequences $\{a_n\}$ and $\{b_n\}$ converge to limits L_1 and L_2, respectively, and c is a constant. Then:*

(a) $\lim_{n \to +\infty} c = c$

(b) $\lim_{n \to +\infty} ca_n = c \lim_{n \to +\infty} a_n = cL_1$

(c) $\lim_{n \to +\infty} (a_n + b_n) = \lim_{n \to +\infty} a_n + \lim_{n \to +\infty} b_n = L_1 + L_2$

(d) $\lim_{n \to +\infty} (a_n - b_n) = \lim_{n \to +\infty} a_n - \lim_{n \to +\infty} b_n = L_1 - L_2$

(e) $\lim_{n \to +\infty} (a_n b_n) = \lim_{n \to +\infty} a_n \cdot \lim_{n \to +\infty} b_n = L_1 L_2$

(f) $\lim_{n \to +\infty} \left(\dfrac{a_n}{b_n}\right) = \dfrac{\lim_{n \to +\infty} a_n}{\lim_{n \to +\infty} b_n} = \dfrac{L_1}{L_2}$ (if $L_2 \neq 0$)

Additional limit properties follow from those in Theorem 10.1.3. For example, use part (e) to show that if $a_n \to L$ and m is a positive integer, then

$$\lim_{n \to +\infty} (a_n)^m = L^m$$

▶ **Example 3** In each part, determine whether the sequence converges or diverges. If it converges, find the limit.

$$\text{(a)}\ \left\{\frac{n}{2n+1}\right\}_{n=1}^{+\infty} \qquad \text{(b)}\ \left\{(-1)^{n+1}\frac{n}{2n+1}\right\}_{n=1}^{+\infty}$$

$$\text{(c)}\ \left\{(-1)^{n+1}\frac{1}{n}\right\}_{n=1}^{+\infty} \qquad \text{(d)}\ \{8-2n\}_{n=1}^{+\infty}$$

Solution (a). Dividing numerator and denominator by n yields

$$\lim_{n\to+\infty}\frac{n}{2n+1} = \lim_{n\to+\infty}\frac{1}{2+1/n} = \frac{\lim\limits_{n\to+\infty} 1}{\lim\limits_{n\to+\infty}(2+1/n)} = \frac{\lim\limits_{n\to+\infty} 1}{\lim\limits_{n\to+\infty} 2 + \lim\limits_{n\to+\infty} 1/n}$$

$$= \frac{1}{2+0} = \frac{1}{2}$$

Thus, the sequence converges to $\frac{1}{2}$.

Solution (b). This sequence is the same as that in part (a), except for the factor of $(-1)^{n+1}$, which oscillates between $+1$ and -1. Thus, the terms in this sequence oscillate between positive and negative values, with the odd-numbered terms being identical to those in part (a) and the even-numbered terms being the negatives of those in part (a). Since the sequence in part (a) has a limit of $\frac{1}{2}$, it follows that the odd-numbered terms in this sequence approach $\frac{1}{2}$, and the even-numbered terms approach $-\frac{1}{2}$. Therefore, this sequence has no limit—it diverges.

Solution (c). Since $\lim_{n\to+\infty} 1/n = 0$, the product $(-1)^{n+1}(1/n)$ oscillates between positive and negative values, with the odd-numbered terms approaching 0 through positive values and the even-numbered terms approaching 0 through negative values. Thus,

$$\lim_{n\to+\infty}(-1)^{n+1}\frac{1}{n} = 0$$

so the sequence converges to 0.

Solution (d). $\lim\limits_{n\to+\infty}(8-2n) = -\infty$, so the sequence $\{8-2n\}_{n=1}^{+\infty}$ diverges. ◀

If the general term of a sequence is $f(n)$, where $f(x)$ is a function defined on the entire interval $[1, +\infty)$, then the values of $f(n)$ can be viewed as "sample values" of $f(x)$ taken at the positive integers. Thus,

if $f(x)\to L$ as $x\to+\infty$, then $f(n)\to L$ as $n\to+\infty$

(Figure 10.1.4*a*). However, the converse is not true; that is, one cannot infer that $f(x)\to L$ as $x\to+\infty$ from the fact that $f(n)\to L$ as $n\to+\infty$ (Figure 10.1.4*b*).

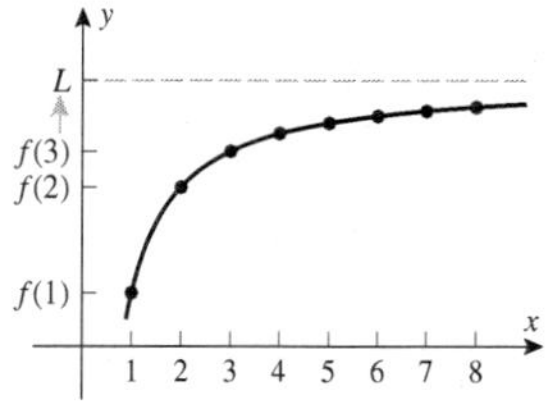

If $f(x)\to L$ as $x\to+\infty$, then $f(n)\to L$ as $n\to+\infty$.

(*a*)

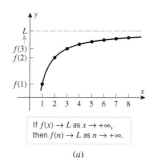

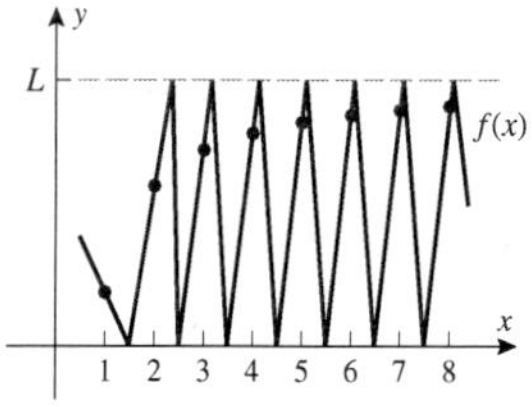

$f(n)\to L$ as $n\to+\infty$, but $f(x)$ diverges by oscillation as $x\to+\infty$.

(*b*)

Figure 10.1.4

▶ **Example 4** In each part, determine whether the sequence converges, and if so, find its limit.

$$\text{(a)}\ 1, \frac{1}{2}, \frac{1}{2^2}, \frac{1}{2^3}, \ldots, \frac{1}{2^n}, \ldots \qquad \text{(b)}\ 1, 2, 2^2, 2^3, \ldots, 2^n, \ldots$$

Solution. Replacing n by x in the first sequence produces the power function $(1/2)^x$, and replacing n by x in the second sequence produces the power function 2^x. Now recall that if

$0 < b < 1$, then $b^x \to 0$ as $x \to +\infty$, and if $b > 1$, then $b^x \to +\infty$ as $x \to +\infty$ (Figure 7.1.1). Thus,

$$\lim_{n\to+\infty} \frac{1}{2^n} = 0 \quad \text{and} \quad \lim_{n\to+\infty} 2^n = +\infty \ ◀$$

▶ **Example 5** Find the limit of the sequence $\left\{\dfrac{n}{e^n}\right\}_{n=1}^{+\infty}$.

Solution. The expression n/e^n is an indeterminate form of type ∞/∞ as $n \to +\infty$, so L'Hôpital's rule is indicated. However, we cannot apply this rule directly to n/e^n because the functions n and e^n have been defined here only at the positive integers, and hence are not differentiable functions. To circumvent this problem we extend the domains of these functions to all real numbers, here implied by replacing n by x, and apply L'Hôpital's rule to the limit of the quotient x/e^x. This yields

$$\lim_{x\to+\infty} \frac{x}{e^x} = \lim_{x\to+\infty} \frac{1}{e^x} = 0$$

from which we can conclude that

$$\lim_{n\to+\infty} \frac{n}{e^n} = 0 \ ◀$$

▶ **Example 6** Show that $\lim_{n\to+\infty} \sqrt[n]{n} = 1$.

Solution.

$$\lim_{n\to+\infty} \sqrt[n]{n} = \lim_{n\to+\infty} n^{1/n} = \lim_{n\to+\infty} e^{(1/n)\ln n} = e^0 = 1$$

By L'Hôpital's rule applied to $(1/x)\ln x$ ◀

Sometimes the even-numbered and odd-numbered terms of a sequence behave sufficiently differently that it is desirable to investigate their convergence separately. The following theorem, whose proof is omitted, is helpful for that purpose.

10.1.4 THEOREM. *A sequence converges to a limit L if and only if the sequences of even-numbered terms and odd-numbered terms both converge to L.*

▶ **Example 7** The sequence

$$\frac{1}{2}, \frac{1}{3}, \frac{1}{2^2}, \frac{1}{3^2}, \frac{1}{2^3}, \frac{1}{3^3}, \ldots$$

converges to 0, since the even-numbered terms and the odd-numbered terms both converge to 0, and the sequence

$$1, \tfrac{1}{2}, 1, \tfrac{1}{3}, 1, \tfrac{1}{4}, \ldots$$

diverges, since the odd-numbered terms converge to 1 and the even-numbered terms converge to 0. ◀

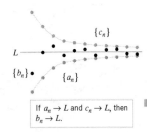

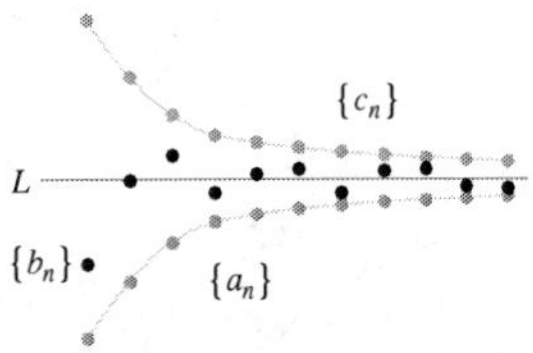

If $a_n \to L$ and $c_n \to L$, then $b_n \to L$.

Figure 10.1.5

THE SQUEEZING THEOREM FOR SEQUENCES

The following theorem, illustrated in Figure 10.1.5, is an adaptation of the Squeezing Theorem (2.6.2) to sequences. This theorem will be useful for finding limits of sequences that cannot be obtained directly. The proof is omitted.

10.1.5 THEOREM (*The Squeezing Theorem for Sequences*). *Let* $\{a_n\}$, $\{b_n\}$, *and* $\{c_n\}$ *be sequences such that*

$$a_n \le b_n \le c_n \quad (\textit{for all values of } n \textit{ beyond some index } N)$$

If the sequences $\{a_n\}$ *and* $\{c_n\}$ *have a common limit* L *as* $n \to +\infty$, *then* $\{b_n\}$ *also has the limit* L *as* $n \to +\infty$.

Recall that if n is a positive integer, then $n!$ (read "n factorial") is the product of the first n positive integers. In addition, it is convenient to define $0! = 1$.

▶ **Example 8** Use numerical evidence to make a conjecture about the limit of the sequence

$$\left\{\frac{n!}{n^n}\right\}_{n=1}^{+\infty}$$

and then confirm that your conjecture is correct.

Table 10.1.4

n	$\dfrac{n!}{n^n}$
1	1.0000000000
2	0.5000000000
3	0.2222222222
4	0.0937500000
5	0.0384000000
6	0.0154320988
7	0.0061198990
8	0.0024032593
9	0.0009366567
10	0.0003628800
11	0.0001399059
12	0.0000537232

Solution. Table 10.1.4, which was obtained with a calculating utility, suggests that the limit of the sequence may be 0. To confirm this we need to examine the limit of

$$a_n = \frac{n!}{n^n}$$

as $n \to +\infty$. Although this is an indeterminate form of type ∞/∞, L'Hôpital's rule is not helpful because we have no definition of $x!$ for values of x that are not integers. However, let us write out some of the initial terms and the general term in the sequence:

$$a_1 = 1, \quad a_2 = \frac{1 \cdot 2}{2 \cdot 2}, \quad a_3 = \frac{1 \cdot 2 \cdot 3}{3 \cdot 3 \cdot 3}, \ldots, \quad a_n = \frac{1 \cdot 2 \cdot 3 \cdots n}{n \cdot n \cdot n \cdots n}, \ldots$$

We can rewrite the general term as

$$a_n = \frac{1}{n}\left(\frac{2 \cdot 3 \cdots n}{n \cdot n \cdots n}\right)$$

from which it is evident that

$$0 \le a_n \le \frac{1}{n}$$

However, the two outside expressions have a limit of 0 as $n \to +\infty$; thus, the Squeezing Theorem for Sequences implies that $a_n \to 0$ as $n \to +\infty$, which confirms our conjecture. ◀

The following theorem is often useful for finding the limit of a sequence with both positive and negative terms—it states that if the sequence $\{|a_n|\}$ that is obtained by taking the absolute value of each term in the sequence $\{a_n\}$ converges to 0, then $\{a_n\}$ also converges to 0.

10.1.6 THEOREM. *If* $\lim_{n \to +\infty} |a_n| = 0$, *then* $\lim_{n \to +\infty} a_n = 0$.

PROOF. Depending on the sign of a_n, either $a_n = |a_n|$ or $a_n = -|a_n|$. Thus, in all cases we have

$$-|a_n| \le a_n \le |a_n|$$

However, the limit of the two outside terms is 0, and hence the limit of a_n is 0 by the Squeezing Theorem for Sequences. ■

▶ **Example 9** Consider the sequence

$$1, -\frac{1}{2}, \frac{1}{2^2}, -\frac{1}{2^3}, \ldots, (-1)^n\frac{1}{2^n}, \ldots$$

If we take the absolute value of each term, we obtain the sequence

$$1, \frac{1}{2}, \frac{1}{2^2}, \frac{1}{2^3}, \ldots, \frac{1}{2^n}, \ldots$$

which, as shown in Example 4, converges to 0. Thus, from Theorem 10.1.6 we have

$$\lim_{n\to+\infty}\left[(-1)^n\frac{1}{2^n}\right] = 0 \blacktriangleleft$$

SEQUENCES DEFINED RECURSIVELY

Some sequences do not arise from a formula for the general term, but rather from a formula or set of formulas that specify how to generate each term in the sequence from terms that precede it; such sequences are said to be defined ***recursively***, and the defining formulas are called ***recursion formulas***. A good example is the mechanic's rule for approximating square roots. In Exercise 19 of Section 4.6 you were asked to show that

$$x_1 = 1, \quad x_{n+1} = \frac{1}{2}\left(x_n + \frac{a}{x_n}\right) \tag{2}$$

describes the sequence produced by Newton's Method to approximate $\sqrt{a}$ as a zero of the function $f(x) = x^2 - a$. Table 10.1.5 shows the first five terms in an application of the mechanic's rule to approximate $\sqrt{2}$.

Table 10.1.5

n	$x_1 = 1, \quad x_{n+1} = \frac{1}{2}\left(x_n + \frac{2}{x_n}\right)$	DECIMAL APPROXIMATION
	$x_1 = 1$ (Starting value)	1.00000000000
1	$x_2 = \frac{1}{2}\left[1 + \frac{2}{1}\right] = \frac{3}{2}$	1.50000000000
2	$x_3 = \frac{1}{2}\left[\frac{3}{2} + \frac{2}{3/2}\right] = \frac{17}{12}$	1.41666666667
3	$x_4 = \frac{1}{2}\left[\frac{17}{12} + \frac{2}{17/12}\right] = \frac{577}{408}$	1.41421568627
4	$x_5 = \frac{1}{2}\left[\frac{577}{408} + \frac{2}{577/408}\right] = \frac{665{,}857}{470{,}832}$	1.41421356237
5	$x_6 = \frac{1}{2}\left[\frac{665{,}857}{470{,}832} + \frac{2}{665{,}857/470{,}832}\right] = \frac{886{,}731{,}088{,}897}{627{,}013{,}566{,}048}$	1.41421356237

It would take us too far afield to investigate the convergence of sequences defined recursively, but we will conclude this section with a useful technique that can sometimes be used to compute limits of such sequences.

► **Example 10** Assuming that the sequence in Table 10.1.5 converges, show that the limit is $\sqrt{2}$.

Solution. Assume that $x_n \to L$, where L is to be determined. Since $n+1 \to +\infty$ as $n \to +\infty$, it is also true that $x_{n+1} \to L$ as $n \to +\infty$. Thus, if we take the limit of the expression

$$x_{n+1} = \frac{1}{2}\left(x_n + \frac{2}{x_n}\right)$$

as $n \to +\infty$, we obtain

$$L = \frac{1}{2}\left(L + \frac{2}{L}\right)$$

which can be rewritten as $L^2 = 2$. The negative solution of this equation is extraneous because $x_n > 0$ for all n, so $L = \sqrt{2}$. ◄

✔QUICK CHECK EXERCISES 10.1 *(See page 635 for answers.)*

1. Consider the sequence 4, 6, 8, 10, 12, ...
   (a) If $\{a_n\}_{n=1}^{+\infty}$ denotes this sequence, then $a_1 =$ ______, $a_4 =$ ______, and $a_7 =$ ______. The general term is $a_n =$ ______.
   (b) If $\{b_n\}_{n=0}^{+\infty}$ denotes this sequence, then $b_0 =$ ______, $b_4 =$ ______, and $b_8 =$ ______. The general term is $b_n =$ ______.
2. What does it mean to say that a sequence $\{a_n\}$ *converges*?
3. Consider the sequences $\{a_n\}$ and $\{b_n\}$, where
$$a_n = \frac{n(2n+1)}{n^2} \quad \text{and} \quad b_n = \frac{(-1)^n}{5}$$
Determine which of the following sequences converge and which diverge. If a sequence converges, indicate its limit.
   (a) $\{a_n\}$ (b) $\{b_n\}$ (c) $\{3a_n - 1\}$ (d) $\{b_n^2\}$
   (e) $\{a_n + b_n\}$ (f) $\{1/a_n\}$ (g) $\{a_n/b_n\}$
4. Let f be the function $f(x) = \cos\left(\frac{\pi}{2}x\right)$ and define sequences $\{a_n\}$ and $\{b_n\}$ by $a_n = f(2n)$ and $b_n = f(2n+1)$.
   (a) Does $\lim_{x\to+\infty} f(x)$ exist?
   (b) $a_1 =$ ______, $a_2 =$ ______, $a_3 =$ ______, $a_4 =$ ______.
   (c) Does $\{a_n\}$ converge?
   (d) $b_1 =$ ______, $b_2 =$ ______, $b_3 =$ ______, $b_4 =$ ______.
   (e) Does $\{b_n\}$ converge?
   (f) Does the sequence $\{f(n)\}$ converge?
5. Suppose that $\{a_n\}$, $\{b_n\}$, and $\{c_n\}$ are sequences such that $a_n \le b_n \le c_n$ for all $n \ge 10$, and that $\{a_n\}$ and $\{c_n\}$ both converge to 12. Then the ______ Theorem for Sequences implies that $\{b_n\}$ converges to ______.

EXERCISE SET 10.1 Graphing Utility C CAS

1. In each part, find a formula for the general term of the sequence, starting with $n = 1$.
   (a) $1, \frac{1}{3}, \frac{1}{9}, \frac{1}{27}, \ldots$ (b) $1, -\frac{1}{3}, \frac{1}{9}, -\frac{1}{27}, \ldots$
   (c) $\frac{1}{2}, \frac{3}{4}, \frac{5}{6}, \frac{7}{8}, \ldots$ (d) $\frac{1}{\sqrt{\pi}}, \frac{4}{\sqrt[3]{\pi}}, \frac{9}{\sqrt[4]{\pi}}, \frac{16}{\sqrt[5]{\pi}}, \ldots$
2. In each part, find two formulas for the general term of the sequence, one starting with $n = 1$ and the other with $n = 0$.
   (a) $1, -r, r^2, -r^3, \ldots$ (b) $r, -r^2, r^3, -r^4, \ldots$
3. (a) Write out the first four terms of the sequence $\{1 + (-1)^n\}$, starting with $n = 0$.
   (b) Write out the first four terms of the sequence $\{\cos n\pi\}$, starting with $n = 0$.
   (c) Use the results in parts (a) and (b) to express the general term of the sequence 4, 0, 4, 0, ... in two different ways, starting with $n = 0$.
4. In each part, find a formula for the general term using factorials and starting with $n = 1$.
   (a) $1\cdot 2, 1\cdot 2\cdot 3\cdot 4, 1\cdot 2\cdot 3\cdot 4\cdot 5\cdot 6, 1\cdot 2\cdot 3\cdot 4\cdot 5\cdot 6\cdot 7\cdot 8, \ldots$
   (b) $1, 1\cdot 2\cdot 3, 1\cdot 2\cdot 3\cdot 4\cdot 5, 1\cdot 2\cdot 3\cdot 4\cdot 5\cdot 6\cdot 7, \ldots$

5–22 Write out the first five terms of the sequence, determine whether the sequence converges, and if so find its limit.

5. $\left\{\frac{n}{n+2}\right\}_{n=1}^{+\infty}$ 6. $\left\{\frac{n^2}{2n+1}\right\}_{n=1}^{+\infty}$ 7. $\{2\}_{n=1}^{+\infty}$
8. $\left\{\ln\left(\frac{1}{n}\right)\right\}_{n=1}^{+\infty}$ 9. $\left\{\frac{\ln n}{n}\right\}_{n=1}^{+\infty}$ 10. $\left\{n \sin\frac{\pi}{n}\right\}_{n=1}^{+\infty}$
11. $\{1 + (-1)^n\}_{n=1}^{+\infty}$ 12. $\left\{\frac{(-1)^{n+1}}{n^2}\right\}_{n=1}^{+\infty}$

13. $\left\{(-1)^n \dfrac{2n^3}{n^3+1}\right\}_{n=1}^{+\infty}$

14. $\left\{\dfrac{n}{2^n}\right\}_{n=1}^{+\infty}$

15. $\left\{\dfrac{(n+1)(n+2)}{2n^2}\right\}_{n=1}^{+\infty}$

16. $\left\{\dfrac{\pi^n}{4^n}\right\}_{n=1}^{+\infty}$

17. $\left\{\cos \dfrac{3}{n}\right\}_{n=1}^{+\infty}$

18. $\left\{\cos \dfrac{\pi n}{2}\right\}_{n=1}^{+\infty}$

19. $\{n^2 e^{-n}\}_{n=1}^{+\infty}$

20. $\{\sqrt{n^2+3n} - n\}_{n=1}^{+\infty}$

21. $\left\{\left(\dfrac{n+3}{n+1}\right)^n\right\}_{n=1}^{+\infty}$

22. $\left\{\left(1-\dfrac{2}{n}\right)^n\right\}_{n=1}^{+\infty}$

23–30 Find the general term of the sequence, starting with $n = 1$, determine whether the sequence converges, and if so find its limit.

23. $\dfrac{1}{2}, \dfrac{3}{4}, \dfrac{5}{6}, \dfrac{7}{8}, \ldots$

24. $0, \dfrac{1}{2^2}, \dfrac{2}{3^2}, \dfrac{3}{4^2}, \ldots$

25. $\dfrac{1}{3}, -\dfrac{1}{9}, \dfrac{1}{27}, -\dfrac{1}{81}, \ldots$

26. $-1, 2, -3, 4, -5, \ldots$

27. $\left(1-\dfrac{1}{2}\right), \left(\dfrac{1}{3}-\dfrac{1}{2}\right), \left(\dfrac{1}{3}-\dfrac{1}{4}\right), \left(\dfrac{1}{5}-\dfrac{1}{4}\right), \ldots$

28. $3, \dfrac{3}{2}, \dfrac{3}{2^2}, \dfrac{3}{2^3}, \ldots$

29. $(\sqrt{2}-\sqrt{3}), (\sqrt{3}-\sqrt{4}), (\sqrt{4}-\sqrt{5}), \ldots$

30. $\dfrac{1}{3^5}, -\dfrac{1}{3^6}, \dfrac{1}{3^7}, -\dfrac{1}{3^8}, \ldots$

FOCUS ON CONCEPTS

31. Give two examples of sequences, all of whose terms are between -10 and 10, that do not converge. Use graphs of your sequences to explain their properties.

32. (a) Suppose that f satisfies $\lim_{x\to 0^+} f(x) = +\infty$. Is it possible that the sequence $\{f(1/n)\}$ converges? Explain.
(b) Find a function f such that $\lim_{x\to 0^+} f(x)$ does not exist but the sequence $\{f(1/n)\}$ converges.

33. (a) Starting with $n = 1$, write out the first six terms of the sequence $\{a_n\}$, where

$$a_n = \begin{cases} 1, & \text{if } n \text{ is odd} \\ n, & \text{if } n \text{ is even} \end{cases}$$

(b) Starting with $n = 1$, and considering the even and odd terms separately, find a formula for the general term of the sequence

$$1, \frac{1}{2^2}, 3, \frac{1}{2^4}, 5, \frac{1}{2^6}, \ldots$$

(c) Starting with $n = 1$, and considering the even and odd terms separately, find a formula for the general term of the sequence

$$1, \frac{1}{3}, \frac{1}{3}, \frac{1}{5}, \frac{1}{5}, \frac{1}{7}, \frac{1}{7}, \frac{1}{9}, \frac{1}{9}, \ldots$$

(d) Determine whether the sequences in parts (a), (b), and (c) converge. For those that do, find the limit.

34. For what positive values of b does the sequence $b, 0, b^2, 0, b^3, 0, b^4, \ldots$ converge? Justify your answer.

C 35. (a) Use numerical evidence to make a conjecture about the limit of the sequence $\{\sqrt[n]{n^3}\}_{n=2}^{+\infty}$.
(b) Use a CAS to confirm your conjecture.

C 36. (a) Use numerical evidence to make a conjecture about the limit of the sequence $\{\sqrt[n]{3^n + n^3}\}_{n=2}^{+\infty}$.
(b) Use a CAS to confirm your conjecture.

37. Assuming that the sequence given in Formula (2) of this section converges, use the method of Example 10 to show that the limit of this sequence is $\sqrt{a}$.

38. Consider the sequence

$$a_1 = \sqrt{6}$$
$$a_2 = \sqrt{6+\sqrt{6}}$$
$$a_3 = \sqrt{6+\sqrt{6+\sqrt{6}}}$$
$$a_4 = \sqrt{6+\sqrt{6+\sqrt{6+\sqrt{6}}}}$$
$$\vdots$$

(a) Find a recursion formula for a_{n+1}.
(b) Assuming that the sequence converges, use the method of Example 10 to find the limit.

39. Consider the sequence $\{a_n\}_{n=1}^{+\infty}$, where

$$a_n = \frac{1}{n^2} + \frac{2}{n^2} + \cdots + \frac{n}{n^2}$$

(a) Find a_1, a_2, a_3, and a_4.
(b) Use numerical evidence to make a conjecture about the limit of the sequence.
(c) Confirm your conjecture by expressing a_n in closed form and calculating the limit.

40. Follow the directions in Exercise 39 with

$$a_n = \frac{1^2}{n^3} + \frac{2^2}{n^3} + \cdots + \frac{n^2}{n^3}$$

41–42 Use numerical evidence to make a conjecture about the limit of the sequence, and then use the Squeezing Theorem for Sequences (Theorem 10.1.5) to confirm that your conjecture is correct.

41. $\displaystyle\lim_{n\to+\infty} \frac{\sin^2 n}{n}$

42. $\displaystyle\lim_{n\to+\infty} \left(\frac{1+n}{2n}\right)^n$

43. (a) A bored student enters the number 0.5 in a calculator display and then repeatedly computes the square of the number in the display. Taking $a_0 = 0.5$, find a formula for the general term of the sequence $\{a_n\}$ of numbers that appear in the display.

(b) Try this with a calculator and make a conjecture about the limit of a_n.
(c) Confirm your conjecture by finding the limit of a_n.
(d) For what values of a_0 will this procedure produce a convergent sequence?

44. Let
$$f(x) = \begin{cases} 2x, & 0 \le x < 0.5 \\ 2x - 1, & 0.5 \le x < 1 \end{cases}$$
Does the sequence $f(0.2),\ f(f(0.2)),\ f(f(f(0.2))), \ldots$ converge? Justify your reasoning.

45. (a) Use a graphing utility to generate the graph of the equation $y = (2^x + 3^x)^{1/x}$, and then use the graph to make a conjecture about the limit of the sequence
$$\{(2^n + 3^n)^{1/n}\}_{n=1}^{+\infty}$$
(b) Confirm your conjecture by calculating the limit.

46. Consider the sequence $\{a_n\}_{n=1}^{+\infty}$ whose nth term is
$$a_n = \frac{1}{n}\sum_{k=1}^{n} \frac{1}{1 + (k/n)}$$
Show that $\lim_{n\to+\infty} a_n = \ln 2$ by interpreting a_n as the Riemann sum of a definite integral.

47. Let a_n be the average value of $f(x) = 1/x$ over the interval $[1, n]$. Determine whether the sequence $\{a_n\}$ converges, and if so find its limit.

48. The sequence whose terms are $1, 1, 2, 3, 5, 8, 13, 21, \ldots$ is called the ***Fibonacci sequence*** in honor of the Italian mathematician Leonardo ("Fibonacci") da Pisa (c. 1170–1250). This sequence has the property that after starting with two 1's, each term is the sum of the preceding two.
(a) Denoting the sequence by $\{a_n\}$ and starting with $a_1 = 1$ and $a_2 = 1$, show that
$$\frac{a_{n+2}}{a_{n+1}} = 1 + \frac{a_n}{a_{n+1}} \quad \text{if } n \ge 1$$
(b) Give a reasonable informal argument to show that if the sequence $\{a_{n+1}/a_n\}$ converges to some limit L, then the sequence $\{a_{n+2}/a_{n+1}\}$ must also converge to L.
(c) Assuming that the sequence $\{a_{n+1}/a_n\}$ converges, show that its limit is $(1 + \sqrt{5})/2$.

49. If we accept the fact that the sequence $\{1/n\}_{n=1}^{+\infty}$ converges to the limit $L = 0$, then according to Definition 10.1.2, for every $\epsilon > 0$, there exists a positive integer N such that $|a_n - L| = |(1/n) - 0| < \epsilon$ when $n \ge N$. In each part, find the smallest possible value of N for the given value of ϵ.
(a) $\epsilon = 0.5$ (b) $\epsilon = 0.1$ (c) $\epsilon = 0.001$

50. If we accept the fact that the sequence
$$\left\{\frac{n}{n+1}\right\}_{n=1}^{+\infty}$$
converges to the limit $L = 1$, then according to Definition 10.1.2, for every $\epsilon > 0$ there exists an integer N such that
$$|a_n - L| = \left|\frac{n}{n+1} - 1\right| < \epsilon$$
when $n \ge N$. In each part, find the smallest value of N for the given value of ϵ.
(a) $\epsilon = 0.25$ (b) $\epsilon = 0.1$ (c) $\epsilon = 0.001$

51. Use Definition 10.1.2 to prove that
(a) the sequence $\{1/n\}_{n=1}^{+\infty}$ converges to 0
(b) the sequence $\left\{\dfrac{n}{n+1}\right\}_{n=1}^{+\infty}$ converges to 1.

52. Find $\lim_{n\to+\infty} r^n$, where r is a real number. [*Hint:* Consider the cases $|r| < 1$, $|r| > 1$, $r = 1$, and $r = -1$ separately.]

✔ QUICK CHECK ANSWERS 10.1

1. (a) 4; 10; 16; $2n+2$ (b) 4; 12; 20; $2n+4$ **2.** $\lim_{n\to+\infty} a_n$ exists **3.** (a) converges to 2 (b) diverges (c) converges to 5 (d) converges to $\frac{1}{25}$ (e) diverges (f) converges to $\frac{1}{2}$ (g) diverges **4.** (a) no (b) -1; 1; -1; 1 (c) no (d) 0; 0; 0; 0 (e) yes (to 0) (f) no **5.** Squeezing; 12

10.2 MONOTONE SEQUENCES

There are many situations in which it is important to know whether a sequence converges, but the value of the limit is not relevant to the problem at hand. In this section we will study several techniques that can be used to determine whether a sequence converges.

TERMINOLOGY

We begin with some terminology.

Note that an increasing sequence need not be strictly increasing, and a decreasing sequence need not be strictly decreasing.

10.2.1 DEFINITION. A sequence $\{a_n\}_{n=1}^{+\infty}$ is called

strictly increasing if $a_1 < a_2 < a_3 < \cdots < a_n < \cdots$

increasing if $a_1 \le a_2 \le a_3 \le \cdots \le a_n \le \cdots$

strictly decreasing if $a_1 > a_2 > a_3 > \cdots > a_n > \cdots$

decreasing if $a_1 \ge a_2 \ge a_3 \ge \cdots \ge a_n \ge \cdots$

A sequence that is either increasing or decreasing is said to be ***monotone***, and a sequence that is either strictly increasing or strictly decreasing is said to be ***strictly monotone***.

Some examples are given in Table 10.2.1 and their corresponding graphs are shown in Figure 10.2.1.

Table 10.2.1

SEQUENCE	DESCRIPTION
$\frac{1}{2}, \frac{2}{3}, \frac{3}{4}, \ldots, \frac{n}{n+1}, \ldots$	Strictly increasing
$1, \frac{1}{2}, \frac{1}{3}, \ldots, \frac{1}{n}, \ldots$	Strictly decreasing
$1, 1, 2, 2, 3, 3, \ldots$	Increasing; not strictly increasing
$1, 1, \frac{1}{2}, \frac{1}{2}, \frac{1}{3}, \frac{1}{3}, \ldots$	Decreasing; not strictly decreasing
$1, -\frac{1}{2}, \frac{1}{3}, -\frac{1}{4}, \ldots, (-1)^{n+1}\frac{1}{n}, \ldots$	Neither increasing nor decreasing

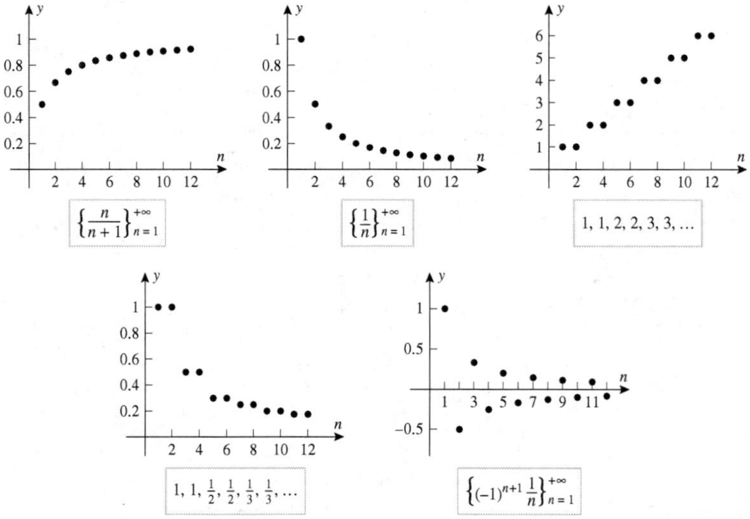

Figure 10.2.1

Can a sequence be both increasing and decreasing? Explain.

The first and second sequences in Table 10.2.1 are strictly monotone; the third and fourth sequences are monotone but not strictly monotone; and the fifth sequence is neither strictly monotone nor monotone.

TESTING FOR MONOTONICITY

Frequently, one can *guess* whether a sequence is monotone or strictly monotone by writing out some of the initial terms. However, to be certain that the guess is correct, one must give a precise mathematical argument. Table 10.2.2 provides two ways of doing this, one based on differences of successive terms and the other on ratios of successive terms. It is assumed in the latter case that the terms are positive. One must show that the specified conditions hold for *all* pairs of successive terms.

Table 10.2.2

DIFFERENCE BETWEEN SUCCESSIVE TERMS	RATIO OF SUCCESSIVE TERMS	CONCLUSION
$a_{n+1} - a_n > 0$	$a_{n+1}/a_n > 1$	Strictly increasing
$a_{n+1} - a_n < 0$	$a_{n+1}/a_n < 1$	Strictly decreasing
$a_{n+1} - a_n \geq 0$	$a_{n+1}/a_n \geq 1$	Increasing
$a_{n+1} - a_n \leq 0$	$a_{n+1}/a_n \leq 1$	Decreasing

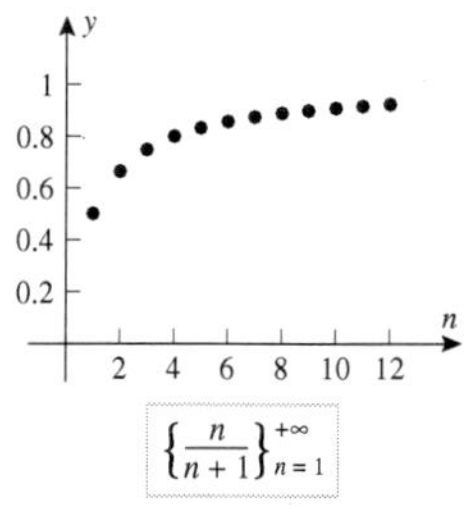

Figure 10.2.2

▶ **Example 1** Use differences of successive terms to show that

$$\frac{1}{2}, \frac{2}{3}, \frac{3}{4}, \ldots, \frac{n}{n+1}, \ldots$$

(Figure 10.2.2) is a strictly increasing sequence.

Solution. The pattern of the initial terms suggests that the sequence is strictly increasing. To prove that this is so, let

$$a_n = \frac{n}{n+1}$$

We can obtain a_{n+1} by replacing n by $n+1$ in this formula. This yields

$$a_{n+1} = \frac{n+1}{(n+1)+1} = \frac{n+1}{n+2}$$

Thus, for $n \geq 1$

$$a_{n+1} - a_n = \frac{n+1}{n+2} - \frac{n}{n+1} = \frac{n^2+2n+1-n^2-2n}{(n+1)(n+2)} = \frac{1}{(n+1)(n+2)} > 0$$

which proves that the sequence is strictly increasing. ◀

▶ **Example 2** Use ratios of successive terms to show that the sequence in Example 1 is strictly increasing.

Solution. As shown in the solution of Example 1,

$$a_n = \frac{n}{n+1} \quad \text{and} \quad a_{n+1} = \frac{n+1}{n+2}$$

Thus,

$$\frac{a_{n+1}}{a_n} = \frac{(n+1)/(n+2)}{n/(n+1)} = \frac{n+1}{n+2} \cdot \frac{n+1}{n} = \frac{n^2+2n+1}{n^2+2n} \tag{1}$$

Since the numerator in (1) exceeds the denominator, it follows that $a_{n+1}/a_n > 1$ for $n \geq 1$. This proves that the sequence is strictly increasing. ◀

The following example illustrates still a third technique for determining whether a sequence is strictly monotone.

▶ **Example 3** In Examples 1 and 2 we proved that the sequence

$$\frac{1}{2}, \frac{2}{3}, \frac{3}{4}, \ldots, \frac{n}{n+1}, \ldots$$

is strictly increasing by considering the difference and ratio of successive terms. Alternatively, we can proceed as follows. Let

$$f(x) = \frac{x}{x+1}$$

so that the nth term in the given sequence is $a_n = f(n)$. The function f is increasing for $x \geq 1$ since

$$f'(x) = \frac{(x+1)(1) - x(1)}{(x+1)^2} = \frac{1}{(x+1)^2} > 0$$

Thus,

$$a_n = f(n) < f(n+1) = a_{n+1}$$

which proves that the given sequence is strictly increasing. ◀

Table 10.2.3

DERIVATIVE OF f FOR $x \geq 1$	CONCLUSION FOR THE SEQUENCE WITH $a_n = f(n)$
$f'(x) > 0$	Strictly increasing
$f'(x) < 0$	Strictly decreasing
$f'(x) \geq 0$	Increasing
$f'(x) \leq 0$	Decreasing

In general, if $f(n) = a_n$ is the nth term of a sequence, and if f is differentiable for $x \geq 1$, then the results in Table 10.2.3 can be used to investigate the monotonicity of the sequence.

PROPERTIES THAT HOLD EVENTUALLY

Sometimes a sequence will behave erratically at first and then settle down into a definite pattern. For example, the sequence

$$9, -8, -17, 12, 1, 2, 3, 4, \ldots \quad (2)$$

is strictly increasing from the fifth term on, but the sequence as a whole cannot be classified as strictly increasing because of the erratic behavior of the first four terms. To describe such sequences, we introduce the following terminology.

> **10.2.2 DEFINITION.** If discarding finitely many terms from the beginning of a sequence produces a sequence with a certain property, then the original sequence is said to have that property ***eventually***.

For example, although we cannot say that sequence (2) is strictly increasing, we can say that it is eventually strictly increasing.

▶ **Example 4** Show that the sequence $\left\{\dfrac{10^n}{n!}\right\}_{n=1}^{+\infty}$ is eventually strictly decreasing.

Solution. We have

$$a_n = \frac{10^n}{n!} \quad \text{and} \quad a_{n+1} = \frac{10^{n+1}}{(n+1)!}$$

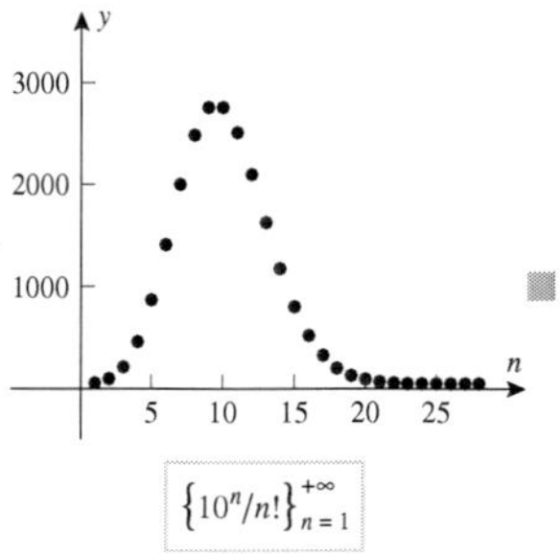

Figure 10.2.3

so

$$\frac{a_{n+1}}{a_n} = \frac{10^{n+1}/(n+1)!}{10^n/n!} = \frac{10^{n+1}n!}{10^n(n+1)!} = 10\frac{n!}{(n+1)n!} = \frac{10}{n+1} \tag{3}$$

From (3), $a_{n+1}/a_n < 1$ for all $n \geq 10$, so the sequence is eventually strictly decreasing, as confirmed by the graph in Figure 10.2.3. ◄

AN INTUITIVE VIEW OF CONVERGENCE

Informally stated, the convergence or divergence of a sequence does not depend on the behavior of its *initial terms*, but rather on how the terms behave *eventually*. For example, the sequence

$$3, -9, -13, 17, 1, \frac{1}{2}, \frac{1}{3}, \frac{1}{4}, \ldots$$

eventually behaves like the sequence

$$1, \frac{1}{2}, \frac{1}{3}, \ldots, \frac{1}{n}, \ldots$$

and hence has a limit of 0.

CONVERGENCE OF MONOTONE SEQUENCES

The following two theorems, whose proofs are discussed at the end of this section, show that a monotone sequence either converges or becomes infinite—divergence by oscillation cannot occur.

10.2.3 THEOREM. *If a sequence $\{a_n\}$ is eventually increasing, then there are two possibilities:*

(*a*) *There is a constant M, called an* **upper bound** *for the sequence, such that $a_n \leq M$ for all n, in which case the sequence converges to a limit L satisfying $L \leq M$.*

(*b*) *No upper bound exists, in which case* $\lim_{n\to+\infty} a_n = +\infty$.

10.2.4 THEOREM. *If a sequence $\{a_n\}$ is eventually decreasing, then there are two possibilities:*

(*a*) *There is a constant M, called a* **lower bound** *for the sequence, such that $a_n \geq M$ for all n, in which case the sequence converges to a limit L satisfying $L \geq M$.*

(*b*) *No lower bound exists, in which case* $\lim_{n\to+\infty} a_n = -\infty$.

Theorems 10.2.3 and 10.2.4 are examples of ***existence theorems***; they tell us whether a limit exists, but they do not provide a method for finding it.

► **Example 5** Show that the sequence $\left\{\dfrac{10^n}{n!}\right\}_{n=1}^{+\infty}$ converges and find its limit.

Solution. We showed in Example 4 that the sequence is eventually strictly decreasing. Since all terms in the sequence are positive, it is bounded below by $M = 0$, and hence Theorem 10.2.4 guarantees that it converges to a nonnegative limit L. However, the limit is not evident directly from the formula $10^n/n!$ for the nth term, so we will need some ingenuity to obtain it.

Recall from Formula (3) of Example 4 that successive terms in the given sequence are related by the recursion formula

$$a_{n+1} = \frac{10}{n+1}a_n \tag{4}$$

where $a_n = 10^n/n!$. We will take the limit as $n \to +\infty$ of both sides of (4) and use the fact that

$$\lim_{n \to +\infty} a_{n+1} = \lim_{n \to +\infty} a_n = L$$

We obtain

$$L = \lim_{n \to +\infty} a_{n+1} = \lim_{n \to +\infty} \left(\frac{10}{n+1} a_n \right) = \lim_{n \to +\infty} \frac{10}{n+1} \lim_{n \to +\infty} a_n = 0 \cdot L = 0$$

so that

$$L = \lim_{n \to +\infty} \frac{10^n}{n!} = 0 \blacktriangleleft$$

In the exercises we will show that the technique illustrated in the last example can be adapted to obtain

$$\lim_{n \to +\infty} \frac{x^n}{n!} = 0 \tag{5}$$

for any real value of x (Exercise 27). This result will be useful in our later work.

THE COMPLETENESS AXIOM

In this text we have accepted the familiar properties of real numbers without proof, and indeed, we have not even attempted to define the term *real number*. Although this is sufficient for many purposes, it was recognized by the late nineteenth century that the study of limits and functions in calculus requires a precise axiomatic formulation of the real numbers analogous to the axiomatic development of Euclidean geometry. Although we will not attempt to pursue this development, we will need to discuss one of the axioms about real numbers in order to prove Theorems 10.2.3 and 10.2.4. But first we will introduce some terminology.

If S is a nonempty set of real numbers, then we call u an ***upper bound*** for S if u is greater than or equal to every number in S, and we call l a ***lower bound*** for S if l is smaller than or equal to every number in S. For example, if S is the set of numbers in the interval $(1, 3)$, then $u = 4$, 10, and 100 are upper bounds for S and $l = -10$, 0, and $\frac{1}{2}$ are lower bounds for S. Observe also that $u = 3$ is the smallest of all upper bounds and $l = 1$ is the largest of all lower bounds. The existence of a smallest upper bound and a greatest lower bound for S is not accidental; it is a consequence of the following axiom.

10.2.5 AXIOM (*The Completeness Axiom*). *If a nonempty set S of real numbers has an upper bound, then it has a smallest upper bound (called the* ***least upper bound****), and if a nonempty set S of real numbers has a lower bound, then it has a largest lower bound (called the* ***greatest lower bound****).*

PROOF OF THEOREM 10.2.3.

(a) We will prove the result for increasing sequences, and leave it for the reader to adapt the argument to sequences that are eventually increasing. Assume there exists a number M such that $a_n \le M$ for $n = 1, 2, \ldots$. Then M is an upper bound for the set of terms in the sequence. By the Completeness Axiom there is a least upper bound for the terms; call it L. Now let ϵ be any positive number. Since L is the least upper bound for the terms, $L - \epsilon$ is not an upper bound for the terms, which means that there is at least one term a_N such that

$$a_N > L - \epsilon$$

Moreover, since $\{a_n\}$ is an increasing sequence, we must have

$$a_n \geq a_N > L - \epsilon \tag{6}$$

when $n \geq N$. But a_n cannot exceed L since L is an upper bound for the terms. This observation together with (6) tells us that $L \geq a_n > L - \epsilon$ for $n \geq N$, so all terms from the Nth on are within ϵ units of L. This is exactly the requirement to have

$$\lim_{n \to +\infty} a_n = L$$

Finally, $L \leq M$ since M is an upper bound for the terms and L is the least upper bound. This proves part (a).

(b) If there is no number M such that $a_n \leq M$ for $n = 1, 2, \ldots$, then no matter how large we choose M, there is a term a_N such that

$$a_N > M$$

and, since the sequence is increasing,

$$a_n \geq a_N > M$$

when $n \geq N$. Thus, the terms in the sequence become arbitrarily large as n increases. That is,

$$\lim_{n \to +\infty} a_n = +\infty$$

■

The proof of Theorem 10.2.4 will be omitted since it is similar to that of 10.2.3.

✔ QUICK CHECK EXERCISES 10.2 *(See page 642 for answers.)*

1. Classify each sequence as (I) increasing, (D) decreasing, or (N) neither increasing nor decreasing.
   _______ $\{2n\}$ _______ $\{2^{-n}\}$
   _______ $\left\{\frac{5-n}{n^2}\right\}$ _______ $\left\{\frac{-1}{n^2}\right\}$
   _______ $\left\{\frac{(-1)^n}{n^2}\right\}$

2. Classify each sequence as (M) monotonic, (S) strictly monotonic, or (N) not monotonic.
   _______ $\{n + (-1)^n\}$ _______ $\{2n + (-1)^n\}$
   _______ $\{3n + (-1)^n\}$

3. Since
   $$\frac{n/[2(n+1)]}{(n-1)/(2n)} = \frac{n^2}{n^2-1} > \underline{\qquad}$$
   the sequence $\{(n-1)/(2n)\}$ is strictly _______.

4. Since
   $$\frac{d}{dx}[(x-8)^2] > 0 \text{ for } x > \underline{\qquad}$$
   the sequence $\{(n-8)^2\}$ is _______ strictly _______.

EXERCISE SET 10.2

1–6 Use $a_{n+1} - a_n$ to show that the given sequence $\{a_n\}$ is strictly increasing or strictly decreasing.

1. $\left\{\frac{1}{n}\right\}_{n=1}^{+\infty}$ **2.** $\left\{1 - \frac{1}{n}\right\}_{n=1}^{+\infty}$ **3.** $\left\{\frac{n}{2n+1}\right\}_{n=1}^{+\infty}$

4. $\left\{\frac{n}{4n-1}\right\}_{n=1}^{+\infty}$ **5.** $\{n - 2^n\}_{n=1}^{+\infty}$ **6.** $\{n - n^2\}_{n=1}^{+\infty}$

7–12 Use a_{n+1}/a_n to show that the given sequence $\{a_n\}$ is strictly increasing or strictly decreasing.

7. $\left\{\frac{n}{2n+1}\right\}_{n=1}^{+\infty}$ **8.** $\left\{\frac{2^n}{1+2^n}\right\}_{n=1}^{+\infty}$ **9.** $\{ne^{-n}\}_{n=1}^{+\infty}$

10. $\left\{\frac{10^n}{(2n)!}\right\}_{n=1}^{+\infty}$ **11.** $\left\{\frac{n^n}{n!}\right\}_{n=1}^{+\infty}$ **12.** $\left\{\frac{5^n}{2^{(n^2)}}\right\}_{n=1}^{+\infty}$

13–18 Use differentiation to show that the given sequence is strictly increasing or strictly decreasing.

13. $\left\{\dfrac{n}{2n+1}\right\}_{n=1}^{+\infty}$ **14.** $\left\{3-\dfrac{1}{n}\right\}_{n=1}^{+\infty}$

15. $\left\{\dfrac{1}{n+\ln n}\right\}_{n=1}^{+\infty}$ **16.** $\{ne^{-2n}\}_{n=1}^{+\infty}$

17. $\left\{\dfrac{\ln(n+2)}{n+2}\right\}_{n=1}^{+\infty}$ **18.** $\{\tan^{-1} n\}_{n=1}^{+\infty}$

19–24 Show that the given sequence is eventually strictly increasing or eventually strictly decreasing.

19. $\{2n^2-7n\}_{n=1}^{+\infty}$ **20.** $\{n^3-4n^2\}_{n=1}^{+\infty}$

21. $\left\{\dfrac{n}{n^2+10}\right\}_{n=1}^{+\infty}$ **22.** $\left\{n+\dfrac{17}{n}\right\}_{n=1}^{+\infty}$

23. $\left\{\dfrac{n!}{3^n}\right\}_{n=1}^{+\infty}$ **24.** $\{n^5e^{-n}\}_{n=1}^{+\infty}$

FOCUS ON CONCEPTS

25. (a) Suppose that $\{a_n\}$ is a monotone sequence such that $1 \le a_n \le 2$ for all n. Must the sequence converge? If so, what can you say about the limit?
(b) Suppose that $\{a_n\}$ is a monotone sequence such that $a_n \le 2$ for all n. Must the sequence converge? If so, what can you say about the limit?

26. Give an example of a monotone sequence that is not eventually strictly monotone. What must be true of such a sequence?

27. The goal in this exercise is to prove Formula (5) in this section. The case where $x = 0$ is obvious, so we will focus on the case where $x \ne 0$.
(a) Let $a_n = |x|^n/n!$. Show that

$$a_{n+1} = \frac{|x|}{n+1}a_n$$

(b) Show that the sequence $\{a_n\}$ is eventually strictly decreasing.
(c) Show that the sequence $\{a_n\}$ converges.
(d) Use the results in parts (a) and (c) to show that $a_n \to 0$ as $n \to +\infty$.
(e) Obtain Formula (5) from the result in part (d).

28. Let $\{a_n\}$ be the sequence defined recursively by $a_1 = \sqrt{2}$ and $a_{n+1} = \sqrt{2+a_n}$ for $n \ge 1$.
(a) List the first three terms of the sequence.
(b) Show that $a_n < 2$ for $n \ge 1$.
(c) Show that $a_{n+1}^2 - a_n^2 = (2-a_n)(1+a_n)$ for $n \ge 1$.
(d) Use the results in parts (b) and (c) to show that $\{a_n\}$ is a strictly increasing sequence. [*Hint:* If x and y are positive real numbers such that $x^2 - y^2 > 0$, then it follows by factoring that $x - y > 0$.]
(e) Show that $\{a_n\}$ converges and find its limit L.

29. Let $\{a_n\}$ be the sequence defined recursively by $a_1 = 1$ and $a_{n+1} = \frac{1}{2}[a_n + (3/a_n)]$ for $n \ge 1$.
(a) Show that $a_n \ge \sqrt{3}$ for $n \ge 2$. [*Hint:* What is the minimum value of $\frac{1}{2}[x + (3/x)]$ for $x > 0$?]
(b) Show that $\{a_n\}$ is eventually decreasing. [*Hint:* Examine $a_{n+1} - a_n$ or a_{n+1}/a_n and use the result in part (a).]
(c) Show that $\{a_n\}$ converges and find its limit L.

30. (a) Compare appropriate areas in the accompanying figure to deduce the following inequalities for $n \ge 2$:

$$\int_1^n \ln x\,dx < \ln n! < \int_1^{n+1} \ln x\,dx$$

(b) Use the result in part (a) to show that

$$\frac{n^n}{e^{n-1}} < n! < \frac{(n+1)^{n+1}}{e^n}, \quad n > 1$$

(c) Use the Squeezing Theorem for Sequences (Theorem 10.1.5) and the result in part (b) to show that

$$\lim_{n\to+\infty} \frac{\sqrt[n]{n!}}{n} = \frac{1}{e}$$

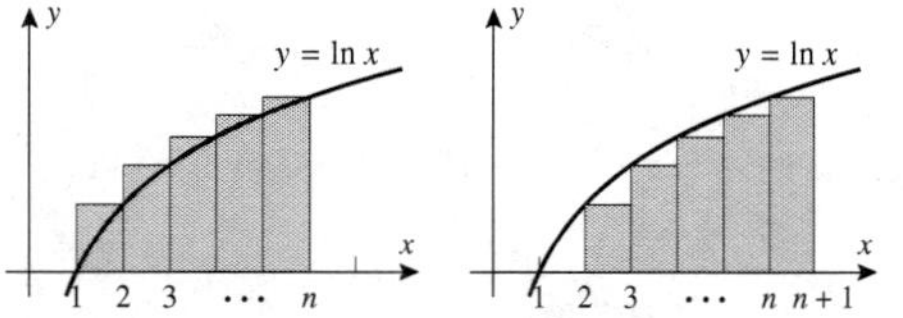

Figure Ex-30

31. Use the left inequality in Exercise 30(b) to show that

$$\lim_{n\to+\infty} \sqrt[n]{n!} = +\infty$$

✔QUICK CHECK ANSWERS 10.2

1. I; D; N; I; N **2.** N; M; S **3.** 1; increasing **4.** 8; eventually; increasing

10.3 INFINITE SERIES

The purpose of this section is to discuss sums that contain infinitely many terms. The most familiar examples of such sums occur in the decimal representations of real numbers. For example, when we write $\frac{1}{3}$ in the decimal form $\frac{1}{3} = 0.3333\ldots$, we mean

$$\frac{1}{3} = 0.3 + 0.03 + 0.003 + 0.0003 + \cdots$$

which suggests that the decimal representation of $\frac{1}{3}$ can be viewed as a sum of infinitely many real numbers.

SUMS OF INFINITE SERIES

Our first objective is to define what is meant by the "sum" of infinitely many real numbers. We begin with some terminology.

10.3.1 DEFINITION. An ***infinite series*** is an expression that can be written in the form

$$\sum_{k=1}^{\infty} u_k = u_1 + u_2 + u_3 + \cdots + u_k + \cdots$$

The numbers $u_1, u_2, u_3, \ldots$ are called the ***terms*** of the series.

Since it is impossible to add infinitely many numbers together directly, sums of infinite series are defined and computed by an indirect limiting process. To motivate the basic idea, consider the decimal

$$0.3333\ldots \tag{1}$$

This can be viewed as the infinite series

$$0.3 + 0.03 + 0.003 + 0.0003 + \cdots$$

or, equivalently,

$$\frac{3}{10} + \frac{3}{10^2} + \frac{3}{10^3} + \frac{3}{10^4} + \cdots \tag{2}$$

Since (1) is the decimal expansion of $\frac{1}{3}$, any reasonable definition for the sum of an infinite series should yield $\frac{1}{3}$ for the sum of (2). To obtain such a definition, consider the following sequence of (finite) sums:

$$s_1 = \frac{3}{10} = 0.3$$

$$s_2 = \frac{3}{10} + \frac{3}{10^2} = 0.33$$

$$s_3 = \frac{3}{10} + \frac{3}{10^2} + \frac{3}{10^3} = 0.333$$

$$s_4 = \frac{3}{10} + \frac{3}{10^2} + \frac{3}{10^3} + \frac{3}{10^4} = 0.3333$$

$$\vdots$$

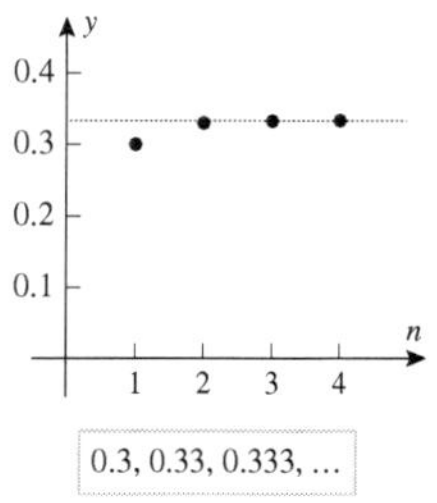

Figure 10.3.1

The sequence of numbers $s_1, s_2, s_3, s_4, \ldots$ (Figure 10.3.1) can be viewed as a succession of approximations to the "sum" of the infinite series, which we want to be $\frac{1}{3}$. As we progress through the sequence, more and more terms of the infinite series are used, and the approximations get better and better, suggesting that the desired sum of $\frac{1}{3}$ might be the *limit* of this sequence of approximations. To see that this is so, we must calculate the limit of the general term in the sequence of approximations, namely,

$$s_n = \frac{3}{10} + \frac{3}{10^2} + \cdots + \frac{3}{10^n} \tag{3}$$

The problem of calculating

$$\lim_{n\to+\infty} s_n = \lim_{n\to+\infty} \left(\frac{3}{10} + \frac{3}{10^2} + \cdots + \frac{3}{10^n} \right)$$

is complicated by the fact that both the last term and the number of terms in the sum change with n. It is best to rewrite such limits in a closed form in which the number of terms does not vary, if possible. (See the discussion of closed form and open form following Example 3 in Section 5.4.) To do this, we multiply both sides of (3) by $\frac{1}{10}$ to obtain

$$\frac{1}{10}s_n = \frac{3}{10^2} + \frac{3}{10^3} + \cdots + \frac{3}{10^n} + \frac{3}{10^{n+1}} \tag{4}$$

and then subtract (4) from (3) to obtain

$$\begin{aligned} s_n - \frac{1}{10}s_n &= \frac{3}{10} - \frac{3}{10^{n+1}} \\ \frac{9}{10}s_n &= \frac{3}{10}\left(1 - \frac{1}{10^n}\right) \\ s_n &= \frac{1}{3}\left(1 - \frac{1}{10^n}\right) \end{aligned}$$

Since $1/10^n \to 0$ as $n \to +\infty$, it follows that

$$\lim_{n\to+\infty} s_n = \lim_{n\to+\infty} \frac{1}{3}\left(1 - \frac{1}{10^n}\right) = \frac{1}{3}$$

which we denote by writing

$$\frac{1}{3} = \frac{3}{10} + \frac{3}{10^2} + \frac{3}{10^3} + \cdots + \frac{3}{10^n} + \cdots$$

Motivated by the preceding example, we are now ready to define the general concept of the "sum" of an infinite series

$$u_1 + u_2 + u_3 + \cdots + u_k + \cdots$$

We begin with some terminology: Let s_n denote the sum of the initial terms of the series, up to and including the term with index n. Thus,

$$\begin{aligned} s_1 &= u_1 \\ s_2 &= u_1 + u_2 \\ s_3 &= u_1 + u_2 + u_3 \\ &\vdots \\ s_n &= u_1 + u_2 + u_3 + \cdots + u_n = \sum_{k=1}^{n} u_k \end{aligned}$$

The number s_n is called the ***nth partial sum*** of the series and the sequence $\{s_n\}_{n=1}^{+\infty}$ is called the ***sequence of partial sums***.

As n increases, the partial sum $s_n = u_1 + u_2 + \cdots + u_n$ includes more and more terms of the series. Thus, if s_n tends toward a limit as $n \to +\infty$, it is reasonable to view this limit as the sum of *all* the terms in the series. This suggests the following definition.

10.3.2 DEFINITION. Let $\{s_n\}$ be the sequence of partial sums of the series

$$u_1 + u_2 + u_3 + \cdots + u_k + \cdots$$

If the sequence $\{s_n\}$ converges to a limit S, then the series is said to ***converge*** to S, and S is called the ***sum*** of the series. We denote this by writing

$$S = \sum_{k=1}^{\infty} u_k$$

If the sequence of partial sums diverges, then the series is said to ***diverge***. A divergent series has no sum.

WARNING

In everyday language the words "sequence" and "series" are often used interchangeably. However, in mathematics there is a difference between the two terms—a sequence is a *succession* whereas a series is a *sum*. It is essential that you keep this distinction in mind.

▶ **Example 1** Determine whether the series

$$1 - 1 + 1 - 1 + 1 - 1 + \cdots$$

converges or diverges. If it converges, find the sum.

Solution. It is tempting to conclude that the sum of the series is zero by arguing that the positive and negative terms cancel one another. However, this is *not correct*; the problem is that algebraic operations that hold for finite sums do not carry over to infinite series in all cases. Later, we will discuss conditions under which familiar algebraic operations can be applied to infinite series, but for this example we turn directly to Definition 10.3.2. The partial sums are

$$s_1 = 1$$
$$s_2 = 1 - 1 = 0$$
$$s_3 = 1 - 1 + 1 = 1$$
$$s_4 = 1 - 1 + 1 - 1 = 0$$

and so forth. Thus, the sequence of partial sums is

$$1, 0, 1, 0, 1, 0, \ldots$$

(Figure 10.3.2). Since this is a divergent sequence, the given series diverges and consequently has no sum. ◀

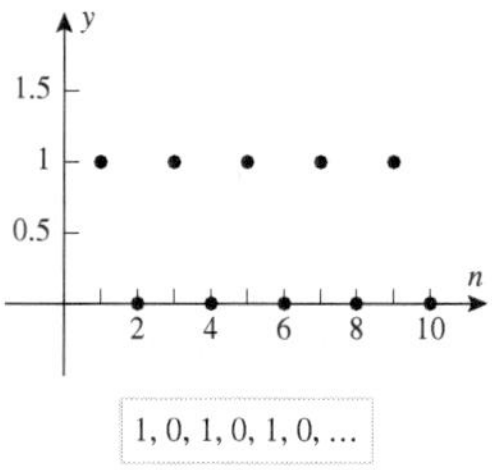

Figure 10.3.2

GEOMETRIC SERIES

In many important series, each term is obtained by multiplying the preceding term by some fixed constant. Thus, if the initial term of the series is a and each term is obtained by multiplying the preceding term by r, then the series has the form

$$\sum_{k=0}^{\infty} ar^k = a + ar + ar^2 + ar^3 + \cdots + ar^k + \cdots \quad (a \neq 0) \tag{5}$$

Such series are called ***geometric series***, and the number r is called the ***ratio*** for the series. Here are some examples:

$$1+2+4+8+\cdots+2^k+\cdots \qquad a=1, r=2$$

$$\frac{3}{10}+\frac{3}{10^2}+\frac{3}{10^3}+\cdots+\frac{3}{10^k}+\cdots \qquad a=\tfrac{3}{10}, r=\tfrac{1}{10}$$

$$\frac{1}{2}-\frac{1}{4}+\frac{1}{8}-\frac{1}{16}+\cdots+(-1)^{k+1}\frac{1}{2^k}+\cdots \qquad a=\tfrac{1}{2}, r=-\tfrac{1}{2}$$

$$1+1+1+\cdots+1+\cdots \qquad a=1, r=1$$

$$1-1+1-1+\cdots+(-1)^{k+1}+\cdots \qquad a=1, r=-1$$

$$1+x+x^2+x^3+\cdots+x^k+\cdots \qquad a=1, r=x$$

The following theorem is the fundamental result on convergence of geometric series.

Sometimes it is desirable to start the index of summation of an infinite series at $k=0$ rather than $k=1$, in which case we would call u_0 the *zeroth term* and $s_0=u_0$ the *zeroth partial sum*. One can prove that changing the starting value for the index of summation of an infinite series has no effect on the convergence, the divergence, or the sum. In the case of (5), the general term of the series would have been more complicated had we started the index at $k=1$. What would it have been?

10.3.3 THEOREM. *A geometric series*

$$\sum_{k=0}^{\infty} ar^k = a+ar+ar^2+\cdots+ar^k+\cdots \qquad (a\neq 0)$$

converges if $|r|<1$ *and diverges if* $|r|\geq 1$. *If the series converges, then the sum is*

$$\sum_{k=0}^{\infty} ar^k = \frac{a}{1-r}$$

PROOF. Let us treat the case $|r|=1$ first. If $r=1$, then the series is

$$a+a+a+a+\cdots$$

so the nth partial sum is $s_n=(n+1)a$ and $\lim_{n\to+\infty} s_n = \lim_{n\to+\infty}(n+1)a=\pm\infty$ (the sign depending on whether a is positive or negative). This proves divergence. If $r=-1$, the series is

$$a-a+a-a+\cdots$$

so the sequence of partial sums is

$$a, 0, a, 0, a, 0, \ldots$$

which diverges.

Now let us consider the case where $|r|\neq 1$. The nth partial sum of the series is

$$s_n = a+ar+ar^2+\cdots+ar^n \tag{6}$$

Multiplying both sides of (6) by r yields

$$rs_n = ar+ar^2+\cdots+ar^n+ar^{n+1} \tag{7}$$

and subtracting (7) from (6) gives

$$s_n - rs_n = a - ar^{n+1}$$

or

$$(1-r)s_n = a-ar^{n+1} \tag{8}$$

Since $r\neq 1$ in the case we are considering, this can be rewritten as

$$s_n = \frac{a-ar^{n+1}}{1-r} = \frac{a}{1-r}(1-r^{n+1}) \tag{9}$$

If $|r| < 1$, then $\lim_{n \to +\infty} r^{n+1} = 0$ (can you see why?), so $\{s_n\}$ converges. From (9)

$$\lim_{n \to +\infty} s_n = \frac{a}{1-r}$$

If $|r| > 1$, then either $r > 1$ or $r < -1$. In the case $r > 1$, $\lim_{n \to +\infty} r^{n+1} = +\infty$, and in the case $r < -1$, r^{n+1} oscillates between positive and negative values that grow in magnitude, so $\{s_n\}$ diverges in both cases. ■

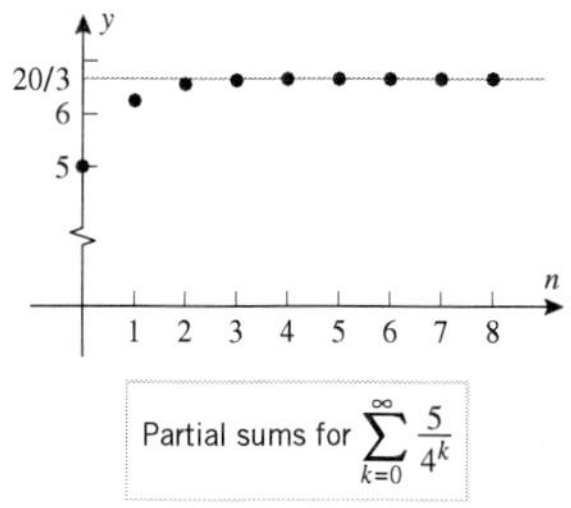

Figure 10.3.3

► **Example 2** The series

$$\sum_{k=0}^{\infty} \frac{5}{4^k} = 5 + \frac{5}{4} + \frac{5}{4^2} + \cdots + \frac{5}{4^k} + \cdots$$

is a geometric series with $a = 5$ and $r = \frac{1}{4}$. Since $|r| = \frac{1}{4} < 1$, the series converges and the sum is

$$\frac{a}{1-r} = \frac{5}{1-\frac{1}{4}} = \frac{20}{3}$$

(Figure 10.3.3). ◄

TECHNOLOGY MASTERY

Computer algebra systems have commands for finding sums of convergent series. If you have a CAS, use it to compute the sums in Examples 2 and 3. Also, see what happens when you try to compute the sum in Example 4(a).

► **Example 3** Find the rational number represented by the repeating decimal

$$0.784784784\ldots$$

Solution. We can write

$$0.784784784\ldots = 0.784 + 0.000784 + 0.000000784 + \cdots$$

so the given decimal is the sum of a geometric series with $a = 0.784$ and $r = 0.001$. Thus,

$$0.784784784\ldots = \frac{a}{1-r} = \frac{0.784}{1-0.001} = \frac{0.784}{0.999} = \frac{784}{999} \quad ◄$$

► **Example 4** In each part, determine whether the series converges, and if so find its sum.

$$\text{(a)} \sum_{k=1}^{\infty} 3^{2k} 5^{1-k} \qquad \text{(b)} \sum_{k=0}^{\infty} x^k$$

Solution (a). This is a geometric series in a concealed form, since we can rewrite it as

$$\sum_{k=1}^{\infty} 3^{2k} 5^{1-k} = \sum_{k=1}^{\infty} \frac{9^k}{5^{k-1}} = \sum_{k=1}^{\infty} 9\left(\frac{9}{5}\right)^{k-1}$$

Since $r = \frac{9}{5} > 1$, the series diverges.

Solution (b). The expanded form of the series is

$$\sum_{k=0}^{\infty} x^k = 1 + x + x^2 + \cdots + x^k + \cdots$$

The series is a geometric series with $a = 1$ and $r = x$, so it converges if $|x| < 1$ and diverges otherwise. When the series converges its sum is

$$\sum_{k=0}^{\infty} x^k = \frac{1}{1-x} \quad ◄$$

TELESCOPING SUMS

► Example 5 Determine whether the series

$$\sum_{k=1}^{\infty} \frac{1}{k(k+1)} = \frac{1}{1\cdot 2} + \frac{1}{2\cdot 3} + \frac{1}{3\cdot 4} + \frac{1}{4\cdot 5} + \cdots$$

converges or diverges. If it converges, find the sum.

Solution. The nth partial sum of the series is

$$s_n = \sum_{k=1}^{n} \frac{1}{k(k+1)} = \frac{1}{1\cdot 2} + \frac{1}{2\cdot 3} + \frac{1}{3\cdot 4} + \cdots + \frac{1}{n(n+1)}$$

To calculate $\lim_{n\to+\infty} s_n$ we will rewrite s_n in closed form. This can be accomplished by using the method of partial fractions to obtain (verify)

$$\frac{1}{k(k+1)} = \frac{1}{k} - \frac{1}{k+1}$$

from which we obtain the sum

$$s_n = \sum_{k=1}^{n}\left(\frac{1}{k} - \frac{1}{k+1}\right)$$

$$= \left(1 - \frac{1}{2}\right) + \left(\frac{1}{2} - \frac{1}{3}\right) + \left(\frac{1}{3} - \frac{1}{4}\right) + \cdots + \left(\frac{1}{n} - \frac{1}{n+1}\right)$$

$$= 1 + \left(-\frac{1}{2} + \frac{1}{2}\right) + \left(-\frac{1}{3} + \frac{1}{3}\right) + \cdots + \left(-\frac{1}{n} + \frac{1}{n}\right) - \frac{1}{n+1}$$

$$= 1 - \frac{1}{n+1} \tag{10}$$

The sum in (10) is an example of a ***telescoping sum.*** The name is derived from the fact that in simplifying the sum, one term in each parenthetical expression cancels one term in the next parenthetical expression, until the entire sum collapses (like a folding telescope) into just two terms.

so

$$\sum_{k=1}^{\infty} \frac{1}{k(k+1)} = \lim_{n\to+\infty} s_n = \lim_{n\to+\infty}\left(1 - \frac{1}{n+1}\right) = 1 \blacktriangleleft$$

HARMONIC SERIES

One of the most important of all diverging series is the ***harmonic series***,

$$\sum_{k=1}^{\infty} \frac{1}{k} = 1 + \frac{1}{2} + \frac{1}{3} + \frac{1}{4} + \frac{1}{5} + \cdots$$

which arises in connection with the overtones produced by a vibrating musical string. It is not immediately evident that this series diverges. However, the divergence will become apparent when we examine the partial sums in detail. Because the terms in the series are all positive, the partial sums

$$s_1 = 1,\quad s_2 = 1 + \frac{1}{2},\quad s_3 = 1 + \frac{1}{2} + \frac{1}{3},\quad s_4 = 1 + \frac{1}{2} + \frac{1}{3} + \frac{1}{4}, \ldots$$

form a strictly increasing sequence

$$s_1 < s_2 < s_3 < \cdots < s_n < \cdots$$

(Figure 10.3.4*a*). Thus, by Theorem 10.2.3 we can prove divergence by demonstrating that there is no constant M that is greater than or equal to *every* partial sum. To this end, we will consider some selected partial sums, namely, $s_2, s_4, s_8, s_{16}, s_{32}, \ldots$. Note that the

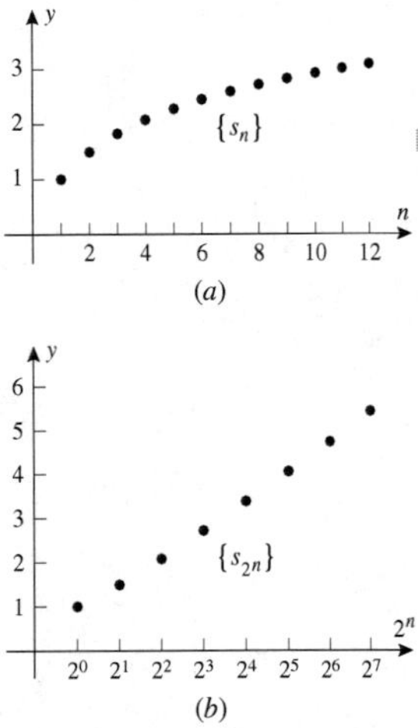

Partial sums for the harmonic series

Figure 10.3.4

subscripts are successive powers of 2, so that these are the partial sums of the form s_{2^n} (Figure 10.3.4*b*). These partial sums satisfy the inequalities

$$s_2 = 1 + \tfrac{1}{2} > \tfrac{1}{2} + \tfrac{1}{2} = \tfrac{2}{2}$$

$$s_4 = s_2 + \tfrac{1}{3} + \tfrac{1}{4} > s_2 + \left(\tfrac{1}{4} + \tfrac{1}{4}\right) = s_2 + \tfrac{1}{2} > \tfrac{3}{2}$$

$$s_8 = s_4 + \tfrac{1}{5} + \tfrac{1}{6} + \tfrac{1}{7} + \tfrac{1}{8} > s_4 + \left(\tfrac{1}{8} + \tfrac{1}{8} + \tfrac{1}{8} + \tfrac{1}{8}\right) = s_4 + \tfrac{1}{2} > \tfrac{4}{2}$$

$$s_{16} = s_8 + \tfrac{1}{9} + \tfrac{1}{10} + \tfrac{1}{11} + \tfrac{1}{12} + \tfrac{1}{13} + \tfrac{1}{14} + \tfrac{1}{15} + \tfrac{1}{16}$$

$$> s_8 + \left(\tfrac{1}{16} + \tfrac{1}{16} + \tfrac{1}{16} + \tfrac{1}{16} + \tfrac{1}{16} + \tfrac{1}{16} + \tfrac{1}{16} + \tfrac{1}{16}\right) = s_8 + \tfrac{1}{2} > \tfrac{5}{2}$$

$$\vdots$$

$$s_{2^n} > \frac{n+1}{2}$$

XVI. *Summa seriei infinitæ harmonicè progressionalium*, $\frac{1}{1}+\frac{1}{2}+\frac{1}{3}+\frac{1}{4}+\frac{1}{5}$ *&c. est infinita.*

Id primus deprehendit Frater: inventa namque per præced. summa seriei $\frac{1}{2}+\frac{1}{6}+\frac{1}{12}+\frac{1}{20}+\frac{1}{30}$, &c. visurus porrò, quid emergeret ex ista serie, $\frac{1}{2}+\frac{2}{6}+\frac{3}{12}+\frac{4}{20}+\frac{5}{30}$, &c. si resolveretur methodo Prop. XIV. collegit propositionis veritatem ex absurditate manifesta, quæ sequeretur, si summa seriei harmonicæ finita statueretur. Animadvertit enim,

Seriem A, $\frac{1}{2}+\frac{1}{3}+\frac{1}{4}+\frac{1}{5}+\frac{1}{6}+\frac{1}{7}$, &c. ∞ (fractionibus singulis in alias, quarum numeratores sunt 1, 2, 3, 4, &c. transmutatis)

seriei B, $\frac{1}{2}+\frac{2}{6}+\frac{3}{12}+\frac{4}{20}+\frac{5}{30}+\frac{6}{42}$, &c. ∞ C+D+E+F, &c.

C. $\frac{1}{2}+\frac{1}{6}+\frac{1}{12}+\frac{1}{20}+\frac{1}{30}+\frac{1}{42}$, &c. ∞ per præc. $\frac{1}{1}$
D. .. $+\frac{1}{6}+\frac{1}{12}+\frac{1}{20}+\frac{1}{30}+\frac{1}{42}$, &c. ∞ C − $\frac{1}{2}$ ∞ $\frac{1}{2}$
E. . . . $+\frac{1}{12}+\frac{1}{20}+\frac{1}{30}+\frac{1}{42}$, &c. ∞ D − $\frac{1}{6}$ ∞ $\frac{1}{3}$
F. . . . . . . $+\frac{1}{20}+\frac{1}{30}+\frac{1}{42}$, &c. ∞ E − $\frac{1}{12}$ ∞ $\frac{1}{4}$
&c. ∞ &c.
} ∞ G; unde sequitur, seriem G ∞ A, totum parti, si summa finita esset.

Ego

This is a proof of the divergence of the harmonic series, as it appeared in an appendix of Jakob Bernoulli's posthumous publication, *Ars Conjectandi*, which appeared in 1713.

If M is any constant, we can find a positive integer n such that $(n+1)/2 > M$. But for this n

$$s_{2^n} > \frac{n+1}{2} > M$$

so that no constant M is greater than or equal to *every* partial sum of the harmonic series. This proves divergence.

This divergence proof, which predates the discovery of calculus, is due to a French bishop and teacher, Nicole Oresme (1323–1382). This series eventually attracted the interest of Johann and Jakob Bernoulli (p. 92) and led them to begin thinking about the general concept of convergence, which was a new idea at that time.

✔QUICK CHECK EXERCISES 10.3 *(See page 652 for answers.)*

1. In mathematics, the terms "sequence" and "series" have different meanings: a ______ is a succession, whereas a ______ is a sum.

2. Consider the series

$$\sum_{k=1}^{\infty} \frac{1}{2^k}$$

If $\{s_n\}$ is the sequence of partial sums for this series, then $s_1 =$ ______, $s_2 =$ ______, $s_3 =$ ______, $s_4 =$ ______, and $s_n =$ ______.

3. What does it mean to say that a series $\sum u_k$ *converges*?

4. A geometric series is a series of the form

$$\sum_{k=0}^{\infty} \text{\_\_\_\_\_\_}$$

This series converges to ______ if ______. This series diverges if ______.

5. The harmonic series has the form

$$\sum_{k=1}^{\infty} \text{\_\_\_\_\_\_}$$

Does the harmonic series converge or diverge?

EXERCISE SET 10.3 [C] CAS

1–2 In each part, find exact values for the first four partial sums, find a closed form for the nth partial sum, and determine whether the series converges by calculating the limit of the nth partial sum. If the series converges, then state its sum.

1. (a) $2 + \dfrac{2}{5} + \dfrac{2}{5^2} + \cdots + \dfrac{2}{5^{k-1}} + \cdots$

(b) $\dfrac{1}{4} + \dfrac{2}{4} + \dfrac{2^2}{4} + \cdots + \dfrac{2^{k-1}}{4} + \cdots$

(c) $\dfrac{1}{2\cdot 3} + \dfrac{1}{3\cdot 4} + \dfrac{1}{4\cdot 5} + \cdots + \dfrac{1}{(k+1)(k+2)} + \cdots$

2. (a) $\displaystyle\sum_{k=1}^{\infty}\left(\frac{1}{4}\right)^k$ (b) $\displaystyle\sum_{k=1}^{\infty} 4^{k-1}$ (c) $\displaystyle\sum_{k=1}^{\infty}\left(\frac{1}{k+3} - \frac{1}{k+4}\right)$

3–14 Determine whether the series converges, and if so find its sum.

3. $\sum_{k=1}^{\infty}\left(-\frac{3}{4}\right)^{k-1}$

4. $\sum_{k=1}^{\infty}\left(\frac{2}{3}\right)^{k+2}$

5. $\sum_{k=1}^{\infty}(-1)^{k-1}\frac{7}{6^{k-1}}$

6. $\sum_{k=1}^{\infty}\left(-\frac{3}{2}\right)^{k+1}$

7. $\sum_{k=1}^{\infty}\frac{1}{(k+2)(k+3)}$

8. $\sum_{k=1}^{\infty}\left(\frac{1}{2^k}-\frac{1}{2^{k+1}}\right)$

9. $\sum_{k=1}^{\infty}\frac{1}{9k^2+3k-2}$

10. $\sum_{k=2}^{\infty}\frac{1}{k^2-1}$

11. $\sum_{k=3}^{\infty}\frac{1}{k-2}$

12. $\sum_{k=5}^{\infty}\left(\frac{e}{\pi}\right)^{k-1}$

13. $\sum_{k=1}^{\infty}\frac{4^{k+2}}{7^{k-1}}$

14. $\sum_{k=1}^{\infty}5^{3k}7^{1-k}$

15. Match a series from one of Exercises 3, 5, 7, or 9 with the graph of its sequence of partial sums.

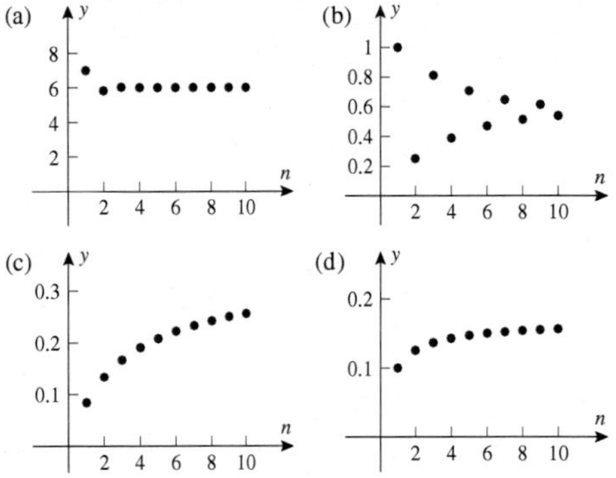

16. Match a series from one of Exercises 4, 6, 8, or 10 with the graph of its sequence of partial sums.

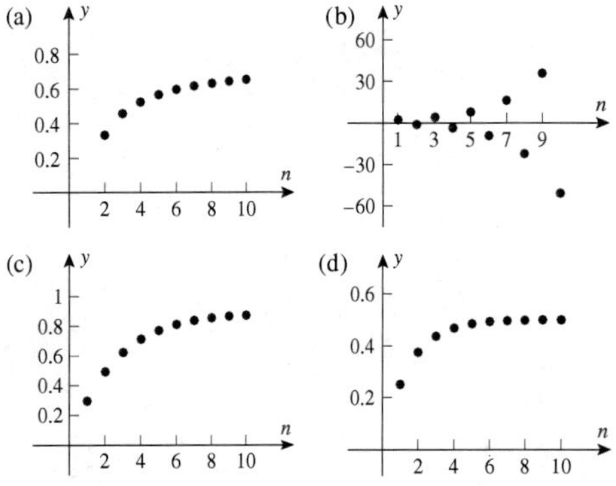

17–20 Express the repeating decimal as a fraction.

17. 0.4444...

18. 0.9999...

19. 5.373737...

20. 0.451141414...

21. Recall that a *terminating decimal* is a decimal whose digits are all 0 from some point on ($0.5 = 0.50000\ldots$, for example). Show that a decimal of the form $0.a_1a_2\ldots a_n9999\ldots$, where $a_n \neq 9$, can be expressed as a terminating decimal.

FOCUS ON CONCEPTS

22. The great Swiss mathematician Leonhard Euler (biography on p. 3) sometimes reached incorrect conclusions in his pioneering work on infinite series. For example, Euler deduced that

$$\tfrac{1}{2} = 1 - 1 + 1 - 1 + \cdots$$

and

$$-1 = 1 + 2 + 4 + 8 + \cdots$$

by substituting $x = -1$ and $x = 2$ in the formula

$$\frac{1}{1-x} = 1 + x + x^2 + x^3 + \cdots$$

What was the problem with his reasoning?

23. A ball is dropped from a height of 10 m. Each time it strikes the ground it bounces vertically to a height that is $\frac{3}{4}$ of the preceding height. Find the total distance the ball will travel if it is assumed to bounce infinitely often.

24. The accompanying figure shows an "infinite staircase" constructed from cubes. Find the total volume of the staircase, given that the largest cube has a side of length 1 and each successive cube has a side whose length is half that of the preceding cube.

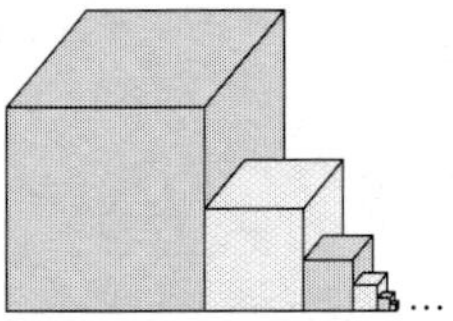

Figure Ex-24

25. In each part, find a closed form for the nth partial sum of the series, and determine whether the series converges. If so, find its sum.

(a) $\ln\frac{1}{2} + \ln\frac{2}{3} + \ln\frac{3}{4} + \cdots + \ln\frac{k}{k+1} + \cdots$

(b) $\ln\left(1-\frac{1}{4}\right) + \ln\left(1-\frac{1}{9}\right) + \ln\left(1-\frac{1}{16}\right) + \cdots + \ln\left(1-\frac{1}{(k+1)^2}\right) + \cdots$

26. Use geometric series to show that

(a) $\sum_{k=0}^{\infty}(-1)^k x^k = \frac{1}{1+x}$ if $-1 < x < 1$

(b) $\displaystyle\sum_{k=0}^{\infty}(x-3)^k = \frac{1}{4-x}$ if $2 < x < 4$

(c) $\displaystyle\sum_{k=0}^{\infty}(-1)^k x^{2k} = \frac{1}{1+x^2}$ if $-1 < x < 1$.

27. In each part, find all values of x for which the series converges, and find the sum of the series for those values of x.
(a) $x - x^3 + x^5 - x^7 + x^9 - \cdots$
(b) $\dfrac{1}{x^2} + \dfrac{2}{x^3} + \dfrac{4}{x^4} + \dfrac{8}{x^5} + \dfrac{16}{x^6} + \cdots$
(c) $e^{-x} + e^{-2x} + e^{-3x} + e^{-4x} + e^{-5x} + \cdots$

28. Show that for all real values of x

$$\sin x - \frac{1}{2}\sin^2 x + \frac{1}{4}\sin^3 x - \frac{1}{8}\sin^4 x + \cdots = \frac{2\sin x}{2+\sin x}$$

29. Let a_1 be any real number, and let $\{a_n\}$ be the sequence defined recursively by

$$a_{n+1} = \tfrac{1}{2}(a_n + 1)$$

Make a conjecture about the limit of the sequence, and confirm your conjecture by expressing a_n in terms of a_1 and taking the limit.

30. Show: $\displaystyle\sum_{k=1}^{\infty}\frac{\sqrt{k+1}-\sqrt{k}}{\sqrt{k^2+k}} = 1$.

31. Show: $\displaystyle\sum_{k=1}^{\infty}\left(\frac{1}{k} - \frac{1}{k+2}\right) = \frac{3}{2}$.

32. Show: $\dfrac{1}{1\cdot 3} + \dfrac{1}{2\cdot 4} + \dfrac{1}{3\cdot 5} + \cdots = \dfrac{3}{4}$.

33. Show: $\dfrac{1}{1\cdot 3} + \dfrac{1}{3\cdot 5} + \dfrac{1}{5\cdot 7} + \cdots = \dfrac{1}{2}$.

34. As shown in the accompanying figure, suppose that lines L_1 and L_2 form an angle θ, $0 < \theta < \pi/2$, at their point of intersection P. A point P_0 is chosen that is on L_1 and a units from P. Starting from P_0 a zig-zag path is constructed by successively going back and forth between L_1 and L_2 along a perpendicular from one line to the other. Find the following sums in terms of θ and a.
(a) $P_0P_1 + P_1P_2 + P_2P_3 + \cdots$
(b) $P_0P_1 + P_2P_3 + P_4P_5 + \cdots$
(c) $P_1P_2 + P_3P_4 + P_5P_6 + \cdots$

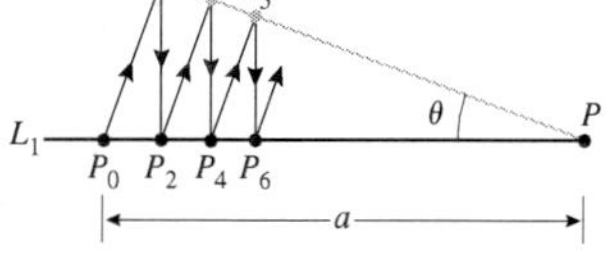

Figure Ex-34

35. As shown in the accompanying figure, suppose that an angle θ is bisected using a straightedge and compass to produce ray R_1, then the angle between R_1 and the initial side is bisected to produce ray R_2. Thereafter, rays $R_3, R_4, R_5, \ldots$ are constructed in succession by bisecting the angle between the preceding two rays. Show that the sequence of angles that these rays make with the initial side has a limit of $\theta/3$.

Source: This problem is based on *Trisection of an Angle in an Infinite Number of Steps* by Eric Kincannon, which appeared in *The College Mathematics Journal*, Vol. 21, No. 5, November 1990.

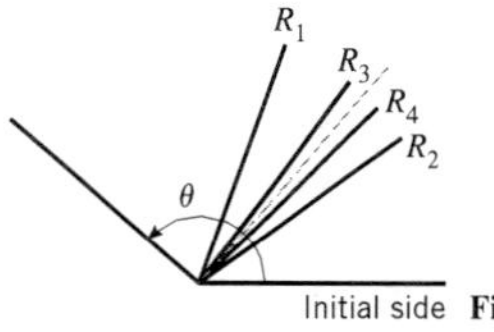

Figure Ex-35

36. In his *Treatise on the Configurations of Qualities and Motions* (written in the 1350s), the French Bishop of Lisieux, Nicole Oresme, used a geometric method to find the sum of the series

$$\sum_{k=1}^{\infty}\frac{k}{2^k} = \frac{1}{2} + \frac{2}{4} + \frac{3}{8} + \frac{4}{16} + \cdots$$

In part (a) of the accompanying figure, each term in the series is represented by the area of a rectangle, and in part (b) the configuration in part (a) has been divided into rectangles with areas $A_1, A_2, A_3, \ldots$. Find the sum $A_1 + A_2 + A_3 + \cdots$.

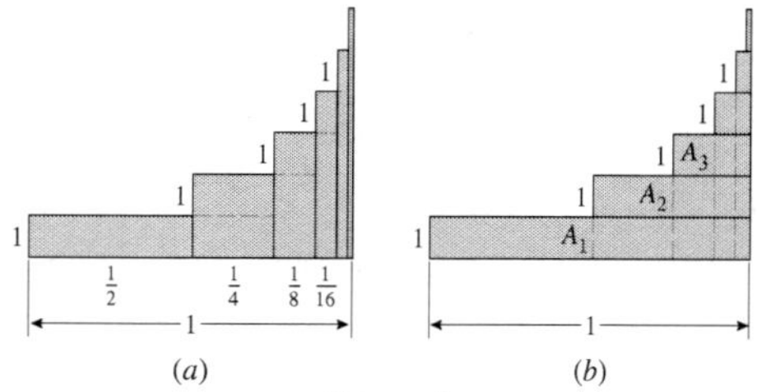

Not to scale

Figure Ex-36

C **37.** (a) See if your CAS can find the sum of the series

$$\sum_{k=1}^{\infty}\frac{6^k}{(3^{k+1}-2^{k+1})(3^k-2^k)}$$

(b) Find A and B such that

$$\frac{6^k}{(3^{k+1}-2^{k+1})(3^k-2^k)} = \frac{2^k A}{3^k-2^k} + \frac{2^k B}{3^{k+1}-2^{k+1}}$$

(c) Use the result in part (b) to find a closed form for the nth partial sum, and then find the sum of the series.

Source: This exercise is adapted from a problem that appeared in the Forty-Fifth Annual William Lowell Putnam Competition.

C **38.** In each part, use a CAS to find the sum of the series if it converges, and then confirm the result by hand calculation.

(a) $\displaystyle\sum_{k=1}^{\infty}(-1)^{k+1}2^k 3^{2-k}$ (b) $\displaystyle\sum_{k=1}^{\infty}\frac{3^{3k}}{5^{k-1}}$ (c) $\displaystyle\sum_{k=1}^{\infty}\frac{1}{4k^2-1}$

✔QUICK CHECK ANSWERS 10.3

1. sequence; series **2.** $\frac{1}{2}; \frac{3}{4}; \frac{7}{8}; \frac{15}{16}; 1 - \frac{1}{2^n}$ **3.** The sequence of partial sums converges.
4. ar^k $(a \neq 0)$; $\frac{a}{1-r}$; $|r| < 1$; $|r| \geq 1$ **5.** $\frac{1}{k}$; diverge

10.4 CONVERGENCE TESTS

In the last section we showed how to find the sum of a series by finding a closed form for the nth partial sum and taking its limit. However, it is relatively rare that one can find a closed form for the nth partial sum of a series, so alternative methods are needed for finding the sum of a series. One possibility is to prove that the series converges, and then to approximate the sum by a partial sum with sufficiently many terms to achieve the desired degree of accuracy. In this section we will develop various tests that can be used to determine whether a given series converges or diverges.

THE DIVERGENCE TEST

In stating general results about convergence or divergence of series, it is convenient to use the notation $\sum u_k$ as a generic template for a series, thus avoiding the issue of whether the sum begins with $k = 0$ or $k = 1$ or some other value. Indeed, we will see shortly that the starting index value is irrelevant to the issue of convergence. The kth term in an infinite series $\sum u_k$ is called the ***general term*** of the series. The following theorem establishes a relationship between the limit of the general term and the convergence properties of a series.

10.4.1 THEOREM (*The Divergence Test*).

(a) If $\lim_{k\to+\infty} u_k \neq 0$, *then the series* $\sum u_k$ *diverges.*

(b) If $\lim_{k\to+\infty} u_k = 0$, *then the series* $\sum u_k$ *may either converge or diverge.*

PROOF (*a*). To prove this result, it suffices to show that if the series converges, then $\lim_{k\to+\infty} u_k = 0$ (why?). We will prove this alternative form of (*a*).

Let us assume that the series converges. The general term u_k can be written as

$$u_k = s_k - s_{k-1} \tag{1}$$

where s_k is the sum of the terms through u_k and s_{k-1} is the sum of the terms through u_{k-1}. If S denotes the sum of the series, then $\lim_{k\to+\infty} s_k = S$, and since $(k-1) \to +\infty$ as $k \to +\infty$, we also have $\lim_{k\to+\infty} s_{k-1} = S$. Thus, from (1)

$$\lim_{k\to+\infty} u_k = \lim_{k\to+\infty} (s_k - s_{k-1}) = S - S = 0$$

PROOF (*b*). To prove this result, it suffices to produce both a convergent series and a divergent series for which $\lim_{k\to+\infty} u_k = 0$. The following series both have this property:

$$\frac{1}{2} + \frac{1}{2^2} + \cdots + \frac{1}{2^k} + \cdots \quad \text{and} \quad 1 + \frac{1}{2} + \frac{1}{3} + \cdots + \frac{1}{k} + \cdots$$

The first is a convergent geometric series and the second is the divergent harmonic series. ■

WARNING

The converse of Theorem 10.4.2 is false. Showing that

$$\lim_{k\to+\infty} u_k = 0$$

does not prove that $\sum u_k$ converges, since this property may hold for divergent as well as convergent series. This is illustrated in the proof of part (*b*) of Theorem 10.4.1.

The alternative form of part (a) given in the preceding proof is sufficiently important that we state it separately for future reference.

10.4.2 THEOREM. *If the series $\sum u_k$ converges, then $\lim_{k\to+\infty} u_k = 0$.*

► **Example 1** The series

$$\sum_{k=1}^{\infty} \frac{k}{k+1} = \frac{1}{2} + \frac{2}{3} + \frac{3}{4} + \cdots + \frac{k}{k+1} + \cdots$$

diverges since

$$\lim_{k\to+\infty} \frac{k}{k+1} = \lim_{k\to+\infty} \frac{1}{1+1/k} = 1 \neq 0 \blacktriangleleft$$

ALGEBRAIC PROPERTIES OF INFINITE SERIES

For brevity, the proof of the following result is omitted.

10.4.3 THEOREM.

(a) *If $\sum u_k$ and $\sum v_k$ are convergent series, then $\sum(u_k + v_k)$ and $\sum(u_k - v_k)$ are convergent series and the sums of these series are related by*

$$\sum_{k=1}^{\infty}(u_k + v_k) = \sum_{k=1}^{\infty} u_k + \sum_{k=1}^{\infty} v_k$$

$$\sum_{k=1}^{\infty}(u_k - v_k) = \sum_{k=1}^{\infty} u_k - \sum_{k=1}^{\infty} v_k$$

(b) *If c is a nonzero constant, then the series $\sum u_k$ and $\sum cu_k$ both converge or both diverge. In the case of convergence, the sums are related by*

$$\sum_{k=1}^{\infty} cu_k = c\sum_{k=1}^{\infty} u_k$$

(c) *Convergence or divergence is unaffected by deleting a finite number of terms from a series; in particular, for any positive integer K, the series*

$$\sum_{k=1}^{\infty} u_k = u_1 + u_2 + u_3 + \cdots$$

$$\sum_{k=K}^{\infty} u_k = u_K + u_{K+1} + u_{K+2} + \cdots$$

both converge or both diverge.

WARNING

Do not read too much into part (c) of Theorem 10.4.3. Although convergence is not affected when finitely many terms are deleted from the beginning of a convergent series, the *sum* of the series is changed by the removal of those terms.

▶ Example 2 Find the sum of the series

$$\sum_{k=1}^{\infty}\left(\frac{3}{4^k}-\frac{2}{5^{k-1}}\right)$$

Solution. The series

$$\sum_{k=1}^{\infty}\frac{3}{4^k}=\frac{3}{4}+\frac{3}{4^2}+\frac{3}{4^3}+\cdots$$

is a convergent geometric series $\left(a=\frac{3}{4}, r=\frac{1}{4}\right)$, and the series

$$\sum_{k=1}^{\infty}\frac{2}{5^{k-1}}=2+\frac{2}{5}+\frac{2}{5^2}+\frac{2}{5^3}+\cdots$$

is also a convergent geometric series $\left(a=2, r=\frac{1}{5}\right)$. Thus, from Theorems 10.4.3(*a*) and 10.3.3 the given series converges and

$$\sum_{k=1}^{\infty}\left(\frac{3}{4^k}-\frac{2}{5^{k-1}}\right)=\sum_{k=1}^{\infty}\frac{3}{4^k}-\sum_{k=1}^{\infty}\frac{2}{5^{k-1}}$$

$$=\frac{\frac{3}{4}}{1-\frac{1}{4}}-\frac{2}{1-\frac{1}{5}}=-\frac{3}{2} \blacktriangleleft$$

▶ Example 3 Determine whether the following series converge or diverge.

(a) $\displaystyle\sum_{k=1}^{\infty}\frac{5}{k}=5+\frac{5}{2}+\frac{5}{3}+\cdots+\frac{5}{k}+\cdots$ (b) $\displaystyle\sum_{k=10}^{\infty}\frac{1}{k}=\frac{1}{10}+\frac{1}{11}+\frac{1}{12}+\cdots$

Solution. The first series is a constant times the divergent harmonic series, and hence diverges by part (*b*) of Theorem 10.4.3. The second series results by deleting the first nine terms from the divergent harmonic series, and hence diverges by part (*c*) of Theorem 10.4.3. ◀

THE INTEGRAL TEST

The expressions

$$\sum_{k=1}^{\infty}\frac{1}{k^2} \quad\text{and}\quad \int_1^{+\infty}\frac{1}{x^2}\,dx$$

are related in that the integrand in the improper integral results when the index k in the general term of the series is replaced by x and the limits of summation in the series are replaced by the corresponding limits of integration. The following theorem shows that there is a relationship between the convergence of the series and the integral.

10.4.4 THEOREM (*The Integral Test*). *Let $\sum u_k$ be a series with positive terms. If f is a function that is decreasing and continuous on an interval $[a,+\infty)$ and such that $u_k=f(k)$ for all $k\ge a$, then*

$$\sum_{k=1}^{\infty}u_k \quad\textit{and}\quad \int_a^{+\infty}f(x)\,dx$$

both converge or both diverge.

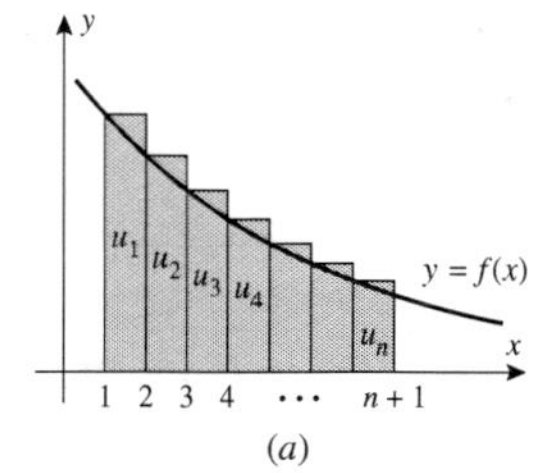

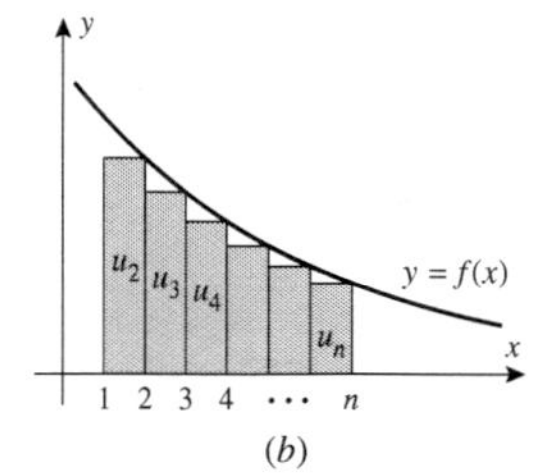

Figure 10.4.1

The proof of the integral test is deferred to the end of this section. However, the gist of the proof is captured in Figure 10.4.1: if the integral diverges, then so does the series (Figure 10.4.1a), and if the integral converges, then so does the series (Figure 10.4.1b).

▶ Example 4 Use the integral test to determine whether the following series converge or diverge.

$$\text{(a)}\ \sum_{k=1}^{\infty} \frac{1}{k} \qquad \text{(b)}\ \sum_{k=1}^{\infty} \frac{1}{k^2}$$

Solution (a). We already know that this is the divergent harmonic series, so the integral test will simply provide another way of establishing the divergence. If we replace k by x in the general term $1/k$, we obtain the function $f(x) = 1/x$, which is decreasing and continuous for $x \geq 1$ (as required to apply the integral test with $a = 1$). Since

$$\int_1^{+\infty} \frac{1}{x}\,dx = \lim_{b\to+\infty} \int_1^b \frac{1}{x}\,dx = \lim_{b\to+\infty} [\ln b - \ln 1] = +\infty$$

the integral diverges and consequently so does the series.

Solution (b). If we replace k by x in the general term $1/k^2$, we obtain the function $f(x) = 1/x^2$, which is decreasing and continuous for $x \geq 1$. Since

$$\int_1^{+\infty} \frac{1}{x^2}\,dx = \lim_{b\to+\infty} \int_1^b \frac{dx}{x^2} = \lim_{b\to+\infty} \left[-\frac{1}{x}\right]_1^b = \lim_{b\to+\infty} \left[1 - \frac{1}{b}\right] = 1$$

the integral converges and consequently the series converges by the integral test with $a = 1$. ◀

p-SERIES

The series in Example 4 are special cases of a class of series called ***p-series*** or ***hyperharmonic series***. A p-series is an infinite series of the form

$$\sum_{k=1}^{\infty} \frac{1}{k^p} = 1 + \frac{1}{2^p} + \frac{1}{3^p} + \cdots + \frac{1}{k^p} + \cdots$$

where $p > 0$. Examples of p-series are

$$\sum_{k=1}^{\infty} \frac{1}{k} = 1 + \frac{1}{2} + \frac{1}{3} + \cdots + \frac{1}{k} + \cdots \qquad p = 1$$

$$\sum_{k=1}^{\infty} \frac{1}{k^2} = 1 + \frac{1}{2^2} + \frac{1}{3^2} + \cdots + \frac{1}{k^2} + \cdots \qquad p = 2$$

$$\sum_{k=1}^{\infty} \frac{1}{\sqrt{k}} = 1 + \frac{1}{\sqrt{2}} + \frac{1}{\sqrt{3}} + \cdots + \frac{1}{\sqrt{k}} + \cdots \qquad p = \tfrac{1}{2}$$

The following theorem tells when a p-series converges.

WARNING

In part (b) of Example 4, do not erroneously conclude that the sum of the series is 1 because the value of the corresponding integral is 1. You can see that this is not so since the sum of the first two terms alone exceeds 1. Later, we will see that the sum of the series is actually $\pi^2/6$.

10.4.5 THEOREM (*Convergence of p-Series*).

$$\sum_{k=1}^{\infty} \frac{1}{k^p} = 1 + \frac{1}{2^p} + \frac{1}{3^p} + \cdots + \frac{1}{k^p} + \cdots$$

converges if $p > 1$ and diverges if $0 < p \leq 1$.

PROOF. To establish this result when $p \neq 1$, we will use the integral test.

$$\int_1^{+\infty} \frac{1}{x^p}\,dx = \lim_{b\to+\infty} \int_1^b x^{-p}\,dx = \lim_{b\to+\infty} \left.\frac{x^{1-p}}{1-p}\right]_1^b = \lim_{b\to+\infty} \left[\frac{b^{1-p}}{1-p} - \frac{1}{1-p}\right]$$

If $p > 1$, then $1 - p < 0$, so $b^{1-p} \to 0$ as $b \to +\infty$. Thus, the integral converges [its value is $-1/(1-p)$] and consequently the series also converges. For $0 < p < 1$, it follows that $1 - p > 0$ and $b^{1-p} \to +\infty$ as $b \to +\infty$, so the integral and the series diverge. The case $p = 1$ is the harmonic series, which was previously shown to diverge. ■

▶ Example 5

$$1 + \frac{1}{\sqrt[3]{2}} + \frac{1}{\sqrt[3]{3}} + \cdots + \frac{1}{\sqrt[3]{k}} + \cdots$$

diverges since it is a p-series with $p = \frac{1}{3} < 1$. ◀

PROOF OF THE INTEGRAL TEST

Before we can prove the integral test, we need a basic result about convergence of series with *nonnegative* terms. If $u_1 + u_2 + u_3 + \cdots + u_k + \cdots$ is such a series, then its sequence of partial sums is increasing, that is,

$$s_1 \le s_2 \le s_3 \le \cdots \le s_n \le \cdots$$

Thus, from Theorem 10.2.3 the sequence of partial sums converges to a limit S if and only if it has some upper bound M, in which case $S \le M$. If no upper bound exists, then the sequence of partial sums diverges. Since convergence of the sequence of partial sums corresponds to convergence of the series, we have the following theorem.

10.4.6 THEOREM. *If $\sum u_k$ is a series with nonnegative terms, and if there is a constant M such that*

$$s_n = u_1 + u_2 + \cdots + u_n \le M$$

for every n, then the series converges and the sum S satisfies $S \le M$. If no such M exists, then the series diverges.

In words, this theorem implies that *a series with nonnegative terms converges if and only if its sequence of partial sums is bounded above.*

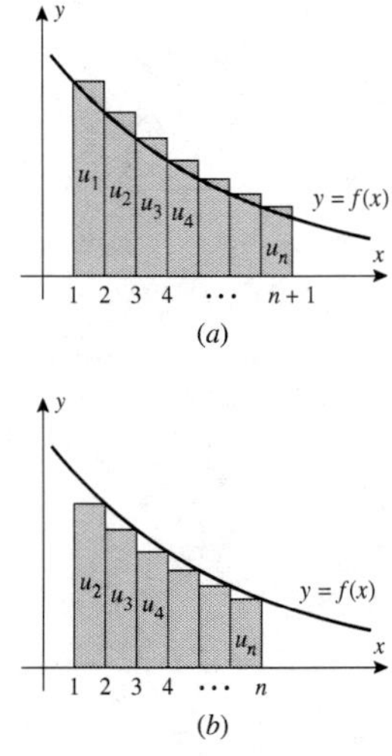

Figure 10.4.2

PROOF OF THEOREM 10.4.4. We need only show that the series converges when the integral converges and that the series diverges when the integral diverges. For simplicity, we will limit the proof to the case where $a = 1$. Assume that $f(x)$ satisfies the hypotheses of the theorem for $x \ge 1$. Since

$$f(1) = u_1,\ f(2) = u_2, \ldots,\ f(n) = u_n, \ldots$$

the values of $u_1, u_2, \ldots, u_n, \ldots$ can be interpreted as the areas of the rectangles shown in Figure 10.4.2.

The following inequalities result by comparing the areas under the curve $y = f(x)$ to the areas of the rectangles in Figure 10.4.2 for $n > 1$:

$$\int_1^{n+1} f(x)\,dx < u_1 + u_2 + \cdots + u_n = s_n \quad \text{Figure 10.4.2}a$$

$$s_n - u_1 = u_2 + u_3 + \cdots + u_n < \int_1^n f(x)\,dx \quad \text{Figure 10.4.2}b$$

These inequalities can be combined as

$$\int_1^{n+1} f(x)\,dx < s_n < u_1 + \int_1^n f(x)\,dx \tag{2}$$

If the integral $\int_1^{\infty} f(x)\,dx$ converges to a finite value L, then from the right-hand inequality in (2)

$$s_n < u_1 + \int_1^n f(x)\,dx < u_1 + \int_1^{+\infty} f(x)\,dx = u_1 + L$$

Thus, each partial sum is less than the finite constant $u_1 + L$, and the series converges by Theorem 10.4.6. On the other hand, if the integral $\int_1^{+\infty} f(x)\,dx$ diverges, then

$$\lim_{n\to+\infty} \int_1^{n+1} f(x)\,dx = +\infty$$

so that from the left-hand inequality in (2), $\lim_{n\to+\infty} s_n = +\infty$. This implies that the series also diverges. ■

✔QUICK CHECK EXERCISES 10.4 *(See page 659 for answers.)*

1. The divergence test says that if ______ $\neq 0$, then the series $\sum u_k$ diverges.
2. Given that

$$\sum_{k=1}^{\infty} \frac{1}{k(k+1)} = 1 \quad \text{and} \quad \sum_{k=1}^{\infty} \frac{1}{6^k} = \frac{1}{5}$$

it follows that

$$\sum_{k=1}^{\infty} \frac{7}{6^{k-1}} = \underline{\qquad}$$

and

$$\sum_{k=1}^{\infty} \left(\frac{1}{2k(k+1)} - \frac{1}{6^k}\right) = \underline{\qquad}$$

3. Since $\int_1^{+\infty}(1/\sqrt{x})\,dx = +\infty$, the ______ test applied to the series $\sum_{k=1}^{\infty}$ ______ shows that this series ______.
4. A p-series is a series of the form

$$\sum_{k=1}^{\infty} \underline{\qquad}$$

This series converges if ______. This series diverges if ______.

EXERCISE SET 10.4 Graphing Utility [C] CAS

1. Use Theorem 10.4.3 to find the sum of each series.
   (a) $\left(\frac{1}{2}+\frac{1}{4}\right)+\left(\frac{1}{2^2}+\frac{1}{4^2}\right)+\cdots+\left(\frac{1}{2^k}+\frac{1}{4^k}\right)+\cdots$
   (b) $\sum_{k=1}^{\infty}\left(\frac{1}{5^k}-\frac{1}{k(k+1)}\right)$
2. Use Theorem 10.4.3 to find the sum of each series.
   (a) $\sum_{k=2}^{\infty}\left[\frac{1}{k^2-1}-\frac{7}{10^{k-1}}\right]$ (b) $\sum_{k=1}^{\infty}\left[7^{-k}3^{k+1}-\frac{2^{k+1}}{5^k}\right]$

3–4 For each given p-series, identify p and determine whether the series converges.

3. (a) $\sum_{k=1}^{\infty}\frac{1}{k^3}$ (b) $\sum_{k=1}^{\infty}\frac{1}{\sqrt{k}}$ (c) $\sum_{k=1}^{\infty}k^{-1}$ (d) $\sum_{k=1}^{\infty}k^{-2/3}$
4. (a) $\sum_{k=1}^{\infty}k^{-4/3}$ (b) $\sum_{k=1}^{\infty}\frac{1}{\sqrt[4]{k}}$ (c) $\sum_{k=1}^{\infty}\frac{1}{\sqrt[3]{k^5}}$ (d) $\sum_{k=1}^{\infty}\frac{1}{k^{\pi}}$

5–6 Apply the divergence test and state what it tells you about the series.

5. (a) $\sum_{k=1}^{\infty}\frac{k^2+k+3}{2k^2+1}$ (b) $\sum_{k=1}^{\infty}\left(1+\frac{1}{k}\right)^k$
   (c) $\sum_{k=1}^{\infty}\cos k\pi$ (d) $\sum_{k=1}^{\infty}\frac{1}{k!}$
6. (a) $\sum_{k=1}^{\infty}\frac{k}{e^k}$ (b) $\sum_{k=1}^{\infty}\ln k$
   (c) $\sum_{k=1}^{\infty}\frac{1}{\sqrt{k}}$ (d) $\sum_{k=1}^{\infty}\frac{\sqrt{k}}{\sqrt{k}+3}$

7–8 Confirm that the integral test is applicable and use it to determine whether the series converges.

7. (a) $\sum_{k=1}^{\infty}\frac{1}{5k+2}$ (b) $\sum_{k=1}^{\infty}\frac{1}{1+9k^2}$

8. (a) $\sum_{k=1}^{\infty} \frac{k}{1+k^2}$ (b) $\sum_{k=1}^{\infty} \frac{1}{(4+2k)^{3/2}}$

9–24 Determine whether the series converges.

9. $\sum_{k=1}^{\infty} \frac{1}{k+6}$ **10.** $\sum_{k=1}^{\infty} \frac{3}{5k}$ **11.** $\sum_{k=1}^{\infty} \frac{1}{\sqrt{k+5}}$

12. $\sum_{k=1}^{\infty} \frac{1}{\sqrt[k]{e}}$ **13.** $\sum_{k=1}^{\infty} \frac{1}{\sqrt[3]{2k-1}}$ **14.** $\sum_{k=3}^{\infty} \frac{\ln k}{k}$

15. $\sum_{k=1}^{\infty} \frac{k}{\ln(k+1)}$ **16.** $\sum_{k=1}^{\infty} ke^{-k^2}$ **17.** $\sum_{k=1}^{\infty} \left(1+\frac{1}{k}\right)^{-k}$

18. $\sum_{k=1}^{\infty} \frac{k^2+1}{k^2+3}$ **19.** $\sum_{k=1}^{\infty} \frac{\tan^{-1} k}{1+k^2}$ **20.** $\sum_{k=1}^{\infty} \frac{1}{\sqrt{k^2+1}}$

21. $\sum_{k=1}^{\infty} k^2 \sin^2\left(\frac{1}{k}\right)$ **22.** $\sum_{k=1}^{\infty} k^2 e^{-k^3}$

23. $\sum_{k=5}^{\infty} 7k^{-1.01}$ **24.** $\sum_{k=1}^{\infty} \operatorname{sech}^2 k$

25–26 Use the integral test to investigate the relationship between the value of p and the convergence of the series.

25. $\sum_{k=2}^{\infty} \frac{1}{k(\ln k)^p}$ **26.** $\sum_{k=3}^{\infty} \frac{1}{k(\ln k)[\ln(\ln k)]^p}$

FOCUS ON CONCEPTS

27. Suppose that the series $\sum u_k$ converges and the series $\sum v_k$ diverges. Show that the series $\sum(u_k+v_k)$ and $\sum(u_k-v_k)$ both diverge. [*Hint:* Assume that $\sum(u_k+v_k)$ converges and use Theorem 10.4.3 to obtain a contradiction.]

28. Find examples to show that if the series $\sum u_k$ and $\sum v_k$ both diverge, then the series $\sum(u_k+v_k)$ and $\sum(u_k-v_k)$ may either converge or diverge.

29–30 Use the results of Exercises 27 and 28, if needed, to determine whether each series converges or diverges.

29. (a) $\sum_{k=1}^{\infty} \left[\left(\frac{2}{3}\right)^{k-1} + \frac{1}{k}\right]$ (b) $\sum_{k=1}^{\infty} \left[\frac{1}{3k+2} - \frac{1}{k^{3/2}}\right]$

30. (a) $\sum_{k=2}^{\infty} \left[\frac{1}{k(\ln k)^2} - \frac{1}{k^2}\right]$ (b) $\sum_{k=2}^{\infty} \left[ke^{-k^2} + \frac{1}{k \ln k}\right]$

C **31.** Use a CAS to confirm that

$$\sum_{k=1}^{\infty} \frac{1}{k^2} = \frac{\pi^2}{6} \quad \text{and} \quad \sum_{k=1}^{\infty} \frac{1}{k^4} = \frac{\pi^4}{90}$$

and then use these results in each part to find the sum of the series.

(a) $\sum_{k=1}^{\infty} \frac{3k^2-1}{k^4}$ (b) $\sum_{k=3}^{\infty} \frac{1}{k^2}$ (c) $\sum_{k=2}^{\infty} \frac{1}{(k-1)^4}$

32–37 Exercise 32 will show how a partial sum can be used to obtain upper and lower bounds on the sum of a series when the hypotheses of the integral test are satisfied. This result will be needed in Exercises 33–37.

32. (a) Let $\sum_{k=1}^{\infty} u_k$ be a convergent series with positive terms, and let f be a function that is decreasing and continuous on $[n, +\infty)$ and such that $u_k = f(k)$ for $k \geq n$. Use an area argument and the accompanying figure to show that

$$\int_{n+1}^{+\infty} f(x)\,dx < \sum_{k=n+1}^{\infty} u_k < \int_{n}^{+\infty} f(x)\,dx$$

(b) Show that if S is the sum of the series $\sum_{k=1}^{\infty} u_k$ and s_n is the nth partial sum, then

$$s_n + \int_{n+1}^{+\infty} f(x)\,dx < S < s_n + \int_{n}^{+\infty} f(x)\,dx$$

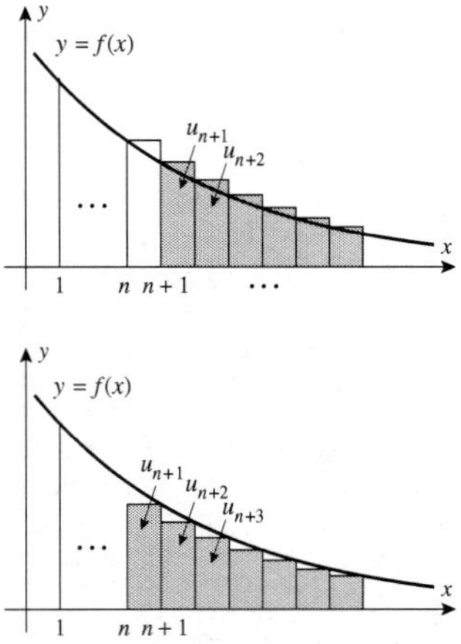

Figure Ex-32

33. (a) It was stated in Exercise 31 that

$$\sum_{k=1}^{\infty} \frac{1}{k^2} = \frac{\pi^2}{6}$$

Show that if s_n is the nth partial sum of this series, then

$$s_n + \frac{1}{n+1} < \frac{\pi^2}{6} < s_n + \frac{1}{n}$$

(b) Calculate s_3 exactly, and then use the result in part (a) to show that

$$\frac{29}{18} < \frac{\pi^2}{6} < \frac{61}{36}$$

(c) Use a calculating utility to confirm that the inequalities in part (b) are correct.

(d) Find upper and lower bounds on the error that results if the sum of the series is approximated by the 10th partial sum.

34. In each part, find upper and lower bounds on the error that results if the sum of the series is approximated by the 10th partial sum.

(a) $\sum_{k=1}^{\infty} \frac{1}{(2k+1)^2}$ (b) $\sum_{k=1}^{\infty} \frac{1}{k^2+1}$ (c) $\sum_{k=1}^{\infty} \frac{k}{e^k}$

35. Our objective in this problem is to approximate the sum of the series $\sum_{k=1}^{\infty} 1/k^3$ to two decimal-place accuracy.

(a) Show that if S is the sum of the series and s_n is the nth partial sum, then

$$s_n + \frac{1}{2(n+1)^2} < S < s_n + \frac{1}{2n^2}$$

(b) For two decimal-place accuracy, the error must be less than 0.005 (see Table 2.5.1 on p. 132). We can achieve this by finding an interval of length 0.01 (or less) that contains S and approximating S by the midpoint of that interval. Find the smallest value of n such that the interval containing S in part (a) has a length of 0.01 or less.

(c) Approximate S to two decimal-place accuracy.

36. (a) Use the method of Exercise 35 to approximate the sum of the series $\sum_{k=1}^{\infty} 1/k^4$ to two decimal-place accuracy.

(b) It was stated in Exercise 31 that the sum of this series is $\pi^4/90$. Use a calculating utility to confirm that your answer in part (a) is accurate to two decimal places.

37. We showed in Section 10.3 that the harmonic series $\sum_{k=1}^{\infty} 1/k$ diverges. Our objective in this problem is to demonstrate that although the partial sums of this series approach $+\infty$, they increase extremely slowly.

(a) Use inequality (2) to show that for $n \geq 2$

$$\ln(n+1) < s_n < 1 + \ln n$$

(b) Use the inequalities in part (a) to find upper and lower bounds on the sum of the first million terms in the series.

(c) Show that the sum of the first billion terms in the series is less than 22.

(d) Find a value of n so that the sum of the first n terms is greater than 100.

38. Investigate the relationship between the value of a and the convergence of the series $\sum_{k=1}^{\infty} k^{-\ln a}$.

39. Use a graphing utility to confirm that the integral test applies to the series $\sum_{k=1}^{\infty} k^2 e^{-k}$, and then determine whether the series converges.

C **40.** (a) Show that the hypotheses of the integral test are satisfied by the series $\sum_{k=1}^{\infty} 1/(k^3+1)$.

(b) Use a CAS and the integral test to confirm that the series converges.

(c) Construct a table of partial sums for $n = 10, 20, 30, \ldots, 100$, showing at least six decimal places.

(d) Based on your table, make a conjecture about the sum of the series to three decimal-place accuracy.

(e) Use part (b) of Exercise 32 to check your conjecture.

✔QUICK CHECK ANSWERS 10.4

1. $\lim_{k\to+\infty} u_k$ **2.** $\frac{42}{5}; \frac{3}{10}$ **3.** integral; $\frac{1}{\sqrt{k}}$; diverges **4.** $\frac{1}{k^p}$; $p > 1$; $0 < p \leq 1$

10.5 THE COMPARISON, RATIO, AND ROOT TESTS

In this section we will develop some more basic convergence tests for series with nonnegative terms. Later, we will use some of these tests to study the convergence of Taylor series.

THE COMPARISON TEST

We will begin with a test that is useful in its own right and is also the building block for other important convergence tests. The underlying idea of this test is to use the known convergence or divergence of a series to deduce the convergence or divergence of another series.

10.5.1 THEOREM (*The Comparison Test*). *Let $\sum_{k=1}^{\infty} a_k$ and $\sum_{k=1}^{\infty} b_k$ be series with nonnegative terms and suppose that*

$$a_1 \leq b_1,\ a_2 \leq b_2,\ a_3 \leq b_3, \ldots, a_k \leq b_k, \ldots$$

(*a*) *If the "bigger series" Σb_k converges, then the "smaller series" Σa_k also converges.*

(*b*) *If the "smaller series" Σa_k diverges, then the "bigger series" Σb_k also diverges.*

It is not essential in Theorem 10.5.1 that the condition $a_k \leq b_k$ hold for all k, as stated; the conclusions of the theorem remain true if this condition is eventually true.

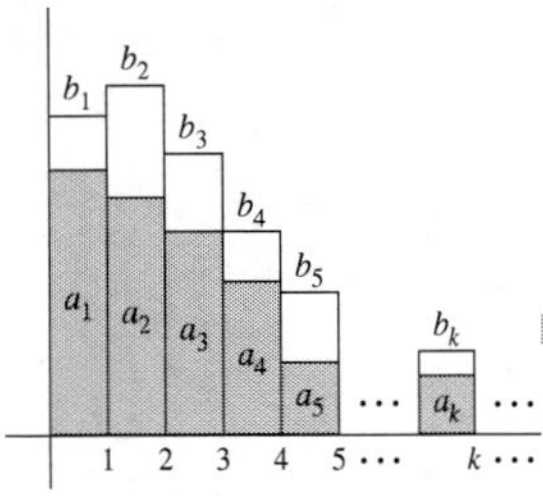

For each rectangle, b_k is the entire area and a_k is the area of the blue portion.

Figure 10.5.1

We have left the proof of this theorem for the exercises; however, it is easy to visualize why the theorem is true by interpreting the terms in the series as areas of rectangles (Figure 10.5.1). The comparison test states that if the total area $\sum b_k$ is finite, then the total area $\sum a_k$ must also be finite; and if the total area $\sum a_k$ is infinite, then the total area $\sum b_k$ must also be infinite.

USING THE COMPARISON TEST

There are two steps required for using the comparison test to determine whether a series $\sum u_k$ with positive terms converges:

- Guess at whether the series $\sum u_k$ converges or diverges.
- Find a series that proves the guess to be correct. That is, if the guess is divergence, we must find a divergent series whose terms are "smaller" than the corresponding terms of $\sum u_k$, and if the guess is convergence, we must find a convergent series whose terms are "bigger" than the corresponding terms of $\sum u_k$.

In most cases, the series $\sum u_k$ being considered will have its general term u_k expressed as a fraction. To help with the guessing process in the first step, we have formulated two principles that are based on the form of the denominator for u_k. These principles sometimes *suggest* whether a series is likely to converge or diverge. We have called these "informal principles" because they are not intended as formal theorems. In fact, we will not guarantee that they *always* work. However, they work often enough to be useful.

10.5.2 INFORMAL PRINCIPLE. *Constant terms in the denominator of u_k can usually be deleted without affecting the convergence or divergence of the series.*

10.5.3 INFORMAL PRINCIPLE. *If a polynomial in k appears as a factor in the numerator or denominator of u_k, all but the leading term in the polynomial can usually be discarded without affecting the convergence or divergence of the series.*

▶ Example 1 Use the comparison test to determine whether the following series converge or diverge.

$$\text{(a)} \sum_{k=1}^{\infty} \frac{1}{\sqrt{k}-\frac{1}{2}} \qquad \text{(b)} \sum_{k=1}^{\infty} \frac{1}{2k^2+k}$$

Solution (a). According to Principle 10.5.2, we should be able to drop the constant in the denominator without affecting the convergence or divergence. Thus, the given series is likely to behave like

$$\sum_{k=1}^{\infty} \frac{1}{\sqrt{k}} \tag{1}$$

which is a divergent p-series $\left(p=\frac{1}{2}\right)$. Thus, we will guess that the given series diverges and try to prove this by finding a divergent series that is "smaller" than the given series. However, series (1) does the trick since

$$\frac{1}{\sqrt{k}-\frac{1}{2}} > \frac{1}{\sqrt{k}} \quad \text{for } k=1,2,\ldots$$

Thus, we have proved that the given series diverges.

Solution (b). According to Principle 10.5.3, we should be able to discard all but the leading term in the polynomial without affecting the convergence or divergence. Thus, the given series is likely to behave like

$$\sum_{k=1}^{\infty} \frac{1}{2k^2} = \frac{1}{2} \sum_{k=1}^{\infty} \frac{1}{k^2} \tag{2}$$

which converges since it is a constant times a convergent p-series ($p = 2$). Thus, we will guess that the given series converges and try to prove this by finding a convergent series that is "bigger" than the given series. However, series (2) does the trick since

$$\frac{1}{2k^2 + k} < \frac{1}{2k^2} \quad \text{for } k = 1, 2, \ldots$$

Thus, we have proved that the given series converges. ◀

THE LIMIT COMPARISON TEST

In the last example, Principles 10.5.2 and 10.5.3 provided the guess about convergence or divergence as well as the series needed to apply the comparison test. Unfortunately, it is not always so straightforward to find the series required for comparison, so we will now consider an alternative to the comparison test that is usually easier to apply. The proof is given in Appendix C.

10.5.4 THEOREM (*The Limit Comparison Test*). *Let $\sum a_k$ and $\sum b_k$ be series with positive terms and suppose that*

$$\rho = \lim_{k \to +\infty} \frac{a_k}{b_k}$$

If ρ is finite and $\rho > 0$, then the series both converge or both diverge.

The cases where $\rho = 0$ or $\rho = +\infty$ are discussed in the exercises (Exercise 54).

▶ **Example 2** Use the limit comparison test to determine whether the following series converge or diverge.

$$\text{(a)} \sum_{k=1}^{\infty} \frac{1}{\sqrt{k} + 1} \qquad \text{(b)} \sum_{k=1}^{\infty} \frac{1}{2k^2 + k} \qquad \text{(c)} \sum_{k=1}^{\infty} \frac{3k^3 - 2k^2 + 4}{k^7 - k^3 + 2}$$

Solution (a). As in Example 1, Principle 10.5.2 suggests that the series is likely to behave like the divergent p-series (1). To prove that the given series diverges, we will apply the limit comparison test with

$$a_k = \frac{1}{\sqrt{k} + 1} \quad \text{and} \quad b_k = \frac{1}{\sqrt{k}}$$

We obtain

$$\rho = \lim_{k \to +\infty} \frac{a_k}{b_k} = \lim_{k \to +\infty} \frac{\sqrt{k}}{\sqrt{k} + 1} = \lim_{k \to +\infty} \frac{1}{1 + \dfrac{1}{\sqrt{k}}} = 1$$

Since ρ is finite and positive, it follows from Theorem 10.5.4 that the given series diverges.

Solution (b). As in Example 1, Principle 10.5.3 suggests that the series is likely to behave like the convergent series (2). To prove that the given series converges, we will apply the limit comparison test with

$$a_k = \frac{1}{2k^2 + k} \quad \text{and} \quad b_k = \frac{1}{2k^2}$$

We obtain

$$\rho = \lim_{k \to +\infty} \frac{a_k}{b_k} = \lim_{k \to +\infty} \frac{2k^2}{2k^2 + k} = \lim_{k \to +\infty} \frac{2}{2 + \dfrac{1}{k}} = 1$$

Since ρ is finite and positive, it follows from Theorem 10.5.4 that the given series converges, which agrees with the conclusion reached in Example 1 using the comparison test.

Solution (c). From Principle 10.5.3, the series is likely to behave like

$$\sum_{k=1}^{\infty} \frac{3k^3}{k^7} = \sum_{k=1}^{\infty} \frac{3}{k^4} \tag{3}$$

which converges since it is a constant times a convergent p-series. Thus, the given series is likely to converge. To prove this, we will apply the limit comparison test to series (3) and the given series. We obtain

$$\rho = \lim_{k \to +\infty} \frac{\dfrac{3k^3 - 2k^2 + 4}{k^7 - k^3 + 2}}{\dfrac{3}{k^4}} = \lim_{k \to +\infty} \frac{3k^7 - 2k^6 + 4k^4}{3k^7 - 3k^3 + 6} = 1$$

Since ρ is finite and nonzero, it follows from Theorem 10.5.4 that the given series converges, since (3) converges. ◀

THE RATIO TEST

The comparison test and the limit comparison test hinge on first making a guess about convergence and then finding an appropriate series for comparison, both of which can be difficult tasks in cases where Principles 10.5.2 and 10.5.3 cannot be applied. In such cases the next test can often be used, since it works exclusively with the terms of the given series—it requires neither an initial guess about convergence nor the discovery of a series for comparison. Its proof is given in Appendix C.

10.5.5 THEOREM (*The Ratio Test*). *Let $\sum u_k$ be a series with positive terms and suppose that*

$$\rho = \lim_{k \to +\infty} \frac{u_{k+1}}{u_k}$$

(a) *If $\rho < 1$, the series converges.*

(b) *If $\rho > 1$ or $\rho = +\infty$, the series diverges.*

(c) *If $\rho = 1$, the series may converge or diverge, so that another test must be tried.*

▶ **Example 3** Use the ratio test to determine whether the following series converge or diverge.

(a) $\displaystyle\sum_{k=1}^{\infty} \frac{1}{k!}$ (b) $\displaystyle\sum_{k=1}^{\infty} \frac{k}{2^k}$ (c) $\displaystyle\sum_{k=1}^{\infty} \frac{k^k}{k!}$ (d) $\displaystyle\sum_{k=3}^{\infty} \frac{(2k)!}{4^k}$ (e) $\displaystyle\sum_{k=1}^{\infty} \frac{1}{2k-1}$

Solution (a). The series converges, since

$$\rho = \lim_{k\to+\infty} \frac{u_{k+1}}{u_k} = \lim_{k\to+\infty} \frac{1/(k+1)!}{1/k!} = \lim_{k\to+\infty} \frac{k!}{(k+1)!} = \lim_{k\to+\infty} \frac{1}{k+1} = 0 < 1$$

Solution (b). The series converges, since

$$\rho = \lim_{k\to+\infty} \frac{u_{k+1}}{u_k} = \lim_{k\to+\infty} \frac{k+1}{2^{k+1}} \cdot \frac{2^k}{k} = \frac{1}{2} \lim_{k\to+\infty} \frac{k+1}{k} = \frac{1}{2} < 1$$

Solution (c). The series diverges, since

$$\rho = \lim_{k\to+\infty} \frac{u_{k+1}}{u_k} = \lim_{k\to+\infty} \frac{(k+1)^{k+1}}{(k+1)!} \cdot \frac{k!}{k^k} = \lim_{k\to+\infty} \frac{(k+1)^k}{k^k} = \lim_{k\to+\infty} \left(1+\frac{1}{k}\right)^k = e > 1$$

See Formula (3) of Section 7.1

Solution (d). The series diverges, since

$$\rho = \lim_{k\to+\infty} \frac{u_{k+1}}{u_k} = \lim_{k\to+\infty} \frac{[2(k+1)]!}{4^{k+1}} \cdot \frac{4^k}{(2k)!} = \lim_{k\to+\infty} \left(\frac{(2k+2)!}{(2k)!} \cdot \frac{1}{4}\right)$$

$$= \frac{1}{4} \lim_{k\to+\infty} (2k+2)(2k+1) = +\infty$$

Solution (e). The ratio test is of no help since

$$\rho = \lim_{k\to+\infty} \frac{u_{k+1}}{u_k} = \lim_{k\to+\infty} \frac{1}{2(k+1)-1} \cdot \frac{2k-1}{1} = \lim_{k\to+\infty} \frac{2k-1}{2k+1} = 1$$

However, the integral test proves that the series diverges since

$$\int_1^{+\infty} \frac{dx}{2x-1} = \lim_{b\to+\infty} \int_1^b \frac{dx}{2x-1} = \lim_{b\to+\infty} \frac{1}{2} \ln(2x-1) \Big]_1^b = +\infty$$

Both the comparison test and the limit comparison test would also have worked here (verify). ◀

THE ROOT TEST

In cases where it is difficult or inconvenient to find the limit required for the ratio test, the next test is sometimes useful. Since its proof is similar to the proof of the ratio test, we will omit it.

10.5.6 THEOREM (*The Root Test*). *Let $\sum u_k$ be a series with positive terms and suppose that*

$$\rho = \lim_{k\to+\infty} \sqrt[k]{u_k} = \lim_{k\to+\infty} (u_k)^{1/k}$$

(*a*) *If $\rho < 1$, the series converges.*

(*b*) *If $\rho > 1$ or $\rho = +\infty$, the series diverges.*

(*c*) *If $\rho = 1$, the series may converge or diverge, so that another test must be tried.*

▶ **Example 4** Use the root test to determine whether the following series converge or diverge.

$$\text{(a)} \sum_{k=2}^{\infty}\left(\frac{4k-5}{2k+1}\right)^k \qquad \text{(b)} \sum_{k=1}^{\infty}\frac{1}{(\ln(k+1))^k}$$

Solution (a). The series diverges, since

$$\rho = \lim_{k\to+\infty}(u_k)^{1/k} = \lim_{k\to+\infty}\frac{4k-5}{2k+1} = 2 > 1$$

Solution (b). The series converges, since

$$\rho = \lim_{k\to+\infty}(u_k)^{1/k} = \lim_{k\to+\infty}\frac{1}{\ln(k+1)} = 0 < 1 \blacktriangleleft$$

✔QUICK CHECK EXERCISES 10.5 (See page 665 for answers.)

1–4 Select between *converges* or *diverges* to fill the first blank.

1. The series

$$\sum_{k=1}^{\infty}\frac{2k^2+1}{2k^{8/3}-1}$$

_______ by comparison with the p-series $\sum_{k=1}^{\infty}$ _______.

2. Since

$$\lim_{k\to+\infty}\frac{(k+1)^3/3^{k+1}}{k^3/3^k} = \lim_{k\to+\infty}\frac{\left(1+\dfrac{1}{k}\right)^3}{3} = \frac{1}{3}$$

the series $\sum_{k=1}^{\infty}k^3/3^k$ _______ by the _______ test.

3. Since

$$\lim_{k\to+\infty}\frac{(k+1)!/3^{k+1}}{k!/3^k} = \lim_{k\to+\infty}\frac{k+1}{3} = +\infty$$

the series $\sum_{k=1}^{\infty}k!/3^k$ _______ by the _______ test.

4. Since

$$\lim_{k\to+\infty}\left(\frac{1}{k^{k/2}}\right)^{1/k} = \lim_{k\to+\infty}\frac{1}{k^{1/2}} = 0$$

the series $\sum_{k=1}^{\infty}1/k^{k/2}$ _______ by the _______ test.

EXERCISE SET 10.5 [C] CAS

1–2 Make a guess about the convergence or divergence of the series, and confirm your guess using the comparison test.

1. (a) $\displaystyle\sum_{k=1}^{\infty}\frac{1}{5k^2-k}$ (b) $\displaystyle\sum_{k=1}^{\infty}\frac{3}{k-\frac{1}{4}}$

2. (a) $\displaystyle\sum_{k=2}^{\infty}\frac{k+1}{k^2-k}$ (b) $\displaystyle\sum_{k=1}^{\infty}\frac{2}{k^4+k}$

3. In each part, use the comparison test to show that the series converges.

(a) $\displaystyle\sum_{k=1}^{\infty}\frac{1}{3^k+5}$ (b) $\displaystyle\sum_{k=1}^{\infty}\frac{5\sin^2 k}{k!}$

4. In each part, use the comparison test to show that the series diverges.

(a) $\displaystyle\sum_{k=1}^{\infty}\frac{\ln k}{k}$ (b) $\displaystyle\sum_{k=1}^{\infty}\frac{k}{k^{3/2}-\frac{1}{2}}$

5–10 Use the limit comparison test to determine whether the series converges.

5. $\displaystyle\sum_{k=1}^{\infty}\frac{4k^2-2k+6}{8k^7+k-8}$ **6.** $\displaystyle\sum_{k=1}^{\infty}\frac{1}{9k+6}$

7. $\displaystyle\sum_{k=1}^{\infty}\frac{5}{3^k+1}$ **8.** $\displaystyle\sum_{k=1}^{\infty}\frac{k(k+3)}{(k+1)(k+2)(k+5)}$

9. $\displaystyle\sum_{k=1}^{\infty}\frac{1}{\sqrt[3]{8k^2-3k}}$ **10.** $\displaystyle\sum_{k=1}^{\infty}\frac{1}{(2k+3)^{17}}$

11–16 Use the ratio test to determine whether the series converges. If the test is inconclusive, then say so.

11. $\displaystyle\sum_{k=1}^{\infty}\frac{3^k}{k!}$ **12.** $\displaystyle\sum_{k=1}^{\infty}\frac{4^k}{k^2}$ **13.** $\displaystyle\sum_{k=1}^{\infty}\frac{1}{5k}$

14. $\displaystyle\sum_{k=1}^{\infty}k\left(\frac{1}{2}\right)^k$ **15.** $\displaystyle\sum_{k=1}^{\infty}\frac{k!}{k^3}$ **16.** $\displaystyle\sum_{k=1}^{\infty}\frac{k}{k^2+1}$

17–20 Use the root test to determine whether the series converges. If the test is inconclusive, then say so.

17. $\sum_{k=1}^{\infty}\left(\frac{3k+2}{2k-1}\right)^k$ **18.** $\sum_{k=1}^{\infty}\left(\frac{k}{100}\right)^k$

19. $\sum_{k=1}^{\infty}\frac{k}{5^k}$ **20.** $\sum_{k=1}^{\infty}(1-e^{-k})^k$

21–44 Use any method to determine whether the series converges.

21. $\sum_{k=0}^{\infty}\frac{7^k}{k!}$ **22.** $\sum_{k=1}^{\infty}\frac{1}{2k+1}$ **23.** $\sum_{k=1}^{\infty}\frac{k^2}{5^k}$

24. $\sum_{k=1}^{\infty}\frac{k!10^k}{3^k}$ **25.** $\sum_{k=1}^{\infty}k^{50}e^{-k}$ **26.** $\sum_{k=1}^{\infty}\frac{k^2}{k^3+1}$

27. $\sum_{k=1}^{\infty}\frac{\sqrt{k}}{k^3+1}$ **28.** $\sum_{k=1}^{\infty}\frac{4}{2+3^k k}$

29. $\sum_{k=1}^{\infty}\frac{1}{\sqrt{k(k+1)}}$ **30.** $\sum_{k=1}^{\infty}\frac{2+(-1)^k}{5^k}$

31. $\sum_{k=1}^{\infty}\frac{2+\sqrt{k}}{(k+1)^3-1}$ **32.** $\sum_{k=1}^{\infty}\frac{4+|\cos k|}{k^3}$

33. $\sum_{k=1}^{\infty}\frac{1}{1+\sqrt{k}}$ **34.** $\sum_{k=1}^{\infty}\frac{k!}{k^k}$ **35.** $\sum_{k=1}^{\infty}\frac{\ln k}{e^k}$

36. $\sum_{k=1}^{\infty}\frac{k!}{e^{k^2}}$ **37.** $\sum_{k=0}^{\infty}\frac{(k+4)!}{4!k!4^k}$ **38.** $\sum_{k=1}^{\infty}\left(\frac{k}{k+1}\right)^{k^2}$

39. $\sum_{k=1}^{\infty}\frac{1}{4+2^{-k}}$ **40.** $\sum_{k=1}^{\infty}\frac{\sqrt{k}\ln k}{k^3+1}$ **41.** $\sum_{k=1}^{\infty}\frac{\tan^{-1}k}{k^2}$

42. $\sum_{k=1}^{\infty}\frac{5^k+k}{k!+3}$ **43.** $\sum_{k=0}^{\infty}\frac{(k!)^2}{(2k)!}$ **44.** $\sum_{k=1}^{\infty}\frac{(k!)^2 2^k}{(2k+2)!}$

45–46 Find the general term of the series and use the ratio test to show that the series converges.

45. $1+\frac{1\cdot 2}{1\cdot 3}+\frac{1\cdot 2\cdot 3}{1\cdot 3\cdot 5}+\frac{1\cdot 2\cdot 3\cdot 4}{1\cdot 3\cdot 5\cdot 7}+\cdots$

46. $1+\frac{1\cdot 3}{3!}+\frac{1\cdot 3\cdot 5}{5!}+\frac{1\cdot 3\cdot 5\cdot 7}{7!}+\cdots$

47–48 Use a CAS to investigate the convergence of the series.

C **47.** $\sum_{k=1}^{\infty}\frac{\ln k}{3^k}$ C **48.** $\sum_{k=1}^{\infty}\frac{[\pi(k+1)]^k}{k^{k+1}}$

FOCUS ON CONCEPTS

49. (a) Make a conjecture about the convergence of the series $\sum_{k=1}^{\infty}\sin(\pi/k)$ by considering the local linear approximation of $\sin x$ at $x=0$.
(b) Try to confirm your conjecture using the limit comparison test.

50. (a) We will see later that the polynomial $1-x^2/2$ is the "local quadratic" approximation for $\cos x$ at $x=0$. Make a conjecture about the convergence of the series
$$\sum_{k=1}^{\infty}\left[1-\cos\left(\frac{1}{k}\right)\right]$$
by considering this approximation.
(b) Try to confirm your conjecture using the limit comparison test.

51. Show that $\ln x < \sqrt{x}$ if $x>0$, and use this result to investigate the convergence of
(a) $\sum_{k=1}^{\infty}\frac{\ln k}{k^2}$ (b) $\sum_{k=2}^{\infty}\frac{1}{(\ln k)^2}$

52. For which positive values of α does the series $\sum_{k=1}^{\infty}(\alpha^k/k^\alpha)$ converge?

53. Use Theorem 10.4.6 to prove the comparison test (Theorem 10.5.1).

54. Let $\sum a_k$ and $\sum b_k$ be series with positive terms. Prove:
(a) If $\lim_{k\to+\infty}(a_k/b_k)=0$ and $\sum b_k$ converges, then $\sum a_k$ converges.
(b) If $\lim_{k\to+\infty}(a_k/b_k)=+\infty$ and $\sum b_k$ diverges, then $\sum a_k$ diverges.

✔ QUICK CHECK ANSWERS 10.5

1. diverges; $1/k^{2/3}$ **2.** converges; ratio **3.** diverges; ratio **4.** converges; root

10.6 ALTERNATING SERIES; CONDITIONAL CONVERGENCE

Up to now we have focused exclusively on series with nonnegative terms. In this section we will discuss series that contain both positive and negative terms.

ALTERNATING SERIES

Series whose terms alternate between positive and negative, called ***alternating series***, are of special importance. Some examples are

$$\sum_{k=1}^{\infty}(-1)^{k+1}\frac{1}{k} = 1 - \frac{1}{2} + \frac{1}{3} - \frac{1}{4} + \frac{1}{5} - \cdots$$

$$\sum_{k=1}^{\infty}(-1)^{k}\frac{1}{k} = -1 + \frac{1}{2} - \frac{1}{3} + \frac{1}{4} - \frac{1}{5} + \cdots$$

In general, an alternating series has one of the following two forms:

$$\sum_{k=1}^{\infty}(-1)^{k+1}a_k = a_1 - a_2 + a_3 - a_4 + \cdots \tag{1}$$

$$\sum_{k=1}^{\infty}(-1)^{k}a_k = -a_1 + a_2 - a_3 + a_4 - \cdots \tag{2}$$

where the a_k's are assumed to be positive in both cases.

The following theorem is the key result on convergence of alternating series.

10.6.1 THEOREM (*Alternating Series Test*). *An alternating series of either form* (1) *or form* (2) *converges if the following two conditions are satisfied:*

(*a*) $a_1 \geq a_2 \geq a_3 \geq \cdots \geq a_k \geq \cdots$

(*b*) $\lim_{k\to+\infty} a_k = 0$

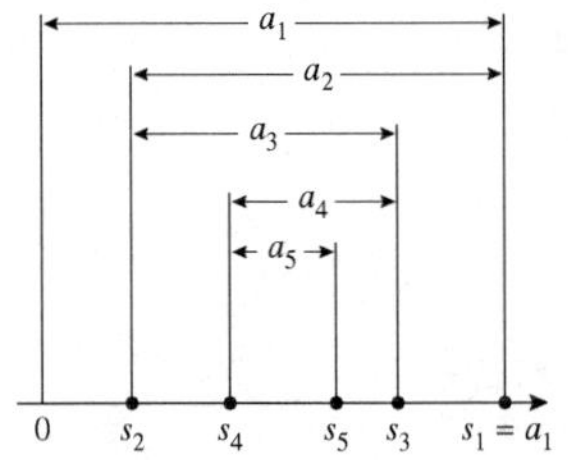

Figure 10.6.1

It is not essential for condition (*a*) in Theorem 10.6.1 to hold for all terms; an alternating series will converge if condition (*b*) is true and condition (*a*) holds eventually.

PROOF. We will consider only alternating series of form (1). The idea of the proof is to show that if conditions (*a*) and (*b*) hold, then the sequences of even-numbered and odd-numbered partial sums converge to a common limit S. It will then follow from Theorem 10.1.4 that the entire sequence of partial sums converges to S.

Figure 10.6.1 shows how successive partial sums satisfying conditions (*a*) and (*b*) appear when plotted on a horizontal axis. The even-numbered partial sums

$$s_2, s_4, s_6, s_8, \ldots, s_{2n}, \ldots$$

form an increasing sequence bounded above by a_1, and the odd-numbered partial sums

$$s_1, s_3, s_5, \ldots, s_{2n-1}, \ldots$$

form a decreasing sequence bounded below by 0. Thus, by Theorems 10.2.3 and 10.2.4, the even-numbered partial sums converge to some limit S_E and the odd-numbered partial sums converge to some limit S_O. To complete the proof we must show that $S_E = S_O$. But the $(2n)$-th term in the series is $-a_{2n}$, so that $s_{2n} - s_{2n-1} = -a_{2n}$, which can be written as

$$s_{2n-1} = s_{2n} + a_{2n}$$

However, $2n \to +\infty$ and $2n - 1 \to +\infty$ as $n \to +\infty$, so that

$$S_O = \lim_{n\to+\infty} s_{2n-1} = \lim_{n\to+\infty}(s_{2n} + a_{2n}) = S_E + 0 = S_E$$

which completes the proof. ■

▶ **Example 1** Use the alternating series test to show that the following series converge.

$$\text{(a)} \sum_{k=1}^{\infty}(-1)^{k+1}\frac{1}{k} \qquad \text{(b)} \sum_{k=1}^{\infty}(-1)^{k+1}\frac{k+3}{k(k+1)}$$

The series in part (a) of Example 1 is called the ***alternating harmonic series***. Note that this series converges, whereas the harmonic series diverges.

Solution (a). The two conditions in the alternating series test are satisfied since

$$a_k = \frac{1}{k} > \frac{1}{k+1} = a_{k+1} \quad \text{and} \quad \lim_{k\to+\infty} a_k = \lim_{k\to+\infty}\frac{1}{k} = 0$$

Solution (b). The two conditions in the alternating series test are satisfied since

$$\frac{a_{k+1}}{a_k} = \frac{k+4}{(k+1)(k+2)} \cdot \frac{k(k+1)}{k+3} = \frac{k^2+4k}{k^2+5k+6} = \frac{k^2+4k}{(k^2+4k)+(k+6)} < 1$$

so

$$a_k > a_{k+1}$$

and

$$\lim_{k\to+\infty} a_k = \lim_{k\to+\infty}\frac{k+3}{k(k+1)} = \lim_{k\to+\infty}\frac{\dfrac{1}{k}+\dfrac{3}{k^2}}{1+\dfrac{1}{k}} = 0 \;◀$$

APPROXIMATING SUMS OF ALTERNATING SERIES

The following theorem is concerned with the error that results when the sum of an alternating series is approximated by a partial sum.

10.6.2 THEOREM. *If an alternating series satisfies the hypotheses of the alternating series test, and if S is the sum of the series, then:*

(*a*) *S lies between any two successive partial sums; that is, either*

$$s_n \le S \le s_{n+1} \quad or \quad s_{n+1} \le S \le s_n \tag{3}$$

depending on which partial sum is larger.

(*b*) *If S is approximated by s_n, then the absolute error $|S - s_n|$ satisfies*

$$|S - s_n| \le a_{n+1} \tag{4}$$

Moreover, the sign of the error $S - s_n$ is the same as that of the coefficient of a_{n+1}.

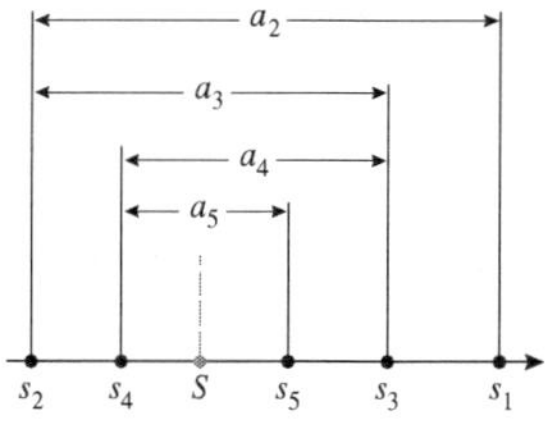

Figure 10.6.2

PROOF. We will prove the theorem for series of form (1). Referring to Figure 10.6.2 and keeping in mind our observation in the proof of Theorem 10.6.1 that the odd-numbered partial sums form a decreasing sequence converging to S and the even-numbered partial sums form an increasing sequence converging to S, we see that successive partial sums oscillate from one side of S to the other in smaller and smaller steps with the odd-numbered partial sums being larger than S and the even-numbered partial sums being smaller than S. Thus, depending on whether n is even or odd, we have

$$s_n \le S \le s_{n+1} \quad \text{or} \quad s_{n+1} \le S \le s_n$$

which proves (3). Moreover, in either case we have

$$|S - s_n| \le |s_{n+1} - s_n| \tag{5}$$

But $s_{n+1} - s_n = \pm a_{n+1}$ (the sign depending on whether n is even or odd). Thus, it follows from (5) that $|S - s_n| \le a_{n+1}$, which proves (4). Finally, since the odd-numbered partial sums are larger than S and the even-numbered partial sums are smaller than S, it follows that $S - s_n$ has the same sign as the coefficient of a_{n+1} (verify). ■

In words, inequality (4) states that for a series satisfying the hypotheses of the alternating series test, the magnitude of the error that results from approximating S by s_n is at most that of the first term that is *not* included in the partial sum. Also, note that if $a_1 > a_2 > \cdots > a_k > \cdots$, then inequality (4) can be strengthened to $|s - s_n| < a_{n+1}$.

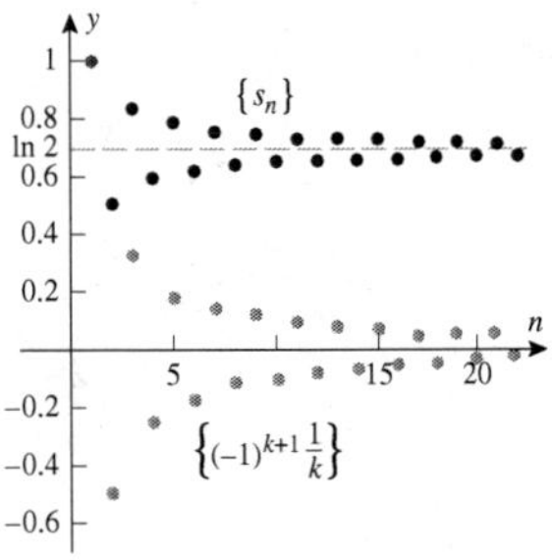

Graph of the sequences of terms and nth partial sums for the alternating harmonic series

Figure 10.6.3

▶ Example 2 Later in this chapter we will show that the sum of the alternating harmonic series is

$$\ln 2 = 1 - \frac{1}{2} + \frac{1}{3} - \frac{1}{4} + \cdots + (-1)^{k+1}\frac{1}{k} + \cdots$$

This is illustrated in Figure 10.6.3.

(a) Accepting this to be so, find an upper bound on the magnitude of the error that results if $\ln 2$ is approximated by the sum of the first eight terms in the series.

(b) Find a partial sum that approximates $\ln 2$ to one decimal-place accuracy (the nearest tenth).

Solution (a). It follows from the strengthened form of (4) that

$$|\ln 2 - s_8| < a_9 = \frac{1}{9} < 0.12 \tag{6}$$

As a check, let us compute s_8 exactly. We obtain

$$s_8 = 1 - \frac{1}{2} + \frac{1}{3} - \frac{1}{4} + \frac{1}{5} - \frac{1}{6} + \frac{1}{7} - \frac{1}{8} = \frac{533}{840}$$

Thus, with the help of a calculator

$$|\ln 2 - s_8| = \left|\ln 2 - \frac{533}{840}\right| \approx 0.059$$

This shows that the error is well under the estimate provided by upper bound (6).

Solution (b). For one decimal-place accuracy, we must choose a value of n for which $|\ln 2 - s_n| \le 0.05$. However, it follows from the strengthened form of (4) that

$$|\ln 2 - s_n| < a_{n+1}$$

so it suffices to choose n so that $a_{n+1} \le 0.05$.

One way to find n is to use a calculating utility to obtain numerical values for $a_1, a_2, a_3, \ldots$ until you encounter the first value that is less than or equal to 0.05. If you do this, you will find that it is $a_{20} = 0.05$; this tells us that partial sum s_{19} will provide the desired accuracy. Another way to find n is to solve the inequality

$$\frac{1}{n+1} \le 0.05$$

algebraically. We can do this by taking reciprocals, reversing the sense of the inequality, and then simplifying to obtain $n \ge 19$. Thus, s_{19} will provide the required accuracy, which is consistent with the previous result.

With the help of a calculating utility, the value of s_{19} is approximately $s_{19} \approx 0.7$ and the value of $\ln 2$ obtained directly is approximately $\ln 2 \approx 0.69$, which agrees with s_{19} when rounded to one decimal place. ◀

As Example 2 illustrates, the alternating harmonic series does not provide an efficient way to approximate ln 2, since too many terms and hence too much computation is required to achieve reasonable accuracy. Later, we will develop better ways to approximate logarithms.

ABSOLUTE CONVERGENCE

The series

$$1 - \frac{1}{2} - \frac{1}{2^2} + \frac{1}{2^3} + \frac{1}{2^4} - \frac{1}{2^5} - \frac{1}{2^6} + \cdots$$

does not fit in any of the categories studied so far—it has mixed signs but is not alternating. We will now develop some convergence tests that can be applied to such series.

10.6.3 DEFINITION. A series

$$\sum_{k=1}^{\infty} u_k = u_1 + u_2 + \cdots + u_k + \cdots$$

is said to ***converge absolutely*** if the series of absolute values

$$\sum_{k=1}^{\infty} |u_k| = |u_1| + |u_2| + \cdots + |u_k| + \cdots$$

converges and is said to ***diverge absolutely*** if the series of absolute values diverges.

► **Example 3** Determine whether the following series converge absolutely.

(a) $1 - \frac{1}{2} - \frac{1}{2^2} + \frac{1}{2^3} + \frac{1}{2^4} - \frac{1}{2^5} - \cdots$ (b) $1 - \frac{1}{2} + \frac{1}{3} - \frac{1}{4} + \frac{1}{5} - \cdots$

Solution (a). The series of absolute values is the convergent geometric series

$$1 + \frac{1}{2} + \frac{1}{2^2} + \frac{1}{2^3} + \frac{1}{2^4} + \frac{1}{2^5} + \cdots$$

so the given series converges absolutely.

Solution (b). The series of absolute values is the divergent harmonic series

$$1 + \frac{1}{2} + \frac{1}{3} + \frac{1}{4} + \frac{1}{5} + \cdots$$

so the given series diverges absolutely. ◄

It is important to distinguish between the notions of convergence and absolute convergence. For example, the series in part (b) of Example 3 converges, since it is the alternating harmonic series, yet we demonstrated that it does not converge absolutely. However, the following theorem shows that *if a series converges absolutely, then it converges.*

10.6.4 THEOREM. *If the series*

$$\sum_{k=1}^{\infty} |u_k| = |u_1| + |u_2| + \cdots + |u_k| + \cdots$$

converges, then so does the series

$$\sum_{k=1}^{\infty} u_k = u_1 + u_2 + \cdots + u_k + \cdots$$

Theorem 10.6.4 provides a way of inferring convergence of a series with positive and negative terms from a related series with nonnegative terms (the series of absolute values). This is important because most of the convergence tests that we have developed apply only to series with nonnegative terms.

PROOF. We will write the series $\sum u_k$ as

$$\sum_{k=1}^{\infty} u_k = \sum_{k=1}^{\infty} [(u_k + |u_k|) - |u_k|] \tag{7}$$

We are assuming that $\sum |u_k|$ converges, so that if we can show that $\sum(u_k + |u_k|)$ converges, then it will follow from (7) and Theorem 10.4.3(a) that $\sum u_k$ converges. However, the value of $u_k + |u_k|$ is either 0 or $2|u_k|$, depending on the sign of u_k. Thus, in all cases it is true that

$$0 \le u_k + |u_k| \le 2|u_k|$$

But $\sum 2|u_k|$ converges, since it is a constant times the convergent series $\sum |u_k|$; hence $\sum(u_k + |u_k|)$ converges by the comparison test. ■

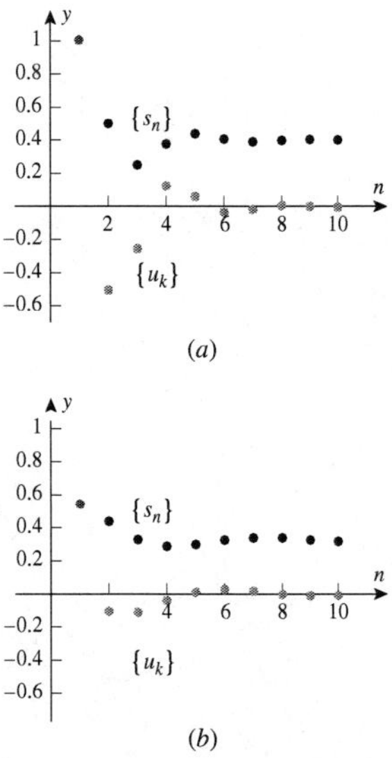

Graphs of the sequences of terms and nth partial sums for the series in Example 4

Figure 10.6.4

► **Example 4** Show that the following series converge.

$$\text{(a) } 1 - \frac{1}{2} - \frac{1}{2^2} + \frac{1}{2^3} + \frac{1}{2^4} - \frac{1}{2^5} - \frac{1}{2^6} + \cdots \qquad \text{(b) } \sum_{k=1}^{\infty} \frac{\cos k}{k^2}$$

Solution (a). Observe that this is not an alternating series because the signs alternate in pairs after the first term. Thus, we have no convergence test that can be applied directly. However, we showed in Example 3(a) that the series converges absolutely, so Theorem 10.6.4 implies that it converges (Figure 10.6.4a).

Solution (b). With the help of a calculating utility, you will be able to verify that the signs of the terms in this series vary irregularly. Thus, we will test for absolute convergence. The series of absolute values is

$$\sum_{k=1}^{\infty} \left| \frac{\cos k}{k^2} \right|$$

However,

$$\left| \frac{\cos k}{k^2} \right| \le \frac{1}{k^2}$$

But $\sum 1/k^2$ is a convergent p-series ($p = 2$), so the series of absolute values converges by the comparison test. Thus, the given series converges absolutely and hence converges (Figure 10.6.4b). ◄

CONDITIONAL CONVERGENCE

Although Theorem 10.6.4 is a useful tool for series that converge absolutely, it provides no information about the convergence or divergence of a series that diverges absolutely. For example, consider the two series

$$1 - \frac{1}{2} + \frac{1}{3} - \frac{1}{4} + \cdots + (-1)^{k+1}\frac{1}{k} + \cdots \tag{8}$$

$$-1 - \frac{1}{2} - \frac{1}{3} - \frac{1}{4} - \cdots - \frac{1}{k} - \cdots \tag{9}$$

Both of these series diverge absolutely, since in each case the series of absolute values is the divergent harmonic series

$$1 + \frac{1}{2} + \frac{1}{3} + \cdots + \frac{1}{k} + \cdots$$

However, series (8) converges, since it is the alternating harmonic series, and series (9) diverges, since it is a constant times the divergent harmonic series. As a matter of terminology, a series that converges but diverges absolutely is said to ***converge conditionally*** (or to be ***conditionally convergent***). Thus, (8) is a conditionally convergent series.

THE RATIO TEST FOR ABSOLUTE CONVERGENCE

Although one cannot generally infer convergence or divergence of a series from absolute divergence, the following variation of the ratio test provides a way of deducing divergence from absolute divergence in certain situations. We omit the proof.

10.6.5 THEOREM (*Ratio Test for Absolute Convergence*). *Let $\sum u_k$ be a series with nonzero terms and suppose that*

$$\rho = \lim_{k \to +\infty} \frac{|u_{k+1}|}{|u_k|}$$

(*a*) *If $\rho < 1$, then the series $\sum u_k$ converges absolutely and therefore converges.*

(*b*) *If $\rho > 1$ or if $\rho = +\infty$, then the series $\sum u_k$ diverges.*

(*c*) *If $\rho = 1$, no conclusion about convergence or absolute convergence can be drawn from this test.*

▶ Example 5 Use the ratio test for absolute convergence to determine whether the series converges.

$$\text{(a)} \sum_{k=1}^{\infty} (-1)^k \frac{2^k}{k!} \qquad \text{(b)} \sum_{k=1}^{\infty} (-1)^k \frac{(2k-1)!}{3^k}$$

Solution (a). Taking the absolute value of the general term u_k we obtain

$$|u_k| = \left|(-1)^k \frac{2^k}{k!}\right| = \frac{2^k}{k!}$$

Thus,

$$\rho = \lim_{k \to +\infty} \frac{|u_{k+1}|}{|u_k|} = \lim_{k \to +\infty} \frac{2^{k+1}}{(k+1)!} \cdot \frac{k!}{2^k} = \lim_{k \to +\infty} \frac{2}{k+1} = 0 < 1$$

which implies that the series converges absolutely and therefore converges.

Solution (b). Taking the absolute value of the general term u_k we obtain

$$|u_k| = \left|(-1)^k \frac{(2k-1)!}{3^k}\right| = \frac{(2k-1)!}{3^k}$$

Thus,

$$\rho = \lim_{k \to +\infty} \frac{|u_{k+1}|}{|u_k|} = \lim_{k \to +\infty} \frac{[2(k+1)-1]!}{3^{k+1}} \cdot \frac{3^k}{(2k-1)!}$$

$$= \lim_{k \to +\infty} \frac{1}{3} \cdot \frac{(2k+1)!}{(2k-1)!} = \frac{1}{3} \lim_{k \to +\infty} (2k)(2k+1) = +\infty$$

which implies that the series diverges. ◀

SUMMARY OF CONVERGENCE TESTS

We conclude this section with a summary of convergence tests that can be used for reference. The skill of selecting a good test is developed through lots of practice. In some instances a test may be inconclusive, so another test must be tried.

Summary of Convergence Tests

NAME	STATEMENT	COMMENTS
Divergence Test (10.4.1)	If $\lim_{k\to+\infty} u_k \neq 0$, then $\sum u_k$ diverges.	If $\lim_{k\to+\infty} u_k = 0$, then $\sum u_k$ may or may not converge.
Integral Test (10.4.4)	Let $\sum u_k$ be a series with positive terms. If f is a function that is decreasing and continuous on an interval $[a, +\infty)$ and such that $u_k = f(k)$ for all $k \geq a$, then $\sum_{k=1}^{\infty} u_k$ and $\int_a^{+\infty} f(x)\,dx$ both converge or both diverge.	This test only applies to series that have positive terms. Try this test when $f(x)$ is easy to integrate.
Comparison Test (10.5.1)	Let $\sum_{k=1}^{\infty} a_k$ and $\sum_{k=1}^{\infty} b_k$ be series with nonnegative terms such that $a_1 \leq b_1, a_2 \leq b_2, \ldots, a_k \leq b_k, \ldots$ If $\sum b_k$ converges, then $\sum a_k$ converges, and if $\sum a_k$ diverges, then $\sum b_k$ diverges.	This test only applies to series with nonnegative terms. Try this test as a last resort; other tests are often easier to apply.
Limit Comparison Test (10.5.4)	Let $\sum a_k$ and $\sum b_k$ be series with positive terms and let $\rho = \lim_{k\to+\infty} \frac{a_k}{b_k}$ If $0 < \rho < +\infty$, then both series converge or both diverge.	This is easier to apply than the comparison test, but still requires some skill in choosing the series $\sum b_k$ for comparison.
Ratio Test (10.5.5)	Let $\sum u_k$ be a series with positive terms and suppose that $\rho = \lim_{k\to+\infty} \frac{u_{k+1}}{u_k}$ (a) Series converges if $\rho < 1$. (b) Series diverges if $\rho > 1$ or $\rho = +\infty$. (c) The test is inconclusive if $\rho = 1$.	Try this test when u_k involves factorials or kth powers.
Root Test (10.5.6)	Let $\sum u_k$ be a series with positive terms and suppose that $\rho = \lim_{k\to+\infty} \sqrt[k]{u_k}$ (a) The series converges if $\rho < 1$. (b) The series diverges if $\rho > 1$ or $\rho = +\infty$. (c) The test is inconclusive if $\rho = 1$.	Try this test when u_k involves kth powers.
Alternating Series Test (10.6.1)	If $a_k > 0$ for $k = 1, 2, 3, \ldots$, then the series $a_1 - a_2 + a_3 - a_4 + \cdots$ $-a_1 + a_2 - a_3 + a_4 - \cdots$ converge if the following conditions hold: (a) $a_1 \geq a_2 \geq a_3 \geq \cdots$ (b) $\lim_{k\to+\infty} a_k = 0$	This test applies only to alternating series.
Ratio Test for Absolute Convergence (10.6.5)	Let $\sum u_k$ be a series with nonzero terms and suppose that $\rho = \lim_{k\to+\infty} \frac{\lvert u_{k+1}\rvert}{\lvert u_k\rvert}$ (a) The series converges absolutely if $\rho < 1$. (b) The series diverges if $\rho > 1$ or $\rho = +\infty$. (c) The test is inconclusive if $\rho = 1$.	The series need not have positive terms and need not be alternating to use this test.

✔QUICK CHECK EXERCISES 10.6 *(See page 675 for answers.)*

1. What characterizes an *alternating* series?
2. (a) The series
$$\sum_{k=1}^{\infty} \frac{(-1)^{k+1}}{k^2}$$
converges by the alternating series test since ______ and ______.
(b) If
$$S = \sum_{k=1}^{\infty} \frac{(-1)^{k+1}}{k^2} \quad \text{and} \quad s_9 = \sum_{k=1}^{9} \frac{(-1)^{k+1}}{k^2}$$
then $|S - s_9| <$ ______.
3. Classify each sequence as conditionally convergent, absolutely convergent, or divergent.
(a) $\sum_{k=1}^{\infty} (-1)^{k+1} \frac{1}{k}$: ______
(b) $\sum_{k=1}^{\infty} (-1)^k \frac{3k-1}{9k+15}$: ______
(c) $\sum_{k=1}^{\infty} (-1)^k \frac{1}{k(k+2)}$: ______
(d) $\sum_{k=1}^{\infty} (-1)^{k+1} \frac{1}{\sqrt[4]{k^3}}$: ______
4. Given that
$$\lim_{k\to+\infty} \frac{(k+1)^4/4^{k+1}}{k^4/4^k} = \lim_{k\to+\infty} \frac{\left(1+\frac{1}{k}\right)^4}{4} = \frac{1}{4}$$
is the series $\sum_{k=1}^{\infty} (-1)^k k^4/4^k$ conditionally convergent, absolutely convergent, or divergent?

EXERCISE SET 10.6 ◻ Graphing Utility ◻C CAS

1–2 Show that the series converges by confirming that it satisfies the hypotheses of the alternating series test (Theorem 10.6.1).

1. $\sum_{k=1}^{\infty} \frac{(-1)^{k+1}}{2k+1}$
2. $\sum_{k=1}^{\infty} (-1)^{k+1} \frac{k}{3^k}$

3–6 Determine whether the alternating series converges; justify your answer.

3. $\sum_{k=1}^{\infty} (-1)^{k+1} \frac{k+1}{3k+1}$
4. $\sum_{k=1}^{\infty} (-1)^{k+1} \frac{k+1}{\sqrt{k}+1}$
5. $\sum_{k=1}^{\infty} (-1)^{k+1} e^{-k}$
6. $\sum_{k=3}^{\infty} (-1)^k \frac{\ln k}{k}$

7–12 Use the ratio test for absolute convergence (Theorem 10.6.5) to determine whether the series converges or diverges. If the test is inconclusive, say so.

7. $\sum_{k=1}^{\infty} \left(-\frac{3}{5}\right)^k$
8. $\sum_{k=1}^{\infty} (-1)^{k+1} \frac{2^k}{k!}$
9. $\sum_{k=1}^{\infty} (-1)^{k+1} \frac{3^k}{k^2}$
10. $\sum_{k=1}^{\infty} (-1)^k \frac{k}{5^k}$
11. $\sum_{k=1}^{\infty} (-1)^k \frac{k^3}{e^k}$
12. $\sum_{k=1}^{\infty} (-1)^{k+1} \frac{k^k}{k!}$

13–30 Classify each series as absolutely convergent, conditionally convergent, or divergent.

13. $\sum_{k=1}^{\infty} \frac{(-1)^{k+1}}{3k}$
14. $\sum_{k=1}^{\infty} \frac{(-1)^{k+1}}{k^{4/3}}$
15. $\sum_{k=1}^{\infty} \frac{(-4)^k}{k^2}$
16. $\sum_{k=1}^{\infty} \frac{(-1)^{k+1}}{k!}$
17. $\sum_{k=1}^{\infty} \frac{\cos k\pi}{k}$
18. $\sum_{k=3}^{\infty} \frac{(-1)^k \ln k}{k}$
19. $\sum_{k=1}^{\infty} (-1)^{k+1} \frac{k+2}{k(k+3)}$
20. $\sum_{k=1}^{\infty} \frac{(-1)^{k+1} k^2}{k^3+1}$
21. $\sum_{k=1}^{\infty} \sin \frac{k\pi}{2}$
22. $\sum_{k=1}^{\infty} \frac{\sin k}{k^3}$
23. $\sum_{k=2}^{\infty} \frac{(-1)^k}{k \ln k}$
24. $\sum_{k=1}^{\infty} \frac{(-1)^k}{\sqrt{k(k+1)}}$
25. $\sum_{k=2}^{\infty} \left(-\frac{1}{\ln k}\right)^k$
26. $\sum_{k=1}^{\infty} \frac{(-1)^{k+1}}{\sqrt{k+1}+\sqrt{k}}$
27. $\sum_{k=2}^{\infty} \frac{(-1)^k (k^2+1)}{k^3+2}$
28. $\sum_{k=1}^{\infty} \frac{k \cos k\pi}{k^2+1}$
29. $\sum_{k=1}^{\infty} \frac{(-1)^{k+1} k!}{(2k-1)!}$
30. $\sum_{k=1}^{\infty} (-1)^{k+1} \frac{3^{2k-1}}{k^2+1}$

31–34 Each series satisfies the hypotheses of the alternating series test. For the stated value of n, find an upper bound on the absolute error that results if the sum of the series is approximated by the nth partial sum.

31. $\sum_{k=1}^{\infty} \frac{(-1)^{k+1}}{k}$; $n = 7$
32. $\sum_{k=1}^{\infty} \frac{(-1)^{k+1}}{k!}$; $n = 5$

33. $\sum_{k=1}^{\infty} \frac{(-1)^{k+1}}{\sqrt{k}}$; $n = 99$

34. $\sum_{k=1}^{\infty} \frac{(-1)^{k+1}}{(k+1)\ln(k+1)}$; $n = 3$

35–38 Each series satisfies the hypotheses of the alternating series test. Find a value of n for which the nth partial sum is ensured to approximate the sum of the series to the stated accuracy.

35. $\sum_{k=1}^{\infty} \frac{(-1)^{k+1}}{k}$; $|\text{error}| < 0.0001$

36. $\sum_{k=1}^{\infty} \frac{(-1)^{k+1}}{k!}$; $|\text{error}| < 0.00001$

37. $\sum_{k=1}^{\infty} \frac{(-1)^{k+1}}{\sqrt{k}}$; two decimal places

38. $\sum_{k=1}^{\infty} \frac{(-1)^{k+1}}{(k+1)\ln(k+1)}$; one decimal place

39–40 Find an upper bound on the absolute error that results if s_{10} is used to approximate the sum of the given *geometric* series. Compute s_{10} rounded to four decimal places and compare this value with the exact sum of the series.

39. $\frac{3}{4} - \frac{3}{8} + \frac{3}{16} - \frac{3}{32} + \cdots$ **40.** $1 - \frac{2}{3} + \frac{4}{9} - \frac{8}{27} + \cdots$

41–44 Each series satisfies the hypotheses of the alternating series test. Approximate the sum of the series to two decimal-place accuracy.

41. $1 - \frac{1}{3!} + \frac{1}{5!} - \frac{1}{7!} + \cdots$ **42.** $1 - \frac{1}{2!} + \frac{1}{4!} - \frac{1}{6!} + \cdots$

43. $\frac{1}{1 \cdot 2} - \frac{1}{2 \cdot 2^2} + \frac{1}{3 \cdot 2^3} - \frac{1}{4 \cdot 2^4} + \cdots$

44. $\frac{1}{1^5 + 4 \cdot 1} - \frac{1}{3^5 + 4 \cdot 3} + \frac{1}{5^5 + 4 \cdot 5} - \frac{1}{7^5 + 4 \cdot 7} + \cdots$

FOCUS ON CONCEPTS

C **45.** The purpose of this exercise is to show that the error bound in part (*b*) of Theorem 10.6.2 can be overly conservative in certain cases.

(a) Use a CAS to confirm that

$$\frac{\pi}{4} = 1 - \frac{1}{3} + \frac{1}{5} - \frac{1}{7} + \cdots$$

(b) Use the CAS to show that $|(\pi/4) - s_{25}| < 10^{-2}$.

(c) According to the error bound in part (*b*) of Theorem 10.6.2, what value of n is required to ensure that $|(\pi/4) - s_n| < 10^{-2}$?

46. Prove: If a series $\sum a_k$ converges absolutely, then the series $\sum a_k^2$ converges.

47. (a) Find examples to show that if $\sum a_k$ converges, then $\sum a_k^2$ may diverge or converge.

(b) Find examples to show that if $\sum a_k^2$ converges, then $\sum a_k$ may diverge or converge.

48. Show that the alternating p-series

$$1 - \frac{1}{2^p} + \frac{1}{3^p} - \frac{1}{4^p} + \cdots + (-1)^{k+1}\frac{1}{k^p} + \cdots$$

converges absolutely if $p > 1$, converges conditionally if $0 < p \le 1$, and diverges if $p \le 0$.

49–51 It can be proved that any series that is constructed from an absolutely convergent series by rearranging the terms is absolutely convergent and has the same sum as the original series. Use this fact together with parts (*a*) and (*b*) of Theorem 10.4.3 in these exercises.

49. It was stated in Exercise 31 of Section 10.4 that

$$\frac{\pi^2}{6} = 1 + \frac{1}{2^2} + \frac{1}{3^2} + \frac{1}{4^2} + \cdots$$

Use this to show that

$$\frac{\pi^2}{8} = 1 + \frac{1}{3^2} + \frac{1}{5^2} + \frac{1}{7^2} + \cdots$$

50. Use the series for $\pi^2/6$ given in the preceding exercise to show that

$$\frac{\pi^2}{12} = 1 - \frac{1}{2^2} + \frac{1}{3^2} - \frac{1}{4^2} + \cdots$$

51. It was stated in Exercise 31 of Section 10.4 that

$$\frac{\pi^4}{90} = 1 + \frac{1}{2^4} + \frac{1}{3^4} + \frac{1}{4^4} + \cdots$$

Use this to show that

$$\frac{\pi^4}{96} = 1 + \frac{1}{3^4} + \frac{1}{5^4} + \frac{1}{7^4} + \cdots$$

FOCUS ON CONCEPTS

52. It can be proved that the terms of any conditionally convergent series can be rearranged to give either a divergent series or a conditionally convergent series whose sum is any given number S. For example, we stated in Example 2 that

$$\ln 2 = 1 - \frac{1}{2} + \frac{1}{3} - \frac{1}{4} + \frac{1}{5} - \frac{1}{6} + \cdots$$

Show that we can rearrange this series so that its sum is $\frac{1}{2}\ln 2$ by rewriting it as

$$\left(1 - \frac{1}{2} - \frac{1}{4}\right) + \left(\frac{1}{3} - \frac{1}{6} - \frac{1}{8}\right) + \left(\frac{1}{5} - \frac{1}{10} - \frac{1}{12}\right) + \cdots$$

[*Hint:* Add the first two terms in each grouping.]

53. Consider the series

$$1 - \frac{1}{2} + \frac{2}{3} - \frac{1}{3} + \frac{2}{4} - \frac{1}{4} + \frac{2}{5} - \frac{1}{5} + \cdots$$

(a) Show that this series diverges.

(b) Explain why the alternating series test does not apply to this series.

54. (a) Use a graphing utility to graph

$$f(x) = \frac{4x - 1}{4x^2 - 2x}, \quad x \geq 1$$

(b) Based on your graph, do you think that the series

$$\sum_{k=1}^{\infty} (-1)^{k+1} \frac{4k - 1}{4k^2 - 2k}$$

converges? Explain your reasoning.

55. As illustrated in the accompanying figure, a bug, starting at point A on a 180-cm wire, walks the length of the wire, stops and walks in the opposite direction for half the length of the wire, stops again and walks in the opposite direction for one-third the length of the wire, stops again and walks in the opposite direction for one-fourth the length of the wire, and so forth until it stops for the 1000th time.

(a) Give upper and lower bounds on the distance between the bug and point A when it finally stops. [*Hint:* As stated in Example 2, assume that the sum of the alternating harmonic series is ln 2.]

(b) Give upper and lower bounds on the total distance that the bug has traveled when it finally stops. [*Hint:* Use inequality (2) of Section 10.4.]

Figure Ex-55

✔QUICK CHECK ANSWERS 10.6

1. Terms alternate between positive and negative. **2.** (a) $1 \geq \frac{1}{4} \geq \frac{1}{9} \geq \cdots \geq \frac{1}{k^2} \geq \frac{1}{(k+1)^2} \geq \cdots$; $\lim_{k \to +\infty} \frac{1}{k^2} = 0$ (b) $\frac{1}{100}$ **3.** (a) conditionally convergent (b) divergent (c) absolutely convergent (d) conditionally convergent **4.** absolutely convergent

10.7 MACLAURIN AND TAYLOR POLYNOMIALS

In a local linear approximation the tangent line to the graph of a function is used to obtain a linear approximation of the function near the point of tangency. In this section we will consider how one might improve on the accuracy of local linear approximations by using higher-order polynomials as approximations functions. We will also investigate the error associated with such approximations.

LOCAL QUADRATIC APPROXIMATIONS

Recall from Formula (1) in Section 3.9 that the local linear approximation of a function f at x_0 is

$$f(x) \approx f(x_0) + f'(x_0)(x - x_0) \tag{1}$$

In this formula, the approximating function

$$p(x) = f(x_0) + f'(x_0)(x - x_0)$$

is a first-degree polynomial satisfying $p(x_0) = f(x_0)$ and $p'(x_0) = f'(x_0)$ (verify). Thus, the local linear approximation of f at x_0 has the property that its value and the value of its first derivative match those of f at x_0.

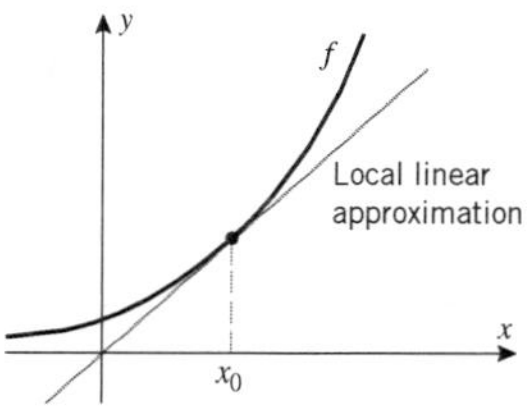

Figure 10.7.1

If the graph of a function f has a pronounced "bend" at x_0, then we can expect that the accuracy of the local linear approximation of f at x_0 will decrease rapidly as we progress away from x_0 (Figure 10.7.1). One way to deal with this problem is to approximate the function f at x_0 by a polynomial p of degree 2 with the property that the value of p and the values of its first two derivatives match those of f at x_0. This ensures that the graphs of f and p not only have the same tangent line at x_0, but they also bend in the same direction at x_0 (both concave up or concave down). As a result, we can expect that the graph of p will remain close to the graph of f over a larger interval around x_0 than the graph of the local linear approximation. The polynomial p is called the ***local quadratic approximation of f at $x = x_0$***.

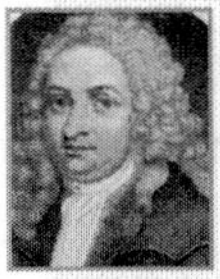

Colin Maclaurin (1698–1746) Scottish mathematician. Maclaurin's father, a minister, died when the boy was only six months old, and his mother when he was nine years old. He was then raised by an uncle who was also a minister. Maclaurin entered Glasgow University as a divinity student but switched to mathematics after one year. He received his Master's degree at age 17 and, in spite of his youth, began teaching at Marischal College in Aberdeen, Scotland. He met Isaac Newton during a visit to London in 1719 and from that time on became Newton's disciple. During that era, some of Newton's analytic methods were bitterly attacked by major mathematicians and much of Maclaurin's important mathematical work resulted from his efforts to defend Newton's ideas geometrically. Maclaurin's work, *A Treatise of Fluxions* (1742), was the first systematic formulation of Newton's methods. The treatise was so carefully done that it was a standard of mathematical rigor in calculus until the work of Cauchy in 1821. Maclaurin was also an outstanding experimentalist; he devised numerous ingenious mechanical devices, made important astronomical observations, performed actuarial computations for insurance societies, and helped to improve maps of the islands around Scotland.

To illustrate this idea, let us try to find a formula for the local quadratic approximation of a function f at $x = 0$. This approximation has the form

$$f(x) \approx c_0 + c_1 x + c_2 x^2 \tag{2}$$

where c_0, c_1, and c_2 must be chosen so that the values of

$$p(x) = c_0 + c_1 x + c_2 x^2$$

and its first two derivatives match those of f at 0. Thus, we want

$$p(0) = f(0), \quad p'(0) = f'(0), \quad p''(0) = f''(0) \tag{3}$$

But the values of $p(0)$, $p'(0)$, and $p''(0)$ are as follows:

$$\begin{aligned} p(x) &= c_0 + c_1 x + c_2 x^2 \qquad & p(0) &= c_0 \\ p'(x) &= c_1 + 2c_2 x & p'(0) &= c_1 \\ p''(x) &= 2c_2 & p''(0) &= 2c_2 \end{aligned}$$

Thus, it follows from (3) that

$$c_0 = f(0), \quad c_1 = f'(0), \quad c_2 = \frac{f''(0)}{2}$$

and substituting these in (2) yields the following formula for the local quadratic approximation of f at $x = 0$:

$$f(x) \approx f(0) + f'(0)x + \frac{f''(0)}{2}x^2 \tag{4}$$

▶ **Example 1** Find the local linear and quadratic approximations of e^x at $x = 0$, and graph e^x and the two approximations together.

Solution. If we let $f(x) = e^x$, then $f'(x) = f''(x) = e^x$; and hence

$$f(0) = f'(0) = f''(0) = e^0 = 1$$

Thus, from (4) the local quadratic approximation of e^x at $x = 0$ is

$$e^x \approx 1 + x + \frac{x^2}{2}$$

and the local linear approximation (which is the linear part of the local quadratic approximation) is

$$e^x \approx 1 + x$$

The graphs of e^x and the two approximations are shown in Figure 10.7.2. As expected, the local quadratic approximation is more accurate than the local linear approximation near $x = 0$. ◀

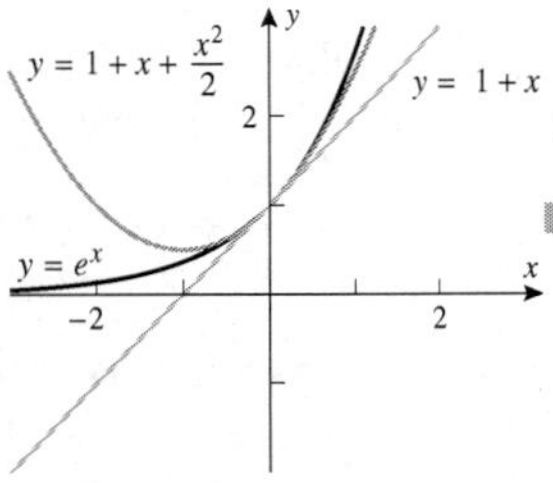

Figure 10.7.2

MACLAURIN POLYNOMIALS

It is natural to ask whether one can improve on the accuracy of a local quadratic approximation by using a polynomial of degree 3. Specifically, one might look for a polynomial of degree 3 with the property that its value and the values of its first three derivatives match those of f at a point; and if this provides an improvement in accuracy, why not go on to polynomials of even higher degree? Thus, we are led to consider the following general problem.

10.7.1 PROBLEM. Given a function f that can be differentiated n times at $x = x_0$, find a polynomial p of degree n with the property that the value of p and the values of its first n derivatives match those of f at x_0.

We will begin by solving this problem in the case where $x_0 = 0$. Thus, we want a polynomial

$$p(x) = c_0 + c_1x + c_2x^2 + c_3x^3 + \cdots + c_nx^n \tag{5}$$

such that

$$f(0) = p(0), \quad f'(0) = p'(0), \quad f''(0) = p''(0), \ldots, \quad f^{(n)}(0) = p^{(n)}(0) \tag{6}$$

But

$$\begin{aligned} p(x) &= c_0 + c_1x + c_2x^2 + c_3x^3 + \cdots + c_nx^n \\ p'(x) &= c_1 + 2c_2x + 3c_3x^2 + \cdots + nc_nx^{n-1} \\ p''(x) &= 2c_2 + 3 \cdot 2c_3x + \cdots + n(n-1)c_nx^{n-2} \\ p'''(x) &= 3 \cdot 2c_3 + \cdots + n(n-1)(n-2)c_nx^{n-3} \\ &\vdots \\ p^{(n)}(x) &= n(n-1)(n-2)\cdots(1)c_n \end{aligned}$$

Thus, to satisfy (6) we must have

$$\begin{aligned} f(0) &= p(0) &&= c_0 \\ f'(0) &= p'(0) &&= c_1 \\ f''(0) &= p''(0) &&= 2c_2 = 2!c_2 \\ f'''(0) &= p'''(0) &&= 3 \cdot 2c_3 = 3!c_3 \\ &\vdots \\ f^{(n)}(0) &= p^{(n)}(0) = n(n-1)(n-2)\cdots(1)c_n = n!c_n \end{aligned}$$

which yields the following values for the coefficients of $p(x)$:

$$c_0 = f(0), \quad c_1 = f'(0), \quad c_2 = \frac{f''(0)}{2!}, \quad c_3 = \frac{f'''(0)}{3!}, \ldots, \quad c_n = \frac{f^{(n)}(0)}{n!}$$

The polynomial that results by using these coefficients in (5) is called the *nth Maclaurin polynomial for f*.

Augustin Louis Cauchy (1789–1857) French mathematician. Cauchy's early education was acquired from his father, a barrister and master of the classics. Cauchy entered L'Ecole Polytechnique in 1805 to study engineering, but because of poor health, was advised to concentrate on mathematics. His major mathematical work began in 1811 with a series of brilliant solutions to some difficult outstanding problems. In 1814 he wrote a treatise on integrals that was to become the basis for modern complex variable theory; in 1816 there followed a classic paper on wave propagation in liquids that won a prize from the French Academy; and in 1822 he wrote a paper that formed the basis of modern elasticity theory. Cauchy's mathematical contributions for the next 35 years were brilliant and staggering in quantity, over 700 papers filling 26 modern volumes. Cauchy's work initiated the era of modern analysis. He brought to mathematics standards of precision and rigor undreamed of by Leibniz and Newton.

Cauchy's life was inextricably tied to the political upheavals of the time. A strong partisan of the Bourbons, he left his wife and children in 1830 to follow the Bourbon king Charles X into exile. For his loyalty he was made a baron by the ex-king. Cauchy eventually returned to France, but refused to accept a university position until the government waived its requirement that he take a loyalty oath.

It is difficult to get a clear picture of the man. Devoutly Catholic, he sponsored charitable work for unwed mothers, criminals, and relief for Ireland. Yet other aspects of his life cast him in an unfavorable light. The Norwegian mathematician Abel described him as, "mad, infinitely Catholic, and bigoted." Some writers praise his teaching, yet others say he rambled incoherently and, according to a report of the day, he once devoted an entire lecture to extracting the square root of seventeen to ten decimal places by a method well known to his students. In any event, Cauchy is undeniably one of the greatest minds in the history of science.

Verify that $f(x) \approx p_1(x)$ is the local linear approximation of f at $x = 0$, and $f(x) \approx p_2(x)$ is the local quadratic approximation at $x = 0$. Thus, the polynomials in these approximations are special cases of the Maclaurin polynomials for f.

10.7.2 DEFINITION. If f can be differentiated n times at 0, then we define the ***nth Maclaurin polynomial for f*** to be

$$p_n(x) = f(0) + f'(0)x + \frac{f''(0)}{2!}x^2 + \frac{f'''(0)}{3!}x^3 + \cdots + \frac{f^{(n)}(0)}{n!}x^n \tag{7}$$

This polynomial has the property that its value and the values of its first n derivatives match the values of f and its first n derivatives at $x = 0$.

Example 2 Find the Maclaurin polynomials p_0, p_1, p_2, p_3, and p_n for e^x.

Solution. Let $f(x) = e^x$. Thus,

$$f'(x) = f''(x) = f'''(x) = \cdots = f^{(n)}(x) = e^x$$

and

$$f(0) = f'(0) = f''(0) = f'''(0) = \cdots = f^{(n)}(0) = e^0 = 1$$

Therefore,

$$\begin{aligned}
p_0(x) &= f(0) = 1 \\
p_1(x) &= f(0) + f'(0)x = 1 + x \\
p_2(x) &= f(0) + f'(0)x + \frac{f''(0)}{2!}x^2 = 1 + x + \frac{x^2}{2!} = 1 + x + \frac{1}{2}x^2 \\
p_3(x) &= f(0) + f'(0)x + \frac{f''(0)}{2!}x^2 + \frac{f'''(0)}{3!}x^3 \\
&= 1 + x + \frac{x^2}{2!} + \frac{x^3}{3!} = 1 + x + \frac{1}{2}x^2 + \frac{1}{6}x^3 \\
p_n(x) &= f(0) + f'(0)x + \frac{f''(0)}{2!}x^2 + \cdots + \frac{f^{(n)}(0)}{n!}x^n \\
&= 1 + x + \frac{x^2}{2!} + \cdots + \frac{x^n}{n!}
\end{aligned}$$

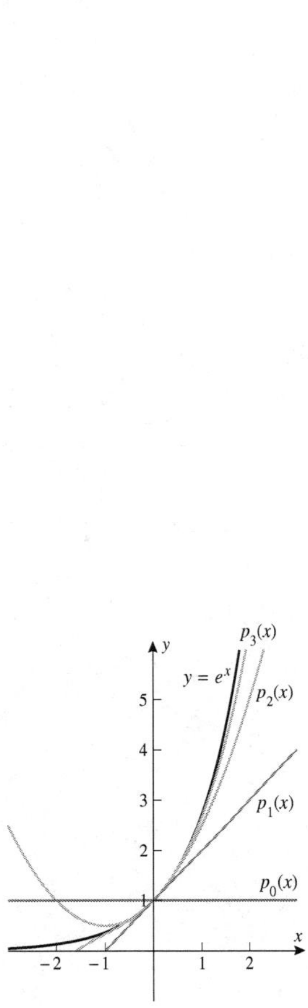

Figure 10.7.3

Figure 10.7.3 shows the graph of e^x (in blue) and the graph of the first four Maclaurin polynomials. Note that the graphs of $p_1(x)$, $p_2(x)$, and $p_3(x)$ are virtually indistinguishable from the graph of e^x near $x = 0$, so these polynomials are good approximations of e^x for x near 0. However, the farther x is from 0, the poorer these approximations become. This is typical of the Maclaurin polynomials for a function $f(x)$; they provide good approximations of $f(x)$ near 0, but the accuracy diminishes as x progresses away from 0. It is usually the case that the higher the degree of the polynomial, the larger the interval on which it provides a specified accuracy. Accuracy issues will be investigated later.

TAYLOR POLYNOMIALS

Up to now we have focused on approximating a function f in the vicinity of $x = 0$. Now we will consider the more general case of approximating f in the vicinity of an arbitrary domain value x_0. The basic idea is the same as before; we want to find an nth-degree polynomial p with the property that its value and the values of its first n derivatives match those of f at x_0. However, rather than expressing $p(x)$ in powers of x, it will simplify the computations if we express it in powers of $x - x_0$; that is,

$$p(x) = c_0 + c_1(x - x_0) + c_2(x - x_0)^2 + \cdots + c_n(x - x_0)^n \tag{8}$$

Verify that $f(x) \approx p_1(x)$ is the local linear approximation of f at $x = x_0$, and $f(x) \approx p_2(x)$ is the local quadratic approximation at $x = x_0$. Thus, the polynomials in these approximations are special cases of the Taylor polynomials for f at $x = x_0$.

We will leave it as an exercise for you to imitate the computations used in the case where $x_0 = 0$ to show that

$$c_0 = f(x_0), \quad c_1 = f'(x_0), \quad c_2 = \frac{f''(x_0)}{2!}, \quad c_3 = \frac{f'''(x_0)}{3!}, \ldots, \quad c_n = \frac{f^{(n)}(x_0)}{n!}$$

Substituting these values in (8) we obtain a polynomial called the *nth Taylor polynomial about $x = x_0$ for f.*

10.7.3 DEFINITION. If f can be differentiated n times at x_0, then we define the ***nth Taylor polynomial for f about x = x₀*** to be

$$p_n(x) = f(x_0) + f'(x_0)(x - x_0) + \frac{f''(x_0)}{2!}(x - x_0)^2 + \frac{f'''(x_0)}{3!}(x - x_0)^3 + \cdots + \frac{f^{(n)}(x_0)}{n!}(x - x_0)^n \tag{9}$$

▶ **Example 3** Find the first four Taylor polynomials for $\ln x$ about $x = 2$.

The Maclaurin polynomials are the special cases of the Taylor polynomials in which $x_0 = 0$. Thus, theorems about Taylor polynomials also apply to Maclaurin polynomials.

Solution. Let $f(x) = \ln x$. Thus,

$$\begin{aligned} f(x) &= \ln x & f(2) &= \ln 2 \\ f'(x) &= 1/x & f'(2) &= 1/2 \\ f''(x) &= -1/x^2 & f''(2) &= -1/4 \\ f'''(x) &= 2/x^3 & f'''(2) &= 1/4 \end{aligned}$$

Substituting in (9) with $x_0 = 2$ yields

$$p_0(x) = f(2) = \ln 2$$

$$p_1(x) = f(2) + f'(2)(x - 2) = \ln 2 + \tfrac{1}{2}(x - 2)$$

$$p_2(x) = f(2) + f'(2)(x - 2) + \frac{f''(2)}{2!}(x - 2)^2 = \ln 2 + \tfrac{1}{2}(x - 2) - \tfrac{1}{8}(x - 2)^2$$

$$\begin{aligned} p_3(x) &= f(2) + f'(2)(x - 2) + \frac{f''(2)}{2!}(x - 2)^2 + \frac{f'''(2)}{3!}(x - 2)^3 \\ &= \ln 2 + \tfrac{1}{2}(x - 2) - \tfrac{1}{8}(x - 2)^2 + \tfrac{1}{24}(x - 2)^3 \end{aligned}$$

Brook Taylor (1685–1731) English mathematician. Taylor was born of well-to-do parents. Musicians and artists were entertained frequently in the Taylor home, which undoubtedly had a lasting influence on him. In later years, Taylor published a definitive work on the mathematical theory of perspective and obtained major mathematical results about the vibrations of strings. There also exists an unpublished work, *On Musick*, that was intended to be part of a joint paper with Isaac Newton. Taylor's life was scarred with unhappiness, illness, and tragedy. Because his first wife was not rich enough to suit his father, the two men argued bitterly and parted ways. Subsequently, his wife died in childbirth. Then, after he remarried, his second wife also died in childbirth, though his daughter survived. Taylor's most productive period was from 1714 to 1719, during which time he wrote on a wide range of subjects—magnetism, capillary action, thermometers, perspective, and calculus. In his final years, Taylor devoted his writing efforts to religion and philosophy. According to Taylor, the results that bear his name were motivated by coffeehouse conversations about works of Newton on planetary motion and works of Halley ("Halley's comet") on roots of polynomials. Unfortunately, Taylor's writing style was so terse and hard to understand that he never received credit for many of his innovations.

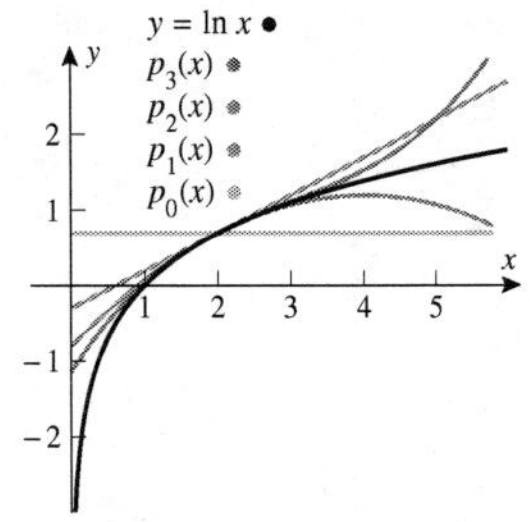

Figure 10.7.4

The graph of $\ln x$ (in blue) and its first four Taylor polynomials about $x = 2$ are shown in Figure 10.7.4. As expected, these polynomials produce their best approximations of $\ln x$ near 2. ◀

SIGMA NOTATION FOR TAYLOR AND MACLAURIN POLYNOMIALS

Frequently, we will want to express Formula (9) in sigma notation. To do this, we use the notation $f^{(k)}(x_0)$ to denote the kth derivative of f at $x = x_0$, and we make the convention that $f^{(0)}(x_0)$ denotes $f(x_0)$. This enables us to write

$$\sum_{k=0}^{n} \frac{f^{(k)}(x_0)}{k!}(x - x_0)^k = f(x_0) + f'(x_0)(x - x_0) + \frac{f''(x_0)}{2!}(x - x_0)^2 + \cdots + \frac{f^{(n)}(x_0)}{n!}(x - x_0)^n \tag{10}$$

In particular, we can write the nth Maclaurin polynomial for $f(x)$ as

$$\sum_{k=0}^{n} \frac{f^{(k)}(0)}{k!}x^k = f(0) + f'(0)x + \frac{f''(0)}{2!}x^2 + \cdots + \frac{f^{(n)}(0)}{n!}x^n \tag{11}$$

▶ **Example 4** Find the nth Maclaurin polynomials for

(a) $\sin x$ (b) $\cos x$ (c) $\dfrac{1}{1-x}$

Solution (a). In the Maclaurin polynomials for $\sin x$, only the odd powers of x appear explicitly. To see this, let $f(x) = \sin x$; thus,

$$\begin{aligned} f(x) &= \sin x & f(0) &= 0 \\ f'(x) &= \cos x & f'(0) &= 1 \\ f''(x) &= -\sin x & f''(0) &= 0 \\ f'''(x) &= -\cos x & f'''(0) &= -1 \end{aligned}$$

Since $f^{(4)}(x) = \sin x = f(x)$, the pattern $0, 1, 0, -1$ will repeat as we evaluate successive derivatives at 0. Therefore, the successive Maclaurin polynomials for $\sin x$ are

$$\begin{aligned} p_0(x) &= 0 \\ p_1(x) &= 0 + x \\ p_2(x) &= 0 + x + 0 \\ p_3(x) &= 0 + x + 0 - \frac{x^3}{3!} \\ p_4(x) &= 0 + x + 0 - \frac{x^3}{3!} + 0 \\ p_5(x) &= 0 + x + 0 - \frac{x^3}{3!} + 0 + \frac{x^5}{5!} \\ p_6(x) &= 0 + x + 0 - \frac{x^3}{3!} + 0 + \frac{x^5}{5!} + 0 \\ p_7(x) &= 0 + x + 0 - \frac{x^3}{3!} + 0 + \frac{x^5}{5!} + 0 - \frac{x^7}{7!} \end{aligned}$$

Because of the zero terms, each even-order Maclaurin polynomial [after $p_0(x)$] is the same as the preceding odd-order Maclaurin polynomial. That is,

$$p_{2k+1}(x) = p_{2k+2}(x) = x - \frac{x^3}{3!} + \frac{x^5}{5!} - \frac{x^7}{7!} + \cdots + (-1)^k \frac{x^{2k+1}}{(2k+1)!} \quad (k = 0, 1, 2, \ldots)$$

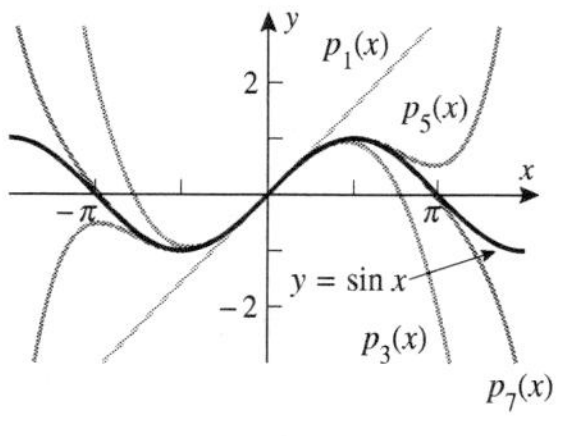

Figure 10.7.5

The graphs of $\sin x$, $p_1(x)$, $p_3(x)$, $p_5(x)$, and $p_7(x)$ are shown in Figure 10.7.5.

Solution (b). In the Maclaurin polynomials for $\cos x$, only the even powers of x appear explicitly; the computations are similar to those in part (a). The reader should be able to show that

$$p_0(x) = p_1(x) = 1$$

$$p_2(x) = p_3(x) = 1 - \frac{x^2}{2!}$$

$$p_4(x) = p_5(x) = 1 - \frac{x^2}{2!} + \frac{x^4}{4!}$$

$$p_6(x) = p_7(x) = 1 - \frac{x^2}{2!} + \frac{x^4}{4!} - \frac{x^6}{6!}$$

In general, the Maclaurin polynomials for $\cos x$ are given by

$$p_{2k}(x) = p_{2k+1}(x) = 1 - \frac{x^2}{2!} + \frac{x^4}{4!} - \frac{x^6}{6!} + \cdots + (-1)^k \frac{x^{2k}}{(2k)!} \quad (k = 0, 1, 2, \ldots)$$

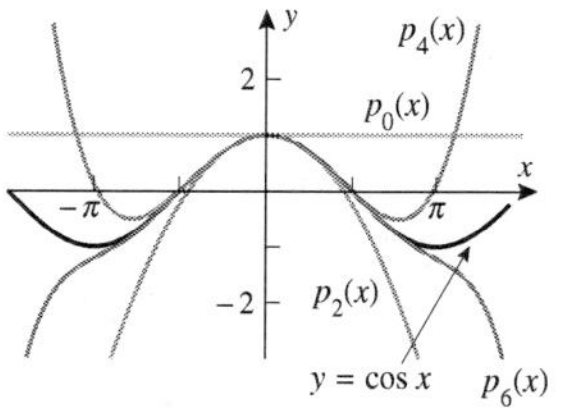

Figure 10.7.6

The graphs of $\cos x$, $p_0(x)$, $p_2(x)$, $p_4(x)$, and $p_6(x)$ are shown in Figure 10.7.6.

Solution (c). Let $f(x) = 1/(1-x)$. The values of f and its first k derivatives at $x = 0$ are as follows:

$$\begin{aligned} f(x) &= \frac{1}{1-x} & f(0) &= 1 = 0! \\ f'(x) &= \frac{1}{(1-x)^2} & f'(0) &= 1 = 1! \\ f''(x) &= \frac{2}{(1-x)^3} & f''(0) &= 2 = 2! \\ f'''(x) &= \frac{3 \cdot 2}{(1-x)^4} & f'''(0) &= 3! \\ f^{(4)}(x) &= \frac{4 \cdot 3 \cdot 2}{(1-x)^5} & f^{(4)}(0) &= 4! \\ &\vdots & &\vdots \\ f^{(k)}(x) &= \frac{k!}{(1-x)^{k+1}} & f^{(k)}(0) &= k! \end{aligned}$$

TECHNOLOGY MASTERY

Computer algebra systems have commands for generating Taylor polynomials of any specified degree. If you have a CAS, use it to find some of the Maclaurin and Taylor polynomials in Examples 3 and 4.

Thus, substituting $f^{(k)}(0) = k!$ into Formula (11) yields the nth Maclaurin polynomial for $1/(1-x)$:

$$p_n(x) = \sum_{k=0}^{n} x^k = 1 + x + x^2 + \cdots + x^n \quad (n = 0, 1, 2, \ldots) ◀$$

▶ **Example 5** Find the nth Taylor polynomial for $1/x$ about $x = 1$.

Solution. Let $f(x) = 1/x$. The computations are similar to those in part (c) of Example 4. We leave it for you to show that

$$f(1) = 1, \quad f'(1) = -1, \quad f''(1) = 2!, \quad f'''(1) = -3!,$$
$$f^{(4)}(1) = 4!, \ldots, \quad f^{(k)}(1) = (-1)^k k!$$

Thus, substituting $f^{(k)}(1) = (-1)^k k!$ into Formula (10) with $x_0 = 1$ yields the nth Taylor polynomial for $1/x$:

$$\sum_{k=0}^{n}(-1)^k(x-1)^k = 1 - (x-1) + (x-1)^2 - (x-1)^3 + \cdots + (-1)^n(x-1)^n \quad ◀$$

THE *n*TH REMAINDER

It will be convenient to have a notation for the error in the approximation $f(x) \approx p_n(x)$. Accordingly, we will let $R_n(x)$ denote the difference between $f(x)$ and its nth Taylor polynomial; that is,

$$R_n(x) = f(x) - p_n(x) = f(x) - \sum_{k=0}^{n} \frac{f^{(k)}(x_0)}{k!}(x-x_0)^k \tag{12}$$

This can also be written as

$$f(x) = p_n(x) + R_n(x) = \sum_{k=0}^{n} \frac{f^{(k)}(x_0)}{k!}(x-x_0)^k + R_n(x) \tag{13}$$

The function $R_n(x)$ is called the ***nth remainder*** for the Taylor series of f, and Formula (13) is called ***Taylor's formula with remainder***.

Finding a bound for $R_n(x)$ gives an indication of the accuracy of the approximation $p_n(x) \approx f(x)$. The following theorem, which is proved in Appendix C, provides such a bound.

10.7.4 THEOREM (*The Remainder Estimation Theorem*). *If the function f can be differentiated $n+1$ times on an interval I containing the number x_0, and if M is an upper bound for $|f^{(n+1)}(x)|$ on I, that is, $|f^{(n+1)}(x)| \le M$ for all x in I, then*

$$|R_n(x)| \le \frac{M}{(n+1)!}|x-x_0|^{n+1} \tag{14}$$

for all x in I.

The bound for $|R_n(x)|$ in (14) is called the ***Lagrange error bound***.

▶ **Example 6** Use an nth Maclaurin polynomial for e^x to approximate e to five decimal-place accuracy.

Solution. We note first that the exponential function e^x has derivatives of all orders for every real number x. From Example 2, the nth Maclaurin polynomial for e^x is

$$\sum_{k=0}^{n} \frac{x^k}{k!} = 1 + x + \frac{x^2}{2!} + \cdots + \frac{x^n}{n!}$$

from which we have

$$e = e^1 \approx \sum_{k=0}^{n} \frac{1^k}{k!} = 1 + 1 + \frac{1}{2!} + \cdots + \frac{1}{n!}$$

Thus, our problem is to determine how many terms to include in a Maclaurin polynomial for e^x to achieve five decimal-place accuracy; that is, we want to choose n so that the absolute value of the nth remainder at $x = 1$ satisfies

$$|R_n(1)| \leq 0.000005$$

To determine n we use the Remainder Estimation Theorem with $f(x) = e^x$, $x = 1$, $x_0 = 0$, and I being the interval $[0, 1]$. In this case it follows from Formula (14) that

$$|R_n(1)| \leq \frac{M}{(n+1)!} \tag{15}$$

where M is an upper bound on the value of $f^{(n+1)}(x) = e^x$ for x in the interval $[0, 1]$. However, e^x is an increasing function, so its maximum value on the interval $[0, 1]$ occurs at $x = 1$; that is, $e^x \leq e$ on this interval. Thus, we can take $M = e$ in (15) to obtain

$$|R_n(1)| \leq \frac{e}{(n+1)!} \tag{16}$$

Unfortunately, this inequality is not very useful because it involves e, which is the very quantity we are trying to approximate. However, if we accept that $e < 3$, then we can replace (16) with the following less precise, but more easily applied, inequality:

$$|R_n(1)| \leq \frac{3}{(n+1)!}$$

Thus, we can achieve five decimal-place accuracy by choosing n so that

$$\frac{3}{(n+1)!} \leq 0.000005 \quad \text{or} \quad (n+1)! \geq 600{,}000$$

Since $9! = 362{,}880$ and $10! = 3{,}628{,}800$, the smallest value of n that meets this criterion is $n = 9$. Thus, to five decimal-place accuracy

$$e \approx 1 + 1 + \frac{1}{2!} + \frac{1}{3!} + \frac{1}{4!} + \frac{1}{5!} + \frac{1}{6!} + \frac{1}{7!} + \frac{1}{8!} + \frac{1}{9!} \approx 2.71828$$

As a check, a calculator's 12-digit representation of e is $e \approx 2.71828182846$, which agrees with the preceding approximation when rounded to five decimal places. ◀

✔ QUICK CHECK EXERCISES 10.7 *(See page 685 for answers.)*

1. If f can be differentiated three times at 0, then the third Maclaurin polynomial for f is $p_3(x) =$ ______.
2. The third Maclaurin polynomial for $f(x) = e^{2x}$ is

$$p_3(x) = \underline{\quad\quad} + \underline{\quad\quad}\, x + \underline{\quad\quad}\, x^2 + \underline{\quad\quad}\, x^3$$

3. If $f(2) = 3$, $f'(2) = -4$, and $f''(2) = 10$, then the second Taylor polynomial for f about $x = 2$ is $p_2(x) =$ ______.
4. The third Taylor polynomial for $f(x) = x^5$ about $x = -1$ is

$$p_3(x) = \underline{\quad\quad} + \underline{\quad\quad}\,(x+1) + \underline{\quad\quad}\,(x+1)^2 + \underline{\quad\quad}\,(x+1)^3$$

5. (a) If a function f has nth Taylor polynomial $p_n(x)$ about $x = x_0$, then the nth remainder $R_n(x)$ is defined by $R_n(x) =$ ______.
   (b) Suppose that a function f can be differentiated five times on an interval I containing $x_0 = 2$ and that $|f^{(5)}(x)| \leq 20$ for all x in I. Then $|R_4(x)| \leq$ ______ for all x in I.

EXERCISE SET 10.7 Graphing Utility C CAS

1. In each part, find the local quadratic approximation of f at $x = x_0$, and use that approximation to find the local linear approximation of f at x_0. Use a graphing utility to graph f and the two approximations on the same screen.
(a) $f(x) = e^{-x}$; $x_0 = 0$ (b) $f(x) = \cos x$; $x_0 = 0$
(c) $f(x) = \sin x$; $x_0 = \pi/2$ (d) $f(x) = \sqrt{x}$; $x_0 = 1$

2. In each part, use a CAS to find the local quadratic approximation of f at $x = x_0$, and use that approximation to find the local linear approximation of f at $x = x_0$.
(a) $f(x) = e^{\sin x}$; $x_0 = 0$ (b) $f(x) = \sqrt{x}$; $x_0 = 9$
(c) $f(x) = \sec^{-1} x$; $x_0 = 2$
(d) $f(x) = \sin^{-1} x$; $x_0 = 0$

3. (a) Find the local quadratic approximation of $\sqrt{x}$ at $x_0 = 1$.
(b) Use the result obtained in part (a) to approximate $\sqrt{1.1}$, and compare your approximation to that produced directly by your calculating utility. [See Example 1 of Section 3.9.]

4. (a) Find the local quadratic approximation of $\cos x$ at $x_0 = 0$.
(b) Use the result obtained in part (a) to approximate $\cos 2°$, and compare the approximation to that produced directly by your calculating utility.

5. Use an appropriate local quadratic approximation to approximate $\tan 61°$, and compare the result to that produced directly by your calculating utility.

6. Use an appropriate local quadratic approximation to approximate $\sqrt{36.03}$, and compare the result to that produced directly by your calculating utility.

7–16 Find the Maclaurin polynomials of orders $n = 0, 1, 2, 3$, and 4, and then find the nth Maclaurin polynomials for the function in sigma notation.

7. e^{-x}
8. e^{ax}
9. $\cos \pi x$
10. $\sin \pi x$
11. $\ln(1+x)$
12. $\dfrac{1}{1+x}$
13. $\cosh x$
14. $\sinh x$
15. $x \sin x$
16. xe^x

17–24 Find the Taylor polynomials of orders $n = 0, 1, 2, 3$, and 4 about $x = x_0$, and then find the nth Taylor polynomial for the function in sigma notation.

17. e^x; $x_0 = 1$
18. e^{-x}; $x_0 = \ln 2$
19. $\dfrac{1}{x}$; $x_0 = -1$
20. $\dfrac{1}{x+2}$; $x_0 = 3$
21. $\sin \pi x$; $x_0 = \dfrac{1}{2}$
22. $\cos x$; $x_0 = \dfrac{\pi}{2}$
23. $\ln x$; $x_0 = 1$
24. $\ln x$; $x_0 = e$

25. (a) Find the third Maclaurin polynomial for
$$f(x) = 1 + 2x - x^2 + x^3$$
(b) Find the third Taylor polynomial about $x = 1$ for
$$f(x) = 1 + 2(x-1) - (x-1)^2 + (x-1)^3$$

26. (a) Find the nth Maclaurin polynomial for
$$f(x) = c_0 + c_1 x + c_2 x^2 + \cdots + c_n x^n$$
(b) Find the nth Taylor polynomial about $x = 1$ for
$$f(x) = c_0 + c_1(x-1) + c_2(x-1)^2 + \cdots + c_n(x-1)^n$$

27–30 Find the first four distinct Taylor polynomials about $x = x_0$, and use a graphing utility to graph the given function and the Taylor polynomials on the same screen.

27. $f(x) = e^{-2x}$; $x_0 = 0$
28. $f(x) = \sin x$; $x_0 = \pi/2$
29. $f(x) = \cos x$; $x_0 = \pi$
30. $\ln(x+1)$; $x_0 = 0$

31. Use the method of Example 6 to approximate $\sqrt{e}$ to four decimal-place accuracy, and check your work by comparing your answer to that produced directly by your calculating utility. [*Suggestion:* Write $\sqrt{e}$ as $e^{0.5}$.]

32. Use the method of Example 6 to approximate $1/e$ to three decimal-place accuracy, and check your work by comparing your answer to that produced directly by your calculating utility.

33. Show that the nth Taylor polynomial for $\sinh x$ about $x = \ln 4$ is
$$\sum_{k=0}^{n} \frac{16 - (-1)^k}{8k!}(x - \ln 4)^k$$

34. (a) The accompanying figure shows a sector of radius r and central angle 2α. Assuming that the angle α is small, use the local quadratic approximation of $\cos \alpha$ at $\alpha = 0$ to show that $x \approx r\alpha^2/2$.
(b) Assuming that the Earth is a sphere of radius 4000 mi, use the result in part (a) to approximate the maximum amount by which a 100-mi arc along the equator will diverge from its chord.

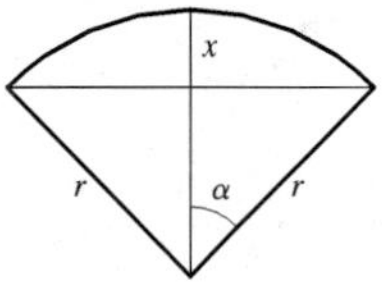

Figure Ex-34

FOCUS ON CONCEPTS

35. Which of the functions graphed in the following figure is most likely to have $p(x) = 1 - x + 2x^2$ as its second-order Maclaurin polynomial? Explain your reasoning.

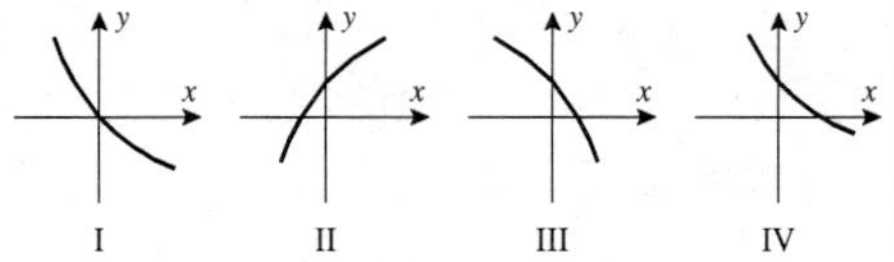

36. Suppose that the values of a function f and its first three derivatives at $x = 1$ are

$$f(1) = 2, \quad f'(1) = -3, \quad f''(1) = 0, \quad f'''(1) = 6$$

Find as many Taylor polynomials for f as you can about $x = 1$.

37. Let $p_1(x)$ and $p_2(x)$ be the local linear and local quadratic approximations of $f(x) = e^{\sin x}$ at $x = 0$.

(a) Use a graphing utility to generate the graphs of $f(x)$, $p_1(x)$, and $p_2(x)$ on the same screen for $-1 \le x \le 1$.

(b) Construct a table of values of $f(x)$, $p_1(x)$, and $p_2(x)$ for $x = -1.00, -0.75, -0.50, -0.25, 0, 0.25, 0.50, 0.75, 1.00$. Round the values to three decimal places.

(c) Generate the graph of $|f(x) - p_1(x)|$, and use the graph to determine an interval on which $p_1(x)$ approximates $f(x)$ with an error of at most ± 0.01. [*Suggestion:* Review the discussion relating to Figure 3.8.4.]

(d) Generate the graph of $|f(x) - p_2(x)|$, and use the graph to determine an interval on which $p_2(x)$ approximates $f(x)$ with an error of at most ± 0.01.

38. (a) Find an interval $[0, b]$ over which e^x can be approximated by $1 + x + (x^2/2!)$ to three decimal-place accuracy throughout the interval.

(b) Check your answer in part (a) by graphing

$$\left| e^x - \left(1 + x + \frac{x^2}{2!}\right) \right|$$

over the interval you obtained.

39–42 Use the Remainder Estimation Theorem to find an interval containing $x = 0$ over which $f(x)$ can be approximated by $p(x)$ to three decimal-place accuracy throughout the interval. Check your answer by graphing $|f(x) - p(x)|$ over the interval you obtained.

39. $f(x) = \sin x;\ p(x) = x - \dfrac{x^3}{3!}$

40. $f(x) = \cos x;\ p(x) = 1 - \dfrac{x^2}{2!} + \dfrac{x^4}{4!}$

41. $f(x) = \dfrac{1}{1 + x^2};\ p(x) = 1 - x^2 + x^4$

42. $f(x) = \ln(1 + x);\ p(x) = x - \dfrac{x^2}{2} + \dfrac{x^3}{3}$

✔ QUICK CHECK ANSWERS 10.7

1. $f(0) + f'(0)x + \dfrac{f''(0)}{2!}x^2 + \dfrac{f'''(0)}{3!}x^3$ **2.** $1; 2; 2; \frac{4}{3}$ **3.** $3 - 4(x - 2) + 5(x - 2)^2$ **4.** $-1; 5; -10; 10$
5. (a) $f(x) - p_n(x)$ (b) $\frac{1}{6}|x - 2|^5$

10.8 MACLAURIN AND TAYLOR SERIES; POWER SERIES

Recall from the last section that the nth Taylor polynomial $p_n(x)$ at $x = x_0$ for a function f was defined so its value and the values of its first n derivatives match those of f at x_0. This being the case, it is reasonable to expect that for values of x near x_0 the values of $p_n(x)$ will become better and better approximations of $f(x)$ as n increases, and may possibly converge to $f(x)$ as $n \to +\infty$. We will explore this idea in this section.

MACLAURIN AND TAYLOR SERIES

In Section 10.7 we defined the nth Maclaurin polynomial for a function f as

$$\sum_{k=0}^{n} \frac{f^{(k)}(0)}{k!}x^k = f(0) + f'(0)x + \frac{f''(0)}{2!}x^2 + \cdots + \frac{f^{(n)}(0)}{n!}x^n$$

and the nth Taylor polynomial for f about $x = x_0$ as

$$\sum_{k=0}^{n} \frac{f^{(k)}(x_0)}{k!}(x - x_0)^k = f(x_0) + f'(x_0)(x - x_0) + \frac{f''(x_0)}{2!}(x - x_0)^2 + \cdots + \frac{f^{(n)}(x_0)}{n!}(x - x_0)^n$$

It is not a big step to extend the notions of Maclaurin and Taylor polynomials to series by not stopping the summation index at n. Thus, we have the following definition.

10.8.1 DEFINITION. If f has derivatives of all orders at x_0, then we call the series

$$\sum_{k=0}^{\infty} \frac{f^{(k)}(x_0)}{k!}(x-x_0)^k = f(x_0) + f'(x_0)(x-x_0) + \frac{f''(x_0)}{2!}(x-x_0)^2 + \cdots + \frac{f^{(k)}(x_0)}{k!}(x-x_0)^k + \cdots \quad (1)$$

the ***Taylor series for f about x = x₀***. In the special case where $x_0 = 0$, this series becomes

$$\sum_{k=0}^{\infty} \frac{f^{(k)}(0)}{k!}x^k = f(0) + f'(0)x + \frac{f''(0)}{2!}x^2 + \cdots + \frac{f^{(k)}(0)}{k!}x^k + \cdots \quad (2)$$

in which case we call it the ***Maclaurin series for f***.

Note that the nth Maclaurin and Taylor polynomials are the nth partial sums for the corresponding Maclaurin and Taylor series.

▶ Example 1 Find the Maclaurin series for

(a) e^x (b) $\sin x$ (c) $\cos x$ (d) $\dfrac{1}{1-x}$

Solution (a). In Example 2 of Section 10.7 we found that the nth Maclaurin polynomial for e^x is

$$p_n(x) = \sum_{k=0}^{n} \frac{x^k}{k!} = 1 + x + \frac{x^2}{2!} + \cdots + \frac{x^n}{n!}$$

Thus, the Maclaurin series for e^x is

$$\sum_{k=0}^{\infty} \frac{x^k}{k!} = 1 + x + \frac{x^2}{2!} + \cdots + \frac{x^k}{k!} + \cdots$$

Solution (b). In Example 4(a) of Section 10.7 we found that the Maclaurin polynomials for $\sin x$ are given by

$$p_{2k+1}(x) = p_{2k+2}(x) = x - \frac{x^3}{3!} + \frac{x^5}{5!} - \frac{x^7}{7!} + \cdots + (-1)^k \frac{x^{2k+1}}{(2k+1)!} \quad (k = 0, 1, 2, \ldots)$$

Thus, the Maclaurin series for $\sin x$ is

$$\sum_{k=0}^{\infty} (-1)^k \frac{x^{2k+1}}{(2k+1)!} = x - \frac{x^3}{3!} + \frac{x^5}{5!} - \frac{x^7}{7!} + \cdots + (-1)^k \frac{x^{2k+1}}{(2k+1)!} + \cdots$$

Solution (c). In Example 4(b) of Section 10.7 we found that the Maclaurin polynomials for $\cos x$ are given by

$$p_{2k}(x) = p_{2k+1}(x) = 1 - \frac{x^2}{2!} + \frac{x^4}{4!} - \frac{x^6}{6!} + \cdots + (-1)^k \frac{x^{2k}}{(2k)!} \quad (k = 0, 1, 2, \ldots)$$

Thus, the Maclaurin series for $\cos x$ is

$$\sum_{k=0}^{\infty}(-1)^k \frac{x^{2k}}{(2k)!} = 1 - \frac{x^2}{2!} + \frac{x^4}{4!} - \frac{x^6}{6!} + \cdots + (-1)^k \frac{x^{2k}}{(2k)!} + \cdots$$

Solution (d). In Example 4(c) of Section 10.7 we found that the nth Maclaurin polynomial for $1/(1-x)$ is

$$p_n(x) = \sum_{k=0}^{n} x^k = 1 + x + x^2 + \cdots + x^n \quad (n = 0, 1, 2, \ldots)$$

Thus, the Maclaurin series for $1/(1-x)$ is

$$\sum_{k=0}^{\infty} x^k = 1 + x + x^2 + \cdots + x^k + \cdots \blacktriangleleft$$

▶ **Example 2** Find the Taylor series for $1/x$ about $x = 1$.

Solution. In Example 5 of Section 10.7 we found that the nth Taylor polynomial for $1/x$ about $x = 1$ is

$$\sum_{k=0}^{n}(-1)^k(x-1)^k = 1 - (x-1) + (x-1)^2 - (x-1)^3 + \cdots + (-1)^n(x-1)^n$$

Thus, the Taylor series for $1/x$ about $x = 1$ is

$$\sum_{k=0}^{\infty}(-1)^k(x-1)^k = 1 - (x-1) + (x-1)^2 - (x-1)^3 + \cdots + (-1)^k(x-1)^k + \cdots \blacktriangleleft$$

POWER SERIES IN *x*

Maclaurin and Taylor series differ from the series that we have considered in Sections 10.3 to 10.6 in that their terms are not merely constants, but instead involve a variable. These are examples of *power series*, which we now define.

If $c_0, c_1, c_2, \ldots$ are constants and x is a variable, then a series of the form

$$\sum_{k=0}^{\infty} c_k x^k = c_0 + c_1 x + c_2 x^2 + \cdots + c_k x^k + \cdots \tag{3}$$

is called a ***power series in x***. Some examples are

$$\sum_{k=0}^{\infty} x^k = 1 + x + x^2 + x^3 + \cdots$$

$$\sum_{k=0}^{\infty} \frac{x^k}{k!} = 1 + x + \frac{x^2}{2!} + \frac{x^3}{3!} + \cdots$$

$$\sum_{k=0}^{\infty}(-1)^k \frac{x^{2k}}{(2k)!} = 1 - \frac{x^2}{2!} + \frac{x^4}{4!} - \frac{x^6}{6!} + \cdots$$

From Example 1, these are the Maclaurin series for the functions $1/(1-x)$, e^x, and $\cos x$, respectively. Indeed, every Maclaurin series

$$\sum_{k=0}^{\infty} \frac{f^{(k)}(0)}{k!} x^k = f(0) + f'(0)x + \frac{f''(0)}{2!}x^2 + \cdots + \frac{f^{(k)}(0)}{k!}x^k + \cdots$$

is a power series in x.

RADIUS AND INTERVAL OF CONVERGENCE

If a numerical value is substituted for x in a power series $\sum c_k x^k$, then the resulting series of numbers may either converge or diverge. This leads to the problem of determining the set of x-values for which a given power series converges; this is called its ***convergence set***.

Observe that every power series in x converges at $x = 0$, since substituting this value in (3) produces the series

$$c_0 + 0 + 0 + 0 + \cdots + 0 + \cdots$$

whose sum is c_0. In some cases $x = 0$ may be the only number in the convergence set; in other cases the convergence set is some finite or infinite interval containing $x = 0$. This is the content of the following theorem, whose proof will be omitted.

10.8.2 THEOREM. *For any power series in x, exactly one of the following is true:*

(*a*) *The series converges only for $x = 0$.*

(*b*) *The series converges absolutely (and hence converges) for all real values of x.*

(*c*) *The series converges absolutely (and hence converges) for all x in some finite open interval $(-R, R)$ and diverges if $x < -R$ or $x > R$. At either of the values $x = R$ or $x = -R$, the series may converge absolutely, converge conditionally, or diverge, depending on the particular series.*

This theorem states that the convergence set for a power series in x is always an interval centered at $x = 0$ (possibly just the value $x = 0$ itself or possibly infinite). For this reason, the convergence set of a power series in x is called the ***interval of convergence***. In the case where the convergence set is the single value $x = 0$ we say that the series has ***radius of convergence* 0**, in the case where the convergence set is $(-\infty, +\infty)$ we say that the series has ***radius of convergence* $+\infty$**, and in the case where the convergence set extends between $-R$ and R we say that the series has ***radius of convergence R*** (Figure 10.8.1).

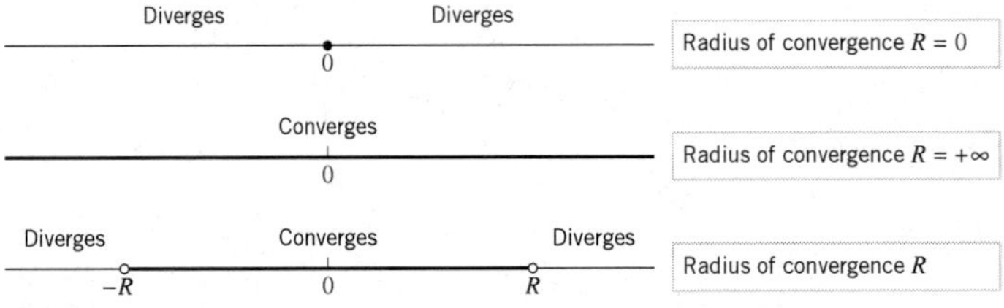

Figure 10.8.1

FINDING THE INTERVAL OF CONVERGENCE

The usual procedure for finding the interval of convergence of a power series is to apply the ratio test for absolute convergence (Theorem 10.6.5). The following example illustrates how this works.

▶ **Example 3** Find the interval of convergence and radius of convergence of the following power series.

(a) $\displaystyle\sum_{k=0}^{\infty} x^k$ (b) $\displaystyle\sum_{k=0}^{\infty} \frac{x^k}{k!}$ (c) $\displaystyle\sum_{k=0}^{\infty} k!x^k$ (d) $\displaystyle\sum_{k=0}^{\infty} \frac{(-1)^k x^k}{3^k(k+1)}$

Solution (a). We apply the ratio test for absolute convergence. We have

$$\rho = \lim_{k\to+\infty}\left|\frac{u_{k+1}}{u_k}\right| = \lim_{k\to+\infty}\left|\frac{x^{k+1}}{x^k}\right| = \lim_{k\to+\infty}|x| = |x|$$

so the series converges absolutely if $\rho = |x| < 1$ and diverges if $\rho = |x| > 1$. The test is inconclusive if $|x| = 1$ (i.e., if $x = 1$ or $x = -1$), which means that we will have to investigate convergence at these values separately. At these values the series becomes

$$\sum_{k=0}^{\infty} 1^k = 1 + 1 + 1 + 1 + \cdots \qquad \boxed{x = 1}$$

$$\sum_{k=0}^{\infty} (-1)^k = 1 - 1 + 1 - 1 + \cdots \qquad \boxed{x = -1}$$

both of which diverge; thus, the interval of convergence for the given power series is $(-1, 1)$, and the radius of convergence is $R = 1$.

Solution (b). Applying the ratio test for absolute convergence, we obtain

$$\rho = \lim_{k\to+\infty}\left|\frac{u_{k+1}}{u_k}\right| = \lim_{k\to+\infty}\left|\frac{x^{k+1}}{(k+1)!}\cdot\frac{k!}{x^k}\right| = \lim_{k\to+\infty}\left|\frac{x}{k+1}\right| = 0$$

Since $\rho < 1$ for all x, the series converges absolutely for all x. Thus, the interval of convergence is $(-\infty, +\infty)$ and the radius of convergence is $R = +\infty$.

Solution (c). If $x \neq 0$, then the ratio test for absolute convergence yields

$$\rho = \lim_{k\to+\infty}\left|\frac{u_{k+1}}{u_k}\right| = \lim_{k\to+\infty}\left|\frac{(k+1)!x^{k+1}}{k!x^k}\right| = \lim_{k\to+\infty}|(k+1)x| = +\infty$$

Therefore, the series diverges for all nonzero values of x. Thus, the interval of convergence is the single value $x = 0$ and the radius of convergence is $R = 0$.

Solution (d). Since $|(-1)^k| = |(-1)^{k+1}| = 1$, we obtain

$$\begin{aligned}\rho = \lim_{k\to+\infty}\left|\frac{u_{k+1}}{u_k}\right| &= \lim_{k\to+\infty}\left|\frac{x^{k+1}}{3^{k+1}(k+2)}\cdot\frac{3^k(k+1)}{x^k}\right| \\ &= \lim_{k\to+\infty}\left[\frac{|x|}{3}\cdot\left(\frac{k+1}{k+2}\right)\right] \\ &= \frac{|x|}{3}\lim_{k\to+\infty}\left(\frac{1+(1/k)}{1+(2/k)}\right) = \frac{|x|}{3}\end{aligned}$$

The ratio test for absolute convergence implies that the series converges absolutely if $|x| < 3$ and diverges if $|x| > 3$. The ratio test fails to provide any information when $|x| = 3$, so the cases $x = -3$ and $x = 3$ need separate analyses. Substituting $x = -3$ in the given series yields

$$\sum_{k=0}^{\infty}\frac{(-1)^k(-3)^k}{3^k(k+1)} = \sum_{k=0}^{\infty}\frac{(-1)^k(-1)^k3^k}{3^k(k+1)} = \sum_{k=0}^{\infty}\frac{1}{k+1}$$

which is the divergent harmonic series $1 + \frac{1}{2} + \frac{1}{3} + \frac{1}{4} + \cdots$. Substituting $x = 3$ in the given series yields

$$\sum_{k=0}^{\infty}\frac{(-1)^k3^k}{3^k(k+1)} = \sum_{k=0}^{\infty}\frac{(-1)^k}{k+1} = 1 - \frac{1}{2} + \frac{1}{3} - \frac{1}{4} + \cdots$$

which is the conditionally convergent alternating harmonic series. Thus, the interval of convergence for the given series is $(-3, 3]$ and the radius of convergence is $R = 3$. ◀

POWER SERIES IN $x - x_0$

If x_0 is a constant, and if x is replaced by $x - x_0$ in (3), then the resulting series has the form

$$\sum_{k=0}^{\infty} c_k(x-x_0)^k = c_0 + c_1(x-x_0) + c_2(x-x_0)^2 + \cdots + c_k(x-x_0)^k + \cdots$$

This is called a ***power series in*** $\boldsymbol{x - x_0}$. Some examples are

$$\sum_{k=0}^{\infty} \frac{(x-1)^k}{k+1} = 1 + \frac{(x-1)}{2} + \frac{(x-1)^2}{3} + \frac{(x-1)^3}{4} + \cdots \qquad \boxed{x_0 = 1}$$

$$\sum_{k=0}^{\infty} \frac{(-1)^k(x+3)^k}{k!} = 1 - (x+3) + \frac{(x+3)^2}{2!} - \frac{(x+3)^3}{3!} + \cdots \qquad \boxed{x_0 = -3}$$

The first of these is a power series in $x - 1$ and the second is a power series in $x + 3$. Note that a power series in x is a power series in $x - x_0$ in which $x_0 = 0$. More generally, the Taylor series

$$\sum_{k=0}^{\infty} \frac{f^{(k)}(x_0)}{k!}(x-x_0)^k$$

is a power series in $x - x_0$.

The main result on convergence of a power series in $x - x_0$ can be obtained by substituting $x - x_0$ for x in Theorem 10.8.2. This leads to the following theorem.

10.8.3 THEOREM. *For a power series $\sum c_k(x-x_0)^k$, exactly one of the following statements is true:*

(*a*) *The series converges only for $x = x_0$.*

(*b*) *The series converges absolutely (and hence converges) for all real values of x.*

(*c*) *The series converges absolutely (and hence converges) for all x in some finite open interval $(x_0 - R, x_0 + R)$ and diverges if $x < x_0 - R$ or $x > x_0 + R$. At either of the values $x = x_0 - R$ or $x = x_0 + R$, the series may converge absolutely, converge conditionally, or diverge, depending on the particular series.*

It follows from this theorem that the set of values for which a power series in $x - x_0$ converges is always an interval centered at $x = x_0$; we call this the ***interval of convergence*** (Figure 10.8.2). In part (*a*) of Theorem 10.8.3 the interval of convergence reduces to the single value $x = x_0$, in which case we say that the series has ***radius of convergence*** $\boldsymbol{R = 0}$; in part (*b*) the interval of convergence is infinite (the entire real line), in which case we say that the series has ***radius of convergence*** $\boldsymbol{R = +\infty}$; and in part (*c*) the interval extends between $x_0 - R$ and $x_0 + R$, in which case we say that the series has ***radius of convergence*** $\boldsymbol{R}$.

▶ **Example 4** Find the interval of convergence and radius of convergence of the series

$$\sum_{k=1}^{\infty} \frac{(x-5)^k}{k^2}$$

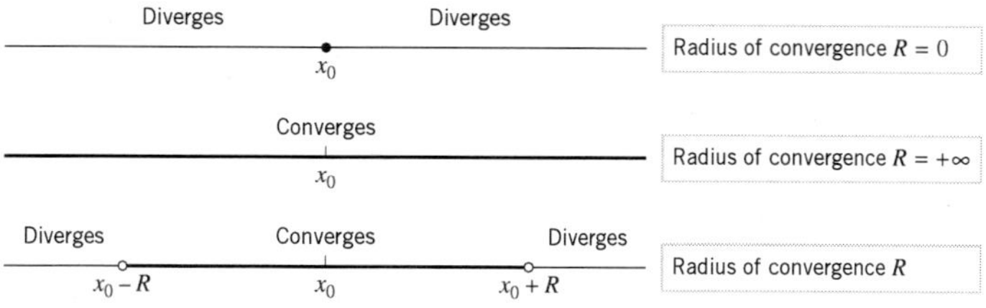

Figure 10.8.2

Solution. We apply the ratio test for absolute convergence.

$$\begin{aligned}\rho = \lim_{k\to+\infty}\left|\frac{u_{k+1}}{u_k}\right| &= \lim_{k\to+\infty}\left|\frac{(x-5)^{k+1}}{(k+1)^2}\cdot\frac{k^2}{(x-5)^k}\right| \\ &= \lim_{k\to+\infty}\left[|x-5|\left(\frac{k}{k+1}\right)^2\right] \\ &= |x-5|\lim_{k\to+\infty}\left(\frac{1}{1+(1/k)}\right)^2 = |x-5|\end{aligned}$$

Thus, the series converges absolutely if $|x-5| < 1$, or $-1 < x-5 < 1$, or $4 < x < 6$. The series diverges if $x < 4$ or $x > 6$.

To determine the convergence behavior at the endpoints $x = 4$ and $x = 6$, we substitute these values in the given series. If $x = 6$, the series becomes

$$\sum_{k=1}^{\infty}\frac{1^k}{k^2} = \sum_{k=1}^{\infty}\frac{1}{k^2} = 1 + \frac{1}{2^2} + \frac{1}{3^2} + \frac{1}{4^2} + \cdots$$

which is a convergent p-series ($p = 2$). If $x = 4$, the series becomes

$$\sum_{k=1}^{\infty}\frac{(-1)^k}{k^2} = -1 + \frac{1}{2^2} - \frac{1}{3^2} + \frac{1}{4^2} - \cdots$$

Since this series converges absolutely, the interval of convergence for the given series is $[4, 6]$. The radius of convergence is $R = 1$ (Figure 10.8.3). ◂

It will always be a waste of time to test for convergence at the endpoints of the interval of convergence using the ratio test, since ρ will always be 1 at those points if

$$\rho = \lim_{n\to+\infty}|a_{n+1}/a_n|$$

exists. Explain why this must be so.

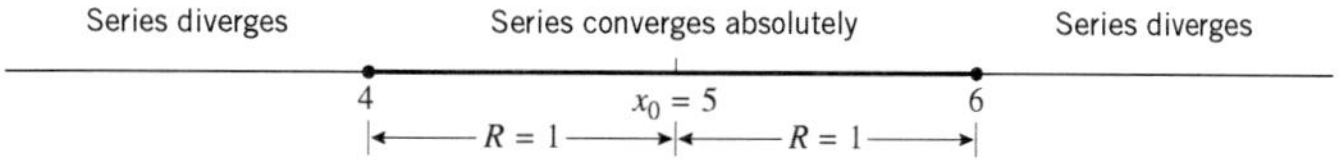

Figure 10.8.3

FUNCTIONS DEFINED BY POWER SERIES

If a function f is expressed as a power series on some interval, then we say that the power series ***represents*** f on that interval. For example, we saw in Example 4(b) of Section 10.3 that

$$\frac{1}{1-x} = \sum_{k=0}^{\infty} x^k$$

if $|x| < 1$, so this power series represents the function $1/(1-x)$ on the interval $-1 < x < 1$.

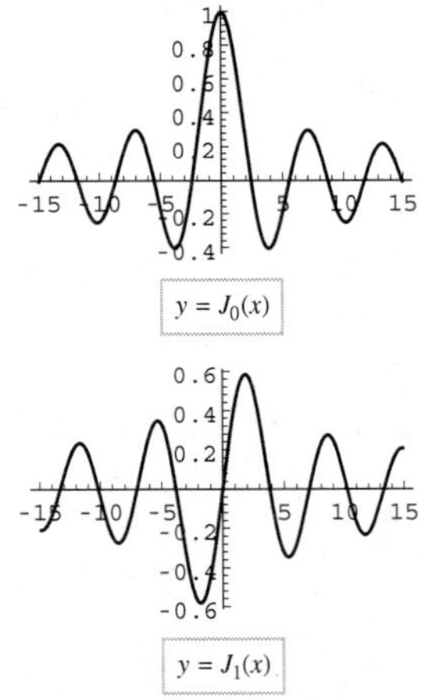

Figure 10.8.4

Sometimes new functions actually originate as power series, and the properties of the functions are developed by working with their power series representations. For example, the functions

$$J_0(x) = \sum_{k=0}^{\infty} \frac{(-1)^k x^{2k}}{2^{2k}(k!)^2} = 1 - \frac{x^2}{2^2(1!)^2} + \frac{x^4}{2^4(2!)^2} - \frac{x^6}{2^6(3!)^2} + \cdots \tag{4}$$

and

$$J_1(x) = \sum_{k=0}^{\infty} \frac{(-1)^k x^{2k+1}}{2^{2k+1}(k!)(k+1)!} = \frac{x}{2} - \frac{x^3}{2^3(1!)(2!)} + \frac{x^5}{2^5(2!)(3!)} - \cdots \tag{5}$$

which are called ***Bessel functions*** in honor of the German mathematician and astronomer Friedrich Wilhelm Bessel (1784–1846), arise naturally in the study of planetary motion and in various problems that involve heat flow.

To find the domains of these functions, we must determine where their defining power series converge. For example, in the case of $J_0(x)$ we have

$$\rho = \lim_{k\to+\infty}\left|\frac{u_{k+1}}{u_k}\right| = \lim_{k\to+\infty}\left|\frac{x^{2(k+1)}}{2^{2(k+1)}[(k+1)!]^2}\cdot\frac{2^{2k}(k!)^2}{x^{2k}}\right|$$

$$= \lim_{k\to+\infty}\left|\frac{x^2}{4(k+1)^2}\right| = 0 < 1$$

so the series converges for all x; that is, the domain of $J_0(x)$ is $(-\infty, +\infty)$. We leave it as an exercise to show that the power series for $J_1(x)$ also converges for all x. Computer-generated graphs of $J_0(x)$ and $J_1(x)$ are shown in Figure 10.8.4.

TECHNOLOGY MASTERY

Many computer algebra systems have the Bessel functions as part of their libraries. If you have a CAS with Bessel functions, use it to generate the graphs in Figure 10.8.4.

✔QUICK CHECK EXERCISES 10.8 *(See page 694 for answers.)*

1. If f has derivatives of all orders at x_0, then the Taylor series for f about $x = x_0$ is defined to be

$$\sum_{k=0}^{\infty} \_\_\_\_\_\_$$

2. Since

$$\lim_{k\to+\infty}\left|\frac{2^{k+1}x^{k+1}}{2^k x^k}\right| = 2|x|$$

the radius of convergence for the infinite series $\sum_{k=0}^{\infty} 2^k x^k$ is ______.

3. Since

$$\lim_{k\to+\infty}\left|\frac{(3^{k+1}x^{k+1})/(k+1)!}{(3^k x^k)/k!}\right| = \lim_{k\to+\infty}\left|\frac{3x}{k+1}\right| = 0$$

the interval of convergence for the series $\sum_{k=0}^{\infty}(3^k/k!)x^k$ is ______.

4. (a) Since

$$\lim_{k\to+\infty}\left|\frac{(x-4)^{k+1}/\sqrt{k+1}}{(x-4)^k/\sqrt{k}}\right| = \lim_{k\to+\infty}\left|\sqrt{\frac{k}{k+1}}(x-4)\right|$$

$$= |x-4|$$

the radius of convergence for the infinite series $\sum_{k=1}^{\infty}(1/\sqrt{k})(x-4)^k$ is ______.

(b) When $x = 3$,

$$\sum_{k=1}^{\infty}\frac{1}{\sqrt{k}}(x-4)^k = \sum_{k=1}^{\infty}\frac{1}{\sqrt{k}}(-1)^k$$

Does this series converge or diverge?

(c) When $x = 5$,

$$\sum_{k=1}^{\infty}\frac{1}{\sqrt{k}}(x-4)^k = \sum_{k=1}^{\infty}\frac{1}{\sqrt{k}}$$

Does this series converge or diverge?

(d) The interval of convergence for the infinite series $\sum_{k=1}^{\infty}(1/\sqrt{k})(x-4)^k$ is ______.

EXERCISE SET 10.8 Graphing Utility C CAS

1–10 Use sigma notation to write the Maclaurin series for the function.

1. e^{-x} **2.** e^{ax} **3.** $\cos \pi x$ **4.** $\sin \pi x$

5. $\ln(1+x)$ **6.** $\dfrac{1}{1+x}$ **7.** $\cosh x$

8. $\sinh x$ **9.** $x \sin x$ **10.** xe^x

11–18 Use sigma notation to write the Taylor series about $x = x_0$ for the function.

11. e^x; $x_0 = 1$ **12.** e^{-x}; $x_0 = \ln 2$

13. $\dfrac{1}{x}$; $x_0 = -1$ **14.** $\dfrac{1}{x+2}$; $x_0 = 3$

15. $\sin \pi x$; $x_0 = \dfrac{1}{2}$ **16.** $\cos x$; $x_0 = \dfrac{\pi}{2}$

17. $\ln x$; $x_0 = 1$ **18.** $\ln x$; $x_0 = e$

19–22 Find the interval of convergence of the power series, and find a familiar function that is represented by the power series on that interval.

19. $1 - x + x^2 - x^3 + \cdots + (-1)^k x^k + \cdots$

20. $1 + x^2 + x^4 + \cdots + x^{2k} + \cdots$

21. $1 + (x-2) + (x-2)^2 + \cdots + (x-2)^k + \cdots$

22. $1 - (x+3) + (x+3)^2 - (x+3)^3 + \cdots + (-1)^k (x+3)^k | \ldots$

23. Suppose that the function f is represented by the power series

$$f(x) = 1 - \frac{x}{2} + \frac{x^2}{4} - \frac{x^3}{8} + \cdots + (-1)^k \frac{x^k}{2^k} + \cdots$$

(a) Find the domain of f. (b) Find $f(0)$ and $f(1)$.

24. Suppose that the function f is represented by the power series

$$f(x) = 1 - \frac{x-5}{3} + \frac{(x-5)^2}{3^2} - \frac{(x-5)^3}{3^3} + \cdots$$

(a) Find the domain of f. (b) Find $f(3)$ and $f(6)$.

25–48 Find the radius of convergence and the interval of convergence.

25. $\displaystyle\sum_{k=0}^{\infty} \frac{x^k}{k+1}$ **26.** $\displaystyle\sum_{k=0}^{\infty} 3^k x^k$ **27.** $\displaystyle\sum_{k=0}^{\infty} \frac{(-1)^k x^k}{k!}$

28. $\displaystyle\sum_{k=0}^{\infty} \frac{k!}{2^k} x^k$ **29.** $\displaystyle\sum_{k=1}^{\infty} \frac{5^k}{k^2} x^k$ **30.** $\displaystyle\sum_{k=2}^{\infty} \frac{x^k}{\ln k}$

31. $\displaystyle\sum_{k=1}^{\infty} \frac{x^k}{k(k+1)}$ **32.** $\displaystyle\sum_{k=0}^{\infty} \frac{(-2)^k x^{k+1}}{k+1}$

33. $\displaystyle\sum_{k=1}^{\infty} (-1)^{k-1} \frac{x^k}{\sqrt{k}}$ **34.** $\displaystyle\sum_{k=0}^{\infty} \frac{(-1)^k x^{2k}}{(2k)!}$

35. $\displaystyle\sum_{k=0}^{\infty} (-1)^k \frac{x^{2k+1}}{(2k+1)!}$ **36.** $\displaystyle\sum_{k=1}^{\infty} (-1)^k \frac{x^{3k}}{k^{3/2}}$

37. $\displaystyle\sum_{k=0}^{\infty} \frac{3^k}{k!} x^k$ **38.** $\displaystyle\sum_{k=2}^{\infty} (-1)^{k+1} \frac{x^k}{k(\ln k)^2}$

39. $\displaystyle\sum_{k=0}^{\infty} \frac{x^k}{1+k^2}$ **40.** $\displaystyle\sum_{k=0}^{\infty} \frac{(x-3)^k}{2^k}$

41. $\displaystyle\sum_{k=1}^{\infty} (-1)^{k+1} \frac{(x+1)^k}{k}$ **42.** $\displaystyle\sum_{k=0}^{\infty} (-1)^k \frac{(x-4)^k}{(k+1)^2}$

43. $\displaystyle\sum_{k=0}^{\infty} \left(\frac{3}{4}\right)^k (x+5)^k$ **44.** $\displaystyle\sum_{k=1}^{\infty} \frac{(2k+1)!}{k^3} (x-2)^k$

45. $\displaystyle\sum_{k=1}^{\infty} (-1)^k \frac{(x+1)^{2k+1}}{k^2+4}$ **46.** $\displaystyle\sum_{k=1}^{\infty} \frac{(\ln k)(x-3)^k}{k}$

47. $\displaystyle\sum_{k=0}^{\infty} \frac{\pi^k (x-1)^{2k}}{(2k+1)!}$ **48.** $\displaystyle\sum_{k=0}^{\infty} \frac{(2x-3)^k}{4^{2k}}$

49. Use the root test to find the interval of convergence of

$$\sum_{k=2}^{\infty} \frac{x^k}{(\ln k)^k}$$

50. Find the domain of the function

$$f(x) = \sum_{k=1}^{\infty} \frac{1 \cdot 3 \cdot 5 \cdots (2k-1)}{(2k-2)!} x^k$$

51. Show that the series

$$1 - \frac{x}{2!} + \frac{x^2}{4!} - \frac{x^3}{6!} + \cdots$$

is the Maclaurin series for the function

$$f(x) = \begin{cases} \cos \sqrt{x}, & x \ge 0 \\ \cosh \sqrt{-x}, & x < 0 \end{cases}$$

[*Hint:* Use the Maclaurin series for $\cos x$ and $\cosh x$ to obtain series for $\cos \sqrt{x}$, where $x \ge 0$, and $\cosh \sqrt{-x}$, where $x \le 0$.]

FOCUS ON CONCEPTS

52. If a function f is represented by a power series on an interval, then the graphs of the partial sums can be used as approximations to the graph of f.

(a) Use a graphing utility to generate the graph of $1/(1-x)$ together with the graphs of the first four partial sums of its Maclaurin series over the interval $(-1, 1)$.

(b) In general terms, where are the graphs of the partial sums the most accurate?

53. Prove:

(a) If f is an even function, then all odd powers of x in its Maclaurin series have coefficient 0.

(b) If f is an odd function, then all even powers of x in its Maclaurin series have coefficient 0.

54. Suppose that the power series $\sum c_k(x-x_0)^k$ has radius of convergence R and p is a nonzero constant. What can you say about the radius of convergence of the power series $\sum pc_k(x-x_0)^k$? Explain your reasoning. [*Hint:* See Theorem 10.4.3.]

55. Suppose that the power series $\sum c_k(x-x_0)^k$ has a finite radius of convergence R, and the power series $\sum d_k(x-x_0)^k$ has a radius of convergence of $+\infty$. What can you say about the radius of convergence of $\sum(c_k+d_k)(x-x_0)^k$? Explain your reasoning.

56. Suppose that the power series $\sum c_k(x-x_0)^k$ has a finite radius of convergence R_1 and the power series $\sum d_k(x-x_0)^k$ has a finite radius of convergence R_2. What can you say about the radius of convergence of $\sum(c_k+d_k)(x-x_0)^k$? Explain your reasoning. [*Hint:* The case $R_1=R_2$ requires special attention.]

57. Show that if p is a positive integer, then the power series

$$\sum_{k=0}^{\infty}\frac{(pk)!}{(k!)^p}x^k$$

has a radius of convergence of $1/p^p$.

58. Show that if p and q are positive integers, then the power series

$$\sum_{k=0}^{\infty}\frac{(k+p)!}{k!(k+q)!}x^k$$

has a radius of convergence of $+\infty$.

59. Show that the power series representation of the Bessel function $J_1(x)$ converges for all x [Formula (5)].

C **60.** If the constant p in the general p-series is replaced by a variable x for $x>1$, then the resulting function is called the ***Riemann zeta function*** and is denoted by

$$\zeta(x)=\sum_{k=1}^{\infty}\frac{1}{k^x}$$

(a) Let s_n be the nth partial sum of the series for $\zeta(3.7)$. Find n such that s_n approximates $\zeta(3.7)$ to two decimal-place accuracy, and calculate s_n using this value of n. [*Hint:* Use the right inequality in Exercise 32(b) of Section 10.4 with $f(x)=1/x^{3.7}$.]

(b) Determine whether your CAS can evaluate the Riemann zeta function directly. If so, compare the value produced by the CAS to the value of s_n obtained in part (a).

61. Prove: If $\lim_{k\to+\infty}|c_k|^{1/k}=L$, where $L\neq 0$, then $1/L$ is the radius of convergence of the power series $\sum_{k=0}^{\infty}c_kx^k$.

62. Prove: If the power series $\sum_{k=0}^{\infty}c_kx^k$ has radius of convergence R, then the series $\sum_{k=0}^{\infty}c_kx^{2k}$ has radius of convergence $\sqrt{R}$.

63. Prove: If the interval of convergence of the series $\sum_{k=0}^{\infty}c_k(x-x_0)^k$ is $(x_0-R, x_0+R]$, then the series converges conditionally at x_0+R.

✔ QUICK CHECK ANSWERS 10.8

1. $\dfrac{f^{(k)}(x_0)}{k!}(x-x_0)^k$ **2.** $\dfrac{1}{2}$ **3.** $(-\infty,+\infty)$ **4.** (a) 1 (b) converges (c) diverges (d) $[3,5)$

10.9 CONVERGENCE OF TAYLOR SERIES

In this section we will investigate when a Taylor series for a function converges to that function on some interval, and we will consider how Taylor series can be used to approximate values of trigonometric, exponential, and logarithmic functions.

THE CONVERGENCE PROBLEM FOR TAYLOR SERIES

Recall that the nth Taylor polynomial for a function f about $x=x_0$ has the property that its value and the values of its first n derivatives match those of f at x_0. As n increases, more and more derivatives match up, so it is reasonable to hope that for values of x near x_0 the values of the Taylor polynomials might converge to the value of $f(x)$; that is,

$$f(x)=\lim_{n\to+\infty}\sum_{k=0}^{n}\frac{f^{(k)}(x_0)}{k!}(x-x_0)^k \tag{1}$$

However, the nth Taylor polynomial for f is the nth partial sum of the Taylor series for f, so (1) is equivalent to stating that the Taylor series for f converges at x, and its sum is $f(x)$. Thus, we are led to consider the following problem.

It is important to understand that Problem 10.9.1 is concerned with more than just convergence of the Taylor series for f; it is concerned with whether the series converges to the function f itself. Indeed, it is possible for a Taylor series of a function f to converge to values different from $f(x)$ for certain values of x (Exercise 14).

10.9.1 PROBLEM. Given a function f that has derivatives of all orders at $x = x_0$, determine whether there is an open interval containing x_0 such that $f(x)$ is the sum of its Taylor series about $x = x_0$ at each point in the interval; that is,

$$f(x) = \sum_{k=0}^{\infty} \frac{f^{(k)}(x_0)}{k!}(x - x_0)^k \tag{2}$$

for all values of x in the interval.

One way to show that (1) holds is to show that

$$\lim_{n\to+\infty}\left[f(x) - \sum_{k=0}^{n} \frac{f^{(k)}(x_0)}{k!}(x - x_0)^k\right] = 0$$

However, the difference appearing on the left side of this equation is the nth remainder for the Taylor series [Formula (12) of Section 10.7]. Thus, we have the following result.

10.9.2 THEOREM. *The equality*

$$f(x) = \sum_{k=0}^{\infty} \frac{f^{(k)}(x_0)}{k!}(x - x_0)^k$$

holds at a point x if and only if $\lim_{n\to+\infty} R_n(x) = 0$.

ESTIMATING THE nTH REMAINDER

It is relatively rare that one can prove directly that $R_n(x)\to 0$ as $n\to+\infty$. Usually, this is proved indirectly by finding appropriate bounds on $|R_n(x)|$ and applying the Squeezing Theorem for Sequences. The Remainder Estimation Theorem (Theorem 10.7.4) provides a useful bound for this purpose. Recall that this theorem asserts that if M is an upper bound for $|f^{(n+1)}(x)|$ on an interval I containing x_0, then

$$|R_n(x)| \le \frac{M}{(n+1)!}|x - x_0|^{n+1} \tag{3}$$

for all x in I.

The following example illustrates how the Remainder Estimation Theorem is applied.

► **Example 1** Show that the Maclaurin series for $\cos x$ converges to $\cos x$ for all x; that is,

$$\cos x = \sum_{k=0}^{\infty}(-1)^k \frac{x^{2k}}{(2k)!} = 1 - \frac{x^2}{2!} + \frac{x^4}{4!} - \frac{x^6}{6!} + \cdots \qquad (-\infty < x < +\infty)$$

Solution. From Theorem 10.9.2 we must show that $R_n(x)\to 0$ for all x as $n\to+\infty$. For this purpose let $f(x) = \cos x$, so that for all x we have

$$f^{(n+1)}(x) = \pm\cos x \quad \text{or} \quad f^{(n+1)}(x) = \pm\sin x$$

In all cases we have $|f^{(n+1)}(x)| \le 1$, so we can apply (3) with $M = 1$ and $x_0 = 0$ to conclude that

$$0 \le |R_n(x)| \le \frac{|x|^{n+1}}{(n+1)!} \tag{4}$$

The method of Example 1 can be easily modified to prove that the Taylor series for $\sin x$ and $\cos x$ about any point $x = x_0$ converge to $\sin x$ and $\cos x$ respectively, for all x (Exercises 21 and 22). For reference, some of the most important Maclaurin series are listed in Table 10.9.1 at the end of this section.

However, it follows from Formula (5) of Section 10.2 with $n+1$ in place of n and $|x|$ in place of x that

$$\lim_{n\to+\infty} \frac{|x|^{n+1}}{(n+1)!} = 0 \tag{5}$$

Using this result and the Squeezing Theorem for Sequences (Theorem 10.1.5), it follows from (4) that $|R_n(x)| \to 0$ and hence that $R_n(x) \to 0$ as $n \to +\infty$ (Theorem 10.1.6). Since this is true for all x, we have proved that the Maclaurin series for $\cos x$ converges to $\cos x$ for all x. This is illustrated in Figure 10.9.1, where we can see how successive partial sums approximate the cosine curve more and more closely. ◀

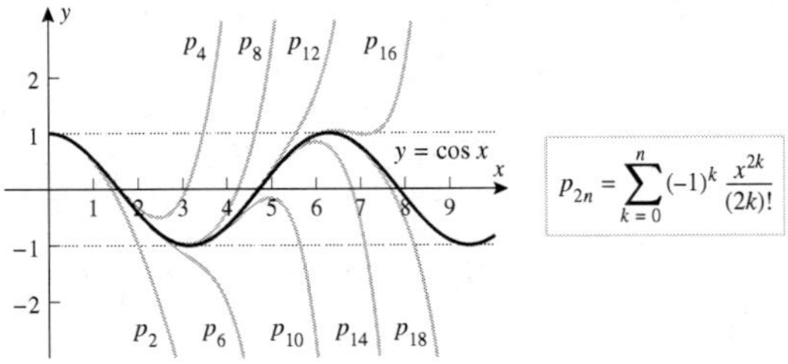

Figure 10.9.1

APPROXIMATING TRIGONOMETRIC FUNCTIONS

In general, to approximate the value of a function f at a point x using a Taylor series, there are two basic questions that must be answered:

- About what point x_0 should the Taylor series be expanded?
- How many terms in the series should be used to achieve the desired accuracy?

In response to the first question, x_0 needs to be a point at which the derivatives of f can be evaluated easily, since these values are needed for the coefficients in the Taylor series. Furthermore, if the function f is being evaluated at x, then x_0 should be chosen as close as possible to x, since Taylor series tend to converge more rapidly near x_0. For example, to approximate $\sin 3^\circ$ $(= \pi/60$ radians), it would be reasonable to take $x_0 = 0$, since $\pi/60$ is close to 0 and the derivatives of $\sin x$ are easy to evaluate at 0. On the other hand, to approximate $\sin 85^\circ$ $(= 17\pi/36$ radians), it would be more natural to take $x_0 = \pi/2$, since $17\pi/36$ is close to $\pi/2$ and the derivatives of $\sin x$ are easy to evaluate at $\pi/2$.

In response to the second question posed above, the number of terms required to achieve a specific accuracy needs to be determined on a problem-by-problem basis. The next example gives two methods for doing this.

▶ **Example 2** Use the Maclaurin series for $\sin x$ to approximate $\sin 3^\circ$ to five decimal-place accuracy.

Solution. In the Maclaurin series

$$\sin x = \sum_{k=0}^{\infty} (-1)^k \frac{x^{2k+1}}{(2k+1)!} = x - \frac{x^3}{3!} + \frac{x^5}{5!} - \frac{x^7}{7!} + \cdots \tag{6}$$

the angle x is assumed to be in radians (because the differentiation formulas for the trigonometric functions were derived with this assumption). Since $3^\circ = \pi/60$ radians, it follows from (6) that

$$\sin 3^\circ = \sin \frac{\pi}{60} = \left(\frac{\pi}{60}\right) - \frac{(\pi/60)^3}{3!} + \frac{(\pi/60)^5}{5!} - \frac{(\pi/60)^7}{7!} + \cdots \tag{7}$$

We must now determine how many terms in the series are required to achieve five decimal-place accuracy. We will consider two possible approaches, one using the Remainder Estimation Theorem (Theorem 10.7.4) and the other using the fact that (7) satisfies the hypotheses of the alternating series test (Theorem 10.6.1).

Method 1. (*The Remainder Estimation Theorem*)

Since we want to achieve five decimal-place accuracy, our goal is to choose n so that the absolute value of the nth remainder at $x = \pi/60$ does not exceed $0.000005 = 5 \times 10^{-6}$; that is,

$$\left|R_n\left(\frac{\pi}{60}\right)\right| \le 0.000005 \tag{8}$$

However, if we let $f(x) = \sin x$, then $f^{(n+1)}(x)$ is either $\pm \sin x$ or $\pm \cos x$, and in either case $|f^{(n+1)}(x)| \le 1$ for all x. Thus, it follows from the Remainder Estimation Theorem with $M = 1$, $x_0 = 0$, and $x = \pi/60$ that

$$\left|R_n\left(\frac{\pi}{60}\right)\right| \le \frac{(\pi/60)^{n+1}}{(n+1)!}$$

Thus, we can satisfy (8) by choosing n so that

$$\frac{(\pi/60)^{n+1}}{(n+1)!} \le 0.000005$$

With the help of a calculating utility you can verify that the smallest value of n that meets this criterion is $n = 3$. Thus, to achieve five decimal-place accuracy we need only keep terms up to the third power in (7). This yields

$$\sin 3^\circ \approx \left(\frac{\pi}{60}\right) - \frac{(\pi/60)^3}{3!} \approx 0.05234 \tag{9}$$

(verify). As a check, a calculator gives $\sin 3^\circ \approx 0.05233595624$, which agrees with (9) when rounded to five decimal places.

Method 2. (*The Alternating Series Test*)

We leave it for you to check that (7) satisfies the hypotheses of the alternating series test (Theorem 10.6.1).

Let s_n denote the sum of the terms in (7) up to and including the nth power of $\pi/60$. Since the exponents in the series are odd integers, the integer n must be odd, and the exponent of the first term *not* included in the sum s_n must be $n + 2$. Thus, it follows from part (*b*) of Theorem 10.6.2 that

$$|\sin 3^\circ - s_n| < \frac{(\pi/60)^{n+2}}{(n+2)!}$$

This means that for five decimal-place accuracy we must look for the first positive odd integer n such that

$$\frac{(\pi/60)^{n+2}}{(n+2)!} \le 0.000005$$

With the help of a calculating utility you can verify that the smallest value of n that meets this criterion is $n = 3$. This agrees with the result obtained above using the Remainder Estimation Theorem and hence leads to approximation (9) as before. ◀

ROUNDOFF AND TRUNCATION ERROR

There are two types of errors that occur when computing with series. The first, called ***truncation error***, is the error that results when a series is approximated by a partial sum; and the second, called ***roundoff error***, is the error that arises from approximations in numerical computations. For example, in our derivation of (9) we took $n = 3$ to keep the truncation error below 0.000005. However, to evaluate the partial sum we had to approximate π, thereby introducing roundoff error. Had we not exercised some care in choosing this approximation, the roundoff error could easily have degraded the final result.

Methods for estimating and controlling roundoff error are studied in a branch of mathematics called ***numerical analysis***. However, as a rule of thumb, to achieve n decimal-place accuracy in a final result, all intermediate calculations must be accurate to at least $n + 1$ decimal places. Thus, in (9) at least six decimal-place accuracy in π is required to achieve the five decimal-place accuracy in the final numerical result. As a practical matter, a good working procedure is to perform all intermediate computations with the maximum number of digits that your calculating utility can handle and then round at the end.

APPROXIMATING EXPONENTIAL FUNCTIONS

▶ **Example 3** Show that the Maclaurin series for e^x converges to e^x for all x; that is,

$$e^x = \sum_{k=0}^{\infty} \frac{x^k}{k!} = 1 + x + \frac{x^2}{2!} + \frac{x^3}{3!} + \cdots + \frac{x^k}{k!} + \cdots \qquad (-\infty < x < +\infty)$$

Solution. Let $f(x) = e^x$, so that

$$f^{(n+1)}(x) = e^x$$

We want to show that $R_n(x) \to 0$ as $n \to +\infty$ for all x in the interval $-\infty < x < +\infty$. However, it will be helpful here to consider the cases $x \le 0$ and $x > 0$ separately. If $x \le 0$, then we will take the interval I in the Remainder Estimation Theorem (Theorem 10.7.4) to be $[x, 0]$, and if $x > 0$, then we will take it to be $[0, x]$. Since $f^{(n+1)}(x) = e^x$ is an increasing function, it follows that if c is in the interval $[x, 0]$, then

$$|f^{(n+1)}(c)| \le |f^{(n+1)}(0)| = e^0 = 1$$

and if c is in the interval $[0, x]$, then

$$|f^{(n+1)}(c)| \le |f^{(n+1)}(x)| = e^x$$

Thus, we can apply Theorem 10.7.4 with $M = 1$ in the case where $x \le 0$ and with $M = e^x$ in the case where $x > 0$. This yields

$$0 \le |R_n(x)| \le \frac{|x|^{n+1}}{(n+1)!} \qquad \text{if } x \le 0$$

$$0 \le |R_n(x)| \le e^x \frac{|x|^{n+1}}{(n+1)!} \qquad \text{if } x > 0$$

Thus, in both cases it follows from (5) and the Squeezing Theorem for Sequences that $|R_n(x)| \to 0$ as $n \to +\infty$, which in turn implies that $R_n(x) \to 0$ as $n \to +\infty$. Since this is true for all x, we have proved that the Maclaurin series for e^x converges to e^x for all x. ◀

Since the Maclaurin series for e^x converges to e^x for all x, we can use partial sums of the Maclaurin series to approximate powers of e to arbitrary precision. Recall that in Example 6 of Section 10.7 we were able to use the Remainder Estimation Theorem to determine that evaluating the ninth Maclaurin polynomial for e^x at $x = 1$ yields an approximation for e with five decimal-place accuracy:

$$e \approx 1 + 1 + \frac{1}{2!} + \frac{1}{3!} + \frac{1}{4!} + \frac{1}{5!} + \frac{1}{6!} + \frac{1}{7!} + \frac{1}{8!} + \frac{1}{9!} \approx 2.71828$$

APPROXIMATING LOGARITHMS

The Maclaurin series

$$\ln(1+x) = x - \frac{x^2}{2} + \frac{x^3}{3} - \frac{x^4}{4} + \cdots \qquad (-1 < x \le 1) \tag{10}$$

is the starting point for the approximation of natural logarithms. Unfortunately, the usefulness of this series is limited because of its slow convergence and the restriction $-1 < x \le 1$. However, if we replace x by $-x$ in this series, we obtain

$$\ln(1-x) = -x - \frac{x^2}{2} - \frac{x^3}{3} - \frac{x^4}{4} - \cdots \qquad (-1 \le x < 1) \tag{11}$$

and on subtracting (11) from (10) we obtain

$$\ln\left(\frac{1+x}{1-x}\right) = 2\left(x + \frac{x^3}{3} + \frac{x^5}{5} + \frac{x^7}{7} + \cdots\right) \qquad (-1 < x < 1) \tag{12}$$

In Example 2 of Section 10.6, we stated without proof that

$$\ln 2 = 1 - \frac{1}{2} + \frac{1}{3} - \frac{1}{4} + \frac{1}{5} - \cdots$$

This result can be obtained by letting $x = 1$ in (10), but as indicated in the text discussion, this series converges too slowly to be of practical use.

Series (12), first obtained by James Gregory in 1668, can be used to compute the natural logarithm of any positive number y by letting

$$y = \frac{1+x}{1-x}$$

or, equivalently,

$$x = \frac{y-1}{y+1} \tag{13}$$

and noting that $-1 < x < 1$. For example, to compute $\ln 2$ we let $y = 2$ in (13), which yields $x = \frac{1}{3}$. Substituting this value in (12) gives

$$\ln 2 = 2\left[\frac{1}{3} + \frac{\left(\frac{1}{3}\right)^3}{3} + \frac{\left(\frac{1}{3}\right)^5}{5} + \frac{\left(\frac{1}{3}\right)^7}{7} + \cdots\right] \tag{14}$$

In Exercise 19 we will ask you to show that five decimal-place accuracy can be achieved using the partial sum with terms up to and including the 13th power of $\frac{1}{3}$. Thus, to five decimal-place accuracy

$$\ln 2 \approx 2\left[\frac{1}{3} + \frac{\left(\frac{1}{3}\right)^3}{3} + \frac{\left(\frac{1}{3}\right)^5}{5} + \frac{\left(\frac{1}{3}\right)^7}{7} + \cdots + \frac{\left(\frac{1}{3}\right)^{13}}{13}\right] \approx 0.69315$$

(verify). As a check, a calculator gives $\ln 2 \approx 0.69314718056$, which agrees with the preceding approximation when rounded to five decimal places.

James Gregory (1638–1675) Scottish mathematician and astronomer. Gregory, the son of a minister, was famous in his time as the inventor of the Gregorian reflecting telescope, so named in his honor. Although he is not generally ranked with the great mathematicians, much of his work relating to calculus was studied by Leibniz and Newton and undoubtedly influenced some of their discoveries. There is a manuscript, discovered posthumously, which shows that Gregory had anticipated Taylor series well before Taylor.

APPROXIMATING π

In the next section we will show that

$$\tan^{-1} x = x - \frac{x^3}{3} + \frac{x^5}{5} - \frac{x^7}{7} + \cdots \qquad (-1 \le x \le 1) \tag{15}$$

Letting $x = 1$, we obtain

$$\frac{\pi}{4} = \tan^{-1} 1 = 1 - \frac{1}{3} + \frac{1}{5} - \frac{1}{7} + \cdots$$

or

$$\pi = 4\left[1 - \frac{1}{3} + \frac{1}{5} - \frac{1}{7} + \cdots\right]$$

This famous series, obtained by Leibniz in 1674, converges too slowly to be of computational value. A more practical procedure for approximating π uses the identity

$$\frac{\pi}{4} = \tan^{-1}\frac{1}{2} + \tan^{-1}\frac{1}{3} \tag{16}$$

which was derived in Exercise 89 of Section 7.7. By using this identity and series (15) to approximate $\tan^{-1}\frac{1}{2}$ and $\tan^{-1}\frac{1}{3}$, the value of π can be approximated efficiently to any degree of accuracy.

BINOMIAL SERIES

If m is a real number, then the Maclaurin series for $(1+x)^m$ is called the ***binomial series***; it is given by

$$1 + mx + \frac{m(m-1)}{2!}x^2 + \frac{m(m-1)(m-2)}{3!}x^3 + \cdots + \frac{m(m-1)\cdots(m-k+1)}{k!}x^k + \cdots$$

In the case where m is a nonnegative integer, the function $f(x) = (1+x)^m$ is a polynomial of degree m, so

$$f^{(m+1)}(0) = f^{(m+2)}(0) = f^{(m+3)}(0) = \cdots = 0$$

and the binomial series reduces to the familiar binomial expansion

$$(1+x)^m = 1 + mx + \frac{m(m-1)}{2!}x^2 + \frac{m(m-1)(m-2)}{3!}x^3 + \cdots + x^m$$

which is valid for $-\infty < x < +\infty$.

Let $f(x) = (1+x)^m$. Verify that

$$f(0) = 1$$
$$f'(0) = m$$
$$f''(0) = m(m-1)$$
$$f'''(0) = m(m-1)(m-2)$$
$$\vdots$$
$$f^{(k)}(0) = m(m-1)\cdots(m-k+1)$$

It can be proved that if m is not a nonnegative integer, then the binomial series converges to $(1+x)^m$ if $|x| < 1$. Thus, for such values of x

$$(1+x)^m = 1 + mx + \frac{m(m-1)}{2!}x^2 + \cdots + \frac{m(m-1)\cdots(m-k+1)}{k!}x^k + \cdots \tag{17}$$

or in sigma notation,

$$(1+x)^m = 1 + \sum_{k=1}^{\infty} \frac{m(m-1)\cdots(m-k+1)}{k!}x^k \quad \text{if } |x| < 1 \tag{18}$$

▶ **Example 4** Find binomial series for

$$\text{(a) } \frac{1}{(1+x)^2} \qquad \text{(b) } \frac{1}{\sqrt{1+x}}$$

Solution (a). Since the general term of the binomial series is complicated, you may find it helpful to write out some of the beginning terms of the series, as in Formula (17), to see developing patterns. Substituting $m = -2$ in this formula yields

$$\begin{aligned}
\frac{1}{(1+x)^2} = (1+x)^{-2} &= 1 + (-2)x + \frac{(-2)(-3)}{2!}x^2 \\
&\quad + \frac{(-2)(-3)(-4)}{3!}x^3 + \frac{(-2)(-3)(-4)(-5)}{4!}x^4 + \cdots \\
&= 1 - 2x + \frac{3!}{2!}x^2 - \frac{4!}{3!}x^3 + \frac{5!}{4!}x^4 - \cdots \\
&= 1 - 2x + 3x^2 - 4x^3 + 5x^4 - \cdots \\
&= \sum_{k=0}^{\infty}(-1)^k(k+1)x^k
\end{aligned}$$

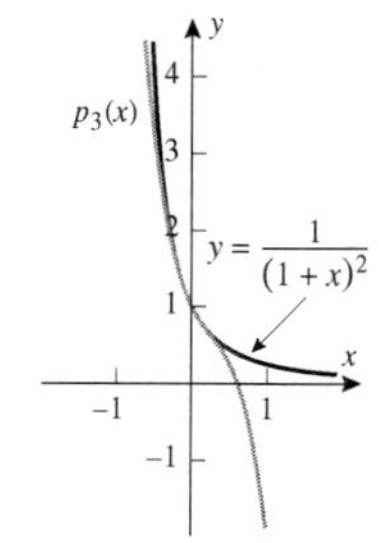

$p_3(x) = 1 - 2x + 3x^2 - 4x^3$

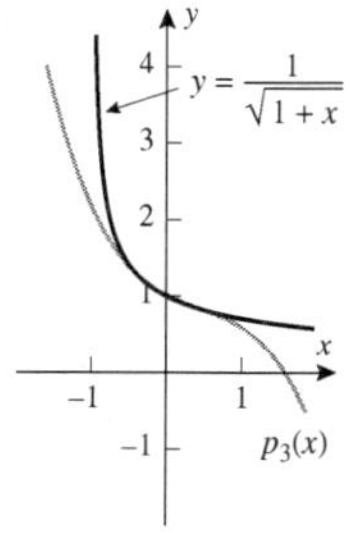

$p_3(x) = 1 - \frac{1}{2}x + \frac{3}{8}x^2 - \frac{5}{16}x^3$

Figure 10.9.2

Solution (b). Substituting $m = -\frac{1}{2}$ in (17) yields

$$\frac{1}{\sqrt{1+x}} = 1 - \frac{1}{2}x + \frac{\left(-\frac{1}{2}\right)\left(-\frac{1}{2}-1\right)}{2!}x^2 + \frac{\left(-\frac{1}{2}\right)\left(-\frac{1}{2}-1\right)\left(-\frac{1}{2}-2\right)}{3!}x^3 + \cdots$$

$$= 1 - \frac{1}{2}x + \frac{1\cdot 3}{2^2\cdot 2!}x^2 - \frac{1\cdot 3\cdot 5}{2^3\cdot 3!}x^3 + \cdots$$

$$= 1 + \sum_{k=1}^{\infty}(-1)^k\frac{1\cdot 3\cdot 5\cdots(2k-1)}{2^k k!}x^k \blacktriangleleft$$

Figure 10.9.2 shows the graphs of the functions in Example 4 compared to their third-degree Maclaurin polynomials.

SOME IMPORTANT MACLAURIN SERIES

For reference, Table 10.9.1 lists the Maclaurin series for some of the most important functions, together with a specification of the intervals over which the Maclaurin series converge to those functions. Some of these results are derived in the exercises and others will be derived in the next section using some special techniques that we will develop.

Table 10.9.1

MACLAURIN SERIES	INTERVAL OF CONVERGENCE
$\frac{1}{1-x} = \sum_{k=0}^{\infty} x^k = 1 + x + x^2 + x^3 + \cdots$	$-1 < x < 1$
$\frac{1}{1+x^2} = \sum_{k=0}^{\infty} (-1)^k x^{2k} = 1 - x^2 + x^4 - x^6 + \cdots$	$-1 < x < 1$
$e^x = \sum_{k=0}^{\infty} \frac{x^k}{k!} = 1 + x + \frac{x^2}{2!} + \frac{x^3}{3!} + \frac{x^4}{4!} + \cdots$	$-\infty < x < +\infty$
$\sin x = \sum_{k=0}^{\infty} (-1)^k \frac{x^{2k+1}}{(2k+1)!} = x - \frac{x^3}{3!} + \frac{x^5}{5!} - \frac{x^7}{7!} + \cdots$	$-\infty < x < +\infty$
$\cos x = \sum_{k=0}^{\infty} (-1)^k \frac{x^{2k}}{(2k)!} = 1 - \frac{x^2}{2!} + \frac{x^4}{4!} - \frac{x^6}{6!} + \cdots$	$-\infty < x < +\infty$
$\ln(1+x) = \sum_{k=1}^{\infty} (-1)^{k+1} \frac{x^k}{k} = x - \frac{x^2}{2} + \frac{x^3}{3} - \frac{x^4}{4} + \cdots$	$-1 < x \le 1$
$\tan^{-1} x = \sum_{k=0}^{\infty} (-1)^k \frac{x^{2k+1}}{2k+1} = x - \frac{x^3}{3} + \frac{x^5}{5} - \frac{x^7}{7} + \cdots$	$-1 \le x \le 1$
$\sinh x = \sum_{k=0}^{\infty} \frac{x^{2k+1}}{(2k+1)!} = x + \frac{x^3}{3!} + \frac{x^5}{5!} + \frac{x^7}{7!} + \cdots$	$-\infty < x < +\infty$
$\cosh x = \sum_{k=0}^{\infty} \frac{x^{2k}}{(2k)!} = 1 + \frac{x^2}{2!} + \frac{x^4}{4!} + \frac{x^6}{6!} + \cdots$	$-\infty < x < +\infty$
$(1+x)^m = 1 + \sum_{k=1}^{\infty} \frac{m(m-1)\cdots(m-k+1)}{k!}x^k$	$-1 < x < 1$* ($m \ne 0, 1, 2, \ldots$)

*The behavior at the endpoints depends on m: For $m > 0$ the series converges absolutely at both endpoints; for $m \le -1$ the series diverges at both endpoints; and for $-1 < m < 0$ the series converges conditionally at $x = 1$ and diverges at $x = -1$.

✔ QUICK CHECK EXERCISES 10.9

(See page 704 for answers.)

1. $\cos x = \sum_{k=0}^{\infty}$ ______

2. $e^x = \sum_{k=0}^{\infty}$ ______

3. $\ln(1+x) = \sum_{k=1}^{\infty}$ ______ for x in the interval ______.

4. If m is a real number but not a nonnegative integer, the *binomial series*
$$1 + \sum_{k=1}^{\infty} \underline{\qquad}$$
converges to $(1+x)^m$ if $|x| <$ ______.

EXERCISE SET 10.9

Graphing Utility | C CAS

1. Use both of the methods given in Example 2 to approximate $\sin 4°$ to five decimal-place accuracy, and check your work by comparing your answer to that produced directly by your calculating utility.

2. Use both of the methods given in Example 2 to approximate $\cos 3°$ to three decimal-place accuracy, and check your work by comparing your answer to that produced directly by your calculating utility.

3. Use the Maclaurin series for $\cos x$ to approximate $\cos 0.1$ to five decimal-place accuracy, and check your work by comparing your answer to that produced directly by your calculating utility.

4. Use the Maclaurin series for $\tan^{-1} x$ to approximate $\tan^{-1} 0.1$ to three decimal-place accuracy, and check your work by comparing your answer to that produced directly by your calculating utility.

5. Use an appropriate Taylor series to approximate $\sin 85°$ to four decimal-place accuracy, and check your work by comparing your answer to that produced directly by your calculating utility.

6. Use a Taylor series to approximate $\cos(-175°)$ to four decimal-place accuracy, and check your work by comparing your answer to that produced directly by your calculating utility.

7. Use the Maclaurin series for $\sinh x$ to approximate $\sinh 0.5$ to three decimal-place accuracy. Check your work by computing $\sinh 0.5$ with a calculating utility.

8. Use the Maclaurin series for $\cosh x$ to approximate $\cosh 0.1$ to three decimal-place accuracy. Check your work by computing $\cosh 0.1$ with a calculating utility.

9. Use the Remainder Estimation Theorem and the method of Example 1 to prove that the Taylor series for $\sin x$ about $x = \pi/4$ converges to $\sin x$ for all x.

10. Use the Remainder Estimation Theorem and the method of Example 3 to prove that the Taylor series for e^x about $x = 1$ converges to e^x for all x.

11. (a) Use Formula (12) in the text to find a series that converges to $\ln 1.25$.
    (b) Approximate $\ln 1.25$ using the first two terms of the series. Round your answer to three decimal places, and compare the result to that produced directly by your calculating utility.

12. (a) Use Formula (12) to find a series that converges to $\ln 3$.
    (b) Approximate $\ln 3$ using the first two terms of the series. Round your answer to three decimal places, and compare the result to that produced directly by your calculating utility.

FOCUS ON CONCEPTS

13. (a) Use the Maclaurin series for $\tan^{-1} x$ to approximate $\tan^{-1} \frac{1}{2}$ and $\tan^{-1} \frac{1}{3}$ to three decimal-place accuracy.
    (b) Use the results in part (a) and Formula (16) to approximate π.
    (c) Would you be willing to guarantee that your answer in part (b) is accurate to three decimal places? Explain your reasoning.
    (d) Compare your answer in part (b) to that produced by your calculating utility.

14. The purpose of this exercise is to show that the Taylor series of a function f may possibly converge to a value different from $f(x)$ for certain values of x. Let
$$f(x) = \begin{cases} e^{-1/x^2}, & x \neq 0 \\ 0, & x = 0 \end{cases}$$
    (a) Use the definition of a derivative to show that $f'(0) = 0$.
    (b) With some difficulty it can be shown that if $n \geq 2$ then $f^{(n)}(0) = 0$. Accepting this fact, show that the Maclaurin series of f converges for all x, but converges to $f(x)$ only at $x = 0$.

15. (a) Find an upper bound on the error that can result if $\cos x$ is approximated by $1 - (x^2/2!) + (x^4/4!)$ over the interval $[-0.2, 0.2]$.

(b) Check your answer in part (a) by graphing

$$\left|\cos x - \left(1 - \frac{x^2}{2!} + \frac{x^4}{4!}\right)\right|$$

over the interval.

16. (a) Find an upper bound on the error that can result if $\ln(1+x)$ is approximated by x over the interval $[-0.01, 0.01]$.

(b) Check your answer in part (a) by graphing

$$|\ln(1+x) - x|$$

over the interval.

17. Use Formula (17) for the binomial series to obtain the Maclaurin series for

(a) $\dfrac{1}{1+x}$ (b) $\sqrt[3]{1+x}$ (c) $\dfrac{1}{(1+x)^3}$.

18. If m is any real number, and k is a nonnegative integer, then we define the ***binomial coefficient***

$$\binom{m}{k} \text{ by the formulas } \binom{m}{0} = 1 \text{ and}$$

$$\binom{m}{k} = \frac{m(m-1)(m-2)\cdots(m-k+1)}{k!}$$

for $k \geq 1$. Express Formula (17) in the text in terms of binomial coefficients.

19. In this exercise we will use the Remainder Estimation Theorem to determine the number of terms that are required in Formula (14) to approximate ln 2 to five decimal-place accuracy. For this purpose let

$$f(x) = \ln\frac{1+x}{1-x} = \ln(1+x) - \ln(1-x) \quad (-1 < x < 1)$$

(a) Show that

$$f^{(n+1)}(x) = n!\left[\frac{(-1)^n}{(1+x)^{n+1}} + \frac{1}{(1-x)^{n+1}}\right]$$

(b) Use the triangle inequality [Theorem 1.1.4(*d*)] to show that

$$|f^{(n+1)}(x)| \leq n!\left[\frac{1}{(1+x)^{n+1}} + \frac{1}{(1-x)^{n+1}}\right]$$

(c) Since we want to achieve five decimal-place accuracy, our goal is to choose n so that the absolute value of the nth remainder at $x = \frac{1}{3}$ does not exceed the value $0.000005 = 0.5 \times 10^{-5}$; that is, $\left|R_n\left(\frac{1}{3}\right)\right| \leq 0.000005$. Use the Remainder Estimation Theorem to show that this condition will be satisfied if n is chosen so that

$$\frac{M}{(n+1)!}\left(\frac{1}{3}\right)^{n+1} \leq 0.000005$$

where $|f^{(n+1)}(x)| \leq M$ on the interval $\left[0, \frac{1}{3}\right]$.

(d) Use the result in part (b) to show that M can be taken as

$$M = n!\left[1 + \frac{1}{\left(\frac{2}{3}\right)^{n+1}}\right]$$

(e) Use the results in parts (c) and (d) to show that five decimal-place accuracy will be achieved if n satisfies

$$\frac{1}{n+1}\left[\left(\frac{1}{3}\right)^{n+1} + \left(\frac{1}{2}\right)^{n+1}\right] \leq 0.000005$$

and then show that the smallest value of n that satisfies this condition is $n = 13$.

20. Use Formula (12) and the method of Exercise 19 to approximate $\ln\left(\frac{5}{3}\right)$ to five decimal-place accuracy. Then check your work by comparing your answer to that produced directly by your calculating utility.

21. Prove: The Taylor series for $\cos x$ about any value $x = x_0$ converges to $\cos x$ for all x.

22. Prove: The Taylor series for $\sin x$ about any value $x = x_0$ converges to $\sin x$ for all x.

23. Research has shown that the proportion p of the population with IQs (intelligence quotients) between α and β is approximately

$$p = \frac{1}{16\sqrt{2\pi}}\int_\alpha^\beta e^{-\frac{1}{2}\left(\frac{x-100}{16}\right)^2}\,dx$$

Use the first three terms of an appropriate Maclaurin series to estimate the proportion of the population that has IQs between 100 and 110.

24. In Section 6.7 we defined the kinetic energy K of a particle with mass m and velocity v to be $K = \frac{1}{2}mv^2$ [see Formula (6) of that section]. In this formula the mass m is assumed to be constant, and K is called the ***Newtonian kinetic energy***. However, in Albert Einstein's relativity theory the mass m increases with the velocity and the kinetic energy K is given by the formula

$$K = m_0c^2\left[\frac{1}{\sqrt{1-(v/c)^2}} - 1\right]$$

in which m_0 is the mass of the particle when its velocity is zero, and c is the speed of light. This is called the ***relativistic kinetic energy***. Use an appropriate binomial series to show that if the velocity is small compared to the speed of light (i.e., $v/c \approx 0$), then the Newtonian and relativistic kinetic energies are in close agreement.

C **25.** (a) In 1706 the British astronomer and mathematician John Machin discovered the following formula for $\pi/4$, called ***Machin's formula***:

$$\frac{\pi}{4} = 4\tan^{-1}\frac{1}{5} - \tan^{-1}\frac{1}{239}$$

Use a CAS to approximate $\pi/4$ using Machin's formula to 25 decimal places.

(b) In 1914 the brilliant Indian mathematician Srinivasa Ramanujan (1887–1920) showed that

$$\frac{1}{\pi} = \frac{\sqrt{8}}{9801}\sum_{k=0}^{\infty}\frac{(4k)!(1103 + 26{,}390k)}{(k!)^4 396^{4k}}$$

Use a CAS to compute the first four partial sums in ***Ramanujan's formula***.

✔ QUICK CHECK ANSWERS 10.9

1. $(-1)^k \dfrac{x^{2k}}{(2k)!}$ **2.** $\dfrac{x^k}{k!}$ **3.** $(-1)^{k+1}\dfrac{x^k}{k}$; $(-1, 1]$ **4.** $\dfrac{m(m-1)\cdots(m-k+1)}{k!}x^k$; 1

10.10 DIFFERENTIATING AND INTEGRATING POWER SERIES; MODELING WITH TAYLOR SERIES

In this section we will discuss methods for finding power series for derivatives and integrals of functions, and we will discuss some practical methods for finding Taylor series that can be used in situations where it is difficult or impossible to find the series directly.

DIFFERENTIATING POWER SERIES

We begin by considering the following problem.

10.10.1 PROBLEM. Suppose that a function f is represented by a power series on an open interval. How can we use the power series to find the derivative of f on that interval?

The solution to this problem can be motivated by considering the Maclaurin series for $\sin x$:

$$\sin x = x - \frac{x^3}{3!} + \frac{x^5}{5!} - \frac{x^7}{7!} + \cdots \qquad (-\infty < x < +\infty)$$

Of course, we already know that the derivative of $\sin x$ is $\cos x$; however, we are concerned here with using the Maclaurin series to deduce this. The solution is easy—all we need to do is differentiate the Maclaurin series term by term and observe that the resulting series is the Maclaurin series for $\cos x$:

$$\frac{d}{dx}\left[x - \frac{x^3}{3!} + \frac{x^5}{5!} - \frac{x^7}{7!} + \cdots\right] = 1 - 3\frac{x^2}{3!} + 5\frac{x^4}{5!} - 7\frac{x^6}{7!} + \cdots$$
$$= 1 - \frac{x^2}{2!} + \frac{x^4}{4!} - \frac{x^6}{6!} + \cdots = \cos x$$

Here is another example.

$$\frac{d}{dx}[e^x] = \frac{d}{dx}\left[1 + x + \frac{x^2}{2!} + \frac{x^3}{3!} + \frac{x^4}{4!} + \cdots\right]$$
$$= 1 + 2\frac{x}{2!} + 3\frac{x^2}{3!} + 4\frac{x^3}{4!} + \cdots = 1 + x + \frac{x^2}{2!} + \frac{x^3}{3!} + \cdots = e^x$$

The preceding computations suggest that if a function f is represented by a power series on an open interval, then a power series representation of f' on that interval can be obtained by differentiating the power series for f term by term. This is stated more precisely in the following theorem, which we give without proof.

10.10.2 THEOREM (*Differentiation of Power Series*). *Suppose that a function f is represented by a power series in $x - x_0$ that has a nonzero radius of convergence R; that is,*

$$f(x) = \sum_{k=0}^{\infty} c_k(x - x_0)^k \qquad (x_0 - R < x < x_0 + R)$$

Then:

(*a*) *The function f is differentiable on the interval $(x_0 - R, x_0 + R)$.*

(*b*) *If the power series representation for f is differentiated term by term, then the resulting series has radius of convergence R and converges to f' on the interval $(x_0 - R, x_0 + R)$; that is,*

$$f'(x) = \sum_{k=0}^{\infty} \frac{d}{dx}[c_k(x - x_0)^k] \qquad (x_0 - R < x < x_0 + R)$$

This theorem has an important implication about the differentiability of functions that are represented by power series. According to the theorem, the power series for f' has the same radius of convergence as the power series for f, and this means that the theorem can be applied to f' as well as f. However, if we do this, then we conclude that f' is differentiable on the interval $(x_0 - R, x_0 + R)$, and the power series for f'' has the same radius of convergence as the power series for f and f'. We can now repeat this process ad infinitum, applying the theorem successively to $f'', f''', \ldots, f^{(n)}, \ldots$ to conclude that f has derivatives of all orders on the interval $(x_0 - R, x_0 + R)$. Thus, we have established the following result.

10.10.3 THEOREM. *If a function f can be represented by a power series in $x - x_0$ with a nonzero radius of convergence R, then f has derivatives of all orders on the interval $(x_0 - R, x_0 + R)$.*

In short, it is only the most "well-behaved" functions that can be represented by power series; that is, if a function f does not possess derivatives of all orders on an interval $(x_0 - R, x_0 + R)$, then it cannot be represented by a power series in $x - x_0$ on that interval.

▶ **Example 1** In Section 10.8, we showed that the Bessel function $J_0(x)$, represented by the power series

$$J_0(x) = \sum_{k=0}^{\infty} \frac{(-1)^k x^{2k}}{2^{2k}(k!)^2} \qquad (1)$$

has radius of convergence $+\infty$ [see Formula (4) of that section and the related discussion]. Thus, $J_0(x)$ has derivatives of all orders on the interval $(-\infty, +\infty)$, and these can be obtained by differentiating the series term by term. For example, if we write (1) as

$$J_0(x) = 1 + \sum_{k=1}^{\infty} \frac{(-1)^k x^{2k}}{2^{2k}(k!)^2}$$

and differentiate term by term, we obtain

See Exercise 44 for a relationship between $J_0'(x)$ and $J_1(x)$.

$$J_0'(x) = \sum_{k=1}^{\infty} \frac{(-1)^k (2k) x^{2k-1}}{2^{2k}(k!)^2} = \sum_{k=1}^{\infty} \frac{(-1)^k x^{2k-1}}{2^{2k-1} k!(k-1)!} \quad ◀$$

The computations in this example use some techniques that are worth noting. First, when a power series is expressed in sigma notation, the formula for the general term of the series will often not be of a form that can be used for differentiating the constant term. Thus, if the series has a nonzero constant term, as here, it is usually a good idea to split it off from the summation before differentiating. Second, observe how we simplified the final formula by canceling the factor k from one of the factorials in the denominator. This is a standard simplification technique.

INTEGRATING POWER SERIES

Since the derivative of a function that is represented by a power series can be obtained by differentiating the series term by term, it should not be surprising that an antiderivative of a function represented by a power series can be obtained by integrating the series term by term. For example, we know that $\sin x$ is an antiderivative of $\cos x$. Here is how this result can be obtained by integrating the Maclaurin series for $\cos x$ term by term:

$$\begin{aligned}\int \cos x\,dx &= \int \left[1 - \frac{x^2}{2!} + \frac{x^4}{4!} - \frac{x^6}{6!} + \cdots\right] dx \\ &= \left[x - \frac{x^3}{3(2!)} + \frac{x^5}{5(4!)} - \frac{x^7}{7(6!)} + \cdots\right] + C \\ &= \left[x - \frac{x^3}{3!} + \frac{x^5}{5!} - \frac{x^7}{7!} + \cdots\right] + C = \sin x + C\end{aligned}$$

The same idea applies to definite integrals. For example, by direct integration we have

$$\int_0^1 \frac{dx}{1+x^2} = \tan^{-1} x\bigg]_0^1 = \tan^{-1} 1 - \tan 0 = \frac{\pi}{4} - 0 = \frac{\pi}{4}$$

and we will show later in this section that

$$\frac{\pi}{4} = 1 - \frac{1}{3} + \frac{1}{5} - \frac{1}{7} + \cdots \tag{2}$$

Thus,

$$\int_0^1 \frac{dx}{1+x^2} = 1 - \frac{1}{3} + \frac{1}{5} - \frac{1}{7} + \cdots$$

Here is how this result can be obtained by integrating the Maclaurin series for $1/(1+x^2)$ term by term (see Table 10.9.1):

$$\begin{aligned}\int_0^1 \frac{dx}{1+x^2} &= \int_0^1 [1 - x^2 + x^4 - x^6 + \cdots]\,dx \\ &= x - \frac{x^3}{3} + \frac{x^5}{5} - \frac{x^7}{7} + \cdots\bigg]_0^1 = 1 - \frac{1}{3} + \frac{1}{5} - \frac{1}{7} + \cdots\end{aligned}$$

The preceding computations are justified by the following theorem, which we give without proof.

10.10.4 THEOREM (*Integration of Power Series*). *Suppose that a function f is represented by a power series in $x - x_0$ that has a nonzero radius of convergence R; that is,*

$$f(x) = \sum_{k=0}^{\infty} c_k(x - x_0)^k \qquad (x_0 - R < x < x_0 + R)$$

(*a*) *If the power series representation of f is integrated term by term, then the resulting series has radius of convergence R and converges to an antiderivative for $f(x)$ on the interval $(x_0 - R, x_0 + R)$; that is,*

$$\int f(x)\,dx = \sum_{k=0}^{\infty}\left[\frac{c_k}{k+1}(x - x_0)^{k+1}\right] + C \qquad (x_0 - R < x < x_0 + R)$$

(*b*) *If α and β are points in the interval $(x_0 - R, x_0 + R)$, and if the power series representation of f is integrated term by term from α to β, then the resulting series converges absolutely on the interval $(x_0 - R, x_0 + R)$ and*

$$\int_\alpha^\beta f(x)\,dx = \sum_{k=0}^{\infty}\left[\int_\alpha^\beta c_k(x - x_0)^k\,dx\right]$$

POWER SERIES REPRESENTATIONS MUST BE TAYLOR SERIES

For many functions it is difficult or impossible to find the derivatives that are required to obtain a Taylor series. For example, to find the Maclaurin series for $1/(1+x^2)$ directly would require some tedious derivative computations (try it). A more practical approach is to substitute $-x^2$ for x in the geometric series

$$\frac{1}{1-x} = 1 + x + x^2 + x^3 + x^4 + \cdots \qquad (-1 < x < 1)$$

to obtain

$$\frac{1}{1+x^2} = 1 - x^2 + x^4 - x^6 + x^8 - \cdots$$

However, there are two questions of concern with this procedure:

- Where does the power series that we obtained for $1/(1+x^2)$ actually converge to $1/(1+x^2)$?
- How do we know that the power series we have obtained is actually the Maclaurin series for $1/(1+x^2)$?

The first question is easy to resolve. Since the geometric series converges to $1/(1-x)$ if $|x| < 1$, the second series will converge to $1/(1+x^2)$ if $|-x^2| < 1$ or $|x^2| < 1$. However, this is true if and only if $|x| < 1$, so the power series we obtained for the function $1/(1+x^2)$ converges to this function if $-1 < x < 1$.

The second question is more difficult to answer and leads us to the following general problem.

10.10.5 PROBLEM. Suppose that a function f is represented by a power series in $x - x_0$ that has a nonzero radius of convergence. What relationship exists between the given power series and the Taylor series for f about $x = x_0$?

The answer is that they are the same; and here is the theorem that proves it.

Theorem 10.10.6 tells us that no matter how we arrive at a power series representation of a function f, be it by substitution, by differentiation, by integration, or by some algebraic process, that series will be the Taylor series for f about $x = x_0$, provided the series converges to f on some open interval containing x_0.

10.10.6 THEOREM. *If a function f is represented by a power series in $x - x_0$ on some open interval containing x_0, then that power series is the Taylor series for f about $x = x_0$.*

PROOF. Suppose that

$$f(x) = c_0 + c_1(x - x_0) + c_2(x - x_0)^2 + \cdots + c_k(x - x_0)^k + \cdots$$

for all x in some open interval containing x_0. To prove that this is the Taylor series for f about $x = x_0$, we must show that

$$c_k = \frac{f^{(k)}(x_0)}{k!} \quad \text{for} \quad k = 0, 1, 2, 3, \ldots$$

However, the assumption that the series converges to $f(x)$ on an open interval containing x_0 ensures that it has a nonzero radius of convergence R; hence we can differentiate term by term in accordance with Theorem 10.10.2. Thus,

$$
\begin{aligned}
f(x) &= c_0 + c_1(x - x_0) + c_2(x - x_0)^2 + c_3(x - x_0)^3 + c_4(x - x_0)^4 + \cdots \\
f'(x) &= c_1 + 2c_2(x - x_0) + 3c_3(x - x_0)^2 + 4c_4(x - x_0)^3 + \cdots \\
f''(x) &= 2!c_2 + (3 \cdot 2)c_3(x - x_0) + (4 \cdot 3)c_4(x - x_0)^2 + \cdots \\
f'''(x) &= 3!c_3 + (4 \cdot 3 \cdot 2)c_4(x - x_0) + \cdots \\
&\vdots
\end{aligned}
$$

On substituting $x = x_0$, all the powers of $x - x_0$ drop out, leaving

$$f(x_0) = c_0, \quad f'(x_0) = c_1, \quad f''(x_0) = 2!c_2, \quad f'''(x_0) = 3!c_3, \ldots$$

from which we obtain

$$c_0 = f(x_0), \quad c_1 = f'(x_0), \quad c_2 = \frac{f''(x_0)}{2!}, \quad c_3 = \frac{f'''(x_0)}{3!}, \ldots$$

which shows that the coefficients $c_0, c_1, c_2, c_3, \ldots$ are precisely the coefficients in the Taylor series about x_0 for $f(x)$. ■

SOME PRACTICAL WAYS TO FIND TAYLOR SERIES

▶ **Example 2** Find the Maclaurin series for $\tan^{-1} x$.

Solution. It would be tedious to find the Maclaurin series directly. A better approach is to start with the formula

$$\int \frac{1}{1+x^2}\,dx = \tan^{-1} x + C$$

and integrate the Maclaurin series

$$\frac{1}{1+x^2} = 1 - x^2 + x^4 - x^6 + x^8 - \cdots \qquad (-1 < x < 1)$$

term by term. This yields

$$\tan^{-1} x + C = \int \frac{1}{1+x^2}\,dx = \int [1 - x^2 + x^4 - x^6 + x^8 - \cdots]\,dx$$

or

$$\tan^{-1} x = \left[x - \frac{x^3}{3} + \frac{x^5}{5} - \frac{x^7}{7} + \frac{x^9}{9} - \cdots\right] - C$$

The constant of integration can be evaluated by substituting $x = 0$ and using the condition $\tan^{-1} 0 = 0$. This gives $C = 0$, so that

$$\tan^{-1} x = x - \frac{x^3}{3} + \frac{x^5}{5} - \frac{x^7}{7} + \frac{x^9}{9} - \cdots \qquad (-1 < x < 1) \tag{3}$$

◀

Observe that neither Theorem 10.10.2 nor Theorem 10.10.3 addresses what happens at the endpoints of the interval of convergence. However, it can be proved that if the Taylor series for f about $x = x_0$ converges to $f(x)$ for all x in the interval $(x_0 - R, x_0 + R)$, and if the Taylor series converges at the right endpoint $x_0 + R$, then the value that it converges to at that point is the limit of $f(x)$ as $x \to x_0 + R$ from the left; and if the Taylor series converges at the left endpoint $x_0 - R$, then the value that it converges to at that point is the limit of $f(x)$ as $x \to x_0 - R$ from the right.

For example, the Maclaurin series for $\tan^{-1} x$ given in (3) converges at both $x = -1$ and $x = 1$, since the hypotheses of the alternating series test (Theorem 10.6.1) are satisfied at those points. Thus, the continuity of $\tan^{-1} x$ on the interval $[-1, 1]$ implies that at $x = 1$ the Maclaurin series converges to

$$\lim_{x\to 1^-} \tan^{-1} x = \tan^{-1} 1 = \frac{\pi}{4}$$

and at $x = -1$ it converges to

$$\lim_{x\to -1^+} \tan^{-1} x = \tan^{-1}(-1) = -\frac{\pi}{4}$$

This shows that the Maclaurin series for $\tan^{-1} x$ actually converges to $\tan^{-1} x$ on the closed interval $-1 \le x \le 1$. Moreover, the convergence at $x = 1$ establishes Formula (2).

Taylor series provide an alternative to Simpson's rule and other numerical methods for approximating definite integrals.

▶ **Example 3** Approximate the integral

$$\int_0^1 e^{-x^2}\,dx$$

to three decimal-place accuracy by expanding the integrand in a Maclaurin series and integrating term by term.

Solution. The simplest way to obtain the Maclaurin series for e^{-x^2} is to replace x by $-x^2$ in the Maclaurin series

$$e^x = 1 + x + \frac{x^2}{2!} + \frac{x^3}{3!} + \frac{x^4}{4!} + \cdots$$

to obtain

$$e^{-x^2} = 1 - x^2 + \frac{x^4}{2!} - \frac{x^6}{3!} + \frac{x^8}{4!} - \cdots$$

Therefore,

$$\begin{aligned}\int_0^1 e^{-x^2}\,dx &= \int_0^1 \left[1 - x^2 + \frac{x^4}{2!} - \frac{x^6}{3!} + \frac{x^8}{4!} - \cdots\right] dx\\ &= \left[x - \frac{x^3}{3} + \frac{x^5}{5(2!)} - \frac{x^7}{7(3!)} + \frac{x^9}{9(4!)} - \cdots\right]_0^1\\ &= 1 - \frac{1}{3} + \frac{1}{5\cdot 2!} - \frac{1}{7\cdot 3!} + \frac{1}{9\cdot 4!} - \cdots\\ &= \sum_{k=0}^{\infty} \frac{(-1)^k}{(2k+1)k!}\end{aligned}$$

Since this series clearly satisfies the hypotheses of the alternating series test (Theorem 10.6.1), it follows from Theorem 10.6.2 that if we approximate the integral by s_n (the nth partial sum of the series), then

$$\left|\int_0^1 e^{-x^2}\,dx - s_n\right| < \frac{1}{[2(n+1)+1](n+1)!} = \frac{1}{(2n+3)(n+1)!}$$

Thus, for three decimal-place accuracy we must choose n such that

$$\frac{1}{(2n+3)(n+1)!} \le 0.0005 = 5 \times 10^{-4}$$

With the help of a calculating utility you can show that the smallest value of n that satisfies this condition is $n = 5$. Thus, the value of the integral to three decimal-place accuracy is

$$\int_0^1 e^{-x^2}\,dx \approx 1 - \frac{1}{3} + \frac{1}{5\cdot 2!} - \frac{1}{7\cdot 3!} + \frac{1}{9\cdot 4!} - \frac{1}{11\cdot 5!} \approx 0.747$$

As a check, a calculator with a built-in numerical integration capability produced the approximation 0.746824, which agrees with our result when rounded to three decimal places. ◀

What advantages does the method of Example 3 have over Simpson's rule? What are its disadvantages?

FINDING MACLAURIN SERIES BY MULTIPLICATION AND DIVISION

The following examples illustrate some algebraic techniques that are sometimes useful for finding Taylor series.

$$\begin{array}{r} 1 - x^2 + \frac{x^4}{2} - \cdots \\ \times \quad x - \frac{x^3}{3} + \frac{x^5}{5} - \cdots \\ \hline x - x^3 + \frac{x^5}{2} - \cdots \\ -\frac{x^3}{3} + \frac{x^5}{3} - \frac{x^7}{6} + \cdots \\ \frac{x^5}{5} - \frac{x^7}{5} + \cdots \\ \hline x - \frac{4}{3}x^3 + \frac{31}{30}x^5 - \cdots \end{array}$$

▶ Example 4 Find the first three nonzero terms in the Maclaurin series for the function $f(x) = e^{-x^2}\tan^{-1} x$.

Solution. Using the series for e^{-x^2} and $\tan^{-1} x$ obtained in Examples 2 and 3 gives

$$e^{-x^2}\tan^{-1} x = \left(1 - x^2 + \frac{x^4}{2} - \cdots\right)\left(x - \frac{x^3}{3} + \frac{x^5}{5} - \cdots\right)$$

Multiplying, as shown in the margin, we obtain

$$e^{-x^2}\tan^{-1} x = x - \frac{4}{3}x^3 + \frac{31}{30}x^5 - \cdots$$

More terms in the series can be obtained by including more terms in the factors. Moreover, one can prove that a series obtained by this method converges at each point in the intersection of the intervals of convergence of the factors (and possibly on a larger interval). Thus, we can be certain that the series we have obtained converges for all x in the interval $-1 \le x \le 1$ (why?). ◀

$$\begin{array}{r|l} & x + \frac{x^3}{3} + \frac{2x^5}{15} + \cdots \\ \hline 1 - \frac{x^2}{2} + \frac{x^4}{24} - \cdots & x - \frac{x^3}{6} + \frac{x^5}{120} - \cdots \\ & x - \frac{x^3}{2} + \frac{x^5}{24} - \cdots \\ \hline & \frac{x^3}{3} - \frac{x^5}{30} + \cdots \\ & \frac{x^3}{3} - \frac{x^5}{6} + \cdots \\ \hline & \frac{2x^5}{15} + \cdots \end{array}$$

▶ Example 5 Find the first three nonzero terms in the Maclaurin series for $\tan x$.

Solution. Using the first three terms in the Maclaurin series for $\sin x$ and $\cos x$, we can express $\tan x$ as

$$\tan x = \frac{\sin x}{\cos x} = \frac{x - \frac{x^3}{3!} + \frac{x^5}{5!} - \cdots}{1 - \frac{x^2}{2!} + \frac{x^4}{4!} - \cdots}$$

Dividing, as shown in the margin, we obtain

$$\tan x = x + \frac{x^3}{3} + \frac{2x^5}{15} + \cdots$$ ◀

TECHNOLOGY MASTERY

If you have a CAS, use its capability for multiplying and dividing polynomials to perform the computations in Examples 4 and 5.

MODELING PHYSICAL LAWS WITH TAYLOR SERIES

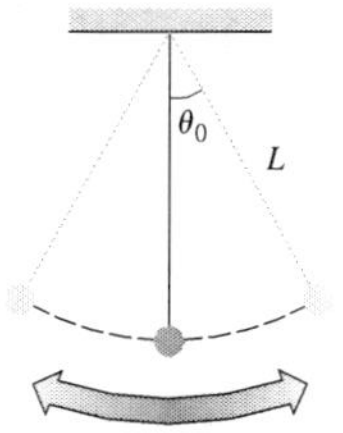

Figure 10.10.1

Taylor series provide an important way of modeling physical laws. To illustrate the idea we will consider the problem of modeling the period of a simple pendulum (Figure 10.10.1). As explained in Exercise 69 of Section 8.8, the period T of such a pendulum is given by

$$T = 4\sqrt{\frac{L}{g}}\int_0^{\pi/2} \frac{1}{\sqrt{1-k^2\sin^2\phi}}\,d\phi \tag{4}$$

where

L = length of the supporting rod
g = acceleration due to gravity
$k = \sin(\theta_0/2)$, where θ_0 is the initial angle of displacement from the vertical

The integral, which is called a ***complete elliptic integral of the first kind***, cannot be expressed in terms of elementary functions and is often approximated by numerical methods. Unfortunately, numerical values are so specific that they often give little insight into general physical principles. However, if we expand the integrand of (4) in a series and integrate term by term, then we can generate an infinite series that can be used to construct various mathematical models for the period T that give a deeper understanding of the behavior of the pendulum.

To obtain a series for the integrand, we will substitute $-k^2\sin^2\phi$ for x in the binomial series for $1/\sqrt{1+x}$ that we derived in Example 4 of Section 10.9. If we do this, then we can rewrite (4) as

$$T = 4\sqrt{\frac{L}{g}}\int_0^{\pi/2}\left[1 + \frac{1}{2}k^2\sin^2\phi + \frac{1\cdot 3}{2^2 2!}k^4\sin^4\phi + \frac{1\cdot 3\cdot 5}{2^3 3!}k^6\sin^6\phi + \cdots\right]d\phi \tag{5}$$

If we integrate term by term, then we can produce a series that converges to the period T. However, one of the most important cases of pendulum motion occurs when the initial displacement is small, in which case all subsequent displacements are small, and we can assume that $k = \sin(\theta_0/2) \approx 0$. In this case we expect the convergence of the series for T to be rapid, and we can approximate the sum of the series by dropping all but the constant term in (5). This yields

$$T = 2\pi\sqrt{\frac{L}{g}} \tag{6}$$

which is called the ***first-order model*** of T or the model for ***small vibrations***. This model can be improved on by using more terms in the series. For example, if we use the first two terms in the series, we obtain the ***second-order model***

$$T = 2\pi\sqrt{\frac{L}{g}}\left(1 + \frac{k^2}{4}\right) \tag{7}$$

(verify).

✔QUICK CHECK EXERCISES 10.10 *(See page 714 for answers.)*

1. The Maclaurin series for e^{-x^2} obtained by substituting $-x^2$ for x in the series

$$e^x = \sum_{k=0}^{\infty}\frac{x^k}{k!}$$

is $e^{-x^2} = \sum_{k=0}^{\infty}$ ______.

2. $\dfrac{d}{dx}\left[\sum_{k=1}^{\infty}(-1)^{k+1}\dfrac{x^k}{k}\right]$ = ______ + ______ x + ______ x^2 + ______ $x^3 + \cdots$ $= \sum_{k=0}^{\infty}$ ______

3. $\left(\sum_{k=0}^{\infty}\frac{x^k}{k!}\right)\left(\sum_{k=0}^{\infty}\frac{x^k}{k+1}\right)$

$= \left(1+x+\frac{x^2}{2!}+\cdots\right)\left(1+\frac{x}{2}+\frac{x^2}{3}+\cdots\right)$

$=$ ______ $+$ ______ $x+$ ______ $x^2+\cdots$

4. Suppose that $f(1)=4$ and $f'(x)=\sum_{k=0}^{\infty}\frac{(-1)^k}{(k+1)!}(x-1)^k$

(a) $f''(1)=$ ______

(b) $f(x)=$ ______ $+$ ______ $(x-1)$

$+$ ______ $(x-1)^2+$ ______ $(x-1)^3+\cdots$

$=$ ______ $+\sum_{k=1}^{\infty}$ ______

EXERCISE SET 10.10 C CAS

1. In each part, obtain the Maclaurin series for the function by making an appropriate substitution in the Maclaurin series for $1/(1-x)$. Include the general term in your answer, and state the radius of convergence of the series.
(a) $\frac{1}{1+x}$ (b) $\frac{1}{1-x^2}$ (c) $\frac{1}{1-2x}$ (d) $\frac{1}{2-x}$

2. In each part, obtain the Maclaurin series for the function by making an appropriate substitution in the Maclaurin series for $\ln(1+x)$. Include the general term in your answer, and state the radius of convergence of the series.
(a) $\ln(1-x)$ (b) $\ln(1+x^2)$
(c) $\ln(1+2x)$ (d) $\ln(2+x)$

3. In each part, obtain the first four nonzero terms of the Maclaurin series for the function by making an appropriate substitution in one of the binomial series obtained in Example 4 of Section 10.9.
(a) $(2+x)^{-1/2}$ (b) $(1-x^2)^{-2}$

4. (a) Use the Maclaurin series for $1/(1-x)$ to find the Maclaurin series for $1/(a-x)$, where $a \neq 0$, and state the radius of convergence of the series.
(b) Use the binomial series for $1/(1+x)^2$ obtained in Example 4 of Section 10.9 to find the first four nonzero terms in the Maclaurin series for $1/(a+x)^2$, where $a \neq 0$, and state the radius of convergence of the series.

5–8 Find the first four nonzero terms of the Maclaurin series for the function by making an appropriate substitution in a known Maclaurin series and performing any algebraic operations that are required. State the radius of convergence of the series.

5. (a) $\sin 2x$ (b) e^{-2x} (c) e^{x^2} (d) $x^2\cos \pi x$

6. (a) $\cos 2x$ (b) x^2e^x (c) xe^{-x} (d) $\sin(x^2)$

7. (a) $\frac{x^2}{1+3x}$ (b) $x\sinh 2x$ (c) $x(1-x^2)^{3/2}$

8. (a) $\frac{x}{x-1}$ (b) $3\cosh(x^2)$ (c) $\frac{x}{(1+2x)^3}$

9–10 Find the first four nonzero terms of the Maclaurin series for the function by using an appropriate trigonometric identity or property of logarithms and then substituting in a known Maclaurin series.

9. (a) $\sin^2 x$ (b) $\ln[(1+x^3)^{12}]$

10. (a) $\cos^2 x$ (b) $\ln\left(\frac{1-x}{1+x}\right)$

11. (a) Use a known Maclaurin series to find the Taylor series of $1/x$ about $x=1$ by expressing this function as

$$\frac{1}{x}=\frac{1}{1-(1-x)}$$

(b) Find the interval of convergence of the Taylor series.

12. Use the method of Exercise 11 to find the Taylor series of $1/x$ about $x=x_0$, and state the interval of convergence of the Taylor series.

13–14 Find the first four nonzero terms of the Maclaurin series for the function by multiplying the Maclaurin series of the factors.

13. (a) $e^x \sin x$ (b) $\sqrt{1+x}\ln(1+x)$

14. (a) $e^{-x^2}\cos x$ (b) $(1+x^2)^{4/3}(1+x)^{1/3}$

15–16 Find the first four nonzero terms of the Maclaurin series for the function by dividing appropriate Maclaurin series.

15. (a) $\sec x \left(=\frac{1}{\cos x}\right)$ (b) $\frac{\sin x}{e^x}$

16. (a) $\frac{\tan^{-1} x}{1+x}$ (b) $\frac{\ln(1+x)}{1-x}$

17. Use the Maclaurin series for e^x and e^{-x} to derive the Maclaurin series for $\sinh x$ and $\cosh x$. Include the general terms in your answers and state the radius of convergence of each series.

18. Use the Maclaurin series for $\sinh x$ and $\cosh x$ to obtain the first four nonzero terms in the Maclaurin series for $\tanh x$.

19–20 Find the first five nonzero terms of the Maclaurin series for the function by using partial fractions and a known Maclaurin series.

19. $\dfrac{4x-2}{x^2-1}$ **20.** $\dfrac{x^3+x^2+2x-2}{x^2-1}$

21–22 Confirm the derivative formula by differentiating the appropriate Maclaurin series term by term.

21. (a) $\dfrac{d}{dx}[\cos x] = -\sin x$ (b) $\dfrac{d}{dx}[\ln(1+x)] = \dfrac{1}{1+x}$

22. (a) $\dfrac{d}{dx}[\sinh x] = \cosh x$ (b) $\dfrac{d}{dx}[\tan^{-1} x] = \dfrac{1}{1+x^2}$

23–24 Confirm the integration formula by integrating the appropriate Maclaurin series term by term.

23. (a) $\int e^x\,dx = e^x + C$

(b) $\int \sinh x\,dx = \cosh x + C$

24. (a) $\int \sin x\,dx = -\cos x + C$

(b) $\int \dfrac{1}{1+x}\,dx = \ln(1+x) + C$

25. (a) Use the Maclaurin series for $1/(1-x)$ to find the Maclaurin series for

$$f(x) = \frac{x}{1-x^2}$$

(b) Use the Maclaurin series obtained in part (a) to find $f^{(5)}(0)$ and $f^{(6)}(0)$.

(c) What can you say about the value of $f^{(n)}(0)$?

26. Let $f(x) = x^2\cos 2x$. Use the method of Exercise 25 to find $f^{(99)}(0)$.

27–28 The limit of an indeterminate form as $x \to x_0$ can sometimes be found by expanding the functions involved in Taylor series about $x = x_0$ and taking the limit of the series term by term. Use this method to find the limits in these exercises.

27. (a) $\lim\limits_{x\to 0} \dfrac{\sin x}{x}$ (b) $\lim\limits_{x\to 0} \dfrac{\tan^{-1} x - x}{x^3}$

28. (a) $\lim\limits_{x\to 0} \dfrac{1-\cos x}{\sin x}$ (b) $\lim\limits_{x\to 0} \dfrac{\ln\sqrt{1+x} - \sin 2x}{x}$

29–32 Use Maclaurin series to approximate the integral to three decimal-place accuracy.

29. $\int_0^1 \sin(x^2)\,dx$ **30.** $\int_0^{1/2} \tan^{-1}(2x^2)\,dx$

31. $\int_0^{0.2} \sqrt[3]{1+x^4}\,dx$ **32.** $\int_0^{1/2} \dfrac{dx}{\sqrt[4]{x^2+1}}$

FOCUS ON CONCEPTS

33. (a) Find the Maclaurin series for e^{x^4}. What is the radius of convergence?

(b) Explain two different ways to use the Maclaurin series for e^{x^4} to find a series for $x^3e^{x^4}$. Confirm that both methods produce the same series.

34. (a) Differentiate the Maclaurin series for $1/(1-x)$, and use the result to show that

$$\sum_{k=1}^{\infty} kx^k = \frac{x}{(1-x)^2} \quad \text{for } -1 < x < 1$$

(b) Integrate the Maclaurin series for $1/(1-x)$, and use the result to show that

$$\sum_{k=1}^{\infty} \frac{x^k}{k} = -\ln(1-x) \quad \text{for } -1 < x < 1$$

(c) Use the result in part (b) to show that

$$\sum_{k=1}^{\infty} (-1)^{k+1}\frac{x^k}{k} = \ln(1+x) \quad \text{for } -1 < x < 1$$

(d) Show that the series in part (c) converges if $x = 1$.

(e) Use the remark following Example 2 to show that

$$\sum_{k=1}^{\infty} (-1)^{k+1}\frac{x^k}{k} = \ln(1+x) \quad \text{for } -1 < x \le 1$$

35. Use the results in Exercise 34 to find the sum of the series.

(a) $\displaystyle\sum_{k=1}^{\infty} \frac{k}{3^k} = \frac{1}{3} + \frac{2}{3^2} + \frac{3}{3^3} + \frac{4}{3^4} + \cdots$

(b) $\displaystyle\sum_{k=1}^{\infty} \frac{1}{k(4^k)} = \frac{1}{4} + \frac{1}{2(4^2)} + \frac{1}{3(4^3)} + \frac{1}{4(4^4)} + \cdots$

36. Use the results in Exercise 34 to find the sum of each series.

(a) $\displaystyle\sum_{k=1}^{\infty} (-1)^{k+1}\frac{1}{k} = 1 - \frac{1}{2} + \frac{1}{3} - \frac{1}{4} + \cdots$

(b) $\displaystyle\sum_{k=1}^{\infty} \frac{(e-1)^k}{ke^k} = \frac{e-1}{e} + \frac{(e-1)^2}{2(e^2)} - \frac{(e-1)^3}{3(e^3)} + \cdots$

37. (a) Use the relationship

$$\int \frac{1}{\sqrt{1+x^2}}\,dx = \sinh^{-1} x + C$$

to find the first four nonzero terms in the Maclaurin series for $\sinh^{-1} x$.

(b) Express the series in sigma notation.

(c) What is the radius of convergence?

38. (a) Use the relationship

$$\int \frac{1}{\sqrt{1-x^2}}\,dx = \sin^{-1} x + C$$

to find the first four nonzero terms in the Maclaurin series for $\sin^{-1} x$.

(b) Express the series in sigma notation.

(c) What is the radius of convergence?

39. We showed by Formula (11) of Section 9.3 that if there are y_0 units of radioactive carbon-14 present at time $t = 0$, then the number of units present t years later is

$$y(t) = y_0e^{-0.000121t}$$

(a) Express $y(t)$ as a Maclaurin series.
(b) Use the first two terms in the series to show that the number of units present after 1 year is approximately $(0.999879)y_0$.
(c) Compare this to the value produced by the formula for $y(t)$.

40. In Section 9.1 we studied the motion of a falling object that has mass m and is retarded by air resistance. We showed that if the initial velocity is v_0 and the drag force F_R is proportional to the velocity, that is, $F_R = -cv$, then the velocity of the object at time t is

$$v(t) = e^{-ct/m}\left(v_0 + \frac{mg}{c}\right) - \frac{mg}{c}$$

where g is the acceleration due to gravity [see Formula (27) of Section 9.1].
(a) Use a Maclaurin series to show that if $ct/m \approx 0$, then the velocity can be approximated as

$$v(t) \approx v_0 - \left(\frac{cv_0}{m} + g\right)t$$

(b) Improve on the approximation in part (a).

C **41.** Suppose that a simple pendulum with a length of $L = 1$ meter is given an initial displacement of $\theta_0 = 5°$ from the vertical.
(a) Approximate the period of the pendulum using Formula (6) for the first-order model. [Take $g = 9.8$ m/s^2.]
(b) Approximate the period of the pendulum using Formula (7) for the second-order model.
(c) Use the numerical integration capability of a CAS to approximate the period of the pendulum from Formula (4), and compare it to the values obtained in parts (a) and (b).

42. Use the first three nonzero terms in Formula (5) and the Wallis sine formula in the Endpaper Integral Table (Formula 122) to obtain a model for the period of a simple pendulum.

43. Recall that the gravitational force exerted by the Earth on an object is called the object's *weight* (or more precisely, its *Earth weight*). If an object of mass m is on the surface of the Earth (mean sea level), then the magnitude of its weight is mg, where g is the acceleration due to gravity at the Earth's surface. A more general formula for the magnitude of the gravitational force that the Earth exerts on an object of mass m is

$$F = \frac{mgR^2}{(R+h)^2}$$

where R is the radius of the Earth and h is the height of the object above the Earth's surface.
(a) Use the binomial series for $1/(1+x)^2$ obtained in Example 4 of Section 10.9 to express F as a Maclaurin series in powers of h/R.
(b) Show that if $h = 0$, then $F = mg$.
(c) Show that if $h/R \approx 0$, then $F \approx mg - (2mgh/R)$. [*Note:* The quantity $2mgh/R$ can be thought of as a "correction term" for the weight that takes the object's height above the Earth's surface into account.]
(d) If we assume that the Earth is a sphere of radius $R = 4000$ mi at mean sea level, by approximately what percentage does a person's weight change in going from mean sea level to the top of Mt. Everest (29,028 ft)?

44. (a) Show that the Bessel function $J_0(x)$ given by Formula (4) of Section 10.8 satisfies the differential equation $xy'' + y' + xy = 0$. (This is called the ***Bessel equation of order zero***.)
(b) Show that the Bessel function $J_1(x)$ given by Formula (5) of Section 10.8 satisfies the differential equation $x^2y'' + xy' + (x^2 - 1)y = 0$. (This is called the ***Bessel equation of order one***.)
(c) Show that $J_0'(x) = -J_1(x)$.

45. Prove: If the power series $\sum_{k=0}^{\infty} a_k x^k$ and $\sum_{k=0}^{\infty} b_k x^k$ have the same sum on an interval $(-r, r)$, then $a_k = b_k$ for all values of k.

✔ QUICK CHECK ANSWERS 10.10

1. $(-1)^k \dfrac{x^{2k}}{k!}$ **2.** $1; -1; 1; -1; (-1)^k x^k$ **3.** $1; \dfrac{3}{2}; \dfrac{4}{3}$ **4.** (a) $-\dfrac{1}{2}$ (b) $4; 1; -\dfrac{1}{4}; \dfrac{1}{18}; 4; (-1)^{k+1}\dfrac{(x-1)^k}{k\cdot(k!)}$

CHAPTER REVIEW EXERCISES

1. What is the difference between an infinite sequence and an infinite series?

2. What is meant by the sum of an infinite series?

3. (a) What is a geometric series? Give some examples of convergent and divergent geometric series.
(b) What is a p-series? Give some examples of convergent and divergent p-series.

4. State conditions under which an alternating series is guaranteed to converge.

5. (a) What does it mean to say that an infinite series converges absolutely?
(b) What relationship exists between convergence and absolute convergence of an infinite series?

6. State the Remainder Estimation Theorem, and describe some of its uses.

7. If a power series in $x - x_0$ has radius of convergence R, what can you say about the set of x-values at which the series converges?

8. (a) Write down the formula for the Maclaurin series for f in sigma notation.
(b) Write down the formula for the Taylor series for f about $x = x_0$ in sigma notation.

9. Are the following statements true or false? If true, state a theorem to justify your conclusion; if false, then give a counterexample.
(a) If $\sum u_k$ converges, then $u_k \to 0$ as $k \to +\infty$.
(b) If $u_k \to 0$ as $k \to +\infty$, then $\sum u_k$ converges.
(c) If $f(n) = a_n$ for $n = 1, 2, 3, \ldots$, and if $a_n \to L$ as $n \to +\infty$, then $f(x) \to L$ as $x \to +\infty$.
(d) If $f(n) = a_n$ for $n = 1, 2, 3, \ldots$, and if $f(x) \to L$ as $x \to +\infty$, then $a_n \to L$ as $n \to +\infty$.
(e) If $0 < a_n < 1$, then $\{a_n\}$ converges.
(f) If $0 < u_k < 1$, then $\sum u_k$ converges.
(g) If $\sum u_k$ and $\sum v_k$ converge, then $\sum (u_k + v_k)$ diverges.
(h) If $\sum u_k$ and $\sum v_k$ diverge, then $\sum (u_k - v_k)$ converges.
(i) If $0 \le u_k \le v_k$ and $\sum v_k$ converges, then $\sum u_k$ converges.
(j) If $0 \le u_k \le v_k$ and $\sum u_k$ diverges, then $\sum v_k$ diverges.
(k) If an infinite series converges, then it converges absolutely.
(l) If an infinite series diverges absolutely, then it diverges.

10. State whether each of the following is true or false. Justify your answers.
(a) The function $f(x) = x^{1/3}$ has a Maclaurin series.
(b) $1 + \frac{1}{2} - \frac{1}{2} + \frac{1}{3} - \frac{1}{3} + \frac{1}{4} - \frac{1}{4} + \cdots = 1$
(c) $1 + \frac{1}{2} - \frac{1}{2} + \frac{1}{2} - \frac{1}{2} + \frac{1}{2} - \frac{1}{2} + \cdots = 1$

11. Find the general term of the sequence, starting with $n = 1$, determine whether the sequence converges, and if so find its limit.
(a) $\dfrac{3}{2^2 - 1^2}, \dfrac{4}{3^2 - 2^2}, \dfrac{5}{4^2 - 3^2}, \ldots$
(b) $\dfrac{1}{3}, -\dfrac{2}{5}, \dfrac{3}{7}, -\dfrac{4}{9}, \ldots$

12. Suppose that the sequence $\{a_k\}$ is defined recursively by

$$a_0 = c, \quad a_{k+1} = \sqrt{a_k}$$

Assuming that the sequence converges, find its limit if
(a) $c = \frac{1}{2}$ (b) $c = \frac{3}{2}$.

13. Show that the sequence is eventually strictly monotone.
(a) $\{(n - 10)^4\}_{n=0}^{+\infty}$ (b) $\left\{\dfrac{100^n}{(2n)!(n!)}\right\}_{n=1}^{+\infty}$

14. (a) Give an example of a bounded sequence that diverges.
(b) Give an example of a monotonic sequence that diverges.

15–20 Use any method to determine whether the series converge.

15. (a) $\displaystyle\sum_{k=1}^{\infty} \frac{1}{5^k}$ (b) $\displaystyle\sum_{k=1}^{\infty} \frac{1}{5^k + 1}$

16. (a) $\displaystyle\sum_{k=1}^{\infty} (-1)^k \frac{k+4}{k^2+k}$ (b) $\displaystyle\sum_{k=1}^{\infty} (-1)^{k+1} \left(\frac{k+2}{3k-1}\right)^k$

17. (a) $\displaystyle\sum_{k=1}^{\infty} \frac{1}{k^3 + 2k + 1}$ (b) $\displaystyle\sum_{k=1}^{\infty} \frac{1}{(3+k)^{2/5}}$

18. (a) $\displaystyle\sum_{k=1}^{\infty} \frac{\ln k}{k\sqrt{k}}$ (b) $\displaystyle\sum_{k=1}^{\infty} \frac{k^{4/3}}{8k^2 + 5k + 1}$

19. (a) $\displaystyle\sum_{k=1}^{\infty} \frac{9}{\sqrt{k} + 1}$ (b) $\displaystyle\sum_{k=1}^{\infty} \frac{\cos(1/k)}{k^2}$

20. (a) $\displaystyle\sum_{k=1}^{\infty} \frac{k^{-1/2}}{2 + \sin^2 k}$ (b) $\displaystyle\sum_{k=1}^{\infty} \frac{(-1)^{k+1}}{k^2 + 1}$

21. Find a formula for the exact error that results when the sum of the geometric series $\sum_{k=0}^{\infty} (1/5)^k$ is approximated by the sum of the first 100 terms in the series.

22. Suppose that $\displaystyle\sum_{k=1}^{n} u_k = 2 - \frac{1}{n}$. Find

(a) u_{100} (b) $\displaystyle\lim_{k \to +\infty} u_k$ (c) $\displaystyle\sum_{k=1}^{\infty} u_k$.

23. In each part, determine whether the series converges; if so, find its sum.

(a) $\displaystyle\sum_{k=1}^{\infty} \left(\frac{3}{2^k} - \frac{2}{3^k}\right)$ (b) $\displaystyle\sum_{k=1}^{\infty} [\ln(k+1) - \ln k]$

(c) $\displaystyle\sum_{k=1}^{\infty} \frac{1}{k(k+2)}$ (d) $\displaystyle\sum_{k=1}^{\infty} [\tan^{-1}(k+1) - \tan^{-1} k]$

24. It can be proved that

$$\lim_{n \to +\infty} \sqrt[n]{n!} = +\infty \quad \text{and} \quad \lim_{n \to +\infty} \frac{\sqrt[n]{n!}}{n} = \frac{1}{e}$$

In each part, use these limits and the root test to determine whether the series converges.

(a) $\displaystyle\sum_{k=0}^{\infty} \frac{2^k}{k!}$ (b) $\displaystyle\sum_{k=0}^{\infty} \frac{k^k}{k!}$

25. Let a, b, and p be positive constants. For which values of p does the series $\displaystyle\sum_{k=1}^{\infty} \frac{1}{(a + bk)^p}$ converge?

26. Find the interval of convergence of

$$\sum_{k=0}^{\infty} \frac{(x - x_0)^k}{b^k} \quad (b > 0)$$

27. (a) Show that $k^k \ge k!$.
(b) Use the comparison test to show that $\displaystyle\sum_{k=1}^{\infty} k^{-k}$ converges.
(c) Use the root test to show that the series converges.

28. Does the series $1 - \frac{2}{3} + \frac{3}{5} - \frac{4}{7} + \frac{5}{9} + \cdots$ converge? Justify your answer.

29. (a) Find the first five Maclaurin polynomials of the function $p(x) = 1 - 7x + 5x^2 + 4x^3$.
(b) Make a general statement about the Maclaurin polynomials of a polynomial of degree n.

30. Show that the approximation

$$\sin x \approx x - \frac{x^3}{3!} + \frac{x^5}{5!}$$

is accurate to four decimal places if $0 \le x \le \pi/4$.

31. Use a Maclaurin series and properties of alternating series to show that $|\ln(1+x) - x| \le x^2/2$ if $0 < x < 1$.

32. Use Maclaurin series to approximate the integral

$$\int_0^1 \frac{1 - \cos x}{x}\, dx$$

to three decimal-place accuracy.

33. In parts (a)–(d), find the sum of the series by associating it with some Maclaurin series.

(a) $2 + \dfrac{4}{2!} + \dfrac{8}{3!} + \dfrac{16}{4!} + \cdots$

(b) $\pi - \dfrac{\pi^3}{3!} + \dfrac{\pi^5}{5!} - \dfrac{\pi^7}{7!} + \cdots$

(c) $1 - \dfrac{e^2}{2!} + \dfrac{e^4}{4!} - \dfrac{e^6}{6!} + \cdots$

(d) $1 - \ln 3 + \dfrac{(\ln 3)^2}{2!} - \dfrac{(\ln 3)^3}{3!} + \cdots$

34. In each part, write out the first four terms of the series, and then find the radius of convergence.

(a) $\displaystyle\sum_{k=1}^{\infty} \frac{1 \cdot 2 \cdot 3 \cdots k}{1 \cdot 4 \cdot 7 \cdots (3k-2)} x^k$

(b) $\displaystyle\sum_{k=1}^{\infty} (-1)^k \frac{1 \cdot 2 \cdot 3 \cdots k}{1 \cdot 3 \cdot 5 \cdots (2k-1)} x^{2k+1}$

35. Use an appropriate Taylor series for $\sqrt[3]{x}$ to approximate $\sqrt[3]{28}$ to three decimal-place accuracy, and check your answer by comparing it to that produced directly by your calculating utility.

36. Differentiate the Maclaurin series for xe^x and use the result to show that

$$\sum_{k=0}^{\infty} \frac{k+1}{k!} = 2e$$

37. Use the supplied Maclaurin series for $\sin x$ and $\cos x$ to find the first four nonzero terms of the Maclaurin series for the given functions.

$$\sin x = \sum_{k=0}^{\infty} (-1)^k \frac{x^{2k+1}}{(2k+1)!}$$

$$\cos x = \sum_{k=0}^{\infty} (-1)^k \frac{x^{2k}}{(2k)!}$$

(a) $\sin x \cos x$ (b) $\frac{1}{2} \sin 2x$

9TH EDITION

Advanced Engineering Mathematics

ERWIN KREYSZIG

Professor of Mathematics
Ohio State University
Columbus, Ohio

WILEY JOHN WILEY & SONS, INC.

Vice President and Publisher: Laurie Rosatone
Editorial Assistant: Daniel Grace
Associate Production Director: Lucille Buonocore
Senior Production Editor: Ken Santor
Media Editor: Stefanie Liebman
Cover Designer: Madelyn Lesure
Cover Photo: © John Sohm/Chromosohm/Photo Researchers

This book was set in Times Roman by GGS Information Services
and printed and bound by Von Hoffmann, Inc.
The cover was printed by Von Hoffmann, Inc.

This book is printed on acid-free paper. ♾™

Kreyszig, Erwin.
Advanced engineering mathematics / Erwin Kreyszig.—9th ed.
p. cm.
Accompanied by instructor's manual.
Includes bibliographical references and index.
ISBN 0-471-48885-2 (cloth : acid-free paper)
1. Mathematical physics. 2. Engineering mathematics. 1. Title.
ISBN-13: 978-0-471-48885-9
ISBN-10: 0-471-48885-2

Printed in the United States of America

10 9 8 7 6 5 4 3 2 1

PART C
Fourier Analysis. Partial Differential Equations

Fourier analysis concerns **periodic phenomena,** as they occur quite frequently in engineering and elsewhere—think of rotating parts of machines, alternating electric currents, or the motion of planets. Related periodic functions may be complicated. This situation poses the important practical task of representing these complicated functions in terms of simple periodic functions, namely, cosines and sines. These representations will be infinite series, called **Fourier series.**[1]

The creation of these series was one of the most path-breaking events in applied mathematics, and we mention that it also had considerable influence on mathematics as a whole, on the concept of a function, on integration theory, on convergence theory for series, and so on (see Ref. [GR7] in App. 1).

Chapter 11 is concerned mainly with Fourier series. However, the underlying ideas can also be extended to *nonperiodic* phenomena. This leads to *Fourier integrals* and *transforms.* A common name for the whole area is **Fourier analysis.**

Chapter 12 deals with the most important **partial differential equations (PDEs)** of physics and engineering. This is the area in which Fourier analysis has its most basic applications, related to boundary and initial value problems of mechanics, heat flow, electrostatics, and other fields.

---

[1]JEAN-BAPTISTE JOSEPH FOURIER (1768–1830), French physicist and mathematician, lived and taught in Paris, accompanied Napoléon in the Egyptian War, and was later made prefect of Grenoble. The beginnings on Fourier series can be found in works by Euler and by Daniel Bernoulli, but it was Fourier who employed them in a systematic and general manner in his main work, *Théorie analytique de la chaleur* (*Analytic Theory of Heat,* Paris, 1822), in which he developed the theory of heat conduction (heat equation; see Sec. 12.5), making these series a most important tool in applied mathematics.

CHAPTER 11

Fourier Series, Integrals, and Transforms

Fourier series (Sec. 11.1) are infinite series designed to represent general periodic functions in terms of simple ones, namely, cosines and sines. They constitute a very important tool, in particular in solving problems that involve ODEs and PDEs.

In this chapter we discuss Fourier series and their engineering use from a practical point of view, in connection with ODEs and with the approximation of periodic functions. Application to PDEs follows in Chap. 12.

The *theory* of Fourier series is complicated, but we shall see that the *application* of these series is rather simple. Fourier series are in a certain sense more universal than the familiar Taylor series in calculus because many *discontinuous* periodic functions of practical interest can be developed in Fourier series but, of course, do not have Taylor series representations.

In the last sections (11.7–11.9) we consider **Fourier integrals** and **Fourier transforms,** which extend the ideas and techniques of Fourier series to nonperiodic functions and have basic applications to PDEs (to be shown in the next chapter).

Prerequisite: Elementary integral calculus (needed for Fourier coefficients)
Sections that may be omitted in a shorter course: 11.4–11.9
References and Answers to Problems: App. 1 Part C, App. 2.

11.1 Fourier Series

Fourier series are the basic tool for representing periodic functions, which play an important role in applications. A function $f(x)$ is called a **periodic function** if $f(x)$ is defined for all real x (perhaps except at some points, such as $x = \pm\pi/2, \pm 3\pi/2, \cdots$ for $\tan x$) and if there is some positive number p, called a **period** of $f(x)$, such that

$$f(x + p) = f(x) \qquad \text{for all } x. \tag{1}$$

The graph of such a function is obtained by periodic repetition of its graph in any interval of length p (Fig. 255).

Familiar periodic functions are the cosine and sine functions. Examples of functions that are not periodic are $x, x^2, x^3, e^x, \cosh x$, and $\ln x$, to mention just a few.

If $f(x)$ has period p, it also has the period $2p$ because (1) implies
$f(x + 2p) = f([x + p] + p) = f(x + p) = f(x)$, etc.; thus for any integer $n = 1, 2, 3, \cdots$,

$$f(x + np) = f(x) \qquad \text{for all } x. \tag{2}$$

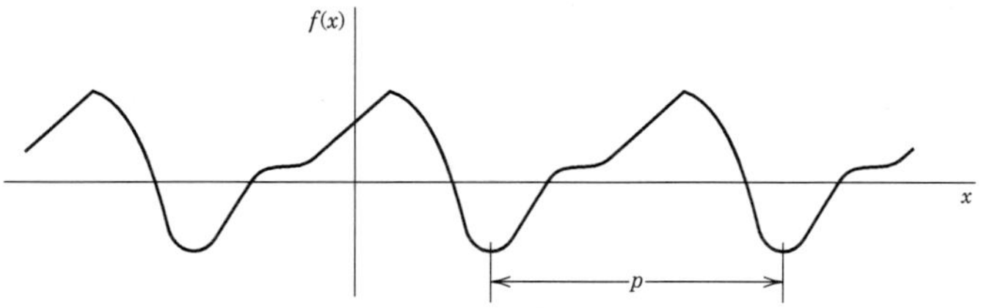

Fig. 255. Periodic function

Furthermore if $f(x)$ and $g(x)$ have period p, then $af(x) + bg(x)$ with any constants a and b also has the period p.

Our problem in the first few sections of this chapter will be the representation of various ***functions*** $f(x)$ ***of period*** 2π in terms of the simple functions

$$1, \quad \cos x, \quad \sin x, \quad \cos 2x, \quad \sin 2x, \cdots, \quad \cos nx, \quad \sin nx, \cdots. \tag{3}$$

All these functions have the period 2π. They form the so-called **trigonometric system.** Figure 256 shows the first few of them (except for the constant 1, which is periodic with any period).

The series to be obtained will be a **trigonometric series,** that is, a series of the form

$$\begin{aligned} &a_0 + a_1 \cos x + b_1 \sin x + a_2 \cos 2x + b_2 \sin 2x + \cdots \\ &= a_0 + \sum_{n=1}^{\infty} (a_n \cos nx + b_n \sin nx). \end{aligned} \tag{4}$$

$a_0, a_1, b_1, a_2, b_2, \cdots$ are constants, called the **coefficients** of the series. We see that each term has the period 2π. Hence *if the coefficients are such that the series converges, its sum will be a function of period* 2π.

It can be shown that if the series on the left side of (4) converges, then inserting parentheses on the right gives a series that converges and has the same sum as the series on the left. This justifies the equality in (4).

Now suppose that $f(x)$ is a given function of period 2π and is such that it can be **represented** by a series (4), that is, (4) converges and, moreover, has the sum $f(x)$. Then, using the equality sign, we write

$$f(x) = a_0 + \sum_{n=1}^{\infty} (a_n \cos nx + b_n \sin nx) \tag{5}$$

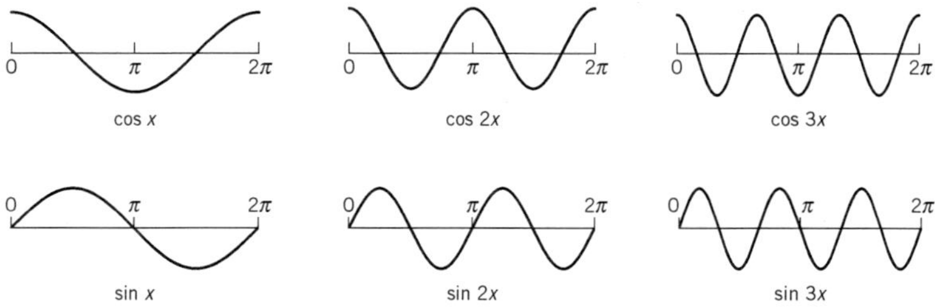

Fig. 256. Cosine and sine functions having the period 2π

and call (5) the **Fourier series** of $f(x)$. We shall prove that in this case the coefficients of (5) are the so-called **Fourier coefficients** of $f(x)$, given by the **Euler formulas**

$$\text{(6)}\qquad \begin{aligned} &\text{(a)} && a_0 = \frac{1}{2\pi}\int_{-\pi}^{\pi} f(x)\,dx \\ &\text{(b)} && a_n = \frac{1}{\pi}\int_{-\pi}^{\pi} f(x)\cos nx\,dx && n = 1, 2, \cdots \\ &\text{(c)} && b_n = \frac{1}{\pi}\int_{-\pi}^{\pi} f(x)\sin nx\,dx && n = 1, 2, \cdots . \end{aligned}$$

The name "Fourier series" is sometimes also used in the exceptional case that (5) with coefficients (6) does not converge or does not have the sum $f(x)$—this may happen but is merely of theoretical interest. (For Euler see footnote 4 in Sec. 2.5.)

A Basic Example

Before we derive the Euler formulas (6), let us become familiar with the application of (5) and (6) in the case of an important example. Since your work for other functions will be quite similar, try to fully understand every detail of the integrations, which because of the n involved differ somewhat from what you have practiced in calculus. Do not just routinely use your software, but make observations: How are continuous functions (cosines and sines) able to represent a given discontinuous function? How does the quality of the approximation increase if you take more and more terms of the series? Why are the approximating functions, called the **partial sums** of the series, always zero at 0 and π? Why is the factor $1/n$ (obtained in the integration) important?

EXAMPLE 1 **Periodic Rectangular Wave (Fig. 257a)**

Find the Fourier coefficients of the periodic function $f(x)$ in Fig. 257a. The formula is

$$\text{(7)}\qquad f(x) = \begin{cases} -k & \text{if } -\pi < x < 0 \\ k & \text{if } 0 < x < \pi \end{cases} \qquad \text{and} \qquad f(x + 2\pi) = f(x).$$

Functions of this kind occur as external forces acting on mechanical systems, electromotive forces in electric circuits, etc. (The value of $f(x)$ at a single point does not affect the integral; hence we can leave $f(x)$ undefined at $x = 0$ and $x = \pm\pi$.)

Solution. From (6a) we obtain $a_0 = 0$. This can also be seen without integration, since the area under the curve of $f(x)$ between $-\pi$ and π is zero. From (6b),

$$a_n = \frac{1}{\pi}\int_{-\pi}^{\pi} f(x)\cos nx\,dx = \frac{1}{\pi}\left[\int_{-\pi}^{0}(-k)\cos nx\,dx + \int_{0}^{\pi} k\cos nx\,dx\right]$$

$$= \frac{1}{\pi}\left[-k\,\frac{\sin nx}{n}\Bigg|_{-\pi}^{0} + k\,\frac{\sin nx}{n}\Bigg|_{0}^{\pi}\right] = 0$$

because $\sin nx = 0$ at $-\pi$, 0, and π for all $n = 1, 2, \cdots$. Similarly, from (6c) we obtain

$$b_n = \frac{1}{\pi}\int_{-\pi}^{\pi} f(x)\sin nx\,dx = \frac{1}{\pi}\left[\int_{-\pi}^{0}(-k)\sin nx\,dx + \int_{0}^{\pi} k\sin nx\,dx\right]$$

$$= \frac{1}{\pi}\left[k\,\frac{\cos nx}{n}\Bigg|_{-\pi}^{0} - k\,\frac{\cos nx}{n}\Bigg|_{0}^{\pi}\right].$$

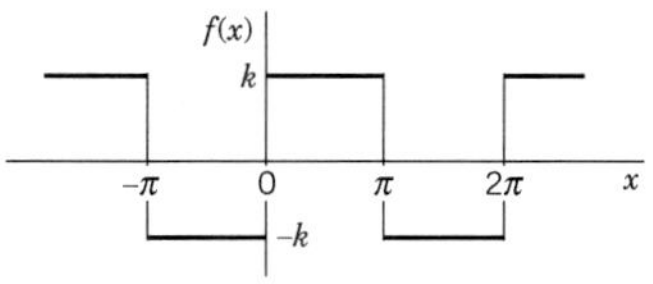

(a) The given function $f(x)$ (Periodic rectangular wave)

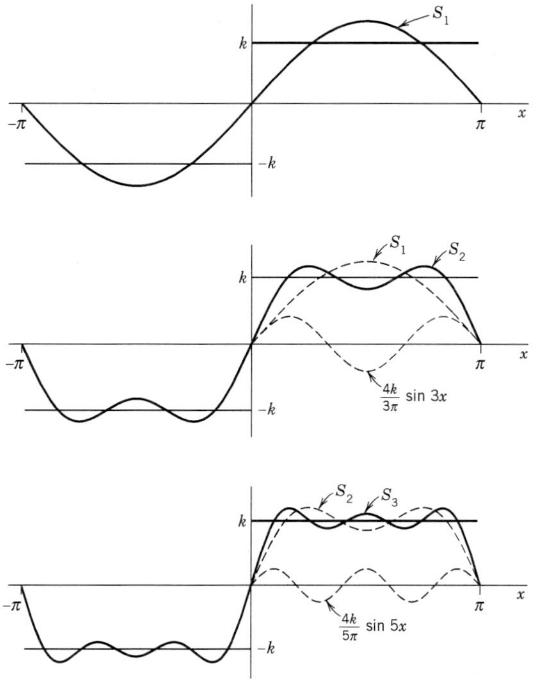

(b) The first three partial sums of the corresponding Fourier series

Fig. 257. Eample 1

Since $\cos(-\alpha) = \cos \alpha$ and $\cos 0 = 1$, this yields

$$b_n = \frac{k}{n\pi}[\cos 0 - \cos(-n\pi) - \cos n\pi + \cos 0] = \frac{2k}{n\pi}(1 - \cos n\pi).$$

Now, $\cos \pi = -1$, $\cos 2\pi = 1$, $\cos 3\pi = -1$, etc.; in general,

$$\cos n\pi = \begin{cases} -1 & \text{for odd } n, \\ 1 & \text{for even } n, \end{cases} \qquad \text{and thus} \qquad 1 - \cos n\pi = \begin{cases} 2 & \text{for odd } n, \\ 0 & \text{for even } n. \end{cases}$$

Hence the Fourier coefficients b_n of our function are

$$b_1 = \frac{4k}{\pi}, \qquad b_2 = 0, \qquad b_3 = \frac{4k}{3\pi}, \qquad b_4 = 0, \qquad b_5 = \frac{4k}{5\pi}, \cdots .$$

Since the a_n are zero, the Fourier series of $f(x)$ is

$$\frac{4k}{\pi}\left(\sin x + \frac{1}{3}\sin 3x + \frac{1}{5}\sin 5x + \cdots\right). \tag{8}$$

The partial sums are

$$S_1 = \frac{4k}{\pi}\sin x, \qquad S_2 = \frac{4k}{\pi}\left(\sin x + \frac{1}{3}\sin 3x\right), \qquad \text{etc.,}$$

Their graphs in Fig. 257 seem to indicate that the series is convergent and has the sum $f(x)$, the given function. We notice that at $x = 0$ and $x = \pi$, the points of discontinuity of $f(x)$, all partial sums have the value zero, the arithmetic mean of the limits $-k$ and k of our function, at these points.

Furthermore, assuming that $f(x)$ is the sum of the series and setting $x = \pi/2$, we have

$$f\left(\frac{\pi}{2}\right) = k = \frac{4k}{\pi}\left(1 - \frac{1}{3} + \frac{1}{5} - + \cdots\right).$$

thus

$$1 - \frac{1}{3} + \frac{1}{5} - \frac{1}{7} + - \cdots = \frac{\pi}{4}.$$

This is a famous result obtained by Leibniz in 1673 from geometric considerations. It illustrates that the values of various series with constant terms can be obtained by evaluating Fourier series at specific points. ■

Derivation of the Euler Formulas (6)

The key to the Euler formulas (6) is the **orthogonality** of (3), a concept of basic importance, as follows.

THEOREM 1

Orthogonality of the Trigonometric System (3)

The trigonometric system (3) *is orthogonal on the interval* $-\pi \leqq x \leqq \pi$ (hence also on $0 \leqq x \leqq 2\pi$ or any other interval of length 2π because of periodicity); *that is, the integral of the product of any two functions in* (3) *over that interval is* 0, *so that for any integers n and m,*

$$\text{(a)} \quad \int_{-\pi}^{\pi} \cos nx \cos mx \, dx = 0 \qquad (n \neq m)$$

$$\text{(b)} \quad \int_{-\pi}^{\pi} \sin nx \sin mx \, dx = 0 \qquad (n \neq m) \tag{9}$$

$$\text{(c)} \quad \int_{-\pi}^{\pi} \sin nx \cos mx \, dx = 0 \qquad (n \neq m \text{ or } n = m).$$

PROOF This follows simply by transforming the integrands trigonometrically from products into sums. In (9a) and (9b), by (11) in App. A3.1,

$$\int_{-\pi}^{\pi} \cos nx \cos mx \, dx = \frac{1}{2}\int_{-\pi}^{\pi} \cos (n + m)x \, dx + \frac{1}{2}\int_{-\pi}^{\pi} \cos (n - m)x \, dx$$

$$\int_{-\pi}^{\pi} \sin nx \sin mx \, dx = \frac{1}{2}\int_{-\pi}^{\pi} \cos (n - m)x \, dx - \frac{1}{2}\int_{-\pi}^{\pi} \cos (n + m)x \, dx.$$

Since $m \neq n$ (integer!), the integrals on the right are all 0. Similarly, in (9c), for all integer m and n (without exception; do you see why?)

$$\int_{-\pi}^{\pi} \sin nx \cos mx \, dx = \frac{1}{2} \int_{-\pi}^{\pi} \sin (n + m)x \, dx + \frac{1}{2} \int_{-\pi}^{\pi} \sin (n - m)x \, dx = 0 + 0. \quad \blacksquare$$

Application of Theorem 1 to the Fourier Series (5)

We prove (6a). Integrating on both sides of (5) from $-\pi$ to π, we get

$$\int_{-\pi}^{\pi} f(x) \, dx = \int_{-\pi}^{\pi} \left[a_0 + \sum_{n=1}^{\infty} (a_n \cos nx + b_n \sin nx) \right] dx.$$

We now assume that termwise integration is allowed. (We shall say in the proof of Theorem 2 when this is true.) Then we obtain

$$\int_{-\pi}^{\pi} f(x) \, dx = a_0 \int_{-\pi}^{\pi} dx + \sum_{n=1}^{\infty} \left(a_n \int_{-\pi}^{\pi} \cos nx \, dx + b_n \int_{-\pi}^{\pi} \sin nx \, dx \right).$$

The first term on the right equals $2\pi a_0$. Integration shows that all the other integrals are 0. Hence division by 2π gives (6a).

We prove (6b). Multiplying (5) on both sides by $\cos mx$ with any ***fixed*** positive integer m and integrating from $-\pi$ to π, we have

$$(10) \qquad \int_{-\pi}^{\pi} f(x) \cos mx \, dx = \int_{-\pi}^{\pi} \left[a_0 + \sum_{n=1}^{\infty} (a_n \cos nx + b_n \sin nx) \right] \cos mx \, dx.$$

We now integrate term by term. Then on the right we obtain an integral of $a_0 \cos mx$, which is 0; an integral of $a_n \cos nx \cos mx$, which is $a_m \pi$ for $n = m$ and 0 for $n \neq m$ by (9a); and an integral of $b_n \sin nx \cos mx$, which is 0 for all n and m by (9c). Hence the right side of (10) equals $a_m \pi$. Division by π gives (6b) (with m instead of n).

We finally prove (6c). Multiplying (5) on both sides by $\sin mx$ with any ***fixed*** positive integer m and integrating from $-\pi$ to π, we get

$$(11) \qquad \int_{-\pi}^{\pi} f(x) \sin mx \, dx = \int_{-\pi}^{\pi} \left[a_0 + \sum_{n=1}^{\infty} (a_n \cos nx + b_n \sin nx) \right] \sin mx \, dx.$$

Integrating term by term, we obtain on the right an integral of $a_0 \sin mx$, which is 0; an integral of $a_n \cos nx \sin mx$, which is 0 by (9c); and an integral of $b_n \sin nx \sin mx$, which is $b_m \pi$ if $n = m$ and 0 if $n \neq m$, by (9b). This implies (6c) (with n denoted by m). This completes the proof of the Euler formulas (6) for the Fourier coefficients. $\blacksquare$

Convergence and Sum of a Fourier Series

The class of functions that can be represented by Fourier series is surprisingly large and general. Sufficient conditions valid in most applications are as follows.

THEOREM 2

Representation by a Fourier Series

Let $f(x)$ be periodic with period 2π and piecewise continuous (see Sec. 6.1) *in the interval $-\pi \leqq x \leqq \pi$. Furthermore, let $f(x)$ have a left-hand derivative and a right-hand derivative at each point of that interval. Then the Fourier series* (5) *of $f(x)$* [with coefficients (6)] *converges. Its sum is $f(x)$, except at points x_0 where $f(x)$ is discontinuous. There the sum of the series is the average of the left- and right-hand limits*[2] *of $f(x)$ at x_0.*

PROOF We prove convergence in Theorem 2. We prove convergence for a continuous function $f(x)$ having continuous first and second derivatives. Integrating (6b) by parts, we obtain.

$$a_n = \frac{1}{\pi}\int_{-\pi}^{\pi} f(x)\cos nx\,dx = \frac{f(x)\sin nx}{n\pi}\bigg|_{-\pi}^{\pi} - \frac{1}{n\pi}\int_{-\pi}^{\pi} f'(x)\sin nx\,dx.$$

The first term on the right is zero. Another integration by parts gives

$$a_n = \frac{f'(x)\cos nx}{n^2\pi}\bigg|_{-\pi}^{\pi} - \frac{1}{n^2\pi}\int_{-\pi}^{\pi} f''(x)\cos nx\,dx.$$

The first term on the right is zero because of the periodicity and continuity of $f'(x)$. Since f'' is continuous in the interval of integration, we have

$$|f''(x)| < M$$

for an appropriate constant M. Furthermore, $|\cos nx| \leqq 1$. It follows that

$$|a_n| = \frac{1}{n^2\pi}\left|\int_{-\pi}^{\pi} f''(x)\cos nx\,dx\right| < \frac{1}{n^2\pi}\int_{-\pi}^{\pi} M\,dx = \frac{2M}{n^2}.$$

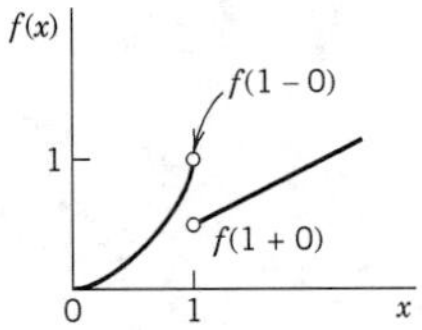

Fig. 258. Left- and right-hand limits

$f(1 - 0) = 1,$

$f(1 + 0) = \frac{1}{2}$

of the function

$$f(x) = \begin{cases} x^2 & \text{if } x < 1 \\ x/2 & \end{cases}$$

[2]The **left-hand limit** of $f(x)$ at x_0 is defined as the limit of $f(x)$ as x approaches x_0 from the left and is commonly denoted by $f(x_0 - 0)$. Thus

$$f(x_0 - 0) = \lim_{h \to 0} f(x_0 - h) \text{ as } h \to 0 \text{ through positive values.}$$

The **right-hand limit** is denoted by $f(x_0 + 0)$ and

$$f(x_0 + 0) = \lim_{h \to 0} f(x_0 + h) \text{ as } h \to 0 \text{ through positive values.}$$

The **left-** and **right-hand derivatives** of $f(x)$ at x_0 are defined as the limits of

$$\frac{f(x_0 - h) - f(x_0 - 0)}{-h} \quad \text{and} \quad \frac{f(x_0 + h) - f(x_0 + 0)}{h},$$

respectively, as $h \to 0$ through positive values. Of course if $f(x)$ is continuous at x_0, the last term in both numerators is simply $f(x_0)$.

Similarly, $|b_n| < 2\,M/n^2$ for all n. Hence the absolute value of each term of the Fourier series of $f(x)$ is at most equal to the corresponding term of the series

$$|a_0| + 2M\left(1 + 1 + \frac{1}{2^2} + \frac{1}{2^2} + \frac{1}{3^2} + \frac{1}{3^2} + \cdots\right)$$

which is convergent. Hence that Fourier series converges and the proof is complete. (Readers already familiar with uniform convergence will see that, by the Weierstrass test in Sec. 15.5, under our present assumptions the Fourier series converges uniformly, and our derivation of (6) by integrating term by term is then justified by Theorem 3 of Sec. 15.5.)

The proof of convergence in the case of a piecewise continuous function $f(x)$ and the proof that under the assumptions in the theorem the Fourier series (5) with coefficients (6) represents $f(x)$ are substantially more complicated; see, for instance, Ref. [C12]. ■

EXAMPLE 2 Convergence at a Jump as Indicated in Theorem 2

The rectangular wave in Example 1 has a jump at $x = 0$. Its left-hand limit there is $-k$ and its right-hand limit is k (Fig. 257). Hence the average of these limits is 0. The Fourier series (8) of the wave does indeed converge to this value when $x = 0$ because then all its terms are 0. Similarly for the other jumps. This is in agreement with Theorem 2. ■

Summary. A Fourier series of a given function $f(x)$ of period 2π is a series of the form (5) with coefficients given by the Euler formulas (6). Theorem 2 gives conditions that are sufficient for this series to converge and at each x to have the value $f(x)$, except at discontinuities of $f(x)$, where the series equals the arithmetic mean of the left-hand and right-hand limits of $f(x)$ at that point.

PROBLEM SET 11.1

1. **(Calculus review)** Review integration techniques for integrals as they are likely to arise from the Euler formulas, for instance, definite integrals of $x \cos nx$, $x^2 \sin nx$, $e^{-2x} \cos nx$, etc.

2–3 FUNDAMENTAL PERIOD

The *fundamental period* is the smallest positive period. Find it for

2. $\cos x$, $\sin x$, $\cos 2x$, $\sin 2x$, $\cos \pi x$, $\sin \pi x$, $\cos 2\pi x$, $\sin 2\pi x$

3. $\cos nx$, $\sin nx$, $\cos \dfrac{2\pi x}{k}$, $\sin \dfrac{2\pi x}{k}$, $\cos \dfrac{2\pi n x}{k}$, $\sin \dfrac{2\pi n x}{k}$

4. Show that $f = const$ is periodic with any period but has no fundamental period.

5. If $f(x)$ and $g(x)$ have period p, show that $h(x) = af(x) + bg(x)$ (a, b, constant) has the period p. Thus all functions of period p form a **vector space.**

6. **(Change of scale)** If $f(x)$ has period p, show that $f(ax)$, $a \neq 0$, and $f(x/b)$, $b \neq 0$, are periodic functions of x of periods p/a and bp, respectively. Give examples.

7–12 GRAPHS OF 2π-PERIODIC FUNCTIONS

Sketch or graph $f(x)$, of period 2π, which for $-\pi < x < \pi$ is given as follows.

7. $f(x) = x$

8. $f(x) = e^{-|x|}$

9. $f(x) = \pi - |x|$

10. $f(x) = |\sin 2x|$

11. $f(x) = \begin{cases} -x^3 & \text{if } -\pi < x < 0 \\ x^3 & \text{if } \ 0 < x < \pi \end{cases}$

12. $f(x) = \begin{cases} 1 & \text{if } -\pi < x < 0 \\ \cos \frac{1}{2}x & \text{if } \ 0 < x < \pi \end{cases}$

13–24 FOURIER SERIES

Showing the details of your work, find the Fourier series of the given $f(x)$, which is assumed to have the period 2π. Sketch or graph the partial sums up to that including $\cos 5x$ and $\sin 5x$.

13.

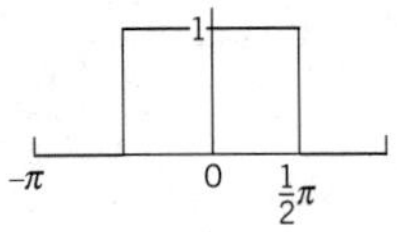

14.

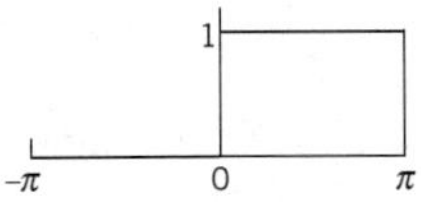

15.

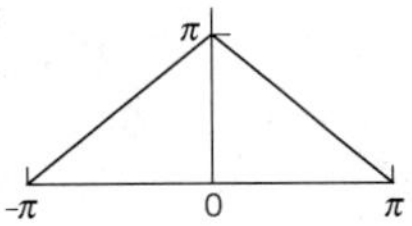

16.

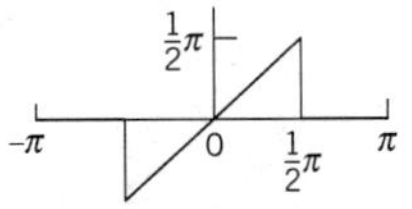

17.

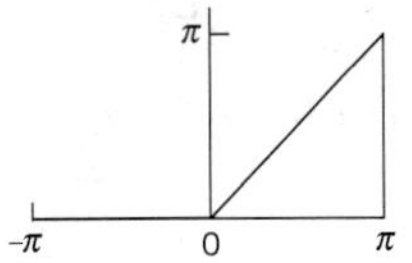

18.

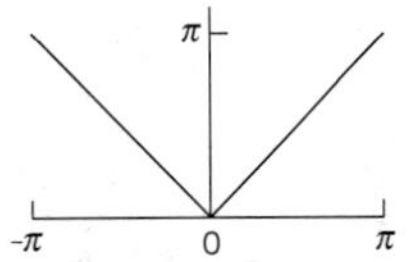

19.

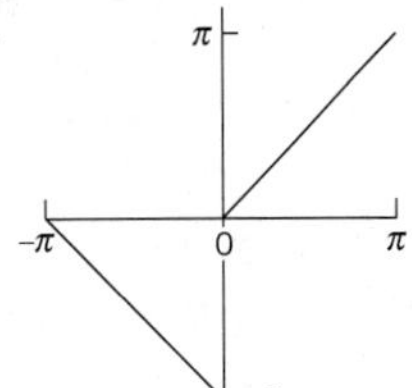

20.

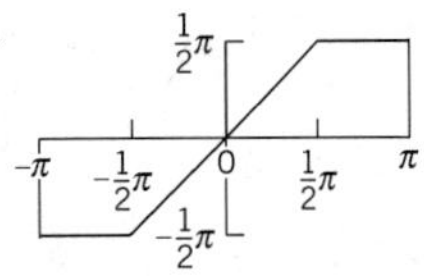

21. $f(x) = x^2 \ (-\pi < x < \pi)$

22. $f(x) = x^2 \ (0 < x < 2\pi)$

23. $f(x) = \begin{cases} x^2 & \text{if } -\frac{1}{2}\pi < x < \frac{1}{2}\pi \\ \frac{1}{4}\pi^2 & \text{if } \frac{1}{2}\pi < x < \frac{3}{2}\pi \end{cases}$

24. $f(x) = \begin{cases} -4x & \text{if } -\pi < x < 0 \\ 4x & \text{if } 0 < x < \pi \end{cases}$

25. (Discontinuities) Verify the last statement in Theorem 2 for the discontinuities of $f(x)$ in Prob. 13.

26. CAS EXPERIMENT. Graphing. Write a program for graphing partial sums of the following series. Guess from the graph what $f(x)$ the series may represent. Confirm or disprove your guess by using the Euler formulas.

(a) $2(\sin x + \frac{1}{3}\sin 3x + \frac{1}{5}\sin 5x + \cdots)$

$- 2(\frac{1}{2}\sin 2x + \frac{1}{4}\sin 4x + \frac{1}{6}\sin 6x \cdots)$

(b) $\frac{1}{2} + \frac{4}{\pi^2}(\cos x + \frac{1}{9}\cos 3x + \frac{1}{25}\cos 5x + \cdots)$

(c) $\frac{2}{3}\pi^2 + 4(\cos x - \frac{1}{4}\cos 2x + \frac{1}{9}\cos 3x - \frac{1}{16}\cos 4x + - \cdots)$

27. CAS EXPERIMENT. Order of Fourier Coefficients. The order seems to be $1/n$ if f is discontinous, and $1/n^2$ if f is continuous but $f' = df/dx$ is discontinuous, $1/n^3$ if f and f' are continuous but f'' is discontinuous, etc. Try to verify this for examples. Try to prove it by integrating the Euler formulas by parts. What is the practical significance of this?

28. PROJECT. Euler Formulas in Terms of Jumps Without Integration. Show that for a function whose third derivative is identically zero,

$$a_n = \frac{1}{n\pi}\left[-\sum j_s \sin nx_s - \frac{1}{n}\sum j_s' \cos nx_s + \frac{1}{n^2}\sum j_s'' \sin nx_s\right]$$

$$b_n = \frac{1}{n\pi}\left[\sum j_s \cos nx_s - \frac{1}{n}\sum j_s' \sin nx_s - \frac{1}{n^2}\sum j_s'' \cos nx_s\right]$$

where $n = 1, 2, \cdots$ and we sum over all the jumps j_s, j_s', j_s'' of f, f', f', respectively, located at x_s.

29. Apply the formulas in Project 28 to the function in Prob. 21 and compare the results.

30. CAS EXPERIMENT. Orthogonality. Integrate and graph the integral of the product $\cos mx \cos nx$ (with various integer m and n of your choice) from $-a$ to a as a function of a and conclude orthogonality of $\cos mx$ and $\cos nx$ $(m \neq n)$ for $a = \pi$ from the graph. For what m and n will you get orthogonality for $a = \pi/2$, $\pi/3$, $\pi/4$? Other a? Extend the experiment to $\cos mx \sin nx$ and $\sin mx \sin nx$.

11.2 Functions of Any Period $p = 2L$

The functions considered so far had period 2π, for the simplicity of the formulas. Of course, periodic functions in applications will generally have other periods. However, we now show that the transition from period $p = 2\pi$ to a period $2L$ is quite simple. The notation $p = 2L$ is practical because L will be the length of a violin string (Sec. 12.2) or the length of a rod in heat conduction (Sec. 12.5), and so on.

The idea is simply to find and use a ***change of scale*** that gives from a function $g(v)$ of period 2π a function of period $2L$. Now from (5) and (6) in the last section with $g(v)$ instead of $f(x)$ we have the Fourier series

$$g(v) = a_0 + \sum_{n=1}^{\infty} (a_n \cos nv + b_n \sin nv) \tag{1}$$

with coefficients

$$\begin{aligned} a_0 &= \frac{1}{2\pi} \int_{-\pi}^{\pi} g(v)\, dv \\ a_n &= \frac{1}{\pi} \int_{-\pi}^{\pi} g(v) \cos nv\, dv \\ b_n &= \frac{1}{\pi} \int_{-\pi}^{\pi} g(v) \sin nv\, dv. \end{aligned} \tag{2}$$

We can now write the change of scale as $v = kx$ with k such that the old period $v = 2\pi$ gives for the new variable x the new period $x = 2L$. Thus, $2\pi = k2L$. Hence $k = \pi/L$ and

$$v = kx = \pi x/L. \tag{3}$$

This implies $dv = (\pi/L)\, dx$, which upon substitution into (2) cancels $1/2\pi$ and $1/\pi$ and gives instead the factors $1/2L$ and $1/L$. Writing

$$g(v) = f(x), \tag{4}$$

we thus obtain from (1) the **Fourier series** of the function $f(x)$ of period $2L$

$$f(x) = a_0 + \sum_{n=1}^{\infty} \left(a_n \cos \frac{n\pi}{L} x + b_n \sin \frac{n\pi}{L} x\right) \tag{5}$$

with the **Fourier coefficients** of $f(x)$ given by the **Euler formulas**

$$\begin{aligned} &\textbf{(a)} \quad a_0 = \frac{1}{2L} \int_{-L}^{L} f(x)\, dx \\ &\textbf{(b)} \quad a_n = \frac{1}{L} \int_{-L}^{L} f(x) \cos \frac{n\pi x}{L}\, dx \qquad n = 1, 2, \cdots \\ &\textbf{(c)} \quad b_n = \frac{1}{L} \int_{-L}^{L} f(x) \sin \frac{n\pi x}{L}\, dx \qquad n = 1, 2, \cdots. \end{aligned} \tag{6}$$

Just as in Sec. 11.1, we continue to call (5) with any coefficients a **trigonometric series.** And we can integrate from 0 to $2L$ or over any other interval of length $p = 2L$.

EXAMPLE 1 Periodic Rectangular Wave

Find the Fourier series of the function (Fig. 259)

$$f(x) = \begin{cases} 0 & \text{if} \quad -2 < x < -1 \\ k & \text{if} \quad -1 < x < 1 \\ 0 & \text{if} \quad 1 < x < 2 \end{cases} \qquad p = 2L = 4, \quad L = 2.$$

Solution. From (6a) we obtain $a_0 = k/2$ (verify!). From (6b) we obtain

$$a_n = \frac{1}{2}\int_{-2}^{2} f(x)\cos\frac{n\pi x}{2}\,dx = \frac{1}{2}\int_{-1}^{1} k\cos\frac{n\pi x}{2}\,dx = \frac{2k}{n\pi}\sin\frac{n\pi}{2}.$$

Thus $a_n = 0$ if n is even and

$$a_n = 2k/n\pi \quad \text{if} \quad n = 1, 5, 9, \cdots, \qquad a_n = -2k/n\pi \quad \text{if} \quad n = 3, 7, 11, \cdots.$$

From (6c) we find that $b_n = 0$ for $n = 1, 2, \cdots$. Hence the Fourier series is

$$f(x) = \frac{k}{2} + \frac{2k}{\pi}\left(\cos\frac{\pi}{2}x - \frac{1}{3}\cos\frac{3\pi}{2}x + \frac{1}{5}\cos\frac{5\pi}{2}x - + \cdots\right).$$ ■

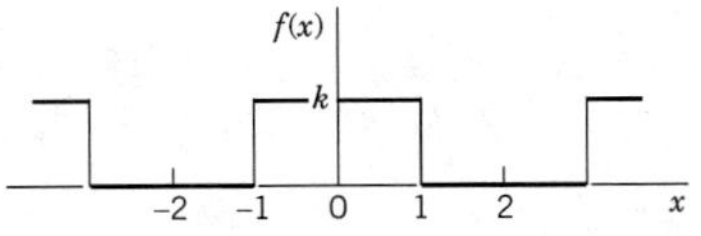

Fig. 259. Example 1

EXAMPLE 2 Periodic Rectangular Wave

Find the Fourier series of the function (Fig. 260)

$$f(x) = \begin{cases} -k & \text{if} \quad -2 < x < 0 \\ k & \text{if} \quad 0 < x < 2 \end{cases} \qquad p = 2L = 4, \qquad L = 2.$$

Solution. $a_0 = 0$ from (6a). From (6b), with $1/L = 1/2$,

$$a_n = \frac{1}{2}\left[\int_{-2}^{0}(-k)\cos\frac{n\pi x}{2}\,dx + \int_{0}^{2} k\cos\frac{n\pi x}{2}\,dx\right]$$

$$= \frac{1}{2}\left[-\frac{2k}{n\pi}\sin\frac{n\pi x}{2}\Big|_{-2}^{0} + \frac{2k}{n\pi}\sin\frac{n\pi x}{2}\Big|_{0}^{2}\right] = 0,$$

so that the Fourier series has no cosine terms. From (6c),

$$b_n = \frac{1}{2}\left[\frac{2k}{n\pi}\cos\frac{n\pi x}{2}\Big|_{-2}^{0} - \frac{2k}{n\pi}\cos\frac{n\pi x}{2}\Big|_{0}^{2}\right]$$

$$= \frac{k}{n\pi}(1 - \cos n\pi - \cos n\pi + 1) = \begin{cases} 4k/n\pi & \text{if} \quad n = 1, 3, \cdots \\ 0 & \text{if} \quad n = 2, 4, \cdots. \end{cases}$$

Hence the Fourier series of $f(x)$ is

$$f(x) = \frac{4k}{\pi}\left(\sin\frac{\pi}{2}x + \frac{1}{3}\sin\frac{3\pi}{2}x + \frac{1}{5}\sin\frac{5\pi}{2}x + \cdots\right).$$

It is interesting that we could have derived this from (8) in Sec. 11.1, namely, by the scale change (3). Indeed, writing v instead of x, we have in (8), Sec. 11.1,

$$\frac{4k}{\pi}\left(\sin v + \frac{1}{3}\sin 3v + \frac{1}{5}\sin 5v + \cdots\right).$$

Since the period 2π in v corresponds to $2L = 4$, we have $k = \pi/L = \pi/2$ and $v = kx = \pi x/2$ in (3); hence we obtain the Fourier series of $f(x)$, as before. ■

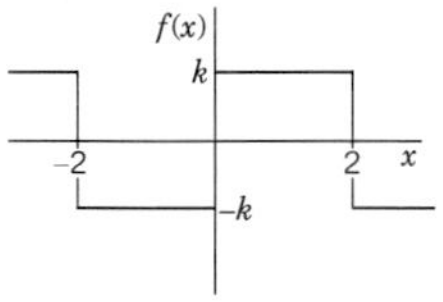

Fig. 260. Example 2

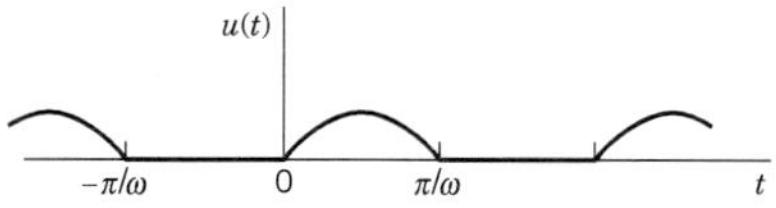

Fig. 261. Half-wave rectifier

EXAMPLE 3 Half-Wave Rectifier

A sinusoidal voltage $E \sin \omega t$, where t is time, is passed through a half-wave rectifier that clips the negative portion of the wave (Fig. 261). Find the Fourier series of the resulting periodic function

$$u(t) = \begin{cases} 0 & \text{if } -L < t < 0, \\ E\sin\omega t & \text{if } \quad 0 < t < L \end{cases} \qquad p = 2L = \frac{2\pi}{\omega}, \qquad L = \frac{\pi}{\omega}.$$

Solution. Since $u = 0$ when $-L < t < 0$, we obtain from (6a), with t instead of x,

$$a_0 = \frac{\omega}{2\pi}\int_0^{\pi/\omega} E\sin\omega t\,dt = \frac{E}{\pi}$$

and from (6b), by using formula (11) in App. A3.1 with $x = \omega t$ and $y = n\omega t$,

$$a_n = \frac{\omega}{\pi}\int_0^{\pi/\omega} E\sin\omega t\cos n\omega t\,dt = \frac{\omega E}{2\pi}\int_0^{\pi/\omega}[\sin(1+n)\omega t + \sin(1-n)\omega t]\,dt.$$

If $n = 1$, the integral on the right is zero, and if $n = 2, 3, \cdots$, we readily obtain

$$a_n = \frac{\omega E}{2\pi}\left[-\frac{\cos(1+n)\omega t}{(1+n)\omega} - \frac{\cos(1-n)\omega t}{(1-n)\omega}\right]_0^{\pi/\omega}$$

$$= \frac{E}{2\pi}\left(\frac{-\cos(1+n)\pi + 1}{1+n} + \frac{-\cos(1-n)\pi + 1}{1-n}\right).$$

If n is odd, this is equal to zero, and for even n we have

$$a_n = \frac{E}{2\pi}\left(\frac{2}{1+n} + \frac{2}{1-n}\right) = -\frac{2E}{(n-1)(n+1)\pi} \qquad (n = 2, 4, \cdots).$$

In a similar fashion we find from (6c) that $b_1 = E/2$ and $b_n = 0$ for $n = 2, 3, \cdots$. Consequently,

$$u(t) = \frac{E}{\pi} + \frac{E}{2}\sin\omega t - \frac{2E}{\pi}\left(\frac{1}{1\cdot 3}\cos 2\omega t + \frac{1}{3\cdot 5}\cos 4\omega t + \cdots\right).$$ ■

PROBLEM SET 11.2

1–11 FOURIER SERIES FOR PERIOD $p = 2L$

Find the Fourier series of the function $f(x)$, of period $p = 2L$, and sketch or graph the first three partial sums. (Show the details of your work.)

1. $f(x) = -1\ (-2 < x < 0),\ f(x) = 1\ (0 < x < 2),\ p = 4$
2. $f(x) = 0\ (-2 < x < 0),\ f(x) = 4\ (0 < x < 2),\ p = 4$
3. $f(x) = x^2 \quad (-1 < x < 1), \quad p = 2$
4. $f(x) = \pi x^3/2 \quad (-1 < x < 1), \quad p = 2$
5. $f(x) = \sin \pi x \quad (0 < x < 1), \quad p = 1$
6. $f(x) = \cos \pi x \quad (-\frac{1}{2} < x < \frac{1}{2}), \quad p = 1$
7. $f(x) = |x| \quad (-1 < x < 1), \quad p = 2$
8. $f(x) = \begin{cases} 1 + x \text{ if } -1 < x < 0 \\ 1 - x \text{ if } \quad 0 < x < 1, \end{cases} \quad p = 2$
9. $f(x) = 1 - x^2 \quad (-1 < x < 1), \quad p = 2$
10. $f(x) = 0\ (-2 < x < 0),\ f(x) = x\ (0 < x < 2),\ p = 4$
11. $f(x) = -x \quad (-1 < x < 0), \quad f(x) = x \quad (0 < x < 1),$ $f(x) = 1 \quad (1 < x < 3), \quad p = 4$
12. **(Rectifier)** Find the Fourier series of the function obtained by passing the voltage $v(t) = V_0 \cos 100\pi t$ through a half-wave rectifier.
13. Show that the familiar identities $\cos^3 x = \frac{3}{4} \cos x + \frac{1}{4} \cos 3x$ and $\sin^3 x = \frac{3}{4} \sin x - \frac{1}{4} \sin 3x$ can be interpreted as Fourier series expansions. Develop $\cos^4 x$.
14. Obtain the series in Prob. 7 from that in Prob. 8.
15. Obtain the series in Prob. 6 from that in Prob. 5.
16. Obtain the series in Prob. 3 from that in Prob. 21 of Problem Set 11.1.
17. Using Prob. 3, show that $1 - \frac{1}{4} + \frac{1}{9} - \frac{1}{16} + - \cdots = \frac{1}{12}\pi^2$.
18. Show that $1 + \frac{1}{4} + \frac{1}{9} + \frac{1}{16} + \cdots = \frac{1}{6}\pi^2$.
19. **CAS PROJECT. Fourier Series of $2L$-Periodic Functions. (a)** Write a program for obtaining partial sums of a Fourier series (1).

    (b) Apply the program to Probs. 2–5, graphing the first few partial sums of each of the four series on common axes. Choose the first five or more partial sums until they approximate the given function reasonably well. Compare and comment.
20. **CAS EXPERIMENT. Gibbs Phenomenon.** The partial sums $s_n(x)$ of a Fourier series show oscillations near a discontinuity point. These oscillations do not disappear as n increases but instead become sharp "spikes." They were explained mathematically by J. W. Gibbs[3]. Graph $s_n(x)$ in Prob. 10. When $n = 50$, say, you will see those oscillations quite distinctly. Consider other Fourier series of your choice in a similar way. Compare.

11.3 Even and Odd Functions. Half-Range Expansions

The function in Example 1, Sec. 11.2, is *even,* and its Fourier series has only *cosine* terms. The function in Example 2, Sec. 11.2, is *odd*, and its Fourier series has only *sine* terms.

Recall that g is **even** if $g(-x) = g(x)$, so that its graph is symmetric with respect to the vertical axis (Fig. 262). A function h is **odd** if $h(-x) = -h(x)$ (Fig. 263).

Now the cosine terms in the Fourier series (5), Sec. 11.2, are even and the sine terms are odd. So it should not be a surprise that an even function is given by a series of cosine terms and an odd function by a series of sine terms. Indeed, the following holds.

[3]JOSIAH WILLARD GIBBS (1839–1903), American mathematician, professor of mathematical physics at Yale from 1871 on, one of the founders of vector calculus [another being O. Heaviside (see Sec. 6.1)], mathematical thermodynamics, and statistical mechanics. His work was of great importance to the development of mathematical physics.

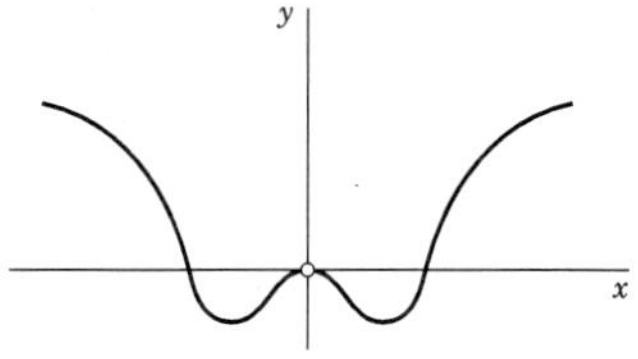

Fig. 262. Even function

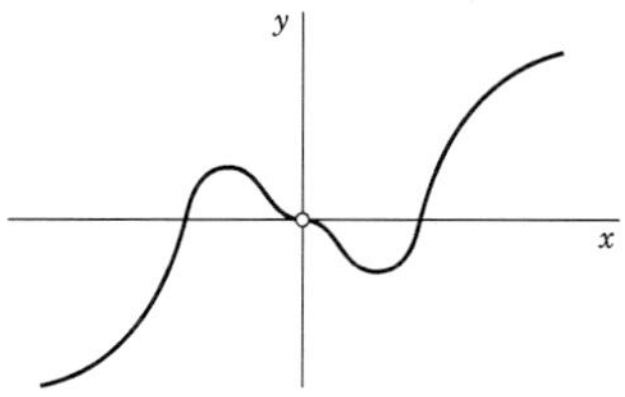

Fig. 263. Odd function

THEOREM 1

Fourier Cosine Series, Fourier Sine Series

The Fourier series of an **even** *function of period* $2L$ *is a* **"Fourier cosine series"**

$$f(x) = a_0 + \sum_{n=1}^{\infty} a_n \cos \frac{n\pi}{L} x \qquad (f \text{ even}) \tag{1}$$

with coefficients (note: integration from 0 to L only!)

$$a_0 = \frac{1}{L}\int_0^L f(x)\,dx, \qquad a_n = \frac{2}{L}\int_0^L f(x)\cos\frac{n\pi x}{L}\,dx, \qquad n = 1, 2, \cdots. \tag{2}$$

The Fourier series of an **odd** *function of period* $2L$ *is a* **"Fourier sine series"**

$$f(x) = \sum_{n=1}^{\infty} b_n \sin \frac{n\pi}{L} x \qquad (f \text{ odd}) \tag{3}$$

with coefficients

$$b_n = \frac{2}{L}\int_0^L f(x)\sin\frac{n\pi x}{L}\,dx. \tag{4}$$

PROOF Since the definite integral of a function gives the area under the curve of the function between the limits of integration, we have

$$\int_{-L}^{L} g(x)\,dx = 2\int_0^L g(x)\,dx \qquad \text{for even } g$$

$$\int_{-L}^{L} h(x)\,dx = 0 \qquad \text{for odd } h$$

as is obvious from the graphs of g and h. (Give a formal proof.) Now let f be even. Then (6a), Sec. 11.2, gives a_0 in (2). Also, the integrand in (6b), Sec. 11.2, is even (a product of even functions is even), so that (6b) gives a_n in (2). Furthermore, the integrand in (6c), Sec. 11.2, is the even f times the odd sine, so that the integrand (the product) is odd, the integral is zero, and there are no sine terms in (1).

Similarly, if f is odd, the integrals for a_0 and a_n in (6a) and (6b), Sec. 11.2, are zero, f times the sine in (6c) is even, (6c) implies (4), and there are no cosine terms in (3). ■

The Case of Period 2π. If $L = \pi$, then $f(x) = a_0 + \sum_{n=1}^{\infty} a_n \cos nx$ (f even) with coefficients

$$(2^*) \qquad a_0 = \frac{1}{\pi}\int_0^{\pi} f(x)\,dx, \qquad a_n = \frac{2}{\pi}\int_0^{\pi} f(x)\cos nx\,dx, \qquad n = 1, 2, \cdots$$

and $f(x) = \sum_{n=1}^{\infty} b_n \sin nx$ (f odd) with coefficients

$$(4^*) \qquad b_n = \frac{2}{\pi}\int_0^{\pi} f(x)\sin nx\,dx, \qquad n = 1, 2, \cdots.$$

For instance, $f(x)$ in Example 1, Sec. 11.1, is odd and is represented by a Fourier sine series.

Further simplifications result from the following property, whose very simple proof is left to the student.

THEOREM 2

Sum and Scalar Multiple

The Fourier coefficients of a sum $f_1 + f_2$ are the sums of the corresponding Fourier coefficients of f_1 and f_2.

The Fourier coefficients of cf are c times the corresponding Fourier coefficients of f.

EXAMPLE 1 **Rectangular Pulse**

The function $f^*(x)$ in Fig. 264 is the sum of the function $f(x)$ in Example 1 of Sec 11.1 and the constant k. Hence, from that example and Theorem 2 we conclude that

$$f^*(x) = k + \frac{4k}{\pi}\left(\sin x + \frac{1}{3}\sin 3x + \frac{1}{5}\sin 5x + \cdots\right). \qquad ■$$

EXAMPLE 2 **Half-Wave Rectifier**

The function $u(t)$ in Example 3 of Sec. 11.2 has a Fourier cosine series plus a single term $v(t) = (E/2)\sin \omega t$. We conclude from this and Theorem 2 that $u(t) - v(t)$ must be an even function. Verify this graphically. (See Fig. 265.) ■

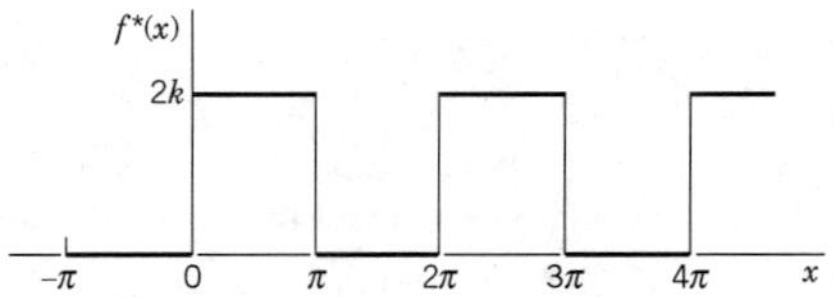

Fig. 264. Example 1

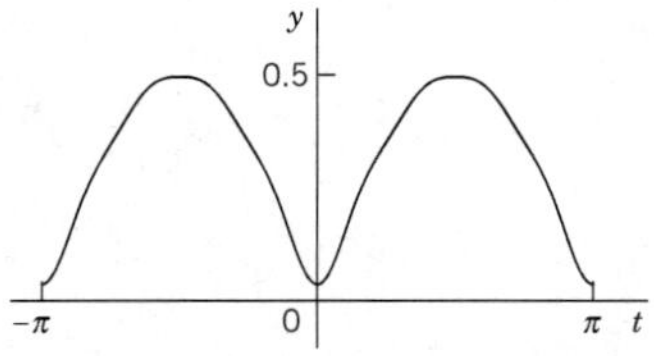

Fig. 265. $u(t) - v(t)$ with $E = 1$, $\omega = 1$

EXAMPLE 3 Sawtooth Wave

Find the Fourier series of the function (Fig. 266)

$$f(x) = x + \pi \quad \text{if} \quad -\pi < x < \pi \qquad \text{and} \qquad f(x + 2\pi) = f(x).$$

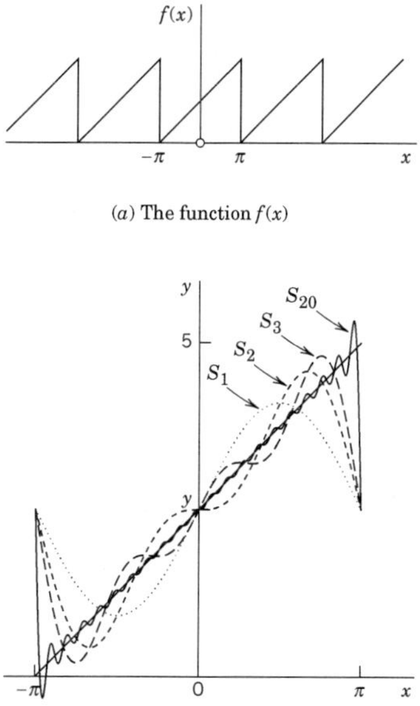

(a) The function $f(x)$

(b) Partial sums S_1, S_2, S_3, S_{20}

Fig. 266. Example 3

Solution. We have $f = f_1 + f_2$, where $f_1 = x$ and $f_2 = \pi$. The Fourier coefficients of f_2 are zero, except for the first one (the constant term), which is π. Hence, by Theorem 2, the Fourier coefficients a_n, b_n are those of f_1, except for a_0, which is π. Since f_1 is odd, $a_n = 0$ for $n = 1, 2, \cdots$, and

$$b_n = \frac{2}{\pi}\int_0^{\pi} f_1(x) \sin nx \, dx = \frac{2}{\pi}\int_0^{\pi} x \sin nx \, dx.$$

Integrating by parts, we obtain

$$b_n = \frac{2}{\pi}\left[\frac{-x \cos nx}{n}\bigg|_0^{\pi} + \frac{1}{n}\int_0^{\pi} \cos nx \, dx\right] = -\frac{2}{n}\cos n\pi.$$

Hence $b_1 = 2$, $b_2 = -2/2$, $b_3 = 2/3$, $b_4 = -2/4, \cdots$, and the Fourier series of $f(x)$ is

$$f(x) = \pi + 2\left(\sin x - \frac{1}{2}\sin 2x + \frac{1}{3}\sin 3x - + \cdots\right). \qquad \blacksquare$$

Half-Range Expansions

Half-range expansions are Fourier series. The idea is simple and useful. Figure 267 explains it. We want to represent $f(x)$ in Fig. 267a by a Fourier series, where $f(x)$ may be the shape of a distorted violin string or the temperature in a metal bar of length L, for example. (Corresponding problems will be discussed in Chap. 12.) Now comes the idea.

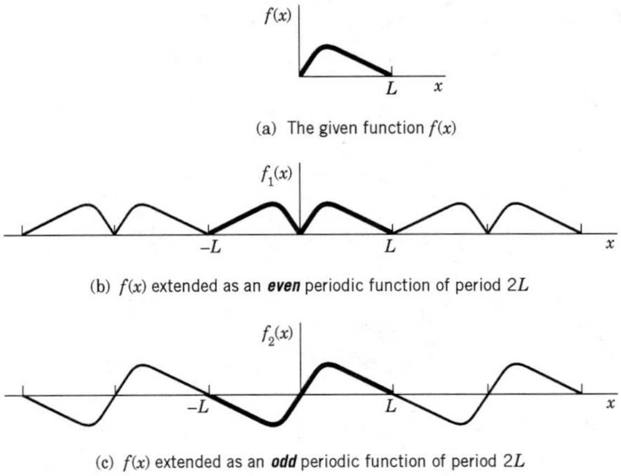

Fig. 267. (a) Function $f(x)$ given on an interval $0 \leqq x \leqq L$

(b) **Even extension** to the full "range" (interval) $-L \leqq x \leqq L$ (heavy curve) and the periodic extension of period $2L$ to the x-axis

(c) **Odd extension** to $-L \leqq x \leqq L$ (heavy curve) and the periodic extension of period $2L$ to the x-axis

We could extend $f(x)$ as a function of period L and develop the extended function into a Fourier series. But this series would in general contain *both* cosine *and* sine terms. We can do better and get simpler series. Indeed, for our given f we can calculate Fourier coefficients from (2) or from (4) in Theorem 1. And we have a choice and can take what seems more practical. If we use (2), we get (1). This is the **even periodic extension** f_1 of f in Fig. 267b. If we choose (4) instead, we get (3), the **odd periodic extension** f_2 of f in Fig. 267c.

Both extensions have period $2L$. This motivates the name **half-range expansions:** f is given (and of physical interest) only on half the range, half the interval of periodicity of length $2L$.

Let us illustrate these ideas with an example that we shall also need in Chap. 12.

EXAMPLE 4 **"Triangle" and Its Half-Range Expansions**

Find the two half-range expansions of the function (Fig. 268)

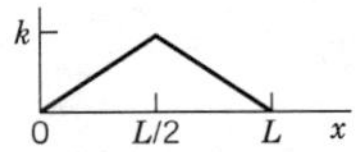

Fig. 268. The given function in Example 4

$$f(x) = \begin{cases} \dfrac{2k}{L}x & \text{if } 0 < x < \dfrac{L}{2} \\ \dfrac{2k}{L}(L - x) & \text{if } \dfrac{L}{2} < x < L. \end{cases}$$

Solution. ***(a) Even periodic extension.*** From (2) we obtain

$$a_0 = \frac{1}{L}\left[\frac{2k}{L}\int_0^{L/2} x\,dx + \frac{2k}{L}\int_{L/2}^{L}(L - x)\,dx\right] = \frac{k}{2},$$

$$a_n = \frac{2}{L}\left[\frac{2k}{L}\int_0^{L/2} x\cos\frac{n\pi}{L}x\,dx + \frac{2k}{L}\int_{L/2}^{L}(L - x)\cos\frac{n\pi}{L}x\,dx\right].$$

We consider a_n. For the first integral we obtain by integration by parts

$$\int_0^{L/2} x \cos \frac{n\pi}{L} x \, dx = \frac{Lx}{n\pi} \sin \frac{n\pi}{L} x \Big|_0^{L/2} - \frac{L}{n\pi} \int_0^{L/2} \sin \frac{n\pi}{L} x \, dx$$

$$= \frac{L^2}{2n\pi} \sin \frac{n\pi}{2} + \frac{L^2}{n^2\pi^2} \left(\cos \frac{n\pi}{2} - 1 \right).$$

Similarly, for the second integral we obtain

$$\int_{L/2}^{L} (L - x) \cos \frac{n\pi}{L} x \, dx = \frac{L}{n\pi} (L - x) \sin \frac{n\pi}{L} x \Big|_{L/2}^{L} + \frac{L}{n\pi} \int_{L/2}^{L} \sin \frac{n\pi}{L} x \, dx$$

$$= \left(0 - \frac{L}{n\pi} \left(L - \frac{L}{2} \right) \sin \frac{n\pi}{2} \right) - \frac{L^2}{n^2\pi^2} \left(\cos n\pi - \cos \frac{n\pi}{2} \right).$$

We insert these two results into the formula for a_n. The sine terms cancel and so does a factor L^2. This gives

$$a_n = \frac{4k}{n^2\pi^2} \left(2 \cos \frac{n\pi}{2} - \cos n\pi - 1 \right).$$

Thus,

$$a_2 = -16k/(2^2\pi^2), \qquad a_6 = -16k/(6^2\pi^2), \qquad a_{10} = -16k/(10^2\pi^2), \cdots$$

and $a_n = 0$ if $n \neq 2, 6, 10, 14, \cdots$. Hence the first half-range expansion of $f(x)$ is (Fig. 269a)

$$f(x) = \frac{k}{2} - \frac{16k}{\pi^2} \left(\frac{1}{2^2} \cos \frac{2\pi}{L} x + \frac{1}{6^2} \cos \frac{6\pi}{L} x + \cdots \right).$$

This Fourier cosine series represents the even periodic extension of the given function $f(x)$, of period $2L$.

(b) ***Odd periodic extension.*** Similarly, from (4) we obtain

$$b_n = \frac{8k}{n^2\pi^2} \sin \frac{n\pi}{2}. \tag{5}$$

Hence the other half-range expansion of $f(x)$ is (Fig. 269b)

$$f(x) = \frac{8k}{\pi^2} \left(\frac{1}{1^2} \sin \frac{\pi}{L} x - \frac{1}{3^2} \sin \frac{3\pi}{L} x + \frac{1}{5^2} \sin \frac{5\pi}{L} x - + \cdots \right).$$

This series represents the odd periodic extension of $f(x)$, of period $2L$.

Basic applications of these results will be shown in Secs. 12.3 and 12.5. ■

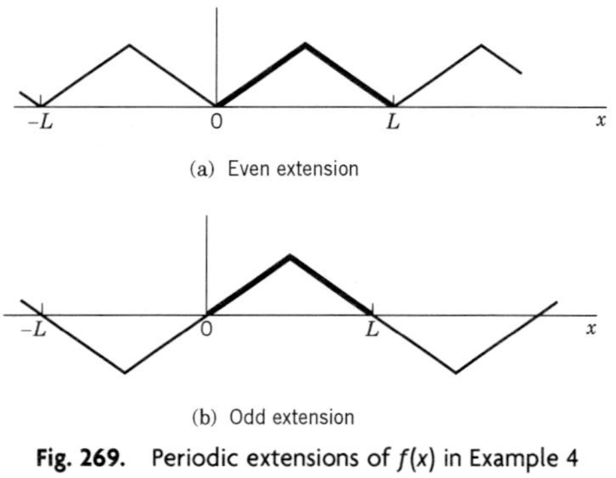

Fig. 269. Periodic extensions of $f(x)$ in Example 4

PROBLEM SET 11.3

1–9 EVEN AND ODD FUNCTIONS

Are the following functions even, odd, or neither even nor odd?

1. $|x|$, $x^2 \sin nx$, $x + x^2$, $e^{-|x|}$, $\ln x$, $x \cosh x$

2. $\sin(x^2)$, $\sin^2 x$, $x \sinh x$, $|x^3|$, $e^{\pi x}$, xe^x, $\tan 2x$, $x/(1 + x^2)$

Are the following functions, which are assumed to be periodic of period 2π, even, odd, or neither even nor odd?

3. $f(x) = x^3 \quad (-\pi < x < \pi)$

4. $f(x) = x^2 \quad (-\pi/2 < x < 3\pi/2)$

5. $f(x) = e^{-4x} \quad (-\pi < x < \pi)$

6. $f(x) = x^3 \sin x \quad (-\pi < x < \pi)$

7. $f(x) = x|x| - x^3 \quad (-\pi < x < \pi)$

8. $f(x) = 1 - x + x^3 - x^5 \quad (-\pi < x < \pi)$

9. $f(x) = 1/(1 + x^2)$ if $-\pi < x < 0$, $f(x) = -1/(1 + x^2)$ if $0 < x < \pi$

10. PROJECT. Even and Odd Functions. (a) Are the following expressions even or odd? Sums and products of even functions and of odd functions. Products of even times odd functions. Absolute values of odd functions. $f(x) + f(-x)$ and $f(x) - f(-x)$ for arbitrary $f(x)$.

(b) Write e^{kx}, $1/(1 - x)$, $\sin(x + k)$, $\cosh(x + k)$ as sums of an even and an odd function.

(c) Find all functions that are both even and odd.

(d) Is $\cos^3 x$ even or odd? $\sin^3 x$? Find the Fourier series of these functions. Do you recognize familiar identities?

11–16 FOURIER SERIES OF EVEN AND ODD FUNCTIONS

Is the given function even or odd? Find its Fourier series. Sketch or graph the function and some partial sums. (Show the details of your work.)

11. $f(x) = \pi - |x| \quad (-\pi < x < \pi)$

12. $f(x) = 2x|x| \quad (-1 < x < 1)$

13. $f(x) = \begin{cases} x & \text{if } -\pi/2 < x < \pi/2 \\ \pi - x & \text{if } \pi/2 < x < 3\pi/2 \end{cases}$

14. $f(x) = \begin{cases} \pi e^{-x} & \text{if } -\pi < x < 0 \\ \pi e^{x} & \text{if } 0 < x < \pi \end{cases}$

15. $f(x) = \begin{cases} 2 & \text{if } -2 < x < 0 \\ 0 & \text{if } 0 < x < 2 \end{cases}$

16. $f(x) = \begin{cases} 1 - \tfrac{1}{2}|x| & \text{if } -2 < x < 2 \\ 0 & \text{if } 2 < x < 6 \end{cases} \quad (p = 8)$

17–25 HALF-RANGE EXPANSIONS

Find (a) the Fourier cosine series, (b) the Fourier sine series. Sketch $f(x)$ and its two periodic extensions. (Show the details of your work.)

17. $f(x) = 1 \quad (0 < x < 2)$

18. $f(x) = x \quad (0 < x < \tfrac{1}{2})$

19. $f(x) = 2 - x \quad (0 < x < 2)$

20. $f(x) = \begin{cases} 0 & (0 < x < 2) \\ 1 & (2 < x < 4) \end{cases}$

21. $f(x) = \begin{cases} 1 & (0 < x < 1) \\ 2 & (1 < x < 2) \end{cases}$

22. $f(x) = \begin{cases} x & (0 < x < \pi/2) \\ \pi/2 & (\pi/2 < x < \pi) \end{cases}$

23. $f(x) = x \quad (0 < x < L)$

24. $f(x) = x^2 \quad (0 < x < L)$

25. $f(x) = \pi - x \quad (0 < x < \pi)$

26. Illustrate the formulas in the proof of Theorem 1 with examples. Prove the formulas.

11.4 Complex Fourier Series. *Optional*

In this optional section we show that the Fourier series

$$f(x) = a_0 + \sum_{n=1}^{\infty} (a_n \cos nx + b_n \sin nx) \tag{1}$$

can be written in complex form, which sometimes simplifies calculations (see Example 1, on page 498). This complex form can be obtained because in complex, the exponential function e^{it} and $\cos t$ and $\sin t$ are related by the basic **Euler formula** (see (11) in Sec. 2.2)

$$e^{it} = \cos t + i \sin t. \qquad \text{Thus} \qquad e^{-it} = \cos t - i \sin t. \tag{2}$$

Conversely, by adding and subtracting these two formulas, we obtain

$$(3) \qquad \text{(a)} \quad \cos t = \frac{1}{2}(e^{it} + e^{-it}), \qquad \text{(b)} \quad \sin t = \frac{1}{2i}(e^{it} - e^{-it}).$$

From (3), using $1/i = -i$ in $\sin t$ and setting $t = nx$ in both formulas, we get

$$\begin{aligned} a_n \cos nx + b_n \sin nx &= \frac{1}{2}\, a_n(e^{inx} + e^{-inx}) + \frac{1}{2i}\, b_n(e^{inx} - e^{-inx}) \\ &= \frac{1}{2}\,(a_n - ib_n)e^{inx} + \frac{1}{2}\,(a_n + ib_n)e^{-inx}. \end{aligned}$$

We insert this into (1). Writing $a_0 = c_0$, $\frac{1}{2}(a_n - ib_n) = c_n$, and $\frac{1}{2}(a_n + ib_n) = k_n$, we get from (1)

$$(4) \qquad f(x) = c_0 + \sum_{n=1}^{\infty} (c_n e^{inx} + k_n e^{-inx}).$$

The coefficients $c_1, c_2, \cdots$, and $k_1, k_2, \cdots$ are obtained from (6b), (6c) in Sec. 11.1 and then (2) above with $t = nx$,

$$(5) \qquad \begin{aligned} c_n &= \frac{1}{2}\,(a_n - ib_n) = \frac{1}{2\pi}\int_{-\pi}^{\pi} f(x)(\cos nx - i \sin nx)\, dx = \frac{1}{2\pi}\int_{-\pi}^{\pi} f(x)e^{-inx}\, dx \\ k_n &= \frac{1}{2}\,(a_n + ib_n) = \frac{1}{2\pi}\int_{-\pi}^{\pi} f(x)(\cos nx + i \sin nx)\, dx = \frac{1}{2\pi}\int_{-\pi}^{\pi} f(x)e^{inx}\, dx. \end{aligned}$$

Finally, we can combine (5) into a single formula by the trick of writing $k_n = c_{-n}$. Then (4), (5), and $c_0 = a_0$ in (6a) of Sec. 11.1 give (summation from $-\infty$!)

$$(6) \qquad \begin{aligned} f(x) &= \sum_{n=-\infty}^{\infty} c_n e^{inx}, \\ c_n &= \frac{1}{2\pi}\int_{-\pi}^{\pi} f(x)e^{-inx}\, dx, \qquad n = 0, \pm 1, \pm 2, \cdots. \end{aligned}$$

This is the so-called *complex form of the Fourier series* or, more briefly, the **complex Fourier series,** of $f(x)$. The c_n are called the **complex Fourier coefficients** of $f(x)$.

For a function of period $2L$ our reasoning gives the **complex Fourier series**

$$(7) \qquad \begin{aligned} f(x) &= \sum_{n=-\infty}^{\infty} c_n e^{in\pi x/L}, \\ c_n &= \frac{1}{2L}\int_{-L}^{L} f(x)e^{-in\pi x/L}\, dx, \qquad n = 0, \pm 1, \pm 2, \cdots. \end{aligned}$$

EXAMPLE 1 Complex Fourier Series

Find the complex Fourier series of $f(x) = e^x$ if $-\pi < x < \pi$ and $f(x + 2\pi) = f(x)$ and obtain from it the usual Fourier series.

Solution. Since $\sin n\pi = 0$ for integer n, we have

$$e^{\pm in\pi} = \cos n\pi \pm i \sin n\pi = \cos n\pi = (-1)^n.$$

With this we obtain from (6) by integration

$$c_n = \frac{1}{2\pi}\int_{-\pi}^{\pi} e^x e^{-inx}\,dx = \frac{1}{2\pi}\,\frac{1}{1-in}\,e^{x-inx}\Big|_{x=-\pi}^{\pi} = \frac{1}{2\pi}\,\frac{1}{1-in}\,(e^{\pi} - e^{-\pi})(-1)^n.$$

On the right,

$$\frac{1}{1-in} = \frac{1+in}{(1-in)(1+in)} = \frac{1+in}{1+n^2} \qquad \text{and} \qquad e^{\pi} - e^{-\pi} = 2\sinh\pi.$$

Hence the complex Fourier series is

$$e^x = \frac{\sinh\pi}{\pi}\sum_{n=-\infty}^{\infty}(-1)^n\frac{1+in}{1+n^2}\,e^{inx} \qquad (-\pi < x < \pi). \tag{8}$$

From this let us derive the real Fourier series. Using (2) with $t = nx$ and $i^2 = -1$, we have in (8)

$$(1 + in)e^{inx} = (1 + in)(\cos nx + i \sin nx) = (\cos nx - n \sin nx) + i(n \cos nx + \sin nx).$$

Now (8) also has a corresponding term with $-n$ instead of n. Since $\cos(-nx) = \cos nx$ and $\sin(-nx) = -\sin nx$, we obtain in this term

$$(1 - in)e^{-inx} = (1 - in)(\cos nx - i \sin nx) = (\cos nx - n \sin nx) - i(n \cos nx + \sin nx).$$

If we add these two expressions, the imaginary parts cancel. Hence their sum is

$$2(\cos nx - n \sin nx), \qquad n = 1, 2, \cdots.$$

For $n = 0$ we get 1 (not 2) because there is only one term. Hence the real Fourier series is

$$e^x = \frac{2\sinh\pi}{\pi}\left[\frac{1}{2} - \frac{1}{1+1^2}(\cos x - \sin x) + \frac{1}{1+2^2}(\cos 2x - 2\sin 2x) - + \cdots\right]. \tag{9}$$

In Fig. 270 the poor approximation near the jumps at $\pm\pi$ is a case of the Gibbs phenomenon (see CAS Experiment 20 in Problem Set 11.2). ■

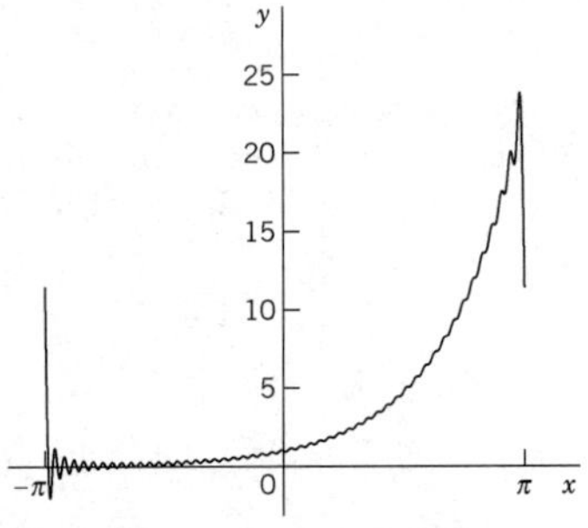

Fig. 270. Partial sum of (9), terms from $n = 0$ to 50

PROBLEM SET 11.4

1. **(Calculus review)** Review complex numbers.
2. **(Even and odd functions)** Show that the complex Fourier coefficients of an even function are real and those of an odd function are pure imaginary.
3. **(Fourier coefficients)** Show that
$a_0 = c_0,\ a_n = c_n + c_{-n},\ b_n = i(c_n - c_{-n})$.
4. Verify the calculations in Example 1.
5. Find further terms in (9) and graph partial sums with your CAS.
6. Obtain the real series in Example 1 directly from the Euler formulas in Sec. 11.

7–13 COMPLEX FOURIER SERIES

Find the complex Fourier series of the following functions. (Show the details of your work.)

7. $f(x) = -1$ if $-\pi < x < 0$, $f(x) = 1$ if $0 < x < \pi$
8. Convert the series in Prob. 7 to real form.
9. $f(x) = x \quad (-\pi < x < \pi)$
10. Convert the series in Prob. 9 to real form.
11. $f(x) = x^2 \quad (-\pi < x < \pi)$
12. Convert the series in Prob. 11 to real form.
13. $f(x) = x \quad (0 < x < 2\pi)$
14. **PROJECT. Complex Fourier Coefficients.** It is very interesting that the c_n in (6) can be derived directly by a method similar to that for a_n and b_n in Sec. 11.1. For this, multiply the series in (6) by e^{-imx} with fixed integer m, and integrate termwise from $-\pi$ to π on both sides (allowed, for instance, in the case of uniform convergence) to get

$$\int_{-\pi}^{\pi} f(x)e^{-imx}\,dx = \sum_{n=-\infty}^{\infty} c_n \int_{-\pi}^{\pi} e^{i(n-m)x}\,dx.$$

Show that the integral on the right equals 2π when $n = m$ and 0 when $n \neq m$ [use (3b)], so that you get the coefficient formula in (6).

11.5 Forced Oscillations

Fourier series have important applications in connection with ODEs and PDEs. We show this for a basic problem modeled by an ODE. Various applications to PDEs will follow in Chap. 12. This will show the enormous usefulness of Euler's and Fourier's ingenious idea of splitting up periodic functions into the simplest ones possible.

From Sec. 2.8 we know that forced oscillations of a body of mass m on a spring of modulus k are governed by the ODE

$$my'' + cy' + ky = r(t) \tag{1}$$

where $y = y(t)$ is the displacement from rest, c the damping constant, k the spring constant (spring modulus), and $r(t)$ the external force depending on time t. Figure 271 shows the model and Fig. 272 its electrical analog, an RLC-circuit governed by

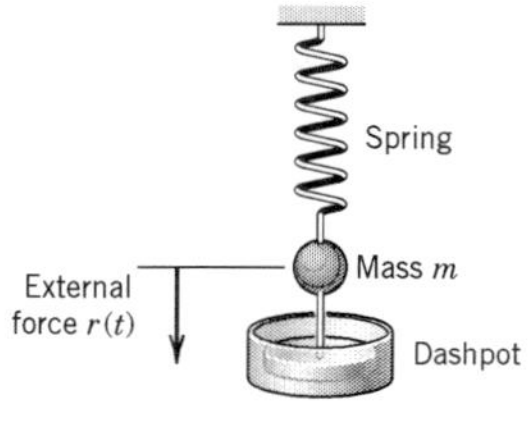

Fig. 271. Vibrating system under consideration

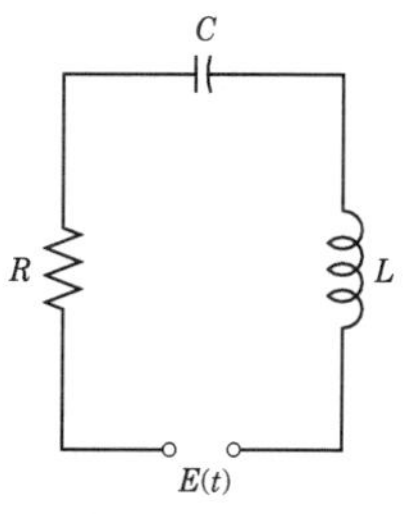

Fig. 272. Electrical analog of the system in Fig. 271 (RLC-circuit)

(1*) $$LI'' + RI' + \frac{1}{C} I = E'(t)$$ (Sec. 2.9).

We consider (1). If $r(t)$ is a sine or cosine function and if there is damping ($c > 0$), then the steady-state solution is a harmonic oscillation with frequency equal to that of $r(t)$. However, if $r(t)$ is not a pure sine or cosine function but is any other periodic function, then the steady-state solution will be a superposition of harmonic oscillations with frequencies equal to that of $r(t)$ and integer multiples of the latter. And if one of these frequencies is close to the (practical) resonant frequency of the vibrating system (see Sec. 2.8), then the corresponding oscillation may be the dominant part of the response of the system to the external force. This is what the use of Fourier series will show us. Of course, this is quite surprising to an observer unfamiliar with Fourier series, which are highly important in the study of vibrating systems and resonance. Let us discuss the entire situation in terms of a typical example.

EXAMPLE 1 Forced Oscillations under a Nonsinusoidal Periodic Driving Force

In (1), let $m = 1$ (gm), $c = 0.05$ (gm/sec), and $k = 25$ (gm/sec^2), so that (1) becomes

(2) $$y'' + 0.05y' + 25y = r(t)$$

where $r(t)$ is measured in gm · cm/sec^2. Let (Fig. 273)

$$r(t) = \begin{cases} t + \dfrac{\pi}{2} & \text{if } -\pi < t < 0, \\ -t + \dfrac{\pi}{2} & \text{if } \quad 0 < t < \pi, \end{cases} \qquad r(t + 2\pi) = r(t).$$

Find the steady-state solution $y(t)$.

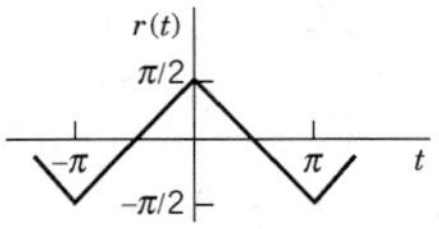

Fig. 273. Force in Example 1

Solution. We represent $r(t)$ by a Fourier series, finding

(3) $$r(t) = \frac{4}{\pi}\left(\cos t + \frac{1}{3^2}\cos 3t + \frac{1}{5^2}\cos 5t + \cdots\right)$$

(take the answer to Prob. 11 in Problem Set 11.3 minus $\frac{1}{2}\pi$ and write t for x). Then we consider the ODE

(4) $$y'' + 0.05y' + 25y = \frac{4}{n^2\pi}\cos nt \qquad (n = 1, 3, \cdots)$$

whose right side is a single term of the series (3). From Sec. 2.8 we know that the steady-state solution $y_n(t)$ of (4) is of the form

(5) $$y_n = A_n \cos nt + B_n \sin nt.$$

By substituting this into (4) we find that

$$(6) \qquad A_n = \frac{4(25 - n^2)}{n^2 \pi D_n}, \qquad B_n = \frac{0.2}{n \pi D_n}, \qquad \text{where} \qquad D_n = (25 - n^2)^2 + (0.05n)^2.$$

Since the ODE (2) is linear, we may expect the steady-state solution to be

$$(7) \qquad y = y_1 + y_3 + y_5 + \cdots$$

where y_n is given by (5) and (6). In fact, this follows readily by substituting (7) into (2) and using the Fourier series of $r(t)$, provided that termwise differentiation of (7) is permissible. (Readers already familiar with the notion of uniform convergence [Sec. 15.5] may prove that (7) may be differentiated term by term.)

From (6) we find that the amplitude of (5) is (a factor $\sqrt{D_n}$ cancels out)

$$C_n = \sqrt{A_n^2 + B_n^2} = \frac{4}{n^2 \pi \sqrt{D_n}}.$$

Numeric values are

$$C_1 = 0.0531$$
$$C_3 = 0.0088$$
$$C_5 = 0.2037$$
$$C_7 = 0.0011$$
$$C_9 = 0.0003.$$

Figure 274 shows the input (multiplied by 0.1) and the output. For $n = 5$ the quantity D_n is very small, the denominator of C_5 is small, and C_5 is so large that y_5 is the dominating term in (7). Hence the output is almost a harmonic oscillation of five times the frequency of the driving force, a little distorted due to the term y_1, whose amplitude is about 25% of that of y_5. You could make the situation still more extreme by decreasing the damping constant c. Try it. ∎

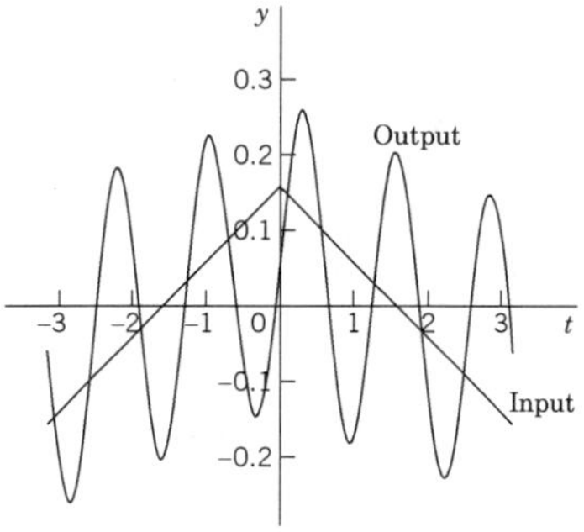

Fig. 274. Input and steady-state output in Example 1

PROBLEM SET 11.5

1. **(Coefficients)** Derive the formula for C_n from A_n and B_n.
2. **(Spring constant)** What would happen to the amplitudes C_n in Example 1 (and thus to the form of the vibration) if we changed the spring constant to the value 9? If we took a stiffer spring with $k = 81$? First guess.
3. **(Damping)** In Example 1 change c to 0.02 and discuss how this changes the output.
4. **(Input)** What would happen in Example 1 if we replaced $r(t)$ with its derivative (the rectangular wave)? What is the ratio of the new C_n to the old ones?

5–11 GENERAL SOLUTION

Find a general solution of the ODE $y'' + \omega^2 y = r(t)$ with $r(t)$ as given. (Show the details of your work.)

5. $r(t) = \cos \omega t, \ \omega = 0.5, 0.8, 1.1, 1.5, 5.0, 10.0$

6. $r(t) = \cos \omega_1 t + \cos \omega_2 t \quad (\omega^2 \neq \omega_1{}^2, \omega_2{}^2)$

7. $r(t) = \sum_{n=1}^{N} a_n \cos nt, \quad |\omega| \neq 1, 2, \cdots, N$

8. $r(t) = \sin t + \frac{1}{3} \sin 3t + \frac{1}{5} \sin 5t + \frac{1}{7} \sin 7t$

9. $r(t) = \begin{cases} t + \pi & \text{if} \quad -\pi < t < 0 \\ -t + \pi & \text{if} \quad 0 < t < \pi \end{cases}$

and $r(t + 2\pi) = r(t), |\omega| \neq 0, 1, 3, \cdots$

10. $r(t) = \begin{cases} t & \text{if} \quad -\pi/2 < t < \pi/2 \\ \pi - t & \text{if} \quad \pi/2 < t < 3\pi/2 \end{cases}$

and $r(t + 2\pi) = r(t), \quad |\omega| \neq 1, 3, 5, \cdots$

11. $r(t) = \dfrac{\pi}{4} |\sin t|$ if $-\pi < t < \pi$ and

$r(t + 2\pi) = r(t), \quad |\omega| \neq 0, 2, 4, \cdots$

12. **(CAS Program)** Write a program for solving the ODE just considered and for jointly graphing input and output of an initial value problem involving that ODE. Apply the program to Probs. 5 and 9 with initial values of your choice.

13. **(Sign of coefficients)** Some A_n in Example 1 are positive and some negative. Is this physically understandable?

14–17 STEADY-STATE DAMPED OSCILLATIONS

Find the steady-state oscillation of $y'' + cy' + y = r(t)$ with $c > 0$ and $r(t)$ as given. (Show the details of your work.)

14. $r(t) = a_n \cos nt$

15. $r(t) = \sin 3t$

16. $r(t) = \begin{cases} \pi t & \text{if} \quad -\pi/2 < t < \pi/2 \\ \pi(\pi - t) & \text{if} \quad \pi/2 < t < 3\pi/2 \end{cases}$

and $r(t + 2\pi) = r(t)$

17. $r(t) = \sum_{n=1}^{N} b_n \sin nt$

18. **CAS EXPERIMENT. Maximum Output Term.** Graph and discuss outputs of $y'' + cy' + ky = r(t)$ with $r(t)$ as in Example 1 for various c and k with emphasis on the maximum C_n and its ratio to the second largest $|C_n|$.

19–20 *RLC*-CIRCUIT

Find the steady-state current $I(t)$ in the *RLC*-circuit in Fig. 272, where $R = 100\ \Omega$, $L = 10$ H, $C = 10^{-2}$ F and $E(t)$ V as follows and periodic with period 2π. Sketch or graph the first four partial sums. Note that the coefficients of the solution decrease rapidly.

19. $E(t) = 200t(\pi^2 - t^2) \ (-\pi < t < \pi)$

20. $E(t) = \begin{cases} 100(\pi t + t^2) & \text{if} \quad -\pi < t < 0 \\ 100(\pi t - t^2) & \text{if} \quad 0 < t < \pi \end{cases}$

11.6 Approximation by Trigonometric Polynomials

Fourier series play a prominent role in differential equations. Another field in which they have major applications is **approximation theory,** which concerns the approximation of functions by other (usually simpler) functions. In connection with Fourier series the idea is as follows.

Let $f(x)$ be a function on the interval $-\pi \leqq x \leqq \pi$ that can be represented on this interval by a Fourier series. Then the ***N*th partial sum** of the series

$$f(x) \approx a_0 + \sum_{n=1}^{N} (a_n \cos nx + b_n \sin nx) \tag{1}$$

is an approximation of the given $f(x)$. It is natural to ask whether (1) is the "best" approximation of f by a **trigonometric polynomial of degree *N*,** that is, by a function of the form

$$F(x) = A_0 + \sum_{n=1}^{N} (A_n \cos nx + B_n \sin nx) \qquad (N \text{ fixed}) \tag{2}$$

where "best" means that the "error" of the approximation is as small as possible.

Of course, we must first define what we mean by the **error** E of such an approximation. We could choose the maximum of $|f - F|$. But in connection with Fourier series it is better to choose a definition that measures the goodness of agreement between f and F *on the whole interval* $-\pi \leqq x \leqq \pi$. This seems preferable, in particular if f has jumps: F in Fig. 275 is a good overall approximation of f, but the maximum of $|f - F|$ (more precisely, the *supremum*) is large (it equals at least half the jump of f at x_0). We choose

$$E = \int_{-\pi}^{\pi} (f - F)^2 \, dx. \tag{3}$$

This is called the **square error** of F relative to the function f on the interval $-\pi \leqq x \leqq \pi$. Clearly, $E \geqq 0$.

N being fixed, we want to determine the coefficients in (2) such that E is minimum. Since $(f - F)^2 = f^2 - 2fF + F^2$, we have

$$E = \int_{-\pi}^{\pi} f^2 \, dx - 2 \int_{-\pi}^{\pi} fF \, dx + \int_{-\pi}^{\pi} F^2 \, dx. \tag{4}$$

We square (2), insert it into the last integral in (4), and evaluate the occurring integrals. This gives integrals of $\cos^2 nx$ and $\sin^2 nx$ ($n \geqq 1$), which equal π, and integrals of $\cos nx$, $\sin nx$, and $(\cos nx)(\sin mx)$, which are zero (just as in Sec. 11.1). Thus

$$\int_{-\pi}^{\pi} F^2 \, dx = \int_{-\pi}^{\pi} \left[A_0 + \sum_{n=1}^{N} (A_n \cos nx + B_n \sin nx) \right]^2 dx$$

$$= \pi(2A_0^2 + A_1^2 + \cdots + A_N^2 + B_1^2 + \cdots + B_N^2).$$

We now insert (2) into the integral of fF in (4). This gives integrals of $f \cos nx$ as well as $f \sin nx$, just as in Euler's formulas, Sec. 11.1, for a_n and b_n (each multiplied by A_n or B_n). Hence

$$\int_{-\pi}^{\pi} fF \, dx = \pi(2A_0a_0 + A_1a_1 + \cdots + A_Na_N + B_1b_1 + \cdots + B_Nb_N).$$

With these expressions, (4) becomes

$$\begin{aligned} E = \int_{-\pi}^{\pi} f^2 \, dx &- 2\pi \left[2A_0a_0 + \sum_{n=1}^{N} (A_na_n + B_nb_n) \right] \\ &+ \pi \left[2A_0^2 + \sum_{n=1}^{N} (A_n^2 + B_n^2) \right]. \end{aligned} \tag{5}$$

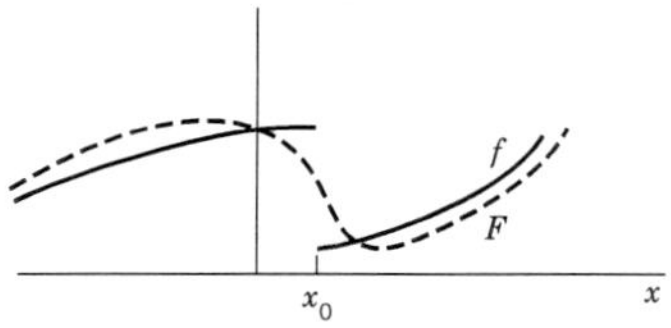

Fig. 275. Error of approximation

We now take $A_n = a_n$ and $B_n = b_n$ in (2). Then in (5) the second line cancels half of the integral-free expression in the first line. Hence for this choice of the coefficients of F the square error, call it E^*, is

$$E^* = \int_{-\pi}^{\pi} f^2\,dx - \pi\left[2a_0^2 + \sum_{n=1}^{N} (a_n^2 + b_n^2)\right]. \tag{6}$$

We finally subtract (6) from (5). Then the integrals drop out and we get terms $A_n^2 - 2A_n a_n + a_n^2 = (A_n - a_n)^2$ and similar terms $(B_n - b_n)^2$:

$$E - E^* = \pi\left\{2(A_0 - a_0)^2 + \sum_{n=1}^{N} [(A_n - a_n)^2 + (B_n - b_n)^2]\right\}.$$

Since the sum of squares of real numbers on the right cannot be negative,

$$E - E^* \geqq 0, \qquad \text{thus} \qquad E \geqq E^*,$$

and $E = E^*$ if and only if $A_0 = a_0, \cdots, B_N = b_N$. This proves the following fundamental minimum property of the partial sums of Fourier series.

THEOREM 1

Minimum Square Error

The square error of F in (2) *(with fixed N) relative to f on the interval $-\pi \leqq x \leqq \pi$ is minimum if and only if the coefficients of F in* (2) *are the Fourier coefficients of f. This minimum value E^* is given by* (6).

From (6) we see that E^* cannot increase as N increases, but may decrease. Hence *with increasing N the partial sums of the Fourier series of f yield better and better approximations to f*, considered from the viewpoint of the square error.

Since $E^* \geqq 0$ and (6) holds for every N, we obtain from (6) the important **Bessel's inequality**

$$2a_0^2 + \sum_{n=1}^{\infty} (a_n^2 + b_n^2) \leqq \frac{1}{\pi}\int_{-\pi}^{\pi} f(x)^2\,dx \tag{7}$$

for the Fourier coefficients of any function f for which integral on the right exists. (For F. W. Bessel see Sec. 5.5.)

It can be shown (see [C12] in App. 1) that for such a function f, **Parseval's theorem** holds; that is, formula (7) holds with the equality sign, so that it becomes **Parseval's identity**[4]

$$2a_0^2 + \sum_{n=1}^{\infty} (a_n^2 + b_n^2) = \frac{1}{\pi}\int_{-\pi}^{\pi} f(x)^2\,dx. \tag{8}$$

[4]MARC ANTOINE PARSEVAL (1755–1836), French mathematician. A physical interpretation of the identity follows in the next section.

EXAMPLE 1 **Minimum Square Error for the Sawtooth Wave**

Compute the minimum square error E^* of $F(x)$ with $N = 1, 2, \cdots, 10, 20, \cdots, 100$ and 1000 relative to

$$f(x) = x + \pi \qquad (-\pi < x < \pi)$$

on the interval $-\pi \leqq x \leqq \pi$.

Solution. $F(x) = \pi + 2(\sin x - \frac{1}{2}\sin 2x + \frac{1}{3}\sin 3x - + \cdots + \frac{(-1)^{N+1}}{N}\sin Nx)$ by Example 3 in Sec. 11.3. From this and (6),

$$E^* = \int_{-\pi}^{\pi} (x + \pi)^2\, dx - \pi\left(2\pi^2 + 4\sum_{n=1}^{N}\frac{1}{n^2}\right).$$

Numeric values are:

N	E^*	N	E^*	N	E^*	N	E^*
1	8.1045	6	1.9295	20	0.6129	70	0.1782
2	4.9629	7	1.6730	30	0.4120	80	0.1561
3	3.5666	8	1.4767	40	0.3103	90	0.1389
4	2.7812	9	1.3216	50	0.2488	100	0.1250
5	2.2786	10	1.1959	60	0.2077	1000	0.0126

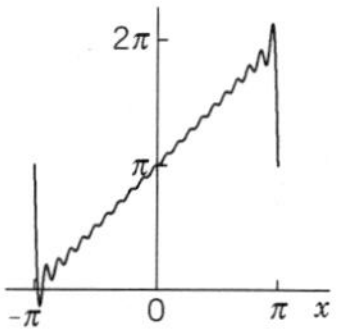

Fig. 276. F with $N = 20$ in Example 1

$F = S_1, S_2, S_3$ are shown in Fig. 266 in Sec. 11.3, and $F = S_{20}$ is shown in Fig. 276. Although $|f(x) - F(x)|$ is large at $\pm\pi$ (how large?), where f is discontinuous, F approximates f quite well on the whole interval, except near $\pm\pi$, where "waves" remain owing to the Gibbs phenomenon (see CAS Experiment 20 in Problem Set 11.2).

Can you think of functions f for which E^* decreases more quickly with increasing N? ■

This is the end of our discussion of Fourier series, which has emphasized the practical aspects of these series, as needed in applications. In the last three sections of this chapter we show how ideas and techniques in Fourier series can be extended to ***nonperiodic*** functions.

PROBLEM SET 11.6

1–9 **MINIMUM SQUARE ERROR**

Find the trigonometric polynomial $F(x)$ of the form (2) for which the square error with respect to the given $f(x)$ on the interval $-\pi \leqq x \leqq \pi$ is minimum, and compute the minimum value for $N = 1, 2, \cdots, 5$ (or also for larger values if you have a CAS).

1. $f(x) = x \; (-\pi < x < \pi)$
2. $f(x) = x^2 \; (-\pi < x < \pi)$
3. $f(x) = |x| \; (-\pi < x < \pi)$
4. $f(x) = x^3 \; (-\pi < x < \pi)$
5. $f(x) = |\sin x| \; (-\pi < x < \pi)$
6. $f(x) = e^{-|x|} \; (-\pi < x < \pi)$
7. $f(x) = \begin{cases} -1 & \text{if } -\pi < x < 0 \\ 1 & \text{if } 0 < x < \pi \end{cases}$
8. $f(x) = \begin{cases} x & \text{if } -\frac{1}{2}\pi < x < \frac{1}{2}\pi \\ 0 & \text{if } \frac{1}{2}\pi < x < \frac{3}{2}\pi \end{cases}$
9. $f(x) = x(x + \pi)$ if $-\pi < x < 0$, $f(x) = x(-x + \pi)$ if $0 < x < \pi$
10. **CAS EXPERIMENT. Size and Decrease of E^*.** Compare the size of the minimum square error E^* for functions of your choice. Find experimentally the factors on which the decrease of E^* with N depends. For each function considered find the smallest N such that $E^* < 0.1$.
11. **(Monotonicity)** Show that the minimum square error (6) is a monotone decreasing function of N. How can you use this in practice?

12–16 PARSEVAL'S IDENTITY

Using Parseval's identity, prove that the series have the indicated sums. Compute the first few partial sums to see that the convergence is rapid.

12. $1 + \frac{1}{3^4} + \frac{1}{5^4} + \frac{1}{7^4} + \cdots = \frac{\pi^4}{96} = 1.01467\,8032$

(Use Prob. 15 in Sec. 11.1.)

13. $1 + \frac{1}{3^2} + \frac{1}{5^2} + \cdots = \frac{\pi^2}{8} = 1.23370\,0550$

(Use Prob. 13 in Sec. 11.1.)

14. $\frac{1}{1^2 \cdot 3^2} + \frac{1}{3^2 \cdot 5^2} + \frac{1}{5^2 \cdot 7^2} + \cdots$

$= \frac{\pi^2}{16} - \frac{1}{2} = 0.11685\,0275$

(Use Prob. 5, this set.)

15. $1 + \frac{1}{2^4} + \frac{1}{3^4} + \cdots = \frac{\pi^4}{90} = 1.08232\,3234$

(Use Prob. 21 in Sec. 11.1.)

16. $1 + \frac{1}{3^6} + \frac{1}{5^6} + \frac{1}{7^6} + \cdots = \frac{\pi^6}{960} = 1.00144\,7078$

(Use Prob. 9, this set.)

11.7 Fourier Integral

Fourier series are powerful tools for problems involving functions that are periodic or are of interest on a finite interval only. Sections 11.3 and 11.5 first illustrated this, and various further applications follow in Chap. 12. Since, of course, many problems involve functions that are **nonperiodic** *and are of interest on the whole x-axis,* we ask what can be done to extend the method of Fourier series to such functions. This idea will lead to "Fourier integrals."

In Example 1 we start from a special function f_L of period $2L$ and see what happens to its Fourier series if we let $L \to \infty$. Then we do the same for an *arbitrary* function f_L of period $2L$. This will motivate and suggest the main result of this section, which is an integral representation given in Theorem 1 (below).

EXAMPLE 1 Rectangular Wave

Consider the periodic rectangular wave $f_L(x)$ of period $2L > 2$ given by

$$f_L(x) = \begin{cases} 0 & \text{if} \quad -L < x < -1 \\ 1 & \text{if} \quad -1 < x < \ \ 1 \\ 0 & \text{if} \quad \ \ 1 < x < \ L. \end{cases}$$

The left part of Fig. 277 shows this function for $2L = 4, 8, 16$ as well as the nonperiodic function $f(x)$, which we obtain from f_L if we let $L \to \infty$,

$$f(x) = \lim_{L\to\infty} f_L(x) = \begin{cases} 1 & \text{if } -1 < x < 1 \\ 0 & \text{otherwise.} \end{cases}$$

We now explore what happens to the Fourier coefficients of f_L as L increases. Since f_L is even, $b_n = 0$ for all n. For a_n the Euler formulas (6), Sec. 11.2, give

$$a_0 = \frac{1}{2L}\int_{-1}^{1} dx = \frac{1}{L}, \qquad a_n = \frac{1}{L}\int_{-1}^{1} \cos\frac{n\pi x}{L}\,dx = \frac{2}{L}\int_{0}^{1} \cos\frac{n\pi x}{L}\,dx = \frac{2}{L}\,\frac{\sin(n\pi/L)}{n\pi/L}.$$

This sequence of Fourier coefficients is called the **amplitude spectrum** of f_L because $|a_n|$ is the maximum amplitude of the wave $a_n \cos(n\pi x/L)$. Figure 277 shows this spectrum for the periods $2L = 4, 8, 16$. We see that for increasing L these amplitudes become more and more dense on the positive w_n-axis, where $w_n = n\pi/L$. Indeed, for $2L = 4, 8, 16$ we have 1, 3, 7 amplitudes per "half-wave" of the function $(2 \sin w_n)/(Lw_n)$ (dashed in the figure). Hence for $2L = 2^k$ we have $2^{k-1} - 1$ amplitudes per half-wave, so that these amplitudes will eventually be everywhere dense on the positive w_n-axis (and will decrease to zero).

The outcome of this example gives an intuitive impression of what about to expect if we turn from our special function to an arbitrary one, as we shall do next. ■

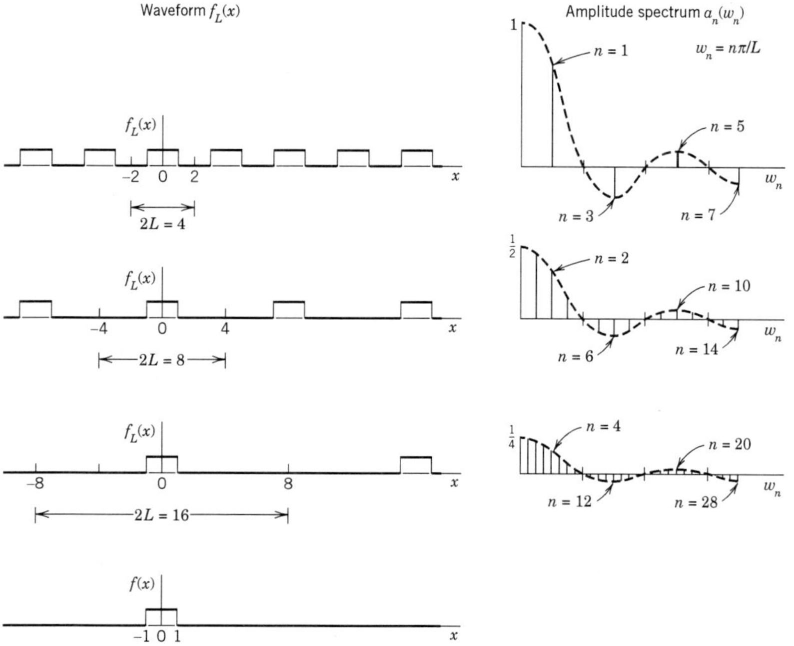

Fig. 277. Waveforms and amplitude spectra in Example 1

From Fourier Series to Fourier Integral

We now consider any periodic function $f_L(x)$ of period $2L$ that can be represented by a Fourier series

$$f_L(x) = a_0 + \sum_{n=1}^{\infty} (a_n \cos w_n x + b_n \sin w_n x), \qquad w_n = \frac{n\pi}{L}$$

and find out what happens if we let $L \to \infty$. Together with Example 1 the present calculation will suggest that we should expect an integral (instead of a series) involving $\cos wx$ and $\sin wx$ with w no longer restricted to integer multiples $w = w_n = n\pi/L$ of π/L but taking *all* values. We shall also see what form such an integral might have.

If we insert a_n and b_n from the Euler formulas (6), Sec. 11.2, and denote the variable of integration by v, the Fourier series of $f_L(x)$ becomes

$$f_L(x) = \frac{1}{2L} \int_{-L}^{L} f_L(v)\, dv + \frac{1}{L} \sum_{n=1}^{\infty} \left[\cos w_n x \int_{-L}^{L} f_L(v) \cos w_n v\, dv + \sin w_n x \int_{-L}^{L} f_L(v) \sin w_n v\, dv \right].$$

We now set

$$\Delta w = w_{n+1} - w_n = \frac{(n+1)\pi}{L} - \frac{n\pi}{L} = \frac{\pi}{L}.$$

Then $1/L = \Delta w/\pi$, and we may write the Fourier series in the form

$$(1)\quad f_L(x) = \frac{1}{2L}\int_{-L}^{L} f_L(v)\,dv + \frac{1}{\pi}\sum_{n=1}^{\infty}\left[(\cos w_n x)\,\Delta w \int_{-L}^{L} f_L(v)\cos w_n v\,dv + (\sin w_n x)\,\Delta w \int_{-L}^{L} f_L(v)\sin w_n v\,dv\right].$$

This representation is valid for any fixed L, arbitrarily large, but finite.

We now let $L \to \infty$ and assume that the resulting nonperiodic function

$$f(x) = \lim_{L\to\infty} f_L(x)$$

is **absolutely integrable** on the x-axis; that is, the following (finite!) limits exist:

$$(2)\quad \lim_{a\to-\infty}\int_a^0 |f(x)|\,dx + \lim_{b\to\infty}\int_0^b |f(x)|\,dx \quad \left(\text{written } \int_{-\infty}^{\infty} |f(x)|\,dx\right).$$

Then $1/L \to 0$, and the value of the first term on the right side of (1) approaches zero. Also $\Delta w = \pi/L \to 0$ and it seems ***plausible*** that the infinite series in (1) becomes an integral from 0 to ∞, which represents $f(x)$, namely,

$$(3)\quad f(x) = \frac{1}{\pi}\int_0^{\infty}\left[\cos wx \int_{-\infty}^{\infty} f(v)\cos wv\,dv + \sin wx \int_{-\infty}^{\infty} f(v)\sin wv\,dv\right]dw.$$

If we introduce the notations

$$\textbf{(4)}\quad A(w) = \frac{1}{\pi}\int_{-\infty}^{\infty} f(v)\cos wv\,dv, \qquad B(w) = \frac{1}{\pi}\int_{-\infty}^{\infty} f(v)\sin wv\,dv$$

we can write this in the form

$$\textbf{(5)}\quad f(x) = \int_0^{\infty} [A(w)\cos wx + B(w)\sin wx]\,dw.$$

This is called a representation of $f(x)$ by a **Fourier integral.**

It is clear that our naive approach merely ***suggests*** the representation (5), but by no means establishes it; in fact, the limit of the series in (1) as Δw approaches zero is not the definition of the integral (3). Sufficient conditions for the validity of (5) are as follows.

THEOREM 1

Fourier Integral

If $f(x)$ is piecewise continuous (see Sec. 6.1) *in every finite interval and has a right-hand derivative and a left-hand derivative at every point* (see Sec 11.1) *and if the integral* (2) *exists, then $f(x)$ can be represented by a Fourier integral* (5) *with A and B given by* (4). *At a point where $f(x)$ is discontinuous the value of the Fourier integral equals the average of the left- and right-hand limits of $f(x)$ at that point* (see Sec. 11.1). (Proof in Ref. [C12]; see App. 1.)

Applications of Fourier Integrals

The main application of Fourier integrals is in solving ODEs and PDEs, as we shall see for PDEs in Sec. 12.6. However, we can also use Fourier integrals in integration and in discussing functions defined by integrals, as the next examples (2 and 3) illustrate.

EXAMPLE 2 Single Pulse, Sine Integral

Find the Fourier integral representation of the function

$$f(x) = \begin{cases} 1 & \text{if} \quad |x| < 1 \\ 0 & \text{if} \quad |x| > 1 \end{cases} \qquad \text{(Fig. 278).}$$

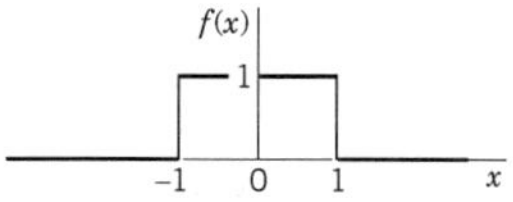

Fig. 278. Example 2

Solution. From (4) we obtain

$$A(w) = \frac{1}{\pi}\int_{-\infty}^{\infty} f(v)\cos wv\, dv = \frac{1}{\pi}\int_{-1}^{1}\cos wv\, dv = \frac{\sin wv}{\pi w}\bigg|_{-1}^{1} = \frac{2\sin w}{\pi w}$$

$$B(w) = \frac{1}{\pi}\int_{-1}^{1}\sin wv\, dv = 0$$

and (5) gives the answer

$$f(x) = \frac{2}{\pi}\int_{0}^{\infty}\frac{\cos wx \sin w}{w}\, dw. \tag{6}$$

The average of the left- and right-hand limits of $f(x)$ at $x = 1$ is equal to $(1 + 0)/2$, that is, $1/2$.

Furthermore, from (6) and Theorem 1 we obtain (multiply by $\pi/2$)

$$\int_{0}^{\infty}\frac{\cos wx \sin w}{w}\, dw = \begin{cases} \pi/2 & \text{if} \quad 0 \leqq x < 1, \\ \pi/4 & \text{if} \quad x = 1, \\ 0 & \text{if} \quad x > 1. \end{cases} \tag{7}$$

We mention that this integral is called **Dirichlet's discontinous factor.** (For P. L. Dirichlet see Sec. 10.8.)

The case $x = 0$ is of particular interest. If $x = 0$, then (7) gives

$$\int_{0}^{\infty}\frac{\sin w}{w}\, dw = \frac{\pi}{2}. \tag{8*}$$

We see that this integral is the limit of the so-called **sine integral**

$$\mathrm{Si}(u) = \int_{0}^{u}\frac{\sin w}{w}\, dw \tag{8}$$

as $u \to \infty$. The graphs of $\mathrm{Si}(u)$ and of the integrand are shown in Fig. 279.

In the case of a Fourier series the graphs of the partial sums are approximation curves of the curve of the periodic function represented by the series. Similarly, in the case of the Fourier integral (5), approximations are obtained by replacing ∞ by numbers a. Hence the integral

$$\frac{2}{\pi}\int_{0}^{a}\frac{\cos wx \sin w}{w}\, dw \tag{9}$$

approximates the right side in (6) and therefore $f(x)$.

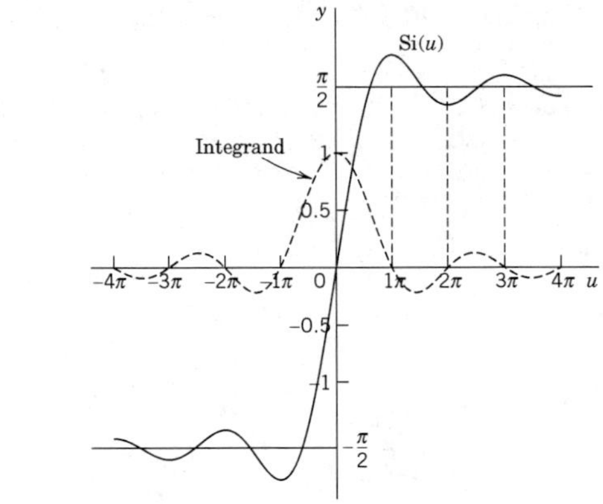

Fig. 279. Sine integral Si(u) and integrand

Figure 280 shows oscillations near the points of discontinuity of $f(x)$. We might expect that these oscillations disappear as a approaches infinity. But this is not true; with increasing a, they are shifted closer to the points $x = \pm 1$. This unexpected behavior, which also occurs in connection with Fourier series, is known as the **Gibbs phenomenon.** (See also Problem Set 11.2.) We can explain it by representing (9) in terms of sine integrals as follows. Using (11) in App. A3.1, we have

$$\frac{2}{\pi}\int_0^a \frac{\cos wx \sin w}{w}\, dw = \frac{1}{\pi}\int_0^a \frac{\sin(w + wx)}{w}\, dw + \frac{1}{\pi}\int_0^a \frac{\sin(w - wx)}{w}\, dw.$$

In the first integral on the right we set $w + wx = t$. Then $dw/w = dt/t$, and $0 \leqq w \leqq a$ corresponds to $0 \leqq t \leqq (x + 1)a$. In the last integral we set $w - wx = -t$. Then $dw/w = dt/t$, and $0 \leqq w \leqq a$ corresponds to $0 \leqq t \leqq (x - 1)a$. Since $\sin(-t) = -\sin t$, we thus obtain

$$\frac{2}{\pi}\int_0^a \frac{\cos wx \sin w}{w}\, dw = \frac{1}{\pi}\int_0^{(x+1)a} \frac{\sin t}{t}\, dt - \frac{1}{\pi}\int_0^{(x-1)a} \frac{\sin t}{t}\, dt.$$

From this and (8) we see that our integral (9) equals

$$\frac{1}{\pi}\,\mathrm{Si}(a[x + 1]) - \frac{1}{\pi}\,\mathrm{Si}(a[x - 1])$$

and the oscillations in Fig. 280 result from those in Fig. 279. The increase of a amounts to a transformation of the scale on the axis and causes the shift of the oscillations (the waves) toward the points of discontinuity -1 and 1. ■

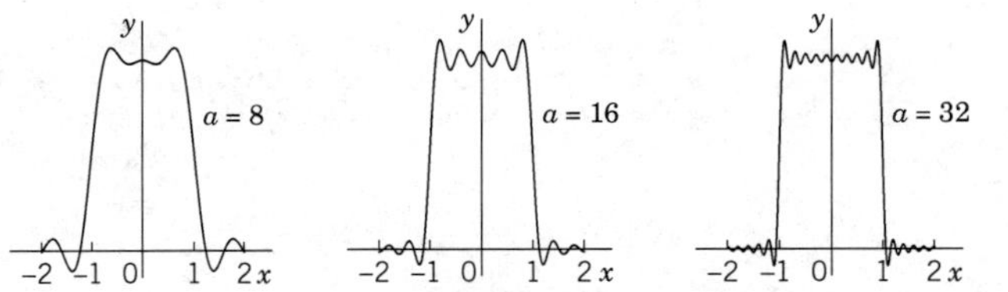

Fig. 280. The integral (9) for a = 8, 16, and 32

Fourier Cosine Integral and Fourier Sine Integral

For an even or odd function the Fourier integral becomes simpler. Just as in the case of Fourier series (Sec. 11.3), this is of practical interest in saving work and avoiding errors. The simplifications follow immediately from the formulas just obtained.

Indeed, if $f(x)$ is an ***even*** function, then $B(w) = 0$ in (4) and

$$A(w) = \frac{2}{\pi}\int_0^\infty f(v)\cos wv\, dv. \tag{10}$$

The Fourier integral (5) then reduces to the **Fourier cosine integral**

$$f(x) = \int_0^\infty A(w)\cos wx\, dw \qquad (f \text{ even}). \tag{11}$$

Similarly, if $f(x)$ is ***odd,*** then in (4) we have $A(w) = 0$ and

$$B(w) = \frac{2}{\pi}\int_0^\infty f(v)\sin wv\, dv. \tag{12}$$

The Fourier integral (5) then reduces to the **Fourier sine integral**

$$f(x) = \int_0^\infty B(w)\sin wx\, dw \qquad (f \text{ odd}). \tag{13}$$

Evaluation of Integrals

Earlier in this section we pointed out that the main application of the Fourier integral is in differential equations but that Fourier integral representations also help in evaluating certain integrals. To see this, we show the method for an important case, the Laplace integrals.

EXAMPLE 3 Laplace Integrals

We shall derive the Fourier cosine and Fourier sine integrals of $f(x) = e^{-kx}$, where $x > 0$ and $k > 0$ (Fig. 281). The result will be used to evaluate the so-called Laplace integrals.

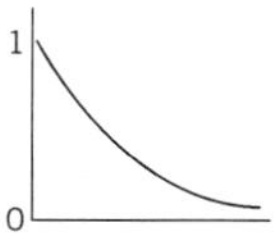

Fig. 281. $f(x)$ in Example 3

Solution. (a) From (10) we have $A(w) = \frac{2}{\pi}\int_0^\infty e^{-kv}\cos wv\, dv$. Now, by integration by parts,

$$\int e^{-kv}\cos wv\, dv = -\frac{k}{k^2+w^2}\, e^{-kv}\left(-\frac{w}{k}\sin wv + \cos wv\right).$$

If $v = 0$, the expression on the right equals $-k/(k^2 + w^2)$. If v approaches infinity, that expression approaches zero because of the exponential factor. Thus

$$A(w) = \frac{2k/\pi}{k^2+w^2}. \tag{14}$$

By substituting this into (11) we thus obtain the Fourier cosine integral representation

$$f(x) = e^{-kx} = \frac{2k}{\pi}\int_0^\infty \frac{\cos wx}{k^2+w^2}\, dw \qquad (x > 0,\quad k > 0).$$

From this representation we see that

$$\int_0^\infty \frac{\cos wx}{k^2 + w^2}\, dw = \frac{\pi}{2k} e^{-kx} \qquad (x > 0,\quad k > 0). \tag{15}$$

(b) Similarly, from (12) we have $B(w) = \frac{2}{\pi}\int_0^\infty e^{-kv} \sin wv\, dv$. By integration by parts,

$$\int e^{-kv} \sin wv\, dv = -\frac{w}{k^2 + w^2} e^{-kv}\left(\frac{k}{w}\sin wv + \cos wv\right).$$

This equals $-w/(k^2 + w^2)$ if $v = 0$, and approaches 0 as $v \to \infty$. Thus

$$B(w) = \frac{2w/\pi}{k^2 + w^2}. \tag{16}$$

From (13) we thus obtain the Fourier sine integral representation

$$f(x) = e^{-kx} = \frac{2}{\pi}\int_0^\infty \frac{w \sin wx}{k^2 + w^2}\, dw.$$

From this we see that

$$\int_0^\infty \frac{w \sin wx}{k^2 + w^2}\, dw = \frac{\pi}{2} e^{-kx} \qquad (x > 0,\quad k > 0). \tag{17}$$

The integrals (15) and (17) are called the **Laplace integrals.** ■

PROBLEM SET 11.7

1–6 EVALUATION OF INTEGRALS

Show that the given integral represents the indicated function. *Hint.* Use (5), (11), or (13); the integral tells you which one, and its value tells you what function to consider. (Show the details of your work.)

1. $$\int_0^\infty \frac{\cos xw + w \sin xw}{1 + w^2}\, dw = \begin{cases} 0 & \text{if } x < 0 \\ \pi/2 & \text{if } x = 0 \\ \pi e^{-x} & \text{if } x > 0 \end{cases}$$

2. $$\int_0^\infty \frac{\sin w - w \cos w}{w^2} \sin xw\, dw = \begin{cases} \pi x/2 & \text{if } 0 < x < 1 \\ \pi/4 & \text{if } x = 1 \\ 0 & \text{if } x > 1 \end{cases}$$

3. $$\int_0^\infty \frac{\cos xw}{1 + w^2}\, dw = \frac{\pi}{2} e^{-x} \text{ if } x > 0$$

4. $$\int_0^\infty \frac{\sin w}{w} \cos xw\, dw = \begin{cases} \pi/2 & \text{if } 0 \leqq x < 1 \\ \pi/4 & \text{if } x = 1 \\ 0 & \text{if } x > 1 \end{cases}$$

5. $$\int_0^\infty \frac{\cos(\pi w/2)}{1 - w^2} \cos xw\, dw = \begin{cases} \frac{\pi}{2}\cos x & \text{if } 0 < |x| < \pi/2 \\ 0 & \text{if } |x| \geqq \pi/2 \end{cases}$$

6. $$\int_0^\infty \frac{\sin \pi w \sin xw}{1 - w^2}\, dw = \begin{cases} \frac{\pi}{2}\sin x & \text{if } 0 \leqq x \leqq \pi \\ 0 & \text{if } x > \pi \end{cases}$$

7–12 FOURIER COSINE INTEGRAL REPRESENTATIONS

Represent $f(x)$ as an integral (11).

7. $$f(x) = \begin{cases} 1 & \text{if } 0 < x < a \\ 0 & \text{if } x > a \end{cases}$$

8. $f(x) = \begin{cases} x^2 & \text{if } 0 < x < a \\ 0 & \text{if } x > a \end{cases}$

9. $f(x) = \begin{cases} x & \text{if } 0 < x < 1 \\ 0 & \text{if } x > 1 \end{cases}$

10. $f(x) = \begin{cases} x/2 & \text{if } 0 < x < 1 \\ 1 - x/2 & \text{if } 1 < x < 2 \\ 0 & \text{if } x > 2 \end{cases}$

11. $f(x) = \begin{cases} \sin x & \text{if } 0 < x < \pi \\ 0 & \text{if } x > \pi \end{cases}$

12. $f(x) = \begin{cases} e^{-x} & \text{if } 0 < x < a \\ 0 & \text{if } x > a \end{cases}$

13. CAS EXPERIMENT. Approximate Fourier Cosine Integrals. Graph the integrals in Prob. 7, 9, and 11 as functions of x. Graph approximations obtained by replacing ∞ with finite upper limits of your choice. Compare the quality of the approximations. Write a short report on your empirical results and observations.

14–19 FOURIER SINE INTEGRAL REPRESENTATIONS

Represent $f(x)$ as an integral (13).

14. $f(x) = \begin{cases} 1 & \text{if } 0 < x < a \\ 0 & \text{if } x > a \end{cases}$

15. $f(x) = \begin{cases} \sin x & \text{if } 0 < x < \pi \\ 0 & \text{if } x > \pi \end{cases}$

16. $f(x) = \begin{cases} 1 - x^2 & \text{if } 0 < x < 1 \\ 0 & \text{if } x > 1 \end{cases}$

17. $f(x) = \begin{cases} \pi - x & \text{if } 0 < x < \pi \\ 0 & \text{if } x > \pi \end{cases}$

18. $f(x) = \begin{cases} \cos x & \text{if } 0 < x < \pi \\ 0 & \text{if } x > \pi \end{cases}$

19. $f(x) = \begin{cases} a - x & \text{if } 0 < x < a \\ 0 & \text{if } x > a \end{cases}$

20. PROJECT. Properties of Fourier Integrals
(a) Fourier cosine integral. Show that (11) implies

$$\text{(a1)} \quad f(ax) = \frac{1}{a}\int_0^\infty A\left(\frac{w}{a}\right) \cos xw \, dw$$

$(a > 0)$ *(Scale change)*

$$\text{(a2)} \quad xf(x) = \int_0^\infty B^*(w) \sin xw \, dw,$$

$$B^* = -\frac{dA}{dw}, \quad A \text{ as in (10)}$$

$$\text{(a3)} \quad x^2 f(x) = \int_0^\infty A^*(w) \cos xw \, dw,$$

$$A^* = -\frac{d^2A}{dw^2}.$$

(b) Solve Prob. 8 by applying (a3) to the result of Prob. 7.

(c) Verify (a2) for $f(x) = 1$ if $0 < x < a$ and $f(x) = 0$ if $x > a$.

(d) Fourier sine integral. Find formulas for the Fourier sine integral similar to those in (a).

11.8 Fourier Cosine and Sine Transforms

An **integral transform** is a transformation in the form of an integral that produces from given functions new functions depending on a different variable. These transformations are of interest mainly as tools for solving ODEs, PDEs, and integral equations, and they often also help in handling and applying special functions. The **Laplace transform** (Chap. 6) is of this kind and is by far the most important integral transform in engineering.

The next in order of importance are Fourier transforms. We shall see that these transforms can be obtained from the Fourier integral in Sec. 11.7 in a rather simple fashion. In this section we consider two of them, which are real, and in the next section a third one that is complex.

Fourier Cosine Transform

For an ***even*** function $f(x)$, the Fourier integral is the Fourier cosine integral

$$(1)\quad \text{(a)}\quad f(x) = \int_0^\infty A(w)\cos wx\, dw, \qquad \text{where} \qquad \text{(b)}\quad A(w) = \frac{2}{\pi}\int_0^\infty f(v)\cos wv\, dv$$

[see (10), (11), Sec. 11.7]. We now set $A(w) = \sqrt{2/\pi}\, \hat{f}_c(w)$, where c suggests "cosine." Then from (1b), writing $v = x$, we have

$$\textbf{(2)}\qquad \hat{f}_c(w) = \sqrt{\frac{2}{\pi}}\int_0^\infty f(x)\cos wx\, dx$$

and from (1a),

$$\textbf{(3)}\qquad f(x) = \sqrt{\frac{2}{\pi}}\int_0^\infty \hat{f}_c(w)\cos wx\, dw.$$

ATTENTION! In (2) we integrate with respect to x and in (3) with respect to w. Formula (2) gives from $f(x)$ a new function $\hat{f}_c(w)$, called the **Fourier cosine transform** of $f(x)$. Formula (3) gives us back $f(x)$ from $\hat{f}_c(w)$, and we therefore call $f(x)$ the **inverse Fourier cosine transform** of $\hat{f}_c(w)$.

The process of obtaining the transform $\hat{f}_c$ from a given f is also called the **Fourier cosine transform** or the *Fourier cosine transform method.*

Fourier Sine Transform

Similarly, for an ***odd*** function $f(x)$, the Fourier integral is the Fourier sine integral [see (12), (13), Sec. 11.7]

$$(4)\quad \text{(a)}\quad f(x) = \int_0^\infty B(w)\sin wx\, dw, \qquad \text{where} \qquad \text{(b)}\quad B(w) = \frac{2}{\pi}\int_0^\infty f(v)\sin wv\, dv.$$

We now set $B(w) = \sqrt{2/\pi}\, \hat{f}_s(w)$, where s suggests "sine." Then from (4b), writing $v = x$, we have

$$\textbf{(5)}\qquad \hat{f}_s(w) = \sqrt{\frac{2}{\pi}}\int_0^\infty f(x)\sin wx\, dx.$$

This is called the **Fourier sine transform** of $f(x)$. Similarly, from (4a) we have

$$\textbf{(6)}\qquad f(x) = \sqrt{\frac{2}{\pi}}\int_0^\infty \hat{f}_s(w)\sin wx\, dw.$$

This is called the **inverse Fourier sine transform** of $\hat{f}_s(w)$. The process of obtaining $\hat{f}_s(w)$ from $f(x)$ is also called the **Fourier sine transform** or the *Fourier sine transform method.*

Other notations are

$$\mathcal{F}_c(f) = \hat{f}_c, \qquad \mathcal{F}_s(f) = \hat{f}_s$$

and $\mathcal{F}_c^{-1}$ and $\mathcal{F}_s^{-1}$ for the inverses of $\mathcal{F}_c$ and $\mathcal{F}_s$, respectively.

EXAMPLE 1 Fourier Cosine and Fourier Sine Transforms

Find the Fourier cosine and Fourier sine transforms of the function

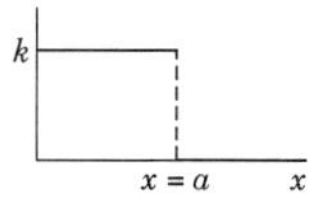

Fig. 282. $f(x)$ in Example 1

$$f(x) = \begin{cases} k & \text{if } 0 < x < a \\ 0 & \text{if } x > a \end{cases} \qquad \text{(Fig. 282).}$$

Solution. From the definitions (2) and (5) we obtain by integration

$$\hat{f}_c(w) = \sqrt{\frac{2}{\pi}}\, k \int_0^a \cos wx\, dx = \sqrt{\frac{2}{\pi}}\, k \left(\frac{\sin aw}{w}\right)$$

$$\hat{f}_s(w) = \sqrt{\frac{2}{\pi}}\, k \int_0^a \sin wx\, dx = \sqrt{\frac{2}{\pi}}\, k \left(\frac{1 - \cos aw}{w}\right).$$

This agrees with formulas 1 in the first two tables in Sec. 11.10 (where $k = 1$).

Note that for $f(x) = k = const$ $(0 < x < \infty)$, these transforms do not exist. (Why?) ■

EXAMPLE 2 Fourier Cosine Transform of the Exponential Function

Find $\mathcal{F}_c(e^{-x})$.

Solution. By integration by parts and recursion,

$$\mathcal{F}_c(e^{-x}) = \sqrt{\frac{2}{\pi}} \int_0^\infty e^{-x} \cos wx\, dx = \sqrt{\frac{2}{\pi}}\, \frac{e^{-x}}{1 + w^2}(-\cos wx + w \sin wx)\bigg|_0^\infty = \frac{\sqrt{2/\pi}}{1 + w^2}.$$

This agrees with formula 3 in Table I, Sec. 11.10, with $a = 1$. See also the next example. ■

What did we do to introduce the two integral transforms under consideration? Actually not much: We changed the notations A and B to get a "symmetric" distribution of the constant $2/\pi$ in the original formulas (10)–(13), Sec. 11.7. This redistribution is a standard convenience, but it is not essential. One could do without it.

What have we gained? We show next that these transforms have operational properties that permit them to convert differentiations into algebraic operations (just as the Laplace transform does). This is the key to their application in solving differential equations.

Linearity, Transforms of Derivatives

If $f(x)$ is absolutely integrable (see Sec. 11.7) on the positive x-axis and piecewise continuous (see Sec. 6.1) on every finite interval, then the Fourier cosine and sine transforms of f exist.

Furthermore, if f and g have Fourier cosine and sine transforms, so does $af + bg$ for any constants a and b, and by (2),

$$\begin{aligned}\mathcal{F}_c(af + bg) &= \sqrt{\frac{2}{\pi}} \int_0^\infty [af(x) + bg(x)] \cos wx\, dx \\ &= a\sqrt{\frac{2}{\pi}} \int_0^\infty f(x) \cos wx\, dx + b\sqrt{\frac{2}{\pi}} \int_0^\infty g(x) \cos wx\, dx.\end{aligned}$$

The right side is $a\mathcal{F}_c(f) + b\mathcal{F}_c(g)$. Similarly for $\mathcal{F}_s$, by (5). This shows that the Fourier cosine and sine transforms are **linear operations,**

$$\text{(7)} \qquad \begin{aligned} &\text{(a)} \quad \mathcal{F}_c(af + bg) = a\mathcal{F}_c(f) + b\mathcal{F}_c(g), \\ &\text{(b)} \quad \mathcal{F}_s(af + bg) = a\mathcal{F}_s(f) + b\mathcal{F}_s(g).\end{aligned}$$

THEOREM 1

Cosine and Sine Transforms of Derivatives

Let $f(x)$ be continuous and absolutely integrable on the x-axis, let $f'(x)$ be piecewise continuous on every finite interval, and let let $f(x) \to 0$ *as* $x \to \infty$. *Then*

$$\text{(8)} \quad \begin{aligned} &\text{(a)} \quad \mathcal{F}_c\{f'(x)\} = w\mathcal{F}_s\{f(x)\} - \sqrt{\frac{2}{\pi}}\, f(0), \\ &\text{(b)} \quad \mathcal{F}_s\{f'(x)\} = -w\mathcal{F}_c\{f(x)\}. \end{aligned}$$

PROOF This follows from the definitions by integration by parts, namely,

$$\begin{aligned} \mathcal{F}_c\{f'(x)\} &= \sqrt{\frac{2}{\pi}} \int_0^\infty f'(x) \cos wx \, dx \\ &= \sqrt{\frac{2}{\pi}} \left[f(x) \cos wx \Big|_0^\infty + w \int_0^\infty f(x) \sin wx \, dx \right] \\ &= -\sqrt{\frac{2}{\pi}}\, f(0) + w\mathcal{F}_s\{f(x)\}; \end{aligned}$$

and similarly,

$$\begin{aligned} \mathcal{F}_s\{f'(x)\} &= \sqrt{\frac{2}{\pi}} \int_0^\infty f'(x) \sin wx \, dx \\ &= \sqrt{\frac{2}{\pi}} \left[f(x) \sin wx \Big|_0^\infty - w \int_0^\infty f(x) \cos wx \, dx \right] \\ &= 0 - w\mathcal{F}_c\{f(x)\}. \end{aligned}$$

■

Formula (8a) with f' instead of f gives (when f', f'' satisfy the respective assumptions for f, f' in Theorem 1)

$$\mathcal{F}_c\{f''(x)\} = w\mathcal{F}_s\{f'(x)\} - \sqrt{\frac{2}{\pi}}\, f'(0);$$

hence by (8b)

$$\text{(9a)} \qquad \mathcal{F}_c\{f''(x)\} = -w^2\mathcal{F}_c\{f(x)\} - \sqrt{\frac{2}{\pi}}\, f'(0).$$

Similarly,

$$\text{(9b)} \qquad \mathcal{F}_s\{f''(x)\} = -w^2\mathcal{F}_s\{f(x)\} + \sqrt{\frac{2}{\pi}}\, wf(0).$$

A basic application of (9) to PDEs will be given in Sec. 12.6. For the time being we show how (9) can be used for deriving transforms.

EXAMPLE 3 An Application of the Operational Formula (9)

Find the Fourier cosine transform $\mathcal{F}_c(e^{-ax})$ of $f(x) = e^{-ax}$, where $a > 0$.

Solution. By differentiation, $(e^{-ax})'' = a^2 e^{-ax}$; thus

$$a^2 f(x) = f''(x).$$

From this, (9a), and the linearity (7a),

$$\begin{aligned} a^2 \mathcal{F}_c(f) &= \mathcal{F}_c(f'') \\ &= -w^2 \mathcal{F}_c(f) - \sqrt{\frac{2}{\pi}}\, f'(0) \\ &= -w^2 \mathcal{F}_c(f) + a\sqrt{\frac{2}{\pi}}\,. \end{aligned}$$

Hence

$$(a^2 + w^2)\mathcal{F}_c(f) = a\sqrt{2/\pi}.$$

The *answer* is (see Table I, Sec. 11.10)

$$\mathcal{F}_c(e^{-ax}) = \sqrt{\frac{2}{\pi}}\left(\frac{a}{a^2 + w^2}\right) \qquad (a > 0).$$ ■

Tables of Fourier cosine and sine transforms are included in Sec. 11.10.

PROBLEM SET 11.8

1–10 FOURIER COSINE TRANSFORM

1. Let $f(x) = -1$ if $0 < x < 1$, $f(x) = 1$ if $1 < x < 2$, $f(x) = 0$ if $x > 2$. Find $\hat{f}_c(w)$.
2. Let $f(x) = x$ if $0 < x < k$, $f(x) = 0$ if $x > k$. Find $\hat{f}_c(w)$.
3. Derive formula 3 in Table I of Sec. 11.10 by integration.
4. Find the inverse Fourier cosine transform $f(x)$ from the answer to Prob. 1. *Hint.* Use Prob. 4 in Sec. 11.7.
5. Obtain $\mathcal{F}_c^{-1}(1/(1 + w^2))$ from Prob. 3 in Sec. 11.7.
6. Obtain $\mathcal{F}_c^{-1}(e^{-w})$ by integration.
7. Find $\mathcal{F}_c((1 - x^2)^{-1} \cos(\pi x/2))$. *Hint.* Use Prob. 5 in Sec. 11.7.
8. Let $f(x) = x^2$ if $0 < x < 1$ and 0 if $x > 1$. Find $\mathcal{F}_c(f)$.
9. Does the Fourier cosine transform of $x^{-1} \sin x$ exist? Of $x^{-1} \cos x$? Give reasons.
10. $f(x) = 1$ $(0 < x < \infty)$ has no Fourier cosine or sine transform. Give reasons.

11–20 FOURIER SINE TRANSFORM

11. Find $\mathcal{F}_s(e^{-\pi x})$ by integration.
12. Find the answer to Prob. 11 from (9b).
13. Obtain formula 8 in Table II of Sec. 11.11 from (8b) and a suitable formula in Table I.
14. Let $f(x) = \sin x$ if $0 < x < \pi$ and 0 if $x > \pi$. Find $\mathcal{F}_s(f)$. Compare with Prob. 6 in Sec. 11.7. Comment.
15. In Table II of Sec. 11.10 obtain formula 2 from formula 4, using $\Gamma(\frac{1}{2}) = \sqrt{\pi}$ [(30) in App. 3.1].
16. Show that $\mathcal{F}_s(x^{-1/2}) = w^{-1/2}$ by setting $wx = t^2$ and using $S(\infty) = \sqrt{\pi/8}$ in (38) of App. 3.1.
17. Obtain $\mathcal{F}_s(e^{-ax})$ from (8a) and formula 3 in Table I of Sec. 11.10.
18. Show that $\mathcal{F}_s(x^{-3/2}) = 2w^{1/2}$. *Hint.* Set $wx = t^2$, integrate by parts, and use $C(\infty) = \sqrt{\pi/8}$ in (38) of App. 3.1.
19. **(Scale change)** Using the notation of (5), show that $f(ax)$ has the Fourier sine transform $(1/a)\hat{f}_s(w/a)$.
20. **WRITING PROJECT. Obtaining Fourier Cosine and Sine Transforms.** Write a short report on ways of obtaining these transforms, giving illustrations with examples of your own.

11.9 Fourier Transform. Discrete and Fast Fourier Transforms

The two transforms in the last section are real. We now consider a third one, called the **Fourier transform,** which is complex. We shall obtain this transform from the complex Fourier integral, which we explain first.

Complex Form of the Fourier Integral

The (real) Fourier integral is [see (4), (5), Sec. 11.7]

$$f(x) = \int_0^\infty [A(w) \cos wx + B(w) \sin wx]\, dw$$

where

$$A(w) = \frac{1}{\pi} \int_{-\infty}^{\infty} f(v) \cos wv\, dv, \qquad B(w) = \frac{1}{\pi} \int_{-\infty}^{\infty} f(v) \sin wv\, dv.$$

Substituting A and B into the integral for f, we have

$$f(x) = \frac{1}{\pi} \int_0^\infty \int_{-\infty}^{\infty} f(v) [\cos wv \cos wx + \sin wv \sin wx]\, dv\, dw.$$

By the addition formula for the cosine [(6) in App. A3.1] the expression in the brackets $[\cdots]$ equals $\cos (wv - wx)$ or, since the cosine is even, $\cos (wx - wv)$. We thus obtain

$$\text{(1*)} \qquad f(x) = \frac{1}{\pi} \int_0^\infty \left[\int_{-\infty}^{\infty} f(v) \cos (wx - wv)\, dv \right] dw.$$

The integral in brackets is an *even* function of w, call it $F(w)$, because $\cos (wx - wv)$ is an even function of w, the function f does not depend on w, and we integrate with respect to v (not w). Hence the integral of $F(w)$ from $w = 0$ to ∞ is 1/2 times the integral of $F(w)$ from $-\infty$ to ∞. Thus (note the change of the integration limit!)

$$\text{(1)} \qquad f(x) = \frac{1}{2\pi} \int_{-\infty}^{\infty} \left[\int_{-\infty}^{\infty} f(v) \cos (wx - wv)\, dv \right] dw.$$

We claim that the integral of the form (1) with sin instead of cos is zero:

$$\text{(2)} \qquad \frac{1}{2\pi} \int_{-\infty}^{\infty} \left[\int_{-\infty}^{\infty} f(v) \sin (wx - wv)\, dv \right] dw = 0.$$

This is true since $\sin (wx - wv)$ is an odd function of w, which makes the integral in brackets an odd function of w, call it $G(w)$. Hence the integral of $G(w)$ from $-\infty$ to ∞ is zero, as claimed.

We now take the integrand of (1) plus $i \; (= \sqrt{-1})$ times the integrand of (2) and use the **Euler formula** [(11) in Sec. 2.2]

$$\text{(3)} \qquad e^{ix} = \cos x + i \sin x.$$

Taking $wx - wv$ instead of x in (3) and multiplying by $f(v)$ gives

$$f(v) \cos (wx - wv) + if(v) \sin (wx - wv) = f(v)e^{i(wx-wv)}.$$

Hence the result of adding (1) plus i times (2), called the **complex Fourier integral,** is

$$f(x) = \frac{1}{2\pi} \int_{-\infty}^{\infty} \int_{-\infty}^{\infty} f(v)e^{iw(x-v)}\, dv\, dw \qquad (i = \sqrt{-1}). \tag{4}$$

It is now only a very short step to our present goal, the Fourier transform.

Fourier Transform and Its Inverse

Writing the exponential function in (4) as a product of exponential functions, we have

$$f(x) = \frac{1}{\sqrt{2\pi}} \int_{-\infty}^{\infty} \left[\frac{1}{\sqrt{2\pi}} \int_{-\infty}^{\infty} f(v)e^{-iwv}\, dv \right] e^{iwx}\, dw. \tag{5}$$

The expression in brackets is a function of w, is denoted by $\hat{f}(w)$, and is called the **Fourier transform** of f; writing $v = x$, we have

$$\hat{f}(w) = \frac{1}{\sqrt{2\pi}} \int_{-\infty}^{\infty} f(x)e^{-iwx}\, dx. \tag{6}$$

With this, (5) becomes

$$f(x) = \frac{1}{\sqrt{2\pi}} \int_{-\infty}^{\infty} \hat{f}(w)e^{iwx}\, dw \tag{7}$$

and is called the **inverse Fourier transform** of $\hat{f}(w)$.

Another notation for the Fourier transform is

$$\hat{f} = \mathcal{F}(f),$$

so that

$$f = \mathcal{F}^{-1}(\hat{f}).$$

The process of obtaining the Fourier transform $\mathcal{F}(f) = \hat{f}$ from a given f is also called the **Fourier transform** or the *Fourier transform method.*

Conditions sufficient for the existence of the Fourier transform (involving concepts defined in Secs. 6.1 and 11.7) are as follows, as we state without proof.

THEOREM 1

Existence of the Fourier Transform

If $f(x)$ is absolutely integrable on the x-axis and piecewise continuous on every finite interval, then the Fourier transform $\hat{f}(w)$ of $f(x)$ given by (6) exists.

EXAMPLE 1 **Fourier Transform**

Find the Fourier transform of $f(x) = 1$ if $|x| < 1$ and $f(x) = 0$ otherwise.

Solution. Using (6) and integrating, we obtain

$$\hat{f}(w) = \frac{1}{\sqrt{2\pi}} \int_{-1}^{1} e^{-iwx}\, dx = \frac{1}{\sqrt{2\pi}} \cdot \frac{e^{-iwx}}{-iw} \Bigg|_{-1}^{1} = \frac{1}{-iw\sqrt{2\pi}} (e^{-iw} - e^{iw}).$$

As in (3) we have $e^{iw} = \cos w + i \sin w$, $e^{-iw} = \cos w - i \sin w$, and by subtraction

$$e^{iw} - e^{-iw} = 2i \sin w.$$

Substituting this in the previous formula on the right, we see that i drops out and we obtain the answer

$$\hat{f}(w) = \sqrt{\frac{\pi}{2}}\, \frac{\sin w}{w}.$$ ■

EXAMPLE 2 **Fourier Transform**

Find the Fourier transform $\mathcal{F}(e^{-ax})$ of $f(x) = e^{-ax}$ if $x > 0$ and $f(x) = 0$ if $x < 0$; here $a > 0$.

Solution. From the definition (6) we obtain by integration

$$\mathcal{F}(e^{-ax}) = \frac{1}{\sqrt{2\pi}} \int_{0}^{\infty} e^{-ax} e^{-iwx}\, dx$$

$$= \frac{1}{\sqrt{2\pi}}\, \frac{e^{-(a+iw)x}}{-(a+iw)} \Bigg|_{x=0}^{\infty} = \frac{1}{\sqrt{2\pi}\,(a + iw)}.$$

This proves formula 5 of Table III in Sec. 11.10. ■

Physical Interpretation: Spectrum

The nature of the representation (7) of $f(x)$ becomes clear if we think of it as a superposition of sinusoidal oscillations of all possible frequencies, called a **spectral representation.** This name is suggested by optics, where light is such a superposition of colors (frequencies). In (7), the **"spectral density"** $\hat{f}(w)$ measures the intensity of $f(x)$ in the frequency interval between w and $w + \Delta w$ (Δw small, fixed). We claim that in connection with vibrations, the integral

$$\int_{-\infty}^{\infty} |\hat{f}(w)|^2\, dw$$

can be interpreted as the **total energy** of the physical system. Hence an integral of $|\hat{f}(w)|^2$ from a to b gives the contribution of the frequencies w between a and b to the total energy.

To make this plausible, we begin with a mechanical system giving a single frequency, namely, the harmonic oscillator (mass on a spring, Sec. 2.4)

$$my'' + ky = 0.$$

Here we denote time t by x. Multiplication by y' gives $my'y'' + ky'y = 0$. By integration,

$$\tfrac{1}{2}mv^2 + \tfrac{1}{2}ky^2 = E_0 = const$$

where $v = y'$ is the velocity. The first term is the kinetic energy, the second the potential energy, and E_0 the total energy of the system. Now a general solution is (use (3) in Sec. 11.4 with $t = x$)

$$y = a_1 \cos w_0 x + b_1 \sin w_0 x = c_1 e^{iw_0 x} + c_{-1} e^{-iw_0 x}, \qquad w_0^2 = k/m$$

where $c_1 = (a_1 - ib_1)/2$, $c_{-1} = \bar{c}_1 = (a_1 + ib_1)/2$. We write simply $A = c_1 e^{iw_0 x}$, $B = c_{-1} e^{-iw_0 x}$. Then $y = A + B$. By differentiation, $v = y' = A' + B' = iw_0(A - B)$. Substitution of v and y on the left side of the equation for E_0 gives

$$E_0 = \tfrac{1}{2}mv^2 + \tfrac{1}{2}ky^2 = \tfrac{1}{2}m(iw_0)^2(A - B)^2 + \tfrac{1}{2}k(A + B)^2.$$

Here $w_0^2 = k/m$, as just stated; hence $mw_0^2 = k$. Also $i^2 = -1$, so that

$$E_0 = \tfrac{1}{2}k[-(A - B)^2 + (A + B)^2] = 2kAB = 2kc_1 e^{iw_0 x} c_{-1} e^{-iw_0 x} = 2kc_1 c_{-1} = 2k|c_1|^2.$$

Hence *the energy is proportional to the square of the amplitude* $|c_1|$.

As the next step, if a more complicated system leads to a periodic solution $y = f(x)$ that can be represented by a Fourier series, then instead of the single energy term $|c_1|^2$ we get a series of squares $|c_n|^2$ of Fourier coefficients c_n given by (6), Sec. 11.4. In this case we have a **"discrete spectrum"** (or **"point spectrum"**) consisting of countably many isolated frequencies (infinitely many, in general), the corresponding $|c_n|^2$ being the contributions to the total energy.

Finally, a system whose solution can be represented by an integral (7) leads to the above integral for the energy, as is plausible from the cases just discussed.

Linearity. Fourier Transform of Derivatives

New transforms can be obtained from given ones by

THEOREM 2

Linearity of the Fourier Transform

The Fourier transform is a ***linear operation;*** *that is, for any functions $f(x)$ and $g(x)$ whose Fourier transforms exist and any constants a and b, the Fourier transform of $af + bg$ exists, and*

$$\mathcal{F}(af + bg) = a\mathcal{F}(f) + b\mathcal{F}(g). \tag{8}$$

PROOF This is true because integration is a linear operation, so that (6) gives

$$\begin{aligned}\mathcal{F}\{af(x) + bg(x)\} &= \frac{1}{\sqrt{2\pi}} \int_{-\infty}^{\infty} [af(x) + bg(x)]e^{-iwx}\, dx \\ &= a \frac{1}{\sqrt{2\pi}} \int_{-\infty}^{\infty} f(x)e^{-iwx}\, dx + b \frac{1}{\sqrt{2\pi}} \int_{-\infty}^{\infty} g(x)e^{-iwx}\, dx \\ &= a\mathcal{F}\{f(x)\} + b\mathcal{F}\{g(x)\}.\end{aligned}$$

■

In applying the Fourier transform to differential equations, the key property is that differentiation of functions corresponds to multiplication of transforms by iw:

THEOREM 3 **Fourier Transform of the Derivative of $f(x)$**

Let $f(x)$ be continuous on the x-axis and $f(x) \to 0$ as $|x| \to \infty$. Furthermore, let $f'(x)$ be absolutely integrable on the x-axis. Then

$$\mathcal{F}\{f'(x)\} = iw\mathcal{F}\{f(x)\}. \tag{9}$$

PROOF From the definition of the Fourier transform we have

$$\mathcal{F}\{f'(x)\} = \frac{1}{\sqrt{2\pi}} \int_{-\infty}^{\infty} f'(x)e^{-iwx}\,dx.$$

Integrating by parts, we obtain

$$\mathcal{F}\{f'(x)\} = \frac{1}{\sqrt{2\pi}} \left[f(x)e^{-iwx} \Big|_{-\infty}^{\infty} - (-iw)\int_{-\infty}^{\infty} f(x)e^{-iwx}\,dx \right].$$

Since $f(x) \to 0$ as $|x| \to \infty$, the desired result follows, namely,

$$\mathcal{F}\{f'(x)\} = 0 + iw\mathcal{F}\{f(x)\}.$$ ■

Two successive applications of (9) give

$$\mathcal{F}(f'') = iw\mathcal{F}(f') = (iw)^2\mathcal{F}(f).$$

Since $(iw)^2 = -w^2$, we have for the transform of the second derivative of f

$$\mathcal{F}\{f''(x)\} = -w^2\mathcal{F}\{f(x)\}. \tag{10}$$

Similarly for higher derivatives.

An application of (10) to differential equations will be given in Sec. 12.6. For the time being we show how (9) can be used to derive transforms.

EXAMPLE 3 **Application of the Operational Formula (9)**

Find the Fourier transform of xe^{-x^2} from Table III, Sec 11.10.

Solution. We use (9). By formula 9 in Table III.

$$\begin{aligned}
\mathcal{F}(xe^{-x^2}) &= \mathcal{F}\left\{-\frac{1}{2}\left(e^{-x^2}\right)'\right\} \\
&= -\frac{1}{2}\mathcal{F}\left\{\left(e^{-x^2}\right)'\right\} \\
&= -\frac{1}{2}\, iw\mathcal{F}(e^{-x^2}) \\
&= -\frac{1}{2}\, iw \frac{1}{\sqrt{2}}\, e^{-w^2/4} \\
&= -\frac{iw}{2\sqrt{2}}\, e^{-w^2/4}.
\end{aligned}$$ ■

Convolution

The **convolution** $f * g$ of functions f and g is defined by

(11) $$h(x) = (f * g)(x) = \int_{-\infty}^{\infty} f(p)g(x-p)\,dp = \int_{-\infty}^{\infty} f(x-p)g(p)\,dp.$$

The purpose is the same as in the case of Laplace transforms (Sec. 6.5): taking the convolution of two functions and then taking the transform of the convolution is the same as multiplying the transforms of these functions (and multiplying them by $\sqrt{2\pi}$):

THEOREM 4

Convolution Theorem

Suppose that $f(x)$ and $g(x)$ are piecewise continuous, bounded, and absolutely integrable on the x-axis. Then

(12) $$\mathcal{F}(f * g) = \sqrt{2\pi}\,\mathcal{F}(f)\mathcal{F}(g).$$

PROOF By the definition,

$$\mathcal{F}(f * g) = \frac{1}{\sqrt{2\pi}} \int_{-\infty}^{\infty} \int_{-\infty}^{\infty} f(p)g(x-p)\,dp\; e^{-iwx}\,dx.$$

An interchange of the order of integration gives

$$\mathcal{F}(f * g) = \frac{1}{\sqrt{2\pi}} \int_{-\infty}^{\infty} \int_{-\infty}^{\infty} f(p)g(x-p)e^{-iwx}\,dx\,dp.$$

Instead of x we now take $x - p = q$ as a new variable of integration. Then $x = p + q$ and

$$\mathcal{F}(f * g) = \frac{1}{\sqrt{2\pi}} \int_{-\infty}^{\infty} \int_{-\infty}^{\infty} f(p)g(q)e^{-iw(p+q)}\,dq\,dp.$$

This double integral can be written as a product of two integrals and gives the desired result

$$\mathcal{F}(f * g) = \frac{1}{\sqrt{2\pi}} \int_{-\infty}^{\infty} f(p)e^{-iwp}\,dp \int_{-\infty}^{\infty} g(q)e^{-iwq}\,dq$$

$$= \frac{1}{\sqrt{2\pi}} [\sqrt{2\pi}\,\mathcal{F}(f)][\sqrt{2\pi}\,\mathcal{F}(g)] = \sqrt{2\pi}\,\mathcal{F}(f)\mathcal{F}(g). \qquad \blacksquare$$

By taking the inverse Fourier transform on both sides of (12), writing $\hat{f} = \mathcal{F}(f)$ and $\hat{g} = \mathcal{F}(g)$ as before, and noting that $\sqrt{2\pi}$ and $1/\sqrt{2\pi}$ in (12) and (7) cancel each other, we obtain

(13) $$(f * g)(x) = \int_{-\infty}^{\infty} \hat{f}(w)\hat{g}(w)e^{iwx}\,dw,$$

a formula that will help us in solving partial differential equations (Sec. 12.6).

Discrete Fourier Transform (DFT), Fast Fourier Transform (FFT)

In using Fourier series, Fourier transforms, and trigonometric approximations (Sec. 11.6) we have to assume that a function $f(x)$, to be developed or transformed, is given on some interval, over which we integrate in the Euler formulas, etc. Now very often a function $f(x)$ is given only in terms of values at finitely many points, and one is interested in extending Fourier analysis to this case. The main application of such a "discrete Fourier analysis" concerns large amounts of equally spaced data, as they occur in telecommunication, time series analysis, and various simulation problems. In these situations, dealing with sampled values rather than with functions, we can replace the Fourier transform by the so-called **discrete Fourier transform (DFT)** as follows.

Let $f(x)$ be periodic, for simplicity of period 2π. We assume that N measurements of $f(x)$ are taken over the interval $0 \leqq x \leqq 2\pi$ at regularly spaced points

$$x_k = \frac{2\pi k}{N}, \qquad k = 0, 1, \cdots, N-1. \tag{14}$$

We also say that $f(x)$ is being **sampled** at these points. We now want to determine a **complex trigonometric polynomial**

$$q(x) = \sum_{n=0}^{N-1} c_n e^{inx_k} \tag{15}$$

that **interpolates** $f(x)$ at the nodes (14), that is, $q(x_k) = f(x_k)$, written out, with f_k denoting $f(x_k)$,

$$f_k = f(x_k) = q(x_k) = \sum_{n=0}^{N-1} c_n e^{inx_k}, \qquad k = 0, 1, \cdots, N-1. \tag{16}$$

Hence we must determine the coefficients $c_0, \cdots, c_{N-1}$ such that (16) holds. We do this by an idea similar to that in Sec. 11.1 for deriving the Fourier coefficients by using the orthogonality of the trigonometric system. Instead of integrals we now take sums. Namely, we multiply (16) by e^{-imx_k} (note the minus!) and sum over k from 0 to $N-1$. Then we interchange the order of the two summations and insert x_k from (14). This gives

$$\sum_{k=0}^{N-1} f_k e^{-imx_k} = \sum_{k=0}^{N-1}\sum_{n=0}^{N-1} c_n e^{i(n-m)x_k} = \sum_{n=0}^{N-1} c_n \sum_{k=0}^{N-1} e^{i(n-m)2\pi k/N}. \tag{17}$$

Now

$$e^{i(n-m)2\pi k/N} = \left[e^{i(n-m)2\pi/N}\right]^k.$$

We donote $[\cdots]$ by r. For $n = m$ we have $r = e^0 = 1$. The sum of *these* terms over k equals N, the number of these terms. For $n \neq m$ we have $r \neq 1$ and by the formula for a geometric sum [(6) in Sec. 15.1 with $q = r$ and $n = N-1$]

$$\sum_{k=0}^{N-1} r^k = \frac{1 - r^N}{1 - r} = 0$$

because $r^N = 1$; indeed, since k, m, and n are integers,

$$r^N = e^{i(n-m)2\pi k} = \cos 2\pi k(n - m) + i \sin 2\pi k(n - m) = 1 + 0 = 1.$$

This shows that the right side of (17) equals $c_m N$. Writing n for m and dividing by N, we thus obtain the desired coefficient formula

$$\textbf{(18*)} \qquad c_n = \frac{1}{N} \sum_{k=0}^{N-1} f_k e^{-inx_k} \qquad f_k = f(x_k), \quad n = 0, 1, \cdots, N - 1.$$

Since computation of the c_n (by the fast Fourier transform, below) involves successive halfing of the problem size N, it is practical to drop the factor $1/N$ from c_n and define the **discrete Fourier transform** of the given signal $\mathbf{f} = [f_0 \quad \cdots \quad f_{N-1}]^\mathsf{T}$ to be the vector $\hat{\mathbf{f}} = [\hat{f}_0 \quad \cdots \quad \hat{f}_{N-1}]$ with components

$$\textbf{(18)} \qquad \hat{f}_n = Nc_n = \sum_{k=0}^{N-1} f_k e^{-inx_k}, \qquad f_k = f(x_k), \quad n = 0, \cdots, N - 1.$$

This is the frequency spectrum of the signal.

In vector notation, $\hat{\mathbf{f}} = \mathbf{F}_N \mathbf{f}$, where the $N \times N$ **Fourier matrix** $\mathbf{F}_N = [e_{nk}]$ has the entries [given in (18)]

$$(19) \qquad e_{nk} = e^{-inx_k} = e^{-2\pi ink/N} = w^{nk}, \qquad w = w_N = e^{-2\pi i/N},$$

where $n, k = 0, \cdots, N - 1$.

EXAMPLE 4 Discrete Fourier Transform (DFT). Sample of $N = 4$ Values

Let $N = 4$ measurements (sample values) be given. Then $w = e^{-2\pi i/N} = e^{-\pi i/2} = -i$ and thus $w^{nk} = (-i)^{nk}$. Let the sample values be, say $\mathbf{f} = [0 \quad 1 \quad 4 \quad 9]^\mathsf{T}$. Then by (18) and (19),

$$(20) \qquad \hat{\mathbf{f}} = \mathbf{F}_4 \mathbf{f} = \begin{bmatrix} w^0 & w^0 & w^0 & w^0 \\ w^0 & w^1 & w^2 & w^3 \\ w^0 & w^2 & w^4 & w^6 \\ w^0 & w^3 & w^6 & w^9 \end{bmatrix} \mathbf{f} = \begin{bmatrix} 1 & 1 & 1 & 1 \\ 1 & -i & -1 & i \\ 1 & -1 & 1 & -1 \\ 1 & i & -1 & -i \end{bmatrix} \begin{bmatrix} 0 \\ 1 \\ 4 \\ 9 \end{bmatrix} = \begin{bmatrix} 14 \\ -4 + 8i \\ -6 \\ -4 - 8i \end{bmatrix}.$$

From the first matrix in (20) it is easy to infer what $\mathbf{F}_N$ looks like for arbitrary N, which in practice may be 1000 or more, for reasons given below. ■

From the DFT (the frequency spectrum) $\hat{\mathbf{f}} = \mathbf{F}_N \mathbf{f}$ we can recreate the given signal $\mathbf{f} = \mathbf{F}_N^{-1} \hat{\mathbf{f}}$, as we shall now prove. Here $\mathbf{F}_N$ and its complex conjugate $\overline{\mathbf{F}}_N = \frac{1}{N} [\overline{w}^{nk}]$ satisfy

$$(21a) \qquad \overline{\mathbf{F}}_N \mathbf{F}_N = \mathbf{F}_N \overline{\mathbf{F}}_N = N\mathbf{I}$$

where $\mathbf{I}$ is the $N \times N$ unit matrix; hence $\mathbf{F}_N$ has the inverse

$$(21b) \qquad \mathbf{F}_N^{-1} = \frac{1}{N} \overline{\mathbf{F}}_N.$$

PROOF We prove (21). By the multiplication rule (row times column) the product matrix $\mathbf{G}_N = \overline{\mathbf{F}}_N \mathbf{F}_N = [g_{jk}]$ in (21a) has the entries g_{jk} = *Row* j *of* $\overline{\mathbf{F}}_N$ *times Column* k *of* $\mathbf{F}_N$. That is, writing $W = \overline{w}^j w^k$, we prove that

$$g_{jk} = (\overline{w}^j w^k)^0 + (\overline{w}^j w^k)^1 + \cdots + (\overline{w}^j w^k)^{N-1}$$

$$= W^0 + W^1 + \cdots + W^{N-1} = \begin{cases} 0 & \text{if} \quad j \neq k \\ N & \text{if} \quad j = k. \end{cases}$$

Indeed, when $j = k$, then $\overline{w}^k w^k = (\overline{w}w)^k = (e^{2\pi i/N}e^{-2\pi i/N})^k = 1^k = 1$, so that the sum of *these* N terms equals N; these are the diagonal entries of $\mathbf{G}_N$. Also, when $j \neq k$, then $W \neq 1$ and we have a geometric sum (whose value is given by (6) in Sec. 15.1 with $q = W$ and $n = N - 1$)

$$W^0 + W^1 + \cdots + W^{N-1} = \frac{1 - W^N}{1 - W} = 0$$

because $W^N = (\overline{w}^j w^k)^N = (e^{2\pi i})^j (e^{-2\pi i})^k = 1^j \cdot 1^k = 1$. ■

We have seen that $\hat{\mathbf{f}}$ is the frequency spectrum of the signal $f(x)$. Thus the components $\hat{f}_n$ of $\hat{\mathbf{f}}$ give a resolution of the 2π-periodic function $f(x)$ into simple (complex) harmonics. Here one should use only n's that are much smaller than $N/2$, to avoid **aliasing.** By this we mean the effect caused by sampling at too few (equally spaced) points, so that, for instance, in a motion picture, rotating wheels appear as rotating too slowly or even in the wrong sense. Hence in applications, N is usually large. But this poses a problem. Eq. (18) requires $O(N)$ operations for any particular n, hence $O(N^2)$ operations for, say, all $n < N/2$. Thus, already for 1000 sample points the straightforward calculation would involve millions of operations. However, this difficulty can be overcome by the so called **fast Fourier transform (FFT),** for which codes are readily available (e.g. in Maple). The FFT is a computational method for the DFT that needs only $O(N) \log_2 N$ operations instead of $O(N^2)$. It makes the DFT a practical tool for large N. Here one chooses $N = 2^p$ (p integer) and uses the special form of the Fourier matrix to break down the given problem into smaller problems. For instance, when $N = 1000$, those operations are reduced by a factor $1000/\log_2 1000 \approx 100$.

The breakdown produces two problems of size $M = N/2$. This breakdown is possible because for $N = 2M$ we have in (19)

$$w_N{}^2 = w_{2M}{}^2 = (e^{-2\pi i/N})^2 = e^{-4\pi i/(2M)} = e^{-2\pi i/M} = w_M.$$

The given vector $\mathbf{f} = [f_0 \quad \cdots \quad f_{N-1}]^\mathsf{T}$ is split into two vectors with M components each, namely, $\mathbf{f}_{\text{ev}} = [f_0 \quad f_2 \quad \cdots \quad f_{N-2}]^\mathsf{T}$ containing the even components of $\mathbf{f}$, and $\mathbf{f}_{\text{od}} = [f_1 \quad f_3 \quad \cdots \quad f_{N-1}]^\mathsf{T}$ containing the odd components of $\mathbf{f}$. For $\mathbf{f}_{\text{ev}}$ and $\mathbf{f}_{\text{od}}$ we determine the DFTs

$$\hat{\mathbf{f}}_{\text{ev}} = \left[\hat{f}_{\text{ev},0} \quad \hat{f}_{\text{ev},2} \quad \cdots \quad \hat{f}_{\text{ev},N-2}\right]^\mathsf{T} = \mathbf{F}_M \mathbf{f}_{\text{ev}}$$

and

$$\hat{\mathbf{f}}_{\text{od}} = \left[\hat{f}_{\text{od},1} \quad \hat{f}_{\text{od},3} \quad \cdots \quad \hat{f}_{\text{od},N-1}\right]^\mathsf{T} = \mathbf{F}_M \mathbf{f}_{\text{od}}$$

involving the same $M \times M$ matrix $\mathbf{F}_M$. From these vectors we obtain the components of the DFT of the given vector $\mathbf{f}$ by the formulas

$$\text{(22)} \qquad \begin{aligned} &\text{(a)} \quad \hat{f}_n = \hat{f}_{\text{ev},n} + w_N{}^n \hat{f}_{\text{od},n} && n = 0, \cdots, M - 1 \\ &\text{(b)} \quad \hat{f}_{n+M} = \hat{f}_{\text{ev},n} - w_N{}^n \hat{f}_{\text{od},n} && n = 0, \cdots, M - 1. \end{aligned}$$

For $N = 2^p$ this breakdown can be repeated $p - 1$ times in order to finally arrive at $N/2$ problems of size 2 each, so that the number of multiplications is reduced as indicated above.

We show the reduction from $N = 4$ to $M = N/2 = 2$ and then prove (22).

EXAMPLE 5 Fast Fourier Transform (FFT). Sample of $N = 4$ Values

When $N = 4$, then $w = w_N = -i$ as in Example 4 and $M = N/2 = 2$, hence $w = w_M = e^{-2\pi i/2} = e^{-\pi i} = -1$. Consequently,

$$\hat{\mathbf{f}}_{\text{ev}} = \begin{bmatrix} \hat{f}_0 \\ \hat{f}_2 \end{bmatrix} = \mathbf{F}_2 \mathbf{f}_{\text{ev}} = \begin{bmatrix} 1 & 1 \\ 1 & -1 \end{bmatrix} \begin{bmatrix} f_0 \\ f_2 \end{bmatrix} = \begin{bmatrix} f_0 + f_2 \\ f_0 - f_2 \end{bmatrix}$$

$$\hat{\mathbf{f}}_{\text{od}} = \begin{bmatrix} \hat{f}_1 \\ \hat{f}_3 \end{bmatrix} = \mathbf{F}_2 \mathbf{f}_{\text{od}} = \begin{bmatrix} 1 & 1 \\ 1 & -1 \end{bmatrix} \begin{bmatrix} f_1 \\ f_3 \end{bmatrix} = \begin{bmatrix} f_1 + f_3 \\ f_1 - f_3 \end{bmatrix}.$$

From this and (22a) we obtain

$$\hat{f}_0 = \hat{f}_{\text{ev},0} + w_N{}^0 \hat{f}_{\text{od},0} = (f_0 + f_2) + (f_1 + f_3) = f_0 + f_1 + f_2 + f_3$$

$$\hat{f}_1 = \hat{f}_{\text{ev},1} + w_N{}^1 \hat{f}_{\text{od},1} = (f_0 - f_2) - i(f_1 + f_3) = f_0 - if_1 - f_2 + if_3.$$

Similarly, by (22b),

$$\hat{f}_2 = \hat{f}_{\text{ev},0} - w_N{}^0 \hat{f}_{\text{od},0} = (f_0 + f_2) - (f_1 + f_3) \qquad = f_0 - f_1 + f_2 - f_3$$

$$\hat{f}_3 = \hat{f}_{\text{ev},1} - w_N{}^1 \hat{f}_{\text{od},1} = (f_0 - f_2) - (-i)(f_1 - f_3) = f_0 + if_1 - f_2 - if_3.$$

This agrees with Example 4, as can be seen by replacing 0, 1, 4, 9 with f_0, f_1, f_2, f_3. ■

We prove (22). From (18) and (19) we have for the components of the DFT

$$\hat{f}_n = \sum_{k=0}^{N-1} w_N^{kn} f_k.$$

Splitting into two sums of $M = N/2$ terms each gives

$$\hat{f}_n = \sum_{k=0}^{M-1} w_N^{2kn} f_{2k} + \sum_{k=0}^{M-1} w_N^{(2k+1)n} f_{2k+1}.$$

We now use $w_N{}^2 = w_M$ and pull out $w_N{}^n$ from under the second sum, obtaining

$$\hat{f}_n = \sum_{k=0}^{M-1} w_M^{kn} f_{\text{ev},k} + w_N{}^n \sum_{k=0}^{M-1} w_M^{kn} f_{\text{od},k}. \tag{23}$$

The two sums are $f_{\text{ev},n}$ and $f_{\text{od},n}$, the components of the "half-size" transforms $\mathbf{F}\,\mathbf{f}_{\text{ev}}$ and $\mathbf{F}\,\mathbf{f}_{\text{od}}$.

Formula (22a) is the same as (23). In (22b) we have $n + M$ instead of n. This causes a sign change in (23), namely $-w_N{}^n$ before the second sum because

$$w_N{}^M = e^{-2\pi iM/N} = e^{-2\pi i/2} = e^{-\pi i} = -1.$$

This gives the minus in (22b) and completes the proof. ■

PROBLEM SET 11.9

1. **(Review)** Show that $1/i = -i$, $e^{ix} + e^{-ix} = 2\cos x$, $e^{ix} - e^{-ix} = 2i \sin x$.

2–9 FOURIER TRANSFORMS BY INTEGRATION

Find the Fourier transform of $f(x)$ (without using Table III in Sec. 11.10). Show the details.

2. $f(x) = \begin{cases} e^{kx} & \text{if } x < 0 \quad (k > 0) \\ 0 & \text{if } x > 0 \end{cases}$

3. $f(x) = \begin{cases} k & \text{if } 0 < x < b \\ 0 & \text{otherwise} \end{cases}$

4. $f(x) = \begin{cases} e^{2ix} & \text{if } -1 < x < 1 \\ 0 & \text{otherwise} \end{cases}$

5. $f(x) = \begin{cases} k & \text{if } -1 < x < 1 \\ 0 & \text{otherwise} \end{cases}$

6. $f(x) = \begin{cases} x & \text{if } -1 < x < 1 \\ 0 & \text{otherwise} \end{cases}$

7. $f(x) = \begin{cases} x & \text{if } 0 < x < 1 \\ 0 & \text{otherwise} \end{cases}$

8. $f(x) = \begin{cases} xe^{-x} & \text{if } -1 < x < 0 \\ 0 & \text{otherwise} \end{cases}$

9. $f(x) = \begin{cases} -1 & \text{if } -1 < x < 0 \\ 1 & \text{if } 0 < x < 1 \\ 0 & \text{otherwise} \end{cases}$

OTHER METHODS

10. Find the Fourier transform of $f(x) = xe^{-x}$ if $x > 0$ and 0 if $x < 0$ from formula 5 in Table III and (9) in the text. *Hint:* Consider xe^{-x} and e^{-x}.

11. Obtain $\mathcal{F}(e^{-x^2/2})$ from formula 9 in Table III.

12. Obtain formula 7 in Table III from formula 8.

13. Obtain formula 1 in Table III from formula 2.

14. **TEAM PROJECT. Shifting. (a)** Show that if $f(x)$ has a Fourier transform, so does $f(x - a)$, and $\mathcal{F}\{f(x - a)\} = e^{-iwa}\mathcal{F}\{f(x)\}$.

    (b) Using (a), obtain formula 1 in Table III, Sec. 11.10, from formula 2.

    (c) Shifting on the w-Axis. Show that if $\hat{f}(w)$ is the Fourier transform of $f(x)$, then $\hat{f}(w - a)$ is the Fourier transform of $e^{iax}f(x)$.

    (d) Using (c), obtain formula 7 in Table III from 1 and formula 8 from 2.

11.10 Tables of Transforms

Table I. Fourier Cosine Transforms

See (2) in Sec. 11.8.

	$f(x)$	$\hat{f}_c(w) = \mathcal{F}_c(f)$
1	$\begin{cases} 1 & \text{if } 0 < x < a \\ 0 & \text{otherwise} \end{cases}$	$\sqrt{\frac{2}{\pi}} \frac{\sin aw}{w}$
2	x^{a-1} $(0 < a < 1)$	$\sqrt{\frac{2}{\pi}} \frac{\Gamma(a)}{w^a} \cos \frac{a\pi}{2}$ ($\Gamma(a)$ see App. A3.1.)
3	e^{-ax} $(a > 0)$	$\sqrt{\frac{2}{\pi}} \left(\frac{a}{a^2 + w^2}\right)$
4	$e^{-x^2/2}$	$e^{-w^2/2}$
5	e^{-ax^2} $(a > 0)$	$\frac{1}{\sqrt{2a}} e^{-w^2/(4a)}$
6	$x^n e^{-ax}$ $(a > 0)$	$\sqrt{\frac{2}{\pi}} \frac{n!}{(a^2 + w^2)^{n+1}} \text{Re}\,(a + iw)^{n+1}$ Re = Real part
7	$\begin{cases} \cos x & \text{if } 0 < x < a \\ 0 & \text{otherwise} \end{cases}$	$\frac{1}{\sqrt{2\pi}} \left[\frac{\sin a(1 - w)}{1 - w} + \frac{\sin a(1 + w)}{1 + w}\right]$
8	$\cos (ax^2)$ $(a > 0)$	$\frac{1}{\sqrt{2a}} \cos \left(\frac{w^2}{4a} - \frac{\pi}{4}\right)$
9	$\sin (ax^2)$ $(a > 0)$	$\frac{1}{\sqrt{2a}} \cos \left(\frac{w^2}{4a} + \frac{\pi}{4}\right)$
10	$\frac{\sin ax}{x}$ $(a > 0)$	$\sqrt{\frac{\pi}{2}} (1 - u(w - a))$ (See Sec. 6.3.)
11	$\frac{e^{-x} \sin x}{x}$	$\frac{1}{\sqrt{2\pi}} \arctan \frac{2}{w^2}$
12	$J_0(ax)$ $(a > 0)$	$\sqrt{\frac{2}{\pi}} \frac{1}{\sqrt{a^2 - w^2}} (1 - u(w - a))$ (See Secs. 5.5, 6.3.)

Table II. Fourier Sine Transforms

See (5) in Sec. 11.8.

	$f(x)$	$\hat{f}_s(w) = \mathcal{F}_s(f)$
1	$\begin{cases} 1 & \text{if } 0 < x < a \\ 0 & \text{otherwise} \end{cases}$	$\sqrt{\frac{2}{\pi}} \left[\frac{1 - \cos aw}{w} \right]$
2	$1/\sqrt{x}$	$1/\sqrt{w}$
3	$1/x^{3/2}$	$2\sqrt{w}$
4	x^{a-1} $(0 < a < 1)$	$\sqrt{\frac{2}{\pi}} \frac{\Gamma(a)}{w^a} \sin \frac{a\pi}{2}$ ($\Gamma(a)$ see App. A3.1.)
5	e^{-ax} $(a > 0)$	$\sqrt{\frac{2}{\pi}} \left(\frac{w}{a^2 + w^2} \right)$
6	$\frac{e^{-ax}}{x}$ $(a > 0)$	$\sqrt{\frac{2}{\pi}} \arctan \frac{w}{a}$
7	$x^n e^{-ax}$ $(a > 0)$	$\sqrt{\frac{2}{\pi}} \frac{n!}{(a^2 + w^2)^{n+1}} \operatorname{Im} (a + iw)^{n+1}$ Im = Imaginary part
8	$xe^{-x^2/2}$	$we^{-w^2/2}$
9	xe^{-ax^2} $(a > 0)$	$\frac{w}{(2a)^{3/2}} e^{-w^2/4a}$
10	$\begin{cases} \sin x & \text{if } 0 < x < a \\ 0 & \text{otherwise} \end{cases}$	$\frac{1}{\sqrt{2\pi}} \left[\frac{\sin a(1 - w)}{1 - w} - \frac{\sin a(1 + w)}{1 + w} \right]$
11	$\frac{\cos ax}{x}$ $(a > 0)$	$\sqrt{\frac{\pi}{2}} u(w - a)$ (See Sec. 6.3.)
12	$\arctan \frac{2a}{x}$ $(a > 0)$	$\sqrt{2\pi} \frac{\sinh aw}{w} e^{-aw}$

Table III. Fourier Transforms

See (6) in Sec. 11.9.

	$f(x)$	$\hat{f}(w) = \mathcal{F}(f)$
1	$\begin{cases} 1 & \text{if } -b < x < b \\ 0 & \text{otherwise} \end{cases}$	$\sqrt{\frac{2}{\pi}}\,\frac{\sin bw}{w}$
2	$\begin{cases} 1 & \text{if } b < x < c \\ 0 & \text{otherwise} \end{cases}$	$\frac{e^{-ibw} - e^{-icw}}{iw\sqrt{2\pi}}$
3	$\frac{1}{x^2 + a^2} \quad (a > 0)$	$\sqrt{\frac{\pi}{2}}\,\frac{e^{-a\lvert w\rvert}}{a}$
4	$\begin{cases} x & \text{if } 0 < x < b \\ 2x - b & \text{if } b < x < 2b \\ 0 & \text{otherwise} \end{cases}$	$\frac{-1 + 2e^{ibw} - e^{-2ibw}}{\sqrt{2\pi}\, w^2}$
5	$\begin{cases} e^{-ax} & \text{if } x > 0 \\ 0 & \text{otherwise} \end{cases} \quad (a > 0)$	$\frac{1}{\sqrt{2\pi}(a + iw)}$
6	$\begin{cases} e^{ax} & \text{if } b < x < c \\ 0 & \text{otherwise} \end{cases}$	$\frac{e^{(a-iw)c} - e^{(a-iw)b}}{\sqrt{2\pi}(a - iw)}$
7	$\begin{cases} e^{iax} & \text{if } -b < x < b \\ 0 & \text{otherwise} \end{cases}$	$\sqrt{\frac{2}{\pi}}\,\frac{\sin b(w - a)}{w - a}$
8	$\begin{cases} e^{iax} & \text{if } b < x < c \\ 0 & \text{otherwise} \end{cases}$	$\frac{i}{\sqrt{2\pi}}\,\frac{e^{ib(a-w)} - e^{ic(a-w)}}{a - w}$
9	$e^{-ax^2} \quad (a > 0)$	$\frac{1}{\sqrt{2a}}\, e^{-w^2/4a}$
10	$\frac{\sin ax}{x} \quad (a > 0)$	$\sqrt{\frac{\pi}{2}}$ if $\lvert w\rvert < a$; 0 if $\lvert w\rvert > a$

CHAPTER 11 REVIEW QUESTIONS AND PROBLEMS

1. What is a Fourier series? A Fourier sine series? A half-range expansion?
2. Can a discontinuous function have a Fourier series? A Taylor series? Explain.
3. Why did we start with period 2π? How did we proceed to functions of any period p?
4. What is the trigonometric system? Its main property by which we obtained the Euler formulas?
5. What do you know about the convergence of a Fourier series?
6. What is the Gibbs phenomenon?
7. What is approximation by trigonometric polynomials? The minimum square error?
8. What is remarkable about the response of a vibrating system to an *arbitrary* periodic force?
9. What do you know about the Fourier integral? Its applications?
10. What is the Fourier sine transform? Give examples.

11–20 **FOURIER SERIES**

Find the Fourier series of $f(x)$ as given over one period. Sketch $f(x)$. (Show the details of your work.)

11. $f(x) = \begin{cases} -k & \text{if } -1 < x < 0 \\ k & \text{if } \quad 0 < x < 1 \end{cases}$

12. $f(x) = \begin{cases} 0 & \text{if } -\pi/2 < x < \pi/2 \\ 1 & \text{if } \quad \pi/2 < x < 3\pi/2 \end{cases}$

13. $f(x) = x \quad (-2\pi < x < 2\pi)$

14. $f(x) = |x| \quad (-2 < x < 2)$

15. $f(x) = \begin{cases} x & \text{if } -1 < x < 1 \\ 2 - x & \text{if } \quad 1 < x < 3 \end{cases}$

16. $f(x) = \begin{cases} -1 - x & \text{if } -1 < x < 0 \\ 1 - x & \text{if } \quad 0 < x < 1 \end{cases}$

17. $f(x) = |\sin 8\pi x| \ (-1/8 < x < 1/8)$

18. $f(x) = e^x \ (-\pi < x < \pi)$

19. $f(x) = x^2 \ (-\pi/2 < x < \pi/2)$

20. $f(x) = x \ (0 < x < 2\pi)$

21–23 Using the answers to suitable odd-numbered problems, find the sum of

21. $1 - \frac{1}{3} + \frac{1}{5} - \frac{1}{7} + - \cdots$

22. $\dfrac{1}{1 \cdot 3} + \dfrac{1}{3 \cdot 5} + \dfrac{1}{5 \cdot 7} + \cdots$

23. $1 + \frac{1}{9} + \frac{1}{25} + \cdots$

24. **(Parseval's identity)** Obtain the result of Prob. 23 by applying Parseval's identity to Prob. 12.

25. What are the sum of the cosine terms and the sum of the sine terms in a Fourier series whose sum is $f(x)$? Give two examples.

26. **(Half-range expansion)** Find the half-range sine series of $f(x) = 0$ if $0 < x < \pi/2$, $f(x) = 1$ if $\pi/2 < x < \pi$. Compare with Prob. 12.

27. **(Half-range cosine series)** Find the half-range cosine series of $f(x) = x \ (0 < x < 2\pi)$. Compare with Prob. 20.

28–29 **MINIMUM SQUARE ERROR**

Compute the minimum square errors for the trigonometric polynomials of degree $N = 1, \cdots, 8$:

28. For $f(x)$ in Prob. 12.

29. For $f(x) = x \ (-\pi < x < \pi)$.

30–31 **GENERAL SOLUTION**

Solve $y'' + \omega^2 y = r(t)$, where $|\omega| \neq 0, 1, 2, \cdots, r(t)$ is 2π-periodic and:

30. $r(t) = t(\pi^2 - t^2) \quad (-\pi < t < \pi)$

31. $r(t) = t^2 \quad (-\pi < t < \pi)$

32–37 **FOURIER INTEGRALS AND TRANSFORMS**

Sketch the given function and represent it as indicated. If you have a CAS, graph approximate curves obtained by replacing ∞ with finite limits; also look for Gibbs phenomena.

32. $f(x) = 1$ if $1 < x < 2$ and 0 otherwise, by a Fourier integral

33. $f(x) = x$ if $0 < x < 1$ and 0 otherwise, by a Fourier integral

34. $f(x) = 1 + x/2$ if $-2 < x < 0$, $f(x) = 1 - x/2$ if $0 < x < 2$, $f(x) = 0$ otherwise, by a Fourier cosine integral

35. $f(x) = -1 - x/2$ if $-2 < x < 0$, $f(x) = 1 - x/2$ if $0 < x < 2$, $f(x) = 0$ otherwise, by a Fourier sine integral

36. $f(x) = -4 + x^2$ if $-2 < x < 0$, $f(x) = 4 - x^2$ if $0 < x < 2$, $f(x) = 0$ otherwise, by a Fourier sine integral

37. $f(x) = 4 - x^2$ if $-2 < x < 2$, $f(x) = 0$ otherwise, by a Fourier cosine integral

38. Find the Fourier transform of $f(x) = k$ if $a < x < b$, $f(x) = 0$ otherwise.

39. Find the Fourier cosine transform of $f(x) = e^{-2x}$ if $x > 0$, $f(x) = 0$ if $x < 0$.

40. Find $\mathcal{F}_c(e^{-2x})$ and $\mathcal{F}_s(e^{-2x})$ by formulas involving second derivatives.

SUMMARY OF CHAPTER 11
Fourier Series, Integrals, Transforms

Fourier series concern **periodic functions** $f(x)$ of period $p = 2L$, that is, by definition $f(x + p) = f(x)$ for all x and some fixed $p > 0$; thus, $f(x + np) = f(x)$ for any integer n. These series are of the form

$$f(x) = a_0 + \sum_{n=1}^{\infty} \left(a_n \cos \frac{n\pi}{L} x + b_n \sin \frac{n\pi}{L} x \right) \qquad \text{(1)} \quad \text{(Sec. 11.2)}$$

with coefficients, called the **Fourier coefficients** of $f(x)$, given by the Euler formulas (Sec. 11.2)

$$\text{(2)} \qquad a_0 = \frac{1}{2L} \int_{-L}^{L} f(x)\, dx, \qquad a_n = \frac{1}{L} \int_{-L}^{L} f(x) \cos \frac{n\pi x}{L}\, dx$$
$$b_n = \frac{1}{L} \int_{-L}^{L} f(x) \sin \frac{n\pi x}{L}\, dx$$

where $n = 1, 2, \cdots$. For period 2π we simply have (Sec. 11.1)

$$\text{(1*)} \qquad f(x) = a_0 + \sum_{n=1}^{\infty} (a_n \cos nx + b_n \sin nx)$$

with the *Fourier coefficients* of $f(x)$ (Sec. 11.1)

$$a_0 = \frac{1}{2\pi} \int_{-\pi}^{\pi} f(x)\, dx, \quad a_n = \frac{1}{\pi} \int_{-\pi}^{\pi} f(x) \cos nx\, dx, \quad b_n = \frac{1}{\pi} \int_{-\pi}^{\pi} f(x) \sin nx\, dx.$$

Fourier series are fundamental in connection with periodic phenomena, particularly in models involving differential equations (Sec. 11.5, Chap. 12). If $f(x)$ is even $[f(-x) = f(x)]$ or odd $[f(-x) = -f(x)]$, they reduce to **Fourier cosine** or **Fourier sine series,** respectively (Sec. 11.3). If $f(x)$ is given for $0 \leqq x \leqq L$ only, it has two **half-range expansions** of period $2L$, namely, a cosine and a sine series (Sec. 11.3).

The set of cosine and sine functions in (1) is called the **trigonometric system.** Its most basic property is its **orthogonality** on an interval of length $2L$; that is, for all integers m and $n \neq m$ we have

$$\int_{-L}^{L} \cos \frac{m\pi x}{L} \cos \frac{n\pi x}{L}\, dx = 0, \qquad \int_{-L}^{L} \sin \frac{m\pi x}{L} \sin \frac{n\pi x}{L}\, dx = 0$$

and for all integers m and n,

$$\int_{-L}^{L} \cos \frac{m\pi x}{L} \sin \frac{n\pi x}{L}\, dx = 0.$$

This orthogonality was crucial in deriving the Euler formulas (2).

Partial sums of Fourier series minimize the **square error** (Sec. 11.6).

Ideas and techniques of Fourier series extend to nonperiodic functions $f(x)$ defined on the entire real line; this leads to the **Fourier integral**

(3) $$f(x) = \int_0^{\infty} [A(w) \cos wx + B(w) \sin wx]\, dw \qquad \text{(Sec. 11.7)}$$

where

(4) $$A(w) = \frac{1}{\pi} \int_{-\infty}^{\infty} f(v) \cos wv\, dv, \qquad B(w) = \frac{1}{\pi} \int_{-\infty}^{\infty} f(v) \sin wv\, dv$$

or, in complex form (Sec. 11.9),

(5) $$f(x) = \frac{1}{\sqrt{2\pi}} \int_{-\infty}^{\infty} \hat{f}(w) e^{iwx}\, dw \qquad (i = \sqrt{-1})$$

where

(6) $$\hat{f}(w) = \frac{1}{\sqrt{2\pi}} \int_{-\infty}^{\infty} f(x) e^{-iwx}\, dx.$$

Formula (6) transforms $f(x)$ into its **Fourier transform** $\hat{f}(w)$, and (5) is the inverse transform.

Related to this are the **Fourier cosine transform** (Sec. 11.8)

(7) $$\hat{f}_c(w) = \sqrt{\frac{2}{\pi}} \int_0^{\infty} f(x) \cos wx\, dx$$

and the **Fourier sine transform** (Sec. 11.8)

(8) $$\hat{f}_s(w) = \sqrt{\frac{2}{\pi}} \int_0^{\infty} f(x) \sin wx\, dx.$$

The **discrete Fourier transform (DFT)** and a practical method of computing it, called the **fast Fourier transform (FFT),** are discussed in Sec. 11.9.

CHAPTER 12

Partial Differential Equations (PDEs)

PDEs are models of various physical and geometrical problems, arising when the unknown functions (the solutions) depend on two or more variables, usually on time t and one or several space variables. It is fair to say that only the simplest physical systems can be modeled by ODEs, whereas most problems in dynamics, elasticity, heat transfer, electromagnetic theory, and quantum mechanics require PDEs. Indeed, the range of applications of PDEs is enormous, compared to that of ODEs.

In this chapter we concentrate on the most important PDEs of applied mathematics, the wave equations governing the vibrating string (Sec. 12.2) and the vibrating membrane (Sec. 12.7), the heat equation (Sec. 12.5), and the Laplace equation (Secs. 12.5, 12.10). We derive these PDEs from physics and consider methods for solving **initial and boundary value problems,** that is, methods of obtaining solutions satisfying conditions that are given by the physical situation.

In Secs. 12.6 and 12.11 we show that PDEs can also be solved by Fourier and Laplace transform methods.

COMMENT. *Numerics for PDEs* is explained in Secs. 21.4–21.7.

Prerequisites: Linear ODEs (Chap. 2), Fourier series (Chap. 11)
Sections that may be omitted in a shorter course: 12.6, 12.9–12.11
References and Answers to Problems: App. 1 Part C, App. 2

12.1 Basic Concepts

A **partial differential equation (PDE)** is an equation involving one or more partial derivatives of an (unknown) function, call it u, that depends on two or more variables, often time t and one or several variables in space. The order of the highest derivative is called the **order** of the PDE. As for ODEs, second-order PDEs will be the most important ones in applications.

Just as for ordinary differential equations (ODEs) we say that a PDE is **linear** if it is of the first degree in the unknown function u and its partial derivatives. Otherwise we call it **nonlinear.** Thus, all the equations in Example 1 on p. 536 are linear. We call a *linear* PDE **homogeneous** if each of its terms contains either u or one of its partial derivatives. Otherwise we call the equation **nonhomogeneous.** Thus, (4) in Example 1 (with f not identically zero) is nonhomogeneous, whereas the other equations are homogeneous.

EXAMPLE 1 Important Second-Order PDEs

(1) $$\frac{\partial^2 u}{\partial t^2} = c^2 \frac{\partial^2 u}{\partial x^2}$$ *One-dimensional wave equation*

(2) $$\frac{\partial u}{\partial t} = c^2 \frac{\partial^2 u}{\partial x^2}$$ *One-dimensional heat equation*

(3) $$\frac{\partial^2 u}{\partial x^2} + \frac{\partial^2 u}{\partial y^2} = 0$$ *Two-dimensional Laplace equation*

(4) $$\frac{\partial^2 u}{\partial x^2} + \frac{\partial^2 u}{\partial y^2} = f(x, y)$$ *Two-dimensional Poisson equation*

(5) $$\frac{\partial^2 u}{\partial t^2} = c^2 \left(\frac{\partial^2 u}{\partial x^2} + \frac{\partial^2 u}{\partial y^2}\right)$$ *Two-dimensional wave equation*

(6) $$\frac{\partial^2 u}{\partial x^2} + \frac{\partial^2 u}{\partial y^2} + \frac{\partial^2 u}{\partial z^2} = 0$$ *Three-dimensional Laplace equation*

Here c is a positive constant, t is time, x, y, z are Cartesian coordinates, and *dimension* is the number of these coordinates in the equation. ■

A **solution** of a PDE in some region R of the space of the independent variables is a function that has all the partial derivatives appearing in the PDE in some domain D (definition in Sec. 9.6) containing R, and satisfies the PDE everywhere in R.

Often one merely requires that the function is continuous on the boundary of R, has those derivatives in the interior of R, and satisfies the PDE in the interior of R. Letting R lie in D simplifies the situation regarding derivatives on the boundary of R, which is then the same on the boundary as it is in the interior of R.

In general, the totality of solutions of a PDE is very large. For example, the functions

(7) $$u = x^2 - y^2, \qquad u = e^x \cos y, \qquad u = \sin x \cosh y, \qquad u = \ln (x^2 + y^2)$$

which are entirely different from each other, are solutions of (3), as you may verify. We shall see later that the unique solution of a PDE corresponding to a given physical problem will be obtained by the use of **additional conditions** arising from the problem. For instance, this may be the condition that the solution u assume given values on the boundary of the region R **("boundary conditions").** Or, when time t is one of the variables, u (or $u_t = \partial u/\partial t$ or both) may be prescribed at $t = 0$ **("initial conditions").**

We know that if an ODE is linear and homogeneous, then from known solutions we can obtain further solutions by superposition. For PDEs the situation is quite similar:

THEOREM 1

Fundamental Theorem on Superposition

If u_1 and u_2 are solutions of a ***homogeneous linear*** *PDE in some region R, then*

$$u = c_1 u_1 + c_2 u_2$$

with any constants c_1 and c_2 is also a solution of that PDE in the region R.

The simple proof of this important theorem is quite similar to that of Theorem 1 in Sec. 2.1 and is left to the student.

Verification of solutions in Probs. 14–25 proceeds as for ODEs. Problems 1–12 concern PDEs solvable like ODEs. To help the student with them, we consider two typical examples.

EXAMPLE 2 Solving $u_{xx} - u = 0$ Like an ODE

Find solutions u of the PDE $u_{xx} - u = 0$ depending on x and y.

Solution. Since no y-derivatives occur, we can solve this PDE like $u'' - u = 0$. In Sec. 2.2 we would have obtained $u = Ae^{x} + Be^{-x}$ with constant A and B. Here A and B may be functions of y, so that the answer is

$$u(x, y) = A(y)e^{x} + B(y)e^{-x}$$

with arbitrary functions A and B. We thus have a great variety of solutions. Check the result by differentiation. ■

EXAMPLE 3 Solving $u_{xy} = -u_x$ Like an ODE

Find solutions $u = u(x, y)$ of this PDE.

Solution. Setting $u_x = p$, we have $p_y = -p$, $p_y/p = -1$, $\ln p = -y + \tilde{c}(x)$, $p = c(x)e^{-y}$ and by integration with respect to x,

$$u(x, y) = f(x)e^{-y} + g(y) \qquad \text{where} \qquad f(x) = \int c(x)\, dx;$$

here, $f(x)$ and $g(y)$ are arbitrary. ■

PROBLEM SET 12.1

1–12 PDEs SOLVABLE AS ODEs

This happens if a PDE involves derivatives with respect to one variable only (or can be transformed to such a form), so that the other variable(s) can be treated as parameter(s). Solve for $u = u(x, y)$:

1. $u_{yy} + 16u = 0$
2. $u_{xx} = u$
3. $u_{yy} = 0$
4. $u_y + 2yu = 0$
5. $u_y + u = e^{xy}$
6. $u_{xx} = 4y^2 u$
7. $u_y = (\cosh x)yu$
8. $u_y = 2xyu$
9. $y^2 u_{yy} + 2yu_y - 2u = 0$
10. $u_{yy} = 4xu_y$
11. $u_{xy} = u_x$
12. $u_{yy} + 10u_y + 25u = e^{-5y}$

13. **(Fundamental Theorem)** Prove Fundamental Theorem 1 for second-order PDEs in two and three independent variables.

14–25 VERIFICATION OF SOLUTIONS

Verify (by substitution) that the given function is a solution of the indicated PDE. Sketch or graph the solution as a surface in space.

14–17 Wave Equation (1) with suitable c

14. $u = 4x^2 + t^2$
15. $u = \sin 8x \cos 2t$
16. $u = \sin 3x \sin 18t$
17. $u = \sin kx \cos kct$

18–21 Heat Equation (2) with suitable c

18. $u = e^{-2kt} \cos 8x$
19. $u = e^{-\pi^2 t} \sin 4x$
20. $u = e^{-4\omega^2 t} \sin \omega x$
21. $u = e^{-\omega^2 c^2 t} \cos \omega x$

22–25 Laplace Equation (3)

22. u in (7) in the text
23. $u = \cos 2y \sinh 2x$
24. $u = \arctan (y/x)$
25. $u = e^{x^2 - y^2} \sin 2xy$

26. **TEAM PROJECT. Verification of Solutions**
(a) Wave equation. Verify that $u(x, t) = v(x + ct) + w(x - ct)$ with any twice differentiable functions v and w satisfies (1).
(b) Poisson equation. Verify that each u satisfies (4) with $f(x, y)$ as indicated.

$$u = x^4 + y^4 \qquad f = 12(x^2 + y^2)$$

$$u = \cos x \sin y \qquad f = -2 \cos x \sin y$$

$$u = y/x \qquad f = 2y/x^3$$

(c) Laplace equation. Verify that $u = 1/\sqrt{x^2 + y^2 + z^2}$ satisfies (6) and $u = \ln (x^2 + y^2)$ satisfies (3). Is $u = 1/\sqrt{x^2 + y^2}$ a solution of (3)? Of what Poisson equation?

(d) Verify that u with any (sufficiently often differentiable) v and w satisfies the given PDE.

$u = v(x) + w(y)$ $\qquad u_{xy} = 0$

$u = v(x)w(y)$ $\qquad uu_{xy} = u_x u_y$

$u = v(x + 3t) + w(x - 3t)$ $\qquad u_{tt} = 9u_{xx}$

27. **(Boundary value problem)** Verify that the function $u(x, y) = a \ln(x^2 + y^2) + b$ satisfies Laplace's equation (3) and determine a and b so that u satisfies the boundary conditions $u = 110$ on the circle $x^2 + y^2 = 1$ and $u = 0$ on the circle $x^2 + y^2 = 100$.

28–30 **SYSTEMS OF PDEs**

Solve

28. $u_x = 0,\ u_y = 0$

29. $u_{xx} = 0,\ u_{xy} = 0$

30. $u_{xx} = 0,\ u_{yy} = 0$

12.2 Modeling: Vibrating String, Wave Equation

As a first important PDE let us derive the equation modeling small transverse vibrations of an elastic string, such as a violin string. We place the string along the x-axis, stretch it to length L, and fasten it at the ends $x = 0$ and $x = L$. We then distort the string, and at some instant, call it $t = 0$, we release it and allow it to vibrate. The problem is to determine the vibrations of the string, that is, to find its deflection $u(x, t)$ at any point x and at any time $t > 0$; see Fig. 283.

$u(x, t)$ will be the solution of a PDE that is the model of our physical system to be derived. This PDE should not be too complicated, so that we can solve it. Reasonable simplifying assumptions (just as for ODEs modeling vibrations in Chap. 2) are as follows.

Physical Assumptions

1. The mass of the string per unit length is constant ("homogeneous string"). The string is perfectly elastic and does not offer any resistance to bending.
2. The tension caused by stretching the string before fastening it at the ends is so large that the action of the gravitational force on the string (trying to pull the string down a little) can be neglected.
3. The string performs small transverse motions in a vertical plane; that is, every particle of the string moves strictly vertically and so that the deflection and the slope at every point of the string always remain small in absolute value.

Under these assumptions we may expect solutions $u(x, t)$ that describe the physical reality sufficiently well.

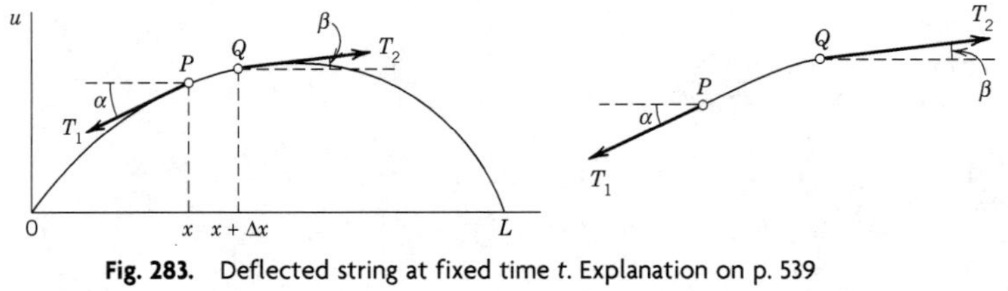

Fig. 283. Deflected string at fixed time t. Explanation on p. 539

Derivation of the PDE of the Model ("Wave Equation") from Forces

The model of the vibrating string will consist of a PDE ("wave equation") and additional conditions. To obtain the PDE, we consider the ***forces acting on a small portion of the string*** (Fig. 283). This method is typical of modeling in mechanics and elsewhere.

Since the string offers no resistance to bending, the tension is tangential to the curve of the string at each point. Let T_1 and T_2 be the tension at the endpoints P and Q of that portion. Since the points of the string move vertically, there is no motion in the horizontal direction. Hence the horizontal components of the tension must be constant. Using the notation shown in Fig. 283, we thus obtain

$$T_1 \cos \alpha = T_2 \cos \beta = T = \mathit{const.} \tag{1}$$

In the vertical direction we have two forces, namely, the vertical components $-T_1 \sin \alpha$ and $T_2 \sin \beta$ of T_1 and T_2; here the minus sign appears because the component at P is directed downward. By **Newton's second law** the resultant of these two forces is equal to the mass $\rho \, \Delta x$ of the portion times the acceleration $\partial^2 u/\partial t^2$, evaluated at some point between x and $x + \Delta x$; here ρ is the mass of the undeflected string per unit length, and Δx is the length of the portion of the undeflected string. (Δ is generally used to denote small quantities; this has nothing to do with the Laplacian ∇^2, which is sometimes also denoted by Δ.) Hence

$$T_2 \sin \beta - T_1 \sin \alpha = \rho \, \Delta x \, \frac{\partial^2 u}{\partial t^2} \, .$$

Using (1), we can divide this by $T_2 \cos \beta = T_1 \cos \alpha = T$, obtaining

$$\frac{T_2 \sin \beta}{T_2 \cos \beta} - \frac{T_1 \sin \alpha}{T_1 \cos \alpha} = \tan \beta - \tan \alpha = \frac{\rho \, \Delta x}{T} \frac{\partial^2 u}{\partial t^2} \, . \tag{2}$$

Now $\tan \alpha$ and $\tan \beta$ are the slopes of the string at x and $x + \Delta x$:

$$\tan \alpha = \left(\frac{\partial u}{\partial x}\right)\bigg|_{x} \qquad \text{and} \qquad \tan \beta = \left(\frac{\partial u}{\partial x}\right)\bigg|_{x+\Delta x} .$$

Here we have to write *partial* derivatives because u depends also on time t. Dividing (2) by Δx, we thus have

$$\frac{1}{\Delta x}\left[\left(\frac{\partial u}{\partial x}\right)\bigg|_{x+\Delta x} - \left(\frac{\partial u}{\partial x}\right)\bigg|_{x}\right] = \frac{\rho}{T} \frac{\partial^2 u}{\partial t^2} \, .$$

If we let Δx approach zero, we obtain the linear PDE

$$\frac{\partial^2 u}{\partial t^2} = c^2 \frac{\partial^2 u}{\partial x^2} \, , \qquad c^2 = \frac{T}{\rho} \, . \tag{3}$$

This is called the **one-dimensional wave equation.** We see that it is homogeneous and of the second order. The physical constant T/ρ is denoted by c^2 (instead of c) to indicate that this constant is *positive,* a fact that will be essential to the form of the solutions. "One-dimensional" means that the equation involves only one space variable, x. In the next section we shall complete setting up the model and then show how to solve it by a general method that is probably the most important one for PDEs in engineering mathematics.

12.3 Solution by Separating Variables. Use of Fourier Series

The model of a vibrating elastic string (a violin string, for instance) consists of the **one-dimensional wave equation**

$$\frac{\partial^2 u}{\partial t^2} = c^2 \frac{\partial^2 u}{\partial x^2} \qquad c^2 = \frac{T}{\rho} \tag{1}$$

for the unknown deflection $u(x, t)$ of the string, a PDE that we have just obtained, and some ***additional conditions,*** which we shall now derive.

Since the string is fastened at the ends $x = 0$ and $x = L$ (see Sec. 12.2), we have the two **boundary conditions**

$$\text{(a)} \quad u(0, t) = 0, \qquad \text{(b)} \quad u(L, t) = 0 \qquad \text{for all } t. \tag{2}$$

Furthermore, the form of the motion of the string will depend on its *initial deflection* (deflection at time $t = 0$), call it $f(x)$, and on its *initial velocity* (velocity at $t = 0$), call it $g(x)$. We thus have the two **initial conditions**

$$\text{(a)} \quad u(x, 0) = f(x), \qquad \text{(b)} \quad u_t(x, 0) = g(x) \qquad (0 \leqq x \leqq L) \tag{3}$$

where $u_t = \partial u/\partial t$. We now have to find a solution of the PDE (1) satisfying the conditions (2) and (3). This will be the solution of our problem. We shall do this in three steps, as follows.

Step 1. By the **"method of separating variables"** or *product method,* setting $u(x, t) = F(x)G(t)$, we obtain from (1) two ODEs, one for $F(x)$ and the other one for $G(t)$.

Step 2. We determine solutions of these ODEs that satisfy the boundary conditions (2).

Step 3. Finally, using **Fourier series,** we compose the solutions gained in Step 2 to obtain a solution of (1) satisfying both (2) and (3), that is, the solution of our model of the vibrating string.

Step 1. Two ODEs from the Wave Equation (1)

In the **method of separating variables,** or *product method,* we determine solutions of the wave equation (1) of the form

$$u(x, t) = F(x)G(t) \tag{4}$$

which are a product of two functions, each depending only on one of the variables x and t. This is a powerful general method that has various applications in engineering mathematics, as we shall see in this chapter. Differentiating (4), we obtain

$$\frac{\partial^2 u}{\partial t^2} = F\ddot{G} \qquad \text{and} \qquad \frac{\partial^2 u}{\partial x^2} = F''G$$

where dots denote derivatives with respect to t and primes derivatives with respect to x. By inserting this into the wave equation (1) we have

$$F\ddot{G} = c^2 F''G.$$

Dividing by c^2FG and simplifying gives

$$\frac{\ddot{G}}{c^2 G} = \frac{F''}{F}.$$

The variables are now separated, the left side depending only on t and the right side only on x. Hence both sides must be constant because if they were variable, then changing t or x would affect only one side, leaving the other unaltered. Thus, say,

$$\frac{\ddot{G}}{c^2 G} = \frac{F''}{F} = k.$$

Multiplying by the denominators gives immediately two ***ordinary*** DEs

$$\textbf{(5)} \qquad F'' - kF = 0$$

and

$$\textbf{(6)} \qquad \ddot{G} - c^2 kG = 0.$$

Here, the **separation constant** k is still arbitrary.

Step 2. Satisfying the Boundary Conditions (2)

We now determine solutions F and G of (5) and (6) so that $u = FG$ satisfies the boundary conditions (2), that is,

$$(7) \qquad u(0, t) = F(0)G(t) = 0, \qquad u(L, t) = F(L)G(t) = 0 \qquad \text{for all } t.$$

We first solve (5). If $G \equiv 0$, then $u = FG \equiv 0$, which is of no interest. Hence $G \not\equiv 0$ and then by (7),

$$(8) \qquad \text{(a)} \quad F(0) = 0, \qquad \text{(b)} \quad F(L) = 0.$$

We show that k must be negative. For $k = 0$ the general solution of (5) is $F = ax + b$, and from (8) we obtain $a = b = 0$, so that $F \equiv 0$ and $u = FG \equiv 0$, which is of no interest. For positive $k = \mu^2$ a general solution of (5) is

$$F = Ae^{\mu x} + Be^{-\mu x}$$

and from (8) we obtain $F \equiv 0$ as before (verify!). Hence we are left with the possibility of choosing k negative, say, $k = -p^2$. Then (5) becomes $F'' + p^2F = 0$ and has as a general solution

$$F(x) = A \cos px + B \sin px.$$

From this and (8) we have

$$F(0) = A = 0 \qquad \text{and then} \qquad F(L) = B \sin pL = 0.$$

We must take $B \neq 0$ since otherwise $F \equiv 0$. Hence $\sin pL = 0$. Thus

$$(9) \qquad pL = n\pi, \qquad \text{so that} \qquad p = \frac{n\pi}{L} \qquad (n \text{ integer}).$$

Setting $B = 1$, we thus obtain infinitely many solutions $F(x) = F_n(x)$, where

$$(10) \qquad F_n(x) = \sin \frac{n\pi}{L} x \qquad (n = 1, 2, \cdots).$$

These solutions satisfy (8). [For negative integer n we obtain essentially the same solutions, except for a minus sign, because $\sin(-\alpha) = -\sin \alpha$.]

We now solve (6) with $k = -p^2 = -(n\pi/L)^2$ resulting from (9), that is,

$$(11^*) \qquad \ddot{G} + \lambda_n{}^2 G = 0 \qquad \text{where} \qquad \lambda_n = cp = \frac{cn\pi}{L}.$$

A general solution is

$$G_n(t) = B_n \cos \lambda_n t + B_n{}^* \sin \lambda_n t.$$

Hence solutions of (1) satisfying (2) are $u_n(x, t) = F_n(x)G_n(t) = G_n(t)F_n(x)$, written out

$$\textbf{(11)} \qquad u_n(x, t) = (B_n \cos \lambda_n t + B_n{}^* \sin \lambda_n t) \sin \frac{n\pi}{L} x \qquad (n = 1, 2, \cdots).$$

These functions are called the **eigenfunctions,** or *characteristic functions,* and the values $\lambda_n = cn\pi/L$ are called the **eigenvalues,** or *characteristic values,* of the vibrating string. The set $\{\lambda_1, \lambda_2, \cdots\}$ is called the **spectrum.**

Discussion of Eigenfunctions. We see that each u_n represents a harmonic motion having the **frequency** $\lambda_n/2\pi = cn/2L$ cycles per unit time. This motion is called the nth **normal mode** of the string. The first normal mode is known as the *fundamental mode* ($n = 1$), and the others are known as *overtones;* musically they give the octave, octave plus fifth, etc. Since in (11)

$$\sin \frac{n\pi x}{L} = 0 \qquad \text{at} \qquad x = \frac{L}{n}, \frac{2L}{n}, \cdots, \frac{n-1}{n} L,$$

the nth normal mode has $n - 1$ **nodes,** that is, points of the string that do not move (in addition to the fixed endpoints); see Fig. 284.

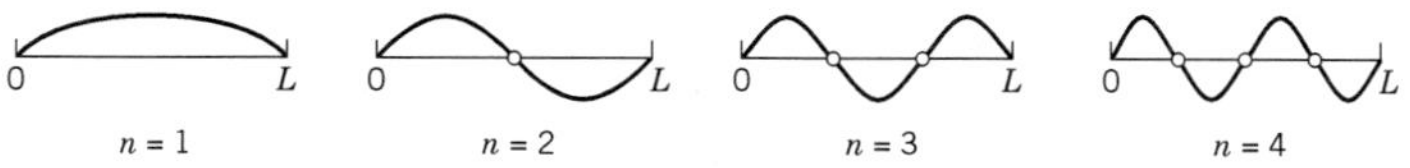

Fig. 284. Normal modes of the vibrating string

Figure 285 shows the second normal mode for various values of t. At any instant the string has the form of a sine wave. When the left part of the string is moving down, the other half is moving up, and conversely. For the other modes the situation is similar.

Tuning is done by changing the tension T. Our formula for the frequency $\lambda_n/2\pi = cn/2L$ of u_n with $c = \sqrt{T/\rho}$ [see (3), Sec. 12.2] confirms that effect because it shows that the frequency is proportional to the tension. T cannot be increased indefinitely, but can you see what to do to get a string with a high fundamental mode? (Think of both L and ρ.) Why is a violin smaller than a double-bass?

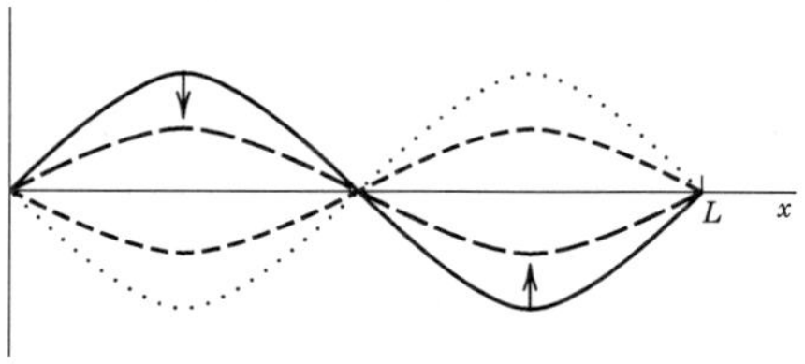

Fig. 285. Second normal mode for various values of t

Step 3. Solution of the Entire Problem. Fourier Series

The eigenfunctions (11) satisfy the wave equation (1) and the boundary conditions (2) (string fixed at the ends). A single u_n will generally not satisfy the initial conditions (3). But since the wave equation (1) is linear and homogeneous, it follows from Fundamental Theorem 1 in Sec. 12.1 that the sum of finitely many solutions u_n is a solution of (1). To obtain a solution that also satisfies the initial conditions (3), we consider the infinite series (with $\lambda_n = cn\pi/L$ as before)

$$u(x, t) = \sum_{n=1}^{\infty} u_n(x, t) = \sum_{n=1}^{\infty} (B_n \cos \lambda_n t + B_n^* \sin \lambda_n t) \sin \frac{n\pi}{L} x. \tag{12}$$

Satisfying Initial Condition (3a) (Given Initial Displacement). From (12) and (3a) we obtain

$$u(x, 0) = \sum_{n=1}^{\infty} B_n \sin \frac{n\pi}{L} x = f(x). \tag{13}$$

Hence we must choose the B_n's so that $u(x, 0)$ becomes the **Fourier sine series** of $f(x)$. Thus, by (4) in Sec. 11.3,

$$B_n = \frac{2}{L} \int_0^L f(x) \sin \frac{n\pi x}{L} dx, \qquad n = 1, 2, \cdots. \tag{14}$$

Satisfying Initial Condition (3b) (Given Initial Velocity). Similarly, by differentiating (12) with respect to t and using (3b), we obtain

$$\left.\frac{\partial u}{\partial t}\right|_{t=0} = \left[\sum_{n=1}^{\infty} (-B_n\lambda_n \sin \lambda_n t + B_n{}^*\lambda_n \cos \lambda_n t) \sin \frac{n\pi x}{L}\right]_{t=0}$$

$$= \sum_{n=1}^{\infty} B_n{}^*\lambda_n \sin \frac{n\pi x}{L} = g(x).$$

Hence we must choose the $B_n{}^*$'s so that for $t = 0$ the derivative $\partial u/\partial t$ becomes the Fourier sine series of $g(x)$. Thus, again by (4) in Sec. 11.3,

$$B_n{}^*\lambda_n = \frac{2}{L}\int_0^L g(x) \sin \frac{n\pi x}{L}\, dx.$$

Since $\lambda_n = cn\pi/L$, we obtain by division

$$B_n{}^* = \frac{2}{cn\pi}\int_0^L g(x) \sin \frac{n\pi x}{L}\, dx, \qquad n = 1, 2, \cdots. \tag{15}$$

Result. Our discussion shows that $u(x, t)$ given by (12) with coefficients (14) and (15) is a solution of (1) that satisfies all the conditions in (2) and (3), provided the series (12) converges and so do the series obtained by differentiating (12) twice termwise with respect to x and t and have the sums $\partial^2 u/\partial x^2$ and $\partial^2 u/\partial t^2$, respectively, which are continuous.

Solution (12) Established. According to our derivation the solution (12) is at first a purely formal expression, but we shall now establish it. For the sake of simplicity we consider only the case when the initial velocity $g(x)$ is identically zero. Then the $B_n{}^*$ are zero, and (12) reduces to

$$u(x, t) = \sum_{n=1}^{\infty} B_n \cos \lambda_n t \sin \frac{n\pi x}{L}, \qquad \lambda_n = \frac{cn\pi}{L}. \tag{16}$$

It is possible to ***sum this series,*** that is, to write the result in a closed or finite form. For this purpose we use the formula [see (11), App. A3.1]

$$\cos \frac{cn\pi}{L}t \sin \frac{n\pi}{L}x = \frac{1}{2}\left[\sin\left\{\frac{n\pi}{L}(x - ct)\right\} + \sin\left\{\frac{n\pi}{L}(x + ct)\right\}\right].$$

Consequently, we may write (16) in the form

$$u(x, t) = \frac{1}{2}\sum_{n=1}^{\infty} B_n \sin\left\{\frac{n\pi}{L}(x - ct)\right\} + \frac{1}{2}\sum_{n=1}^{\infty} B_n \sin\left\{\frac{n\pi}{L}(x + ct)\right\}.$$

These two series are those obtained by substituting $x - ct$ and $x + ct$, respectively, for the variable x in the Fourier sine series (13) for $f(x)$. Thus

$$u(x, t) = \tfrac{1}{2}[f^*(x - ct) + f^*(x + ct)] \tag{17}$$

where f^* is the odd periodic extension of f with the period $2L$ (Fig. 286). Since the initial deflection $f(x)$ is continuous on the interval $0 \leqq x \leqq L$ and zero at the endpoints, it follows from (17) that $u(x, t)$ is a continuous function of both variables x and t for all values of the variables. By differentiating (17) we see that $u(x, t)$ is a solution of (1), provided $f(x)$ is twice differentiable on the interval $0 < x < L$, and has one-sided second derivatives at $x = 0$ and $x = L$, which are zero. Under these conditions $u(x, t)$ is established as a solution of (1), satisfying (2) and (3) with $g(x) \equiv 0$. ■

Fig. 286. Odd periodic extension of $f(x)$

Generalized Solution. If $f'(x)$ and $f''(x)$ are merely piecewise continuous (see Sec. 6.1), or if those one-sided derivatives are not zero, then for each t there will be finitely many values of x at which the second derivatives of u appearing in (1) do not exist. Except at these points the wave equation will still be satisfied. We may then regard $u(x, t)$ as a **"generalized solution,"** as it is called, that is, as a solution in a broader sense. For instance, a triangular initial deflection as in Example 1 (below) leads to a generalized solution.

Physical Interpretation of the Solution (17). The graph of $f^*(x - ct)$ is obtained from the graph of $f^*(x)$ by shifting the latter ct units to the right (Fig. 287). This means that $f^*(x - ct)$ $(c > 0)$ represents a wave that is traveling to the right as t increases. Similarly, $f^*(x + ct)$ represents a wave that is traveling to the left, and $u(x, t)$ is the superposition of these two waves.

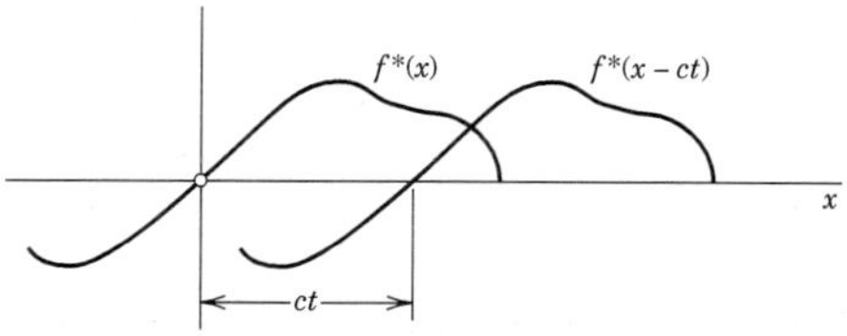

Fig. 287. Interpretation of (17)

EXAMPLE 1 **Vibrating String if the Initial Deflection Is Triangular**

Find the solution of the wave equation (1) corresponding to the triangular initial deflection

$$f(x) = \begin{cases} \dfrac{2k}{L} x & \text{if} \quad 0 < x < \dfrac{L}{2} \\[2mm] \dfrac{2k}{L}(L - x) & \text{if} \quad \dfrac{L}{2} < x < L \end{cases}$$

and initial velocity zero. (Figure 288 shows $f(x) = u(x, 0)$ at the top.)

Solution. Since $g(x) \equiv 0$, we have $B_n^* = 0$ in (12), and from Example 4 in Sec. 11.3 we see that the B_n are given by (5), Sec. 11.3. Thus (12) takes the form

$$u(x, t) = \frac{8k}{\pi^2}\left[\frac{1}{1^2} \sin \frac{\pi}{L} x \cos \frac{\pi c}{L} t - \frac{1}{3^2} \sin \frac{3\pi}{L} x \cos \frac{3\pi c}{L} t + - \cdots\right].$$

For graphing the solution we may use $u(x, 0) = f(x)$ and the above interpretation of the two functions in the representation (17). This leads to the graph shown in Fig. 288. ■

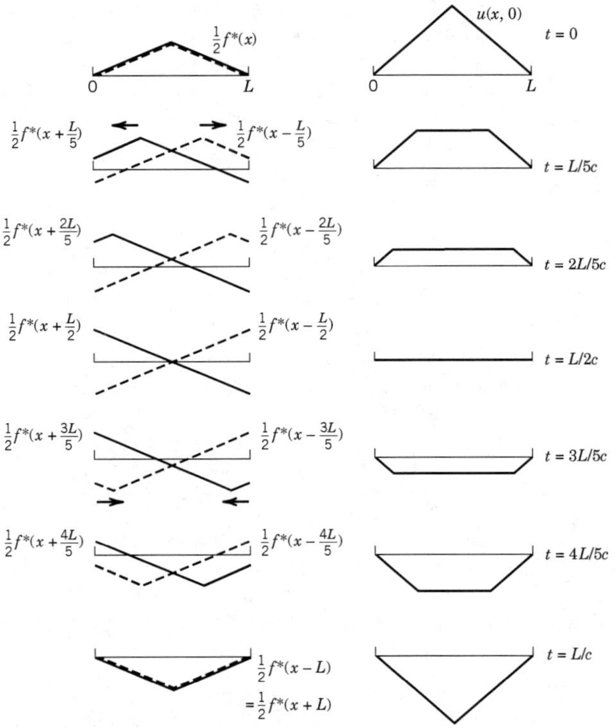

Fig. 288. Solution $u(x, t)$ in Example 1 for various values of t (right part of the figure) obtained as the superposition of a wave traveling to the right (dashed) and a wave traveling to the left (left part of the figure)

PROBLEM SET 12.3

1–10 DEFLECTION OF THE STRING

Find $u(x, t)$ for the string of length $L = 1$ and $c^2 = 1$ when the initial velocity is zero and the initial deflection with small k (say, 0.01) is as follows. Sketch or graph $u(x, t)$ as in Fig. 288.

1. $k \sin 2\pi x$

2. $k(\sin \pi x - \frac{1}{3} \sin 3\pi x)$

3. $kx(1 - x)$

4. $kx(1 - x^2)$

5.

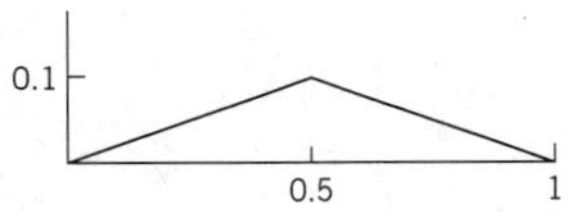

6.

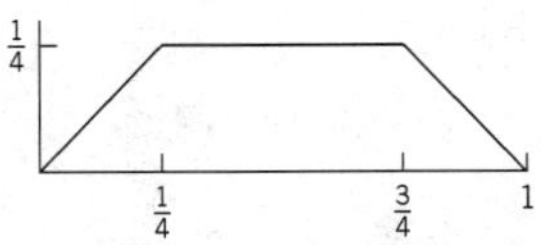

7.

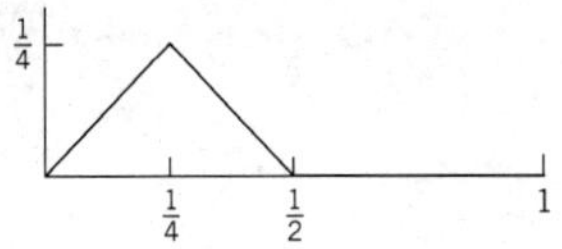

8.

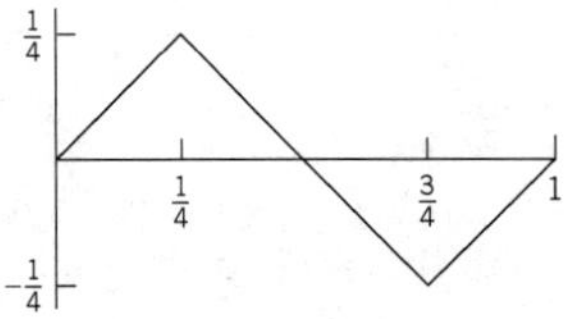

9.

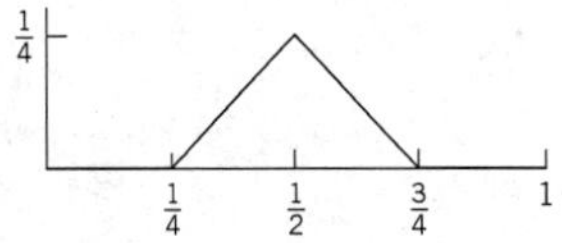

10. 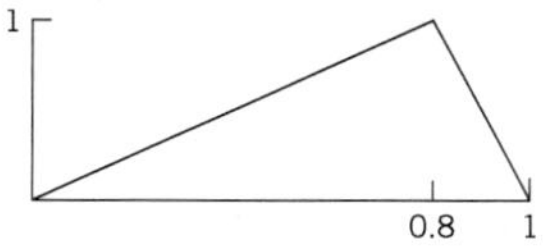

11. (Frequency) How does the frequency of the fundamental mode of the vibrating string depend on the length of the string? On the mass per unit length? What happens to the string if we double the tension? Why is a contrabass larger than a violin?

12. (Nonzero initial velocity) Find the deflection $u(x, t)$ of the string of length $L = \pi$ and $c^2 = 1$ for zero initial displacement and "triangular" initial velocity $u_t(x, 0) = 0.01x$ if $0 \leqq x \leqq \frac{1}{2}\pi$, $u_t(x, 0) = 0.01(\pi - x)$ if $\frac{1}{2}\pi \leqq x \leqq \pi$. (Initial conditions with $u_t(x, 0) \neq 0$ are hard to realize experimentally.)

13. CAS PROJECT. Graphing Normal Modes. Write a program for graphing u_n with $L = \pi$ and c^2 of your choice similarly as in Fig. 284. Apply the program to u_2, u_3, u_4. Also graph these solutions as surfaces over the xt-plane. Explain the connection between these two kinds of graphs.

14. TEAM PROJECT. Forced Vibrations of an Elastic String. Show the following.

(a) Substitution of

$$u(x, t) = \sum_{n=1}^{\infty} G_n(t) \sin \frac{n\pi x}{L} \tag{17}$$

(L = length of the string) into the wave equation (1) governing free vibrations leads to [see (10*)]

$$\ddot{G}_n + \lambda_n^2 G = 0, \qquad \lambda_n = \frac{cn\pi}{L}. \tag{18}$$

(b) Forced vibrations of the string under an external force $P(x, t)$ per unit length acting normal to the string are governed by the PDE

$$u_{tt} = c^2 u_{xx} + \frac{P}{\rho}. \tag{19}$$

(c) For a sinusoidal force $P = A\rho \sin \omega t$ we obtain

$$\frac{P}{\rho} = A \sin \omega t = \sum_{n=1}^{\infty} k_n(t) \sin \frac{n\pi x}{L},$$
$$k_n(t) = \begin{cases} (4A/n\pi) \sin \omega t & (n \text{ odd}) \\ 0 & (n \text{ even}). \end{cases} \tag{20}$$

Substituting (17) and (20) into (19) gives

$$\ddot{G}_n + \lambda_n^2 G_n = \frac{2A}{n\pi}(1 - \cos n\pi) \sin \omega t.$$

If $\lambda_n^2 \neq \omega^2$, the solution is

$$G_n(t) = B_n \cos \lambda_n t + B_n^* \sin \lambda_n t + \frac{2A(1 - \cos n\pi)}{n\pi(\lambda_n^2 - \omega^2)} \sin \omega t.$$

Determine B_n and B_n^* so that u satisfies the initial conditions $u(x, 0) = f(x)$, $u_t(x, 0) = 0$.

(d) (Resonance) Show that if $\lambda_n = \omega$, then

$$G_n(t) = B_n \cos \omega t + B_n^* \sin \omega t - \frac{A}{n\pi\omega}(1 - \cos n\pi) t \cos \omega t.$$

(e) (Reduction of boundary conditions) Show that a problem (1)–(3) with more complicated boundary conditions $u(0, t) = 0$, $u(L, t) = h(t)$, can be reduced to a problem for a new function v satisfying conditions $v(0, t) = v(L, t) = 0$, $v(x, 0) = f_1(x)$, $v_t(x, 0) = g_1(x)$ but a nonhomogeneous wave equation. *Hint:* Set $u = v + w$ and determine w suitably.

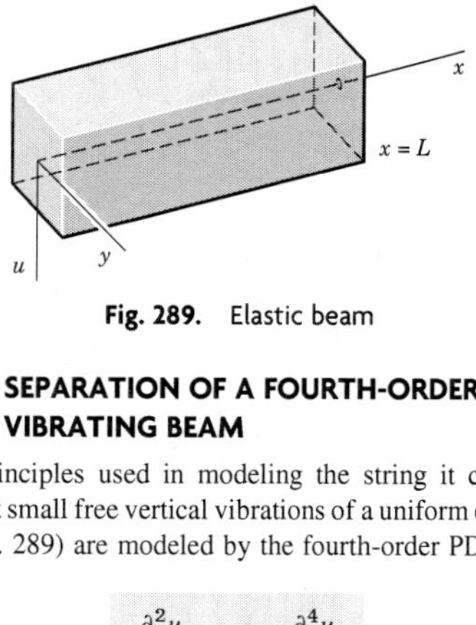

Fig. 289. Elastic beam

15–20 SEPARATION OF A FOURTH-ORDER PDE. VIBRATING BEAM

By the principles used in modeling the string it can be shown that small free vertical vibrations of a uniform elastic beam (Fig. 289) are modeled by the fourth-order PDE

$$\frac{\partial^2 u}{\partial t^2} = -c^2 \frac{\partial^4 u}{\partial x^4} \tag{21}$$

(Ref. [C11])

where $c^2 = EI/\rho A$ (E = Young's modulus of elasticity, I = moment of intertia of the cross section with respect to the y-axis in the figure, ρ = density, A = cross-sectional area). (*Bending* of a beam under a load is discussed in Sec. 3.3.)

15. Substituting $u = F(x)G(t)$ into (21), show that

$$F^{(4)}/F = -\ddot{G}/c^2 G = \beta^4 = const,$$
$$F(x) = A \cos \beta x + B \sin \beta x + C \cosh \beta x + D \sinh \beta x,$$
$$G(t) = a \cos c\beta^2 t + b \sin c\beta^2 t.$$

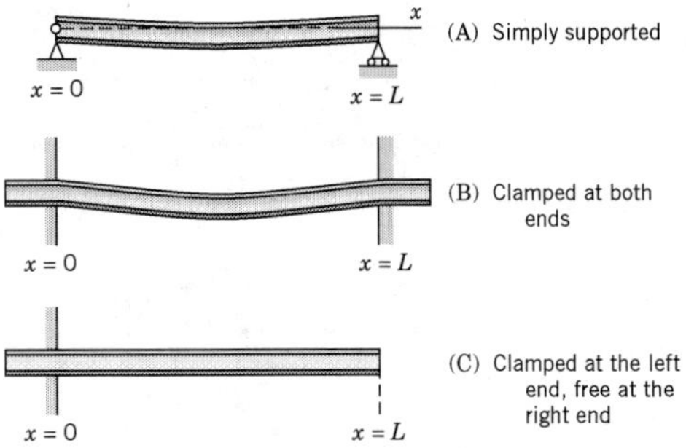

Fig. 290. Supports of a beam

16. (Simply supported beam in Fig. 290A) Find solutions $u_n = F_n(x)G_n(t)$ of (21) corresponding to zero initial velocity and satisfying the boundary conditions (see Fig. 290A)

$$u(0, t) = 0, \; u(L, t) = 0$$
(ends simply supported for all times t),

$$u_{xx}(0, t) = 0, \; u_{xx}(L, t) = 0$$
(zero moments, hence zero curvature, at the ends).

17. Find the solution of (21) that satisfies the conditions in Prob. 16 as well as the initial condition

$$u(x, 0) = f(x) = x(L - x).$$

18. Compare the results of Probs. 17 and 3. What is the basic difference between the frequencies of the normal modes of the vibrating string and the vibrating beam?

19. (Clamped beam in Fig. 290B) What are the boundary conditions for the clamped beam in Fig. 290B? Show that F in Prob. 15 satisfies these conditions if βL is a solution of the equation

$$\cosh \beta L \cos \beta L = 1. \tag{22}$$

Determine approximate solutions of (22), for instance, graphically from the intersections of the curves of $\cos \beta L$ and $1/\cosh \beta L$.

20. (Clamped-free beam in Fig. 290C) If the beam is clamped at the left and free at the right (Fig. 290C), the boundary conditions are

$$u(0, t) = 0, \qquad u_x(0, t) = 0,$$
$$u_{xx}(L, t) = 0, \qquad u_{xxx}(L, t) = 0.$$

Show that F in Prob. 15 satisfies these conditions if βL is a solution of the equation

$$\cosh \beta L \cos \beta L = -1. \tag{23}$$

Find approximate solutions of (18).

12.4 D'Alembert's Solution of the Wave Equation. Characteristics

It is interesting that the solution (17), Sec. 12.3, of the wave equation

$$\frac{\partial^2 u}{\partial t^2} = c^2 \frac{\partial^2 u}{\partial x^2}, \qquad c^2 = \frac{T}{\rho}, \tag{1}$$

can be immediately obtained by transforming (1) in a suitable way, namely, by introducing the new independent variables

$$v = x + ct, \qquad w = x - ct. \tag{2}$$

Then u becomes a function of v and w. The derivatives in (1) can now be expressed in terms of derivatives with respect to v and w by the use of the chain rule in Sec. 9.6. Denoting partial derivatives by subscripts, we see from (2) that $v_x = 1$ and $w_x = 1$. For simplicity let us denote $u(x, t)$, as a function of v and w, by the same letter u. Then

$$u_x = u_v v_x + u_w w_x = u_v + u_w.$$

We now apply the chain rule to the right side of this equation. We assume that all the partial derivatives involved are continuous, so that $u_{wv} = u_{vw}$. Since $v_x = 1$ and $w_x = 1$, we obtain

$$u_{xx} = (u_v + u_w)_x = (u_v + u_w)_v v_x + (u_v + u_w)_w w_x = u_{vv} + 2u_{vw} + u_{ww}.$$

Transforming the other derivative in (1) by the same procedure, we find

$$u_{tt} = c^2(u_{vv} - 2u_{vw} + u_{ww}).$$

By inserting these two results in (1) we get (see footnote 2 in App. A3.2)

$$u_{vw} \equiv \frac{\partial^2 u}{\partial w \, \partial v} = 0. \tag{3}$$

The point of the present method is that (3) can be readily solved by two successive integrations, first with respect to w and then with respect to v. This gives

$$\frac{\partial u}{\partial v} = h(v) \qquad \text{and} \qquad u = \int h(v)\, dv + \psi(w).$$

Here $h(v)$ and $\psi(w)$ are arbitrary functions of v and w, respectively. Since the integral is a function of v, say, $\phi(v)$, the solution is of the form $u = \phi(v) + \psi(w)$. In terms of x and t, by (2), we thus have

$$u(x, t) = \phi(x + ct) + \psi(x - ct). \tag{4}$$

This is known as **d'Alembert's solution**[1] of the wave equation (1).

Its derivation was much more elegant than the method in Sec. 12.3, but d'Alembert's method is special, whereas the use of Fourier series applies to various equations, as we shall see.

D'Alembert's Solution Satisfying the Initial Conditions

$$\text{(a)} \quad u(x, 0) = f(x), \qquad \text{(b)} \quad u_t(x, 0) = g(x). \tag{5}$$

These are the same as (3) in Sec. 12.3. By differentiating (4) we have

$$u_t(x, t) = c\phi'(x + ct) - c\psi'(x - ct) \tag{6}$$

[1]JEAN LE ROND D'ALEMBERT (1717–1783), French mathematician, also known for his important work in mechanics.

We mention that the general theory of PDEs provides a systematic way for finding the transformation (2) that simplifies (1). See Ref. [C8] in App. 1.

where primes denote derivatives with respect to the *entire* arguments $x + ct$ and $x - ct$, respectively, and the minus sign comes from the chain rule. From (4)–(6) we have

$$u(x, 0) = \phi(x) + \psi(x) = f(x), \tag{7}$$

$$u_t(x, 0) = c\phi'(x) - c\psi'(x) = g(x). \tag{8}$$

Dividing (8) by c and integrating with respect to x, we obtain

$$\phi(x) - \psi(x) = k(x_0) + \frac{1}{c}\int_{x_0}^{x} g(s)\, ds, \qquad k(x_0) = \phi(x_0) - \psi(x_0). \tag{9}$$

If we add this to (7), then ψ drops out and division by 2 gives

$$\phi(x) = \frac{1}{2} f(x) + \frac{1}{2c}\int_{x_0}^{x} g(s)\, ds + \frac{1}{2} k(x_0). \tag{10}$$

Similarly, subtraction of (9) from (7) and division by 2 gives

$$\psi(x) = \frac{1}{2} f(x) - \frac{1}{2c}\int_{x_0}^{x} g(s)\, ds - \frac{1}{2} k(x_0). \tag{11}$$

In (10) we replace x by $x + ct$; we then get an integral from x_0 to $x + ct$. In (11) we replace x by $x - ct$ and get minus an integral from x_0 to $x - ct$ or plus an integral from $x - ct$ to x_0. Hence addition of $\phi(x + ct)$ and $\psi(x - ct)$ gives $u(x, t)$ [see (4)] in the form

$$u(x, t) = \frac{1}{2}[f(x + ct) + f(x - ct)] + \frac{1}{2c}\int_{x-ct}^{x+ct} g(s)\, ds. \tag{12}$$

If the initial velocity is zero, we see that this reduces to

$$u(x, t) = \tfrac{1}{2}[f(x + ct) + f(x - ct)], \tag{13}$$

in agreement with (17) in Sec. 12.3. You may show that because of the boundary conditions (2) in that section the function f must be odd and must have the period $2L$.

Our result shows that the two initial conditions [the functions $f(x)$ and $g(x)$ in (5)] determine the solution uniquely.

The solution of the wave equation by the Laplace transform method will be shown in Sec. 12.11.

Characteristics. Types and Normal Forms of PDEs

The idea of d'Alembert's solution is just a special instance of the **method of characteristics.** This concerns PDEs of the form

$$Au_{xx} + 2Bu_{xy} + Cu_{yy} = F(x, y, u, u_x, u_y) \tag{14}$$

(as well as PDEs in more than two variables). Equation (14) is called **quasilinear** because it is linear in the highest derivatives (but may be arbitrary otherwise). There are three types of PDEs (14), depending on the discriminant $AC - B^2$, as follows.

Type	Defining Condition	Example in Sec. 12.1
Hyperbolic	$AC - B^2 < 0$	Wave equation (1)
Parabolic	$AC - B^2 = 0$	Heat equation (2)
Elliptic	$AC - B^2 > 0$	Laplace equation (3)

Note that (1) and (2) in Sec. 12.1 involve t, but to have y as in (14), we set $y = ct$ in (1), obtaining $u_{tt} - c^2u_{xx} = c^2(u_{yy} - u_{xx}) = 0$. And in (2) we set $y = c^2t$, so that $u_t - c^2u_{xx} = c^2(u_y - u_{xx})$.

A, B, C may be functions of x, y, so that a PDE may be **of mixed type,** that is, of different type in different regions of the xy-plane. An important mixed-type PDE is the **Tricomi equation** (see Prob. 10).

Transformation of (14) to Normal Form. The normal forms of (14) and the corresponding transformations depend on the type of the PDE. They are obtained by solving the **characteristic equation** of (14), which is the ODE

$$Ay'^2 - 2By' + C = 0 \quad (15)$$

where $y' = dy/dx$ (note $-2B$, not $+2B$). The solutions of (15) are called the **characteristics** of (14), and we write them in the form $\Phi(x, y) = const$ and $\Psi(x, y) = const$. Then the transformations giving new variables v, w instead of x, y and the normal forms of (14) are as follows.

Type	New Variables		Normal Form
Hyperbolic	$v = \Phi$	$w = \Psi$	$u_{vw} = F_1$
Parabolic	$v = x$	$w = \Phi = \Psi$	$u_{ww} = F_2$
Elliptic	$v = \frac{1}{2}(\Phi + \Psi)$	$w = \frac{1}{2i}(\Phi - \Psi)$	$u_{vv} + u_{ww} = F_3$

Here, $\Phi = \Phi(x, y)$, $\Psi = \Psi(x, y)$, $F_1 = F_1(v, w, u, u_v, u_w)$, etc., and we denote u as function of v, w again by u, for simplicity. We see that the normal form of a hyperbolic PDE is as in d'Alembert's solution. In the parabolic case we get just one family of solutions $\Phi = \Psi$. In the elliptic case, $i = \sqrt{-1}$, and the characteristics are complex and are of minor interest. For derivation, see Ref. [GR3] in App. 1.

EXAMPLE 1 D'Alembert's Solution Obtained Systematically

The theory of characteristics gives d'Alembert's solution in a systematic fashion. To see this, we write the wave equation $u_{tt} - c^2u_{xx} = 0$ in the form (14) by setting $y = ct$. By the chain rule, $u_t = u_y y_t = cu_y$ and $u_{tt} = c^2u_{yy}$. Division by c^2 gives $u_{xx} - u_{yy} = 0$, as stated before. Hence the characteristic equation is $y'^2 - 1 = (y' + 1)(y' - 1) = 0$. The two families of solutions (characteristics) are $\Phi(x, y) = y + x = const$ and $\Psi(x, y) = y - x = const$. This gives the new variables $v = \Phi = y + x = ct + x$ and $w = \Psi = y - x = ct - x$ and d'Alembert's solution $u = f_1(x + ct) + f_2(x - ct)$. ■

PROBLEM SET 12.4

1. Show that c is the speed of each of the two waves given by (4).
2. Show that because of the boundary conditions (2), Sec. 12.3, the function f in (13) of this section must be odd and of period $2L$.
3. If a steel wire 2 m in length weighs 0.9 nt (about 0.20 lb) and is stretched by a tensile force of 300 nt (about 67.4 lb), what is the corresponding speed of transverse waves?
4. What are the frequencies of the eigenfunctions in Prob. 3?
5. **Longitudinal Vibrations of an Elastic Bar or Rod.** These vibrations in the direction of the x-axis are modeled by the wave equation $u_{tt} = c^2 u_{xx}$, $c^2 = E/\rho$ (see Tolstov [C9], p. 275). If the rod is fastened at one end, $x = 0$, and free at the other, $x = L$, we have $u(0, t) = 0$ and $u_x(L, t) = 0$. Show that the motion corresponding to initial displacement $u(x, 0) = f(x)$ and initial velocity zero is

$$u = \sum_{n=0}^{\infty} A_n \sin p_n x \cos p_n ct,$$

$$A_n = \frac{2}{L} \int_0^L f(x) \sin p_n x \, dx, \qquad p_n = \frac{(2n + 1)\pi}{2L}.$$

6–9 GRAPHING SOLUTIONS

Using (13), sketch or graph a figure (similar to Fig. 288 in Sec. 12.3) of the deflection $u(x, t)$ of a vibrating string (length $L = 1$, ends fixed, $c = 1$) starting with initial velocity 0 and initial deflection (k small, say, $k = 0.01$).

6. $f(x) = k \sin \pi x$
7. $f(x) = k(1 - \cos 2\pi x)$
8. $f(x) = kx(1 - x)$
9. $f(x) = k(x - x^3)$

10. **(Tricomi and Airy equations[2])** Show that the *Tricomi equation* $yu_{xx} + u_{yy} = 0$ is of mixed type. Obtain the **Airy equation** $G'' - yG = 0$ from the Tricomi equation by separation. (For solutions, see p. 446 of Ref. [GR1] listed in App. 1.)

11–20 NORMAL FORMS

Find the type, transform to normal form, and solve. (Show the details of your work.)

11. $u_{xy} - u_{yy} = 0$
12. $u_{xx} - 2u_{xy} + u_{yy} = 0$
13. $u_{xx} + 9u_{yy} = 0$
14. $u_{xx} + u_{xy} - 2u_{yy} = 0$
15. $u_{xx} + 2u_{xy} + u_{yy} = 0$
16. $xu_{xy} - yu_{yy} = 0$
17. $u_{xx} - 4u_{xy} + 4u_{yy} = 0$
18. $u_{xx} + 2u_{xy} + 5u_{yy} = 0$
19. $xu_{xx} - yu_{xy} = 0$
20. $u_{xx} - 4u_{xy} + 3u_{yy} = 0$

12.5 Heat Equation: Solution by Fourier Series

From the wave equation we now turn to the next "big" PDE, the **heat equation**

$$\frac{\partial u}{\partial t} = c^2 \nabla^2 u, \qquad c^2 = \frac{K}{\sigma\rho},$$

which gives the temperature $u(x, y, z, t)$ in a body of homogeneous material. Here c^2 is the thermal diffusivity, K the thermal conductivity, σ the specific heat, and ρ the density of the material of the body. $\nabla^2 u$ is the Laplacian of u, and with respect to Cartesian coordinates x, y, z,

$$\nabla^2 u = \frac{\partial^2 u}{\partial x^2} + \frac{\partial^2 u}{\partial y^2} + \frac{\partial^2 u}{\partial z^2}.$$

The heat equation was derived in Sec. 10.8. It is also called the **diffusion equation.**

As an important application, let us first consider the temperature in a long thin metal bar or wire of constant cross section and homogeneous material, which is oriented along the x-axis (Fig. 291) and is perfectly insulated laterally, so that heat flows in the x-direction

[2] SIR GEORGE BIDELL AIRY (1801–1892), English mathematician, known for his work in elasticity. FRANCESCO TRICOMI (1897–1978), Italian mathematician, who worked in integral equations.

Fig. 291. Bar under consideration

only. Then u depends only on x and time t, and the heat equation becomes the **one-dimensional heat equation**

$$\frac{\partial u}{\partial t} = c^2 \frac{\partial^2 u}{\partial x^2}. \tag{1}$$

This seems to differ only very little from the wave equation, which has a term u_{tt} instead of u_t, but we shall see that this will make the solutions of (1) behave quite differently from those of the wave equation.

We shall solve (1) for some important types of boundary and initial conditions. We begin with the case in which the ends $x = 0$ and $x = L$ of the bar are kept at temperature zero, so that we have the **boundary conditions**

$$u(0, t) = 0, \qquad u(L, t) = 0 \qquad \text{for all } t. \tag{2}$$

Furthermore, the initial temperature in the bar at time $t = 0$ is given, say, $f(x)$, so that we have the **initial condition**

$$u(x, 0) = f(x) \qquad [f(x) \text{ given}]. \tag{3}$$

Here we must have $f(0) = 0$ and $f(L) = 0$ because of (2).

We shall determine a solution $u(x, t)$ of (1) satisfying (2) and (3)—one initial condition will be enough, as opposed to two initial conditions for the wave equation. Technically, our method will parallel that for the wave equation in Sec. 12.3: a separation of variables, followed by the use of Fourier series. You may find a step-by-step comparison worthwhile.

***Step 1.* Two ODEs from the heat equation (1).** Substitution of a product $u(x, t) = F(x)G(t)$ into (1) gives $F\dot{G} = c^2F''G$ with $\dot{G} = dG/dt$ and $F'' = d^2F/dx^2$. To separate the variables, we divide by c^2FG, obtaining

$$\frac{\dot{G}}{c^2G} = \frac{F''}{F}. \tag{4}$$

The left side depends only on t and the right side only on x, so that both sides must equal a constant k (as in Sec. 12.3). You may show that for $k = 0$ or $k > 0$ the only solution $u = FG$ satisfying (2) is $u \equiv 0$. For negative $k = -p^2$ we have from (4)

$$\frac{\dot{G}}{c^2G} = \frac{F''}{F} = -p^2.$$

Multiplication by the denominators gives immediately the two ODEs

$$F'' + p^2F = 0 \tag{5}$$

and

$$\dot{G} + c^2p^2G = 0. \tag{6}$$

***Step 2.* Satisfying the boundary conditions (2).** We first solve (5). A general solution is

$$F(x) = A \cos px + B \sin px. \tag{7}$$

From the boundary conditions (2) it follows that

$$u(0, t) = F(0)G(t) = 0 \qquad \text{and} \qquad u(L, t) = F(L)G(t) = 0.$$

Since $G \equiv 0$ would give $u \equiv 0$, we require $F(0) = 0$, $F(L) = 0$ and get $F(0) = A = 0$ by (7) and then $F(L) = B \sin pL = 0$, with $B \neq 0$ (to avoid $F \equiv 0$); thus,

$$\sin pL = 0, \qquad \text{hence} \qquad p = \frac{n\pi}{L}, \qquad n = 1, 2, \cdots.$$

Setting $B = 1$, we thus obtain the following solutions of (5) satisfying (2):

$$F_n(x) = \sin \frac{n\pi x}{L}, \qquad n = 1, 2, \cdots.$$

(As in Sec. 12.3, we need not consider *negative* integral values of n.)

All this was literally the same as in Sec. 12.3. From now on it differs since (6) differs from (6) in Sec. 12.3. We now solve (6). For $p = n\pi/L$, as just obtained, (6) becomes

$$\dot{G} + \lambda_n{}^2G = 0 \qquad \text{where} \qquad \lambda_n = \frac{cn\pi}{L}.$$

It has the general solution

$$G_n(t) = B_n e^{-\lambda_n{}^2 t}, \qquad n = 1, 2, \cdots$$

where B_n is a constant. Hence the functions

$$u_n(x, t) = F_n(x)G_n(t) = B_n \sin \frac{n\pi x}{L} e^{-\lambda_n{}^2 t} \qquad (n = 1, 2, \cdots) \tag{8}$$

are solutions of the heat equation (1), satisfying (2). These are the **eigenfunctions** of the problem, corresponding to the **eigenvalues** $\lambda_n = cn\pi/L$.

***Step 3.* Solution of the entire problem. Fourier series.** So far we have solutions (8) satisfying the boundary conditions (2). To obtain a solution that also satisfies the initial condition (3), we consider a series of these eigenfunctions,

$$u(x, t) = \sum_{n=1}^{\infty} u_n(x, t) = \sum_{n=1}^{\infty} B_n \sin \frac{n\pi x}{L} e^{-\lambda_n{}^2 t} \qquad \left(\lambda_n = \frac{cn\pi}{L}\right). \tag{9}$$

From this and (3) we have

$$u(x, 0) = \sum_{n=1}^{\infty} B_n \sin \frac{n\pi x}{L} = f(x).$$

Hence for (9) to satisfy (3), the B_n's must be the coefficients of the **Fourier sine series,** as given by (4) in Sec. 11.3; thus

$$(10) \qquad B_n = \frac{2}{L} \int_0^L f(x) \sin \frac{n\pi x}{L}\, dx \qquad (n = 1, 2, \cdots).$$

The solution of our problem can be established, assuming that $f(x)$ is piecewise continuous (see Sec. 6.1) on the interval $0 \leqq x \leqq L$ and has one-sided derivatives (see Sec. 11.1) at all interior points of that interval; that is, under these assumptions the series (9) with coefficients (10) is the solution of our physical problem. A proof requires knowledge of uniform convergence and will be given at a later occasion (Probs. 19, 20 in Problem Set 15.5).

Because of the exponential factor, all the terms in (9) approach zero as t approaches infinity. The rate of decay increases with n.

EXAMPLE 1 Sinusoidal Initial Temperature

Find the temperature $u(x, t)$ in a laterally insulated copper bar 80 cm long if the initial temperature is $100 \sin(\pi x/80)$ °C and the ends are kept at 0°C. How long will it take for the maximum temperature in the bar to drop to 50°C? First guess, then calculate. *Physical data for copper:* density 8.92 gm/cm^3, specific heat 0.092 cal/(gm °C), thermal conductivity 0.95 cal/(cm sec °C).

Solution. The initial condition gives

$$u(x, 0) = \sum_{n=1}^{\infty} B_n \sin \frac{n\pi x}{80} = f(x) = 100 \sin \frac{\pi x}{80}.$$

Hence, by inspection or from (9) we get $B_1 = 100$, $B_2 = B_3 = \cdots = 0$. In (9) we need $\lambda_1{}^2 = c^2\pi^2/L^2$, where $c^2 = K/(\sigma\rho) = 0.95/(0.092 \cdot 8.92) = 1.158$ [cm^2/sec]. Hence we obtain

$$\lambda_1{}^2 = 1.158 \cdot 9.870/80^2 = 0.001785 \ [\text{sec}^{-1}].$$

The solution (9) is

$$u(x, t) = 100 \sin \frac{\pi x}{80}\, e^{-0.001785t}.$$

Also, $100e^{-0.001785t} = 50$ when $t = (\ln 0.5)/(-0.001785) = 388$ [sec] ≈ 6.5 [min]. Does your guess, or at least its order of magnitude, agree with this result? ■

EXAMPLE 2 Speed of Decay

Solve the problem in Example 1 when the initial temperature is $100 \sin(3\pi x/80)$ °C and the other data are as before.

Solution. In (9), instead of $n = 1$ we now have $n = 3$, and $\lambda_3{}^2 = 3^2\lambda_1{}^2 = 9 \cdot 0.001785 = 0.01607$, so that the solution now is

$$u(x, t) = 100 \sin \frac{3\pi x}{80}\, e^{-0.01607t}.$$

Hence the maximum temperature drops to 50°C in $t = (\ln 0.5)/(-0.01607) \approx 43$ [seconds], which is much faster (9 times as fast as in Example 1; why?).

Had we chosen a bigger n, the decay would have been still faster, and in a sum or series of such terms, each term has its own rate of decay, and terms with large n are practically 0 after a very short time. Our next example is of this type, and the curve in Fig. 292 corresponding to $t = 0.5$ looks almost like a sine curve; that is, it is practically the graph of the first term of the solution. ■

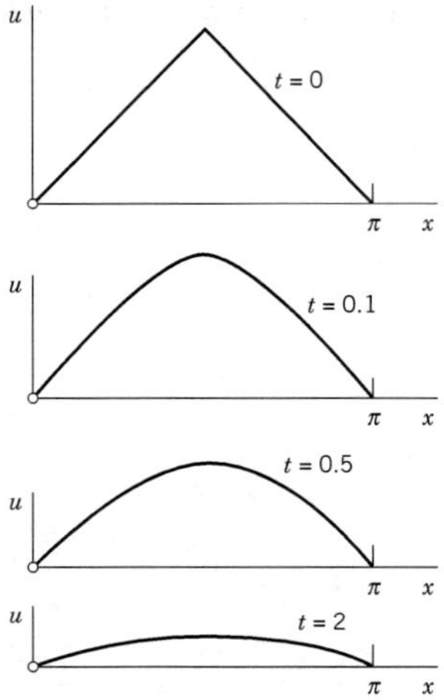

Fig. 292. Example 3. Decrease of temperature with time t for $L = \pi$ and $c = 1$

EXAMPLE 3 "Triangular" Initial Temperature in a Bar

Find the temperature in a laterally insulated bar of length L whose ends are kept at temperature 0, assuming that the initial temperature is

$$f(x) = \begin{cases} x & \text{if} \quad 0 < x < L/2, \\ L - x & \text{if} \quad L/2 < x < L. \end{cases}$$

(The uppermost part of Fig. 292 shows this function for the special $L = \pi$.)

Solution. From (10) we get

$$B_n = \frac{2}{L}\left(\int_0^{L/2} x \sin\frac{n\pi x}{L}\,dx + \int_{L/2}^{L} (L - x)\sin\frac{n\pi x}{L}\,dx\right). \tag{10*}$$

Integration gives $B_n = 0$ if n is even,

$$B_n = \frac{4L}{n^2\pi^2} \quad (n = 1, 5, 9, \cdots) \qquad \text{and} \qquad B_n = -\frac{4L}{n^2\pi^2} \quad (n = 3, 7, 11, \cdots).$$

(see also Example 4 in Sec. 11.3 with $k = L/2$). Hence the solution is

$$u(x, t) = \frac{4L}{\pi^2}\left[\sin\frac{\pi x}{L}\exp\left[-\left(\frac{c\pi}{L}\right)^2 t\right] - \frac{1}{9}\sin\frac{3\pi x}{L}\exp\left[-\left(\frac{3c\pi}{L}\right)^2 t\right] + - \cdots\right].$$

Figure 292 shows that the temperature decreases with increasing t, because of the heat loss due to the cooling of the ends.

Compare Fig. 292 and Fig. 288 in Sec. 12.3 and comment. ■

EXAMPLE 4 Bar with Insulated Ends. Eigenvalue 0

Find a solution formula of (1), (3) with (2) replaced by the condition that both ends of the bar are insulated.

Solution. Physical experiments show that the rate of heat flow is proportional to the gradient of the temperature. Hence if the ends $x = 0$ and $x = L$ of the bar are insulated, so that no heat can flow through the ends, we have grad $u = u_x = \partial u/\partial x$ and the boundary conditions

$$(2^*) \qquad u_x(0, t) = 0, \qquad u_x(L, t) = 0 \qquad \text{for all } t.$$

Since $u(x, t) = F(x)G(t)$, this gives $u_x(0, t) = F'(0)G(t) = 0$ and $u_x(L, t) = F'(L)G(t) = 0$. Differentiating (7), we have $F'(x) = -Ap \sin px + Bp \cos px$, so that

$$F'(0) = Bp = 0 \qquad \text{and then} \qquad F'(L) = -Ap \sin pL = 0.$$

The second of these conditions gives $p = p_n = n\pi/L$, $(n = 0, 1, 2, \cdots)$. From this and (7) with $A = 1$ and $B = 0$ we get $F_n(x) = \cos(n\pi x/L)$, $(n = 0, 1, 2, \cdots)$. With G_n as before, this yields the eigenfunctions

$$(11) \qquad u_n(x, t) = F_n(x)G_n(t) = A_n \cos \frac{n\pi x}{L} e^{-\lambda_n{}^2 t} \qquad (n = 0, 1, \cdots)$$

corresponding to the eigenvalues $\lambda_n = cn\pi/L$. The latter are as before, but we now have the additional eigenvalue $\lambda_0 = 0$ and eigenfunction $u_0 = const$, which is the solution of the problem if the initial temperature $f(x)$ is constant. This shows the remarkable fact that ***a separation constant can very well be zero, and zero can be an eigenvalue.***

Furthermore, whereas (8) gave a Fourier sine series, we now get from (11) a Fourier cosine series

$$\textbf{(12)} \qquad u(x, t) = \sum_{n=0}^{\infty} u_n(x, t) = \sum_{n=0}^{\infty} A_n \cos \frac{n\pi x}{L} e^{-\lambda_n{}^2 t} \qquad \left(\lambda_n = \frac{cn\pi}{L}\right).$$

Its coefficients result from the initial condition (3),

$$u(x, 0) = \sum_{n=0}^{\infty} A_n \cos \frac{n\pi x}{L} = f(x),$$

in the form (2), Sec. 11.3, that is,

$$(13) \qquad A_0 = \frac{1}{L}\int_0^L f(x)\,dx, \qquad A_n = \frac{2}{L}\int_0^L f(x) \cos \frac{n\pi x}{L}\,dx, \qquad n = 1, 2, \cdots. \qquad \blacksquare$$

EXAMPLE 5 "Triangular" Initial Temperature in a Bar with Insulated Ends

Find the temperature in the bar in Example 3, assuming that the ends are insulated (instead of being kept at temperature 0).

Solution. For the triangular initial temperature, (13) gives $A_0 = L/4$ and (see also Example 4 in Sec. 11.3 with $k = L/2$)

$$A_n = \frac{2}{L}\left[\int_0^{L/2} x \cos \frac{n\pi x}{L}\,dx + \int_{L/2}^{L} (L - x) \cos \frac{n\pi x}{L}\,dx\right] = \frac{2L}{n^2\pi^2}\left(2 \cos \frac{n\pi}{2} - \cos n\pi - 1\right).$$

Hence the solution (12) is

$$u(x, t) = \frac{L}{4} - \frac{8L}{\pi^2}\left\{\frac{1}{2^2} \cos \frac{2\pi x}{L} \exp\left[-\left(\frac{2c\pi}{L}\right)^2 t\right] + \frac{1}{6^2} \cos \frac{6\pi x}{L} \exp\left[-\left(\frac{6c\pi}{L}\right)^2 t\right] + \cdots\right\}.$$

We see that the terms decrease with increasing t, and $u \to L/4$ as $t \to \infty$; this is the mean value of the initial temperature. This is plausible because no heat can escape from this totally insulated bar. In contrast, the cooling of the ends in Example 3 led to heat loss and $u \to 0$, the temperature at which the ends were kept. $\blacksquare$

Steady Two-Dimensional Heat Problems. Laplace's Equation

We shall now extend our discussion from one to two space dimensions and consider the two-dimensional heat equation

$$\frac{\partial u}{\partial t} = c^2\nabla^2 u = c^2\left(\frac{\partial^2 u}{\partial x^2} + \frac{\partial^2 u}{\partial y^2}\right)$$

for **steady** (that is, *time-independent*) problems. Then $\partial u/\partial t = 0$ and the heat equation reduces to **Laplace's equation**

$$\nabla^2 u = \frac{\partial^2 u}{\partial x^2} + \frac{\partial^2 u}{\partial y^2} = 0 \tag{14}$$

(which has already occurred in Sec. 10.8 and will be considered further in Secs. 12.7–12.10). A heat problem then consists of this PDE to be considered in some region R of the xy-plane and a given boundary condition on the boundary curve C of R. This is a **boundary value problem (BVP).** One calls it:

First BVP or **Dirichlet Problem** if u is prescribed on C (**"Dirichlet boundary condition"**)

Second BVP or **Neumann Problem** if the normal derivative $u_n = \partial u/\partial n$ is prescribed on C (**"Neumann boundary condition"**)

Third BVP, Mixed BVP, or **Robin Problem** if u is prescribed on a portion of C and u_n on the rest of C (**"Mixed boundary condition").**

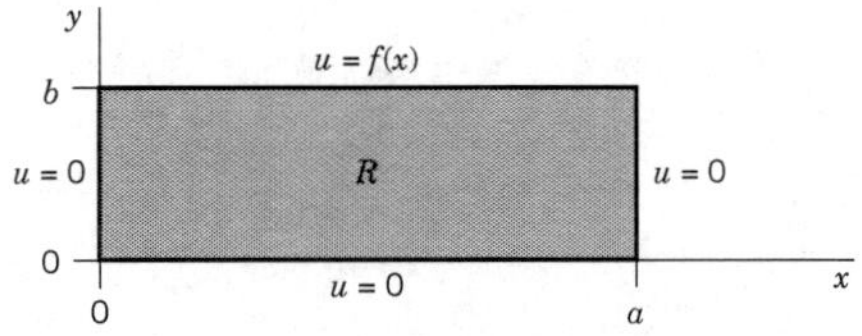

Fig. 293. Rectangle R and given boundary values

Dirichlet Problem in a Rectangle R (Fig. 293). We consider a Dirichlet problem for Laplace's equation (14) in a rectangle R, assuming that the temperature $u(x, y)$ equals a given function $f(x)$ on the upper side and 0 on the other three sides of the rectangle.

We solve this problem by separating variables. Substituting $u(x, y) = F(x)G(y)$ into (14) written as $u_{xx} = -u_{yy}$, dividing by FG, and equating both sides to a negative constant, we obtain

$$\frac{1}{F} \cdot \frac{d^2F}{dx^2} = -\frac{1}{G} \cdot \frac{d^2G}{dy^2} = -k.$$

From this we get

$$\frac{d^2F}{dx^2} + kF = 0,$$

and the left and right boundary conditions imply

$$F(0) = 0, \qquad \text{and} \qquad F(a) = 0.$$

This gives $k = (n\pi/a)^2$ and corresponding nonzero solutions

$$F(x) = F_n(x) = \sin \frac{n\pi}{a} x, \qquad n = 1, 2, \cdots. \tag{15}$$

The ODE for G with $k = (n\pi/a)^2$ then becomes

$$\frac{d^2G}{dy^2} - \left(\frac{n\pi}{a}\right)^2 G = 0.$$

Solutions are

$$G(y) = G_n(y) = A_n e^{n\pi y/a} + B_n e^{-n\pi y/a}.$$

Now the boundary condition $u = 0$ on the lower side of R implies that $G_n(0) = 0$; that is, $G_n(0) = A_n + B_n = 0$ or $B_n = -A_n$. This gives

$$G_n(y) = A_n(e^{n\pi y/a} - e^{-n\pi y/a}) = 2A_n \sinh \frac{n\pi y}{a}.$$

From this and (15), writing $2A_n = A_n^*$, we obtain as the **eigenfunctions** of our problem

$$u_n(x, y) = F_n(x)G_n(y) = A_n^* \sin \frac{n\pi x}{a} \sinh \frac{n\pi y}{a}. \tag{16}$$

These solutions satisfy the boundary condition $u = 0$ on the left, right, and lower sides.

To get a solution also satisfying the boundary condition $u(x, b) = f(x)$ on the upper side, we consider the infinite series

$$u(x, y) = \sum_{n=1}^{\infty} u_n(x, y).$$

From this and (16) with $y = b$ we obtain

$$u(x, b) = f(x) = \sum_{n=1}^{\infty} A_n^* \sin \frac{n\pi x}{a} \sinh \frac{n\pi b}{a}.$$

We can write this in the form

$$u(x, b) = \sum_{n=1}^{\infty} \left(A_n^* \sinh \frac{n\pi b}{a}\right) \sin \frac{n\pi x}{a}.$$

This shows that the expressions in the parentheses must be the Fourier coefficients b_n of $f(x)$; that is, by (4) in Sec. 11.3,

$$b_n = A_n^* \sinh \frac{n\pi b}{a} = \frac{2}{a} \int_0^a f(x) \sin \frac{n\pi x}{a} \, dx.$$

From this and (16) we see that the solution of our problem is

$$u(x, y) = \sum_{n=1}^{\infty} u_n(x, y) = \sum_{n=1}^{\infty} A_n^* \sin \frac{n\pi x}{a} \sinh \frac{n\pi y}{a} \tag{17}$$

where

$$A_n^* = \frac{2}{a \sinh (n\pi b/a)} \int_0^a f(x) \sin \frac{n\pi x}{a} \, dx. \tag{18}$$

We have obtained this solution formally, neither considering convergence nor showing that the series for u, u_{xx}, and u_{yy} have the right sums. This can be proved if one assumes that f and f' are continuous and f'' is piecewise continuous on the interval $0 \leqq x \leqq a$. The proof is somewhat involved and relies on uniform convergence. It can be found in [C4] listed in App. 1.

Unifying Power of Methods. Electrostatics, Elasticity

The Laplace equation (14) also governs the electrostatic potential of electrical charges in any region that is free of these charges. Thus our steady-state heat problem can also be interpreted as an electrostatic potential problem. Then (17), (18) is the potential in the rectangle R when the upper side of R is at potential $f(x)$ and the other three sides are grounded.

Actually, in the steady-state case, the two-dimensional wave equation (to be considered in Secs. 12.7, 12.8) also reduces to (14). Then (17), (18) is the displacement of a rectangular elastic membrane (rubber sheet, drumhead) that is fixed along its boundary, with three sides lying in the xy-plane and the fourth side given the displacement $f(x)$.

This is another impressive demonstration of the ***unifying power*** of mathematics. It illustrates that ***entirely different physical systems may have the same mathematical model*** and can thus be treated by the same mathematical methods.

PROBLEM SET 12.5

1. **WRITING PROJECT. Wave and Heat Equations.** Compare the two PDEs with respect to type, general behavior of eigenfunctions, and kind of boundary and initial conditions and resulting practical problems. Also discuss the difference between Figs. 288 in Sec. 12.3 and 292.
2. **(Eigenfunctions)** Sketch (or graph) and compare the first three eigenfunctions (8) with $B_n = 1$, $c = 1$, $L = \pi$ for $t = 0, 0.2, 0.4, 0.6, 0.8, 1.0$.
3. **(Decay)** How does the rate of decay of (8) with fixed n depend on the specific heat, the density, and the thermal conductivity of the material?

4. If the first eigenfunction (8) of the bar decreases to half its value within 10 sec, what is the value of the diffusivity?

5–9 LATERALLY INSULATED BAR

A laterally insulated bar of length 10 cm and constant cross-sectional area 1 cm^2, of density 10.6 gm/cm^3, thermal conductivity 1.04 cal/(cm sec °C), and specific heat 0.056 cal/(gm °C) (this corresponds to silver, a good heat conductor) has initial temperature $f(x)$ and is kept at 0°C at the ends $x = 0$ and $x = 10$. Find the temperature $u(x, t)$ at later times. Here, $f(x)$ equals:

5. $f(x) = \sin 0.4\pi x$

6. $f(x) = \sin 0.1\pi x + \frac{1}{2}\sin 0.2\pi x$

7. $f(x) = 0.2x$ if $0 < x < 5$ and 0 otherwise

8. $f(x) = 1 - 0.2|x - 5|$

9. $f(x) = x$ if $0 < x < 2.5$, $f(x) = 2.5$ if $2.5 < x < 7.5$, $f(x) = 10 - x$ if $7.5 < x < 10$

10. (Arbitrary temperatures at ends) If the ends $x = 0$ and $x = L$ of the bar in the text are kept at constant temperatures U_1 and U_2, respectively, what is the temperature $u_I(x)$ in the bar after a long time (theoretically, as $t \to \infty$)? First guess, then calculate.

11. In Prob. 10 find the temperature at any time.

12. (Changing end temperatures) Assume that the ends of the bar in Probs. 5–9 have been kept at 100°C for a long time. Then at some instant, call it $t = 0$, the temperature at $x = L$ is suddenly changed to 0°C and kept at 0°C, whereas the temperature at $x = 0$ is kept at 100°C. Find the temperature in the middle of the bar at $t = 1, 2, 3, 10, 50$ sec. First guess, then calculate.

BAR UNDER ADIABATIC CONDITIONS

"Adiabatic" means no heat exchange with the neighborhood, because the bar is completely insulated, also at the ends. *Physical Information:* The heat flux at the ends is proportional to the value of $\partial u/\partial x$ there.

13. Show that for the completely insulated bar, $u_x(0, t) = 0$, $u_x(L, t) = 0$, $u(x, t) = f(x)$ and separation of variables gives the following solution, with A_n given by (2) in Sec. 11.3.

$$u(x, t) = A_0 + \sum_{n=1}^{\infty} A_n \cos \frac{n\pi x}{L} e^{-(cn\pi/L)^2 t}$$

14–19 Find the temperature in Prob. 13 with $L = \pi$, $c = 1$, and

14. $f(x) = x$

15. $f(x) = 1$

16. $f(x) = 0.5 \cos 4x$

17. $f(x) = \pi^2 - x^2$

18. $f(x) = \frac{1}{2}\pi - |x - \frac{1}{2}\pi|$

19. $f(x) = (x - \frac{1}{2}\pi)^2$

20. Find the temperature of the bar in Prob. 13 if the left end is kept at 0°C, the right end is insulated, and the initial temperature is $U_0 = const.$

21. The **boundary condition of heat transfer**

$$-u_x(\pi, t) = k[u(\pi, t) - u_0] \tag{19}$$

applies when a bar of length π with $c = 1$ is laterally insulated, the left end $x = 0$ is kept at 0°C, and at the right end heat is flowing into air of constant temperature u_0. Let $k = 1$ for simplicity, and $u_0 = 0$. Show that a solution is $u(x, t) = \sin px\, e^{-p^2 t}$, where p is a solution of $\tan p\pi = -p$. Show graphically that this equation has infinitely many positive solutions $p_1, p_2, p_3, \cdots$, where $p_n > n - \frac{1}{2}$ and $\lim_{n\to\infty} (p_n - n + \frac{1}{2}) = 0$. (Formula (19) is also known as **radiation boundary condition,** but this is misleading; see Ref. [C3], p. 19.)

22. (Discontinuous f) Solve (1), (2), (3) with $L = \pi$ and $f(x) = U_0 = const\ (\neq 0)$ if $0 < x < \pi/2$, $f(x) = 0$ if $\pi/2 < x < \pi$.

23. (Heat flux) The *heat flux* of a solution $u(x, t)$ across $x = 0$ is defined by $\phi(t) = -Ku_x(0, t)$. Find $\phi(t)$ for the solution (9). Explain the name. Is it physically understandable that ϕ goes to 0 as $t \to \infty$?

OTHER HEAT EQUATIONS

24. (Bar with heat generation) If heat is generated at a constant rate throughout a bar of length $L = \pi$ with initial temperature $f(x)$ and the ends at $x = 0$ and π are kept at temperature 0, the heat equation is $u_t = c^2 u_{xx} + H$ with constant $H > 0$. Solve this problem. *Hint.* Set $u = v - Hx(x - \pi)/(2c^2)$.

25. (Convection) If heat in the bar in the text is free to flow through an end into the surrounding medium kept at 0°C, the PDE becomes $v_t = c^2 v_{xx} - \beta v$. Show that it can be reduced to the form (1) by setting $v(x, t) = u(x, t)w(t)$.

26. Consider $v_t = c^2 v_{xx} - v \quad (0 < x < L, t > 0)$, $v(0, t) = 0$, $v(L, t) = 0$, $v(x, 0) = f(x)$, where the term $-v$ models heat transfer to the surrounding medium kept at temperature 0. Reduce this PDE by setting $v(x, t) = u(x, t)w(t)$ with w such that u is given by (9), (10).

27. (Nonhomogeneous heat equation) Show that the problem modeled by

$$u_t - c^2 u_{xx} = Ne^{-\alpha x}$$

and (2), (3) can be reduced to a problem for the homogeneous heat equation by setting

$$u(x, t) = v(x, t) + w(x)$$

and determining w so that v satisfies the homogeneous PDE and the conditions $v(0, t) = v(L, t) = 0$, $v(x, 0) = f(x) - w(x)$. (The term $Ne^{-\alpha x}$ may represent heat loss due to radioactive decay in the bar.)

28–35 TWO-DIMENSIONAL PROBLEMS

28. (Laplace equation) Find the potential in the rectangle $0 \leqq x \leqq 20$, $0 \leqq y \leqq 40$ whose upper side is kept at potential 220 V and whose other sides are grounded.

29. Find the potential in the square $0 \leqq x \leqq 2$, $0 \leqq y \leqq 2$ if the upper side is kept at the potential $\sin \frac{1}{2}\pi x$ and the other sides are grounded.

30. CAS PROJECT. Isotherms. Find the steady-state solutions (temperatures) in the square plate in Fig. 294 with $a = 2$ satisfying the following boundary conditions. Graph isotherms.

(a) $u = \sin \pi x$ on the upper side, 0 on the others.

(b) $u = 0$ on the vertical sides, assuming that the other sides are perfectly insulated.

(c) Boundary conditions of your choice (such that the solution is not identically zero).

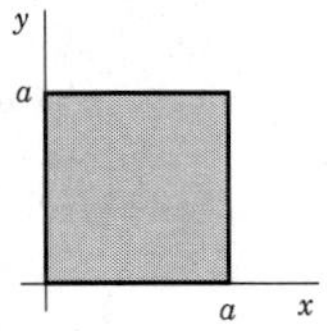

Fig. 294. Square plate

31. (Heat flow in a plate) The faces of the thin square plate in Fig. 294 with side $a = 24$ are perfectly insulated. The upper side is kept at 20°C and the other sides are kept at 0°C. Find the steady-state temperature $u(x, y)$ in the plate.

32. Find the steady-state temperature in the plate in Prob. 31 if the lower side is kept at U_0°C, the upper side at U_1°C, and the other sides are kept at 0°C. *Hint:* Split into two problems in which the boundary temperature is 0 on three sides for each problem.

33. (Mixed boundary value problem) Find the steady-state temperature in the plate in Prob. 31 with the upper and lower sides perfectly insulated, the left side kept at 0°C, and the right side kept at $f(y)$°C.

34. (Radiation) Find steady-state temperatures in the rectangle in Fig. 293 with the upper and left sides perfectly insulated and the right side radiating into a medium at 0°C according to $u_x(a, y) + hu(a, y) = 0$, $h > 0$ constant. (You will get many solutions since no condition on the lower side is given.)

35. Find formulas similar to (17), (18) for the temperature in the rectangle R of the text when the lower side of R is kept at temperature $f(x)$ and the other sides are kept at 0°C.

12.6 Heat Equation: Solution by Fourier Integrals and Transforms

Our discussion of the heat equation

$$\frac{\partial u}{\partial t} = c^2 \frac{\partial^2 u}{\partial x^2} \tag{1}$$

in the last section extends to bars of infinite length, which are good models of very long bars or wires (such as a wire of length, say, 300 ft). Then the role of Fourier series in the solution process will be taken by **Fourier integrals** (Sec. 11.7).

Let us illustrate the method by solving (1) for a bar that extends to infinity on both sides (and is laterally insulated as before). Then we do not have boundary conditions, but only the **initial condition**

$$u(x, 0) = f(x) \qquad (-\infty < x < \infty) \tag{2}$$

where $f(x)$ is the given initial temperature of the bar.

To solve this problem, we start as in the last section, substituting $u(x, t) = F(x)G(t)$ into (1). This gives the two ODEs

$$F'' + p^2 F = 0 \qquad \text{[see (5), Sec. 12.5]} \tag{3}$$

and

$$\dot{G} + c^2p^2G = 0 \qquad \text{[see (6), Sec. 12.5].} \tag{4}$$

Solutions are

$$F(x) = A\cos px + B\sin px \qquad \text{and} \qquad G(t) = e^{-c^2p^2t},$$

respectively, where A and B are any constants. Hence a solution of (1) is

$$u(x, t; p) = FG = (A\cos px + B\sin px)e^{-c^2p^2t}. \tag{5}$$

Here we had to choose the separation constant k negative, $k = -p^2$, because positive values of k would lead to an increasing exponential function in (5), which has no physical meaning.

Use of Fourier Integrals

Any series of functions (5), found in the usual manner by taking p as multiples of a fixed number, would lead to a function that is periodic in x when $t = 0$. However, since $f(x)$ in (2) is not assumed to be periodic, it is natural to use **Fourier integrals** instead of Fourier series. Also, A and B in (5) are arbitrary and we may regard them as functions of p, writing $A = A(p)$ and $B = B(p)$. Now, since the heat equation (1) is linear and homogeneous, the function

$$u(x, t) = \int_0^\infty u(x, t; p)\,dp = \int_0^\infty [A(p)\cos px + B(p)\sin px]\,e^{-c^2p^2t}\,dp \tag{6}$$

is then a solution of (1), provided this integral exists and can be differentiated twice with respect to x and once with respect to t.

Determination of $A(p)$ and $B(p)$ from the Initial Condition. From (6) and (2) we get

$$u(x, 0) = \int_0^\infty [A(p)\cos px + B(p)\sin px]\,dp = f(x). \tag{7}$$

This gives $A(p)$ and $B(p)$ in terms of $f(x)$; indeed, from (4) in Sec. 11.7 we have

$$A(p) = \frac{1}{\pi}\int_{-\infty}^{\infty} f(v)\cos pv\,dv, \qquad B(p) = \frac{1}{\pi}\int_{-\infty}^{\infty} f(v)\sin pv\,dv. \tag{8}$$

According to (1*), Sec. 11.9, our Fourier integral (7) with these $A(p)$ and $B(p)$ can be written

$$u(x, 0) = \frac{1}{\pi}\int_0^\infty \left[\int_{-\infty}^{\infty} f(v)\cos(px - pv)\,dv\right] dp.$$

Similarly, (6) in this section becomes

$$u(x, t) = \frac{1}{\pi}\int_0^\infty \left[\int_{-\infty}^{\infty} f(v)\cos(px - pv)\,e^{-c^2p^2t}\,dv\right] dp.$$

Assuming that we may reverse the order of integration, we obtain

$$u(x, t) = \frac{1}{\pi} \int_{-\infty}^{\infty} f(v) \left[\int_0^{\infty} e^{-c^2p^2t} \cos(px - pv)\, dp \right] dv. \tag{9}$$

Then we can evaluate the inner integral by using the formula

$$\int_0^{\infty} e^{-s^2} \cos 2bs\, ds = \frac{\sqrt{\pi}}{2}\, e^{-b^2}. \tag{10}$$

[A derivation of (10) is given in Problem Set 16.4 (Team Project 28).] This takes the form of our inner integral if we choose $p = s/(c\sqrt{t})$ as a new variable of integration and set

$$b = \frac{x - v}{2c\sqrt{t}}\,.$$

Then $2bs = (x - v)p$ and $ds = c\sqrt{t}\, dp$, so that (10) becomes

$$\int_0^{\infty} e^{-c^2p^2t} \cos(px - pv)\, dp = \frac{\sqrt{\pi}}{2c\sqrt{t}} \exp\left\{ -\frac{(x - v)^2}{4c^2t} \right\}.$$

By inserting this result into (9) we obtain the representation

$$u(x, t) = \frac{1}{2c\sqrt{\pi t}} \int_{-\infty}^{\infty} f(v) \exp\left\{ -\frac{(x - v)^2}{4c^2t} \right\} dv. \tag{11}$$

Taking $z = (v - x)/(2c\sqrt{t})$ as a variable of integration, we get the alternative form

$$u(x, t) = \frac{1}{\sqrt{\pi}} \int_{-\infty}^{\infty} f(x + 2cz\sqrt{t})\, e^{-z^2}\, dz. \tag{12}$$

If $f(x)$ is bounded for all values of x and integrable in every finite interval, it can be shown (see Ref. [C10]) that the function (11) or (12) satisfies (1) and (2). Hence this function is the required solution in the present case.

EXAMPLE 1 Temperature in an Infinite Bar

Find the temperature in the infinite bar if the initial temperature is (Fig. 295)

$$f(x) = \begin{cases} U_0 = \text{const} & \text{if } |x| < 1, \\ 0 & \text{if } |x| > 1. \end{cases}$$

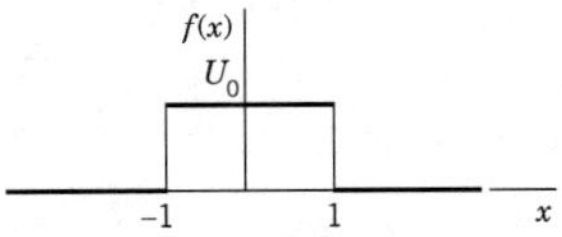

Fig. 295. Initial temperature in Example 1

Solution. From (11) we have

$$u(x, t) = \frac{U_0}{2c\sqrt{\pi t}} \int_{-1}^{1} \exp\left\{-\frac{(x - v)^2}{4c^2 t}\right\} dv.$$

If we introduce the above variable of integration z, then the integration over v from -1 to 1 corresponds to the integration over z from $(-1 - x)/(2c\sqrt{t})$ to $(1 - x)/(2c\sqrt{t})$, and

(13) $$u(x, t) = \frac{U_0}{\sqrt{\pi}} \int_{-(1+x)/(2c\sqrt{t})}^{(1-x)/(2c\sqrt{t})} e^{-z^2} dz \qquad (t > 0).$$

We mention that this integral is not an elementary function, but can be expressed in terms of the error function, whose values have been tabulated. (Table A4 in App. 5 contains a few values; larger tables are listed in Ref. [GR1] in App. 1. See also CAS Project 10, p. 568.) Figure 296 shows $u(x, t)$ for $U_0 = 100°\text{C}$, $c^2 = 1\ \text{cm}^2/\text{sec}$, and several values of t. ■

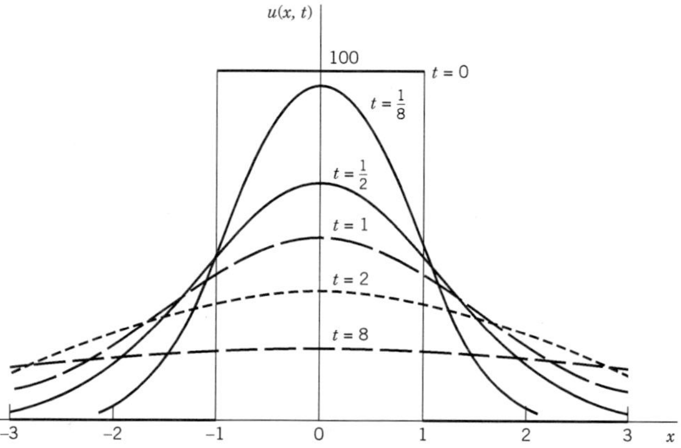

Fig. 296. Solution $u(x, t)$ in Example 1 for $U_0 = 100°\text{C}$, $c^2 = 1\ \text{cm}^2/\text{sec}$, and several values of t

Use of Fourier Transforms

The Fourier transform is closely related to the Fourier integral, from which we obtained the transform in Sec. 11.9. And the transition to the Fourier cosine and sine transform in Sec. 11.8 was even simpler. (You may perhaps wish to review this before going on.) Hence it should not surprise you that we can use these transforms for solving our present or similar problems. The Fourier transform applies to problems concerning the entire axis, and the Fourier cosine and sine transforms to problems involving the positive half-axis. Let us explain these transform methods by typical applications that fit our present discussion.

EXAMPLE 2 Temperature in the Infinite Bar in Example 1

Solve Example 1 using the Fourier transform.

Solution. The problem consists of the heat equation (1) and the initial condition (2), which in this example is

$$f(x) = U_0 = const \quad \text{if } |x| < 1 \qquad \text{and 0 otherwise.}$$

Our strategy is to take the Fourier transform with respect to x and then to solve the resulting ***ordinary*** DE in t. The details are as follows.

Let $\hat{u} = \mathcal{F}(u)$ denote the Fourier transform of u, ***regarded as a function of x.*** From (10) in Sec. 11.9 we see that the heat equation (1) gives

$$\mathcal{F}(u_t) = c^2\mathcal{F}(u_{xx}) = c^2(-w^2)\mathcal{F}(u) = -c^2w^2\hat{u}.$$

On the left, assuming that we may interchange the order of differentiation and integration, we have

$$\mathcal{F}(u_t) = \frac{1}{\sqrt{2\pi}}\int_{-\infty}^{\infty} u_t e^{-iwx}\,dx = \frac{1}{\sqrt{2\pi}}\frac{\partial}{\partial t}\int_{-\infty}^{\infty} ue^{-iwx}\,dx = \frac{\partial \hat{u}}{\partial t}.$$

Thus

$$\frac{\partial \hat{u}}{\partial t} = -c^2w^2\hat{u}.$$

Since this equation involves only a derivative with respect to t but none with respect to w, this is a first-order ***ordinary DE,*** with t as the independent variable and w as a parameter. By separating variables (Sec. 1.3) we get the general solution

$$\hat{u}(w, t) = C(w)e^{-c^2w^2t}$$

with the arbitrary "constant" $C(w)$ depending on the parameter w. The initial condition (2) yields the relationship $\hat{u}(w, 0) = C(w) = \hat{f}(w) = \mathcal{F}(f)$. Our intermediate result is

$$\hat{u}(w, t) = \hat{f}(w)e^{-c^2w^2t}.$$

The inversion formula (7), Sec. 11.9, now gives the solution

$$u(x, t) = \frac{1}{\sqrt{2\pi}}\int_{-\infty}^{\infty} \hat{f}(w)e^{-c^2w^2t}e^{iwx}\,dw. \tag{14}$$

In this solution we may insert the Fourier transform

$$\hat{f}(w) = \frac{1}{\sqrt{2\pi}}\int_{-\infty}^{\infty} f(v)e^{-ivw}\,dv.$$

Assuming that we may invert the order of integration, we then obtain

$$u(x, t) = \frac{1}{2\pi}\int_{-\infty}^{\infty} f(v)\left[\int_{-\infty}^{\infty} e^{-c^2w^2t}e^{i(wx-wv)}\,dw\right]dv.$$

By the Euler formula (3). Sec. 11.9, the integrand of the inner integral equals

$$e^{-c^2w^2t}\cos(wx - wv) + ie^{-c^2w^2t}\sin(wx - wv).$$

We see that its imaginary part is an odd function of w, so that its integral is 0. (More precisely, this is the principal part of the integral; see Sec. 16.4.) The real part is an even function of w, so that its integral from $-\infty$ to ∞ equals twice the integral from 0 to ∞:

$$u(x, t) = \frac{1}{\pi}\int_{-\infty}^{\infty} f(v)\left[\int_{0}^{\infty} e^{-c^2w^2t}\cos(wx - wv)\,dw\right]dv.$$

This agrees with (9) (with $p = w$) and leads to the further formulas (11) and (13). ■

EXAMPLE 3 **Solution in Example 1 by the Method of Convolution**

Solve the heat problem in Example 1 by the method of convolution.

Solution. The beginning is as in Example 2 and leads to (14), that is,

$$u(x, t) = \frac{1}{\sqrt{2\pi}}\int_{-\infty}^{\infty} \hat{f}(w)e^{-c^2w^2t}e^{iwx}\,dw. \tag{15}$$

Now comes the crucial idea. We recognize that this is of the form (13) in Sec. 11.9, that is,

$$u(x, t) = (f * g)(x) = \int_{-\infty}^{\infty} \hat{f}(w)\hat{g}(w)e^{iwx}\, dw \tag{16}$$

where

$$\hat{g}(w) = \frac{1}{\sqrt{2\pi}}\, e^{-c^2w^2t}. \tag{17}$$

Since, by the definition of convolution [(11), Sec. 11.9],

$$(f * g)(x) = \int_{-\infty}^{\infty} f(p)g(x - p)\, dp, \tag{18}$$

as our next and last step we must determine the inverse Fourier transform g of $\hat{g}$. For this we can use formula 9 in Table III of Sec. 11.10,

$$\mathcal{F}(e^{-ax^2}) = \frac{1}{\sqrt{2a}}\, e^{-w^2/(4a)}$$

with a suitable a. With $c^2t = 1/(4a)$ or $a = 1/(4c^2t)$, using (17) we obtain

$$\mathcal{F}(e^{-x^2/(4c^2t)}) = \sqrt{2c^2t}\, e^{-c^2w^2t} = \sqrt{2c^2t}\,\sqrt{2\pi}\,\hat{g}(w).$$

Hence $\hat{g}$ has the inverse

$$\frac{1}{\sqrt{2c^2t}\,\sqrt{2\pi}}\, e^{-x^2/(4c^2t)}.$$

Replacing x with $x - p$ and substituting this into (18) we finally have

$$u(x, t) = (f * g)(x) = \frac{1}{2c\sqrt{\pi t}} \int_{-\infty}^{\infty} f(p) \exp\left\{-\frac{(x - p)^2}{4c^2t}\right\} dp. \tag{19}$$

This solution formula of our problem agrees with (11). We wrote $(f * g)(x)$, without indicating the parameter t with respect to which we did not integrate. ■

EXAMPLE 4 Fourier Sine Transform Applied to the Heat Equation

If a laterally insulated bar extends from $x = 0$ to infinity, we can use the Fourier sine transform. We let the initial temperature be $u(x, 0) = f(x)$ and impose the boundary condition $u(0, t) = 0$. Then from the heat equation and (9b) in Sec. 11.8, since $f(0) = u(0, 0) = 0$, we obtain

$$\mathcal{F}_s(u_t) = \frac{\partial \hat{u}_s}{\partial t} = c^2\mathcal{F}_s(u_{xx}) = -c^2w^2\mathcal{F}_s(u) = -c^2w^2\hat{u}_s(w, t).$$

This is a first-order ODE $\partial\hat{u}_s/\partial t + c^2w^2\hat{u}_s = 0$. Its solution is

$$\hat{u}_s(w, t) = C(w)e^{-c^2w^2t}.$$

From the initial condition $u(x, 0) = f(x)$ we have $\hat{u}_s(w, 0) = \hat{f}_s(w) = C(w)$. Hence

$$\hat{u}_s(w, t) = \hat{f}_s(w)e^{-c^2w^2t}.$$

Taking the inverse Fourier sine transform and substituting

$$\hat{f}_s(w) = \sqrt{\frac{2}{\pi}} \int_0^{\infty} f(p) \sin wp\, dp$$

on the right, we obtain the solution formula

$$u(x, t) = \frac{2}{\pi}\int_0^{\infty}\int_0^{\infty} f(p) \sin wp\, e^{-c^2w^2t} \sin wx\, dp\, dw. \tag{20}$$

Figure 297 shows (20) with $c = 1$ for $f(x) = 1$ if $0 \leqq x \leqq 1$ and 0 otherwise, graphed over the xt-plane for $0 \leqq x \leqq 2$, $0.01 \leqq t \leqq 1.5$. Note that the curves of $u(x, t)$ for constant t resemble those in Fig. 296 on p. 565. ■

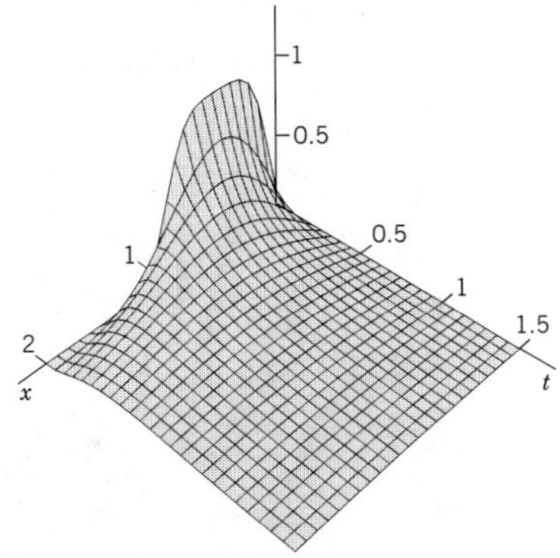

Fig. 297. Solution (20) in Example 4

PROBLEM SET 12.6

1–7 SOLUTION IN INTEGRAL FORM

Using (6), obtain the solution of (1) in integral form satisfying the initial condition $u(x, 0) = f(x)$, where

1. $f(x) = 1$ if $|x| < a$ and 0 otherwise
2. $f(x) = e^{-k|x|}$ $(k > 0)$
3. $f(x) = 1/(1 + x^2)$. [Use (15) in Sec. 11.7.]
4. $f(x) = (\sin x)/x$. [Use Prob. 4 in Sec. 11.7.]
5. $f(x) = (\sin \pi x)/x$. [Use Prob. 4 in Sec. 11.7.]
6. $f(x) = x$ if $|x| < 1$ and 0 otherwise
7. $f(x) = |x|$ if $|x| < 1$ and 0 otherwise.

8. Verify that u in Prob. 5 satisfies the initial condition.
9. **CAS PROJECT. Heat Flow. (a)** Graph the basic Fig. 296.
   (b) In (a) apply animation to "see" the heat flow in terms of the decrease of temperature.
   (c) Graph $u(x, t)$ with $c = 1$ as a surface over the upper xt-half-plane.
10. **CAS PROJECT. Error Function**

$$\text{erf}\, x = \frac{2}{\sqrt{\pi}}\int_0^x e^{-w^2}\, dw \tag{21}$$

This function is important in applied mathematics and physics (probability theory and statistics, thermodynamics, etc.) and fits our present discussion. Regarding it as a typical case of a special function defined by an integral that cannot be evaluated as in elementary calculus, do the following.

(a) Sketch or graph the **bell-shaped curve** [the curve of the integrand in (21)]. Show that erf x is odd. Show that

$$\int_a^b e^{-w^2}\, dw = \frac{\sqrt{\pi}}{2}(\text{erf}\, b - \text{erf}\, a),$$

$$\int_{-b}^b e^{-w^2}\, dw = \sqrt{\pi}\, \text{erf}\, b.$$

(b) Obtain the Maclaurin series of erf x from that of the integrand. Use that series to compute a table of erf x for $x = 0(0.01)3$ (meaning $x = 0, 0.01, 0.02, \cdots, 3$).

(c) Obtain the values required in (b) by an integration command of your CAS. Compare accuracy.

(d) It can be shown that erf $(\infty) = 1$. Confirm this experimentally by computing erf x for large x.

(e) Let $f(x) = 1$ when $x > 0$ and 0 when $x < 0$. Using $\text{erf}(\infty) = 1$, show that (12) then gives

$$u(x, t) = \frac{1}{\sqrt{\pi}} \int_{-x/(2c\sqrt{t})}^{\infty} e^{-z^2}\, dz = \frac{1}{2} - \frac{1}{2}\,\text{erf}\left(-\frac{x}{2c\sqrt{t}}\right) \qquad (t > 0).$$

(f) Express the temperature (13) in terms of the error function.

(g) Show that $\Phi(x) = \frac{1}{\sqrt{2\pi}} \int_{-\infty}^{x} e^{-s^2/2}\, ds$

$$= \frac{1}{2} + \frac{1}{2}\,\text{erf}\left(\frac{x}{\sqrt{2}}\right).$$

Here, the integral is the definition of the "distribution function of the normal distribution" to be discussed in Sec. 24.8.

12.7 Modeling: Membrane, Two-Dimensional Wave Equation

The vibrating string in Sec. 12.2 is a basic one-dimensional vibrational problem. Equally important is its two-dimensional analog, namely, the motion of an elastic membrane, such as a drumhead, that is stretched and then fixed along its edge. Indeed, setting up the model will proceed almost as in Sec. 12.2.

Physical Assumptions

1. The mass of the membrane per unit area is constant ("homogeneous membrane"). The membrane is perfectly flexible and offers no resistance to bending.
2. The membrane is stretched and then fixed along its entire boundary in the xy-plane. The tension per unit length T caused by stretching the membrane is the same at all points and in all directions and does not change during the motion.
3. The deflection $u(x, y, t)$ of the membrane during the motion is small compared to the size of the membrane, and all angles of inclination are small.

Although these assumptions cannot be realized exactly, they hold relatively accurately for small transverse vibrations of a thin elastic membrane, so that we shall obtain a good model, for instance, of a drumhead.

Derivation of the PDE of the Model ("Two-Dimensional Wave Equation") from Forces. As in Sec. 12.2 the model will consist of a PDE and additional conditions. The PDE will be obtained by the same method as in Sec. 12.2, namely, by considering the forces acting on a small portion of the physical system, the membrane in Fig. 298 on the next page, as it is moving up and down.

Since the deflections of the membrane and the angles of inclination are small, the sides of the portion are approximately equal to Δx and Δy. The tension T is the force per unit length. Hence the forces acting on the sides of the portion are approximately $T\,\Delta x$ and $T\,\Delta y$. Since the membrane is perfectly flexible, these forces are tangent to the moving membrane at every instant.

Horizontal Components of the Forces. We first consider the horizontal components of the forces. These components are obtained by multiplying the forces by the cosines of the angles of inclination. Since these angles are small, their cosines are close to 1. Hence

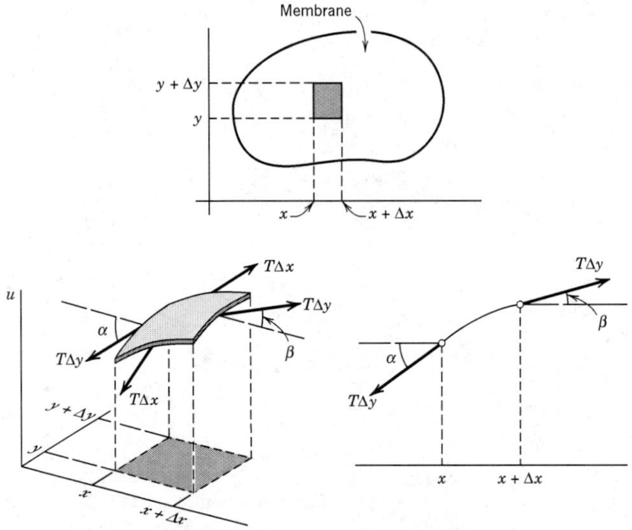

Fig. 298. Vibrating membrane

the horizontal components of the forces at opposite sides are approximately equal. Therefore, the motion of the particles of the membrane in a horizontal direction will be negligibly small. From this we conclude that we may regard the motion of the membrane as transversal; that is, each particle moves vertically.

Vertical Components of the Forces. These components along the right side and the left side are (Fig. 298), respectively,

$$T\,\Delta y \sin\beta \qquad \text{and} \qquad -T\,\Delta y \sin\alpha.$$

Here α and β are the values of the angle of inclination (which varies slightly along the edges) in the middle of the edges, and the minus sign appears because the force on the left side is directed downward. Since the angles are small, we may replace their sines by their tangents. Hence the resultant of those two vertical components is

$$\text{(1)} \qquad \begin{aligned} T\,\Delta y\,(\sin\beta - \sin\alpha) &\approx T\,\Delta y\,(\tan\beta - \tan\alpha) \\ &= T\,\Delta y\,[u_x(x + \Delta x, y_1) - u_x(x, y_2)] \end{aligned}$$

where subscripts x denote partial derivatives and y_1 and y_2 are values between y and $y + \Delta y$. Similarly, the resultant of the vertical components of the forces acting on the other two sides of the portion is

$$\text{(2)} \qquad T\,\Delta x\,[u_y(x_1, y + \Delta y) - u_y(x_2, y)]$$

where x_1 and x_2 are values between x and $x + \Delta x$.

Newton's Second Law Gives the PDE of the Model. By Newton's second law (see Sec. 2.4) the sum of the forces given by (1) and (2) is equal to the mass $\rho\Delta A$ of that small portion times the acceleration $\partial^2 u/\partial t^2$; here ρ is the mass of the undeflected membrane per unit area, and $\Delta A = \Delta x\, \Delta y$ is the area of that portion when it is undeflected. Thus

$$\rho\, \Delta x\, \Delta y\, \frac{\partial^2 u}{\partial t^2} = T\, \Delta y\, [u_x(x + \Delta x, y_1) - u_x(x, y_2)] + T\, \Delta x\, [u_y(x_1, y + \Delta y) - u_y(x_2, y)]$$

where the derivative on the left is evaluated at some suitable point $(\tilde{x}, \tilde{y})$ corresponding to that portion. Division by $\rho\, \Delta x\, \Delta y$ gives

$$\frac{\partial^2 u}{\partial t^2} = \frac{T}{\rho}\left[\frac{u_x(x + \Delta x, y_1) - u_x(x, y_2)}{\Delta x} + \frac{u_y(x_1, y + \Delta y) - u_y(x_2, y)}{\Delta y}\right].$$

If we let Δx and Δy approach zero, we obtain the PDE of the model

$$\text{(3)} \qquad \frac{\partial^2 u}{\partial t^2} = c^2\left(\frac{\partial^2 u}{\partial x^2} + \frac{\partial^2 u}{\partial y^2}\right) \qquad c^2 = \frac{T}{\rho}.$$

This PDE is called the **two-dimensional wave equation.** The expression in parentheses is the Laplacian $\nabla^2 u$ of u (Sec. 10.8). Hence (3) can be written

$$\text{(3')} \qquad \frac{\partial^2 u}{\partial t^2} = c^2 \nabla^2 u.$$

Solutions of the wave equation (3) will be obtained and discussed in the next section.

12.8 Rectangular Membrane. Double Fourier Series

The model of the vibrating membrane for obtaining the displacement $u(x, y, t)$ of a point (x, y) of the membrane from rest ($u = 0$) at time t is

$$\text{(1)} \qquad \frac{\partial^2 u}{\partial t^2} = c^2\left(\frac{\partial^2 u}{\partial x^2} + \frac{\partial^2 u}{\partial y^2}\right)$$

$$\text{(2)} \qquad u = 0 \text{ on the boundary}$$

$$\text{(3a)} \qquad u(x, y, 0) = f(x, y)$$

$$\text{(3b)} \qquad u_t(x, y, 0) = g(x, y).$$

Here (1) is the **two-dimensional wave equation** with $c^2 = T/\rho$ just derived, (2) is the **boundary condition** (membrane fixed along the boundary in the xy-plane for all times $t \geqq 0$), and (3) are the **initial conditions** at $t = 0$, consisting of the given *initial displacement* (initial shape) $f(x, y)$ and the given *initial velocity* $g(x, y)$, where $u_t = \partial u/\partial t$. We see that these conditions are quite similar to those for the string in Sec. 12.2.

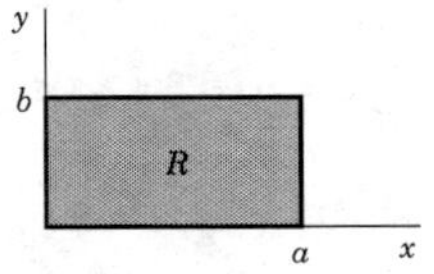

Fig. 299. Rectangular membrane

As a first important model, let us consider the **rectangular membrane** R in Fig. 299, which is simpler than the circular drumhead to follow. Then the boundary in (2) is the rectangle in Fig. 299. We shall solve this problem in three steps:

Step 1. By separating variables, setting $u(x, y, t) = F(x, y)G(t)$ and later $F(x, y) = H(x)Q(y)$ we obtain from (1) an ODE (4) for G and later from a PDE (5) for F two ODEs (6) and (7) for H and Q.

Step 2. From the solutions of those ODEs we determine solutions (13) of (1) (**"eigenfunctions"** u_{mn}) that satisfy the boundary condition (2).

Step 3. We compose the u_{mn} into a double series (14) solving the whole model (1), (2), (3).

Step 1. Three ODEs From the Wave Equation (1)

To obtain ODEs from (1), we apply two successive separations of variables. In the first separation we set $u(x, y, t) = F(x, y)G(t)$. Substitution into (1) gives

$$F\ddot{G} = c^2(F_{xx}G + F_{yy}G)$$

where subscripts denote partial derivatives and dots denote derivatives with respect to t. To separate the variables, we divide both sides by c^2FG:

$$\frac{\ddot{G}}{c^2G} = \frac{1}{F}(F_{xx} + F_{yy}).$$

Since the left side depends only on t, whereas the right side is independent of t, both sides must equal a constant. By a simple investigation we see that only negative values of that constant will lead to solutions that satisfy (2) without being identically zero; this is similar to Sec. 12.3. Denoting that negative constant by $-\nu^2$, we have

$$\frac{\ddot{G}}{c^2G} = \frac{1}{F}(F_{xx} + F_{yy}) = -\nu^2.$$

This gives two equations: for the **"time function"** $G(t)$ we have the ODE

$$\text{(4)} \qquad \ddot{G} + \lambda^2 G = 0 \qquad \text{where } \lambda = c\nu,$$

and for the **"amplitude function"** $F(x, y)$ a PDE, called the *two-dimensional* **Helmholtz**[3] **equation**

$$\text{(5)} \qquad F_{xx} + F_{yy} + \nu^2 F = 0.$$

[3]HERMANN VON HELMHOLTZ (1821–1894), German physicist, known for his basic work in thermodynamics, fluid flow, and acoustics.

Separation of the Helmholtz equation is achieved if we set $F(x, y) = H(x)Q(y)$. By substitution of this into (5) we obtain

$$\frac{d^2H}{dx^2} Q = -\left(H \frac{d^2Q}{dy^2} + \nu^2 HQ\right).$$

To separate the variables, we divide both sides by HQ, finding

$$\frac{1}{H} \frac{d^2H}{dx^2} = -\frac{1}{Q}\left(\frac{d^2Q}{dy^2} + \nu^2 Q\right).$$

Both sides must equal a constant, by the usual argument. This constant must be negative, say, $-k^2$, because only negative values will lead to solutions that satisfy (2) without being identically zero. Thus

$$\frac{1}{H} \frac{d^2H}{dx^2} = -\frac{1}{Q}\left(\frac{d^2Q}{dy^2} + \nu^2 Q\right) = -k^2.$$

This yields two ODEs for H and Q, namely,

$$\frac{d^2H}{dx^2} + k^2 H = 0 \tag{6}$$

and

$$\frac{d^2Q}{dy^2} + p^2 Q = 0 \qquad \text{where } p^2 = \nu^2 - k^2. \tag{7}$$

Step 2. Satisfying the Boundary Condition

General solutions of (6) and (7) are

$$H(x) = A \cos kx + B \sin kx \qquad \text{and} \qquad Q(y) = C \cos py + D \sin py$$

with constant A, B, C, D. From $u = FG$ and (2) it follows that $F = HQ$ must be zero on the boundary, that is, on the edges $x = 0, x = a, y = 0, y = b$; see Fig. 299. This gives the conditions

$$H(0) = 0, \qquad H(a) = 0, \qquad Q(0) = 0, \qquad Q(b) = 0.$$

Hence $H(0) = A = 0$ and then $H(a) = B \sin ka = 0$. Here we must take $B \neq 0$ since otherwise $H(x) \equiv 0$ and $F(x, y) \equiv 0$. Hence $\sin ka = 0$ or $ka = m\pi$, that is,

$$k = \frac{m\pi}{a} \quad (m \text{ integer}).$$

In precisely the same fashion we conclude that $C = 0$ and p must be restricted to the values $p = n\pi/b$ where n is an integer. We thus obtain the solutions $H = H_m$, $Q = Q_n$, where

$$H_m(x) = \sin \frac{m\pi x}{a} \qquad \text{and} \qquad Q_n(y) = \sin \frac{n\pi y}{b}, \qquad \begin{aligned} m &= 1, 2, \cdots, \\ n &= 1, 2, \cdots. \end{aligned}$$

As in the case of the vibrating string, it is not necessary to consider $m, n = -1, -2, \cdots$ since the corresponding solutions are essentially the same as for positive m and n, except for a factor -1. Hence the functions

$$\text{(8)} \qquad F_{mn}(x, y) = H_m(x)Q_n(y) = \sin \frac{m\pi x}{a} \sin \frac{n\pi y}{b}, \qquad \begin{aligned} m &= 1, 2, \cdots, \\ n &= 1, 2, \cdots, \end{aligned}$$

are solutions of the Helmholtz equation (5) that are zero on the boundary of our membrane.

Eigenfunctions and Eigenvalues. Having taken care of (5), we turn to (4). Since $p^2 = \nu^2 - k^2$ in (7) and $\lambda = c\nu$ in (4), we have

$$\lambda = c\sqrt{k^2 + p^2}.$$

Hence to $k = m\pi/a$ and $p = n\pi/b$ there corresponds the value

$$\text{(9)} \qquad \lambda = \lambda_{mn} = c\pi \sqrt{\frac{m^2}{a^2} + \frac{n^2}{b^2}}, \qquad \begin{aligned} m &= 1, 2, \cdots, \\ n &= 1, 2, \cdots, \end{aligned}$$

in the ODE (4). A corresponding general solution of (4) is

$$G_{mn}(t) = B_{mn} \cos \lambda_{mn} t + B^*_{mn} \sin \lambda_{mn} t.$$

It follows that the functions $u_{mn}(x, y, t) = F_{mn}(x, y)G_{mn}(t)$, written out

$$\text{(10)} \qquad u_{mn}(x, y, t) = (B_{mn} \cos \lambda_{mn} t + B^*_{mn} \sin \lambda_{mn} t) \sin \frac{m\pi x}{a} \sin \frac{n\pi y}{b}$$

with λ_{mn} according to (9), are solutions of the wave equation (1) that are zero on the boundary of the rectangular membrane in Fig. 299. These functions are called the **eigenfunctions** or *characteristic functions,* and the numbers λ_{mn} are called the **eigenvalues** or *characteristic values* of the vibrating membrane. The frequency of u_{mn} is $\lambda_{mn}/2\pi$.

Discussion of Eigenfunctions. It is very interesting that, depending on a and b, several functions F_{mn} may correspond to the same eigenvalue. Physically this means that there may exist vibrations having the same frequency but entirely different **nodal lines** (curves of points on the membrane that do not move). Let us illustrate this with the following example.

EXAMPLE 1 Eigenvalues and Eigenfunctions of the Square Membrane

Consider the square membrane with $a = b = 1$. From (9) we obtain its eigenvalues

$$\lambda_{mn} = c\pi\sqrt{m^2 + n^2}. \tag{11}$$

Hence $\lambda_{mn} = \lambda_{nm}$, but for $m \neq n$ the corresponding functions

$$F_{mn} = \sin m\pi x \sin n\pi y \qquad \text{and} \qquad F_{nm} = \sin n\pi x \sin m\pi y$$

are certainly different. For example, to $\lambda_{12} = \lambda_{21} = c\pi\sqrt{5}$ there correspond the two functions

$$F_{12} = \sin \pi x \sin 2\pi y \qquad \text{and} \qquad F_{21} = \sin 2\pi x \sin \pi y.$$

Hence the corresponding solutions

$$u_{12} = (B_{12} \cos c\pi\sqrt{5}t + B_{12}^* \sin c\pi\sqrt{5}t)F_{12} \qquad \text{and} \qquad u_{21} = (B_{21} \cos c\pi\sqrt{5}t + B_{21}^* \sin c\pi\sqrt{5}t)F_{21}$$

have the nodal lines $y = \frac{1}{2}$ and $x = \frac{1}{2}$, respectively (see Fig. 300). Taking $B_{12} = 1$ and $B_{12}^* = B_{21}^* = 0$, we obtain

$$u_{12} + u_{21} = \cos c\pi\sqrt{5}t\,(F_{12} + B_{21}F_{21}) \tag{12}$$

which represents another vibration corresponding to the eigenvalue $c\pi\sqrt{5}$. The nodal line of this function is the solution of the equation

$$F_{12} + B_{21}F_{21} = \sin \pi x \sin 2\pi y + B_{21} \sin 2\pi x \sin \pi y = 0$$

or, since $\sin 2\alpha = 2 \sin \alpha \cos \alpha$,

$$\sin \pi x \sin \pi y\,(\cos \pi y + B_{21} \cos \pi x) = 0. \tag{13}$$

This solution depends on the value of B_{21} (see Fig. 301).

From (11) we see that even more than two functions may correspond to the same numerical value of λ_{mn}. For example, the four functions F_{18}, F_{81}, F_{47}, and F_{74} correspond to the value

$$\lambda_{18} = \lambda_{81} = \lambda_{47} = \lambda_{74} = c\pi\sqrt{65}, \qquad \text{because} \qquad 1^1 + 8^2 = 4^2 + 7^2 = 65.$$

This happens because 65 can be expressed as the sum of two squares of positive integers in several ways. According to a theorem by Gauss, this is the case for every sum of two squares among whose prime factors there are at least two different ones of the form $4n + 1$ where n is a positive integer. In our case we have $65 = 5 \cdot 13 = (4 + 1)(12 + 1)$. ■

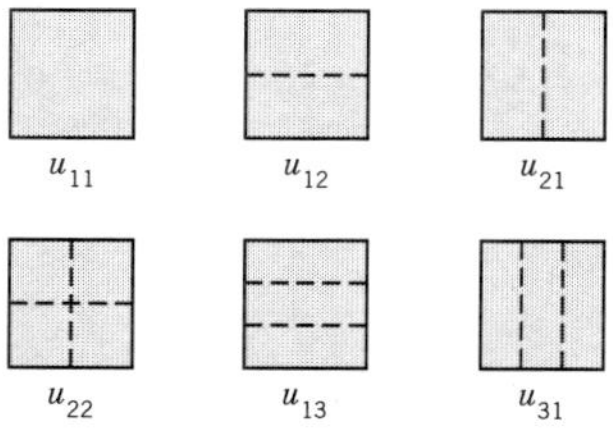

Fig. 300. Nodal lines of the solutions u_{11}, u_{12}, u_{21}, u_{22}, u_{13}, u_{31} in the case of the square membrane

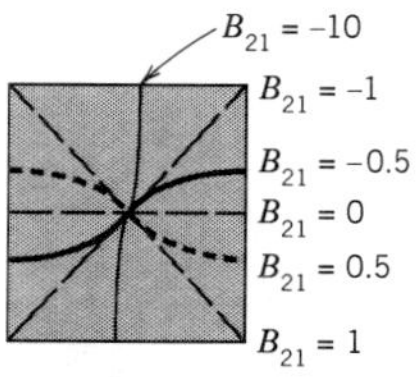

Fig. 301. Nodal lines of the solution (12) for some values of B_{21}

Step 3. Solution of the Model (1), (2), (3). Double Fourier Series

So far we have solutions (10) satisfying (1) and (2) only. To obtain the solution that also satisfies (3), we proceed as in Sec. 12.3. We consider the double series

$$\begin{aligned} u(x, y, t) &= \sum_{m=1}^{\infty} \sum_{n=1}^{\infty} u_{mn}(x, y, t) \\ &= \sum_{m=1}^{\infty} \sum_{n=1}^{\infty} (B_{mn} \cos \lambda_{mn} t + B_{mn}^* \sin \lambda_{mn} t) \sin \frac{m\pi x}{a} \sin \frac{n\pi y}{b} \end{aligned} \tag{14}$$

(without discussing convergence and uniqueness). From (14) and (3a), setting $t = 0$, we have

$$u(x, y, 0) = \sum_{m=1}^{\infty} \sum_{n=1}^{\infty} B_{mn} \sin \frac{m\pi x}{a} \sin \frac{n\pi y}{b} = f(x, y). \tag{15}$$

Suppose that $f(x, y)$ can be represented by (15). (Sufficient for this is the continuity of f, $\partial f/\partial x$, $\partial f/\partial y$, $\partial^2 f/\partial x \partial y$ in R.) Then (15) is called the **double Fourier series** of $f(x, y)$. Its coefficients can be determined as follows. Setting

$$K_m(y) = \sum_{n=1}^{\infty} B_{mn} \sin \frac{n\pi y}{b} \tag{16}$$

we can write (15) in the form

$$f(x, y) = \sum_{m=1}^{\infty} K_m(y) \sin \frac{m\pi x}{a} .$$

For fixed y this is the Fourier sine series of $f(x, y)$, considered as a function of x. From (4) in Sec. 11.3 we see that the coefficients of this expansion are

$$K_m(y) = \frac{2}{a} \int_0^a f(x, y) \sin \frac{m\pi x}{a} \, dx. \tag{17}$$

Furthermore, (16) is the Fourier sine series of $K_m(y)$, and from (4) in Sec. 11.3 it follows that the coefficients are

$$B_{mn} = \frac{2}{b} \int_0^b K_m(y) \sin \frac{n\pi y}{b} \, dy.$$

From this and (17) we obtain the **generalized Euler formula**

$$B_{mn} = \frac{4}{ab} \int_0^b \int_0^a f(x, y) \sin \frac{m\pi x}{a} \sin \frac{n\pi y}{b} \, dx \, dy \qquad \begin{aligned} m &= 1, 2, \cdots \\ n &= 1, 2, \cdots \end{aligned} \tag{18}$$

for the **Fourier coefficients** of $f(x, y)$ in the double Fourier series (15).

The B_{mn} in (14) are now determined in terms of $f(x, y)$. To determine the B^*_{mn}, we differentiate (14) termwise with respect to t; using (3b), we obtain

$$\left.\frac{\partial u}{\partial t}\right|_{t=0} = \sum_{m=1}^{\infty}\sum_{n=1}^{\infty} B^*_{mn}\lambda_{mn} \sin\frac{m\pi x}{a} \sin\frac{n\pi y}{b} = g(x, y).$$

Suppose that $g(x, y)$ can be developed in this double Fourier series. Then, proceeding as before, we find that the coefficients are

$$B^*_{mn} = \frac{4}{ab\lambda_{mn}}\int_0^b\int_0^a g(x, y) \sin\frac{m\pi x}{a} \sin\frac{n\pi y}{b}\, dx\, dy \qquad \begin{matrix} m = 1, 2, \cdots \\ n = 1, 2, \cdots . \end{matrix} \tag{19}$$

Result. *If f and g in* (3) *are such that u can be represented by* (14), *then* (14) *with coefficients* (18) *and* (19) *is the solution of the model* (1), (2), (3).

EXAMPLE 2 Vibration of a Rectangular Membrane

Find the vibrations of a rectangular membrane of sides $a = 4$ ft and $b = 2$ ft (Fig. 302) if the tension is 12.5 lb/ft, the density is 2.5 slugs/ft^2 (as for light rubber), the initial velocity is 0, and the initial displacement is

$$f(x, y) = 0.1(4x - x^2)(2y - y^2) \text{ ft.} \tag{20}$$

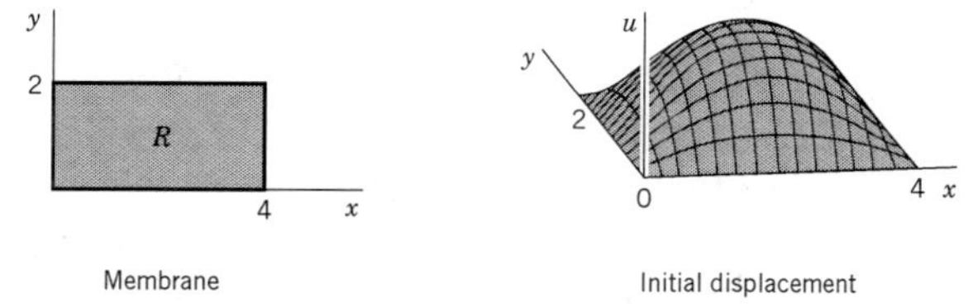

Fig. 302. Example 2

Solution. $c^2 = T/\rho = 12.5/2.5 = 5$ [ft^2/sec^2]. Also, $B^*_{mn} = 0$ from (19). From (18) and (20),

$$B_{mn} = \frac{4}{4\cdot 2}\int_0^2\int_0^4 0.1(4x - x^2)(2y - y^2) \sin\frac{m\pi x}{4} \sin\frac{n\pi y}{2}\, dx\, dy$$

$$= \frac{1}{20}\int_0^4 (4x - x^2) \sin\frac{m\pi x}{4}\, dx \int_0^2 (2y - y^2) \sin\frac{n\pi y}{2}\, dy.$$

Two integrations by parts give for the first integral on the right

$$\frac{128}{m^3\pi^3}[1 - (-1)^m] = \frac{256}{m^3\pi^3} \qquad (m \text{ odd})$$

and for the second integral

$$\frac{16}{n^3\pi^3}[1 - (-1)^n] = \frac{32}{n^3\pi^3} \qquad (n \text{ odd}).$$

For even m or n we get 0. Together with the factor 1/20 we thus have $B_{mn} = 0$ if m or n is even and

$$B_{mn} = \frac{256\cdot 32}{20m^3n^3\pi^6} \approx \frac{0.426\,050}{m^3n^3} \qquad (m \text{ and } n \text{ both odd}).$$

From this, (9), and (14) we obtain the answer

$$u(x, y, t) = 0.426\,050 \sum_{m,n\ \text{odd}} \sum \frac{1}{m^3 n^3} \cos\left(\frac{\sqrt{5}\pi}{4}\sqrt{m^2 + 4n^2}\right)t \sin\frac{m\pi x}{4} \sin\frac{n\pi y}{2}$$

$$(21) \qquad = 0.426\,050 \left(\cos\frac{\sqrt{5}\pi\sqrt{5}}{4}t \sin\frac{\pi x}{4} \sin\frac{\pi y}{2} + \frac{1}{27}\cos\frac{\sqrt{5}\pi\sqrt{37}}{4}t \sin\frac{\pi x}{4} \sin\frac{3\pi y}{2}\right.$$

$$\left. + \frac{1}{27}\cos\frac{\sqrt{5}\pi\sqrt{13}}{4}t \sin\frac{3\pi x}{4} \sin\frac{\pi y}{2} + \frac{1}{729}\cos\frac{\sqrt{5}\pi\sqrt{45}}{4}t \sin\frac{3\pi x}{4} \sin\frac{3\pi y}{2} + \cdots\right).$$

To discuss this solution, we note that the first term is very similar to the initial shape of the membrane, has no nodal lines, and is by far the dominating term because the coefficients of the next terms are much smaller. The second term has two horizontal nodal lines ($y = 2/3, 4/3$), the third term two vertical ones ($x = 4/3, 8/3$), the fourth term two horizontal and two vertical ones, and so on. ■

PROBLEM SET 12.8

1. **(Frequency)** How does the frequency of the eigenfunctions of the rectangular membrane change if (a) we double the tension, (b) we take a membrane of half the mass of the original one, (c) we double the sides of the membrane? (Give reason.)

SQUARE MEMBRANE

2. Determine and sketch the nodal lines of the eigenfunctions of the square membrane for $m = 1, 2, 3, 4$ and $n = 1, 2, 3, 4$.

3–8 **Double Fourier Series.** Represent $f(x, y)$ by a series (15), where $0 < x < 1$, $0 < y < 1$.

3. $f(x, y) = 1$
4. $f(x, y) = x$
5. $f(x, y) = y$
6. $f(x, y) = x + y$
7. $f(x, y) = xy$
8. $f(x, y) = xy(1 - x)(1 - y)$

9. **CAS PROJECT. Double Fourier Series. (a)** Write a program that gives and graphs partial sums of (15). Apply it to Probs. 4 and 5. Do the graphs show that those partial sums satisfy the boundary condition (3a)? Explain why. Why is the convergence rapid?

   (b) Do the tasks in (a) for Prob. 3. Graph a portion, say, $0 < x < \frac{1}{2}$, $0 < y < \frac{1}{2}$, of several partial sums on common axes, so that you can see how they differ. (See Fig. 303.)

   (c) Do the tasks in (b) for functions of your choice.

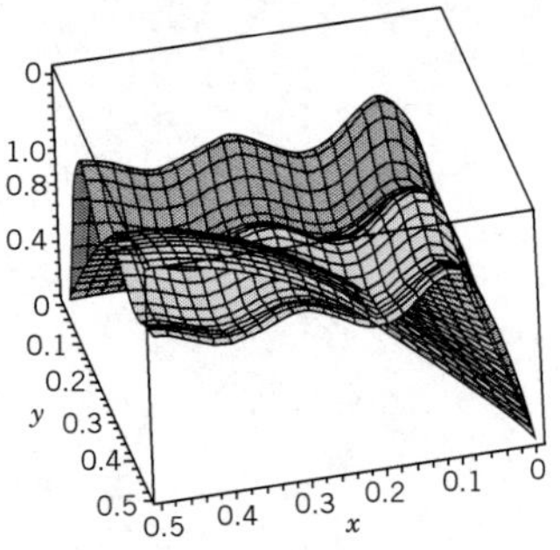

Fig. 303. Partial sums $S_{2,2}$ and $S_{10,10}$ in CAS Project 9b

10. **CAS EXPERIMENT. Quadruples of F_{mn}.** Write a program that gives you four numerically equal λ_{mn} in Example 1, so that four different F_{mn} correspond to it. Sketch the nodal lines of F_{18}, F_{81}, F_{47}, F_{74} in Example 1 and similarly for further F_{mn} that you will find.

11–13 **Deflection.** Find the deflection $u(x, y, t)$ of the square membrane of side π and $c^2 = 1$ if the initial velocity is 0 and the initial deflection is

11. $k \sin 2x \sin 5y$
12. $0.1 \sin x \sin y$
13. $0.1xy(\pi - x)(\pi - y)$

RECTANGULAR MEMBRANE

14. Verify the discussion of the terms of (21) in Example 2.
15. Repeat the task of Prob. 2 when $a = 4$ and $b = 1$.

16. Verify the calculation of B_{mn} in Example 2 by integration by parts.

17. Find eigenvalues of the rectangular membrane of sides $a = 2$ and $b = 1$ to which there correspond two or more different (independent) eigenfunctions.

18. (Minimum property) Show that among all rectangular membranes of the same area $A = ab$ and the same c the square membrane is that for which u_{11} [see (10)] has the lowest frequency.

19–22 **Double Fourier Series.** Represent $f(x, y)$ $(0 < x < a,\ 0 < y < b)$ by a double Fourier series (15).

19. $f(x, y) = k$

20. $f(x, y) = 0.25xy$

21. $f(x, y) = xy(a^2 - x^2)(b^2 - y^2)$

22. $f(x, y) = xy(a - x)(b - y)$

23. (Deflection) Find the deflection of the membrane of sides a and b with $c^2 = 1$ for the initial deflection

$$f(x, y) = \sin\frac{3\pi x}{a}\sin\frac{4\pi y}{b} \quad \text{and initial velocity 0.}$$

24. Repeat the task in Prob. 23 with $c^2 = 1$, for $f(x, y)$ as in Prob. 22 and initial velocity 0.

25. (Forced vibrations) Show that forced vibrations of a membrane are modeled by the PDE $u_{tt} = c^2\nabla^2 u + P/\rho$, where $P(x, y, t)$ is the external force per unit area acting perpendicular to the xy-plane.

12.9 Laplacian in Polar Coordinates. Circular Membrane. Fourier–Bessel Series

In boundary value problems for PDEs it is a ***general principle*** to use coordinates in which the formula for the boundary is as simple as possible. Since we want to discuss circular membranes (drumheads), we first transform the Laplacian in the wave equation (1), Sec. 12.8,

$$u_{tt} = c^2\nabla^2 u = c^2(u_{xx} + u_{yy}) \tag{1}$$

(subscripts denoting partial derivatives) into **polar coordinates**

$$r = \sqrt{x^2 + y^2}, \qquad \theta = \arctan\frac{y}{x}.$$

Hence $x = r\cos\theta$, $y = r\sin\theta$. By the chain rule (Sec. 9.6) we obtain

$$u_x = u_r r_x + u_\theta \theta_x.$$

Differentiating once more with respect to x and using the product rule and then again the chain rule gives

$$\begin{aligned} u_{xx} &= (u_r r_x)_x + (u_\theta\theta_x)_x \\ &= (u_r)_x r_x + u_r r_{xx} + (u_\theta)_x\theta_x + u_\theta\theta_{xx} \\ &= (u_{rr}r_x + u_{r\theta}\theta_x)r_x + u_r r_{xx} + (u_{\theta r}r_x + u_{\theta\theta}\theta_x)\theta_x + u_\theta\theta_{xx}. \end{aligned} \tag{2}$$

Also, by differentiation of r and θ we find

$$r_x = \frac{x}{\sqrt{x^2 + y^2}} = \frac{x}{r}, \qquad \theta_x = \frac{1}{1 + (y/x)^2}\left(-\frac{y}{x^2}\right) = -\frac{y}{r^2}.$$

Differentiating these two formulas again, we obtain

$$r_{xx} = \frac{r - xr_x}{r^2} = \frac{1}{r} - \frac{x^2}{r^3} = \frac{y^2}{r^3}, \qquad \theta_{xx} = -y\left(-\frac{2}{r^3}\right) r_x = \frac{2xy}{r^4}.$$

We substitute all these expressions into (2). Assuming continuity of the first and second partial derivatives, we have $u_{r\theta} = u_{\theta r}$, and by simplifying,

$$u_{xx} = \frac{x^2}{r^2} u_{rr} - 2\frac{xy}{r^3} u_{r\theta} + \frac{y^2}{r^4} u_{\theta\theta} + \frac{y^2}{r^3} u_r + 2\frac{xy}{r^4} u_\theta. \tag{3}$$

In a similar fashion it follows that

$$u_{yy} = \frac{y^2}{r^2} u_{rr} + 2\frac{xy}{r^3} u_{r\theta} + \frac{x^2}{r^4} u_{\theta\theta} + \frac{x^2}{r^3} u_r - 2\frac{xy}{r^4} u_\theta. \tag{4}$$

By adding (3) and (4) we see that the **Laplacian of u in polar coordinates** is

$$\nabla^2 u = \frac{\partial^2 u}{\partial r^2} + \frac{1}{r}\frac{\partial u}{\partial r} + \frac{1}{r^2}\frac{\partial^2 u}{\partial \theta^2}. \tag{5}$$

Circular Membrane

Circular membranes occur in drums, pumps, microphones, telephones, and so on. This accounts for their great importance in engineering. Whenever a circular membrane is plane and its material is elastic, but offers no resistance to bending (this excludes thin metallic membranes!), its vibrations are modeled by the **two-dimensional wave equation in polar coordinates** obtained from (1) with $\nabla^2 u$ given by (5), that is,

$$\frac{\partial^2 u}{\partial t^2} = c^2\left(\frac{\partial^2 u}{\partial r^2} + \frac{1}{r}\frac{\partial u}{\partial r} + \frac{1}{r^2}\frac{\partial^2 u}{\partial \theta^2}\right) \qquad c^2 = \frac{T}{\rho}. \tag{6}$$

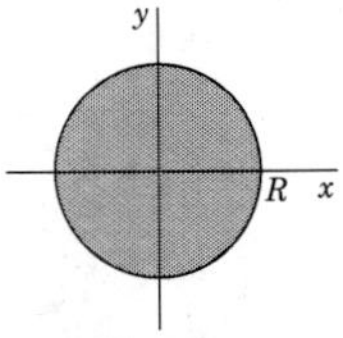

Fig. 304. Circular membrane

We shall consider a membrane of radius R (Fig. 304) and determine solutions $u(r, t)$ that are radially symmetric. (Solutions also depending on the angle θ will be discussed in the problem set.) Then $u_{\theta\theta} = 0$ in (6) and the model of the problem (the analog of (1), (2), (3) in Sec. 12.8) is

$$\frac{\partial^2 u}{\partial t^2} = c^2\left(\frac{\partial^2 u}{\partial r^2} + \frac{1}{r}\frac{\partial u}{\partial r}\right) \tag{7}$$

$$u(R, t) = 0 \text{ for all } t \geqq 0 \tag{8}$$

$$u(r, 0) = f(r) \tag{9a}$$

$$u_t(r, 0) = g(r). \tag{9b}$$

Here (8) means that the membrane is fixed along the boundary circle $r = R$. The initial deflection $f(r)$ and the initial velocity $g(r)$ depend only on r, not on θ, so that we can expect radially symmetric solutions $u(r, t)$.

Step 1. Two ODEs From the Wave Equation (7). Bessel's Equation

Using the **method of separation of variables,** we first determine solutions $u(r, t) = W(r)G(t)$. (We write W, not F because W depends on r, whereas F, used before, depended on x.) Substituting $u = WG$ and its derivatives into (7) and dividing the result by c^2WG, we get

$$\frac{\ddot{G}}{c^2G} = \frac{1}{W}\left(W'' + \frac{1}{r}W'\right)$$

where dots denote derivatives with respect to t and primes denote derivatives with respect to r. The expressions on both sides must equal a constant. This constant must be negative, say, $-k^2$, in order to obtain solutions that satisfy the boundary condition without being identically zero. Thus,

$$\frac{\ddot{G}}{c^2G} = \frac{1}{W}\left(W'' + \frac{1}{r}W'\right) = -k^2.$$

This gives the two linear ODEs

(10) $$\ddot{G} + \lambda^2 G = 0 \qquad \text{where } \lambda = ck$$

and

(11) $$W'' + \frac{1}{r}W' + k^2W = 0.$$

We can reduce (11) to Bessel's equation (Sec. 5.5) if we set $s = kr$. Then $1/r = k/s$ and, retaining the notation W for simplicity, we obtain by the chain rule

$$W' = \frac{dW}{dr} = \frac{dW}{ds}\frac{ds}{dr} = \frac{dW}{ds}k \qquad \text{and} \qquad W'' = \frac{d^2W}{ds^2}k^2.$$

By substituting this into (11) and omitting the common factor k^2 we have

(12) $$\frac{d^2W}{ds^2} + \frac{1}{s}\frac{dW}{ds} + W = 0.$$

This is **Bessel's equation** (1), Sec. 5.5, with parameter $\nu = 0$.

Step 2. Satisfying the Boundary Condition (8)

Solutions of (12) are the Bessel functions J_0 and Y_0 of the first and second kind (see Secs. 5.5, 5.6). But Y_0 becomes infinite at 0, so that we cannot use it because the deflection of the membrane must always remain finite. This leaves us with

(13) $$W(r) = J_0(s) = J_0(kr) \qquad (s = kr).$$

On the boundary $r = R$ we get $W(R) = J_0(kR) = 0$ from (8) (because $G \equiv 0$ would imply $u \equiv 0$). We can satisfy this condition because J_0 has (infinitely many) positive zeros, $s = \alpha_1, \alpha_2, \cdots$ (see Fig. 305), with numerical values

$$\alpha_1 = 2.4048, \quad \alpha_2 = 5.5201, \quad \alpha_3 = 8.6537, \quad \alpha_4 = 11.7915, \quad \alpha_5 = 14.9309$$

and so on. (For further values, consult your CAS or Ref. [GR1] in App. 1.) These zeros are slightly irregularly spaced, as we see. Equation (13) now implies

$$kR = \alpha_m \qquad \text{thus} \qquad k = k_m = \frac{\alpha_m}{R}, \qquad m = 1, 2, \cdots. \tag{14}$$

Hence the functions

$$W_m(r) = J_0(k_m r) = J_0\left(\frac{\alpha_m}{R} r\right), \qquad m = 1, 2, \cdots \tag{15}$$

are solutions of (11) that are zero on the boundary circle $r = R$.

Eigenfunctions and Eigenvalues. For W_m in (15), a corresponding general solution of (10) with $\lambda = \lambda_m = ck_m = c\alpha_m/R$ is

$$G_m(t) = A_m \cos \lambda_m t + B_m \sin \lambda_m t.$$

Hence the functions

$$u_m(r, t) = W_m(r)G_m(t) = (A_m \cos \lambda_m t + B_m \sin \lambda_m t)J_0(k_m r) \tag{16}$$

with $m = 1, 2, \cdots$ are solutions of the wave equation (7) satisfying the boundary condition (8). These are the **eigenfunctions** of our problem. The corresponding **eigenvalues** are λ_m.

The vibration of the membrane corresponding to u_m is called the mth **normal mode;** it has the frequency $\lambda_m/2\pi$ cycles per unit time. Since the zeros of the Bessel function J_0 are not regularly spaced on the axis (in contrast to the zeros of the sine functions appearing in the case of the vibrating string), the sound of a drum is entirely different from that of a violin. The forms of the normal modes can easily be obtained from Fig. 305 and are shown in Fig. 306. For $m = 1$, all the points of the membrane move up (or down) at the same time. For $m = 2$, the situation is as follows. The function $W_2(r) = J_0(\alpha_2 r/R)$ is zero for $\alpha_2 r/R = \alpha_1$, thus $r = \alpha_1 R/\alpha_2$. The circle $r = \alpha_1 R/\alpha_2$ is, therefore, **nodal line,** and when at some instant the central part of the membrane moves up, the outer part ($r > \alpha_1 R/\alpha_2$) moves down, and conversely. The solution $u_m(r, t)$ has $m - 1$ nodal lines, which are circles (Fig. 306).

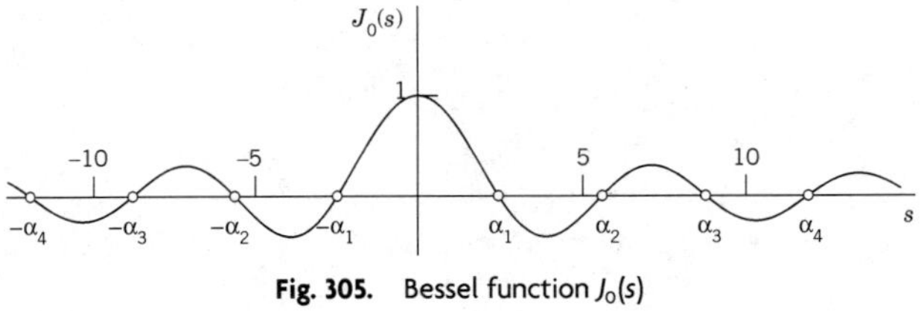

Fig. 305. Bessel function $J_0(s)$

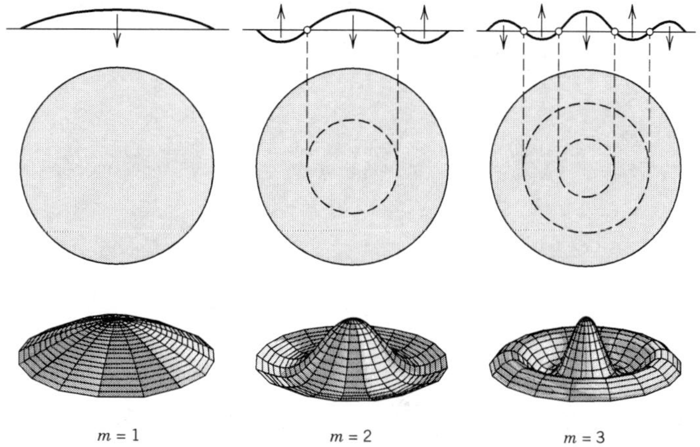

Fig. 306. Normal modes of the circular membrane in the case of vibrations independent of the angle

Step 3. Solution of the Entire Problem

To obtain a solution $u(r, t)$ that also satisfies the initial conditions (9), we may proceed as in the case of the string. That is, we consider the series

$$u(r, t) = \sum_{m=1}^{\infty} W_m(r)G_m(t) = \sum_{m=1}^{\infty} (A_m \cos \lambda_m t + B_m \sin \lambda_m t)\, J_0\left(\frac{\alpha_m}{R} r\right) \tag{17}$$

(leaving aside the problems of convergence and uniqueness). Setting $t = 0$ and using (9a), we obtain

$$u(r, 0) = \sum_{m=1}^{\infty} A_m J_0\left(\frac{\alpha_m}{R} r\right) = f(r). \tag{18}$$

Thus for the series (17) to satisfy the condition (9a), the constants A_m must be the coefficients of the **Fourier–Bessel series** (18) that represents $f(r)$ in terms of $J_0(\alpha_m r/R)$; that is [see (10) in Sec. 5.8 with $n = 0$, $\alpha_{0,m} = \alpha_m$, and $x = r$],

$$A_m = \frac{2}{R^2 J_1^{\,2}(\alpha_m)} \int_0^R r f(r) J_0\left(\frac{\alpha_m}{R} r\right) dr \qquad (m = 1, 2, \cdots). \tag{19}$$

Differentiability of $f(r)$ in the interval $0 \leqq r \leqq R$ is sufficient for the existence of the development (18); see Ref. [A13]. The coefficients B_m in (17) can be determined from (9b) in a similar fashion. Numeric values of A_m and B_m may be obtained from a CAS or by a numeric integration method, using tables of J_0 and J_1. However, numeric integration can sometimes be ***avoided,*** as the following example shows.

EXAMPLE 1 Vibrations of a Circular Membrane

Find the vibrations of a circular drumhead of radius 1 ft and density 2 slugs/ft^2 if the tension is 8 lb/ft, the initial velocity is 0, and the initial displacement is

$$f(r) = 1 - r^2 \text{ [ft]}.$$

Solution. $c^2 = T/\rho = 8/2 = 4\ [\text{ft}^2/\text{sec}^2]$. Also $B_m = 0$, since the initial velocity is 0. From (19) and Example 3 in Sec. 5.8, since $R = 1$, we obtain

$$\begin{aligned} A_m &= \frac{2}{J_1^2(\alpha_m)} \int_0^1 r(1 - r^2)\, J_0(\alpha_m r)\, dr \\ &= \frac{4J_2(\alpha_m)}{\alpha_m^2 J_1^2(\alpha_m)} \\ &= \frac{8}{\alpha_m^3 J_1(\alpha_m)} \end{aligned}$$

where the last equality follows from (24c), Sec. 5.5, with $\nu = 1$, that is,

$$J_2(\alpha_m) = \frac{2}{\alpha_m} J_1(\alpha_m) - J_0(\alpha_m) = \frac{2}{\alpha_m} J_1(\alpha_m).$$

Table 9.5 on p. 409 of [GR1] gives α_m and $J_0'(\alpha_m)$. From this we get $J_1(\alpha_m) = -J_0'(\alpha_m)$ by (24b), Sec. 5.5, with $\nu = 0$, and compute the coefficients A_m:

m	α_m	$J_1(\alpha_m)$	$J_2(\alpha_m)$	A_m
1	2.40483	0.51915	0.43176	1.10801
2	5.52008	−0.34026	−0.12328	−0.13978
3	8.65373	0.27145	0.06274	0.04548
4	11.79153	−0.23246	−0.03943	−0.02099
5	14.93092	0.20655	0.02767	0.01164
6	18.07106	−0.18773	−0.02078	−0.00722
7	21.21164	0.17327	0.01634	0.00484
8	24.35247	−0.16170	−0.01328	−0.00343
9	27.49348	0.15218	0.01107	0.00253
10	30.63461	−0.14417	−0.00941	−0.00193

Thus

$$f(r) = 1.108 J_0(2.4048r) - 0.140 J_0(5.5201r) + 0.045 J_0(8.6537r) - \cdots.$$

We see that the coefficients decrease relatively slowly. The sum of the explicitly given coefficients in the table is 0.99915. The sum of *all* the coefficients should be 1. (Why?) Hence by the Leibniz test in App. A3.3 the partial sum of those terms gives about three correct decimals of the amplitude $f(r)$.

Since

$$\lambda_m = ck_m = c\alpha_m/R = 2\alpha_m,$$

from (17) we thus obtain the solution (with r measured in feet and t in seconds)

$$u(r, t) = 1.108 J_0(2.4048 r) \cos 4.8097t - 0.140 J_0(5.5201r) \cos 11.0402t + 0.045 J_0(8.6537r) \cos 17.3075t - \cdots.$$

In Fig. 306, $m = 1$ gives an idea of the motion of the first term of our series, $m = 2$ of the second term, and $m = 3$ of the third term, so that we can "see" our result about as well as for a violin string in Sec. 12.3. ■

PROBLEM SET 12.9

1. Why did we use polar coordinates in this section?

2. Work out the details of the calculation leading to the Laplacian in polar coordinates.

3. If u is independent of θ, then (5) reduces to $\nabla^2 u = u_{rr} + u_r/r$. Derive this directly from the Laplacian in Cartesian coordinates.

4. An **alternative form** of (5) is $\nabla^2 u = \dfrac{1}{r}\dfrac{\partial}{\partial r}\left(r\dfrac{\partial u}{\partial r}\right) + \dfrac{1}{r^2}\dfrac{\partial^2 u}{\partial \theta^2}$. Derive this from (5).

5. **(Radial solution)** Show that the only solution of $\nabla^2 u = 0$ depending only on $r = \sqrt{x^2 + y^2}$ is $u = a \ln r + b$ with constant a and b.

6. **TEAM PROJECT. Series for Dirichlet and Neumann Problems**

(a) Show that $u_n = r^n \cos n\theta$, $u_n = r^n \sin n\theta$, $n = 0, 1, \cdots$, are solutions of Laplace's equation $\nabla^2 u = 0$ with $\nabla^2 u$ given by (5). (What would u_n be in Cartesian coordinates? Experiment with small n.)

(b) Dirichlet problem (See Sec. 12.5) Assuming that termwise differentiation is permissible, show that a solution of the Laplace equation in the disk $r < R$ satisfying the boundary condition $u(R, \theta) = f(\theta)$ (f given) is

$$u(r, \theta) = a_0 + \sum_{n=1}^{\infty} \left[a_n \left(\frac{r}{R}\right)^n \cos n\theta + b_n \left(\frac{r}{R}\right)^n \sin n\theta \right] \tag{20}$$

where a_n, b_n are the Fourier coefficients of f (see Sec. 11.1).

(c) Dirichlet problem Solve the Dirichlet problem using (20) if $R = 1$ and the boundary values are $u(\theta) = -100$ volts if $-\pi < \theta < 0$, $u(\theta) = 100$ volts if $0 < \theta < \pi$. (Sketch this disk, indicate the boundary values.)

(d) Neumann problem Show that the solution of the Neumann problem $\nabla^2 u = 0$ if $r < R$, $u_N(R, \theta) = f(\theta)$ (where $u_N = \partial u/\partial N$ is the directional derivative in the direction of the outer normal) is

$$u(r, \theta) = A_0 + \sum_{n=1}^{\infty} r^n (A_n \cos n\theta + B_n \sin n\theta)$$

with arbitrary A_0 and

$$A_n = \frac{1}{\pi n R^{n-1}} \int_{-\pi}^{\pi} f(\theta) \cos n\theta \, d\theta,$$

$$B_n = \frac{1}{\pi n R^{n-1}} \int_{-\pi}^{\pi} f(\theta) \sin n\theta \, d\theta.$$

(e) Compatibility condition Show that (9), Sec. 10.4, imposes on $f(\theta)$ in (d) the *"compatibility condition"*

$$\int_{-\pi}^{\pi} f(\theta) \, d\theta = 0.$$

(f) Neumann problem Solve $\nabla^2 u = 0$ in the annulus $1 < r < 3$ if $u_r(1, \theta) = \sin \theta$, $u_r(3, \theta) = 0$.

7–12 ELECTROSTATIC POTENTIAL. STEADY-STATE HEAT PROBLEMS

The electrostatic potential satisfies Laplace's equation $\nabla^2 u = 0$ in any region free of charges. Also the heat equation $u_t = c^2 \nabla^2 u$ (Sec. 12.5) reduces to Laplace's equation if the temperature u is time-independent **("steady-state case")**. Using (20), find the potential (equivalently: the steady-state temperature) in the disk $r < 1$ if the boundary values are (sketch them, to see what is going on).

7. $u(1, \theta) = 40 \cos^3 \theta$
8. $u(1, \theta) = 800 \sin^3 \theta$
9. $u(1, \theta) = 110$ if $-\frac{1}{2}\pi < \theta < \frac{1}{2}\pi$ and 0 otherwise
10. $u(1, \theta) = \theta$ if $-\frac{1}{2}\pi < \theta < \frac{1}{2}\pi$ and 0 otherwise
11. $u(1, \theta) = |\theta|$ if $-\pi < \theta < \pi$
12. $u(1, \theta) = \theta^2$ if $-\pi < \theta < \pi$

13. **CAS EXPERIMENT. Equipotential Lines.** Guess what the equipotential lines $u(r, \theta) = const$ in Probs. 9 and 11 may look like. Then graph some of them, using partial sums of the series.

14. **(Semidisk)** Find the electrostatic potential in the semidisk $r < 1$, $0 < \theta < \pi$ which equals $110\theta(\pi - \theta)$ on the semicircle $r = 1$ and 0 on the segment $-1 < x < 1$.

15. **(Semidisk)** Find the steady-state temperature in a semicircular thin plate $r < a$, $0 < \theta < \pi$ with the semicircle $r = a$ kept at constant temperature u_0 and the segment $-a < x < a$ at 0.

16. **(Invariance)** Show that $\nabla^2 u$ is invariant under translations $x^* = x + a$, $y^* = y + b$ and under rotations $x^* = x \cos \alpha - y \sin \alpha$, $y^* = x \sin \alpha + y \cos \alpha$.

CIRCULAR MEMBRANE

17. (Frequency) What happens to the frequency of an eigenfunction of a drum if you double the tension?

18. (Size of a drum) A small drum should have a higher fundamental frequency than a large one, tension and density being the same. How does this follow from our formulas?

19. (Tension) Find a formula for the tension required to produce a desired fundamental frequency f_1 of a drum.

20. CAS PROJECT. Normal Modes. (a) Graph the normal modes u_4, u_5, u_6 as in Fig. 306.

(b) Write a program for calculating the A_m's in Example 1 and extend the table to $m = 15$. Verify numerically that $\alpha_m \approx (m - \frac{1}{4})\pi$ and compute the error for $m = 1, \cdots, 10$.

(c) Graph the initial deflection $f(r)$ in Example 1 as well as the first three partial sums of the series. Comment on accuracy.

(d) Compute the radii of the nodal lines of u_2, u_3, u_4 when $R = 1$. How do these values compare to those of the nodes of the vibrating string of length 1? Can you establish any empirical laws by experimentation with further u_m?

21. (Nodal lines) Is it possible that for fixed c and R two or more u_m [see (16)] with different nodal lines correspond to the same eigenvalue? (Give a reason.)

22. Why is $A_1 + A_2 + \cdots = 1$ in Example 1? Compute the first few partial sums until you get 3-digit accuracy. What does this problem mean in the field of music?

23. (Nonzero initial velocity) Show that for (17) to satisfy (9b) we must have

$$B_m = \frac{2}{c\alpha_m R J_1{}^2(\alpha_m)} \times \int_0^R rg(r)J_0(\alpha_m r/R)\, dr. \tag{21}$$

VIBRATIONS OF A CIRCULAR MEMBRANE DEPENDING ON BOTH *r* AND *θ*

24. (Separations) Show that substitution of $u = F(r, \theta)G(t)$ into the wave equation (6), that is,

$$u_{tt} = c^2\left(u_{rr} + \frac{1}{r}u_r + \frac{1}{r^2}u_{\theta\theta}\right) \tag{22}$$

gives an ODE and a PDE

$$\ddot{G} + \lambda^2 G = 0, \qquad \text{where } \lambda = ck, \tag{23}$$

$$F_{rr} + \frac{1}{r}F_r + \frac{1}{r^2}F_{\theta\theta} + k^2F = 0. \tag{24}$$

Show that the PDE can now be separated by substituting $F = W(r)Q(\theta)$, giving

$$Q'' + n^2Q = 0, \tag{25}$$

$$r^2W'' + rW' + (k^2r^2 - n^2)W = 0. \tag{26}$$

25. (Periodicity) Show that $Q(\theta)$ must be periodic with period 2π and, therefore, $n = 0, 1, 2, \cdots$ in (25) and (26). Show that this yields the solutions $Q_n = \cos n\theta$, $Q_n{}^* = \sin n\theta$, $W_n = J_n(kr)$, $n = 0, 1, \cdots$.

26. (Boundary condition) Show that the boundary condition

$$u(R, \theta, t) = 0 \tag{27}$$

leads to $k = k_{mn} = \alpha_{mn}/R$, where $s = \alpha_{mn}$ is the mth positive zero of $J_n(s)$.

27. (Solutions depending on both *r* and *θ*) Show that solutions of (22) satisfying (27) are (see Fig. 307)

$$\begin{aligned} u_{mn} &= (A_{mn}\cos ck_{mn}t + B_{mn}\sin ck_{mn}t) \times \\ &\quad \times J_n(k_{mn}r)\cos n\theta \\ u^*_{mn} &= (A^*_{mn}\cos ck_{mn}t + B^*_{mn}\sin ck_{mn}t) \times \\ &\quad \times J_n(k_{mn}r)\sin n\theta \end{aligned} \tag{28}$$

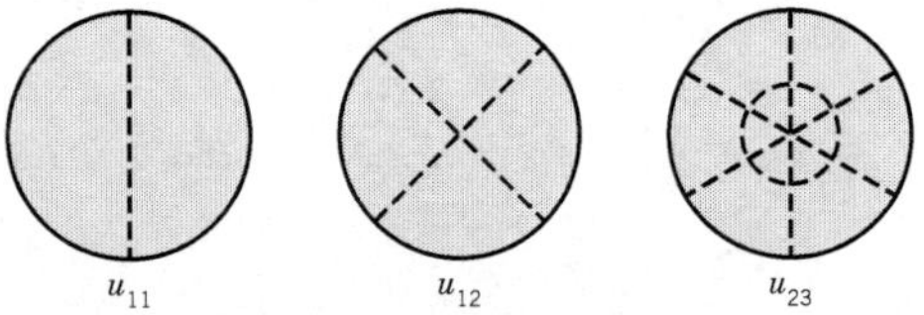

Fig. 307. Nodal lines of some of the solutions (28)

28. (Initial condition) Show that $u_t(r, \theta, 0) = 0$ gives $B_{mn} = 0$, $B^*_{mn} = 0$ in (28).

29. Show that $u^*_{m0} = 0$ and u_{m0} is identical with (16) in the current section.

30. (Semicircular membrane) Show that u_{11} represents the fundamental mode of a semicircular membrane and find the corresponding frequency when $c^2 = 1$ and $R = 1$.

12.10 Laplace's Equation in Cylindrical and Spherical Coordinates. Potential

Laplace's equation

(1) $$\nabla^2 u = u_{xx} + u_{yy} + u_{zz} = 0$$

is one of the most important PDEs in physics and its engineering applications. Here, x, y, z are Cartesian coordinates in space (Fig. 165 in Sec. 9.1), $u_{xx} = \partial^2 u/\partial x^2$, etc. The expression $\nabla^2 u$ is called the **Laplacian** of u. The theory of the solutions of (1) is called **potential theory.** Solutions of (1) that have ***continuous*** second partial derivatives are known as **harmonic functions.**

Laplace's equation occurs mainly in **gravitation, electrostatics** (see Theorem 3, Sec. 9.7). steady-state **heat flow** (Sec. 12.5), and **fluid flow** (to be discussed in Chap. 18.4).

Recall from Sec. 9.7 that the gravitational **potential** $u(x, y, z)$ at a point (x, y, z) resulting from a single mass located at a point (X, Y, Z) is

(2) $$u(x, y, z) = \frac{c}{r} = \frac{c}{\sqrt{(x - X)^2 + (y - Y)^2 + (z - Z)^2}} \qquad (r > 0)$$

and u satisfies (1). Similarly, if mass is distributed in a region T in space with density $\rho(X, Y, Z)$, its potential at a point (x, y, z) not occupied by mass is

(3) $$u(x, y, z) = k \iiint_T \frac{\rho(X, Y, Z)}{r}\, dX\, dY\, dZ.$$

It satisfies (1) because $\nabla^2(1/r) = 0$ (Sec. 9.7) and ρ is not a function of x, y, z.

Practical problems involving Laplace's equation are boundary value problems in a region T in space with boundary surface S. Such a problem is called (see also Sec. 12.5 for the two-dimensional case):

(I) **First boundary value problem** or **Dirichlet problem** if u is prescribed on S.

(II) **Second boundary value problem** or **Neumann problem** if the normal derivative $u_n = \partial u/\partial n$ is prescribed on S.

(III) **Third** or **mixed boundary value problem** or **Robin problem** if u is prescribed on a portion of S and u_n on the remaining portion of S.

Laplacian in Cylindrical Coordinates

The first step in solving a boundary value problem is generally the introduction of coordinates in which the boundary surface S has a simple representation. Cylindrical symmetry (a cylinder as a region T) calls for cylindrical coordinates r, θ, z related to x, y, z by

(4) $$x = r\cos\theta, \qquad y = r\sin\theta, \qquad z = z$$ (Fig. 308, p. 588).

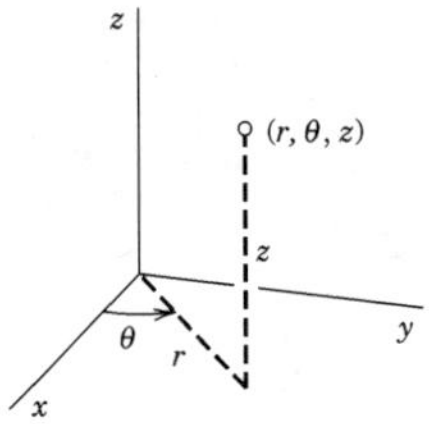

Fig. 308. Cylindrical coordinates

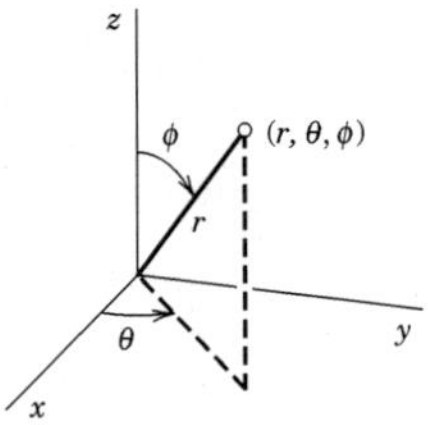

Fig. 309. Spherical coordinates

For these we get $\nabla^2 u$ immediately by adding u_{zz} to (5) in Sec. 12.9; thus,

(5) $$\nabla^2 u = \frac{\partial^2 u}{\partial r^2} + \frac{1}{r}\frac{\partial u}{\partial r} + \frac{1}{r^2}\frac{\partial^2 u}{\partial \theta^2} + \frac{\partial^2 u}{\partial z^2}.$$

Laplacian in Spherical Coordinates

Spherical symmetry (a ball as region T bounded by a sphere S) requires **spherical coordinates** r, θ, ϕ related to x, y, z by

(6) $$x = r\cos\theta\sin\phi, \qquad y = r\sin\theta\sin\phi, \qquad z = r\cos\phi \qquad \text{(Fig. 309)}.$$

Using the chain rule (as in Sec. 12.9), we obtain $\nabla^2 u$ in spherical coordinates

(7) $$\nabla^2 u = \frac{\partial^2 u}{\partial r^2} + \frac{2}{r}\frac{\partial u}{\partial r} + \frac{1}{r^2}\frac{\partial^2 u}{\partial \phi^2} + \frac{\cot\phi}{r^2}\frac{\partial u}{\partial \phi} + \frac{1}{r^2\sin^2\phi}\frac{\partial^2 u}{\partial \theta^2}.$$

We leave the details as an exercise. It is sometimes practical to write (7) in the form

(7′) $$\nabla^2 u = \frac{1}{r^2}\left[\frac{\partial}{\partial r}\left(r^2\frac{\partial u}{\partial r}\right) + \frac{1}{\sin\phi}\frac{\partial}{\partial \phi}\left(\sin\phi\frac{\partial u}{\partial \phi}\right) + \frac{1}{\sin^2\phi}\frac{\partial^2 u}{\partial \theta^2}\right].$$

Remark on Notation. Equation (6) is used in calculus and extends the familiar notation for polar coordinates. Unfortunately, some books use θ and ϕ interchanged, an extension of the notation $x = r\cos\phi$, $y = r\sin\phi$ for polar coordinates (used in some European countries).

Boundary Value Problem in Spherical Coordinates

We shall solve the following **Dirichlet problem** in spherical coordinates:

(8) $$\nabla^2 u = \frac{1}{r^2}\left[\frac{\partial}{\partial r}\left(r^2\frac{\partial u}{\partial r}\right) + \frac{1}{\sin\phi}\frac{\partial}{\partial \phi}\left(\sin\phi\frac{\partial u}{\partial \phi}\right)\right] = 0.$$

(9) $$u(R, \phi) = f(\phi)$$

(10) $$\lim_{r\to\infty} u(r, \phi) = 0.$$

The PDE (8) follows from (7) by assuming that the solution u will not depend on θ because the Dirichlet condition (9) is independent of θ. This may be an electrostatic potential (or a temperature) $f(\phi)$ at which the sphere S: $r = R$ is kept. Condition (10) means that the potential at infinity will be zero.

Separating Variables by substituting $u(r, \phi) = G(r)H(\phi)$ into (8). Multiplying (8) by r^2, making the substitution and then dividing by GH, we obtain

$$\frac{1}{G}\frac{d}{dr}\left(r^2\frac{dG}{dr}\right) = -\frac{1}{H\sin\phi}\frac{d}{d\phi}\left(\sin\phi\frac{dH}{d\phi}\right).$$

By the usual argument both sides must be equal to a constant k. Thus we get the two ODEs

$$\frac{1}{G}\frac{d}{dr}\left(r^2\frac{dG}{dr}\right) = k \qquad \text{or} \qquad r^2\frac{d^2G}{dr^2} + 2r\frac{dG}{dr} = kG \tag{11}$$

and

$$\frac{1}{\sin\phi}\frac{d}{d\phi}\left(\sin\phi\frac{dH}{d\phi}\right) + kH = 0. \tag{12}$$

The solutions of (11) will take a simple form if we set $k = n(n + 1)$. Then, writing $G' = dG/dr$, etc., we obtain

$$r^2G'' + 2rG' - n(n + 1)G = 0. \tag{13}$$

This is an **Euler–Cauchy equation.** From Sec. 2.5 we know that it has solutions $G = r^a$. Substituting this and dropping the common factor r^a gives

$$a(a - 1) + 2a - n(n + 1) = 0. \qquad \text{The roots are} \qquad a = n \quad \text{and} \quad -n - 1.$$

Hence solutions are

$$G_n(r) = r^n \qquad \text{and} \qquad G_n^*(r) = \frac{1}{r^{n+1}}. \tag{14}$$

We now solve (12). Setting $\cos\phi = w$, we have $\sin^2\phi = 1 - w^2$ and

$$\frac{d}{d\phi} = \frac{d}{dw}\frac{dw}{d\phi} = -\sin\phi\frac{d}{dw}.$$

Consequently, (12) with $k = n(n + 1)$ takes the form

$$\frac{d}{dw}\left[(1 - w^2)\frac{dH}{dw}\right] + n(n + 1)H = 0. \tag{15}$$

This is **Legendre's equation** (see Sec. 5.3), written out

$$(1 - w^2)\frac{d^2H}{dw^2} - 2w\frac{dH}{dw} + n(n + 1)H = 0. \tag{15'}$$

For integer $n = 0, 1, \cdots$ the Legendre polynomials

$$H = P_n(w) = P_n(\cos \phi) \qquad n = 0, 1, \cdots,$$

are solutions of Legendre's equation (15). We thus obtain the following two sequences of solution $u = GH$ of Laplace's equation (8), with constant A_n and B_n, where $n = 0, 1, \cdots,$

$$\text{(a)}\quad u_n(r, \phi) = A_n r^n P_n(\cos \phi), \qquad \text{(b)}\quad u_n^*(r, \phi) = \frac{B_n}{r^{n+1}} P_n(\cos \phi) \tag{16}$$

Use of Fourier–Legendre Series

Interior Problem: Potential Within the Sphere S. We consider a series of terms from (16a),

$$u(r, \phi) = \sum_{n=0}^{\infty} A_n r^n P_n(\cos \phi) \qquad (r \leqq R). \tag{17}$$

Since S is given by $r = R$, for (17) to satisfy the Dirichlet condition (9) on the sphere S, we must have

$$u(R, \phi) = \sum_{n=0}^{\infty} A_n R^n P_n(\cos \phi) = f(\phi); \tag{18}$$

that is, (18) must be the **Fourier–Legendre series** of $f(\phi)$. From (7) in Sec. 5.8 we get the coefficients

$$A_n R^n = \frac{2n + 1}{2}\int_{-1}^{1} \tilde{f}(w) P_n(w)\, dw \tag{19*}$$

where $\tilde{f}(w)$ denotes $f(\phi)$ as a function of $w = \cos \phi$. Since $dw = -\sin \phi\, d\phi$, and the limits of integration -1 and 1 correspond to $\phi = \pi$ and $\phi = 0$, respectively, we also obtain

$$A_n = \frac{2n + 1}{2R^n}\int_0^{\pi} f(\phi) P_n(\cos \phi) \sin \phi\, d\phi, \qquad n = 0, 1, \cdots. \tag{19}$$

If $f(\phi)$ and $f'(\phi)$ are piecewise continuous on the interval $0 \leqq \phi \leqq \pi$, then the series (17) with coefficients (19) solves our problem for points inside the sphere because it can be shown that under these continuity assumptions the series (17) with coefficients (19) gives the derivatives occurring in (8) by termwise differentiation, thus justifying our derivation.

Exterior Problem: Potential Outside the Sphere *S*. Outside the sphere we cannot use the functions u_n in (16a) because they do not satisfy (10). But we can use the u_n^* in (16b), which do satisfy (10) (but could not be used inside *S*; why?). Proceeding as before leads to the solution of the exterior problem

$$u(r, \phi) = \sum_{n=0}^{\infty} \frac{B_n}{r^{n+1}} P_n(\cos \phi) \qquad (r \geqq R) \tag{20}$$

satisfying (8), (9), (10), with coefficients

$$B_n = \frac{2n+1}{2} R^{n+1} \int_0^{\pi} f(\phi) P_n(\cos \phi) \sin \phi \, d\phi. \tag{21}$$

The next example illustrates all this for a sphere of radius 1 consisting of two hemispheres that are separated by a small strip of insulating material along the equator, so that these hemispheres can be kept at different potentials (110 V and 0 V).

EXAMPLE 1 Spherical Capacitor

Find the potential inside and outside a spherical capacitor consisting of two metallic hemispheres of radius 1 ft separated by a small slit for reasons of insulation, if the upper hemisphere is kept at 110 V and the lower is grounded (Fig. 310).

Solution. The given boundary condition is (recall Fig. 309)

$$f(\phi) = \begin{cases} 110 & \text{if} \quad 0 \leqq \phi < \pi/2 \\ 0 & \text{if} \quad \pi/2 < \phi \leqq \pi. \end{cases}$$

Since $R = 1$, we thus obtain from (19)

$$A_n = \frac{2n+1}{2} \cdot 110 \int_0^{\pi/2} P_n(\cos \phi) \sin \phi \, d\phi$$

$$= \frac{2n+1}{2} \cdot 110 \int_0^1 P_n(w) \, dw$$

where $w = \cos \phi$. Hence $P_n(\cos \phi) \sin \phi \, d\phi = -P_n(w) \, dw$, we integrate from 1 to 0, and we finally get rid of the minus by integrating from 0 to 1. You can evaluate this integral by your CAS or continue by using (11) in Sec. 5.3, obtaining

$$A_n = 55(2n+1) \sum_{m=0}^{M} (-1)^m \frac{(2n-2m)!}{2^n m! (n-m)! (n-2m)!} \int_0^1 w^{n-2m} \, dw$$

where $M = n/2$ for even n and $M = (n-1)/2$ for odd n. The integral equals $1/(n - 2m + 1)$. Thus

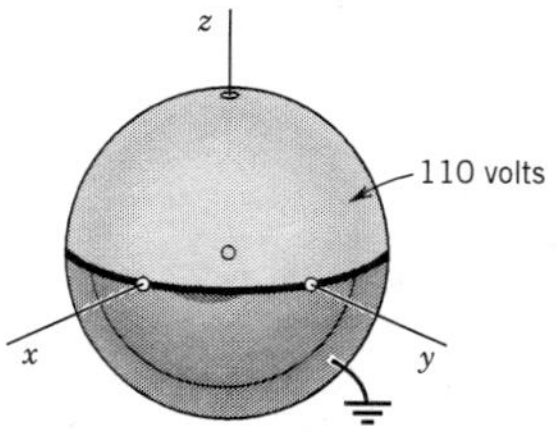

Fig. 310. Spherical capacitor in Example 1

(22)
$$A_n = \frac{55(2n+1)}{2^n} \sum_{m=0}^{M} (-1)^m \frac{(2n-2m)!}{m!(n-m)!(n-2m+1)!}.$$

Taking $n = 0$, we get $A_0 = 55$ (since $0! = 1$). For $n = 1, 2, 3, \cdots$ we get

$$A_1 = \frac{165}{2} \cdot \frac{2!}{0!1!2!} = \frac{165}{2},$$

$$A_2 = \frac{275}{4}\left(\frac{4!}{0!2!3!} - \frac{2!}{1!1!1!}\right) = 0,$$

$$A_3 = \frac{385}{8}\left(\frac{6!}{0!3!4!} - \frac{4!}{1!2!2!}\right) = -\frac{385}{8}, \qquad \text{etc.}$$

Hence the ***potential* (17) *inside the sphere*** is (since $P_0 = 1$)

(23)
$$u(r, \phi) = 55 + \frac{165}{2} r P_1(\cos\phi) - \frac{385}{8} r^3 P_3(\cos\phi) + \cdots \qquad \text{(Fig. 311)}$$

with $P_1, P_3, \cdots$ given by (11′), Sec. 5.3. Since $R = 1$, we see from (19) and (21) in this section that $B_n = A_n$, and (20) thus gives the ***potential outside the sphere***

(24)
$$u(r, \phi) = \frac{55}{r} + \frac{165}{2r^2} P_1(\cos\phi) - \frac{385}{8r^4} P_3(\cos\phi) + \cdots.$$

Partial sums of these series can now be used for computing approximate values of the inner and outer potential. Also, it is interesting to see that far away from the sphere the potential is approximately that of a point charge, namely, $55/r$. (Compare with Theorem 3 in Sec. 9.7.) ■

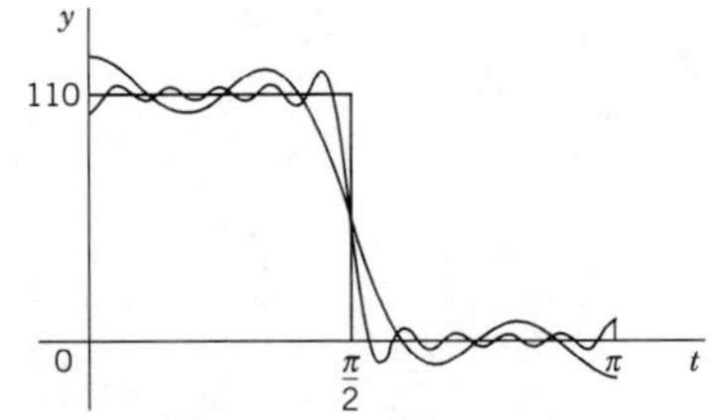

Fig. 311. Partial sums of the first 4, 6, and 11 nonzero terms of (23) for $r = R = 1$

EXAMPLE 2 Simpler Cases. Help with Problems

The technicalities occurring in cases like that of Example 1 can often be avoided. For instance, find the potential inside the sphere S: $r = R = 1$ when S is kept at the potential $f(\phi) = \cos 2\phi$. (Can you see the potential on S? What is it at the North Pole? The equator? The South Pole?)

Solution. $w = \cos\phi$, $\cos 2\phi = 2\cos^2\phi - 1 = 2w^2 - 1 = \frac{4}{3}P_2(w) - \frac{1}{3} = \frac{4}{3}(\frac{3}{2}w^2 - \frac{1}{2}) - \frac{1}{3}$. Hence the potential in the interior of the sphere is

$$u = \tfrac{4}{3}r^2 P_2(w) - \tfrac{1}{3} = \tfrac{4}{3}r^2 P_2(\cos\phi) - \tfrac{1}{3} = \tfrac{2}{3}r^2(3\cos^2\phi - 1) - \tfrac{1}{3}.$$ ■

PROBLEM SET 12.10

1. Derive (7) from $\nabla^2 u$ in Cartesian coordinates. (Show the details.)
2. Find the surfaces on which the functions u_1, u_2, u_3 are zero.
3. Sketch the functions $P_n(\cos\phi)$ for $n = 0, 1, 2$ (see (11′) in Sec. 5.3).
4. Sketch the functions $P_3(\cos\phi)$ and $P_4(\cos\phi)$.
5. Verify that u_n and $u_n{}^*$ in (16) are solutions of (8).

6–11 POTENTIALS DEPENDING ONLY ON r

6. **(Dimension 3)** Show that the only solution of the Laplace equation depending only on $r = \sqrt{x^2 + y^2 + z^2}$ is $u = c/r + k$ with constant c and k.
7. **(Dimension 3)** Verify that $u = c/r$, $r = \sqrt{x^2 + y^2 + z^2}$, satisfies Laplace's equation in spherical coordinates.
8. **(Dirichlet problem).** Find the electrostatic potential between two concentric spheres of radii $r_1 = 10$ cm and $r_2 = 20$ cm kept at potentials $U_1 = 260$ V and $U_2 = 110$ V, respectively.
9. **(Dimension 2, logarithmic potential)** Show that the only solution of the two-dimensional Laplace equation depending only on $r = \sqrt{x^2 + y^2}$ is $u = c \ln r + k$ with constant c and k.
10. **(Logarithmic potential)** Find the electrostatic potential between two coaxial cylinders of radii $r_1 = 10$ cm and $r_2 = 20$ cm kept at potentials $U_1 = 260$ V and $U_2 = 110$ V, respectively. Compare with Prob. 8. Comment.
11. **(Heat problem)** If the surface of the ball $r^2 = x^2 + y^2 + z^2 \leqq R^2$ is kept at temperature zero and the initial temperature in the ball is $f(r)$, show that the temperature $u(r, t)$ in the ball is a solution of $u_t = c^2(u_{rr} + 2u_r/r)$ satisfying the conditions $u(R, t) = 0$, $u(r, 0) = f(r)$. Show that setting $v = ru$ gives $v_t = c^2 v_{rr}$, $v(R, t) = 0$, $v(r, 0) = rf(r)$. Include the condition $v(0, t) = 0$ (which holds because u must be bounded at $r = 0$), and solve the resulting problem by separating variables.
12. **(Two-dimensional potential problems)** Show that the functions $x^2 - y^2$, xy, $x/(x^2 + y^2)$, $e^x \cos y$, $e^x \sin y$, $\cos x \cosh y$, $\ln(x^2 + y^2)$, and $\arctan(y/x)$ satisfy Laplace's equation $u_{xx} + u_{yy} = 0$. (Two-dimensional potential problems are best solved by ***complex analysis***, as we shall see in Chap. 18.)

13–17 BOUNDARY VALUE PROBLEMS IN SPHERICAL COORDINATES r, θ, ϕ

Find the potential in the interior of the sphere $S: r = R = 1$ if this interior is free of charges and the potential on S is:

13. $f(\phi) = 100$
14. $f(\phi) = \cos\phi$
15. $f(\phi) = \cos 3\phi$
16. $f(\phi) = \sin^2\phi$
17. $f(\phi) = 35 \cos 4\phi + 20 \cos 2\phi + 9$
18. Show that in Prob. 13 the potential exterior to the sphere is the same as that of a point charge at the origin. Is this physically plausible?
19. Sketch the intersection of the equipotential surfaces in Prob. 14 with the xz-plane.
20. Find the potential exterior to the sphere in Example 2 of the text and in Prob. 15.
21. What is the temperature in a ball of radius 1 and of homogeneous material if its lower boundary hemisphere is kept at 0°C and its upper at 100°C?
22. **(Reflection in a sphere)** Let r, θ, ϕ be spherical coordinates. If $u(r, \theta, \phi)$ satisfies $\nabla^2 u = 0$, show that $v(r, \theta, \phi) = u(1/r, \theta, \phi)/r$ satisfies $\nabla^2 v = 0$. What does this give for (16)?
23. **(Reflection in a circle)** Let r, θ be polar coordinates. If $u(r, \theta)$ satisfies $\nabla^2 u = 0$, show that the function $v(r, \theta) = u(1/r, \theta)$ satisfies $\nabla^2 v = 0$. What are $u = r\cos\theta$ and v in terms of x and y? Answer the same question for $u = r^2 \cos\theta \sin\theta$ and v.
24. **TEAM PROJECT. Transmission Line and Related PDEs.** Consider a long cable or telephone wire (Fig. 312) that is imperfectly insulated, so that leaks occur along the entire length of the cable. The source S of the current $i(x, t)$ in the cable is at $x = 0$, the receiving end T at $x = l$. The current flows from S to T, through the load, and returns to the ground. Let the constants R, L, C, and G denote the resistance, inductance, capacitance to ground, and conductance to ground, respectively, of the cable per unit length.

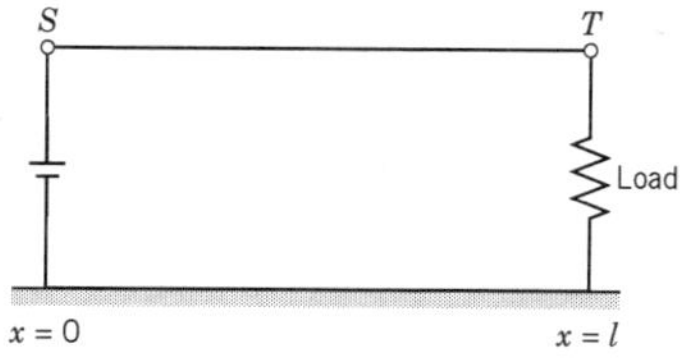

Fig. 312. Transmission line

(a) Show that (**"first transmission line equation"**)

$$-\frac{\partial u}{\partial x} = Ri + L\frac{\partial i}{\partial t}$$

where $u(x, t)$ is the potential in the cable. *Hint:* Apply Kirchhoff's voltage law to a small portion of the cable between x and $x + \Delta x$ (difference of the potentials at x and $x + \Delta x$ = resistive drop + inductive drop).

(b) Show that for the cable in (a) (**"second transmission line equation"**),

$$-\frac{\partial i}{\partial x} = Gu + C\frac{\partial u}{\partial t}$$

Hint: Use Kirchhoff's current law (difference of the currents at x and $x + \Delta x$ = loss due to leakage to ground + capacitive loss).

(c) Second-order PDEs. Show that elimination of i or u from the transmission line equations leads to

$$u_{xx} = LCu_{tt} + (RC + GL)u_t + RGu,$$

$$i_{xx} = LCi_{tt} + (RC + GL)i_t + RGi.$$

(d) Telegraph equations. For a submarine cable, G is negligible and the frequencies are low. Show that this leads to the so-called *submarine cable equations* or **telegraph equations**

$$u_{xx} = RCu_t, \qquad i_{xx} = RCi_t.$$

Find the potential in a submarine cable with ends ($x = 0, x = l$) grounded and initial voltage distribution $U_0 = const.$

(e) High-frequency line equations. Show that in the case of alternating currents of high frequencies the equations in (c) can be approximated by the so-called **high-frequency line equations**

$$u_{xx} = LCu_{tt}, \qquad i_{xx} = LCi_{tt}.$$

Solve the first of them, assuming that the initial potential is

$$U_0 \sin(\pi x/l),$$

and $u_t(x, 0) = 0$ and $u = 0$ at the ends $x = 0$ and $x = l$ for all t.

12.11 Solution of PDEs by Laplace Transforms

Readers familiar with Chap. 6 may wonder whether Laplace transforms can also be used for solving *partial* differential equations. The answer is yes, particularly if one of the independent variables ranges over the positive axis. The steps to obtain a solution are similar to those in Chap. 6. For a PDE in two variables they are as follows.

1. Take the Laplace transform with respect to one of the two variables, usually t. This gives an ***ODE for the transform*** of the unknown function. This is so since the derivatives of this function with respect to the other variable slip into the transformed equation. The latter also incorporates the given boundary and initial conditions.
2. Solving that ODE, obtain the transform of the unknown function.
3. Taking the inverse transform, obtain the solution of the given problem.

If the coefficients of the given equation do not depend on t, the use of Laplace transforms will simplify the problem.

We explain the method in terms of a typical example.

EXAMPLE 1 Semi-Infinite String

Find the displacement $w(x, t)$ of an elastic string subject to the following conditions. (We write w since we need u to denote the unit step function.)

(i) The string is initially at rest on the x-axis from $x = 0$ to ∞ (*"semi-infinite string"*).

(ii) For $t > 0$ the left end of the string ($x = 0$) is moved in a given fashion, namely, according to a single sine wave

$$w(0, t) = f(t) = \begin{cases} \sin t & \text{if } 0 \leqq t \leqq 2\pi \\ 0 & \text{otherwise} \end{cases} \qquad \text{(Fig. 313).}$$

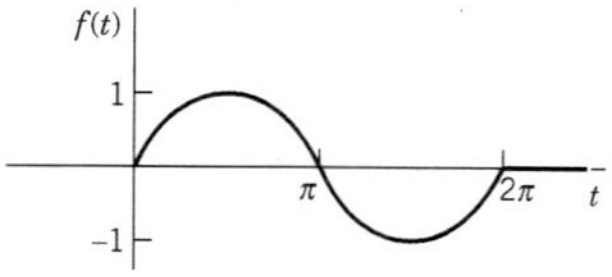

Fig. 313. Motion of the left end of the string in Example 1 as a function of time t

(iii) Furthermore, $\lim_{x\to\infty} w(x, t) = 0$ for $t \geqq 0$.

Of course there is no infinite string, but our model describes a long string or rope (of negligible weight) with its right end fixed far out on the x-axis.

Solution. We have to solve the wave equation (Sec. 12.2)

$$\text{(1)} \qquad \frac{\partial^2 w}{\partial t^2} = c^2 \frac{\partial^2 w}{\partial x^2}, \qquad c^2 = \frac{T}{\rho}$$

for positive x and t, subject to the "boundary conditions"

$$\text{(2)} \qquad w(0, t) = f(t), \qquad \lim_{x\to\infty} w(x, t) = 0 \qquad (t \geqq 0)$$

with f as given above, and the initial conditions

$$\text{(3)} \qquad \text{(a)} \quad w(x, 0) = 0, \qquad \text{(b)} \quad w_t(x, 0) = 0.$$

We take the Laplace transform ***with respect to t.*** By (2) in Sec. 6.2,

$$\mathcal{L}\left\{\frac{\partial^2 w}{\partial t^2}\right\} = s^2\mathcal{L}\{w\} - sw(x, 0) - w_t(x, 0) = c^2\mathcal{L}\left\{\frac{\partial^2 w}{\partial x^2}\right\}.$$

The expression $-sw(x, 0) - w_t(x, 0)$ drops out because of (3). On the right we assume that we may interchange integration and differentiation. Then

$$\mathcal{L}\left\{\frac{\partial^2 w}{\partial x^2}\right\} = \int_0^\infty e^{-st} \frac{\partial^2 w}{\partial x^2}\, dt = \frac{\partial^2}{\partial x^2} \int_0^\infty e^{-st} w(x, t)\, dt = \frac{\partial^2}{\partial x^2} \mathcal{L}\{w(x, t)\}.$$

Writing $W(x, s) = \mathcal{L}\{w(x, t)\}$, we thus obtain

$$s^2 W = c^2 \frac{\partial^2 W}{\partial x^2}, \qquad \text{thus} \qquad \frac{\partial^2 W}{\partial x^2} - \frac{s^2}{c^2} W = 0.$$

Since this equation contains only a derivative with respect to x, it may be regarded as an ***ordinary differential equation*** for $W(x, s)$ considered as a function of x. A general solution is

$$\text{(4)} \qquad W(x, s) = A(s)e^{sx/c} + B(s)e^{-sx/c}.$$

From (2) we obtain, writing $F(s) = \mathcal{L}\{f(t)\}$,

$$W(0, s) = \mathcal{L}\{w(0, t)\} = \mathcal{L}\{f(t)\} = F(s).$$

Assuming that we can interchange integration and taking the limit, we have

$$\lim_{x\to\infty} W(x, s) = \lim_{x\to\infty} \int_0^\infty e^{-st} w(x, t)\, dt = \int_0^\infty e^{-st} \lim_{x\to\infty} w(x, t)\, dt = 0.$$

This implies $A(s) = 0$ in (4) because $c > 0$, so that for every fixed positive s the function $e^{sx/c}$ increases as x increases. Note that we may assume $s > 0$ since a Laplace transform generally exists for *all* s greater than some fixed k (Sec. 6.2). Hence we have

$$W(0, s) = B(s) = F(s),$$

so that (4) becomes

$$W(x, s) = F(s)e^{-sx/c}.$$

From the second shifting theorem (Sec. 6.3) with $a = x/c$ we obtain the inverse transform

$$w(x, t) = f\left(t - \frac{x}{c}\right) u\left(t - \frac{x}{c}\right) \tag{5}$$ (Fig. 314)

that is,

$$w(x, t) = \sin\left(t - \frac{x}{c}\right) \quad \text{if} \quad \frac{x}{c} < t < \frac{x}{c} + 2\pi \quad \text{or} \quad ct > x > (t - 2\pi)c$$

and zero otherwise. This is a single sine wave traveling to the right with speed c. Note that a point x remains at rest until $t = x/c$, the time needed to reach that x if one starts at $t = 0$ (start of the motion of the left end) and travels with speed c. The result agrees with our physical intuition. Since we proceeded formally, we must verify that (5) satisfies the given conditions. We leave this to the student. ∎

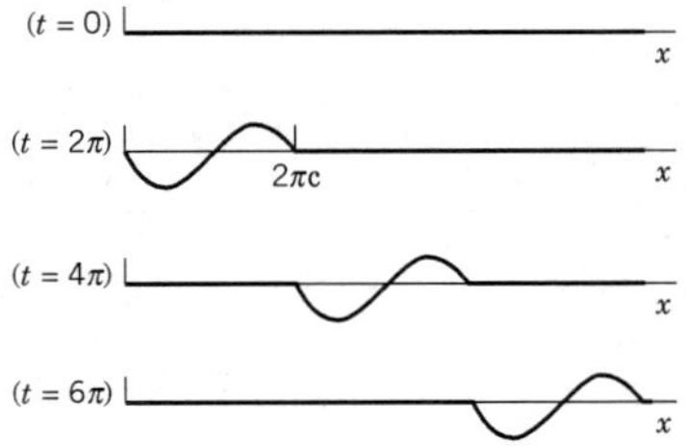

Fig. 314. Traveling wave in Example 1

This is the end of Chap. 12, in which we concentrated on the most important partial differential equations (PDEs) in physics and engineering. This is also the end of Part C on Fourier analysis and PDEs.

We have seen that PDEs have various basic engineering applications, which make them the subject of many ongoing research projects.

Numerics for PDEs follows in Secs. 21.4–21.7, which are independent of the other sections in Part E on numerics.

In the next part, Part D on **complex analysis,** we turn to an area of a different nature that is also highly important to the engineer, as our examples and problems will show. This will include another approach to the (two-dimensional) **Laplace equation** and its applications in Chap. 18.

PROBLEM SET 12.11

1. Sketch a figure similar to Fig. 314 if $c = 1$ and f is "triangular" as in Example 1, Sec. 12.3.
2. How does the speed of the wave in Example 1 depend on the tension and on the mass of the string?
3. Verify the solution in Example 1. What traveling wave do we obtain in Example 1 in the case of a (nonterminating) sinusoidal motion of the left end starting at $t = 0$?

4–6 SOLVE BY LAPLACE TRANSFORMS

4. $\frac{\partial w}{\partial x} + x\frac{\partial w}{\partial t} = x,\ w(x, 0) = 1,\ w(0, t) = 1$

5. $x \dfrac{\partial w}{\partial x} + \dfrac{\partial w}{\partial t} = xt, \quad w(x, 0) = 0$ if $x \geqq 0,$
$w(0, t) = 0$ if $t \geqq 0$

6. $\dfrac{\partial^2 w}{\partial x^2} = 100 \dfrac{\partial^2 w}{\partial t^2} + 100 \dfrac{\partial w}{\partial t} + 25w,$

$w(x, 0) = 0$ if $x \geqq 0, \quad w_t(x, 0) = 0$ if $t \geqq 0,$
$w(0, t) = \sin t$ if $t \geqq 0$

7. Solve Prob. 5 by another method.

8–10 HEAT PROBLEM

Find the temperature $w(x, t)$ in a semi-infinite laterally insulated bar extending from $x = 0$ along the x-axis to infinity, assuming that the initial temperature is 0, $w(x, t) \to 0$ as $x \to \infty$ for every fixed $t \geqq 0$, and $w(0, t) = f(t)$. Proceed as follows.

8. Set up the model and show that the Laplace transform leads to

$$sW = c^2 \frac{\partial^2 W}{\partial x^2} \qquad (W = \mathcal{L}\{w\})$$

and

$$W = F(s)e^{-\sqrt{s}x/c} \qquad (F = \mathcal{L}\{f\}).$$

Applying the convolution theorem, show that

$$w(x, t) = \frac{x}{2c\sqrt{\pi}} \int_0^t f(t - \tau)\tau^{-3/2} e^{-x^2/(4c^2\tau)}\, d\tau.$$

9. Let $w(0, t) = f(t) = u(t)$ (Sec. 6.3). Denote the corresponding w, W, and F by w_0, W_0, and F_0. Show that then in Prob. 8,

$$w_0(x, t) = \frac{x}{2c\sqrt{\pi}} \int_0^t \tau^{-3/2} e^{-x^2/(4c^2\tau)}\, d\tau$$
$$= 1 - \operatorname{erf}\left(\frac{x}{2c\sqrt{t}}\right)$$

with the error function erf as defined in Problem Set 12.6.

10. **(Duhamel's formula[4])** Show that in Prob. 9,

$$W_0(x, s) = \frac{1}{s} e^{-\sqrt{s}x/c}$$

and the convolution theorem gives *Duhamel's formula*

$$w(x, t) = \int_0^t f(t - \tau) \frac{\partial w_0}{\partial \tau}\, d\tau.$$

CHAPTER 12 REVIEW QUESTIONS AND PROBLEMS

1. Write down the three probably most important PDEs from memory and state their main applications.
2. What is the method of separating variables for PDEs? Give an example from memory.
3. What is the superposition principle? Give a typical application.
4. What role did Fourier series play in this chapter? Fourier integrals?
5. What are the eigenfunctions and their frequencies of the vibrating string? Of the heat equation?
6. What additional conditions did we consider for the wave equation? For the heat equation?
7. Name and explain the three kinds of boundary conditions.
8. What do you know about types of PDEs? About transformation to normal forms?
9. What is d'Alembert's method? To what PDE does it apply?
10. When and why did we use polar coordinates? Spherical coordinates?
11. When and why did Legendre's equation occur in this chapter? Bessel's equation?
12. What are the eigenfunctions of the circular membrane? How do their frequencies differ in principle from those of the eigenfunctions of the vibrating string?
13. Explain mathematically (not physically) why we got exponential functions in separating the heat equation, but not for the wave equation.
14. What is the error function? Why did it occur and where?
15. Explain the idea of using Laplace transform methods for PDEs. Give an example from memory.
16. For what k and m are $x^4 + kx^2y^2 + y^4$ and $\sin mx \sinh y$ solutions of Laplace's equation?
17. Verify that $(x^2 - y^2)/(x^2 + y^2)^2$ satisfies Laplace's equation.

18–21 Solve for $u = u(x, y)$:

18. $u_{yy} + 16u = 0$
19. $u_{xx} + u_x - 2u = 0$
20. $u_{xy} + u_y + x + y + 1 = 0$
21. $u_{yy} + u_y = 0,\ u(x, 0) = f(x),\ u_y(x, 0) = g(x)$
22. Find all solution $u(x, y) = F(x)G(y)$ of Laplace's equation in two variables.

[4]JEAN-MARIE CONSTANT DUHAMEL (1797–1872), French mathematician.

23–26 Find and sketch or graph (as in Fig. 285 in Sec. 12.3) the deflection $u(x, t)$ of a vibrating string of length π, extending from $x = 0$ to $x = \pi$, and $c^2 = T/\rho = 1$, starting with velocity 0 and deflection

23. $f(x) = \sin x - \frac{1}{2}\sin 2x$

24. $f(x) = \frac{1}{2}\pi - |x - \frac{1}{2}\pi|$

25. $f(x) = \sin^3 x$

26. $f(x) = x(\pi - x)$

27–30 Find the temperature distribution in a laterally insulated thin copper bar ($c^2 = K/\rho\sigma = 1.158 \text{ cm}^2/\text{sec}$), 50 cm long and of constant cross section with endpoints at $x = 0$ and 50 kept at 0°C and initial temperature

27. $f(x) = \sin(\pi x/50)$

28. $f(x) = x(50 - x)$

29. $f(x) = 25 - |25 - x|$

30. $f(x) = 4\sin^3(\pi x/10)$

31–33 Find the temperature $u(x, t)$ in a laterally insulated bar of length π, extending from $x = 0$ to $x = \pi$, with $c^2 = 1$ for adiabatic boundary condition (see Problem Set 12.5) and initial temperature

31. $100 \cos 4x$

32. $3x^2$

33. $\pi - 2|x - \frac{1}{2}\pi|$

34. Using partial sums, graph $u(x, t)$ in Prob. 33 for several constant t on common axes. Do these graphs agree with your physical intuition?

35. Let $f(x, y) = u(x, y, 0)$ be the initial temperature in a thin square plate of side π with edges kept at 0°C and faces perfectly insulated. Separating variables, obtain from $u_t = c^2\nabla^2 u$ the solution

$$u(x, y, t) = \sum_{m=1}^{\infty}\sum_{n=1}^{\infty} B_{mn} \sin mx \sin ny\, e^{-c^2(m^2+n^2)t}$$

where

$$B_{mn} = \frac{4}{\pi^2}\int_0^{\pi}\int_0^{\pi} f(x, y) \sin mx \sin ny\, dx\, dy.$$

36. Find the temperature in Prob. 35 if $f(x, y) = x(\pi - x)y(\pi - y)$.

37–42 Transform to normal form and solve (showing the details!)

37. $u_{xy} = u_{xx}$

38. $u_{xx} + 4u_{xy} + 4u_{yy} = 0$

39. $u_{xx} + 4u_{yy} = 0$

40. $2u_{xx} + 5u_{xy} + 2u_{yy} = 0$

41. $u_{xx} + 2u_{xy} + u_{yy} = 0$

42. $u_{yy} + u_{xy} - 2u_{xx} = 0$

43–47 Show that the following membranes of area 1 with $c^2 = 1$ have the frequencies of the fundamental mode as given (4-decimal values). Compare.

43. Circle: $\alpha_1/(2\sqrt{\pi}) = 0.6784$

44. Square: $1\sqrt{2} = 0.7071$

45. Rectangle (sides 1:2): $\sqrt{5/8} = 0.7906$

46. Semicircle: $3.832/\sqrt{8\pi} = 0.7644$

47. Quadrant of circle: $\alpha_{12}/(4\sqrt{\pi}) = 0.7244$ ($\alpha_{12} = 5.13562$ = first positive zero of J_2)

48–50 Find the electrostatic potential in the following (charge-free) regions:

48. Between two concentric spheres of radii r_0 and r_1 kept at the potentials u_0 and u_1, respectively.

49. Between two coaxial circular cylinders of radii r_0 and r_1 kept at the potential u_0 and u_1, respectively. (Compare with Prob. 48.)

50. In the interior of a sphere of radius 1 kept at the potential $f(\phi) = \cos 3\phi + 3\cos\phi$ (referred to our usual spherical coordinates).

SUMMARY OF CHAPTER 12
Partial Differential Equations (PDEs)

Whereas ODEs (Chaps. 1–6) serve as models of problems involving only *one* independent variable, problems involving *two or more* independent variables (space variables or time t and one or several space variables) lead to PDEs. This accounts for the enormous importance of PDEs to the engineer and physicist. Most important are:

(1) $u_{tt} = c^2 u_{xx}$ One-dimensional wave equation (Secs. 12.2–12.4)

(2) $u_{tt} = c^2(u_{xx} + u_{yy})$ Two-dimensional wave equation (Secs. 12.7–12.9)

(3) $u_t = c^2 u_{xx}$ One-dimensional heat equation (Secs. 12.5, 12.6)

(4) $\nabla^2 u = u_{xx} + u_{yy} = 0$ Two-dimensional Laplace equation (Secs. 12.5, 12.9)

(5) $\nabla^2 u = u_{xx} + u_{yy} + u_{zz} = 0$ Three-dimensional Laplace equation (Sec. 12.10).

Equations (1) and (2) are hyperbolic, (3) is parabolic, (4) and (5) are elliptic.

In practice, one is interested in obtaining the solution of such an equation in a given region satisfying given additional conditions, such as **initial conditions** (conditions at time $t = 0$) or **boundary conditions** (prescribed values of the solution u or some of its derivatives on the boundary surface S, or boundary curve C, of the region) or both. For (1) and (2) one prescribes two initial conditions (initial displacement and initial velocity). For (3) one prescribes the initial temperature distribution. For (4) and (5) one prescribes a boundary condition and calls the resulting problem a (see Sec. 12.5)

Dirichlet problem if u is prescribed on S,
Neumann problem if $u_n = \partial u/\partial n$ is prescribed on S,
Mixed problem if u is prescribed on one part of S and u_n on the other.

A general method for solving such problems is the method of **separating variables** or **product method,** in which one assumes solutions in the form of products of functions each depending on one variable only. Thus equation (1) is solved by setting $u(x, t) = F(x)G(t)$; see Sec. 12.3; similarly for (3) (see Sec. 12.5). Substitution into the given equation yields ***ordinary*** differential equations for F and G, and from these one gets infinitely many solutions $F = F_n$ and $G = G_n$ such that the corresponding functions

$$u_n(x, t) = F_n(x)G_n(t)$$

are solutions of the PDE satisfying the given boundary conditions. These are the **eigenfunctions** of the problem, and the corresponding **eigenvalues** determine the frequency of the vibration (or the rapidity of the decrease of temperature in the case of the heat equation, etc.). To satisfy also the initial condition (or conditions), one must consider infinite series of the u_n, whose coefficients turn out to be the Fourier coefficients of the functions f and g representing the given initial conditions (Secs. 12.3, 12.5). Hence **Fourier series** (and *Fourier integrals*) are of basic importance here (Secs. 12.3, 12.5, 12.6, 12.8).

Steady-state problems are problems in which the solution does not depend on time t. For these, the heat equation $u_t = c^2 \nabla^2 u$ becomes the Laplace equation.

Before solving an initial or boundary value problem, one often transforms the PDE into coordinates in which the boundary of the region considered is given by simple formulas. Thus in polar coordinates given by $x = r \cos \theta$, $y = r \sin \theta$, the **Laplacian** becomes (Sec. 12.9)

$$\nabla^2 u = u_{rr} + \frac{1}{r} u_r + \frac{1}{r^2} u_{\theta\theta}\,; \tag{6}$$

for spherical coordinates see Sec. 12.10. If one now separates the variables, one gets **Bessel's equation** from (2) and (6) (vibrating circular membrane, Sec. 12.9) and **Legendre's equation** from (5) transformed into spherical coordinates (Sec. 12.10).